Essays on Topology

Louis Funar • Athanase Papadopoulos
Editors

Essays on Topology

Dedicated to Valentin Poénaru

Springer

Editors

Louis Funar
Institut Fourier, UMR 5582
Université Grenoble Alpes, CNRS
CS 40700, 38058 Grenoble cedex 9, France

Athanase Papadopoulos
Institut de Recherche Mathématique
Avancée
CNRS and University of Strasbourg
Strasbourg, France

ISBN 978-3-031-81413-6   ISBN 978-3-031-81414-3 (eBook)
https://doi.org/10.1007/978-3-031-81414-3

Mathematics Subject Classification: 00A30, 46M20, 51F05, 55M05, 55R20, 57M25, 57R90, 20F36, 20F65, 20H05, 30F60, 32G15, 37E30, 51M15, 53C55, 54D35, 55E05, 55Q45, 57R42, 57K10, 58K05, 01A70, 18A10, 20F05, 54E99, 57M27

Cover illustration: Valentin Poénaru, pencil drawing by Milen Poénaru, ca. 1980. ©Milen Poénaru

This Springer imprint is published by the registered company Springer Nature Switzerland AG
The registered company address is: Gewerbestrasse 11, 6330 Cham, Switzerland

Valentin Poénaru at Northeastern University, Boston, 1966. The photo was taken by the students

Valentin Poénaru at Northeastern University, Boston, 1966. The photo was taken by the students

Foreword

Meeting Po, I

One fresh day of September 1990 I passed near an office having the name of Poénaru on the doorplate. I knew this name. Two years before that I discovered some excerpts from the book of Fathi, Laudenbach and Poénaru entitled *Sur les travaux de Thurston* sparking my interest. These days I was mostly spending my time within the library of the Institute of Mathematics, known as chez Pompi, after the name of the librarian. My frustration was great when I realized that the book was missing from the *Astérisque* collection at the library and nobody could find a copy of it. I wanted it so much that I wrote to Valentin Poénaru a letter explaining my grief. Soon after the 1989 fall of communism in Romania, I received a xerox copy of his book which I started immediately reading when preparing the diploma thesis.

As soon as I knocked at the door I heard some voice inviting me in. There was a single person in the office who, when I entered and told my name, greeted me and simply said *I'm Po*. He somewhat knew I had to come, which surprised me at that time, and he asked me directly about the mathematics I'm doing. I took a piece of chalk, and we started a quite long discussion at the blackboard. Po advised me to head to low-dimensional topology, and soon later, he became my PhD advisor. This first meeting with Po will be followed by many others during the next thirty-five years.

I found out much later what role he played in my arrival at Orsay by supporting my application with all his resources. Our common Romanian roots, although living in very different times, may have been one additional link in our longstanding relationship.

There is not enough place here to explain how much I owe to him. I learned mathematics, of course, but also a lot about physics, history, philosophy and many other subjects, which Po addressed with ease and passion. His thoughts about the profession of mathematician had a strong influence on me. His strength and resilience in his work represent a model for me. He is well-known for his kindness and his generosity in sharing ideas and advice. This book is dedicated to Po out of gratitude and great admiration for the man and the mathematician.

Grenoble, France
June 29, 2025

Louis Funar

Meeting Po, II

I first met Valentin Poénaru while I was an undergraduate at Orsay. I followed his course on complex analysis, and that is how I learned topology and hyperbolic geometry. Then, I had him as advisor for what was called in those times *Diplôme d'Études Approfondies*; this was the year that immediately preceded the enrollment in a PhD program. I had already taken the exams (in logic), and I had to write a memoir. He first gave me a reading assignment, the book *Problèmes ergodiques de la mécanique classique* by Arnol'd and Avez, and then asked me to write an essay on Anosov diffeomorphisms. This is how I learned dynamics. Then, I went for PhD with François Laundebach, a colleague and close friend of Poénaru. The first question that François asked me is "Why do you come to me rather than to Poénaru". I don't remember the exact answer I gave him, or whether I really answered his question, but the fact is that I thought I had already learned a lot from Poénaru, and I wanted to learn from someone else. Thinking about this now, I am glad that Po was not my PhD advisor, because in that way we became friends. I think that it is difficult for a relationship with one's PhD advisor to become a true friendship; it very often remains asymmetrical.

Laudenbach gave me a problem on Thurston's pseudo-Anosov mapping classes, as a doctoral research subject. During my two years of PhD, Poénaru has always been in the background, aware of what I was working on. I learned Thurston's theory from the Orsay seminar notes, of which he was the initiator, and later he was the president of my PhD thesis committee. We became friends, meeting whenever it was possible, at our house in Strasbourg and at other places, talking about maths, literature, philosophy, religion, history ... and our own history. Over and above everything I have learned from Poénaru, there is one thing that has particularly affected me, and that is his kindness and his faithfulness in his relations with his friends. I always say that there is a big amount of good mathematicians in the world, but among them, rarely of this kind: modest, gentle and caring. Po reads a lot, and he likes books. I couldn't have responded better to all this, and to everything he has given me, than by dedicating a book to him, and working on it.

Strasbourg, France — Athanase Papadopoulos
June 29, 2025

Contents

Chapter 1
Introduction

Louis Funar and Athanase Papadopoulos

Abstract This is an Introduction to the book *Essays in Geometry*, a collection of articles dedicated to Valentin Poénaru. The Introduction contains an overview of the whole book. It gives an idea of the content of each chapter and at the same time of the spectrum of subjects of interest of Poénaru. The topics discussed include topology, geometry, combinatorial group theory, history and philosophy of mathematics, mathematical physics. Several chapters are of autobiographical nature, written by Poénaru himself.

Keywords Topology · Geometry · Combinatorial group theory · History of mathematics, and philosophy of mathematics · Biographical essays · Poincaré conjecture · Schenflies conjecture · Valentin Poénaru

Two years ago, Valentin Poénaru (*Po*, for his friends), turned ninety. His work had an essential impact on the development of topology in France over the last forty years of the twentieth century. To this day, he is always bubbling with energy, and his mathematics and the rest of his intellectual activities are still fully occupying his mind. At the occasion of his ninetieth birthday, we have decided to offer to him a collection of articles, and we have asked several colleagues and friends to contribute to this project. The result is the essays that are included in the present book, which we collected under the title *Essays on geometry*. The title may seem restrictive, but one can think of the old meaning of the word "géomètre" which, in France, until the end of the XIXth century, meant "mathematician".

It turned out that this collection contains essays on a variety of topics: hyperbolic geometry, 3-manifolds, complex and symplectic geometry, differential topology,

L. Funar
Institut Fourier, University Grenoble Alpes, CNRS, Grenoble, France
e-mail: louis.funar@univ-grenoble-alpes.fr

A. Papadopoulos (✉)
Institut de Recherche Mathématique Avancée, CNRS and University of Strasbourg, Strasbourg, France
e-mail: athanase.papadopoulos@math.unistra.fr

L. Funar, A. Papadopoulos (eds.), *Essays on Topology*,
https://doi.org/10.1007/978-3-031-81414-3_1

combinatorial group theory, PL topology, algebraic geometry, knots and links, homotopy theory, braid groups, phylogenetics, history of geometry, and philosophy of mathematics. Most of these topics are to be found in Po's sphere of interest and their plurality is a sign of his broad mathematical culture and the richness of his work. In fact, Po's work is mentioned several times in the various chapters that constitute this book. To these chapters, contributed by colleagues and friends, we have added some essays composed by Po himself, including biographical sketches of several periods of his life, and reports he wrote on his work, on several occasions.

Let us make now a short but systematic summary of the content of this book.

Chapter 2, by Barry Mazur, is titled *Primes, Knots and Po*. In this chapter, Mazur expresses his admiration and deep friendship for Po. He talks about their common mathematical interests dating back to the time when Po was still living in Romania, more than half a century ago. Then, he leads us into a discussion of the analogy between topological problems and problems in number theory, which are two fields in which he has always been thoroughly involved.

Chapter 3, by Dennis Sullivan, is titled *A short history of open manifolds related to complex analytic coordinates*. In this chapter, Sullivan situates Po's work among a series of outstanding results in topology, in a period that was still part of the "heroic era" of this field, where a large number of fundamental questions were answered one after the other. The results mentioned by Sullivan, besides Po's and his own, are due to Hirsch, Smale, Phillips, Gromov, Haefliger, Landweber, Adachi, Teichner, Vogt, Blank, Bott, Milnor, Quillen, Samuelsen and others.

Chapters 4 and 5 are by Bob Penner. Chapter 4 is titled *Universal Spin Teichmüller Theory, I: The action of P(SL(2,$\mathbb{Z}$)) on $\mathcal{T}ess^+$* and Chap. 5, which is a sequel to Chap. 4, is titled *Universal Spin Teichmüller Theory, II: Finite presentation of P(SL(2,$\mathbb{Z}$))*. These two chapters are a continuation of a series of works that Penner started in the 1980s, on universal Teichmüller theory. The setting is the following: Penner defined a *universal Teichmüller space* as a Fréchet manifold $\mathcal{T}$ equipped with the action of a *universal mapping class group* $\mathcal{P}$ such that (i) the classical mapping class groups inject homomorphically into explicit completions of $\mathcal{P}$; (ii) the classical Teichmüller spaces embed naturally in $\mathcal{T}$ in an equivariant way with respect to the mapping class group actions; (iii) the Weil–Petersson Kähler geometry on each Teichmüller space is obtained by pulling back a universal geometry on $\mathcal{T}$.

In a previous paper titled *Universal constructions in Teichmüller theory* (1993), Penner constructed a beautiful incarnation of the universal Teichmüller space, as a quotient space $\mathcal{T}ess = \mathcal{T}ess'/\mathrm{PSL}(2,\mathbb{R})$ where $\mathcal{T}ess'$ is a space of tesselations of the hyperbolic disc with a distinguished oriented edge. He showed that the auxiliary space $\mathcal{T}ess'$ is homeomorphic to the space of orientation-preserving homeomorphisms of the circle, which naturally appears as the boundary of the Poincaré model of the hyperbolic plane. In previous works, the universal mapping class group was taken to be the group $\mathrm{PPSL}(2,\mathbb{Z})$ of all piecewise $\mathrm{PSL}(2,\mathbb{Z})$ homeomorphisms of the unit circle with finitely many breakpoints contained in the set of rational points. This group is the combinatorial Ptolemy group acting on tesselations of the hyperbolic plane with distinguished oriented edge by flips (Penner 1993 quoted above).

In Chap. 4, Penner introduces several new constructions in universal Teichmüller theory, including a new spin mapping class group P(SL(2,$\mathbb{Z}$)), consisting of the piecewise-constant maps $S^1 \to \mathrm{SL}(2, \mathbb{Z})$ which projectivize to an element of PPSL(2, $\mathbb{Z}$). This new spin mapping class group is defined combinatorially through its action on $\mathcal{T}ess^+$, and it is then identified with the spin mapping class group in the proof of the next result. In Chap. 5, Penner derives a finite presentation of P(SL(2,$\mathbb{Z}$)) from that of PPSL(2,$\mathbb{Z}$) $\approx T$ and from an explicit computation of the orbifold deck group of the spin cover. The finite presentation of PPSL(2, $\mathbb{Z}$) is derived from basic principles of general position as an orbifold fundamental group.

Chapter 6, by Greg McShane and Vlad Sergiescu, is titled *Geometry of Fermat's sum of squares*. The classical Fermat theorem on the sum of two squares says the following: For any prime number p, the equation $x^2+y^2 = p$ has an integer solution if and only if $p = 2$ or $p - 1$ is a multiple of 4. In Chap. 6, McShane and Sergiescu provide a new proof of Fermat's sum of two squares theorem using calculations from hyperbolic geometry and considerations of automorphisms of the 3-punctured sphere.

Chapter 7, by Ken'ichi Ohshika, titled *Ends and compactifications in geometric topology*, is a survey on ends and compactifications of topological spaces, with applications to the topology of manifolds. The author starts by presenting the work of Freudenthal done in the 1930s–1940s on the ends of topological spaces. Based on this concept, the latter introduced a new notion of compactification of general topological spaces, including discrete spaces, which was used later in geometric group theory. Ohshika then reviews more especially the theory of ends of contractible manifolds, considering, after the famous J. H. C. Whitehead construction of a contractible open 3-manifold which is not homeomorphic to Euclidean 3-space, works by several mathematicians on topology in dimension three and then in higher dimensions. This includes works by Stallings, Edwards, Freedman, Donaldson, Browder–Levine–Livesay, Siebenmann, Husch–Price, Tucker, Scott, Marden, Agol, Calegari–Gabai and others, culminating in the work of Thurston on attaching ideal boundaries to manifolds, a major part of his theory of hyperbolic 3-manifolds and Kleinian groups, and, in particular, in his uniformisation theorem for Haken manifolds. In doing so, Ohshika mentions Po's result, obtained after the appearance of Whitehead manifolds, saying that the product of any smooth homotopy 3-sphere with the real line is a 4-manifold which is a union of two standard smooth $\mathbb{R}^4$s. The independent works of Po and Mazur on the existence of compact contractible 4-manifolds whose boundaries are not simply connected are put into context, as well as the independent works of Po and Casson on simply connectedness at infinity of 3-manifolds under some conditions on their fundamental groups.

Chapter 8, by Vladimir Turaev, is titled *Axiomatic phylogenetics*. The topic is motivated by evolutionary biology, more precisely, phylogenetics, a branch of genetics in which one studies genetical modifications in (animal of vegetal) living species and transformations and relations among them. These relations are naturally presented by scientists and by common people in the form of trees, which biologists call phylogenetic trees and which encode the history, evolution or the relation and degree of parentage among individuals. Using the formalism of precategories,

Turaev, in this chapter, sets a general mathematical framework for phylogenetics. The central context in which this mathematical formalism is applied is the consideration, not only of the historical evolution, but of all possible evolutions from primitive beings to complex organisms. From the purely mathematical point of view, Turaev gives examples of phylogenetic precategories arising in set theory, group theory and the theory of metric spaces, in particular, the notion of *trim metric spaces*, a theory which establishes a coherent setting for finite metric spaces, metric graphs and metric trees and which he has previously developed in two papers, *Trimming of finite metric spaces* (2016) and *Trimming of metric spaces and the tight span* (2018).

It is well known that some mathematical methods are used in biology. It is interesting to see that conversely, there are mathematical ideas and new mathematical concepts that are suggested by biology—here, evolutionary biology.

In Chap. 9, titled *A note on the discriminant loci of a hypersurface of bi-degree* (m, n), François Laudenbach gives a new proof of the following theorem: Any two non-vanishing closed 1-forms on $S^2 \times [0, 1]$ that are cohomologous and whose kernels are tangent to the boundary of this manifold are isotopic. This theorem is a particular case of a theorem proved in 1979 by Laudenbach together with Sandy Blank, with the same statement except that the product space $S^2 \times [0, 1]$ is replaced by an arbitrary compact 3-manifold. The new proof of the particular case is less technical than the original proof of the general theorem. This particular case also provides a new proof of a famous theorem of Cerf, known under the name "$\Gamma_4 = 0$", saying that every orientation-preserving diffeomorphism of the 3-sphere is isotopic to the identity. The author describes his proof as a "foliated proof".

Chapter 10, by Liviu Ornea and Misha Verbitsky, is titled *Holomorphic tensors on Vaisman manifolds*. In this chapter, the authors consider a family of complex manifolds called *locally conformally Kähler* (LCK). This is a complex manifold admitting a Hermitian form ω satisfying $d\omega = \omega \wedge \theta$, where θ is a closed 1-form, called the Lee form. An LCK manifold admits a covering equipped with a Kähler metric such that the deck transformations are holomorphic homotheties. An LCK manifold is said to be "with potential" if its universal covering admits a Kähler potential, with deck transformations being homotheties. Examples of the latter are the classical Hopf manifolds and the Vaisman manifolds, that is, the LCK manifolds in which the Lee form is parallel with respect to the Levi-Civita connection. In Chap. 10, the authors prove that any holomorphic tensor on a Vaisman manifold is invariant with respect to the Lee field. They use this result to compute the Kodaira dimension of Vaisman manifolds and they show that the dimension of such a manifold which is the $\mathbb{Z}$-quotient of an algebraic cone over a projective manifold X is equal to the Kodaira dimension of X. The last result is used to prove a deformational stability of the Kodaira dimension of Vaisman manifolds.

Chapter 11, by Mihai Damian, is titled *On the topology of monotone Lagrangians of high dimension*. In this chapter, the author proves that in dimension $n \geq 6$, any closed orientable monotone Lagrangian submanifold of $\mathbb{C}^n$ is the total space of a fibration over the circle, provided some topological condition on the universal cover of the Lagrangian is satisfied. Damian also describes a generalization of this result for an arbitrary symplectic ambient manifold with vanishing second homology group, under the condition that the Lagrangian is displaceable through a Hamiltonian isotopy. Here, a Lagrangian submanifold L of a symplectic manifold X is said to be displaceable if there exists a Hamiltonian isotopy (φ_t) $(0 \leq t \leq 1)$ on X such that $\varphi_1(L) \cap L = \emptyset$. A generalization for monotone Lagrangians in $\mathbf{CP}^n$ is also obtained. The main ingredient the author uses in the proofs is the so-called lifted Floer homology which he has developed in a previous work. Damian formulates several conjectures concerning this theory.

Chapter 12, by Petr M. Akhmetiev, is titled *Kervaire Problems in Stable Homotopy Theory*. Originally, the Kervaire Problem in Stable Homotopy Theory asks for the classification of framed immersions with Arf invariant one. The Arf invariant is an invariant of a quadratic form defined on the homology of the manifold with $\mathbb{Z}/2$ coefficients. The name Kervaire refers to the latter's 1960 article *A manifold which does not admit any differentiable structure*. In Chap. 12, Akhmetiev recalls the history of the Kervaire problem, with several transformations and generalizations it underwent. He describes a series of geometrical constructions around these generalizations and presents a solution of a Generalized Kervaire Problem he has formulated in 2016 in an article titled *Kervaire problems in stable homotopy theory*. Kervaire invariants and Arf–Kervaire invariants naturally appear in this generalized theory. The author mentions a relation with the work of Po, noting that the Kervaire invariants were originally defined by means of self-intersections of immersions that have been investigated by Po in his 1975 paper *Some invariants of generic immersions and their geometric applications*.

Chapter 13, by Sergey A. Melikhov, is titled *Lifting generic maps to embeddings. The double point obstruction*. The subject is the theory of generic PL maps or smooth fold maps $f : N^n \to M^m$ with $m \geq n$ and $2(m+k) \geq 3(n+1)$ for some $k \geq 1$. The author proves that such a map lifts to a PL or to a smooth embedding from N to $M \times \mathbb{R}^k$ if and only if its double point locus $\{(x, y) \in N \times N \mid f(x) = f(y),\ x \neq y\}$ admits an equivariant map to S^{k-1}. A corollary answers a question raised by P. Petersen in 1990. At the same time, Melikhov discusses several criteria for lifting a non-degenerate PL map or C^0-stable smooth map between manifolds $f : N^n \to M^m$ where $m \geq n$, to an embedding into $M^m \times \mathbb{R}$.

The questions addressed in Chap. 13 revive several questions that were at the heart of the research in differential topology developed by Thom, Whitney, Mather and others, and to which Po brought important ideas. This chapter is also an occasion for the author to elaborate on some of the latter's observations from his work *Homotopie régulière et isotopie* (1974). Three valuable appendices supplement the chapter: The first one is on stable smooth maps to which the author adds new results. The second one is on the theory of stable PL maps, a theory which, as the author says, is still at a very primordial stage and which leads to fundamental questions inspired by the theory of smooth stable maps. The third one is on a theory which the author calls extended 2-multi-0-jet transversality which is based on a question raised by C. T. C. Wall in 1971 and which the author also addresses in a very thorough way.

Chapter 14, by Dan Burghelea, is titled *"Barcodes" for continuous maps and a brief introduction to Alternative Morse Theory*. Burghelea first explains the term "barcodes" in the title: In the so-called Alternative Morse Theory, one starts with a smooth manifold M and a continuous map $f : M \to \mathbb{R}$ and then refines the set of critical values of f into a collection of four types of intervals which, in topological persistence homology theory, are referred to as *barcodes*. The barcodes appear here as the support of two vector-space valued maps, the one defined on the Euclidean plane and the other on the "above the diagonal" half-plane of this Euclidean plane. After reviewing the theory of barcodes, Burghelea explains how to recover the Morse complex of a Morse function from them.

Chapter 15, by Francesco Russo, is titled *A note on the quasi simple filtration of finite groups*. In this chapter, the author examines the property called "quasi simple filtration" (QSF) for finitely presented groups, introduced by Brick and Mihalik in their paper *The QSF property for groups and spaces* (1995). The exposition comes with a glimpse into Po's work related to this notion, including his intuitions on the QSF property before it was officially introduced. In describing some steps of the evolution of the QSF property, Russo shows how this notion can be adapted to algebraic fundamental groups in the setting of profinite groups. Among the other important notions that appear in this chapter are the simple connectivity at infinity for a connected locally finite CW-complex X, which is extensively used by Po. This property says that for each compact subset C of X there is a compact set D of X such that each loop in $X \setminus D$ is homotopically trivial in $X \setminus C$. Most of all, this chapter is the occasion for surveying works by Po and some of his students and collaborators, in topology and the theory of discrete groups.

Chapter 16, by Alexey Gorelov, is titled *Lifting maps between graphs to embeddings*. In this chapter, the author studies, for a given, piecewise linear map $f : P \to Q$ between polyhedra, the existence of an embedding $\tilde{f} : P \to Q \times \mathbb{R}$ such that $f = \mathrm{pr}_Q \circ \tilde{f}$. The focus is on non-degenerate maps where, in the case of piecewise linear maps $f : P \to Q$, this means that the inverse image of each point is a finite set. Gorelov introduces combinatorial techniques and establishes necessary and sufficient conditions for the existence of such maps. After a study of the general case, he focuses on the case of maps between graphs. He makes connections between these problems and the question of approximability by embeddings. He also proves that in the setting of generic maps from a tree to a segment, a weaker condition becomes sufficient for the existence of a lifting.

Susumu Tanabé, in Chap. 17, titled *A note on the discriminant loci of a hypersurface of bi-degree* (m, n), presents a topological method for the study of the discriminant locus of an algebraic variety defined in a product of projective spaces. His approach is characterized by the use of a groupoid in the description of the monodromy. As a matter of fact, despite its importance, the fundamental group of the complement of the discriminantal locus of an algebraic variety draws relatively scarce attention by geometers. In Chap. 17, Tanabé proposes a new direction to remedy this unfavourable situation by means of a minute analysis of a concrete family of Calabi–Yau hypersurfaces embedded into a product of projective varieties. This kind of Calabi–Yau varieties appear in homological mirror symmetry and their moduli space has complex dimension larger than two. The monodromy transformation of cycles represents a homomorphism from the fundamental group consisting of loops that avoid the discriminantal variety. In contrast to the cases where discriminantal loci essentially consist of zero-dimensional points, in the bi-degree (m, n) setting, loops avoid a curve that may be reducible with several connected components. The difficulty to give a precise description of the fundamental group, thus of the monodromy of cycles of a bi-degree (m, n) Calabi–Yau hypersurface, produces an obstacle to obtain a clear picture of relations between two mirror symmetric models, namely the correspondence between monodromies of cycles and automorphisms of coherent sheaves of the mirror variety. The entire construction of this chapter is based on the observation that the discriminantal loci admit a covering that transforms them into a complex line arrangement with special symmetries. To describe all loops avoiding this line arrangement, the author slightly generalizes such topological notions as rotation of punctures and annular braids. He decomposes the monodromy action induced by a loop into a composition of groupoid actions. The idea of using the notion of groupoid in the study of the fundamental group of the discriminantal loci was promoted by Alexander Grothendieck—whom, on several occasions, Po tenderly calls "Shourik"— in his *Esquisse d'un programme*.

In Chap. 18, by Benjamin Audoux, Jean-Baptiste Meilhan and Akira Yasuhara, titled *Welded graphs, Wirtinger groups and punctured spheres*, the authors obtain a complete link-homotopy classification of knotted punctured spheres in Euclidean 4-space. They show that every knotted punctured sphere is link-homotopic to a ribbon one. In doing so, they develop a diagrammatic theory of welded graphs and they establish a connection with ribbon surface-links via an extension of the so-called Satoh Tube map from welded graphs to these surface-links. This map was introduced by S. Satoh in the paper *Virtual knot presentation of ribbon torus-knots* (2000). The authors give an interpretation of the theory of welded graphs as a theory of Wirtinger group presentations up to a natural set of transformations which they call self-virtualization. In the course of their arguments, they develop a theory of welded graphs that naturally encompasses all known moves on welded objects for which the Tube map is invariant. We recall that a Wirtinger group is a group that admis a finite presentation as $< x_1, \ldots, x_k \mid R_1, \ldots, R_S >$ where each relation R_i is of the form $\bar{x}_j x_i^w$ where x_i and x_j are two generators and w is an element of the free group on $x_1, \ldots x_k$. Such groups arise in the present chapter as fundamental groups of exteriors of surface-links obtained from a given welded graph by the extended Tube map. The authors also address the injectivity question for the Tube map, identifying a new family of local moves on welded links under which the (non extended) Tube map is invariant. Their construction is based on 4-dimensional Milnor-type invariants they introduced in a previous paper, *Milnor concordance invariant for knotted surfaces and beyond* (2021).

Chapter 19, by Friedrich Hegenbarth and Dušan D. Repovš, is titled *On geometric representation of* $\mathbb{L}$*-homology classes*. In this chapter, the authors address the question of giving a geometric meaning to algebraically defined objects like $\mathbb{L}$-homology classes in the setting of generalized n-manifolds with or without boundary, namely, n-dimensional Euclidean neighborhood retracts with some condition on the local homology at every point.

Chapter 20, by Christine Lescop, is titled *On elementary invariants of genus one knots and Seifert surfaces*. In this chapter, Lescop introduces new invariants of genus one knots in homology 3-spheres. One of these invariants consists of a combination of coefficients of Alexander polynomials of curves of a genus one Seifert surface of the knot. To prove invariance, Lescop uses the properties of an invariant of genus two 3-dimensional handlebodies called the Alexander form. Alexander forms generalize abelian Reidemeister torsion from link complements to more general 3-manifolds. They contain Reidemeister torsions and Alexander polynomials of link exteriors obtained by attaching 2-handles along the boundary of the manifold. Among the related results discussed in this chapter, Lescop exhibits the structural properties of the Alexander forms of genus two rational homology handlebodies. In addition to the result obtained, Chap. 20 contains an introduction to invariants of genus one knots, with complete definitions and background material. Furthermore, the author explores the properties of genus two rational homology handlebodies.

Chapter 21, by Dušan Repovš and Andrei Vesnin, is titled *On a family of hyperbolic Brunnian links and their volumes*. A link L with n components in the 3-sphere is said to be *Brunnian* if it is non-trivial yet every proper sublink of it is trivial. The simplest and best known example of a hyperbolic Brunnian link is the 3-component link known as "Borromean rings". By Thurston's work, the complement of this link admits a hyperbolic structure. In this chapter, Repovš and Vesnin introduce, for $n \geq 2$, an infinite family of Brunnian links $Br(k_1, \dots, k_n)$ having n components with positive integer parameters that generalize examples constructed by Debrunner in his 1964 paper *Links of Brunnian type*. The authors are interested in hyperbolic invariants of the 3-manifold $S^3 \setminus Br(k_1, \dots, k_n)$. They obtain upper bounds for the volumes of the complements of the links they introduce. Their approach is based on Thurston's techniques of Dehn fillings of cusped manifolds and on computations of volumes of ideal right-angled hyperbolic antiprisms.

Chapter 22, by Arkady Plotnitsky, is titled *"All the resources of pure mathematics": Mathematical physics and mathematics as physics*. In this chapter, the author investigates the reciprocal impact of modern mathematics on modern physics. By "modern" mathematics, Plotnitsky means the mathematics that emerged as an abstract science, separate from physics and from daily life occupations, around the year 1800, and that continues until our own time. By "modern" physics, he means the physics that emerged two centuries earlier, at the times of Galileo, which also continues to our own times, and which he calls a "mathematical experimental science". The author argues that in this reciprocal impact, that of modern mathematics on modern physics plays the major role. He develops the idea that modern physics is characterized by the invention of new theories comprised of mathematical concepts that are capable of representing or predicting the physical reality, and that, on the other hand, the invention of new mathematics has most fundamentally defined, and still defines, modern physics. Physics, in this description, becomes a mathematical-experimental science, with a stress on the mathematical component. Plotnitsky defends this idea with a detailed discussion based on arguments from twentieth-century physics, especially relativity and quantum theory. Beyond its physical dimension, the article is an affirmation of the role of mathematical thinking in human thought in general. The author discusses for instance the problem raised from the fact that Hermann Weyl, while he was studying the concept of continuum, was led to emphasize the divorce between modern mathematics and ordinary thinking, including the representation of nature, but at the same time, considered that physics must use the "abstract schemata" of modern mathematics. Plotnitsky also explains how he views quantum mechanics as "a radical transformation of theoretical physics defined by the birth of physics out of the spirit of mathematics." This is the case for the so-called Heisenberg method, where probability theory, infinite-dimensional matrix algebra and a generalization of Fourier series play central roles. Thorough discussions are conducted from a philosophical point of view, on topics like the role of probability theory in the study of quantum objects, or on the Galois group in renormalization theory, and on many others. A fascinating discussion on discreteness and continuity of space and its impact on various theories that were developed by various mathematicians, from Riemann to Grothendieck, can be read as a subtle watermark all throughout the chapter. In some sense, this chapter is an excursion in the realm of the ideas of several generations of mathematicians and physicists, starting with the Greeks (Presocratics, Plato, Aristotle) and up to Einstein, Dirac, Schrödinger, Heisenberg, Weyl and others among the modern, with Riemann in the background, and the latter's question on the "reality underlying space".

Chapter 23, by Victor Pambuccian, is titled *The case for absolute geometry*. The author presents this chapter as a plea to revive a neglected part of absolute geometry, that is, the geometry whose notions and propositions are valid in both the Euclidean and hyperbolic settings. Pambuccian leads us into a reflection on what geometry is, adding the question of whether the study of geometry has anything to do with a quest for understanding nature, or even whether it has any relation with anything real at all. He also leads us to a reflection on the material with which we work every day as geometers: set theory, continuity, the real numbers, postulates and the like. These observations eventually lead him to the question of why we do it. The discussion, in this chapter, combines geometry, logic and philosophy.

Chapter 24, by Stelios Negrepontis and Dimitrios Protopapas, titled *The Reconstruction of Theaetetus' Theory of Ratios of Magnitudes*, brings us to Plato, that is, to the heart of philosophy of mathematics, and maybe of all Western philosophy. The authors present a reconstruction of Theaetetus' theory of ratios of magnitudes (that is, ratios of lines, areas, etc., as opposed to only ratios of integers) based on a definition saying that two ratios of magnitudes are equal if they have equal anthyphairesis. In modern terms, this means that given four (commensurable or incommensurable) magnitudes a, b, c, d, we have the equality $a/b = c/d$ if and only if a/b and c/d have the same continued fraction expansion. As the authors recall, the existence of such a definition, predating Eudoxus' theory of proportion, is attested by Aristotle in the *Topics*. Negrepontis and Protopapas discuss attempts by Becker (1933), van der Waerden (1954) and Knorr (1975) to reconstruct Theaetetus' earlier theory of ratios of magnitudes from Eudoxus' arithmetical theory of ratios presented in Book V of Euclid's *Elements* and which, as is well known, is a precursor of Dedekind's modern theory of real numbers. A common unwelcome feature of these reconstructions has been the anachronistic need of Eudoxus' condition of Definition 4 in Book V of the *Elements*: given magnitudes a and b, there is a natural number n such that $nb > a$. In contrast, Negrepontis and Protopapas argue that Theaetetus' discovery of the quadratic incommensurabilities is not arithmetical but anthyphairetic. They explore the connection between Theaetetus' discovery and Plato's philosophical theory of knowledge of an Intelligible Being contained in his three dialogues *Theaetetus, Sophist*, and *Meno*, a theory based on division and recollection and especially on Theaetetus' quadratic incommensurabilities. At the same time, they provide an anthyphaeretic explanation of some obscure passages on philosophical knowledge in Plato's *Theaetetus*, emitting and supporting the thesis that Theaetetus' anthyphairetic definition of equality of ratios of magnitudes, especially ratios with finite or periodic anthyphairesis, like his whole approach to incommensurabilty, are the basis of Plato's theory of the anthyphairetic nature and structure of the Intelligible Being. Negrepontis and Protopapas emphasize the meaning and the fundamental role played by Logos, suggesting that this mathematical notion is at the foundation of Plato's whole philosophy. The ideas contained in this chapter constitute an important contribution to the history and philosophy of Greek mathematics, the elucidation of the nature of Theaetetus' theory of proportion of magnitudes and the role that this theory played in Plato's philosophy.

In the last part of the book, we have collected some memories and autobiographical articles of Po. There are six chapters: three of them are more on his personal life, and the other three are more specifically on his work in mathematics and physics.

Chapter 25 is a report written by Po in the year 2000, on his mathematical work until that date, a work which started in Bucharest in 1955. He mentions in particular his efforts, over several decades, focused on the proof of the 3-dimensional Poincaré conjecture. Along the way, we meet his mathematical friends and collaborators during all this period.

The next chapter, Chap. 26, is titled *Mathematical Adventures: A letter to Barry Mazur*. This letter is dated September 2016, and it is followed by a short addendum written in 2024. In this letter, Po writes: "This story will start about 1993, at a time when I thought that I had already finished my proof of the Poincaré Conjecture." Later, he tells his correspondent: "this text is supposed to be largely nontechnical, and I will rather stress the human and emotional aspects." All throughout the letter, we get a glimpse of Po's tension towards the proof the Poincaré 3-dimensional conjecture, the effort and energy that he has put into this program during several decades, and the important notions that came out from this attempt. He also tells us about the fate of the writings that came out of this effort, which constituted a sizable portion of his life. At the same time, we get a feeling of his effort towards the proof of the four-dimensional differentiable Schoenflies problem, and of his works in the fields of combinatorial group theory, differential topology, singularity theory and physics.

In discussing these important episodes of his personal history, Po at the same time talks about his friends, Dave Gabai who accompanied him for several years in his work on the Poincaré conjecture, and of course Barry Mazur, the long-standing friend and recipient of the letter. Po mentions the big breakthrough that Mazur made around 1958 when he proved that any smooth four-dimensional Schoenflies ball is homeomorphic to the standard ball by a homeomorphism which is a diffeomorphism except maybe at a boundary point, a work that greatly inspired Po who makes the following comments : "I was obsessed with the issue of smoothing that four-dimensional boundary point, and I went big at it with all kind of techniques, like Whitney's theory of differentiable functions on arbitrary closed sets. But nothing seemed to work." The whole chapter is centered around these ideas, in relation with Po's past and present results and projects. At the same time, we learn some facts about his life, his colleagues and friends, his students, his wife and his children.

In Chap. 27 titled *Adventures in physics*, Po recounts how, in the 1970s, he started working with physicists and, in particular, with Gérard Toulouse to whom is dedicated the chapter. Their first joint work was concerned with the dynamics of defects of ordered media in connection with the homotopy groups of the manifold of internal states. This led Po to further research works with several physicists on related topics which took place especially in the 1970s.

Chapter 28, titled *Dimitru Poénaru, my father*, is a panorama of Po's life in Romania, before he emigrates to the West.

Chapter 29, titled *Father Andrei*, contains some remembrances from Po's life in Romania, and his relation with a monk called Andrei. This chapter is the occasion for Po to talk about his relation with Christian mysticism, which leads him to what he calls the "creative power of the infinite." Rather than asking about the existence or inexistence of God, a question which Po thinks is irrelevant to the reality of our life, he addresses the question of existence restricted to our world, "universe or multiverse", including our mathematical world. This mathematical world is real and objective, says Po, and it is the place where we find big deep beauty and where transcendence and mystical contemplation become accessible.

In the more personal Chap. 30, titled *The magic summer*, Po talks about his wife's family, and her parental relation with Christopher Zeeman. He tells us how he met his wife through Zeeman, and how Barry Mazur and his wife, Milen Gretchen accompanied the Poénaru couple from the very beginning of this encounter. It also appears from this that his marriage with Milen is no stranger to his mathematical discoveries.

After Po's auto-biographical chapters, we have included a personal tribute written by two of his Italian friends, Corrado Tanasi and Franco Marchese.

There is nothing better for understanding Po's approach to mathematics and the impact of his work than listening to his mathematical friends talk about this work and its place in the history of mathematics. We hope that in this introduction, we have shown that this occurs in several chapters of the present book.

Chapter 2
Primes, Knots and Po

Barry Mazur

Abstract In the comparison between prime numbers and hyperbolic knots the present article focuses on the parallels that can be drawn, as well as the distinctions that can be made.

Keywords Hyperbolic knot · Profinite fundamental group · Frobenius conjugacy class · Étale cohomology

MSC 57 M 25, 57 M 27, 11 R 32

These are notes for the conference "Geometry, Topology and Group Theory" in honor of the 80th birthday of Valentin Poenaru held in Autrans July 1st–6th, 2012—slightly up-dated (2023).

Happy 90th birthday Po!

I presented some of this material in 2021 at the conference CHAT: Career, History and Thoughts [19].

B. Mazur (✉)
Harvard University, Cambridge, MA, USA
e-mail: mazur@g.harvard.edu

L. Funar, A. Papadopoulos (eds.), *Essays on Topology*,
https://doi.org/10.1007/978-3-031-81414-3_2

Po and I became instant friends when we met over half a century ago. In the intervening years our friendship has only deepened.

I viewed Po, especially during the time that the photo in front of you was taken, as someone ready—with his own intense energy and centeredness; as someone ready to create—from scratch—his own culture, his own goals, his own mountains to climb, as well as his own language.

Po once commented that to really engage in any aspect of history or science—that is to say, to understand a culture—you're faced with the contradiction of a certain catch22 in that you must manage, somehow, to have already achieved a *critical mass* of prior questions and reflections and knowledge about it, before anything about it comes into focus. Happily, Po is at home in many cultures, many languages, many histories, and he's blessed with this type of intellectual critical mass for so many subjects; physics and mathematics—of course—included.

We first met each other through correspondence—he was in Romania and I in the US—and we were both working on contractible four-manifolds that bound homology three-spheres, each of us having had constructed examples of *nonsimply connected* homology 3-spheres that occur as boundaries of contractible four-manifolds. (Po's article is: [24]).

Both Po's examples (of contractible 4-manifolds that bound non-simply connected manifolds) and mine [18] were obtained by sewing a thickened 2-disc onto the boundary of $D^3 \times S^1$ along a knot in the boundary that winds its way around the S^1 so as to homotopically kill the S^1. The result—given the particular knots we chose—is the construction of contractible 4-manifolds that with non-simply connected boundary. The kinds of 4-manifolds we construct are simply connected, of course, but—as Po mentioned to me in this conference—an (unpublished) result of Casson implies that they are not *geometrically simply connected* in the sense that it cannot be built as handlebodies having *no* handles of index 1.

The concept of a manifold being *simply connected but not geometrically simply connected* is one that is intimately related to Po's work. This turns out to be uniquely a *dimension four* phenomenon; as Po commented, "strange things happen in dimension four." Indeed, as we now know, for $n \neq 4$ all simply connected compact manifolds are geometrically simply connected. For $n \leq 2$ this is classical; for $n = 3$ it is proved by Perelman and for $n \geq 5$ it is true, thanks to Smale.

As for the related question of whether or not a homology $(n - 1)$-sphere actually bounds a smooth contractible n-manifold, again, and for similar reasons, the case $n = 4$ shows itself to be peculiar. For the answer is *yes* if $n \neq 4$; and the answer is *not always* when $n = 4$, thanks again to an invariant of Casson [6]. (See also [1]).

It is curious that even now, half a century after this issue was first broached, there are still open problems in this general domain, not the least of which is the still unresolved *smooth* four-dimensional Schoenflies Problem.

Both Po's examples (of homology 3-spheres that bound contractible manifolds) and mine [18] had the further feature that when you doubled them on their boundary—i.e., put two copies of them together by identifying their boundaries—you got a closed differentiable manifold diffeomorphic to the four-dimensional sphere, which means that there is a smooth involution of S^4 (switching the two copies of the doubled manifold) with non-simply-connected fixed point set, and therefore "exotic" in the sense that the involution is not equivalent to a linear involution. I remember that at the time—the late fifties of the past century—I was very much in awe of the magical construction of R. H. Bing who showed that the double of the closure of the *bad* component of the complement of the Alexander horned sphere in S^3 is again (topologically) S^3 giving, therefore, a thoroughly wild and untamable involution of S^3. I imagine that Po was similarly inspired by Bing.[1]

The surprising re-creation of S^4 by sewing together the boundaries of two manifolds continued as a theme of Po's work, as in the marvelous theorem that he proved with François Laudenbach in the early seventies [16].

But the great and abiding interest of Po centered on the mysteries of three-dimensional topology, even though—very often—his mode of approach was through four dimensions. The Poincaré Conjecture of course was one of his beloved focal points over decades. Now the way in which Po dealt with the prospect of Perelman's wonderful proof, and the way he engaged in appreciating the insight of Perelman, and at the same time adroitly reshaping his own research projects makes Po one of the truly fine models for our profession: fully at home in the world of ideas for their own sake; and devoted to—and deriving inspiration from—the beauty of that world, their depth, its power of explanation.

Knots and their exquisitely idiosyncratic properties, are the vital essence of three-dimensional topology, the DNA that governs the development, and evolution, of that

[1] One might also mention that a related issue in the theory of homology $n - 1$-spheres is the *double suspension conjecture* that asserts that the double suspension of a homology $n - 1$-sphere is homeomorphic to S^{n+1}. This was proved for all n by Cannon [6] building on earlier work of Edwards [8].

field. I know that Po has a special love and affinity for them, and I also think that they form a link to many other—seemingly far-flung—aspects of mathematics. For example, when I was trying to get a feel for number theory, I found that a certain analogy between the knot theory that I knew as a topologist and the phenomenology of prime numbers (that I was trying to become at home with) was exceedingly helpful, as a bridge. I've returned to it often as a learning device and it seems that it might allow two-way traffic, from knots to primes, and from primes to knots. (See my slide presentation *Thoughts about Primes and Knots* [19].) In celebration of Po for this conference, I will be discussing this analogy. For a beautiful introduction to this subject, see M. Morishita's treatise, *Knots and Primes*, [23]; also: [21, 22].

In this conference, after listening to a lecture by Michel Boileau, it occurred to me that it might make sense to be somewhat sharper than one traditionally is, when one frames the basic analogy. Perhaps we should be making the comparison between *prime numbers* and—more specifically—the class of *hyperbolic knots* (which, in contrast to the class of *all* knots have very few members, conjecturally, in each commensurability class.[2]) This choice also has the virtue of allowing us to make use of the hyperbolic volume of the complement of the knot, $vol(K)$, as a ready-made analogue to the logarithm of the norm of the prime.[3] The format of our comparison is then:

$$\begin{array}{rcl} \textbf{Prime Numbers } p & \leftrightarrow & \textbf{Hyperbolic Knots } K \\ \log p & \leftrightarrow & vol(K) \end{array}$$

It is interesting to focus on the parallels that can be drawn, as well as the distinctions that can be made.

2.1 One Knot and One Prime

2.1.1 A Single Knot K Embedded in the Three-Sphere S^3

The separate geometries of the two spaces involved are these: *the ambient three-sphere* S^3 is 2-connected and enjoys a 3-dimensional Poincaré duality with a canonical isomorphism $H^3(S^3; \mathbf{Z}) \simeq \mathbf{Z}$ while *the knot* K is diffeomorphic to S^1. For technical reasons I will always take K to be given with an orientation–i.e., with a canonical isomorphism $H^1(K; \mathbf{Z}) \simeq \mathbf{Z}$—so K is (canonically) a $K(\mathbf{Z}, 1)$-space. Now consider the knot embedded in S^3,

$$K \hookrightarrow S^3,$$

[2] Two knots K, K' are said to be **commensurate** if there are finite covers M, M' of their respective knot complements such that M is homeomorphic to M'.

[3] As Morishita commented on an early draft of these notes, one might also take the closely related *Gromov norm* of the knot complement.

and the knot complement

$$X = X_K := S^3 - K \hookrightarrow S^3.$$

Alexander duality establishes a **Z**-duality between $H^1(X; \mathbf{Z})$ and

$$\partial : H_2(S^3, K; \mathbf{Z}) \xrightarrow{\simeq} H_1(K; \mathbf{Z}) = \mathbf{Z},$$

giving us a canonical isomorphism:

$$H^1(X; \mathbf{Z}) = \mathbf{Z}$$

which tells us that all finite abelian covering spaces of S^3 branched at the knot, but unramified outside it, have *cyclic* groups of deck transformations, that these cyclic groups have canonical compatible generators, and that $X^{\mathrm{ab}} \to X$, the maximal abelian covering space of X, has group of deck transformations Γ canonically isomorphic to **Z**.

Or equivalently, setting

$$\Pi_K := \pi_1(X, x),$$

with suitable base point x—the *fundamental group of the knot*—we have

$$\Pi^{ab} := \Pi/[\Pi, \Pi] \simeq \mathbf{Z}.$$

Up to isotopy, the knot complement X_K may be viewed as compact manifold with torus boundary, $T_K = \partial X_K$, and within that torus—up to homotopy—there's a normal ('meridianal') loop that generates the infinite cyclic subgroup

$$N_K \subset T_K \subset X_K.$$

In anticipation of our comparison we might call

$$\mathcal{D}_K = \pi_1(T_K) = \mathbf{Z} \times \mathbf{Z}$$

the *decomposition group* of the knot, and

$$\mathcal{I}_K = \pi_1(N_K) = \mathbf{Z}$$

the *inertia subgroup*. The fundamental group of the knot then comes with maps

$$\textbf{(1)} \quad \mathcal{I}_K \hookrightarrow \mathcal{D}_K \longrightarrow \Pi_K.$$

A basic theorem gives us that $V = V_K := H_1(X^{\rm ab}; \mathbf{Q})$ is a *finite dimensional* $\mathbf{Q}$*-vector space*. The natural action of the canonical generator of the group of deck transformations $\Gamma \simeq \mathbf{Z}$ on $X^{\rm ab}$ induces an automorphism of V_K whose characteristic polynomial $P_K(T)$ is the *Alexander Polynomial of the knot* K.

There are multiple ways of approaching, and understanding, the information in $P_K(T)$ (e.g., through the combinatorial braid group theory around HOMFLY). Here is an attitude to the *zeroes* of the Alexander Polynomial that is natural enough: for any nonzero complex number z consider the homomorphism $\psi_z : \Pi_K \to \mathbf{C}^*$ that sends the generator of $\Pi^{ab} := \Pi/[\Pi, \Pi]$ to z. This defines a linear system (of complex vector spaces of dimension one) $V(z)$ over X. We have that $\dim_{\mathbf{C}} H_1(X, V(z))$ is equal to the order of vanishing of the Alexander polynomial $P_K(T)$ at $T = z$.

Since the analogue (in number theory) to the topological fundamental group is the *étale fundamental group*—which for a smooth complex variety is the profinite completion of the topological fundamental group—we might prepare for this, in anticipation of our analogy, by defining two knots K, K' to be **profinitely equivalent** if there is an isomorphism between the profinite completions of their basic group diagrams,

$$(\hat{\mathbf{1}})\quad \hat{\mathcal{I}}_K \hookrightarrow \hat{\mathcal{D}}_K \longrightarrow \hat{\Pi}_K.$$

and

$$(\hat{\mathbf{1}}')\quad \hat{\mathcal{I}}_{K'} \hookrightarrow \hat{\mathcal{D}}_{K'} \longrightarrow \hat{\Pi}_{K'};$$

and similarly for links.

This raises two questions:

1. Are profinitely equivalent knots, or links, isomorphic?[4]
2. Let us say, casually—not precisely—that a knot (or link) invariant has a "profinite definition" if it can be computed directly from the profinite completions $(\hat{\mathbf{1}})$. Which of the knot invariants have profinite definitions (and therefore carry over directly to the context of primes numbers) and which do not? For example (see 4.4 in [3]), if two knots are profinitely equivalent and have the property that

[4] As I learned in the conference from Norbert A'Campo and others, there were some—not then published—investigations of this. Editing these notes now over a decade later there is indeed a significant literature about it. First, profinitely trivial knots are trivial. This follows from the fact that knot groups are residually finite (See [11]) so the knot group of any knot K that has the property that its profinite completion is $\hat{\mathbb{Z}}$ is abelian—hence isomorphic to $\mathbb{Z}$; consequently K is the unknot. (See also subsection 4.4 in [3].) For more about profiniteness, see the Appendix (Sect. 2.3.4 below).

neither of their Alexander polynomials have a zero that is a root of unity, then their Alexander polynomials agree (up to sign).
3. For another take on profinite completion see [13].

2.1.2 A Single Prime Number p in the Integers Z

The algebra here is just given by the natural "reduction mod p" homomorphism

$$\mathbf{Z} \to \mathbf{Z}/p\mathbf{Z} = \mathbf{F}_p.$$

We will avoid the prime $p = 2$ since some minor differences would have to be acknowledged at various points otherwise; so *prime* will mean *odd prime* in the discussion below. We will be taking the standard viewpoint of modern algebraic geometry, and think of this surjective homomorphism as giving us an embedding of schemes,

$$\mathrm{Spec}(\mathbf{F}_p) \hookrightarrow \mathrm{Spec}(\mathbf{Z}),$$

and our analogy begins by thinking of $\mathcal{K} := \mathrm{Spec}(\mathbf{F}_p)$ as 'like' the knot K and $\mathcal{S} := \mathrm{Spec}(\mathbf{Z})$ as 'like" the three-dimensional sphere S^3. To understand this, we should examine, first, the separate geometries of the two schemes $\mathrm{Spec}(\mathbf{F}_p)$ and $\mathrm{Spec}(\mathbf{Z})$. The facts of life of the theory of finite fields tells us that for every positive integer n, up to isomorphism, there is a unique field of cardinality p^n, $\mathbf{F}_{p^n}$ given as a field extension $\mathbf{F}_{p^n}/\mathbf{F}_p$ which is Galois, cyclic, and degree n.

Moreover the (cyclic) Galois group of this field extension has a canonical generator: the Frobenius automorphism $x \mapsto x^p$. In a word

$$\mathrm{Spec}(\mathbf{F}_{p^n}) \to \mathbf{F}_p$$

is a cyclic (unramified!) Galois cover with Galois group canonically $\mathbf{Z}/n\mathbf{Z}$. An algebraic closure $\bar{\mathbf{F}}/\mathbf{F}_p$ is an appropriate union of these field extensions, and its Galois group is (canonically) isomorphic to $\hat{\mathbf{Z}}$, the profinite completion of $\mathbf{Z}$. From the étale homotopy perspective, $\mathrm{Spec}(\bar{\mathbf{F}})$ is contractible, and therefore $\mathcal{K} := \mathrm{Spec}(\mathbf{F}_p)$ is homotopically a $K(\hat{\mathbf{Z}}, 1)$-space.

The theory for $\mathcal{S} := \mathrm{Spec}(\mathbf{Z})$ requires some class field theory, as reformulated in the vocabulary of étale (and some other Grothendieckian) cohomology theories. Firstly, $\mathcal{S}$ is simply connected, in the sense that every connected finite cover of $\mathcal{S}$ is ramified. Moreover, $\mathcal{S}$ enjoys a three-dimensional "Poincaré-type" duality theorem for étale and flat cohomology with values in the multiplicative group $\mathbf{G}_m$ in the sense that

- $H^i(\mathcal{S}, \mathbf{G}_m)$ is (canonically) equal to $\{\pm 1\}, 0, 0, \mathbf{Q}/\mathbf{Z}$, and 0 for $i = 0, 1, 2, 3$, and > 3 respective;

- If F is a finite flat group scheme over $\mathcal{S}$ and $F^* := Hom(F, \mathbf{G}_m)$ its (Cartier) dual finite flat group scheme, then cup-product induces a perfect pairing of flat cohomology groups

$$H^i(\mathcal{S}, F) \otimes H^{3-i}(\mathcal{S}, F^*) \longrightarrow H^3(\mathcal{S}, \mathbf{G}_m) = \mathbf{Q}/\mathbf{Z}.$$

In a word, $\mathcal{S}$ is morally 2-connected and enjoys a 3-dimensional Poincaré duality "oriented" by the coefficient sheaf $\mathbf{G}_m$.

Now consider our prime p viewed as 'knot' $\mathcal{K}$ embedded in $\mathcal{S}$,

$$\mathcal{K} \hookrightarrow \mathcal{S},$$

and form the 'knot complement'

$$\mathcal{X} := \mathcal{S} - \mathcal{K} = \mathrm{Spec}(\mathbf{Z}[1/p] \hookrightarrow \mathcal{S}.$$

An argument very akin to Alexander duality (given the cohomological facts we have just recalled) establishes a canonical isomorphism

$$H_1(\mathcal{X}; \mathbf{Z}) \simeq \mathbf{Z}_p^*$$

(where $\mathbf{Z}_p^*$ is the group of units in the ring $\mathbf{Z}_p$ of p-adic integers. If $p > 2$ we can write

$$\mathbf{Z}_p^* \simeq F_p^* \times \Gamma$$

where Γ is the infinite cyclic pro-p-group of 1-units in $\mathbf{Z}_p$ and is generated, for example, by the 1-unit $1 + p$:

$$\Gamma = (1 + p)^{\mathbf{Z}_p}.$$

In particular, all finite abelian covering spaces of $\mathcal{S}$ branched at $\mathcal{K}$—i.e., finite abelian extensions of $\mathbf{Q}$ unramified except at the prime p (and ∞)—have Galois groups that are cyclic, and canonically isomorphic to the finite quotients $(\mathbf{Z}/p^m\mathbf{Z})^*$ of the topological group $\mathbf{Z}_p^*$. In anticipation of things to come, set:

$$\Lambda := \mathbf{Z}_p[[\mathbf{Z}_p^*]]$$

noting that this ring is isomorphic to a direct product of $p - 1$ copies of the power series ring in one variable $\mathbf{Z}_p[[T]]$, where if i is an integer modulo $p - 1$ the i-th factor of Λ is given by the surjective $\mathbf{Z}_p$-algebra homomorphism

$$\chi_i : \Lambda \longrightarrow \mathbf{Z}_p[[T]].$$

This is the unique $\mathbf{Z}_p$-algebra homomorphism that extends the continuous group homomorphism from $\mathbf{Z}_p^* =\simeq F_p^* \times (1+p)^{\mathbf{Z}_p} \subset \Lambda^*$ to $\mathbf{Z}_p[[T]]^*$ obtained by the stipulations that

- $x \in F_p^*$ be sent to

$$(x^i, 1) \in F_p^* \times \Gamma = \mathbf{Z}_p^* \subset \mathbf{Z}_p[[T]]$$

  and
- $(1+p) \in \Gamma$ be sent to $1+T \in \mathbf{Z}_p[[T]]$.

Set

$$\Pi_{\mathcal{K}} := \pi_1^{et}(\mathcal{X}, x),$$

with suitable base point x —the *étale fundamental group of our knot $\mathcal{K}$*—we have that in relatively standard parlance,

$$\Pi_{\mathcal{K}} = G_{\mathbf{Q},\{p,\infty\}},$$

i.e., is the quotient of Gal($\bar{\mathbf{Q}}/\mathbf{Q}$) that is the Galois group of the maximal extension of $\mathbf{Q}$ in an algebraic closure $\bar{\mathbf{Q}}$ that is unramified except at p and ∞.

As with topological knots the "fundamental group of the prime," comes with inertia and decomposition groups

$$\mathcal{I}_{\mathcal{K}} \hookrightarrow \mathcal{D}_{\mathcal{K}} \longrightarrow \Pi_{\mathcal{K}} = G_{\mathbf{Q},\{p,\infty\}}.$$

From our previous discussion,

$$\Pi_{\mathcal{K}}^{ab} := \Pi_{\mathcal{K}}/[\Pi_{\mathcal{K}}, \Pi_{\mathcal{K}}] \simeq \mathbf{Z}_p^*,$$

and if $\mathcal{X}^{\mathrm{ab}} \to \mathcal{X}$ is the maximal unramified abelian (connected) cover, then we can also say

$$\text{“Gal}(\mathcal{X}^{\mathrm{ab}}/\mathcal{X})\text{”} = \Pi_{\mathcal{K}}^{ab} = \mathbf{Z}_p^*.$$

A natural analogue to the finite dimensional $\mathbf{Q}$-vector space $V_K := H_1(X^{\mathrm{ab}}; \mathbf{Q})$ discussed above is the étale 1-st homology group, taken first, with p-adic *integral* coefficients,

$$M_{\mathcal{K}} := H_1^{et}(\mathcal{X}^{\mathrm{ab}}; \mathbf{Z}_p) = \lim_n H_1^{et}(\mathcal{X}^{\mathrm{ab}}; \mathbf{Z}/p^n\mathbf{Z}),$$

or—tensoring with $\mathbf{Q}_p$—we get the vector space

$$V_{\mathcal{K}} := H_1^{et}(\mathcal{X}^{\text{ab}}; \mathbf{Z}_p) \otimes_{\mathbf{Z}_p} \mathbf{Q}_p.$$

The module $M_{\mathcal{K}}$ is naturally a Λ-module, and $V_{\mathcal{K}}$ a $\Lambda \otimes_{\mathbf{Z}_p} \mathbf{Q}_p$-module. Tensoring $M_{\mathcal{K}}$ with the $p-1$ projection operators χ_i described above, we get for each i mod $p-1$ a $\mathbf{Z}_p[[T]]$-module that we'll call $M^i_{\mathcal{K}}$.

The behavior of these modules depends crucially on the parity of i. It is a marvelous theorem in Iwasawa theory that if i is 'odd' (which makes sense since our prime p is not 2) then our module $M^i_{\mathcal{K}}$ is a finitely generated $\mathbf{Z}_p$-module, and therefore $V^i_{\mathcal{K}} = M^i_{\mathcal{K}} \otimes_{\mathbf{Z}_p} \mathbf{Q}_p$ is a finite dimensional $\mathbf{Q}_p$-vector space. By definition–but subject to possibly different normalization—the **Iwasawa polynomial** for the pair (p, i) (i odd, modulo $p-1$) is the characteristic polynomial $g_p(i; T) \in \mathbf{Z}_p[T]$ of the operator T acting on this vector space $V_{\mathcal{K}}$. These polynomials $g_p(i; T)$ or, more precisely, their zeroes are crucial for much number theoretic phenomena. For example, if for a given p and all odd i mod $p-1$, they are all 1—i.e., have no zeroes—the prime p is what is called *regular* and Kummer's relatively easy procedure of proving Fermat's Last Theorem for exponent p can be made to work. In general, by what is known as the "main conjecture" (which is a theorem) the zeroes of $g_p(i; T)$ correspond in a one-one fashion, and in a natural way, to the zeroes of the Leopold-Kubota L-function $L_p(s, \omega^{1-i})$.

2.1.3 Brief Comments on Comparison and Differences

- If by **unknotted** one means that the fundamental group of the knot is abelian, every prime is 'knotted.'
- A serious distinction between knots and primes has to do with what is called *wild inertia* a phenomenon that exists, and is of crucial importance in number theory, but there's no corresponding complexity in our analogous situation in knot theory.
- There is a duality in the structure of the Alexander polynomial (it is invariant under inversion $t \mapsto t^{-1}$; hence if θ is a root, so is θ^{-1}). But there is nothing like that for Iwasawa polynomials.

  Given the G_m-orientation of $\mathcal{S}$, the corresponding duality for the Iwasawa polynomial would—if it existed—send the index i to $j := 1 - i$ and since j would then be even, $g_j(T)$ has not been defined. Of course, you could simply, by fiat, define $g_j(T)$ so that it exhibits the duality, but lacking (yet) any number theoretic motivation, that would be too formal a move to contemplate.

2.2 Two Knots and Two Primes

2.2.1 *A Pair of (Disjoint) Knots K, L Embedded in the Three-Sphere S^3*

Here we can consider the embeddings:

$$K \hookrightarrow X_L := S^3 - L$$

and

$$L \hookrightarrow X_K := S^3 - K$$

Choose arbitrary base points and consider the induced homomorphisms of π_1,

$$\pi_1(K) \longrightarrow \Pi_L.$$

In anticipation of the analogy to come, let

$$\{Frob_K\} \in \Pi_L$$

denote the conjugacy class of the image of the canonical generator of $\pi_1(K)$. This is indeed well-defined, independent of the choice of base points. Similarly we have

$$\{Frob_L\} \in \Pi_K.$$

To be sure, we can't yet compare these conjugacy classes, since they live in different groups. But passing to the abelian quotient groups of Π_K and Π_L, both are canonically isomorphic to $\mathbf{Z}$ we can indeed compare the images of $\{Frob_K\}$ and $\{Frob_L\}$ in

$$\Pi_K^{\mathrm{ab}} = \mathbf{Z} = \Pi_L^{\mathrm{ab}},$$

and those images are given— respectively—by the linking number of K in L and the linking number of L in K, these being *equal* with opposite sign. The proof of this equality is usually given by identifying these numbers with the cup product of the fundamental classes in $H^1(X_K)$ and $H^1(X_L)$ in $H^2(X_{K,L}) = \mathbf{Z}$ where $X_{K,L} := S^3 - \{K \cup L\}$.

2.2.2 A Pair of (Distinct) Primes p, q

In parallel with our previous subsection, let $\mathcal{K} := \text{Spec}(\mathbf{F}_p)$ and $\mathcal{L} := \text{Spec}(\mathbf{F}_q)$. Consider the embeddings:

$$\mathcal{K} \hookrightarrow X_{\mathcal{L}} := \text{Spec}(\mathbf{Z}[1/p])$$

and

$$\mathcal{L} \hookrightarrow X_{\mathcal{K}} := \text{Spec}(\mathbf{Z}[1/q])$$

Choose arbitrary base points and consider the induced homomorphisms of the étale fundamental groups

$$\pi_1^{et}(\mathcal{K}) \longrightarrow \Pi_{\mathcal{L}}.$$

Denote by

$$\{Frob_{\mathcal{K}}\} \in \Pi_{\mathcal{L}}$$

the conjugacy class of the image of the canonical generator of $\pi_1^{et}(\mathcal{K})$ which is independent of the choice of base points. Similarly we have

$$\{Frob_{\mathcal{L}}\} \in \Pi_{\mathcal{K}}.$$

Here again, we can't yet compare these conjugacy classes, since they live in different groups. Even passing to the abelian quotient groups of $\Pi_{\mathcal{K}}$ and $\Pi_{\mathcal{L}}$, which are canonically $\mathbf{Z}_p^*$ and $\mathbf{Z}_q^*$ respectively, and where the image of $\{Frob_{\mathcal{K}}\}$ is the element $p \in \mathbf{Z}_q^*$ and the image of $\{Frob_{\mathcal{L}}\}$ is the element $q \in \mathbf{Z}_p^*$, we simply have elements in different groups and so are not (yet) comparable. In a word, the linking "number" of p with q (in that order) is the element p in $\mathbf{Z}_q^*$, while the linking "number" of q with p (in that order) is the element q in $\mathbf{Z}_p^*$–no clear way to make any correspondence, yet. Nevertheless each of these groups $\mathbf{Z}_p^*$ and $\mathbf{Z}_q^*$ have unique subgroups of index two (consisting of "squares" of elements) and the famous comparison to be made here is to ask whether p being a square in $\mathbf{Z}_q^*$ (or equivalently, mod q) has anything to do with q being a square in $\mathbf{Z}_p^*$ (or equivalently, mod p). Indeed it does, as given by the classical *quadratic reciprocity theorem.* Namely, p is a square mod q if q is a square mod p , except in the case where both p and q are both congruent to -1 mod 4, in which case p is a square mod q if and only if q is not a square mod p. (One of the many proofs of this follows the lines of the proof I hinted at above of skew-symmetry of linking number.[5])

[5] Po raised the question (in this conference) of whether Gauss himself—who, after all, had introduced the integral formula for the linking number—might have seen some analogy between that concept and the structure surrounding the quadratic reciprocity theorem.

2.3 Borromean Primes and "Chebotarev Arrangements"

2.3.1 Borromean Primes

The *Borromean Ring* is that well-known link of three disjoint 'unknots' that has the property that if you ignore any of the three unknots the other two are unlinked, yet the three taken all together are somehow linked. John Milnor defined a class of invariants that serve as obstructions to linkage of the above sort, these being secondary (or higher) linking numbers that can be defined—in analogy with standard linking numbers—as secondary (or higher) cohomology operations related to the vanishing of cup-products, the Massey triple product being the first example of these. The clean general structure corresponds to what is called an A_∞-algebra structure on chain complexes, such as was discussed by François Laudenbach in this conference (he obtained it from Morse functions on the knot manifold with Dirichlet and Neumann conditions on the boundary).

One can establish a striking analogy to this, with prime numbers, obtaining secondary (or higher) versions of the quadratic reciprocity theorem, as is done in the work of Morishita, Redei, and others (cf. [23]). Specifically, given three distinct primes p, q, r all congruent to 1 mod 4 and each a quadratic residue of any of the others, there is a mod 2 invariant which gauges how triply-entangled the three primes are; moreover, as is the case with old-fashioned linking numbers, the natural definition of this invariant is given somewhat asymmetrically in terms of the roles played by p, q and r; yet, the theorem is that the invariant itself is independent of permutation of these.

Here is the description of this invariant,

$$link(p, q, r) \in \{\pm 1\},$$

as given by Redei (cf. section 8 of [23]). Under the assumptions of the previous paragraph there is a nontrivial integral zero (x, y, z) of the quadratic form

$$X^2 - qY^2 - rZ^2$$

and moreover, one can assume that $g.c.d((x, y, z) = 1$, y is even, and $x - y \equiv 1$ mod 4. Now form $\alpha := x + \sqrt{q}y$ and consider the (non-Galois) extension of $\mathbf{Q}$,

$$K := \mathbf{Q}(\sqrt{q}, \sqrt{\alpha}).$$

Then

$$link(p, q, r) = 1 \in \{\pm 1\}$$

if and only if the prime p splits completely in K. Otherwise, $link(p, q, r) = -1$.

An example of linked Borromean triples of primes is given (by D. Vogan—cf. loc.cit.) by

$$(p, q, r) = (13, 61, 937).$$

2.3.2 'Chebotarev Arrangements'

Here is a thought-experiment that I once mused about a long time ago, but will try to sharpen a bit here: I think of it *not at all* as a problem to be resolved[6] but rather as just a somewhat casual way of appreciating *visually* how vastly *entangled* the collection of all primes are.

Imagine choosing one hyperbolic knot in every commensurable equivalence class of hyperbolic knots, and then arranging these knots (up to equivalence) in S^3 so that they form a mutually disjoint ensemble:

$$\mathcal{K} := \sqcup_i K_i \ \subset S^3$$

where we have ordered them compatibly with their hyperbolic volume. By an **admissible Galois cover of** S^3 **(relative to** $\mathcal{K}$**)** let us mean a finite cover $f : M^3 \to S^3$, Galois and ramified over at worst a finite subcollection of knots $\Sigma = K^{(1)} \sqcup K^{(2)} \sqcup \cdots \sqcup K^{(n)}$ of $\mathcal{C}$ in the natural sense; i.e., such that f restricted to $Y := M^3 - f^{-1}\Sigma$ the pullback of $S^3 - \Sigma$ is a locally trivial covering space of $X := S^3 - \Sigma$ with free action of a finite group G on M^3 (the "Galois group" of the cover) such that $Y/G = X$.

A knot in $\mathcal{K}$ which is branched in $M^3 \to S^3$ we say is *ramified in the cover* and if it isn't we say it is *unramified in the cover*. Any unramified knot K in an admissible cover $M^3 \to S^3$ gives rise to a conjugacy class of elements in $G = \mathrm{Gal}(M^3/S^3)$ by the analogue of the Frobenius construction alluded to earlier. Thus, for all but finitely many knots in $\mathcal{K}$ we have a well-defined conjugacy class

$$\{Frob_K(M^3/S^3)\} \subset G.$$

Let us say that the collection $\mathcal{K}$ is a **Chebotarev Arrangement** if the following statistical rule holds for every admissible cover M^3/S^3 and every conjugacy class $\{\mathcal{C}\} \subset G = \mathrm{Gal}(M^3/S^3)$

$$\lim_{k\to\infty} \frac{1}{k}\#\big[K_i, i \leq k \mid \{Frob_{K_i}(M^3/S^3)\} = \{\mathcal{C}\}\big] \ = \ \frac{|\{\mathcal{C}\}|}{|G|},$$

[6] Although the easiest is just to formulate it as a 'question'.

where the limit here is compiled by ordering the knots compatibly with their hyperbolic volume.

In effect, one is asking that—with these conventions–the Frobenius conjugacy classes are uniformly distributed in fundamental groups.

The only reason for my formulating this notion is that there is an important theorem in number theory (The Chebotarev Density Theorem) that makes the closely analogous statement for primes.

2.3.3 Chebotarev Dynamical Systems Containing All Knot Types

A knot or link L in S^3 is called **fibred** if the complement fibers over S^1, the fibers being spanning surfaces. Fixing a vector field that is transverse to all of those spanning surfaces, consider the collection $\mathcal{K}$ of periodic flowlines coming from such a vector field, the elements of $\mathcal{K}$ being viewed as knots, this structure being also described by the monodromy automorphism the vector field induces on any one of the spanning surfaces. Such a fibered knot or link is said to be *universally fibered* ([10] Definition 1.1) if *every* vector field transverse to the fibration possesses all knot and link types.

In response to a much earlier question posed Birman and Williams [2] many examples of fibred knots and links L were constructed in [10] where the corresponding collection $\mathcal{K}$ contains all *knot types*–i.e. are "universally fibred"; a sufficient condition for this to be the case is given in the Main Theorem in [10] by a neat—general condition—on the braid that determines the fibred knot (or link).

How bewilderingly complex, and yet somehow delicately organized, these entangled collections must be, each knot winding about infinitely many others according to various proportions if they are in a Chebotarev arrangement! As I said, I brought this up only to have a visualizable counterpart to the type of entanglement represented by the facts of life for prime numbers; and–of course—to offer a birthday greeting for my friend Po!

As Curt McMullen explained to me, beautiful examples can be gotten by considering the collection of knots given by closed geodesics in the spherical tangent bundles of hyperbolic surfaces. See [20] for:

Theorem 2.1 *Let X be a closed surface of constant negative curvature, and let $K_1, K_2, \cdots \subset M = T_1(X)$ be the closed orbits of the geodesic flow in the spherical tangent bundle of X, ordered by length. Then the ensemble of knots $\mathcal{K} := \{K_i\}$ obeys the Chebotarev law. For related results about Bernoulli flows, see [14, 15]. (and the same result holds for the closed orbits of any topologically mixing pseudo-Anosov flow on a closed 3-manifold M).*

Following [26] if M^3 is a 3-manifold and $\mathcal{K} = \sqcup_i K_i$ a collection of disjoint knots (or links) in M^3, say that this set-up ($\mathcal{K} \subset M^3$) is **generic** if for $L \subset \mathcal{K}$ any finite sublink in $\mathcal{K}$ every conjugacy class of elements in $\pi_1(M^3 - L)$ is in the image of $\pi_1(K_i)$ for some $K_i \in \mathcal{K}$ disjoint from L. The set-up is called **stably generic** if the same condition holds if for all such links L and for all finite covers $N^3 \xrightarrow{h} M^3$ branched at L, taking the collection of disjoint knots (or links) in N^3 to be $h^{-1}\mathcal{K}$.

It is shown in [26] that any system of links $(M^3, \mathcal{K})$ that satisfy the Chebotarev law are stably generic; hence McMullen's examples in [20] do; and moreover, the Chebotarev sequences $K_1, K_2, K_3, \ldots$ that McMullen discusses contain (thanks to results in [10]) every possible knot and link type in S^3 as a subconfiguration.

2.3.4 The "Chebotarev" Property in More General Circumstances

Also, one needn't stick to manifolds: let $f : X \to X$ be any continuous self-mapping of a connected CW-complex X, fixing a basepoint $x_o \in X$. Let $\Pi(X, x_o; f)$ be the quotient group of $\pi_1(X, x_o)$ that *equalizes* f and the identity map; i.e., it is the quotient by the normal subgroup generated by $f(\alpha)\alpha^{-1}$ for all $\alpha \in \pi_1(X, x_o)$.

Put $I := [0, 1]$, and form $X \times I$. Attaching $X \times \{0\}$ to $X \times \{1\}$ by the mapping $(x, 0) \mapsto (f(x), 1)$, construct the space

$$Y := X \times I/\{(x, 0) \sim (f(x), 1); x \in X\}.$$

Identifying $S^1 = \partial D^2$ with the closed loop $C := x_o \times I/\{(x_o, 0) \sim (x_o, 1)\} \subset Y$ (via, say, $\iota : e^{2\pi it} \mapsto (x_o, t) \in Y$ for $t \in I$) form the CW complex Z by adjoining the disc D^2 with the attaching map

$$\partial D^2 \xrightarrow{\iota} C$$

to build the space:

$$Z := Y \cup_\iota D^2.$$

Fix the base point $z_o :=$ the image of $(x_o, 0)$ in Z.

We have (by van Kampen's theorem):

$$\textbf{(1)} \quad \pi_1(Z, z_o) \simeq \Pi(X, x_o; f).$$

Any finite f-orbit in X,

$$\mathcal{O} = \{x, fx, f^2x, \ldots, f^{n-1}x\} \subset X$$

with period n (denote this: "$Per(\mathcal{O}) = n$"—so n is the smallest positive number such that $f^n x = x$) gives rise to a loop, i.e., an oriented closed curve, $\gamma_{\mathcal{O}} \subset Z$, defined to be the image of

$$x \times I \sqcup fx \times I \sqcup f^2 x \times I \sqcup \cdots \sqcup f^{n-1} x \times I \subset X \times I$$

under the natural mapping

$$X \times I \longrightarrow Z.$$

We give $\gamma_{\mathcal{O}}$ the natural orientation (i.e., 'induced' from the natural orientation of I).

Definition 2.1 By the **Frobenius conjugacy class** attached to the finite f-orbit $\mathcal{O} \subset X$ we mean the conjugacy class in $\Pi(X, x_o; f)$ determined by the loop $\gamma_{\mathcal{O}} \subset Z$. Specifically, the image of the canonical generator of the fundamental group of the oriented loop $\gamma_{\mathcal{O}}$ into the fundamental group of Z (the map on fundamental groups being given by choosing compatible base points) determines a well-defined conjugacy class of elements in $\pi_1(Z, z_o)$ which we identify with a conjugacy class of elements in $\Pi(X, x_o; f)$ via the isomorphism **(1)** above. Denote this Frobenius conjugacy class

$$\{Frob_{\mathcal{O}}\} \subset \Pi(X, x_o; f).$$

Thus we have a canonical mapping

$$\begin{aligned} Frob : &\{\text{Finite } f - \text{orbits in } X\} \\ &- - - - - - - \longrightarrow \{\text{Conjugacy classes in } \Pi(X, x_o; f)\}. \end{aligned} \tag{2.1}$$

Now consider the following condition regarding the dynamics of f.

Per The mapping $f : X \to X$ has only finitely many periodic points of any given order, but infinitely many in all.

When $f : X \to X$ is a mapping that satisfies **Per** we order the collection of finite f-orbits in a manner compatible with their periods, and we can begin to ask distribution questions regarding the dynamics of f that are somewhat analogous to the question answered by the Chebotarev Theorem in classical number theory.

The conditions most naturally related to the facts of life in number theory that set the stage for the classical Chebotarev Theorem are the following:

Definition 2.2 Let $\rho : \Pi(X, x_o; f) \longrightarrow G$ be a surjective homomorphism onto a finite group G. Say that $f : X \to X$ is ρ-**Chebotarev** if for any conjugacy class $C \subset G$ we have:

$$\lim_{n\to\infty} \frac{|\{\mathcal{O}\,;\, Per(\mathcal{O}) \le n;\, \rho(Frob_{\mathcal{O}}) = C\}|}{|\{\mathcal{O}\,;\, Per(\mathcal{O}) \le n\}} = \frac{|C|}{|G|}.$$

Remark 2.1 More generally, let G be a compact Lie group, and μ its Haar measure (normalized to have total measure 1). If $\rho : \Pi(X, x_o; f) \longrightarrow G$ is a homomorphism with dense image, say that $f : X \to X$ is ρ-**Chebotarev** if for every measurable subset $\mathcal{N} \subset G$ closed under conjugation by elements of G, we have:

$$\lim_{n\to\infty} \frac{|\{\mathcal{O}\ ;\ Per(\mathcal{O}) \leq n;\ \rho(Frob_{\mathcal{O}}) \subset \mathcal{N}\}|}{|\{\mathcal{O}\ ;\ Per(\mathcal{O}) \leq n\}} \quad = \quad \mu(\mathcal{N}).$$

Acknowledgments With partial support from the NSF Grant DMS 2152149.
Thanks to Louis Funar and Curt McMullen for very helpful comments about the early draft.

Appendix: Profinite Rigidity

I want to thank Louis Funar for sending me the material in this section—namely, references for the following works—most of them having appeared after the conference.

A finitely generated group G is said to be **profinitely rigid** if its isomorphy type is determined by its profinite completion; that is, letting $\hat{H}$ denote the profinite completion of a group H, G is **profinitely rigid** if for G' any finitely generated group, we have that $\hat{G}$ is isomorphic to $\hat{G}'$ (if and) only if $G \simeq G'$.

For a general discussion, see [25] and [7]. For "positive results" regarding profinite rigidly in the context of hyperbolic geometry see [3, 5, 17, 27]. For results about positive rigidity of knot groups—more specifically torus knots and the figure-eight knot—see [3, 4]. For negative results, see [9, 12].

References

1. S. Akbulut, J. McCarthy, *Casson's Invariant for Oriented Homology 3-Spheres* (Princeton University Press, Princeton, 1990)
2. J. Birman, R.F. Williams, Knotted periodic orbits in dynamical systems—I: Lorenz's equation. Topology **22**(1), 47–82 (1983). ISSN 0040–9383. https://doi.org/10.1016/0040-9383(83)90045-9
3. M. Boileau, S. Friedl, The profinite completion of 3-manifold groups, fiberedness and the Thurston norm, in *What's next? The mathematical legacy of William P. Thurston*, ed. by D.P. Thurston. Annals of Mathematics Studies, vol. 205 (Princeton University Press, Princeton, 2020), pp. 21–44
4. M. Bridson, A.W. Reid, H. Wilton, Profinite rigidity and surface bundles over the circle. Bull. Lond. Math. Soc. **49**(5), 831–841 (2017)
5. M. Bridson, D. McReynolds, A. Reid, R. Spitler, Absolute profinite rigidity and hyperbolic geometry. Ann. Math. **192**(3), 679–719 (2020)
6. J.W. Cannon, Shrinking cell-like decompositions of manifolds. Codimension three. Ann. Math. **110**, 83–112 (1979)

7. N. Dunfield, W. Thurston, Finite covers of random 3-manifolds. Invent. Math. **166**(3), 457–521 (2006)
8. R.D. Edwards, Suspensions of homology spheres. http://arXiv:math/0610573v1
9. L. Funar, Torus bundles not distinguished by TQFT invariants. With an appendix by Louis Funar and Andrei Rapinchuk. Geom. Topol. **17**(4), 2289–2344 (2013)
10. R. Ghrist, E. Kin, Flowlines transverse to knot and link fibrations. Pac. J. Math. **217**, 61–86 (2004)
11. J. Hempel, Residual finiteness for 3-manifolds, in *Combinatorial Group Theory and Topology (Alta, Utah, 1984)*. Annals of Mathematics Studies, vol. 111 (Princeton University Press, Princeton, 1987), pp. 379–396
12. J. Hempel, Some 3-manifold groups with the same finite quotients (2014). https://arxiv.org/abs/1409.3509
13. F. Hidekazu, Galois action on knots I: action of the absolute Galois group. Quant. Topol. **8**, 295–360 (2012)
14. H. Hu, Y. Pesin, A,Talitskaya, Every compact manifold carries a hyperbolic Bernoulli flow, in *Modern Dynamical Systems And Applications* (Cambridge Universit Press, Cambridge, 2004), pp. 347–358
15. A. Katok, Bernoulli diffeomorphisms on surfaces. Ann. Math. Second Ser. **110**(3), 529–547 (1979)
16. F. Laudenbach, V. Poénaru, A note on 4-dimensional handlebodies. Bull. Soc. Math. France **100**, 337–344 (1972)
17. Y. Liu, Finite-volume hyperbolic 3-manifolds are almost determined by their finite quotient groups. Invent. Math. **231**(2), 741–804 (2023)
18. B. Mazur, A note on some contractible 4-manifolds. Ann. Math. **73**, 221–228 (1961)
19. B. Mazur, Thoughts about primes and knots (2021). https://www.math.ucla.edu/~cyhsu/seminars/UCLANTS/slides/Mazur.pdf
20. C. McMullen, Knots which behave like the prime numbers. Compos. Math. **149**, 1235–1244 (2013)
21. M. Morishita, Milnor invariants and Massey products for prime numbers. Compos. Math. **140**, 69–83 (2004). https://www.cambridge.org/core/journals/compositio-mathematica/article/milnor-invariants-and-massey-products-for-prime-numbers/83068A12C788910D3D54B9ADEB3905C
22. M. Morishita, Analogies between knots and primes, 3-manifolds and number rings (2009). https://arxiv.org/abs/0904.3399.pdf
23. M. Morishita, *Knots and Primes* (Springer, Berlin, 2011)
24. V. Poénaru, Les décompositions de l'hypercube en produit topologique. Bull. Soc. Math. France **88**, 113–129 (1960)
25. A.W. Reid, Profinite rigidity (in English), in *Proceedings of the International Congress of Mathematicians, ICM 2018, Rio de Janeiro, Brazil, August 1–9, 2018*, ed. by B. Sirakov et al. Invited Lectures, Volume II (World Scientific, Hackensack, 2018), pp. 1193–1216. Rio de Janeiro: Sociedade Brasileira de Matemática (SBM)
26. J. Ueki, Chebotarev links are stably generic. Bull. Lond. Math. Soc. **53**, 82–91 (2021). https://doi.org/10.1112/blms.12400
27. H. Wilton, P. Zalesski, Distinguishing geometries using finite quotients. Geom. Topol. **21**, 345–384 (2017). https://arxiv.org/abs/1411.5212v2

Chapter 3
A Short History of Open Manifolds Related to Complex Analytic Coordinates

Dennis Sullivan

Abstract In this note we give the major steps that show that an open connected oriented four manifold has local holomorphic projective coordinates compatible with a holomorphic foliation obtained by a holomorphic local isomorphism to the complex projective plane minus a point foliated by complex lines through that point. The work of Valentin Poénaru is put in context.

Keywords Immersion · 4-manifold · Complex structure

MSC 57 R 15, 57 K 40, 57 M 50

1. Hirsch and Smale immersions in positive codimension for all manifolds, circa 1960.
2. The equidimensional case for immersions of open manifolds: Valentin Poénaru, early sixties Harvard notes.
3. The submersion case for open manifolds based on Poénaru's proof scheme: Tony Phillips, PhD thesis Princeton University, 1965.
4. Maps transversal to a foliation for open manifolds, independently by Tony Phillips at Berkeley and Misha Gromov in Leningrad, 1967.
5. Haefliger general coordinate structures for open manifolds and an obstruction theory based on the Gromov–Phillips transversality theory, 1969.
6. Landweber's almost complex open $2k$-manifolds on a neighborhood of their k-skeletons have complex coordinates, using Haefliger theory, 1970.

D. Sullivan (✉)
Mathematics Department SUNY, Stony Brook, New York, USA
e-mail: dennis@math.stonybrook.edu

L. Funar, A. Papadopoulos (eds.), *Essays on Topology*,
https://doi.org/10.1007/978-3-031-81414-3_3

7. Gromov wrote and Adachi explained using the complex cell decomposition of the complex Grassmannian that the Landweber result is true also on a neighborhood of their $(k+1)$-skeletons, using Landweber, circa 1980.
Corollary *Any almost complex open four-manifold has holomorphic coordinates.*

8. Any oriented open four-manifold is spin^c, Teichner and Vogt, 1995. This was motivated by the Seiberg–Witten theory for closed manifolds (1995) where the spin^c result was due to Hirzebruch and Hopf (late fifties).

9. An easy obstruction theory argument shows that the tangent bundle of any open connected oriented four-manifold can be induced by the spin^c structure map to $\mathbb{CP}^2$ minus a point, from the tangent bundle of $\mathbb{CP}^2$, see my paper, *All Open connected oriented 4D manifolds are Complex Analytic Surfaces* (2022), to appear in PAMQ.
By Poénaru (**2**) this map is homotopic to a map which is a local diffeomorphism.
This proves, using only (**2**) and (**8**), the following:
Theorem Any open connected oriented four manifold has local holomorphic projective coordinates compatible with a holomorphic foliation obtained by a holomorphic local isomorphism to the complex projective plane minus a point foliated by complex lines through that point.

10. The geometric picture suggested by the theorem in a real version for parallelizable three-manifolds immersing to Euclidean three-space needs to be analyzed pictorially analogously to the beautiful theory of Sandy Blank for punctured two-spheres immersing to the plane, circa 1970. This program was Poénaru's road to conquer the Poincaré conjecture in dimension three.

This path is still open in dimension three and if carried out would motivate a similar study in dimension four where the smooth Poincaré conjecture is still open.

11. By the 1970's, open connected manifolds with complex tangent bundles were known to be complex manifolds with these tangent bundles only in dimensions 2, 4 and 6. Dimensions 8 and 10 were just now achieved, 50 years later, in the PhD thesis of Filip Samuelsen (2025) by vigorous use of the Haefliger pseudogroup theory of the 1970's. Dimensions 12 and higher are still unknown.

12. The Bott vanishing theorem (1970's) inspired the conjecture that the fibre of the Haefliger total space over the infinite Grassmannian of complex n-planes, known to be n-connected, is actually $2n$-connected. The Samuelsen thesis also proves the conditional result that open connected $2n$ manifolds with complex tangent bundles are actually complex manifolds for all n if the groups conjectured to be zero are just known to be finite. This, using the surprising coherence between the group cohomology of $U(n)$ made discrete and the cohomology (both with finite coefficients) of the complex Grassmannian (Friedlander, Milnor, Quillen).

Chapter 4
Universal Spin Teichmüller Theory, I: The Action of P(SL(2,$\mathbb{Z}$)) on $\mathcal{T}$ess$^+$

Robert Penner

Abstract Earlier work took as universal mapping class group the collection PPSL$(2, \mathbb{Z})$ of all piecewise PSL$(2, \mathbb{Z})$ homeomorphisms of the unit circle $S^1 = \partial\mathbb{D}$ with finitely many breakpoints among the rational points in S^1. The spin mapping class group P(SL(2,$\mathbb{Z}$)) introduced here consists of all piecewise-constant maps $S^1 \to \mathrm{SL}(2, \mathbb{Z})$ which projectivize to an element of PPSL$(2, \mathbb{Z})$. We also introduce a spin universal Teichmüller space $\mathcal{T}ess^+$ covering the earlier universal Teichmüller space $\mathcal{T}ess$ of tesselations of $\mathbb{D}$ with fiber the space of $\mathbb{Z}/2$ connections on the graph dual to the tesselation in $\mathbb{D}$. There is a natural action P(SL$(2, \mathbb{Z})$) $\circlearrowright$ $\mathcal{T}ess^+$ which is universal for finite-type hyperbolic surfaces with spin structure in the same sense that PPSL$(2, \mathbb{Z})$ $\circlearrowright$ $\mathcal{T}ess$ is universal for finite-type hyperbolic surfaces. Three explicit elements of P(SL(2,$\mathbb{Z}$)) are defined combinatorially via their actions on $\mathcal{T}ess^+$, and the main new result here is that they generate P(SL(2,$\mathbb{Z}$)). Background, including material on hyperbolic and spin structures on finite-type surfaces, is sketched down to first principles in order to motivate the new constructions and to provide an overall survey. A companion chapter to this one gives a finite presentation of the universal spin mapping class group P(SL$(2, \mathbb{Z})$) introduced here.

Keywords Classical and universal Teichmüller space · Riemann moduli space · Mapping class group · Spin structure · Thompson group T

MSC Codes 30F60 32G15 32C11 37E30

R. Penner (✉)
Institut des Hautes Études Scientifiques, Le Bois Marie, France
e-mail: rpenner@ihes.fr

L. Funar, A. Papadopoulos (eds.), *Essays on Topology*,
https://doi.org/10.1007/978-3-031-81414-3_4

Introduction

I am happy and honored to contribute here to the 90th birthday Festschrift for Valentin Poénaru, that is, for Po. We made our acquaintance only relatively recently with my residence in Bures-sur-Yvette and quickly became fast friends. We have enjoyed wide-ranging discussions across math and physics, in both scientific and sociological detail, as well as our fair share of delectations of fine wines and dinners including Milen and their lovely family. Po and Milen are a magical couple of transcendent love, their lives comfortably merging Po's science with Milen's graphic art, even as this confluence has been passed on to the next generation of their children.

Po is one of the main founders of the Orsay topology group, and thanks to his network of friends and relations, he is the one in that group who was the most open to the world mathematical community, both East and West. He has always worked on the most difficult problems [29], including the Poincaré Conjecture in dimension three and the smooth Schoenflies Problem in dimension four, on which his studies continue today, including the related massive project [28, 30, 31] in geometric group theory.

Along with his lifelong best friend Barry Mazur, independently Po [27] and Barry [19] gave the headline first examples of pseudo 4-cells which are not topological 4-cells, but whose product with the unit interval are topological 5-cells. (A pseudo cell is a contractible compact combinatorial manifold with boundary.) Among other fruitful collaborations including his celebrated Immersion Theorem [11] with André Haefliger, Po's work with Boone and Haken [3] provides a fundamental tool for decision problems in topology,

Meanwhile, Po the man is the gentlest, kindest and warmest person I know, a beacon of calm and clear reason to whom I often turn for guidance both mathematical and otherwise. Here to Po in gratitude, admiration and friendship in this celebratory volume, I offer the universal action $\mathrm{P}(\mathrm{SL}(2,\mathbb{Z})) \circlearrowright \mathcal{T}ess^+$ of the spin mapping class group on its spin Teichmüller space as a healtfelt 90th birthday present. In order to formulate this, I first take two paragraphs to introduce the main new characters and recall their antecedents:

Universal Spaces First define a *tesselation* τ of the Poincaré disk $\mathbb{D}$ to be a locally finite collection of geodesics decomposing $\mathbb{D}$ into ideal triangles. We shall typically choose a distinguished oriented edge or simply *doe* in the tesselation τ, and define the auxiliary space

$$\mathcal{T}ess' = \{\text{tesselations of } \mathbb{D} \text{ with doe}\},$$

which we shall prove later is homeomorphic to the space of orientation-preserving homeomorphisms of the circle $S^1 = \partial\mathbb{D}$. The Möbius group $\mathrm{PSL}(2,\mathbb{R})$ acts on $\mathcal{T}ess'$ with Fréchet-manifold quotient

$$\mathcal{T}ess = \mathcal{T}ess'/\mathrm{PSL}(2,\mathbb{R})$$

the *universal Teichmüller space* of [23]. The spin version is introduced here as the total space of the bundle

$$\mathcal{T}ess^+ \to \mathcal{T}ess$$

of all equivalence classes of finitely supported functions $\tau \to \{0, 1\}$, where the equivalence relation is generated by changing all three values on the boundary of a fixed triangle complementary to τ in $\mathbb{D}$.

Universal Groups Recall that PSL(2,$\mathbb{R}$) denotes the group of fractional linear transformations of the upper half plane, and its double cover SL(2,$\mathbb{R}$) the area-preserving linear mappings of the plane $\mathbb{R}^2$. SL(2,$\mathbb{Z}$) $<$ SL(2,$\mathbb{R}$) is the subgroup that preserves the integral lattice in $\mathbb{R}^2$, and PSL(2,$\mathbb{Z}$) is the projection of SL(2,$\mathbb{Z}$) to PSL(2,$\mathbb{R}$). Riemann surfaces (with known elementary exceptions) are quotients of the upper half plane by torsion free discrete subgroups of PSL(2,$\mathbb{R}$), and lifts of these subgroups to SL(2,$\mathbb{R}$) correspond precisely to spin structures on these Riemann surfaces. In the universal setting, the mapping class group is the collection PPSL(2, $\mathbb{Z}$) of all piecewise PSL(2, $\mathbb{Z}$) homeomorphisms of the unit circle S^1 with finitely many breakpoints among the rational points in S^1. Elements of PPSL(2, $\mathbb{Z}$) are automatically $\mathcal{C}^1$- but never $\mathcal{C}^2$-smooth at minimal breakpoints. Moreover, there is a triple incarnation of this group also as the Richard Thompson group T [5] and the combinatorial Ptolemy group acting on tesselations with doe by flips from [23]. The new spin mapping class group P(SL(2,$\mathbb{Z}$)) consists of all piecewise-constant maps $S^1 \to$ SL(2, $\mathbb{Z}$) which projectivize to an element of PPSL(2, $\mathbb{Z}$). This group is defined combinatorially by its action on $\mathcal{T}ess^+$ and then identified with the spin mapping class group in the proof of the next result.

Here then is the birthday gift for Po:

Main Theorem *The group* P(SL(2, $\mathbb{Z}$)) *generated by the three transformations* α, β, t *illustrated in Fig. 4.1 acting on* $\mathcal{T}ess^+$ *is precisely the group of piecewise-constant maps* $S^1 \to$ SL(2, $\mathbb{Z}$) *which projectivize to homeomorphisms in* PPSL(2, $\mathbb{Z}$).

My goal here is to explain these new constructions with motivations going back to first principles in some detail, which I hope might be useful whether as survey or invitation. The entire story comes together diagrammatically in Fig. 4.1 illustrating the action of the generators of the universal spin mapping class group on its Teichmüller space, a kind of emblem that I hope Milen might even appreciate as design.

A companion chapter [25] to this one derives a finite presentation of P(SL(2,$\mathbb{Z}$)) from that of PPSL(2,$\mathbb{Z}$) $\approx T$. In fact, [25] contains the first complete derivation in the literature of the latter finite presentation as well, though an equivalent

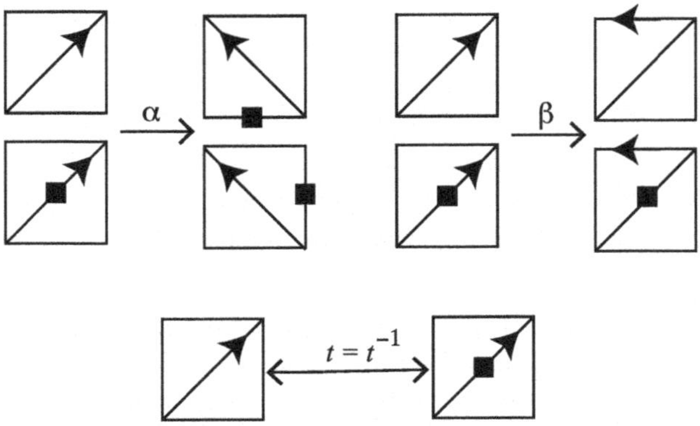

Fig. 4.1 The three generators α, β, t of P(SL(2,$\mathbb{Z}$)) acting on a marked tesselation with distinguished oriented edge in the two cases that the distinguished edge may or may not have a non-zero label in $\mathbb{Z}/2$ as indicated with a box icon. The distinguished oriented edge is indicated with an arrow

presentation was given in [17] seemingly based on [23] and on unpublished notes of Thompson. By construction, the diagram

$$\begin{array}{ccc} \mathrm{P(SL}(2,\mathbb{Z})) \times \mathcal{T}ess^+ & \to & \mathcal{T}ess^+ \\ \downarrow & & \downarrow \\ \mathrm{PPSL}(2,\mathbb{Z}) \times \mathcal{T}ess & \to & \mathcal{T}ess \end{array}$$

commutes, where the vertical maps are the natural forgetful maps.

As I first learned as a graduate student from the Orsay *Travaux de Thurston sur les Surfaces* seminars [8] co-organized by Po, the classical Teichmüller theory originated as a topic in complex analysis and evolved under William Thurston's masterful infusion of hyperbolic geometry. This has led to a combinatorial approach useful more recently for studying the Riemann moduli space and Teichmüller space $T(F)$ of an orientable surface $F = F_g^s$ of genus $g \geq 0$ with $s \geq 1$ punctures, where $2 - 2g - s < 0$.

The key construction, as recalled later, canonically associates a decomposition $\Delta(\Gamma)$ of F into ideal polygons with a hyperbolic structure Γ on F suitably decorated with one horocycle about each of its punctures. Furthermore, *solving the equation* $\Delta(\Gamma) = \Delta$ *in* Γ *for fixed* Δ yields an ideal cell decomposition of the decorated Teichmüller space itself, which is a difficult theorem in [22]. Generically, an ideal triangulation of F is the label for a top-dimensional cell in the decorated Teichmüller space of F. In effect in the universal setting, an ideal triangulation of the surface is replaced by a tesselation of its universal cover.

For a *universal Teichmüller space*, we demand [23] a Fréchet manifold $\mathcal{T}$ supporting the action $\mathcal{P} \circlearrowright \mathcal{T}$ of a *universal mapping class group* $\mathcal{P}$ so that:

- there are injective homomorphisms of the classical mapping class groups $MC(F)$ into explicit completions $\widehat{\mathcal{P}}_F \circlearrowright \mathcal{T}$ of $\mathcal{P}$;

- there are embeddings $T(F) \subset \mathcal{T}$ of the classical Teichmüller spaces which are equivariant for the $MC(F)$-action;
- the Weil–Petersson Kähler geometry on each $T(F)$ is induced by pulling back a universal geometry on $\mathcal{T}$.

Our universal Teichmüller space satisfying these conditions and more is given by the quotient

$$\mathcal{T}ess = \mathcal{T}ess'/\mathrm{PSL}(2,\mathbb{R}).$$

Suitably decorating a tesselation with horocycles at its ideal points, a construction analogous to the classical case for finite-type surfaces (namely, a convex hull construction in Minkowski space, cf. the appendix) provides an ideal polygonal decomposition of $\mathbb{D}$, and the space of decorated tesselations itself again admits a corresponding decomposition by *solving the equation.*

Also from [23], there are global so-called *lambda length coordinates* on $\mathcal{T}ess$ providing the Fréchet structure. In fact, the log lambda length deformations also from [23] lead to an infinite-dimensional Lie algebra of piecewise-constant $\mathfrak{sl}_2$ vector fields on S^1 with breakpoints among the rationals, a kind of completion of the loop algebra of $\mathfrak{sl}_2$, introduced in [18] and more recently studied in [10] as part of a larger program with implications to and from the current chapter. This Lie algebra denoted $\mathfrak{psl}_2$ in [18] (for piecewise-constant $\mathfrak{sl}_2$) was renamed $\mathrm{pp}\mathfrak{sl}_2$ in [10] (to emphasize its relation with PPSL(2, ℤ)) and is now more precisely seen to be *exactly the Lie algebra of* P(SL(2,ℤ)). Lambda lengths and the associated Lie algebra are not further discussed in the current chapter.

The novel contributions of this chapter involve including spin structure in the universal setting. But it is not so much novel as it is the natural universal [23] extension from [26] in the classical case. The latter is reformulated in [13], as we shall explain, with related aspects of GL(1|1)-graph connections also studied in [4]. An overview of the several equivalent formulations of spin structures on finite-type surfaces is given in the appendix. Considerations of universal spin structure are prefatory to any discussion of super universal Teichmüller theory.

The most elegant and immediately useful characterization of spin structures in our setting is due to Natanzon [21] for a hyperbolic structure on F specified by a (conjugacy class of projective) uniformizing representation $\pi_1(F) \to \mathrm{PSL}(2,\mathbb{R})$ of the fundamental group: a *spin structure on* F is a lift $\pi_1(F) \to \mathrm{SL}(2,\mathbb{R})$ of the uniformizing representation from $\mathrm{PSL}(2,\mathbb{R})$ to $\mathrm{SL}(2,\mathbb{R})$.

The natural spin generalization of PPSL(2, ℤ) based on Natanzon's description is our universal spin mapping class group P(SL(2,ℤ)) of all piecewise-constant maps $S^1 \to \mathrm{SL}(2,\mathbb{Z})$ with no conditions at the breakpoints except that the underlying quotient map $S^1 \to \mathrm{PSL}(2,\mathbb{Z})$ lies in PPSL(2, ℤ). Composition in the group P(SL(2,ℤ)) is given by taking common refinements of the piecewise structures and ordinary composition in SL(2, ℤ) on the resulting pieces, just as in PPSL(2, ℤ).

To define the spin universal space, we extend a different characterization of spin structure from [26] given by equivalence classes of orientations $\omega : \tau \to \tilde{\tau}$ on the edges in $\tau \in \mathcal{T}ess$, where $\tilde{\tau}$ denotes the set of oriented edges of τ. The equivalence relation is generated by a move called *(Kastelyn) reflection* on an ideal triangle complementary to τ, which reverses the orientations of all three frontier edges of a fixed complementary triangle.

In fact, the doe e itself also extends to another orientation $\omega_e : \tau \to \tilde{\tau}$ induced by the flow in S^1 from initial to terminal point on its right and the reverse orientation on its left. (Any specific orientation will suffice to specify the "standard" spin structure determined by the doe.) It follows that on a tesselation τ with doe $e \in \tilde{\tau}$, we might specify the orientation ω determining a universal spin structure by instead marking with a box icon each edge $f \in \tau$ where $\omega(f) \neq \omega_e(f)$, called a *marking*. In this notation taken modulo two on each edge, a reflection on a complementary ideal triangle simply adds a mark to each frontier edge of the triangle on which the reflection is performed. Equivalently as in [4, 13] and the Abstract, a spin structure is a $\mathbb{Z}/2$-graph connection on the graph dual to an ideal triangulation of a punctured surface in the classical case, and also now dual to a tesselation with doe in the universal case. In this chapter, we shall primarily stick to the formalism of equivalence classes of $\mathbb{Z}/2$-markings on edges of a tesselation.

Section 4.1 gives an overview of Teichmüller theory, both classical and universal, necessary for a further discussion of our results, with more complete details but few proofs in an appendix. (Let us apologize for pedagogically inevitable minor expositional repetition in slightly different contexts in these two surveys.) Section 4.2 discusses universal spin structure and the proof of the Main Theorem. The reference [24] subsumes the classical [22] and universal [23] decorated Teichmüller theories but does not treat spin structures from [26].

4.1 Overview of Pt $\approx$ PPSL(2, $\mathbb{Z}$) $\circlearrowleft$ $\mathcal{T}ess \approx$ Homeo$_+(S^1)$/ PSL(2, $\mathbb{R}$)

Fix an oriented surface $F = F_g^s$ as in the Introduction and choose a point in its Teichmüller space $T(F)$, i.e., specify some complete finite-area metric on F of constant Gauss curvature -1 modulo push-forward by diffeomorphisms of F isotopic to the identity. The universal setting is intended to provide an infinite-dimensional space $\mathcal{T}$, together with embeddings of each $T(F) \subset \mathcal{T}$ which may depend upon choices, so that both the geometry and mapping class quotient topology of the classical spaces $T(F)$ pull-back corresponding universal structures on $\mathcal{T}$.

It turns out that the further assignment of one real parameter to each puncture of F, which may be interpreted as the hyperbolic length of a specified horocycle about the puncture, is enough to determine a collection in F of pairwise disjointly embedded arcs with endpoints among the punctures and decomposing F into ideal polygons, as illustrated on the top in Fig. 4.2. (This extra decoration is also sufficient to define the global affine coordinates on the decorated bundles given by lambda

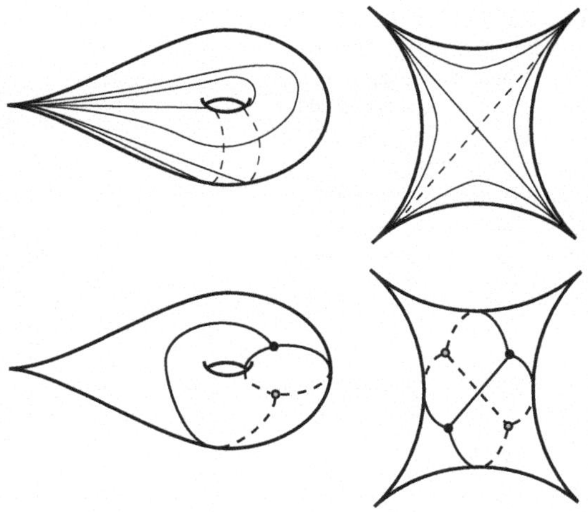

Fig. 4.2 Examples of ideal triangulations in F_1^1 and F_0^4 above and their dual fatgraphs below

lengths.) The existence of such a canonical decomposition in Theorem 4.2 from [22] is the starting point for all the combinatorics.

The Poincaré dual of this cell decomposition in F is a graph G embedded as its deformation retract, as illustrated on the bottom of Fig. 4.2, and it has the extra structure of a cyclic ordering on the half-edges about each vertex coming from an orientation on F, a so-called *fatgraph* $G \subset F$. Generically in Teichmüller space and as in the figure, the decomposition Δ is an *ideal triangulation* with only complementary triangles, and in this case G is a trivalent fatgraph.

In effect, the passage to the universal case entails lifting such an ideal triangulation of F to its universal cover $\mathbb{D}$, namely the open unit disk in the complex plane $\mathbb{C}$ with the Poincaré metric. In the generic case, the ideal triangulation Δ of F lifts to a tesselation τ of $\mathbb{D}$. Of course, such a lift to $\mathbb{D}$ is invariant under the action of the cofinite Fuchsian group uniformizing F. The universal case amounts to keeping the tesselation but dropping the Fuchsian group.

In order to have one natural example in mind and as discussed in detail in the appendix, the *Farey tesselation* τ_* is illustrated in Fig. 4.3 with its *distinguished oriented edge* or *doe* $e_* \in \tilde{\tau}_*$, where $\tilde{\tau}$ denotes the set of orientations on edges in τ. Define

$$\mathcal{T}ess' = \{\text{tesselations with doe of } \mathbb{D}\}.$$

The doe is not intrinsic and its specification as extra data is critical to the universal setting since tesselations with doe are rigid in the following sense: Let $\tau^0 \subset S^1 = \partial\mathbb{D}$ denote the collection of ideal points of edges in τ. Given two tesselations τ_0 and τ with doe, there is a unique $f : \tau_0^0 \to \tau^0$ so that $x, y \in S^1$ are ideal vertices of an edge in τ_0 if and only if $f(x), f(y)$ are likewise for τ. This is easy to see inductively, one triangle at a time starting with the triangle to the

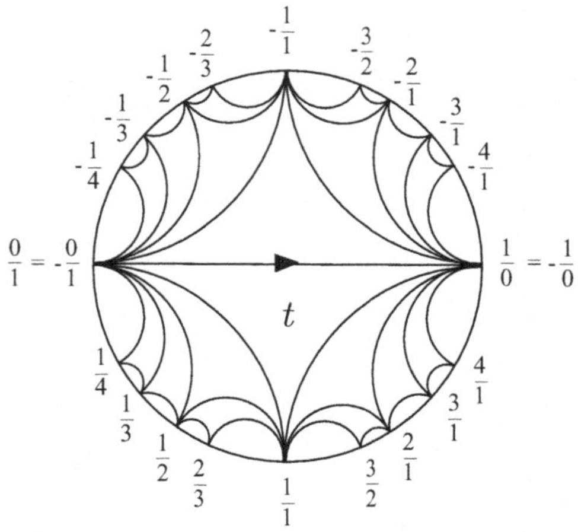

Fig. 4.3 The Farey tesselation τ_* of the Poincaré disk $\mathbb{D}$ with its triangle t to the right of its distinguished oriented edge. The extended rationals $\mathbb{Q} \cup \{\infty\}$ are naturally identified with the ideal vertices of τ_*, as illustrated in low generation

right of the doe. As explained in the appendix, this map f moreover interpolates an orientation-preserving homeomorphism also denoted $f : S^1 \to S^1$.

Taking the Farey tesselation $\tau_0 = \tau_* \in \mathcal{T}ess'$ with doe as a chosen base point in this construction gives the *characteristic map*

$$f_{e \in \tilde{\tau}} \in \text{Homeo}_+(S^1) = \{\text{orientation}-\text{preserving homeomorphisms of } S^1\}$$

canonically associated with $e \in \tilde{\tau}$, and in fact, this association of characteristic map $f_{e \in \tilde{\tau}}$ to a tesselation τ with doe e induces an isomorphism

$$\mathcal{T}ess' \approx \text{Homeo}_+(S^1).$$

Our model of *universal Teichmüller space* is the quotient

$$\mathcal{T}ess = \mathcal{T}ess'/\text{PSL}(2, \mathbb{R}) \approx \text{Homeo}_+(S^1)/\text{PSL}(2, \mathbb{R}),$$

and its *universal mapping class group* is generated by a combinatorial move, called a *flip* defined as follows for any edge $e \in \tau \in \mathcal{T}ess$: remove e from τ so as to produce a complementary quadrilateral and then replace the one diagonal e with the other diagonal f of this quadrilateral, as indicated in Fig. 4.4, along with the dual move on fatgraphs. Using the characteristic map, edges and hence flips can be indexed by elements τ_*, so the collection of all finite sequences[1] of such elements forms a group, the *Ptolemy group* Pt.

[1] The completion $\widehat{\mathcal{P}}_F$ of PPSL(2,$\mathbb{Z}$) in the wish-list for a universal theory in the Introduction arises from finite sequences of simultaneous flips along all the edges in an orbit of the uniformizing Fuchsian group, i.e., the lift to the universal cover of a flip on one edge in the underlying surface.

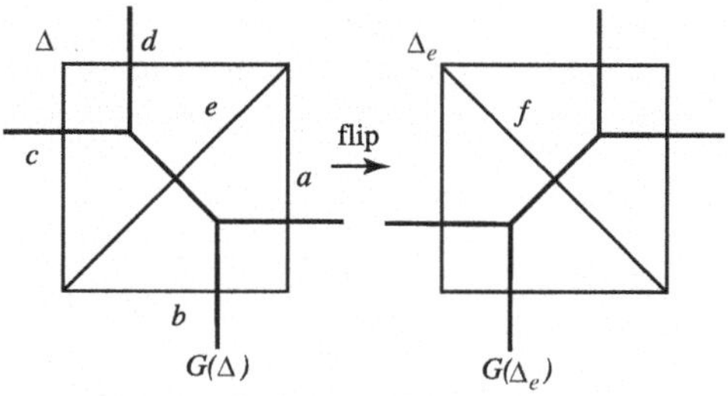

Fig. 4.4 The flip on an edge e in the ideal triangulation Δ produces another ideal triangulation Δ_e, and equivalently the flip in its dual trivalent fatgraph $G(\Delta)$ produces $G(\Delta_e)$

In fact via the characteristic map, the Ptolemy group Pt can be identified with the subgroup $\mathrm{PPSL}(2,\mathbb{Z})$ of $\mathrm{Homeo}_+(S^1)$ consisting of all piecewise-constant $\mathrm{PSL}(2,\mathbb{Z})$ homeomorphisms of S^1 with only finitely many pieces whose endpoints lie in τ_*^0, and this group turns out to furthermore be isomorphic to Richard Thompson's group T, all of which is detailed in the appendix. Thus, our universal mapping class group

$$\text{Thompson group } T \approx \mathrm{PPSL}(2,\mathbb{Z}) \approx \text{Ptolemy group Pt}$$

acts on the universal Teichmüller space $\mathcal{T}ess$ by flips.

4.2 Proof of Main Theorem

One key explanation for the gift/theorem/emblem in the Introduction is that the *enhanced flip on an oriented edge in a tesselation* engendered by the move α in Fig. 4.1 is precisely the dual of the flip transformation on fatgraphs with spin structure discovered and illustrated in Theorem 4.6 from [13, 26].

We have already defined the universal spin mapping class group

$$\mathrm{P}(\mathrm{SL}(2,\mathbb{Z})) = \left\{ \begin{array}{l} \text{piecewise-constant } \phi : S^1 \to \mathrm{SL}(2,\mathbb{Z}) \\ \text{with rational breakpoints so that the} \\ \text{projectivization of } \phi \text{ lies in } \mathrm{PPSL}(2,\mathbb{Z}) \end{array} \right\}$$

and universal spin Teichmüller space $\mathcal{T}ess^+ = \mathcal{T}ess'^+/\mathrm{PSL}(2,\mathbb{R})$, where

$$\mathcal{T}ess'^+ = \{(e \in \tilde{\tau}, \mu) \in \mathcal{T}ess' \times \{0,1\}^\tau : \mu \text{ has finite support}\}/\sim$$

with the equivalence relation $\sim$ on markings μ given by finite compositions of *(Kastelyn) reflections* which change marking on all three frontier edges of any fixed complementary triangle. Just as in [23] for flips, infinite sequences of reflections can also be allowed provided the marking of each edge eventually stabilizes.

For $A = \left(\begin{smallmatrix} a & b \\ c & d \end{smallmatrix}\right) \in \mathrm{SL}(2, \mathbb{R})$, define the sign

$$\mathrm{sgn}(A) = \begin{cases} \mathrm{sgn}(a+d); & a+d \neq 0, \\ \mathrm{sgn}(c); & \text{else}, \end{cases}$$

where $\mathrm{sgn}(x)$ denotes the usual sign of $0 \neq x \in \mathbb{R}$. In particular, $\mathrm{trace}(A) = a+d = 0 = c$ is impossible since $ad - bc = 1$. It follows that a spin hyperbolic structure on a finite-type surface may be thought of as a pair comprised of a projective uniformizing representation

$$\pi_1(F) \to \mathrm{PSL}(2, \mathbb{R})$$

as usual together with an induced function

$$\mathrm{sgn} : \pi_1(F) \to \{\pm 1\}$$

determining which sign to assign to the representing projective matrix, the one of positive trace (for sgn=1) or of negative trace (for sgn=-1) generically, with a similar definition using the sign of the (2,1)-entry of the representing matrix in the traceless case.

Analogously, given $\phi : S^1 \to \mathrm{SL}(2, \mathbb{Z})$ in P(SL(2,$\mathbb{Z}$)), there is by definition the induced projectivization $\bar{\phi} : S^1 \to \mathrm{PSL}(2, \mathbb{Z})$ in PPSL(2, $\mathbb{Z}$). Using the sign function in the previous paragraph, ϕ itself is uniquely determined by $\bar{\phi}$ together with a function $\sigma : \Pi(\bar{\phi}) \to \{\pm 1\}$ establishing the signs, where $\Pi(\bar{\phi}) = \Pi(\phi)$ is the set of pieces of $\bar{\phi}$ or ϕ.

To define σ, we first define an auxiliary function $\sigma : \tau - \{doe\} \to \{\pm 1\}$ as follows. As depicted in Fig. 4.5 in the presence of a doe, another edge $e \in \tau$ determines a point $p_e \in U_e \subset S^1$, where U_e is the component of $S^1 - \partial e$ on the other side from the doe. A horocycle h_e centered at p_e meets a countable collection of edges of τ, some finite number of which have non-zero marking, and we define $\hat{\sigma}(e) = +1$ if this number is even, and otherwise $\hat{\sigma}(e) = -1$, completing the definition of $\hat{\sigma}$. As illustrated in Fig. 4.6, this function $\hat{\sigma} : \tau \to \{\pm 1\}$ is well-defined on reflection equivalence classes. Let us emphasize that this determination according to parity of the number of marked edges meeting a horocycle derives from the classical case of finite-type surfaces. It is not simply invented here for our separate purposes.

Now suppose that $e \in \tilde{\tau}$, where τ agrees with the Farey tesselation τ_* outside of a finite ideal polygon P which moreover contains all edges with non-zero markings as well as the doe. An induction over innermost disks proves that there is a marking equivalent to the given one with all non-zero marked edges lying in the frontier of P, so we may henceforth assume that the marking is of this type.

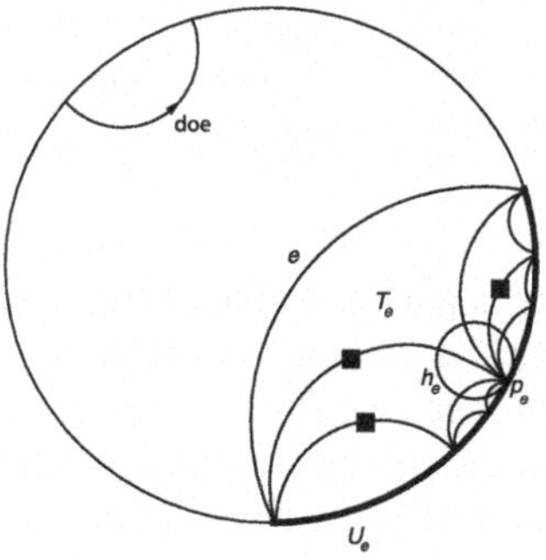

Fig. 4.5 In the presence of the doe, any other edge $e \in \tau$ independently of orientations determines a complementary triangle T_e with its ideal vertex $p_e = \partial T_e - \partial e \in S^1$ in a component U_e of $S^1 - \partial e$ in boldface. Also illustrated is a horocycle h_e centered at p_e. A finite number of nearby edges may have non-zero markings, as indicated with the box icon

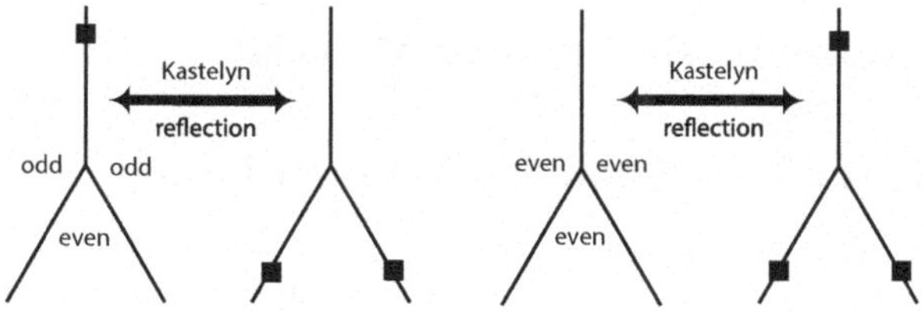

Fig. 4.6 Reflection preserves parity modulo two of the number of marked edges on each nearby horocycle as depicted on the level of dual fatgraphs

The ideal vertices of P contain the breakpoints of $\Pi(\phi)$, and it therefore suffices to define σ on the pieces so determined, and we set $\sigma(I) = \hat{\sigma}(e)$ if the circular interval I comprising a piece in the piecewise structure of ϕ agrees with some U_e. This defines the mapping

$$\langle \alpha, \beta, t \rangle \to \text{P(SL(2,}\mathbb{Z}\text{))}$$

from the group generated by α, β, t to the group of piecewise-constant SL(2,$\mathbb{Z}$)-valued maps which descend to PPSL(2,$\mathbb{Z}$).

It remains to prove that $\langle \alpha, \beta, t \rangle \to$ P(SL(2,$\mathbb{Z}$)) is bijective. We begin with the proof of surjectivity, and to this end claim that there is a finite composition of reflections on triangles in P achieving any specified marking on its frontier, which we take to be an ideal n-gon. To see this, consider the fatgraph G dual to P, which has $V_e = n$ univalent external vertices and $V_i = n - 2$ trivalent internal ones, with n edges incident on the former and $n - 3$ edges with endpoints among the latter, for a total of $E = 2n - 3$, so of course $V_e + V_i - E = 1$ according to Euler characteristic, whence $V_e = 1 + E - V_i$.

Rather than Kastelyn reflect on a complementary ideal triangle in P, let us dually imagine reflecting on an interior vertex of G by changing orientations on each incident edge, and consider the collection of all functions from the set of all edges of G to $\mathbb{Z}/2$, a set with cardinality 2^E. Interior vertex reflections act on this set of functions in the natural way, and there are evidently 2^{V_i} possible such compositions. The simultaneous reflection at all interior vertices of G acts trivially on this set of functions, so only 2^{V_i-1} such compositions may act non-trivially. Insofar as $2^E/2^{V_i-1} = 2^{1+E-V_i}$ and there are $1 + E - V_i$ external edges, the claim follows, proving surjectivity.

As for injectivity of $\langle\alpha, \beta, t\rangle \to$ P(SL(2,$\mathbb{Z}$)), consider an element of the kernel, namely a word w in α, β, t which in particular descends to the identity in PPSL(2,$\mathbb{Z}$) upon setting $t = 1 \in$ P(SL(2,$\mathbb{Z}$)). In other words in the language of the companion chapter [25], w is a t-insertion in a relation in PPSL(2,$\mathbb{Z}$)). Since it is in the kernel, w moreover preserves some marking, and hence any marking since markings, just as spin structures, provide a torsor. This is precisely the characterization of a relation in P(SL(2,$\mathbb{Z}$)) based on orbifold covers of quotients that is exploited in [25] to derive the finite presentation of P(SL(2,$\mathbb{Z}$)). An element in the kernel of $\langle\alpha, \beta, t\rangle \to$ P(SL(2,$\mathbb{Z}$)) is thus a relation in the domain, proving injectivity. □

Appendix: Background Material

Farey Tesselation and Modular Group

Let $\mathbb{Z} \subseteq \mathbb{Q} \subseteq \mathbb{R} \subseteq \mathbb{C}$ denote the integers, rational, real and complex numbers, respectively, and let $i = \sqrt{-1}$. Let ∞ indicate a *point at infinity* and the superscript plus sign denote the one-point completion by ∞, e.g., $\mathbb{Q}^+ = \mathbb{Q} \cup \{\infty\}$. The upper half-plane $\mathcal{U} = \{x + iy \in \mathbb{C} : y > 0\}$ endowed with the metric $ds^2 = \frac{dx^2+dy^2}{y^2}$ provides a standard model for hyperbolic geometry. The *Cayley transform* $C : s \mapsto \frac{s-i}{s+i}$ induces an isomorphism of pairs $(\mathcal{U}, \mathbb{R}^+) \to (\mathbb{D}, S^1)$, where $\mathbb{D} \approx \mathcal{U}$ is the open unit disk in the complex plane with the induced metric, or *Poincaré disk* model of the hyperbolic plane, with its frontier S^1 the *circle at infinity*.

Let t denote the ideal triangle with vertices $\pm 1, -i \in S^1$ as in Fig. 4.1, and consider the group Γ generated by hyperbolic reflections in its frontier edges. Define the *Farey tesselation* τ_* to be the Γ-orbit of the frontier edges of t, regarded as a set of geodesic *edges*, together with its *distinguished oriented edge*, or *doe*, given by the edge from -1 to $+1$. A rational point $\frac{p}{q} \in \mathbb{Q}^+ = \mathbb{Q} \cup \{\infty\}$, as illustrated in Fig. 4.1, is said to be of *generation* g if the radial arc in $\mathbb{D}$ from the origin to $\frac{p}{q}$ meets the interior of $g \geq 0$ distinct complementary triangles.

The *modular group* PSL(2, $\mathbb{Z}$) is the subgroup of Γ consisting of compositions of an even number of reflections, or in other words the group of two-by-two integral matrices A of unit determinant modulo the equivalence relation generated by identifying A with $-A$. This group acts on the left by orientation-preserving hyperbolic isometry on $z \in \mathcal{U}$ by fractional linear transformation $A = \left(\begin{smallmatrix} a & b \\ c & d \end{smallmatrix}\right) : z \mapsto \frac{az+b}{cz+d}$. There is also the action of $A \in$ PSL(2, $\mathbb{Z}$) on the right (following Gauss) on the rational points, for which we introduce the notation A : doe $\mapsto e_A \in \tilde{\tau}$, where

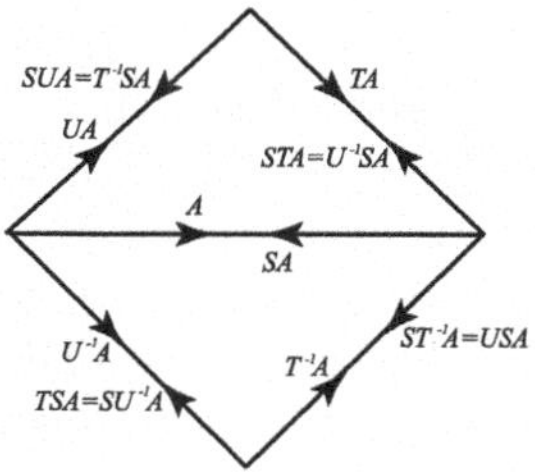

Fig. 4.7 Enumeration of oriented edges near e_A in τ_*

e_A has respective initial and terminal points $\frac{b+ia}{b-ia}$ and $\frac{d+ic}{d-ic}$ in $\mathbb{C}$, that is, respective labels $-\frac{b}{a}$ and $-\frac{d}{c}$ in Fig. 4.3.

The main point for us is that the modular group leaves set-wise invariant the Farey tesselation τ_*, mapping $\cup\,\tau_*$ onto $\cup\,\tau_*$, and any orientation-preserving homeomorphism of the circle leaving τ_* invariant in this manner lies in PSL(2, $\mathbb{Z}$). In fact, the modular group acts simply transitively on the set $\tilde{\tau}_*$ of orientations on edges of τ_*, and a generating set for PSL(2, $\mathbb{Z}$) is given by any two of

$$R = \begin{pmatrix} 0 & -1 \\ 1 & 1 \end{pmatrix},\ S = \begin{pmatrix} 0 & -1 \\ 1 & 0 \end{pmatrix},\ T = \begin{pmatrix} 1 & -1 \\ 0 & 1 \end{pmatrix},\ U = \begin{pmatrix} 1 & 0 \\ 1 & 1 \end{pmatrix}.$$

Moreover, $S^2 = 1 = R^3$ is a complete set of relations in the generators $R = T^{-1}U$ and $S = TU^{-1}T$, so PSL(2, $\mathbb{Z}$) $\approx \mathbb{Z}/2 \star \mathbb{Z}/3$.

Geometrically in Fig. 4.1: the elliptic element S fixes the distinguished edge in τ_* reversing the orientation of the doe; R is the elliptic transformation which counter clockwise cyclically permutes the vertices of the triangle t; and U (respectively T) is the parabolic transformation with the fixed point $\frac{0}{1}$ (respectively $\frac{1}{0}$) which cyclically permutes the incident edges of τ_* in the counter clockwise sense about $\frac{0}{1}$ (respectively the clockwise sense about $\frac{1}{0}$). Typical aspects of this enumeration of oriented edges by elements of PSL(2, $\mathbb{Z}$) are illustrated in Fig. 4.7.

Tesselations of $\mathbb{D}$

An arbitrary tesselation of $\mathbb{D}$ is a collection τ of bi-infinite geodesics locally finite in $\mathbb{D}$ and decomposing it into complementary ideal triangles. Geodesics in τ are called its *edges*, and τ itself is regarded as a set of edges. Let $\tau^0 \subset S^1$ denote the collection of endpoints of all the edges in τ, which is automatically dense in S^1 by local finiteness using that complementary regions are ideal triangles. Let $\tilde{\tau}$ denote the set of oriented edges of τ. A distinguished oriented edge, or simply *doe*, on τ is the specification of an element of $\tilde{\tau}$.

Another canonical tesselation of $\mathbb{D}$ with doe is the dyadic tesselation τ_d, which has the same doe as τ_* as well as the same generation zero vertices, and which is recursively characterized by the property that one vertex of each triangle complementary to τ_d bisects the angle between its other two vertices. Thus, τ_d^0 consists of the points in S^1 with (dyadic) rational arguments, as opposed to $\tau_*^0 \subset S^1$ which is comprised of points with rational rectilinear coordinates.

We again exploit the fact that tesselations with doe are rigid, namely, suppose that τ is any tesselation of $\mathbb{D}$ with doe $e \in \tilde{\tau}$. There is a unique $A \in \mathrm{PSL}(2, \mathbb{R})$ mapping the doe of τ_* to e and the triangle t to its right to the triangle complementary to τ on the right of e. One continues in this manner mapping complementary triangles to induce a map $\phi = \phi_{e \in \tilde{\tau}} : \tau_*^0 \to \tau^0$ called the *characteristic map* of $e \in \tilde{\tau}$.

It is not hard to see that ϕ is a surjection using local finiteness in $\mathbb{D}$, and it is an order-preserving injection by construction. As an order-preserving bijection between dense sets in S^1, the characteristic map interpolates an orientation-preserving homeomorphism $\phi : S^1 \to S^1$ of the same name. Letting $\mathrm{Homeo}_+(S^1)$ denote the topological group of orientation-preserving homeomorphisms of S^1 with the compact-open topology and $\mathcal{T}ess'$ the space of tesselations of $\mathbb{D}$ with doe with the Hausdorff topology on closed subsets of $\mathbb{D}$, we have

Theorem 4.1 ([23]) *The assignment $\phi_{\tau,e}$ of characteristic map to $e \in \tilde{\tau}$ induces a homeomorphism $\mathcal{T}ess' \to \mathrm{Homeo}_+(S^1)$.*

Our model of *universal Teichmüller space* is the quotient

$$\mathcal{T}ess = \mathcal{T}ess'/\mathrm{PSL}(2, \mathbb{R}) \approx \mathrm{Homeo}_+(S^1)/\mathrm{PSL}(2, \mathbb{R}),$$

which can be identified with the collection of all ideal triangulations of $\mathbb{D}$ with the same doe as τ_* and the same triangle t to its right; in effect in $\mathcal{T}ess'$, the doe determines a triangle, which then is normalized in $\mathcal{T}ess$ to t by $\mathrm{PSL}(2, \mathbb{R})$. Bers' universal Teichmüller space [2] is the collection of all quasi-symmetric homeomorphisms of S^1 modulo $\mathrm{PSL}(2, \mathbb{R})$, so our construction naturally generalizes Bers' version by extending to all orientation-preserving homeomorphisms of S^1.

Teichmüller Theory for Finite-Type Surfaces

Consider a connected closed orientable surface with genus $g \geq 0$, and let $F = F_g^s$ be the complement of a finite set of cardinality $s \geq 1$, which are taken to be the punctures of F. Let us tacitly fix an orientation on F and choose a basepoint for its fundamental group.

Provided the Euler characteristic $2-2g-s$ is negative, $F = \mathcal{U}/\Gamma$ is *uniformized by a Fuchsian group*, by which we mean: let $\mathcal{U}$ denote the upper half plane with its Poincaré metric $ds^2 = \frac{dx^2+dy^2}{y^2}$ and its projective matrix group $\mathrm{PSL}(2, \mathbb{R}) = \mathrm{SL}(2, \mathbb{R})/\pm I$ of oriented isometries, where I denotes the identity matrix; upon choosing a basepoint in F, there is a (conjugacy class of) injective homomorphism called the *uniformizing representation* $\pi_1 \to \mathrm{PSL}(2, \mathbb{R})$ of the fundamental group $\pi_1 = \pi_1(F)$ of F onto a discrete *Fuchsian subgroup* $\Gamma < \mathrm{PSL}(2, \mathbb{R})$ so that non-trivial loops about punctures in F are represented by parabolic transformations, namely, those with absolute trace equal to two. See [1, 9, 24] for example.

The *Teichmüller space* of F is

$$T(F) = \mathrm{Hom}'(\pi_1, \mathrm{PSL}(2, \mathbb{R}))/\mathrm{PSL}(2, \mathbb{R}),$$

where the prime indicates Fuchsian representations as just defined and the action of $\mathrm{PSL}(2, \mathbb{R})$ on Hom' is by conjugation. The *decorated Teichmüller space* is simply $\tilde{T}(F) = T(F_g^s) \times \mathbb{R}_+^s$, where the decoration is conveniently regarded as an s-tuple of positive real weights on the punctures. In particular, the *mapping class group* $MC(F)$ of homotopy classes of orientation-preserving homeomorphisms of F acts on $T(F)$ and on $\tilde{T}(F)$ in the natural way by push-forward. $T(F)$ is homeomorphic to an open ball of real dimension $6g - 6 + 2s$.

The homotopy class of a collection Δ of pairwise disjointly embedded essential arcs in F connecting punctures is called an *arc family* in F provided no two distinct arcs in Δ are isotopic. Δ is said to *fill* F provided each component of $F - \Delta$ is simply connected. An *ideal cell decomposition* is a decomposition of a topological space into simplices minus certain of their faces of codimension at least two, e.g., an ideal triangulation Δ filling F decomposes it into triangles with their *ideal vertices at infinity* at the punctures of F.

There is the following foundational result based on a convex hull construction[2] in Minkowski space:

Theorem 4.2 ([22, 24]) *There is an $MC(F)$-invariant smooth ideal cell decomposition $\mathcal{C}(F)$ of $\tilde{T}(F)$ whose simplices are indexed by arc families filling F with faces corresponding to inclusion of arc families.*

A maximal arc family is called an *ideal triangulation* of $F = F_g^s$ and contains $6g - g + 3s$ edges. In fact, the lambda lengths discussed in the Introduction on the edges of any ideal triangulation give global affine coordinates on $\tilde{T}(F)$. By construction, crossing a codimension-one face of $\mathcal{C}(F)$ corresponds to a *flip* on an edge e in an ideal triangulation Δ as illustrated in Fig. 4.4, replacing e by f to produce another ideal triangulaton Δ_e of F.

As also depicted, dual to an arc family Δ filling F, there is a graph $G = G(\Delta) \subset F$ embedded in F as a deformation retract, called a *spine* of F. An orientation on F induces the counter clockwise ordering on the half edges of G incident on each fixed vertex thus giving the abstract graph G the structure of a *fatgraph* (also called a *ribbon graph*). Dual to flipping diagonals of a quadrilateral in an ideal triangulation

---

[2] Uniformize the punctured surface F in Minkowski space by $\Gamma \circlearrowright \mathbb{R}^{2,1}$, where Γ acts by Lorentz isometry on $\mathbb{R}^{2,1}$ and lies in the component of the identity. A choice of horocycles about the punctures in F provides a finite collection of Γ-orbits of points in the positive light-cone in $\mathbb{R}^{2,1}$, the closed convex hull of which provides a Γ-invariant convex body in $\mathbb{R}^3 \approx \mathbb{R}^{2,1}$. The extreme edges of this body project to an ideal cell decomposition of F. This basic convex hull construction is from [22]. The same applies to the analogous end for discrete radially dense subsets of the light-cone in the universal context of [23].

of F, there is the combinatorial move depicted in Fig. 4.4 also called a *flip* on the dual trivalent fatgraph spine $G \subset F$: contract an edge of T with distinct endpoints and then expand the resulting 4-valent vertex in the unique distinct manner in order to produce another trivalent fatgraph.

This leads to the *Ptolemy groupoid* Pt(F) of F whose objects are homotopy classes of ideal triangulations of F, or dually trivalent fatgraph spines, and whose morphisms are compositions of flips. Pt(F) is the fundamental path groupoid of $\tilde{T}(F)$ according to Theorem 4.2. In fact, finite compositions of flips act transitively on homotopy classes of trivalent fatgraph spines. To see this, use the evident path connectivity of $\tilde{T}(F)$ to join by a smooth path cells corresponding to distinct ideal triangulations. Put this path in general position with respect to the ideal cell decomposition $\mathcal{C}(F)$ so that it crosses only codimension-one faces, as required. See [22, 24] for the details, namely, $\mathcal{C}(F)$ is algebraic hence smooth so that general position is viable, and every ideal triangulation of F actually occurs in $\mathcal{C}(F)$.

It follows that flips generate $MC(F)$ in the sense that if $G \subset F$ is a trivalent fatgraph spine and $\varphi \in MC(F)$, then there is a sequence

$$\varphi(G) = G_1 - G_2 - \cdots - G_n = G$$

of trivalent fatgraph spines of F where any consecutive pair differ by a flip, with the similar statement for ideal triangulations. Moreover, the general position argument just given, but now for two-dimensional homotopies of paths, implies

Corollary 4.3 *A complete set of relations in the fundamental path groupoid of $\tilde{T}(F)$ is given by declaring flips idempotent together with the links of codimension-two faces as illustrated in Fig. 4.8, namely: commutativity of flips on disjoint quadrilaterals; and the pentagon relation of five consecutive flips alternating between two distinct edges of a triangle.*

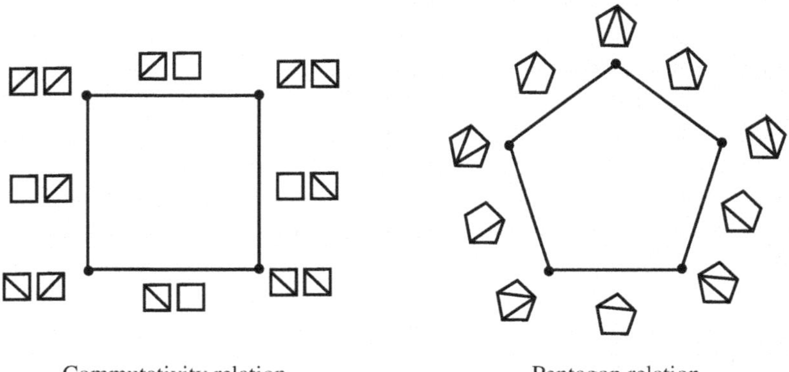

Fig. 4.8 Links of codimension-two cells in the idea cell decomposition of decorated Teichmüller spaces

Ptolemy Group

The *Ptolemy group(oid)* Pt has objects given by tesselations with doe of $\mathbb{D}$ which coincide with τ_* outside of a finite ideal polygon and morphisms given by finite compositions of flips. A typical flip clearly has order two, however, the flip on the doe is defined to produce another doe so that these oriented edges in this order respect the orientation of F, and hence the flip on the doe has order four, cf. the top-left for α in Fig. 4.1 ignoring boxes

Using combinatorial rigidity of tesselations with doe, flips can be labeled by edges of a fixed tesselation, so words in these labels render Pt in fact a group, not just a groupoid as it is for finite-type surfaces F without doe. Furthermore, $\mathrm{PSL}(2, \mathbb{Z})$ sits inside Pt as those tesselations which are identical to τ_* except perhaps for the location of the doe.

Let $\mathrm{PPSL}(2, \mathbb{Z}) < \mathrm{Homeo}_+(S^1)$ denote the collection of all piecewise-$\mathrm{PSL}(2, \mathbb{Z})$ homeomorphisms of S^1 with only finitely many pieces whose endpoints lie in τ_*^0. Since any orientation-preserving homeomophism of $\mathbb{D}$ preserving $\cup\, \tau_*$ must lie in the modular group, it follows that $\mathrm{PPSL}(2, \mathbb{Z})$ can also be described as the collection of characteristic maps of objects in Pt. As pointed out to me by Maxim Kontsevich [16] decades ago now and as exposed by Imbert in [12], by the same logic these two discrete subgroups of $\mathrm{Homeo}_+(S^1)$ coincide under the isomorphism in Theorem 4.1, so we have $\mathrm{PPSL}(2, \mathbb{Z}) \approx \mathrm{Pt}$.

Moreover, it is not difficult to see that the characteristic map $\tau_* \to \tau_d$ of the dyadic tesselation τ_d is precisely the Minkowski ?-function, and it conjugates $\mathrm{PPSL}(2, \mathbb{Z})$ to the celebrated Thompson group T, cf. [5]. We have arrived at the remarkable fact that the three avatars

$$\text{Thompson group } T \approx \mathrm{PPSL}(2, \mathbb{Z}) \approx \text{Ptolemy group Pt}$$

of our *universal mapping class group* Pt act on the universal Teichmüller space $\mathcal{T}ess \approx \mathrm{Homeo}_+(S^1)/\mathrm{PSL}(2, \mathbb{R})$ by flips.

Let α denote the flip on the doe and β denote the move that fixes the tesselation and moves the doe around the triangle to its left, cf. β in Fig. 4.1 again ignoring boxes. Clearly $\alpha^2 \sim S$ and $\beta \sim R$ generate $\mathrm{PSL}(2, \mathbb{Z})$, which acts simply transitively on $\tilde{\tau}_*$.

As explained in [25], there is the following direct consequence of the ideal analogue of Theorem 4.2 from [23]:

Theorem 4.4 $\mathrm{PPSL}(2, \mathbb{Z})$ *is generated by the flip* α *on the doe and the transformation* β *which moves the doe one edge counter clockwise in the triangle to its left. A presentation in these generators is given by the following relations:* α^4, β^3, $(\alpha\beta)^5$ *and the two commutators* $[\beta\alpha\beta, \alpha^2\beta\alpha\beta\alpha^2]$ *and* $[\beta\alpha\beta, \alpha^2\beta^2\alpha^2\ \beta\alpha\beta\ \alpha^2\beta\alpha^2]$.

See [17] for the discussion of an algebraic proof of an equivalent presentation based on unpublished computations of Thompson, and [25] for a complete and self-contained proof.

Spin Structures on Finite-Type Surfaces

Milnor's elegant definition [20] of a *spin structure* on a smooth surface F is a class in the modulo two first cohomology group of the unit tangent bundle of F which is non-zero on the fiber class. An equally elegant definition of immediate utility in our situation due to Natanzon [21] on a uniformized surface F is given by a lift of the uniformizing representation $\pi_1 \to \mathrm{PSL}(2,\mathbb{R})$ to $\pi_1 \to \mathrm{SL}(2,\mathbb{R})$; this immediately leads to P(SL(2,$\mathbb{R}$)) as the universal spin mapping class group.

Our starting point for the universal spin Teichmüller space is Johnson's general formulation [15] as an element of the affine $H^1(F;\mathbb{Z}/2)$-space $\mathcal{Q}(F)$ of quadratic forms, i.e., functions

$$q : H_1 = H_1(F;\mathbb{Z}/2) \to \mathbb{Z}_2$$

which are quadratic with respect to the homology intersection pairing $\cdot : H_1 \otimes H_1 \to \mathbb{Z}_2$ on H_1 in the sense that $q(a+b) = q(a) + q(b) + a \cdot b$, for $a, b \in H_1$. Our combinatorial starting point towards Johnson's definition as an element of $\mathcal{Q}(F)$ is the special case of a general technique from statistical physics by Cimazoni–Reshetihkin in [6, 7]. This is expressed in terms of Kastelyn orientations (which disagree with the orientation as a boundary an odd number of times, cf. the top of Fig. 4.9) and dimer configurations (disjoint unions of closed edges covering the vertices, again cf. Fig. 4.9) on the one-skeleton of a suitable CW decomposition of F:

Theorem 4.5 (Theorem 2.2 of [7]) *Fix a dimer configuration D on a surface graph with boundary $\mathcal{G}$ for the surface Σ and let $\alpha \in H_1(\Sigma;\mathbb{Z}_2)$ be represented by oriented closed curves $C_1, \ldots, C_m \in \bar{\mathcal{G}}$. If K is a Kasteleyn orientation on $\mathcal{G}$, then the function $q_D^K : H_1(\Sigma;\mathbb{Z}_2) \to \mathbb{Z}_2$ given by*

$$q_D^K(\alpha) = \sum_{i<j} C_i \cdot C_j + \sum_{n=1}^{m} (1 + n_{C_i}^K + \ell_{C_i}^D) \pmod 2$$

is a well-defined quadratic form, where ℓ_C^D is the number of edges of D sticking out to the left of C, and n_C^K is the number of edges counted with multiplicity where the orientation of C disagrees with that of K. Moreover for each fixed dimer D, this establishes an isomorphism as affine $H^1(\Sigma;\mathbb{Z}_2)$-space between $\mathcal{Q}(\Sigma)$ and the collection of equivalence classes of Kastelyn orientations on $\mathcal{G}$, with equivalence generated by reversing orientations of all edges incident on some fixed vertex of $\mathcal{G}$.

We actually rely here on another formulation of spin structure from [26] which amounts to choosing a CW decomposition in Theorem 4.5 suited to the tesselation with a special dimer on it, cf. Fig. 4.9. Fix a trivalent fatgraph spine $G \subset F$ and consider the set $\mathcal{O}(G)$ of equivalence classes of all orientations on its edges, where the equivalence relation is generated by *reflections* at vertices v of G by reversing

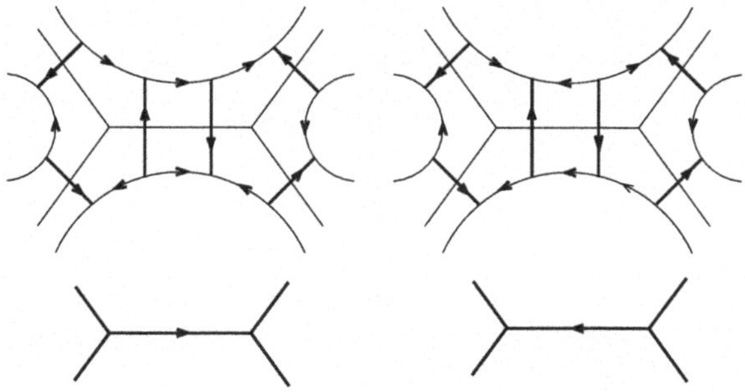

Fig. 4.9 On the top, CW decomposition of a neighborhood of a trivalent fatgraph G with its special dimer indicated in boldface. Orientations on hexagons extend to Kastelyn orientations on the top, and on the bottom is the corresponding orientation on the edge of G in the two cases

the orientation of each edge incident on v. (These reflections on G are compositions of six Kastelyn reflections on $\overline{\mathcal{G}}$ as in Theorem 4.5.)

Here is the combinatorial result from [26] which explains the genesis of the generator α of P(SL(2,$\mathbb{Z}$)):

Theorem 4.6 ([13, 26]) *Suppose $G \subset F = F_g^s$ is a trivalent fatgraph spine. Then $\mathcal{O}(G)$ and $\mathcal{Q}(F)$ are isomorphic as affine $H^1(F;\mathbb{Z}_2)$-spaces. Moreover, the action of the mapping class group $MC(F)$ on $\mathcal{Q}(F)$ lifts to the action of the Ptolemy groupoid on $\mathcal{O}(G)$ described by the following figure*

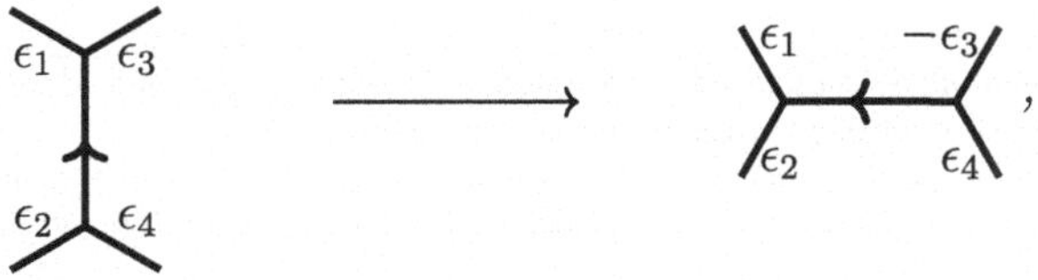

where ϵ_i indicates orientations on edges, and the extra minus sign denotes orientation reversal; in the special case $F = F_1^1$, ϵ_3 replaces $-\epsilon_3$.

Embarrassingly, there were inadvertently two cases missing after a lengthy reduction from 16 cases in [26], but as noted in [13], the two cases are each reflection-equivalent to the diagram in the theorem.

A notion from physics, a spin structure $q \in \mathcal{Q}(F)$ on F distinguishes two types of punctures as follows: If γ_p is a simple loop about the puncture p of F, then p is a *Neveu-Schwarz* puncture if $q([\gamma_p]) = 0$ and otherwise p is a *Ramond* puncture, where bracket denotes homology class.

Theorem 4.7 ([14]) *Consider a fatgraph spine G of F and a simple oriented edge-path γ in G surrounding a fixed puncture p of F. A spin structure as in the previous theorem determined by the equivalence class of an orientation ω on G has p as a Ramond puncture if and only if the length of γ has the same parity modulo two as the number of edges of G where ω agrees with γ.*

Acknowledgments It is a pleasure to thank Igor Frenkel, Athanase Papadopoulos and Anton Zeitlin for discussions and Barry Mazur and Dennis Sullivan for remarks on exposition.

References

1. A. Beardon, *The Geometry of Discrete Groups* (Springer, Berlin, 1983)
2. L. Bers, Universal Teichmüller space, in *Analytic Methods in Mathematical Physics* (Gordon and Breach, New York, 1968), pp. 65–83
3. W.W. Boone, W. Haken, V. Poénaru, *On Recursively Unsolvable Problems in Topology and Their Classification*. Studies in Logic and the Foundations of Mathematics, vol. 50 (Elsevier, Amsterdam, 1968), pp. 37–74
4. A. Bourque, A. Zeitlin, Flat GL(1|1)-connections and fatgraphs (2022, preprint). arXiv:2208.08033
5. J. Cannon, W. Floyd, W. Parry, Introductory remarks on Richard Thompson's groups. L'Enseignement Math. **42**, 215–256 (1996)
6. D. Cimasoni, N. Reshetikhin, Dimers on surface graphs and spin structures. I. Commun. Math. Phys. **275**, 187–208 (2007)
7. D. Cimasoni, N. Reshetikhin, Dimers on surface graphs and spin structures. II. Commun. Math. Phys. **281**, 445–468 (2008)
8. A. Fathi, F. Laudenbach, V. Poénaru, *Travaux de Thurston sur les Surfaces*. Asterisque (Société Mathématique de France, Paris, 1979), pp. 66–67
9. L. Ford, *Automorphic Functions* (American Mathematical Society, Providence, 1951)
10. I. Frenkel, R. Penner, Sketch of a program for automorphic functions from universal Teichmüller theory to capture monstrous moonshine. Commun. Math. Phys. **389**, 1525–1567 (2022)
11. A. Haefliger, V. Poénaru, La classification des immersions combinatoires. Publ. Math. l'IHÉS **23**, 75–91 (1964)
12. M. Imbert, Sur l'isomorphisme du groupe de Richard Thompson avec le groupe de Ptoélmée, in *Geometric Galois Actions II*, ed. by P. Lochak, L. Schneps. London Math Society Lecture Notes, vol. 243 (Cambridge University Press, Cambridge, 1997)
13. I. Ip, R. Penner, A. Zeitlin, N=2 Super-Teichmüller theory. Adv. Math. **336**, 409–454 (2018)
14. I. Ip, R. Penner, A. Zeitlin, On ramond decorations. Commun. Math. Phys. **371**, 145–157 (2019)
15. D. Johnson, Spin structures and quadratic forms on surfaces. J. Lond. Math. Soc.**22**, 365–373 (1980)
16. M. Kontsevich, Private communication (1995)
17. P. Lochak, L. Schneps, The universal Ptolemy-Teichmüller groupoid, in *Geometric Galois Actions II*, ed. by P. Lochak, L. Schneps. London Math Society Lecture Notes, vol. 243 (Cambridge University Press, Cambridge, 1997)
18. F. Malikov, R. Penner, The Lie algebra of homeomorphisms of the circle. Adv. Math. **140**, 282–322 (1999)
19. B. Mazur, A note on some contractible 4-manifolds. Ann. Math. **73**, 221–228 (1961)
20. J. Milnor, Remarks concerning spin manifolds, in *Differential and Combinatorial Topology* (Princeton University Press, Princeton, 1965)
21. S.M. Natanzon, *Moduli of Riemann Surfaces, Real Algebraic Curves, and Their Superanalogs*. Translations of Mathematical Monographs, vol. 225 (American Mathematical Society, Providence, 2004)
22. R.C. Penner, The decorated Teichmüller space of punctured surfaces. Commun. Math. Phys. **113**, 299–339 (1987)
23. R.C. Penner, Universal constructions in Teichmüller theory. Adv. Math. **98**, 143–215 (1993)
24. R.C. Penner, *Decorated Teichmüller Theory* (European Mathematical Society, Helsinki, 2012)

25. R.C. Penner, Universal Spin Teichmüller Theory, II. Finte Presentation of P(SL(2,$\mathbb{Z}$)) (preprint, 2023)
26. R. Penner, A. Zeitlin, Decorated super-Teichmüller space. J. Differ. Geom. **111**, 527–566 (2019)
27. V. Poénaru, La decomposition de l'hypercube en produit topologique. Bull. Soc. Math. France **88**, 113–129 (1960)
28. V. Poénaru, Geometric simple connectivity and low-dimentional topology (2003). http://mathnet.ru/links/f1937b66240c8471d3a2533073eed2f8/tm19.pdf
29. V. Poénaru, On the 3-dimensional Poincaré conjecture and the 4-dimensional smooth schoenflies problem (2006). http://arxiv.org/abs/math/0612554
30. V. Poénaru, Geometric simple connectivity and finitely presented groups (2014). http://arxiv.org/abs/1404.4283
31. V. Poénaru, All finitely presented groups are QSF (2014). http://arxiv.org/abs/1409.7325

Chapter 5
Universal Spin Teichmüller Theory, II: Finite Presentation of P(SL(2,$\mathbb{Z}$))

Robert Penner

For Valentin Poénaru on the happy occasion of his 90th birthday.

Abstract In previous works, the universal mapping class group was taken to be the group PPSL(2, $\mathbb{Z}$) of all piecewise PSL(2, $\mathbb{Z}$) homeomorphisms of the unit circle S^1 with finitely many breakpoints among the rational points in S^1, and in fact, the Thompson group $T \approx$ PPSL(2, $\mathbb{Z}$). The new spin mapping class group P(SL(2,$\mathbb{Z}$)) is given by all piecewise-constant maps $S^1 \to$ SL(2, $\mathbb{Z}$) which projectivize to an element of PPSL(2, $\mathbb{Z}$). We compute a finite presentation of PPSL(2, $\mathbb{Z}$) from basic principles of general position as an orbifold fundamental group. The orbifold deck group of the spin cover is explicitly computed here, from which follows also a finite presentation of P(SL(2, $\mathbb{Z}$)). This is our main new achievement. Certain commutator relations in P(SL(2, $\mathbb{Z}$)) seem to organize according to root lattices, which would be a novel development. We naturally wonder what is the automorphism group of P(SL(2, $\mathbb{Z}$)) and speculate that it is a large sporadic group. There is a companion chapter to this one which explains the topological background from first principles and proves that the group studied here using combinatorial group theory is indeed P(SL(2, $\mathbb{Z}$)).

Keywords Classical and universal Teichmüller space · Riemann moduli space · Mapping class group · Spin structure · Thompson group T

MSC Codes 30F60 32G15 32C11 37E30

R. Penner (✉)
Institut des Hautes Études Scientifiques, Le Bois Marie, France
e-mail: rpenner@ihes.fr

L. Funar, A. Papadopoulos (eds.), *Essays on Topology*,
https://doi.org/10.1007/978-3-031-81414-3_5

Introduction

Consider the space of all tesselations τ of the Poincaré disk $\mathbb{D}$, i.e., τ is a locally-finite collection of geodesics which decompose $\mathbb{D}$ into ideal triangles, and let us choose a distinguished oriented edge or *doe* in τ. Let $\mathcal{T}ess$ be the space of PSL(2,$\mathbb{R}$)-orbits of all tesselations of $\mathbb{D}$ with doe. Consider a finitely supported $\mathbb{Z}/2$-*marking* $\tau \to \{0, 1\}$, and define an equivalence relation on such markings on τ generated by adding unity modulo two to each of the edges in the frontier of some triangle complementary to τ, i.e., change the values of the marking on all three edges of a fixed complementary triangle.

Our new universal spin Teichmüller space $\mathcal{T}ess^+$ is the collection of all equivalence classes of marked tesselations with doe. It can be convenient to think of the marking as determined by finite collections on each edge, where the parity modulo two of the cardinality of the collection determines the $\mathbb{Z}/2$-marking on the edge.

Three combinatorial moves α, β, t on marked tesselations with doe are illustrated in Fig. 5.1, where adding a mark on an edge corresponds to adding unity modulo two to the value of the marking of the edge. The moves on unmarked tesselations with doe underlying α and β generate PPSL(2,$\mathbb{Z}$), as we prove in the appendix and shall recall presently, and there is a commutative diagram

$$\begin{array}{ccc} \mathrm{P(SL(2, \mathbb{Z}))} \times \mathcal{T}ess^+ & \to & \mathcal{T}ess^+ \\ \downarrow & & \downarrow \\ \mathrm{PPSL(2, \mathbb{Z})} \times \mathcal{T}ess & \to & \mathcal{T}ess \end{array}$$

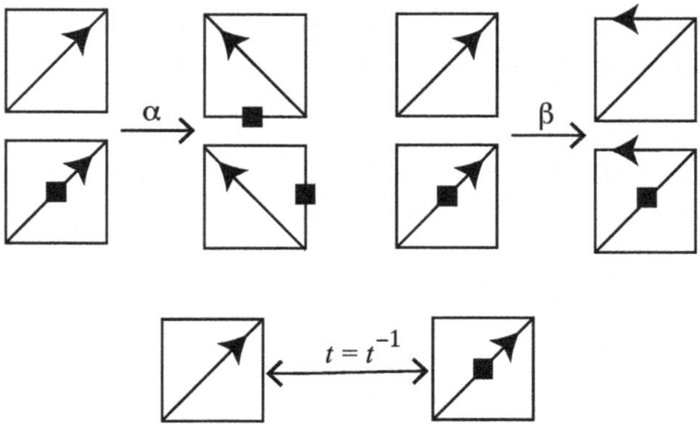

Fig. 5.1 The three generators α, β, t of P(SL(2,$\mathbb{Z}$)) acting on a marked tesselation with doe in the two cases that the doe, which is indicated with an arrow, may or may not have a non-zero marking, as indicated with or without a box icon. It is easy to confirm that these moves are well defined on equivalence classes of markings

where the vertical maps are the natural forgetful maps. PPSL(2,ℤ) is conjugate to Thompson's group T [1] by an orientation-preserving homeomorphism of the circle.

There is a companion [8] to this chapter which explains this isomorphism $T \approx \mathrm{PPSL}(2, \mathbb{Z})$ and gives the proof that the group P(SL(2, ℤ)) generated by α, β, t is precisely the group of piecewise-constant maps $S^1 \to \mathrm{SL}(2,\mathbb{Z})$ which projectivize to homeomorphisms in PPSL(2,ℤ). In both these groups, composition is defined by taking the usual composition on each piece of a common refinement of the two piecewise structures. The companion [8] also surveys the entire construction of P(SL(2,ℤ)) $\circlearrowleft \mathcal{T}ess^+$ back to first principles, for punctured surfaces of finite type from [4] as well as in the universal case from [5], both of which are covered in [6]. In particular as described there, the new move α including markings arises from the treatment of spin structures on finite-type punctured surfaces introduced in [7].

This chapter derives a finite presentation of the universal spin mapping class group P(SL(2, ℤ)), for whose formulation we require a definition: Given a word $w = w_1 \cdots w_n$ in the generators α and β, a *t-insertion* in w is any word $\hat{w} = w_1 t^{\epsilon_1} w_2 t^{\epsilon_2} \cdots t^{\epsilon_{n-1}} w_n$, for some choice of $\epsilon_i \in \{0, 1\}$. Two further notational points are that square brackets in the sequel denote group commutators, and we shall sometimes include extra spaces in group-theoretic words in order to improve readability or emphasize the structure, for example, a conjugate of the flip $\beta\alpha\beta$ or explicitly as a commutator.

Main Theorem 1 *The group* P(SL(2, ℤ)) *generated by* α, β, t *admits a finite presentation with the following relations:*

- ***Power Laws:*** $t^2, \beta^3, \alpha^4, (t\beta)^3, (t\alpha)^4, [t, \alpha^2]$ and $[t, \alpha t\alpha]$;
- ***Pentagon:*** $(\beta\alpha)^5$ and $(\beta t\alpha t)^5$;
- ***Degeneracy:***

$$t = [\alpha, \beta t\beta^2] = [\alpha t, \beta t\beta^2] = [\alpha, t\beta^2 t\beta] = [\alpha t, t\beta^2 t\beta];$$

- ***Insertion:*** $[t, \hat{w}]$ *for any* $\mu, \nu \in \{0, 1\}$ *and any t-insertion* $\hat{w}$, *for* $w \in \{\beta\alpha\beta,\ \beta^{2\mu}\alpha^2\ \beta\alpha\beta\ \alpha^2\beta^{\mu},\ \beta^{2\nu}\alpha^2\beta^{2\mu}\alpha^2\ \beta\alpha\beta\ \alpha^2\beta^{\mu}\alpha^2\beta^{\nu}\}$;
- ***First Commutator:*** *any word of the form* $t^{r_0}\beta\alpha\beta\ \alpha t^{r_4}\alpha\ \beta\alpha\beta\ \alpha t^{r_3}\alpha\ \beta t^{s_4}\beta t^{s_3}\ \alpha^3\ t^{s_2}\beta t^{s_1}\beta\ \alpha t^{r_2}\alpha\ \beta t^{t_4}\beta t^{t_3}\ \alpha^3\ t^{t_2}\beta t^{t_1}\beta\ \alpha t^{r_1}\alpha$, *where the exponents satisfy*

$$\sum_{i=0}^{4} r_i = s_1 + s_3 + t_1 + t_3,$$

$$r_1 + r_2 = s_1 + s_2, \quad r_2 + r_3 = t_3 + t_4,$$

$$r_3 + r_4 = s_3 + s_4, \quad r_1 + r_4 = t_1 + t_2;$$

- ***Second Commutator:*** *any word of the form* $t^{r_0}\beta\alpha\beta\ \alpha t^{r_8}\alpha\ \beta t^{s_5}\beta$ $\alpha t^{r_7}\alpha\ \beta\alpha\beta\ \alpha t^{r_6}\ \alpha\ \beta t^{s_4}\beta t^{s_3}\alpha^3 t^{s_2}\beta t^{s_1}\beta\ \alpha t^{r_4}\alpha\beta t^{t_5}\beta\ \alpha t^{r_3}\alpha\ \beta t^{t_4}\beta t^{t_3}\alpha^3$ $t^{t_2}\beta t^{t_1}\beta\ \alpha t^{r_2}\alpha\beta\ \alpha t^{r_1}\alpha$, *where the exponents satisfy*

$$r_0 = s_1 + s_3 + s_5 + \sum_{i=1}^{5} t_i,$$

$$r_1 + r_4 = s_1 + s_2, \quad r_2 + r_7 = t_1 + t_2,$$

$$r_5 + r_8 = s_3 + s_4, \quad r_3 + r_6 = t_3 + t_4,$$

$$\sum_{i=1}^{8} r_i = s_5 + t_1 + t_3 + t_5.$$

The last two collections of relations arise from t-insertions in the *First* $w_1 = [\beta\alpha\beta, \alpha^2\ \beta\alpha\beta\ \alpha^2]$ and *Second Commutator Relations* $w_2 = [\beta\alpha\beta,\ \alpha^2\beta^2\alpha^2\ \beta\alpha\beta\ \alpha^2\ \beta\alpha^2]$ in PPSL(2,$\mathbb{Z}$), cf. Theorem A below. There may be redundancies among the Commutator Relations in the Main Theorem. Note that β and t generate a dihedral subgroup D_6, and β and α^2 generate a subgroup PSL(2,$\mathbb{Z}$) of P(SL(2,$\mathbb{Z}$)). Moreover, t lies in the first derived subgroup according to the Degeneracy Relations, so P(SL(2,$\mathbb{Z}$)) is perfect since T is.

In the First Commutator Relation, the 8-tuple of s- and t-variables mimics part of the root lattice of E_8 since $\sum_{i=1}^4 s_i = \sum_{i=1}^4 t_i$. In the spirit of [2], it is natural to wonder if the commutator relations of P(SL(2,$\mathbb{Z}$)) are organized according to an interesting lattice and to ask: What is the automorphism group of P(SL(2,$\mathbb{Z}$))?

The basic idea for the proof of the Main Theorem is as follows. There are decorated bundles $\widetilde{\mathcal{T}ess} \to \mathcal{T}ess$ and $\widetilde{\mathcal{T}ess^+} \to \mathcal{T}ess^+$ with fibers given by collections of horocycles, one centered at each ideal point of the tesselation. As in [5], these respective decorated spaces come equipped with *ideal cell decompositions* $\mathcal{C}$ and $\mathcal{C}^+$, namely, decompositions into simplices plus certain of their faces, where $\mathcal{C}$ and $\mathcal{C}^+$ are invariant under the respective actions PPSL(2,$\mathbb{Z}$) $\circlearrowleft$ $\widetilde{\mathcal{T}ess}$ and P(SL(2,$\mathbb{Z}$)) $\circlearrowleft$ $\widetilde{\mathcal{T}ess^+}$.

Recall from [5] that a codimension-one face in $\mathcal{C}$ corresponds to removing one edge of an ideal triangulation. General position of a path in $\widetilde{\mathcal{T}ess}$ with respect to $\mathcal{C}$ therefore shows that the fundamental path groupoid of $\widetilde{\mathcal{T}ess}$ is generated by *flips*, that is, the combinatorial moves underlying α in Fig. 5.1: namely, remove an edge from τ so as to produce a complementary ideal quadrilateral, and replace the removed edge with the other diagonal of this quadrilateral.

General position of a homotopy of paths in $\widetilde{\mathcal{T}ess}$ likewise shows that a complete set of relations for the fundamental path groupoid is provided by the collection of links of codimension-two cells. These links correspond to removing two edges, which may either lie in the frontier of a common triangle, or not, and respectively correspond to the *(Classical) Pentagon* and *Commutativity Relations* illustrated in Fig. 5.2. There is a further relation called *Idempotence* arising from a degenerate

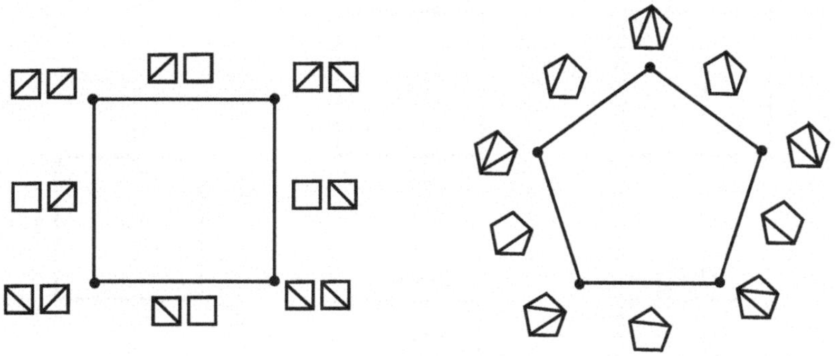

Fig. 5.2 Links of codimension-two cells in the ideal cell decomposition of decorated Teichmüller spaces. If one of the edges happens to be the doe, then the pentagon relation has order ten, since the five moves depicted interchange the two edges, as one can check

codimension-two face corresponding to performing a flip to produce a new edge upon which you then flip.

It is easy and satisfying to check that ignoring markings, the flip on $\beta(doe)$ is given by $\beta\alpha\beta$, and on $\beta^2(doe)$ by its inverse $\beta^2\alpha^3\beta^2$. Instead of flips, the orbifold fundamental group PPSL(2,ℤ) of the quotient can equally well be regarded as generated by these α, β, and there is the following presentation, whose proof is given in the appendix.

Theorem A *PPSL(2,ℤ) is generated by the flip α on the doe and the transformation β which moves the doe one edge counter-clockwise in the triangle to its left. A presentation in these generators is given by the following relations: α^4, β^3, $(\alpha\beta)^5$ and the two commutators $w_1 = [\beta\alpha\beta, \alpha^2\beta\alpha\beta\alpha^2]$ and $w_2 = [\beta\alpha\beta, \alpha^2\beta^2\alpha^2\ \beta\alpha\beta\ \alpha^2\beta\alpha^2]$.*

In fact, the current chapter provides the first complete proof of this result, which is stated in the context of [5] in [3] but depends in [3] upon unpublished computations of Richard Thompson.

Our approach for P(SL(2,ℤ)) is to take each of the defining relations in PPSL(2,ℤ) and consider all of its t-insertions. As we shall explain presently in an example, we can compute those t-insertions which leave invariant one (and hence each) equivalence class of marking. These provide a complete but highly redundant presentation, which effectively describes the deck group of $\widetilde{\mathcal{T}ess}^+/\mathrm{P(SL(2,\mathbb{Z}))} \to \widetilde{\mathcal{T}ess}/\mathrm{PPSL(2,\mathbb{Z})}$.

Example (The Power Law Relations) The fourth power of the generator $\alpha \in$ P(SL(2,ℤ)) returns the doe to its starting position and adds a mark on each edge in the frontier of the quadrilateral near the doe, and likewise for $(t\alpha)^4$. However, this marking is equivalent to the trivial one with which we started, and so $\alpha^4 = 1$ in P(SL(2,ℤ)). More generally, consider words $t^{t_4}\alpha t^{t_3}\alpha t^{t_2}\alpha t^{t_1}\alpha$, where

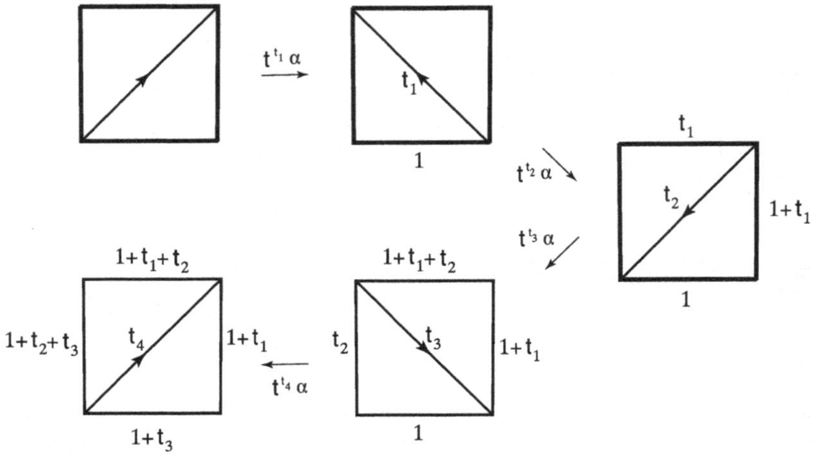

Fig. 5.3 Evolution of marking under the t-insertion $\hat{w} = t^{t_4}\alpha t^{t_3}\alpha t^{t_2}\alpha t^{t_1}\alpha$ in $w = \alpha^4$, where words act from right to left

the $t_i \in \{0, 1\}$. We can compute the effects on markings of these words, with letters applied from right to left, as illustrated in Fig. 5.3. In order that the resulting marking is equivalent to the trivial one with which we began, we must have $t_1 = t_3$ and $t_2 = t_4$, so we find precisely the Power Law Relations on α. The other Power Law Relations $\beta^3 = (t\beta)^3 = 1$ follow similarly.

The overall procedure is thus clear: First, compute the effect on markings of each t-insertion in each relation of PPSL(2,$\mathbb{Z}$) to produce a finite presentation of P(SL(2,$\mathbb{Z}$)) as in the Example; second, find their redundancies and compute a minimal set of relations. The latter step involves combinatorial group-theoretic calculations, which can be quite involved as in the Degeneracy Relations, and have yet to be completed for the full presentation in the Main Theorem.

The Insertion Relations were separately discovered in this way, one at a time, and they are in fact special cases of a general lemma: if a word w in α, β, t leaves the doe invariant, then $[t, w] = 1$. The general proof is clear, or one can check these several cases directly combinatorially as in the Power Law Example.

There thus remain four sections, one to analyze each of the Pentagon, Degeneracy, First and Second Commutator Relations, in each case first simplifying as much as possible using the Insertion and Power Law Relations to reduce the number of variables. The Degeneracy Relation with no t-insertions reads simply $(\beta\alpha\beta)(\beta^2\alpha^3\beta^2)$, which vanishes since $\beta^3 = \alpha^4 = 1$; however with t-insertions, there are the four commutators equal to t in the Degeneracy Relations. The appendix is independent of the body of the chapter and constitutes a fifth and final section which derives the finite presentation of PPSL(2,$\mathbb{Z}$) from general position.

Before diving into these detailed computation in subsequent sections, we collect here several simple algebraic facts to be used in the sequel without further comment, throughout which x, y, t are group elements and $n \in \mathbb{Z}_{>0}$: $(xy)^n = 1$ if and only if $(yx)^n = 1$; if x is finite-order, then the following are equivalent: $[x, y] = 1$,

$[x^{-1}, y] = 1$, $[x, y^{-1}] = 1$, $[x^{-1}, y^{-1}] = 1$; if $t^2 = 1$, then $(txty)^n = 1$ if and only if $(xtyt)^n = 1$, and $(xt)^n = 1$ if and only if $(x^{-1}t)^n = 1$.

5.1 Pentagon Relations

The general insertion of t in $(\beta\alpha)^5$ can be simplified in P(SL(2,ℤ)) using the fact from the Insertion Relations that t commutes with any t-insertion in $\beta\alpha\beta$. (In practice, the pentagon t-insertion relations provided the vehicle for discovering these Insertion Relations.) It therefore suffices to consider only words of the form

$$\hat{w} = \beta\alpha\beta t^{t_4}\alpha\beta t^{t_3}\alpha\beta t^{t_2}\alpha\beta t^{t_1}\alpha t^{t_0}.$$

Computing as in the Power Law Example, we find the change of marking under $\hat{w}$ as in Fig. 5.4, which is equivalent to the trivial marking if and only if $t_1 = t_2 = t_3$ and $t_4 = t_0 + t_3$. The four solutions to this give rise to the relation $(\alpha\beta)^5 = 1$ as usual, plus two copies of $(\alpha\beta)^3 = t(\alpha\beta t)^3$ and one copy of the familiar $[t, \beta\alpha\beta] = 1$. Meanwhile by the Insertion Relations, $(\alpha\beta)^3 = t(\alpha\beta t)^3$ is equivalent to the relation $(\beta t\alpha t)^5 = 1$ given in the Main Theorem.

5.2 Degeneracy Relations

The action on markings of the t-insertion $\hat{w}$ in $w = (\beta\alpha\beta)(\beta^2\alpha^3\beta^2)$ indicated in Fig. 5.5 produces the trivial marking, again as in the Example in the Introduction, if and only if the following linear system holds:

$$t_1 + t_2 = s_1 + s_5,$$
$$t_3 + t_4 = s_2 + s_3,$$
$$t_3 + t_5 = s_1 + s_4.$$

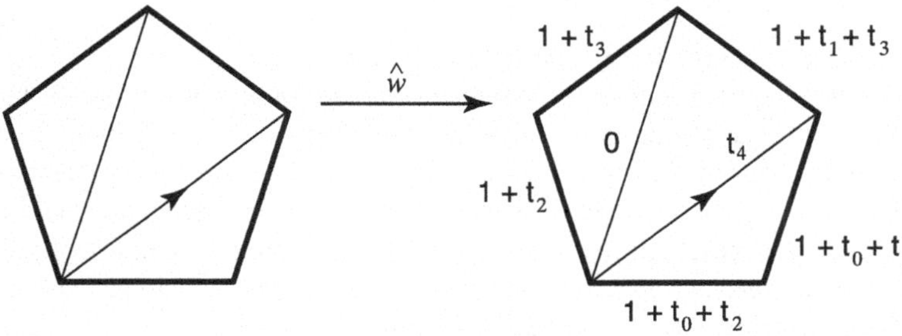

Fig. 5.4 Evolution of marking under the t-insertion $\hat{w} = \beta\alpha\beta t^{t_4}\alpha\beta t^{t_3}\alpha\beta t^{t_2}\alpha\beta t^{t_1}\alpha t^{t_0}$ in $w = (\beta\alpha)^5$

2. Degeneracy Relations

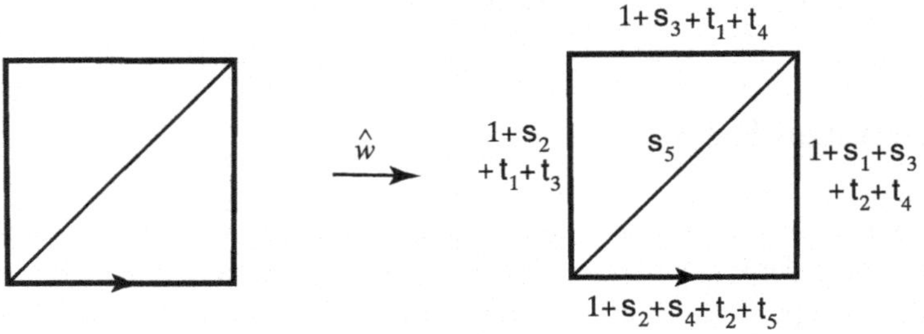

Fig. 5.5 Evolution of marking under the t-insertion $\hat{w} = t^{t_5}\beta t^{s_5}\alpha t^{t_4}\beta t^{s_4}\beta t^{t_3}\beta t^{s_3}\alpha t^{t_2}\alpha t^{s_2}\alpha$ $t^{t_1}\beta t^{s_1}\beta$ in $w = (\beta\alpha\beta)(\beta^2\alpha^3\beta^2)$

In particular, the complement map that adds one modulo two to each variable preserves solutions, so to enumerate all $2^{10-3} = 128$ solutions, as we shall do here in order to extract a minimal set of relations, it suffices first to enumerate only those solutions with at most 5 non-zero variables and then adjoin their complements. We shall in general let K denote the number of non-zero variables of a solution. Define the Boolean predicate $T(r, s, t) = [(r \wedge s) \vee (s \wedge t) \vee (t \wedge r)]$, where $\wedge$ is logical AND and $\vee$ is logical OR. Taking the $\{0, 1\}$-valued variables as truth values, it is not difficult to combinatorially enumerate the non-zero solutions to the linear system above with $K \leq 5$, as follows:

$\underline{K=2}$: $s_4 \wedge t_5$, $\quad T(s_2, s_3, t_4)$, $\quad T(t_1, t_2, s_5)$;

$\underline{K=3}$: $(i)\ s_1 \wedge (s_4 \vee t_5) \wedge (t_1 \vee t_2 \vee s_5)$,

$t_3 \wedge (s_4 \vee t_5) \wedge (s_2 \vee s_3 \vee t_4)$;

$\underline{K=4}$: $T(s_2, s_3, t_4) \wedge T(t_1, t_2, s_5)$,

$(ii)\ s_4 \wedge t_5 \wedge [T(t_1, t_2, s_5) \vee T(s_2, s_3, t_4)]$,;

$(iii)\ s_1 \wedge t_3 \wedge (s_2 \vee s_3 \vee t_4) \wedge (t_1 \vee t_2 \vee s_5)$;

$\underline{K=5}$: $s_1 \wedge (s_4 \vee t_5) \wedge (t_1 \wedge t_2 \wedge s_5)$, $\quad t_3 \wedge (s_4 \vee t_5) \wedge (s_2 \wedge s_3 \wedge t_4)$,

$s_1 \wedge (s_4 \vee t_5) \wedge (t_1 \vee t_2 \vee s_5) \wedge T(s_2, s_3, t_4)$,

$t_3 \wedge (s_4 \vee t_5) \wedge (s_2 \vee s_3 \vee t_4) \wedge T(t_1, t_2, s_5)$,

where we bring cases (i)–(iii) to particular attention.

A tedious group-theoretic computation, which we omit, shows that all but four of the 128 relations arising from these are tautologies assuming these four new relations together with the Power Law Relations and the Insertion Relations for $\beta\alpha\beta$. Two of the four new relations, $t\alpha\beta^2 t\beta t\alpha^3\beta t\beta^2$ and $\alpha t\beta^2 t\beta\alpha^3\beta t\beta^2 t$, arise from (iii), $\alpha\beta t\beta^2\alpha^3 t\beta t\beta^2$ from (i), and $t\beta\alpha t\beta t\beta^2 t\alpha^3$ from (ii). One checks directly that these equations are identical with the Degeneracy Relations expressed as commutators equal to t as in the Main Theorem.

3. First Commutator Relations

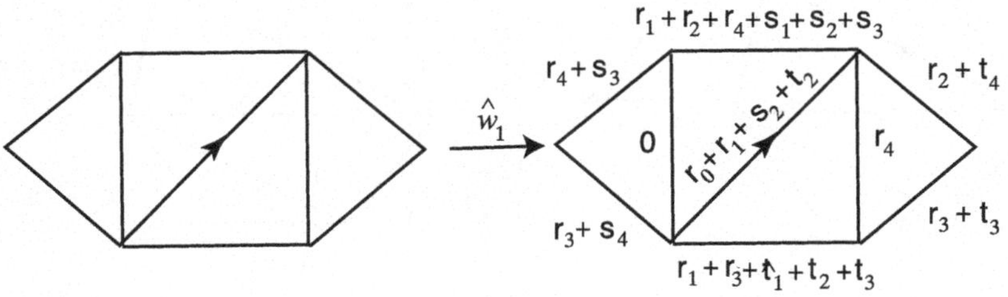

Fig. 5.6 Evolution of marking under the t-insertion $\hat{w}_1$ in the First Commutator w_1 given the text

5.3 First Commutator Relations

We use the Insertion Relation for $\beta\alpha\beta$ and the Power Law Relations for α to reduce the dimension of the general t-insertion in the First Commutator w_1, so as to produce

$$\begin{aligned}\hat{w}_1 = & t^{r_0}\beta\alpha\beta\ \alpha t^{r_4}\alpha\ \beta\alpha\beta\ \alpha t^{r_3}\alpha \\ & \beta t^{s_4}\beta t^{s_3}\ \alpha^3\ t^{s_2}\beta t^{s_1}\beta\ \alpha t^{r_2}\alpha \\ & \beta t^{t_4}\beta t^{t_3}\ \alpha^3\ t^{t_2}\beta t^{t_1}\beta\ \alpha t^{r_1}\alpha\,.\end{aligned}$$

The action of $\hat{w}_1$ on the trivial marking is computed as before with the result illustrated in Fig. 5.6, which is equivalent to the trivial marking if and only if the five equations in the Main Theorem for the First Commutator are satisfied.

5.4 Second Commutator Relations

We again use the Insertion Relation for $\beta\alpha\beta$ and Power Laws for α to reduce the dimension of the general t-insertion in the Second Commutator w_2, so as to produce

$$\begin{aligned}\hat{w}_2 = & t^{r_0}\beta\alpha\beta\ \alpha t^{r_8}\alpha\ \beta t^{s_5}\beta\ \alpha t^{r_7}\alpha\ \beta\alpha\beta\ \alpha t^{r_6}\alpha\ \beta\ \alpha t^{r_5}\alpha \\ & \beta t^{s_4}\beta t^{s_3}\ \alpha^3\ t^{s_2}\beta t^{s_1}\beta\ \alpha t^{r_4}\alpha\ \beta t^{t_5}\beta\ \alpha t^{r_3}\alpha \\ & \beta t^{t_4}\beta t^{t_3}\ \alpha^3\ t^{t_2}\beta t^{t_1}\beta\ \alpha t^{r_2}\alpha\ \beta\alpha t^{r_1}\alpha\,.\end{aligned}$$

The action of $\hat{w}_2$ on the trivial marking is computed as before with the result illustrated in Fig. 5.7, which is equivalent to the trivial marking if and only if the six equations in the Main Theorem for the Second Commutator are satisfied.

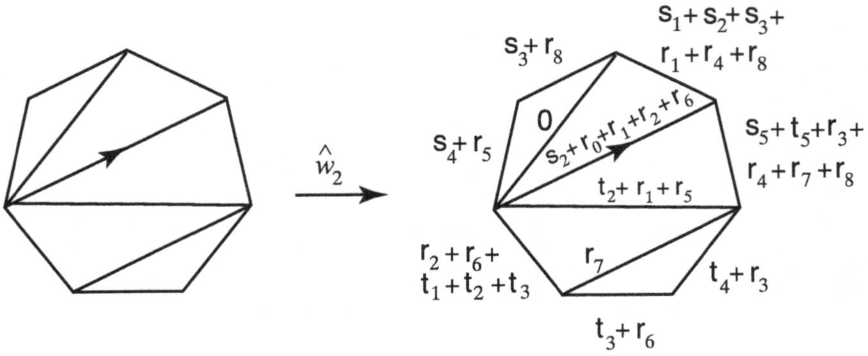

Fig. 5.7 Evolution of marking under the t-insertion $\hat{w}_2$ in the Second Commutator w_2 given in the text

Appendix: Presentation of PPSL(2,ℤ)

This appendix is dedicated to the proof of the following theorem. The symbols α, β here denote the operations induced from Fig. 5.1 on tesselations with doe but without marking.

Theorem 1 *PPSL(2,ℤ) is generated by the flip α on the doe and the transformation β which moves the doe one edge counter-clockwise in the triangle to its left. A presentation in these generators is given by the following relations: α^4, β^3, $(\alpha\beta)^5$ and the two commutators $[\beta\alpha\beta, \alpha^2\beta\alpha\beta\alpha^2]$ and $[\beta\alpha\beta, \alpha^2\beta^2\alpha^2\ \beta\alpha\beta\ \alpha^2\beta\alpha^2]$.*

Proof According to [5], PPSL(2,ℤ) is the orbifold fundamental group of PPSL(2,ℤ) $\circlearrowleft \widetilde{Tess}$. This space admits a natural ideal cell decomposition whose codimension-two skeleton more or less immediately gives this presentation via general position, as discussed in the Introduction. Namely, the geometry gives three classes of relations:

- the pentagon relation $(\alpha\beta)^5 = 1$ when the two edges cobound a single triangle;
- commutativity of two flips supported on disjoint quadrilaterals when the two edges do not cobound a single triangle;

  together with the degenerate codimension-two face of consecutively crossing the same codimension-one face:
- idempotence of flips, i.e., perform a flip to produce a new edge upon which one subsequently flips.

  It is evident that the flip α on the doe has order 4, and that β has order 3, in the orbifold fundamental group. Notice that the flip on $\beta(doe)$ is given by $\beta\alpha\beta$ and on $\beta^2(doe)$ is given by $\beta^2\alpha^3\beta^2$. It is well known that finite words in $\alpha^2 \sim \left(\begin{smallmatrix} 0 & -1 \\ 1 & 0 \end{smallmatrix}\right) \in \mathrm{PSL}(2, \mathbb{Z})$ and $\beta \sim \left(\begin{smallmatrix} 1 & -1 \\ 0 & 1 \end{smallmatrix}\right) \in \mathrm{PSL}(2, \mathbb{Z})$ act simply transitively on the oriented edges of the (Farey) tesselation, so the general flip $\phi(g)$ on the edge $g(doe)$, where

$$g = \beta^{\epsilon_n}\alpha^2\beta^{\epsilon_{n-1}}\alpha^2\cdots\alpha^2\beta^{\epsilon_1} \text{ with } \epsilon_j \in \{1,2\} \text{ for } n \geq j > 1 \text{ and } \epsilon_1 \in \{0,1,2\},$$

can be written

$$\phi(g) = g^{-1}\beta^{2\epsilon_n}\alpha^{2\epsilon_n - 1}g$$
$$= \begin{cases} \beta^{2\epsilon_1}\alpha^2\cdots\alpha^2\beta^{2\epsilon_{n-1}}\alpha^2\ \beta\alpha\beta\ \alpha^2\beta^{\epsilon_{n-1}}\alpha^2\cdots\alpha^2\beta^{\epsilon_1}, \text{if } \epsilon_n = 1; \\ \beta^{2\epsilon_1}\alpha^2\cdots\alpha^2\beta^{2\epsilon_{n-1}}\alpha^2\ \beta^2\alpha^3\beta^2\ \alpha^2\beta^{\epsilon_{n-1}}\alpha^2\cdots\alpha^2\beta^{\epsilon_1}, \text{if } \epsilon_n = 2, \end{cases}$$

applying words from right to left as before. These are simply the conjugates in PSL(2,ℤ) of the flips noted above.

In particular, it follows that α and β indeed generate PPSL(2,ℤ) since the group they generate contains the flips on all edges. Moreover, $[\phi(g)]^{-1} = \phi(\beta^{\epsilon_n}g)$, namely, to invert, change the first (leftmost) exponent $1 \leftrightarrow 2$ of β in g. Idempotence of a flip is simply a conjugate of $(\beta\alpha\beta)(\beta^2\alpha^3\beta^2) = 1$, which is not noteworthy for PPSL(2,ℤ) since it follows from $\beta^3 = 1 = \alpha^4$, but it is of consequence for P(SL(2,ℤ)). As with flips conjugating by elements of PSL(2,ℤ), the one pentagon relation gives rise to all pentagon relations.

It remains only to prove that the two commutators in the theorem imply the commutativity relations for flips on any pair of edges which do not bound a common triangle. This is called a "remarkable fact" in [3], and here follows its proof by induction: First note that quite generally two flips commute if and only if their conjugates commute. Thus, if flips on two respective edges e, f commute, then flips on any two edges in the same relative positions in the tesselation as e, f also commute. The first commutator in the statement of the theorem is thus $[\phi(\beta), \phi(\beta\alpha^2)] = 1$, or equivalently

$$[\alpha^2, \beta\alpha\beta\ \alpha^2\ \beta\alpha\beta] = 1, \tag{*}$$

and this in turn one finds is also equivalent to $[\phi(\beta), \phi(\beta^2\alpha^2)] = 1$; taking inverses one has also $[\phi(\beta^2), \phi(\beta\alpha^2)] = [\phi(\beta^2), \phi(\beta^2\alpha^2)] = 1$. Thus, the flip on any edge commutes with the flips on edges *exactly two triangles away*, i.e., a path in general position between the edges meets precisely three edges in the tesselation; this is equivalent to (*).

The flip α on the doe does not commute with other flips because it engenders a change of notation, cf. Fig. 5.8. Rather, the flip $\phi(g)$ on $g(doe)$ evidently satisfies the following functional equation

$$\text{FE}(g): \qquad \alpha\,\phi(g) = \phi(T(g))\,\alpha,$$

where in the notation for g above

$$T(g) = \begin{cases} g\beta^{\epsilon_1}\alpha^{2\epsilon_1}, & \text{if } \epsilon_1 \neq 0; \\ g\alpha^2\beta^{\epsilon_2}\alpha^{2\epsilon_2}, & \text{if } \epsilon_1 = 0, \end{cases}$$

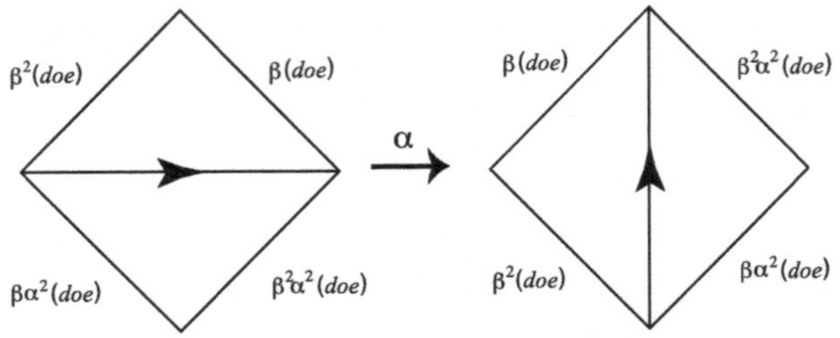

Fig. 5.8 Change of notation under α near the doe

or in other words the transformation T acts on suffixes of g by

$$\beta \mapsto \beta^2\alpha^2 \mapsto \beta\alpha^2 \mapsto \beta^2 \mapsto \beta.$$

In particular, T is a square root of right multiplication by α^2.

The flip on the edge underlying the doe commutes with the flip $\phi(g)$ if and only if g satisfies the functional equation FE(g). Thus, the first commutator in the statement of the theorem is equivalent to FE(g) holding for all $g = \beta^{\epsilon_2}\alpha^2\beta^{\epsilon_1}$ with $\epsilon_1 \neq 0$ and $n = 2$ in our normal form. Moreover if $\epsilon_1(g) \neq 0$, then FE(g) trivially implies FE($g\alpha^2$), so it suffices to assume henceforth that $\epsilon_1 \neq 0$ and proceed by induction on n, as we next undertake. One confirms that the case $n = 3$ of FE is likewise equivalent to the identity

$$[v, \beta v \beta^2] = 1, \text{ where } v = \alpha^2\, \beta\alpha\beta\, \alpha^2, \tag{†}$$

which is evidently equivalent to $[\beta^{\epsilon} v^{\pm 1}\beta^{2\epsilon}, \beta^{\delta} v^{\pm 1}\beta^{2\delta}] = 1$, for any $\epsilon, \delta \in \{1, 2\}$, and finds that (†) is in turn equivalent to the vanishing of the second commutator in the statement of the theorem.

For the inductive step, suppose that $g = \beta^{\epsilon_n}\alpha^2 \cdots \alpha^2\beta^{\epsilon_1}$ with $n > 3$ in our normal form with $\epsilon_1 = \epsilon_1(g) \neq 0$, and set $h = \beta^{\epsilon_n}\alpha^2 \cdots \alpha^2\beta^{\epsilon_2}$. FE($g$) is given by

$$\begin{aligned} \alpha\, \phi(h\alpha^2\beta^{\epsilon_1}) &= \phi(h\alpha^2\beta^{\epsilon_1}\, \beta^{\epsilon_1}\alpha^{2\epsilon_1})\, \alpha \\ &= \alpha^{2\epsilon_1}\, \phi(h\alpha^2\beta^{2\epsilon_1})\, \alpha^{2\epsilon_1+1}, \end{aligned}$$

so for $\epsilon_1(g) \neq 0$, FE(g) reads

$$\alpha^{2\epsilon_1+1}\, \phi(g)\, \alpha^{2\epsilon_1-1} = \beta^{2\epsilon_1}\, \phi(g)\, \beta^{\epsilon_1},$$

which simply means: if $\epsilon_1(g) = 2$, then you can pull α to the right across $\phi(g)$ at the expense of changing this to $\epsilon_1 = 1$, or equivalently to the left again changing the terminal exponent if $\epsilon_1(g) = 1$. In particular for any $N \geq 1$, FE(g) holds for $g = (\beta\alpha^2)^N\beta$ and likewise for βg.

Suppose first that in fact $\epsilon_1(g) = 1$. If g differs from the forms above that automatically satisfy FE, then there is some index $1 < m < n$ with $\epsilon_m = 2$, so that $g = h\alpha^2 k$ where h ends with β^2 and k of course ends with β since we assume that g does, and both h and k satisfy FE by the strong inductive hypothesis. Using the simple description of the FE as right/left commutativity laws for α across flips in the previous paragraph, one finds that FE(g) is equivalent to $[\phi(h'), \phi(k')] = 1$, where h', k' respectively arise from h, k by altering their terminal β-exponents. Since the relative positions of the flipped edges is decreased by unity, FE holds in general by induction. The analogous argument holds for $\epsilon_1(g) = 2$ using the automatic solutions FE($\beta^2(\alpha^2\beta^2)^N\alpha^2$), for $N \geq 1$. □

Notice that according to the proof there are exactly two conjugacy classes in PPSL(2, ℤ) of flips on edges other than the doe, namely, the conjugacy classes of $\beta\alpha\beta$ and $\beta^2\alpha^3\beta^2$. Since these are inverses, the collection of flips on edges other than the doe abelianizes to a cyclic group. Inspection of Fig. 5.2 shows that the five flips of the pentagon relation are comprised of two from one class and three from the other. It follows that the pentagon relations alone imply that flips on edges other than the doe lie in the first derived subgroup.

References

1. J. Cannon, W. Floyd, W. Parry, Introductory remarks on Richard Thompson's groups. L'Enseignement Math. **42**, 215–256 (1996)
2. I. Frenkel, R. Penner, Sketch of a program for automorphic functions from universal Teichmüller theory to capture monstrous moonshine. Commun. Math. Phys. **389**, 1525–1567 (2022)
3. P. Lochak, L. Schneps, The universal Ptolemy-Teichmüller groupoid, in *Geometric Galois Actions II*, ed. by P. Lochak, L. Schneps. London Math Society Lecture Notes, vol. 243 (Cambridge University Press, Cambridge, 1997)
4. R.C. Penner, The decorated Teichmüller space of punctured surfaces. Commun. Math. Phys. **113**, 299–339 (1987)
5. R.C. Penner, Universal constructions in Teichmüller theory. Adv. Math. **98**, 143–215 (1993)
6. R.C. Penner, *Decorated Teichmüller Theory* (European Mathematical Society, Zurich, Switzerland, 2012)
7. R. Penner, A. Zeitlin, Decorated super-Teichmüller space. J. Differ. Geom. **111**, 527–566 (2019)
8. R.C. Penner, Universal Spin Teichmüller Theory, I. The action of P(SL(2,ℤ)) on $\mathcal{T}ess^+$ (2023, preprint)

Chapter 6
Geometry of Fermat's Sum of Squares

Greg McShane and Vlad Sergiescu

Abstract In this paper, we present a novel geometric proof of Fermat's sum of two squares theorem, which states that a prime number p can be expressed as the sum of two squares if and only if $p \equiv 1 \pmod 4$. Our proof relies on classical techniques from hyperbolic geometry, specifically leveraging calculations familiar to most graduate students, and an analysis of the fixed point sets of automorphisms of the three-punctured sphere. We propose a fresh geometric perspective on this centuries-old result, which parallels the key ideas found in Heath-Brown's celebrated proof. As such this approach offers a new connection between number theory and hyperbolic geometry, enriching the understanding of Fermat's theorem and opening potential avenues for further exploration.

Keywords Hyperbolic geometry · Two squares theorem

AMS Codes 11D41, 11D72

6.1 Introduction

Consider the following pair of well-known theorems from elementary number theory:

Theorem 6.1.1 *Let p be a prime then the equation*

$$x^2 = -1$$

admits a solution in $\mathbb{F}_p$ *if and only if* $p = 2$ *or* $p - 1$ *is a multiple of* 4.

G. McShane (✉) · V. Sergiescu
Institut Fourier 100 rue des Maths, St Martin d'Hères Cedex, France
e-mail: mcshane@univ-grenoble-alpes.fr; vlad.sergiescu@univ-grenoble-alpes.fr

L. Funar, A. Papadopoulos (eds.), *Essays on Topology*,
https://doi.org/10.1007/978-3-031-81414-3_6

Theorem 6.1.2 (Fermat) *Let p be a prime then the equation*

$$x^2 + y^2 = p$$

has a solution in integers if and only if $p = 2$ or $p - 1$ is a multiple of 4.

These results are intimately linked and often one deduces the second as a corollary of the first, for example, by using unique factorisation in the Gaussian integers. We present a unified geometric approach to these results using the theory of group actions and in particular an application of Burnsides's lemma.

As in Zagier's remarkable proof [10] (see also [1, 2, 4, 5] for closely related constructions and discussion) both results follow from showing that a certain involution has a fixed point. Amusingly Burnsides's lemma reduces this to showing that another involution has exactly two fixed points:

- In the proof of Theorem 6.1.1 this is a consequence of the fact that a quadratic equation over a field has at most two solutions.
- In the proof of Theorem 6.1.2 this follows from some geometry and the fact that every odd prime can be written in an essentially unique way as the dif and only iference of two squares $a^2 - b^2$.

6.1.1 Organisation and Remarks

In Sect. 6.2 we recall the statement of Burnside's lemma and apply it to a Klein four group generated by involutions of $\mathbb{F}_p^*$ yielding a proof of Theorem 6.1.1. In Sect. 6.3 we introduce $\Gamma(2)$ and the associated Riemann surface $\mathbb{H}/\Gamma(2)$. In Sect. 6.4, for each prime p we study how the automorphisms of $\mathbb{H}/\Gamma(2)$ act on a family of geodesic on this surface obtained in a natural way from the rationals k/p. In particular we show that if p is congruent to 1 modulo 4 there is always an orientation-preserving involution that leaves one of our geodesics invariant and from this we deduce Theorem 6.1.2.

6.1.1.1 Heath-Brown's Proof

In 1984 Heath-Brown published a proof [4] of Theorem 6.1.2 apparently in the journal of the Oxford University undergraduate mathematics society. His proof arose from a study of the account of Liouville's papers on identities for parity functions, presented by Uspensky and Heaslet in the 1970s. Zagier's proof in [10] is a clever reformulation of ths argument.

Like our proof it is based on an action of a Klein four group on a finite set and considerations of parity. To define his set Heath-Brown introduces an auxiliary equation, namely

$$p = 4xy + z^2,$$

whereas in our proof the sum of squares decomposition arises directly as the result of a geometric construction. As such, the motivation for our work is to show that the finite sets involved in the proof can be chosen to be both natural and have a geometric interpretation. For example, in Sect. 6.2 we give a proof of Theorem 6.1.1 using a group generated by

$$\begin{aligned} x &\mapsto -x \\ x &\mapsto 1/x \end{aligned}$$

and in the proof of Theorem 6.1.2 our group is generated by

$$\begin{aligned} z &\mapsto -\bar{z} \\ z &\mapsto 1/\bar{z} \end{aligned}.$$

6.1.1.2 Burnside's Lemma and Signatures

The astute reader will surely realise that Burnside's lemma is not essential to our argument and that one can achieve the same reduction by considering the signature of the permutations associated to the involutions we consider. This approach is closer to the parity arguments in Heath-Brown [4].

6.1.1.3 Farey Tessellation, λ-Lengths

Much of our inspiration for this approach comes from endless contemplation of the *Farey tessellation.*

The idea of associating a length to a geodesic joining cusps (paragraph 6.3.1) appears in Penner's work on moduli [6], see also [8] for a more recent account. He defined the λ-length of a simple bicuspidal geodesic on a punctured surface to be the exponential of half the length of the portion outside of some fixed system of cusp regions. Though not strictly necessary, we will phrase parts of proof in terms of λ-lengths, as this is the origin of the intuition behind our approach. Lemma 6.3.2 shows that in the context we consider, a λ-length is always the determinant of an integer matrix.

In [6] Penner presents an approach to the Markoff equation using λ-lengths. Alternatively, using the existing calculations in Wolpert [9] one can show that, for a suitable choice of cusp region on the modular torus the λ-lengths of arcs coincide with the Markoff numbers. Then, using the fact that each arc is invariant under the elliptic involution one can show, using Lemma 6.3.3, that every Markoff number is the sum of two squares. In fact this was the observation that was the starting point for this paper.

6.1.1.4 Bézouts Lemma

We are implicitly using Bézout's Lemma which states that for any pair of integers a and b, if d is the greatest common divisor of a and b, then there exist integers x and y such that:

$$ax + by = d$$

and in particular if a, b are coprime then

$$ax + by = 1.$$

For example in the proof of Lemma 6.5.1 when we assert that

- $SL(2, \mathbb{Z})$ is transitive on $\mathbb{Q} \cup \infty$ (which is equivalent to Bézout's Theorem.)
- $\Gamma(2)$ has exactly three orbits on $\mathbb{Q} \cup \infty$.

we are relying on Bézout's Lemma. In fact Lemma 6.5.1 can be proved without using our notion of length for a bicuspidal geodesics but instead by studying the action of the lifts U, V of the generators of our group G_4 and applying Bézout's Theorem.

6.1.1.5 References

Almost all of the material in Sects. 6.3 and 6.4 can be found in Serre's book [7] and the reader should not need any other references to understand this paper if they are already familiar with Burnside's lemma.

6.1.2 Thanks

The first author thanks Louis Funar and the second author for many useful conversations over the years concerning this subject. He would also like to thank Xu Binbin for reading early drafts of the manuscript.

6.2 Burnside's Lemma

We give a proof of Theorem 6.1.1 using Burnside's Lemma. Recall that if G is a group acting on a finite set X then Burnside's lemma says

$$|G||X/G| = \sum_g |X^g| \tag{6.1}$$

where, as usual, X^g denotes the set of fixed points of the element g and X/G the orbit space.

Let $p \neq 2$, $X = \mathbb{F}_p^*$ and G be the group generated by the two involutions

$$x \mapsto -x$$
$$x \mapsto 1/x.$$

The group G has exactly four elements namely:

- the trivial element which has $p - 1$ fixed points
- $x \mapsto -x$ which has no fixed points
- $x \mapsto 1/x$ has exactly two fixed points namely 1 and -1.
- $g : x \mapsto -1/x$ is the remaining element, and the theorem is equivalent to the existence of a fixed point for it.

Since $\mathbb{F}_p$ is a an integral domain the equation $x^2 = -1$ has at most two solutions. If there is a solution α then $-\alpha$ is also a solution and these are distinct unless $p = 2$ which we have excluded above so $|X^g| = \sharp\{x^2 = -1,\ x \in \mathbb{F}_p^*\}$ is either 0 or 2. Now for our choice of X and G Eq. (6.2) yields

$$4|X/G| = (p-1) + 2 + |X^g|. \tag{6.2}$$

The LHS is always divisible by 4 so the RHS is too and hence

$$|X^g| = \begin{cases} 0 \ (p-1) = 2 \mod 4 \\ 2 \ (p-1) = 0 \mod 4. \end{cases}$$

This proves Theorem 6.1.1.

Note

As was noted in the introduction one can obtain the same conclusion by calculating the signature of $x \mapsto -1/x$ using the fact that it is the composition of $x \mapsto -x$ and $x \mapsto 1/x$.

6.3 Reciprocals of Sums of Squares and Arcs

This group of integer matrices $\mathrm{SL}(2, \mathbb{Z})$ acts on $\mathbb{H}$ by linear fractional transformations that is:

$$\begin{pmatrix} a & b \\ c & d \end{pmatrix} \in \mathrm{SL}(2, \mathbb{Z}),\ z \in \mathbb{H},\ \begin{pmatrix} a & b \\ c & d \end{pmatrix} \cdot z = \frac{az+b}{cz+d}.$$

The key lemma that relates this $\mathrm{SL}(2, \mathbb{Z})$ action to sums of squares is Lemma 6.3.3.

6.3.1 Ford Circles, Lengths, Midpoints

Lemma 6.3.3 illustrates the connexion between sums of squares, the orbit $\mathrm{SL}(2, \mathbb{Z})\cdot i$ and Poincaré geodesics. We now recall some standard ideas from hyperbolic geometry which in particular will allow us to give an intuitive definition of our set X in the next section. We define an *arc* to be a Poincaré geodesic with endpoints in $\partial\mathbb{H}$ a pair of extended rationals, that is elements of $\mathbb{Q} \cup \infty$.

We denote by F the set $\{z, \operatorname{Im} z > 1\}$ this is a *horoball in* $\mathbb{H}$ centered at ∞. The image of F under the $\mathrm{SL}(2, \mathbb{Z})$ action consists of F and infinitely many disjoint discs, which we will refer to as *Ford circles*, each tangent to the real line at some rational m/n. We adopt the convention that F is also a Ford circle of infinite Euclidean radius tangent to the extended real line at $\infty = 1/0$ (Fig. 6.1).

The following is well known and is easily checked:

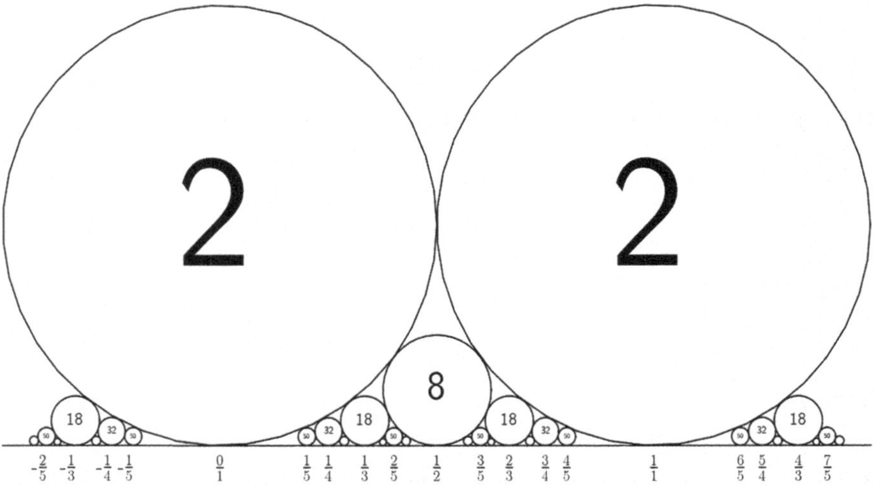

Fig. 6.1 Ford circles with tangent points and curvatures. Recall that the curvature of a euclidean circle is the reciprocal of the square of its radius

Lemma 6.3.1 (1) *The Ford circle tangent to the real line at m/n has Euclidean diameter $1/n^2$.*

(2) *The closures of a pair of distinct Ford circles are either disjoint or meet in a point of the* $\mathrm{SL}(2,\mathbb{Z})$*-orbit of i.*

Let a/c, b/d be a pair of distinct rationals. We define the *length* of the arc joining these rationals to be the length, with respect to the Poincaré metric on $\mathbb{H}$, of the portion of this arc outside of the Ford circles tangent at $a/c, b/d$.

Following Penner [6], see also [8] for a more recent account, we define the λ-length of the arc to be the exponential of half of this length and define its *midpoint* to be the midpoint of this sub arc.

Lemma 6.3.2 *Let $a/c, b/d$ be a pair of distinct extended rationals. Then the λ-length of the arc joining them is the absolute value of the determinant of the matrix*

$$\begin{pmatrix} a & b \\ c & d \end{pmatrix}.$$

Further if $a/c = 1/0$ then the arc is a vertical line whose midpoint has imaginary part equal to $1/d$.

Proof By transitivity of the $\mathrm{SL}(2,\mathbb{Z})$ action on the extended rationals we may suppose $a/b = 1/0 = \infty$. The arc joining $a/c, b/d$ is a vertical line and the determinant of the matrix in the statement is just d. By Lemma 6.3.1 the Ford circle tangent at b/d has diameter $1/d^2$ so the portion of the arc outside F (the Ford circle tangent at ∞) and this circle is a segment between z_1 with imaginary part 1 and z_2 with imaginary part $1/d^2$. A standard calculation using the Poincaré metric $\frac{dz}{\mathrm{Im} z}$ shows that the length of this segment is indeed $\log(d^2)$. The second part of the statement follows from a similar calculation. □

The arcs of λ-length 1 are the edges in the so-called *Farey diagram* (see Fig. 6.2). Recall that the Farey diagram is the graph whose vertices are the rationals and where two vertices are joined by an edge if the determinant of the matrix

$$\begin{pmatrix} a & b \\ c & d \end{pmatrix}$$

is ± 1.

6.3.2 Sums of Squares

Lemma 6.3.3 *Let n be a positive integer. The number of ways of writing n as a sum of squares*

$$n = c^2 + d^2$$

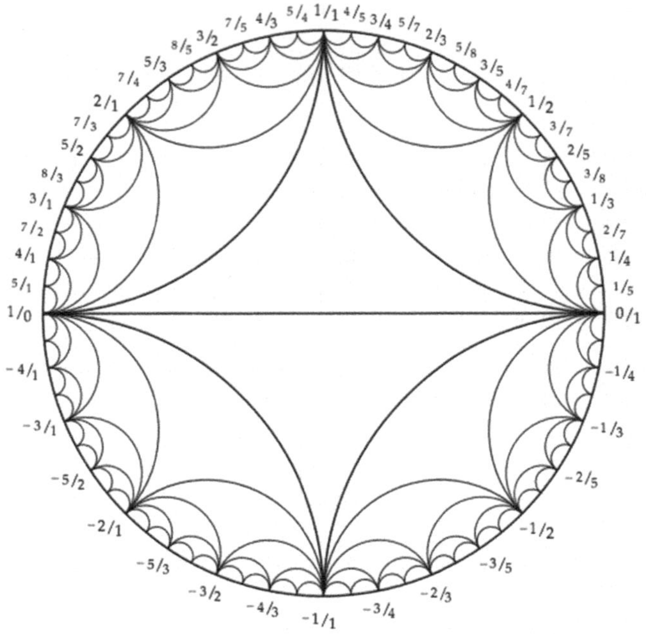

Fig. 6.2 Farey diagram

with c, d coprime positive integers is equal to the number of integers $0 \leq k < n-1$ coprime to n such that the line

$$\{k/n + it,\ t > 0\}$$

contains a point in the $\mathrm{SL}(2, \mathbb{Z})$ *orbit of* i.

Proof Suppose there is such a point which we denote w. The point w is a fixed point of some element of order 2 in $\mathrm{SL}(2, \mathbb{Z})$. Since the Ford circles are $\mathrm{SL}(2, \mathbb{Z})$ invariant this element must permute $F = \{z, \operatorname{Im} z > 1\}$ with the Ford circle tangent to the real line at the real part of w. So, in particular, w is the midpoint of the line that it lies on and by Lemma 6.3.2 one has:

$$\frac{1}{n} = \operatorname{Im} \frac{1}{n}(k+i) = \operatorname{Im} \frac{ai+b}{ci+d} = \frac{\operatorname{Im} i}{c^2+d^2}.$$

Conversely if c, d are coprime integers then there exists a, b such that

$$ad - bc = 1 \Rightarrow \begin{pmatrix} a & b \\ c & d \end{pmatrix} \in \mathrm{SL}(2, \mathbb{Z}).$$

By applying a suitable iterate of the parabolic transformation $z \mapsto z+1$, one can choose w such that $0 \leq \operatorname{Re} w < 1$. So if $n = c^2 + d^2$ then $\frac{ai+b}{ci+d}$ is on one of the lines of the family in the statement. □

6.3.3 Relation with Square Roots of -1

As we saw in the proof of Theorem 6.1.1 under the hypothesis of that p is a prime and $p-1$ is a multiple of 4 there is $\bar{m} \in \mathbb{F}_p$ such that $\bar{m}^2 = -1$. In fact the real part of $\frac{ai+b}{ci+d}$ is related to the square this $\bar{m}$ in a simple way which, although we will not need it later, we explain now. Begin by writing

$$\frac{ai+b}{ci+d} = \frac{m}{p} + \frac{i}{p}$$

then

$$\frac{m^2+1}{p^2} = \left(\frac{ai+b}{ci+d}\right)\overline{\left(\frac{ai+b}{ci+d}\right)} = \frac{a^2+b^2}{c^2+d^2}$$

now since $p = c^2 + d^2$ one has:

$$m^2 + 1 = p(a^2 + b^2). \tag{6.3}$$

Thus m determines a square root of -1 in $\mathbb{F}_p$.

6.4 The Three Punctured Sphere

In this section we are concerned with the geometry of the surface $\mathbb{H}/\Gamma(2)$ associated to $\Gamma(2)$, the principal level 2 congruence subgroup of $\mathrm{SL}(2, \mathbb{Z})$ that is the kernel of the canonical (reduction) homomorphism $\mathrm{SL}(2, \mathbb{Z}) \to \mathrm{SL}(2, \mathbb{Z})/2\mathbb{Z}$. This group acts on $\mathbb{Z}^2$, that is pairs of integers, preserving parity.

It also acts on $\mathbb{H}$ by linear fractional transformations that is:

$$\begin{pmatrix} a & b \\ c & d \end{pmatrix} \in \mathrm{SL}(2, \mathbb{Z}),\ z \in \mathbb{H},\ \begin{pmatrix} a & b \\ c & d \end{pmatrix} \cdot z = \frac{az+b}{cz+d}.$$

The quotient $\mathbb{H}/\Gamma(2)$ is conformally equivalent to the Riemann sphere minus three points which we will refer to as *cusps* (see Fig. 6.4). Following convention we label these cusps $0, 1, \infty$ respectively, corresponding to the three $\Gamma(2)$ orbits of $\mathbb{Q} \cup \infty$. Finally, the *standard fundamental domain* for $\Gamma(2)$ is the convex hull of the points $\infty, -1, 0, 1$. This region can be decomposed into two ideal triangles $\infty, -1, 0$ and $0, 1, \infty$ as in Fig. 6.3. The edges of the ideal triangles project to three disjoint simple geodesics on $\mathbb{H}/\Gamma(2)$ and each edge has a *midpoint* which is a point of the $\mathrm{SL}(2, \mathbb{Z})$ orbit of i (see Figure 6.4).

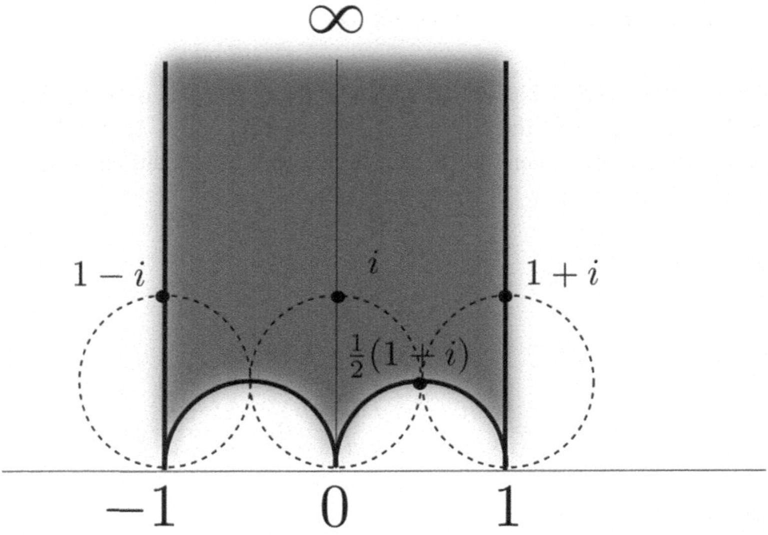

Fig. 6.3 Standard fundamental domain for $\Gamma(2)$ and its decomposition into ideal triangles

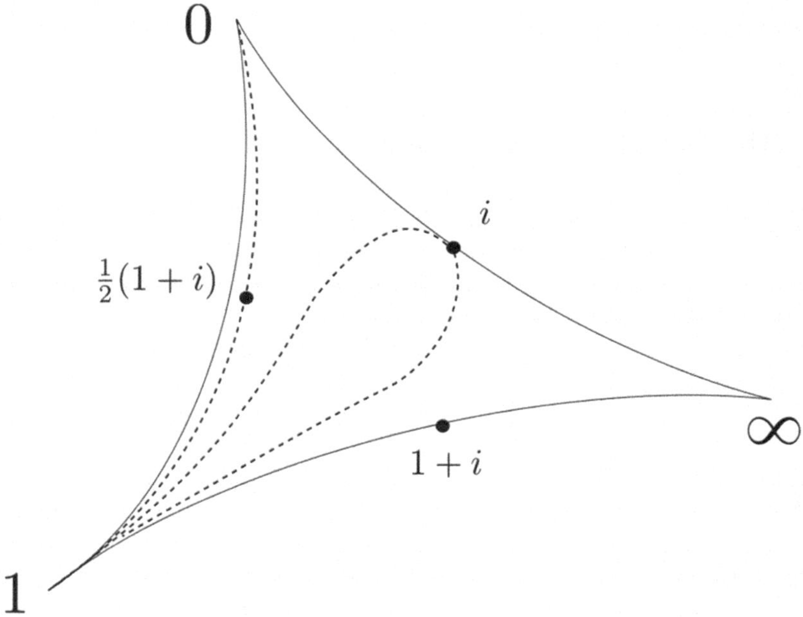

Fig. 6.4 A depiction of the three punctured sphere with cusps and midpoints labelled. The dotted loop is the fixed point set of the automorphism induced by V'

6.4.1 Cusp Regions

The image of a Ford circle on $\mathbb{H}/\Gamma(2)$ is a *cusp region* around one of the three cusps $0, 1, \infty$. Any pair of these cusp regions are tangent are tangent at one of the midpoints labelled i, $1+i$, $\frac{1}{2}(1+i)$. It is not dif and only ificult to see that these cusp regions are permuted by the automorphisms of $\mathbb{H}/\Gamma(2)$. It follows that if an automorphism preserves a geodesic joining cusps on $\mathbb{H}/\Gamma(2)$ then it must permute the Ford regions at each end of a lift to $\mathbb{H}$.

6.4.2 Automorphism Groups of $\mathbb{H}/\Gamma(2)$

From covering theory an isometry of $\mathbb{H}$ induces an automorphism of $\mathbb{H}/\Gamma(2)$ if and only if it normalises the covering group i.e. $\Gamma(2)$. Since $\Gamma(2)$ is a normal subgroup of $\mathrm{SL}(2, \mathbb{Z})$, the quotient group

$$H^+ := \mathrm{SL}(2, \mathbb{Z})/\Gamma(2)$$

acts as a group of (orientation preserving) automorphisms of the surface $\mathbb{H}/\Gamma(2)$. More generally, $\Gamma(2)$ is normal in $\mathrm{GL}(2, \mathbb{Z})$ and

$$H := \mathrm{GL}(2, \mathbb{Z})/\Gamma(2)$$

acts as a group of possibly orientation reversing automorphisms of the surface $\mathbb{H}/\Gamma(2)$.

6.4.3 Orientation Reversing Automorphisms

To prove Theorem 6.1.2 we will have to work with automorphisms that do not preserve the orientation and in particular those induced by the involutions:

$$U : z \mapsto -\bar{z}$$
$$V : z \mapsto 1/\bar{z},$$

generate a group which we denote G_4. Note that the composition of U and V is none other than the involution

$$z \mapsto -1/z,$$

so we have a Klein 4-group as in Sect. 6.2. Both U and V normalise $\Gamma(2)$ so induce automorphisms of $\mathbb{H}/\Gamma(2)$.

Lemma 6.4.1 *The group G_4 descends to a group of automorphisms on the three punctured sphere which preserve the cusp labelled* 1 *(see Fig. 6.4).*

Proof One checks that each of the generators U and V preserve the standard fundamental domain and further they preserve the set $\{1, -1\}$. Note that this pair of points belong to the same $\Gamma(2)$ orbit and so project to the same cusp on $\mathbb{H}/\Gamma(2)$. It follows that the generators preserve this cusp ie the cusp labelled 1. □

6.4.4 Fixed Point Sets

The group generated by U, V which we will denoted G_4 in paragraph 6.4.3. Consider the fixed point sets of the elements:

- U fixes the vertical line $\{it, t > 0\}$;
- V fixes the semi circle joining -1 and 1;
- the composition $U \circ V(z)$ is just $z \mapsto -1/z$ and so fixes i.

From this we may deduce that the automorphisms of $\mathbb{H}/\Gamma(2)$ induced by U and V each fix a pair of lines on the surface. The fixed point set of V projects to a geodesic on $\mathbb{H}/\Gamma(2)$ (depicted as a dotted loop in Fig. 6.4) separating the surface into two pieces which are permuted by the corresponding automorphism, so the fixed point set is exactly this geodesic. For U the fixed point set of the induced automorphism is strictly bigger as it will also fix the images on the surface of $\{1 + it, t \in \mathbb{R}\}$ and the semi circle joining 0 to 1. This is because these arcs are sides of the standard fundamental domain for $\Gamma(2)$:

- $U(1+it) = -1+it = f(1+it)$, where $z \mapsto z-2$ is the side-pairing for these sides.
- U exchanges the semi circle joining 0 to 1 and the semi circle joining 0 to -1 and these sides are paired by $z \mapsto z/(2z+1)$

This proves the first part of:

Lemma 6.4.2 *Any arc of the Farey diagram projects to a component of the fixed point set of the automorphism of $\mathbb{H}/\Gamma(2)$ induced by U.*

The fixed point set of the automorphism induced by $U \circ V$ is exactly the intersection of the fixed point sets of the automorphisms induced by U and V. This is a single point namely the image of i on $\mathbb{H}/\Gamma(2)$

Proof The first part of the statement follows from the discussion above: the automorphism of U exchanges the pair of ideal triangles in the standard decomposition of $\mathbb{H}/\Gamma(2)$ leaving their edges fixed.

The second part of lemma follows from the observation that $U \circ V$ leaves the standard fundamental domain for $\Gamma(2)$ invariant swapping the two ideal triangles in the complement of the fixed point set of U. One sees from this that any fixed point of $U \circ V$ is contained in the fixed point set of V. Further, $U \circ V$ fixes the cusp labeled

1 but exchanges the cusps labeled 0 and 1. It follow that any fixed point of $U \circ V$ is contained in the projection of the vertical line $\{it, t > 0\}$. This line contains i and the lemma is proven. □

6.5 Action on a Family of Arcs

Now G_4 permutes the cusps labelled ∞ and 0 on $\mathbb{H}/\Gamma(2)$ and will obviously permute the geodesics joining them. If γ is such a geodesic then any lift $\hat{\gamma} \subset \mathbb{H}$ is an arc joining a point in the $\Gamma(2)$ orbit of ∞ to another in the $\Gamma(2)$ orbit of 0 and so γ has a well defined λ-length. For any integer n G_4 permutes the set of geodesics joining the cusps labelled ∞ and 0 on $\mathbb{H}/\Gamma(2)$ of λ-length n^2. This will be our set X.

6.5.1 Canonical Lifts

Let n be an integer and N' the set of integers coprime with n. Consider the collection of geodesics of $\mathbb{H}$.

$$\{k/n + it,\ t > 0\},\ k \in N'.$$

The image of this collection on the quotient surface $\mathbb{H}/\Gamma(2)$ is a family of arcs and, since $\Gamma(2)$ preserves parity, these split into two sub families namely:

- those joining the cusps labelled ∞ and 0 that is belonging our set X
- those joining the cusps labelled ∞ and 1.

The first of these sub families consists of projections of the lines

$$\hat{X} := \{2k/n + it,\ t > 0\},\ k \in N'$$

Lemma 6.5.1 *Let p be a prime then the set X consists of $p - 1$ elements.*

Proof The $z \mapsto z + 2$ is a generator of the subgroup of $\Gamma(2)$ that stabilises ∞ so to have a complete set of representatives of $\hat{X}$ we need only consider $2k/n$ with $0 < k < n$. If n is prime then for any such k the integers $2k$ and n are coprime so $\hat{X}$ contains exactly $n - 1$ elements. □

6.6 Proof of Fermat's Theorem

Throughout this section the integer n is a prime which we denote p. Theorem 6.1.2 follows from Lemma 6.3.3 and the following result:

Lemma 6.6.1 *Let p be a prime congruent to 1 or 2 modulo 4. Then there is always a geodesic in the family* $\hat{X}$ *that has as its midpoint a point in the* SL(2, $\mathbb{Z}$) *orbit of* i.

This is equivalent to saying that, on projecting to the surface $\mathbb{H}/\Gamma(2)$, there is always a geodesic in X which passes through the fixed point of the map induced by $U \circ V$. For $p = 2$ this can be done explicitly and for the general case by the argument using Burnside's lemma in Sect. 6.2 it suffices to show that:

- U fixes no element of X
- V fixes at most two.

6.6.1 *The Singular Case of Lemma 6.6.1*

The case $p = 2$ is exceptional and we will deal with it first. From the preceding paragraph there is a single geodesic namely the projection of the line

$$\{1/2 + it,\ t \in \mathbb{R}\}$$

and this contains the point $\frac{1}{2}(1+i)$ Note that one has

$$\begin{pmatrix} 1 & 0 \\ 1 & 1 \end{pmatrix} \in \mathrm{SL}(2,\mathbb{Z}),\ \begin{pmatrix} 1 & 0 \\ 1 & 1 \end{pmatrix} \cdot i = \frac{1}{2}(1+i)$$

so this point is in the SL(2, $\mathbb{Z}$) orbit of i> Then one has as in Lemma 6.3.3:

$$\mathrm{Im}\,\frac{1}{2}(1+i) = \frac{1}{2} = \mathrm{Im}\begin{pmatrix} 1 & 0 \\ 1 & 1 \end{pmatrix} \cdot i = \frac{\mathrm{Im}\, i}{1^2 + 1^2}$$

So, in a rather roundabout way, we obtain 2 as a sum of squares by comparing denominators:

$$2 = 1^2 + 1^2.$$

6.6.2 *Inversions and Fixed Points in X*

We will finish the proof of Lemma 6.6.1 by showing that there is a geodesic invariant by the orientation preserving automorphism in G_4, obtaining the required midpoint as the fixed point of the automorphism induced by $U \circ V$. Our argument is exactly the same as for Theorem 6.1.1. More precisely, we show that, for any $p > 2$:

(1) the automorphism induced by U preserves no geodesic in X
(2) the automorphism induced by V preserves at most two geodesics in X

The first point is rather easy (the automorphism induced by U fixes three disjoint geodesics joining cusps and permutes the pair of ideal triangles in their complement) but the second requires establishing the analogue of the fact that the equation

$$x^2 = 1$$

has at most two solutions in any field or integral domain for that matter.

Let us start by considering the action of V on the set of rationals, recall that

$$V : z \mapsto 1/\bar{z}$$

so that for coprime integers a, b

$$a/b \mapsto b/a.$$

If $a/b \neq \pm 1$ then this map preserves the geodesic joining a/b and b/a and the λ-length of this arc can be computed using the determinant formula:

$$\lambda - \text{length} = \left| \begin{pmatrix} a & b \\ b & a \end{pmatrix} \right| = a^2 - b^2 = \pm p$$

So if this is to be a lift of one of our arcs and since p is prime the only solutions must satisfy (after possibly permuting a, b):

$$a + b = \pm p,\ a - b = \pm 1.$$

Thus up to changing sign

$$a = \pm(p + 1)/2,\ b = (p - 1)/2.$$

Thus we have:

Lemma 6.6.2 *The automorphism induced by V preserves two and exactly two geodesics in X.*

6.6.3 Composite Integers

It is well known that the set of integers n which can be written as a sum of squares $c^2 + d^2$ with c, d coprime is (almost) closed under multiplication: if p, q are sums of squares and at least one is odd then the product is also a sum of squares. For example one has

$$50 = 5^2 \times 2 = |(1 + 2i)^2(1 + i)|^2 = |-7 + i|^2 = 7^2 + 1^2$$

and

$$65 = 5 \times 13 = |(1+2i)(2+3i)|^2 = |-4+7i|^2 = 7^2 + 4^2.$$

but of course

$$20 = 2 \times 10 = |(1+i)(1+3i)|^2 = |-2+4i|^2 = 2^2 + 4^2.$$

Despite this multiplication by 2 seems to have a nice geometric interpretation in the spirit of our analysis of arcs but multiplication by 5 does not seem to have one. Let p be a prime such that $p-1$ is a multiple of 4 and m a square root of -1 then $\frac{m}{p} + \frac{i}{p}$ is a point of the $\Gamma(2)$-orbit of i. The semi circle joining $\frac{m-1}{p}$ to $\frac{m+1}{p}$ has λ-length $2p$ and passes through $\frac{m}{p} + \frac{i}{p}$. So by Lemma 6.3.3 $2p$ is a sum of squares.
It seems dif and only ificult to prove that 65 is a sum of squares directly using our approach with Burnside's lemma and considering the set X of arcs of λ-length 65. Naively applying Burnside's lemma yields:

$$4|X/G| = \phi(65) + |\{x, x^2 = 1\}| + |\{x, x^2 = -1\}| = 48 + 4 + |\{x, x^2 = -1\}|.$$

6.6.3.1 The Ptolemy Identity and Multiplicativity

Looking at this equation modulo 4 is not helpful. However, by using the fact that λ-lengths satisfy the Ptolemy Identity one can prove a weaker form of the closure property quite easily.

The *Ptolemy Identity* concerns the λ-lengths of an ideal quadrilateral see Fig. 6.5. If $X^{\pm}$, $Y^{\pm}$ are the λ-lengths of pairs of opposite sides and $Z^{\pm}$ of the diagonals then

$$X^+X^- + Y^+Y^- = Z^+Z^- \tag{6.4}$$

Now suppose that p, q are distinct integers which can be written as sums of squares, by Lemma 6.3.3 there is a pair of arcs on $\mathbb{H}/\Gamma(2)$ with λ-lengths p and q respectively which meet in the projection of i to the surface. In the upper half

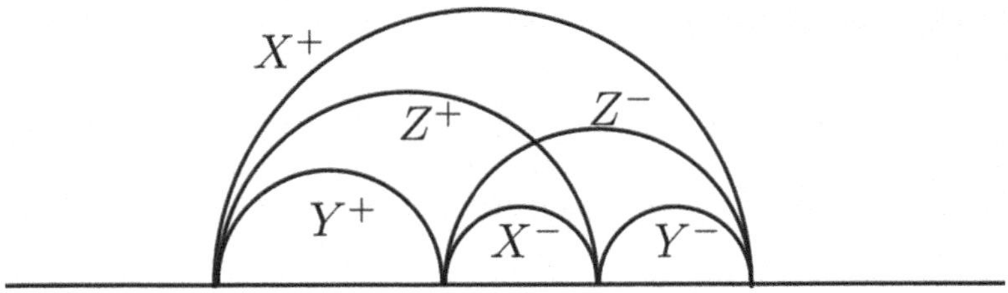

Fig. 6.5 Ptolemy identity diagram

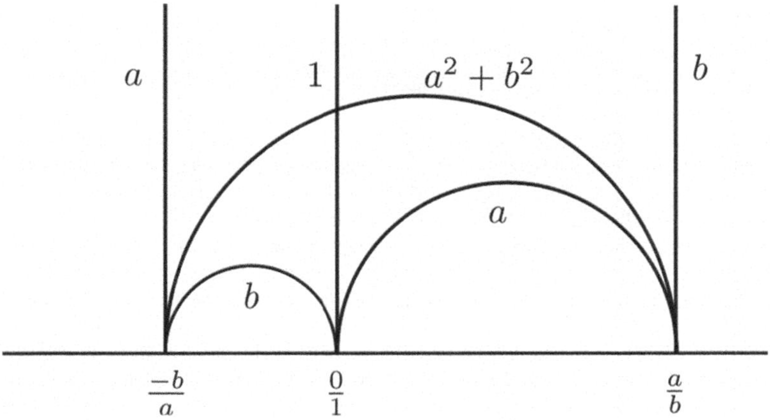

Fig. 6.6 A λ-length parallelogram invariant under $z \mapsto -1/z$. Its vertices are $\infty, -b/a, 0, a/b$. This illustrates a special case of the Ptolemy Identity

space $\mathbb{H}$ there is a corresponding pair of lifts that intersect in i. The convex hull of these arcs is an ideal quadrilateral in $\mathbb{H}$ to which we can apply the Ptolemy Identity. Observe that in addition this quadrilateral is invariant under $z \mapsto -1/z$ so $X^+ = X^-$ and $Y^+ = Y^-$, such a quadrilateral is a *parallelogram* for λ-length, and we have

$$(X^+)^2 + (Y^+)^2 = pq. \tag{6.5}$$

Note that if p, q are square free and coprime then X^+ and Y^+ are automatically coprime though this may not be the case in general so more work is required to prove the strong form of closure under multiplication (Fig. 6.6).

If $q = p > 2$ then find an arc in $\mathbb{H}$ passing through i and of λ-length p. The image of the arc under $z \mapsto \bar{z}$ also passes through i and has λ-length p. Now proceed as before to obtain an ideal parallelogram which represents p^2 as a sum of squares. Note that this fails for 2 as the initial arc joins -1 to 1 and so is invariant under $z \mapsto \bar{z}$.

6.7 Concluding Remarks

We have given a geometric treatment of Fermat's theorem using the automorphisms of the surface $\mathbb{H}/\Gamma(2)$ and Penner's λ-lengths. In another work we will study the relation between the quadratic form $x^2 + xy + y^2$ and $\mathbb{H}/\Gamma(2)$.

References

1. M. Aigner, G.M. Ziegler, Representing numbers as sums of two squares, in *Proofs from the Book* (Springer, Berlin, 2010)
2. C.A. Elsholtz, Combinatorial approach to sums of two squares and related problems, in *Additive Number Theory*, D. Chudnovsky, G. Chudnovsky (Springer, New York, 2010)
3. L.R. Ford, *Automorphic Functions* (New York, McGraw-Hill, 2004). ISBN 978-0-8218-3741-2
4. R. Heath-Brown, *Fermat's Two Squares Theorem* Invariant (1984)
5. S. Northshield, A short proof of fermat's two-square theorem. Am. Math. Month. **127**, 638–638 (2020)
6. R.C. Penner, The decorated Teichmüller space of punctured surfaces. Commun. Math. Phys. **113**, 299–339 (1987)
7. J.P. Serre, *A Course in Arithmetic*. Graduate Texts in Mathematics (Springer, New York, 1973)
8. B. Springborn, The hyperbolic geometry of Markov's theorem on Diophantine approximation and quadratic forms. Enseign. Math. **63**(3–4), 333–373 (2017). https://doi.org/10.4171/LEM/63-3/4-5
9. S. Wolpert, On the Kahler form of the moduli space of once-punctured tori. Comment. Math. Helv. **58**, 246–256 (1983)
10. D. Zagier, A one-sentence proof that every prime p = 1 (mod 4) is a sum of two squares. Am. Math. Month. **97**(2), 144 (1990)

Chapter 7
Ends and Compactifications in Geometric Topology

Ken'ichi Ohshika

Abstract This chapter is a survey on ends and compactification of topological spaces and on their applications in manifold topology. Starting from the basic notion and properties of ends, which were first introduced by Freudenthal, we look at how it was used to deepen understanding of open manifolds. We put a particular emphasis on problems of finding open contractible manifolds different from Euclidean spaces and of attaching boundaries to open 3-manifolds. We also touch upon the special case of hyperbolic 3-manifolds with finitely generated fundamental groups, and explain how such manifolds can be compactified by attaching boundaries.

Keywords Ends · Compactification · Fundamental group · Open 3-manifold

2020 Mathematics Subject Classification 54D35, 57K30, 57K32, 57Q30

7.1 Introduction

Looking at a long history of geometric topology, we realise that compact manifolds were studied earlier and better than non-compact ones. One of the driving forces of research in topology in the twentieth century was the famous Poincaré conjecture, which concerns only closed 3-manifolds. On the other hand it was already recognised in the first half of the twentieth century that open manifolds have more complicated characters. In the 1930s as a matter of fact, Whitehead constructed an open contractible 3-manifold which is not homeomorphic to the Euclidean space.

To fully understand the differences between open manifolds and closed manifolds, it is necessary to use the notion of end. Its definition and basic properties are the topic of Sect. 7.2. The beginning of the notion of end of topological spaces or discrete groups dates back to the famous work of Freudenthal. He defined

K. Ohshika (✉)
Department of Mathematics, Faculty of Science, Gakushuin University, Tokyo, Japan
e-mail: ohshika@math.gakushuin.ac.jp

L. Funar, A. Papadopoulos (eds.), *Essays on Topology*,
https://doi.org/10.1007/978-3-031-81414-3_7

ends as equivalence classes of descending sequences of open sets with some properties. Making use of this notion of end, he defined a compactification of non-compact topological spaces, which is different from the one-point compactification introduced by Alexandroff. His original definition applies only to topological spaces with conditions such as local connectedness. The assumption was loosened a bit later by replacing it with semi-compactness and compactness of component spaces. He further generalised the definition in order to apply it to discrete spaces. This was quite useful for the later development of geometric group theory.

The theory of ends can be applied to open manifolds and enables us to understand them better, especially those having trivial algebraic invariants such as homology groups and homotopy groups. We deal with ends of contractible manifolds in Sect. 7.3. As mentioned above, a contractible open 3-manifold which is not homeomorphic to $\mathbb{R}^3$ was found by Whitehead. By Stallings's theorem, it is known that any contractible open manifold has only one end. We can define simple connectedness at infinity for such manifolds, and it was proved by Edwards that this is in fact a sufficient condition for a contractible open 3-manifold to be homeomorphic to the Euclidean space modulo the Poincaré conjecture which was not yet resolved at that time. In dimension $n \geq 5$, Stallings's theorem says that any contractible open PL n-manifold that is simply connected at infinity is PL homeomorphic to $\mathbb{R}^n$, and the same holds in both the differentiable and the topological categories. In dimension 4, based on celebrated work of Freedman and Donaldson, it was proved that there are uncountably many exotic $\mathbb{R}^4$s, i.e. manifolds homeomorphic but not diffeomorphic (or PL homeomorphic) to $\mathbb{R}^4$. Still in the topological category, thanks to Freedman's work, it turns out that the same result as Stallings's holds.

Studying ends of open manifolds is closely related to a problem of finding their "ideal" boundaries, which we shall deal with in Sect. 7.4. This problem was first studied by Browder–Levine–Levesay in higher-dimension and then improved and refined by Siebenmann in his thesis. Making use of their tools, in particular the fundamental groups at ends, Husch–Price obtained a necessary and sufficient condition for an open irreducible 3-manifold to be homeomorphic to the interior of a compact 3-manifold. The condition was much simplified later by Tucker whose theorem does not refer to the fundamental groups at ends and whose proof makes use of progress of 3-dimensional topology in the 1970s, in particular Scott's theorems.

In the late 1979s, Thurston found that the problem of attaching ideal boundaries to manifolds is very important in his theory of hyperbolic 3-manifolds and Kleinian groups, in particular for his uniformisation theorem of Haken manifolds. We shall look at how this problem was studied and solved in Sect. 7.5. Marden was aware of this problem before Thurston's work appeared, and proved that geometrically finite hyperbolic 3-manifolds can be compactified to compact 3-manifolds with boundaries by attaching ideal boundaries, derived from the ideal boundary of the hyperbolic 3-space. Thurston needed to consider geometrically infinite hyperbolic 3-manifolds for his theory. He introduced the notion of geometric tameness for ends of hyperbolic 3-manifolds, and showed that this implies topological tameness, i.e. the possibility of attaching a boundary at infinity. Making use of tools introduced by

Thurston, and also considering the space of geodesic currents on surfaces, Bonahon proved that if the fundamental group of a hyperbolic 3-manifold is indecomposable with respect to free product, then every end of the manifold is either geometrically finite or geometrically tame in the sense of Thurston. The same was proved in general, without the assumption of indecomposability of fundamental groups with respect to free products, by Agol and Calegari–Gabai independently.

The universal cover of any hyperbolic 3-manifold is homeomorphic to $\mathbb{R}^3$, whereas examples of open 3-manifolds which are not simply connected at infinity constructed by Whitehead and others are contractible, and hence their universal covers are themselves. Therefore, one can naturally ask whether there is an open 3-manifold covered by $\mathbb{R}^3$ which cannot be homeomorphic to the interior of a compact 3-manifold with boundary. Such examples were constructed concretely by Scott and Tucker. Thus, we see that hyperbolic metrics give a constraint for the topology of open 3-manifolds.

In the last section, we shall deal with the question of whether the universal cover of a closed aspherical manifold is always homeomorphic to a Euclidean space. A positive answer to this question for 3-manifolds is one of the important consequences of the resolution of the geometrisation conjecture. On the other hand, in dimension greater than 3, Davis constructed much earlier aspherical manifolds whose universal covers are not simply connected at infinity, based on a construction of Coxeter systems with certain properties. This again demonstrates the peculiarity of dimension 3 in manifold topology.

7.2 Foundation of Ends

The notion of ends was first introduced and studied by Freudenthal in [12] for locally compact, locally connected, connected and second countable Hasdorff spaces, and then developed for manifolds by Hopf in [21] and for discrete groups by Freudenthal again in [14]. In this section, we shall review the original definition of ends and their fundamental properties.

Definition 7.2.1 Let X be a locally compact, connected, locally connected and second countable Hausdorff space. An *endpoint* of X is a sequence of connected open subsets with compact boundaries $U_1 \supset U_2 \supset \dots$ such that $\bigcap_{n=1}^{\infty} U_n = \emptyset$.

An endpoint e represented by $U_1 \supset U_2 \supset \dots$ is said to be contained in an open set O when U_n is contained in O for sufficiently large n. For two endpoints e_1, represented by $U_1 \supset U_2 \supset$, and e_2, we write $e_2 \sim e_1$ when every e_2 is contained in every U_n.

It is easy to verify the following:

Lemma 7.2.2 *The relation* $\sim$ *is an equivalence relation in the set of endpoints of* X.

We call the equivalence classes the *ends* of X, and denote it by $E(X)$.

Remark 7.2.3 In Freudethal's original papers, the term "Endpunkt" is used to refer to an end in our sense. To distinguish a representative of an equivalence class from the equivalence class itself, we reserve the term "endpoint" for representative.

Freudenthal considered compactifying X by taking the disjoint union of X and its ends. The topology is defined as follows.

Definition 7.2.4 For an end e represented by $U_1 \supset U_2 \supset \dots$, a neighbourhood of e is a set of the form $e \cup O$ where O is an open set in X containing the endpoints representing e. We introduce a topology on $X \cup E(X)$ by neighbourhood systems thus defined, and denote $X \cup E(X)$ with this topology by $\bar{X}$.

This gives a compactification of X which is called the Freudenthal compactification.

Theorem 7.2.5 *The following hold for $\bar{X}$.*

(1) $\bar{X}$ is a compact Hausdorff space.
(2) The inclusion from X to $\bar{X}$ is an embedding, i.e. a homeomorphism to its image, whose image is dense in $\bar{X}$.
(3) $\bar{X} \setminus X$ is totally disconnected.
(4) $\bar{X}$ is maximal among compactifications of X satisfying (1)–(3) above.

Sketch of Proof The conditions (1)–(3) are easy to verify. To prove (4), let $\hat{X}$ be another compactification satisfying (1)–(3). From the compactness of $\hat{X}$, it follows that for any sequence of open sets $U_1 \supset U_2 \supset \dots$ corresponding to $x \in E(X)$, there is a point in $\hat{X} \setminus X$ contained in $\cap_{n=1}^{\infty} \overline{U_n}$, where $\overline{U_n}$ denotes the closure of U_n in $\hat{X}$. The condition (2) implies that any point in $\hat{X} \setminus X$ must be contained in such a set. Finally, by the condition (3), we see that there is only one point in $\hat{X} \setminus X$ corresponding to $x \in E(X)$. Therefore, if $\hat{X}$ does not coincide with $\bar{X}$, then it is obtained from $\bar{X}$ as a quotient space by identifying points in $\bar{X} \setminus X = E(X)$. This shows the maximality of $\bar{X}$. □

This characterisation of the Freudenthal compactification implies the following immediately.

Corollary 7.2.6 *Any homeomorphism $h: X \to X$ extends uniquely to a homeomorphism from $\bar{X}$ to $\bar{X}$.*

This in turn gives rise to the following fact on ends of topological groups.

Theorem 7.2.7 *Any locally compact, second countable, locally connected and connected Hausdorff topological group has at most two ends.*

Sketch of Proof Let G be a topological group as in the statement. For any $g \in G$, the action $x \mapsto gx$ is a self-homeomorphism of G. By Corollary 7.2.6, this extends to a self-homeomorphism of $\bar{G}$. Since $E(G)$ is totally disconnected, G is connected, and the identity element fixes all points of $E(G)$, we see that any g fixes $E(G)$ pointwise.

Let e be an end in $E(G)$. Then there is a compact subset K in G separating e from the other ends, i.e. $G \setminus K$ consists of two components U, U^* one of which, say U, contains only one end e. Suppose that there is an end e' distinct from e, and let $V \subset G$ be the intersection of its connected neighbourhood with G. Then we can find an element $g \in G$ such that gK is contained in V. Since V contains only one end, this shows that it is impossible that G contains ends other than e and e'. □

The theory was generalised in [13], changing the assumptions on local connectedness and local compactness in Definition 7.2.1 to semi-compactness and compactness of the component space. The notion of semi-compactness was introduced by Zippin [57]. A Hausdorff space X is said to be semi-compact when every neighbourhood of a point has a sub-neighbourhood whose boundary is compact. The latter condition on compactness of component space is satisfied by definition if X has only finitely many components. Freudenthal gives an explicit example which does not satisfy this condition, as follows. He considers a space $X \subset \mathbb{R}^2$ which is a union of the subsets $\{0\}\times]n, n+1[$ $(n \in \mathbb{Z})$ and the subsets $\{1/n\} \times \mathbb{R}$ $(n \in \mathbb{N})$. In this X, there is a sequence of components not converging to any other component.

In the subsequent paper [14], Freudenthal considers ends of discrete spaces, in particular discrete groups. In a discrete space R, he considers a "neighbouring structure" in the following sense. Suppose that a reflective and symmetric relation $\sim$ is defined on R. (A typical example will be given below for finitely generated groups.) Then we say that y is a neighbour of x when y is related to x.

Now Freudenthal introduced two symbols $\mathfrak{H}$ and $\mathfrak{F}$. For a subset M of R, its hull $\mathfrak{H}(M)$ is defined to be $\{y \in R \mid \exists x \in X \text{ with } y \sim x\}$, and its fringe $\mathfrak{F}(M)$ is defined to be $\mathfrak{H}(M) \setminus M$. In the same way as there were assumptions on topological spaces in [12, 13], here we assume the following two conditions.

(1) For every point $a \in R$, its hull $\mathfrak{H}(a)$ is a finite set.
(2) R is a countable set.

Now an end of R is defined as follows. Let $\mathfrak{Q}$ be the set of subsets X of R such that $\mathfrak{F}(X)$ is a finite set. An end of R is a subset $\mathfrak{e}$ of $\mathfrak{Q}$ satisfying the following four conditions.

(1) $\emptyset$ is not contained in $\mathfrak{e}$.
(2) For two sets $E_1, E_2 \in \mathfrak{e}$, we have $E_1 \cap E_2 \in \mathfrak{e}$.
(3) If M is a finite set, then $R \setminus M$ is always contained in $\mathfrak{e}$.
(4) $\mathfrak{e}$ is a maximal subset of $\mathfrak{Q}$ among those having the three conditions above.

The set of ends $E(X)$ is defined to be the set of all $\mathfrak{e}$ as above, and its topology can also be defined in the same way as before.

A typical and important example of a discrete set with neighbouring structure as above is a group G with a finite generator system S, which we assume to be symmetric and contain the identity element. Then we can define two elements $g, h \in G$ to be related by $\sim$ when there is a generator $s \in S$ with $h = gs$. It is easy to verify that the ends of G defined above coincide with the ends of the Cayley graph $\Gamma(G, S)$ as a topological space, defined at the beginning of this section.

As explained above, Freudenthal considered the ends of topological groups and discrete groups. Hopf studied ends of discrete groups acting as covering translations. He proved the following theorem in [21].

Theorem 7.2.8 *Let $\mathfrak{G}$ be a finitely generated group acting on a connected, locally connected, locally compact and second countable Hausdorff space as a covering translation group with compact quotient. Then the ends of X consist of one point or two points or uncountably many points.*

Sketch of Proof We have only to prove that if $E(X)$ contains more than two points, then for any point $e \in E(X)$, every neighbourhood of e contains a point of $E(X)$ distinct from e.

To prove this, we assume that e has a neighbourhood U containing no other end points, and that X has three distinct ends e_1, e_2 and e_3, one of which may be e. By taking a smaller neighbourhood we can assume that both U and $X \setminus U$ are connected. We can assume (by the definition of ends) that $\operatorname{Fr} U$ is compact, and can take disjoint connected neighbourhoods U_1, U_2 and U_3 of e_1, e_2 and e_3 respectively whose frontiers are all compact. Since $\mathfrak{G}$ acts on X with compact quotient, there is $g \in \mathfrak{G}$ such that $g(\operatorname{Fr} U_1 \cup \operatorname{Fr} U_2 \cup \operatorname{Fr} U_3) \subset U$. By Corollary 7.2.6, the action of g extends to $E(X)$ as a homeomorphism. By our choice of U, it contains at most only one end among $g(e_1), g(e_2), g(e_3)$. Therefore, by renumbering them if necessary, we can assume that U does not contain $g(e_1)$. Since $g(\operatorname{Fr} U_1)$ is contained in U and $X \setminus U$ is connected, we see that $X \setminus U$ is contained in $g(U_1)$. This implies that both e_2 and e_3 are contained in U, contradicting our choice of U. □

7.3 Ends of Contractible Manifolds

The Freudenthal compactification which we explained in the preceding section makes it possible to count the number of ends of a space. This gives a topological invariant for open manifolds in particular. A typical open manifold with one end is the Euclidean space, which is also contractible. Conversely the following was proved by Stallings.

Proposition 7.3.1 (Stallings [50]) *A contractible open manifold of dimension greater than* 1 *has exactly one end.*

Stallings's proof essentially uses the following theorem of Specker.

Theorem 7.3.2 (Specker [49]) *The first cohomology group with compact support of an infinite (locally finite) simplicial complex with n ends has rank $n - 1$. (By convention, we assume that $\infty - 1 = \infty$.)*

The proof of Proposition 7.3.1 is simple once Theorem 7.3.2 is admitted. Indeed, if a contractible open manifold M has n ends with $n > 1$, then Theorem 7.3.2 shows that $H^1_c(M)$ has rank $n - 1 \geq 1$. By the duality theorem, this implies that $H_{n-1}(M)$ is not trivial, contradicting the assumption that M is contractible.

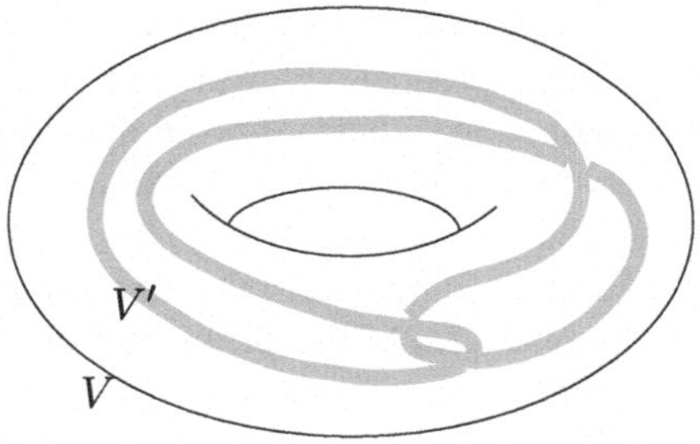

Fig. 7.1 The embedding of a smaller solid torus V' into V

The famous Poincaré conjecture, which was given in [40], asks if a closed simply-connected 3-manifold is always homeomorphic to the 3-sphere. The same kind of problem can be thought of for open manifolds, and a natural question is: Is an open contractible 3-manifold always homeomorphic to the Euclidean space?

It was recognised already in the first part of the twentieth century that the answer is 'no' and that open 3-manifolds are much more complicated. The first example showing this complexity was given by Whitehead [56]. He constructed a contractible open 3-manifold as follows.

Consider a solid torus V embedded in S^3 so that $S^3 \setminus \text{Int}\, V$ is also a solid torus, which we denote by V_0. We next consider a sub-solid torus V' embedded in V in such a way that its core curve is null-homotopic but not ambient isotopic to a core curve in V. (See Fig. 7.1.) We denote the 3-manifold with boundary $V \setminus \text{Int}\, V'$ by W. The boundary ∂W consists of two tori, the one coming from ∂V and the other coming from $\partial V'$. We denote the former by $\partial_e W$ and the latter by $\partial_i W$. Each of them has a marking consisting of a meridian of V (resp. V') and a longitude (which bounds a disc in the complement of) V (resp. V'). We consider countably infinitely many copies W_n ($n \in \mathbb{N}$) of W letting W_1 be W, and paste W_{n+1} to W_n by identifying $\partial_e W_{n+1}$ to $\partial_i W_n$ preserving the markings.

Now we consider the union $M = V_0 \cup \bigcup_{n=1}^{\infty} W_n$, which is a 3-manifold and a subset of S^3. The meridian of V and the longitude of V' form a link called the Whitehead link, and both the meridian and the longitude of ∂V are null-homotopic in $S^3 \setminus V'$. In the same way, both the meridian and the longitude of ∂W_n are null-homotopic in $S^3 \setminus W_{n+1}$. This implies that M is simply connected. Furthermore, since any map from the n-sphere is contained in the union of V_0 and $V_0 \cup_{n=1}^{n_0} W_n$ for some $n_0 \in \mathbb{N}$, we see that the higher homotopy groups all vanish. Therefore M is contractible.

On the other hand, it can be shown that M is not homeomorphic to $\mathbb{R}^3$ as follows. Suppose that M is homeomorphic to $\mathbb{R}^3$. In $\mathbb{R}^3$, every closed curve is contained in a 3-ball. Therefore, there must be a 3-ball B containing the meridian of $\partial V = \partial_e W_1$, which is a longitude of V_0. This is impossible since such a ball in S^3 must intersect the complement of M and cannot lie inside M. Thus M is a contractible open 3-manifold embedded in S^3 which is not homeomorphic to $\mathbb{R}^3$, and is often referred to as the Whitehead manifold.

Looking at this construction, we can see that once we move to dimension four by taking a product with $\mathbb{R}$, then a smaller solid torus embedded in a larger one can be isotoped to a standard position, and hence $M \times \mathbb{R}$ is homeomorphic to $\mathbb{R}^4$. This observation was first given by Shapiro and Glimm [15].

A similar construction can be done based on two-component links other than the Whitehead link, such that one component is null-homotopic in the complement of the other. McMillan Jr. proved that in fact this is the only way to construct a contractible open 3-manifold under some condition on embeddability into $\mathbb{R}^3$. He showed the following.

Theorem 7.3.3 ([32]) *Let M be a contractible open 3-manifold each of whose compact subsets can be embedded in $\mathbb{R}^3$. Then there is an ascending sequence of handlebodies $H_1 \subset H_2 \subset H_3 \ldots$ such that M is homeomorphic to $\cup_{j=1}^{\infty} H_j$.*

As in the case of the Whitehead manifold, we can verify that for any such manifold, its product with $\mathbb{R}$ is homeomorphic to $\mathbb{R}^4$. Making use of such manifolds, Poénaru wrote a short note [42] in which he showed that for any (smooth) homotopy sphere Σ, its product with $\mathbb{R}$ is a 4-manifold which is a union of two standard smooth $\mathbb{R}^4$s.

The Whitehead manifold and its variations have only one end. The property which distinguishes such manifolds from $\mathbb{R}^3$ is the following simple connectedness at infinity. Indeed, we shall see in Theorem 7.3.7 that simple connectedness at infinity is a necessary and sufficient condition for a contractible open 3-manifold to be homeomorphic to $\mathbb{R}^3$.

Definition 7.3.4 Let X be a topological space and e an end of X. We say that X is simply connected at e when for any neighbourhood U of e, there is another neighbourhood V contained in U such that the inclusion $V \subset U$ induces the trivial map from $\pi_1(V)$ to $\pi_1(U)$. If X has only one end e and X is simply connected at e, we say that X is simply connected at infinity.

Although the product of the Whitehead manifold with $\mathbb{R}^n$ is homeomorphic to $\mathbb{R}^{n+3}$, there is a way to construct higher-dimensional contractible manifolds which are not simply connected at infinity. These manifolds are obtained as the interior of compact manifolds whose boundaries are not simply connected. The construction is due to Newman when the dimension is greater than 4.

Theorem 7.3.5 (Newman [35]) *For any $n \geq 5$, there is a contractible open n-manifold in $\mathbb{R}^n$ which is bounded by a closed $n-1$-manifold which is not simply connected.*

In dimension 4, Mazur and Poénaru constructed the same kind of manifolds.

Theorem 7.3.6 (Mazur [30], Poénaru [41]) *There are compact contractible 4-manifolds whose boundaries are not simply connected.*

On the other hand, if we impose the condition of simple connectedness at infinity, we have positive results. The following theorem was proved by Edwards [10] as a three-dimensional analogue of Stallings's theorem, which we shall present as Theorem 7.3.8.

Theorem 7.3.7 *Let M be a contractible open 3-manifold which is simply connected at infinity. Suppose furthermore that any compact subset of M can be embedded in $\mathbb{R}^3$. Then M is homeomorphic to $\mathbb{R}^3$.*

Sketch of Proof Since M has only one end, which is simply connected, for any compact set C in M, there is a 3-submanifold N of M containing C such that $M \setminus \operatorname{Int} N$ is connected and simply connected.

We can show that ∂N is a sphere as follows. Suppose that ∂N is not a sphere. Then there is a pair of simple closed curves a, b on ∂N intersecting at one point. Since $M \setminus \operatorname{Int} N$ is simply connected, a bounds a singular disc in $M \setminus \operatorname{Int} N$, which represents a non-trivial element in $H_2(M \setminus \operatorname{Int} N, \partial N)$. By duality, this implies that b is non-trivial in $H_1(M \setminus \operatorname{Int} N)$, which contradicts the assumption that $M \setminus \operatorname{Int} N$ is simply connected.

Now, by assumption N can be embedded in $\mathbb{R}^3$, and since ∂N is a sphere, it bounds a 3-ball in $\mathbb{R}^3$, which implies that N is a 3-ball. Therefore, we have an exhaustion $N_1 \subset N_2 \subset \dots$ of M by balls $N_1, N_2, \dots$. By Brown's theorem [5], this shows that M is homeomorphic to $\mathbb{R}^3$. □

The assumption that any compact subset can be embedded in $\mathbb{R}^3$ was necessary to avoid a difficulty concerning the Poincaré conjecture, which was unsolved at that time. Today, we can safely remove this assumption thanks to the resolution of the conjecture by Perelman [37–39].

In higher dimensions, a similar argument works in the PL category if we replace the part making use of the classification of closed surfaces with the engulfing theorem. In fact, chronologically, the following theorem of Stallings appeared before Theorem 7.3.7.

Theorem 7.3.8 (Stalling [50]) *Let M be an n-dimensional open contractible PL manifold which is simply connected at infinity. If $n \geq 5$, then M is PL-homeomorphic to $\mathbb{R}^n$.*

Sketch of Proof Since M is simply connected at infinity, for any compact set C, there is a compact n-submanifold $N \supset C$ such that $M \setminus N$ is simply connected. By the homotopy exact sequence for $(M, M \setminus N)$, we have that $\pi_2(M, M \setminus N) = 0$. We consider a triangulation τ of M, and let $\tau^{(2)}$ denote its 2-skeleton. In this situation, we can apply a version of the engulfing theorem (see for instance Hirsch–Zeeman [20]) to conclude that there are a compact set E and a PL homeomorphism $h\colon M \to M$ such that $N \subset E$, $\tau^{(2)} \subset h(M \setminus N)$, and $h \mid (M \setminus E) = id|(M \setminus E)$.

We consider a subcomplex K of τ consisting of $\tau^{(2)}$ and all simplices of τ contained in $M \setminus E$. We define L to be a subcomplex of the barycentric subdivision of τ consisting of all simplices disjoint from K. Then, L has dimension less than $n-2$, and hence by the englulfing theorem again, there is a PL n-cell B in M which contains L. By our construction of L, we can isotope B so that $M = h(M \setminus N) \cup B$, which implies that $h^{-1}(B)$ contains C.

Repeating this argument, we have an exhausting sequence of PL n-balls in M, and by Gugenheim's theorem, which says that all embeddings of (orientation-preserving) PL balls are ambient isotopically equivalent, we obtain a PL-homeomorphism from M to $\mathbb{R}^n$. □

The same holds for the category of differentiable manifolds, which follows from the work of Munkres [33] showing that a differentiable manifold which is PL equivalent to $\mathbb{R}^n$ is diffeomorphic to $\mathbb{R}^n$. In the category of topological manifolds, Luft [28] proved the same. This also follows from the work of Kirby–Siebenmann on the Hauptvermutung [26], combined with Theorem 7.3.8.

The case of dimension 4 is contained in neither Theorem 7.3.7 nor Theorem 7.3.8. In fact, in the PL or the differentiable category, the same kind of theorem does not hold in dimension 4. It is known that there are 'extotic $\mathbb{R}^4$s', i.e. open 4-manifolds which are homeomorphic to but not diffeomorphic to (or PL-homeomorphic to) the standard $\mathbb{R}^4$. This fact was first pointed out by Freedman, as a result obtained by making use of his result [11] combined with Donaldson's revolutionary theorem [9] based on gauge theory (see also Gompf [16].) A bit later, it was found by Gompf [17] that there are infinitely many such manifolds and by Taubes [52] that there are indeed uncountably many. Still in the topological category, an analogous theorem holds, which is due to Freedman.

Theorem 7.3.9 (Freedman [11]) *Any open contractible 4-manifold that is simply connected at infinity is homeomorphic to* $\mathbb{R}^4$.

7.4 Attaching Boundaries

In the preceding section, we saw that there are open contractible manifolds with non-simply connected boundaries. Examples of open contractible manifolds given by Newmann in dimension ≥ 5 and by Mazur and Poénaru in dimension 4 are the interiors of compact manifolds with boundaries. On the other hand, examples in dimension 3 cannot be the interiors of compact 3-manifolds: since any surface other than the sphere cannot bound a contractible 3-manifold, and if the boundary is a sphere, its interior is simply connected at infinity.

In general, to study the topology of an open manifold, it is important to see whether it can be compactified to a manifold with boundary or not. In dimension greater than 4, the problem was first studied in Browder–Levine–Livesay [4], and then generalised and refined by Sibenmann's thesis [46]. Although we are going to focus on the three-dimensional case in this section, we need to borrow tools provided by [46] for higher-dimensions, even in dimension 3. We note that from now on to the end of this chapter, *we always assume that manifolds are orientable*, for simplicity.

Since we have to deal with the situation where ends may not be simply connected, we need to introduce the following notion of the fundamental group at infinity. Let e be an end of a manifold M (or more generally locally compact, connected, locally connected, second countable Hausdorff space as in Sect. 7.2). Let $U_1 \supset U_2 \supset \dots$ be a base of open connected neighbourhoods of e. Taking $x_i \in U_i$, we have a sequence of homomorphisms

$$\pi_1(U_1, x_1) \xleftarrow{f_1} \pi_1(U_2, x_2) \xleftarrow{f_2} \cdots$$
$$\xleftarrow{f_{i-1}} \pi_1(U_i, x_i) \xleftarrow{f_i} \pi_1(U_{i+1}, x_{i+1}) \xleftarrow{f_{i+1}} \cdots.$$

This sequence induces another one

$$\mathrm{Im}(f_1) \xleftarrow{f_1} \mathrm{Im}(f_2) \longleftarrow \cdots$$
$$\xleftarrow{f_{i-1}} \mathrm{Im}(f_i) \xleftarrow{f_i} \mathrm{Im}(f_{i+1}) \xleftarrow{f_{i+1}} \cdots.$$

The original sequence is said to be *stable* if after passing to a subsequence, the homomorphisms f_i in the second sequence are isomorphisms. Although this definition depends on the choice of a neighbourhood system $\{U_i\}$, it turns out that the stability does not depend on the choice of a neighbourhood system, and neither does the projective limit $\varprojlim G_i$ (see Siebenmann [46] and Jackson [23]). Therefore, we can say that $\pi_1^e(M)$ is stable if such a neighbourhood system of e exists. We refer to the projective limit (under the assumption of stability) as the fundamental group at e, and denote it by $\pi_1^e(M)$. In particular, when M has only one end, we call it the fundamental group at infinity, and denote it by $\pi_1^\infty(M)$. It is easy to see that a manifold M with one end is simply connected at infinity if and only if $\pi_1^\infty(M) = \{1\}$.

Suppose that an open manifold M can be compactified to a manifold with boundary $\bar{M}$. Since each boundary component of $\bar{M}$ has a collar neighbourhood, each end e of M has a neighbourhood system whose corresponding projective sequence of fundamental groups is stable.

Now we can state the following powerful result of Husch and Price on compactification of 3-manifolds.

Theorem 7.4.1 (Husch–Price [22]) *Let M be an open orientable irreducible 3-manifold with finitely many ends $e_1, \ldots, e_n$. Suppose that for each end e_j, the fundamental group $\pi_1^{e_j}(M)$ is stable and finitely presented. Then there is a compact 3-manifold $\bar{M}$ whose interior is homeomorphic to M.*

We note that the assumption that $\pi_1^{e_j}(M)$ is finitely presented can be relaxed to the condition that it is finitely generated, by Scott's theorem [44], which appeared later than this theorem of Husch and Price.

Sketch of Proof Since each end should correspond to a boundary component, it suffices to consider the case when M has only one end e. Let $U_1 \supset U_2 \supset \cdots \supset U_i \supset U_{i+1} \supset \ldots$ be an open connected neighbourhood base of e such that the corresponding sequence $\pi_1(U_1, x_1) \xleftarrow{f_1} \pi_1(U_2, x_2) \xleftarrow{f_2} \cdots$ is stable. The stability and the assumption that $\pi_1^e(M)$ is finitely presented implies, just applying van Kampen's theorem, that $\pi_1(U_i, x_i)$ is finitely presented.

We first modify U_1 so that $\mathrm{Fr}\, U_1$ becomes connected, by removing regular neighbourhoods of arcs in U_1 connecting different frontier components if necessary.

Since the arcs are disjoint from U_i for large i, passing to a subsequence, we can assume that this modification does not affect $U_2, U_3, \ldots$. In the same way, passing to a subsequence, we can make all $\mathrm{Fr}\, U_i$ connected without changing the stability. We consider compressing $\mathrm{Fr}\, U_i$ inductively for $i \geq 3$ to make it incompressible both in $U_{i-1} \setminus U_i$ and $\bar{U}_i \setminus U_{i+1}$ keeping it connected. If we simply compress $\mathrm{Fr}\, U_i$ in $U_{i-1} \setminus U_i$ or $\bar{U}_i \setminus U_{i+1}$, there is a possibility that $\mathrm{Fr}\, U_i$ becomes disconnected.

We explain how to deal with the problem up to a subsequence and a modification of U_i keeping the required conditions. Let D be a compressing disc for $\mathrm{Fr}\, U_i$ contained in $U_{i-1} \setminus U_i$. Suppose that $\mathrm{Fr}\, U_i \setminus D$ is disconnected, but that $U_{i-1} \setminus (U_i \cup D)$ is connected. Then there is a loop γ contained in U_i intersecting D at one point. If this occurs for infinitely many i, then $H_1(U_i, \mathbb{Z}_2)$ is infinitely generated, which contradicts the fact $\pi_1(U_i)$ is finitely, generated mentioned above. Therefore, taking a subsequence, we can assume that there is no such compression disc. Next suppose that $U_{i-1} \setminus D$ is also disconnected. Let F_1 and F_2 be the components of the surfaces obtained by compressing $\mathrm{Fr}\, U_i$ along D. Since we assumed that $\mathrm{Fr}\, U_{i-1}$ is connected, either F_1 or F_2 must bound a compact submanifold C contained in U_{i-1}. Then we add C to U_i and eliminate one of F_1 and F_2.

In the same way, if the compressing disc D is contained in $\bar{U}_i \setminus U_{i+1}$ and $\mathrm{Fr}\, U_i \setminus D$ is disconnected, by passing to a subsequence, we can assume that $U_i \setminus D$ is disconnected. Then one of the surfaces F_1, F_2 obtained by compressing $\mathrm{Fr}\, U_i$ along D bound a compact submanifold C in $\bar{U}_i \setminus U_{i+1}$. Then we remove C from U_i. Repeating these, we can assume that $\mathrm{Fr}\, U_i$ is incompressible preserving its connectivity.

Now we consider a compact 3-manifold $K_i := U_i \setminus \mathrm{Int}\, U_{i+1}$ for each i after the above modification. Let $g_i : \pi_1(\mathrm{Fr}\, U_{i+1}) \to \pi_1(U_i)$ be a homomorphism induced by the inclusion. Since we made $\mathrm{Fr}\, U_{i+1}$ incompressible, g_i is injective. On the other hand, the surjectivity of $f_i : \mathrm{Im} f_{i+1} \to \mathrm{Im} f_i$ implies, by way of van Kampen's theorem, that g_i is surjective. Since K_i is irreducible, this is possible only when K_i is homeomorphic to $\mathrm{Fr}\, U_{i+1} \times I$ (see [6] and [19, §10]). Thus we have shown that e has a neighbourhood homeomorphic to $S \times \mathbb{R}$ for a closed surface S, and we are done. □

The assumption of Theorem 7.4.1 corresponds well to that of Siebenmann's theorem for higher dimension. In fact, a much simpler assumption is sufficient for an open manifold to be compactified to a manifold with boundary in dimension 3, as was shown by Tucker's theorem below, which makes use of the development of three-dimensional topology in the 1970s.

Theorem 7.4.2 (Tucker [55]) *Let M be an open irreducible 3-manifold. Then M is homeomorphic to the interior of a compact 3-manifold if and only if for every compact 3-submanifold C of M, the fundamental group of every component of $M \setminus C$ is finitely generated.*

Sketch of Proof We can take a compact 3-submanifold C of M such that each component of $M \setminus C$ has exactly one end. Since $\pi_1(M)$ is finitely generated by assumption, by adding regular neighbourhoods of loops generating $\pi_1(M)$ if necessary, we can assume that the inclusion of C into M induces an epimorphism between the fundamental groups.

If $\operatorname{Fr} C$ is compressible in $M \setminus \operatorname{Int} C$, there is a compressing disc D of $\operatorname{Fr} C$ contained in a component U of $M \setminus \operatorname{Int} C$. Then we can enlarge C again by adding a regular neighbourhood of D, and repeating this finitely many times, we can make every component of $\operatorname{Fr} C$ incompressible in $M \setminus \operatorname{Int} C$. By a standard argument using a continuous map to an aspherical CW-complex, we see that each component U_i of $M \setminus \operatorname{Int} C$ has fundamental group which cannot be decomposed into a non-trivial free product. Then by the work of Scott [44], we can find a compact connected 3-manifold C_i in U_i such that the inclusion of $\operatorname{Fr} C_i$ into $U_i \setminus \operatorname{Int} C_i$ is a homotopy equivalence. This in particular implies that there is a frontier component of C_i which bounds a compact 3-manifold K_i with $\operatorname{Fr} C$. By letting C' be $C \cup (\cup_i (K_i \cup C_i))$, we see that there is a one-to-one correspondence between the components of $\operatorname{Fr} C'$ and the components of $M \setminus C'$ facing them, each of which contains exactly one end, and the inclusion of each component of $\operatorname{Fr} C'$ into the corresponding component of $M \setminus C'$ is a homotopy equivalence. We can repeat the same procedure to construct an exhaustion of M by such compact 3-submanifolds, which gives us a neighbourhood of each end with a product structure. □

To finish this section, we mention that Scott proved in his paper [43] refining [44] that every open 3-manifold with finitely generated fundamental group has a compact 3-submanifold, called a *compact core*, whose inclusion induces a homotopy equivalence. The precise statement is as follows.

Theorem 7.4.3 (Scott [43]) *Every open 3-manifold M with finitely generated fundamental group contains a compact submanifold C whose inclusion induces a homotopy equivalence. Each component of $M \setminus C$ contains exactly one end. In particular M has only finitely many ends.*

Although there is no uniqueness of compact cores up to ambient isotopy, we have the following uniqueness of homeomorphism type.

Theorem 7.4.4 (McCullough–Swarup [31]) *Let M be an open 3-manifold with finitely generated fundamental group, and C_1, C_2 its compact cores with inclusions ι_1, ι_2 to M. Then there is a homeomorphism $h\colon C_1 \to C_2$ such that ι_1 and $\iota_1 \circ h$ are homotopic.*

These results in particular show that if we can attach boundaries to open 3-manifolds, the resulting compact manifolds are unique up to homeomorphisms.

7.5 Ends of Hyperbolic 3-Manifolds

In this section, we shall consider compactification of open hyperbolic 3-manifolds. In this section, *we always assume that the fundamental group of an open hyperbolic 3-manifold we consider is finitely generated.* For simplicity, we shall only deal with hyperbolic 3-manifolds which do not have parabolic cusps. The first work on compactification for such manifolds was given by Marden [29] for geometrically finite hyperbolic 3-manifolds. Here we adopt the following definition for geometric finiteness, which is different from its original definition used by Marden.

Definition 7.5.1 An end e of an open hyperbolic 3-manifold M with finitely generated fundamental group having no parabolic cusps is said to be *geometrically finite* if there is a neighbourhood U of e which intersects no closed geodesics. We say that such M is geometrically finite when all ends of M are geometrically finite.

Theorem 7.5.2 (Marden [29]) *Any geometrically finite open hyperbolic 3-manifold without parabolic cusps is homeomorphic to the interior of a compact 3-manifold with boundary.*

Sketch of Proof Let M be a geometrically finite open hyperbolic 3-manifold without parabolic cusps. Let C be the convex core of M, that is, a minimal convex submanifold containing all closed geodesics. This can be constructed concretely as follows. Consider the action of $\pi_1(M)$ on the universal cover $\mathbb{H}^3$ of M as the covering transformation group. Let Λ be the limit set of the action of $\pi_1(M)$ on the sphere at infinity S^2_∞ of $\mathbb{H}^3$, i.e. the closure of the set consisting of points on S^2_∞ fixed by non-trivial elements of $\pi_1(M)$. Consider the convex hull H of the union of all geodesics whose endpoints at infinity lie on Λ. Then $H/\pi_1(M)$ is the convex core of M.

Since we assumed that every end of e is geometrically finite, we see that e is contained in a component U_e of $M \setminus C$. For each point $x \in U_e$, we define $r(x)$ to be the nearest point of C from x, which is well defined by the convexity of C. This map r gives a homeomorphism from U_e to $\mathrm{Fr}\, U_e \times \mathbb{R}$. Since we assumed that $\pi_1(M)$ is finitely generated, by Theorem 7.4.3, M has only finitely many ends. Since r gives a product structure for every end of M, we see that M is homeomorphic to the interior of a compact 3-manifold with boudary. □

In Thurston's lecture notes where he explained how to prove the uniformisation theorem of Haken manifolds ([54]), the notion of geometric tameness for hyperbolic 3-manifolds was introduced, and it was proved that this condition implies the topological tameness, i.e. the possibility of making them the interiors of compact 3-manifolds with boundaries. To explain what is geometric tameness, we first need to define pleated surfaces.

Definition 7.5.3 Let M be an open hyperbolic 3-manifold without cusps and (S, m) a closed surface with a hyperbolic metric m. A continuous map $f\colon S \to M$ is said to be a *pleated surface* when for each point $x \in S$, there exists a geodesic segment s (with respect to m) containing x in its interior such that $f|s$ is totally geodesic. The subset of S consisting of points where the germs of such geodesic segments are unique is called the *pleating locus* of f.

A pleating locus is a closed set consisting of disjoint complete simple geodesics. Such a set is called a *geodesic lamination*. Now we can define geometric tameness.

Definition 7.5.4 Let e be an end of an open hyperbolic 3-manifold M with finitely generated fundamental group having no parabolic cusps. Let C be a compact core of M. Suppose that an end e is contained in a component E of $M \setminus C$ such that $S := \operatorname{Fr} E$ is incompressible. Then the end e is said to be *geometrically tame* if there is a sequence of pleated surfaces $f_i \colon S \to M$ homotopic to the inclusion which tends to e as $i \to \infty$: i.e. for any neighbourhood U of e, there is $i_0 \in \mathbb{N}$ such that for any $i \geq i_0$, the image $f_i(S)$ is contained in U.

Thurston proved in [54] that the geometric tameness of an open hyperbolic 3-manifold implies that it can be compactified to a 3-manifold with boundary.

Proposition 7.5.5 (Thurston [54]) *Let M be an open hyperbolic 3-manifold with finitely generated fundamental group without parabolic cusps, each of whose ends is either geometrically finite or geometrically tame. Then M is homeomorphic to the interior of a compact 3-manifold with boundary.*

Sketch of Proof Here we just give an outline of the proof given by Thurston in §9 of [54]. Since a neighbourhood of a geometrically finite end can be compactified as shown in Theorem 7.5.2, we have only to deal with geometrically tame ends. Let e be a geometrically tame end, and take a compact core C and its boundary component S as in Definition 7.5.4. Then, by definition, there is a sequence of pleated surfaces $f_i \colon (S, m_i) \to M$ homotopic to the inclusion tending to e. Let λ_i be the pleating locus of f_i on (S, m_i). Fix a hyperbolic metric m on S and isotope λ_i to a geodesic lamination with respect to m, for which we use the same symbol λ_i. Making use of area computation in hyperbolic geometry, we can show that the limit lamination of the λ_i with respect to the Hausdorff topology is uniquely determined up to isolated non-compact geodesics. The limit has a unique minimal component, which is well defined, and it is often referred to as the ending lamination of e.

Let μ_e denote the ending lamination of e. Then, we can show, using the compactness of the set of pleating surfaces intersecting a compact set, that every sequence of pleated surfaces whose pleating loci, regarded as geodesic laminations on (S, m), converge to a geodesic lamination containing μ_e must tend to e. Let $\mathcal{GL}(S)$ denote the set of geodesic laminations. We endow $\mathcal{GL}(S)$ with a topology weaker than the Hausdorff topology: we define that $\mu_i \in \mathcal{GL}(S)$ converges to $\mu \in \mathcal{GL}(S)$ if its Hausdorff limit contains μ. Now we consider the pleating loci of the f_i, and construct an open continuous path $\alpha : [0, \infty) \to \mathcal{GL}(S)$ in the geodesic lamination space $\mathcal{GL}(S)$ which tends to a geodesic lamination μ_e as the parameter goes to ∞. We can perturb the path so that for every geodesic lamination λ on the path, either the components of $S \setminus \lambda$ are all ideal triangles, or they are all ideal triangles except for one, which is an ideal quadrilateral and the path passes by only countably many points corresponding to the latter type of geodesic laminations. We say that the lamination has depth 0 in the former case, and depth 1 in the latter case. We can further assume, by another perturbation, that at a point t_0 where $\alpha(t_0)$ has depth 1, the Hausdorff limits of $\alpha(t)$ as $t \to t_0 - 0$ and $t \to t_0 + 0$ are well defined and both have depth 0.

For each t such that the complement of $\alpha(t)$ has depth 0, we consider a pleated surface f_t homotopic to the inclusion of S whose pleating locus is $\alpha(t)$. Such a pleated surface is unique since $\alpha(t)$ has depth 0. At each point $t_0 \in [0, \infty)$ with $\alpha(t_0)$ having depth 1, we can insert a family of negatively curved surfaces containing the geodesic realisation of $\alpha(t_0)$, which joins $\lim_{t \to t_0 - 0} f_t$ with $\lim_{t \to t_0 + 0} f_t$. This gives rise to a proper map $F: S \times [0, \infty) \to M$ such that $F(S, t)$ tends to e as $t \to \infty$.

Recall that the end e is contained in a component E of $M \setminus C$. By the properness of F, there is $t_1 \in [0, \infty)$ such that $F([t_1, \infty))$ is contained in E. Let Σ_r be the subset of E consisting of points having distance r from C. We can assume that Σ_r is a submanifold in $E \setminus F([0, t_1])$ taking a generic large r. Since C is homotopy equivalent to M, the component E of its complement cannot contain a non-separating surface. Therefore, we can take a component Σ'_r of Σ_r which separates e from C. We perform a compression in $E \setminus F([0, t_1])$ to get a surface $\bar{\Sigma}_r$ which is incompressible in $E \setminus F([0, t_1])$. Again we can assume $\bar{\Sigma}_r$ to be connected by taking a component separating e from C. We make F transverse to Σ'_r by a perturbation, and consider the inverse image $S_r := F^{-1}(\Sigma_r)$. We take a component of S_r which separates $S \times \{\infty\}$ from $S \times \{0\}$ and which is farthest from $S \times \{0\}$ if there are more than one such components, and let it be $\bar{S}_r$. Since $\bar{S}_r$ is homologous to $S \times \{0\}$ and $\bar{\Sigma}_r$ is homologous to S in E, we see that $F|\bar{S}_r$ must have degree 1. If $\bar{S}_r$ is compressible in $S \times [0, \infty)$, then so is it in $S \times [t_1, \infty)$, which contradicts the fact that $\bar{\Sigma}_r$ is incompressible in $E \setminus F([0, t_0])$. Thus we see that $\bar{S}_r$ is incompressible in $S \times [0, \infty)$, and hence is isotopic to $S \times \{t_1\}$. Repeating this construction by letting $t_1 \to \infty$, we see that e has a neighbourhood homeomorphic to $S \times (0, \infty)$. □

In the argument above, the assumption that S is incompressible is essential. In the case when the boundary component S facing e is compressible, we need to add an extra-condition for pleated surfaces tending to e. See Ohshika [36, §4.E] or Souto [48].

Thurston showed in [54] that all ends of hyperbolic 3-manifolds corresponding to (algebraic) limits of quasi-Fuchsian groups are either geometrically finite or geometrically tame. Much more generally, Bonahon proved the following.

Theorem 7.5.6 (Bonahon [2]) *Let M be an open hyperbolic 3-manifold without parabolic cusps such that $\pi_1(M)$ is finitely generated and cannot be decomposed into a non-trivial free product. Then every end of M is either geometrically finite or geometrically tame.*

In fact his argument works for every end that faces an incompressible component of a compact core even when $\pi_1(M)$ can be decomposed into a non-trivial free product.

Sketch of Proof Let C be a compact core of M, and e an end of M, which we assume to face a boundary component S of C, i.e. the component of $M \setminus C$ containing e has S in its closure. Suppose that e is not geometrically finite. We need to show that e is geometrically tame then. Since e is assumed not to be geometrically finite, there is a sequence of closed geodesics c_i such that for any neighbourhood U of e, there exists i_0 such that c_i intersects U for all $i \geq i_0$.

What we really need is a sequence of closed geodesics d_n such that d_n is contained in U for large n for every neighbourhood U. If $\{c_i\}$ does not have this property, then there is a neighbourhood V of e such that $\operatorname{Fr} U$ intersects c_i for all large i. Since c_i travels farther and farther toward e as $i \to \infty$, we can take a component α_i of $V \cap c_i$ also travelling farther and farther as $i \to \infty$, and passing to a subsequence, we can assume that the a_i are all in distinct homotopy classes. Since $\operatorname{Fr} V$ is compact, passing to a subsequence again, we can find a sequence of pairs of distinct arcs α_{i_n} and α_{j_n} in $\{\alpha_i\}$ such that both of the endpoints of α_{i_n} can be connected to those of α_{j_n} respectively by arcs a_n, b_n on $\operatorname{Fr} V$ of length less than $1/n$. Then we construct a closed curve obtained by the concatenation $a_n * \alpha_{j_n} * b_n^{-1} * \alpha_{i_n}$, and let d_n be the closed geodesic freely homotopic to this closed curve. By the fact that both c_{i_n} and c_{j_n} travel farther and farther towards e, elementary area calculation for homotopies implies that for any neighbourhood U, the closed geodesic d_n is contained in U for sufficiently large n.

The next step is to replace d_n with a closed geodesic which is homotopic to a simple closed curve in S keeping the property that it tends to e as $n \to \infty$. Since S is incompressible and the inclusion of C is a homotopy equivalence, d_n is homotopic in U to a closed curve $\bar{d}_n$ on S. We show that the self-intersection of $\bar{d}_n$ grows more slowly than its geodesic length with respect to the fixed hyperbolic metric m. This can be proved by showing that for two closed curves a, b on S which are homtopic to closed geodesics a^*, b^* respectively, their intersection number $i(a, b)$ can be bounded from above by a constant of the order $\exp(-d)\text{length}(a^*)\text{length}(b^*)$, where d denotes the distance from S to $a^* \cup b^*$. In particular, in the case when $a = b = \bar{d}_n$, this implies that $\{d_n^*/\text{length}_m(d_n)\}$ converges in the space of geodesic currents to a "measured lamination" λ, where length_m denotes the length on (S, m). Here a measured lamination is meant to be a geodesic lamination equipped with a transverse measure. Once this is proved, it can be shown that for any sequence of simple closed curves γ_n on S such that $\gamma_n/\text{length}_S(\gamma_n)$ converges to λ, the closed geodesic γ_n^* homotopic to γ_n in M tends to e as $n \to \infty$. By constructing a pleated surface whose pleating locus contains γ_n, we can have a sequence of pleated surfaces tending to e, and see that e is geometrically tame. □

For hyperbolic 3-manifolds whose fundamental groups admit non-trivial decompositions into free products, the same kind of tameness was proved much later by Agol [1] and Calegari–Gabai [7]. A simpler proof somehow combining arguments of these two papers can be found in Soma [47]. An argument distinguishing clearly a part which necessitates hyperbolicity from the other which needs only ordinary three-manifold theory can be found in Bowditch [3].

Theorem 7.5.7 (Agol [1], Calegari–Gabai [7]) *Let M be a hyperbolic 3-manifold with finitely generated fundamental group. Then M is homeomorphic to the interior of a compact 3-manifold with boundary.*

Now we move away from hyperbolic 3-manifolds. Some examples given in Sect. 7.3 show that there are open 3-manifolds which cannot be homeomorphic to any compact 3-manifolds with boundaries. They are all simply connected, and hence their universal covers, which are themselves, cannot be homeomorphic to $\mathbb{R}^3$. Furthermore, Myers proved in [34], without relying on the geometrisation theorem for 3-manifolds, that these manifolds cannot cover closed 3-manifolds. To be more precise, he showed this for any 3-manifold obtained as a union of infinitely many solid tori V_n with the property that V_n is contained in V_{n+1} in such a way that the inclusion is null-homotopic and ∂V_{n+1} is incompressible in $V_{n+1} \setminus V_n$. In particular, for the Whitehead manifold, he also proved that it cannot cover any other 3-manifold non-trivially.

On the other hand, one may also ask whether there exists an open 3-manifold with finitely generated fundamental group whose universal cover is homeomorphic to $\mathbb{R}^3$, but which is not homeomorphic to the interior of any compact 3-manifold with boundary. Such a manifold would elucidate the specialness of hyperbolic 3-manifolds. Scott and Tucker proved in [45] that there are infinitely many such open 3-manifolds.

Their construction is again by taking the union of infinitely many compact 3-manifolds which are homeomorphic to each other. We start from a compact 3-manifold W which is obtained from $S \times I$ for a closed surface S by attaching a solid torus V to $S \times \{1\}$ along a disc on the boundary of V. We consider countably infinitely many copies W_n of W. We express W_n as $S \times [a_n, b_n] \cup V_n$ so that we identify $S \times [a_n, b_n]$ with $S \times I$ and V_n with V. We regard V_n as a regular neighbourhood of the union of a circle C_n and an arc α_n intersecting each other at one point such that a_n is attached to $S \times \{b_n\}$ at one point. We embed W_n to W_{n+1} as follows: we set $a_{n+1} < a_n < b_n < b_{n+1}$, and hence $S \times [a_n, b_n]$ is embedded in the interior of $S \times [a_{n+1}, b_{n+1}]$ naturally. On the other hand, V_n is embedded in $S \times [b_n, b_{n+1}] \cup V_{n+1}$ in such a way that α_n starting from $S \times \{b_n\}$ goes around V_{n+1}, and then comes back to $S \times [b_n, b_{n+1}]$, whereas C_n encircles α_n. (See Fig. 7.2.) We

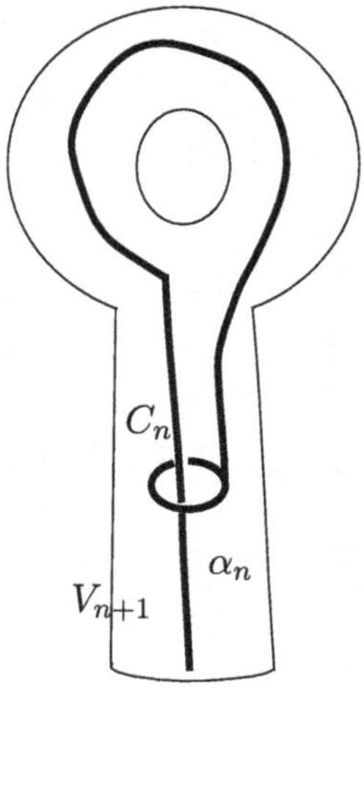

Fig. 7.2 The construction of Scott–Tucker

define W_∞ to be the union $\cup_{n=1}^{\infty} W_n$. We note that W_∞ is homotopy equivalent to S whereas the fundamental group of a component of $W_\infty \setminus W_1$ is infinitely generated. This shows that W_∞ is not homeomorphic to the interior of any compact 3-manifold with boundary by Theorem 7.4.2. On the other hand, it is possible to show that the universal cover of W_∞ is homeomorphic to $\mathbb{R}^3$ drawing a figure concretely.

7.6 Simple Connectedness at Infinity for Universal Covers

The geometrisation theorem for closed 3-manifolds, conjectured by Thurston [53] and proved by Perelman [37–39], implies the following as we mentioned before.

Theorem 7.6.1 *The universal cover of any closed aspherical 3-manifold is homeomorhic to* $\mathbb{R}^3$.

The implication from the geometrisation theorem to Theorem 7.6.1 is roughly as follows. Let M be a closed aspherical 3-manifold. By the prime decomposition theorem due to Kneser [27], M can be decomposed into a connected sum of prime 3-manifolds. Since we assumed that M is aspherical, each summand must be irreducible. If $\pi_1(M)$ is finite, then its universal cover is a homotopy sphere, and hence M cannot be aspherical. Therefore $\pi_1(M)$ is infinite.

Now, by the Jaco–Shalen–Johannson theory [24, 25], M can be decomposed into Seifert fibred manifolds and atoroidal 3-manifolds, (i.e. 3-manifolds not containing non-peripheral incompressible tori,) by cutting it along incompressible tori. The universal cover of M is obtained by pasting universal covers of the pieces together along their plane boundaries, and hence we have only to prove that the universal cover of each piece is a region in $\mathbb{R}^3$ bounded by a properly embedded union of planes. By the geometrisation conjecture, each piece is either a Seifert fibred manifold whose boundary is fibred over a boundary component of the base surface, or a hyperbolic 3-manifold with torus cusp boundaries. It is easy to check that such a manifold always satisfies the required condition.

According to [18, 51], before the geometrisation conjecture was resolved, there were attempts to show a weaker version of Theorem 7.6.1, i.e. simple connectedness at infinity, by Poénaru and Casson independently imposing some extra conditions on fundamental groups.

One can ask naturally whether the same kind of theorem as Theorem 7.6.1 holds for dimensions higher than 3. The answer is negative, as was shown by Davis [8], much prior to the resolution of the 3-dimensional case.

Theorem 7.6.2 (Davis) *In every dimension n greater than 3, there is an n-dimensional closed manifold covered by a contractible open manifold which is not simply connected at infinity.*

His proof is based on the "Coxeter construction". It was shown that there are open contractible (hence one-ended by Proposition 7.3.1) n-manifolds M with infinitely generated π_1^∞ on which the group generated by reflections acts in such a way that there is a finite-index subgroup acting on M freely with compact quotient. The construction uses the fact that there exists a "generalised homology sphere", i.e. a homology manifold with homology groups isomorphic to those of a sphere, in dimension higher than 2.

References

1. I. Agol, Tameness of hyperbolic 3-manifolds (2004). arXiv.org math.GT, 22 pp.
2. F. Bonahon, Bouts des variétés hyperboliques de dimension 3. Ann. Math. 124(1), 71–158 (1986)
3. B.H. Bowdich, Notes on tameness. LEnseign. Mathématique **56**, 229–285 (2010)
4. W. Browder, J. Levine, G.R. Livesay, Finding a boundary for an open manifold. Am. J. Math. 87(4), 1017–1028 (1965)
5. M. Brown, The monotone union of open n-cells is an open n-cell. Proc. Am. Math. Soc. **12**(5), 812–814 (1961)
6. E.M. Brown, Unknotting in $M^2 \times I$. Trans. Am. Math. Soc. **123**(2), 835–847 (1966)
7. D. Calegari, D. Gabai, Shrinkwrapping and the taming of hyperbolic 3-manifolds. J. Am. Math. Soc. **19**(20), 385–446 (2006)
8. M.W. Davis, Groups Generated by reflections and aspherical manifolds not covered by Euclidean space. Ann. Math. **117**(2), 293–324 (1983)
9. S.K. Donaldson, An application of gauge theory to four-dimensional topology. J. Differ. Geom. **18**(2), 279–315 (1983)
10. C.H. Edwards, Open 3-manifolds which are simply connected at infinity. Proc. Am. Math. Soc. **14**(3), 391–395 (1963)
11. M.H. Freedman, The topology of four-dimensional manifolds. J. Differ. Geom. **17**(3), 357–453 (1982)
12. H. Freudenthal, Über die Enden topologischer Räume und Gruppen. Math. Z. **33**(1), 692–713 (1931)
13. H. Freudenthal, Neuaufbau der Endentheorie. Ann. Math. **43**(2), 261–279 (1942)
14. H. Freudenthal, Über die Enden diskreter Räume und Gruppen. Comment. Math. Helv. **17**(1), 1–38 (1944)
15. J. Glimm, Two cartesian products which are euclidean spaces. Bullet. Soc. Math. France **79**, 131–135 (1960)
16. R.E. Gompf, Three exotic $\mathbf{R}^4$'s and other anomalies. J. Differ. Geom. **18**(2), 317–328 (1983)
17. R.E. Gompf, An infinite set of exotic $\mathbf{R}^4$'s. J. Differ. Geom. **21**(2), 283–300 (1985)
18. R. Grimaldi, V. Poénaru, Topologie à l'infini et variétés à géométrie bornée. Bull. Math. Soc. Sci. Math. Roumanie **56**(4), 419–434 (2013)
19. J. Hempel, *3-Manifolds*, vol. 349 (AMS Chelsea Publishing, Providence, 2004). Reprint of the 1976 original
20. M.W. Hirsch, E.C. Zeeman, Engulfing. Bullet. Am. Math. Soc. **72**(1), 113–115 (1966)
21. H. Hopf, Enden offener Räume und unendliche diskontinuierliche Gruppen. Comment. Math. Helv. **16**(1), 81–100 (1943)
22. L.S. Husch, T.M. Price, Finding a boundary for a 3-manifold. Ann. Math. **91**(1), 223 (1970)
23. B. Jackson, End invariants of group extensions. Topology **21**(1), 71–81 (1982)
24. W.H. Jaco, P.B. Shalen, Seifert fibered spaces in 3-manifolds. Memoir. Am. Math. Soc. **21**(220), viii–192 (1979)
25. K. Johannson, *Homotopy Equivalences of 3-Manifolds with Boundaries*, vol. 761 (Springer, Berlin, 1979)

26. R.C. Kirby, L.C. Siebenmann, *Foundational Essays on Topological Manifolds, Smoothings, and Triangulations*. Annals of Mathematics Studies, vol. 88 (Princeton University Press, Princeton, 1977). With Notes by John Milnor and Michael Atiyah
27. H. Kneser, Geschlossene flächen in dreidimensionalen mannigfaltigkeiten. Jahresber. Dtsch. Math.-Ver. **38**, 248–259 (1929)
28. E. Luft, On contractible open topological manifolds. Invent. Math. **4**(3), 192–201 (1967)
29. A. Marden, The geometry of finitely generated kleinian groups. Ann. Math. **99**(3), 383–462 (1974)
30. B. Mazur, A note on some contractible 4-manifolds. Ann. Math. **73**(1), 221 (1961)
31. D. McCullough, A. Miller, G.A. Swarup, Uniqueness of cores of non-compact 3-manifolds. J. Lond. Math. Soc. **s2–32**(3), 548–556 (1985)
32. D.R. McMillan, Cartesian products of contractible open manifolds. Bullet. Am. Math. Soc. **67**(5), 510–514 (1961)
33. J. Munkres, Obstructions to the smoothing of piecewise-differentiable homeomorphisms. Ann. Math. **72**(3), 521 (1960)
34. R. Myers, Contractible open 3-manifolds which are not covering spaces. Topology **27**(1), 27–35 (1988)
35. M.H.A. Newman, Boundaries of ULC sets in euclidean n-space. Proc. Natl. Acad. Sci. **34**(5), 193–196 (1948)
36. K. Ohshika, Kleinian groups which are limits of geometrically finite groups. Mem. Am. Math. Soc. **177**(834), xii–116 (2005)
37. G. Perelman, The entropy formula for the Ricci flow and its geometric applications (2002). arXiv:math/0211159
38. G. Perelman, Finite extinction time for the solutions to the Ricci flow on certain three-manifolds (2003). arXiv:math/0307245
39. G. Perelman, Ricci flow with surgery on three-manifolds (2003). arXiv:math/0303109
40. M.H. Poincaré, Cinquième complément à l'analysis situs. Rend. Circ. Mat. Palermo **18**, 45–110 (1904)
41. V. Poénaru, Les décompositions de l'hypercube en produit topologique. Bullet. Soc. Math. France **79**, 113–129 (1960)
42. V. Poénaru, A remark on simply-connected 3-manifolds. Bullet. Am. Math. Soc. **80**(6), 1203–1204 (1974)
43. G.P. Scott, Compact submanifolds of 3-manifolds. J. Lond. Math. Soc. Ser. **7**(2), 246–250 (1973)
44. G.P. Scott, Finitely generated 3-manifold groups are finitely presented. J. Lond. Math. Soc. **s2–6**(3), 437–440 (1973)
45. P. Scott, T. Tucker, Some examples of exotic non-compact 3-manifolds. Quart. J. Math. **40**(4), 481–499 (1989)
46. L.C. Siebenmann, *The Obstruction to Finding a Boundary for an Open Manifold of Dimension Greater Than Five* (ProQuest LLC, Ann Arbor, 1965). Thesis (Ph.D.)–Princeton University
47. T. Soma, Existence of ruled wrappings in hyperbolic 3-manifolds. Geom. Topol. **10**(2), 1173–1184 (2006)
48. J. Souto, A note on the tameness of hyperbolic 3-manifolds. Topology **44**(2), 459–474 (2005)
49. E. Specker, Die erste Cohomologiegruppe von Überlagerungen und Homotopie-Eigenschaften dreidimensionaler Mannigfaltigkeiten. Comment. Math. Helv. **23**(1), 303–333 (1949)
50. J. Stallings, The piecewise-linear structure of euclidean space. Math. Proc. Cambridge Philos. Soc. **58**(3), 481–488 (1962)
51. J.R. Stallings, S.M. Gersten, Casson's idea about 3-manifolds whose universal cover is $\mathbf{R}^3$. Int. J. Algebra Comput. **1**(4), 395–406 (1991)
52. C.H. Taubes, Gauge theory on asymptotically periodic 4-manifolds. J. Differ. Geom. **25**(3), 363–430 (1987)
53. W.P. Thurston, Three-dimensional manifolds, Kleinian groups and hyperbolic geometry. Am. Math. Soc. Bullet. New Ser. **6**(3), 357–381 (1982)

54. W.P. Thurston, *The Geometry and Topology of Three-Manifolds*, vol. IV (American Mathematical Society, Providence, 2022). Edited and with a preface by Steven P. Kerckhoff and a chapter by J.W. Milnor
55. T.W. Tucker, Non-compact 3-manifolds and the missing-boundary problem. Topology **13**, 267–273 (1974)
56. J.H.C. Whitehead, A certain open manifold whose group is unity. Quart. J. Math. **os-6**(1), 268–279 (1935)
57. L. Zippin, On semicompact spaces. Am. J. Math. **57**(2), 327–341 (1935)

Chapter 8
Axiomatic Phylogenetics

Vladimir Turaev

Abstract We use the language of quivers to formulate a mathematical framework for phylogenetics.

Keywords Philogenetics · Quivers · Evolutionary sequence · Finite metric spaces · Ultrametrics

2010 Mathematics Subject Classification 18A10, 54E99, 92D15

8.1 Introduction

Mathematical methods are commonly used in biology which, in some cases, leads to new mathematical theories, see [1–6]. In this paper we study certain mathematical ideas suggested by evolutionary biology. Biological evolution is described in Wikipedia as the "change in the heritable characteristics of biological populations over successive generations". Evolution creates a sequence of generations of species, each generation arising from the previous one via natural selection and/or genetic drift. Every species produces descendants in the next generation. The branching diagram showing the evolutionary relationships between species is called the evolutionary tree. This tree has a distinguished vertex—a root—as all life on Earth is believed to share a common ancestor known as the last universal common ancestor (LUCA). Phylogenetics studies the evolutionary tree and aims to recover it from the current generation of living organisms. For a review of phylogenetic analysis, see [9]; for mathematical aspects of phylogenetics, see [1, 5].

We introduce here a mathematical formalism for evolution emphasizing its phylogenetic aspects. While it remains to be seen whether this formalism may be of use in theoretical biology, it does suggest new mathematical concepts. Our main idea

V. Turaev (✉)
Department of Mathematics, Indiana University, Bloomington, IN, USA
e-mail: vturaev@yahoo.com

L. Funar, A. Papadopoulos (eds.), *Essays on Topology*,
https://doi.org/10.1007/978-3-031-81414-3_8

is to consider not only the historical evolution but all possible evolutions of primitive beings into complex organisms. To this end we use the language of quivers (directed graphs). We introduce analogues of many key notions of phylogenetics in the setting of quivers. This includes analogues of the notions of species, parents/children, LUCAs, evolutions, generations, etc. We show that under appropriate assumptions on the quiver, it gives rise to an evolutionary tree. We give examples of quivers satisfying these assumptions.

This work was partially supported by the NSF grant DMS-1664358.

8.2 Quivers and Evolutions

8.2.1 *Quivers*

A *quiver* $\mathcal{O}$ is formed by a class of vertices and a collection of sets $\{\mathcal{O}(A, B)\}_{A,B}$ where A, B run over the vertices of $\mathcal{O}$. The elements of the set $\mathcal{O}(A, B)$ are called *edges* from A to B and are represented by arrows $A \to B$ or $B \leftarrow A$. (We allow $A = B$, i.e, the quiver $\mathcal{O}$ may have loops). To indicate that A is a vertex of $\mathcal{O}$ we write $A \in \mathcal{O}$. A biologically minded reader may replace the word vertex with "species". An arrow $B \leftarrow A$ is understood in the sense that B is a "parent" of A.

8.2.2 *Evolutions*

An *evolutionary chain* or, shorter, an *evolution* of length $m \geq 0$ in a quiver $\mathcal{O}$ is a sequence

$$A_0 \longleftarrow A_1 \longleftarrow A_2 \longleftarrow \cdots \longleftarrow A_m \tag{8.2.1}$$

where $A_0, A_1, \ldots, A_m \in \mathcal{O}$ and the arrows are edges of $\mathcal{O}$. We call A_0 the *initial* vertex and A_m the *terminal* vertex of the evolution.

For $A, B \in \mathcal{O}$, we write $A \leq B$ and say that A is an *ancestor* of B and B is a *descendant* of A if there is an evolution in $\mathcal{O}$ leading from A to B, i.e., starting at A and terminating at B.

Lemma 8.2.1 *The relation $\leq$ is reflexive and transitive.*

Proof The reflexivity of $\leq$ means that $A \leq A$ for any $A \in \mathcal{O}$. In other words, every $A \in \mathcal{O}$ is both its own ancestor and its own descendant. This is obtained using the evolution of A to itself of length 0. The transitivity of $\leq$ means that if $A \leq B$ and $B \leq C$, then $A \leq C$ for any $A, B, C \in \mathcal{O}$. This is obtained by concatenating an evolution from A to B with an evolution from B to C. □

8.2.3 Isotypy

We say that vertices A, B of a quiver $\mathcal{O}$ are *isotypic* and write $A \sim B$ if both $A \le B$ and $B \le A$, i.e., if A, B are descendants of each other. The vertices A, B are isotypic iff there is a circle formed by arrows and traversing both A and B:

$$A \longleftarrow \cdots \longleftarrow B \longleftarrow \cdots \longleftarrow A.$$

The relation $\sim$ is an equivalence relation: every vertex is isotypic to itself (the reflexivity); if $A \sim B$, then $B \sim A$ for any $A, B \in \mathcal{O}$ (the symmetry); if $A \sim B \sim C$, then $A \sim C$ for any $A, B, C \in \mathcal{O}$ (the transitivity). The first two properties follow from the definitions. The transitivity holds because if $A \sim B \sim C$, then $A \le B \le C$ and $C \le B \le A$. Therefore $A \le C$ and $C \le A$, i.e., $A \sim C$.

The next lemma shows that isotypic vertices are equivalent from the evolutionary viewpoint and may be considered as inessential variations of each other.

Lemma 8.2.2 *For any $A, B \in \mathcal{O}$ the following five conditions are equivalent:*

(i) A and B are isotypic;
(ii) A is both an ancestor and a descendant of B;
(iii) B is both an ancestor and a descendant of A;
(iv) A and B have the same ancestors;
(v) A and B have the same descendants.

Proof The equivalences $(i) \iff (ii) \iff (iii)$ follow directly from the definitions. If $A \sim B$ and X is an ancestor of A, then $X \le A \le B$. So, $X \le B$, i.e., X is an ancestor of B. Similarly, all ancestors of B are ancestors of A. Thus, $(i) \implies (iv)$. Conversely, if (iv) holds, then A being its own ancestor must be an ancestor of B, i.e., $A \le B$. Analogously, $B \le A$. Thus, $(iv) \implies (i)$ and so $(i) \iff (iv)$. The equivalence $(i) \iff (v)$ is checked similarly. □

Note that all vertices appearing in an evolution between isotypic vertices are isotypic: For any evolution (8.2.1) between isotypic vertices A_0, A_m, we have $A_0 \sim A_k$ for $k = 0, 1, \ldots, m$. Indeed, the existence of the evolution (8.2.1) implies that $A_0 \le A_k \le A_m$. Since A_0, A_m are isotypic, $A_m \le A_0$. Therefore $A_k \le A_0$. Thus, $A_k \sim A_0$.

8.2.4 Hereditary and Anti-hereditary Properties

A property, say, P of vertices of a quiver is *hereditary* if for any vertex having P, all its descendants also have P. Similarly, the property P is *anti-hereditary* if for any vertex having P, all its ancestors also have P. If the property P is hereditary or anti-hereditary, then it is isotypy invariant, i.e., all vertices isotypic to a vertex having P also have P. If a property is hereditary, then its negation is anti-hereditary and vice versa.

8.2.5 Examples

1. Let $\mathcal{SET}$ be the quiver formed by finite non-empty sets as vertices and maps between sets as edges. An evolutionary chain of length $m \geq 0$ in $\mathcal{SET}$ is a sequence of finite non-empty sets $A_0, A_1, \ldots, A_m$ and maps $\{A_k \to A_{k-1}\}_{k=1}^m$. The elements of A_k can be viewed as the species of the k-th generation while the map $A_k \to A_{k-1}$ carries each species to its parent. Such evolutions reflect asexual reproduction: each species has a single parent. (Reproduction involving two parents may be formalized by taking as edges between sets A, B the maps $B \to A \times A$.) For any $A, B \in \mathcal{SET}$, a map carrying B to a single element of A yields a length 1 evolution $A \leftarrow B$. Thus, $A \leq B$. Consequently, all vertices of $\mathcal{SET}$ are isotypic.
2. Let $\mathcal{S}$ be the quiver formed by finite non-empty sets as vertices and surjective maps between sets as edges. It is clear that $A \leq B$ for $A, B \in \mathcal{S}$ if and only if $\mathrm{card}(A) \leq \mathrm{card}(B)$ where card is the number of elements of a set. Therefore, $A \sim B$ if and only if $\mathrm{card}(A) = \mathrm{card}(B)$.

8.3 Primitivity and the Height

8.3.1 Primitive Vertices

The role of LUCAs in evolutionary biology will be played here by so-called primitive vertices. We call a vertex A of a quiver $\mathcal{O}$ *primitive* if all ancestors of A also are descendants of A. Thus, $A \in \mathcal{O}$ is primitive if $B \leq A \Longrightarrow A \leq B$ for all $B \in \mathcal{O}$. It is clear from our definitions that a vertex A is primitive iff all its ancestors are isotypic to A.

A quiver may have no primitive vertices. If it has primitive vertices, then they may be non-isotypic to each other. At the same time, all vertices isotypic to a primitive vertex are primitive, as is clear from the following lemma.

Lemma 8.3.1 *All ancestors of a primitive vertex are primitive.*

Proof Let A be a primitive vertex of a quiver $\mathcal{O}$ and let $B \in \mathcal{O}$ be an ancestor of A. If C is an ancestor of B, then $C \leq B \leq A$ and by transitivity, $C \leq A$. Since A is primitive, we must have $A \leq C$. Since $B \leq A$, the transitivity yields $B \leq C$. Thus, $C \leq B \Longrightarrow B \leq C$, i.e., B is primitive. □

8.3.2 The Height

By a *full evolution* for a vertex $X \in \mathcal{O}$ we mean an evolutionary chain in $\mathcal{O}$ which starts at a primitive vertex and terminates at X. We view such an evolution as an evolutionary history of X. A full evolution for X does not necessarily exist, and if it exists, it is not necessarily unique.

We define the *height* $h(X)$ of $X \in \mathcal{O}$ to be the smallest integer $h \geq 0$ such that there is a full evolution for X of length h. The height measures the evolutionary complexity of X, i.e., the number of steps needed to evolve from a primitive vertex to X. If there are no full evolutions for X, then we set $h(X) = \infty$.

It is clear that a vertex $X \in \mathcal{O}$ has a finite height if and only if X has a primitive ancestor. Concatenating evolutionary chains we conclude that the property of a vertex to have finite height is hereditary. Note also that $h(X) = 0$ if and only if X is primitive, and $h(X) = 1$ if and only if X is not primitive but there is an edge from X to a primitive vertex.

A full evolution for a vertex $X \in \mathcal{O}$ is said to be *short* if its length is the smallest among all full evolutions for X. Such an evolution exists if and only if $h(X) < \infty$ and its length is equal to $h(X)$.

Lemma 8.3.2 *Let $\alpha = (A_0 \leftarrow \cdots \leftarrow A_m)$ be a full evolution for a vertex $A_m \in \mathcal{O}$. Then $h(A_k) \leq k$ for all $k = 0, 1, \ldots, m$. If α is short, then $h(A_k) = k$ for all k.*

Proof For $k = 0, 1, \ldots, m$, let $\alpha_k = (A_0 \leftarrow \cdots \leftarrow A_k)$ be the initial segment of α of length k and let $\alpha^k = (A_k \leftarrow \cdots \leftarrow A_m)$ be the terminal segment of α of length $m - k$. Since α is a full evolution, A_0 is a primitive vertex and so α_k is a full evolution for A_k. Thus, $h(A_k) \leq k$. If there is a full evolution for A_k of length $< k$ then concatenating it with α^k we obtain a full evolution for A_m of length $< k + (m - k) = m$. Therefore, if α is short, then so is α_k and $h(A_k) = k$. □

Lemma 8.3.2 implies that any vertex of finite height $m \geq 1$ has ancestors of heights $0, 1, \ldots, m - 1$.

8.3.3 Examples

1. All vertices of the quiver $\mathcal{SET}$ from Example 8.2.5.1 are primitive and have zero height. Consider next the quiver $\mathcal{S}$ from Example 8.2.5.2. A finite non-empty set $X \in \mathcal{S}$ is primitive if and only if $\text{card}(X) = 1$. Every $X \in \mathcal{S}$ is the terminal vertex of a full evolution: if $\text{card}(X) = 1$, then this is the length 0 evolution (X); if $\text{card}(X) \geq 2$, then this is the length 1 evolution $A \leftarrow X$ where A is a 1-point set and the arrow stands for the only map from X to A. Thus, $h(X) = 0$ if $\text{card}(X) = 1$ and $h(X) = 1$ otherwise.
2. Let Γ be a rooted tree, i.e., a connected graph without cycles and with a distinguished vertex $*$ (the root). We direct all edges of Γ as follows: if an edge of Γ connects vertices A, B and there is a path from $*$ to A missing B, then this edge is directed from B to A. This turns Γ into a quiver. It is easy to check that: different vertices of Γ cannot be isotypic; the vertex $*$ is the only primitive vertex of Γ; every vertex $X \in \Gamma$ is the terminal vertex of a unique full evolution formed

by the vertices and edges of the shortest path from $*$ to X, this full evolution is short.
3. Consider the quiver with three vertices A, B, C and three edges leading from B to A, from B to C, and from C to B. Then $A \leq B \sim C$ and $h(A) = 0, h(B) = 1$, $h(C) = 2$. Thus, isotypic vertices may have different heights.
4. The quiver with vertices $\{A_k\}_{k\in\mathbb{Z}}$ and arrows $\{A_k \leftarrow A_{k+1}\}_{k\in\mathbb{Z}}$ has no primitive vertices and $h(A_k) = \infty$ for all k.

8.4 Universal Evolutions and Phylogenetic Vertices

We introduce so-called universal evolutions and phylogenetic vertices.

8.4.1 Universal Evolutions

Given two evolutionary chains

$$\alpha = (A_0 \leftarrow \cdots \leftarrow A_m) \quad \text{and} \quad \beta = (B_0 \leftarrow \cdots \leftarrow B_n)$$

in a quiver $\mathcal{O}$, we say that α *embeds* in β if $m \leq n$ and there are integers

$$0 \leq r_0 < r_1 < \cdots < r_m \leq n$$

such that $A_k \sim B_{r_k}$ for $k = 0, 1, \ldots, m$. For $m = n$, this condition amounts to $A_k \sim B_k$ for all k. In this case we say that the evolutions α, β are *isotypic*.

A *universal evolution* for a vertex $X \in \mathcal{O}$ is a full evolution for X which embeds in all full evolutions for X. Thus, the vertices of a universal evolution for X must appear (in the same order) in any full evolution for X, at least up to isotypy. A universal evolution for X is necessarily short. So, its length is equal to $h(X)$. A universal evolution for X may exist only if $h(X) < \infty$.

Clearly, any two universal evolutions for a vertex $X \in \mathcal{O}$ are isotypic. Any evolution isotypic to a universal evolution is itself universal. We view a universal evolution for X as a complete evolutionary history of X.

8.4.2 Phylogenetic Vertices

A vertex $X \in \mathcal{O}$ is *phylogenetic* if there is a universal evolution for X in $\mathcal{O}$. For example, for a primitive X, the length zero evolution (X) is universal. Thus, all primitive vertices are phylogenetic.

We state several properties of phylogenetic vertices.

Theorem 8.4.1 *Let $X \in \mathcal{O}$ be a phylogenetic vertex. Then:*

(i) X has a primitive ancestor;
(ii) all primitive ancestors of X are isotypic to each other;
(iii) all short full evolutions for X are universal;
(iv) all vertices appearing in a universal evolution for X are phylogenetic.

Proof Pick a universal evolution $\alpha = (A_0 \leftarrow \cdots \leftarrow A_m)$ for $X = A_m$. Clearly, A_0 is a primitive ancestor of X which yields (i). To prove (ii), consider another primitive ancestor B of X and an evolution $\beta = (B_0 \leftarrow \cdots \leftarrow B_n)$ from $B_0 = B$ to $B_n = X$. Since B is primitive, β is full. Since α is universal, it embeds in β. Consequently, $A_0 \sim B_k$ for some $k \leq n$. Then $B = B_0 \leq B_k \leq A_0$. Hence $B \leq A_0$, and the primitivity of A_0 ensures that $B \sim A_0$. This gives (ii).

To prove (iii), consider a short full evolution $\beta = (B_0 \leftarrow \cdots \leftarrow B_m)$ for $X = B_m$. Since α embeds in β and α, β have the same length, $A_k \sim B_k$ for all $k \leq m$. Therefore β is universal.

For $k = 0, 1, \ldots, m$, let α_k and α^k be the initial and terminal segments of the universal evolution α as defined in the proof of Lemma 8.3.2. Since α is a full evolution for A, α_k is a full evolution for A_k. Any full evolution γ for A_k concatenated with α^k yields a full evolution $\gamma\,\alpha^k$ for A_m. Since the evolution $\alpha_k\,\alpha^k = \alpha$ is universal, it embeds in $\gamma\,\alpha^k$. Therefore α_k embeds in γ. Thus, α_k is a universal evolution for A_k and the vertex A_k is phylogenetic. □

8.4.3 Examples

The full evolutions in Examples 8.3.3.1 and 8.3.3.2 are universal. All vertices of the quivers in these examples are phylogenetic.

8.5 Critical Ancestors and Normality

We further develop our language and, in particular, introduce the notion of a critical ancestor. In this section, $\mathcal{O}$ is an arbitrary quiver.

8.5.1 The Step Inequality

The height of a vertex of $\mathcal{O}$ cannot increase too quickly under evolutions. In fact, for any length 1 evolution (i.e., an edge of our quiver) $A \leftarrow B$ we have the *step inequality*

$$h(B) \leq h(A) + 1. \tag{8.5.1}$$

More precisely, if $h(A) < \infty$, then (8.5.1) holds and, in particular, $h(B) < \infty$. Indeed, concatenating a full evolution for A of length $h(A)$ with the evolution $A \leftarrow B$ we obtain a full evolution for B of length $h(A) + 1$. If $h(A) = \infty$, then the inequality (8.5.1) provides no information on $h(B)$.

8.5.2 Critical Ancestors

A vertex $A \in \mathcal{O}$ is a *critical ancestor* of a vertex $B \in \mathcal{O}$ if $h(A) < \infty$ and there is an evolution $A \leftarrow A_1 \leftarrow \cdots \leftarrow B$ of length ≥ 1 such that $h(A_1) = h(A) + 1$. We view critical ancestors of B as gatekeepers yielding access to higher levels in evolutions from the primitives to B. It follows from Lemma 8.3.2 that any vertex of a finite height $m \geq 1$ has critical ancestors of heights $0, 1, \ldots, m-1$. Primitive vertices have no critical ancestors because all ancestors of a primitive vertex are primitive and have zero height. So, the equality $h(A_1) = h(A) + 1$ above cannot hold.

A critical ancestor of a vertex $B \in \mathcal{O}$ is necessarily a critical ancestor of all descendants of B. Consequently, isotypic vertices have the same critical ancestors.

8.5.3 Normal Vertices

We call a vertex $B \in \mathcal{O}$ *normal* if any two critical ancestors of B of the same height are isotypic. For instance, all primitive vertices are normal simply because they have no critical ancestors.

Lemma 8.5.1 *The normality of a vertex is anti-hereditary. Vertices isotypic to a normal vertex are normal.*

Proof If B is a descendant of A, then all critical ancestors of A are critical ancestors of B. Consequently, if B is normal, then so is A. The second claim of the lemma follows from the first claim. □

Theorem 8.5.2 *Any normal vertex of finite height is phylogenetic.*

Proof Let X be a normal vertex of finite height m. If $m = 0$, then X is primitive and therefore phylogenetic. Assume that $m \geq 1$ and pick a short full evolution $\alpha = (A_0 \leftarrow \cdots \leftarrow A_m)$ for $X = A_m$. We will prove that α is universal, i.e., that α embeds in any full evolution $\beta = (B_0 \leftarrow \cdots \leftarrow B_n)$ for $X = B_n$. Since β starts at a vertex of zero height and terminates at a vertex of height m, the step inequality implies that for each $k = 0, 1, \ldots, m-1$, there is $r \in \{0, 1, \ldots, n-1\}$ such that $h(B_r) = k$ and $h(B_{r+1}) = k + 1$. Let r_k be the smallest such r. Also, set $r_m = n$. Since the segment $B_0 \leftarrow \cdots \leftarrow B_{r_k}$ of β starts from a vertex of zero height and terminates at a vertex of height k, the same argument as above shows that if $k > 0$, then there is an index $r < r_k$ such that $h(B_r) = k - 1$ and $h(B_{r+1}) = k$. On the other hand, r_{k-1} is the smallest such index. Therefore $r_{k-1} \leq r < r_k$. Hence, $r_0 < r_1 < \cdots < r_m$. We claim that $A_k \sim B_{r_k}$ for $k = 0, 1, \ldots, m$. For $k = m$, this is obvious because $A_m = X = B_n = B_{r_m}$. For $k < m$, Lemma 8.3.2 and the definition of r_k imply that both A_k and B_{r_k} are critical ancestors of X of height k. By the normality of X, we have $A_k \sim B_{r_k}$. Therefore, α embeds in β. □

8.5.4 Examples

In Example 8.3.3.3, the vertex A is primitive and has no critical ancestors. The vertex B has A, B as critical ancestors. This is clear from the evolutions $A \leftarrow B$ and $B \leftarrow C \leftarrow B$. The vertex C is not a critical ancestor of B because this quiver does not have a vertex with height $h(C) + 1 = 3$. The vertex C is isotypic to B and therefore has the same critical ancestors A, B. In this example and in Examples 8.2.5.1, 8.2.5.2, and 8.3.3.2 all vertices are normal. The quiver in Example 8.3.3.4 has no normal vertices and its vertices have no critical ancestors.

8.6 Monotonous and Phylogenetic Quivers

We introduce monotonous and phylogenetic quivers.

8.6.1 Monotonous Quivers

The height of a vertex may decrease under certain evolutions, and we view such evolutions as degenerate. We call a quiver $\mathcal{O}$ *monotonous* if it does not have such degenerate evolutions, i.e., if for any edge $A \to B$ in $\mathcal{O}$, we have $h(A) \geq h(B)$. This condition may be reformulated by saying that the descendants of any vertex $B \in \mathcal{O}$ have the height $\geq h(B)$. A useful consequence: in a monotonous quiver, isotypic vertices have the same height. For monotonous quivers we can invert Theorem 8.5.2 as follows.

Theorem 8.6.1 *A vertex of a monotonous quiver is phylogenetic if and only if it is normal and has finite height.*

Proof In view of Theorem 8.5.2, it suffices to prove the "only if" part. Consider a phylogenetic vertex X of a monotonous quiver and a universal evolution $\alpha = (A_0 \leftarrow \cdots \leftarrow A_m)$ for $X = A_m$ where $m = h(X) < \infty$. The evolution α is short, and so, by Lemma 8.3.2, the vertices $A_0, A_1, \ldots, A_{m-1}$ are critical ancestors of X of heights respectively $0, 1, \ldots, m-1$. To prove the normality of X, consider an arbitrary critical ancestor B of X of height $r < \infty$. By the definition of a critical ancestor, there is an evolution $\beta = (B \leftarrow B_1 \leftarrow \cdots \leftarrow X)$ with $h(B_1) = r + 1$. By the monotonicity, all vertices in β except B have the height $\geq h(B_1) > r$. Pick any short full evolution γ for B. By Lemma 8.3.2, all vertices in γ except B have height $< r$. Concatenating γ with β, we obtain a full evolution $\gamma\beta$ for X whose only vertex of height r is B. Since the universal evolution α must embed in $\gamma\beta$, we have $A_r \sim B$. Thus, X is normal. □

It is clear that in a monotonous quiver, the property of a vertex to have finite height is anti-hereditary. Combining this observation with Lemma 8.5.1 and Theorem 8.6.1, we obtain the following.

Corollary 8.6.2 *In a monotonous quiver, the phylogeneticity of a vertex is anti-hereditary.*

Theorem 8.6.1 and Lemma 8.3.2 imply the following claim.

Corollary 8.6.3 *In a monotonous quiver, a phylogenetic vertex X has precisely $h(X)$ isotypy classes of critical ancestors.*

Corollary 8.6.4 *In a monotonous quiver, universal evolutions for isotypic phylogenetic vertices are themselves isotypic.*

Proof Consider universal evolutions $A_0 \leftarrow \cdots \leftarrow A_m$ and $B_0 \leftarrow \cdots \leftarrow B_m$ for isotypic phylogenetic vertices $X = A_m, Y = B_m$ of height m. For any $k = 0, 1, \ldots, m-1$, the critical ancestors A_k of X and B_k of Y have the same height k. Since $X \sim Y$, the vertex B_k is also a critical ancestor of X. By Theorem 8.6.1, X is normal and therefore $A_k \sim B_k$. Also, $A_m = X \sim Y = B_m$. □

8.6.2 Phylogenetic Quivers

A quiver is *phylogenetic* if it is monotonous and all its vertices are phylogenetic. Theorem 8.6.1 shows that a quiver is phylogenetic if and only if it is monotonous and all its vertices are normal and have finite height.

It is easy to check that the quivers $\mathcal{SET}$ and $\mathcal{S}$ from Sect. 8.2.5 as well as the tree quiver $\mathcal{O}(\Gamma)$ from Sect. 8.3.3.2 are phylogenetic. Further examples of phylogenetic quivers can be derived from various algebraic theories involving filtrations. As a specific case, we consider a quiver of finite nilpotent groups. (Similar phylogenetic quivers can be formed from finite solvable groups and from nilpotent/solvable finite-dimensional Lie algebras.) Recall the lower central series $G_0 \supset G_1 \supset \cdots$ of a group G: by definition, $G_0 = G$ and for $n \geq 0$, the group $G_{n+1} \subset G_n$ is generated by the commutators $xyx^{-1}y^{-1}$ with $x \in G_n$ and $y \in G$. The group G is *nilpotent* if $G_n = \{1\}$ for some $n \geq 0$, and the smallest such n is denoted $n(G)$. Let $\mathcal{N}$ be the quiver formed by finite nilpotent groups and group epimorphisms $f : G \to H$ such that $n(G) \geq 1$ and $\operatorname{Ker}(f) \subset G_{n(G)-1}$. It is easy to check that: groups in $\mathcal{N}$ are isotypic if and only if they are isomorphic; a group in $\mathcal{N}$ is primitive as a vertex if and only if it is trivial; for any $G \in \mathcal{N}$, the sequence of quotient groups and projections

$$\{e\} = G/G_0 \longleftarrow G/G_1 \longleftarrow \cdots \longleftarrow G/G_{n(G)} = G$$

is a universal evolution for G. The quiver $\mathcal{N}$ is phylogenetic.

8.6.3 Remark

Any monotonous quiver $\mathcal{O}$ determines a quiver $\mathcal{O}'$ consisting of all phylogenetic vertices of $\mathcal{O}$ and all edges between them in $\mathcal{O}$. It is easy to show using Corollary 8.6.2 that the quiver $\mathcal{O}'$ is phylogenetic.

8.7 The Evolutionary Sequence and the Evolutionary Forest

8.7.1 The Evolutionary Sequence

Here we restrict ourselves to so-called small quivers. We call a quiver *small* if the isotypy classes of its vertices form a set. This condition is satisfied in all our examples.

Consider a small phylogenetic quiver $\mathcal{O}$ and let $\widetilde{\mathcal{O}}$ be the set of isotypy classes of vertices of $\mathcal{O}$. Each vertex $A \in \mathcal{O}$ represents an element $[A]$ of $\widetilde{\mathcal{O}}$. Two vertices $A, B \in \mathcal{O}$ represent the same element of $\widetilde{\mathcal{O}}$ if and only if $A \sim B$. The relation $\leq$ in the class of vertices of $\mathcal{O}$ induces a relation $\leq$ in $\widetilde{\mathcal{O}}$: for $a, b \in \widetilde{\mathcal{O}}$, we set $a \leq b$ if $A \leq B$ for some (and then for all) $A, B \in \mathcal{O}$ representing respectively a and b. The relation $\leq$ in $\widetilde{\mathcal{O}}$ is a partial order, i.e., it is reflexive, transitive, and antisymmetric (if $a \leq b$ and $b \leq a$, then $a = b$). Clearly, $\widetilde{\mathcal{O}} = \amalg_{m \geq 0} \mathcal{O}_m$ where $\mathcal{O}_m$ is the set of isotypy classes of vertices of $\mathcal{O}$ of height m. In particular, $\mathcal{O}_0$ is the set of isotypy classes of primitive vertices of $\mathcal{O}$.

For each $m \geq 1$, we define the *parental map*

$$p = p_m : \mathcal{O}_m \to \mathcal{O}_{m-1} \tag{8.7.1}$$

as follows. For any $A \in \mathcal{O}$ of height $m \geq 1$, consider a universal evolution $A_0 \leftarrow \cdots \leftarrow A_{m-1} \leftarrow A_m = A$ and set $p([A]) = [A_{m-1}]$. By Corollary 8.6.4, this yields a well-defined map (8.7.1). The resulting sequence of sets and maps

$$\mathcal{O}_0 \xleftarrow{p} \mathcal{O}_1 \xleftarrow{p} \mathcal{O}_2 \xleftarrow{p} \mathcal{O}_3 \xleftarrow{p} \cdots \tag{8.7.2}$$

is called the *evolutionary sequence* of $\mathcal{O}$.

Theorem 8.7.1 *Let $a \in \mathcal{O}_m$ and $b \in \mathcal{O}_n$ with $m, n \geq 0$. Then:*

(i) $a \leq b$ if and only if $m \leq n$ and $a \leq p^{n-m}(b) \in \mathcal{O}_m$;
(ii) For $m = n = 0$, we have $a \leq b \iff a = b$;
(iii) For $m = n \geq 1$, if $a \leq b$, then $p(a) = p(b) \in \mathcal{O}_{m-1}$;
(iv) For $m = n - 1$, the equality $a = p(b)$ holds if and only if there is an edge $B \to A$ in $\mathcal{O}$ such that $A, B \in \mathcal{O}$ represent respectively the isotypy classes a, b.

Theorem 8.7.1 is proved below. Note that by Claim (i), the partial order in $\widetilde{\mathcal{O}}$ is fully determined by its restrictions to the sets $\{\mathcal{O}_m\}_m$ and the parental maps.

8.7.2 Proof of Theorem 8.7.1

We begin with a lemma.

Lemma 8.7.2 *Let $\alpha = (A_0 \leftarrow \cdots \leftarrow A_m)$ and $\beta = (B_0 \leftarrow \cdots \leftarrow B_n)$ be universal evolutions for vertices A_m, B_n in a monotonous quiver such that $A_m \leq B_n$. If $m = n$, then $A_{m-1} \sim B_{n-1}$. If $m < n$, then $A_m \leq B_{n-1}$.*

Proof By Lemma 8.3.2, $h(A_k) = k$ for all $k \leq m$ and $h(B_l) = l$ for all $l \leq n$. Since $A_m \leq B_n$, there is an evolution γ from A_m to B_n, and then $\alpha\gamma$ is an evolution from A_0 to B_n. Since β is universal, it embeds in $\alpha\gamma$. So, B_{n-1} is isotypic to a vertex in $\alpha\gamma$ of the same height $n-1$ (here we use the monotonicity of the quiver). Clearly, all vertices in γ have height $\geq h(A_m) = m$. If $m = n$, then the only vertex of height $n - 1 = m - 1$ in $\alpha\gamma$ is A_{m-1}. Thus, $A_{m-1} \sim B_{n-1}$. If $m < n$, then all vertices in α other than A_m have height $< m \leq n-1$. In this case, B_{n-1} has to be isotypic to a vertex, C, appearing in the evolution γ, and so $A_m \leq C \leq B_{n-1}$ and $A_m \leq B_{n-1}$. □

We now prove Theorem 8.7.1. We start with Claim (i). Suppose that $a \leq b$. By the monotonicity, $m \leq n$. If $m = n$, then $a \leq b = p^{n-m}(b)$. If $m < n$, then the second claim of Lemma 8.7.2 implies that $a \leq p(b)$. Iterating, we get $a \leq p^{n-m}(b)$. Conversely, suppose that $m \leq n$ and $a \leq p^{n-m}(b)$. It follows from the definitions that $p(b) \leq b$ for all b. Hence,

$$a \leq p^{n-m}(b) \leq p^{n-m-1}(b) \leq \cdots \leq p(b) \leq b.$$

Claim (ii) holds because any vertex of zero height is primitive and so is isotypic to all its ancestors. Claim (iii) follows from the first claim of Lemma 8.7.2. We prove Claim (iv). If $p(b) = a$, then pick a representative $B \in \mathcal{O}$ of b and a universal evolution $B_0 \leftarrow \cdots \leftarrow B_{n-1} \leftarrow B_n = B$. The edge $B_{n-1} \leftarrow B_n$ satisfies our conditions because $[B_{n-1}] = p(b) = a$. Conversely, suppose that there is an edge $A \leftarrow B$ in $\mathcal{O}$ such that A, B represent respectively a, b. Concatenating a short full evolution for A with the 1-edge evolution $A \leftarrow B$, we obtain a full evolution $\cdots \leftarrow A \leftarrow B$ of length $h(A) + 1 = m + 1 = n = h(B)$. This evolution is short and, by Theorem 8.4.1(iii), universal. Therefore $p(b) = [A] = a$.

8.7.3 The Evolutionary Forest

Any sequence of sets and maps

$$P_0 \xleftarrow{p} P_1 \xleftarrow{p} P_2 \xleftarrow{p} P_3 \xleftarrow{p} \cdots$$

determines a graph Γ: take the disjoint union $\amalg_{m\geq 0} P_m$ as the set of vertices and connect each $a \in P_m$ with $m \geq 1$ to $p(a) \in P_{m-1}$ by an edge. This graph is a forest in the sense that all its components are trees. Clearly, every component of Γ has a unique vertex in P_0. If $\mathrm{card}(P_0) = 1$, then Γ is a tree. In this case we define a metric d in the set $\amalg_{m\geq 0} P_m$: For any elements a, b of this set, the distance $d(a, b)$ is the minimal number of edges in a path in Γ from a to b. Clearly, $d(a, b) = k + l$ where $k, l \geq 0$ are minimal integers such that $p^k(a) = p^l(b)$. Note that for each $m \geq 0$, the restriction of d to the set P_m takes only even values and is an ultrametric (the definition of an ultrametric is recalled in Sect. 8.10.1).

Applying these constructions to the evolutionary sequence (8.7.2), we obtain a forest called the *evolutionary forest* of the quiver $\mathcal{O}$. If $\mathrm{card}(\mathcal{O}_0) = 1$, i.e., if all primitive vertices of $\mathcal{O}$ are isotypic, then we also obtain a metric in the set $\widetilde{\mathcal{O}}$ of isotypy classes of vertices of $\mathcal{O}$.

8.7.4 Examples

The evolutionary forest of the quiver $\mathcal{SET}$ is a single point. The evolutionary forest of the quiver $\mathcal{S}$ from Example 8.2.5.2 is a wedge of a countable number of segments. The evolutionary forest of the tree quiver Γ from Example 8.3.3.2 can be identified with Γ.

8.8 E-sequences and Reconstruction

8.8.1 E-sequences

Axiomatizing the properties of the evolutionary sequences, we define so-called E-sequences. An *E-sequence* consists of partially ordered sets $(P_m, \leq)_{m\geq 0}$ and maps $(p = p_m : P_m \to P_{m-1})_{m\geq 1}$ such that the partial order in P_0 is trivial and for any $a, b \in P_m$ with $m \geq 1$, if $a \leq b$, then $p(a) = p(b)$. We will sometimes use the strict partial order $<$ in P_m defined by $a < b$ if $a \leq b$ and $a \neq b$. Two E-sequences P, P' are *isomorphic* if there are bijections $\{f_m : P_m \to P'_m\}_{m\geq 0}$ such that for all $m \geq 1$, we have $p' f_m = f_{m-1} p : P_m \to P'_{m-1}$ and f_m is order-preserving in the sense that $a \leq b \iff f_m(a) \leq f_m(b)$ for any $a, b \in P_m$.

By Theorem 8.7.1, the evolutionary sequence of a small phylogenetic quiver is an E-sequence. We show now that all E-sequences arise in this way.

Theorem 8.8.1 *Every E-sequence is isomorphic to the evolutionary sequence of a small phylogenetic quiver.*

Proof Given an E-sequence P, we define a quiver $\mathcal{O}$ as follows. The vertices of $\mathcal{O}$ are the elements of the set $\amalg_{m\geq 0} P_m$. For all $a, b \in P_m$ with $m \geq 0$, there is a single edge $a \to b$ if $b < a$ and a single edge $a \to p(a) \in P_{m-1}$ if $m \geq 1$. We claim that the quiver $\mathcal{O}$ is phylogenetic. A typical evolutionary chain in $\mathcal{O}$ starts at some $a_0 \in P_m$ with $m \geq 0$ and consecutively takes bigger and bigger elements $a_0 < a_1 < \cdots$ of P_m. At some step, one either stops at $a_i \in P_m$ or proceeds to the higher level by taking for a_{i+1} any element of the set $p^{-1}(a_i) \subset P_{m+1}$ (if this set is non-void). Then the whole process is repeated starting at a_{i+1}, etc. (One has to stop eventually.) The antisymmetry of the partial order implies that an evolution in $\mathcal{O}$ of non-zero length cannot start and end at the same vertex. Consequently, isotypic vertices of $\mathcal{O}$ must coincide so that $\mathcal{O}$ is small and $\widetilde{\mathcal{O}} = \amalg_{m\geq 0} P_m$. The primitive vertices of $\mathcal{O}$ are the elements of P_0 (here we use that the partial order in P_0 is trivial). All full evolutions for any $a \in P_m$ with $m \geq 1$ start at $p^m(a) \in P_0$ and include the vertices $\{p^k(a)\}_{k=1}^m$. Therefore the full evolution

$$p^m(a) \leftarrow p^{m-1}(a) \leftarrow \cdots \leftarrow p(a) \leftarrow a$$

is universal for a. As a consequence, $h(a) = m$ and $\mathcal{O}_m = P_m$. Therefore the quiver $\mathcal{O}$ is monotonous and all its vertices are phylogenetic. It is clear that the evolutionary sequence of $\mathcal{O}$ is isomorphic to P. □

8.8.2 Reconstruction

The main objective of phylogenetics is to recover the evolutionary tree from the current generation of species. We briefly discuss reconstruction in our context. The reconstruction aims to recover the initial segment

$$P_0 \xleftarrow{p} P_1 \xleftarrow{p} \cdots \xleftarrow{p} P_N \tag{8.8.1}$$

of an E-sequence from the set P_N, eventually endowed with additional data. For simplicity, we assume here that $\mathrm{card}(P_0) = 1$ and all the maps p are surjective. One well-known approach to reconstruction uses the ultrametric ρ in P_N defined by letting the distance $\rho(a, b)$ between any $a, b \in P_N$ to be the minimal integer $k \geq 0$ such that $p^k(a) = p^k(b)$. (Note that $\rho(a, b) = \frac{1}{2}d(a, b)$ where d is the metric defined in Sect. 8.7.3.) The sets and maps (8.8.1) can be fully recovered from the ultrametric space (P_N, ρ). Namely, for $s = 0, 1, \ldots, N$, the elements of P_s can be identified with balls in P_N or radius $N - s$; for $s \geq 1$, the map $p : P_s \to P_{s-1}$ carries a ball $B \subset P_N$ or radius $N - s$ to the unique ball of radius $N - s + 1$ in P_N containing B. Next, we encode in terms of P_N the given strict partial order $<$ in $P_1, \ldots, P_N$. This partial order induces a binary relation $\prec$ in P_N by the rule $a \prec b$ if $a \neq b$ and $p^{k-1}(a) < p^{k-1}(b)$ for $k = \rho(a, b) \geq 1$. Conversely, the strict partial order $<$ in $P_1, \ldots, P_N$ can be fully recovered from $\prec$ and ρ: two balls $B, B' \subset P_N$ of the same radius r satisfy $B < B'$ if and only if $a \prec b$ and $\rho(a, b) = r + 1$ for some (and then for all) $a \in B, b \in B'$. In particular. for $r = 0$, two points $a, b \in P_N$ satisfy $a < b$ if and only if $a \prec b$ and $\rho(a, b) = 1$.

8.8.3 *Remark*

So far we have studied evolutions in a static world in which all vertices (species) coexist together. To relate to the real world, we briefly discuss the timeline. One way to involve time is to accept the following three principles:

(i) (the moment zero) all primitive vertices come to existence at the same moment of time, the moment zero;
(ii) (the molecular clock) the time needed for an accomplishment of an evolutionary chain is equal to a constant coefficient C times the length of the chain;
(iii) (the least wait) every non-primitive vertex comes to existence at the earliest possibility, i.e., at the end of a short full evolution.

These principles ensure that each vertex X of a phylogenetic quiver $\mathcal{O}$ evolves in the moment of time $C \cdot h(X)$. The sequence (8.7.2) is then the sequence of generations: each set $\mathcal{O}_m$ with $m \geq 0$ is the generation of vertices that have evolved at the moment of time Cm.

The principles (i) and (ii) above can be generalized by agreeing that (i)$'$ each primitive vertex comes to existence at a certain moment of time (not necessarily the same) and (ii)$'$ each edge carries a positive length and the length of any evolution is the sum of the lengths of the constituent edges.

8.8.4 *Remark*

A binary relation $\prec$ in an ultrametric space (X, ρ) arises as in Sect. 8.8.2 from the initial segment of length $N \geq 1$ of an E-sequence iff

(i) the ultrametric ρ takes values in the set $\{0, 1, \ldots, N\}$;
(ii) $a \prec b \Longrightarrow b \not\prec a$ for all $a, b \in X$ (in particular, $a \not\prec a$ for all $a \in X$);
(iii) for any distinct $a, b, c \in X$,

- if $a \prec b$ and $\rho(a, c) < \rho(a, b)$, then $c \prec b$;
- if $a \prec b$ and $\rho(b, c) < \rho(a, b)$, then $a \prec c$;
- if $a \prec b \prec c$ and $\rho(a, b) = \rho(a, c) = \rho(b, c)$, then $a \prec c$.

8.9 Clades in Phylogenetic Quivers

8.9.1 *Clades*

Any vertex A of a quiver $\mathcal{O}$ determines a quiver $\mathcal{O}_A$ formed by all descendants of A and all edges between them in $\mathcal{O}$. Following the standard taxonomic terminology, we call $\mathcal{O}_A$ the *clade* of A. Clearly, $A \in \mathcal{O}_A$. We state a few properties of $\mathcal{O}_A$.

Lemma 8.9.1 *The primitive vertices of $\mathcal{O}_A$ are all the vertices of $\mathcal{O}$ isotypic to A. In particular, A is a primitive vertex of $\mathcal{O}_A$.*

Proof If $B \in \mathcal{O}_A$ is primitive in $\mathcal{O}_A$, then the relation $A \leq B$ implies that $B \leq A$, i.e., that A, B are isotypic in $\mathcal{O}$. Conversely, if $B \in \mathcal{O}$ is isotypic to A, then $B \in \mathcal{O}_A$ and for any $C \in \mathcal{O}_A$, we have $B \leq A \leq C$ so that $B \leq C$. Thus, B is primitive as a vertex of $\mathcal{O}_A$. □

Lemma 8.9.2 *All vertices of $\mathcal{O}_A$ have finite height in $\mathcal{O}_A$.*

Proof Since A is primitive in $\mathcal{O}_A$ and all vertices of $\mathcal{O}_A$ are terminal vertices of evolutions starting at A, all vertices of $\mathcal{O}_A$ have final height. □

The heights of vertices in the quivers $\mathcal{O}$ and $\mathcal{O}_A$ will be denoted respectively by h and h_A. The following theorem estimates h_A via h for monotonous $\mathcal{O}$.

Theorem 8.9.3 *If A is a vertex of finite height in a monotonous quiver $\mathcal{O}$, then for all $B \in \mathcal{O}_A$ we have $\infty > h(B) \geq h(A)$ and $h_A(B) \geq h(B) - h(A)$.*

Proof The inequalities $\infty > h(B) \geq h(A)$ follow from the definitions of Sect. 8.3.2 and the monotonicity of $\mathcal{O}$. By Lemmas 8.9.1 and 8.9.2, there is a short full evolution, β, in $\mathcal{O}_A$ starting at a vertex $C \in \mathcal{O}$ isotypic to A and terminating at B. We have $h(C) = h(A) < \infty$ so that there is a short full evolution, γ, for C in $\mathcal{O}$. Then $\gamma\beta$ is a full evolution for B in $\mathcal{O}$ of length $h(C) + h_A(B) = h(A) + h_A(B)$. Therefore $h(A) + h_A(B) \geq h(B)$ and $h_A(B) \geq h(B) - h(A)$. □

8.9.2 Regular Vertices

We call a vertex A of a quiver $\mathcal{O}$ *regular* if for any $B \in \mathcal{O}_A$ with $h(A) = h(B)$, there is an edge $A \leftarrow B$ in $\mathcal{O}$. This condition is a very weak form of a composition law in $\mathcal{O}$: it may be rephrased by saying that if there is a finite chain of edges $A \leftarrow \cdots \leftarrow B$ in $\mathcal{O}$ and $h(A) = h(B)$, then there is an edge $A \leftarrow B$ in $\mathcal{O}$. The reader may check that in our examples of phylogenetic quivers all vertices are regular.

We now state the main result of this section.

Theorem 8.9.4 *The clade of any regular vertex of a small phylogenetic quiver is a small phylogenetic quiver.*

The proof of this theorem occupies the rest of the section.

8.9.3 Lemmas

Consider a vertex A of a small phylogenetic quiver $\mathcal{O}$ and a vertex $B \in \mathcal{O}_A$. Set $m = h(A) \geq 0$ and $n = h(B)$. By Theorem 8.9.3, we have $n \geq m$ and $h_A(B) \geq n - m$. Recall the isotypy classes $[A] \in \mathcal{O}_m$, $[B] \in \mathcal{O}_n$, and the iterated parental map $p^{n-m} : \mathcal{O}_n \to \mathcal{O}_m$.

Lemma 8.9.5 *If $p^{n-m}([B]) = [A]$, then B is a phylogenetic vertex of $\mathcal{O}_A$ and $h_A(B) = n - m$.*

Proof Suppose first that $n = m$. Then the condition $p^{n-m}([B]) = [A]$ means that $B \sim A$. Then B is primitive in $\mathcal{O}_A$ by Lemma 8.9.1. Therefore $h_A(B) = 0 = n - m$ and B is phylogenetic in $\mathcal{O}_A$. Suppose now that $n > m$. Pick universal evolutions α for A and $\beta = (B_0 \leftarrow \cdots \leftarrow B_n = B)$ for B in $\mathcal{O}$. The condition $p^{n-m}([B]) = [A]$ implies that $B_m \sim A$. Then the terminal segment $\beta^m = (B_m \leftarrow \cdots \leftarrow B_n)$ of β is a full evolution in $\mathcal{O}_A$ for B of length $n - m$. For any evolution γ from A to B in $\mathcal{O}$, the universal evolution β must embed in $\alpha\gamma$. Comparing the heights of vertices in $\mathcal{O}$, we observe that such an embedding carries β^m to γ and yields an embedding of β^m in γ. Thus, β^m is a universal evolution for B in $\mathcal{O}_A$. So, B is a phylogenetic vertex of $\mathcal{O}_A$ and $h_A(B) = n - m$. □

Lemma 8.9.6 *If $p^{n-m}([B]) \neq [A]$ and A is regular, then B is a phylogenetic vertex of $\mathcal{O}_A$ and $h_A(B) = n - m + 1$.*

Proof Suppose first that $n = m$. Then the condition $p^{n-m}([B]) \neq [A]$ means that A and B are not isotypic in $\mathcal{O}$. By Lemma 8.9.1, B is not a primitive vertex of $\mathcal{O}_A$ and so $h_A(B) \geq 1$. Since $B \in \mathcal{O}_A$, there is an evolution from A to B in $\mathcal{O}$ of non-zero length. The regularity of A and the assumption $h(B) = n = m = h(A)$ imply that there is an edge $A \leftarrow B$ in $\mathcal{O}$. This edge yields a universal evolution for B in $\mathcal{O}_A$ of length 1. Thus, B is a phylogenetic vertex of $\mathcal{O}_A$ and $h_A(B) = 1 = n - m + 1$.

Suppose now that $n > m$. Pick a universal evolution $\beta = (B_0 \leftarrow \cdots \leftarrow B_n)$ for $B = B_n$ in $\mathcal{O}$ and an arbitrary evolution $\gamma = (A_0 \leftarrow \cdots \leftarrow A_k)$ from $A = A_0$ to $B = A_k$ in $\mathcal{O}$ (for some $k \geq 0$). The same argument as in the proof of the previous lemma shows that the terminal segment $\beta^m = (B_m \leftarrow \cdots \leftarrow B_n)$ of β embeds in γ. In particular, $B_m \sim A_i$ for some i. So $A \leq A_i \leq B_m$ and $B_m \in \mathcal{O}_A$. The condition $p^{n-m}([B]) \neq [A]$ implies that B_m and $A = A_0$ are not isotypic, i.e., $i \neq 0$. Therefore there is an evolution from $A = A_0$ to $B_m \sim A_i$ in $\mathcal{O}$ of length $\geq i \neq 0$. The regularity of A and the equalities $h(B_m) = m = h(A)$ imply that there is an edge $A \leftarrow B_m$ in $\mathcal{O}$. Concatenating this edge with β^m, we obtain an evolution β^m_+ from A to $B = B_n$ of length $n - m + 1$. By the above, β^m_+ embeds in γ. Therefore, β^m_+ is a universal evolution for $B \in \mathcal{O}_A$, B is phylogenetic in $\mathcal{O}_A$, and $h_A(B) = n - m + 1$. □

8.9.4 Proof of Theorem 8.9.4

Let A be a regular vertex of a phylogenetic quiver $\mathcal{O}$. Since $\mathcal{O}$ is small, so is the clade $\mathcal{O}_A$ of A. In view of Lemmas 8.9.5 and 8.9.6, we need only to prove that $\mathcal{O}_A$ is monotonous, i.e., that $h_A(B) \leq h_A(C)$ for any $B, C \in \mathcal{O}_A$ such that there is an edge $B \leftarrow C$ in $\mathcal{O}$. Set $m = h(A)$, $n = h(B)$ and $k = h(C)$. By the monotonicity of $\mathcal{O}$ and the definition of the height,

$$m \leq n \leq k \leq n + 1.$$

We distinguish three cases.

Case $m = n = k$. If $A \sim B$, then $h_A(B) = 0 \leq h_A(C)$. If $A \sim C$, then $A \leq B \leq C \leq A$ so that $A \sim B$ and we proceed as above. If neither B nor C are isotypic to A, then Lemma 8.9.6 applies to B, C and gives $h_A(B) = 1 = h_A(C)$.

Case $m < n = k$. Since there is an edge $B \leftarrow C$ in $\mathcal{O}$ and $h(B) = n = k = h(C)$, Theorem 8.7.1(iii) implies that $p([B]) = p([C])$. Therefore $p^{n-m}([B]) = p^{n-m}([C])$. Then either $p^{n-m}([B]) = [A]$ and Lemma 8.9.5 gives

$$h_A(B) = n - m = k - m = h_A(C)$$

or $p^{n-m}([B]) \neq [A]$ and Lemma 8.9.6 gives

$$h_A(B) = n - m + 1 = k - m + 1 = h_A(C).$$

Case $k = n + 1$. Concatenating a universal evolution for B in $\mathcal{O}$ with the edge $B \leftarrow C$ we obtain a short full evolution for C in $\mathcal{O}$. The latter evolution is universal and so $p([C]) = [B]$. Therefore $p^{n-m}([B]) = p^{n-m+1}([C])$. If $p^{n-m}([B]) = [A]$, then Lemma 8.9.5 gives

$$h_A(B) = n - m \text{ and } h_A(C) = k - m.$$

If $p^{n-m}([B]) \neq [A]$, then Lemma 8.9.6 gives

$$h_A(B) = n - m + 1 \text{ and } h_A(C) = k - m + 1.$$

In both cases, $h_A(B) < h_A(C)$.

8.10 The Quiver of Finite Ultrametric Spaces

We form a phylogenetic quiver from finite ultrametric spaces. We first recall the definition of an ultrametric space.

8.10.1 Ultrametrics and Contractions

A *metric space* is a pair (X, d) consisting of a non-empty set X and a map $d : X \times X \to \mathbb{R}_+ = [0, \infty)$, the *metric*, such that for all $x, y, z \in X$, we have $d(x, y) = d(y, x)$, $d(x, y) = 0 \Longleftrightarrow x = y$, and

$$d(x, y) \leq d(x, z) + d(y, z). \tag{8.10.1}$$

An isometry between metric spaces is a metric-preserving bijection. A metric space (X, d) is *finite* if X is a finite set.

An *ultrametric space* is a metric space (X, d) such that for all $x, y, z \in X$,

$$d(x, y) \leq \max(d(x, z), d(y, z)). \tag{8.10.2}$$

The map d is called then an *ultrametric*. The condition (8.10.2) is stronger than (8.10.1); it implies that for any $x, y, z \in X$, two of the numbers $d(x, y), d(x, z), d(y, z)$ are equal and are greater than or equal to the third number.

For a real number $\epsilon > 0$, we call a map $f : X \to Y$ between metric spaces $X = (X, d)$ and $Y = (Y, \rho)$ an *ϵ-contraction* if $f(X) = Y$ and

$$\rho(f(x), f(y)) = d(x, y) - \varepsilon \tag{8.10.3}$$

for any distinct $x, y \in X$. Various ϵ-contractions with $\epsilon \in (0, \infty)$ are collectively called *contractions*. Contractions are surjective but not necessarily bijective. The composition of two contractions is not necessarily a contraction.

Theorem 8.10.1 *Let $\mathcal{U}$ be the quiver whose vertices are finite ultrametric spaces and whose edges are contractions and isometries. Then:*

(i) Two finite ultrametric spaces are isotypic to each other in $\mathcal{U}$ if and only if they are isometric;
(ii) A finite ultrametric space is a primitive vertex of $\mathcal{U}$ if and only if it consists of a single point;
(iii) The height of a finite ultrametric space $X = (X, d)$ in $\mathcal{U}$ is equal to the number of non-zero elements in the set $d(X \times X) \subset \mathbb{R}_+$;
(iv) The quiver $\mathcal{U}$ is phylogenetic;
(v) All vertices of $\mathcal{U}$ are regular.

The proof of Theorem 8.10.1 occupies the rest of the section. We begin with notation and a lemma. For a finite metric space $X = (X, d)$, set

$$||X|| = \sum_{x,y \in X} d(x, y) = \sum_{x,y \in X, x \neq y} d(x, y).$$

If $\mathrm{card}(X) \geq 2$, then set

$$|X| = \min\{d(x, y) \mid x, y \in X, x \neq y\}.$$

Lemma 8.10.2 *Let $f : X \to Y$ be an ϵ-contraction between finite metric spaces $X = (X, d)$ and $Y = (Y, \rho)$ where $\varepsilon > 0$. If* $\mathrm{card}(Y) \geq 2$, *then* $||Y|| < ||X||$. *If f is not a bijection, then* $\mathrm{card}(X) \geq 2$ *and* $|X| = \varepsilon$.

Proof Since $f(X) = Y$, we can pick for each $a \in Y$ a point $\overline{a} \in f^{-1}(a)$. Then

$$||Y|| = \sum_{a,b \in Y, a \neq b} \rho(a, b) = \sum_{a,b \in Y, a \neq b} (d(\overline{a}, \overline{b}) - \varepsilon) < \sum_{a,b \in Y, a \neq b} d(\overline{a}, \overline{b}) \leq ||X||.$$

If f is non-injective, then $\text{card}(X) \geq \text{card}(Y) + 1 \geq 2$ and $f(a) = f(b)$ for some distinct $a, b \in X$. Formula (8.10.3) implies that $d(a, b) = \varepsilon$ and $d(x, y) \geq \varepsilon$ for any distinct $x, y \in X$. Hence, $|X| = \varepsilon$. □

8.10.2 Proof of Theorem 8.10.1

Since isometries are edges in $\mathcal{U}$ and each edge determines a length 1 evolution, isometric ultrametric spaces are isotypic in $\mathcal{U}$. Conversely, consider isotypic $X, Y \in \mathcal{U}$. If $\text{card}(X) = 1$, then the condition $Y \leq X$ implies that $\text{card}(Y) = 1$. If $\text{card}(Y) = 1$, then the condition $X \leq Y$ implies that $\text{card}(X) = 1$. In both cases, X is isometric to Y. Suppose that $\text{card}(X) \geq 2$ and $\text{card}(Y) \geq 2$. Since $Y \leq X$, there is an evolution $Y \leftarrow \cdots \leftarrow X$ in $\mathcal{U}$. By Lemma 8.10.2, either all edges in this evolution are isometries or $||Y|| < ||X||$. Similarly, the relation $X \leq Y$ implies that either X, Y are isometric or $||X|| < ||Y||$. Since we cannot have $||Y|| < ||X|| < ||Y||$, the only option is that X and Y are isometric. This implies (i).

To proceed, we define certain edges in $\mathcal{U}$. Given $\epsilon \in \mathbb{R}$ and a finite ultrametric space $X = (X, d)$ having at least two points, the formula

$$d_\varepsilon(x, y) = \begin{cases} 0 & \text{if } x = y, \\ d(x, y) - \varepsilon & \text{if } x \neq y \end{cases}$$

defines a map $d_\varepsilon : X \times X \to \mathbb{R}$. If $\epsilon < |X|$, then d_ε is an ultrametric in X and the identity map $\text{id}_X : (X, d) \to (X, d_\varepsilon)$ is a bijective ε-contraction. For $\epsilon = |X|$, the map $d^\bullet = d_{|X|} : X \times X \to \mathbb{R}$ satisfies all requirements on an ultrametric except one: there are distinct $x, y \in X$ with $d^\bullet(x, y) = 0$. We define a relation $\sim_{d^\bullet}$ in X by $x_1 \sim_{d^\bullet} x_2$ if $d^\bullet(x_1, x_2) = 0$. It is straightforward to check that $\sim_{d^\bullet}$ is an equivalence relation. Let $Y = X/{\sim_{d^\bullet}}$ be the quotient set and let $p : X \to Y$ be the projection. Then there is a unique map $\rho : Y \times Y \to \mathbb{R}$ such that $d^\bullet(x, y) = \rho(p(x), p(y))$ for all $x, y \in X$. The map ρ is an ultrametric. We denote the ultrametric space (Y, ρ) by $u(X)$. Clearly, the projection $p : X \to u(X)$ is a non-injective $|X|$-contraction. Applying this construction recursively, we obtain an evolution in $\mathcal{U}$

$$u^m(X) \leftarrow \cdots \leftarrow u^2(X) \leftarrow u(X) \leftarrow X \tag{8.10.4}$$

where m is the smallest integer such that $u^m(X)$ has only one point. Thus, X has a 1-point ancestor. We can now prove Claim (ii) of the theorem. By the definition of $\mathcal{U}$, the only edges from a 1-point ultrametric space A to vertices of $\mathcal{U}$ are isometries. Thus, all ancestors of A are isometric to A and A is primitive. If $X \in \mathcal{U}$ is primitive, then X is isotypic to all its ancestors and, in particular, is isotypic to a 1-point space. By (i), X is a 1-point space. This proves (ii).

We verify now that every vertex $X \in \mathcal{U}$ is phylogenetic. Let α be the full evolution (8.10.4) for X. We claim that α is universal. We must show that α embeds in an arbitrary full evolution for X, say,

$$\beta = (B_0 \xleftarrow{f_1} B_1 \xleftarrow{f_2} \cdots \xleftarrow{f_{n-1}} B_{n-1} \xleftarrow{f_n} B_n = X)$$

where B_0 is a 1-point space. Note that if for some $k = 1, \ldots, n-1$, the edge $f_k : B_k \to B_{k-1}$ in β is an isometry, then we can delete B_k from β and replace f_k, f_{k+1} with their composition $f_k f_{k+1} : B_{k+1} \to B_{k-1}$ which is a contraction or an isometry depending on whether f_{k+1} is a contraction or an isometry. This gives a shorter full evolution β' for X which embeds in β. It suffices to prove that α embeds in β'. Similarly, if the edge $f_n : B_n \to B_{n-1}$ in β is an isometry, then we can delete B_{n-1} from β and replace f_{n-1}, f_n with their composition. Thus, we can reduce ourselves to the case where all edges in β are contractions. In the same way, we can get rid of *bijective* contractions in β using the obvious fact that the composition of a bijective ε-contraction with any ε'-contraction is an $(\varepsilon + \varepsilon')$-contraction. It remains therefore to treat the case where all edges in β are non-bijective contractions. In particular, $f_n : X = B_n \to B_{n-1}$ is a non-bijective ε-contraction for some $\varepsilon > 0$. By Lemma 8.10.2, $\varepsilon = |X|$. Consequently, B_{n-1} is isometric to $u(X)$. Proceeding by induction, we obtain that B_{n-k} is isometric to $u^k(X)$ for all $k \geq 1$. Since m is the smallest integer such that $u^m(X)$ has only one point and since B_0 is the only 1-point space in the evolution β, we conclude that $m = n$ and that the evolutions α and β are isotypic. In particular, α embeds in β. This proves the universality of α.

To prove (iii), let $N(X)$ be the number of non-zero elements in $d(X \times X) \subset \mathbb{R}_+$. By the above, the evolution (8.10.4) is universal and so $h(X) = m$. The values of the ultrametric in $u(X)$ are obtained from those of d by finding the smallest non-zero value of d and subtracting it from all non-zero values of d. Therefore $N(u(X)) = N(X) - 1$. Inductively, $N(u^k(X)) = N(X) - k$ for $k = 1, \ldots, m$. Since $u^m(X)$ is a 1-point set, $0 = N(u^m(X)) = N(X) - m$. Thus, $h(X) = m = N(X)$.

We now prove (iv). That $\mathcal{U}$ is small follows from (i). As we know, all vertices of $\mathcal{U}$ are phylogenetic. For any contraction or isometry $X \to Y$ between finite ultrametric spaces, one easily sees that $N(X) \geq N(Y)$. Therefore $h(X) = N(X) \geq N(Y) = h(Y)$. Thus, $\mathcal{U}$ is monotonous.

We leave it to the reader to check that all vertices of $\mathcal{U}$ are regular.

8.11 The Quiver of Finite Metric Spaces

We form a phylogenetic quiver from finite metric spaces. We start by defining trim metric spaces and drifts following [7, 8].

8.11.1 Trim Metric Spaces

A metric space (X, d) is *trim* if either $\text{card}(X) = 1$ or for each $x \in X$, there are distinct $y, z \in X \setminus \{x\}$ such that

$$d(x, y) + d(x, z) = d(y, z).$$

The latter equality may be expressed by saying that x lies between y and z. The class of finite trim metric spaces is quite narrow. In particular, there are no trim metric spaces having two or three points. A finite subset of a Euclidean space with ≥ 2 points and with the induced metric cannot be trim: such a subset must contain a pair of points lying at the maximal distance; these points cannot lie between other points of the subset. For examples of trim metric spaces, see [7, 8].

With any metric space (X, d), we associate a function $\underline{d} : X \to \mathbb{R}_+$ as follows: if X has only one point, then $\underline{d} = 0$; if X has two points x, y, then $\underline{d}(x) = \underline{d}(y) = d(x, y)/2$; if X has three or more points, then for all $x \in X$,

$$\underline{d}(x) = \inf_{y,z \in X \setminus \{x\}, y \neq z} \frac{d(x, y) + d(x, z) - d(y, z)}{2} \geq 0.$$

It is easy to check that $\underline{d}(x) + \underline{d}(y) \leq d(x, y)$ for any distinct $x, y \in X$, see [7]. It follows from the definitions that $\underline{d} = 0$ if and only if (X, d) is trim.

8.11.2 Drifts

We call a map $f : X \to Y$ between metric spaces $X = (X, d)$ and $Y = (Y, \rho)$ a *drift* if $f(X) = Y$ and for any distinct $x, y \in X$,

$$\rho(f(x), f(y)) = d(x, y) - \underline{d}(x) - \underline{d}(y). \tag{8.11.1}$$

A drift is surjective but not necessarily bijective. If X is trim, then all drifts $X \to Y$ are isometries. We state a version of Theorem 8.10.1 for metric spaces.

Theorem 8.11.1 *Let $\mathcal{M}$ be the quiver whose vertices are finite metric spaces and whose edges are drifts and isometries. Then:*

(i) Two finite metric spaces are isotypic in $\mathcal{M}$ if and only if they are isometric;
(ii) A finite metric space is a primitive vertex of $\mathcal{M}$ if and only if it is trim;
(iii) The quiver $\mathcal{M}$ is phylogenetic;
(iv) All vertices of $\mathcal{M}$ are regular.

Proof Recall from Sect. 8.10.1 the isometry invariant $||X||$ of a finite metric space X. It is clear that for any edge $f : X \to Y$ in $\mathcal{M}$, either $||X|| > ||Y||$ or f is an isometry. If $X, Y \in \mathcal{M}$ are isotypic, then applying this argument to the edges in an evolution $Y \leftarrow \cdots \leftarrow X$, we obtain that either $||X|| > ||Y||$ or all these edges are isometries. Similarly, either $||Y|| > ||X||$ or all edges in an evolution $X \leftarrow \cdots \leftarrow Y$ are isometries. This implies (i).

To proceed, consider a finite metric space $X = (X, d)$ with at least two points. The formula

$$d^\bullet(x, y) = \begin{cases} 0 & \text{if } x = y, \\ d(x, y) - \underline{d}(x) - \underline{d}(y) & \text{if } x \neq y \end{cases}$$

defines a map $d^\bullet : X \times X \to \mathbb{R}$ which satisfies all requirements on a metric except, possibly, one: there may be distinct $x, y \in X$ with $d^\bullet(x, y) = 0$. We define an equivalence relation $\sim_{d^\bullet}$ in X by $x_1 \sim_{d^\bullet} x_2$ if $d^\bullet(x_1, x_2) = 0$. Let $Y = X/{\sim_{d^\bullet}}$ be the quotient set and let $p : X \to Y$ be the projection. Then there is a unique map $\rho : Y \times Y \to \mathbb{R}$ such that $d^\bullet(x, y) = \rho(p(x), p(y))$ for all $x, y \in X$. The map ρ is a metric in Y, and we denote the metric space (Y, ρ) by $\upsilon(X)$. Clearly, the projection $X \to \upsilon(X)$ is a drift. Applying this procedure recursively, we obtain an evolution in $\mathcal{M}$

$$\upsilon^m(X) \leftarrow \cdots \leftarrow \upsilon^2(X) \leftarrow \upsilon(X) \leftarrow X \tag{8.11.2}$$

where $m \geq 0$ is the smallest integer such that $\upsilon^m(X)$ is trim. (The existence of such an m follows from the fact that if a drift $X \to Y$ is bijective, then Y is trim, cf. [7], Lemma 2.1). We conclude that each vertex of $\mathcal{M}$ has a trim ancestor.

We can now prove Claim (ii) of the theorem. By the definition of $\mathcal{M}$, the only edges from a trim metric space X to vertices of $\mathcal{M}$ are isometries. Thus, all ancestors of X are isometric to X, and so X is primitive. Conversely, if $X \in \mathcal{M}$ is primitive, then X is isotypic to all its ancestors. By the above, X is isotypic to a trim metric space. By (i), X is itself a trim metric space.

As in the proof of Theorem 8.10.1, the evolution (8.11.2) is universal, and so all vertices of $\mathcal{M}$ are phylogenetic. We leave it to the reader to check that $\mathcal{M}$ is monotonous and all its vertices are regular. □

References

1. A. Dress, K. Huber, J. Koolen, V. Moulton, A. Spillner, *Basic Phylogenetic Combinatorics* (Cambridge University Press, Cambridge, 2012), xii+264 pp.
2. N. Eriksson, K. Ranestad, B. Sturmfels, S. Sullivant, Phylogenetic algebraic geometry, in *Projective Varieties with Unexpected Properties* (de Gruyter, Berlin, 2005), pp. 237–255
3. M. Gromov, Mendelian dynamics and Sturtevant's paradigm, in *Geometric and Probabilistic Structures in Dynamics*, Contemp. Math., vol. 469 (Amer. Math. Soc., Providence, RI, 2008), pp. 227–242
4. M.L. Reed, Algebraic structure of genetic inheritance. Bull. Amer. Math. Soc. (N.S.) **34**, 107–130 (1997)
5. C. Semple, M. Steel, *Phylogenetics*, Oxford Lecture Series in Mathematics and Its Applications, vol. 24 (Oxford University Press, Oxford, 2003)

6. J.P. Tian, *Evolution Algebras and Their Applications*, Lecture Notes in Mathematics, 1921 (Springer, Berlin, 2008)
7. V. Turaev, Trimming of finite metric spaces, arXiv:1612.06760
8. V. Turaev, Trimming of metric spaces and the tight span. Discrete Math. **341**, 2925–2937 (2018)
9. Z. Yang, B. Rannala, Molecular phylogenetics: principles and practice. Nat. Rev. Genet. **13**, 303–314 (2012)

Chapter 9
About the Diffeomorphisms of the 3-Sphere and a Famous Theorem of Cerf ($\Gamma_4 = 0$)

François Laudenbach

To my old friend Valentin Poénaru—who prefers his nickname Po—on the occasion of his ninetieth birthday, Po from whom I learnt so much maths on some long and freezing chair lifts in ski periods, and
To the memory of my friend and co-author Sandy Blank to whom Po introduced me in the mid sixties.

Abstract At the Golden Age of Differential Topology, for every dimension, the question of *exotic spheres* arose. A sub-question emerged: does there exist in dimension n an exotic sphere just obtained by gluing two n-balls along their boundaries? Such spheres, up to diffeomorphism, form a group for the connected sum denoted by Γ_n.

The first discovery of this type of exotic spheres goes back to J. Milnor (1956): he proved that Γ_7 is not trivial. By proving $\Gamma_4 = 0$, J. Cerf (1968) stated that the gluing of two four-dimensional balls always gives rise to the standard S^4. Actually, Cerf proved that every diffeomorphism of S^3 is isotopic to a linear diffeomorphism. In this chapter we present a *foliated* proof of Cerf's theorem.

Keywords Diffeomorphisms of the 3-sphere · Closed 1-forms · Dehn twist · Dehn modification

2000 Mathematics Subject Classification 57R19

The famous theorem of Jean Cerf in question in the title states the following.

> *Every orientation-preserving diffeomorphism of the 3-sphere is isotopic to the identity.*

F. Laudenbach (✉)
Laboratoire de mathématiques Jean Leray, UMR 6629 du CNRS, Faculté des Sciences et Techniques, Université de Nantes, Nantes Cedex 3, France
e-mail: francois.laudenbach@univ-nantes.fr

L. Funar, A. Papadopoulos (eds.), *Essays on Topology*,
https://doi.org/10.1007/978-3-031-81414-3_9

It follows that every diffeomorphism of S^3 extends to the 4-ball. Then, $\Gamma_4 := \mathrm{Diff}S^3/\rho(\mathrm{Diff}D^4)$ is a trivial group; here, ρ stands for the restriction morphism to the boundary of D^4. So, $\Gamma_4 = 0$ is a surname for the real theorem of Cerf. Actually, in 1992, using recent results at the time in 3-dimensional contact geometry and the theory of holomorphic discs, Y. Eliashberg gave a direct proof of $\Gamma_4 = 0$ [2, section 6], avoiding Cerf's theorem.

In 1979, I solved in [5], with the help of Samuel Blank, a problem raised by J. Moser [7]:

(A) *On a compact 3-dimensional manifold, two non-vanishingclosedone-forms that are cohomologous and whose kernels are tangent to the boundary are isotopic.*

Applied to $S^2 \times [0, 1]$, this immediately implies the original theorem of Cerf, not just $\Gamma_4 = 0$.[1] Nevertheless, I should confess that the proof of this isotopy theorem of 1-forms is somehow technical. A simpler proof of (A) is given by N.V. Quê and R. Roussarie [9], but depending on Cerf's theorem.

We name Theorem (A') the particular case of Theorem (A) where the ambient manifold is $S^2 \times [0, 1]$. The present paper aims to give a less technical proof of Theorem (A') than [5], and hence of Cerf's Theorem.

9.1 Preliminaries and Plan of Proof

In the rest of the paper, the ambient manifold will be $S^2 \times [0, 1]$ and z will denote the variable in its second factor.

9.1.1 Every non-vanishing closed one-form ω on $S^2 \times [0, 1]$ that is *tangent to the boundary*—meaning that $\ker\omega$ is tangent to $\partial(S^2 \times [0, 1])$—reads $\varphi^* dz$ for some diffeomorphism φ that is equal to the identity on $S^2 \times \{0\}$. This follows from the fact that every orbit of a vector field X, that fulfills $\omega(X) > 0$ everywhere, goes from $S^2 \times \{0\}$ to $S^2 \times \{1\}$. Moreover, if $\varphi^* dz = dz$ then φ is isotopic to the identity through diffeomorphisms of the same type.[2]

9.1.2 Moser's Method. Given two non-vanishing closed 1-forms ω_0 and ω_1 on $S^2 \times [0, 1]$, tangent to the boundary, a point $a \in S^2 \times [0, 1]$ is said to be a *negative contact* (resp. a *positive contact*) between these forms if there exists $\mu > 0$ (resp. $\mu < 0$) such that $\omega_0(a) + \mu\,\omega_1(a) = 0$. In other words, the kernels of $\omega_0(a)$ and $\omega_1(a)$ coincide but their co-orientations are opposite (resp. equal). The locus of negative (resp. positive) contacts will be denoted by C_- (resp. C_+).

[1] For completeness, one should apply some classical fibration theorems also due to Cerf [1, Appendice].

[2] These facts hold true on $S^n \times [0, 1]$ for every dimension.

Using the exactness of the two forms under consideration and setting $\omega_i = df_i$, $i = 0, 1$, allow us to characterize the contact points—forgetting the sign of the contact. A point a is a *contact point* if and only if a is a critical point of the restriction of f_0 to the sphere $\{f_1 = f_1(a)\}$ (or the same when exchanging the roles of f_0 and f_1).

Lemma 9.1.3 *If the two non-vanishing closed one-forms are cohomologous and their mutual negative contact locus C_- is empty, then the two forms are isotopic.*

Indeed, the barycentric combination $\omega_t := (1-t)\omega_0 + t\omega_1$ provides a path $(\omega_t)_{t\in[0,1]}$ of closed 1-forms, from ω_0 to ω_1, tangent to the boundary and nowhere singular. In that case, Moser's *isotopy theorem* applies, as in the case of volume forms [7]. Namely, there exists an autonomous flow $(\psi_t)_{t\in[0,1]}$ that conjugates ω_t to ω_0 for every $t \in [0, 1]$.[3]

Therefore, our method to prove the isotopy Theorem (A') on $S^2 \times [0, 1]$ will be to cancel the locus C_- of negative contacts between $\omega_0 := dz$ and $\omega_1 := \varphi^* dz$.

9.1.4 Reduction to a More Special Case. Let f be the function which is equal to the projection onto the factor $[0, 1]$ and set $\ell := \varphi^* f = f \circ \varphi$. From now on, the two forms ω_0 and ω_1 will be respectively denoted by $\omega_{\mathcal{F}}$ and $\omega_{\mathcal{L}}$; and the foliations they define will be respectively named $\mathcal{F}$ and $\mathcal{L}$. The leaves of $\omega_{\mathcal{F}}$ will be named the *level sets* (implied by f) while the level sets of ℓ will continue to be named the *leaves* (underlying $\mathcal{L}$).

Without loss of generality, to prove Theorem (A') we may assume the following in the rest of the chapter:

$$\left.\begin{array}{l}\text{The restriction of } f \text{ to every leaf of } \mathcal{L}\text{, close enough to}\\ \text{the boundary } S^2 \times \{0, 1\}\text{, has only two critical points,}\\ \text{a maximum and a minimum, and these belong to } C_+.\end{array}\right\} \qquad (9.1.1)$$

9.1.5 Plan of the Proof of Theorem (A'). Generically, the contact points (positive or negative) have a Morse index: 0, 1 or 2. The contact points of index 1 are named *saddles*. They have also a *type*: Y, λ, or X. A saddle s on a leaf $L \in \mathcal{L}$ is said to be of type λ (resp. Y) if the connected component of $L \cap [f(s) - \varepsilon, f(s) + \varepsilon]$ which contains s looks like a *pair-of-pants* (resp. a reversed one). A saddle is said to be of type X if it is the common limit of two sequences of saddles, one of λ-saddles and the other of Y-saddles, the pair (f, ℓ) being fixed (Definition 9.3.12). This necessarily involves a pair of saddles both of type X in the same connected component of level curves (Lemma 9.3.14). One will speak of a λ-, Y-, and X-saddle respectively.

Section 9.2 is devoted to basics on generic properties. Section 9.3 reviews elementary isotopies whose effect is to simplify C_-. One of them will allow us to kill all the finitely many X-saddles. Typically, this phenomenon occurs in the setting *saddle-center-saddle* (Definition 9.3.4).

[3] In the case of non-vanishing closed one-forms, a very simple proof is given in [5, Appendice I].

The next operation will consist of pushing the saddles of type Y to levels higher than all saddles of type λ. Unfortunately, this operation is obstructed by connecting orbits $Y \to \lambda$ of the $\mathcal{L}$-gradient of f, a phenomenon that generically appears finitely many times. Here, the $\mathcal{L}$-gradient of f is the orthogonal projection of $\nabla f = \partial_z$ onto the leaves of ℓ with respect to a given Riemannian metric; it will be denoted by $\nabla_{\mathcal{L}} f$.

These obstructions will be destroyed thanks to the main operation of the paper, namely the *Dehn modification*. Section 9.4 will be devoted to this type of 3-dimensional surgery considered in our specific setting. Section 9.5 makes use of all these tools to prove the main result, namely Theorem (A').

9.2 Basics on Pairs of Non-vanishing Closed One-Forms

9.2.1 Regularity. We set $C = C_- \cup C_+$. With local coordinates (x, y) on S^2, the contact locus is defined by the following system:

$$C = \left\{ \frac{\partial \ell}{\partial x}(x, y, z) = 0,\ \frac{\partial \ell}{\partial y}(x, y, z) = 0 \right\}. \tag{9.2.1}$$

The subsets C_+ (resp. C_-) are defined by adding the inequation $\dfrac{\partial \ell}{\partial z}(x, y, z) > 0$ (resp. < 0). By the assumption (9.1.1), C_- does not approach the boundary and hence is compact.

By Thom's transversality theorem of in jet spaces, some approximation of φ makes maximal the rank of the linearized system associated with system (9.2.1) at every point of $S^2 \times]0, 1[$. In this case the pair $(\omega_{\mathcal{F}}, \omega_{\mathcal{L}})$—or (f, ℓ)—is said to be *regular* and C is a smooth curve. The open subset of points in C where

$$\Delta := \frac{\partial^2 \ell}{\partial x^2} \frac{\partial^2 \ell}{\partial y^2} - \left(\frac{\partial^2 \ell}{\partial x \partial y} \right)^2$$

is non-zero is the locus where C is transverse to both foliations $\mathcal{F}$ and $\mathcal{L}$; at such a point, their contact is quadratic. If a is such a point and L is the leaf of a, this point has a Morse index in $\{0, 1, 2\}$ as a critical point of the function f_L, the restriction of f to L. The critical points of f_L are named *minimum, saddle, maximum,* depending on their Morse index.

The remaining points of C are called *inflection points*. Generically, the equation $\Delta(x, y, z) = 0$ is regular and hence, the inflections are isolated in C and do not approach the boundary. So, there are finitely many of them. At an inflection, the tangency of both $\mathcal{F}$ and $\mathcal{L}$ with C is quadratic while the mutual tangency of $\mathcal{F}$ and $\mathcal{L}$ is cubic.

By the *versal unfolding* theory [6, 8], there are coordinates in a neighborhood of an inflection where the two functions f and ℓ read

$$f(x, y, z) = z \text{ and } \ell(x, y, z) = x^3 \pm y^2 + \lambda(z)x + \mu(z). \tag{9.2.2}$$

The sign in the above formula depends on the Morse indices of nearby contact points. The Hessian of f_L at an inflection $I \in L$ has index 0 (resp. 1), and is said to be a *saddle-min* (resp. *saddle-max*) inflection. If the restriction f_C of f to C is locally minimal (resp. maximal) at $p \in C$, p is said to be a *birth* (resp. *cancellation*) inflection.

Since C_- is generically a closed—in general non-connected—curve, any of its connected component carries at least two inflections, one maximum and one minimum of the restriction of f to C_-.

The Morse index of a quadratic contact is constant on each arc of C ending at an inflection (or at a point in the boundary). When two arcs in C of quadratic contact have a common inflection point in their closure, then their indices differ by 1. So, the Morse index is constantly equal to 1 along one of these two arcs; every contact point of this arc is a saddle and the other arc has the index of an f_L-extremum.

9.2.2 Excellence. The pair of functions (f, ℓ) is said to be *excellent* if the following conditions are fulfilled.

(E1) Let a and b be two distinct quadratic contact points at the same level and in the same leaf; let $(a(t), b(t))$ be a pair of local parametrizations of the respective contact arcs such that $a(0) = a,\ b(0) = b$ and $\ell(a(t)) = \ell(b(t))$ for every t close to 0. Then we have $\dfrac{d}{dt}[f(a(t)) - f(b(t))] \neq 0$ at $t = 0$.

(E2) There are neither three contact points in the same leaf and in the same level set nor two pairs of contact points in the same leaf at two distinct levels.

(E3) Let σ be an inflection point. The leaf of σ (resp. its level set) contains neither another inflection point nor a pair of contacts at the same level (resp. in the same leaf) as σ.

Lemma 9.2.3 *In the space of smooth functions $\ell : S^2 \times [0, 1] \to [0, 1]$ such that the pair (f, ℓ) fulfils the assumption* (9.1.1), *the subspace such that the pair (f, ℓ) is regular and excellent is an open dense set.*

Proof Such a statement is well known to the topologists when one speaks of a generic path of functions, the variable z being the time. But every path of functions is not the primitive of a non-vanishing closed one-form $\omega_{\mathcal{L}}$ tangent to the boundary. The main difference comes from the points where $\dfrac{\partial \ell}{\partial z} = 0$. In our setting, $\dfrac{\partial \ell}{\partial z}$ is non-zero at every point where the two other partial derivatives vanish. That means that we work in an open set of the general space of smooth real-valued functions. So, Thom's transversality theorems in bi-one-jet[4] spaces of real-valued functions apply. □

[4] A bi-one-jet is just a pair of two one-jets.

9.3 Elementary Isotopies and Application

Here is the list of the isotopies of ℓ that we consider as *elementary*: cancellation—or creation—of a *simple loop* in the contact locus C; bypass of the *cusp* singularity x^4; exchange (that is, another way of bypassing a cusp); and cut. To this list, we add the *isotopies along a saturated set*, though they are not related to singularity theory.

As announced, these techniques will allow us to cancel the saddles of type X (in the sense of Sect. 9.1.5). Before starting with the description of these isotopies, some notation is needed.

9.3.1 Given a leaf $L \in \mathcal{L}$ and an extremum m of f_L, the *cone* of m, denoted by $C_{\mathcal{L}}(m)$, is the largest open disc in L—if it exists—that contains m and is bounded by a *singular* simple closed curve $\partial C_{\mathcal{L}}(m)$ in a level set of f such that:

(i) $\partial C_{\mathcal{L}}(m)$ contains a saddle (or an inflection point) and bounds a disc $\Delta(m)$ in its level set;
(ii) $\Delta(m) \cup C_{\mathcal{L}}(m)$ bounds a (topological) ball $B_{\mathcal{L}}(m)$ in the ambient manifold.

Note that if L is close to the boundary of $S^2 \times [0, 1]$, by (9.1.1) the cone of the minimum (resp. maximum) is not defined since the leaf contains no saddles .

Definition 9.3.2 (Simple Loop) A connected component Γ of the contact locus C of the pair (f, ℓ) of smooth functions on $S^2 \times [0, 1]$ is said to be a *simple loop* if the following conditions are fulfilled.

(1) Γ is made of two open arcs, α and β, of quadratic contact points of the same sign and two inflection points, I_0 and I_1, which make the topological closure of each of the arcs α and β; say α is of Morse index 1; the index of β, 0 or 2, depends on the sign in front of y^2 in Formula *(9.2.2)*.
(2) Every leaf L that crosses α at a saddle s_L[5] also crosses β at a unique extremum m_L and conversely. Moreover, s_L belongs to $\partial C_{\mathcal{L}}(m_L)$.
(3) There is a unique $\mathcal{L}$-gradient line[6] γ_L joining s_L to m_L on every leaf L meeting $\alpha \cup \beta$.
(4) The only arc of definite quadratic contacts located in the ball $B_{\mathcal{L}}(m_L)$ is $\beta \cap B_{\mathcal{L}}(m_L)$.

By Formula (9.2.2), if s_L goes to I_j $(j = 0$ or $1)$ then γ_L with its field of unit tangent vectors goes to I_j with $\pm\,\partial_x$. Note that, by permuting f and ℓ, there are $\mathcal{F}$-gradient lines connecting pair of points in $\alpha \times \beta$ lying in the same level set of f.

Proposition 9.3.3 *Let* Γ *be a* simple loop *in the contact locus of the pair* (f, ℓ). *Then, there exists a smooth closed 3-ball* N *which is an arbitrary small neighbourhood of the union* $\bigcup_{m_L \in \beta} B_{\mathcal{L}}(m_L)$ *satisfying the following two properties.*

[5] By the Stokes formula this point is unique when it exists.

[6] See the definition in Sect. 9.1.5.

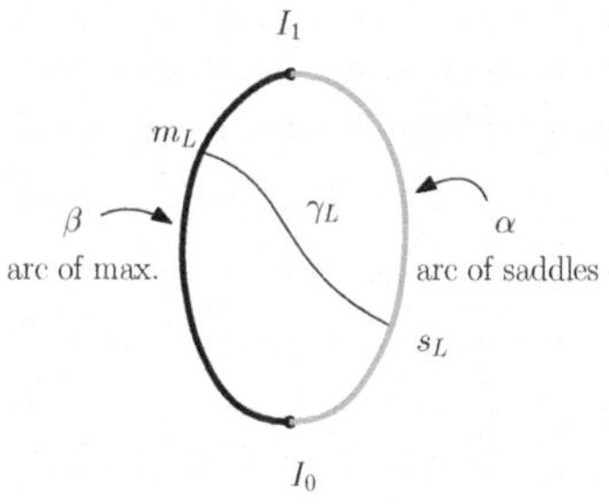

Fig. 9.1 A max-saddle simple loop

(1) *The boundary is foliated by smooth closed curves drawn by* $\mathcal{L}\cap\partial N$, *with exactly two singularities at* $\max \ell_{\partial N}$ *and* $\min \ell_{\partial N}$.[7]
(2) *The germ of* ℓ *along* ∂N *extends to* N *as a smooth function* ℓ' *without contact point with* f *in* N. *Moreover,* ℓ' *is isotopic to* ℓ *by an isotopy supported in the interior of* N.

Proof We may only consider the case where β is an arc of maxima. Let ε be a small positive number. For every $m_L \in \beta$, consider the stable manifold $W^s(m_L, \nabla_{\mathcal{L}} f)$, truncated at the level $f = f(s_L) - \varepsilon$. Then slightly enlarge this surface to a disc $D_{m_L} \subset L$ in order to have a smooth boundary and to contain an ε-neighbourhood of the truncated stable manifold of s_L. If ε is small enough, D_{m_L} avoids all other connected components of the contact locus (Fig. 9.1). If this construction is performed smoothly with respect to m_L with a fixed ε, the desired N is $\cup_{m\in\beta} D_m$.

There exists a proper disc $S \subset N$, transverse both to the leaves of ℓ and the level sets of f, that contains all segments γ_L (item (3) in Definition 9.3.2). This S can be endowed with coordinates (x, z) where $f(x, z) = z$ and x allows us to use the normal form of inflections. Finally, one finds the y-coordinate by slicing N transversely to S so that the maximum of $f_{|\{x=x_0, \ell(x_0,y,z)=\ell_0\}}$ is quadratic non-degenerate and located in $\{y = 0\}$. In Morse theory, such extension of coordinates is named a *suspension* [3].

We are able to solve the problem of isotopy for a pair of functions like $(f_{|S}, \ell_{|S})$ since the two-dimensional problem is known by the contractibility of $\mathrm{Diff}(D^2 \text{ rel. } S^1)$ [10]. An isotopy of S extends to N, relatively to a small neighbourhood of ∂N by suspension. Nothing is changed in $(S^2 \times [0, 1]) \smallsetminus N$. The proof is now complete.[8] □

Conversely, creating a simple loop requires no condition. It can be realized in any open set of the ambient manifold where the two functions f and ℓ have no contact points. The details are left to the reader.

---

[7] These two points are close to I_0 and I_1 respectively, depending on the distance of ∂N to $\cup_{m_L\in\beta} B_{\mathcal{L}}(m_L)$.

[8] This proof is inspired by my proof of a theorem, initially due to Morse, about the cancellation of a pair of critical points of a real-valued function [4].

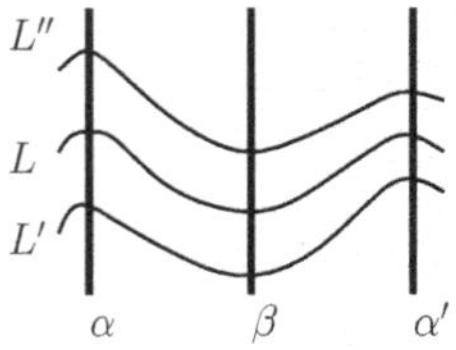

Fig. 9.2 The leaves are denoted on the left. Vertically are the three contact arcs

The next elementary isotopy will follow the same idea of suspension. This deals with the configuration of contacts named *saddle-center-saddle*. Here, *center* stands for minimum or maximum; here, we only consider the case of a minimum. The case where the center is a maximum is similar.

Definition 9.3.4 The saddle-center-saddle configuration is the following. The three contact points are in the same leaf L of $\mathcal{L}$; the center is a minimum m_L whose cone $C_{\mathcal{L}}(m_L)$ contains two saddles s_L and s'_L in its boundary—so, $f(s_L) = f(s'_L)$; there is exactly one $\mathcal{L}$-gradient line from m_L to s_L, and similarly from m_L to s'_L.

Let α, β, α' denote the contact arcs containing s_L, m_L, s'_L respectively. Here are the main requirements:

(1) Let L' be any leaf close to L crossing β at $m_{L'}$ with $f(m_{L'}) < f(m_L)$. Then $f(s_{L'}) < f(s'_{L'})$.
(2) Let L'' be any leaf close to L crossing β at $m_{L''}$ with $f(m_{L''}) > f(m_L)$. Then $f(s_{L''}) > f(s'_{L''})$.
(3) The interior of the ball $B_{\mathcal{L}}(m_L)$ meets no other contact arc than an arc of β.[9]

Note that, up permuting α and α', (1) and (2) from Definition 9.3.4 follow from the excellence of the pair (f, ℓ). Moreover, it is easily seen by (3) that s_L, m_L, and s'_L have the same sign as contact points. The planar picture of this configuration is represented in Fig. 9.2. One recognizes the graphs of real functions of one-variables in a one-parameter family belonging to the *versal unfolding*—up to sign—of the singularity x^4, named *cusp*.[10] Its *versal unfolding*—up to conjugation by the group Diff(source) × Diff(target)—reads

$$(x, \lambda, \mu) \mapsto -\left(\frac{x^4}{4} + \frac{\lambda x^2}{2} + \mu x\right)$$

where λ and μ are two real parameters. As m_L is a minimum, the so-called *suspension* consists of adding a *positive square* in the added variable y. Hence, it reads

$$(x, y, \lambda, \mu) \mapsto -\left(\frac{x^4}{4} + \frac{\lambda x^2}{2} + \mu x\right) + y^2.$$

[9] See Sect. 9.3.1 for notation $B_{\mathcal{L}}(m_L)$.

[10] The terminology comes from the fact that, in singularity theory, an additive constant has no interest. Only the derivatives are important.

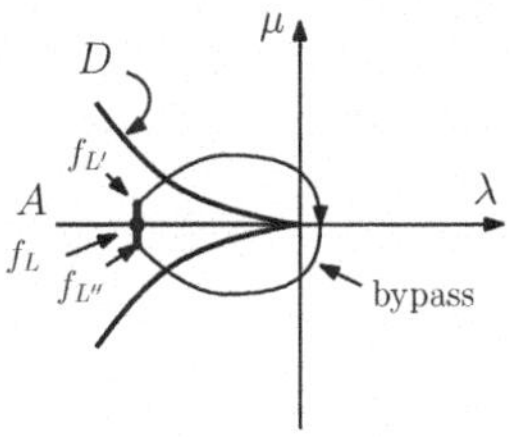

Fig. 9.3 Discriminant locus of the singular function x^4 or $x^4 + y^2$

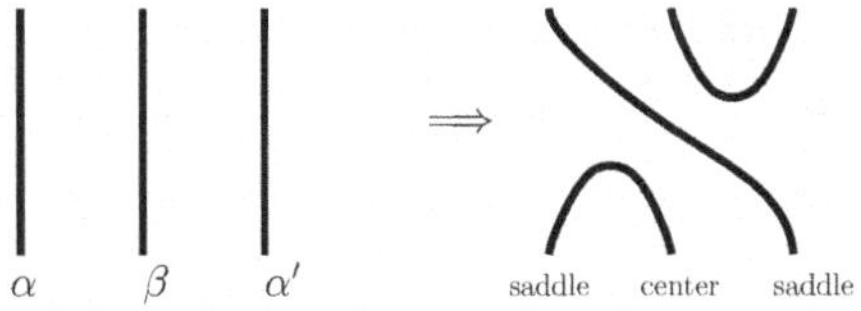

Fig. 9.4 Bypass in configuration saddle-center-saddle

The critical points of f restricted to a leaf close to L are given by $y = 0,\ x^3 + \lambda x + \mu = 0$. The equation of this discriminant locus reads

$$4\lambda^3 + 27\mu^2 = 0.$$

The (λ, μ)-space, denoted by $\mathbb{R}_{(\lambda,\mu)}$, is stratified, apart from the origin, by the discriminant locus D of the *fold*, and the half axis $A = \{\lambda < 0\}$, the set of parameters for which the corresponding function has two critical points with the same value. This is exactly the case of f_L. The deformation from L' to L'' corresponds to a path in $\mathbb{R}_{(\lambda,\mu)}$ crossing A transversely (Fig. 9.3).

In the setting saddle-center-saddle, we have the 3-ball $B_{\mathcal{L}}(m_L)$ that we can extend up to the level $\{f = f(s_L) + \varepsilon\}$. The closure of this extension contains a piece of the unstable manifolds $W^u(s_L, \nabla_{\mathcal{L}} f)$ and $W^u(s'_L, \nabla_{\mathcal{L}} f)$. Then, we take a neighbourhood N of this closure. There are coordinates (x, y, z) on N such that the loci of contact points in N are contained in the surface $S := \{y = 0\}$ and have (locally, since there are three arcs) constant x-coordinates.

Definition 9.3.5 A *bypass* of the cusp singularity is any path of functions as in Fig. 9.3.

Figure 9.4 represents the modified contact locus along such a bypass. The proof of the next proposition is analogous to that of Proposition 9.3.3, by suspending a planar configuration. Namely, one first constructs the desired isotopy in the plane of Fig. 9.2 so that, at time one, the contact locus is as shown in the left part of Fig. 9.4. Then one suspends by adding $+\,y^2$ transversely.

Proposition 9.3.6 *In the above-mentioned setting, there is an isotopy supported in N that, at time 1, carries $\mathcal{E}$ to a function $\mathcal{E}'$ whose one-parameter family of leaves realizes a bypass of the cusp singularity.*

After these first two elementary isotopies, we present more briefly one useful variation of the latter.

Fig. 9.5 Another way of bypassing the cusp singularity

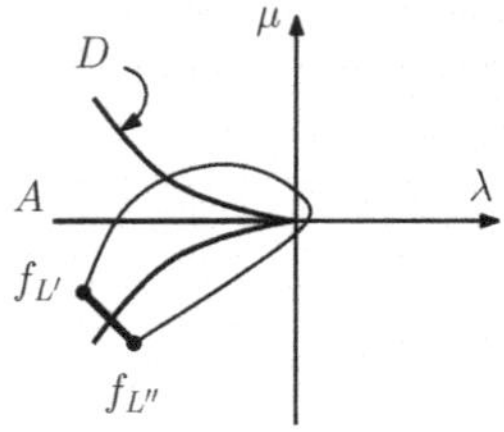

Fig. 9.6 Min-saddle-min configuration

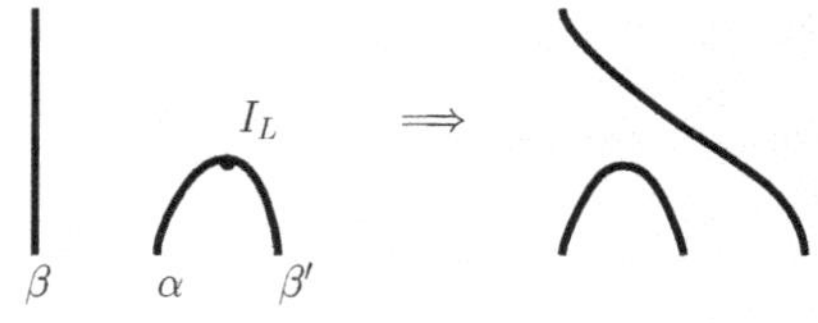

Definition 9.3.7 The configuration *min-saddle-min* is the following. A quadratic contact m_L of index 0, located in the leaf L, has a cone $C_{\mathcal{L}}(m_L)$ whose boundary contains an inflection I_L of the same sign as m_L. Let β be the contact curve containing m_L. It is assumed that:

(1) The ball $B_{\mathcal{L}}(m_L)$ contains no other contact arcs but a sub-arc of β.
(2) I_L is the cancellation point of a pair saddle-min, that is the common upper bound of an arc α of saddles and an arc β' of minima, distinct from β (Fig. 9.6, left-hand side).

Therefore, there is an $\mathcal{L}$-gradient line from m_L to I_L. So, if $m_{L'}$ is just below m_L on β, then there is a saddle $s_{L'}$ connected to $m_{L'}$ by an $\mathcal{L}$-gradient line, and hence, the other branch of the stable manifold $W^s(s_{L'}, \nabla_{\mathcal{L}} f)$ comes from a minimum $m'_{L'}$ distinct from $m_{L'}$.

If one forgets the suspension by $+\,y^2$, it is clear that the new configuration deals with the cusp singularity x^4. Instead of crossing the equality of two critical values, one crosses the cancellation of a saddle with a center (Fig. 9.5).

In this setting, bypassing the cusp singularity has the following effect on the contact curves, as shown in Fig. 9.6.

Proposition 9.3.8 *In the configuration min-saddle-min, there is an isotopy supported in a neighbourhood N of the ball $B_{\mathcal{L}}(m_L)$ that, at time 1, carries ℓ to a function whose contact curves in N have the behaviour presented in the right hand side of* Fig. 9.6.

We now turn to *isotopies along saturated sets.*

Definition 9.3.9 Let $x \in S^2 \times [0, 1]$ and $z_0 > f(x)$. The *ascending saturated set* of x up to the level z_0, denoted by $S_{z_0}(x)$, is the minimal closed subset of $\{f \le z_0\}$ that contains x and satisfies:

(i) For every $y \in S_{z_0}(x)$, the positive orbit of y under $\nabla_{\mathcal{L}} f$, truncated at the level z_0, is included in $S_{z_0}(x)$.

(ii) Let C denote the contact locus of $(\mathcal{F}, \mathcal{L})$. If $y \in C \cap S_{z_0}(x)$, every arc of C ascending from y and truncated at the level z_0 is contained in $S_{z_0}(x)$.

Of course, there is an analogous definition of a *descending satured set*. Such a subset is stratified in an obvious way. By the excellence assumption (Sect. 9.2.2), $S_{z_0}(x)$ contains finitely many *accidents*, meaning level sets where the stratified type changes. From one accident to the next one, one proves that $S_{z_0}(x)$ is surrounded by a fundamental system of *collapsible domains* that, in our setting, are defined as follows.

Definition 9.3.10 A *collapsible domain* K is a 3-ball whose boundary ∂K is the angular union of two discs, the *lower* boundary $\partial_{lo} K$ and the *horizontal* boundary $\partial_h K$, fulfilling the following conditions:

(i) $\partial_h K$ is contained in $\{f = z_0\}$.
(ii) $\partial_{lo} K$ (resp. its interior) is contained in $\{f \leq z_0\}$ (resp. $\{f < z_0\}$).
(iii) There exists a field of directions on K transverse to the f-level sets and also transverse to $\partial_{lo} K$.

Lemma 9.3.11 *Let K be a collapsible domain which contains $S_{z_0}(x)$ and whose lower boundary is disjoint from $S_{z_0}(x)$. For every $\varepsilon > 0$, there exists an isotopy $(\Phi^t)_{t\in[0,1]}$, supported in the interior of K, fulfilling the following conditions.*

(1) $\Phi^0 = Id$ *and* $\Phi^1(S_{z_0}(x))$ *is contained in* $\{z_0 - \varepsilon < f \leq z_0\}$.
(2) *For every* $t \in [0, 1]$ *and every contact point* y *of the pair* (f, ℓ), *its image* $\Phi^t(y)$ *is a contact point of the pair* $(f, \ell \circ (\Phi^t)^{-1})$, *with the same sign and the same Morse index.*
(3) *If* γ *is an* $\mathcal{L}$*-gradient arc, then* $\Phi^t(\gamma)$ *is transverse to the* f*-level sets.*[11]

Such an isotopy is said to be an *isotopy along a saturated set*.

9.3.12 Application. In this subsection, we give an application of Proposition 9.3.6, namely, the cancellation of a pair of saddles that are of type X (Proposition 9.3.15). We specify the definition of an X-saddle when the pair (f, ℓ) is excellent.

Definition 9.3.13 A saddle s in a contact arc α is said to be of type X, or an *X-saddle*, if s is the upper bound on α of an interval of saddles of type λ (resp. Y) and the lower bound of an interval of saddles of type Y (resp. λ).

The following facts are easily checkable when the pair $(\mathcal{F}, \mathcal{L})$ is excellent and the leaves are 2-spheres.

Lemma 9.3.14

(1) *If s is an X-saddle there is another X-saddle s' in some contact arc α' interacting with s, meaning that the two saddles are located in the same*

[11] See more details in [5, chap. III §4].

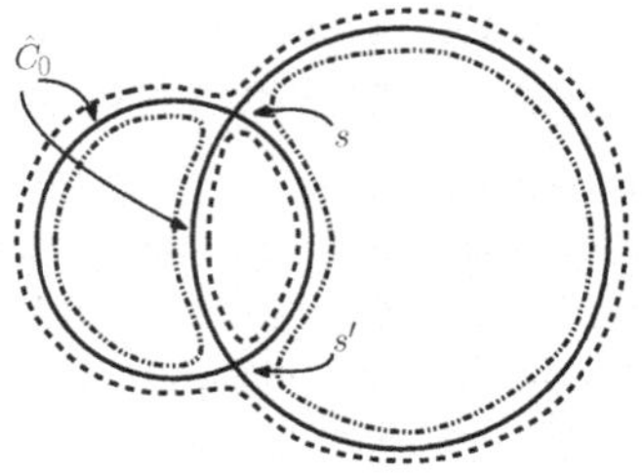

Fig. 9.7 $\hat{C}_0$ denotes the connected component of s in the level curve $F_0 \cap L_0$. The dashed (resp. dotted) curves lie in L_0 and in level sets above (resp. below) z_0

connected component $\hat{C}_0$ of the intersection $L_0 \cap F_0$ of a leaf $L_0 := \{\ell = u_0\}$ and a level set $F_0 := \{f = z_0\}$.

(2) *Both saddles have the same sign and $\hat{C}_0$ looks like the bold line on Fig. 9.7.*

(3) *$\hat{C}_0$ is made of two closed smooth curves C_0 and C_0' in F_0 crossing in s and s' only.*[12]

Proof It is clear enough that a second saddle s' in $\hat{C}_0$ is necessary. Indeed, if not, the germ of $\hat{C}_0$ in $\mathcal{L}$ is stable in the sense that, up to isotopy and identification of close leaves, it does not depend on small variation of z_0. Therefore, s could not be of type X.

By excellence, $\hat{C}_0$ cannot have three saddles (or double points). Then, it contains exactly two saddles. Up to isotopy, there are very few possible configurations for $\hat{C}_0$:

(a) The one represented in Fig. 9.7.
(b) A second one which is the image—named a *chain*—of a smooth immersion $S^1 \to F_0$ with two double points.

Consider case (b) and the levels $z_0 \pm \varepsilon$. One checks that the type of the saddles in the contact arc α between these two levels is constant; and similarly for s'. Therefore, the saddles s and s' are not of type X. It remains to show that s and s' do have the same sign in case (a).

We prove this by considering two leaves L_+ and L_- of $\mathcal{L}$ on both sides of L_0 and close to it. They cross α (resp. α') at two saddles s_+ and s_- (resp. s'_+ and s'_-). One finds two simple closed curves γ_+ and γ_- that fulfill the following: $\gamma_\pm$ is contained in $L_\pm$ and passes through both saddles $s_\pm$ and $s'_\pm$. Let D denote some minimal closed disc in F_0 that contains $\hat{C}_0$. If the signs of s and s' are opposite the two loops we are considering are linked in $D \times [0, 1]$, which is impossible since they lie in disjoint spheres. □

Let $s(t)$ and $s'(t)$ be two local parametrizations of α and α', with $s(0) = s$, $s'(0) = s'$, such that $\ell(s(t)) = \ell(s'(t))$ for every t close to 0. Say $f(s(t))$ and $f(s'(t))$ increase with t. Since the pair (f, ℓ) is excellent, the two f-velocities

[12] There are two possibilities for $C_0 \cup C_0'$: either it is the limit of curves like the dotted curve from Fig. 9.7 or it is the limit of curves like the dashed curve.

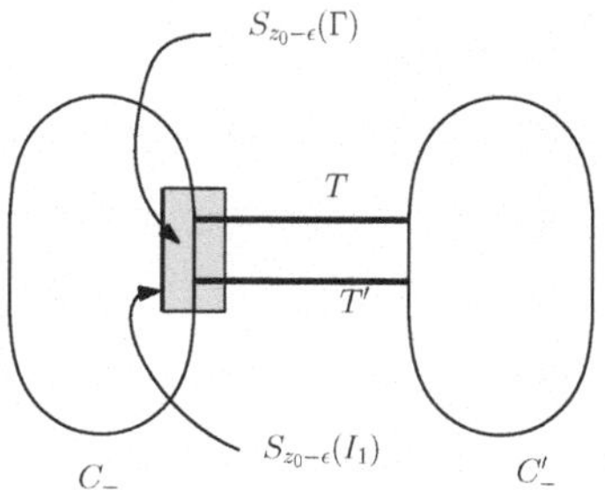

Fig. 9.8 The gray rectangle represents the domain marked by the ascending saturated set of Γ in the level set $\{f = z_0 - \varepsilon\}$

at $t = 0$ are different. We continue this analysis by specifying that the two contacts are positive. The case of two negative contacts would be similar.

For ε small enough, let C_- and C'_- be the two disjoint smooth closed curves in the level set $\{f = z_0 - \varepsilon\}$ whose union goes to $C_0 \cup C'_0$ by $\mathcal{L}$-gradient lines in the leaf L_0. Let T and T' be the lower boundaries of the saturated sets descending from $s(0)$ and $s'(0)$ respectively (Fig. 9.8).

Proposition 9.3.15 *There is an isotopy supported in* $\{z_0 - 2\varepsilon \leq f \leq z_0 + \varepsilon\}$ *whose effect on* ℓ *is to decrease by one the number of pairs of contact points with* $\mathcal{F}$ *that are of type* X.

Proof The idea is to create a configuration saddle-center-saddle which contains $s(0)$ and $s'(0)$. Since the center is missing, one starts by introducing a simple loop saddle-min, noted Γ, whose inflection points I_0 and I_1 satisfy $z_0 - 2\varepsilon < f(I_0) < f(I_1) < z_0 - \varepsilon$ and so that the upper boundary of $S_{z_0-\varepsilon}(\Gamma)$ is in a position as shown in Fig. 9.8. Note the important fact for the proof: $S_{z_0-\varepsilon}(I_1) \cap (T \cup T') = \emptyset$.[13] So, the saturated set $S_{z_0+\varepsilon}(I_1)$ avoids $T \cup T'$.

The next operation is an isotopy along $S_{z_0+\varepsilon}(I_1)$. By application of Lemma 9.3.11, one can collapse this domain above z_0. After this isotopy, still naming ℓ and Γ the carried function and simple loop, one sees a configuration saddle-center-saddle as shown in Fig. 9.9. Then we may use Proposition 9.3.6 to get the desired cancellation of a pair of X-saddles.

There is a new arc α'' of positive saddles which crosses $\{f = z_0 - \varepsilon\}$; this is the arc of saddles in the simple loop Γ after the isotopy just above mentioned. Its arc of minima is denoted by β. One checks that α'' contains only saddles of type λ. Now we can apply Proposition 9.3.6. This cancels the pair (s, s') of X-saddles without creating a new one and Proposition 9.3.15 is proved. □

The outcome of Proposition 9.3.15 is the cancellation of every X-saddle.

[13] The set $S_{z_0-\varepsilon}(I_1)$ is just an interval in the boundary of the gray rectangle of Fig. 9.8.

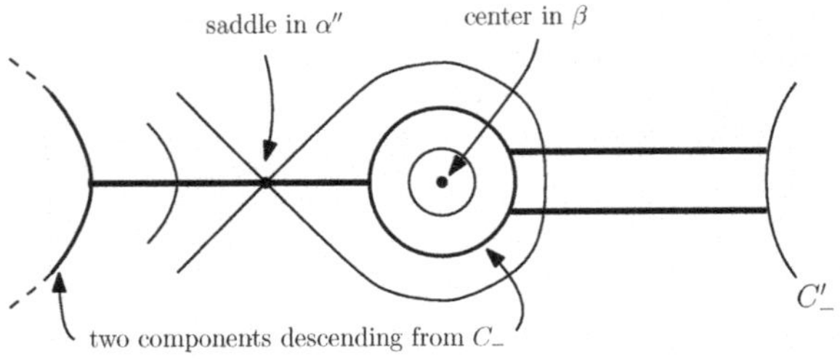

Fig. 9.9 The very thick lines are lower boundaries of descending saturated sets of saddles $s(0)$, $s'(0)$ and $\alpha'' \cap \{f = z_0 - \varepsilon\}$

9.4 Dehn Modification

After Thom's transversality arguments that generate generic properties, the Dehn modification, also called Dehn surgery, is the main tool to prove Theorem (A), and this is also the case for its 1-connected version Theorem (A'). We are going to speak of this notion in our setting only: the ambient manifold is $M := S^2 \times [0, 1]$, provided with two foliations, respectively as the level sets of two functions, f and ℓ, that are constant on each boundary component and whose critical sets are empty.

As an application, we prove that a suitable series of Dehn modifications allows us to move the saddles of type Y above all saddles of type λ.

Definition 9.4.1 (Dehn Twist and Dehn Modification)

(1) A *right Dehn twist* on the standard annulus $\mathbb{A} := S^1 \times [1, 2]$ is a diffeomorphism τ that, in polar coordinates, reads

$$(\theta, r) \overset{\tau}{\longmapsto} (\theta + \varphi(r - 1), r) \tag{9.4.1}$$

where φ is a non-decreasing smooth function from $\varphi([0, \varepsilon]) = 0$ to $\varphi([1 - \varepsilon, 1]) = 2\pi$.
The *left* Dehn twist is the inverse τ^{-1} of the right Dehn twist.

(2) Let A be an annulus embedded in M and parametrized by $\mathbb{A}$; hence, $\partial_i A$ is parametrized by $\partial_i \mathbb{A} := S^1 \times \{i\}$. Let $\check{M}$ be the manifold obtained by cutting M along the interior of A; it is provided with a singular boundary made of two lips A^+ and A^- corresponding to the co-orientation of A. The *Dehn modification* of M along A is the manifold M_τ obtained from $\check{M}$ by gluing $x \in A^+$ to $\tau(x) \in A^-$.

In our setting, A is an annulus, embedded in a level set F_0 of f, transversely to $\mathcal{L}$; the leaves of $\mathcal{L} \cap A$ are closed curves isotopic to each component of ∂A. The identification with $\mathbb{A}$ is chosen so that these curves are parametrized by the circles of $\mathbb{A}$. In this setting we shall say that the Dehn modification is *adapted* to the pair (f, ℓ) or $(\mathcal{F}, \mathcal{L})$. By construction, the functions f and ℓ are mapped to some uniquely defined functions f_τ and ℓ_τ (Fig. 9.10).

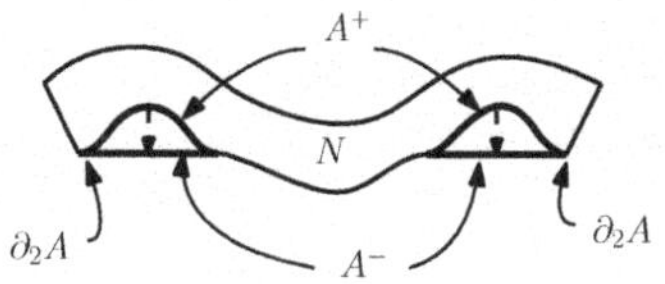

Fig. 9.10 The two dotted vertical arrows indicate the gluing that makes M_τ

Important Remark 9.4.2 *In the previous setting, a Dehn modification along A maps the contact locus of the pair (f, ℓ) to the one of (f_τ, ℓ_τ).* Indeed, the support of this modification is away from the contact locus.

Definition 9.4.3 Let A be an annulus, parametrized by $\mathbb{A}$, in a level set $F_0 := \{f = z_0\}$. Let $p_A \in S^2$ such that the vertical arc $\{p_A\} \times [0, 1] \subset S^2 \times [0, 1]$ avoids A and intersects the unique disc in F_0 that contains A and is bounded by $\partial_2 A$.

Let $\mathcal{D}_A$ denote the space of smooth oriented 2-discs embedded in M that contain A with the right orientation, have algebraic intersection $+1$ with $\{p_A\} \times [0, 1]$ and have $\partial_2 A$ as a boundary.

Proposition 9.4.4 *With this notation, the following holds.*

(1) *Every $D \in \mathcal{D}_A$ defines a unique diffeomorphism $\psi_D : M_\tau \to M$, up to isotopy.*
(2) *For every pair (D_0, D_1) of elements in $\mathcal{D}_A$, the diffeomorphism ψ_{D_1} is isotopic to ψ_{D_0}.*
(3) *There are two discs D_0 and D_1, elements of $\mathcal{D}_A$, such that $(\psi_{D_0})^* df = df_\tau$, and $(\psi_{D_1})^* d\ell = d\ell_\tau$.*
(4) *The functions f and ℓ are isotopic if and only if the functions f_τ and ℓ_τ are isotopic.*

Proof

(1) Since A is an imposed collar of ∂D for every $D \in \mathcal{D}_A$, it is natural to say that, abstractly, D is the disc of radius 2 in the Euclidean plane. This disc may be considered as embedded in M or in M_τ as well. First, we consider D in M_τ. One endows D with a 3-dimensional collar $N \cong D \times [0, 1]$ on the side of A^+, that is $A^+ \subset D \times \{0\}$. So, N already existed in $\check{M}$ and M.

On defines a diffeomorphism $\rho : D \to D$ by $\rho(x) = \tau(x)$ if $\|x\| \in [1, 2]$ and $\rho(x) = x$ if $\|x\| \leq 1$. One chooses an isotopy $(h_t)_{t \in [0,1]}$ among the diffeomorphisms of D from ρ for $t = 0$ to the identity of D for $t = 1$, with an extra requirement:

$$\|h_t(x)\| = \|x\| \text{ for every } t \in [0, 1]. \tag{9.4.2}$$

Then, on defines $\psi_D : M_\tau \to M$ by

$$\begin{cases} \psi_D(y) &= y \quad \text{if } y \in M \smallsetminus N \\ \psi_D(x, t) &= (h_t(x), t) \quad \text{if } (x, t) \in N. \end{cases} \tag{9.4.3}$$

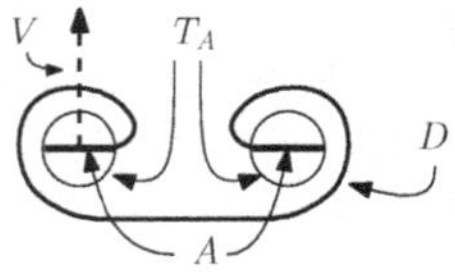

Fig. 9.11 Example of meridional twist. Here, $w_A(D) = -1$

Since all choices are made in contractible spaces, ψ_D is uniquely defined up to isotopy.

(2) Usual tools of 3-dimensional topology yield the following topological result.[14] *Each 2-disc $D \in \mathcal{D}_A$ has a winding number around A, $w_A(D) \in \mathbb{Z}$, that is the algebraic intersection of D with a* vertical *arc V in $S^2 \times [0, 1]$ starting from the interior of A and ending in $S^2 \times \{1\}$ (Fig. 9.11). Moreover, w_A provides a bijection from $\pi_0(\mathcal{D}_A)$ onto $\mathbb{Z}$.*

Here, the linking condition imposed to D for being in $\mathcal{D}_A$ plays an important role. The same result holds true in M_τ. Moreover, one can think of A as a flat annulus in a solid torus $T_A \cong S^1 \times D^2$, each ray of A being a diameter of $\{\theta\} \times D^2$. An exterior collar of ∂T_A has coordinates (θ, x, t) with $(x, t) \in \partial D^2 \times [0, 1]$. This allows one to define a *meridional Dehn twist* in the coordinates (x, t) independent of θ.[15] Let us denote it by μ. For $D \in \mathcal{D}_A$, one has

$$w_A(\mu(D)) = w_A(D) + 1. \tag{9.4.4}$$

This holds true on M and M_τ as well because the support of τ and μ are disjoint.

Using the fact that translations on the torus commute, one has for every $D \in \mathcal{D}_A$:

$$\psi_{\mu(D)} = \mu \circ \psi_D \circ \mu^{-1}. \tag{9.4.5}$$

Moreover, μ is isotopic to the identity if A is let to rotate freely in T_A around its core. Rotating A does not change M_τ. Therefore, $\psi_{\mu(D)}$ is isotopic to ψ_D which proves what was desired.

(3) The disc D_0 is the unique disc in the level set F_0 fulfilling the requirement $\partial D_0 = \partial_2 A$. One can choose its collar $N \cong D \times [0, 1]$ to be foliated by level sets of f. Then, Formula (9.4.3) shows that $(\psi_{D_0})^* f$ is sensitive neither to τ nor to the isotopy (h_t). So, $(\psi_{D_0})^* f = f$ is clear.

For D_1, this is slightly more subtle. There is a disc Δ which is bounded by $\partial_1 A$ in some $\mathcal{L}$-leaf and intersects $\{p_A\} \times [0, 1]$ positively. The union $A \cup \Delta$ is angular along $\partial_1 A$. But its smoothing is easy: Take a collar N_Δ of Δ in the direction outward to A along $\partial_1 A$ and foliated by $\mathcal{L}$. Then, cap $\partial_1 A$ in N_Δ with

[14] Unfortunately, I have no reference for this exercise. One has to use the classical method of the *innermost* intersection curve for two discs in $\mathcal{D}_A$ with the same winding number w_A.

[15] Use formula (9.4.1) replacing (θ, r) with (x, t).

a disc $\tilde{\Delta}$ tangent to A along $\partial_1 A$ and so that $\mathcal{L}$ induces a foliation by circles with one center. Now, we take $D_1 = A \cup \tilde{\Delta}$.

The last step consists of choosing a collar $N_1 = D_1 \times [0, 1]$ of D_1 in M. Such a collar exists in $\check{M}$ (on the side of A^+) so that $\mathcal{L}$ induces on each $D_1 \times \{t\}$ a foliation by circles. Therefore, by formula (9.4.3) the isotopy is compatible with $\mathcal{L}$. And hence, $\psi_{D_1}(\mathcal{L}_\tau) = \mathcal{L}$.

(4) This is a formal consequence of items (2) and (3).

□

We have seen that a Dehn modification adapted to $(\mathcal{F}, \mathcal{L})$ allows us to induce some structures, like f or ℓ, from M to M_τ. But an isotopy, for instance along a saturated set, cannot be induced to M_τ. Fortunately, the last item of Proposition 9.4.4 allows us to induce the property that the pair $(\mathcal{F}, \mathcal{L})$ is made of isotopic foliations. This is the reason why we introduce the notion of *weak isotopy*.

Definition 9.4.5 A *weak isotopy* of $\mathcal{L}$ to $\mathcal{F}$ in M is an alternating finite sequence of Dehn modifications—right or left—and isotopies, $\mathcal{M}_1, (\psi_1^t), \dots, \mathcal{M}_k, (\psi_k^t)$, $t \in [0, 1]$, and a sequence of pairs of foliations, $(\mathcal{F}, \mathcal{L}), (\mathcal{F}_1, \mathcal{L}_1), \dots, (\mathcal{F}_k, \mathcal{L}_k)$, fulfilling the following recursive conditions for every integer $j \in [1, k]$.

(i) $\mathcal{M}_j$ is a Dehn modification adapted to the pair of foliations $\big(\mathcal{F}_{j-1}, \psi_{j-1}^1(\mathcal{L}_{j-1})\big)$;[16]

(ii) $\mathcal{M}_j$ carries $\mathcal{F}_{j-1}$ to $\mathcal{F}_j$ and $\psi_{j-1}^1(\mathcal{L}_{j-1})$ to $\mathcal{L}_j$;

(iii) $\mathcal{F}_k = \psi_k^1(\mathcal{L}_k)$.

In this language, Theorem (A') reduces to: $\mathcal{F}$ and $\mathcal{L}$ are weakly isotopic.

Now comes the announced application of Dehn modifications dealing with connecting orbits of saddles. One should first mention that, as for a one-parameter family of functions on a Riemannian manifold, generically with respect to the metric, the vector field $\nabla_{\mathcal{L}} f$ has finitely many orbits connecting two saddles; moreover there are connecting orbits neither from a max-saddle inflection to a λ-saddle nor from a Y-saddle to a min-saddle inflection. We now include this property in the definition of *excellence* and assume it in the remainder of this paper. We recall that, by Proposition 9.3.15, we may assume that *no saddle is of type X*.

Proposition 9.4.6 *In this setting, there exists a* weak isotopy *of ℓ to a new function ℓ' in M such that all its Y-saddles are located above its λ-saddles, with respect to the order of their f-values.*

Proof Without X-saddles the type of saddles is constant on every saddle arc. The two ends of such an arc are inflections of the same Morse index:[17] if this index is 0 (resp. 1) the saddles are of type λ (resp. Y) as it can be seen near the inflection.

[16] Here, $(\mathcal{F}_0, \mathcal{L}_0) = (\mathcal{F}, \mathcal{L})$.

[17] The index of an inflection p is the index of the maximal non-degenerate quadratic part in the Hessian of f_L at p.

The matter is to destroy the $\mathcal{L}$-gradient connecting orbits from a Y-saddle to a λ-saddle. Indeed, assume there are no such connections. Consider a level z_0 above all λ-saddles and a contact arc α of Y-saddles. By the assumption on the metric, the ascending saturated set $S_{z_0}(\alpha)$ avoids every closure of λ-saddle arcs. Of course, $S_{z_0}(\alpha)$ could meet maxima or inflections of Morse index 1. This does not matter, such inflections and intervals of maxima are incorporated to $S_{z_0}(\alpha)$. An isotopy along $S_{z_0}(\alpha)$ realizes what is wished for α, and so on for the other Y-saddle arcs.

To bypass the connecting orbits we use Dehn modifications. At the beginning, we have finitely many connecting orbits from Y-saddle to λ-saddle lying in distinct leaves $L_1, L_2, \ldots, L_n$ where each $L_i = \ell^{-1}(y_i)$ is a generic leaf. There exists $\varepsilon > 0$ such that $\hat{L}_i := \ell^{-1}([y_i - \varepsilon, y_i + \varepsilon])$ is still made of generic leaves.

The connecting orbit on L_i goes from the Y-saddle s_i to the λ-saddle s'_i. Denote the arc of contact points that contains s_i (resp. s'_i). Set $\hat{\alpha}_i := \hat{L}_i \cap \alpha_i$ and $\hat{\alpha}'_i := \hat{L}_i \cap \alpha'_i$; all are named *saddle intervals*. If ε is small enough there exists η such that every interval in the collection $\left\{ f(\hat{\alpha}_i), f(\hat{\alpha}'_j) \right\}_{i,j\in[1,n]}$ has a length less than η and any two of them have a mutual distance larger than η. In addition, one may require that, for every $j \in [1, n]$ and every pair (a, b) of distinct saddle points in a same leaf from $\hat{L}_j$,

$$|f(a) - f(b)| > 2\eta. \tag{9.4.6}$$

Say $f(s_1) > f(s_i)$ for every $i > 1$. Cut $\hat{L}_1$ at the level $z_1 := f(s'_1) - \eta$. By condition (9.4.6), this level set of $\hat{L}_1$ is a finite collection of annuli. Among them, call A_1 the one which meets the connecting orbit $s_1 \to s'_1$. The foliation $\mathcal{L}$ induces on A_1 a foliation by circles; the central circle lies in the leaf L_1. Knowing that s_1 is of type Y and s'_1 of type λ one sees in A_1 two arcs T and T' joining the two boundary circles: $T = A_1 \cap S_{z_1}(\hat{\alpha}_1)$ and $T' = A_1 \cap S_{z_1}(\hat{\alpha}'_1)$. Here, the first satured set is ascending and the second one is descending. Moreover, $T \cap T'$ is the point where the connecting orbit crosses A_1—because the connecting orbits are on isolated leaves.

Applying a Dehn twist τ to T'—left or right depending on the sign of the intersection $\langle T, T'\rangle$—together with the corresponding Dehn modification to M kills the connecting orbit since the intersection $\tau(T') \cap T$ is empty.

However, new connecting orbits could have been created in the foliated domain $\hat{L}_1$ by this first Dehn modification (this is the case in Fig. 9.12 due to $\tau(T'')\cap T \neq \emptyset$). One solves this new difficulty as follows.

Look at all λ-saddle intervals lying in $\hat{L}_1$ and at level higher than z_1. Name them $\hat{\beta}_1, \ldots, \hat{\beta}_k$ such that $f(\hat{\beta}_1) < f(\hat{\beta}_2) < \cdots < f(\hat{\beta}_k)$.[18] Their number is not affected by the forthcoming Dehn modifications, it is kept invariant (Remark 9.4.2).

Say that $\hat{\beta}_1 \subset \hat{L}_1$ is the lowest λ-saddle interval which is connected to $\hat{\alpha}_1$ by a $\mathcal{L}_\tau$-gradient line; if there is no such a connecting orbit, consider $\hat{\beta}_2$, and so on. Take

[18] Inequalities of intervals mean that their are disjoint and ordered as it is written.

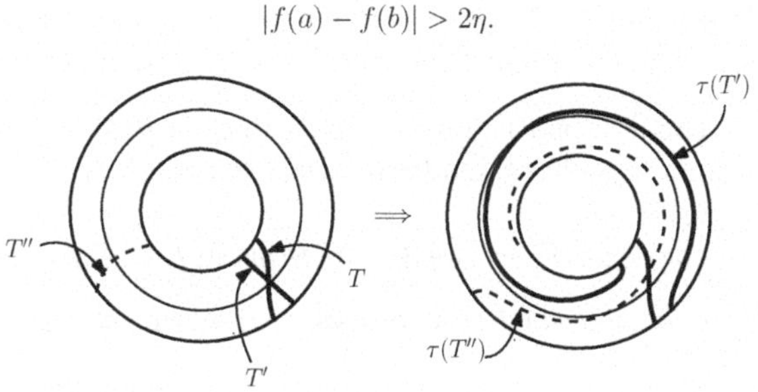

Fig. 9.12 T'' is marked by $S_{z_1}(\hat{\alpha}'')$ where $\hat{\alpha}''$ is a saddle interval of type λ in $\hat{L}_1$ and $f(\hat{\alpha}'') > f(\hat{\alpha}'_1)$

a proper annulus $A'_1 \subset \hat{L}_1$, at the level $z'_1 := f(\hat{\beta}_1) - \eta$, that crosses a connecting orbit from $\hat{\alpha}_1$ to $\hat{\beta}_1$.

The figure is similar to Fig. 9.12, but $S_{z'_1}(\hat{\beta}_1)$ marks a short arc $T'(\hat{\beta}_1)$ joining the two boundary components of A'_1 while $S_{z'_1}(\hat{\alpha}_1)$ marks a long arc like $\tau^{-1}(T)$. These two arcs are transverse to the foliation $\mathcal{L}_\tau \cap A'_1$ and, up to isotopy, there is only one intersection point between them. One can repeat the trick of Dehn modification to cancel the connection $\hat{\alpha}_1 \to \hat{\beta}_1$. Of course, new connecting orbits are created, but only from $\hat{\alpha}_1$ to $\hat{\beta}_j$, $j > 1$. So, the process goes on, without increasing the complexity.

One can destroy all connecting orbits from $\hat{\alpha}_1$ to $\hat{\beta}_1 \cup \cdots \cup \hat{\beta}_k$. As a result, after this long series of Dehn modifications, the ascending saturated set $S_{z_0}(\hat{\alpha}_1)$ avoids every λ-saddle, where z_0 has been chosen at the very beginning of the proof. If the saddles $(s_i)_{i=1}^n$, from which one connecting orbit goes up in the leaf L_i, are ranked in the decreasing order of their f-values $(f(s_i))_1^n$, the process continues with the interval $\hat{\alpha}_2$ and so on.

□

9.5 Proof of Theorem (A')

9.5.1 So far, we got that every saddle is of type λ or Y—that is, no X-saddle— and, up to a weak isotopy applied to the function ℓ (Definition 9.4.5,) every Y-saddle is located at a level higher than all λ-saddles. Recall that an inflection adhering to a contact arc of λ-saddles (resp. Y-saddles) is of type saddle-min (resp. saddle-max). Say that the λ-saddles are located in $\{0 < f < 1/2\}$ and the Y-saddles are located in $\{1/2 < f < 1\}$.

A way to achieve the proof of Theorem (A') is to kill every negative contact of the pair (f, ℓ) and then to apply Lemma 9.1.3. This requires some algorithm since the leaves of $\mathcal{L}$ could have a very tricky topology as embedded surfaces in $S^2 \times [0, 1]$. In [5], this algorithm was based on words in an alphabet with four letters. Here, it is much simpler; it is based on two topological concepts: *basin* and *co-basin* that we are going to define.

Definition 9.5.2 The *basin* of $L := \{\ell = u_0\}$ determined by a *negative* minimum $m \in L$ is the 3-ball $B(m)$ in $\{\ell \leq u_0\}$ whose boundary is made of two parts:

- The first part is a horizontal[19] disc $\partial_{\mathcal{F}} B(m)$ whose boundary curve has exactly one *negative* λ-saddle if L is a generic leaf or two saddles, one of them being negative.
- The second part is a disc $\partial_{\mathcal{L}} B(m)$ in L without negative saddle in its interior.

The *threshold* of $B(m)$ consists of all negative saddles at the top of $\partial_{\mathcal{L}} B(m)$.

Note that the basin of every negative minimum in β is uniquely defined.

Definition 9.5.3 A basin $B(m)$ being given as above, a *joint co-basin* is a maximal ball B^* whose boundary is made of two parts: the first one is a horizontal disc $\partial_{\mathcal{F}} B^*$—let z_0 denote its level—and the second one is a disc $\partial_{\mathcal{L}} B^*$ contained in the $\mathcal{L}$-boundary of $B(m)$. Its *threshold* s is the, actually unique, λ-saddle which has only one descending separatrix lying in $\partial_{\mathcal{L}} B^*$ and the other in the pathwise connected component of $\left(\partial_{\mathcal{L}} B(m) \cap \{f \leq z_0\}\right) \smallsetminus \{s\}$ as m.

We see that the germ of B^* along $\partial_{\mathcal{L}} B^*$ is contained in $\{\ell \geq u_0\}$ while $B(m)$ is contained in $\{\ell \leq u_0\}$. Every positive minimum in $\partial_{\mathcal{L}} B(m)$ belongs to a unique co-basin.

Somehow, a co-basin is like a pocket with respect to a basin. The uniqueness of its threshold holds since there are no X-saddles. Many co-basins could be glued to a given basin. In our setting where the Y-saddles are above all λ-saddles, $\partial_{\mathcal{L}} B^*$ cannot have maxima.

9.5.4 Lower Complexity. Given the pair (f, ℓ) of functions without critical point as considered above, we define its *lower complexity* $\kappa^-(f, \ell)$ as the pair (μ, ν) of non-negative integers arranged in lexicographical order and defined in the following way.[20]

The first entry μ is the number of contact arcs of negative minima.

Let I_0 be the upper, unique by excellence, negative inflection which is the birth of an arc β of negative minima. Every minimum $m \in \beta$ defines a unique basin. A minimum m, distinct from a cancellation inflection, iis said to be *accidental* if one of the following cases happens:

- the boundary of $\partial_{\mathcal{F}} B(m)$ contains either two saddles or a negative inflection—necessarily a cancellation inflection by definition of I_0;
- $\partial_{\mathcal{L}} B(m)$ contains an inflection in its interior.[21]

By definition, the entry ν of the lower complexity is the number of accidental minima in β.

[19] A horizontal disc is contained in an $\mathcal{F}$-leaf.

[20] This pair is meant to measure the topological complexity of $\mathcal{L}$ in the domain $\{0 < f < 1/2\}$.

[21] This inflection should be positive or a negative cancellation inflection not ending β, by definition of I_0.

If $\mu = 0$ then there are neither negative minima nor negative λ-saddles anymore and, by convention, $\kappa^-(f, \ell)$ is equal to $(0, 0)$.

Similarly, by the symmetry $z \mapsto 1 - z$, some *upper complexity* $\kappa^+(f, \ell)$ is defined that measures the topological complexity of $\mathcal{L}$ in the domain $\{1/2 < f < 1\}$. Its vanishing means that there are no negative contacts in this domain. By symmetry, any result about negative λ-saddles/minima applies to negative Y-saddles/maxima—and conversely. Therefore we focused on $\{0 < f < 1/2\}$.

Lemma 9.5.5 *Let $B(m)$ be the basin defined by some $m \in \beta$ and let B^* be a joint co-basin. The following holds:*

(1) *The minimum of $f_{|B(m)}$ is realized in m.*
(2) *Any contact arc descending from a positive contact point in $\partial_{\mathcal{L}} B(m)$ descends to an inflection in the interior of $B(m)$.*
(3) *The $\mathcal{L}$-boundary of B^* contains no negative minimum.*

Proof

(1) Obviously, such a minimum cannot be either realized in the interior of $B(m)$ or in a positive contact in $\partial_{\mathcal{L}} B(m)$. So, it should be realized in a negative contact point of the $\mathcal{L}$-boundary of $B(m)$. Since this disc contains m as a unique negative contact in its interior, m has to be the minimum in question.
(2) By Stokes' formula, such an arc cannot cross $\partial_{\mathcal{L}} B(M)$ twice. Hence its lower point is an inflection in the interior of $B(M)$.
(3) Assume $\partial_{\mathcal{L}} B^*$ contains a negative minimum contact m'. By the inclusion $\partial_{\mathcal{L}} B^* \subset \partial_{\mathcal{L}} B(m)$, we get $m' \in int\partial_{\mathcal{L}} B(m)$ which is impossible by the definition of a basin.

□

The topic of the next lemma is to clean up the basins that we are going to deal with in the proof of Theorem (A'), namely to make sure that no co-basins are joint to them. Again, we are in the setting generated by the *upper* birth negative saddle-min inflection I_0 and the contact arc β of minima that it generates. Every basin whose negative minimum belongs to β will be named an I_0-basin.

Lemma 9.5.6 *In this setting, let I' be a* positive *birth saddle-min inflection and let $m \in \beta$ such that $\partial_{\mathcal{L}} B(m)$ contains I'. Then there exists a* contact conjugating[22] *isotopy Φ_t, $t \in [0, 1]$, fulfilling the following:*

- *The isotopy is supported in $\{0 < f < 1/2\}$ and keeps the negative contacts fixed.*
- *$\Phi_1(I')$ is away from every so-called I_0-basin.*

After having iterated such isotopies, every I_0-basin is free from positive contact.

[22] A diffeomorphism Φ of $S^2 \times [0, 1]$ is said to be contact conjugating if Φ maps every contact point of the pair $(\mathcal{F}, \mathcal{L})$ to a contact point of the pair $(\mathcal{F}, \Phi(\mathcal{L}))$ and conversely.

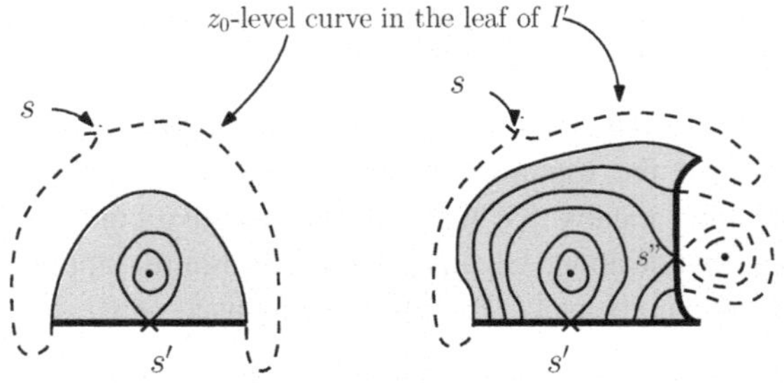

Fig. 9.13 In gray is the upper boundary of the ascending satured set $\Sigma_{z_0}(I')$. The thick lines are drawn by the unstable manifolds of saddles in Σ'. The other lines are curves in leaves of $\mathcal{L}$ at level z_0. On the left, there is the elementary case. On the right, there is a connecting orbit from the saddle arc emanating from I' to another positive saddle (compare to a handle slide)

Proof By excellence, the threshold of the basin $B(m)$ is made of a unique saddle s; let z_0 be its level. Let $\Sigma' := \Sigma_{z_0}(I')$ be the ascending saturated set of $I' \in \partial_{\mathcal{L}} B(m)$ up to level z_0. Let $(m_t)_{t\in[0,1]}$ be a descending parametrization of β from $m_0 = m$. Let γ be the arc of negative thresholds descending from s and parametrized by $(s_t)_{t\in[0,1]}$ where s_t is the threshold of $B(m_t)$. This γ is continuous as there are no X-saddles. Finally, let s'_t be a threshold of any co-basin joint to $B(m_t)$; the three contact points m_t, s_t, s'_t are in the same leaf of $\mathcal{L}$ (Fig. 9.13).

We recall that, when the negative threshold of a basin goes down on its contact arc, meanwhile every positive threshold of a joint co-basin goes up on its own positive contact arc. So, their heights should cross. The time $t = 1$ is chosen so that $f(s_1) = f(s'_1)$ where s'_1 is the lower positive threshold among all co-basins joint to $B(m_1)$.

Note that, by Lemma 9.5.5, the arc γ cannot enter any co-basin.

Claim For a convenient choice of the $\mathcal{L}$-gradient, Σ' does not approach the negative contacts.

Proof of the Claim One chooses the $\mathcal{L}$-gradient $\nabla_{\mathcal{L}}$ so that the separatrix $\sigma(s_0)$ of s_0 descending in $\partial_{\mathcal{L}} B(m)$ does not enter the unstable manifold of I' truncated at level $f(s_0) - \varepsilon$ (a half-disc in its leaf). Assume that the claim does not hold. Then there is a first time $t_0 > 0$ where the separatrix $\sigma(s_{t_0})$ enters Σ' truncated at level $f(s_{t_0}) - \varepsilon$. Since we are sure that s_1 is not confined in any co-basin, this one-parameter family of $\mathcal{L}$-gradients may be seen as a path in the $\mathcal{L}$-gradient space with a given origin and a free extremity. So, we are allowed to retract it by deformation onto its beginning segment.

One can argue similarly for any negative saddle arc with respect to Σ'. □

Then, we apply Lemma 9.3.11 to $\Sigma_{z_0+\varepsilon}$. The outcome is to clean $B(m)$ of the co-basin generated by I'. The same process can be applied to every positive birth inflection in the basin of I_0. Since there are finitely may such inflections the process stops. The order of the positive inflections is not relevant. □

Now we are ready for the decisive part of the proof of Theorem (A'). It will consists of decreasing the lower complexity $\kappa^-(f,\ell)$ until it vanishes. By the symmetry $z \mapsto 1-z$, the same holds true about the upper complexity.

Proposition 9.5.7 *If the set of negative minima contacts of the pair (f,ℓ) is non-empty there exists an isotopy supported in $\{0 < f < 1/2\}$ that carries ℓ to a function ℓ' such that*

$$\kappa^-(f,\ell') < \kappa^-(f,\ell). \tag{9.5.1}$$

Proof There are several cases depending on how the complexity is topologically made.

(1) Assume $\mu > 0$ and $\nu = 0$. We continue with the upper negative birth saddle-min inflection I_0 and the arc β of index 0 that it generates. By Lemma 9.5.6, every basin $B(m)$, $m \in \beta$, is free from positive contact. In this case, for every minimum $m \in \beta$ the cone $C_{\mathcal{L}}(m)$ is standard.[23] meaning that $\mathcal{F}$ induces a smooth foliation by closed curves on its interior except in m.

If the negative λ-saddle arc α and the minimum arc β emanating from I_0 close up in a common cancellation inflection point then the union $\alpha \cup \beta$ forms a *simple loop* in the sense of Definition 9.3.2. By Proposition 9.3.3 there is an isotopy supported in a neighbourhood of $\cup_{m\in\beta} C_{\mathcal{L}}(m)$ that eliminates this simple loop and decreases μ by 1.

Let us show that there are no other configurations with $\nu = 0$. Let J_1 be the cancellation inflection ending α;[24] let I_1, distinct from J_1, be the inflection ending β and let I_2 be the birth inflection of the λ-arc, named α', descending from I_1. We have $f(I_0) > f(I_2)$ by definition of I_0 (Fig. 9.14). Moreover, $f(J_1) > f(I_1)$; if not, the unique separatrix converging to J_1 starts from a minimum $m_1 \in \beta$, and hence $\nu > 0$.

When $\nu = 0$, and hence $f(J_1) > f(I_1)$, every $m \in \beta$ ends one separatrix descending from some saddle $s(m) \in \alpha$ and one separatrix descending from some saddle $s'(m) \in \alpha'$. The last two claims are proved similarly: the existence of $s(m)$ is clear near the inflection I_0 and extends to β since no accident happens along this arc; the existence of $s'(m)$ is clear near I_1 and also extends to β for the same reason.

When m is close to I_0 one has $f(s'(m)) > f(s(m))$ and when $s'(m)$ is close to I_1 one has $f(s'(m)) < f(s(m))$. Then there is $m_0 \in \beta$ such that

$$f(s(m_0)) = f(s'(m_0)). \tag{9.5.2}$$

Hence, $C_{\mathcal{L}}(m_0)$ is in the *saddle-center-saddle* configuration (Definition 9.3.4) and ν is positive. Contradiction.

[23] See Sect. 9.3.1 for notation.

[24] Recall that each contact arc goes increasingly from an inflection point, called a birth inflection, to another inflection point, called a cancellation inflection.

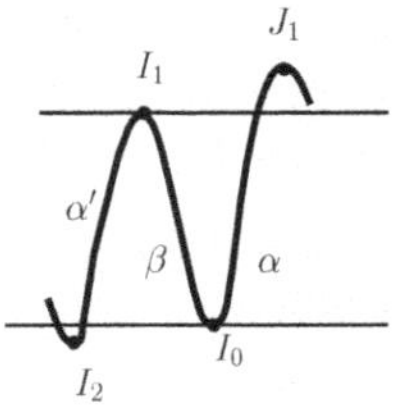

Fig. 9.14 The horizontal lines stand for level sets

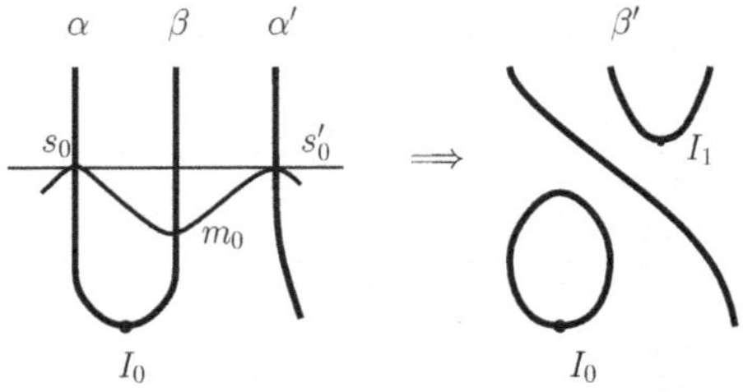

Fig. 9.15 Bypass creating a simple loop

(2) We assume $\nu > 0$ in the rest of the proof. Consider the first accidental minimum m_0 in β when coming from I_0 and assume that the curve $\partial C_{\mathcal{L}}(m_0)$ contains two saddles s_0 and s'_0, necessarily negative by the "cleaning" Lemma 9.5.6. Then we have the saddle-center-saddle configuration; item (3) from Definition 9.3.4 is fulfilled by the choice of I_0 and the cleaning of the I_0-basins.

Let α and α' denote the contact arcs containing s_0 and s'_0 respectively; let s_t, s'_t and m_t denote local parametrizations of α, α' and β respectively near $t = 0$ such that these three contact points are in the same leaf for every t close to 0. If α comes from I_0 and knowing that m_0 is the first accidental point on β, one has $f(m_t) < f(s_t) < f(s'_t)$ for every $t < 0$. In other words, the f-values of s_t and s'_t are ordered as in Fig. 9.2. Thus every saddle in the sub-arc $(I_0, s_0) \subset \alpha$ is cancellable with the corresponding minimum in β; for every small $t > 0$, the pair (s'_t, m_t) can be cancelled. Proposition 9.3.6 applies and its effect is shown in Fig. 9.4. In particular, this *bypass* creates a simple loop which contains I_0 and a new birth inflection I_1 with the following properties:

- $f(I_1) > f(I_0)$.
- The minimum arc β' emanating from I_1 has less accidental minima than β since it coincides with β in $\{f > f(I_1) + \varepsilon\}$ for ε small enough (Fig. 9.15).

Once the simple loop has been cancelled the number of birth inflections remains equal to the initial μ but the entry ν of the lower complexity decreases by 1: the arc β' has one less accidental minimum than β. Moreover, the arc β' is free from positive contact as the initial β was.

(3) With the same notation, we assume $\nu > 0$ and the first accidental minimum $m_1 \in \beta$ has a cone with an inflection point J_1 in its boundary. This inflection cannot be a birth since it lies at a higher level than I_0. So, this is a cancellation inflection, which is a negative contact due to the cleaning Lemma 9.5.6 applied to the basins generated by I_0. Let β denote the index-0 arc starting from I_0 and let I_1 be its

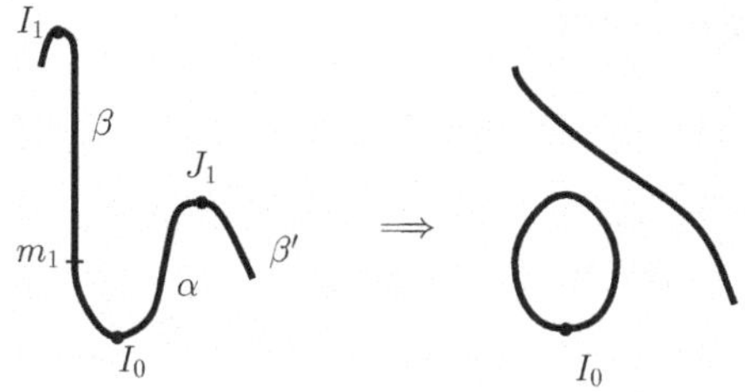

Fig. 9.16 Bypass in configuration min-saddle-min

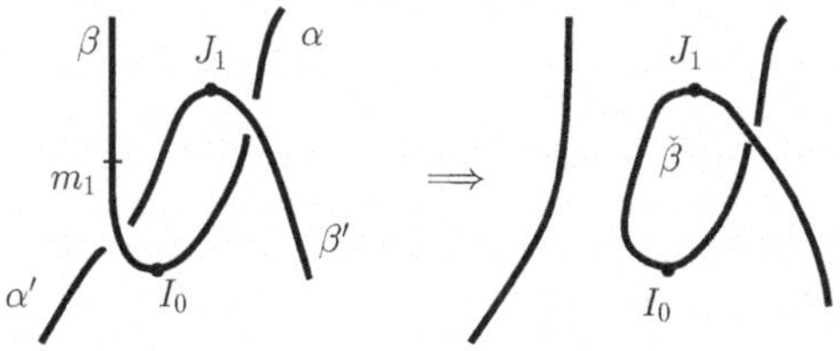

Fig. 9.17 Variant of Fig. 9.16

upper end. We have $I_1 \neq J_1$ since the arc β that is transverse to $\mathcal{L}$—except at end points—cannot meet the same leaf twice. Here there are two cases.

(3-1) Assume that the saddle arc α starting from I_0 ends at J_1. Here, we are in the configuration min-saddle-min (Definition 9.3.7). Let β' be the arc of index 0 descending from J_1. The isotopy from Proposition 9.3.8 modifies the contact arcs as it is shown in Fig. 9.6, including their names. The outcome is a simple loop containing I_0 and an arc of minima. After the simple loop has been cancelled, the number of negative birth saddle-min inflections has decreased by one and hence the complexity decreases (Fig. 9.16).

(3-2) Assume that α does not end at J_1. Let α' and β' respectively denote the saddle arc and the minimum arc ending at J_1. The arcs (β, α', β') and the leaf L that contains $m_0 \in \beta$ slightly below m_1 also presents a configuration min-saddle-min. The effect of the isotopy from Proposition 9.3.8 does not produce any simple loop that could be cancelled. But it keeps I_0 as the upper negative birth inflection and the index-0 arc $\check{\beta}$ emanating from I_0 contains no accidental minimum. Therefore, the complexity decreases from (μ, ν) to $(\mu, 0)$ (Fig. 9.17). □

When the two complexities vanish, no negative contacts are remaining and the Moser trick (Lemma 9.1.3) completes the proof of Theorem (A'). □

Acknowledgments I am grateful to Robert Roussarie for his reading of a preliminary version of this text, his advice about the saddles of type X and other comments. I would like to thank the referee for his careful reading of this text and also the project COSY ANR-21-CE40-0002 for its support.

References

1. J. Cerf, *Sur les difféomorphismes de la sphère de dimension trois* (Γ_4=0), Lect. Notes in Math. 53 (Springer, 1968), xii+133 pp.
2. Y. Eliashberg, Contact 3-manifolds twenty years since J. Martinet's work. Ann. Inst. Fourier **42**(1–2), 165–192 (1992)
3. A. Hatcher, J. Wagoner, *Pseudo-Isotopies of Compact Manifolds*, Astérisque 6 (Soc. Math. France, Paris, 1973)
4. F. Laudenbach, A proof of Morse's theorem about the cancellation of critical points. C. R. Math. Acad. Sci. Paris **351**(11–12), 483–488 (2013)
5. F. Laudenbach, S. Blank, *Isotopie de formes fermées en dimension trois*. Invent. Math. **54**(2), 103–177 (1979)
6. J. Mather, Stability of C^∞ mappings VI, in *Proc. Liverpool Singularities Symp. I.*, ed. by C.T.C. Wall, Lect. Notes in Math. 192 (Springer, 1971)
7. J. Moser, On the volume elements on a manifold. Trans. Amer. Math. Soc. **120**, 286–294 (1965)
8. V. Poénaru, Singularities of analytical spaces, in *CIME 1974 ciclo II* (Cremonese, 1975)
9. N.V. Quê, R. Roussarie, Sur l'isotopie des formes fermées en dimension 3. Invent. Math. **64**, 69–88 (1981)
10. S. Smale, Diffeomorphisms of the 2-sphere. Proc. Amer. Math. Soc. **10**, 621–626 (1959)

Chapter 10
Holomorphic Tensors on Vaisman Manifolds

Liviu Ornea and Misha Verbitsky

Dedicated to Professor Valentin Poénaru at his ninetieth birthday

Abstract An LCK (locally conformally Kähler) manifold is a complex manifold admitting a Hermitian form ω which satisfies $d\omega = \omega \wedge \theta$, where θ is a closed 1-form, called **the Lee form**. An LCK manifold is called **Vaisman** if the Lee form is parallel with respect to the Levi-Civita connection. The dual vector field, called **the Lee field**, is holomorphic and Killing. We prove that any holomorphic tensor on a Vaisman manifold is invariant with respect to the Lee field. This is used to compute the Kodaira dimension of Vaisman manifolds. We prove that the Kodaira dimension of a Vaisman manifold obtained as a $\mathbb{Z}$-quotient of an algebraic cone over a projective manifold X is equal to the Kodaira dimension of X. This can be applied to prove the deformational stability of the Kodaira dimension of Vaisman manifolds.

Keywords Locally conformally Kähler · LCK potential · Algebraic group · Zariski closure

2010 Mathematics Subject Classification 53C55, 32G05

Liviu Ornea is partially supported by Romanian Ministry of Education and Research, Program PN-III, Project number PN-III-P4-ID-PCE-2020-0025, Contract 30/04.02.2021 and by PNRR-III-C9-2023-I8 grant CF 149/31.07.2023 Conformal Aspects of Geometry and Dynamics.
Misha Verbitsky is partially supported by FAPERJ SEI-260003/000410/2023 and CNPq - Process 310952/2021-2.

L. Ornea (✉)
University of Bucharest, Faculty of Mathematics and Informatics, Bucharest, Romania

Institute of Mathematics "Simion Stoilow" of the Romanian Academy, Bucharest, Romania
e-mail: lornea@fmi.unibuc.ro; liviu.ornea@imar.ro

M. Verbitsky
Instituto Nacional de Matemática Pura e Aplicada (IMPA), Rio de Janeiro, RJ, Brasil
e-mail: verbit@impa.br

L. Funar, A. Papadopoulos (eds.), *Essays on Topology*,
https://doi.org/10.1007/978-3-031-81414-3_10

10.1 Introduction

Locally conformally Kähler geometry is probably the simplest complex geometry beyond the Kähler which admits many examples, explicit and non-explicit. A manifold is called **LCK (locally conformally Kähler)** if it admits a Kähler covering, with the deck group acting by holomorphic homotheties (Definition 10.2.1). Every Hopf manifold is LCK, because its universal cover admits a Kähler potential, with the deck transform multiplying the potential by a number. A manifold with this property is called **an LCK manifold with potential**; an LCK manifold admits a potential if and only if it can be realized as a complex submanifold in a Hopf manifold [14].

Recall that **a diagonal Hopf manifold** is a manifold of the form $\frac{\mathbb{C}^n\backslash 0}{\langle A\rangle}$, where A is a linear contraction which can be diagonalized in an appropriate basis. A more special class of LCK manifolds with potential are **Vaisman manifolds**, Sect. 10.2.2.2, which can be characterized as complex manifolds admitting a holomorphic embedding to a diagonal Hopf manifold [20, Theorem 4.9].

Every Vaisman manifold M is equipped with an action of the Lee and anti-Lee fields, which are Killing and holomorphic (Proposition 10.2.17). In [30], it was shown that any stable holomorphic bundle on M is equivariant with respect to this action. From Theorem 10.2.18, it follows that any complex subvariety is invariant under this action as well.

Let G be the group of holomorphic isometries of a Vaisman manifold M obtained as the closure of the group generated by the Lee and the anti-Lee flow. In this paper, we prove that any holomorphic tensor field on M is G-invariant.

For holomorphic vector fields and differential forms this result was obtained by K. Tsukada in [25, Theorem 3.3, Theorem 4.2].

We prove a similar result for LCK manifolds with potential. Let M be an LCK manifold with potential, and $\tilde{M}$ its Kähler $\mathbb{Z}$-cover. We assume that $\tilde{M}$ is an open algebraic cone (Definition 10.2.10); this is true automatically when $\dim_{\mathbb{C}} M \geqslant 3$ [19]. When $\dim_{\mathbb{C}} M = 2$, this would follow if we assume the GSS conjecture [18, Section 25.6]. Since $\tilde{M}$ is algebraic, it makes sense to speak of the Zariski closure of the $\mathbb{Z}$-action on $\tilde{M}$. We consider the Zariski closure $\mathcal{G}$ of the $\mathbb{Z}$-action as the smallest algebraic subgroup of $\operatorname{Aut}(\tilde{M})$ containing this $\mathbb{Z}$-action; this group is clearly commutative and positive-dimensional. Since the action of $\mathcal{G}$ commutes with the action of the deck transform group $\mathbb{Z} \subset \mathcal{G}$ on $\tilde{M}$, the quotient $\mathcal{G}/\mathbb{Z}$ naturally acts on M.

We prove that any holomorphic tensor field on an LCK manifold with potential is $\mathcal{G}$-invariant (Proposition 10.3.3).

Let $H = \frac{\mathbb{C}^n\backslash 0}{\langle A\rangle}$ be a diagonal Hopf manifold, and $\mathcal{G}$ the Zariski closure of $\langle A\rangle$. Let α_i be the eigenvalues of A, and A_1 an operator which has the eigenvalues $|\alpha_i|$ in the same basis. In Theorem 10.4.3, we show that the linear field $\log A_1$ belongs to the Zariski closure of $\langle A\rangle$, and in Theorem 10.5.1, step 3, we prove that $\log A_1$ can be obtained as the Lee field of a Vaisman structure on H.

This implies that the Lee and anti-Lee field of a Vaisman manifold belong to the Lie algebra of $\mathcal{G}$ (the Zariski closure of the $\mathbb{Z}$-action). Therefore, any holomorphic tensor field is Lee and anti-Lee invariant.

10.2 Preliminaries

For the material in this section, we refer the reader to [18].

10.2.1 Locally Conformally Kähler Manifolds

Definition 10.2.1 A complex manifold (M, I) is called **locally conformally Kähler** (LCK, for short) if it admits a covering $(\tilde{M}, I)$ equipped with a Kähler metric $\tilde{\omega}$ such that the deck group of the cover acts on $(\tilde{M}, \tilde{\omega})$ by holomorphic homotheties. An **LCK metric** on an LCK manifold is an Hermitian metric on (M, I) such that its pullback to $\tilde{M}$ is conformal with $\tilde{\omega}$.

Proposition 10.2.2 *A Hermitian manifold (M, I, g) is LCK if and only if there exists a closed 1-form θ (called* **the Lee form***) such that the fundamental form $\omega(x, y) := g(Ix, y)$ satisfies $d\omega = \theta \wedge \omega$.*

Definition 10.2.3 An LCK structure (ω, θ) on a complex manifold (M, I) is called **strict** if the Lee form is not exact. When the Lee form is exact, the structure (ω, θ) is called **globally conformally Kähler** (GCK).

Example 10.2.4 Let $H := (\mathbb{C}^n \setminus 0)/\langle A \rangle$, $A = \lambda \operatorname{Id}$, $|\lambda| > 1$, be a **classical Hopf manifold**. It is LCK because the flat Kähler metric $\tilde{g}_0 = \sum dz_i \otimes d\bar{z}_i$ is multiplied by λ by the deck group $\mathbb{Z}$. The Lee form on the covering $\mathbb{C}^n \setminus 0$ is $\theta = -d \log |z|^2$. Since the Hopf manifold is diffeomorphic to $S^1 \times S^{2n-1}$, the LCK structure is strict.

Remark 10.2.5 All complex submanifolds of an LCK manifold are still LCK (but not necessarily strict).

10.2.2 LCK Manifolds with Potential

Definition 10.2.6 An LCK manifold has **LCK potential** if it admits a Kähler covering on which the Kähler metric has a global and positive potential function

ψ such that the deck group multiplies ψ by a constant.[1] In this case, M is called **LCK manifold with potential**.

Example 10.2.7 The diagonal Hopf manifold in Example 10.2.4 is LCK with potential because $\tilde{g}_0$ has the global automorphic potential $\psi := \sum |z_i|^2$.

By [14], compact LCK manifolds with potential are stable to small deformations of the complex structure. This implies that all **linear Hopf manifolds** $(\mathbb{C}^n \setminus 0)/\langle A \rangle$, $A \in \mathrm{GL}(n, \mathbb{C})$, with eigenvalues of absolute value > 1, are LCK with potential.

Remark 10.2.8 One can easily prove that all complex submanifolds of an LCK manifold with potential are still LCK with potential.

One of the most important properties of LCK manifolds with potential is the following Kodaira-type embedding result:

Theorem 10.2.9 ([14]) *Let M be a compact LCK manifold with potential of complex dimension at least 3. Then M admits a holomorphic embedding into a linear Hopf manifold.*

The idea of the proof is that the Kähler cover of an LCK manifold with potential is not a cone in the Riemannian sense (it lacks the homothety), but is very close to the idea of cone, because it can be completed with just one (singular) point. This completion can be realized using a theorem by Rossi and Andreotti–Siu which only works in dimension greater than 3. In dimension 2, the result is true, if one assumes the GSS conjecture [17, Theorem 5.6], which is a long-standing conjecture, often assumed to be true.

Below, we give a precise description of the Kähler cover of an LCK manifold with potential.

10.2.2.1 Algebraic Cones

Definition 10.2.10 ([15]) A **closed algebraic cone** is an affine variety $\mathcal{C}$ admitting an algebraic $\mathbb{C}^*$-action ρ with a unique fixed point x_0, called **the origin** or **the apex**, and satisfying the following:

(i) $\mathcal{C}$ is smooth outside of x_0, and $\mathbb{C}^*$ acts freely on $\mathcal{C}\backslash x_0$.
(ii) ρ acts on the Zariski tangent space $T_{x_0}\mathcal{C}$ with all eigenvalues $|\alpha_i| < 1$.

An **open algebraic cone** is a complex manifold which is biholomorphic to a closed algebraic cone without the origin. ■

Theorem 10.2.11 ([19]) *Let M be an LCK manifold with proper potential, and $\tilde{M}$ its Kähler $\mathbb{Z}$-cover. Then $\tilde{M}$ is an open algebraic cone.*

[1] In the sequel, a differential form which is multiplied by a constant factor by the action of the deck group is called **automorphic**.

Remark 10.2.12 The restriction to proper potentials is not significant, since any LCK metric with potential can be approximated in the C^∞-topology (while keeping the complex structure fixed) by an LCK metric with proper potential [13, 16].

As a topological space, the closed algebraic cone $\tilde{M}_c$ is obtained as a one-point completion of an open algebraic cone $\tilde{M}$. However, the complex structure on a closed algebraic cone is not uniquely determined by $\tilde{M}$. We refer the reader to [19] for this intricate and fascinating subject.

The manifold $\tilde{M}$ does not define the LCK manifold M uniquely: $M = \frac{\tilde{M}}{\mathbb{Z}}$ is determined by the choice of the $\mathbb{Z}$-action which may vary.

In [19], we proved that an algebraic cone of an LCK manifold with potential is isomorphic to an affine cone over a projective orbifold. This nearly transforms the geometry of LCK manifolds with potential into a chapter of algebraic geometry.

We end these preliminaries by citing a result which gives a method of constructing LCK manifolds with potential once the algebraic cone which will be the covering is known. We first need the following notion.

Definition 10.2.13 Let $\tilde{M}$ be an open algebraic cone, $\tilde{M}_c$ the corresponding closed cone, and $\vec{r} \in T\tilde{M}_c$ a holomorphic vector field such that for all $t > 0$ the diffeomorphism $e^{t\vec{r}}$ is a holomorphic contraction of $\tilde{M}_c$ to the origin. In this situation, the 1-parameter family $e^{t\vec{r}}$ is called **the flow of contractions** (or **contraction flow**). A strictly pseudoconvex hypersurface $S \subset \tilde{M}$ is called a **pseudoconvex shell** if S intersects each orbit of $e^{t\vec{r}}$, $t \in \mathbb{R}$, exactly once.

Pseudoconvex shells are used to obtain automorphic plurisubharmonic functions, as follows.

Theorem 10.2.14 ([15]) *Let $\tilde{M}$ be an algebraic cone, $e^{t\vec{r}}$ a contraction flow, and $S \subset \tilde{M}$ a pseudoconvex shell. Then for each $\lambda \in \mathbb{R}$ there exists a unique function φ_λ such that* $\operatorname{Lie}_{\vec{r}} \varphi = \lambda\varphi$ *and* $\varphi_\lambda\big|_S = 1$. *Moreover, such φ_λ is strictly plurisubharmonic when $\lambda \gg 0$.* ■

10.2.2.2 Vaisman Manifolds

Among the LCK manifolds with potential, the best understood subclass is formed by the **Vaisman manifolds**. These are LCK manifolds with the Lee form parallel with respect to the Levi-Civita connection of the LCK metric.

Let (M, I, ω, θ) be a Vaisman manifold and $\pi : \tilde{M} \to M$ a Kähler cover on which $\pi^*\theta$ is exact. Then it can be seen that the $\tilde{\omega}$-squared norm of $\pi^*\theta$ is a positive, automorphic potential for the Kähler metric $\tilde{\omega}$ [14].

Example 10.2.15 Some examples of Vaisman manifolds:

(i) Diagonal Hopf manifolds $(\mathbb{C}^n \backslash 0)/\langle A \rangle$ where A is semi-simple and with eigenvalues α_i of absolute value > 1, [6, 15].

(ii) Elliptic complex surfaces (see [3] for the complete classification of Vaisman compact surfaces; see also [31]).
(iii) All compact submanifolds of a Vaisman manifold [29].

Remark 10.2.16 The class of Vaisman manifolds is strict: neither the LCK Inoue surfaces, nor the non-diagonal Hopf manifolds can bear Vaisman metrics [3, 15].

The metric dual $\theta^\sharp$ of the Lee form is called **the Lee field**.

Proposition 10.2.17 ([27, 28]) *Let (M, g, I, θ) be a Vaisman manifold. Then the Lee field $\theta^\sharp$ is Killing and holomorphic; moreover, it commutes with $I\theta^\sharp$.* ■

Denote by Σ the holomorphic 1-dimensional foliation generated by $\theta^\sharp$ and $I\theta^\sharp$. It is called **the canonical foliation** (the motivation is given in the next theorem).

Theorem 10.2.18 *Let M be a compact Vaisman manifold, and $\Sigma \subset TM$ its canonical foliation. Then:*

(i) *Σ is independent from the choice of the Vaisman metric [26].*
(ii) *$d^c\theta = \omega - \theta \wedge I\theta$ [28] and the exact (1,1)-form $\omega_0 := d^c\theta$ is semi-positive [29]. Therefore, $\Sigma = \ker \omega_0$, and ω_0 is transversally Kähler with respect to Σ.*

The following result is almost obvious. We give its proof for consistency.

Lemma 10.2.19 *Let M be a compact Vaisman manifold, and G the closure of the Lie group generated by the action of the Lee and the anti-Lee fields on M. Then G is a compact torus.*

Proof Since the Lee and anti-Lee fields act on M by isometries (Proposition 10.2.17), the group G is a closed subgroup of the isometry group of M, and the latter is compact. Then G is compact; since it is the closure of an abelian group, it is abelian. ■

10.3 Holomorphic Tensors on LCK Manifolds with Potential

The purpose of this section is to prove Proposition 10.3.3. For its proof, we shall recall some facts related to coherent sheaves.

10.3.1 Extension of Reflexive Coherent Sheaves

Given a coherent sheaf $\mathcal{F}$ over X, let $\mathcal{F}^* := \operatorname{Hom}(\mathcal{F}, \mathcal{O}_X)$ be **the dual sheaf**, that is, the sheaf of module homomorphisms to the ring of regular functions. The natural morphism of sheaves $\mathcal{F} \longrightarrow \mathcal{F}^{**}$ does not need to be an isomorphism: for example, its kernel contains the torsion of $\mathcal{F}$. A coherent sheaf is called **reflexive** if the natural map $\mathcal{F} \longrightarrow \mathcal{F}^{**}$ is an isomorphism. For an introduction to reflexive sheaves, see [11]. Here we state some results which are relevant to our work.

First of all, notice that the sheaf $\mathcal{F}^*$ is already reflexive [11, Ch. II, Lemma 1.1.8]. Moreover, the singular set of a reflexive sheaf over a normal variety has codimension $\geqslant 3$; in particular, a reflexive sheaf over a complex surface is locally free [11, Ch. II, 1.1.10]. Also, a reflexive sheaf of rank one over a smooth manifold is locally free [11, Ch. II, Lemma 1.1.15].

The most important property of reflexive sheaves is **normality**. Let $Z \subset X$ be a subvariety of a normal complex variety, $\operatorname{codim}_X Z \geqslant 2$. Consider the open embedding map $j : X\backslash Z \hookrightarrow X$, and let j^* and j_* be the sheaf pullback and pushforward, For any sheaf $\mathcal{F}$, there exists a natural sheaf morphism $\mathcal{F} \longrightarrow j_*j^*\mathcal{F}$ taking a section of $\mathcal{F}$ to its restriction to $X\backslash Z$. A coherent sheaf $\mathcal{F}$ is **normal** if the natural map $\mathcal{F} \longrightarrow j_*j^*\mathcal{F}$ is an isomorphism, for any subvariety $Z \subset X$ of codimension $\geqslant 2$.

The following theorem can be understood as a sheaf version of the Hartogs extension theorem.

Theorem 10.3.1 *A coherent sheaf $\mathcal{F}$ over a normal complex variety is normal if and only if it is reflexive.*

Proof [23, Proposition 7]. ■

Theorem 10.3.2 ([18, Theorem 27.6]) *Let M be a normal complex variety with isolated singularities, $\dim_{\mathbb{C}} M \geqslant 3$, $x \in M$ a point, and $M_0 := M\backslash\{x\} \stackrel{j}{\hookrightarrow} M$. Consider a reflexive sheaf $\mathcal{F}$ on M_0. Then $j_*\mathcal{F}$ is a reflexive coherent sheaf. Moreover, $j_*\mathcal{F}$ is the unique coherent reflexive sheaf on M which is isomorphic to $\mathcal{F}$ on M_0.*

Proof Uniqueness of a reflexive extension follows from the fact that a coherent sheaf $\mathcal{F}$ over a normal variety is normal if and only if it is reflexive (10.3.1). The existence of coherent extension of $\mathcal{F}$ follows from [1, Proposition 6.1]. ■

10.3.2 Invariance of Holomorphic Tensor Fields

We are now ready to prove the main result of this section.

Proposition 10.3.3 *Let M be a compact LCK manifold with potential, $\tilde{M}$ its Kähler $\mathbb{Z}$-cover, considered as an open algebraic cone, and $\Phi \in H^0(M, B)$ be a holomorphic tensor field on M, where $B = (\Omega^1 M)^{\otimes k} \otimes TM^{\otimes l}$. Denote by $\tilde{\Phi}$ its lift to $\tilde{M}$, and let $\mathcal{G}$ be the Zariski closure of the $\mathbb{Z}$-action on $\tilde{M}$. Then $\tilde{\Phi}$ is $\mathcal{G}$-invariant.*

Proof **Step 1:** Consider a $\mathbb{Z}$-action on a finite-dimensional space W. Then any $\mathbb{Z}$-invariant vector $w \in W$ is invariant under the Zariski closure of $\mathbb{Z}$. This is why the statement of Proposition 10.3.3 is not surprising. However, the space of tensor fields on $\tilde{M}$ is not finite-dimensional, which makes Proposition 10.3.3 non-trivial.

Denote by $\mathfrak{m}$ the maximal ideal of the origin in the closed algebraic cone $\tilde{M}_c$. We think of the quotients $\mathcal{O}_{\tilde{M}_c}/\mathfrak{m}^k$ as of the spaces of jets of holomorphic (or algebraic) functions. As in [19], we choose $\tilde{M}_c$ normal, so that the $\mathbb{Z}$-action is extended to $\tilde{M}_c$. Then any $\mathbb{Z}$-invariant jet $u \in \mathcal{O}_{\tilde{M}_c}/\mathfrak{m}^k$ is also $\mathcal{G}$-invariant.

Step 2: From now on, we use the same letter Φ to denote the lift of Φ to $\tilde{M}$. Consider Φ as a $\mathbb{Z}$-invariant section of the appropriate tensor bundle $B = (\Omega^1 M)^{\otimes k} \otimes TM^{\otimes l}$, which is by construction $\mathbb{Z}$-equivariant. Using Theorem 10.3.2, we extend B to a reflexive coherent sheaf B_c on $\tilde{M}_c$. Since coherent reflexive sheaves are normal ([11, Ch. II, Lemma 1.1.12], [23, Proposition 7]), the section Φ admits a holomorphic extension to $\tilde{M}_c$, denoted as $\Phi_c \in H^0(\tilde{M}_c, B_c)$. Consider the space $J_B^k := \frac{B_c}{\mathfrak{m}^{k+1} B_c}$ of k-jets of B_c. The $\mathbb{Z}$-action on $H^0(\tilde{M}_c, B_c)$ preserves $H^0(\tilde{M}_c, \mathfrak{m}^k B_c) \subset H^0(\tilde{M}_c, B_c)$. Let $\Phi_c^k \in J_B^k$ be the k-jet of $\tilde{\Phi}_c$. Since Φ_c is $\mathbb{Z}$-invariant, and the space of k-jets is finite-dimensional, the k-jet Φ_c^k is $\mathcal{G}$-invariant. Let $\hat{B}_c := \varprojlim J_B^k$ be the $\mathfrak{m}$-adic completion of B_c, and $\hat{\Phi}_c$ the image of Φ_c in this completion. Since all k-jets Φ_c^k are $\mathcal{G}$-invariant, the completion $\hat{\Phi}_c$ is also $\mathcal{G}$-invariant. Since the ring of germs of holomorphic functions on $\tilde{M}_c$ in c is Noetherian [7, Chapter II, Theorem B.9], and B_c, being reflexive, is torsion-free, the completion map $H^0(\tilde{M}_c, B_c) \longrightarrow \hat{B}_c$ is injective [2, Theorem 10.17]. Therefore, $\Phi \in H^0(\tilde{M}_c, B_c)$ is also $\mathcal{G}$-invariant. ■

10.4 Zariski Closures and the Chevalley Theorem

We recall the following famous theorem due to Chevalley.

Let $G \subset \mathrm{GL}(V)$ be an algebraic group, and $W = V^{\otimes k} \otimes (V^*)^{\otimes l}$ a tensor representation of G. A G-invariant vector $v \in W$ is called **a tensor invariant of** G. A point $x \in \mathbb{P}W$ is called **a projective tensor invariant of** G if it is G-invariant.

Theorem 10.4.1 (Chevalley Theorem [10]) *An algebraic group is uniquely determined by the set of its projective tensor invariants.* ■

When an algebraic group is *reductive* (over $\mathbb{C}$, a group is reductive if and only if it has a compact real form), a stronger version of Chevalley theorem is available.

Theorem 10.4.2 *A reductive algebraic group is uniquely determined by the set of its tensor invariants.*

Proof [5, Proposition 3.1 (c)]. ■

We are going to prove the following result, used further on to determine the Lee field action on a Vaisman manifold.

Theorem 10.4.3 *Let $A \in \mathrm{GL}(n, \mathbb{C})$ be a diagonal linear operator with the eigenvalues $\alpha_1, \ldots, \alpha_n$, and A_1 an operator which is diagonal in the same basis,* *with the eigenvalues* $|\alpha_1|, \ldots, |\alpha_n|$. *Denote by* $\mathcal{G}$ *the Zariski closure of the group* $\langle A\rangle$ *in* $\mathrm{GL}(n, \mathbb{C})$. *Then* $\mathcal{G}$ *contains* A_1.

Proof Since the operator A is diagonalizable, the algebraic closure of $\langle A\rangle$ is a commutative group which contains only semisimple elements. By [4, Proposition 1.5], the connected component of $\mathcal{G}$ is isomorphic to $(\mathbb{C}^*)^d$, hence $\mathcal{G}$ is reductive. Then Theorem 10.4.2 implies that $\mathcal{G}$ is the set of all $g \in \mathrm{GL}(n, \mathbb{C})$ which preserve all A-invariant vectors $w \in W = V^{\otimes k} \otimes (V^*)^{\otimes l}$. The eigenvalues of A on W are products of k instances of α_i and l instances of α_i^{-1}. The eigenvectors $w \in W$ are $w = \bigotimes_{i=1}^k z_{m_i} \otimes \bigotimes_{i=1}^l \zeta_{n_i}$, where z_{n_i} are the eigenvectors in V with the eigenvalues α_{n_i} and ζ_{m_i} are the eigenvectors in V^* with the eigenvalues $\alpha_{m_i}^{-1}$. Then $A(w) = \prod_{i=1}^k \alpha_{n_i} \prod_{i=1}^l \alpha_{m_i}^{-1}$. The eigenvector w is A-invariant if and only if $\prod_{i=1}^k \alpha_{n_i} \prod_{i=1}^l \alpha_{m_i}^{-1} = 1$. This implies $\prod_{i=1}^k |\alpha_{n_i}| \prod_{i=1}^l |\alpha_{m_i}^{-1}| = 1$, hence w is A_1-invariant as well. ■

10.5 Holomorphic Tensors on Vaisman Manifolds

Let M be a Vaisman manifold, and $\theta^\sharp$ its Lee field. Then $\theta^\sharp$ and $I(\theta^\sharp)$ are holomorphic Killing fields (Proposition 10.2.17). The closure G of the group $e^{t_1\theta^\sharp + t_2 I(\theta^\sharp)}$ in the group of isometries of M is compact and commutative, hence it is a compact torus (Lemma 10.2.19).

Theorem 10.5.1 *Let* M *be a compact Vaisman manifold,* $\Phi \in H^0(M, B)$ *a holomorphic tensor, where* $B = (\Omega^1 M)^{\otimes k} \otimes TM^{\otimes l}$, *and* G *the smallest closed Lie group containing the flows generated by the Lee and the anti-Lee fields. Then* Φ *is* G*-invariant.*

Proof **Step 1:** Choose a Vaisman metric on M with LCK rank 1, and let $\tilde{M}$ be the corresponding Kähler $\mathbb{Z}$-cover of M, which is considered as an open algebraic cone. Let $\mathcal{G} \subset \mathrm{Aut}(\tilde{M})$ be the Zariski closure of the $\mathbb{Z}$-action. Then Φ is $\mathcal{G}$-invariant by Proposition 10.3.3. To finish the proof of Theorem 10.5.1, it would suffice to show that $G \subset \mathcal{G}$; since $\mathcal{G}$ is closed, this would follow if we prove that the Lee field $\theta^\sharp$ is tangent to $\mathcal{G}$.

Step 2: Let $M \longrightarrow H$ be a holomorphic embedding of M to a diagonal Hopf manifold. By Theorem 10.2.18, this embedding commutes with the group generated by the Lee and the anti-Lee flow, hence it would suffice to show that $\theta^\sharp$ is tangent to $\mathcal{G}$ when $M = H$. It remains to prove Theorem 10.5.1 assuming that M is a Hopf manifold. For this purpose, we compute the groups G and $\mathcal{G}$ for a diagonal Hopf manifold. This can be done explicitly using the expression for the Vaisman metric on a Hopf manifold obtained in Theorem 10.2.14 via the pseudoconvex shells.

Step 3: The rank 2 algebra generated by the Lee and anti-Lee flows on a Vaisman manifold M is uniquely determined by the complex structure on M (Theorem 10.2.18). We express the Lee flow on a Hopf manifold using a Vaisman structure we shall construct explicitly, and prove that it belongs to $\mathrm{Lie}(\mathcal{G})$. This would imply that the original Lee flow also belongs to $\mathrm{Lie}(\mathcal{G})$.

Let $H = \frac{\mathbb{C}^n \backslash 0}{\langle A \rangle}$, where $A \in \mathrm{GL}(n, \mathbb{C})$ is a diagonal linear contraction with eigenvalues $\alpha_1, \ldots, \alpha_n$. Let A_1 be the matrix which is diagonal in the same basis as A, and has eigenvalues $|\alpha_i|$. Then $\vec{r} := \log A_1$ is a holomorphic contraction vector field, and $e^{t\vec{r}}$ is an A-invariant contraction flow. The sphere S^{2n-1} is a pseudoconvex shell compatible with this contraction flow. By [15], we obtain a $e^{t \log A_1}$-automorphic plurisubharmonic function φ. We are going to prove that φ is A-automorphic.

Consider the diagonal $\mathrm{U}(1)^n$-action on $\mathbb{C}^n$. The plurisubharmonic function φ is $\mathrm{U}(1)^n$-invariant because $\vec{r}$ and S^{2n-1} are $\mathrm{U}(1)^n$-invariant. Since $A \in \mathrm{U}(1)^n \cdot A_1$, and φ is A_1-automorphic, it is A-automorphic. Therefore, $\frac{dd^c\varphi}{\varphi}$ defines a Vaisman metric on H. By definition, the Lee flow maps the level sets of φ to the level sets of φ, with $\varphi(e^{t\theta^\sharp} z) = t\varphi(z)$. By construction of φ, the same is true for $e^{-t \log A_1}$, that is, $\varphi(e^{-t \log A_1} z) = t\varphi(z)$. Therefore, $\theta^\sharp = -\log A_1$, where $\log A_1$ is the diagonal matrix with the eigenvalues $\log|\alpha_i|$ in the same basis. This is the Lee flow for the Vaisman metric $\frac{dd^c\varphi}{\varphi}$.

The vector field $\log A_1$ belongs to $\mathrm{Lie}(\mathcal{G})$, because any power of A_1 fixes all tensor invariants of $\mathcal{G}$ by Theorem 10.4.3, hence the corresponding 1-parametric subgroup also belongs to $\mathrm{Lie}(\mathcal{G})$; this gives $\log A_1 \in \mathrm{Lie}(\mathcal{G})$.

10.6 An Application: The Kodaira Dimension of Vaisman Manifolds

Definition 10.6.1 Let M be a compact complex manifold. Its **pluricanonical bundle** is the tensor power $K_M^n = K_M^{\otimes n}$, $n \geqslant 0$. The **Kodaira dimension** $\kappa(M)$ is defined as $\kappa(M) := \limsup_n \frac{\log(\dim H^0(K_M^n))}{\log n}$.

Remark 10.6.2 The Kodaira dimension is equal to $-\infty$ if $H^0(K_M^n) = 0$ for almost all $n > 0$, and to 0 if $\dim H^0(K_M^n)$ is bounded, but non-zero for infinitely many n. If the function $n \mapsto \dim H^0(K_M^n)$ grows as a polynomial of degree d, the Kodaira dimension of M is d.

Let M be a Vaisman manifold, G the closure of the group generated by the Lee and the anti-Lee action, $\tilde{M}$ a Kähler $\mathbb{Z}$-covering of M, and $\tilde{G}$ the lift of G to $\tilde{M}$. Denote by $\mathfrak{G}$ the Zariski closure of $\tilde{G}$ in $\mathrm{Aut}(\tilde{M})$, considered as an algebraic variety [19].

Since $\tilde{G}$ acts on $\tilde{M}$ by homotheties, and contains contractions, there is an open set $U \subset \mathfrak{G}$ in the connected component of $\mathfrak{G}$ such that for all $\gamma' \in U$, the quotient $M' := \frac{\tilde{M}}{\langle \gamma' \rangle}$ is LCK [19]. The automorphism γ commutes with the action of the group generated by the Lee and the anti-Lee field, hence by [8] M' is Vaisman.[2] By [4, Proposition 1.5], the connected component of $\mathfrak{G}$ is isomorphic to $(\mathbb{C}^*)^k$. It is not hard to see that the union of all one-dimensional closed subgroups is dense in any algebraic group, hence U contains a dense subset of γ' which are contained in a subgroup $\mathbb{C}^* \subset \mathfrak{G}$. Since $\mathbb{C}^*/\langle\gamma'\rangle$ is an elliptic curve, the Vaisman manifold $M' = \frac{\tilde{M}}{\langle\gamma'\rangle}$ is elliptically fibered.[3] As shown in [12], the leaf space X is a projective orbifold.

Theorem 10.6.3 *Let M be a Vaisman manifold, M' its quasi-regular deformation obtained above, and X the corresponding leaf space, which is a projective orbifold. Then $\kappa(M) = \kappa(M') = \kappa(X)$, where κ denotes the Kodaira dimension.*

Proof **Step 1:** The equality $\kappa(M') = \kappa(X)$ is clear from the adjunction formula: since $\pi : M' \longrightarrow X$ is an elliptic fibration, one has $H^0(K^k_{M'}) = H^0(K^k_X)$ for all k.

Step 2: By Theorem 10.5.1, any section h of the pluricanonical bundle K^k_M on M is $\mathfrak{G}$-invariant. Then, for any $\gamma' \in U$, the lift of h to $\tilde{M}$ is γ'-invariant, hence can be obtained as a pullback of a section of $K^k_{M'}$ under the quotient map $\tilde{M} \longrightarrow M'$. This defines an injective map $\Psi : H^0(K^k_M) \longrightarrow H^0(K^k_{M'})$.

Step 3: To finish the proof of Theorem 10.6.3 it remains to show that $\Psi : H^0(K^k_M) \longrightarrow H^0(K^k_{M'})$ is surjective. This would follow if we show that any section $h' \in H^0(K^k_{M'})$ is $\mathfrak{G}$-invariant. Since $H^0(K^k_{M'}) = H^0(K^k_X)$ (Step 1), the $\mathfrak{G}$-invariance of h' would follow if we prove that the corresponding section of K^k_X is invariant under the natural action of $\mathfrak{G}$ on X.

Recall that **the pluricanonical representation** is the natural action of $\mathrm{Aut}(X)$ on $H^0(K^k_X)$. Since X is an orbifold, its singularities are klt [21, Proposition 1.2.1]. By [9, Theorem 10.61], the image of the pluricanonical representation is finite; since $\mathfrak{G}$ is connected, it acts on $H^0(K^k_X)$ trivially. This gives $H^0(K^k_M) = H^0(K^k_{M'}) = H^0(K^k_X)$. ■

As an application, we outline the proof of the deformational stability of Kodaira dimension for Vaisman manifolds.

Conjecture 10.6.4 Let $M_t, t \in \mathbb{R}$, be a smooth family of compact Vaisman manifolds. Then the Kodaira dimension of M_t is constant.

Sketch of a Proof We sketch a proof of this conjecture, reducing it to Conjecture 10.6.5 below.

Let $\tilde{M}_t$ be the family of open algebraic cones obtained from each M_t by taking a Kähler $\mathbb{Z}$-cover; this is possible to do by [12]. Each $\tilde{M}_t$ is equipped with a natural algebraic structure by [19].

[2] As shown in [8], an LCK manifold which admits a holomorphic conformal $\mathbb{C}$-action is Vaisman, assuming that this action cannot be lifted to an isometry of the Kähler cover.

[3] Elliptically fibered Vaisman manifolds are called **quasi-regular**.

What follows is a conjecture, which most likely can be proven by the methods used in [19].

Let $\tilde{M}_t$, $t \in \mathbb{R}$, be a smooth family of open algebraic cones. As shown in [19], for an appropriate choice of the $\mathbb{C}^*$-action on $\tilde{M}_t$ the quotient $\tilde{M}_t/\mathbb{C}^*$ is isomorphic to a complex projective orbifold X_t; this isomorphism is extended to an isomorphism between $\tilde{M}_t$ and $\mathrm{Tot}^\circ(L_t)$, where L_t is an ample line bundle over X_t.

Conjecture 10.6.5 Let $\tilde{M}_t$, $t \in \mathbb{R}$, be a smooth family of open algebraic cones. Then for any sufficiently small open subset $U \subset \mathbb{R}$, the choice of $\mathbb{C}^*$-action as above can be performed in such a way that the family (X_t, L_t) depends smoothly on $t \in U$.

Now, by Theorem 10.6.3, the Kodaira dimension of M_t is equal to the Kodaira dimension of X_t, which is constant, as shown by Siu in [24].[4] ■

Acknowledgments We are grateful to Nicolina Istrati and Victor Vuletescu for their interest and valuable advice.

References

1. A. Andreotti, Y.T. Siu, Projective embeddings of pseudoconcave spaces. Ann. Scuola Norm. Sup. Pisa **24**, 231–278 (1970)
2. M.F. Atiyah, I.G. MacDonald, *Introduction to Commutative Algebra* (Addison-Wesley, 1969)
3. F.A. Belgun, On the metric structure of non-Kähler complex surfaces. Math. Ann. **317**, 1–40 (2000)
4. A. Borel, J. Tits, Groupes réductifs. Publ. Math. l'IHÉS **27**, 55–151 (1965)
5. P. Deligne, Hodge cycles on abelian varieties (Notes by J. Milne.), in *Hodge Cycles, Motives, and Shimura Varieties*, ed. by P. Deligne et al. Lecture Notes in Mathematics, vol. 900 (Springer, Berlin, 1982)
6. P. Gauduchon, L. Ornea, Locally conformally Kähler metrics on Hopf surfaces. Ann. Inst. Fourier (Grenoble) **48**, 1107–1128 (1998)
7. R.C. Gunning, H. Rossi, *Analytic Functions of Several Complex Variables*, Reprint of the 1965 original (AMS Chelsea Publishing, Providence, RI, 2009)
8. Y. Kamishima, L. Ornea, Geometric flow on compact locally conformally Kähler manifolds. Tohoku Math. J. **57**(2), 201–221 (2005)
9. J. Kollár, with collaboration of S. Kovács, *Singularities of the Minimal Model Program* (Cambridge University Press, 2013)
10. D.W. Morris, *Ratner's Theorems on Unipotent Flows* (Univ. of Chicago Press, Chicago, 2005)
11. C. Okonek, M. Schneider, H. Spindler, *Vector Bundles on Complex Projective Spaces*, Corrected reprint of the 1988 edition. With an appendix by S. I. Gelfand. Modern Birkhäuser Classics (Birkhäuser/Springer Basel AG, Basel, 2011)
12. L. Ornea, M. Verbitsky, An immersion theorem for compact Vaisman manifolds. Math. Ann. **332**, 121–143 (2005)
13. L. Ornea, M. Verbitsky, Morse-Novikov cohomology of locally conformally Kähler manifolds. J. Geom. Phys. **59**(3), 295–305 (2009)

[4] Siu's result was proven for a projective family of varieties; for a complex-analytic family of projective varieties, the invariance of plurigenera follows from [22]. We are grateful to Runze Zhang for giving us this reference.

14. L. Ornea, M. Verbitsky, Locally conformal Kähler manifolds with potential. Math. Ann. **348**, 25–33 (2010)
15. L. Ornea, M. Verbitsky, Locally conformally Kähler metrics obtained from pseudoconvex shells. Proc. Amer. Math. Soc. **144**, 325–335 (2016)
16. L. Ornea, M. Verbitsky, LCK rank of locally conformally Kähler manifolds with potential. J. Geom. Phys. **107**, 92–98 (2016)
17. L. Ornea, M. Verbitsky, Embedding of LCK manifolds with potential into Hopf manifolds using Riesz-Schauder theorem, in *Complex and Symplectic Geometry*, Springer INdAM Series (2017), pp. 137–148
18. L. Ornea, M. Verbitsky, Principles of locally conformally Kähler geometry. Progress Math. Birkhauser, **353** (2024)
19. L. Ornea, M. Verbitsky, *Algebraic cones on LCK manifolds with potential*, J. Geom. Phys. **198** 105103 (2024). https://doi.org/10.1016/j.geomphys.2024.105103.
20. L. Ornea, M. Verbitsky, *Lee classes on LCK manifolds with potential*, Tohoku Math. J **76**(1), 105–125 (2024)
21. Y.G. Prokhorov, *Lectures on Complements on Log Surfaces*, MSJ Memoirs, 10 (Mathematical Society of Japan, Tokyo, 2001)
22. S. Rao, I-H. Tsai, *Invariance of plurigenera and Chow-type lemma*, arXiv:2011.03306
23. J.-P. Serre, Prolongement de faisceaux analytiques cohérents. Ann. Inst. Fourier **16**(1), 363–374 (1966)
24. Y.-T. Siu, Extension of twisted pluricanonical sections with plurisubharmonic weight and invariance of semipositively twisted plurigenera for manifolds not necessarily of general type, in *Complex Geometry (Göttingen, 2000)* (Springer, Berlin, 2002), pp. 223–277
25. K. Tsukada, Holomorphic forms and holomorphic vector fields on compact generalized Hopf manifolds. Compos. Math. **93**(1), 1–22 (1994)
26. K. Tsukada, The canonical foliation of a compact generalized Hopf manifold. Differential Geom. Appl. **11**(1), 13–28 (1999)
27. I. Vaisman, Locally conformal Kähler manifolds with parallel Lee form. Rend. Mat. (6) **12**(2), 263–284 (1979)
28. I. Vaisman, Generalized Hopf manifolds. Geom. Dedicata **13**, 231–255 (1982)
29. M. Verbitsky, Theorems on the vanishing of cohomology for locally conformally hyper-Kähler manifolds. Proc. Steklov Inst. Math. no. 3 **246**, 54–78 (2004)
30. M. Verbitsky, Stable bundles on positive principal elliptic fibrations. Math. Res. Lett. **12**(2–3), 251–264 (2005)
31. M. Verbitsky, V. Vuletescu, L. Ornea, Classification of non-Kähler surfaces and locally conformally Kähler geometry. Russian Math. Surv. **76**, 261–290 (2021). arxiv:1810.05768

Chapter 11
On the Topology of Monotone Lagrangians of High Dimension

Mihai Damian

Dedicated to Professor Valentin Poénaru on the occasion of his 90th birthday.

Abstract In my previous work (M. Damian, Ann Sci Ec Nom Sup 4ème série 48(1):237–252, 2015) I conjectured that any closed orientable monotone Lagrangian submanifold in $\mathbf{C}^n$ is the total space of a fibration over the circle. In the present paper I prove that this statement is true in dimension greater than 6 under some topological hypothesis on the universal cover of the Lagrangian. More generally, the proof is valid for any symplectic ambient manifold with vanishing second homology group, provided that the Lagrangian is displaceable through a Hamiltonian isotopy. A generalisation for monotone Lagrangians in $\mathbf{CP}^n$ is also proved.

Keywords Lagrangian submanifolds · Floer homology · Novikov homology

AMS Classification 53D12, 53D40, 57R19

11.1 Introduction and Main Results

11.1.1 Some Conjectures on Lagrangian Submanifolds

Let (X^{2n}, ω) be a symplectic manifold. A Lagrangian submanifold $L^n \subset X$ is a submanifold of maximal dimension on which the symplectic form vanishes. The present paper studies the topological properties of closed Lagrangians. A special case is the one of the so-called local Lagrangians $L \subset \mathbf{C}^n$; by Darboux's theorem these manifolds admit Lagrangian embeddings in any symplectic manifold X. This

M. Damian (✉)
IRMA, Université de Strasbourg, Strasbourg, France
e-mail: damian@math.unistra.fr

L. Funar, A. Papadopoulos (eds.), *Essays on Topology*,
https://doi.org/10.1007/978-3-031-81414-3_11

particular case is already interesting. Let us recall some results about the topology of the local Lagrangians. We begin with a celebrated theorem of M. Gromov [11].

Theorem 11.1.1 *There is no closed simply connected Lagrangian in* $\mathbf{C}^n$.

In [1] Michèle Audin pointed out that for any Lagrangian immersion $L \hookrightarrow \mathbf{C}^n$ there exists a Lagrangian embedding $L \times \mathbf{S}^1 \hookrightarrow \mathbf{C}^{n+1}$. Since the Lagrangian immersions are governed by the h-principle [12], this provides a large class of examples of local Lagrangian submanifolds. Some years after, L. Polterovich used Lagrangian immersions to produce another class of examples which are not $\mathbf{S}^1$ products [16]. His result states

Theorem 11.1.2 *For n odd, given an immersion* $L \hookrightarrow \mathbf{C}^n$ *with k double points, there is a Lagrangian embedding of the connected sum of L with the k-fold connected sum of* $\mathbf{S}^{n-1} \times \mathbf{S}^1$

$$L\#^k(\mathbf{S}^{n-1} \times \mathbf{S}^1) \hookrightarrow \mathbf{C}^n.$$

More recently, T. Ekholm et al. [7] refined this theorem in the case $n = 3$.

Theorem 11.1.3 *For any closed orientable* 3*-manifold L there is a Lagrangian embedding*

$$L\#(\mathbf{S}^2 \times \mathbf{S}^1) \hookrightarrow \mathbf{C}^3.$$

Therefore there is a lot of flexibility for the topology of a closed Lagrangian submanifold especially in dimension 3, even in the local case. In order to recover some topological constraints we have to impose more hypotheses on L. A natural one would be L *prime*, meaning that L is not a non-trivial connected sum. Under this hypothesis K. Fukaya [10] sketched a proof of the following result:

Theorem 11.1.4 *If a Lagrangian* $L \subset \mathbf{C}^3$ *is closed orientable and prime then it is diffeomorphic to an* $\mathbf{S}^1$*-product* $\mathbf{S}^1 \times \Sigma$.

Fukaya's proof was later completed by K. Irie [13]. Another option is to work under the hypothesis L *monotone* (see the next subsection for a brief definition). In a previous work [6], I established the same result in this case (see also J. Evans and J. Kedra [8] for a similar statement proved by a different method):

Theorem 11.1.5 *Any monotone closed orientable Lagrangian* $L \subset \mathbf{C}^3$ *is an* $\mathbf{S}^1$*-product* $\mathbf{S}^1 \times \Sigma$.

In the same paper I stated the following

Conjecture 11.1.1 Any monotone closed orientable Lagrangian $L \subset \mathbf{C}^n$ is the total space of a fibration over the circle.

Actually, Theorem 11.1.5 is valid in the more general setting of a general monotone ambient symplectic manifold X with $H_2(X, \mathbf{Z}) = 0$, provided that L is *displaceable*:

Definition 11.1.6 A Lagrangian submanifold $L \subset X$ is called displaceable if there is a Hamiltonian isotopy (φ_t) on X such that $\varphi_1(L) \cap L = \emptyset$.

An example is $X = \mathbf{C} \times T^*\Sigma$ where Σ is different from $\mathbf{S}^2$ and in fact the argument could be adapted to $\mathbf{S}^2$ too; the hypothesis L displaceable is of course fulfilled here. It is tempting to formulate the above conjecture in this more general setting. Then one may also try to go beyond the case where L is displaceable. Consider for instance the case $X = \mathbf{CP}^n$. We know that $L = \mathbf{RP}^n$ is a monotone Lagrangian which does not fiber over the circle, so we need to rule out this example by hypothesis. A reasonable conjecture would be

Conjecture 11.1.2 Any monotone closed orientable Lagrangian $L \subset \mathbf{CP}^n$ such that $H_1(L, \mathbf{Z})$ is not torsion is the total space of a fibration over the circle.

The purpose of the present chapter is to present the closest results to these conjectures which I am able to prove at this moment. They will be stated in the next section.

11.1.2 Monotone Lagrangians and the Main Results

We start by recalling the definition of a monotone Lagrangian submanifold. Given a Lagrangian $L \subset X$ one can define two morphisms $\pi_2(X, L) \to \mathbb{R}$. The first one I_ω is simply the integration of the symplectic form $I_\omega(u) = \int_D u^*\omega$ for $u : (D, \partial D) \to (X, L)$ representing a given homotopy class in $\pi_2(M, L)$. To define the second one, $I_\mu : \pi_2(X, L) \to \mathbf{Z}$, called the Maslov morphism, start by trivialising the symplectic bundle u^*TX. From the tangent spaces of L along $u|_{\partial D}$ we thus get a loop of Lagrangian subspaces of $\mathbf{C}^n$, i.e. an application $\Gamma_u : \mathbf{S}^1 \to \Lambda(n)$ with values in the Grassmanian Lagrangian. Using the identification $\Lambda(n) = U(n)/O(n)$, we consider the complex determinant $\det_{\mathbf{C}} : \Lambda(n) \to \mathbf{S}^1$ and then we define

$$I_\mu(u) = \deg(\det_{\mathbf{C}} \circ \Gamma_u)$$

Note that for an element u of the image $\pi_2(X) \to \pi_2(X, L)$ the Maslov morphism equals twice $c_1(u)$ where c_1 is the Chern class associated with an almost complex structure J compatible with ω (meaning that $\omega(\cdot J(\cdot))$ is a Riemannian metric).

We define the *Maslov number* $N_L \in \mathbf{N}$ by

$$\mathrm{Im}(I_\mu) = N_L \cdot \mathbf{Z} \tag{11.1}$$

and the monotone Lagrangians as follows

Definition 11.1.7 A Lagrangian $L \subset X$ is called monotone if

$$I_\mu = \lambda \cdot I_\omega$$

for some $\lambda > 0$.

Monotone Lagrangians (tori) naturally appear in the context of toric symplectic manifolds. Another source of examples is the construction of P. Biran [2] which will be discussed later on in the paper, see Theorem 11.4.4. On the other hand, I proved in [6] under some hypothesis that non-trivial connected sums cannot be monotone displaceable Lagrangians and in particular that the flexibility results of [16] and [7] do not provide *monotone* examples. Monotone Lagrangians seem rather to satisfy some topological rigidity and the results we are able to prove go in the direction of the conjectures above. Here are the statements:

Theorem 11.1.8 *Let $n \geq 3$ and X^{2n} a symplectic manifold which satisfies $H_2(X, \mathbf{Z}) = 0$. Suppose that $L \subset X$ is a closed, orientable and monotone Lagrangian which is displaceable. Denote by $\widetilde{L}$ the universal cover of L and assume that*

(i) the homology $H_i(\widetilde{L}, \mathbf{Z})$ is finitely generated,
(ii) the Euler characteristic

$$\chi(\widetilde{L}) = \sum_i (-1)^i \dim(H_i(\widetilde{L}, \mathbf{Z}_2))$$

satisfies $\chi(\widetilde{L}) \neq 0$.

Then $N_L = 2$ and $\mathbf{S}^3 \times L$ is the total space of a fibration over the circle. Moreover, if $n \geq 6$ and if the Whitehead group $\mathrm{Wh}_1(L)$ *vanishes, then L is the total space of a fibration over the circle.*

Example 11.1.9 Let $L = L_0 \times \mathbf{S}^{2k_1} \times \cdots \times \mathbf{S}^{2k_r} \times \mathbf{CP}^{l_1} \times \cdots \times \mathbf{CP}^{l_s}$ be the product of a closed orientable manifold of negative scalar curvature L_0 with some arbitrary number (including zero) of spheres and complex projective spaces. Then L satisfies the hypothesis of the theorem and therefore it is the total space of a fibration over the circle provided that it is Lagrangian monotone of dimension $\dim(L) \geq 6$ and displaceable. It is indeed known that $\mathrm{Wh}_1(L_0) = 0$ for any closed manifold L_0 which is negatively curved [9].

Remark 11.1.10 The first condition on the homology groups of the universal cover $\widetilde{L}$ may be stated in terms of homotopy groups. Namely the homology $H_*(\widetilde{L}, \mathbf{Z})$ is finitely generated if and only if $\pi_i(L)$ is finitely generated for any $i \geq 2$, according to [19].

Remark 11.1.11 In a previous work we proved that an n-dimensional displaceable monotone Lagrangian L is the total space of a fibration over the circle with fiber homeomorphic to the $(n-1)$-sphere, provided that $n \geq 6$ and $N_L = n$ ([5], Theorem1.7).

Our other main result deals with monotone Lagrangians which are not assumed to be displaceable:

Theorem 11.1.12 *Let* $L \subset \mathbf{CP}^n$ *be a closed orientable monotone Lagrangian. Suppose moreover that the conditions (i) and (ii) from Theorem 11.1.8 are satisfied and that moreover* $H_1(L, \mathbf{Z})$ *has no* $(n+1)$*-torsion elements. Then the same conclusion as in Theorem 11.1.8 is valid.*

Remark 11.1.13 This result is valid for a larger class of symplectic manifolds which contains $\mathbf{CP}^n$: hypersurfaces X which are subcritical polarisations (see the paper [3] of P. Biran and K. Cieliebak for definition and more details), with hypothesis (iii) replaced by the following condition on the Chern number of X: (iii) $H_1(L, \mathbf{Z})$ has no element of N_X-torsion.

11.1.3 Structure of the Paper

In the next section we briefly recall the lifted Floer homology from our previous paper [5]: it will be the main ingredient of our proofs. Then, in the third section, we present some conditions which guarantee the existence of a fibration over the circle, mostly coming from F. Latour's paper [14]. In the last section we combine the symplectic arguments from Sect. 11.2 with the topological arguments from Sect. 11.3 to complete the proofs of our two theorems.

11.2 Lifted Floer Homology

We recall the definition from [5]. Let $L \subset X$ be a closed monotone Lagrangian with $N_L \geq 2$. Consider a Hamiltonian isotopy $(\varphi_t)_{t\in[0,1]}$ on X such that $\varphi_1(L) \pitchfork L$. Denote $L_1 = \varphi_1(L)$. If $p : \widetilde{L} \to L$ denotes the projection, then the lifted Floer complex is the free complex over $\mathbf{Z}_2$ spanned by $p^{-1}(L \cap L_1)$; it is therefore infinite dimensional whenever $\pi_1(L)$ is infinite. To define its differential we take a J-almost complex structure on X which is compatible with the symplectic form and then we consider the moduli space of J-holomorphic strips with boundary in $L \cup L_1$ as in the case of usual Floer homology: for any $x, y \in L \cap L_1$ we define

$$\mathcal{M}(x, y) = \left\{ v : \mathbb{R} \times [0,1] \to X \;\middle|\; \begin{array}{c} \frac{\partial v}{\partial s} + J\frac{\partial v}{\partial t} = 0 \\ \\ v(-\infty, t) = x, \quad v(+\infty, t) = y \\ \\ v(s, 0) \in L, \quad v(s, 1) \in L_1 \\ \\ E(v) := \int v^*\omega < +\infty \end{array} \right\}$$

and

$$\mathcal{L}(x, y) = \mathcal{M}(x, y)/\mathbb{R},$$

the quotient by the action of $\mathbb{R}$ given by $\sigma \cdot v(s, t) = v(s+\sigma, t)$. We know from usual Floer theory that, for a generic choice of the Hamiltonian and of the almost complex structure, $\mathcal{L}(x, y)$ is a manifold and moreover, under the hypothesis $N_L \geq 2$, its 0-dimensional component $\mathcal{L}^0(x, y)$ is finite. For $\overline{x} \in p^{-1}(x)$ and $\overline{y} \in p^{-1}(y)$, denote by $\mathcal{L}^0(\overline{x}, \overline{y})$ the subset of $\mathcal{L}^0(x, y)$ formed by the strips v with the property that the path $s \mapsto v(s, 0) \in L$ lifts to a path between $\overline{x}$ and $\overline{y}$ in $\widetilde{L}$.

On the preferred basis of the lifted Floer complex the differential is defined as follows: for $\overline{x} \in p^{-1}(x)$

$$\partial(\overline{x}) = \sum_{y \in L \cap L_1} \#_2 \mathcal{L}^0(\overline{x}, \overline{y})\, \overline{y},$$

where $\#_2$ denotes the number modulo 2.

The crucial observation of [5] is that unlike in usual Floer homology it is not always true that $\partial^2 = 0$ and in this situation some useful properties can be inferred. Namely, summarising the results Theorem 2.1 of [5] and Theorem 3.1, Proposition 3.2 of [6] we have

Theorem 11.2.1 *If $\partial^2 \neq 0$ then $N_L = 2$ and moreover there is some J-holomorphic disk $w : (D, \partial D) \to (X, L)$ such that*

(a) $I_\mu(w) = 2$ and
(b) If $g \in \pi_1(L)$ denotes the class $[w|_{\partial D}]$ then the centralizer subgroup

$$C(g) = \{\gamma \in \pi_1(L) \mid \gamma g = g\gamma\}$$

is of finite index in $\pi_1(L)$.

Remark 11.2.2 In fact, more is true: There is a finite number of disks $w_1, w_2, \ldots, w_k$ as above such that if $\mathcal{S}_k$ is the permutation group of the elements $g_i = [w_i|_{\partial D}]$, then

$$\Phi(\gamma) = (g_i \mapsto \gamma g_i \gamma^{-1})$$

defines a morphism from $\pi_1(L)$ to $\mathcal{S}_k$. This is obviously stronger than (b) since $\mathrm{Ker}(\Phi)$ has finite index and is contained in $C(g_i)$.

We will not use this stronger result in the sequel. On the other hand, Theorem 11.2.1 will be the main ingredient of our proofs, so in order to apply it let us turn our attention to the situation when its hypothesis *is not fulfilled*, i.e. $\partial^2 = 0$. In this case the lifted Floer complex $CF(\widetilde{L}) = (\mathbf{Z}_2\langle p^{-1}(L \cap L_1)\rangle, \partial)$ is well defined and its homology $HF(\widetilde{L})$ does not depend on (H, J), as pointed out in [5]. There is also a relation between $HF(\widetilde{L})$ and the singular homology $H(\widetilde{L}, \mathbf{Z}/2)$ which can be briefly described as follows:

Proposition 11.2.3 *There exists a spectral sequence E^r_{pq} which converges to $HF(\widetilde{L})$ and whose second page is defined using the singular homology of the universal cover $H(\widetilde{L}; \mathbf{Z}/2)$.*

This spectral sequence is associated with a filtered complex which is built from the lifted Floer complex associated with some particular choice of (H, J), by adding some extra variable. The precise formulation is Theorem 2.9 of [5]. In this paper we only use the following feature of E^r_{pq} which was proved in [5] (Theorem 3.3).

Theorem 11.2.4 *Let $L \subset X$ be a closed monotone orientable Lagrangian which satisfies the conditions (i) and (ii) of Theorem 11.1.8. Then for some p, q we have $E^{\infty}_{pq} \neq 0$ and in particular $HF(\widetilde{L}) \neq 0$.*

We give he idea of the proof. Since L is orientable, its Maslov number N_L is even and this is used to show that the non-vanishing of the Euler characteristic of the usual homology ($\chi \neq 0$) propagates through all the pages of the spectral sequence preventing the limit page to vanish.

On the other hand when L is displaceable through a Hamiltonian isotopy, the corresponding lifted Floer complex is trivial when well defined and therefore $HF(\widetilde{L}) = 0$. Arguing by contradiction we infer from Theorem 11.2.4:

Corollary 11.2.5 *If the Lagrangian $L \subset X$ is closed, monotone, orientable, displaceable and moreover satisfies the hypothesis (i) and (ii) of Theorem 11.1.8 then the lifted Floer homology of L is not defined (i.e. $\partial^2 \neq 0$) and therefore the conclusions of Theorem 11.2.1 are valid.*

It is the part (b) of Theorem 11.2.1 which will be crucial in the proof of the fact that L is the total space of a fibration over the circle. To explain why, we need first to recall some facts about Morse-Novikov theory, mostly from Latour's paper [14].

11.3 Non-singular Closed One-Forms

The existence of a fibration homotopic to some map $f : X \to \mathbf{S}^1$ is obviously equivalent to the existence of a non-singular form in the 1-cohomology class $[f^*d\theta] \in H^1(X; \mathbf{Z})$. The main result of the aforementioned paper [14] is the following algebraic characterisation in high dimension of the existence of such fibrations:

Theorem 11.3.1 *Let L be a closed manifold of dimension $n \geq 6$ and $u \in H^1(L; \mathbb{R})$ a 1-cohomology class. Then there exists a non-singular 1-form $\alpha \in u$ if and only if:*

1. *The Novikov homology vanishes: $H_*(L; u) = 0$.*
2. *The Whitehead torsion vanishes: $\tau(L; u) = 0$.*
3. *The kernel $\mathrm{Ker}(u)$ of $u : \pi_1(L) \to \mathbb{R}$ is finitely presented.*

Recall (from [20] or from [14] for instance) that the Novikov homology $H_*(L; u)$ is the homology of the Morse-type complex defined by a Morse closed 1-form $\alpha \in u$ and a negative generic gradient ξ. The mentioned complex $C_*(\alpha, \xi)$ is free over the Novikov ring

$$\Lambda_u = \left\{ \sum_{i \geq 0} n_i g_i \mid n_i \in \mathbf{Z},\ g_i \in \pi_1(L),\ \lim_{i \to \infty} u(g_i) = +\infty \right\}$$

and its homology does not depend on (α, ξ); actually it is isomorphic to the homology of the tensor product $C_*(\widetilde{L}) \otimes_{\mathbf{Z}[\pi_1(L)]} \Lambda_u$, where $C_*(\widetilde{L})$ is the CW-complex of a cell decomposition of the universal cover $\widetilde{L}$ which is the lift of a CW-decomposition of L; the complex $C_*(\widetilde{L})$ is therefore finite-dimensional and free over $\mathbf{Z}[\pi_1(L)]$. This is proved in [14], Theorem 2.18.

When the Novikov homology vanishes we may define a Whitehead torsion $\tau(L; u) \in \mathrm{Wh}(\pi_1(L; u))$. The group $\mathrm{Wh}(\pi_1(L; u))$ was defined in [14] by

$$\mathrm{Wh}(\pi_1(L; u)) = \mathrm{GL}(\Lambda_u)/\langle [\mathrm{GL}(\Lambda_u), \mathrm{GL}(\Lambda_u)], \Gamma_u \rangle$$

where $\mathrm{GL}(\Lambda_u)$ is the direct limit of $\mathrm{GL}_m(\Lambda_u)$ and $\Gamma_u \leq \mathrm{GL}_1(\Lambda_u)$ represents the subgroup of *trivial units* defined by

$$\Gamma_u = \left\{ \pm g \big(1 + \sum_{i \geq 0} n_i g_i\big) \in \Lambda_u \mid u(g_i) > 0, g \in \pi_1(X) \right\}.$$

The group $\mathrm{Wh}(\pi_1(L; u))$ is thus defined in a similar way to the classical Whitehead group $\mathrm{Wh}(\pi_1(L))$ the difference being that for the latter one takes the quotient of the abelianization of $\mathrm{GL}(\mathbf{Z}[\pi_1(L)])$ by $\{\pm g \mid g \in \pi_1(X)\}$ (seen as a subgroup of $\mathrm{GL}_1(\mathbf{Z}[\pi_1(L)])$. Relating the two groups is in general quite difficult; however the following was established by A. Ranicki in [17]:

Theorem 11.3.2 *If* $\mathrm{Wh}(\pi_1(L)) = 0$ *then* $\mathrm{Wh}(\pi_1(L); u) = 0$ *for any* $u \in H^1(L; \mathbf{Z})$.

The Whitehead torsion $\tau(X; u)$ is also defined in an analogous way to the classical Whitehead torsion of an acyclic complex (C_*, ∂) endowed with a preferred basis. For this general definition [15] one considers a homotopy $s : C_* \to C_{*+1}$ between the identity and the 0-map and then remarks that $(\partial + s\partial s)^2 = \mathrm{Id}$. The torsion $\tau(C_*)$ is then defined as the image in the Whitehead group of the matrix defined by the isomorphism

$$\partial + s\partial s : C_{\mathrm{even}} \to C_{\mathrm{odd}}.$$

In particular, for a complex of the form $C_* \oplus C_{*+2k+1}$ this matrix is of the form

$$\begin{pmatrix} A & 0 \\ 0 & A^{-1} \end{pmatrix}$$

where A is the matrix of C_* and therefore the Whitehead torsion vanishes. Therefore the following is true

Proposition 11.3.3 *For any $k \geq 1$ and for any $u \in H^1(X, \mathbb{R})$ we have*

$$\tau(X \times \mathbf{S}^{2k+1}; u) = 0.$$

11.4 Proofs

11.4.1 Proof of Theorem 11.1.8

Let L be a Lagrangian which satisfies the hypothesis of the theorem. We start by applying Corollary 11.2.5 and infer that the lifted Floer homology $HF(\widetilde{L})$ is not well defined. Therefore we may apply Theorem 11.2.1. The holomorphic disk w : $(D, \partial D) \to (X, L)$ provided by this theorem defines an element $[w] \in H_2(X, L, \mathbf{Z})$ which is non-torsion since $I_\mu([w]) = 2$. Denote by $g \in \pi_1(L)$ the homotopy class of $[\partial w]$ and by $[g]$ the image of g in $H_1(L, \mathbf{Z})$. As $H_2(X, \mathbf{Z}) = 0$ and $\mathrm{Im}(I_\mu) = N_L \cdot \mathbf{Z}$. with N_L even (because L is orientable), the morphism

$$\frac{1}{2} I_\mu : \pi_2(X, L) \to \mathbf{Z}$$

may be extended to $H_2(X, L, \mathbf{Z})$ and then to some morphism $[u] : H_1(L) \to \mathbf{Z}$ such that $u([g]) = 1$. Therefore there exists $u : \pi_1(L) \to \mathbf{Z}$ satisfying $u(g) = 1$, which we see as a cohomology class in $H^1(L, \mathbf{Z})$. We will use Theorem 11.3.1 to prove that this class contains a non-singular one-form, finishing thus the proof of Theorem 11.1.8.

There are three conditions on u to be verified in order to apply Theorem 11.3.1. The second one, on the Whitehead torsion, is automatically fulfilled under the hypothesis of Theorem 11.1.8: on $\mathbf{S}^3 \times L$ using Proposition 11.3.3 and on L when the Whitehead group $\mathrm{Wh}_1(L)$ vanishes, using Theorem 11.3.2. Thus we only have to check conditions 1 and 3. Let us start with the last one, the fact that $\mathrm{Ker}(u)$ is finitely presented. Recall that $g = [\partial w]$ satisfies the conclusion of Theorem 11.2.1. We start with an elementary lemma:

Lemma 11.4.1 *Let G be a finitely presented group and $g \in G$ which satisfies $C(g) = G$, i.e. $\gamma g = g\gamma$ for all $\gamma \in G$. Then, if $u : G \to \mathbf{Z}$ is such that $u(g) = 1$ then* $\mathrm{Ker}(u)$ *is finitely presented.*

Proof Let $G = \langle x_1, x_2, \ldots x_p, g \mid r_1, r_2, \ldots, r_q \rangle$ be a finite presentation of G with g in the set of generators. We derive an explicit finite presentation

$$\mathrm{Ker}(u) = \langle y_1, y_2, \ldots y_p \mid R_1, R_2, \ldots, R_q \rangle$$

with $y_i = x_i g^{-u(x_i)}$ and $R_j = \Pi y_i^{k_i}$ if $r_j = g^{l_j} \Pi x_i^{k_i}$. The details are easy and left to the reader.

□

We know that $C(g)$ is of finite index in $\pi_1(L)$ and therefore it is finitely presented. The morphism $u : C(g) \to \mathbf{Z}$ satisfies $u(g) = 1$ and therefore by the above lemma $\mathrm{Ker}(u|_{C(g)}) = \mathrm{Ker}(u) \cap C(g)$ is finitely presented. Now $\mathrm{Ker}(u) \cap C(g) \subset \mathrm{Ker}(u)$ is obviously a subgroup of finite index, so the latter is also finitely presented, as claimed. Indeed if G is a group admitting a finitely presented group H of finite index we may explicitly construct a finite presentation of G from a finite presentation of H and a representative in each coset of G/H. Details are left to the reader.

Let us now show that the Novikov homology $H_*(L; u)$ vanishes. To this end we consider $\overline{\pi} : \overline{L} \to L$, the finite cover of L whose fundamental group is the centralizer $C(g)$. We use the following statement from [4]:

Proposition 11.4.2 *The Novikov homology $H_*(\overline{L}; u \circ \overline{\pi})$ vanishes.*

Sketch of the Proof Denote $\overline{u} = u \circ \overline{\pi}$. Consider the associated Novikov ring $\Lambda_{\overline{u}}$ and the complex $\overline{C}_* = C_*(\widetilde{L}) \otimes_{\mathbf{Z}[\pi_1(\overline{L})]} \Lambda_{\overline{u}}$ whose homology is the Novikov homology $H_*(\overline{L}; \overline{u})$. We will define an isomorphism of complexes $\Phi : \overline{C}_* \to \overline{C}_*$ which induces zero in homology, showing thus that $H_*(\overline{C}) = 0$.

We have $\pi_1(\overline{L}) = C(g)$ and we know that $\overline{u}(g) = u(g) = 1$. Since g is in the center of $\pi_1(\overline{L})$ the multiplication by g defines a morphism of $\mathbf{Z}[\pi_1(\overline{L})]$-complexes between $C_*(\widetilde{L})$ and itself. In a similar way, any polynomial $P \in \mathbf{Z}[X]$ defines a morphism $P(g) : C_*(\widetilde{L}) \to C_*(\widetilde{L})$. Note that if P is of the form $\pm 1 + XQ(X)$ then $P(g) \in \mathbf{Z}[\pi_1(\overline{L})]$ is invertible in $\Lambda_{\overline{u}}$; its inverse is

$$\lambda_0 = \pm \left(\sum_k 1 + (\pm 1)^k g^k Q^k(g) \right).$$

It follows that

$$m \otimes \lambda \mapsto m \otimes \lambda_0 \cdot \lambda$$

defines an inverse to $P(g) \otimes \mathrm{Id} : \overline{C} \to \overline{C}$. To finish the proof we have to choose $P = \pm 1 + XQ(X)$ such that $P(g) \otimes \mathrm{Id}$ is zero in homology. At this point we use the hypothesis of the finite generation of $H_*(\widetilde{L})$: we take for P the characteristic polynomial of the isomorphism of $H_*(\widetilde{L}, \mathbf{Z})$ induced by the multiplication by g. Therefore $P(g) : H_*(\widetilde{L}, \mathbf{Z}) \to H_*(\widetilde{L}, \mathbf{Z})$ is zero. Finally the Novikov homology

$H_*(\overline{C})$ is the limit of the spectral sequence of change of coefficients whose second page is

$$E^2_{pq} = \mathrm{Tor}_p^{\mathbf{Z}[\pi_1(\overline{L})]}(H_q(\widetilde{L}, \mathbf{Z}), \Lambda_{\overline{u}})$$

and $\Psi = P(g) \otimes \mathrm{Id}$ induces a morphism of spectral sequence which vanishes at the second page. It follows as in [4] that $\Psi_* : H_*(\overline{C}) \to H_*(\overline{C})$ is nilpotent and therefore for some large enough k the morphism $\Phi = \Psi^k$ is invertible and induces zero in homology. □

In order to prove that $H_*(L; u) = 0$ we use another description of Novikov homology due to Jean-Claude Sikorav [20]:

Proposition 11.4.3 *Let $u \in H^1(L, \mathbb{R})$. Consider a one-form $\alpha \in u$, the universal cover $\pi : \widetilde{L} \to L$ and $\tilde{f} : \widetilde{L} \to \mathbb{R}$ such that $d\tilde{f} = \pi^*\alpha$. Then we have*

$$H_*(L; u) = H_*(\varprojlim C_*(\widetilde{L}, \tilde{f} \geq c)).$$

Note that this description only involves a primitive $\tilde{f}$ of some $\alpha \in u$ on the universal cover. We may obviously pick the same primitive $\tilde{f} : \widetilde{L} \to \mathbb{R}$ for $\alpha \in u$ and for $\overline{\pi}^*\alpha \in \overline{u}$. This implies

$$H_*(L; u) = 0 \quad \Longleftrightarrow \quad H_*(\overline{L}; \overline{u}) = 0$$

and the proof of Theorem 11.1.8 is finished using Proposition 11.4.2.

11.4.2 Proof of Theorem 11.1.12

Let $L \subset \mathbf{CP}^n$ be a monotone Lagrangian as in the statement of Theorem 11.1.12. The proof relies on two main ingredients. The first is a construction of Paul Biran [2] of a monotone Lagrangian $\Gamma_L \subset \mathbf{C}^{n+1}$ with the following properties:

Theorem 11.4.4 *Let $L \subset \mathbf{CP}^n$ be a closed monotone Lagrangian submanifold. Then there exists a monotone Lagrangian $\Gamma_L \subset \mathbf{C}^{n+1}$ such that:*

(a) The Maslov numbers are equal: $N_{\Gamma_L} = N_L$.
(b) Γ_L is a circle fibration over L and moreover the universal cover $\widetilde{\Gamma}_L$ is diffeomorphic to $\widetilde{L} \times \mathbb{R}$.

Since Γ_L is obviously displaceable, orientable (since $N_{\Gamma_L} = N_L$ is even) and satisfies the hypothesis (i) and (ii) of Theorem 11.1.8 by (b) above, we may apply Corollary 11.2.5 and infer that the lifted Floer homology $HF(\widetilde{\Gamma}_L)$ is not well defined. We now use the second main ingredient, the following theorem of Simon Schatz [18]:

Theorem 11.4.5 *With the above notation, the lifted Floer homology $HF(\widetilde{\Gamma}_L)$ is well defined if and only if $HF(\widetilde{L})$ is well defined.*

Therefore $HF(\widetilde{L})$ is not well defined as well and we may apply Theorem 11.2.1. Let $g \in \pi_1(L)$ given by this theorem. We prove the following

Lemma 11.4.6 *Suppose as in the statement of Theorem 11.1.12 that the group $H_1(L, \mathbf{Z})$ has no $(n+1)$-torsion elements. Then there is a morphism $u : \pi_1(L) \to \mathbf{Z}$ such that $u(g) \neq 0$.*

Proof Let $[g]$ be the image of g in $H_2(\mathbf{CP}^n, L, \mathbf{Z})$. We consider the following commutative diagram:

$$\begin{array}{ccccc} H_2(\mathbf{CP}^n, \mathbf{Z}) & \xrightarrow{p} & H_2(\mathbf{CP}^n, L, \mathbf{Z}) & \xrightarrow{\delta} & H_1(L, \mathbf{Z}) \\ & \searrow^{2c_1} & \downarrow I_\mu & & \\ & & \mathbf{Z} & & \end{array}$$

Note that we only defined I_μ on the homotopy group $\pi_2(X, L)$; it is actually defined on $H_2(X, L)$ but for our purpose only its definition on the image of $\pi_2(X, L)$ is needed. It suffices to show that $\delta([g])$ is non-torsion in $H_1(L, \mathbf{Z})$. Suppose the contrary: $\delta(m \cdot [g]) = 0$ for some integer m. It follows that $m \cdot [g] = p([h])$ for some $[h] \in H_2(\mathbf{CP}^n, \mathbf{Z})$. Using that $I_\mu([g]) = 2$ and that the Chern number of $\mathbf{CP}^n$ is $n + 1$ we get the relation

$$2m = 2k(n + 1)$$

for some integer k. So m is a multiple of $n + 1$ which contradicts the fact that $H_1(L, \mathbf{Z})$ does not have $(n + 1)$-torsion. □

To finish the proof of Theorem 11.1.12 we point out that the proofs we gave in the preceding section of the facts that $H_*(L; u) = 0$ and $\mathrm{Ker}(u)$ is finitely presented still work under the milder assumption $u(g) \neq 0$ instead of $u(g) = 1$. Actually for the vanishing of the Novikov homology the argument is the same. For the finite presentation of the kernel, if $u(g) = k$, consider the normal subgroup of $\pi_1(L)$

$$H = \{h \in \pi_1(L) \mid u(h) \in k\mathbf{Z}\}.$$

Obviously H is of finite index (since $u : G/H \to \mathbf{Z}/k\mathbf{Z}$ is mono) and therefore H is finitely presented. So the argument in the proof of Lemma 11.4.1 applies to H and implies that $\mathrm{Ker}(u|_H) = \mathrm{Ker}(u)$ is finitely presented.

The proof of Theorem 11.1.12 is now complete.

References

1. M. Audin, Fibrés normaux d'immersions en dimension double, points doubles d'immersions lagrangiennes et plongements totalement réels. Comment. Math. Helv. **63**, 593–623 (1988)
2. P. Biran, Lagrangian non-intersections. Geom. Funct. Anal. **16**, 279–326 (2006)
3. P. Biran, K. Cieliebak, Symplectic topology on subcritical manifolds. Comment. Math. Helv. **76**, 712–753 (2001)
4. M. Damian, On the higher homotopy groups of a finite CW-complex. Topol. Appl. **149**, 273–284 (2005)
5. M. Damian, Floer homology on the universal cover, Audin's conjecture and other constraints on Lagrangian submanifolds. Comment. Math. Helv. **87**(2), 433–462 (2012)
6. M. Damian, On the topology of monotone Lagrangian submanifolds. Ann. Sci. Ec. Nom. Sup. 4ème série **48**(1), 237–252 (2015)
7. T. Ekholm, Y. Eliahsberg, E. Murphy, I. Smith, Constructing exact lagrangian immersions with few double points. Geom. Funct. Anal. **23**, 1772–1803 (2013)
8. J. Evans, J. Kedra, Remarks on monotone lagrangians in $\mathbf{C}^n$. Math. Res. Lett. **21**(6), 1241–1255 (2014)
9. F.T. Farell, L.E. Jones, Stable pseudo-isotopy spaces of non-positively curved manifolds. J. Differ. Geom. **38**, 769–834 (1991)
10. K. Fukaya, Application of Floer homology of Langrangian submanifolds to symplectic topology, in *Morse Theoretic Methods in Nonlinear Analysis and in Symplectic Topology*, volume 217 of NATO Sci. Ser. II Math. Phys. Chem. (Springer, Dordrecht, 2006), pp. 231–276
11. M. Gromov, Pseudo-holomorphic curves in symplectic manifolds. Invent. Math. **82**, 307–347 (1985)
12. M. Gromov, *Partial Differential Relations*, volume 9 of Ergebnisse der Mathematik und Ihrer Grenzgebiete (Springer, Berlin, 1986)
13. K. Irie, Chain level loop bracket and pseudo-holomorphic disks. J. Topol. **13**(2), 870–938 (2020)
14. F. Latour, Existence de 1-formes fermées non singulières dans une classe de cohomologie de de Rham. Inst. Hautes Études Sci. Publ. Math. **80**, 135–194 1994 (1995)
15. S. Maumary, *Type Simple d'homotopie*, volume 48 of Lecture Notes in Mathematics (Springer, Berlin, 1967)
16. L. Polterovich, The surgery of lagrangian manifolds. Geom. Funct. Anal. **1**, 198–210 (1991)
17. A. Ranicki, *High-dimensional Knot Theory* (Springer, Berlin, 1998). Translated from the 2010 French original by Reinie Erné
18. S. Schatz, A topological constraint for monotone lagrangians in hypersurfaces of kähler manifolds. Math. Z. **281**(3), 877–892 (2015)
19. J.-P. Serre, Groupes d'homotopie et classes de groupes abeliens. Ann. Math. (2) **58**, 258–294 (1953)
20. J.-C. Sikorav, Points fixes de difféomorphismes symplectiques, intersections de sous-variétés lagrangiennes et singularités de un-formes fermées. PhD thesis, Université Paris-Sud, Orsay, France, 1987

Chapter 12
Kervaire Problems in Stable Homotopy Theory

Petr M. Akhmetiev

Abstract The Kervaire Problem in Stable Homotopy Theory is a problem to classify framed immersions with Arf-invariant one. The first interesting example is in dimension 30. There exists a closed stably-parallelizable manifold $\tilde{M}^{30}$ with Arf–Kervaire invariant 1. The goal of the chapter is to recall various geometrical constructions concerning the Kervaire Problem. In particular, we give a self-contained description of the Mahowald element in the stable homotopy group of spheres $\Pi_{2^l}, l \geq 3$. Using this construction, we prove that the Generalized Kervaire Problem, formulated by the author in Akhmetiev (Kervaire Problems in Stable Homotopy Theory (2016). arXiv:1608.01206v1 [math.AT]), is solved positively.

Keywords Skew-framed immersions · Hirsch control principle · Stably-framed immersions · The Mahowald elements

1980 AMS Subject Classification 55Q45, 57R90

12.1 Part I: Introduction

Kervaire invariants are defined by means of self-intersections of immersions as was investigated by V. Poénaru in [30]. The Kervaire Problem is an unsolved problem in stable homotopy theory. Take a closed stably-parallizable manifold M^{4k+2} of dimension $dim(M) = 4k + 2$ and assume that this manifold is $2k$-connected. On the homology group the following function:

$$q : H_{2k+1}(M^{4k+2}; \mathbb{Z}/2) \to \mathbb{Z}/2$$

P. M. Akhmetiev (✉)
HSE Tikhonov Moscow Institute of Electronics and Mathematics (MIEM HSE), Moscow, Russia

Pushkov Institute of Terrestrial Magnetism, Ionosphere and Radio Wave Propagation Russian Academy of Sciences Russian Federation, 108840, Moscow, Troitsk, Kaluzhskoe Hwy 4, IZMIRAN
e-mail: pmakhmet@mail.ru

L. Funar, A. Papadopoulos (eds.), *Essays on Topology*,
https://doi.org/10.1007/978-3-031-81414-3_12

is well defined as follows. Take an embedding

$$F : M^{4k+2} \subset \mathbb{R}^{4k+2+N},$$

where N is a very big positive integer. By assumption M^{4k+2} is stably parallelizable and the normal bundle $\nu(F)$ of the embedding F is the trivial bundle: there is a framing $\Xi : \nu(F) \cong N\varepsilon$.

Define $q(x)$, $x \in H_{2k+1}(M^{4k+2}; \mathbb{Z}/2)$. We may assume (because M^{4k+2} is $2k$-connected) that the cycle x is the fundamental class of an embedded standard sphere $\varphi : S^{2k+1} \subset M^{4k+2}$. Take an embedding $G = F \circ \varphi : S^{2n-1} \subset \mathbb{R}^{4k+2+N}$. This embedding is N-framed, because F is framed. Take a regular homotopy $G(t)$ such that $G(0) = G$ and $G(1)$ is an immersion into the standard hyperspace $\mathbb{R}^{4k+2} \subset \mathbb{R}^{4k+2+N}$. Moreover, we assume that the N-framing of $G(1) : S^{2k+1} \looparrowright \mathbb{R}^{4k+2} \subset \mathbb{R}^{4k+2+N}$ coincides with the restriction of the coordinate framing of the hyperspace $\mathbb{R}^{4k+2} \subset \mathbb{R}^{4k+2+N}$. Then define $x = 1$ if and only if $G(1)$ in $\mathbb{R}^{4k+2}$ self-intersects in an odd number of points, and define $x = 0$, if $G(1)$ self-intersects in an even number of points.

The function q is a quadratic form, because this function satisfies the following equation:

$$\langle x, y \rangle = q(x) + q(y) + q(x + y) \pmod 2,$$

where $\langle \dots, \dots \rangle$ is the symmetric intersection form on $H_{2k+1}(M^{4k+2}; \mathbb{Z}/2)$.

The Arf-invariant $Arf(q)$ is defined as follows. Take a Hamiltonian base in $H_{2k+1}(M^{4k+2}; \mathbb{Z}/2)$: $\mathbf{e}_i, \mathbf{f}_i$, $i = 1, \dots, k$ (this terminology arises from the paper [29] and the definition of the quadratic form $q(F)$ for a framed immersion is a reformulation of [28]).

Define

$$Arf(q) = \sum_{i=1}^{k} q(\mathbf{e}_i) q(\mathbf{f}_i) \in \mathbb{Z}/2.$$

The framed cobordism class of F contains a homotopy sphere if and only if $Arf(q) = 0$.

12.1.1 Pontryagin–Thom Construction

Let us recall the Pontryagin Construction [31] to describe homotopy groups of spheres, generalized by R. Thom. Take a framed embedding $f : M^m \subset S^{m+N}$, with the frame $\Xi : U(f) = M^m \times D^N$, and take a collapse $S^{m+N} \to S^N$ as $M^m \times D^N \to D^N \subset S^N$, and $S^{m+N} \setminus U(f)$ is mapped into the marked point $\infty = S^n \setminus D^N$ on S^N. The regular cobordism class of framed embeddings f is in one-to-one correspondence with a homotopy classes of mappings $(S^{m+N}, pt) \rightarrow (S^N, \infty)$. The Arf invariant is the homomorphism

$$\Theta_{4k+2} : \pi_{4k+2+N}(S^N) = \Pi_{4k+2} \rightarrow \mathbb{Z}/2,$$

which is defined as the $Arf(q)$ for a corresponding framed embedding F, which represents an element in $\Pi_{4k+2} = \pi_{4k+2+N}(S^N), \quad N \geq 4k+4$.

12.1.2 Kervaire Problem

For What are dimensions of the form $4k+2$ for which Θ_{4k+2} is onto?

This problem has a complicated history, see http://www.math.rochester.edu/people/faculty/doug, [33]. Examples below give particular positive solutions of the Kervaire Problem. The Theorem in [22] shows that the Kervaire Problem, in general, has a negative solution and this result is generalized by Brown and Peterson and by Browder [12]. A result by Barratt, Mahowald and Tangora, used in [12] and [20], shows that the mentioned positive solutions are not unique. We recall the Jones construction [20], a framed 30-manifold with Arf-invariant 1. This result is the most interesting, because an element in Π_{15} with Hopf invariant one does not exist [1], but the square of the two Hopf invariant one elements in Π_{30} exists!

At first glance, a reasonable conjecture would be the following: in all dimensions $2^l - 2$, $l \geq 5$ there exists a framed closed smooth manifold, representing a non-trivial element in Π_{2^l-2} with Arf-invariant 1. In [11] results toward an infinite family of positive solutions of Kervaire Problem is proposed, this result is related with a closed Strong Kervaire Problem, which is also motivated by results in [19]. The Mahowald elements in Π_{2^l}, $l \geq 3$, constructed in [26], give an analogous infinite family. But by Hill–Hopkins–Ravenel Theorem, we get a negative solution in all dimensions, probably, except the dimension 126. In dimension 126, the Kervaire Problem is open, see Conjecture 12.1 below. The result in [27] claims a positive solution of the Kervaire Problem in dimension 126.

In this chapter we investigate a problem, called Generalized Kervaire Problem, which generalizes the Kervaire Problem and the Strong Kervaire Problem. The Generalized Kervaire Problem was proposed by the author in [2]. Briefly, the Kervaire invariant is reformulated in the most general situation and the problem asks if whether the generalized Kervaire invariant is realizable in an infinite sequence of dimensions. The main result proves that the Generalized Kervaire Problem has a positive solution. The classical Kervaire Problem which has only finite positive solutions is deformed and a more general problem gets an infinite number of positive solutions. One question is: "where do infinitely many solutions appear by the deformation"?

The Generalized Kervaire Problem is related by the Pontryagin–Thom Construction with the concept of stably-framed immersions, see Sect. 12.2.9. The Arf–

Kervaire invariant for framed immersions is generalized as twisted Arf-invariant for stably-framed immersions and admits a geometrical visualization called the Browder–Eccles invariant.

12.1.3 Examples

Take the Hopf generator $\eta_l \in \Pi_{2^l-1}$, $l = 1, 2, 3$. Assume η_l is represented by an embedding framed sphere $F_l : S^{2^l-1} \subset \mathbb{S}^{2^l-1+N}$. Take the corresponding framed embedding torus

$$G_{2^{l+1}-2} F_l \times F_l : S^{2^l-1} \times S^{2^l-1} \subset \mathbb{R}^{2^l-1+N} \times \mathbb{R}^{2^l-1+N} \subset S^{2^{l+1}-2+2N}.$$

The embedding $G_{2^{l+1}-2}$ represents an element $\theta_{2^{l+1}-2} \in \Pi_{2^{l+1}-2}$. By construction we get $\Theta_{2^{l+1}-2}(\theta_{2^{l+1}-2}) = 1$ and the Kervaire Problem in the dimension 2, 6, 14 is solved positively.

12.1.4 Theorem by M. Kervaire (1960) [22]

The homomorphism

$$\Theta_{10} : \Pi_{10} \to \mathbb{Z}/2$$

is trivial.

Proof *(See [8] for a Similar Reasoning in a More General Case,* Θ_{4k+2}*,* $4k+2 \neq 2^l - 2$*)*
For the regular cobordism groups of immersions and embeddings with prescribed normal bundle structure, we use notations by A. Szucs and his approach started in [34]. We briefly recall the definition of regular cobordism groups in Sect. 12.2.2. For a more detailed definition see [7]. By the Hirsch compression principle [18] the following isomorphism is well defined:

$$\mathrm{Emb}^{\mathrm{fr}}(m, N) \to \mathrm{Imm}^{\mathrm{fr}}(m, 1).$$

The following homomorphism (self-intersection of the framed codimension 1 embedding):

$$\delta : \mathrm{Imm}^{\mathrm{fr}}(m, 1) \to \mathrm{Imm}^{\mathrm{sf}}(m - 1, 1)$$

is well defined. Take $x \in \mathrm{Imm}^{\mathrm{fr}}(10, 1)$. Take $y = \delta(x) \in \mathrm{Imm}^{\mathrm{sf}}(9, 1)$. Take an immersion $f : N^9 \looparrowright \mathbb{R}^{10}$ which represents y and let $g : L^8 \looparrowright \mathbb{R}^{10}$ be the self-intersection immersion of f. By the Browder Theorem [12] and Eccles Theorem [16], see Sect. 12.2.7, the Kervaire invariant $\Theta_{10}(x)$ is given by the following characteristic class:

$$\langle w_2^4(L^8); [L^8]\rangle \in \mathbb{Z}/2,$$

where w_2 is the second normal Stiefel–Whitney classes of L^8, which is the Euler class of the normal 2-bundle η over L^8.

Take a submanifold $K^4 \subset L^8$, which is dual to w_2^2 (the Whitney sum of the two isomorphic 2-bundles determines the Euler class). Then

$$\langle w_2^4(L^8); [L^8]\rangle = \langle w_2^2(K^4); [K^4]\rangle.$$

We get for the top class

$$w_4(K^4) = w_2^2 + w_1^2 w_2.$$

Because the top 4-class on the 4-manifold K^4 is trivial, we get:

$$\Theta_{10}(x) = \langle w_2^2(K^4); [K^4]\rangle = w_1^2 w_2.$$

Take the submanifold

$$Q^2 \subset K^4,$$

which is dual to w_2 on K^4. The top normal class

$$\bar{w}_2(Q^2) = w_1^2, \quad w_1 = w_1|_{Q\subset K}.$$

We get:

$$\langle w_2 w_1^2(K^4); [K^4]\rangle = \langle w_1^2(Q^2); [Q^2]\rangle = 0.$$

Because Q_2 is an oriented surface, we have $w_1(Q^2) = w_1(4\eta) = 0$, and for an arbitrary class $\xi \in H^1(Q^2)$ we get $\xi^2 = 0$. □

12.1.5 The Strong Kervaire Invariant Problem

Take the Whitehead product

$$\omega_{k-1} \in S^{2k-1}(S^k),$$

assuming $k-1$ is even. The Strong Kervaire Problem, as in [19] is the following: for which $k-1$ the product $\omega_{k-1} \in \pi_{2k-1}(S^k)$ is halved (that is, divisible by 2)? In the case $k-1 = 0, 2, 6$ $\omega_{k-1} = 0$ and is halved. By the Toda Theorem $\omega_{14} \in S^{29}(S^{15})$ is halved.

A positive solution of the Strong Kervaire Problem (ω_{k-1} is halved) implies a positive solution of the Kervaire Problem in Π_{k-1} [11]. It was assumed then a positive solution of Strong Kervaire Problem in ω_{2^l-2} implies a positive solution of Kervaire Problem in dimension $2^{l+1}-2$. This is not possible, because by [17] for an arbitrary dimension not less then 254, Kervaire Problem has a negative solution.

Let $F : M^{4l-2} \looparrowright \mathbb{R}^{6l-3}$ is an immersion, which is stably framed. This means that the immersion is framed after codimension 1 stabilization in $\mathbb{R}^{6l+4}$. The homology modulo 2 $H_{2l-1}(M^{4l-2};\mathbb{Z}/2)$ is equipped with an even symmetric bilinear form $\langle x, y\rangle \in \mathbb{Z}/2$, $\langle x, x\rangle = 0$. The equation $\langle x, x\rangle = 0$ is satisfied for the following geometrical reason: a generic intersection of a cycle $c_x \cap c'_x$, where c'_x is a small alteration of c_x, even c_x is a non-oriented manifold, is an integer number. This number $[c_x \cap c'_x]$ changes sign, when we take the permutation $c_x \leftrightarrow c'_x$. But, obviously $[c_x \cap c'_x] = [c'_x \cap c_x]$. This proves $[c_x \cap c'_x] = 0 \pmod{\mathbb{Z}}$ and, in particular, $[c_x \cap c'_x]$ is an even number.

The quadratic form

$$q : H_{2l-1}(M^{4l-2};\mathbb{Z}/2) \to \mathbb{Z}/2$$

is well defined as above (we will not assume that M^{4k+2} is $2k$-connected) and also a linear form (assuming $2k-1 \geq 7$)

$$\xi : H_{2l-1}(M^{4l-2};\mathbb{Z}/2) \to \mathbb{Z}/2$$

is well defined as the total class for trivialization of the normal bundle, restricted on a cycle in $H_{2l-1}(M^{4l-2};\mathbb{Z}/2)$. The twisted quadratic form $q^{tw} = q + \xi$ and the $Arf(q^{tw})$ is called the Browder–Eccles invariant of the stably-framed immersion F.

Theorem (See Corollary 12.1, Sect. 12.2.9) *For* $4l+2 = 2^{\lambda}-2 = n$, $\lambda \geq 4$, *there exists a stable framed immersion with Browder–Eccles invariant* 1.

12.2 Part II: Survey

12.2.1 The Kervaire Problem

Let $f : M^{n-1} \looparrowright \mathbb{R}^n$ be an immersion of a closed, in general non-orientable manifold in codimension 1, in $n \equiv 2 \pmod 4$. Let us consider the self-intersection manifold of f. Denote this manifold by N^{n-2}. Consider the characteristic class $w_2 \in H^2(N^{n-2}; \mathbb{Z}/2)$ and denote the characteristic number

$$\langle w_2^{\frac{n-2}{2}}; [N^{n-2}]\rangle \in \mathbb{Z}/2$$

by $\Theta(f)$. The most interesting case is $n = 2^l - 2, l \in \mathbb{N}, l \geq 2$.

Problem 12.1 (Kervaire Problem)

For which n the following equation $\Theta(f) = 1$ for an immersion $f : M^{n-1} \looparrowright \mathbb{R}^n$ is possible? Let us say that the Kervaire Problem has a positive solution if and only if there exists an immersion f with $\Theta(f) = 1$.

In the case $n \neq 2^l - 2$, the Kervaire Problem is solved negatively by [22], [12].

12.2.2 The Generalized Kervaire Problem

Let $g : M^{n-k} \looparrowright \mathbb{R}^n$ be an immersion of a closed non-oriented manifold of dimension $n - k$, where $n \equiv 2 \pmod 4$, $n > k$. Denote by $\kappa : M^k \looparrowright \mathbb{RP}^\infty$ a prescribed characteristic class. In the case $n - k \equiv 1 \pmod 2$ the characteristic class κ is the oriented class over M^{n-k}: $\kappa = w_1(M^{n-k})$. The line bundle $\kappa^*(\gamma)$ (γ is the only non-trivial bundle over $\mathbb{RP}^\infty$) is also denoted by κ, for short. Assume that the following isomorphism $\Psi : \nu(g) \cong k\kappa$ is well defined, where $\nu(g)$ is the normal bundle of g, $k\kappa$ is the Whitney sum of k copies of the bundle κ. The triple (g, κ, Ψ) represents an element in the cobordism group $\mathrm{Imm}^{\mathrm{sf}}(n-k, k)$. For a detailed definition and elementary properties see [7].

Assume that $k = \frac{n}{2}$ and define the characteristic number $\Theta_{\frac{n}{2}}(g, \kappa, \Psi) \in \mathbb{Z}/2$ as the parity of self-intersection points of the immersion g (we assume that g is self-transversal). In fact, θ does not depends on (κ, Ψ) and $\Theta_{\frac{n}{2}}(g, \kappa, \Psi) = \Theta_{\frac{n}{2}}(g)$.

For given k_-, k_+ satisfying the inequalities $n > k_+ > \frac{n}{2} > k_- \geq -1$, the following sequence of homomorphisms is well defined:

$$\mathrm{Imm}^{\mathrm{sf}}(k_+, n-k_+) \xrightarrow{J^{k_+}_{\frac{n}{2}}} \mathrm{Imm}^{\mathrm{sf}}(\frac{n}{2}, \frac{n}{2}) \xrightarrow{J^{\frac{n}{2}}_{k_-}} \mathrm{Imm}^{\mathrm{sf}}(k_-, n-k_-). \quad (12.1)$$

By the group $\mathrm{Imm}^{\mathrm{sf}}(-1, n+1)$ one means the one-element group.

Definition 12.1 For a prescribed n, $n \equiv 2 \pmod 4$, denote the subgroup $Im(J^{k_+}_{\frac{n}{2}}) \cap Ker(J^{\frac{n}{2}}_{k_-}) \subset \mathrm{Imm}^{\mathrm{sf}}(\frac{n}{2}, \frac{n}{2})$ by $K^n_{k_+,k_-} = K_{k_+,k_-} \subset \mathrm{Imm}^{\mathrm{sf}}(\frac{n}{2}, \frac{n}{2})$. The 2-parameter family of subgroups determines the double filtration of the group $\mathrm{Imm}^{\mathrm{sf}}(\frac{n}{2}, \frac{n}{2})$. (Note that $K_{\frac{n}{2},-1}$ coincides with the whole group $\mathrm{Imm}^{\mathrm{sf}}(\frac{n}{2}, \frac{n}{2})$.)

The characteristic class Θ is defined as a homomorphism:

$$\mathrm{Imm}^{\mathrm{sf}}(n-1,1) \stackrel{J^{n-1}_{\frac{n}{2}}}{\longrightarrow} \mathrm{Imm}^{\mathrm{sf}}(\frac{n}{2},\frac{n}{2}) \stackrel{\Theta^{\mathbf{D}}}{\longrightarrow} \mathbb{Z}/2,$$

where $\Theta^{\mathbf{D}} : \mathrm{Imm}^{\mathrm{sf}}(\frac{n}{2},\frac{n}{2}) \longrightarrow \mathbb{Z}/2$ is the number of self-intersection points of a generic immersion, which represents an element in $\mathrm{Imm}^{\mathrm{sf}}(\frac{n}{2},\frac{n}{2})$.

Problem 12.2 (Generalized Kervaire Problem)

1. For what (k_+, k_-) is the restriction of the homomorphism Θ on the subgroup $K^n_{k_+,k_-}$ non-trivial? (if there exists $[(g,\kappa,\Psi)] \in K_{k_+,k_-} \subset \mathrm{Imm}^{\mathrm{sf}}(\frac{n}{2},\frac{n}{2})$, for which $\Theta_{\frac{n}{2}}(g) = 1$?)
2. In particular, are there infinitely many n such that $K_{\frac{n}{2},-1}$ contains an element x with $\Theta_{\frac{n}{2}}(x) = 1$?

Remark 12.1 A positive solution of Statement 2 of Problem 12.2 is presented in Part 2 of the chapter.

Let us consider the following diagram:

$$\begin{array}{ccccc}
\mathrm{Imm}^{\mathrm{sf}}(n-1,1) & \stackrel{\delta^{\mathbf{D}}}{\longrightarrow} & \mathrm{Imm}^{\mathbf{D}}(n-2,2) & \stackrel{\Theta^{\mathbf{D}}}{\longrightarrow} & \mathbb{Z}/2 \\
\downarrow J^{\mathrm{sf}}_{k_+} & & \downarrow J^{\mathbf{D}}_{k_+} & & || \\
\mathrm{Imm}^{\mathrm{sf}}(n-k_+,k_+) & \stackrel{\delta^{\mathbf{D}}_{k_+}}{\longrightarrow} & \mathrm{Imm}^{\mathbf{D}}(n-2k_+,2k_+) & \stackrel{\Theta^{\mathbf{D}}_{k_+}}{\longrightarrow} & \mathbb{Z}/2. \\
\downarrow J^{\mathrm{sf}}_{k_+,k_-} & & & & \\
\mathrm{Imm}^{\mathrm{sf}}(k_-,n-k_-) & & & &
\end{array} \tag{12.2}$$

Proposition 12.1 *The positive answer to Problem 12.2, Statement 1 is equivalent to the following.*

a. There exists an element $x \in \mathrm{Imm}^{\mathrm{sf}}(n-k_+,k_+)$ with $\Theta_{k_+}(x) = 1$, such that $J^{\mathrm{sf}}_{k_+,k_-} = 0$.
b. Moreover, the statement a. implies that the element x is represented by a skew-framed immersion (f,Ψ), $f : M^{n-k_+} \looparrowright \mathbb{R}^n$, for which the cohomology class $\kappa : M^{n-k_+} \to \mathbb{RP}^\infty$ is represented by a composition $M^{n-k_+} \to \mathbb{RP}^{n-k_+-k_-} \subset \mathbb{RP}^\infty$.

Proof of Proposition 12.1 Statement a. is clear. Let us prove Statement b. Assume that $M_1^{k_-} \subset M^{n-k_+}$ is dual to $\kappa^{n-k_+-k_-}$. Let $(W_1^{k_-},\Psi_{W_1})$, $\partial(W_1^{k_-+1}) = M_1^{k_-}$, be a skew-framed cobordism, which exists because (M_1,Ψ_{M_1}) is a skew-boundary in the codimension $n-k_-$. Take a thickning $W \mapsto W_1$ of W as the codimension $n-k_+-k_-$. The cobordism W_1 is the result of a gluing $M^{n-k_+} \times [0,1] \cup U((n-k_+-k_-)\kappa_{W_1})$, where $U((n-k_+-k_-)\kappa_{W_1})$ is the disk bundle associated with the first $n-k_+-k_-$ vectors of the skew-framing Ψ_{W_1} over W_1. The cobordism W_1 is skew-framed in codimension k_+ from (M,Ψ_M) to (M',Ψ'), where the manifold M' is equipped with an immersion $f' : M' \looparrowright \mathbb{R}^n$, which is skew-framed in codimension k_+. Additionally, the cohomology class $\kappa_{M'}$ is represented by a mapping. $M' \to \mathbb{RP}^{n-k_+-k_-} \subset \mathbb{RP}^\infty$. Proposition 12.1 is proved. □

Proposition 12.2 *A positive solution to the Kervaire Problem for* $n = 2^l - 2$ *is equivalent to a positive solution of the Generalized Kervaire Problem for* $K_{n-1,-1}$.

Proof of Proposition 12.2

Consider an immersion $f : M^{n-1} \looparrowright \mathbb{R}^n$. Consider a submanifold $M_1^{\frac{n}{2}} \subset M^{n-1}$, which is dual to the characteristic class $\kappa^{\frac{n}{2}-1} \in H^{\frac{n}{2}-1}(M^{n-1}; \mathbb{Z}/2)$. In this case $\kappa : M^{n-1} \to \mathbb{RP}^\infty$ is the oriented class. (Below we use $\mathbb{Z}/2$-coefficients, in case the coefficients are omitted.) Consider the immersion $g : M_1^{\frac{n}{2}} \looparrowright \mathbb{R}^n$, $g = f|_{M_1^{\frac{n}{2}}}$. Define the characteristic class $\kappa|_{M_1^{\frac{n}{2}}}$, which is not re-denoted. By construction, a skew-framing $\Psi : \nu_g \cong \frac{n}{2}\kappa$ is well defined. The triple (g, κ, Ψ) determines an element in the group $\mathrm{Imm}^{\mathrm{sf}}(\frac{n}{2}, \frac{n}{2})$. The following equation is satisfied: $\Theta(f) = \Theta_{\frac{n}{2}}(g)$. By construction, $(g, \kappa, \Psi) \in K_{\frac{n}{2},-1}$. If the Kervaire Problem admits a positive solution, then there exists f such that $\Theta(f) = 1$. Then $\Theta_{\frac{n}{2}}(g) = 1$ and the Generalized Kervaire Problem for the term $K_{n-1,-1}$ admits a positive solution.

Conversely, assume $\Theta_{\frac{n}{2}}(g, \kappa, \Psi) = 1$. The inverse image of the subgroup $K_{\frac{n}{2},-1}$ in $\mathrm{Imm}^{\mathrm{sf}}(n-1, 1)$, by the homomorphism $J^{n-1}_{\frac{n}{2}}$, contains an element $[f]$ with $\Theta(f) = \Theta_{\frac{n}{2}}([g, \kappa, \Psi]) = 1$. This proves that the Kervaire Problem has the positive solution. □

Proposition 12.3 *The Generalized Kervaire Problem for the term* $K_{\frac{n}{2},\frac{n}{2}-1}$ *is positively solved if and only if* $n = 2, 6, 14$.

Proof of Proposition 12.3

The group $K_{\frac{n}{2},\frac{n}{2}-1}$ is represented by a triple $(g, \Xi, \kappa = \varepsilon)$, where $g : M_1^{\frac{n}{2}} \looparrowright \mathbb{R}^n$, Ξ is a trivialization of the normal bundle of g: $\Xi : \nu_g \cong \frac{n+2}{2}\varepsilon$. Therefore, by the Hirsch principle [18], (g, Ξ, ε) represents an element in the stable homotopy group of the sphere $\mathrm{Imm}^{fr}(\frac{n}{2}, 1) = \Pi_{\frac{n}{2}}$. The value $\Theta_{\frac{n}{2}}(g, \Xi, \varepsilon)$ coincides with the stable Hopf invariant $h([g, \Xi])$. By the Adams theorem, $h([g, \Xi, \varepsilon]) = 0$, if $l \geq 4$; the homomorphism $h : \Pi_{\frac{n}{2}} \to \mathbb{Z}/2$ is onto, if and only if $n = 2, 6, 14$. □

12.2.3 Strong Kervaire Problem

Define the element $\omega_{k-1} \in \pi_{2k-1}(S^k)$ by the Whitehead square of the generator $i_k \in \pi_k(S^k)$, $k - 1 \equiv 2 \pmod 4$. We use an alternative notation for the Whitehead product ω_{k-1} instead of w_k as in [19].

Problem 12.3 (Strong Kervaire Problem)
Assume $k = 2^l - 1$, and let $n = 2k = 2^{l+1} - 2$. For what l does there exist an element $y \in \pi_{2k-1}(S^k)$ such that the following condition in $\pi_{2k-1}(S^k)$ is satisfied:

$$\omega_{k-1} = 2y? \tag{12.3}$$

In this case we will call that the element ω_{k-1} is halved.

Theorem 12.1 *The Strong Kervaire Problem is positively solved (for $2k = n = 2^{l+1} - 2$ the element ω_{k-1} is halved) if and only if the Generalized Kervaire Problem for the term $K^n_{k,k-2}$ is positively solved.*

Proof of Theorem 12.1
Let us prove that if ω_{k-1} is halved, then the Strong Kervaire Problem is positively solved. By the Pontryagin Theorem, the element in the homotopy group $\pi_{2k-1}(S^k)$ is represented by a framed submanifold $(P^{k-1}; \Xi_P)$, $P^{k-1} \subset S^{2k-1}$, $\Xi : \nu_P \cong k\varepsilon$. It is known that the element $\omega_{k-1} \in \pi_{2k-1}(S^k)$ is represented by the Hopf link of the 2 standard framed spheres $(S^{k-1} \cup S^{k-1}) \subset \mathbb{R}^{2k-1} \subset S^{2k-1}$. The Hopf link is the result of the intersection of the two coordinate subspaces $\mathbb{R}^k \vee \mathbb{R}^k \subset \mathbb{R}^{2k}$ with the unit sphere $S^{2k-1} \subset \mathbb{R}^{2k}$. The framing of the Hopf link is defined as the standard coordinate framing of the two subspaces.

Consider the standard (internal) embedding $I_1 : S^{2k-1} \subset \mathbb{R}^n$, $n = 2k$ and the concentric external embedding $I_2 : S^{2k-1} \subset \mathbb{R}^n$. By assumption, ω_k is even, therefore there exists a framed embedded cobordism from the internal Hopf link in $I_1(S^{2k-1})$ to an external framed submanifold (Q^{k-1}, Ξ_Q) in the sphere $I_2(S^{2k-1})$, which is the union of two disjoint copies of a framed embedding.

Define a skew-framed immersion in codimension k of a non-oriented manifold $g : N^k \subset \mathbb{R}^n$ with the only one self-intersection point inside the sphere $I_1(S^{2k-1})$. The immersion g coincides with the Hopf link of the two standard framed spheres in the hypersphere $I_1(S^{2k-1})$, and coincides with (Q^{k-1}, Ξ_Q) in the hypersphere $I_1(S^{2k-1})$. In the exterior domain and in the middle domain $\mathbb{R}^n \setminus \{I_1(S^{2k-1}) \cup I_2(S^{2k-1})\}$, the immersion is a framed embedding. In the interior ball, the immersion is given by the pair of disks with self-intersection point in the center of the ball. In the middle domain the framed embedding is well defined by the assumption that ω_{k-1} is halved. In the exterior domain the skew-framed embedding is given by a cylinder, which is bounded by the two parallel copies of the embedded framed submanifold with opposite framings in the odd codimension k.

By construction, the triple (g, κ_N, Ψ_N) represents an element in $\mathrm{Imm}^{\mathrm{sf}}(k, k)$ and the orienting class $\kappa_N : N^k \to \mathbb{RP}^\infty$ is an integer, $[\kappa_N] \in H^1(N^k; \mathbb{Z})$. We get: $\Psi_N : \nu_g \cong k\kappa_N$. The triple (g, κ_N, Ψ_N) represents an element in the subrgoup $K^n_{\frac{n}{2},\frac{n}{2}-2}$, and $\Theta_{\frac{n}{2}}(g) = 1$.

Assume that the Strong Kervaire Problem is solved positively, and let us prove that w_k is halved. An element in $K^n_{\frac{n}{2},\frac{n}{2}-2}$ is represented by a triple (g, κ_N, Ψ_N), for which the characteristic class κ_N is an integer. If $\Theta_{\frac{n}{2}} = 1$, the immersion g is self-intersected in an odd number of points. There exists an immersion with an integer class κ with an odd number of self-intersection points. By the Hirsch dense control principle, because κ_N is represented by a mapping into the circle, there exists an immersion into a thin regular neighborhood of the standard circle $U \subset \mathbb{R}^n$ with an odd number of self-intersection points.

By the Whitney trick we may assume that there is only one self-intersection point of the immersion g, and that the immersed submanifold $g(N^k)$ is divided by the spheres $I_1(S^{2k-1})$, $I_2(S^{2k-1})$ into three submanifolds with boundaries. The internal boundary is represented by the framed embedding Hopf link in $I_1(S^{2k-1})$; the external boundary is represented by the two-component framed embedding (Q^{k-1}, Ξ_Q); the middle circle of the solid torus U is divided by the external sphere into two n-balls with boundaries n − 1-disks on $I_2(S^{2k-1})$ and the components of (Q^{k-1}, Ξ_Q) are embedded into the disks. The framed embeddings are joined inside the spherical layer between $I_1(S^{2k-1})$, $I_2(S^{2k-1})$ by a skew-framed embedding cobordism. The external cobordism outside $I_2(S^{2k-1})$ proves that the components of (Q^{k-1}, Ξ_Q) are equal, and the element ω_{k-1} is represented by the framed embedding (Q^{k-1}, Ξ_Q). Theorem 12.1 is proved. □

12.2.4 Brown–Peterson Theorem; James Theorem; Hill–Hopkins–Ravenel Theorem; Snaith Conjecture

Let us formulate results concerning the Strong Kervaire Problem. The following theorem is formulated in [11] as the Brown–Peterson Theorem (1964).

Theorem 12.2 *Assume that for a given $n = 2^{l+1} - 2$, $l \geq 4$. The Strong Kervaire Problem is positively solved (recall that this means that the homomorphism Θ_{2^l-1} : $K^{2^{l+1}-2}_{2^l-1,2^l-3} \to \mathbb{Z}/2$ is onto, or, equivalently, by Theorem 12.1 ω_{2^l-2} is halved) if and only if the Kervaire Problem in dimension $2^l - 2$ is positively solved (this means that the homomorphism $\Theta : \mathrm{Imm}^{\mathrm{sf}}(2^l-3, 1) \to \mathbb{Z}/2$ is onto) and, additionally, there exist an element $x \in \mathrm{Imm}^{\mathrm{sf}}(2^l-3, 1)$, $\Theta(x) = 1$.*

Proof of Theorem 12.2 [The Necessary Part]

We present a new a proof of this theorem. This proof is based on the Hirsch control principle [18] and is a simple illustration of an approach in [6].

Let $g_M : M^{k-1} \looparrowright D^k$ be a framed immersion which is regular cobordant to its mirror image and with $\Theta([f]) = 1$. This means that for the self-intersection immersion $h_K : K^{k-2} \looparrowright D^{k-1}$, $[h_K] \in \mathrm{Imm}^{\mathrm{sf}}(k-2, 1)$, the equation $\Theta([h_K]) = 1$ is satisfied. Let us consider a skew-framed codimension 1 immersion $g_N : N^k \looparrowright D^k \times S^1$, with an integer class κ of its skew-frame (there is an integer lift of the orienting class $w_1(N^k)$) over the middle circle of the target solid torus, such that in the transversal cross-section we get the immersion g_M.

Let us consider a skew-framed codimension k immersion $f : N^k \looparrowright D^{2k-1} \times S^1$, $k = \frac{n}{2}$, $n = 2^{l+1} - 2$, which is the vertical generic lift of the immersion g_N. The normal bundle of f is split as follows: $\nu(f) \cong \kappa \oplus (k-1)\varepsilon$.

We will use the formula

$$(k-1)\kappa_N \mapsto (k-1)\varepsilon, \tag{12.4}$$

which is possible for an arbitrary even positive $k-1$. By the same reason one may assume that the normal bundle of a skew-framed immersion f is split as following: $\nu(f) = k\kappa_N$. Arguments of Theorem 12.1 show that we have to prove that f has an odd number (only one) self-intersection points.

Assume that f is an embedding, and let us prove that this contradicts $\Theta([h_K]) = 1$. Take a twisted submanifold (the product of the Moebius band with D^k) $\tilde{V}^{tw} = D^k \tilde{\times} S^1 \subset V = D^{2k-1} \times S^1$. Take a compression of f into $g_N^{tw} : N^k \looparrowright D^k \tilde{\times} S^1$. This immersion is 1-framed, because κ_N is an integer. Take the self-intersection points immersion $h_L^{tw} : L^{tw;k-1} \looparrowright D^k \tilde{\times} S^1$. The normal bundle of the immersion h_L^{tw} is split: $\nu(h_{L^{tw}}) = p \oplus \varepsilon$, where $p \in H^1(L^{tw;k-1})$ is the class of the canonical covering over a self-intersection manifold. The number of self-intersection of f is calculated as following:

$$\langle p^{k-2}\rho; [L^{tw;k-1}]\rangle = \langle p^{k-1}; [L^{tw;k-1}]\rangle = 0, \tag{12.5}$$

where $\rho = p \otimes \kappa$. The left hand-side number is clear geometrically, because a vertical lift over the vertical line of the Moebius band is required. In the right-hand side of the formula we replace $\rho \mapsto p$. The curve $l \subset L^{tw;k-1}$, which is dual to the class p^{k-2}, cannot represents the fundamental class of the middle circle $S^1 \subset \hat{V}^{tw}$, because $k-1$ is even, l is a cooriented curve in $\hat{V}$, and $\hat{V}$ is non-oriented over S^1. This proves the left-hand side of the formula.

The right hand side of (12.5) is equivalent to the following: a characteristic class $p : L^{tw;k-1} \to \mathbb{RP}^\infty$ is compressed as $p : L^{tw;k-1} \to \mathbb{RP}^{k-2} \subset \mathbb{RP}^\infty$. The characteristic number $\langle p^{k-1}; [L^{tw;k-1}]\rangle$ is a degree of the classifying mapping for p into the skeleton $\mathbb{RP}^{k-1} \subset \mathbb{RP}^\infty$. This degree is even. Preimages of the marked point $\mathbb{RP}^0 \subset \mathbb{RP}^{k-1}$ admit no sign, because we get a not-oriented class κ on $L^{tw;k-1}$.

Take the submanifold $Q^{\frac{k+1}{2}} \subset L^{tw;k-1}$ of the even codimension $\frac{k-3}{2} = 2^{l-1} - 2$, which is dual to $p^{\frac{k-3}{2}}$ (recall, because $\frac{k-3}{2}$ is even, see (12.4), $\rho^{\frac{k-3}{2}} = p^{\frac{k-3}{2}}$). Take the restriction h_Q of h_L^{tw} to $Q^{\frac{k+1}{2}}$. Denote by l the self-intersection curve of h_Q, and assume this curve is connected. The canonical covering $\bar{l}$ over the self-intersection curve l^1 cannot be arbitrary.

A priori there are two cases.

Case I: λ is not the (non-oriented) fundamental class of the middle line S^1. This case is not possible by the assumption $\Theta(x) = 1$.

Case II: λ is the (non-oriented) fundamental class of the middle line S^1. This case corresponds to the assumption $\Theta(x) = 1$. In this case there are two subcases:

Case IIA: the canonical covering $\bar{l} \to l$ is connected;

Case IIB: the canonical covering $\bar{l} \to l$ is disconnected.

Let us prove that in the case $k = 2^l - 1 \geq 15$, Case IIA is not possible; Case IIB is not possible, additionally in the case when the Eq. (12.5) is satisfied.

Let us consider the following diagram:

$$\begin{array}{ccccc} & Q^{\frac{k+1}{2}} & & & \\ & \downarrow \hat{h}_Q & \searrow h_Q & & \\ D^{k-3} \times S^1 = W \subset & D^{k-1}\tilde{\times} S^1 = \hat{V}^{tw} & \subset & D^k \tilde{\times} S^1 = V^{tw} & \\ \cup I \times Id & & & & \\ \mathbb{RP}^{\frac{k-1}{2}} \times S^1 & & & & . \end{array} \qquad (12.6)$$

In this diagram, $\hat{h}_Q : Q^{\frac{k+1}{2}} \to D^{k-1}\tilde{\times} S^1$ is an immersion, because h_Q with a cross-section ε, which is vertical by the Hirsch compression with respect to the normal $\vec{n}$ of the cooriented hypersurface $D^{k-1}\tilde{\times} S^1 \subset D^k \tilde{\times} S^1$. The embedding $I : \mathbb{RP}^{\frac{k-1}{2}} \subset D^{k-3}$ (see the left bottom inclusion in the diagram) in the case $k = 2^l - 1$ exists: by Mahowald (1964) if n=3 mod 4 and n>3, then $\mathbb{RP}^n$ is embeddable in $\mathbb{R}^{2n-2}$ [14].

By the Hirsch control principle we may assume that the cohomology class p for the immersed manifold $\hat{h}_Q$ is controlled over $\mathbb{RP}^{\frac{k-1}{2}}$ (see (12.5)); additionally the controlled embedding is inside the horizontal $D^{k-3} \times S^1 \subset D^{k-2}\tilde{\times} S^1 \subset D^{k-1}\tilde{\times} S^1$. Denote the oriented axis of this hyperband by ε_1.

Take the restriction of the immersion $\hat{h}_Q$ on the hypersection inside $D^{k-1} \times c \subset D^{k-1}\tilde{\times} S^1$, $c \in S^1$. Denote this restriction by $\hat{h}_{Q;c} : Q_c^{\frac{k-1}{2}} \looparrowright D^{k-1} \times \{c\}$. The immersion $\hat{h}_{Q;c}$ for an arbitrary generic c admits an odd number of self-intersection points $\{z_i\}$. For simplicity we may assume that there is the only point $\{z_1(c) = z(c)\}$ for an arbitrary $c \in S^1$. This point is equipped with a sign, assuming the horizontal disk $D^{k-1} \subset D^{k-1}\tilde{\times} S^1$ is oriented. Because of the control, for an arbitrary $z(c)$, we may take a path on $Q_c^{\frac{k+1}{2}}$ with the projection of $\mathbb{RP}^{\frac{k-1}{2}}$ homotopic to a contractible loop λ.

Using this path, two preimages $\{\bar{z}(c)\}$ of $z(c)$ admit normal orientations of immersed manifold up to a simultaneously antipodal transformation, because the control projection $Q_c^{\frac{k+1}{2}} \to \mathbb{RP}^{\frac{k-1}{2}}$ corresponds to the orienting class p of immersion $\hat{h}_{Q;c}$.

Over a point on S^1, the cross-section disk D^{k-2} and the twisted disk D^1 are inside the corresponding cross-section disk $D^{k-1} \times \{c\} \subset \hat{V}^{tw} = D^{k-1}\tilde{\times} S^1$ in the semi-direct product. The monodromy over the middle circle $S^1 \subset \hat{V}$ changes orientation on D^{k-1} and on twisted D^1, and preserves the orientation on D^{k-2}. This means that a co-orientation at $\{\bar{z}(c)\}$ is well defined only up to the monodromy transformation over S^1 generator, which changes this sign for the both preimages.

Assume we get Case IIA. In this case the vertical lift of the preimages $\{\bar{z}(c)\}$ of the only point $z(c) \in l$ is fixed and the order of the preimages is fixed. We have a contradiction, because the monodromy over S^1 has to change the sign of $z(c)$. But this is not possible, because the monodromy is continuous over l.

Assume we get Case IIB. In this case the vertical lift h_Q has a unique self-intersection point $z(c_0) \in l$ on the self-intersection of the immersion $\hat{h}_{Q;c_0}$, and this lift changes the preimages $\bar{z}_0(c_0), \bar{z}_1(c_0)$ near the point $z(c_0) \in l$, because $z(c_0)$ is the only self-intersection point of h_Q.

The canonical double covering $\bar{l}$ over the self-intersection curve l of $\hat{h}_Q$ is a curve of even degree with respect to the S^1-projection. This curve admits a prescribed orientation, if we fix an orientation of the hyperdisk at a marked point on the canonical covering $\bar{l}$. The cycle $[l] \in H_1(S^1; \mathbb{Z}/4)$ by this curve is well defined. Let us call this cycle a self-intersection index of $\hat{h}_Q$. This index is an invariant of a horizontal control generic homotopy of $\hat{h}_Q^{up}$. Let us prove that the self-intersection index is equal to zero.

Take a hyperprojection of $\hat{h}_Q$ on the codimension 2 horizontal hyperplane along the horizontal control axis ε_1. We get a cusps curve Σ. On obscuration by Szucs, this curve is a boundary of the canonical double covering $\bar{l}$. The cobordism between $\bar{l}$ and Σ is defined by the canonical covering over the self-intersection surface Π of the ε_1-projection of $\hat{h}_Q$. We may assume that Π is connected. The canonical covering $\bar{\Pi}$ is equipped with a prescribed orientation outside $\bar{l} \subset \bar{\Pi}$ analogously with $\bar{l}$: two points $\bar{y}_1$, $\bar{y}_2$ on $\bar{\Pi}$ over a common self-intersection point $y \in \Pi$ are joined by an arc λ, which is generic in $H_1(S^1)$. Because of a horizontal control, we get a well defined pair of coorientation on $\hat{h}_Q$ at y and the coorientation of Π at y is well defined. Take a path λ_1, which joins y with a marked point $y_0 \in \bar{l} \subset \Pi$. The path λ_1 is well defined up to a conjugation with an element in $\pi_1(S^1)$. The conjugation by a generic in $H_1(S^1)$ path λ_0 changes the side of $\bar{l}$ in Π. Because λ_0 and λ_0^{-1} are on a common side of y_0, the closed loop $\lambda_0 \circ \lambda_1 \circ \lambda_0^{-1}$ has to preserves the orientation on Π. This implies that the loop λ is not generic. This proves that $\Pi \setminus \bar{\lambda}$ is oriented. In particular, the curve Σ, the second component of the boundary of Π, is canonically oriented.

We get an orientation on Σ and in Case IIB the fundamental class $[\Sigma]$ has to be an even non-trivial element in $H_1(S^1; \mathbb{Z}/4)$. This contradicts the following fact : Take a hypersection immersed manifold $\hat{h}_Q$ inside the horizontal hyperdisk D^{k-1}. The normal bundle of the immersion $\hat{h}_Q$ is $(k-1)p_Q$. The restriction to the hypersection admits a cross-section, because the Euler class of this cross-section is an integer trivial zero-cycle. By the Whitney trick this cycle is empty. The cross-section by the Hirsch compression for $\hat{h}_Q$ inside the hypersection disk is ε_1-vertical by a regular homotopy. This implies that the hypersection ε_1-hyperprojection is regular and has no cusps. The cycle $[\Sigma] \in H_1(S^1; \mathbb{Z}/4)$ is a boundary.

With this fact, Case IIB is not possible. Theorem 12.2 is proved. □

12.2.4.1 James Theorem and Strong Kervaire Problem

Let us formulate the James Theorem [19]. Denote by $V_{k,2}$ the Stiefel manifold $V_{k,2}$ of 2-uple frames in $\mathbb{R}^k$; the most interesting case is $k = 2^l - 1, l \geq 2$. Denote by $I : V_{k,2} \to V_{k,2}$ the involution by the formula: $\{e_1, e_2\} \mapsto \{e_1, -e_2\}$.

Definition 12.2 Let us say that the manifold $V_{k,2}$ is neutral, if the involution I is homotopic to the identity.

Theorem 12.3 (James Theorem) *The manifold $V_{k,2}$, $k = 2^l - 1$, $l \geq 2$ is neutral, if and only if the element $w_{k-1} \in \pi_{2k-1}(S^k)$ is halved (see (12.3)). This condition is equivalent to a positive solution of the Strong Kervaire Problem for $n = 2^l - 2 = k - 1$.*

Proof of Theorem 12.3
Theorem 22.6, pp. 139–141 [19].

Example 12.1 The manifold $V_{k,2}$ is neutral for $k = 7$.

Proof of Example 12.3
Define a homotopy $F(t) : V_{k,2} \to V_{k,2}$, $F(0) = Id$, $F(1) = I$. Consider the standard inclusion $\mathbb{R}^7 \subset \mathbb{R}^8$ which is orthogonal to the base vector $\mathbf{1} \in \mathbb{R}^8$. Let $\mathbf{f}_1 \in \mathbb{R}^7 \subset \mathbb{R}^8$ be a base vector which is orthogonal to the vector $\mathbf{1} \neq \mathbf{f}_1$. Denote by $\mathbf{f}_2$ a vector in the orthogonal complement to the vector $\mathbf{f}_1$ in the subspace $\mathbb{R}^7$. By construction, $\mathbf{f}_2 \in (\mathbf{f}_1)^{\perp}$.

A one-parameter family of orthogonal transformations $G(t, \mathbf{f}_1) : \mathbb{R}^7 \to \mathbb{R}^7$ which transforms the subspace $(\mathbf{f}_1)^{\perp}$ to itself is defined by multiplication of vectors from $(\mathbf{f}_1)^{\perp}$ by the unit $\mathbf{1}\cos(t\pi) + \mathbf{f}_1 \sin(t\pi)$. Denote by $F(t, \mathbf{f}_1) : \mathbb{R}^7 \to \mathbb{R}^7$ the transformation which is the identity in the subspace generated by the vector $\mathbf{f}_1$ and which coincides to $G(t, \mathbf{f}_1)$ on the subspace $(\mathbf{f}_1)^{\perp}$. The homotopy $F(t) = G(t, \mathbf{f}_1)$ is required. □

Example 12.2 The manifold $V_{k,2}$ is neutral for $k = 15$.

Proof of Example 12.2
Let us define the manifold $V_{15,2}$ as a subspace of the space of the pairs $\{(\mathbf{f}_1, \mathbf{f}_2)\}$ of pure imaginary unit sedenions $\mathbb{S}$:

$$f_j = \sum_{i=1}^{i=15} z_{j,i}\mathbf{e}_i \quad j = 1, 2.$$

Let us define the submanifold of zero-divisors:

$$G_2 = \{Z_1, Z_2\} \subset \Im(\mathbb{S}) \times \Im(\mathbb{S}), \quad Z_1 \cdot Z_2 = 0.$$

- For an arbitrary $\{Z_1, Z_2\} \in G_2$, $Z_1 \perp Z_2$, and $G_2 \subset V_{k,2} \subset \Im(\mathbb{S}) \times \Im(\mathbb{S})$. One gets $Z_1 \neq \mathbf{e}_8$, $Z_2 \neq \mathbf{e}_8$, where $\mathbf{e}_8$ is the base imaginary sedenion.

The proof is the following. Assume we get a non-zero coefficient with the vector $\mathbf{e}_8$ in the base decomposition of an imaginary sedenion Z in G_2:

$$\mathbf{f}_1 = A\mathbf{e}_8 + \sum_{l=1, i \neq 8}^{15} z_{1,l}\mathbf{e}_l.$$

Take the (right) product $\mathbf{e}_8 \cdot Z$. This product admits the only coefficient A in the base decomposition of the real component. This proves teht $Z_2 \neq \mathbf{e}_8$. The fact that $Z_1 \perp \mathbf{e}_8$ is analogous.

Let us define a (discontinuous) homotopy $F_1(\theta) : V_{k,2} \setminus G_2 \to V_{15,2} \setminus G_2$; it is not continuous on $G_2 \subset V_{15,2}$, by the formula:

$$Z_2(\theta) = Z_2 \cdot Z_1(\theta),$$

as in Example 12.3. By the property • the vector Z_2 determines a cross-section ξ over G_2 and this cross-section is transversal to the vector $\mathbf{e}_8$. Let us extend the cross-section ξ into a small regular neighborhood of G_2.

Let us modify $F_1(t) \mapsto F_2(t)$ to a homotopy into $\bar{V}_{15,2}$, where the target $\bar{V}_{15,2}$ is the space of all noncollinear pairs of vectors. Take the projection $p : V_{15,2} \to S^{14}$ $p(\mathbf{e}_1, \mathbf{e}_2) = \mathbf{e}_2$. By •, we get: $\mathbf{e}_8 \notin p(G_2)$. Take the cross-section ξ and define $F_2(t) : V_{k,2} \to \bar{V}_{k,2}$ by a linear interpolation as a continuous homotopy near the singular locus G_2 for $t = \frac{1}{2} - 0$, namely, replace the vector $\mathbf{f}_2(\theta)$, $\theta \in (\frac{\pi}{2} - \varepsilon, \frac{\pi}{2})$, $\varepsilon \to 0+$, by a sum with a vector parallel to $\mathbf{e}_8$. By this homotopy, the short arc of the great circle is rotated in the plane $Lin(\mathbf{f}_2(\theta), \mathbf{e}_8)$ and $\mathbf{f}_1$ is outside $Lin(\mathbf{f}_2(\theta), \mathbf{e}_8)$. SInce the product $\mathbf{f}_1 \cdot x$ is not real, $x \in Lin(\mathbf{f}_2(\theta), \mathbf{e}_8)$ is an arbitrary non-zero vector. We get $-F_1(\mathbf{f}_2(\frac{1}{2} - \varepsilon)) = F_1(\mathbf{f}_2(\frac{1}{2} + \varepsilon))$ and the vectors $F_2(\mathbf{f}_2(\theta))$, $\theta \in (\frac{\pi}{2}, \frac{\pi}{2} + \varepsilon)$, $\varepsilon \to 0+$ are defined by the relation $-F_2(\mathbf{f}_2(\frac{1}{2} - \varepsilon)) = F_2(\mathbf{f}_2(\frac{1}{2} + \varepsilon))$.

Because the inclusion $V_{15,2} \subset \bar{V}_{15,2}$ is a homotopy equivalence, by an orthogonalization $F_2(t) \mapsto F(t)$, we get the required homotopy.

The manifold G_2 is well-know. This manifold is a 14-dimensional simply-connected Lie group. The codimension of G_2 equals to $14 + 13 - 14 = 13$. □

It is known that Kervaire's Problem is solved by the positive for $n = 2, 6, 14, 30, 62$. In the case $n \geq 254$ Kervaire's Problem is solved by the negative the Hill–Hopkins–Ravenel Theorem, [17]. A negative solution of Kervaire's Problem except, probably, for a number of finite values of n, is proved by the author in [6], using different geometrical arguments.

The following conjecture [33] is open, but the opposite result to the following is announced in [27].

Conjecture 12.1 (Snaith Conjecture) Kervaire's Problem is negatively solved for $n = 126$.

12.2.5 Solution of Strong Kervaire Problem for $\frac{n}{2} = 15$

Define a closed manifold $M^{15} = S^7 \times S^7 \rtimes S^1$ as a semi-direct product of $S^7 \times S^7$ and the circle S^1 which is fibered over S^1 with $S^7 \times S^7$ fiber. The identification of the fiber $S^7 \times S^7$ along S^1 is given by the involution $Perm : S^7 \times S^7 \to S^7 \times S^7$, which permutes the factors. The orientation class $w_1(M^{15}) = \kappa \in H^1(M^{15}; \mathbb{Z}/2)$ is induced from the generator $H^1(S^1)$ by the projection $M^{15} \to S^1$.

Let us define a generic immersion $\hat{\varphi} : M^{15} \looparrowright \mathbb{R}^{16}$, which is skew-framed by $\Psi : \nu_{\hat{\varphi}} \cong \kappa$. Consider the manifold $K^{15} = \mathbb{RP}^7 \times \mathbb{RP}^7 \tilde{\times} S^1$ which is fibered over S^1 with fiber $\mathbb{RP}^7 \times \mathbb{RP}^7$. The identification of the fiber $\mathbb{RP}^7 \times \mathbb{RP}^7$ along S^1 is given by the involution $\mathbb{RP}^7 \times \mathbb{RP}^7 \to \mathbb{RP}^7 \times \mathbb{RP}^7$, which permutes the factors.

Take a generic vertical lift $\varphi : K^{15} \looparrowright \mathbb{R}^{30}$. The normal bundle $\nu_{\hat{\varphi}}$ is isomorphic to the Whitney sum of an odd number $s = 15$ of copies of the line orienting bundle $\nu_{\hat{\varphi}} = 15\kappa$. We may assume that $s = 15$. Since $2\hat{\kappa} \cong 2\varepsilon$ over $\nu_{\hat{K}}$, the normal bundle can be replaced by $\nu_{\hat{K}} \cong (15 - 2r)\hat{\kappa} \oplus 2r\varepsilon$, where r is an arbitrary integer, $1 \le r \le 7$.

Consider the standard 4-sheeted covering $p : M^{15} = (S^7 \times S^7)\tilde{\times} S^1 \to \hat{K}^{15}$, which is induced by the pair of coverings $S^7 \to \mathbb{RP}^7$. The immersion φ is well defined by the formula $\varphi = \hat{\varphi} \circ p$. A skew-framing Ψ is well defined by the formula: $\Psi = \varphi^*(\hat{\Psi})$. The value $\Theta_{\frac{n}{2}}(\varphi, \kappa, \Psi)$ is calculated as a parity of self-intersection points of the immersion $\varphi = \hat{\varphi} \circ p$, which is deformed $\varphi \mapsto \varphi_1$ into $\mathbb{R}^{30}$ by a small perturbation. Let us prove the formula

$$\Theta^{30}_{15}(\varphi, \kappa, \Psi) = \Theta^{30}_{15}(\varphi) = 1. \tag{12.7}$$

To calculate $\Theta_{15} \in \mathbb{Z}/2$ by (12.7), we describe a self-intersection manifold of the (non-generic) immersion $\varphi : M^{15} \looparrowright \mathbb{R}^{30}$. The self-intersection manifold contains 3 components. A component N_1^{15} consists of points $[(x_1, x_2; y_1, y_2)]$, where $x_1 = x_2$, $y_1 = Ty_2$, or $Tx_1 = x_2$, $y_1 = y_2$, where $T : S^7 \to S^7$ is the standard involution on the factors, x_1, x_2 are points on the same factor in the (local) coordinates system $M^{15} \cong S^7 \times S^7 \times z$, $z \in S^1$. The component N_1^{15} is a semi-direct product of a non-connected manifold L_1^{14} with the circle. Obviously, after a regular generic perturbation of $\varphi \mapsto \varphi_1$, the component gives no contribution to the value Θ_{15}.

To prove this, take the restriction of the line bundle κ_N over N_1^{15} to be the trivial bundle. The normal bundle ν_{N_1} over N_1^{15} in $\mathbb{R}^{30}$ is trivial. After a small generic vertical alteration of the canonical covering $\bar{N}_1^{15} \to N_1^{15}$ in the bundle $15\varepsilon \to N_1^{15}$ we get an embedding $\bar{N}_1^{15} \subset \mathbb{R}^{15} \times N_1^{15}$, because $\langle \kappa_N^{15}; [N_1^{15}] \rangle = 1$.

The second component, denoted by N_2^{15}, comes from the self-intersection of the immersion φ. Obviously, this component gives no contribution to the formula (12.9) below.

The last component is denote by N^{15}. This component consists of points $[(x_1, x_2; y_1, y_2)]$, where $x_1 = Tx_2$, $y_1 = Ty_2$. The manifold is also a semi-direct product of a 14-manifold, denoted by L^{14}, with the circle S^1. The characteristic class of the canonical covering

$$p : \bar{N}^{15} \to N^{15} \tag{12.8}$$

over a self-intersection manifold is described by the characteristic class $p_N \in H_1(N^{15})$. The normal bundle ν_N over N^{15} in $\mathbb{R}^{30}$ is isomorphic to the Whitney sum $\kappa_N \oplus 14\varepsilon$, where κ_N is the line bundle with an integer class. This bundle is a pull-back of the non-trivial line bundle over the circle S^1. To calculate Θ, which is defined by the characteristic number in the left-hand side of the following equation, let us prove that

$$\langle p_N^{15}; [N] \rangle = \langle p_N^{14}(p_N + \kappa_N); [N] \rangle = 1. \tag{12.9}$$

Let us describe explicitly the manifold N^{15} and the submanifold $L^{14} \subset N^{15}$. Take the sphere S^{15} with the standard involution, denoted by $T : S^{15} \to S^{15}$. The factor space S^{15}/T is the standard projective space $\mathbb{RP}^{15}$. Take the sphere S^{15} as the join of the two spheres $S^7 * S^7 = S^{15}$. The involution T is defined as the diagonal involution on the factors $I_j : S^7 \subset S^{15}$, $j = 1, 2$.

The middle solid torus $R^{14} = S^7 \times S^7 \subset S^{15}$ is a T-equivariant submanifold, the restriction of the involution T on R^{14} is defined as the diagonal involution on the factors. Recall that $R^{14}/T = L^{14}$, looks like a T-equivariant thin regular neighborhood of $R^{14} \subset S^{15}$ as $R^{14} \times I$, $I = [1, 2]$. The canonical covering over the manifold N^{15} is defined by the identification of the boundary components $R^{14} \times \{0\} \cup R^{14} \times \{1\}$ of the manifold $R^{14} \times I$ by the permutation involution $Perm$. This manifold is equipped the involution T, because T on each boundary component commutes with the permutation involution $Perm$.

Consider the T-equivariant mapping $H : R^{14} \times I \to S^{15}$ which maps the components of the boundary $R^{14} \times \{1\} \cup R^{14} \times \{2\}$ into the T-equivariant diametral spheres $i_j : S^7 \subset S^{15}$, $j = 1, 2$, which are central manifolds of neighborhoods $S^{15} \setminus R^{14}$. This mapping is extended into a T-equivariant mapping over a thin neighborhood of the boundary of $R^{14} \times I$. This extension has an 8-dimensional support, which is a result of an arbitrary equivariant isotopy of the sphere i_1 to the sphere i_2. The mapping H on the component $R^{14} \times \{1\}$ is equivariant and is given by the composition of the standard projection on the first factor of the product $R^{14} \to S^7$ with the standard inclusion on the first middle submanifod of the solid torus $S^{15} \setminus R^{14}$. The mapping H on the component $R^{14} \times \{1\}$ is equivariant and is given by the composition of the standard projection on the first factor of the product $R^{14} \to S^7$ with the standard inclusion on the second middle submanifod of the solid torus $S^{15} \setminus R^{14}$. This proves that the mapping H extends to a T-equivariant mapping on N^{15} of degree 1. The formula (12.9) is proved.

An analogue of the construction exists also in dimension $\frac{n}{2} = 7$. In this case we get the manifold $N^7 = S^3 \times S^3 \rtimes S^1$ and the immersion $\tilde{\varphi} : N^7 \looparrowright \mathbb{R}^8$, which is defined as an element

$$\eta_3^{\mathrm{sf}} \in \mathrm{Imm}^{\mathrm{sf}}(7, 1). \tag{12.10}$$

The manifold N^7 is included into the following tower of double coverings:

$$N^7 \longrightarrow \hat{N}^7 \longrightarrow \tilde{N}^7. \tag{12.11}$$

We may replace the codimension-6 framing of $\tilde{\varphi}$ by a skew-framing, because the orienting class $\kappa_{\tilde{N}}$ is an integer and is represented by a classifying mapping into the circle. This gives the element $\zeta_3^{\mathrm{sf}} = (\tilde{\varphi}_7, \kappa_7, \Psi_7)$ in the group $\mathrm{Imm}^{\mathrm{sf}}(7, 7)$ with $\Theta_7^{14}(\zeta_3^{\mathrm{sf}}) = 1$. Briefly speaking, the reframing $\eta_3^{\mathrm{sf}} \mapsto \zeta_3^{\mathrm{sf}}$ transforms the Hopf invariant on the group $\mathrm{Imm}^{\mathrm{sf}}(7, 1)$ into the Kervaire invariant on the group $\mathrm{Imm}^{\mathrm{sf}}(7, 7)$.

12.2.6 The Jones Manifold

In the paper [20] a framed closed manifold $\tilde{M}^{30}$ of dimension 30 with Arf-invariant 1 is constructed.

Definition of the Quadratic Form

$$q : H_{2k+1}(\tilde{M}^{4k+2}) \to \mathbb{Z}/2 \tag{12.12}$$

for a framed $4k + 2$-manifold

Let $(\tilde{M}^{4k+2}, \Xi)$ be an arbitrary closed manifold with a prescribed trivialization of the stable normal bundle $\Xi : \nu_{\tilde{M}} \cong r\varepsilon, r - 2 \geq 4k + 2 = dim(\tilde{M})$. The framing Ξ determines up to regular concordance an immersion $\varphi : (\tilde{M}^{4k+2} \setminus pt) \looparrowright \mathbb{R}^{4k+2}$. Let $x \in H_{2k+1}(\tilde{M}^{4k+2})$ be a homology class. By Thom's Theorem, the class x is represented by an embedding $l : K^{2k+1} \subset \tilde{M}^{4k+2}$, where K^{2k+1} is a closed, in general, non-oriented closed $2k + 1$-manifold.

Consider the immersion $\varphi \circ l : K^{2k+1} \subset (\tilde{M}^{4k+2} \setminus pt) \looparrowright \mathbb{R}^{4k+2}$. Define $q(x, l)$ as the parity of self-intersection points of the immersion $\varphi \circ l$. The value $q : H_{2k+1}(\tilde{M}^{4k+2}; \mathbb{Z}/2) \to \mathbb{Z}/2$ depends only on x and not on the cycle $l : K^{2k+1} \subset \tilde{M}^{4k+2}$ which represents x. The form q is quadratic and not degenerate, because this form is associated with the intersection of $2k + 1$-cycles on $\tilde{M}^{4k+2}$. Define the Arf-invariant of the form q by $Arf(q) \in \mathbb{Z}/2$. It is not hard to show that $Arf(q)$ is well defined and also is invariant with respect to framed surgery of the framed manifold $(\tilde{M}^{4k+2}, \Xi)$. By Pontryagin's theorem [31] $Arf(q)$ determines a homomorphism $\Pi_{30} \to \mathbb{Z}/2$.

Assume that the manifold $\tilde{M}^{4k+2}$ is $2k-1$-connected and $k \cong 3 \pmod 4$, $4k+2 \geq 22$. In this case the definition of $q : H_{2k+1}(\tilde{M}^{4k+2}) \to \mathbb{Z}/2$ coincides with a well-known definition from [12]. Namely, assume that the homology class x is represented by an embedded sphere $l : S^{2k+1} \subset M^{4k+2}$. One defines $q(x) = 1$ if the normal bundle $\nu(l)$ of the embedding l is non-trivial, and $q(x) = 0$ if the normal bundle of l is trivial. Let us deduce this definition from the definition of q, given above. The normal bundle $\nu(l)$ is isomorphic to the normal bundle of the immersion $\varphi \circ l : S^{2k+1} \looparrowright \mathbb{R}^{4k+2}$ and the number of self-intersection points of this immersion is odd if and only if the bundle $\nu(l)$ is non-trivial.

12.2.6.1 Construction

Assume $4k+2 = 30$. Let us construct a stably parallelized manifold $\tilde{M}^{30}$, which is called the Jones Manifold [20]. Denote by P^2 the closed non-orientable surface of Euler characteristic $+3$ (the connected sum of the projective plane and the torus). Consider an arbitrary embedding $i_P : P^2 \subset \mathbb{R}^{35}$. Denote by ν_P the normal bundle of the embedding i_P, $\dim(\nu_P) = 33$. Denote by $\nu_0 \subset \nu_P$ a trivial subbundle of the dimension 32, $\nu_0 \cong 32\varepsilon$. Let us write $\nu_P \cong \hat{\nu}_P \oplus \nu_0$, where $\hat{\nu}_P$ is the line orienting normal bundle over P. Decompose the fiber of the bundle ν_0 over the marked point $pt \in P^2$ into 4 blocks and denote these blocks by A, B, C, D. Each block is 8-dimensional.

Consider the decomposition of the surface $P^2 \cong Q^2 \cup_S R^2$, where Q^2 is the projective plane with one hole, R^2 is the torus with one hole, the two surfaces Q^2, R^2 being glued along the common boundary circle S. Denote by $a \in H_1(Q^2)$ the generator on the projective plane and by $b_1, b_2 \in H_1(P^2)$ generators on the torus. Define a flat 4-bundle $\mu : E(\mu) \to P$, the structure group of this bundle being given by the permutation of the 1 and 3 base vectors along the loop on Q^2, represented by a; by the permutation $(1,2)(3,4)$ of the base vectors along the loop, represented by b_1; and by the permutation $(2,3)(4,1)$ of the base vectors along the loop, represented by b_2. The bundle ν_0 is the Whitney sum $\nu_0 = 8\mu$, where μ is the $O(4)$ bundle with the dihedral group $\mathbf{D}$ structure. This bundle is well defined by a representation $\mu : \pi_1(P^2) \to O(4)$, described above using generators.

Consider the bundle $8\varepsilon \otimes \mu$, which is isomorphic to the Whitney sum 8μ of copies of the bundle μ. The bundle 8μ over P^2 is trivial, and the corresponding isomorphism $8\mu \cong \nu_0$ is well defined. The fibers of the bundle 8μ (in, particular, the fiber over the marked point $pt \in P$) are decomposed into 8-bundles as follows: $A \oplus B \oplus C \oplus D$, where A is the 8-dimensional fiber $\{e_1 \otimes \mathbb{R}^8\}$ associated with the first base vector $\{e_1\}$ of μ. Obviously, $\mu \cong \varepsilon \oplus \hat{\mu}$, where $\dim(\hat{\mu}) = 3$. It is not hard to prove that $w_1(\hat{\mu}) \in H^1(P^2)$ is dual to a. The element $w_2(\hat{\mu}) \in H^2(P^2)$ represents the fundamental class on P^2. Therefore we have $\hat{\Xi} : \hat{\nu}_P \oplus \hat{\mu} \cong 4\varepsilon$ and $\Xi : \hat{\nu}_P \oplus \mu \cong \hat{\nu}_P \oplus \hat{\mu} \oplus \varepsilon = 5\varepsilon$. In this formula, $\hat{\Xi}$ can be an arbitrary.

Let us consider the fibre $A \oplus B \oplus C \oplus D$ of the bundle ν_0 over the marked point $pt \in P^2$ and consider the Hopf immersion $h : S^7 \looparrowright \mathbb{R}^8$, which determines a generator of $[\Pi_7]_{(2)} \cong \mathbb{Z}/16$. Consider the Cartesian product $h^4 : (S^7)^4 \looparrowright A \oplus B \oplus C \oplus D$ of 4 copies of the immersion h. Define the required manifold $\tilde{M}^{30}$ to be the semi-direct product $P^2 \tilde{\times} (S^7)^4$, where the dihedral group $\mathbf{D}$ transforms the fiber $(S^7)^4$ by the formulas of μ for the generators a, b_1, b_2. By construction, $\mathbf{D}$ keeps the immersion h^4 and keeps the framing $\hat{\mu} \oplus \varepsilon$ of this immersion.

The immersion $\tilde{M}^{30} \looparrowright E(8\mu)$ into the total space of the bundle 8μ is well defined. Because $E(8\mu)$ is diffeomorphic to the total space of the bundle ν_0, the immersion $f : \tilde{M}^{30} \looparrowright \mathbb{R}^{35}$ is well defined. Obviously, the normal bundle ν_f of the immersion f is isomorphic to the Whitney sum $\hat{\nu}_P \oplus \mu$. By the isomorphism Ξ we get the trivial normal bundle ν_f. The manifold $\tilde{M}^{30}$ is framed and f is a framed immersion, which determines an element in Π_{30}. The Jones Manifold is well defined.

12.2.6.2 Calculation of Hamiltonian Base in the Middle Homology of the Jones Manifold

Let us prove that the vector space $H_{15}(\tilde{M}^{30})$ is 8-dimensional, then determine the Hamiltonian base of this space with respect to the standard bilinear product. Take the marked point $pt \in P^2$ and mark the 7-cycles in the fiber $(S^7)^4$ of the projection $p : \tilde{M}^{30} \to P^2$ by A, B, C, D correspondingly with the subfibers of the vector bundle ν_0. The cycles $A \otimes B$, $A \otimes C$, $A \otimes D$, $B \otimes C$, $B \otimes D$, $C \otimes D$ in $H_{14}((S^7)^4)$ are the base cycles in the fibre $(S^7)^4 = p^{-1}(pt)$, represented by products of two spheres. Denote $H_{14}((S^7)^4)$ by Ω for short.

The punctured surface $P^2 \setminus pt$ is homotopy equivalent to a bouquet of 3 circles: $P^2 \setminus pt \cong S^1_a \vee S^1_{b_1} \vee S^1_{b_2}$, where the circles S^1_a, $S^1_{b_1}$, $S^1_{b_2}$ are marked correspondingly to cycles in $H_1(P^2)$. Along each circle the monodromy is well defined: $\theta_a : \Omega \to \Omega$, $\theta_{b_1} : \Omega \to \Omega$, $\theta_{b_2} : \Omega \to \Omega$. The monodromy θ_a admits a 2-dimensional fixed subspace $A \oplus C$. To the complement $B \oplus D$ of this subspace the monodromy θ_a is represented by the permutation of base vectors $B \leftrightarrow D$. The monodromy θ_{b_1} is represented by the pair of permutations $A \leftrightarrow B$, $C \leftrightarrow D$. The monodromy θ_{b_2} is represented by the pair of permutations $A \leftrightarrow D$, $B \leftrightarrow C$.

We may take the denotations of the cycles such that $[A \times B \tilde{\times} b_1]$, $[C \times D \tilde{\times} b_1]$ are closed 15-cycles over b_1; $[B \times C \tilde{\times} b_2]$, $[D \times A \tilde{\times} b_2]$ are closed 15-cycles over b_2. This means that the monodromy along b_1 of the cycle $[A \times B \tilde{\times} b_1]$ transposes the cycles A and B, etc. Therefore along b_1 there exist extra two invariant cycles $A \otimes C + B \otimes D$, $A \otimes D + B \otimes C$. The subspace of Ω generated by the cycles over b_1 is split into the direct sum of the line invariant subspaces $\{A \otimes B\}$ and $\{C \otimes D\}$. The 4-dimensional complement is generated by the two pairs of cycles $A \otimes C$, $A \otimes D$; $B \otimes C$, $B \otimes D$. The monodromy is split into two isomorphic invariant subspaces: $\{A \otimes C, B \otimes D\}$, $\{A \otimes D, B \otimes C\}$. Each subspace admits the only invariant line described above. This proves that $\dim(H_{15}(\Omega \tilde{\times} S^1_{b_1})) = 4$. Analogously, the monodromy along b_2 is described.

Assume that 7-cycles A, C are invariant with respect to the monodromy θ_a along a. Therefore, the cycles $A \otimes C$, $B \otimes D$, $A \otimes B + A \otimes D$, $C \otimes B + C \otimes D$ are invariant. The 2 pairs of cycles ($A \otimes B$ and $A \otimes C$), ($C \otimes B$ and $C \otimes D$) are transposed by θ_a.

This gives the dimension of the $\mathbb{Z}/2$–vector space: $\dim H_{15}(\tilde{M}^{30} \setminus p^{-1}(U(pt))) = [H_{15}(\Omega \tilde{\times} S^1_{b_1}) \oplus H_{15}(\Omega \tilde{\times} S^1_{b_2}) \oplus H_{15}(\Omega \tilde{\times} S^1_{a})] = 12$. Then let us calculate the dimension of the image of the homomorphism $\partial_* : H_{15}(\partial U(pt)) \to H_{15}(\tilde{M}^{30} \setminus p^{-1}(U(pt)))$, which is induced by the inclusion of the boundary into the manifold with boundary. It is not hard to prove that the image of this homomorphism is 4-dimensional. The dimension $\dim(H_{15}(\partial U(pt)))$ is equals to 6, the 2 base cycles are in the kernel $Ker(\partial_*)$, namely, the cycles $A \otimes B + C \otimes D \otimes [\partial U]$, $B \otimes C + C \otimes D \otimes [\partial U]$. The cycles $A \otimes C \times [\partial U]$, $B \otimes D \times [\partial U]$ are in $Ker(\partial_*)$. Therefore, we get $\dim(H_{15}(\tilde{M}^{30})) = 8$.

Let us calculate the Hamiltonian base in $H_{15}(\tilde{M}^{30})$. This base contains 4 pairs of cycles, where cycles in each pair are dual, cycles from different pairs are orthogonal. By construction, the projection $p : \tilde{M}^{30} \to P^2$ on the base with fiber $(S^7)^4$ is well defined. Denote by $\hat{Q} \subset \tilde{M}$ (over the torus with a hull), $\hat{R} \subset \tilde{M}$ (over the projective space with a hull), $\hat{S} \subset \tilde{M}$ (over the complement $\hat{P} \setminus (\hat{S} \cup \hat{R})$ which is a annulus) preimages by p of corresponding surfaces with common boundary. Take a marked point $pt \in S \subset Q^2$.

Consider the cycle $c_{1,1} = [A \otimes B \tilde{\times} b_1]$ as the preimage of b_1. To get the cycle c_1 closed, we assume that over the marked point pt the cycle b_1 is fibered by $A \otimes B$, because θ_{b_1} permutes the cycles A, B. The cycle $A \otimes B$ is an invariant (non-oriented) cycle and c_1 is a closed non-oriented cycle. Define a cycle $c_{1,2} = [B \otimes C \tilde{\times} b_2]'$ as the corresponding cycle in the preimage of the curve ab_2ab_1, covered by $B \otimes C$ over the started marked point pt. The cycles b_1, ab_2ab_1 have an intersection point y. We assume that this point y is on c_2 and the marked point pt is on b_2 and is between the two copies of the segments a.

We get the following sequence of monodromy transformations:

$$B \otimes C \xrightarrow{\theta_a} D \otimes C \xrightarrow{\theta_{b_2}} A \otimes B \xrightarrow{\theta_a} A \otimes D \xrightarrow{\theta_{b_1}} B \otimes C.$$

Over the point y we get the 14-dimensional cycle $A \otimes B$ on $c_{1,1}$ and $D \otimes C$ on $c_{1,2}$. With this assumption, $c_{1,2}$ over the point y is represented by $C \otimes D$ and the cap-product $c_{1,1} \cap c_{1,2}$ is equal to 1. Cycles $c_{1,1} = [A \otimes B \tilde{\times} b_1]$, $c_{1,2} = [B \otimes C \times b_2]'$ determine a Hamiltonian pair.

Analogously, the pair $c_{2,1} = [B \otimes C \tilde{\times} b_2]$, $c_{2,2} = [A \otimes B \times b_1]'$ is well defined. The cycle $[B \times C \times b_2]$ is well defined, because the cycle $B \otimes C$ is invariant along the curve b_2. The cycle $[A \otimes B \times b_1]'$ is the result of an extension of $A \otimes B$ along the curve ab_1ab_2 from pt. The cycles $c_{2,1} = [B \otimes C \tilde{\times} b_2]$, $c_{2,2} = [A \otimes B \times b_1]'$ determine a Hamiltonian pair.

Analogously, the pair $c_{3,1} = [C \otimes D \tilde{\times} b_1]$, $c_{3,2} = [D \otimes A \times b_2]'$ is well defined. The cycle $[A \otimes D \times b_2]'$ is defined as a cycle in the preimage of the curve ab_2ab_1, which coincides with the product $A \otimes B$ over $y = b_1 \cap ab_2ab_1$. As in the previous example, we get $c_{3,1} \cap c_{3,2} = 1$.

Finally, the last Hamiltonian pair is determined by the following cycles: $c_{4,1} = [A \otimes C \times a]$ and $c_{4,2} = [B \otimes D \tilde{\times} a]$.

It is not hard to prove that the intersection numbers between cycles of different pairs equal to zero. For example, the curve b_1, the projection of the first cycle $c_{1,1} = [A \otimes B \tilde{\times} b_1]$, intersects the curve ab_2ab_1, which is the projection of the cycle $c_{2,2} = [A \otimes D \times b_2]'$ at a point $y = b_1 \cap b_2$. Over this point the cycle $c_{1,1}$ is given by $A \otimes B$, the cycle $c_{2,2}$ is given by $A \otimes B$ and we get empty generic intersection, because $\langle A \otimes B, A \otimes B \rangle = 0$. The equation $\langle c_{1,1}, c_{2,1} \rangle = 0$ is evident.

The curve b_2, the projection of the cycle $c_{2,1} = [B \otimes D \tilde{\times} b_2]$, intersects the curve ab_2ab_1, which is projection of the cycle $c_{1,2} = [B \otimes C \tilde{\times} b_2]'$ at a point $y = b_1 \cap b_2$. Over this point the cycle $c_{2,1}$ is given by $B \otimes D$, the cycle $c_{1,2}$ is given by $A \otimes D$ and we get the empty generic intersection of this cycles, because $\langle A \otimes D, B \otimes D \rangle = 0$. The equation $\langle c_{1,2}, c_{2,2} \rangle = 0$ is evident.

The curve b_1, the projection of the cycle $c_{1,1} = [A \otimes B \tilde{\times} b_1]$, intersects the curve ab_1ab_2, which is the projection of the cycle $c_{3,2} = [D \otimes A \times b_2]'$ at a point $y = b_1 \cap b_2$. Over this point the cycle $c_{1,1}$ is given by $A \otimes B$, the cycle $c_{3,2}$ is given by $A \otimes B$ and we get empty generic intersection, because $\langle A \otimes B, A \otimes B \rangle = 0$. The equation $\langle c_{1,1}, c_{3,1} \rangle = 0$ is evident.

The curve b_1, the projection of the cycle $c_{3,1} = [C \otimes D \tilde{\times} b_1]$ intersects the curve ab_2ab_1, which is the projection of the cycle $c_{1,2} = [B \otimes C \times b_2]'$ at a point $y = b_1 \cap b_2$, over this point the cycle $c_{3,1}$ is given by $C \otimes D$, the cycle $c_{1,2}$ is given by $C \otimes D$ and we get the empty generic intersection of this cycles, because $\langle C \otimes D, C \otimes D \rangle = 0$. The equation $\langle c_{1,2}, c_{3,2} \rangle = 0$ is evident.

The curve b_2, the projection of the cycle $c_{2,1} = [B \otimes C \tilde{\times} b_2]$, intersects the curve ab_2ab_1, which is the projection of the cycle $c_{3,2} = [D \otimes A \times b_2]'$ at a point $y = b_1 \cap b_2$, over this point the cycle $c_{2,1}$ is given by $B \otimes C$, the cycle $c_{3,2}$ is given by $B \otimes C$ and we get the empty generic intersection, because $\langle B \otimes C, B \otimes C \rangle = 0$. The equation $\langle c_{2,1}, c_{3,1} \rangle = 0$ is evident.

The curve b_21, the projection of the cycle $c_{3,1} = [C \otimes D \tilde{\times} b_1]$, intersects the curve ab_1ab_2, which is the projection of the cycle $c_{2,2} = [A \otimes B \times b_1]'$ at a point $y = b_1 \cap b_2$. Over this point the cycle $c_{3,1}$ is given by $C \otimes D$, the cycle $c_{2,2}$ is given by $C \otimes D$ and we get the empty generic intersection of this cycles, because $\langle C \otimes D, C \otimes D \rangle = 0$. The equation $\langle c_{1,2}, c_{3,2} \rangle = 0$ is evident. Evidently, the cycle $c_{4,1}, c_{4,2}$ determines a Hamiltonian pair and vectors of the pairs are orthogonal to vectors $c_{i,j}, i = 1, 2, 3, j = 1, 2$, constructed above.

The quadratic form on $H_{15}(\tilde{M}^{30})$ is well defined. This form is split into a pairwise orthogonal direct sum of the 4 Hamiltonian 2-dimensional forms over $c_{1,1} \oplus c_{1,2}$, over $c_{2,1} \oplus c_{2,2}$, and over $c_{3,1} \oplus c_{3,2}$ over $c_{4,1} \oplus c_{4,2}$.

Calculation of the Arf-invariant of the Jones Manifold

Let us calculate the quadratic form q for vectors in the Hamiltonian base of $H_{15}(\tilde{M}^{30})$. One can prove that $q([A \times C \times a]) = 0$; $q([A \times C \times b_1]') = 1$. This Hamiltonian pair gives no contribution to the Arf-invariant.

The equation $q([A \times C \times a]) = 0$ is satisfied, because the cycle is represented by a framed 15-manifold, and $q([A \times C \times a])$ coincides with the Hopf invariant. By the Toda Theorem [35] the Hopf invariant in dimension 15 is trivial. The equation $q([A \times C \times b_1]') = 1$ is proved using results of Sect. 12.2.5. The cycle considered is represented by the manifold $(S^7 \times S^7)\tilde{\times} S^1$. This manifold is diffeomorphic to the manifold $\tilde{M}^{15}$, this manifold if immersed into $\mathbb{R}^{30}$ by $\varphi : \tilde{M}^{15} \looparrowright \mathbb{R}^{30}$. By construction, the restriction of the immersion φ to the submanifold $S^7 \times S^7 \subset (S^7 \times S^7)\tilde{\times} S^1$ coincides with the immersion of the submanifold $\tilde{L}^{14}$, as in the example. This immersion represents the element $(h_7)^2 \in \Pi_{14}$, where $h_7 \in \Pi_7$ is a generator.

Using analogous consideration one may prove that Arf-invariant for each last Hamiltonian pair equals to 1. As a result, we get that the Jones framed manifold $(\tilde{M}^{30}, \Xi)$ has Arf-invariant 1.

12.2.7 Browder Theorem, Eccles Theorem: The Positive Solution of Kervaire Problem in Dimension n = 30

An element α in the stable homotopy group of spheres $\Pi_{30} = \pi_{n+30}(S^n)$, $n \geq 32$, is represented by a framed manifold $(\tilde{M}^{30}, \Xi)$. By the Browder theorem [12] a framed manifold $(\tilde{M}^{30}, \Xi)$ has Arf-invariant one iff the element α in Π_{30} is detected by a secondary cohomological operation, which is based on the Adem relation in the Steenrod algebra:

$$Sq^{2^j} Sq^{2^j} + \sum_{i=1}^{j-1} Sq^{2^{j+1}-2^i} Sq^{2^i} = 0, \quad j = 4. \tag{12.13}$$

By Eccles's Theorem [16] the element α is detected by the considered cohomology operation, if and only if the element $\beta = \lambda_*(\alpha) \in \pi_{29+n}(\Sigma^n(K(\mathbb{Z}/2, 1)))$, which is associated with α by the Khan–Priddy mapping $\lambda : Q\mathbb{RP}^\infty \to Q(S^0)$, is detected by the operation Sq^{16} in the cone of the mapping β using the generic cohomology class in $H_{n+15}(\Sigma^n(\mathbb{RP}^\infty))$. By the Pontyagin–Thom Construction, the element β is represented by an immersion $f : M^{29} \looparrowright \mathbb{R}^{30}$ in codimension 1 of a closed, in general, non-orientable manifold. Moreover, the characteristic number $\Theta(f) = 1$, if and only if Sq^{16} detects β. In the presented form the statement is used in [8].

Because the Jones Manifold is a framed manifold of the Arf-invariant one, there exists an immersion $F : M^{39} \looparrowright \mathbb{R}^{30}$ with $\Theta(f) = 1$. Therefore the Kervaire Problem is positively solved in dimension $n = 30$.

12.2.8 Eccles-Wood Filtration: Estimation of the Filtration for a Standard Kervaire 30-Manifold

The fundamental paper [22] is dedicated to the applications of homotopy theory in differential topology. In this paper, Kervaire constructed a closed 4-connected PL-manifold of dimension 10, which admits no smooth structure. In the dimension 30 the analogous 14-connected manifold is smooth, and is called a standard Kervaire manifold. A standard Kervaire manifold is defined as a result of a normal surgery of the Jones manifold $\tilde{M}^{30}$ into a 14-connected manifold N^{30}, $dim(H_{15}(\tilde{N}^{30};\mathbb{Z}/2)) = 2$. A standard Kervaire manifold is well defined up to a connected sum with a homotopy 30-sphere. By construction, N^{30} is framed. It is well-known, that $\tilde{N}^{30}$ admits a PL-embedding into $\mathbb{R}^{32}$.

Theorem 12.4 *There is no smooth embedding $N^{30} \times D^{14} \subset \mathbb{R}^{46}$, where N^{30} is a standard Kervaire manifold, and D^{14} is the standard disk.*

To prove the theorem one has to estimate the Eccles–Wood filtration [15] for a standard Kervaire manifold. Let N^n be a closed framed manifold (let us restrict to the case $n \equiv 0 \pmod 2$), with a prescribed trivialization of the stable normal bundle: $\Xi : \nu_N \cong r\varepsilon$ of dimension r, $r \geq n+2$. One says that (N^n, Ψ) admits the Eccles-Wood filtration, at least (k_+, k_-), $n > k_+ > \frac{n}{2} \geq k_- > 0$, if there exists an embedding $N^n \subset \mathbb{R}^{2n-k_-}$ with the normal vector field ψ of dimension $k_+ - k_-$ (of codimension $n - k_+$).

The following statement is equivalent to Theorem 12.4.

Proposition 12.4 *The standard Kervaire manifold (N^{30}, Ξ) does not admit a Eccles–Wood filtration* (28, 14).

Remark

In [9] a stronger result is proved: the standard Kervaire manifold (N^{30}, Ξ) does not admit a Eccles–Wood filtration (23, 14).

Proof of Proposition 12.4
Assume, there exists an embedding $f : N^{30} \subset \mathbb{R}^{46}$, for which the normal bundle ν_f admits 14 linear independent sections. Prove that the immersion f cannot be an embedding. By assumption, an isomorphism of the normal bundle $\nu_f \cong \eta \oplus 14\varepsilon$, where η is a real 2–bundle, is well defined. Because $\pi_1(N^{30}) = \pi_2(N^{30}) = 0$, the bundle η is a trivial bundle. Therefore a framed immersion (f, Ξ), $\Xi : \nu_f \cong 16\varepsilon$ is well defined. It is well known that the Arf-invariant for a framed manifold $(N^{30}, \Xi) \in \Pi_{30}$ does not depend on a framing [21]. Therefore for the framed immersion (f, Ξ) we get $Arf(f, \Xi) = 1$.

Let us consider a hyperprojection $\pi : \mathbb{R}^{46} \to \mathbb{R}^{45}$ and consider the mapping $g = \pi \circ f$. By the compression theorem [32] this mapping could be defined as a framed immersion (by a framing $\hat{\Xi}$, where $\hat{\Xi} \oplus Id_\varepsilon \cong \Xi$), which represents an element $[(g, \Psi)])$, in the cobordism group $\mathrm{Imm}^{\mathrm{sf}}(15, 15) \cong \Pi_{15}$.

Consider the self-intersection manifold of the framed immersion $g : N^{30} \looparrowright \mathbb{R}^{45}$. By construction, $[(g, \Psi)] \in K^{30}_{29,14}$ (a priori we have $[(g, \Psi)] \in K^{30}_{28,14}$). The first index k_1 of the filtration is maximal, therefore by Browder and Eccles Theorems the equation $\Theta([(g, \Psi)]) = 1$ is satisfied, because $Arf(g, \hat{\Xi}) = 1$. The second index of the filtration shows that Ψ is a frame, and by Proposition 12.3 we get $\Theta([(g, \Psi)]) = 0$. Therefore an embedding $N^{30} \times D^{14} \subset \mathbb{R}^{46}$ cannot exists. □

12.2.9 Stably-Framed Immersions

In [3] the cobordism group of stably-framed immersions $\mathrm{Imm}^{st-fr}(n-k, k)$ for a non-negative integer $n-k$ and a positive integer k is introduced. An element of the group $\mathrm{Imm}^{st-fr}(n-k, k)$ is represented by a pair $(f, \tilde{\Xi})$, where $f : M^{n-k} \looparrowright \mathbb{R}^n$ is an immersion, $\tilde{\Xi} : \nu(I \circ f) \cong (k+N)\varepsilon$, N is a stable framing, i.e. a prescribed isomorphism of the normal bundle of the immersion $I \circ f : M^{n-k} \looparrowright \mathbb{R}^n \subset \mathbb{R}^{n+N-k}$ with the trivial N-dimensional bundle, where N is a great positive integer. The cobordism relation between pairs of stable framed immersions is a standard equivalece relation on the set of stably-framed immersions. The equivalence classes equipped with the operation of disjoint summation determines an abelian group, which was denoted above.

By the Pontryagin–Thom construction, there exists a natural isomorphism

$$\mathrm{Imm}^{st-fr}(n-k, k) = \pi_{N+n}(V_{n+N,n+N-k})$$

and the Hurewicz homomorphism:

$$\Gamma : \mathrm{Imm}^{st-fr}(n-k, k) \mapsto H_{n+N}(V_{N,N-k}\mathbb{Z}),$$

where $N >> 1$ as above and $V_{N,N-k}$ is the Stiefel manifold of $N-k$ framings in the Euclidean space $\mathbb{R}^N$. The homomorphism Γ is an isomorphism over the rationals Q and the homology groups $H_{n+N}(V_{N,N-k}\mathbb{Z})$ are well-known. There is a generator for $k = n = 2l$ and for $n = 4l - 1$, $2k + 1 < n$.

If we assume in the definition $\mathrm{Imm}^{\mathrm{st-fr}}(n-k, k)$ that an immersion f is an embedding, we obtain the cobordism group of stably-framed embeddings, denoted by $\mathrm{Emb}^{\mathrm{st-fr}}(n-k, k)$. The following theorem [3] describes a relationship between the two groups, which are related by a natural homomorphism:

$$forg : \mathrm{Emb}^{\mathrm{sf-fr}}(n-k, k) \to \mathrm{Imm}^{\mathrm{st-fr}}(n-k, k).$$

Theorem 12.5

1. For an arbitrary n, k, except the case $n = k = 2l$, the homomorphism

$$forg^{\mathbf{Q}}\mathrm{Emb}^{\mathrm{st-fr}}(n-k, k) \otimes \mathbf{Q} \longrightarrow \mathrm{Imm}^{\mathrm{st-fr}}(n-k, k) \otimes \mathbf{Q}$$

is an isomorphism.

2. *The homomorphism*

$$forg^{\mathbf{Q}} : \mathrm{Emb}^{\mathrm{st-fr}}(2l, 2l) \otimes \mathbf{Q} \longrightarrow \mathrm{Imm}^{\mathrm{st-fr}}(2l, 2l) \otimes \mathbf{Q}$$

is trivial with kernel **Q**.

A relationship between skew-framed immersions $\mathrm{Imm}^{\mathrm{sf}}(n-k, k)$ and stably-framed immersions $\mathrm{Imm}^{\mathrm{sf-fr}}(n, k)$ is given by the generalized Khan–Priddy transfer construction, which is recalled. The generalized Khan–Priddy transfer is a homomorphism

$$\lambda : \mathrm{Imm}^{\mathrm{sf}}(n-k, k) \longrightarrow \mathrm{Imm}^{\mathrm{st-fr}}(n, k), \qquad (12.14)$$

which is defined as follows. Take the triple (g, κ, Ψ), $g : M^{n-k} \looparrowright \mathbb{R}^n$, $\Psi : \nu(g) \cong k\kappa$, which represents an element in $\mathrm{Imm}^{\mathrm{sf}}(n-k, k)$. The construction is a generalization of the Koschorke construction of the Khan–Priddy homomorphism, because for $k = 1$ the composition of the homomorphism λ with the forgetful homomorphism $\mathrm{Imm}^{st-fr}(n, 1) \to \mathrm{Imm}^{fr}(n, 1)$ coincides with the Koschorke construction.

Take the codimension k stabilization $\tilde{g} : M^{n-k} \looparrowright \mathbb{R}^{n+k}$, $\tilde{g} = I \circ g$, $I : \mathbb{R}^n \subset \mathbb{R}^{n+k}$. There is an isomorphism $\nu(\tilde{g}) \cong k\kappa \oplus k\varepsilon$. Take a point $z \in M^{n-k}$ and consider the fiber $\nu(\tilde{g})_z$, which is represented by the ordered Whitney sum $\zeta(z) \oplus \zeta^{tw}(z)$, where $\zeta(z)$ is the fiber of the bundle $k\varepsilon$ over z and $\zeta^{tw}(z)$ is the bundle $k\kappa$ over z. Take another decomposition of the fiber into a non-ordered Whitney sum of two isomorphic k-dimensional bundles, $\eta_a(z) \oplus \eta_b(z)$, where $\zeta(z)$ is the bisector bundle of this decomposition, formed by the vectors $x_a \oplus x_b$, with respect to the isomorphism $\eta_a(z) \to \eta_b(z)$. Then $\zeta^{tw}(z)$ is the orthogonal bundle to $\zeta(z)$, formed by vectors $x_a \oplus -x_b$.

Take a disk bundle $D_a \rtimes D_b$, associated with the vector bundle $\eta_a \oplus \eta_b$ over M^{n-k}. In each fiber $[D_a \rtimes D_b](z)$ take a sphere $S^k(z) \subset [D_a \rtimes D_b](z)$, which is defined as the join of the two $k-1$-dimensional spheres $\partial D_a(z)$, $\partial D_b(z)$, the boundaries of the disks in the fiber over a point z. The sphere $S^k(z)$ is embedded into $\zeta(z) \oplus \zeta^{tw}(z)$ as the PL-sphere. We do a smoothing of the corners over $\partial D_a(z) \cup \partial D_b(z)$ to get a smooth immersed sphere with only self-intersection points at the origin of the fiber. The collection of spheres $S^k(z)$ determines a bundle over M^{n-k} with the fiber S^{n-k}. The total space of this bundle is denoted by $\tilde{M}^n$. Let us prove that the immersion $\tilde{f} : \tilde{M}^n \looparrowright \mathbb{R}^{n+k}$ is equipped with a stable trivialization of the normal bundle, namely,

$$\tilde{\Xi}^1 : \nu(\tilde{f}) \oplus \varepsilon \cong (k+1)\varepsilon. \qquad (12.15)$$

Let us investigate the normal bundle $\nu(\tilde{f})$ of the immersion $\tilde{f}$. The normal bundle over a sphere $S^k(z)$ is naturally isomorphic to the tangent bundle $T(S^k(z))$ and the isomorphism $\tilde{\Xi(z)} : T(S^k(z)) \oplus \varepsilon \cong \nu(\tilde{f}(z)) \oplus \varepsilon \longrightarrow (k+1)\varepsilon$ is well defined. Let us construct the total family $\tilde{\Xi}(z)$ over the base M^{n-k}. Take one of the two self-intersection points at the origin $w_a \in S^k(z)$ in the fiber $\zeta(z) \oplus \zeta^{tw}(z)$. This choice corresponds to a prescribed coordinate $\zeta^{tw}_a(z)$ in the fiber $\zeta^{tw}(z)$, or, to an order of fibers $\eta_a(z)$, $\eta_b(z)$. The bundle $\nu(\tilde{f})(z)$ over the point $w_a \in S^k(z)$ is isomorphic to $\zeta^{tw}(z)_a$ and the fiber $\nu(\tilde{f})|_{w_a}$ is described by this isomorphism. In particular, over the antipodal point $w_b = -w_a \in S^k(z)$ the fiber $\nu(\tilde{f})|_{w_b}$ is described by $-\zeta^{tw}_a$. This implies that the two fibers of the bundle $\nu(\tilde{f}) \oplus \varepsilon$ over w_a and w_b, which a priori are related by the antipodal transformation, admit antipodal coordinates description as disk bundles, associated with the vector bundle ζ^{tw}. The isomorphism (12.15) is well defined.

Take the stabilization

$$\tilde{\Xi} : \nu(\tilde{f}) \oplus N\varepsilon \cong (k+N)\varepsilon. \tag{12.16}$$

of the stable framing (12.15). The correspondence $(g, \kappa, \Psi) \mapsto (\tilde{f}, \tilde{\Xi})$ determines the homomorphism (12.14).

The most interesting case of the homomorphism (12.14) concerns $n = 4l + 2$, $k = 2l + 1$. In this case we get:

$$\lambda : \mathrm{Imm}^{\mathrm{sf}}(2l+1, 2l+1) \longrightarrow \mathrm{Imm}^{st-fr}(4l+2, 2l+1). \tag{12.17}$$

There is a homomorphism

$$\Theta^{tw} : \mathrm{Imm}^{\mathrm{st-sf}}(4l+2, 2l+1) \to \mathbb{Z}/2,$$

which is called the Browder–Eccles invariant. This invariant is included into the following commutative diagram:

$$\begin{array}{ccc} \lambda : \mathrm{Imm}^{\mathrm{fr}}(4l+2, 1) & \longrightarrow & \mathrm{Imm}^{\mathrm{st-sf}}(4l+2, 2l+1) \\ \Theta \searrow & & \swarrow \Theta^{tw} \\ & \mathbb{Z}/2, & \end{array} \tag{12.18}$$

where $\Theta : \mathrm{Imm}^{\mathrm{fr}}(4l+1, 1) \to \mathbb{Z}/2$ is the Kervaire invariant. The Browder–Eccles invariant is defined as follows.

Let $(f, \tilde{\Xi})$, $f : M^{4l+2} \looparrowright \mathbb{R}^{6l+3}$ be an immersion, $\Xi : \nu(\tilde{f}) \oplus N\varepsilon \cong (2l+1+N)\varepsilon$ be a stable framing of $\tilde{f} = I \circ f : M^{4l+2} \looparrowright \mathbb{R}^{6l+3} \subset \mathbb{R}^{6l+3+N}$, and assume $2l+1 \geq 9$. Let q be the quadratic form for the framing in the codimension $2l+1+N$, defined by (12.12). Let us denote by

$$\xi : H_{2l+1}(\tilde{M}^{4l+2}; \mathbb{Z}/2) \to \mathbb{Z}/2 \tag{12.19}$$

a linear form defined as follows. Take a cycle $[x] \in H_{2l+1}(\tilde{M}^{4l+2}; \mathbb{Z}/2)$ and represent this cycle by the image of the fundamental class by the mapping $x : S^{2l+1} \longrightarrow \tilde{M}^{4l+2}$ of a closed manifold. The restriction of the normal bundle of $\tilde{f}$ over the submanifold $x(S^{2l+1}) \subset \tilde{M}^{4l+2}$ is a stable-trivial $2l+1$-dimensional bundle. Up to an isomorphism, over a $2l+1$-dimensional manifold there are only two stable-trivial $2l+1$-dimensional bundles. Let us define $\xi(x)=1$ if the bundle $\nu(\tilde{f})|_{x(S^{2l+1})}$ is non-trivial and define $\xi(x)=0$, if this bundle is trivial. The mapping (12.19) is well defined. This mapping is a homomorphism.

Definition 12.3 Define the twisted quadratic form

$$q^{tw}: H_{2l+1}(\tilde{M}^{4l+2}) \to \mathbb{Z}/2$$

by the formula:

$$q^{tw} = q + \xi$$

(because q is a quadratic form and ξ is a linear form, $q+\xi$ is a quadratic form). The Browder–Eccles invariant $Arf^{tw}(f, \tilde{\Xi})$ is defined as the Arf-invariant of q^{tw}.

Example 12.3 Let $i : S^{2k+1} \subset \mathbb{R}^{6k+3}$, $2k+1 \geq 9$ be the standard inclusion, the normal bundle of this inclusion may be represented as the Whitney sum of two isomorphic copies of $2k+1$dimensional bundle ξ_S, each copy is a stably-trivial bundle, but not the trivial bundle. Let us consider two bisector bundles ξ_1, ξ_2, each bundle is isomorphic to ξ_S, we get: $\xi_1 \oplus \xi_2 \cong \xi_{S,1} \oplus \xi_{S,2}$. Let us take two disk bundles $\bar{\xi}_1, \bar{\xi}_2$, which are associated with ξ_1, ξ_2. The cylinder over the boundary determines the fiberized immersion $S^{2k+1}_s \looparrowright (\xi_1 \oplus \xi_2)_s$ over $s \in S^{2k+1}$. Denote this immersion by $I : S^{2k+1} \times S^{2k+1} \looparrowright \mathbb{R}^{6k+3}$, this is a 1-stably-framed immersion. The following formula is satisfied:

$$Arf^{tw}(i, \tilde{\Xi}) = 0. \tag{12.20}$$

Proof of Example 12.3
Take a Hamiltonian basis of cycles on $T^{4k+2} = S^{2k+1} \times S^{2k+1}$, which is defined by the parallel x (the horizontal cycle, which is parallel to $i(S^{2k+1})$) and the meridian y (the vertical cycle in the normal section) of the torus. Let us prove (12.20).

The cycle $x \times \{pt\} \subset S^{2k+1} \times S^{2k+1}$ is defined as the result of a shift of the central manifold $i(S^{2k+1})$ along a generic section of the bundle ξ_1 onto the immersed sphere $I(S^{2k+1} \times \{pt\})$ in the corresponding fiber. The normal bundle $\nu_I|_x$ is isomorphic to the pull-back $\pi^*(\xi_2)$ by the projection $\pi : I(T^{4k+2}) \to i(S^{2k+1})$, therefore $\xi(x) = 1$. The normal bundle of this cycle inside T^{4k+2} is isomorphic to $\pi^*(\xi_1)$, therefore $\theta(x) = 1$. As a result, the form $q^{tw}(I, \tilde{\xi})([]x]) = 0$, this proves (12.20).

□

In [9] the following theorem is proved.

Theorem 12.6 *For an arbitrary positive integer l, there exists a homomorphism*

$$\sigma : \mathrm{Imm}^{\mathrm{st-fr}}(4l+2, 2l+1) \longrightarrow \mathrm{Imm}^{\mathrm{sf}}(2l+1, 2l+1),$$

such that the following diagrams are commutative:

$$\begin{array}{ccc} \sigma : \mathrm{Imm}^{\mathrm{st-fr}}(4l+2,1) & \longrightarrow & \mathrm{Imm}^{\mathrm{sf}}(2l+1,2l+1) \\ \Theta^{tw} \searrow & & \swarrow \Theta^{\mathbf{D}} \\ & \mathbb{Z}/2, & \end{array} \tag{12.21}$$

$$\begin{array}{ccc} \lambda : \mathrm{Imm}^{\mathrm{sf}}(2l+1,2l+1) & \longrightarrow & \mathrm{Imm}^{\mathrm{st-fr}}(4l+2,2l+1) \\ \Theta^{\mathbf{D}} \searrow & & \swarrow \Theta^{tw} \\ & \mathbb{Z}/2. & \end{array} \tag{12.22}$$

Corollary 12.1

1. *There exists a stable-framed immersion* $\tilde{f} : \tilde{M}^{30} \looparrowright \mathbb{R}^{45}$ *with Browder–Eccles invariant* 1.
2. *Assume Problem 12.2 is positively solved. Then there exists infinitely many dimensions* $n = 4l + 2$ *for which there exists a stably-framed immersions* $(\tilde{f}, \tilde{\Xi})$ *with Browder–Eccles invariant* 1.

Statement 2 of Corollary 12.1 is completely opposite to the case of framed immersion, described by the Hill–Hopkins–Ravenel Theorem.

Proof of Corollary 12.1
Statement 1 can be proved by the following different two ways a and b.

a. Take the framed Jones manifold, constructed in Sect. 12.2.6. By the diagram (12.35) this manifold determines a stable-framed immersion with Browder–Eccles invariant 1.
b. Take the skew-framed immersion, constructed in Sect. 12.2.5, see the formula (12.7), which defines an element in $\mathrm{Imm}^{\mathrm{sf}}(15, 15)$ with Kervaire invariant 1. By the diagram (12.38) the image of this skew-framed immersion by the homomorphism λ determines a stable-framed immersion with Browder–Eccles invariant 1.

Statement 2 is proved analogously with b. □

12.2.10 Applications: Spaces of Functions with Moderate Singularities; Perfect Immersions

Let us consider the space F of functions on the real line, fixed at infinity and with no singularities of type A_4 (only Morse type A_2 and birth-death type A_3). By [10], [36] ch.3 Lemma 5, there is a natural homomorphism

$$\pi_2(F) \to \pi_3(S^2) = \mathbb{Z}.$$

A Cerf diagram is an immersed surface (with cusp curve) in the 3D-space of values+parameters. A Cerf diagram determines an element in $y \in \Pi_2$. A negative cusp curve determines an element $x \in \Pi_1$.

Conjecture The product of y with the Hopf generator $t \in \Pi_1$ gives the following relation in Π_2:

$$x \circ t = y. \tag{12.23}$$

In [4], the following is proved. Let us take $M \times I$, where M is a closed manifold of sufficiently large dimension ($\dim(M) \geq 5$ is sufficient). Let G be the space of functions $M \times I \rightarrow I_Y$, which are with A_2, A_3 type singularities and which map the bottom part of the boundary $M \times \{0\}$ into $\{0\} \in I_Y$, the lower part of the boundary $M \times \{1\}$ into $\{1\} \in I_Y$. Assume that the families of stable and unstable separatrices admits an orientation everywhere, outside the cusp curve on the Cerf diagram. Then $x \in \Pi_1$ and $y \in \Pi_2$ are well defined and Eq. (12.23) is satisfied.

In particular, the Konstantinov torus is a surface immersed in $\mathbb{R}^3$ with $Arf = 1$ and, because each cusp of the projection on $\mathbb{R}^2$ is regular homotopic to the standard embedding circle, we get $x \in \Pi_1$ is trivial. The Konstantinov torus cannot be a Cerf diagram of an element $\pi_2(F)$.

Assume that an immersion of the torus T^2 into $\mathbb{R}^3$ relates with the generator of the steam Π_2 [13]. The Konstantinov torus is an immersion, which determines the generator in Π_2, and is a ∞-perfect immersed surface (each component of self-intersection curve on the torus is null-homotopic). This surface is not regular homotopic to its mirror image as a 2-perfect immersion (an arbitrary regular homotopy from the Konstantinov torus to its mirror image contains a generic self-intersection immersion with self-intersection curve of this immersion determining a non-trivial homology class in $H_1(T^2; \mathbb{Z}_2)$) and is not regular homotopic to an inside-out immersion as a 2-perfect immersion (an arbitrary regular homotopy from the Konstantinov torus to its inside-out image contains a surface with generic self-intersection curve with a non-trivial homology class in $H_1(T^2; \mathbb{Z}_4)$).

12.3 Part III: Solution of the Generalized Kervaire Problem

12.3.1 Generalized Kervaire Problem

This part concerns a positive solution of Problem 12.2, Statement 2. The result of the chapter is the following:

Theorem 12.7 *A generalized Kervaire Problem has a positive solution for all dimensions $n = 2^{l+1} - 2$, $l \geq 1$: the homomorphism*

$$\Theta_{2^l-1} : \mathrm{Imm}^{\mathrm{sf}}(2^l - 1, 2^l - 1) \longrightarrow \mathbb{Z}/2$$

is an epimorphism.

The proof of Theorem 12.7 is based on a result by Mahowald [26]. The Mahowald element $\eta_i^{\rm fr} \in \Pi_{2^i}$ is defined using the element $\eta_i^{\rm sf}$, constructed below, by the Khan–Priddy homomorphism :

$$\lambda : \mathrm{Imm}^{st}(2^i - 1, 1) \to \Pi_{2^i} = \mathrm{Imm}^{\rm fr}(2^l, 1), \qquad (12.24)$$

[23]. The homomorphism (12.24) is a particular case of the homomorphism (12.14).

In this section we construct the Mahowald element $\eta_l^{\rm sf} \in \mathrm{Imm}^{\rm sf}(2^l - 1, 1)$, which determines the base of the induction over the parameter $l \geq 3$. The construction explains a geometrical meaning of the following in [26]: the element hh_l is spherical in ASS.

Because our approach is different from the classical approach, even at a level of statements of the results, we will give a self-contained proof of the Mahowald Theorem. Theorem 12.7 is proved geometrically by a short construction using a reformulation of the definition of stable framing into stable skew-framing. Recall, see Sect. 12.2.5, that Theorem 12.7 for $l = 4$ is proved. This result gives the base of the induction. In the case $l = 1, 2, 3$ the proof is evident: the elements in the image $\mathrm{Imm}^{\rm fr}(2^l - 1, 1) \longrightarrow \mathrm{Imm}^{\rm sf}(2^l - 1, 2^l - 1)$ are elements with Θ_{2^l-1} invariant 1.

The Mahowald element $\eta_i^{\rm fr} \in \Pi_{2^i}$ is defined using the element $\eta_i^{\rm sf}$, constructed below, by the Khan–Priddy homomorphism (12.24). This part of the Mahowald construction is discussed in Sect. 12.3.4.

12.3.2 The Mahowald Elements $\eta_i^{\rm sf}$, $i > 3$

In this section let us consider a step of the induction and construct an immersion $\phi : N^{15} \looparrowright \mathbb{R}^{16}$, which represents the element $[\phi] = \eta_4^{\rm sf}$ in the group $\mathrm{Imm}^{\rm sf}(15, 1)$, using the element $\eta_3^{\rm sf} \in \mathrm{Imm}^{\rm sf}(7, 1)$, constructed by (12.10).

Let us repeat a reasoning of Sect. 12.2.5 and define $\tilde{M}^{15} = (\hat{N}^7 \times \hat{N}^7) \tilde{\times} S^1$. The difference is that we previously used the manifold $\mathbb{RP}^7$ instead of $\tilde{N}^7$.

As a component of this semi-direct product, let us take the manifold $\hat{N}^7$, which is constructed in Sect. 12.2.5. Then let us define the tower of 2-sheeted coverings:

$$M^{15} \stackrel{p}{\longrightarrow} \hat{M}^{15} \stackrel{\hat{p}}{\longrightarrow} \tilde{M}^{15}, \qquad (12.25)$$

which is the analogue of the tower (12.11). The bottom covering $\hat{p}$ in (12.25) is called a primary covering, the top covering p is called a secondary covering. We indicate the base of the covering in the notation $p_{\hat{M}}$, etc. The following formula, which is an analogue of the formula (12.9), is satisfied:

$$\langle p^{2^i-1}; [\hat{M}^{2^i-1}] \rangle = h(\phi), \quad i = 3. \qquad (12.26)$$

In the proof we take the (T_1, T_2)-equivariant composition: $\bar{p}_N \times \bar{p}_N : \bar{N}^7 \times \bar{N}^7 \to S^7 \times S^7$, where the factor is defined as the equivariant classifiyng mapping of the covering (12.8) from the previous step of the construction. The T-equivariant mapping is defined from (12.26) by the diagonal involution.

Then we repeat arguments from the previous step of the calculation of the degree of p. Because $deg(\bar{p}_N) = 1$, we get the formula (12.26).

The difference between the constructions (12.25) and (12.11) is the following. The orienting class κ_M in (12.25) is not an integer class. Then, starting by the manifold $\tilde{M}^{15}$, the required manifold $\hat{N}^{15}$, equipped with the covering $p : N^{15} \to \hat{N}^{15}$, and by the immersion $\varphi = \hat{\varphi} \circ p : N^{15} \looparrowright \mathbb{R}^{15}$ is constructed below. We use the same notation for the corresponding manifold in Sect. 12.2.5, but the construction is different. This immersion represents the described element $\eta_4^{\mathrm{sf}} \in \mathrm{Imm}^{\mathrm{sf}}(15, 1)$ with Hopf invariant 1.

Let us consider the Cartesian product $\hat{\varphi}_7 \times \hat{\varphi}_7 : \hat{N}^7 \times \hat{N}^7 \looparrowright \mathbb{R}^8 \times \mathbb{R}^8$ of the two copies of the immersion. The normal bundle $\nu(\hat{\varphi}_7 \times \hat{\varphi}_7)$ of this immersion is 2-dimensional and is represented by the Whitney sum of the two copies of the line bundle $\nu(\hat{\varphi})$.

Let us consider the Cartesian product $D^8 \times D^8$ of two open 8-disks and the open manifold $\tilde{K}^{17} = (D^8 \times D^8)\tilde{\times} S^1$, where the semi-direct product is defined by the monodromy $m : D^8 \times D^8 \to D^8 \times D^8$ along the generator S^1 by the formula: $m(x, y) = (y, x)$. Evidently, $\tilde{K}^{17}$ is diffeomorphic to $D^8 \times D^8 \times S^1$ and there exists an embedding $i_{\tilde{K}} : \tilde{K}^{17} \subset \mathbb{R}^{17}$ (the construction below does not depend on a choice of the embedding). Let us consider the manifold $\tilde{M}^{15} = (\hat{N}^7 \times \hat{N}^7)\tilde{\times} S^1$ (which is analogous to the manifold $(\mathbb{RP}^3 \times \mathbb{RP}^3)\tilde{\times} S^1$ constructed in (12.10)). Let us consider an immersion $j : (\hat{N}^7 \times \hat{N}^7)\tilde{\times} S^1 \looparrowright \tilde{K}^{17}$ and the composition $\tilde{\psi} = i_{\tilde{K}} \circ j : (\hat{N}^7 \times \hat{N}^7)\tilde{\times} S^1 \looparrowright \mathbb{R}^{17}$. The normal bundle $\nu_{\tilde{\psi}}$ of the immersion $\tilde{\psi}$ admits a reduction of the structural group $O(2)$ to the discrete subgroup $\mathbf{D}$.

The dihedral group $\mathbf{D}$ contains 8 elements and has the corepresentation $\{a, b | a^2 = b^4 = 1; [a, b] = b^2\}$. The element $b^2 \in \mathbf{D}$ is also denoted by $-1 \in \mathbf{D}$. The group D is represented in $O(2)$ as follows: the element b is represented by the rotation of the plane through the angle $\frac{\pi}{2}$, and the element a is represented by the symmetry with respect to the bisector of the first-third coordinate quarter. Let us denote $A = a \circ b = ab$. This element is represented by the symmetry with respect to the horizontal coordinate axis. Let us denote $\dot{A} = b^2 A = -A$, $\dot{a} = b^2 a = -a$.

Over the manifold $\tilde{M}^{15}$ the following tower of 2-sheeted coverings (12.25) is well defined. This tower corresponds to the following system of subgroups:

$$\mathbf{I}_a \subset \mathbf{I}_{a,\dot{a}} \subset \mathbf{D}. \tag{12.27}$$

The structured group of the normal bundle $\nu_{\tilde{\psi}}$ of the immersion $\tilde{M}^{15} \looparrowright \mathbb{R}^{17}$ is also isomorphic to the group $\mathbf{D}$. To distinguish the two isomorphic sequence of subgroups: the sequence of coverings over the manifold $\tilde{M}^{15}$ and the sequence of reductions of the structured group of the normal bundle, the subgroup of coverings tower is denoted by (12.27), a subgroup of the structured group are denoted by the corresponding collection of elements. In particular, the sequence of subgroups (12.27) of the covering corresponds to the following sequence of monodromy groups:

$$\{a\} \subset \{a, \dot{a}\} \subset \{a, b\} \simeq \mathbf{D}. \tag{12.28}$$

The element A of the residue class of the subgroup $\{a, \dot{a}\} \subset \mathbf{D}$ acts by an involution on the covering manifold $\hat{M}^{15}$ of the bottom covering $\hat{p}$, this action is denoted by T_A. The element $-1 = b^2$ in the residue class of the subgroup $\{a\} \subset \{a, \dot{a}\}$ acts on M^{15} by a covering translation $M^{15} \to \hat{M}^{15}$. This action is denoted by T_{b^2}. The restriction of T_A on the submanifold

$$M^{15} \supset (N^7 \times N^7)/\{b^2\} \tag{12.29}$$

is defined by the formula: $(n_1, n_2) \mapsto (T_N(n_1), n_2)$, where T_N is the involution in the top covering (12.11).

Let us consider the subgroups $\{a\} \subset \{a, b\}$, $\mathbf{I}_a \subset \mathbf{D}$, which are not normal subgroups. The normalizer of the subgroup $\{a\}$ is the subgroup $\{a, \dot{a}\} \subset \{a, b\}$. Analogously, the normalizer of $\mathbf{I}_a$ is the subgroup $\mathbf{I}_{a,\dot{a}} \subset \mathbf{D}$, recall that $\mathbf{D}$ is the structured group of the bundle $\nu_{\tilde{\psi}}$.

The subgroup $\mathbf{I}_a \subset \mathbf{D}$ is a subgroup of the covering $\hat{p} \circ p : M^{15} \to \tilde{M}^{15}$, see (12.25), which is not a covering with a normal subgroup. The subgroup $\mathbf{I}_{A,\dot{A}} \subset \mathbf{D}$ is visualized as a translation group of the covering $M^{15} \to \tilde{M}^{15}$, restricted to the submanifold $\hat{N}^7 \times \hat{N}^7 \subset \tilde{M}^{15}$. This group defines transformations of sheets of the covering.

Let us define the manifold $\hat{N}^{15}$. Let us consider the vector bundle $\nu_{\tilde{\psi}}$ over $\tilde{M}^{15}$ and denote

$$\nu_{\tilde{\psi}}^{tw} = \nu_{\tilde{\psi}} \otimes \lambda, \tag{12.30}$$

where $\lambda = \lambda(\hat{p})$ is a line bundle over $\tilde{M}^{15}$, associated with the covering $\hat{p} : \hat{M}^{15} \to \tilde{M}^{15}$. Denote by $\tilde{L}^{13} \subset \tilde{M}^{15}$ the Euler class of the bundle $\nu_{\tilde{\psi}}^{tw}$. (Recall that the Euler class of a vector bundle over a base manifold is defined as a submanifold of zero sections of the vector bundle. In particular the normal bundle of the Euler class submanifold is isomorphic to the bundle itself, where $\tilde{K}^{14} \subset \tilde{M}^{15}$ is the boundary of the disk-bundle associated with the vector bundle (12.56), which is the total space of the corresponding S^1-bundle over $\tilde{L}^{13}$.) For the formula (12.56) of the Euler class of the twisted section see [24].

The total space D_λ^{16} of the disk bundle, which is associated with the bundle λ, is a 16-dimensional manifold with boundary $\hat{M}^{15} = \partial(D_\lambda^{16})$. In the manifold with boundary D_λ^{16} let us define a subspace $\hat{N}^{15} \subset D_\lambda^{16}$, which is the required closed 15-dimensional manifold.

Take the submanifold $\tilde{L}^{13}$ and take its preimage by the projection $D^{16}_{\lambda} \to \tilde{M}^{15}$ on the base of the bundle. Let us cut-out fibers over the regular neighborhood $U(\hat{L}^{14})$ in D^{16}_{λ}. The boundary of the manifold obtained consists of a lateral surface of the handle and the horizontal bottom of the handle. The horizontal bottom is the double covering over $\hat{N}^{15}_{pred} \subset D^{16}_{\lambda} \to \tilde{M}^{15}$. Denote this closed manifold by $\hat{N}^{15}$ (this manifold is the required manifold). The decomposition of this manifold as the lateral surface and the bottom of the handle is denoted by

$$\hat{N}^{15} = \hat{N}^{15}_{pred} \cup \hat{K}^{15}. \tag{12.31}$$

Let us consider the projection $q : \hat{N}^{15}_{pred} \to \tilde{M}^{15} \setminus U(\tilde{L}^{13})$, as the restriction of the projection $D^{16}_{\lambda} \to \tilde{M}^{15}$ on the submanifold. By this projection the image $r(\hat{K}^{15})$ coincides with $\partial(U(\tilde{L}^{13}))$, and the image $r(\hat{N}^{15})$ coincides with the interior $\tilde{M}^{15} \setminus U(\tilde{L}^{13})$.

Let us consider the lateral surface of the handle $\hat{K}^{15}$. The middle submanifold of the lateral surface $\hat{K}^{15}$ is denoted by $\hat{K}^{14}_m \subset \hat{K}^{15}$. The natural projection $\hat{K}^{15} \to \hat{K}^{14}_m$ is a D^1-bundle. The restriction of this projection on the boundary $\partial(\hat{K}^{15})$, which is a common boundary of the lateral handle and of the horizontal bottom is the double covering over $\hat{K}^{14}_m$. Let us denote the covering space $\partial(\hat{K}^{15})$ by K^{14}_m. This covering $K^{14}_m \to \hat{K}^{14}_m$ coincides with the pull-back $\lambda \mapsto \lambda^!$, $\lambda^! \simeq \varepsilon$, by the bottom covering in (12.25) over the spherization $S(\nu_{\hat{L}} \otimes \lambda)$ of the normal bundle of the submanifold $\hat{L}^{13} \subset \hat{M}^{15}$, which is a copy the restriction of the normal bundle $\nu_{\hat{M}}$ over its own Euler class. By construction, there is a cohomology class:

$$p : K^{14}_m \to \mathbb{RP}^{\infty}, \tag{12.32}$$

which is the restriction of the top class in (12.25).

We may identify $\hat{K}^{14}_m$ with $S(\nu_{\tilde{L}}) \otimes \lambda$, i.e. with the spherization of the normal Euler class of the bundle (12.56). Obviously,

$$S(\nu_{\tilde{L}} \otimes \lambda) = S(\nu_{\tilde{L}}) \otimes \lambda. \tag{12.33}$$

The left hand side of the formula (12.33) looks more natural. The covering K^{14}_m (12.32) is defined as $S(\nu_{\hat{L}} \otimes \lambda^!)$.

On the manifold $\hat{K}^{14}_m$ the antipodal free involution is well defined:

$$J_{\hat{K}_m} : \hat{K}^{14}_m \to \hat{K}^{14}_m.$$

The quotient $\hat{K}^{14}_m / J_{\hat{K}_m}$ is naturally identified with the projectivization $P(\nu_{\tilde{L}} \otimes \lambda)$ corresponded with the bundle (12.33). Let us denote this manifold by $\tilde{K}^{14}_m$.

Let us consider the following diagram of 2-sheeted coverings:

$$\begin{array}{ccccc} S(\nu_{\tilde{L}} \otimes \lambda) = & \hat{K}^{14}_m & \xrightarrow{\bar{\pi}} & \tilde{K}^{14}_m & = P(\nu_{\tilde{L}} \otimes \lambda) \\ & \hat{q} \uparrow & & \uparrow \tilde{q} & \\ S(\nu_{\hat{L}} \otimes \lambda^!) = & K^{14}_m & \xrightarrow{\hat{\pi}} & K^{14}_m / J_{K_m} & = P(\nu_{\hat{L}} \otimes \lambda^!). \end{array} \qquad (12.34)$$

Lemma 12.1 *There exists a cohomology class $\alpha : K^{14}_m \longrightarrow \mathbb{RP}^{\infty}$, which is a pull-back of p, defined by (12.32), by the projection $K^{14}_m \to \hat{L}^{13} \subset \hat{M}^{15}$.*

Corollary 12.2 *There exist a cohomology class $\hat{\beta} : \hat{K}^{14}_m \longrightarrow \mathbb{RP}^{\infty}$ whose pull-back by the covering $\hat{q}$ coincides with the class $\alpha : K^{14}_m \longrightarrow \mathbb{RP}^{\infty}$.*

Corollary 12.3 *There exist a cohomology class $\tilde{\beta} : \tilde{K}^{14}_m \longrightarrow \mathbb{RP}^{\infty}$ whose pull-back by the covering $\bar{\pi}$ coincides with the class $\hat{\beta} : \hat{K}^{14}_m \longrightarrow \mathbb{RP}^{\infty}$.*

Proof of Lemma 12.1

Take the following diagram, which contains the diagram (12.34) as the Euler class:

$$\begin{array}{ccccc} S(\nu_{\tilde{M}} \otimes \lambda) = & \hat{Q}^{16}_m & \xrightarrow{\bar{\pi}} & \tilde{Q}^{16}_m & = P(\nu_{\tilde{M}} \otimes \lambda) \\ & \hat{q} \uparrow & & \uparrow \tilde{q} & \\ S(\nu_{\hat{M}} \otimes \lambda^!) = & Q^{16}_m & \xrightarrow{\hat{\pi}} & Q^{16}_m / J_{Q_m} & = P(\nu_{\hat{M}} \otimes \lambda^!) \end{array} \qquad (12.35)$$

In the diagram (12.35) spaces and coverings are defined analogously with (12.34), instead of the manifold $\tilde{L}^{13}$, the common base of the covering. We get the manifold $\tilde{M}^{15}$ equipped with the bundles described by the formula (12.56).

Let us consider the diagram of spaces which are double coverings over the corresponding spaces of the diagram (12.35):

$$\begin{array}{ccc} \hat{R}^{16}_m & \xrightarrow{\bar{\pi}} & \tilde{R}^{16}_m \\ \hat{q} \uparrow & & \uparrow \tilde{q} \\ R^{16}_m & \xrightarrow{\hat{\pi}} & R^{16}_m / J_{R_m}, \end{array} \qquad (12.36)$$

which is defined as follows.

Recall that $\tilde{M}^{15} = \hat{N}^7 \times \hat{N}^7 \rtimes_{\tilde{T}} S^1$. The submanifold $\hat{N}^7 \times \hat{N}^7 \subset \tilde{M}^{15}$ is the order 2-points configuration space of $\hat{N}^7$. The involution $\tilde{T} : \hat{N}^7 \times \hat{N}^7 \to \hat{N}^7 \times \hat{N}^7$ permutes points in pairs. The covering $\hat{M}^{15} \to \tilde{M}^{15}$ is described in (12.25), using the diagram (12.27). We get $\hat{M}^{15} = (N^7 \times N^7)/T_{diag} \rtimes_{\hat{T}} S^1$, where, T_{diag} is the diagonal of the involution on the Cartesian product of the two involutions $N^7 \to N^7$, which is defined in the covering space of the top covering (12.11). The involution T_{diag} corresponds to the element $b^2 \in \{a, b\}$. The involution $\hat{T}$ permutes points in pairs. The covering $\hat{N}^7 \times \hat{N}^7 \times S^1 \to \tilde{M}^{15}$, which is the covering with the subgroup $\mathbf{I}_{A \times \dot{A}} \subset \mathbf{D}$ with the involution denoted by T_R, induces the coverings of spaces of (12.36) over corresponding spaces of (12.35).

Let us describe the cohomology classes $\alpha_R : R^{16}_m \to \mathbb{RP}^\infty$. Then let us prove the following formula:

$$\alpha_R = T_R \circ \alpha_R, \tag{12.37}$$

where T_R is the involution of the covering $R^{16}_m \to Q^{16}_m$.

Define the class α_R as the pull-back of the class $[T_{diag}]^* \in H^1((N^7 \times N^7)/T_{diag})$ by the projection $R^{16}_m \to (N^7 \times N^7)/T_{diag} \times S^1 \to (N^7 \times N^7)/T_{diag}$. The fact (12.37) is a corollary of the fact that α_R coincides to the standard $\mathbf{I}_{\dot{a}}$-character (it is invariant with respect to $\mathbf{I}_a$-transformation). The cohomology class $\alpha_Q \in H^1(Q^{16}_m)$ is well defined. The class α is the pull-back of the class α_Q by the inclusion of the diagram (12.34) in (12.35). Lemma 12.1 is proved. □

Proof of Corollary 12.2

Let us collect spaces from the left of the diagrams (12.34), (12.35), (12.36) in a common diagram. There are inclusions: $\hat{K}^{14}_m \subset \tilde{M}^{15} \subset \hat{Q}^{16}_m$. The restriction of the normal S^1-fibration over a regular neighborhood of $\hat{K}^{14}_m \subset \tilde{M}^{15}$ is referred to as the regular domain $\hat{Q}^{16}_{reg} \subset \hat{Q}^{16}_m$. The preimages in $\hat{R}^{16}_m$, R^{16}_m of the regular domain are denoted by $\hat{R}^{16}_{reg}$, R^{16}_{reg} correspondingly. We use only regular domains in the following diagram:

$$\begin{array}{ccccc} \hat{K}^{14}_m & \subset & \hat{Q}^{16}_{reg} & \longleftarrow & \hat{R}^{16}_{reg} \\ \hat{q} \uparrow & & \hat{q}_Q \uparrow & & \uparrow \hat{q}_R \\ K^{14}_m & \subset & Q^{16}_{reg} & \longleftarrow & R^{16}_{reg}. \end{array} \tag{12.38}$$

Take the left-bottom manifold K^{14}_m in (12.34) and consider the section $\varepsilon = \lambda^!$ as the pull-back of the section λ of the bundle $\nu_{\hat{L}} \otimes \lambda$ over the left-upper manifold in the diagram. Because the pull-back $(\nu_{\hat{L}} \otimes \lambda)^!$ is canonically isomorphic to $\nu_{\hat{L}}$, the section $\lambda^!$ looks as a section of the bundle $\nu_{\hat{L}}$. The bundle $\nu_{\hat{L}} \otimes \lambda$ and the section λ are pull-backs of the corresponding bundle and section over the middle bottom manifold of the diagram (we use the same notation).

Let us denote by $H^+(X) \subset H_1(X)$, for a non-orientable closed manifold X, the index 2 subgroup, which is the kernel of the orientation homomorphism: $Ker(w) : H_1(X) \to \mathbb{Z}/2$. Analogously, define the non-trivial residue class of the orientation homomorphism by $H^-(X)$. Obviously, $H^1(X) = H^+(X) \cup H^-(X)$. Denote by

$$H^{\oplus}(\hat{Q}^{16}_{reg}) = [H^+(\hat{Q}^{16}_{reg}) \cap Ker(\mathbf{I}_a)] \cup [H^-(\hat{Q}^{16}_{reg}) \cap \overline{Ker}(\mathbf{I}_a)], \tag{12.39}$$

where $\mathbf{I}_a$ is the homomorphism $H_1(\hat{Q}^{16}_{reg}) \to H_1(S^1)$, which corresponds to the right upper horizontal covering in the diagram (12.38). This notation $H^{\oplus}(\hat{Q}^{16}_{reg})$ means that we take in $H_1(\hat{Q}^{16}_{reg})$ the kernel of the homomorphism $A + \dot{A}$:

$\mathbf{D} = \{a, b\} \rightarrow \mathbb{Z}/2$, described using the normal bundle structure. Namely, the group (12.39) coincides with the kernel of the homomorphism:

$$H_1(\hat{Q}^{16}_{reg}) \longrightarrow \mathbf{D}/[\mathbf{D}, \mathbf{D}] = \{A \oplus \dot{A}\} \rightarrow \{A\} + \{\dot{A}\}. \qquad (12.40)$$

The required cohomology class $\hat{\beta} : \hat{Q}^{16}_{reg} \longrightarrow \mathbb{RP}^{\infty}$ is well defined as an extension of a homomorphism $\hat{\beta}^{\oplus} : H^{\oplus}(\hat{Q}^{16}_{reg}) \rightarrow \mathbb{Z}/2$ on the subgroup (12.39), which is constructed below.

Briefly, the definition of $\hat{\beta}^{\oplus}$ is the following. On the subgroup $H_1^{\oplus}(Q^{16}_{reg}) \subset H_1^{\oplus}(\hat{Q}^{16}_{reg})$ (the middle bottom space of the diagram (12.38)), the class $\beta^{\oplus}$ is defined as the homomorphism

$$H^{\oplus}(Q^{16}_{reg}) \subset H_1(Q^{16}_{reg}) \rightarrow \mathbf{I}_A \oplus \mathbf{I}_{\dot{A}} \rightarrow \mathbf{I}_A$$

(on the space Q^{16}_{reg}, the homomorphisms $\mathbf{I}_A$, $\mathbf{I}_{\dot{A}}$ coincides). This homomorphism is not extended to $H_1(\hat{Q}^{16}_{reg})$, because the left bottom covering in (12.38) does not commute with this $\mathbf{I}_A$-homomorphism. Using the monodromy group $\{a, b\}$ we will define a mapping

$$\sharp : H^{\oplus}(Q^{16}_{reg}) \rightarrow \mathbb{Z}/2 \qquad (12.41)$$

which satisfies the formulas (12.43), (12.46). We get the mapping (12.45), which is a well-defined homomorphism. This mapping admits an extension:

$$\hat{\sharp} : H^{\oplus}(\hat{Q}^{16}_{reg}) \rightarrow \mathbb{Z}/2. \qquad (12.42)$$

Then we prove that the required class $\hat{\beta}$ on the left-top space in the diagram (12.38) is related to the class α.

Definition of the Mapping (12.41)

Consider an arbitrary homology class $[l]_* \in H^{\oplus}(Q^{16}_{reg})$, let it be represented by a connected loop l_Q and consider the double covering l_R over the loop in R^{16}_{reg} (the upper-left space in (12.36)). The loop l_R is visualized as a two-component closed loop [Case I, the standard $\mathbf{I}_a$-character on l_Q is trivial and the covering l_R over l_Q is disconnected]. Or, l_R is represented by a single connected loop (Case II, the $\mathbf{I}_a$-character on l_Q is non-trivial). In Case II, we may see l_R as a segment on R^{16}_{reg} with end-points related by the involution by $T_{R_{reg}}$. Moreover, in the case the class $[l_Q]$ is represented by a segment l_R, such that the end-points of the segment are $T_{R_{reg}}$-involutive, or, the projection of l_R on the circle fiber are $\mathbf{I}_a$-involutive, we assume that the end-points of l_R are not on the section. In Case I components of the loop l_R are ordered, but the numbers defined below do not depend on an order. In Case II segments are also ordered, the numbers depends of the order.

Denote by $\sharp(l_Q) \in \{0, 1\}$ the number of intersection points of the first component of l_R with the positive base vector of cross-section $\lambda^!$. Take the involution $T_{\hat{q}}$ in the covering by the central vertical arrow of (12.38) and consider the lift of $T_{\hat{q}}$ as an involution $T_{\hat{q};R_{reg}}$ on R^{16}_{reg} inside the regular domain. Let us prove that the numbers $\sharp(l)$, $\sharp(T_{\hat{q};R_{reg}}(l))$ coincide in Case I. Recall $[l] \in H^+(R^{16}_{reg})$, see the formula (12.39). In Case II, the numbers $\sharp(l_R)$, $\sharp(T_{\hat{q};R_{reg}}(l_R))$ are different and $[l]_* \in H^-(Q^{16}_{reg})$. In both cases we calculate the indexes for components (segments) with a common number.

Let us consider Case I. Recall that $l \in H^+(R^{16}_{m;reg})$. The bundles κ_A, $\kappa_{\dot{A}}$ over l_R are trivial (Subcase 1), or, each bundle is non-trivial (Subcase 2) simultaneously. Take a coordinate system at the initial point of the path. The section ε over each component of l_R projects on the spherical fiber, which is the projection on the circle in Subcase 1. The number $\sharp(l_R)$ for the first component depends on a number of rotation of the section, but the number $\sharp(T_{\hat{q};R_{reg}}(l_R))$ as well, because the sections are antipodal.

In Subcase 1 assume that the section $\lambda^!$ is represented by a section projected on the circle fiber to the constant point $\mathbf{e}_1$ (to make arguments clear). Then the involution $T_{\hat{q};R_{m;reg}}$ transforms the section into its antipodal point $-\mathbf{e}_1$ on the circle. In Subcase 2, we will see a closed loop as a twisted cylinder, where "twisted" means a rotation of its generator through the angle $\pm\pi$ with respect to the coordinates in $S(\kappa \oplus \kappa_{\dot{a}})$. In this case the section is translated by $T_{\hat{q};R_{reg}}$ into the antipodal twisted cylinder.

Assume that a component l_R is represented by a curve at the point $\mathbf{e}_1$ (the base vector of the bundle $S(\nu_{\hat{L}} \otimes \lambda^!)$). Take the involution $T_{\hat{q};R_{reg}}$ and take the images of l_R and of the section $\lambda^!$ by the involution. The curve $T_{\hat{q};R_{reg}}(l_R)$ is constant at the point $\mathbf{e}_1$, and the section $T_{\hat{q};R_{reg}}(\lambda^!)$ is a constant at the point $-\mathbf{e}_1$, (recall that the bundle $\nu_{\hat{M}} \otimes \lambda^!$ is twisted by the involution $T_{\hat{q}}$). For the curves in the example, we get $\sharp(l_R) = 0$; $\sharp(T_{\hat{q};R_{reg}}(l_R)) = 0$. For an arbitrary l_R we have $\sharp(l_R) = 1$; $\sharp(T_{\hat{q};R_{reg}}(l_R)) = 1$ is also possible.

In Case I, Subcase 1, we get:

$$\sharp(l_R) = \sharp(T_{\hat{q};R_{reg}}(l_R)). \tag{12.43}$$

Case I, Subcase 2 is analogous.

Let us consider the Case II. We may assume that both standard A, $\dot{A}$-characters on l_R are trivial (this case is analogous to Subcase 1 of Case I; the case corresponds to the condition $[l]_* \in H^{\oplus}(Q^{16}_{reg})$, see the formula (12.39); the case when both characters are non-trivial is analogous to the Subcase 2, Case I). The curve l_R is represented by a constant segment at the point $\mathbf{e}_1 + \mathbf{e}_2$, the section $\lambda^!$ is represented by a constant at the point $\mathbf{e}_1 + \mathbf{e}_2$. The curve $T_{\hat{q};R_{reg}}(l_R)$ is a constant at the point $\mathbf{e}_1 + \mathbf{e}_2$. The section $T_{\hat{q};R_{reg}}(\lambda^!)$ is a constant at the point $-\mathbf{e}_1 - \mathbf{e}_2$. A generic alteration gets: $\sharp(l_R) = 1$; $\sharp(T_{\hat{q};R_{reg}}(l_R)) = 0$. As a conclusion, in the Case II we get:

$$\sharp(l_R) = \sharp(T_{\hat{q};R_{reg}}(l_R)) + 1. \tag{12.44}$$

The calculation above means that $\sharp : H^+(R^{16}_{reg}) \to \mathbb{Z}/2$ is a homomorphism and is not a homomorphism on $H^{\oplus}(Q^{16}_{reg})$. Obviously, the standard character $\mathbf{I}_A$ on $H^{\oplus}(R^{14}_{reg})$ admits the same property by the Eqs. (12.43), (12.46). This implies that the homomorphism

$$\sharp + \mathbf{I}_A : H^{\oplus}(R^{16}_{reg}) \to \mathbb{Z}/2 \tag{12.45}$$

is invariant with respect to $T_{\hat{q}} : H^{\oplus}(R^{16}_{reg}) \to H^{\oplus}(R^{16}_{reg})$. This homomorphism determines a homomorphism on $H^{\oplus}(Q^{16}_{reg})$.

As a result, there exists a homomorphism $\hat{\beta} : H_1(\hat{K}^{14}_m) \to \mathbb{Z}/2$ which is defined as the canonical extension of $\sharp + \mathbf{I}_A$ by (12.45). An extension is defined in two possible ways. Let us take $\hat{\beta}$ with a condition, using α, described below.

Take the pull-back $\alpha = \hat{q}^*(\hat{\beta})$ of $\hat{\beta}$ by the covering $\hat{q}$. The restriction of the cocycle $\sharp$ on $H^{\oplus}(K^{14}_m)$ is extended to the cocycle on $H^{\oplus}(\hat{M}^{15} \setminus U(\hat{L}^{13}))$, because the section $\lambda^!$ on this manifold $\hat{M}^{15} \setminus U(\hat{L}^{13})$ is well defined. To calculate the pull-back $\sharp^!$ of $\sharp$ by the covering $\hat{q}$ take a generic intersection of the section on $\hat{M}^{15} \setminus U(\hat{L}^{13})$ with itself. Obviously, the extension $\sharp^!$ coincides with $w_1(\hat{M}^{15} \setminus U(\hat{L}^{13}))$, which is trivial on $H^+(\hat{M}^{15} \setminus U(\hat{L}^{13}))$ and is non-trivial on $H^-(\hat{M}^{15} \setminus U(\hat{L}^{13}))$. The section $\lambda^!$, restricted to $\hat{M}^{15} \setminus U(\hat{L}^{13})$, is generic with respect to itself, and the intersection number of the section with a generic alteration over a homology class $[l]_* \in H_1(\hat{M}^{15} \setminus U(\hat{L}^{13}))$ represents the number $\sharp^!(l)$. One gets $\sharp^!(l) = 1$ when the bundle $\nu_{\hat{M}}$ over l is non-trivial. This exactly means that $\hat{M}$ over l is non-oriented. Because of the assumption $[l]_* \in H^{\oplus}(K^{14}_m)$, or, equivalently, $[l]_* \in H^{\oplus}(\hat{M}^{15} \setminus U(\hat{L}^{13}))$, the condition $\sharp^!(l) = 1$ if and only if $\mathbf{I}_a(l) = 1$, or, equivalently, if and only if $[l]_* \in H^{\oplus}(\hat{M}^{15} \setminus U(\hat{L}^{13}))$. This means that on $H^{\oplus}(\hat{M}^{15} \setminus U(\hat{L}^{13}))$ we get: $\sharp^! = \mathbf{I}_a + \mathbf{I}_{\dot{a}}$.

An extension of a cocycle from $H^{\oplus}(\hat{K}^{14}_m)$ to $H_1(\hat{K}^{14}_m)$, say, the cocycle α, determines the only corresponding extension of $\hat{\beta}$ (a value $\hat{\beta}(t)$, $t \in H_1(\hat{K}^{14}_m) \setminus H^{\oplus}(\hat{K}^{14}_m)$ determines the homomorphism $\hat{\beta}$ on $H_1(\hat{K}^{14}_m)$, which coincides with the prescribed lift $\hat{\beta}$ on the codimension 2 subgroup $H^{\oplus}(\hat{K}^{14}_m)$, where the character $A + \dot{A}$ is trivial). Because

$$\alpha = \sharp^! + \mathbf{I}_a = \mathbf{I}_{\dot{a}} \tag{12.46}$$

on $H^{\oplus}(\hat{M}^{15} \setminus U(\hat{L}^{13}))$, the cocycle $\hat{\beta}$ is well defined with the condition $\hat{q}^*(\hat{\beta}) = \alpha$. Corollary 12.2 is proved. □

Proof of Corollary 12.3
Take the involution in the covering $\tilde{\pi}$ and denote this involution by

$$T_{\tilde{\pi}} : \hat{K}^{14}_m \to \hat{K}^{14}_m.$$

The cocycle $\hat{\beta}$ is invariant with respect to this involution. Let us prove this fact for the pull-back by $\tilde{q}$. Both terms in (12.46) are invariant. For the second term this is obvious, for the term $\sharp^!$ the statement is correct by definition: the involution $T_{\tilde{\pi}}$ is antipodal in S^1-fiber (12.33), and also the orientation of the line bundle $\lambda^!$ is changed into the antipodal. The formula is invariant with respect to the involution in the covering $\tilde{q}$. Corollary 12.3 is proved. □

To get a construction of η_4^{sf} one may start with a simplification, that the Euler class $\tilde{L}^{13}$ of $\nu_{\tilde{M}}$ is empty. With this assumption we may see that $\hat{M}^{15} = \hat{N}^{15}$ is a boundary of the total space of the disk bundle associated with λ (looks like a generalized Möbious band), immersed into $\mathbb{R}^{16}$, and therefore the immersion $g : \hat{M}^{15} \looparrowright \mathbb{R}^{17}$ with the image inside the band admits a cross-section. This gives the required element $\eta_4^{\mathrm{sf}} \in \mathrm{Imm}^{\mathrm{sf}}(15, 1)$ by the composition $M^{15} \to \hat{M}^{15} \looparrowright \mathbb{R}^{17}$. After the Hirsch trick this immersion is compressed into codimension 1 immersion, analogously with a similar construction in Sect. 12.2.5.

Let us consider a general case. The manifold $\hat{N}^{15}$, (12.31), which is equipped with the characteristic class $p_{\hat{N}} \in H^1(\hat{N}^{15})$ by Lemma 12.2 is well defined. Let us prove that $p_{\hat{N}}^{15}$ is the fundamental class of this closed manifold. By the construction, $\hat{N}^{15}$ (see the formula (12.31)) is the result of a surgery of $\hat{M}^{15}$ by a generalized handle with a 14-dimensional bottom. The bottom of the lateral surface of the handle is equipped with the class $p_{\hat{K}}$, which is the pull-back of a class α by the projection of the lateral surface onto the middle submanifold $\hat{K}^{14}_m$, see Lemma 12.2, see (12.46). Let us use the following statement, which is proved using elementary obstruction theory.

Lemma 12.2 *Let K^{14} be an arbitrary closed 14-dimensional non-oriented manifold, which is equipped with a cohomology class $p : K^{14} \to \mathbb{RP}^\infty$, such that the kernel of the induced homomorphism $p_* : H_1(K^{14} \to \mathbb{Z}/2)$ contains non-oriented loops. Consider two mappings $p_i : K^{14} \to \mathbb{RP}^{14}$, $i = 1, 2$ in the prescribed cohomology class p by the composition $I \circ p_i : K^{14} \to \mathbb{RP}^{14} \subset \mathbb{RP}^\infty$, $I \circ p_1 = I \circ p_2 = p$. Then for an arbitrary double covering $s : \bar{K}^{14} \to K^{14}$ wich is not the covering with the class $w_1(K^{14}$, mappings $p_1 \circ s$ and $p_2 \circ s$ are homotopic.*

In a special case, when the covering s coincides with p (the covering p is defined by the class p), the Lemma is deduced from the following: an arbitrary even-degree mapping of a non-oriented $\bar{K}^{14}$ into the sphere S^{14} is homotopic to the constant.

We use the lemma for $K = \hat{K}^{14}_m$ and classes $p_1 = p_{\hat{K}}$, p_2, where p_2 is the composition of the projection of the bottom surface of the handle to the central manifold of the D^2-bundle and the class into the 13-skeleton $\mathbb{RP}^{13}$. By

Corollary 12.3 the classes are compositions with a common double covering $\tilde{q}$. Lemma 12.2 implies that the classes are homotopic inside $\mathbb{RP}^{14}$.

By the surgery along the handle the manifold $\hat{M}^{15}$ is split into two component. The component $\hat{N}^{15}$ is outside the handle and the component $\hat{N}_0^{15}$ is inside the handle. Both components are equipped with the cohomology class p and the total degree of p into the skeleton $\mathbb{RP}^{15}$ coincides with the characteristic number (12.26). The degree of p for the component $\hat{N}_0^{15}$ is equal to zero, because the image of this component by p is inside 14-dimensional skeleton $\mathbb{RP}^{14} \subset \mathbb{RP}^{\infty}$.

The degree of the characteristic mapping into the 15-skeleton for the class of the mapping $p_{\hat{M}} \mapsto p_{\hat{N}}$ by the surgery $\hat{M}^{15} \mapsto \hat{N}^{15}$ remains fixed, and the characteristic number of $p_{\hat{N}}$ and of $p_{\hat{M}}$ coincide. The required equation

$$\langle p_{\hat{N}}^{2^l-1}; [\hat{N}^{2^l-1}]\rangle = 1 \tag{12.47}$$

for $i = 4$ is deduced from (12.26).

An immersion $\phi_{\hat{N}} : \hat{N}^{15} \looparrowright \mathbb{R}^{16}$ is well defined, using $p_{\hat{N}}$ with the Eq. (12.47) and determines the required element in $\mathrm{Imm}^{\mathrm{sf}}(15, 1)$, which will be denoted by η_4^{sf}. Let us repeat the reasoning from Sect. 12.2.5, formula (12.11), and prove that the element η_4^{sf} has non-trivial Hopf invariant.

The construction of the elements $\eta_l^{\mathrm{sf}} \in \mathrm{Imm}^{\mathrm{sf}}(2^l - 1, 1)$, $l \geq 5$ is completely analogous to the case $l = 4$.

Remark

The Mahowald element $\eta_4^{\mathrm{fr}} \in \mathrm{Imm}^{\mathrm{fr}}(16, 1)$ is well defined. It is not difficult to prove, using the calculation of the secondary cohomology operation by the Khan–Priddy transfer, that the element η_4^{fr} is detected by the corresponding secondary cohomology operation, described in [26].

12.3.3 *The Kervaire Invariant* 1 *Element* $\zeta_l^{\mathrm{sf}} \in \mathrm{Imm}^{\mathrm{sf}}(2^l - 1, 2^l - 1)$

Let us prove Theorem 12.7. Consider an immersion $\hat{\phi} : \hat{N}^{2^l-1} \looparrowright \mathbb{R}^{2^l}$, where $p \in H^1(\hat{N}^{2^l-1})$, $\langle p^{2^l-1}; [\hat{N}]\rangle = 1$, $l \geq 3$ (see the formula (12.47)). Such an immersion is constructed in Sect. 12.3.2. In our proof the explicit construction of $\hat{N}^{2^l-1}$ is not unrequired. We will only use the immersion $\hat{\phi}$ and the formula (12.47). Using this data we construct by a regular surgery a self-transversal immersion $\phi_2 : N_2^{2^l-1} \looparrowright \mathbb{R}^{2^{l+1}-2}$ with an odd number of self-intersection points and a stable skew-framing $\Psi_2^{st} : \nu_{\varphi_2} \oplus n\kappa \simeq (2^l - 1 + n)\kappa$, where κ is the line orienting bundle over $N_2^{2^l-1}$. Then we calculate the obstruction to reduce the stable skew-framing Ψ_2^{st} to a skew-framing Ψ_3 and we shall prove that there exists a modification of $(\phi_2, \Psi_2^{st}) \mapsto (\phi_3, \Psi_3)$ of stably skew-framed immersions into a skew-framed immersion. From this data the immersion ϕ_3 is skew-framed by Ψ_3 and has an odd number of self-intersection points, because the surgery preserves self-intersection points of ϕ_2. This proves the theorem.

Let us consider the characteristic mappings $p : \hat{N}^{2^l-1} \to \mathbb{RP}^\infty$ and $\hat{\kappa} : \hat{N}^{2^l-1} \to \mathbb{RP}^\infty$, where $\hat{\kappa}$ is the orienting cohomology class of the manifold $\hat{N}^{2^l-1}$, p a classifying mapping of the given cohomology class. Let us denote by

$$\hat{\mu} : \hat{N}^{2^l-1} \to \mathbb{RP}^\infty \tag{12.48}$$

a classifying mapping with the characteristic class $\hat{\mu} = \hat{\kappa} + p$. The characteristic class $\hat{\mu}$ classifies the line bundle $\hat{\kappa} \otimes p$ over $\hat{N}^{2^l-1}$.

Consider a submanifold $\hat{K}^{2^l-3} \subset \hat{N}^{2^l-1}$, which is defined as the regular preimage by $\hat{\mu}$ of the standard codimension 2 submanifold: $\mathbb{RP}^{\infty-2} \subset \mathbb{RP}^\infty$. The manifold $\hat{K}^{2^l-3}$ is assumed to be connected. Denote by $V(\hat{K}^{2^l-3})$ a thin regular neighborhood of this manifold in $\hat{N}^{2^l-1}$ and by $U(\hat{K}) \subset V(\hat{K}^{2^l-3})$ the boundary of the neighborhood.

Take the immersion

$$\hat{\phi}_1 = I \circ \hat{\phi} : \hat{N}^{2^l-1} \looparrowright \mathbb{R}^{2^{l+1}-2}, \tag{12.49}$$

where $\hat{\phi} : \hat{N}^{2^l-1} \looparrowright \mathbb{R}^{2^l}$ is given and $I : \mathbb{R}^{2^l} \subset \mathbb{R}^{2^{l+1}-2}$ the standard embedding. Let us define the Hurwitz number for $n = 2^l - 1$ by C_n. This is a smallest power of 2, such that the bundle $C_n\gamma$ over $\mathbb{RP}^n$ is trivial, where γ is the canonical line bundle over $\mathbb{RP}^n$. Take the immersion

$$\hat{\phi}_1^{st} = J \circ \hat{\phi} : \hat{N}^{2^l-1} \looparrowright \mathbb{R}^{2^{l+1}-2} \subset \mathbb{R}^{2^l+C_n}, \tag{12.50}$$

where $\hat{\phi} : \hat{N}^{2^l-1} \looparrowright \mathbb{R}^{2^l}$ is given and $J : \mathbb{R}^{2^{l+1}-2} \subset \mathbb{R}^{2^l+C_n}$ the standard embedding. We assume that (12.49) is generic, and, by (12.47) has only one self-intersection point. We do not assume that (12.50) is generic, but the skew-framing, constructed below, is not an extension of a horizontal skew-framing of $\hat{\phi}_1$.

Denote the manifold $\hat{N}^{2^l-1} \setminus V\hat{K}^{2^2-3}$ with boundary $U(\hat{K})$ by $(\hat{N}_2, U(\hat{K}))$. Define a regular surgery

$$p \circ \hat{\phi}_1 \mapsto \phi_2, \tag{12.51}$$

using a generalized handle W, which is defined as the line disk bundle $D(p_K) = W$, $\partial W = U(K)$, over the manifold $U(\hat{K})$, associated with the covering $p|_{U(\hat{K})}$. The manifold N_2 is well defined, the regular surgery (12.51) is defined below.

From this construction, we will get a skew-framing:

$$\Psi_2^{st} : \nu(\phi_2^{st}) \cong (C_{2^l-1} + 1)\mu, \tag{12.52}$$

where

$$\mu = \pi^*(\hat{\mu}), \quad \pi : N_2^{2^l-1} \to \hat{N}^{2^l-1}. \tag{12.53}$$

The immersion $\hat{\phi}_1^{st}$, (12.50) is skew-framed in the ambient space $\mathbb{R}^{2^l+C_n}$:

$$\hat{\Psi}_1^{st} : \nu(\hat{\phi}_1^{st}) \cong (C_n + 1)\hat{\kappa}. \tag{12.54}$$

We may use a general position argument, that the first of the two skew-sections of (12.54) over $\hat{K}^{2^l-3} \subset \hat{N}^{2^l-1}$ is horizontal in $\mathbb{R}^{2^{l+1}-2}$ (the first skew-section is the skew-framing of $\hat{\phi}$ and is horizontal in $\mathbb{R}^{2^l}$.) The 2-skewframing is denoted by

$$\hat{\Psi}_1^{st} \supset \hat{\psi}_2 : 2\hat{\kappa} \subset \mathbb{R}^{2^{l+1}-2} \subset \mathbb{R}^{2^l+C_n}. \tag{12.55}$$

Because the class $\hat{\mu}$ is an integer over $\hat{N}^{2^l-1} \setminus V(\hat{K})$, the following isomorphism

$$2p \cong 2\hat{\kappa} \tag{12.56}$$

is well defined. In particular, over $\hat{N}^{2^l-1} \setminus V(\hat{K})$ there exists a fibrewized monomorphism of the line bundle into the 2-bundle: $i_P \; : \; p \subset \hat{\psi}_2(2\hat{\kappa})$. The image of the orthogonal monomorphism $i_P^{\perp}$ determines segments of the lateral handle W (recall, the lateral handle is the disk bundle of p over $U(\hat{K})$). The handle of the surgery admits a regular codimension 2 projection on the total space of the subbundle (12.55). Let us construct a line subbundle in $\nu(\phi_2)$ inside (12.55). This subbubdle is described as the line bundle $i_P \otimes \kappa = \mu$

The manifolds $\hat{N} \setminus V\hat{K}$, N_2 admit the characteristic mappings

$$\hat{\kappa} : \hat{N} \setminus V\hat{K} \to \mathbb{RP}^{\infty}, \quad \kappa = \hat{\kappa} \circ \pi : N_2 \to \hat{N} \setminus V\hat{K} \to \mathbb{RP}^{\infty}.$$

The manifold $\hat{N}_2 \setminus V\hat{K}$ admits the characteristic mapping $\hat{\mu} : \hat{N}_2 \setminus V\hat{K} \to \mathbb{RP}^1$ and the manifold N_2 admits the characteristic mapping

$$\mu = \hat{\mu} \circ \pi : N_2 \to \hat{N} \setminus V\hat{K} \to \mathbb{RP}^1. \tag{12.57}$$

The submanifold $\hat{\mu}^{-1}(\mathbb{RP}^0) \subset \hat{N} \setminus V\hat{K}$ is of codimension 1. Over this submanifold we get the isomorphism $\hat{\kappa} \cong p$. We assume that over $\hat{\mu}^{-1}(\mathbb{RP}^0)$ the bundle monomorphism i_P is the inclusion in the first factor of $\hat{\psi}_2$.

Take the bundle $\hat{\psi}_2$ over the manifold $U(\hat{K})$ with the left hand structure (12.56) over $\partial(U(\hat{K}))$ and take the reflection $R : \hat{\psi}_2 \to \hat{\psi}_2$ with respect to the fixed first line subbundle $i_p(p) \subset \hat{\psi}_2$. The pair κ, $R(\kappa)$ are two line subunles in $\hat{\psi}_2$ over $U\hat{K} \subset N_2^{2^l-1}$. By a geometrical reason, the subbundles κ, $R(\kappa)$ are line subbundles inside the normal bundle of the inclusion

$$N_2 \setminus V\hat{K} \subset \hat{\psi}_2. \tag{12.58}$$

The line subbundles are identified by the R-permutation.

Let us denote the handle as the disk bundle, associated with $i_P^{\perp}$. The handle is glued to get the immersion (12.51) along the common boundary. We get an immersion with angles, the required regular immersion is the result of a smoothing of the angles.

The skew-framing (12.54) is defined as following. By the surgery we get

$$w_1(N_2^{2^l-1}) = \mu. \tag{12.59}$$

The normal bundle of $N_2^{2^l-1}$ inside (12.58) is split into the Whitney sum of the two line bundles: $\zeta_1 \oplus \zeta_2$, one of the line bundle ζ_1 is a bundle with the class κ, this bundle is parallel to the bisector of the pair κ, $R\kappa$. Over a hypersurface, where the bundles κ, $R\kappa$ are antiparallel and coincide with the line $i_P^{\perp}$, a fiber of ζ_1 is degenerate along the fiber of i_P. We deformed this singular projection parallel to the normal direction of $U(\hat{K}) \subset \hat{N}^{2^l-1}$.

The line bundle ζ_2 is a bundle over $N_2^{2^l-1}$. From a calculation (12.59) we get $\zeta_2 \cong \mu$, after the isomorphism $C(l)\kappa \cong C(l)\mu$, we get the formula (12.54).

Because the class μ over $N_2^{2^l-1}$ is an integer, the bundle ν_a, the normal bundle of the coordinate inclusion (12.50) is a skew-framed bundle with the class μ by:

$$\Psi_0 : \nu_a \cong (C_n - 2^{n+1} - 2)\mu. \tag{12.60}$$

The pair (Ψ_2^{st}, Ψ_0) by the Eqs. (12.54), (12.60) is called a stable skew-framing of the immersion ϕ_2^{st}. This notion is studied in [9]. Let us calculate the total obstruction $o(\Psi_2^{st}, \Psi_0)$ of a destabilization of the stable skew-framing and let us define a skew-framing Ψ_2 of the normal bundle of the immersion $\phi_2 : N^{2^l-1} \looparrowright \mathbb{R}^{2^{l+1}-2}$ in the codimension $2^l - 1$, the top unstable dimension for a normal bundle.

The total obstruction o of the reduction is an integer. Let us recall the definition of the total obstruction $o(\Psi_2^{st}, \Psi_0)$ in the specific case for the immersion ϕ_2 and its stable skew-framing (12.54),(12.60).

Let us consider the following isotopy F_t, $t \in [0, 1]$ of the skew-framing (12.60) for $t = 0$, into the following spherical skew-framing for $t = 1$. The i-th vector of $\tilde{\Psi}_0$ is equivariantly deformed into the i-th vector of Ψ_2^{st} by an orthogonal translation. This is possible by general position arguments for all vector of $\tilde{\Psi}_0$, but the last vector with the number $i_{last} = C_n - 2^{l+1} - 2$. The result $\tilde{F}_1$ is a spherical skew-framing, where the first $i_{last} - 1$ vectors are ono-to-one.

Denote by $T_\pi : N_2^{2^l-1} \to N_2^{2^l-1}$ the involution with a fold $U(\hat{K}) \subset N_2^{2^l-1}$ of the projection $\pi : N_2^{2^l-1} \to \hat{N}^{2^l-1}$. Denote by $\tilde{T}_\mu : \tilde{N}_2^{2^l-1} \to \tilde{N}_2^{2^l-1}$ the involution by the class μ, denote by $\tilde{T}_\pi : \tilde{N}_2^{2^l-1} \to \tilde{N}_2^{2^l-1}$ the lift of T_π.

The skew-framing (12.60) is a $\tilde{T}_\mu$-equivariant mapping:

$$\tilde{G} : \tilde{N}^{2^l-1} \to S^{2^l-1}. \tag{12.61}$$

This mapping is $\tilde{T}_\pi$-equivariant mapping with a fixed point fold of codimension 1. The total obstruction $o(\Psi_2^{st}, \Psi_0)$ is equal to the degree of the mapping (12.61). This integer is changed into its opposite by the involution $\tilde{T}_\mu$. Let us prove that

$$o(\Psi_2^{st}, \Psi_0) \cong 0 \pmod 4. \tag{12.62}$$

The involution $\tilde{T}_\pi$ is generic with respect to i_{last}- skew-section. This proves that this skew-section is the π-pull-back of a generic skew-section over $\hat{N}^{2^l-1}$ of the bundle $(2^l - 1)\hat{\mu}$. Evidently, the skew-section on $\hat{N}^{2^l-1}$ has an even number of zeros, each is outside $V(\hat{\Sigma}) \subset \hat{N}^{2^l-1}$. This proves (12.62).

Consider the standard immersion $f : S^{2^l-1} \looparrowright \mathbb{R}^{2^{l+1}-2}$ with only one self-intersection point. For the immersion f the standard stable framing Ξ^{st} is well defined and this stable skew-framing can be considered as the corresponding stable skew-framing. For this skew-framing Ξ^{st} we get: $o(\Xi^{st}) = \pm 2$, where a sign is related of a coorientation of the standard immersion f. Take even $-2k$ algebraic copies f_j, $j = 0 \pmod 2$ of the skew-framed immersion (f, Ξ^{st}), join these copies to the immersion (ϕ_2, Ψ_2) by paths, we get a stably skew-framed immersion $(\phi_3, \Psi_3^{st}, \Psi_3)$ with $o(\Psi_3^{st}, \Psi_3) = 0$. The immersion ψ_3 has an even number of self-intersection points: the self-intersection point of ϕ_2 and additionally an even number of self-intersection points from the f_j. As a result, we get a skew-framed immersion $\phi_3 : N_3^{2^l-1} \looparrowright \mathbb{R}^{2^{l+1}-2}$ with an odd number of self-intersection points. Theorem 12.7 is proved.

12.3.4 Discussion

In Sect. 12.3.2, an element $\eta_l^{\mathrm{sf}} \in \mathrm{Imm}^{\mathrm{sf}}(2^l-1, 1)$, $l \geq 3$ is constructed. This element is associated by the Khan–Priddy transfer with the element $\eta_l \in \pi_{2^{l+1}+2}(S^{2^l+2}) \simeq \Pi_{2^l} = \mathrm{Imm}^{\mathrm{fr}}(2^l, 1)$, which is not desuspended 5 times (which is not compressed 3 times in the terminology [5]) from the stable range into the unstable range.

At the same time, for an arbitrary $d \geq 0$, there exists $l_0 = l_0(d)$ such that for an arbitrary $l \geq l_0$ an arbitrary element in the stable homotopy group of spheres Π_{2^l-2} admits a d-fold desuspension, as it is proved in [5]. In [25] it is proved that if the group Π_{126} there exists an element with Kervaire invariant 1 then in the residue class of this element there exists an element which is desuspended 11 times (is compressed 9 times) into the unstable range.

The author is grateful to S.A. Melikhov for discussions on Part 2, to Th.Yu. Popelenskii for discussions on Part 1, and to A.S. Mishchenko for discussions on Part 1 and Part 2.

The author is grateful to Athanase Papadopoulos for English and editorial corrections.

References

1. J.F. Adams, On the non-existence of elements of Hopf-invariant one. Ann Math. Ser. 2 **72**, 20–104 (1960)
2. P.M. Akhmetiev, *Kervaire Problems in Stable Homotopy Theory* (2016). arXiv:1608.01206v1 [math.AT]
3. P.M. Akhmetiev, On the properties of the cobordism group of stably-framed immersions in codimension k. Chebyshevskii Sb. **21**(2), 26–36 (2020)
4. P.M. Akhmetiev, Embedding of compacta, stable homotopy groups of spheres, and singularity theory. Uspekhi Mat. Nauk **55**(3) (333), 3–62 (2000). Russian Math. Surveys **55**(3), 405–462 (2000)
5. P.M. Akhmetiev, *A Geometrical Approach Toward Stable Homotopy Groups of Spheres. A Desuspension Theorem* (Revisited). arXiv:2304.03485
6. P.M. Akhmetiev, *Geometric Approach to Stable Homotopy Groups of Spheres II; Arf–Kervaire Invariants* (v6 to appear). arXiv:2301.08889
7. P.M. Akhmetiev, O.D. Frolkina, On properties of skew-framed immersions cobordism groups. Fundam. Prikl. Mat. **21**(5), 19–46 (2016)
8. P.M. Akhmetiev, P.J. Eccles, A geometric proof of Browder's result on the vanishing of the Kervaire invariant. The report at the Conference on 60-th Birthday of S. P. Novikov. Proc. Steklov Math. Inst. **225**, 40–44 (1999)
9. P.M. Akhmetiev, M. Cencelj, D. Repovs, *'The Arf–Kervaire invariant of framed manifolds as an obstruction to embeddability*. Atti della Accademia Perloritana dei Pericolanti. Classe di Scienze Fisiche, Matematiche e Naturali **100**(2) (2022). arXiv:0804.3164v3 [math.AT] 17 Aug 2010
10. V.I. Arnol'd, Spaces of functions with moderate singularities. Funktsional. Anal. i Prilozhen. **23**(3), 1–10, 96 (1989)
11. M.G. Barratt, J.D.S. Jones, M.E. Mahowald, The Kervaire invariant problem, in *Proceedings of the Northwestern Homotopy Theory Conference*, ed. by H.R. Miller, S.B. Priddy. Contemporary Mathematics, vol. 19 (American Mathematical Society, Providence, 1983), pp. 9–22
12. W. Browder, The Kervaire invariant of framed manifolds and its generalization. Ann. Math. Second Series **90**(1), 157–186 (1969)
13. O. Cépas, P. Akhmetiev, Topological interpretation of color exchange invariants: hexagonal lattice on a torus. Scipost Phys. **10**, 1–20 (2021)
14. M. Mahowald, On the embeddability of the real projective spaces, Proc. Amer. Math. Soc. **13**, 763–764 (1962)
15. P.J. Eccles, Filtering framed bordism by embedding codimension. J. Lond. Math. Soc. (2) **19**, 163–169 (1979)
16. P.J. Eccles, Codimension one immersions and the Kervaire invariant one problem. Math. Proc. Camb. Phil. Soc. **90**, 483 (1981)
17. M.A. Hill, M.J. Hopkins, D.C. Ravenel, On the non-existence of elements of Kervaire invariant one. Ann. Math. (2) **184**, 1–262 (2016). arXiv:0908.3724
18. M.W. Hirsch, Immersions of manifolds. Trans. Am. Math. Soc. **93**, 242–276 (1960)
19. I.M. James, *The Topology of Stiefel Manifolds*. Lecture Notes Series, vol. 24 (Cambridge University Press/London Mathematical Society, Cambridge/London, 1976).
20. J.D.S. Jones, The Kervaire invariant of extended power manifolds. Topology **17**, 249–266 (1978)
21. J. Jones, E. Rees, Kervaire's invariant for framed manifolds. *Algebraic and Geometric Topology, Proceedings of Symposia in Pure Mathematics*, vol. 32 (American Mathematical Society, Providence, 1978), pp. 141–147
22. M. Kervaire, A manifold which does not admit any differentiable structure. Commun. Math. Helv. **34**, 256–270 (1960)
23. U. Koschorke, Multiple points of immersions and the Khan-Priddy theorem. Math. Z. **169**, 223–236 (1979)

24. U. Koschorke, Nonstable and stable monomorphisms of vector bundles. Topol. Appl. **75**, 261–28 (1987)
25. K.Y. Lam, D. Randall, Block bundle obstruction to Kervaire invariant one. Contemp. Math. **407**, 163–171 (2006)
26. M. Mahowald, A new infinite family in ${}_2\pi_*^S$. Topology **16**, 249–256 (1977)
27. H. Minami, *Note on the Divisibility of the Whitehead Square*. http://arxiv.org/abs/2106.14604v3
28. S.P. Novikov, Homotopically equivalent smooth manifolds. I. Izv. Akad. Nauk SSSR Ser. Mat. **28**(2), 365–474 (1964)
29. S.P. Novikov, Algebraic construction and properties of hermitian analogs of K-theory over rings with involution from the viewpoint of hamiltonian formalism. applications to differential topology and the theory of characteristic classes. I. Izv. Akad. Nauk SSSR Ser. Mat. **34**(2), 253–288 (1970). Math. USSR-Izv. **4**(2), 257–292 (1970)
30. V. Poénaru, Some invariants of generic immersions and their geometric applications. Bull. Am. Math. Soc. **81**, 1079–1082 (1975)
31. L.S. Pontryagin, *Smooth Manifolds and Their Applications to Homotopy Theory*, 2nd edn. (Nauka, 1976). English transl.: Amer. Math. Soc. Transl. (2) 11 (1959) 1–114; reprinted (1964); L. S. Pontryagin, Selected Works, Vol. 3, Algebraic and Differential topology, pp. 115–252, Gordon and Breach Scientific Publishers (1986)
32. C.P. Rourke, B.J. Sanderson, The compression theorem, I.. Geom. Topol. **5**, 399–429 (2001)
33. V.P. Snaith, *Stable Homotopy Around the Arf–Kervaire Invariant* (Birkhäuser Verlag AG Basel/Boston/Berlin, 2009)
34. A. Szücs. *Geometrical Proof of a Theorem on Suspension in Homotopy Groups of Spheres*. Janos Bolyai 55 Topology Pecs (Colloquia Mathematica Society, Hungary, 1989), pp. 501–505
35. H. Toda, *Composition Methods in the Homotopy Groups of Spheres*, vol. 49 (Annals of Mathematics Studies, Princeton, 1962)
36. V.A. Vassiliev, *Complements of Discriminants of Smooth Maps: Topology and Applications*. Translations of Mathematical Monographs, vol. 98, , 2-d extended edn. (AMS, Providence, 1994), 268 pp.

Chapter 13
Lifting Generic Maps to Embeddings. The Double Point Obstruction

Sergey A. Melikhov

Abstract Given a generic PL map or a generic smooth fold map $f\colon N^n \to M^m$, where $m \geq n$ and $2(m+k) \geq 3(n+1)$, we prove that f lifts to a PL or smooth embedding $N \to M \times \mathbb{R}^k$ if and only if its double point locus $\{(x, y) \in N \times N \mid f(x) = f(y),\ x \neq y\}$ admits an equivariant map to S^{k-1}. As a corollary we answer a 1990 question of P. Petersen and obtain some other applications.

We also discuss several criteria for lifting a non-degenerate PL map or a C^0-stable smooth map $f\colon N^n \to M^m$, where $m \geq n$, to an embedding in $M \times \mathbb{R}$, elaborating on V. Poénaru's observations. In particular, the existence of such a lift is determined by the equivariant homotopy type of the diagram consisting of the three projections from the triple point locus $\{(x, y, z) \in N \times N \times N \mid f(x) = f(y) = f(z),\ x \neq y \neq z \neq x\}$ to the double point locus.

The three Appendices, which can be read independently of the rest of this chapter, are devoted to stable and generic maps. Appendix B introduces an elementary theory of stable PL maps. Appendix C extends the 2-multi-0-jet transversality theorem over the usual compactification of $M \times M \setminus \Delta_M$.

Keywords Projected embedding · Stable PL map · Whitney trick · Penrose staircase · Multijet transversality · Simple fold map

AMS Codes 57Q35, 57Q65, 57R35, 57R40

Disclaimer: Some published papers which refer to the present paper use the author's numbering of theorems and other statements, which is quite different from the Publisher's numbering found below. Please refer to the arXiv version of the present paper (arXiv:1711.03518) for the author's numbering.

S. A. Melikhov (✉)
Steklov Mathematical Institute of Russian Academy of Sciences, Moscow, Russia
e-mail: melikhov@mi-ras.ru

L. Funar, A. Papadopoulos (eds.), *Essays on Topology*,
https://doi.org/10.1007/978-3-031-81414-3_13

13.1 Introduction

A continuous/PL/smooth map $f\colon N \to M$ is called a topological/PL/smooth *k-prem* (from "**pr**ojected **em**bedding") if there exists a map $g\colon N \to \mathbb{R}^k$ such that $f \times g\colon N \to M \times \mathbb{R}^k$ is a topological/PL/smooth embedding. When the choice of a category is irrelevant, we will speak simply of "k-prems". A discussion of this definition, including a comparison of the three categories, as well as a number of references on k-prems can be found in the companion paper [39].

The main objective of the present chapter is to determine algebraically, under some reasonable hypotheses, whether a given map f is a k-prem. The main theorem of the present chapter is applied in [7] and was originally motivated by that application.

For a space N let $\Delta_N = \{(x, x) \in N \times N\}$ and $\tilde{N} = N \times N \setminus \Delta_N$. We endow $\tilde{N}$ with the factor exchanging involution, and we also endow each sphere S^n with the antipodal involution. By a "map" we will mean a continuous map. Given a map $f\colon N \to M$, let $\Delta_f = \{(x, y) \in \tilde{N} \mid f(x) = f(y)\}$.

A necessary condition for $f\colon N \to M$ to be a k-prem is the existence of an equivariant map $\tilde{g}\colon \Delta_f \to S^{k-1}$. Namely, $\tilde{g}(x, y) = \frac{g(y)-g(x)}{\|g(y)-g(x)\|}$, where $g\colon N \to \mathbb{R}^k$ is a map such that $f \times g\colon N \to M \times \mathbb{R}^k$ is an embedding. In general, this condition is not sufficient.

Example 13.1.1 The 3-fold covering $f\colon S^1 \to S^1$ is not a 1-prem, but there exists an equivariant map $\Delta_f \to S^0$. Indeed, Δ_f is homeomorphic to $S^1 \sqcup S^1$, where the involution permutes the two components.

Also there exists a non-degenerate[1] PL map f from S^1 to the triod (=the cone over the three-point set) which is not a 1-prem even though there exists an equivariant map $\Delta_f \to S^0$ [24, Example 5]. Namely, f is the composition of the 3-fold covering $S^1 \to S^1$ and the unique simplicial surjection from the boundary of the hexagon onto the triod.

However, if f is a stable[2] PL map of a graph to $\mathbb{R}^1$, then f is a 1-prem if and only if there exists an equivariant map $\Delta_f \to S^0$ [24, Theorem 11] (this is deduced in [24] from the result of [18]; in connection with the latter see also [54], [9]). Let us note that there exists a stable PL map of a trivalent *tree* to $\mathbb{R}^1$ which is not a 1-prem [52] (see also [8, §3.1]).

More examples will be discussed toward the end of this introduction.

Theorem 13.1 *Suppose that $m \geq n$ and $2(m + k) \geq 3(n + 1)$. Let $f\colon N^n \to M^m$, where N is compact, be one of the following:*

[1] A PL map is called *non-degenerate* if it has no point-inverses of dimension > 0.

[2] Stable PL maps are defined in Appendix B. A PL map f of a graph in $\mathbb{R}^1$ is stable if and only if every point-inverse $f^{-1}(y)$ contains at most one point x such that f does not restrict to a homeomorphism between a neighborhood of x and a neighborhood of y.

(a) a generic PL map of a polyhedron to a PL manifold;
(b) a generic smooth fold map[3] *between smooth manifolds;*
(c) a generic smooth map between smooth manifolds, where $3n - 2m \leq k$.

If there exists an equivariant map $\Delta_f \to S^{k-1}$, *then* f *is a PL* k*-prem in the case of (a) and a smooth* k*-prem in the cases of (b) and (c).*

In this statement and in what follows we assume the following conventions.

(1) The dimension of a manifold or a polyhedron is often indicated by a superscript.
(2) "Generic" is understood as "satisfying a certain fixed set of conditions that determine an open and dense set of maps" (see Sect. 13.2 for the details). This is a slight variation of Gromov's terminology [25, 1.3.2(B)] (what he calls "generic" will be called "weakly generic" in Appendix C).

In order to be able to apply Theorem 13.1 to specific maps, one may want to know explicitly the dense open sets whose existence it asserts. It is not hard to see (see Propositions 13.2.1 and 13.2.2) that parts (a) and (b) are equivalent to their versions with "generic" replaced by "stable".[4] Also, it is easy to see that part (c) implies its version with "generic" replaced by "stable" (see the proof of Proposition 13.2.1). The converse implication holds in a wide range of dimensions (in particular, for all $n \leq 8$), since by Mather's theorem [32] generic smooth maps $N^n \to M^m$, where $m \geq n$, are stable if either $6m \geq 7n - 7$, or $6m \geq 7n - 8$ and $m \leq n + 3$.

Corollary 13.1.2 *Theorem 13.1 is true if "generic" is replaced by "stable".*

Remark 13.1.3 In Theorem 13.1(a) one may also replace "generic" by another explicit condition, which is less restrictive and easier to check than stability (see Remark 13.3.7, which uses the notation introduced in the first few lines of Sect. 13.3.2).

As for part (c) of Theorem 13.1, it is reduced to a relative version of (b) by a half-page argument (see Sect. 13.3.1), and we leave it to the interested reader to formulate an explicit list of genericity conditions that are needed for that argument. That list is not going to be short, although it might be possible to make it shorter by extra work. However, the author is not convinced that part (c) is a definitive result; it seems more likely that the real theorem in this direction has yet to be proved (see Remark 13.1.7).

Remark 13.1.4 Theorem 13.1 is obvious for immersions without triple points. In this case, the projection $\pi : N \times N \to N$ embeds Δ_f. Then any extension of the composition $\pi(\Delta_f) \xrightarrow{\pi^{-1}} \Delta_f \xrightarrow{\varphi} S^{k-1} \subset \mathbb{R}^k$ to a PL or smooth map $g : N \to \mathbb{R}^k$ yields a PL or smooth embedding $f \times g : N \hookrightarrow M \times \mathbb{R}^k$.

[3] A *smooth fold map* is a smooth map whose only singularities are of the fold type. In particular, every smooth immersion is trivially a smooth fold map.

[4] Stable smooth maps are discussed in a number of textbooks, including [23], [10]. See also Appendix A.

It is not hard to elaborate on this construction so as to prove Theorem 13.1 for generic maps without triple points (see Proposition 13.3.6 for the PL case). Let us note that a generic smooth map without triple points is a smooth fold map.

Remark 13.1.5 If $k = 1$, then generic maps $N^n \to M^m$, where $2(m + k) \geq 3(n + 1)$, have no triple points. Thus the case $k = 1$ in Theorem 13.1 is also quite easy.

Remark 13.1.6 A natural approach to proving Theorem 13.1 is by trying to adapt Haefliger's generalized Whitney trick (see [1, §VII.4]; see also [55] for an alternative proof). In fact, our proof below can be used to embed a core part of Haefliger's "standard model" into a generic lift of f in a way that agrees with the projection $M \times \mathbb{R}^k \to M$. However, it is far from clear how one could possibly construct appropriate global Haefliger-style framings especially when Δ_f is not immersed in N.

We prove Theorem 13.1 by what can be called a new kind of generalized Whitney trick. It contrasts with the Whitney–Haefliger(–Koschorke) approach in that ours describes the desired homotopy essentially by an explicit formula. (The formula is only aware of pairs of points, and may overlook triple-point information; because of this, the homotopy given by the formula has to be slightly perturbed.) Haefliger's construction is less explicit in that it depends on the choice of an embedding of the "standard model", whose existence is proved using obstruction theory. However, it is difficult to compare the two versions of generalized Whitney trick directly because they apply under incompatible hypotheses.

Remark 13.1.7 Theorem 13.1(c) falls short of the announcements in [34, third remark after Theorem 5] and [36, §1], where it was conjectured to hold under the weaker restriction $4n - 3m \leq k$. (See Remark 13.4.2 below for an explanation.) The main obstacle in extending Theorem 13.1(b) to maps with more general singularities is that already in the presence of cusps (i.e., singularities of type $\Sigma^{1,1,0}$) it is no longer easy to maintain the connection between the geometric maps and the configuration space data throughout the proof.

Proposition 13.1.8 *Every stable smooth (PL) map $f: N^n \to M^{2n+1-k}$ is a smooth (PL) k-prem.*

Proof We will only discuss the smooth case here; the PL case is proved similarly. Let $g: N \to M \times \mathbb{R}^k$ be a smooth embedding that is sufficiently C^∞-close to the composition $N \xrightarrow{f} M \xrightarrow{\iota} M \times \mathbb{R}^k$, where ι is the inclusion onto $M \times \{0\}$. Since f is C^∞-stable, we may assume that the composition $N \xrightarrow{g} M \times \mathbb{R}^k \xrightarrow{\pi} M$, where π is the projection, is C^∞-left-right-equivalent to f. But then clearly g itself lifts to a smooth embedding. □

In the PL case, one can actually say more:

Theorem 13.2 *Every non-degenerate PL map of compact n-polyhedra is a PL $(n+1)$-prem.*

Proof The given PL map is triangulated by a simplicial map $f\colon K \to L$. This in turn yields a simplicial map $K' \to L'$ between the barycentric subdivisions. For each i-simplex τ of L let us embed the preimage $f^{-1}(\hat\tau)$ of its barycenter $\hat\tau$ into the $(i+1)$st coordinate axis $\langle e_{i+1}\rangle$ of $\mathbb{R}^{n+1}$. Let us define $g\colon K \to \mathbb{R}^{n+1}$ by these embeddings on the vertices of K' and by extending linearly to each simplex of K'. It remains to show that $f \times g\colon |K| \to |L| \times \mathbb{R}^{n+1}$ is injective.

Given distinct points $x, y \in |K|$ such that $f(x) = f(y)$, they lie in the interiors of some distinct simplexes s, t of K' such that $f(s) = f(t)$. Now s is a join $\hat\sigma^{i_0} * \cdots * \hat\sigma^{i_m}$, where $\sigma^{i_0} \subsetneq \cdots \subsetneq \sigma^{i_m}$ are simplexes of K. Moreover, each $f(\sigma^{i_k}) = f(\tau^{i_k})$, where τ^{i_k} is the corresponding factor of the similar decomposition $t = \hat\tau^{i_0} * \cdots * \hat\tau^{i_m}$. Since $s \neq t$, we have $\sigma^{i_m} \neq \tau^{i_m}$ and consequently $\hat\sigma^{i_m} \neq \hat\tau^{i_m}$. Then also $g(\hat\sigma^{i_m}) \neq g(\hat\tau^{i_m})$.

Writing $r = f(s) = f(t)$, we similarly have $r = \hat\rho^{i_0} * \cdots * \hat\rho^{i_m}$. Thinking of r as a subset of $\mathbb{R}^m$, each line $L_k := \hat\rho^{i_k} \times \langle e_{i_k+1}\rangle$ is skew with respect to the affine subspace A_k of $\mathbb{R}^m \times \mathbb{R}^{n+1}$ spanned by $L_0, \dots, L_{k-1}$ (that is, L_k and A_k are disjoint and contain no parallel vectors). Therefore the join $L_0 * \cdots * L_m$ is contained in $\mathbb{R}^m \times \mathbb{R}^{n+1}$, and in fact in $r \times \mathbb{R}^{n+1}$. Since $f \times g$ sends $\hat\sigma^{i_m}$ and $\hat\tau^{i_m}$ into distinct points of L_m, it must also send x and y to distinct points of $L_0 * \cdots * L_m \subset r \times \mathbb{R}^{n+1}$. □

Remark 13.1.9 Theorem 13.2 yields a simple explicit construction of an embedding of a given n-polyhedron in $\mathbb{R}^{2n+1}$ (see [40, 1.2] for another simple explicit construction). Indeed, the barycentric subdivision K' of a given n-dimensional simplicial complex K admits a non-degenerate simplicial map onto the n-simplex Δ^n (by sending the barycenter of each i-simplex of K to the $(i+1)$st vertex of Δ^n and extending linearly).

The following application of Theorem 13.1 is deduced from it in Sect. 13.6 by elaborating on some arguments due to P. M. Akhmetiev.

Theorem 13.3 *Let $f\colon N \to M$ be a stable smooth map between stably parallelizable smooth n-manifolds, where $n \neq 1, 2, 3, 7$. Then f is an n-prem.*

Akhmetiev himself proved that every f as above is *n-realizable* in the sense that the composition $N \xrightarrow{f} M \xrightarrow{\iota} M \times \mathbb{R}^n$ with the inclusion ι onto $M \times \{0\}$ is C^0-approximable by smooth embeddings [3, 4] (see also [34, §5] and [5] for alternative proofs). It is possible to deduce Theorem 13.3 from the statement of this original result of Akhmetiev (although the direct proof of Theorem 13.3 given in Sect. 13.6 is simpler than the one given by this deduction). Namely, it is proved in [7] using Theorem 13.1 of the present chapter that if $f\colon N \to M$ is a stable smooth map between smooth n-manifolds (resp. a stable PL map from an n-polyhedron to a PL n-manifold), where $n \geq 3$, then f is n-realizable if and only if it is a smooth (resp. PL) n-prem. In fact, the equivalence of k-realizability and the k-prem property is proved in [7] under much weaker hypotheses, which are however strictly stronger than those of Theorem 13.1. Moreover, there are explicit examples in [7] showing that a map f as in Theorem 13.1 can be k-realizable without being a k-prem.

Theorem 13.3 fails for $n = 1, 3, 7$ by considering e.g. the double cover $S^n \to \mathbb{R}P^n$ (cf. [34, Example 3]). It is unknown if it holds for $n = 2$. Strong partial results on this problem were obtained by L. Funar and P. Pagotto [20]; see also [36] for another approach.

Here is another application of Theorem 13.1 (the proof is in Sect. 13.6).

Theorem 13.4 *Let N be a PL (smooth) $\mathbb{Z}/2$-homology n-sphere, M an orientable PL (smooth) n-manifold, $n > 2$. A stable PL (smooth) map $f: N \to M$ is a PL (smooth) n-prem if either*

(a) $\deg(f)$ *is zero or odd; or*
(b) $f_*: \pi_1(N) \to \pi_1(M)$ *is onto.*

By a PL $\mathbb{Z}/2$-homology n-sphere we mean a PL manifold (not just a polyhedral homology manifold) with the same mod 2 homology as S^n.

Theorem 13.4(b) implies, in particular, the following:

Corollary 13.1.10 *Stable smooth maps $S^n \to S^n$ are smooth n-prems for $n > 2$.*

For $n = 2$ this is also true [34, Proof of the Yamamoto–Akhmetiev Theorem]. In the cases $n = 3, 7$ Corollary 13.1.10 is a rather delicate result, in the sense that some closely related assertions are false:

- the universal covering map of the Poincaré homology sphere is not a 3-prem [34, Example 3];
- a certain stable self-map of the Poincaré homology sphere is not a 3-prem [34, Remark to Theorem 4];
- for $n = 3, 7$ there exists an equivariant map $F: S^n \times S^n \to S^n \times S^n$ such that $F|_{\tilde{S}^n}$ is transverse to Δ_{S^n} and $(F|_{\tilde{S}^n})^{-1}(\Delta_{S^n})$ admits no equivariant map to S^{n-1} [6] (see also Lemma 13.6.3 below).[5]

Corollary 13.1.10 is also a consequence of the author's previous result that stable smooth maps $S^n \to S^n$ are n-realizable for $n > 2$ [34] and the aforementioned result of [7], proved using Theorem 13.1 of the present chapter.

Theorem 13.4(a) yields a negative solution to Petersen's problem [43, end of §3]: does there exist a 3-dimensional lens space $L(p, q)$ with p odd whose universal covering is not a 3-prem? (The condition "with p odd" is not explicitly stated in Petersen's question, but is implied by the preceding discussion and certainly by his observation on the preceding page that no even degree covering $S^3 \to M^3$ is a 3-prem.)

Corollary 13.1.11 *The universal covering of every 3-dimensional lens space $L(p, q)$ with p odd is a 3-prem.*

[5] It is also asserted in the main result of [6] that such equivariant maps $S^n \times S^n \to S^n \times S^n$ do not exist for $n \neq 1, 3, 7$. A proof of this assertion can be found in the present chapter, in the proof of Theorem 13.3 in Sect. 13.6; see also [5] for a different approach.

The following example shows that the restriction $2(m + k) \geq 3(n + 1)$ in Theorem 13.1 cannot be relaxed.

Example 13.1.12 It is easy to construct a self-transverse immersion f of the Möbius band in $\mathbb{R}^3$ such that Δ_f is homeomorphic to S^1 with the antipodal involution. On the other hand, if f is a self-transverse immersion of an orientable 2-manifold N in $\mathbb{R}^3$, it is easy to see that Δ_f admits an equivariant map to S^0. Nevertheless, there exist self-transverse immersions of orientable 2-manifolds in $\mathbb{R}^3$ that are not 1-prems [22, §2], [2] (see also [16], [50], [8, §3.3]). Below we describe a variant of this construction, which is particularly simple in that it has only one triple point in the case of a surface with boundary, and simultaneously extend it to higher dimensions.

(a) Let us first describe an immersion $\varphi\colon G \to H$ between graphs with only one triple point such that Δ_φ admits an equivariant map to S^0 but φ is not a 1-prem.

Our graphs will have loops and multiple edges. Let H consist of one vertex v (of degree 6) and three edges A, B, C. Let G consist of three vertices a, b, c (each of degree 4) and the following six edges:

- A' and B'' between a and b;
- B' and C'' between b and c;
- C' and A'' between c and a.

We define φ by $\varphi(a) = \varphi(b) = \varphi(c) = v$, $\varphi(A') = \varphi(A'') = A$, $\varphi(B') = \varphi(B'') = B$ and $\varphi(C') = \varphi(C'') = C$.[6] It is easy to see that Δ_φ admits an equivariant map to S^0. However, the pairs (a, b), (b, c) and (c, a) lie in the same component of Δ_φ.[7] Hence if $g\colon G \to \mathbb{R}$ is a map such that $\varphi \times g\colon G \to H \times \mathbb{R}$ is an embedding, then $g(a) - g(b)$, $g(b) - g(c)$ and $g(c) - g(a)$ must all have the same sign. However, this contradicts the fact that their sum is 0.[8] Thus φ is not a 1-prem.

(b) Next we construct an orientable $2n$-manifold M with boundary and a self-transverse immersion $f\colon M \to \mathbb{R}^{3n}$ with only one triple point such that Δ_f admits an equivariant map to S^0 but f is not a 1-prem.

This f will be modeled on φ, in the sense that G and H will appear as the images of Δ_f in M and in $\mathbb{R}^3$, respectively, and φ will appear as the restriction of f.

Let $I = [-1, 1]$, and let us consider three $2n$-disks $D_{xy} = I^n \times I^n \times 0$, $D_{yz} = 0 \times I^n \times I^n$ and $D_{zx} = I^n \times 0 \times I^n$ in the three coordinate $2n$-planes in $\mathbb{R}^{3n}$. (These disks are going to be neighborhoods of the vertices a, b, c respectively, or rather their images under the immersion.) Let $p_\pm = (\pm 1, 0, \dots, 0) \in I^n$ and let $x_\pm$, $y_\pm$ and $z_\pm$ be the images of $p_\pm$ in $I^n \times 0 \times 0$, $0 \times I^n \times 0$ and $0 \times 0 \times I^n$.

---

6 Let us note that if $g\colon \mathbb{R}P^2 \to \mathbb{R}^3$ is the Boy's surface, then the restriction of g to the image of Δ_g in $\mathbb{R}P^2$ can be described similarly, except that A'' will connect c to itself, B'' will connect a to itself, and C'' will connect b to itself.

7 In fact, this remains true even if the edges C, C' and C'' are deleted from G and H.

8 In fact, $\{(a, b, c)\}$ is the simplest example of what is called a Penrose staircase in Theorem 13.5.

Let us consider three arcs with interiors outside I^{3n}: an arc J_{xy} connecting x_+ with y_-; an arc J_{yz} connecting y_+ with z_-; and an arc J_{zx} connecting z_+ with x_-. (These arcs are going to lie in the edges A, B, C respectively.) Let e_i be the standard ith basis vector in $\mathbb{R}^n$, and let v_i^x, v_i^y and v_i^z be its images in $\mathbb{R}^n \times 0 \times 0$, $0 \times \mathbb{R}^n \times 0$ and $0 \times 0 \times \mathbb{R}^n$.

We may assume that J_{xy} lies in the 2-plane through 0 spanned by v_1^x and v_1^y, and similarly for J_{yz} and J_{zx}. The normal bundle λ of J_{xy} in this plane is a line subbundle in the normal bundle ν of J_{xy} in $\mathbb{R}^{3n}$. Let us note that λ is spanned by (a translate of) v_1^y over x_+ and by v_1^x over y_-. Another line subbundle λ' in ν is spanned by v_1^z. The 2-bundle $\lambda \oplus \lambda'$ can also be represented as $\lambda_1 \oplus \lambda_1'$, where λ_1 and λ_1' coincide with λ and λ' over x_+, and then start rotating away from them, eventually (i.e. over y_-) rotating through an angle of $\pi/2$. If we do the rotation counterclockwise, then the frame (v_1^y, v_1^z) over x_+ is transported along $\lambda_1 \oplus \lambda_1'$ into the frame $(v_1^z, -v_1^x)$ over y_-.

For each $i > 1$ the vectors v_i^x, v_i^y and v_i^z also span line subbundles μ_x, μ_y and μ_z of ν. The 3-bundle $\mu_x \oplus \mu_y \oplus \mu_z$ can also be represented as $\mu_i \oplus \mu_i' \oplus \mu_i''$, where μ_i, μ_i' and μ_i'' coincide with μ_x, μ_y and μ_z over x_+, and then start rotating around the vector $v_i^x + v_i^y + v_i^z$, so that eventually the three vectors exchange cyclically according to the permutation (xyz). In other words the frame (v_i^x, v_i^y, v_i^z) over x_+ is transported along $\mu_i \oplus \mu_i' \oplus \mu_i''$ into the frame (v_i^y, v_i^z, v_i^x) over y_-.

Let us now connect D_{xy} with D_{yz} by the total space of the $(2n-1)$-disk bundle over J_{xy} associated with $\lambda_1 \oplus \mu_2 \cdots \oplus \mu_n \oplus \mu_2' \oplus \cdots \oplus \mu_n'$. (The union of this total space with D_{xy} and D_{yz} is going to be a neighborhood of the edge A', or rather its image under the immersion.) We also connect D_{zx} with D_{xy} by the total space of the $(2n-1)$-disk bundle over J_{xy} associated with $\lambda_1' \oplus \mu_2 \oplus \cdots \oplus \mu_n \oplus \mu_2'' \cdots \oplus \mu_n''$. (This is going to be an immersed neighborhood of A''.) We similarly connect D_{xy} with D_{yz} and D_{yz} with D_{zx} along J_{yz}. (These are going to be immersed neighborhoods of B' and B'', respectively.) Also we connect D_{yz} with D_{zx} and D_{zx} with D_{xy} along J_{zx}. (These are going to be immersed neighborhoods of C' and C'', respectively.)

Thus we obtain an orientable manifold with boundary M^{2n} and a self-transverse immersion $f\colon M \to \mathbb{R}^{3n}$ with one triple point which extends φ so that $\Delta_f = \Delta_\varphi$. Hence Δ_f admits an equivariant map to S^0 but f is not a 1-prem.

(c) Finally we show that the previous construction can be improved so as to get rid of the boundary. Thus we will construct a closed orientable $2n$-manifold W and a self-transverse immersion $F\colon W \to \mathbb{R}^{3n}$ (with several triple points) such that Δ_F admits an equivariant map to S^0 but F is not a 1-prem.

First we note that in the previous construction, M is a $2n$-dimensional thickening of the graph G, that is, a handlebody with three 0-handles and six 1-handles; and f factors through an immersion $f\colon (M, \partial M) \to (N, \partial N)$, where N is a $3n$-dimensional thickening of the graph H, that is, a handlebody with one 0-handle h and three 1-handles h_i.

Let B be obtained from N by attaching three 2-handles h_i' so that they cancel geometrically the 1-handles h_i. (Thus B is homeomorphic to the $3n$-

ball.) The 2-handles are attached to N along thin tubular neighborhoods T_i of three embedded circles $C_i \subset \partial N$, and it is not hard to ensure that each $C_i \cap h_i$ is an arc disjoint from $f(\partial M)$, and each $C_i \cap h$ is an arc which meets $f(\partial M)$ transversely (if at all). Thus $f(\partial M)$ lies entirely in ∂B and in the $(3n-1)$-disks $Q_i := T_i \cap h$. Moreover, each $f(\partial M) \cap Q_i$ consists of parallel copies of an unknotted $(2n-1)$-disk in the $(3n-1)$-ball Q_i (since the tubular neighborhood T_i is thin and the intersections of its core circle C_i with $f(\partial M)$ are transversal).

Now the double $2M := M \cup_{\partial M = \partial M} M$ is immersed by $2f$ into $S^{3n} = B \cup_{\partial B = \partial B} B$, but still has boundary, which is embedded by $2f$ into the annuli $2Q_i \cong D^1 \times S^{3n-2}$ which in turn lie in the tori $2T_i \cong S^1 \times S^{3n-2}$. These tori bound the solid tori $2h'_i \cong D^2 \times S^{3n-2}$; and to these solid tori there are attached the $(3n-1)$-handles $2h_i \cong D^1 \times D^{3n-1}$. The unions $U_i := 2h'_i \cup 2h_i \cong D^1 \times D^{3n-1}$ can also be viewed as $(3n-1)$-handles, whose bases are our annuli $2Q_i \cong D^1 \times S^{3n-2}$ containing the image of $\partial(2M)$. Now the image of $\partial(2M)$ in each $2Q_i$ consists of parallel copies of an embedded $(2n-1)$-sphere, which is unknotted in the $(3n-1)$-sphere ∂U_i; consequently they bound parallel $2n$-disks in the $3n$-ball U_i. The desired immersion F is the extension of $2f$ by means of these embedded disks. Although the interior of each of these disks intersects the image of $2f(M)$, the intersection corresponds to two new components of Δ_F which are interchanged by the involution, so the equivariant map $\Delta_{2f} \to S^0$ yields an equivariant map $\Delta_F \to S^0$.

For a space X let $\Delta_X^{(3)} = \{(x, x, x) \in X^3\}$ and $X^{(3)} = \{(x, y, z) \in X^3 \mid x \neq y \neq z \neq x\}$, and for a map $f\colon X \to Y$ let $\Delta_f^{(3)} = \{(x, y, z) \in X^{(3)} \mid f(x) = f(y) = f(z)\}$.

Theorem 13.5 *Let $f\colon X \to Y$ be a non-degenerate PL map between compact polyhedra. The following are equivalent:*

(1) *f is a PL 1-prem;*
(2) *there exist an equivariant map $\varphi\colon \Delta_f \to S^0 = \{1, -1\}$ and for each $y \in Y$ a total ordering of $f^{-1}(y)$ such that $x < x'$ if and only if $\varphi(x, x') = 1$;*
(3) *there exists an equivariant map $\varphi\colon \Delta_f \to S^0$ such that $\varphi^3\colon (\Delta_f)^3 \to (S^0)^3$ sends every triple of the form $\big((x, y), (y, z), (z, x)\big)$ into the complement of $\Delta_{S^0}^{(3)}$;*
(4) *f has no Penrose staircases, where a* Penrose staircase *for f is a finite subset $F \subset \Delta_f^{(3)}$ such that for every equivariant map $\varphi\colon \Delta_f \to S^0$ there exists an $(x, y, z) \in F$ such that $\varphi(x, y) = \varphi(y, z) = \varphi(z, x)$;*
(5) *there exist an S_2-equivariant map $\Phi\colon \Delta_f \to \tilde{\mathbb{R}}$ and an S_3-equivariant[9] map $\Psi\colon \Delta_f^{(3)} \to \mathbb{R}^{(3)}$ that, given any choice of an embedding $j\colon \{1, 2\} \to \{1, 2, 3\}$, commute with the corresponding projections $\pi_j\colon \Delta_f^{(3)} \to \Delta_f$ and $\Pi_j\colon \mathbb{R}^{(3)} \to \tilde{\mathbb{R}}$.*

[9] S_i denotes the symmetric group.

Theorem 13.5 elaborates on some results found in the literature, most notably in Poénaru's paper [44]. The implication (3)⇒(2) is implicit in the proof of [44, Lemme 1.5]. The implication (2)⇒(1) for generic smooth immersions between manifolds is implicit in [44, Lemme 1.4] (whose proof is left to the reader); it is proved in [22, Proposition 4] for generic immersions of orientable 2-manifolds in $\mathbb{R}^3$ and in [24, Theorem 2] in the general case. See also [17] for a related result.

Yet another condition equivalent to those of Theorem 13.5 is contained in A. Gorelov's recent paper [24, Theorem 4]. His approach is somewhat different (as compared to conditions (4) and (5)), and can be regarded as a natural correction of the approach of [44] (whose main result is incorrect, as shown in [24, Theorem 5]).

Proof The implications (1)⇒(2) ⇒(3) ⇒(4) are trivial.

(2)⇒(1). Let K and L triangulate X and Y so that f is triangulated by a simplicial map $K \to L$. For each vertex v of L let us pick a monotone embedding of $f^{-1}(v)$ in $\mathbb{R}$. This defines a map $g\colon X \to \mathbb{R}$ on the vertices of K, and we extend it linearly to each simplex of K. If $(x, y) \in \Delta_f$, then $x = \sum_i \lambda_i v_i$ and $y = \sum_i \lambda_i w_i$ for some $\lambda_i > 0$ such that $\sum_i \lambda_i = 1$ and some vertices v_i and w_i of K such that $f(v_i) = f(w_i)$ for each i, and $v_i \neq w_i$ for at least one i. Clearly, each pair (v_i, w_i) lies either in Δ_X or in the same component of Δ_f as (x, y). In the latter case $\varphi(v_i, w_i) = \varphi(x, y)$ for each i, and therefore each $g(v_i) - g(w_i)$ is of the same sign as $\varphi(x, y) \in S^0 = \{1, -1\}$. Then $g(x) - g(y) = \sum_i \lambda_i \big(g(v_i) - g(w_i)\big)$ is also of that sign, and in particular $g(x) \neq g(y)$. Hence $f \times g\colon X \to Y \times \mathbb{R}$ is injective.

(3)⇒(2). Given an $y \in Y$, let us define a binary relation $<$ on $f^{-1}(y)$ by $x < x'$ if $x \neq x'$ and $\varphi(x, x') = 1$. By (3) $<$ is transitive. Hence it is a strict total order.

(4)⇒(3). Clearly, Δ_f has only finitely many connected components. Hence there exist only finitely many distinct equivariant maps $\Delta_f \to S^0$. Let us denote them by $\varphi_1, \dots, \varphi_n$. If (3) does not hold, then each φ_i sends some triple $\big((x_i, y_i), (y_i, z_i), (z_i, x_i)\big)$ into $\Delta^{(3)}_{S^0}$. Then the triples (x_i, y_i, z_i), $i = 1, \dots, n$, form a Penrose staircase.

(5)⇒(3). The natural embedding $e\colon X^{(3)} \to (\tilde{X})^3$, $e(x, y, z) = \big((y, z), (z, x), (x, y)\big)$, yields embeddings $e_f\colon \Delta^{(3)}_f \to (\Delta_f)^3$ and $e_{\mathbb{R}}\colon \mathbb{R}^{(3)} \to (\tilde{\mathbb{R}})^3$. Clearly, Φ and Ψ commute with the projections π_j and Π_j for each j if and only if Φ^3 and Ψ commute with e_f and $e_{\mathbb{R}}$. On the other hand, if $h\colon \tilde{\mathbb{R}} \to S^0$ is a homotopy equivalence, then $h^3\colon (\tilde{\mathbb{R}})^3 \to (S^0)^3$ sends the image of $e_{\mathbb{R}}$ onto the complement of $\Delta^{(3)}_{S^0}$.

(3)⇒(5). In the notation of the proof of (5)⇒(3), let us note that e is equivariant with respect to the action of S_3 on $(\tilde{X})^3$ given by $\sigma(p_1, p_2, p_3) = (t^\sigma p_{\sigma(1)}, t^\sigma p_{\sigma(2)}, t^\sigma p_{\sigma(3)})$, where $t^\sigma\colon \tilde{X} \to \tilde{X}$ is the factor exchanging involution if σ is odd and the identity if σ is even. Hence the composition $\Delta^{(3)}_f \xrightarrow{e_f} (\Delta_f)^3 \xrightarrow{\varphi^3} (S^0)^3 \setminus \Delta^{(3)}_{S^0}$ is equivariant with respect to the free action of S_3 on $(S^0)^3 \setminus \Delta^{(3)}_{S^0}$ given by $\sigma(\epsilon_1, \epsilon_2, \epsilon_3) = \big((-1)^\sigma \epsilon_{\sigma(1)}, (-1)^\sigma \epsilon_{\sigma(2)}, (-1)^\sigma \epsilon_{\sigma(3)}\big)$. Also, the composition $\mathbb{R}^{(3)} \xrightarrow{e_{\mathbb{R}}} (\tilde{\mathbb{R}})^3 \xrightarrow{h^3} (S^0)^3 \setminus \Delta^{(3)}_{S^0}$ is an S_3-equivariant homotopy equivalence. Hence we obtain the desired S_3-map Ψ. □

Corollary 13.1.13 *Let $f: X^n \to Y^m$ be a C^0-stable smooth map between compact smooth manifolds, where $n \leq m$. Then the assertion of Theorem 13.5 with "PL 1-prem" replaced by "smooth 1-prem" holds for f.*

Proof By Verona's theorem (see Theorem A.1) f is C^0-left-right equivalent to a PL map f'. Since $n \leq m$, both maps are non-degenerate. By [39, Theorem B] f is a smooth 1-prem if and only if it is a topological 1-prem. The latter is equivalent to f' being a topological 1-prem. Now the assertion follows from Theorem 13.5. □

Problem 13.1.14 Let $f: X \to Y$ be a finite-to-one map between compacta. Does the assertion of Theorem 13.5 with "PL 1-prem" replaced by "topological 1-prem" hold for f?

13.2 Generic Maps

13.2.1 Generic Smooth Maps

In the present chapter we adhere to the following understanding of the term *generic*. It can be used only in assertions of certain types. First we consider the smooth case. Let N and M be smooth manifolds, where N is compact.[10] The assertion "every generic smooth map $N \to M$ satisfies property P" (or any logically equivalent assertion) means "$C^\infty(N, M)$ contains a dense open subset whose elements satisfy property P". The assertion "every generic smooth fold map $N \to M$ satisfies property P" (or any logically equivalent assertion) means "the subspace of fold maps in $C^\infty(N, M)$ contains a dense open subset whose elements satisfy property P". Here "fold" can be replaced by any other adjective or adjective phrase, and one can similarly use adjectives or adjective phrases in the PL case, which will be discussed below.

Let us note that, according to our convention, the word "generic" is not an adjective, but a more complex part of speech (which modifies not only a noun phrase but the whole sentence, or even several sentences). One can also understand "generic" as an adjective, whose meaning e.g. in case of "generic smooth maps" would be "satisfying a certain fixed set of conditions that determine an open and dense set of maps in $C^\infty(N, M)$". A drawback of this reading is that the scope of the implicit existential quantifier in "a certain fixed set" is not clearly specified, which may lead to ambiguity.

Proposition 13.2.1 *Theorem 13.1(b) is equivalent to its modified version where "generic" is replaced by "stable".*

[10] The case of non-compact N is discussed in Appendix C but is not needed in the main part of the present chapter.

Proof Let $F, S, A \subset C^\infty(N, M)$ consist of all fold maps, of all stable maps and of all corank one maps, respectively. Since S is open, $S \cap F$ is open in F. Since $S \cap A$ is dense in A (see Theorem A.2) and F is open, $S \cap F$ is dense in F. Thus the modified version of Theorem 13.1(b) implies the original version.

Conversely, let G be the dense open subset of F given by the statement of Theorem 13.1(b), and let $f: N \to M$ be a stable smooth fold map. Since G is dense in F, there exists a $g \in G$ that is C^∞-left-right equivalent to f. Then f is a smooth k-prem if and only if g is a smooth k-prem; also, there exists an equivariant map $\Delta_f \to S^{k-1}$ if and only if there exists an equivariant map $\Delta_g \to S^{k-1}$. Hence Theorem 13.1(b) implies its modified version. □

13.2.2 Generic PL Maps

Here we assume familiarity with Appendix B.

Let N and M be polyhedra, where N is compact. The assertion "every generic PL map $N \to M$ satisfies property P" (or any logically equivalent assertion) means "for each triangulations K, L of N, M and every stratification map φ between their face posets, $C(\varphi)$ contains an open dense subset whose elements satisfy property P".

The proof of Theorem 13.1 will also utilize the following definition. Let N_0 be a closed subpolyhedron of N. The assertion "every PL map $N \to M$, generic on $N \setminus N_0$, satisfies property P" (or any logically equivalent assertion) means "for each triangulations (K, K_0) of (N, N_0) and L of M and every stratification map φ between the face posets of K and L, the space $C(\varphi)$ contains a subset whose elements satisfy property P and whose intersection with each point-inverse F of the restriction map $C(\varphi) \to C(\varphi_0)$ is dense and open in F, where φ_0 denotes the restriction of φ to the face poset of K_0".

Proposition 13.2.2 *Theorem 13.1(a) is equivalent to its modified version where "generic" is replaced by "stable".*

Proof Theorem B.1 implies that generic PL maps $N \to M$ are stable. (In more detail, let φ be a stratification map between the face posets of some triangulations K, L of N, M. By Theorem B.1 $S(\varphi)$ is open and dense in $C(\varphi)$. But every element of $S(\varphi)$ is a stable PL map, by using the same triangulations.) Hence the modified version of Theorem 13.1(a) implies the original version.

Conversely, let $f: N \to M$ be a stable PL map. Then there exist triangulations K, L of P, Q and a stratification map φ between the face posets of K and L such that $f \in S(\varphi)$. Assuming the assertion of Theorem 13.1(a), we get an open dense subset G of $C(\varphi)$ such that if $g \in G$ and there exists an equivariant map $\Delta_g \to S^{k-1}$, then g is a PL k-prem. Since G is dense and f is PL-left-right equivalent to some $f' \in S(\varphi)$, some $g \in G$ is PL-left-right equivalent to f' and hence to f. Then f is a PL k-prem if and only if g is a PL k-prem; also, there exists an equivariant map $\Delta_f \to S^{k-1}$ if and only if there exists an equivariant map $\Delta_g \to S^{k-1}$. Hence Theorem 13.1(a) implies its modified version. □

13.3 Proof of Theorem 13.1 Modulo Main Lemma

13.3.1 Proof of Theorem 13.1 Modulo Main Theorem

The following theorem will be proved later in this section modulo a certain Main Lemma.

Theorem 13.3.1 (Main Theorem) *Suppose that $m \geq n$ and $2(m+k) \geq 3(n+1)$.*

(a) Let M be a PL m-manifold and N a compact n-polyhedron, and let $M_\star$, $N_\star$ be their closed subpolyhedra. Let $f\colon N \to M$ be a PL map such that $f^{-1}(M_\star) = N_\star$ and f is generic on $N \setminus N_\star$, and let $f_\star = f|_{N_\star}$.

Suppose that $e_\star\colon N_\star \to \mathbb{R}^k$ is a PL map such that $f_\star \times e_\star\colon N_\star \to M_\star \times \mathbb{R}^k$ is an embedding and $\tilde{e}_\star\colon \Delta_{f_\star} \to S^{k-1}$ extends to an equivariant map $\alpha\colon \Delta_f \to S^{k-1}$.

Then $e_\star$ extends to a PL map $e\colon N \to \mathbb{R}^k$ such that $f \times e\colon N \to M \times \mathbb{R}^k$ is an embedding and $\tilde{e}\colon \Delta_f \to S^{k-1}$ is equivariantly homotopic to α keeping $\Delta_{f_\star}$ fixed.

(b) Let M be a smooth m-manifold and N a compact smooth n-manifold, and let $M_\star$ and $N_\star$ be codimension zero closed submanifolds of M and N respectively. Let $f\colon N \to M$ be a generic smooth fold map such that $f^{-1}(M_\star) = N_\star$, and let $f_\star = f|_{N_\star}$.

Suppose that $e_\star\colon N_\star \to \mathbb{R}^k$ is a map such that $f_\star \times e_\star\colon N_\star \to M_\star \times \mathbb{R}^k$ is a smooth embedding and $\tilde{e}_\star\colon \Delta_{f_\star} \to S^{k-1}$ extends to an equivariant map $\alpha\colon \Delta_f \to S^{k-1}$.

Then $e_\star$ extends to a map $e\colon N \to \mathbb{R}^k$ such that $f \times e\colon N \to M \times \mathbb{R}^k$ is a smooth embedding and $\tilde{e}\colon \Delta_f \to S^{k-1}$ is equivariantly homotopic to α keeping $\Delta_{f_\star}$ fixed.

Theorem 13.1(a) is an immediate consequence of the case $M_\star = \varnothing$ of Theorem 13.3.1(a). The case $M_\star \neq \varnothing$ will arise from the induction step in the proof of the case $M_\star = \varnothing$.

Theorem 13.1(b) is an immediate consequence of the case $M_\star = \varnothing$ of Theorem 13.3.1(b). The case $M_\star \neq \varnothing$ is needed to deduce Theorem 13.1(c). Before we go into this deduction, let us introduce some notation.

13.3.2 Immersions and Fold Maps

For a space N let $\Delta_N^{(3)} = \{(x, x, x) \in N^3\}$ and for a map $f\colon N \to M$ let

$$\Delta_f^{(3)} = \{(x, y, z) \in N^3 \mid f(x) = f(y) = f(z),\ x \neq y,\ y \neq z,\ z \neq x\}.$$

Let $\bar{\Delta}_f$ be the closure of Δ_f in N^2 and let $\bar{\Delta}_f^{(3)}$ be the closure of $\Delta_f^{(3)}$ in N^3. Let Σ_f be the set of points $p \in N$ such that every neighborhood of p contains distinct points x, y satisfying $f(x) = f(y)$. Let $\Sigma_f^{(3)}$ be the set of points $p \in N$ such that every neighborhood of p contains pairwise distinct points x, y, z satisfying $f(x) = f(y) = f(z)$. Thus $\Delta_{\Sigma_f} = \bar{\Delta}_f \cap \Delta_N$ and $\Delta^{(3)}_{\Sigma_f^{(3)}} = \bar{\Delta}_f^{(3)} \cap \Delta_N^{(3)}$.

A map $f: N \to M$ is called a (smooth) *immersion* if every point of N has a neighborhood that is (smoothly) embedded by f. Clearly, f is an immersion if and only if $\Sigma_f = \varnothing$. By a *fold map* we mean a map $f: N \to M$ such that $\Sigma_f^{(3)} = \varnothing$; this includes *smooth fold maps*, whose only singularities are of the fold type (= of type $\Sigma^{1,0}$, cf. [23]).

Proof of Theorem 13.1(c) This proof essentially uses Appendix C. Let $X = f^{-1}\big(f(\Sigma_f^{(3)})\big)$. Since f is generic, we may assume that X is a compact polyhedron (see Theorem A.1) and that $\dim X = \dim \Sigma_f^{(3)} \le 3n - 2m - 2$. Let $g: N \to \mathbb{R}^k$ be a generic smooth map and let $h = f \times g: N \to M \times \mathbb{R}^k$. Since $k \ge \dim X + 1$, by Theorem C.12 we may assume that Σ_h is disjoint from X. Thus h embeds some closed neighborhood $N_\star$ of X in N such that $N_\star$ is a manifold with boundary, $N_\star$ deformation retracts onto X, and $N_\star = f^{-1}(M_\star)$ for some codimension zero submanifold $M_\star \subset M$. Let $f_\star = f|_{N_\star}$ and $e_\star = g|_{N_\star}$. We may further assume that $\Delta_{f_\star}$ equivariantly deformation retracts onto $Y := \Delta_{f|X}$.

We have $\dim Y \le \dim X \le 3n - 2m - 2$. Since $3n - 2m - 2 < k - 1$, the map $\tilde{g}|_Y: Y \to S^{k-1}$ is equivariantly homotopic to $\varphi|_Y$, where $\varphi: \Delta_f \to S^{k-1}$ is the given equivariant map. It follows that $\tilde{e}_\star: \Delta_{f_\star} \to S^{k-1}$ is equivariantly homotopic to $\varphi|_{\Delta_{f_\star}}$, and consequently extends to an equivariant map $\alpha: \Delta_f \to S^{k-1}$ which is homotopic to φ. Let N' be a codimension zero closed submanifold of N containing $N \setminus N_\star$ and disjoint from X. (If $N_\star \cap \partial N = \varnothing$, we may take N' to be simply the closure of $N \setminus N_\star$; otherwise N' has to be slightly larger in order for $\partial N'$ to be a smooth manifold.) Since f is a fold map on N', we can now apply Theorem 13.3.1(b) to obtain an extension of $e_\star$ to a map $e: N \to \mathbb{R}^k$ such that $f \times e: N \to M \times \mathbb{R}^k$ is a smooth embedding (and, in fact, $\tilde{e}$ is equivariantly homotopic to φ). □

13.3.3 Simple Fold Maps

By a *(smooth) simple fold map* we mean a (smooth) fold map $f: N \to M$ that does not intersect its own folds, that is, $(N \times \Sigma_f) \cap \Delta_f = \varnothing$. Clearly, (smooth) immersions are (smooth) simple fold maps, and generic PL (smooth) triple point free maps are (smooth) simple fold maps. Generic PL (smooth) maps $N^n \to M^m$

- have no triple points when $2m \ge 3n + 1$;
- are simple (smooth) fold maps when $2m \ge 3n$;
- are (smooth) fold maps when $2m \ge 3n - 1$, provided that M and N are PL manifolds rather than polyhedra in the PL case.

Let E_f be the set of distinct pairs $(p, q) \in N \times N$ such that every neighborhood of (p, q) contains pairs (x, y) and (x, z) such that x, y, z are pairwise distinct and $f(x) = f(y) = f(z)$. Let $\bar{E}_f$ be the closure of E_f in N^2. Thus $\bar{\Delta}_f^{(3)} \setminus \Delta_f^{(3)}$ is the union of the images of $\bar{E}_f$ under the three embeddings of $N \times N$ onto the thick diagonals of N^3. Clearly, $\bar{E}_f \setminus E_f = \Delta_{\Sigma_f^{(3)}}$ and $\bar{E}_f \subset N \times \Sigma_f$. It is easy to see that if f is a generic (PL or smooth) map, then $E_f = (N \times \Sigma_f) \cap \Delta_f$. Thus a generic PL (smooth) map is a (smooth) simple fold map if and only if $\bar{E}_f = \varnothing$.

Remark 13.3.2 Smooth simple fold maps have been studied by A. Szűcs. He calls them "maps of singular multiplicity 2" in [56] and "Σ^{1_1}-maps" (not to be confused with "Σ^1-maps") in [58] (but "Σ^{1_1}-maps" have a different meaning in a paper by R. Rimányi and A. Szűcs published in the same year and in later papers by A. Szűcs). Also they would be called "simple Σ^1-singular maps" in direct analogy with the terminology of [57].

Lemma 13.3.3 (Main Lemma)

(a) Theorem 13.3.1(a) holds if $f|_{N \setminus N_\star}$ is a simple fold map.
(b) Theorem 13.3.1(b) holds if f is a smooth simple fold map.

Lemma 13.3.3(a) is proved in Sect. 13.4 and Lemma 13.3.3(b) is proved in Sect. 13.5 by building on the proof of Lemma 13.3.3(a). A sketch of the proof, which applies to both PL and smooth cases, is given in Sect. 13.3.8.

Remark 13.3.4 The full strength of Lemma 13.3.3 will be needed to prove Theorem 13.3.1(b). To prove Theorem 13.3.1(a), we need Lemma 13.3.3(a) only in the cases where $f|_{N \setminus N_\star}$ is either (i) a triple-point free map or (ii) a trivial covering. It is not hard to prove the case (i) directly, which we will do (see Proposition 13.3.6) since it helps to understand the general position assumptions that are really needed for Theorem 13.3.1(a) (see Remark 13.3.7). However, the case (ii) is not so far from the general case (of simple fold maps), at least from the conceptual viewpoint. From the technical viewpoint, the verification of "being not so far" does take some efforts (which are not in vain since they are anyway needed for the smooth case). One reason to be careful about this verification is that f may have complicated behavior near $N_\star$ (since it is assumed to be a simple fold map only over $N \setminus N_\star$), but that is also an issue in the case (ii) (since in that case f is assumed to be a covering only over $N \setminus N_\star$) and so in the case (ii) one would anyway need to be careful about some verification of this sort.

13.3.4 A Lemma on Isovariant Maps

A triangulation of a polyhedron P with a PL action of a finite group G is called *equivariant* if the action is by simplicial maps and the stabilizer of each simplex equals the stabilizer of each interior point of this simplex; this implies, in particular, that the fixed point set P^G is triangulated by a subcomplex.

An *isovariant* map is an equivariant map $f\colon X \to Y$ which does not increase stabilizers of points, i.e. the stabilizer of each $x \in X$ contains the stabilizer of $f(x)$. (Let us note that the reverse inclusion also holds since f is equivariant.) In particular, an equivariant map φ of a $\mathbb{Z}/2$-space X into $\mathbb{R}^k$ with $\mathbb{Z}/2$ acting by the central symmetry $x \mapsto -x$ is isovariant if and only if $\varphi^{-1}(0)$ coincides with the fixed point set $X^{\mathbb{Z}/2}$.

Lemma 13.3.5 *Let P be a compact $\mathbb{Z}/2$-polyhedron and let Q be an invariant closed subpolyhedron of P. Let $\beta\colon P \to \mathbb{R}^k$ be an isovariant map such that $\beta|_Q$ is a PL map. Then β is isovariantly homotopic keeping Q fixed to a PL map.*

Proof Let A be an equivariant triangulation of P such that Q is triangulated by a subcomplex of A. Let B be the union of all closed simplexes of A that do not intersect $P^{\mathbb{Z}/2}$. Let B' be an equivariant subdivision of B into sufficiently small simplexes, and let us extend it to an equivariant subdivision A' of A without adding new vertices. Let C be the simplicial neighborhood of $P^{\mathbb{Z}/2}$ in A'. Let us define φ to coincide with β on the vertices of A' and by extending linearly to the simplexes of A'. Then it is easy to see that φ is isovariant. (Every simplex of C is a join of a simplex S of $P^{\mathbb{Z}/2}$ and a simplex T of B'. Then $\varphi(S) = 0$ and $\varphi(T) \subset \mathbb{R}^k \setminus 0$ by our choice of B'. Hence $\varphi^{-1}(0) \cap (S * T) = S$.) Also, φ is "conical" on C in the sense that it is the composition of the quotient map $C \to C/P^{\mathbb{Z}/2}$ and a conical map of the cone $C/P^{\mathbb{Z}/2}$ over ∂C into the infinite cone $\mathbb{R}^k$.

Now $\beta|_B$ is isovariantly homotopic to $\varphi|_B$ by the linear homotopy $(1-t)\beta|_B + t\varphi|_B$, which keeps Q fixed. Therefore β is isovariantly homotopic keeping Q fixed to a map γ such that $\gamma|_B = \varphi|_B$. Finally, similarly to the Alexander trick, γ is isovariantly homotopic to φ by a homotopy γ_t, $t \in [0,1]$, keeping $B \cup Q$ fixed and such that each γ_t coincides with γ on $C \setminus tC$ for a certain smaller neighborhood tC of $P^{\mathbb{Z}/2}$, and is "conical" on tC. □

13.3.5 The Triple Point Free Case

In the case where $f|_{N\setminus N_\star}$ is a simple fold map *without triple points*, Theorem 13.3.1(a) can be proved by a much simpler construction, which works without any hypotheses on dimensions and genericity:

Proposition 13.3.6 *Let M^m be a PL manifold and N^n a compact polyhedron, and let $M_\star$, $N_\star$ be closed subpolyhedra of M and N respectively. Let $f\colon N \to M$ be a PL map such that $f^{-1}(M_\star) = N_\star$ and $f|_{N\setminus N_\star}$ is a simple fold map without triple points, and let $f_\star = f|_{N_\star}$.*

Suppose that $e_\star\colon N_\star \to \mathbb{R}^k$ is a PL map such that $f_\star \times e_\star\colon N_\star \to M_\star \times \mathbb{R}^k$ is an embedding and $\tilde{e}_\star\colon \Delta_{f_\star} \to S^{k-1}$ extends to an equivariant map $\alpha\colon \Delta_f \to S^{k-1}$.

Then $e_\star$ extends to a PL map $e\colon N \to \mathbb{R}^k$ such that $f \times e\colon N \to M \times \mathbb{R}^k$ is an embedding and $\tilde{e}\colon \Delta_f \to S^{k-1}$ is equivariantly homotopic to α keeping $\Delta_{f_\star}$ fixed.

Let us emphasize that f is not assumed to be generic here, and m, n and k are arbitrary. Let us note that a generic PL map without triple points is automatically a simple fold map.

Proof We are given $e_\star$ and α and we need to construct e. Let $f_\circ = f|_{N\setminus N_\star}$. Let us note that $\Delta_{f_\star} = \Delta_f \setminus \Delta_{f_\circ}$, but $\bar{\Delta}^\star_f := \bar{\Delta}_f \setminus \bar{\Delta}_{f_\circ}$ may be larger than $\bar{\Delta}_{f_\star}$. Clearly, $\bar{\Delta}^\star_f = \bar{\Delta}_f \cap (N_\star \times N_\star)$.

Let $S_\star\colon \bar{\Delta}^\star_f \to \mathbb{R}^k$ be defined by $S_\star(x, y) = \frac{e_\star(x)+e_\star(y)}{2}$, and let $S\colon \bar{\Delta}_f \to \mathbb{R}^k$ be an arbitrary PL map extending $S_\star$ and satisfying $S(x, y) = S(y, x)$. Let $A_\star\colon \bar{\Delta}^\star_f \to \mathbb{R}^k$ and $a_\star\colon \bar{\Delta}^\star_f \to [0, \infty)$ be defined by $A_\star(x, y) = \frac{e_\star(y)-e_\star(x)}{2}$ and $a_\star(x, y) = ||A_\star(x, y)||$, and let $a\colon \bar{\Delta}_f \to [0, \infty)$ be an arbitrary PL map extending $a_\star$ and satisfying $a(x, y) = a(y, x)$ and $a(x, y) = 0 \Leftrightarrow x = y$. Let us define $A\colon \bar{\Delta}_f \to \mathbb{R}^k$ by $A(x, y) = \alpha(x, y) \cdot a(x, y)$ for $x \neq y$ and by $A(x, x) = 0$. Then A is isovariant and $A(x, y) = \frac{e_\star(y)-e_\star(x)}{||e_\star(y)-e_\star(x)||} \cdot \frac{||e_\star(y)-e_\star(x)||}{2} = A_\star(x, y)$ for all $(x, y) \in \Delta_{f_\star}$. Since $A(x, x) = 0$, we have $A|_{\bar{\Delta}^\star_f} = A_\star$. Since $A_\star$ is a PL map, by Lemma 13.3.5 A is isovariantly homotopic keeping $\bar{\Delta}^\star_f$ fixed to a PL map $A'\colon \bar{\Delta}_f \to \mathbb{R}^k$.

Let us define $\varphi\colon \bar{\Delta}_f \to \mathbb{R}^k$ by $\varphi(x, y) = S(x, y) + A'(x, y)$. Then φ is a PL map and $\varphi(x, y) = S_\star(x, y) + A_\star(x, y) = e_\star(y)$ for $(x, y) \in \bar{\Delta}^\star_f$. Let $\pi\colon N \times N \to N$ be the projection onto the second factor. Since $f_\circ$ is a simple fold map without triple points, π embeds $\bar{\Delta}_{f_\circ}$. Let us define $e_1\colon N_\star \cup \pi(\bar{\Delta}_f) \to \mathbb{R}^k$ by $e_1(y) = e_\star(y)$ for $y \in N_\star$ and by $e_1(y) = \varphi(x, y)$ for $(x, y) \in \bar{\Delta}_f$. Let us note that the latter definition agrees with the former when $(x, y) \in \bar{\Delta}^\star_f$, and it follows that e_1 is well-defined. Let $e\colon N \to \mathbb{R}^k$ be any PL map extending e_1. Since $f \times e\colon N \to M \times \mathbb{R}^k$ embeds $N_\star$ and $\pi(\Delta_{f_\circ})$, it is an embedding. We have

$$\tilde{e}(x, y) = \frac{e(y) - e(x)}{||e(y) - e(x)||} = \frac{\varphi(x, y) - \varphi(y, x)}{||\varphi(x, y) - \varphi(y, x)||} = \frac{A'(x, y) - A'(y, x)}{||A'(x, y) - A'(y, x)||}$$

and

$$\frac{A(x, y) - A(y, x)}{||A(x, y) - A(y, x)||} = \frac{\alpha(x, y) - \alpha(y, x)}{||\alpha(x, y) - \alpha(y, x)||} = \frac{2\alpha(x, y)}{||2\alpha(x, y)||} = \alpha(x, y)$$

for all $(x, y) \in \Delta_f$. Since A is isovariantly homotopic to A' keeping $\bar{\Delta}^\star_f$ fixed, it follows that $\tilde{e}$ is equivariantly homotopic to α keeping $\Delta_{f_\star}$ fixed. □

13.3.6 Proof of Main Theorem modulo Main Lemma: PL Case

Proof of Theorem 13.3.1(a) Let K and L be triangulations of N and M such that $f\colon K \to L$ is simplicial and $N_\star$ and $M_\star$ are triangulated by subcomplexes. Let $\bar{D}_3$ be the image of the composition $\bar{\Delta}_f^{(3)} \subset N^3 \to N$ of the inclusion and the projection onto the first factor. Let E be the image of the composition $(\Sigma_f \times N) \cap \Delta_f \subset N^2 \to N$ of the inclusion and the projection onto the first factor. Since f is generic on $N \setminus N_\star$, we may assume that $f|_{N\setminus N_\star}$ is non-degenerate, that $\bar{D}_3 \cap (N \setminus N_\star)$ is of dimension $\leq 3n - 2m$ and that $E \cap (N \setminus N_\star)$ is of dimension $\leq 3n - 2m - 1$ (in fact, $\leq 3n - 2m$ would already suffice for our purposes).

Let $\sigma_1, \dots, \sigma_r$ be the simplexes of $f(\bar{D}_3 \cup E)$ not contained in $M_\star$, arranged in an order of increasing dimension. Let $L_i = M_\star \cup \sigma_1 \cup \dots \cup \sigma_i$, and let $K_i = f^{-1}(L_i)$. Let $f_i\colon K_i \to L_i$ be the restriction of f. Thus $f_0 = f_\star$.

Let $e_0 = e_\star$. Suppose more generally that $e_i\colon K_i \to \mathbb{R}^k$ is a PL map such that $f_i \times e_i\colon K_i \to L_i \times \mathbb{R}^k$ is an embedding and $\tilde{e}_i\colon \Delta_{f_i} \to S^{k-1}$ is equivariantly homotopic to $\alpha|_{\Delta_{f_i}}$ by a homotopy $h_i\colon \Delta_{f_i} \times I \to S^{k-1}$ that keeps $\Delta_{f_\star}$ fixed.

Let $(Q, Q_\star) = f^{-1}(\sigma_{i+1}, \partial\sigma_{i+1})$ and let $(F, F_\star) = f|_{(Q,Q_\star)}$ and $E_\star = e_i|_{Q_\star}$. Since $f|_{N\setminus N_\star}$ is non-degenerate, f sends every simplex of K not contained in $N_\star$ homeomorphically onto some simplex of L. Hence the preimage of every open k-simplex of L not contained in $M_\star$ is a disjoint union of open k-simplexes of K. In particular, F restricts to a trivial finite-fold (possibly 0-fold) covering $Q \setminus Q_\star = K_{i+1} \setminus K_i \to \sigma_{i+1} \setminus \partial\sigma_{i+1} = L_{i+1} \setminus L_i$. The map $F_\star \times E_\star\colon Q_\star \to \partial\sigma_{i+1} \times \mathbb{R}^k$ is an embedding, and $\tilde{E}_\star\colon \Delta_{F_\star} \to S^{k-1}$ extends to an equivariant map $\Phi\colon \Delta_F \to S^{k-1}$ given by $\alpha|_{\Delta_F}$ along with $H_\star := h_i|_{\Delta_{F_\star} \times I}$, using that the inclusion $\Delta_{F_\star} \subset \Delta_F$ is a cofibration.[11] Since every covering is an immersion, and hence a simple fold map, and since every s such that $s \leq 3n - 2m$ satisfies $s + k \geq \frac{3}{2}(s+1)$, by Lemma 13.3.3(a) $E_\star$ extends to a PL map $E\colon Q \to \mathbb{R}^k$ such that $F \times E\colon Q \to \sigma_{i+1} \times \mathbb{R}^k$ is an embedding and $\tilde{E}\colon \Delta_F \to S^{k-1}$ is equivariantly homotopic to Φ keeping $\Delta_{F_\star}$ fixed. Then $\tilde{E}$ is equivariantly homotopic to $\alpha|_{\Delta_F}$ by a homotopy H that extends $H_\star$.

Clearly, $\Delta_{f_{i+1}} = \Delta_{f_i} \cup_{\Delta_{F_\star}} \Delta_F$. Thus $e_{i+1} := e_i \cup_{E_\star} E$ is a PL map $K_{i+1} \to \mathbb{R}^k$ such that $f_{i+1} \times e_{i+1}\colon K_{i+1} \to L_{i+1} \times \mathbb{R}^k$ is an embedding and $\tilde{e}_{i+1}\colon \Delta_{f_{i+1}} \to S^{k-1}$ is equivariantly homotopic to $\alpha|_{\Delta_{f_{i+1}}}$ by the homotopy $h_{i+1} := h_i \cup_{H_\star} H$ which obviously keeps $\Delta_{f_\star}$ fixed.

Finally, we have $L_r = M_\star \cup f(\bar{D}_3 \cup E)$ and consequently $K_r \supset N_\star \cup \bar{D}_3 \cup E$. Hence $f|_{N\setminus K_r}$ is a simple fold map without triple points, and we can apply Proposition 13.3.6. □

[11] If $F\colon Q \to X$ is a PL map, and R is a closed subpolyhedron of Q, then $\tilde{R}$ is a closed subpolyhedron of $\tilde{Q}$, and consequently $\Delta_{F|_R} = (F \times F)^{-1}(\Delta_X) \cap \tilde{R}$ is a closed subpolyhedron of $\Delta_F = (F \times F)^{-1}(\Delta_X) \cap \tilde{Q}$.

Remark 13.3.7 It is clear from the proof of Theorem 13.3.1(a) that the hypothesis "f is generic" in Theorem 13.1(a) can be replaced by the conjunction of the following conditions:

(1) f is non-degenerate;
(2) $\dim \bar{\Delta}_f^{(3)} \le 3n - 2m$;
(3) $\dim(\Sigma_f \times N) \cap \Delta_f \le 3n - 2m - 1$.

The same conditions but with $f|_{N \setminus N_\star}$ in place of f work to replace the hypothesis "f is generic on $N \setminus N_\star$" in Theorem 13.3.1(a).

Remark 13.3.8 The proof of Theorem 13.3.1(a) modulo Lemma 13.3.3(a) almost looks like an induction on Thom–Boardman (or at least Morin) strata (which gets stuck at the penultimate step). Nevertheless, it remains a challenge to make such an induction work in the smooth case.

13.3.7 Proof of Main Theorem Modulo Main Lemma: Smooth Case

In order to prove the smooth case of the Main Theorem, we will actually need a slightly strengthened version of the smooth case of the Main Lemma. It is also convenient to state the analogous strengthening of the PL case.

Addendum 13.3.9 (To Main Lemma)

(a) Let M, N, f, $M_\star$, $N_\star$ and $f_\star$ be as in Theorem 13.3.1(a). Let X be the union of $\Delta_{f_\star}$ and $(\Sigma_f \times N \cup N \times \Sigma_f) \cap \Delta_f$. Suppose that $g\colon N \to \mathbb{R}^k$ is a PL map, generic on $N \setminus N_\star$ and such that $f_\star \times (g|_{N_\star})\colon N_\star \to M_\star \times \mathbb{R}^k$ is an embedding, $\Delta_{f\times g}$ is disjoint from X, and the map $\tilde{g}\colon X \to S^{k-1}$ defined by $\tilde{g}(x, y) = \frac{g(y)-g(x)}{||g(y)-g(x)||}$ extends to an equivariant map $\alpha\colon \Delta_f \to S^{k-1}$. Then g is homotopic keeping $N_\star$ fixed to a PL map $e\colon N \to \mathbb{R}^k$ such that $f \times e\colon N \to M \times \mathbb{R}^k$ is an embedding and $\tilde{e}\colon \Delta_f \to S^{k-1}$ is equivariantly homotopic to α keeping $\Delta_{f_\star}$ fixed.

(b) Let M, N, f, $M_\star$, $N_\star$ and $f_\star$ be as in Theorem 13.3.1(b). Let X be the union of $\Delta_{f_\star}$ and $(\Sigma_f \times N \cup N \times \Sigma_f) \cap \Delta_f$. Suppose that $g\colon N \to \mathbb{R}^k$ is a generic smooth map such that $f_\star \times (g|_{N_\star})\colon N_\star \to M_\star \times \mathbb{R}^k$ is a smooth embedding, $\Delta_{f\times g}$ is disjoint from X, and the map $\tilde{g}\colon X \to S^{k-1}$ defined by $\tilde{g}(x, y) = \frac{g(y)-g(x)}{||g(y)-g(x)||}$ extends to an equivariant map $\alpha\colon \Delta_f \to S^{k-1}$. Then g is homotopic keeping $N_\star$ fixed to a map $e\colon N \to \mathbb{R}^k$ such that $f \times e\colon N \to M \times \mathbb{R}^k$ is a smooth embedding and $\tilde{e}\colon \Delta_f \to S^{k-1}$ is equivariantly homotopic to α keeping $\Delta_{f_\star}$ fixed.

Let us note that Lemma 13.3.3 is a trivial consequence of Addendum 13.3.9: if f is a simple fold map, then $(\Sigma_f \times N) \cap \Delta_f = \varnothing$, so $X = \Delta_{f_\star}$.

Proof of Theorem 13.3.1(b) Since f is a fold map, $S := (\Sigma_f \times N) \cap \Delta_f$ is a compact subset of $N \times N \setminus \Delta_N$. Let π be the projection of $N \times N$ to the first factor and t be the factor exchanging involution of $N \times N$. Let $N_0 = \pi\big(S \cup t(S)\big) \cup N_\star$ and $f_0 = f|_{N_0}$.

Since f is generic, $\dim S \leq 3m - 2n$ (in fact, $\dim S \leq 3m - 2n - 1$, but we do not need this). Then it follows from Theorem 13.3.1(a) and the triangulation theorem (see Theorem A.1) that the given smooth map $e_\star\colon N_\star \to \mathbb{R}^k$ extends to a continuous map $e_0\colon N_0 \to \mathbb{R}^k$ such that $f_0 \times e_0\colon N_0 \to M \times \mathbb{R}^k$ is a topological embedding and $\tilde{e}_0\colon \Delta_{f_0} \to S^{k-1}$ is equivariantly homotopic to $\alpha|_{\Delta_{f_0}}$ keeping $\Delta_{f_\star}$ fixed.

Let $g\colon N \to \mathbb{R}^k$ be a generic smooth map such that $g|_{N_\star} = e_\star$ and $g|_{N_0}$ is sufficiently C^0-close to e_0, and let $h = f \times g\colon N \to M \times \mathbb{R}^k$. Since e_0 is injective and S is a compact subset of $N \times N \setminus \Delta_N$, we may assume that $\bar{\Delta}_h$ is disjoint from $X := S \cup t(S) \cup \Delta_{f_\star}$ and that $\tilde{g}\colon X \to S^{k-1}$ is equivariantly homotopic to $\alpha|_X$ keeping $\Delta_{f_\star}$ fixed. Then $\tilde{g}$ extends to an equivariant map $\alpha'\colon \Delta_f \to S^{k-1}$ which is equivariantly homotopic to α keeping $\Delta_{f_\star}$ fixed.

Then by Addendum 13.3.9(b) g is homotopic keeping $N_\star$ fixed to a map $e\colon N \to \mathbb{R}^k$ such that $f \times e\colon N \to M \times \mathbb{R}^k$ is a smooth embedding and $\tilde{e}\colon \Delta_f \to S^{k-1}$ is equivariantly homotopic to α keeping $\Delta_{f_\star}$ fixed. □

13.3.8 Proof of Main Lemma: Sketch

Here is an outline of the proof of Lemma 13.3.3. In this outline, we will assume that $M_\star = \varnothing$ in order to simplify notation.

Let us pick a generic lift $h = f \times g\colon N \to M \times \mathbb{R}^k$ of the given map $f\colon N \to M$. Let us define $\bar{g}\colon \Delta_f \to \mathbb{R}^k$ by $(x, y) \mapsto g(y) - g(x)$. Using the hypothesis, it is not hard to construct a generic equivariant homotopy Φ from $\bar{g}$ to a map into $\mathbb{R}^k \setminus \{0\}$. Let $D_\circledcirc$ be the image of $\Phi^{-1}(0) \subset \Delta_f \times I$ under the projection $\Delta_f \times I \to \Delta_f$ and let D be the closure of $D_\circledcirc$ in $\bar{\Delta}_f$. Then D is of a small dimension (namely, $2n+1-m-k$, which is less than $n/2$) and therefore the projection $\pi\colon \bar{\Delta}_f \subset N \times N \to N$ embeds D. Also, since f is a simple fold map, π is an immersion, and hence embeds a closed invariant neighborhood U of D. Our plan is to homotope h to an embedding by a homotopy lying over f, with support in $\pi(U)$, using the configuration level homotopy Φ (restricted over U) and the homeomorphism $\pi|_U$ between U and $\pi(U)$. A preliminary version $\xi_t = f \times e_t$ of the desired homotopy is given by an explicit formula involving $\xi_0 = h|_{\pi(U)}$ and Φ, and does eliminate the existing double points of h; thus ξ_1 embeds $\pi(U)$. But, of course, this formula is unaware that f may have triple and 10-tuple points, and because of this ξ_t may accidentally create new double points between $\pi(U)$ and $N \setminus \pi(U)$. However, since D is of small dimension, by a slight perturbation of ξ_t we can at least ensure that new double points do not occur between $\pi(D)$ and $N \setminus \pi(D)$. Then they also do not occur between a small neighborhood $\pi(V^\sharp)$ of $\pi(D)$ in $\pi(U)$ and $N \setminus \pi(V^\sharp)$, and (using again that f is a simple fold map) even between a neighborhood $W^\sharp$ of $\pi(D)$ in N and $N \setminus W^\sharp$. With enough care this remains true (even with the same $V^\sharp$ and $W^\sharp$, so that we don't get a circularity!) after we amend Φ so that it has support in $V^\sharp$.

Remark 13.3.10 One naturally wonders if it is not excessive to first define a preliminary version ξ_t of the homotopy and then amend[12] it (along with Φ): isn't it possible to define ξ_t just once, after an appropriate preparation? The amendment is based on the fact that the images of $\pi(D)$ (of dimension $< n/2$) and $N \setminus \pi(D)$ (of dimension n) in $M \times \mathbb{R}^k$ (of dimension $m + k \geq 3(n + 1)/2$) do not intersect in general position (while lying over f). If they happen to remain disjoint in general position even after projecting to M, then we could indeed define ξ_t just once; but this assumption (namely, $m \geq 3n/2$) would be too restrictive.

We could actually do as follows: first define ξ_t on $\pi(D)$ (by explicit formula), then perturb it into general position (along with Φ), and finally extend it over $\pi(U)$ (by explicit formula). While this seems to be the most logical order of actions, it would require analyzing properties of the construction by explicit formula two times (separately for $\pi(D)$ and for $\pi(U)$), even though the formula is the same. So it is really in order to avoid this repeated analysis that we define ξ_t twice on the entire $\pi(U)$.

13.4 Proof of Main Lemma: PL Case

Here is the proof of Lemma 13.3.3(a), split into a number of steps.

13.4.1 Construction of D

Let us write $N_\circ = N \setminus N_\star$, $M_\circ = M \setminus M_\star$ and $f_\circ = f|_{N_\circ}$. Since f is generic on $N_\circ$, we may assume that $\dim \Delta_{f_\circ} \leq 2n - m$. Let $d = 2n - m - k + 1$. Let $g\colon N \to \mathbb{R}^k$ be a PL map extending $e_\star$ and generic on $N_\circ$, and let $h = f \times g\colon N \to M \times \mathbb{R}^k$. Since $2n - m < k + d$, g lifts (with respect to the projection $\mathbb{R}^{k+d} \to \mathbb{R}^d$) to a PL map $G\colon N \to \mathbb{R}^{k+d}$ extending the composition $N_\star \xrightarrow{e_\star} \mathbb{R}^k \subset \mathbb{R}^{k+d}$ and such that $f \times G\colon N \to M \times \mathbb{R}^{k+d}$ is an embedding. Since g is generic on $N_\circ$ and $g|_{N_\star} = e_\star$, we may assume that Δ_h is a subpolyhedron of codimension $\geq k$ in $\Delta_{f_\circ}$.

Since f is PL, $\bar{\Delta}_f$ is a closed subpolyhedron of $N \times N$ and Σ_f is a closed subpolyhedron of N. Let us note that $\bar{\Delta}^\star_f := \bar{\Delta}_f \setminus \bar{\Delta}_{f_\circ}$ may be larger than $\bar{\Delta}_{f_\star}$, even though $\Delta_{f_\star} = \Delta_f \setminus \Delta_{f_\circ}$. (Clearly, $\bar{\Delta}^\star_f = \bar{\Delta}_f \cap (N_\star \times N_\star)$.) Let us define an isovariant map $\beta\colon \bar{\Delta}_f \to \mathbb{R}^k$ by sending Δ_{Σ_f} to 0 and by $(x, y) \mapsto \alpha(x, y) \cdot ||G(y) - G(x)||$

---

12 Here we are concerned only with amendments that are global, at least on some scale; perturbations into general position do not count as "amendments" here.

for $x \neq y$. Let us note that $\beta(x,y) = \frac{g(y)-g(x)}{||g(y)-g(x)||} \cdot ||g(y)-g(x)|| = g(y)-g(x)$ for $(x,y) \in \bar\Delta^\star_f$. In particular, $\beta|_{\bar\Delta^\star_f}$ is a PL map.

By Lemma 13.3.5 β is isovariantly homotopic keeping $\bar\Delta^\star_f$ fixed to a PL map $\varphi\colon \bar\Delta_f \to \mathbb{R}^k$. By equivariantly perturbing the images of the vertices of an equivariant triangulation of $\bar\Delta_f$ such that φ is linear on its simplexes we may assume that φ is generic on $\bar\Delta_{f_\circ}$ (as an equivariant map). The equivariant map $\bar\Delta_f \to \mathbb{R}^k$ defined by $(x,y) \mapsto g(y)-g(x)$ is equivariantly homotopic to φ by the linear homotopy $\Phi\colon \bar\Delta_f \times I \to \mathbb{R}^k$, which keeps $\bar\Delta^\star_f$ fixed. Thus we have $\Phi(x,y,t) = g(y)-g(x)$ for all $(x,y,t) \in \bar\Delta_f \times \{0\} \cup \bar\Delta^\star_f \times I$ and $\Phi(\Delta_f \times \{1\} \cup \Delta_{f_\star} \times I) \subset \mathbb{R}^k \setminus \{0\}$. Hence $\nabla_\circledcirc := \Phi^{-1}(0) \setminus (\Delta_N \times I)$ lies in $\Delta_{f_\circ} \times [0,1)$, also $\nabla_\circledcirc \cap (\Delta_{f_\circ} \times \{0\}) = \Delta_h$. Let $\nabla_\circ$ be the closure of $\nabla_\circledcirc$ in $\bar\Delta_{f_\circ} \times I$, and let ∇ be its closure in $\bar\Delta_f \times I$. Let $D_\circledcirc$, $D_\circ$ and D be the images of $\nabla_\circledcirc$, $\nabla_\circ$ and ∇ under the projection $\bar\Delta_f \times I \to \bar\Delta_f$. Since Φ is PL, ∇ is a closed subpolyhedron of $\bar\Delta_f \times I$ and consequently D is a closed subpolyhedron of $\bar\Delta_f$. Since Φ is generic on $\bar\Delta_{f_\circ} \times I$, we may assume that $\nabla_\circledcirc$ is of codimension $\geq k$ in $\Delta_{f_\circ} \times I$, hence of dimension $\leq d$. Consequently ∇ and D are also of dimension $\leq d$.

Let $\pi : \bar\Delta_f \subset N \times N \to N$ be the composition of the inclusion and the projection onto the first factor and let $\pi_\circ = \pi|_{\bar\Delta_{f_\circ}}$ and $\pi_\circledcirc = \pi|_{\Delta_{f_\circ}}$.

Lemma 13.4.1

(a) *$\Delta_{\pi_\circledcirc}$ is homeomorphic to $\Delta^{(3)}_{f_\circ}$.*
(b) *$\Delta_{\pi_\circ} \setminus \Delta_{\pi_\circledcirc}$ is a union of two homeomorphic copies of $(\Sigma_{f_\circ} \times N) \cap \Delta_{f_\circ}$.*
(c) *$\Sigma_{\pi_\circledcirc}$ coincides with $(N \times \Sigma_{f_\circ}) \cap \Delta_{f_\circ}$ as long as f is generic on $N_\circ$.*
(d) *$\Sigma_{\pi_\circ} \setminus \Sigma_{\pi_\circledcirc}$ coincides with $\Delta_{\Sigma^{(3)}_{f_\circ}}$ as long as f is generic on $N_\circ$.*

Proof

(a) Every point of $\Delta_{\pi_\circledcirc}$ is of the form $\big((x,y),(x,z)\big)$, where (x,y) and (x,z) belong to $\Delta_{f_\circ}$ and $y \neq z$. Thus the projection $N^4 \to N^3$, $\big((x,y),(x,z)\big) \mapsto (x,y,z)$, sends $\Delta_{\pi_\circledcirc}$ homeomorphically onto $\Delta^{(3)}_{f_\circ}$. □

(b) Clearly, $\Delta_{\pi_\circ} \setminus \Delta_{\pi_\circledcirc}$ is the union of the images of $(\Sigma_{f_\circ} \times N) \cap \Delta_{f_\circ}$ under the two embeddings $N^2 \to N^4$ given by $(x,y) \mapsto \big((x,x),(x,y)\big)$ and $(x,y) \mapsto \big((x,y),(x,x)\big)$. □

(c) It is easy to see that $\Sigma_{\pi_\circledcirc} \subset (N \times \Sigma_{f_\circ}) \cap \Delta_{f_\circ}$. In more detail, by definition, $\Sigma_{\pi_\circledcirc}$ consists of pairs $(p,q) \in \Delta_{f_\circ}$ whose arbitrary neighborhood in $\Delta_{f_\circ}$ contains distinct pairs (x,y) and (x,z). Here $f_\circ(y) = f_\circ(x) = f_\circ(z)$, so $q \in \Sigma_{f_\circ}$. Thus $\Sigma_{\pi_\circledcirc} \subset (N \times \Sigma_{f_\circ}) \cap \Delta_{f_\circ}$.

The reverse inclusion $(N \times \Sigma_{f_\circ}) \cap \Delta_{f_\circ} \subset \Sigma_{\pi_\circledcirc}$ needs the hypothesis that f is generic on $N_\circ$. Assuming this hypothesis, we actually have $\Sigma_{\pi_\circledcirc} = E_{f_\circ} = (N \times \Sigma_{f_\circ}) \cap \Delta_{f_\circ}$ (see Sect. 13.3.3 concerning $E_{f_\circ}$). □

(d) Let $\Sigma'_{f_\circ}$ denote the set of points $p \in N_\circ$ such that every neighborhood of p contains distinct points x and y such that $x \in \Sigma_{f_\circ}$ and $f_\circ(x) = f_\circ(y)$; thus $\Delta_{\Sigma'_{f_\circ}}$ is the intersection of Δ_N with the closure of $(N \times \Sigma_{f_\circ}) \cap \Delta_{f_\circ}$.

It is easy to see that $\Sigma_{\pi_\circ} \setminus \Sigma_{\pi_\circledcirc}$ contains $\Delta_{\Sigma^{(3)}_{f_\circ}}$ and is contained in $\Delta_{\Sigma^{(3)}_{f_\circ}} \cup \Delta_{\Sigma'_{f_\circ}}$; actually, all three sets coincide as long as f is generic on $N_\circ$.

In more detail, clearly $\Sigma_{\pi_\circ} \setminus \Sigma_{\pi_\circledcirc}$ coincides with $\Sigma_{\pi_\circ} \cap \Delta_N$. By definition, the latter consists of pairs $(p, p) \in \bar\Delta_{f_\circ} \cap \Delta_N$ whose arbitrary neighborhood in $\bar\Delta_{f_\circ}$ contains distinct pairs (x, y) and (x, z). The case where x, y, z can be chosen to be pairwise distinct amounts to $p \in \Sigma^{(3)}_{f_\circ}$, and the case where x and y can be chosen to be equal amounts to $p \in \Sigma'_{f_\circ}$. This proves the two inclusions.

Finally, as long as f is generic on $N_\circ$, $(x, y) \in (N \times \Sigma_{f_\circ}) \cap \Delta_{f_\circ}$ implies $(x, y, y) \in \bar\Delta^{(3)}_{f_\circ}$ and therefore $\Sigma'_{f_\circ} \subset \Sigma^{(3)}_{f_\circ}$.

□

13.4.2 *π Embeds D*

Since f is generic on $N_\circ$, we may assume that $\dim \Delta^{(3)}_{f_\circ} \leq 3n - 2m$ and that $\dim(\Sigma_{f_\circ} \times N) \cap \Delta_{f_\circ} \leq 3n - 2m - 1$. Hence by Lemma 13.4.1(a,b) $\bar\Delta_{\pi_\circ}$ is of dimension at most $3n - 2m$. Therefore so is its closure $\bar\Delta_{\pi_\circ}$.

Let K be an equivariant triangulation of $\bar\Delta_f$ and L a triangulation of N such that $\pi\colon K \to L$ is simplicial and $\bar\Delta^\star_f$ and $N_\star$ are triangulated by subcomplexes. Since $f_\circ$ is non-degenerate, so is $\pi_\circ$. Therefore if σ_1, σ_2 are simplexes of K not in $\bar\Delta^\star_f$ such that $\pi(\sigma_1) = \pi(\sigma_2)$, then they have the same dimension s. Moreover, $s \leq \dim \bar\Delta_{\pi_\circ} \leq 3n - 2m$. Since Φ is generic on $\bar\Delta_{f_\circ} \times I$, we may assume that $\nabla \cap \sigma_i \times I$ is of dimension at most $s - (k - 1)$, and therefore so is $D \cap \sigma_i$. Since $2\big(s - (k - 1)\big) - s \leq 3n + 2 - 2m - 2k < 0$, we may assume that π embeds D.

13.4.3 *Construction of U*

Since $f_\circ$ is a simple fold map, we conclude from Lemma 13.4.1(c,d) that $\Sigma_{\pi_\circ} = \varnothing$. Thus $\pi_\circ\colon \bar\Delta_{f_\circ} \to N$ is an immersion.

Since $\pi_\circ$ immerses $\bar\Delta_{f_\circ}$ and embeds $D_\circ$, it must in fact embed some $\mathbb{Z}/2$-invariant closed neighborhood $U_\circ$ of $D_\circ$ in $\bar\Delta_{f_\circ}$.[13] It is not hard to choose $U_\circ$ so that its closure U in $\bar\Delta_f$ is a closed subpolyhedron of $N \times N$, and $U_\star := U \setminus U_\circ$ coincides with $D \setminus D_\circ$. Since π embeds D and $U_\circ$, and $\pi(U_\star) \cap \pi(U_\circ) \subset N_\star \cap N_\circ = \varnothing$, we get that π embeds U. (It would in fact suffice for our purposes to know that it embeds $U_\circ$, but to stay within the PL category it helps to deal with compact polyhedra.)

[13] Indeed, since $D_\circ \times D_\circ$ and $\bar\Delta_{\pi_\circ}$ are disjoint closed subsets of $\bar\Delta_{f_\circ} \times \bar\Delta_{f_\circ}$, there exists an open neighborhood $U'_\circ$ of $D_\circ$ such that $U'_\circ \times U'_\circ$ is disjoint from $\bar\Delta_{\pi_\circ}$. Since $D_\circ$ is invariant, $U'_\circ$ contains an invariant closed neighborhood $U_\circ$ of $D_\circ$.

13.4.4 Construction of e_t

Let us construct a PL homotopy $e_t : \pi(U) \to \mathbb{R}^k$ keeping $\pi(U_\star)$ fixed and such that $e_0 = g|_{\pi(U)}$ and $e_t(y) - e_t(x) = \Phi(x, y, t)$ for all $(x, y) \in U$. (We do not need to worry if e_t keeps the entire frontier of $\pi(U)$ fixed, because we will eventually use only the restriction of e_t to a small neighborhood of $\pi(D)$ in $\pi(U)$.)

For each $x \in \pi(U)$ there is a unique $y = y(x) \in \pi(U)$ such that $(x, y) \in U$. The vector $e_t(y) - e_t(x) := \Phi(x, y, t)$ is given for each t, and we have some freedom in choosing its endpoints $e_t(x)$ and $e_t(y)$. We may, for instance, endow every point $x \in \pi(U)$ with a "mass" continuously depending on x and choose the endpoints $e_t(x)$ and $e_t(y)$ so that their "center of gravity" does not depend on t. For our purposes, it suffices to consider the constant mass function, so that the center of gravity is the midpoint of the vector. The requirement that this midpoint be fixed under the homotopy can be expressed by

$$\frac{e_t(x) + e_t(y)}{2} = \frac{g(x) + g(y)}{2},$$

where the left hand side can be rewritten as

$$e_t(x) + \tfrac{1}{2}\big(e_t(y) - e_t(x)\big) = e_t(x) + \tfrac{1}{2}\Phi(x, y, t).$$

Thus we define $e_t : \pi(U) \to \mathbb{R}^k$ by

$$\begin{aligned} e_t(x) &= \tfrac{1}{2}\big(g(x) + g(y) - \Phi(x, y, t)\big) \\ &= g(x) + \tfrac{1}{2}\big(g(y) - g(x) - \Phi(x, y, t)\big), \end{aligned}$$

where $y \in \pi(U)$ is the unique point such that $(x, y) \in U$. Clearly, e_t is piecewise linear, e_0 is the restriction of g, and $e_t(y) - e_t(x) = \Phi(x, y, t)$ for all $(x, y) \in U$; the latter can also be verified directly:

$$e_t(y) - e_t(x) = g(y) - g(x) - (\tfrac{1}{2} + \tfrac{1}{2})\big(g(y) - g(x) - \Phi(x, y, t)\big) = \Phi(x, y, t).$$

Finally, since $U_\star \subset \bar{\Delta}^\star_f$ and $\Phi(x, y, t) = g(y) - g(x)$ for $(x, y) \in \bar{\Delta}^\star_f$, we have $e_t(x) = g(x)$ for all $x \in \pi(U_\star)$. Let us note that since $e_t(y) - e_t(x) = \Phi(x, y, t)$ for $(x, y) \in U$ and $\Phi(\Delta_f \times \{1\}) \subset \mathbb{R}^k \setminus \{0\}$, $e_1(x) \neq e_1(y)$ for $(x, y) \in U$.

Remark 13.4.2 By considering a non-constant mass function, it is not hard to generalize the above construction of e_t to the case where f embeds Σ_f but is not necessarily a simple fold map. However, in the case where f is a simple fold map, below we extend e_t to a neighborhood of $\pi(U)$ in N without creating new double points; this seems to be very difficult to achieve when f is not a simple fold map. The latter issue was overlooked by the author at the time of the announcements of

Theorem 13.1(c) in [34, third remark after Theorem 5] and [36, §1], which is why those announcements contain the restriction $4n - 3m \leq k$ (instead of $3n - 2m \leq k$).

13.4.5 Perturbation of e_t

The constraint $e_t(y) - e_t(x) = \Phi(x, y, t)$ for $(x, y) \in U$ can be used to reconstruct Φ from e_t. More precisely, if we amend e_t into a PL homotopy g'_t by an amendment with support in U, then this constraint yields a new isovariant PL homotopy $\Phi'\colon \bar{\Delta}_f \times I \to \mathbb{R}^k$ which coincides with Φ outside U. Moreover, if g'_t satisfies $g'_0 = g|_{\pi(U)}$ and $g'_1(x) \neq g'_1(y)$ for $(x, y) \in U$, then Φ' satisfies $\Phi'(x, y, 0) = \Phi(x, y, 0)$ and $\Phi'(\Delta_f \times \{1\}) \subset \mathbb{R}^k \setminus \{0\}$. Also, as long as the amendment preserves the midpoints, that is, $\frac{1}{2}\big(g'_t(x)+g'_t(y)\big) = \frac{1}{2}\big(g(x)+g(y)\big)$ for all $(x, y) \in U$, then it is compatible with the definition of e_t; that is, if we repeat the definition of e_t with Φ' in place of Φ, we will get nothing but g'_t. Thus we are free to perturb e_t keeping the midpoints fixed. Then by arguments similar to equivariant general position we may assume e_t to be generic on $\pi(U_\circ) \times (0, 1]$.

Let us define $\Xi\colon \pi(U) \times I \to M \times \mathbb{R}^k$ by $\Xi(x, t) = \big(f(x), e_t(x)\big)$. We have $\Xi(x, 0) = h(x)$ for $x \in \pi(U)$, where $h = f \times g\colon N \to M \times \mathbb{R}^k$ is our original generic lift of f. Since $e_1(x) \neq e_1(y)$ for $(x, y) \in U$, we get that Ξ embeds $\pi(U) \times \{1\}$. However, since f may have triple points, $\Xi\big(\pi(U) \times \{1\}\big)$ may intersect $h\big(N \setminus \pi(U)\big)$.

Lemma 13.4.3 *By perturbing e_t we may make $\Xi\big(\pi(D_\circ) \times I\big)$ disjoint from $h\big(N_\circ \setminus \pi(D_\circ)\big)$.*

Proof Let T and T^* be the images of $(D_\circ \times N_\circ) \cap \bar{\Delta}^{(3)}_{f_\circ}$ under the projections of $D_\circ \times N_\circ$ to its factors. Then $\pi(T)$ and T^* are closed subsets of $N_\circ$ and $\dim T = \dim T^* \leq d+n-m = 3n-2m-k+1 < k-1$. Let us note that since $f_\circ$ is a simple fold map and f is generic on $N_\circ$, we have $\bar{\Delta}^{(3)}_{f_\circ} = \Delta^{(3)}_{f_\circ}$; consequently $T \subset D_\circledcirc$. Since π embeds $D_\circ$, T^* is disjoint from $\pi(D_\circ)$. Then, since $\Delta_h \subset D_\circ$, $h(T^*)$ is disjoint from $h\big(\pi(D_\circ)\big)$, and in particular from $h\big(\pi(T)\big)$. Since $f|_{\pi(T)\cup T^*}$ is non-degenerate and e_t is generic on $\pi(U_\circ) \times (0, 1]$, we may assume that $\Xi\big(\pi(T) \times I\big)$ is disjoint from $h(T^*)$ in $M \times \mathbb{R}^k$.

Indeed, let Q and R be triangulations of N and M such that $f\colon Q \to R$ is simplicial, and let $\sigma \subset \pi(T)$ and $\tau \subset T^*$ be simplexes of Q. If $\Xi(x \times I)$ is not disjoint from $h(y)$, then $f(x) = f(y)$. Hence if $x \in \sigma$ and $y \in \tau$, then $f(\sigma) = f(\tau)$ and σ, τ have the same dimension $s < k - 1$. Since $(s + 1) + s < s + k$, we may assume $\Xi(\sigma \times I)$ to be disjoint from $h(\tau)$ in $f(\sigma) \times \mathbb{R}^k$. Indeed, it is easy to see that a generic $(s + 1)$-plane in $\mathbb{R}^{s+k}$, where $k > s + 1$, is disjoint from $\mathbb{R}^s \times \{0\}$ and at the same time maps surjectively onto $\mathbb{R}^s$ under the vertical projection.

Therefore $\Xi\big(\pi(D_\circ) \times I\big)$ is disjoint from $h\big(N_\circ \setminus \pi(D_\circ)\big)$. □

Lemma 13.4.4 *There is a closed neighborhood $W^\sharp_\circ$ of $\pi(D_\circ)$ in $N_\circ$ such that $\Delta_{f|_{W^\sharp_\circ}} \subset U$.*

It is clear from the proof that the closure $W^\sharp$ of $W^\sharp_\circ$ in N may be assumed to be a subpolyhedron of N.

Proof Indeed, let $D^*_\circ = \pi^{-1}\big(\pi(D_\circ)\big) \setminus D_\circ \subset \bar\Delta_f$. Since $\pi(D_\circ) \subset N_\circ$, we have $D^*_\circ \subset \bar\Delta_{f_\circ}$, and since $f_\circ$ is a simple fold map, $D^*_\circ \subset \Delta_{f_\circ}$. Let T and T^* be as in the proof of Lemma 13.4.3. Then $\pi(D^*_\circ) = \pi(T)$ and $\pi^*(D^*_\circ) = T^*$, where $\pi^*\colon \bar\Delta_f \subset N \times N \to N$ is the composition of the inclusion and the projection onto the second factor. Let O and O^* be disjoint open neighborhoods of $\pi^*(D_\circ) = \pi(D_\circ)$ and of T^* in $N_\circ$. We may assume that their closures in N are subpolyhedra of N. Clearly, $(\pi^*)^{-1}(O)$ and $(\pi^*)^{-1}(O^*)$ are disjoint open neighborhoods of $D_\circ$ and $D^*_\circ$ in $\bar\Delta_{f_\circ}$. Then $J := \big((\pi^*)^{-1}(O) \cap \operatorname{Int} U\big) \cup (\pi^*)^{-1}(O^*)$ is an open neighborhood of $D_\circ \cup D^*_\circ = \pi^{-1}\big(\pi(D_\circ)\big)$ in $\bar\Delta_{f_\circ}$. Then J is also open in $\bar\Delta_f$, so $\bar\Delta_f \setminus J$ is compact, and consequently so is its image $\pi(\bar\Delta_f \setminus J)$. Since $\bar\Delta_{f_\circ} \setminus J$ is disjoint from $\pi^{-1}\big(\pi(D_\circ)\big)$, so is $\bar\Delta_f \setminus J$, and consequently $\pi(\bar\Delta_f \setminus J)$ is disjoint from $\pi(D_\circ)$. Hence $O' := N_\circ \setminus \pi(\bar\Delta_f \setminus J)$ is an open neighborhood of $\pi(D_\circ)$ in $N_\circ$ such that $\pi^{-1}(O') \subset J$. Let $W^\sharp_\circ$ be a closed neighborhood of $\pi(D_\circ)$ in $O \cap O'$. Then $W^\sharp_\circ$ is disjoint from O^* and $\pi^{-1}(W^\sharp_\circ) \subset J \subset U \cup (\pi^*)^{-1}(O^*)$. Hence $\pi^{-1}(W^\sharp_\circ) \cap (\pi^*)^{-1}(W^\sharp_\circ) \subset U$; in other words, $\Delta_{f|_{W^\sharp_\circ}} \subset U$. □

13.4.6 Construction of W and W$^\flat$

Let $Z^\sharp_\circ$ be the closure of $N_\circ \setminus W^\sharp_\circ$ in $N_\circ$. Since $\Xi\big(\pi(D_\circ) \times I\big)$ is disjoint from $h\big(N_\circ \setminus \pi(D_\circ)\big)$, it is disjoint from $h(Z^\sharp_\circ)$. Let us recall that $\Xi(x, t) = \big(f(x), \frac{1}{2}g(x) + \frac{1}{2}g(y) - \frac{1}{2}\Phi(x, y, t)\big)$, where $y \in \pi(U)$ is the unique point such that $(x, y) \in U$. Since $\Xi\big(\pi(D_\circ) \times I\big)$ and $h(Z^\sharp_\circ)$ are disjoint, where $\pi(D_\circ)$ and $Z^\sharp_\circ$ are disjoint closed subsets of $N_\circ$ whose closures in N are subpolyhedra of N, there exists a PL function $\epsilon\colon N \to [0, \infty)$ such that $\epsilon^{-1}(0) = N_\star$ and for each $x \in \pi(D_\circ)$ and $t \in I$ we have $\big(f(x'), \frac{1}{2}g(x') + \frac{1}{2}g(y') - \frac{1}{2}\Phi(x'', y'', t)\big) \notin h(Z^\sharp_\circ)$ whenever x', x'' are $\epsilon(x)$-close to x and y', y'' are $\epsilon(y)$-close to the unique point $y \in \pi(U)$ such that $(x, y) \in U$.

Let $W_\circ$ be a closed neighborhood of $\pi(D_\circ)$ in $N_\circ$ which is contained in the open ϵ-neighborhood of $\pi(D_\circ)$ (that is, the union of the open balls $B_{\epsilon(x)}(x)$ for all $x \in \pi(D_\circ)$). We may assume that $W_\circ \subset \operatorname{Int} W^\sharp_\circ$ and that the closure W of $W_\circ$ in N is a subpolyhedron of N. Let $W^\flat_\circ$ be a closed neighborhood of $\pi(D_\circ)$ in $\operatorname{Int} W_\circ$. We may assume that the closure $W^\flat$ of $W^\flat_\circ$ in N is a subpolyhedron of N.

Let $V^\flat = \Delta_{f|_{W^\flat}}$ and $V = \Delta_{f|_W}$, and let $\Lambda^\flat$ and Λ be the closures of $\Delta_f \setminus V^\flat$ and $\Delta_f \setminus V$ in Δ_f. Similarly, let $\bar V^\flat = \bar\Delta_{f|_{W^\flat}}$ and $\bar V = \bar\Delta_{f|_W}$, and let $\bar\Lambda^\flat$ and $\bar\Lambda$ be the closures of $\bar{\Delta}_f \setminus \bar{V}^\flat$ and $\bar{\Delta}_f \setminus \bar{V}$ in $\bar{\Delta}_f$. Let us note that $\pi(\bar{V}) \subset W$, and by our choice of ϵ (using $x'' = x'$ and $y'' = y'$), $\Xi\big(\pi(\bar{V}) \times I\big)$ is disjoint from $h(Z_\circ^\sharp)$.

13.4.7 Construction of Φ^+

Let Φ^+ be the composition

$$\bar{\Delta}_f \times I \xrightarrow{r} \bar{\Delta}_f \times \{0\} \cup (\bar{V}^\flat \cup \bar{\Lambda}) \times I \xrightarrow{\Psi} \mathbb{R}^k,$$

where r is an isovariant PL retraction such that $r(x, y, t) = (x', y', t')$ implies $d(x, x') < \epsilon(x')$ and $d(y, y') < \epsilon(y')$ and Ψ is defined by

$$\Psi(x, y, t) = \begin{cases} \Phi(x, y, t), & \text{if } (x, y) \in \bar{V}^\flat; \\ g(y) - g(x), & \text{if } (x, y) \in \bar{\Lambda}; \\ g(y) - g(x), & \text{if } t = 0. \end{cases}$$

Then Φ^+ is an isovariant PL homotopy satisfying $\Phi^+|_{\bar{V}^\flat \times I} = \Phi|_{\bar{V}^\flat \times I}$ and $\Phi^+(x, y, t) = g(y) - g(x)$ for all $(x, y, t) \in \bar{\Delta}_f \times \{0\} \cup \bar{\Lambda} \times I$. Also, since $\Phi^+(V^\flat \times \{1\}) \subset \Phi(\Delta_f \times \{1\}) \subset \mathbb{R}^k \setminus \{0\}$ and $\Phi^+(\Lambda^\flat \times I) \subset \Phi(\Delta_f \times I \setminus D_\circledcirc \times I) \subset \mathbb{R}^k \setminus \{0\}$, we have $\Phi^+(\Delta_f \times \{1\}) \subset \mathbb{R}^k \setminus \{0\}$.

13.4.8 Construction of e_t^+

Let $e_t^+ : \pi(U) \to \mathbb{R}^k$ be defined similarly to e_t but using Φ^+ in place of Φ:

$$e_t^+(x) = g(x) + \tfrac{1}{2}\big(g(y) - g(x) - \Phi^+(x, y, t)\big),$$

where $y \in \pi(U)$ is the unique point such that $(x, y) \in U$. Then like before, e_t^+ keeps $\pi(U_\star)$ fixed, $e_0^+ = g|_{\pi(U)}$ and $e_t^+(y) - e_t^+(x) = \Phi^+(x, y, t)$ for all $(x, y) \in U$. Also, $e_t^+(x) = e_t(x)$ for all $x \in \bar{V}^\flat$ and $e_t^+(x) = g(x)$ for all $x \in \pi(U \cap \bar{\Lambda})$.

Let us define $\Xi^+ : \pi(U) \times I \to M \times \mathbb{R}^k$ by $\Xi^+(x, t) = \big(f(x), e_t^+(x)\big)$. Then $\Xi^+(x, t) = h(x)$ for all $(x, t) \in \pi(U) \times \{0\} \cup \pi(U \cap \bar{\Lambda}) \times I$ and $\Xi^+(x, t) = \Xi(x, t)$ for all $x \in \pi(\bar{V}^\flat)$. Also, since $e_t^+(x) - e_t^+(y) = \Phi^+(x, y, t)$ for all $(x, y) \in U$ and $\Phi^+(\Delta_f \times \{1\}) \subset \mathbb{R}^k \setminus \{0\}$, we get that Ξ^+ embeds $\pi(U) \times \{1\}$. Since $\Xi^+(x, t) = \Xi(x, t)$ for all $x \in \pi(\bar{V}^\flat)$, where $\bar{V}^\flat \subset \bar{V}$ and $\Xi(\pi(\bar{V}) \times I)$ is disjoint from $h(Z_\circ^\sharp)$, we also get that $\Xi^+\big(\pi(\bar{V}^\flat) \times I\big)$ is disjoint from $h(Z_\circ^\sharp)$. Moreover, in fact, $\Xi^+\big(\pi(\bar{V}) \times I\big)$ is disjoint from $h(Z_\circ^\sharp)$ due to our choice of ϵ (this time using $x'' \neq x'$ and $y'' \neq y'$).

13.4.9 Construction of h_t

Let Z be the closure of $N \setminus W_\circ$ in N. Let Γ be the composition

$$N \times I \xrightarrow{R} N \times \{0\} \cup \big(\pi(U) \cup Z\big) \times I \xrightarrow{\Theta} \mathbb{R}^k,$$

where R is a PL retraction such that $r(x,t) = (x', t')$ implies $d(x, x') < \epsilon(x')$ and Θ is defined by

$$\Theta(x,t) = \begin{cases} e_t^+(x,t), & \text{if } (x,y) \in \pi(U); \\ g(x), & \text{if } (x,y) \in Z; \\ g(x), & \text{if } t = 0. \end{cases}$$

Let us note that the three cases agree on overlaps, including $\pi(U) \cap Z = \pi(U \cap \bar{\Lambda})$.

Clearly, $f \times \Gamma\colon N \times I \to M \times \mathbb{R}^k$ is an extension of Ξ^+. Let us define $g_t\colon N \to \mathbb{R}^k$ by $g_t(x) = \Gamma(x,t)$, and let $h_t = f \times g_t\colon N \to M \times \mathbb{R}^k$. Then $h_0 = h$, $h_t|_Z = h|_Z$ and $h_t(x) = \Xi(x,t)$ for all $x \in \pi(\bar{V}^\flat)$. Also, h_1 embeds $\pi(U)$, and $h_t\big(\pi(\bar{V})\big)$ is disjoint from $h(Z_\circ^\sharp)$ for each $t \in I$. Moreover, due to our choice of ϵ, in fact, $h_t(W)$ is disjoint from $h(Z_\circ^\sharp)$ for each $t \in I$. In other words, $h_t(W_\circ)$ is disjoint from $h(Z^\sharp)$ for each $t \in I$, where $Z^\sharp$ is the closure of $N \setminus W_\circ^\sharp$ in N.

13.4.10 Verification

Since $\pi(\Delta_h) \subset \pi(D_\circledcirc) \subset \pi(D)$ and $h\big(\pi(D)\big)$ is disjoint from $h(Z)$, h embeds Z. Since $h_t|_Z = h|_Z$, so does h_t for each $t \in I$. In particular, h_t embeds $Z^\sharp$, and $h_t(Z \setminus Z^\sharp)$ is disjoint from $h_t(Z^\sharp)$. On the other hand, since $N \setminus Z \subset W_\circ$, where $h_t(W_\circ)$ is disjoint from $h(Z^\sharp)$ and $h|_{Z^\sharp} = h_t|_{Z^\sharp}$ for each $t \in I$, we get that $h_t(N \setminus Z)$ is disjoint from $h_t(Z^\sharp)$. Thus $h_t(N \setminus Z^\sharp)$ is disjoint from $h_t(Z^\sharp)$. Since h_t also embeds $Z^\sharp$, we obtain that $\Delta_{h_t} \cap (N \times Z^\sharp \cup Z^\sharp \times N) = \varnothing$ for each $t \in I$. In particular, $\Delta_{h_t} \subset W^\sharp \times W^\sharp$. Finally, since h_1 embeds $\pi(\bar{V}^\sharp)$, where $\bar{V}^\sharp = \bar{\Delta}_{f|_{W^\sharp}}$, it also embeds $W^\sharp$. Consequently, h_1 is an embedding.

It remains to verify that $\tilde{g}_1$ is equivariantly homotopic to the given map $\alpha\colon \Delta_f \to S^{k-1}$. Let $V^\sharp = \Delta_{f|_{W^\sharp}}$ and let $\Lambda^\sharp$ be the closure of $\Delta_f \setminus V^\sharp$ in Δ_f. Since $\Delta_{h_t} \cap (N \times Z^\sharp \cup Z^\sharp \times N) = \varnothing$ for each $t \in I$, so in particular $\Delta_{h_t} \cap \Lambda^\sharp = \varnothing$, there is an equivariant homotopy $\psi_t\colon \Lambda^\sharp \to S^{k-1}$, defined by $\psi_t(x,y) = \frac{g_t(y) - g_t(x)}{||g_t(y) - g_t(x)||}$, such that $\psi_1 = \tilde{g}_1|_{\Lambda^\sharp}$. Since ψ_t keeps $U \cap \Lambda^\sharp$ fixed (where $U \cap \Lambda^\sharp$ contains $U \setminus \operatorname{Int} U$), it extends to an equivariant homotopy $\psi_t^+\colon \Delta_f \to S^{k-1}$ between $\psi_0^+ := (\tilde{g}_1|_{V^\sharp}) \cup \psi_0$ and $\psi_1^+ := \tilde{g}_1$. But it is easy to see (using that $V \subset V^\sharp \subset U$) that ψ_0^+ coincides with the composition $\Delta_f \xrightarrow{\Phi^+|_{\Delta_f \times \{1\}}} \mathbb{R}^k \setminus \{0\} \to S^{k-1}$. It follows from the definition of Φ^+ that the latter is equivariantly homotopic to the composition $\Delta_f \xrightarrow{\Phi|_{\Delta_f\times\{1\}}} \mathbb{R}^k \setminus \{0\} \to S^{k-1}$. But the latter is in turn equivariantly homotopic to α by the construction of Φ. □

13.4.11 *Proof of Addendum*

To prove Addendum 13.3.9(a), we proceed as in the above proof of Lemma 13.3.3(a), with the following modifications.

The map g is now not chosen at random but given. By the hypothesis Δ_h is disjoint from X. Also, since α extends $\tilde{g}$, we may assume that Φ keeps X fixed. Then D will be disjoint from X. Then we may also choose U to be disjoint from X. Then, although $f_\circ$ is no longer assumed to be a simple fold map, the same constructions still work. □

13.5 Proof of Main Lemma: Smooth Case

In this section we will assume familiarity with Appendix C (with the exception of Theorem C.9 and its consequences, which are not needed now).

The proof of Lemma 13.3.3(b) and Addendum 13.3.9(b) is generally similar to the proof of Lemma 13.3.3(a) and Addendum 13.3.9(a), with many straightforward modifications. We will discuss only substantial modifications.

13.5.1 *Boundary Constraints*

Compared to the PL case, f is assumed to be generic not just on $N \setminus N_\star$, but on the entire N. This means, in particular, that f may be assumed to be transverse to $\partial M_\star$. Then whatever is given on $N_\star$ can be extended over a small neighborhood of $N_\star$ in N. Due to this, $\pi(D)$ will be entirely contained in $N \setminus N_\star$, and so we no longer need to care about the intersection of $\pi(U)$ with $N_\star$. (The solution we used in the PL case, to keep this intersection to a minimum, would actually not suffice for the smooth case, had we still assumed f to be generic only on $N \setminus N_\star$.) Due to this simplification, we will no longer discuss the behaviour at the boundary in what follows.

13.5.2 *Making φ Smooth*

The construction of the isovariant map $\varphi\colon \bar{\Delta}_f \to \mathbb{R}^k$ is modified as follows. Since the manifold with boundary $\check{\Delta}_f$ is equivariantly homotopy equivalent to its interior Δ_f, the given map $\alpha\colon \Delta_f \to S^{k-1}$ is equivariantly homotopic to the restriction of a smooth map $\check{\alpha}\colon \check{\Delta}_f \to S^{k-1}$. On the other hand, let $\check{G}\colon \check{\Delta}_f \to S^{k+d-1} \times [0,\infty)$ be as in Lemma C.8, and let $\kappa\colon \check{\Delta}_f \to [0,\infty)$ be the composition of $\check{G}$ and the projection onto $[0,\infty)$. Since κ is smooth, so is $\check{\varphi} := \check{\alpha} \times \kappa\colon \check{\Delta}_f \to S^{k-1} \times [0,\infty)$.

Let $\mathring{\mathbb{R}}^k$ be the blowup of $\mathbb{R}^k$ at 0; thus the map $S^{k-1} \times [0,\infty) \to \mathbb{R}^k$, $(x,s) \mapsto sx$, factors as a composition $S^{k-1} \times [0,\infty) \xrightarrow{Q} \mathring{\mathbb{R}}^k \xrightarrow{R} \mathbb{R}^k$. Also let $\hat{\Delta}_f$ be the image of $\check{\Delta}_f$ in the blowup of $N \times N$ along Δ_N. Then the projection $\check{\Delta}_f \to \bar{\Delta}_f$ factors as a composition $\check{\Delta}_f \xrightarrow{q} \hat{\Delta}_f \xrightarrow{r} \bar{\Delta}_f$. Since $\check{\varphi}$ is equivariant, it descends to a map $\hat{\varphi}\colon \hat{\Delta}_f \to \mathring{\mathbb{R}}^k$, which is easily seen to be smooth. Since f is a corank one map, r is a diffeomorphism, and consequently $\check{\varphi}$ descends to a smooth map $\varphi\colon \bar{\Delta}_f \to \mathbb{R}^k$, as illustrated in the diagram:

$$\begin{array}{ccc} \check{\Delta}_f & \xrightarrow{\check{\varphi}} & S^{k-1} \times [0,\infty) \\ \downarrow{\scriptstyle q} & & \downarrow{\scriptstyle Q} \\ \hat{\Delta}_f & \xrightarrow{\hat{\varphi}} & \mathring{\mathbb{R}}^k \\ \downarrow{\scriptstyle r} & & \downarrow{\scriptstyle R} \\ \bar{\Delta}_f & \xrightarrow{\varphi} & \mathbb{R}^k. \end{array}$$

13.5.3 Analyzing φ at Σ_f

Let us compute $d\varphi_{(x,x)}(v)$ for each $x \in \Sigma_f$ and each unit vector $v \in T_xN$ in the kernel of df_x, where the tangent bundle $\mathcal{T}_N\colon TN \to N$ is identified with the normal bundle ν of $\Delta_N \subset N^2$. By Lemma C.7 there exists a smooth curve $\delta\colon \mathbb{R} \to N$ such that $\delta(0) = x$, $\delta'(0) = v$ and $f\big(\delta(t)\big) = f\big(\delta(-t)\big)$ for each $t \in \mathbb{R}$. Let $\gamma = \delta \times \delta\colon \mathbb{R} \to N \times N$ and let $\check{\gamma}\colon \mathbb{R} \to \check{N}$ be the lift of γ. Then

$$\begin{aligned} d\varphi_{(x,x)}(v) &= d\varphi_{(x,x)}\big(\gamma'(0)\big) = (\varphi\gamma)'(0) \\ &= \lim_{t\to 0^+} \frac{(\varphi\gamma)(t) - 0}{t} = \lim_{t \to 0} \frac{\check{\varphi}\big(\delta(t), \delta(-t)\big)}{t} \\ &= \lim_{t\to 0^+} \frac{\big|\big|G\delta(-t) - G\delta(t)\big|\big|\, \check{\alpha}\big(\delta(t), \delta(-t)\big)}{t} \\ &= \left|\left| \lim_{t\to 0} \frac{G\delta(-t) - G\delta(t)}{t} \right|\right| \check{\alpha}\Big(\lim_{t\to 0^+} \gamma(t)\Big) \\ &= \big|\big|-2(G\delta)'(0)\big|\big|\, \check{\alpha}\big(\check{\gamma}(0)\big) = -2\big|\big|dG_x\big(\delta'(0)\big)\big|\big|\, \check{\alpha}(x,v) \\ &= -2\big|\big|dG_x(v)\big|\big|\check{\alpha}(x,v). \end{aligned}$$

Here $dG_x(v) \neq 0$ since G is a smooth embedding. Hence $d\varphi_{(x,x)}(v) \neq 0$.

13.5.4 Perturbing Φ, Φ^+ and g_t

When perturbing φ along with the map $e_1 : \pi(U) \to \mathbb{R}^k$ so that $\varphi(x, y) = e_1(y) - e_1(x)$ for $(x, y) \in U$, the following conditions must be preserved: φ is smooth and isovariant, and the restriction of $d\varphi$ to $\ker df$ is a monomorphism of bundles. It is not hard to ensure that $\varphi^+ := \Phi^+|_{\bar{\Delta}_f \times \{1\}}$ also satisfies these conditions by slightly perturbing it keeping $\bar{V}^\flat \cup \bar{\Lambda}$ fixed. Also, it is not hard to ensure that g_1 is smooth by slightly perturbing it keeping $\pi(U) \cup Z$ fixed.

13.5.5 Injectivity of the Differential

Let us compute $d(g_1)_x(v)$, where $x \in \Sigma_f \cap \pi(U)$ and $v \in T_x N$ is a unit vector in the kernel of df_x. We may assume that $\gamma\big((0, 1)\big) \subset U$ and so $\delta\big((-1, 1)\big) \subset \pi(U)$, so that we can use the constraint $g_1(y) - g_1(x) = \varphi^+(x, y)$ for $(x, y) \in U$.

$$
\begin{aligned}
d(g_1)_x(v) &= d(g_1)_x\big(\delta'(0)\big) = (g_1\delta)'(0) \\
&= \lim_{t\to 0} \frac{g_1\delta(t) - g_1\delta(-t)}{2t} = -\lim_{t\to 0} \frac{\varphi^+\big(\delta(t), \delta(-t)\big)}{2t} \\
&= -\frac{1}{2} \lim_{t\to 0^+} \frac{\varphi^+\gamma(t) - 0}{t} = -\frac{1}{2}(\varphi^+\gamma)'(0) = -\frac{1}{2} d\varphi^+_{(x,x)}\big(\gamma'(0)\big) \\
&= -\frac{1}{2} d\varphi^+_{(x,x)}(v).
\end{aligned}
$$

But $d\varphi^+_{(x,x)}(v) \neq 0$ since the restriction of $d\varphi^+$ to $\ker df$ is a monomorphism. Hence $d(g_1)_x(v) \neq 0$. Thus $f \times g_1$ is a smooth embedding. $\square$

13.6 Proofs of Theorems 13.3 and 13.4

Proof of Theorem 13.3 Since N is stably parallelizable, the tangent bundle of $\tilde{N}$ is isomorphic to $n\lambda$, where λ is the line bundle associated with the two-fold cover $\tilde{N} \to \tilde{N}/t$ (see Lemma 13.6.1). On the other hand, since M is stably parallelizable, the normal bundle of Δ_f/t in $\tilde{N}/t$ is stably isomorphic to $n\lambda_f$, where λ_f is the restriction of λ over Δ_f/t (see Lemma 13.6.2). Therefore Δ_f/t is stably parallelizable.

By Proposition 13.1.8 f lifts to an embedding $g : M \to N \times \mathbb{R}^{n+1}$. This yields an equivariant map $\tilde{g} : \Delta_f \to S^n$. We may assume that $\tilde{g}$ is transverse to the equatorial S^{n-1}. Let F be the composition $M \xrightarrow{g} N \times \mathbb{R}^{n+1} \xrightarrow{\pi} N \times \mathbb{R}$, where π is the projection onto the vertical line. Then $\Delta_F = \tilde{g}^{-1}(S^{n-1})$. Consequently

$\Delta_F/t = \bar{g}^{-1}(\mathbb{R}P^{n-1})$, where $\bar{g}\colon \Delta_f/t \to \mathbb{R}P^n$ is determined by $\tilde{g}$ and therefore is the classifying map of the bundle λ_f. Hence λ_f is the pullback of the tautological line bundle γ over $\mathbb{R}P^n$. Since $[\mathbb{R}P^{n-1}]$ is Poincaré dual to $w_1(\gamma)$, we obtain that $[\Delta_F/t]$ is Poincaré dual to $w_1(\lambda_f)$.

On the other hand, since the normal bundle of $\mathbb{R}P^{n-1}$ in $\mathbb{R}P^n$ is isomorphic to γ, the normal bundle of Δ_F/t in Δ_f/t is isomorphic to λ. Since Δ_f/t is stably parallelizable, we get that $\tau \oplus \lambda$ is stably trivial, where τ is the tangent bundle of Δ_F/t. Hence by [29, Lemma 3.5] $\tau \oplus \lambda$ is trivial. Then by the Hirsch theorem (see [49], [48]) Δ_F/t immerses in $\mathbb{R}^n$. Since $n \neq 1, 3, 7$, this implies that $w_1^{n-1}(\Delta_F/t) = 0$ (see [34, Lemma 9]). Also we have $w_1(\Delta_F/t) = w_1(\lambda_F)$, where λ_F is λ restricted over Δ_F/t, so $w_1^{n-1}(\lambda_F) = 0$. Since $[\Delta_F/t]$ is Poincaré dual to $w_1(\lambda_f)$, it follows that $w_1(\lambda_f)^n = 0$. Then Δ_f admits an equivariant map to S^{n-1} (see Lemma 13.6.3). □

Lemma 13.6.1 *If N is a stably parallelizable n-manifold, then the tangent bundle of $\tilde{N}/t$ is stably isomorphic to $n\lambda$, where λ is the line bundle associated with the double cover $\tilde{N} \to \tilde{N}/t$.*

The case $N = S^n$ is discussed in [5, Lemma 3.1].

Proof Since N is stably parallelizable, $Q := N \times \mathbb{R}$ is parallelizable [29, proof of Lemma 3.4]. The total space $T_{\tilde{Q}/t}$ of the tangent bundle $\tau_{\tilde{Q}/t}$ is naturally identified with a submanifold of $\widetilde{T_Q}/t$. The points of the latter are of the form $\{(x, y, u, v), (y, x, v, u)\}$, where $u \in T_x Q$ and $v \in T_y Q$. If $e_0, \dots, e_n$ are the basis vectors of a framing of τ_Q, then $\tau_{\tilde{Q}/t}$ is the Whitney sum $\epsilon_0 \oplus \dots \oplus \epsilon_n \oplus \lambda_0 \oplus \dots \oplus \lambda_n$ of its line subbundles ϵ_i with basis vectors $\{(x, y, e_i, e_i), (y, x, e_i, e_i)\}$ and line subbundles λ_i with basis vectors $\{(x, y, e_i, -e_i), (y, x, -e_i, e_i)\}$. Clearly, each ϵ_i is trivial and each λ_i is isomorphic to the line bundle associated with the double cover $\tilde{Q} \to \tilde{Q}/t$. It is also easy to see, by a similar construction, that the normal bundle of $\tilde{N}/t$ in $\tilde{Q}/t$ is isomorphic to $\epsilon \oplus \lambda$. Hence $\tau_{\tilde{N}/t} \oplus \epsilon \oplus \lambda \simeq (n+1)\epsilon \oplus (n+1)\lambda$. By adding a bundle $\lambda^\perp$ such that $\lambda \oplus \lambda^\perp$ is trivial to both sides of the equation, we conclude that $\tau_{\tilde{N}/t}$ is stably isomorphic to $n\lambda$. □

Lemma 13.6.2 *Let M be a stably parallelizable smooth n-manifold, N a smooth n-manifold and $f\colon N \to M$ a self-transverse map. Then the normal bundle ν_f of Δ_f/t in $\tilde{N}$ is stably isomorphic to $n\lambda$, where λ is the line bundle associated with the double cover $\Delta_f \to \Delta_f/t$.*

The case $M = \mathbb{R}^n$ is discussed in [5, §3] (see also [7, §4]).

Proof Since N is stably parallelizable, $Q := N \times \mathbb{R}$ is parallelizable [29, proof of Lemma 3.4]. Let $R = N \times \mathbb{R}$ and let $F = f \times \mathrm{id}_{\mathbb{R}}\colon R \to Q$. Let $e_0, \dots, e_n$ be the basis vectors of a framing of the normal bundle of Δ_Q in $Q \times Q$ such that the differential of the factor exchanging involution of $Q \times Q$ sends each e_i to $-e_i$. Then by pulling back the e_i we obtain an equivariant framing of the normal bundle ν_F of Δ_F in $\tilde{R}$; that is, an isomorphism $\nu_F \simeq (n+1)\epsilon$ which is equivariant with respect to the differential of the factor exchanging involution on $\tilde{R}$ and the sign involution $v \mapsto -v$ on $(n+1)\epsilon$. It follows that the normal bundle of Δ_F/t in $\tilde{R}/t$ splits as $\lambda_0 \oplus \cdots \oplus \lambda_n$, where each λ_i is isomorphic to the line bundle associated with the double cover $\tilde{Q} \to \tilde{Q}/t$. On the other hand, it is easy to see, by a similar construction, that the normal bundle of Δ_f/t in Δ_F/t is isomorphic to λ. Hence $\nu_f \oplus \lambda \simeq (n+1)\lambda$. By adding a bundle $\lambda^\perp$ such that $\lambda \oplus \lambda^\perp$ is trivial to both sides of the equation, we conclude that ν_f is stably isomorphic to $n\lambda$. □

Let Δ be a space with a free involution t. Its *Yang index* is the maximal k such that $w_1(\lambda)^k \neq 0$, where λ is the line bundle associated with the 2-covering $\Delta \to \Delta/t$.

Lemma 13.6.3 *If Δ is an n-manifold with a free involution t, it admits an equivariant map to S^{n-1} with the antipodal involution if and only if it has Yang index $< n$.*

Proof The "only if" part is trivial, since S^{n-1} has Yang index $< n$.

The first obstruction to the existence of an equivariant map $\Delta \to S^{n-1}$ is $e(\lambda)^n \in H^n(\Delta/t; \mathbb{Z}_\lambda^{\otimes n})$, where $\mathbb{Z}_\lambda$ is the integral local coefficient system corresponding to λ, and $e(\lambda) \in H^1(\Delta/t; \mathbb{Z}_\lambda)$ is the Euler class of λ, which is induced from the generator of $H^1(\mathbb{R}P^\infty; \mathbb{Z}_\lambda) \simeq \mathbb{Z}/2$ under a classifying map of λ; this obstruction is complete (see [35, §2]). Since Δ is a manifold, $H^n(C; \mathbb{Z}_\lambda^{\otimes n})$ is either 0 or $\mathbb{Z}$ or $\mathbb{Z}/2$ for each component C of Δ. So $H^n(\Delta/t; \mathbb{Z}_\lambda^{\otimes n})$ contains no elements of order 4. Hence $e(\lambda)^n = 0$ iff its mod 2 reduction $w_1(\lambda)^n = 0$. □

Lemma 13.6.4 ([34, proof of Corollary to Theorem 5]) *Let N be a $\mathbb{Z}/2$-homology n-sphere and $\Delta \subset \tilde{N}$ an n-dimensional closed $\mathbb{Z}/2$-invariant submanifold. Δ has Yang index $< n$ if and only if every (compact) $\mathbb{Z}/2$-invariant component of Δ projects with an odd degree to the first factor of $N \times N$.*

All results of [34] referred to in this section are proved there in the smooth category, but remain true in the PL category by the same arguments, without substantial modifications (apart from using PL transversality in place of smooth transversality).

Proof of Theorem 13.4 By Corollary 13.1.2 and Lemmas 13.6.3 and 13.6.4, we only need to show that every $\mathbb{Z}/2$-invariant component of Δ_f projects with an even degree to the first factor of $N \times N$. Under the assumptions of part (a), this follows immediately from [34, Theorem 2] and [34, diagram (∗) in the introduction].

For (b), let $f_0\colon N_0 \to M_0$ be the map whose mapping cylinder is the universal covering of the mapping cylinder of f. (So $\pi_1(M_0) = 1$ and $\pi_1(N_0) = \ker(f_*)$.)

Let us first consider the case where $\pi_1(M)$ is finite. Each component C of Δ_f is covered by some component C_0 of Δ_{f_0} with finite degree. By [34, Theorem 3], C_0 projects with zero degree to the (first, say) factor of $N_0 \times N_0$. Hence C projects with zero degree to the first factor of $N \times N$.

Now if $\pi_1(M)$ is infinite, let C be a component of Δ_f. Let p be a regular point of f and let b be some point of $S := f^{-1}(p) \cap C$. If $\ell \subset C$ is a path with endpoints in S, let $\alpha_\ell \in \pi_1(M, p)$ be the class of its image under the composition $\Delta_f \subset N \times N \xrightarrow{\pi} N \xrightarrow{f} M$. Let S_0 be the set of those points of S that can be joined with

b by a path $\ell \subset C$ such that $\alpha_\ell = 1$. If $g_1, \dots, g_i \in \pi_1(M, p)$ and $S_i \subset S$ are defined and $S_i \neq S$, pick a point $q \in S \setminus S_i$ and a path $\ell \subset C$ joining q with b; set $g_{i+1} = \alpha_\ell$ and define S_{i+1} to consist of those points of S that can be joined with b by a path ℓ such that $\alpha_\ell \in \{1, g_1, \dots, g_{i+1}\}$. This process terminates at some finite stage $r \leq |S|$.

The outcome is that $S = \widetilde{f^{-1}(p)} \cap C$ is now in bijection with $\bigsqcup_{i \leq r} \widetilde{f_0^{-1}(g_i \hat{p})} \cap \hat{C}_0$, where $\hat{p}$ is a lift of p, and $\hat{C}_0$ is the component of Δ_{f_0} that covers C and contains the point $\hat{b}$ that corresponds to b under the identification $f^{-1}(p) = f_0^{-1}(\hat{p})$. Indeed, the points of $g_{i+1}(\hat{S}_{i+1} \setminus \hat{S}_i)$ are in $\hat{C}_0$ and those of $g_{i+1}(\hat{S} \setminus \hat{S}_{i+1})$ are not in $\hat{C}_0$ by the construction of S_{i+1}. If a point $g_{i+1}(\hat{q})$ of $g_{i+1}(\hat{S}_i)$ is in $\hat{C}_0$, then q can be joined with b by a path $\ell_{qb} \subset C$ such that $\alpha_{\ell_{qb}} = g_{i+1}$. Hence every point of $S_{i+1} \setminus S_i$ can be joined with q by a path ℓ going first to b, then to q via the reverse of ℓ_{qb}, and finally back to b so that $\alpha_\ell = g_{i+1} g_{i+1}^{-1} g_j$ for some $j \leq i$, which is a contradiction.

As a consequence, the degree of the composition $C \to N \to M$, as computed at b, equals r times the degree of the projection $C_0 \to N_0 \to M_0$, as computed at $\hat{b}, g_1\hat{b}, \dots, g_r\hat{b}$. Here we understand that if C_0 and M_0 are non-compact, but the composition is a proper map, the degree is defined via cohomology with compact support (or locally-finite homology) and so may be nonzero. The assertion of [34, Theorem 3] was stated in the case where N_0 is compact, but the proof works for proper maps of non-compact manifolds as well. Thus $C_0 \to N_0$ has degree 0, and therefore so does either $C \to N$ or $N \to M$. In the former case the proof is completed similarly to the case of finite $\pi_1(M)$, and in the latter we refer to part (a). □

Appendix A. Stable Smooth Maps

Continuous maps $f, g \colon N \to M$ between C^r-manifolds are called *C^r-left-right equivalent*, where $r \in \{0, 1, \dots, \infty\}$, if there exist C^r-self-homeomorphisms φ of N and ψ of M such that the following diagram commutes:

$$\begin{array}{ccc} N & \xrightarrow{\varphi} & N \\ {\scriptstyle f}\downarrow & & \downarrow{\scriptstyle g} \\ M & \xrightarrow{\psi} & M. \end{array}$$

A smooth (i.e. C^∞) map $f \colon N \to M$ between smooth (i.e. C^∞) manifolds is called *C^r-stable* if it has a neighborhood in $C^\infty(N, M)$ whose every member is C^r-left-right equivalent to f. By a stable smooth map we mean a C^∞-stable one.

Theorem A.1 (Triangulation Theorem) *Let M and N be smooth manifolds, where N is compact. Then there exists a dense open set $S \subset C^\infty(N, M)$ such that every $f \in S$ is C^0-stable and C^0-left-right equivalent to a PL map (with respect to some smooth triangulations of M and N).*

R. Thom and J. Mather proved that the set of C^0-stable smooth maps $N \to M$ contains a dense open subset of $C^\infty(N, M)$ (see [21]), and A. Verona proved that C^0-stable smooth maps are triangulable [59]. Here is another approach:

Proof By Shiota's theorem [51] a smooth map $f: N \to M$ is C^0-left-right equivalent to a PL map if it is Thom stratified. By [21, IV.3.3] f is Thom stratified if it belongs to the set S of smooth maps that are multi-transverse to a certain Whitney stratification of a suitable jet space. By [21, IV.1.1 and IV.4.1] S is open and dense in $C^\infty(N, M)$. Also, all members of S are C^0-stable [21, IV.4.4]. □

A smooth map $f: N^n \to M^m$, $n \le m$, is called a *corank one* map, if $\dim(\ker df_x) \le 1$ at every point $x \in N$. In particular, every smooth fold map is a corank one map. The set of corank one maps is open in $C^\infty(N, M)$.[14] If $2m \ge 3(n-1)$, corank one maps are also dense in $C^\infty(N, M)$ (see [23, VI.5.2]).

Theorem A.2 (Corank One Stability Theorem) *Let M^m and N^n be smooth manifolds, where N is compact, $m \ge n$. Let A be the set of all corank one maps $N \to M$ and S be the set of all C^∞-stable maps $N \to M$. Then $S \cap A$ is open and dense in A.*

This result is well-known (see [28, §2.1], [23, VII.6.4]), but I did not find a conclusive writeup of the proof in the literature.

Proof Let T be the set of all Thom–Boardman maps $N \to M$ that are self-transverse (="have normal crossings" in the terminology of [23]). Then T is dense in $C^\infty(N, M)$ [23, VI.5.2]. Since A is open, $T \cap A$ is dense in A. For each $f \in T \cap A$ we have $\Sigma_f^{i_1,\dots,i_k} = \varnothing$ if $i_1 > 1$, and hence (see [14, 2.18]) also if some $i_j > 1$. Hence each $x \in N$ belongs to some $\Sigma_f^{1,\dots,1,0}$ (which includes Σ_f^0). Then by Morin's theorem [41] f has stable germs at all $x \in N$. In particular, they are infinitesimally stable (see definition in [23]). Since f is also self-transverse, it has infinitesimally stable multi-germs at all $y \in M$ [31, 1.6]. Hence f is infinitesimally stable (see e.g. [23, V.1.5 and V.1.6]) and therefore stable (see [23]). Thus $T \cap A \subset S$. Since $T \cap A$ is dense in A, so is $S \cap A$. Clearly, S is open in $C^\infty(N, M)$, so $S \cap A$ is open in A. □

Appendix B. Stable PL Maps

B.1 PL Transversality

A subpolyhedron Y of a polyhedron X is said to be *collared* in X if some neighborhood of Y in X is homeomorphic to $Y \times [0, 1]$ by a PL homeomorphism that extends $\mathrm{id}: Y \to Y \times \{0\}$.

[14] Indeed, f is a corank one map if and only if $j^1 f(N)$ is disjoint from the closed subset $\bigcup_{i \ge 2} \Sigma^i$ of $J^1(N, M)$ (see [23]).

A PL map $f: P \to Q$ between polyhedra is said to be *PL transverse* to a triangulation L of Q if $f^{-1}(\partial\sigma)$ is collared in $f^{-1}(\sigma)$ for each simplex σ of L. The map f is called *PL transverse* to a subpolyhedron R of Q if f is PL transverse to some triangulation L of Q such that R is triangulated by a subcomplex of L.

Let K and L be simplicial complexes. A *semi-linear map* $f: K \to L$ is a PL map $|K| \to |L|$ between their underlying polyhedra that sends every simplex of K into some simplex of L by an affine map. Every semi-linear map $f: K \to L$ determines a monotone map $[f]$ between the face posets[15] of K and L, defined by sending every simplex σ of K to the minimal simplex of L containing $f(\sigma)$. Two semi-linear maps $f, g: K \to L$ will be called *combinatorially equivalent* if $[f] = [g]$, or in other words if $f^{-1}(\sigma) = g^{-1}(\sigma)$ for every simplex σ of L.

If $f, g: K \to L$ are combinatorially equivalent semi-linear maps, then f is PL transverse to L if and only if g is PL transverse to L. In this case the monotone map $[f]$ between the face posets of K and L is called a *stratification map*. If L' is a simplicial subdivision of a simplicial complex L, and $\mathrm{id}: |L'| \to |L|$ is regarded as a semi-linear map $s: L' \to L$, then the monotone map $[s]$ is a stratification map.

Since composition of stratification maps is a stratification map [37, 13.4] (see also [15, "Amalgamation" on p. 23 and "Extension to polyhedra" on p. 35]), a PL map $P \to |L|$ that is transverse to L' must also be transverse to L. Conversely, if a PL map $f: P \to |L|$ is transverse to L, it is PL-left-right equivalent[16] to a PL map that is transverse to L' (see [15, Theorem II.2.1 and "Extension to polyhedra" on p. 35]); moreover, the equivalence is via PL homeomorphisms $|L| \to |L|$ and $P \to P$ that preserve the simplexes of L and their preimages. Using a generalization of these results from simplicial complexes to cone complexes (see [15, 37]) and a result of M. M. Cohen[17] it is not hard to show (see [15, §II.4]) that every PL map $f: P \to |L|$ is PL-left-right equivalent to a PL map that is transverse to L.

B.2 Stable PL Maps

If K is a simplicial complex, a *linear map* $f: K \to \mathbb{R}^m$ is a PL map $|K| \to \mathbb{R}^m$ whose restriction to every simplex of K is the restriction of an affine map. Let $C(K, \mathbb{R}^m)$ be the subspace of $C^0(|K|, \mathbb{R}^m)$ consisting of all linear maps. Let $S(K, \mathbb{R}^m)$ be the set of all linear maps $f: K \to \mathbb{R}^m$ such that f has a neighborhood in $C(K, \mathbb{R}^m)$ whose every member is PL-left-right equivalent to f.

More generally, given simplicial complexes K and L and a monotone map φ between their face posets, let $C(\varphi)$ be the subspace of $C^0(|K|, |L|)$ consisting of

[15] By "face poset" we mean the poset of all nonempty faces.

[16] The definition of PL-left-right equivalence repeats that of C^r-left-right equivalence (see Appendix A), with "C^r" replaced by "PL" throughout.

[17] If $f: K_1 \to K_2$ is a simplicial map between simplicial complexes, then $|f|: |K_1| \to |K_2|$ is transverse to the dual cone complex K_2^* (see [37, 16.2]).

all semi-linear maps $f\colon K \to L$ such that $[f] = \varphi$. Let $S(\varphi)$ be the set of all semi-linear maps $f\colon K \to L$ such that f has a neighborhood in $C(\varphi)$ whose every member is PL-left-right equivalent to f. If L triangulates $\mathbb{R}^m$ and φ is a constant map onto some m-simplex, then $C(\varphi)$ is an open subspace of $C(K, \mathbb{R}^m)$, and $S(\varphi) = S(K, \mathbb{R}^m) \cap C(\varphi)$.

Theorem B.1 *Let K and L be simplicial complexes, where K is finite, and φ be a monotone map between their face posets. Then $S(\varphi)$ is open and dense in $C(\varphi)$.*

Proof The definition of $S(\varphi)$ implies that it is open in $C(\varphi)$.

Let us call a map ν from a finite set F to an affine space V a *general position map* if for each $G \subset F$ the affine subspace of V spanned by $\nu(G)$ is of dimension $\min(\#G-1, \dim V)$. In other words, ν is required to be injective, unless $\dim V = 0$; not to send any three points into the same affine line, unless $\dim V \leq 1$; not to send any four points into the same affine plane, unless $\dim V \leq 2$; and so on. Each of these conditions determines an open and dense subset of $C^0(F, V)$, and hence their intersection, which is the set of all general position maps $F \to V$, is also open and dense in $C^0(F, V)$.

Let $G(\varphi)$ be the subset of $C(\varphi)$ consisting of all semi-linear maps $f\colon K \to L$ such that for each simplex σ of L the restriction of f to set of vertices of the subcomplex $f^{-1}(\sigma)$ of K is a general position map into the affine space spanned by σ. Since K is finite, it is easy to see that $G(\varphi)$ is open and dense in $C(\varphi)$.

For every semi-linear map $f\colon K \to L$ there is a standard construction yielding subdivisions K'_f, L'_f of K, L with respect to which f is simplicial (see [62]). It is not hard to see that if $f \in G(\varphi)$, then there is a neighborhood U of f in $C(\varphi)$ such that K'_f and L'_f are isomorphic (as simplicial complexes) to K'_g and L'_g for every $g \in U$. Then g is PL-left-right equivalent to f. Hence $G(\varphi) \subset S(\varphi)$, and therefore $S(\varphi)$ is dense in $C(\varphi)$. □

Corollary B.2 *If K is a simplicial complex, $S(K, \mathbb{R}^m)$ is open and dense in $C(K, \mathbb{R}^m)$.*

A PL map $f\colon P \to Q$ between polyhedra will be called *stable* if there exist triangulations K, L of P, Q and a stratification map φ between their face posets such that f is PL-left-right equivalent to a member of $S(\varphi)$. In particular, stable PL maps $|K| \to \mathbb{R}^m$ include all members of $S(K, \mathbb{R}^m)$.

Remark B.3

(a) There is an alternative approach to stable PL maps. By using the C^1 topology on semi-linear maps (see [42]) one can do without fixing a triangulation of the domain. However, a triangulation of the target still needs to be fixed, and hence PL transversality still needs to be used in this approach.
(b) A classical approach to general position arguments for PL maps from a compact polyhedron to a PL manifold M is to cover M by coordinate charts, and achieve desired general position properties separately in each chart (see [62]). Since the transition maps are not linear but PL, it seems to be difficult to formulate this approach in invariant terms (such as stability), even for a fixed atlas.

B.3 Examples of Stable PL Maps

Proposition B.4 *Every PL embedding $f: P \to Q$ between polyhedra, where P is compact, is stable.*

Proof Upon replacing f by a PL-left-right equivalent embedding we may assume that it is PL transverse to some triangulation L of Q. Let K be a triangulation of P such that f is a semi-linear map $K \to L$. Let $g: K \to L$ be a semi-linear map with $[g] = [f]$ that is ϵ-close to f in the sup metric. Since K is finite and f sends disjoint simplexes of K to disjoint subsets of Q, so does g, as long as ϵ is sufficiently small. But if $g(p) = g(q)$, where the minimal simplexes σ, τ of K containing p and q share a common face $\rho = \sigma \cap \tau$, then we have $\sigma = \rho * \sigma'$ and $\tau = \rho * \tau'$ and it is easy to see that either $g(\sigma)$ meets $g(\tau')$ or $g(\sigma')$ meets $g(\tau)$. Thus g is an embedding. By a similar argument, the linear homotopy between f and g is an isotopy. Moreover, if h_i denotes the semi-linear map $K \to L$ which agrees with g on the first i vertices of K and with f on the remaining ones, so that $h_0 = f$ and $h_k = g$, where k is the number of vertices of K, then by similar arguments each h_i is an embedding, and the linear homotopy between h_i and h_{i+1} is an isotopy. Each $[h_i] = [f]$, hence every such isotopy is covered by a PL ambient isotopy (even if Q is not a manifold). Thus g is PL-left-right equivalent to f. □

Example B.5 Let N^n and M^m be PL manifolds, where N is compact and $m \geq n$, and let $f: N \to M$ a PL map. As long as f is non-degenerate, for each $x \in N$ there is a PL map $\mathrm{lk}(x, f): \mathrm{lk}(x, N) \to \mathrm{lk}\big(f(x), M\big)$, which is well-defined up to PL-left-right equivalence; and for each $y \in M$ there is a PL map $\mathrm{lk}(y, f): \bigsqcup_{f(x)=y} \mathrm{lk}(x, N) \to \mathrm{lk}(y, M)$, which is also well-defined up to PL-left-right equivalence.

(a) If $m \geq 2n + 1$, it follows from Proposition B.4 that f is stable if and only if it is an embedding.
(b) If $m = 2n$, it follows by similar arguments that f is stable if and only if it is an immersion (i.e., locally injective) with a finite set Δ of transverse double points (i.e. points $y \in M$ such that $\mathrm{lk}(y, f)$ is PL-left-right equivalent, not necessarily preserving the orientations, to the Hopf link $\partial I^n \times \{0\} \sqcup \{0\} \times \partial I^n \subset \partial(I^n \times I^n)$, where $I = [-1, 1]$). Let us note that stable maps $N^2 \to M^4$ may be locally knotted at finitely many points of $N \setminus f^{-1}(\Delta)$. Nevertheless, stable maps $N^2 \to \mathbb{R}^4$ have a normal Euler class [12].
(c) If $m = 2n - 1$, $n > 2$, similar techniques work to show that f is stable if and only if M contains a finite subset Σ such that $\mathrm{lk}(y, f)$ is a stable PL map $S^{n-1} \to S^{2n-2}$ (see (b)) for each $y \in \Sigma$, and $f|_{\dots}: N \setminus f^{-1}(\Sigma) \to M \setminus \Sigma$ is an immersion with an embedded curve Δ of transverse double points (i.e. points $y \in M$ such that $\mathrm{lk}(y, f)$ is PL-left-right equivalent to the suspension over the Hopf link, $S^0 * S^{n-2} \sqcup S^0 * S^{n-2} \to S^0 * S^{2n-3}$). Let us note that stable PL maps $N^3 \to M^5$ may be locally knotted at points of an embedded finite graph $G \subset N \setminus f^{-1}(\Delta \cup \Sigma)$.

(c′) If $(n, m) = (2, 3)$, similarly f is stable if and only if M contains disjoint finite subsets Σ and T such that $\operatorname{lk}(y, f)$ is a stable PL map $S^1 \to S^2$ for each $y \in \Sigma$ and is PL-left-right equivalent to the Borromean ornament (see [38]) for each $y \in T$; and $f|_{\dots}: N \setminus f^{-1}(\Sigma \cup T) \to M \setminus (\Sigma \cup T)$ is an immersion with an embedded curve Δ of transverse double points.

(c″) If $(n, m) = (1, 1)$, it is easy to see that f is stable if and only if N contains a finite subset S such that $f|_S$ is an embedding and $f|_{N\setminus S}$ is an immersion.

(d) If $(n, m) = (2, 2)$, it is not hard to see that f is stable if and only if N contains an embedded finite graph G with vertex set V such that $f|_V$ is an embedding, $f|_{G\setminus V}$ is a stable PL map into $M \setminus f(V)$ (see (b)), $f|_{N\setminus G}$ is an immersion, and $\operatorname{lk}(x, f)$ is a stable PL map $S^1 \to S^1$ (see (c″)) for each $x \in V$ and is PL-left-right equivalent to the suspension over some map $S^0 \to S^0$ for each $x \in G \setminus V$.

Example B.6 Let P^n be a polyhedron and $f: P \to \mathbb{R}$ a PL map. For each $x \in P$ let $L^+(x) = \operatorname{lk}\big(x, f^{-1}([y, \infty))\big)$ and $L^-(x) = \operatorname{lk}\big(x, f^{-1}((-\infty, y])\big)$, where $y = f(x)$. Thus $L^+(x) \cup L^-(x) = \operatorname{lk}(x, P)$ and $L^+(x) \cap L^-(x)$ coincides with $L^0(x) := \operatorname{lk}\big(x, f^{-1}(y)\big)$. Let us call f *link-regular* at x if $L^0(x)$ is collared in $L^+(x)$ and in $L^-(x)$, or in other words if $f|_{\operatorname{lk}(x,P)}$ is PL transverse to $\{f(x)\}$. Let us call f *regular* at x if each of $L^+(x)$ and $L^-(x)$ is PL homeomorphic to the cone over $L^0(x)$ keeping $L^0(x)$ fixed, or equivalently (see [37, 12.3]) if $f|_{\operatorname{st}(x,P)}$ is PL transverse to $\{f(x)\}$. Let us call f a *link-submersion* (resp. a *PL submersion*) if it is link-regular (resp. regular) at all points $x \in P$.

It is not hard to see that the following conditions are equivalent for a PL map $P \to \mathbb{R}$, where P is a compact polyhedron:

(1) f is stable;
(2) f is a link-submersion and P contains a finite set S such that $f|_S$ is an embedding and $f|_{P\setminus S}$ is a PL submersion;
(3) f is PL-left-right equivalent to a linear map $K \to \mathbb{R}$ that embeds every 1-simplex.

Stable maps $P \to \mathbb{R}$ have been used in discrete differential geometry [11] and in discrete Morse theory [13].

Remark B.7 The theory of stable PL maps, as outlined above, is clearly in a very early stage of development. In particular, implicit in the definition of a stable PL map are several questions about the status of the obvious modifications of this definition. One may also wonder about a multi-0-jet transversality theorem (cf. Sect. C.2 below) for stable PL maps. Presumably this situation has some similarity with the state of the theory of stable smooth maps in the times of Whitney and Thom, i.e. before Boardman, Mather and others who brought more clarity to that field.

Appendix C. Extended 2-Multi-0-Jet Transversality

The main result of this Appendix, Corollary C.10, addresses the case $s = 2, r = 0$ of the informal problem raised by C. T. C. Wall in 1971: "Is there any natural way to 'fill the hole along the diagonal' [in the s-multi-r-jet transversality theorem] and get a transversality theorem in the completed space (so one could, for example, combine injectivity [of a generic smooth map $N^n \to M^m$ in the range $m \geq 2n + 1$] with the immersion result [in the range $m \geq 2n$])?" [60, p. 192].

One special case of our Corollary C.10 was essentially proved by F. Ronga, but our proof of this case is somewhat different; in fact, we give two proofs somewhat different from each other and from Ronga's (see Theorem C.1 and subsequent remarks).

An application of another special case of Corollary C.10 is obtained in Sect. C.10, and is one of the results of Appendix C that are applied in the main part of the present chapter.

The general case of Wall's problem will hopefully be addressed in a future paper by the author. A related problem is also mentioned in a 1976 paper by Wall [61, p. 761].

C.1 Weakly Generic Maps

A subset of a topological space X is called *massive* if it is a countable intersection of dense open sets. By Baire's category theorem, a massive subset of a complete metric space is dense. In particular, since $C^\infty(N, M)$ with the metric (compact-open-like) topology is completely metrizable (see [26, discussion following 3.4.4]), its massive subsets are dense in it. Although $C^\infty(N, M)$ with the strong topology (also known as the Whitney topology or the Mather topology) is not metrizable if N is non-compact, its massive subsets are also dense in it (see [26, discussion following 3.4.4], [23, II.3.3]). Let us note that for compact N, the two topologies on $C^\infty(N, M)$ coincide (see [26]). A key technical advantage of the strong topology is that if W is a closed subset of M, then the set of smooth maps $N \to M$ that are transverse to W is open and dense in the strong topology, but only massive in the metric topology (see [26, 3.2.1]). The same applies to jet transversality (see [26, 3.2.8 and Exercise 3.8(b)]).

The assertion "every generic smooth map $N \to M$ satisfies property P" (or any logically equivalent assertion) will mean "$C^\infty(N, M)$ with the strong topology contains an open dense subset whose elements satisfy property P". The assertion "every weakly generic smooth map $N \to M$ satisfies property P" (or any logically equivalent assertion) will mean "$C^\infty(N, M)$ with the strong topology contains a massive subset whose elements satisfy property P".[18] The choice of strong topology has only technical significance, because in the end we are only interested in the case of compact N.

C.2 2-Multi-0-Jet Transversality Theorem

Let N be a closed smooth n-manifold, M a smooth m-manifold, $m \geq n$, and let $f\colon N \to M$ be a smooth map. The graph $\Gamma_f\colon N \to N \times M$, defined by $x \mapsto \big(x, f(x)\big)$, is an embedding. Therefore the diagonal $\Delta_N = \{(x,x) \mid x \in N\}$ is the preimage of $\Delta_{N\times M}$ under $\Gamma_f \times \Gamma_f\colon N^2 \to (N \times M)^2$. Let $\tilde\Gamma_f\colon \tilde N \to \widetilde{N \times M}$ be the restriction of $\Gamma_f \times \Gamma_f$ to the deleted product $\tilde N = N \times N \setminus \Delta_N$. Clearly, the image of $\tilde\Gamma_f$ lies in $\tilde N \times M \times M = \widetilde{N \times M} \setminus \Delta_N \times \tilde M$.

Let us recall the 2-multi-0-jet Transversality Theorem [23, II.4.13]: if L is a smooth submanifold of $\tilde N \times M \times M$ and $f\colon N \to M$ is a weakly[19] generic smooth map, then $\tilde\Gamma_f$ is transverse to L. An immediate consequence of this theorem is that Δ_f is a smooth $(2n-m)$-manifold; indeed, $\Delta_f = \tilde\Gamma_f^{-1}(L)$, where $L = \tilde N \times \Delta_M$. Moreover, f is *self-transverse*, that is, the restriction of $f \times f\colon N^2 \to M^2$ to $\tilde N$ is transverse to Δ_M, or, equivalently, for any two distinct points $x, y \in N$ with $f(x) = f(y)$ the tangent space $T_{f(x)}M$ is generated by $df_x(T_xN)$ and $df_y(T_yN)$.

It is necessary to restrict $\Gamma_f \times \Gamma_f$ to $\tilde N$ in the 2-multi-0-jet Transversality Theorem. Indeed, $\Gamma_f \times \Gamma_f$ is not transverse to $\bar L := N^2 \times \Delta_M$ unless f is an immersion, since $(\Gamma_f \times \Gamma_f)^{-1}(\bar L) = \Delta_f \cup \Delta_N$ is not a manifold unless f is an immersion. (Here $\Delta_f = \{(x,y) \in \tilde N \mid f(x) = f(y)\}$.) One way of explaining this failure is that N^2 is, in a sense, a wrong compactification of $\tilde N$.

C.3 Fulton–MacPherson and Axelrod–Singer Compactifications of $\tilde N$

Let N be a closed smooth manifold and let $\tau\colon N^2 \to N^2$ be the factor-exchanging involution. Let R be a τ-invariant tubular neighborhood of Δ_N in N^2, and let $\nu\colon R \to \Delta_N$ be an equivariant normal bundle projection. The closure Q of $N^2 \setminus R$ in N^2 is a manifold with boundary, and ν extends to an equivariant smooth map $\varphi\colon N^2 \to N^2$ that restricts to a homeomorphism between $Q \setminus \partial Q$ and $\tilde N$. Hence

[18] "Weakly generic" corresponds to "generic" in the terminology of Gromov's book [25, 1.3.2(B)].

[19] If L is closed and $L \subset K \times M$, where $K \subset \tilde N$ is compact, then "weakly" can be omitted (see [23, Proof of Lemma II.4.14]). For every compact N (in which case the metric and strong topologies coincide) it is easy to construct an L (with "non-asymptotic" behavior near $\Delta_N \times M$) such that the set of all $f\colon N \to M$ for which $\tilde\Gamma_f$ is transverse to L is not open in $C^\infty(N, M)$.

$\tilde{N}$ is the interior of a $\mathbb{Z}/2$-manifold $(\check{N}, \check{t})$ which is equivariantly homeomorphic to Q. The quotient of $\check{N}$ by the restriction of $\check{t}$ to $\partial\check{N}$ is a closed manifold $\hat{N}$.

The manifolds $\check{N}$ and $\hat{N}$ are well-defined up to equivariant homeomorphism keeping $\tilde{N}$ fixed (indeed, $\hat{N}$ is nothing but the blowup of N^2 along Δ_N) and are special cases of the Axelrod–Singer and Fulton–MacPherson compactifications of configuration spaces (see [19], [53]); in our case of interest they were known long before (see [46]). Since ν is isomorphic to the tangent disk bundle of N, the Axelrod–Singer corona $\check{N} \setminus \tilde{N}$ is homeomorphic to the total space SN of the spherical tangent bundle of N, and the Fulton–MacPherson corona $\hat{N} \setminus \tilde{N}$ is homeomorphic to the total space PN of the projective tangent bundle of N. In fact, using φ, we obtain the commutative diagram

$$\begin{array}{ccc} SN & \longrightarrow & \check{N} \\ \downarrow & & \downarrow \\ PN & \longrightarrow & \hat{N} \\ \downarrow & & \downarrow \\ N & \longrightarrow & N^2. \end{array}$$

To avoid excessive notation, we will identify $\check{N} \setminus \tilde{N}$ with SN and $\hat{N} \setminus \tilde{N}$ with PN.

A continuous curve $\gamma\colon [-1, 1] \to N^2$ with $\gamma^{-1}(\Delta_N) = 0$ lifts to a continuous curve $\hat{\gamma}\colon [-1, 1] \to \hat{N}$ if and only if γ is differentiable at 0; and $\hat{\gamma}(0) = \big(\gamma(0), \langle d\gamma_0(1)\rangle\big) \in PN$.

If N is a non-compact smooth manifold without boundary, all of the above applies, except that $\check{N}$ and $\hat{N}$ cannot be called "compactifications" (but they can be called *completions*, as long as we fix some complete Riemannian metric on N or at least a uniform equivalence class of such metrics). If N is a smooth manifold with boundary, one can define $\check{N}$ and $\hat{N}$ by considering the double of N, that is, $N \cup_{\partial N} N$.

C.4 Complete Self-transversality

Given a smooth map $f\colon N \to M$ between smooth manifolds, let $\hat{\Gamma}_f\colon \hat{N} \to \widehat{N \times M}$ be the extension of $\tilde{\Gamma}_f\colon \tilde{N} \to \widetilde{N \times M}$ by means of the "projective differential" $PN \to P(N \times M)$, given by

$$(x, \ell) \mapsto \Big(\big(x, f(x)\big), \big\langle v + df_x(v) \mid v \in \ell\big\rangle\Big), \qquad (*)$$

where $x \in N$ and ℓ is a 1-subspace of T_xN.

The closure of our old friend $\tilde{N} \times \Delta_M$ in $\widehat{N \times M}$ can be written as $\hat{N} \times \Delta_M$ and intersects $P(N \times M)$ in $PN \times M$. The preimage $\hat{\Gamma}_f^{-1}(PN \times M)$ coincides with the following subset of PN,

$$\hat{\Sigma}_f := \{ \langle v \rangle \subset T_x N \mid x \in N,\ v \in \ker df_x \}.$$

Consequently $\hat{\Gamma}_f^{-1}(\hat{N} \times \Delta_M)$ coincides with the following subset of $\hat{N}$,

$$\hat{\Delta}_f := \Delta_f \cup \hat{\Sigma}_f.$$

The map $f : N \to M$ is called *completely self-transverse* if it is smooth and $\hat{\Gamma}_f$ is transverse to $\hat{N} \times \Delta_M$. (The word "completely" may remind us of the completions.) Thus if f is completely self-transverse, $\hat{\Delta}_f$ is a submanifold of $\hat{N}$.

Theorem C.1 *Let N^n and M^m be smooth manifolds, where $m \geq n$ and N is compact. Then the set of completely self-transverse maps is open and dense in $C^\infty(N, M)$.*

Theorem C.1 is an easy consequence of a result by Ronga [46, 2.5(i)] (see also [47]). As the present author was unaware of Ronga's work, Theorem C.1 was also announced in [34, comments after Proposition 1]. Its proof given below is extracted from an unfinished 2004 manuscript by the author and is somewhat different from Ronga's proof (Ronga's proof is more explicit, whereas ours is coordinate free), although both proofs quote the same cases of the Jet Transversality Theorem.

In Sect. C.1 below we also prove a generalization of Theorem C.1, whose proof on the one hand reuses much of the proof of Theorem C.1, but on the other hand provides alternative arguments elsewhere. This resulting third proof of Theorem C.1 is logically shorter in that it builds directly on the proof of the Jet Transversality Theorem instead of being content with sorting out its consequences.

C.5 Full 1-Transversality

Let $\mathcal{P}_N : PN \to N$ denote the projectivization of the tangent bundle $\mathcal{T}_N : TN \to N$, and let γ be the tautological line bundle over PN, which is associated with its double covering by the total space SN of the spherical tangent bundle. Then γ can be viewed as a subbundle of $\mathcal{P}_N^* \mathcal{T}_N$, and the differential $df : \mathcal{T}_N \to \mathcal{T}_M$ yields a map of bundles $df^{PN} : \mathcal{P}_N^* \mathcal{T}_N \to \mathcal{P}_N^* f^* \mathcal{T}_M$ over PN. The restriction of df^{PN} to γ can be regarded as a section of the Hom-bundle $\eta : \mathrm{Hom}(\gamma, \mathcal{P}_N^* f^* \mathcal{T}_M) \to PN$. The zero set of this section $s_f : PN \to E(\eta)$ clearly coincides with $\hat{\Sigma}_f$. If s_f is transverse to the zero section, then the map f is called *fully* 1*-transverse*, following Porteous [45]. Thus if f is fully 1-transverse, then $\hat{\Sigma}_f$ is a submanifold of PN.

Lemma C.2 *f is completely self-transverse if and only if it is self-transverse and fully* 1*-transverse.*

Proof Clearly, f is self-transverse if and only if $\hat{\Gamma}_f$ is transverse to $\tilde{N} \times \Delta_M$.

On the other hand, let p_1, p_2 denote the projections onto the factors of $N \times M$. Since $p_1\Gamma_f = \mathrm{id}_N$ and $p_2\Gamma_f = f$, we have $\Gamma_f^* p_1^* \mathcal{P}_N = \mathcal{P}_N$ and $\Gamma_f^* p_2^* \mathcal{P}_M = f^*\mathcal{P}_M$. With these identifications, $U := E(\Gamma_f^*\mathcal{P}_{N\times M})\backslash E(f^*\mathcal{P}_M)$ is an open tubular neighborhood of $E(\mathcal{P}_N)$ in $E(\Gamma_f^*\mathcal{P}_{N\times M})$. Since the normal bundle of $E(\mathcal{T}_N)$ in $E(\Gamma_f^*\mathcal{T}_{N\times M}) = E(\mathcal{T}_N \oplus f^*\mathcal{T}_M)$ is isomorphic to $E(\mathcal{T}_N^* f^*\mathcal{T}_M)$, and the normal bundle of $\mathbb{R}P^n$ in $\mathbb{R}P^{n+m}$ at $\ell \in \mathbb{R}P^n$ is canonically identified with $\mathrm{Hom}(\ell, \mathbb{R}^m)$, the normal bundle $\nu\colon U \to PN$ of $E(\mathcal{P}_N)$ in $E(\Gamma_f^*\mathcal{P}_{N\times M})$ can be identified with the Hom-bundle $\eta\colon\ \mathrm{Hom}(\gamma, \mathcal{P}_N^* f^*\mathcal{T}_M) \to PN$.

The bundle projection $\nu\colon U \to PN$, which discards the M-component of the tangent line, has the zero section PN as well as the section $\hat{\Gamma}_f|_{PN}\colon PN \to i^*\big(P(N \times M) \setminus p_2^*PM\big)$, where $i\colon \Gamma_f(N) \to N \times M$ is the inclusion. Under the above identification of ν with η this section $\hat{\Gamma}_f|_{PN}$, which is given by the formula (∗), gets identified with s_f. Thus f is fully 1-transverse if and only if $\hat{\Gamma}_f|_{PN}\colon PN \to i^*P(N \times M)$ is transverse to $i^*p_1^*PN$. Since $\hat{N} \times \Delta_M$ meets $P(N \times M)$ transversely in $PN \times M$, which in turn meets $i^*P(N \times M)$ transversely in $i^*p_1^*PN$, the latter is equivalent to saying that $\hat{\Gamma}_f$ is transverse to $\hat{N} \times \Delta_M$ at each point of $PN \times M$. □

C.6 1-Jet Transversality Theorem

The space $J^1(N, M)$ of 1-jets from N to M is the total space of the vector bundle $\mathcal{J}^1(N, M)\colon\ \mathrm{Hom}(p_1^*\mathcal{T}_N, p_2^*\mathcal{T}_M) \to N \times M$, where p_1, p_2 denote the projections onto the factors of $N \times M$. The differential df can be viewed as a section $s_{df}\colon N \to J^1(N, M)$ of the bundle $J^1(N, M) \to N \times M \to N$. The 1-jet Transversality Theorem (see [26, 3.2.8], [23, II.4.9]) says that if L is a smooth submanifold of $J^1(N, M)$ and f is a weakly[20] generic smooth map $N \to M$, then s_{df} is transverse to L. An immediate consequence of this theorem is that for a weakly generic $f\colon N \to M$, the set $\Sigma_f^i := \{x \in N \mid \dim(\ker df_x) = i\}$ is a (not necessarily closed) smooth submanifold in N of codimension $i(m - n + i)$, as long as $m \geq n$. Indeed, $\Sigma_f^i = s_{df}^{-1}(\Sigma^i)$, where $\Sigma^i \subset J^1(N, M)$ is the (generally non-closed) submanifold of all linear maps $L\colon T_xN \to T_yM$, $(x, y) \in N \times M$, of rank $r := n - i$; and a fiber of the normal bundle of Σ^i at some $L \in \Sigma^i \cap \mathrm{Hom}(T_xN, T_yM)$, can be identified (see [23, proof of II.5.3]) with the coset $C_L := L + \mathrm{Hom}(\ker L, \mathrm{coker}\, L) \subset \mathrm{Hom}(T_xN, T_yM)$, which has dimension $(n - r)(m - r)$. Following Porteous [45], we will say that f is *i-transverse* if s_{df} is transverse to Σ^i. Note that this condition is only non-vacuous for finitely many values of i, namely when $i(m - n + i) \leq n$.

Lemma C.3 ([46, 2.2]) *If f is i-transverse for all i, then it is fully 1-transverse.*

[20] If L is closed, then "weakly" can be dropped (see [26, 3.2.8], [23, II.3.4 and II.4.5]).

Lemma C.3 and the 1-jet Transversality Theorem imply that weakly generic maps are fully 1-transverse. One application of Lemma C.3 is discussed in [33].

Proof The section $s_f\colon PN \to \operatorname{Hom}(\gamma, \mathcal{P}_N^* f^*\mathcal{T}_M)$ is transverse to the zero section if and only if the section $\hat{s}_f\colon PN \to \operatorname{Hom}(\mathcal{P}_N^*\mathcal{T}_N, \mathcal{P}_N^* f^*\mathcal{T}_M)$, given by df^{PN}, is transverse to the subbundle $\Pi := \operatorname{Hom}(\mathcal{P}_N^*\mathcal{T}_N/\gamma, \mathcal{P}_N^* f^*\mathcal{T}_M)$. On the other hand, $s_{df}\colon N \to J^1(N, M)$ factors through a section of the bundle $\Gamma_f^*\mathcal{J}^1(N, M)\colon \operatorname{Hom}(\mathcal{T}_N, f^*\mathcal{T}_M) \to N$. Now $\hat{s}_f$ is the induced section of the induced bundle $\mathcal{P}_N^* \operatorname{Hom}(\mathcal{T}_N, f^*\mathcal{T}_M) \to PN$, and it follows that s_{df} is transverse to Σ^i if and only if $\hat{s}_f$ is transverse to $\hat{\Sigma}^i := \mathcal{P}_N^*\Gamma_f^*(\Sigma^i)$.

We have $\Pi \subset \bigcup_i \hat{\Sigma}^i$, and each $\Pi^i := \Pi \cap \hat{\Sigma}^i$ is a (not necessarily closed) submanifold of $\mathcal{P}_N^* \operatorname{Hom}(\mathcal{T}_N, f^*\mathcal{T}_M)$. By the definition of transversality, to prove that $\hat{s}_f$ is transverse to Π it suffices to show that $\hat{s}_f$ is transverse to each Π^i. By the above, $\hat{s}_f$ is transverse to each $\hat{\Sigma}^i$ since f is assumed to be i-transverse. So it remains to show that $\hat{s}_f$ restricted to each $\hat{\Sigma}_f^i := \hat{s}_f^{-1}(\hat{\Sigma}^i) = \mathcal{P}_N^{-1}(\Sigma_f^i)$ is transverse to Π^i (as a map into $\hat{\Sigma}^i$).

Let ℓ be a 1-subspace of T_xN for some $x \in N$, and let $L = \hat{s}_f(\ell)$. Assuming that $L \in \Pi^i$, we have $\dim(\ker df_x) = i$ and $\ell \subset \ker df_x$. Let K_ℓ denote the tangent space at ℓ to the fiber of $\mathcal{P}_N$ over x, that is, K_ℓ is the kernel of $d(\mathcal{P}_N)_\ell\colon T_\ell(PN) \to T_xN$. We claim that $d(\hat{s}_f)_\ell(K_\ell)$ and $T_L\Pi^i$ span $T_L\hat{\Sigma}^i$. Indeed, direct computation shows that Π^i has codimension r in $\hat{\Sigma}^i$.[21] On the other hand, K_ℓ, which is a tangent space of $\mathbb{R}P^{n-1}$, can be identified with $\operatorname{Hom}(\ell, T_xN/\ell)$. If $v_1, \dots, v_r \in K_\ell$ are representatives of a basis for $K_\ell/\operatorname{Hom}(\ell, \ker df_x/\ell)$, where $r = n - i$ is the rank of df_x, we claim that the r vectors $d(\hat{s}_f)_\ell(v_i) + T_L\Pi^i$ are linearly independent in $T_L\hat{\Sigma}^i/T_L\Pi^i$. Indeed, every nontrivial linear combination $\sum_j \alpha_j d(\hat{s}_f)_\ell(v_j)$ can be written as $d(\hat{s}_f)_\ell(v)$, where $v = \sum_j \alpha_j v_j \neq 0$. This v is the tangent vector to a great circle arc from ℓ to a nearby point $\ell' \in \mathbb{R}P^{n-1}$ such that $\ell' \not\subset \ker df_x$. Then $\hat{s}_f(\ell') \notin \Pi^i$, and it follows that $d(\hat{s}_f)_\ell(v) \notin T_L\Pi_i$. □

Proof of Theorem C.1 Let S be the set of all maps $N \to M$ that are self-transverse and i-transverse for all i such that $i(m-n+i) \leq n$. By the 1-Jet Transversality and the 2-Multi-0-Jet Transversality theorems (see [23]) S is massive, and in particular dense in $C^\infty(N, M)$. By Lemmas C.2 and C.3 S lies in the set T of all completely self-transverse maps $N \to M$. Hence T is also dense in $C^\infty(N, M)$.

On the other hand, the set U of all maps $\hat{N} \to \widehat{N \times M}$ that are transverse to $\hat{N} \times \Delta_M$ is open in $C^\infty(\hat{N}, \widehat{N \times M})$ (see e.g. [23, Proposition II.4.5]). By definition, $T = \varphi^{-1}(U)$, where $\varphi\colon C^\infty(N, M) \to C^\infty(\hat{N}, \widehat{N \times M})$ is given by $f \mapsto \hat{\Gamma}_f$. It is easy to see that φ is continuous. Hence T is open. □

[21] Namely, the fiber of $\hat{\Sigma}^i$ over $(x, \ell) \in PN$ consists of $n \times m$ matrices of rank r, and so has dimension $mn - (m-r)(n-r)$. The fiber of Π^i over (x, ℓ) consists of $(n-1) \times m$ matrices of rank r, and so has dimension $m(n-1) - (m-r)(n-1-r) = mn - (m-r)(n-r) - r$.

C.7 Behavior of $\hat{\Delta}_f$ at $\hat{\Sigma}_f$

If f is a completely self-transverse map, then $\hat{\Delta}_f$ is a submanifold of $\hat{N}$; and also its intersection $\hat{\Sigma}_f$ with PN is a submanifold of PN by Lemma C.2. The following lemma shows that this is not merely a coincidence.

Lemma C.4 *$\hat{\Gamma}_f$ is transverse to $P(N \times M)$ for every smooth map $f : N \to M$.*

Proof We have $\hat{\Gamma}_f^{-1}\big(P(N \times M)\big) = PN$. For each $(x, \ell) \in PN$ we can find a smooth curve $\gamma : \mathbb{R} \to N$ such that $\gamma(0) = x$ and $\langle\gamma'(0)\rangle = \ell$. Then $\delta : \mathbb{R} \to N^2$ given by $\delta(t) = \big(\gamma(t), \gamma(-t)\big)$ is a smooth curve which lifts to a smooth curve $\hat{\delta} : \mathbb{R} \to \hat{N}$ with $\hat{\delta}(0) = (x, \ell)$. Since δ is not tangent to Δ_N at $\delta(0)$ and $\delta = p\Gamma_f^2\delta$, where $p : (N \times M)^2 \to N^2$ is the projection, $\Gamma_f^2\delta$ is not tangent to $\Delta_{N\times M}$ at $\Gamma_f^2\delta(0)$. Consequently its lift $\widehat{\Gamma_f\delta} = \hat{\Gamma}_f\hat{\delta} : \mathbb{R} \to \widehat{N \times M}$ is not tangent to $P(N \times M)$ at $\hat{\Gamma}_f\hat{\delta}(0)$. This suffices, since $P(N \times M)$ has codimension one in $\widehat{N \times M}$. □

Corollary C.5 *If $f : N \to M$ is a completely self-transverse map, then*

(a) $\hat{\Delta}_f$ intersects PN transversely;
(b) $\hat{\Delta}_f$ coincides with the closure of Δ_f in $\hat{N}$;
(c) the image of $\hat{\Delta}_f$ in N^2 coincides with $\bar{\Delta}_f$;
(d) the image of $\hat{\Sigma}_f$ in N coincides with Σ_f.

Here $\bar{\Delta}_f$ is the closure of Δ_f in $N \times N$.

Proof Obviously, Lemma C.4 $\Rightarrow$ (a) $\Rightarrow$ (b) $\Rightarrow$ (c) $\Rightarrow$ (d). □

Corollary C.6 *If $f : N \to M$ is a corank one completely self-transverse map, then $\bar{\Delta}_f$ is a smooth submanifold of $N \times N$ and Σ_f is a smooth submanifold of N.*

Let us note that although $\bar{\Delta}_f$ intersects Δ_N in the submanifold Δ_{Σ_f}, this intersection is never transverse for $n > 1$, since $\hat{\Sigma}_f$ has codimension one in $\hat{\Delta}_f$.

It is a well-known folklore result that $\bar{\Delta}_f$ is a smooth submanifold of $N \times N$ and Σ_f is a smooth submanifold of N for every stable[22] corank one map f. It can be derived from Morin's canonical forms [41] for stable corank one germs (cf. [27, §9]); see also [30, 2.14(i) and 2.16] for a complex algebraic version with $m > n$.

Proof Since f is completely self-transverse, $\hat{\Delta}_f$ is a smooth submanifold of $\hat{N}$. Also, Lemma C.2 implies that $\hat{\Sigma}_f$ is a smooth submanifold of PN and $\mathcal{P}_N : PN \to N$ restricts to a submersion on $\hat{\Sigma}_f$. On the other hand, since f is a corank one map, $\mathcal{P}_N : PN \to N$ is injective on $\hat{\Sigma}_f$, and consequently the blowdown map $\pi : \hat{N} \to N^2$ is injective on $\hat{\Delta}_f$. It follows that $\mathcal{P}_N$ restricts to a smooth embedding

[22] Let us note that stable maps are completely self-transverse, since the latter are dense in C^∞ by Theorem C.1 and self-transversality is preserved by C^∞-left-right equivalence.

on $\hat{\Sigma}_f$, and, using Corollary C.5(a), that π restricts to a smooth embedding on $\hat{\Delta}_f$. Finally, by Corollary C.5(c,d) the images of these two embeddings coincide with Σ_f and $\bar{\Delta}_f$. □

Lemma C.7 *If $f: N \to M$ is a completely self-transverse map, then for each pair $(x, \ell) \in \hat{\Sigma}_f$ there exists a smooth curve $\gamma: \mathbb{R} \to N$ such that $\gamma(0) = x$, $\langle \gamma'(0) \rangle = \ell$ and $f(\gamma(t)) = f(\gamma(-t))$ for each $t \in \mathbb{R}$.*

Proof $\hat{\Sigma}_f$ is the fixed point set of the smooth involution on $\hat{\Delta}_f$, which at each point of $\hat{\Sigma}_f$ is locally equivalent to the orthogonal reflection $(x_1, x_2, \dots, x_k) \mapsto (-x_1, x_2, \dots, x_k)$ in $\mathbb{R}^k$, where $k = 2n - m$. Hence there exists an equivariant (with respect to the involution $x \mapsto -x$ on $\mathbb{R}$) smooth curve $\beta: \mathbb{R} \to \hat{\Delta}_f$ such that $\beta(0) = (x, \ell)$. Let $\pi: \hat{N} \to N^2$ be the blowdown map and $p: N^2 \to N$ the projection onto the first factor. Let $\gamma = p\pi\beta$. Then $\pi\beta(t) = \big(\gamma(t), \gamma(-t)\big) \in \bar{\Delta}_f$, so $f\big(\gamma(t)\big) = f\big(\gamma(-t)\big)$ for each $t \in \mathbb{R}$. Also $\ell = \left\langle \lim_{t\to 0} \frac{\gamma(t)-\gamma(-t)}{2t} \right\rangle = \langle \gamma'(0) \rangle$. □

C.8 Extended Gauss Map

If $f: N \to M$ is a completely self-transverse map between manifolds without boundary, then clearly $\check{\Delta}_f$ is a smooth manifold whose boundary is a double cover of $\hat{\Sigma}_f$.

Lemma C.8 *Let N, M be smooth manifolds, $f: N \to M$ a completely self-transverse map and $G: N \to \mathbb{R}^k$ a smooth map such that $f \times G: N \to M \times \mathbb{R}^k$ is a smooth embedding. Then $\check{G}: \check{\Delta}_f \to S^{k-1} \times [0, \infty)$, defined by $\check{G}(x, y) = \left(\frac{G(y)-G(x)}{||G(y)-G(x)||}, ||G(y) - G(x)||\right)$ for $(x, y) \in \Delta_f$ and by $\check{G}(x, v) = \left(\frac{dG_x(v)}{||dG_x(v)||}, 0\right)$ for $(x, v) \in \check{\Sigma}_f$, is a smooth map.*

Proof Let us first consider the case where $N = M \times \mathbb{R}^k$ and f and G are the projections $p: M \times \mathbb{R}^k \to M$ and $q: M \times \mathbb{R}^k \to \mathbb{R}^k$ so that $f \times G = \mathrm{id}_{M\times\mathbb{R}^k}$. It is easy to see that $\check{q}$ (defined similarly to $\check{G}$) is the composition of the natural map $\check{\Delta}_p \to \check{\Delta}_c$, where $c: \mathbb{R}^k \to \{0\}$ is the constant map, and the obvious retraction of $(\mathbb{R}^k)\check{}$ onto its anti-diagonal. In the general case, $f \times G$ induces a smooth embedding $\check{N} \hookrightarrow (M \times \mathbb{R}^k)\check{}$, which in turn restricts to a smooth embedding $G_*: \check{\Delta}_f \to \check{\Delta}_p$. Now G factors into the composition $\check{\Delta}_f \xrightarrow{G_*} \check{\Delta}_p \xrightarrow{\check{q}} S^{k-1} \times [0, \infty)$. □

C.9 The Extended 2-Multi-0-Jet Transversality Theorem

Theorem C.9 *Let N and M be smooth manifolds and let L be a smooth submanifold of $P(N \times M) \setminus N \times PM$. Let X_L be the set of smooth maps $f : N \to M$ whose "projective differential" $\hat{\Gamma}_f|_{PN} : PN \to P(N \times M) \setminus N \times PM$ is transverse to L. Then X_L is massive in $C^\infty(N, M)$ with the strong topology; if L is closed, then "massive" can be replaced with "open and dense".*

Proof By the proof of Lemma C.2, $\hat{\Gamma}_f|_{PN}$ is identified with the section s_f of the bundle $\mathrm{Hom}(\gamma, \mathcal{P}_N^* f^* \mathcal{T}_M) \to PN$ given by the composition $\gamma \subset \mathcal{P}_N^* \mathcal{T}_N \xrightarrow{df^{PN}} \mathcal{P}_N^* f^* \mathcal{T}_M$. In particular, L is identified with a submanifold of the total space $\mathrm{Hom}(\gamma, \mathcal{P}_N^* f^* \mathcal{T}_M)$. Let $\bar{L}$ be the preimage of L in $\mathrm{Hom}(\mathcal{P}_N^* \mathcal{T}_N, \mathcal{P}_N^* f^* \mathcal{T}_M)$ under the restricting map. Clearly, s_f is transverse to L if and only if the section $\hat{s}_f : PN \to \mathrm{Hom}(\mathcal{P}_N^* \mathcal{T}_N, \mathcal{P}_N^* f^* \mathcal{T}_M)$, given by df^{PN}, is transverse to $\bar{L}$. On the other hand, $s_{df} : N \to J^1(N, M)$ can be identified with a section of the bundle $\Gamma_f^* \mathcal{J}^1(N, M) : \mathrm{Hom}(\mathcal{T}_N, f^* \mathcal{T}_M) \to N$, and $\hat{s}_f$ is the induced section of the induced bundle $\mathcal{P}_N^* \mathrm{Hom}(\mathcal{T}_N, f^* \mathcal{T}_M) \to PN$.

By the proof of the Jet Transversality theorem [23, II.4.9] (see also [26, 3.2.8]), for each $w \in N$ there exists a neighborhood B of the origin in the vector space $\mathrm{Hom}(\mathbb{R}^n, \mathbb{R}^m)$ of all linear maps $\mathbb{R}^n \to \mathbb{R}^m$ and a smooth homotopy $g_b : N \to M$, $b \in B$, such that $g_0 = f$, each $g_b|_{N \setminus U} = f|_{N \setminus U}$ for some fixed compact neighborhood U of w, and the smooth map $\Phi : N \times B \to \mathrm{Hom}(\mathcal{T}_N, f^* \mathcal{T}_M)$ defined by $\Phi(x, b) = (dg_b)_x$ sends some neighborhood of $(w, 0)$ by a diffeomorphism onto some neighborhood of df_w. Then the induced map $P\Phi : PN \times B \to \mathcal{P}_N^* \mathrm{Hom}(\mathcal{T}_N, f^* \mathcal{T}_M)$, defined by $P\Phi(x, \ell, b) = \big(x, \ell, \Phi(x, b)\big) = (dg_b)_{x,\ell}^{PN}$, sends some neighborhood of $(w, \ell, 0)$ by a diffeomorphism onto some neighborhood of $df_{w,\ell}^{PN}$ for each $\ell \in P_w N$. In particular, $P\Phi$ is transverse to $\bar{L}$ at $df_{w,\ell}^{PN}$ for each $\ell \in P_w N$. Let us note that $s_{dg_b}(x) = \Phi(x, b)$ and $\hat{s}_{g_b}(x, \ell) = P\Phi(x, \ell, b)$. By [23, II.4.7] we get that B contains a dense subset B_0 such that $\hat{s}_{g_b}$ is transverse to $\bar{L}$ for each $b \in B_0$. Let $b_1, b_2, \dots \in B_0$ converge to 0, and set $f_i = g_{b_i}$. Then each $\hat{s}_{f_i}$ is transverse to $\bar{L}$ and $f_i \to f$ in the strong topology due to $f_i|_{N \setminus U} = f|_{N \setminus U}$. Thus X_L is dense.

Let $S \subset C^\infty\big(N, \mathrm{Hom}(\mathcal{T}_N, f^* \mathcal{T}_M)\big)$ be the subspace consisting of all sections of the bundle $\mathrm{Hom}(\mathcal{T}_N, f^* \mathcal{T}_M) \to N$. The map $C^\infty(N, M) \to S$ given by $f \mapsto s_{df}$ is continuous (see [23, II.3.4]), and one can check that so is the map $S \to C^\infty\big(PN, \mathcal{P}_N^* \mathrm{Hom}(\mathcal{T}_N, f^* \mathcal{T}_M)\big)$, given by sending every section of $\mathrm{Hom}(\mathcal{T}_N, f^* \mathcal{T}_M) \to N$ to the induced section of the induced bundle $\mathcal{P}_N^* \mathrm{Hom}(\mathcal{T}_N, f^* \mathcal{T}_M) \to PN$. The subset of $C^\infty\big(PN, \mathcal{P}_N^* \mathrm{Hom}(\mathcal{T}_N, f^* \mathcal{T}_M)\big)$ consisting of all maps that are transverse to $\bar{L}$ is open if $\bar{L}$ is closed (see [23, II.4.5]), and is massive in general (see [23, proof of II.4.9]). Hence the set of smooth maps $f : N \to M$ such that $\hat{s}_f$ is transverse to $\bar{L}$ is open in $C^\infty(N, M)$ if $\bar{L}$ is closed, and massive in general. As noted above, this set coincides with X_L; and $\bar{L}$ is closed if L is closed. □

From Theorem C.9, Lemma C.4 and the 2-Multi-0-Jet Transversality Theorem we immediately obtain

Corollary C.10 *Let N and M be smooth manifolds and let L be a smooth submanifold of $\widehat{N \times M} \setminus \Delta_N \times \hat{M}$. Let X_L be the set of smooth maps $f\colon N \to M$ such that $\hat{\Gamma}_f$ is transverse to L. Then X_L is massive in $C^\infty(N, M)$ with the strong topology; if L is closed, then "massive" can be replaced with "open and dense".*

The case $L = \hat{N} \times \Delta_M$ was already covered in Theorem C.1. Next we note an application which needs a different L.

C.10 Taking Σ_f Off a Polyhedron

Corollary C.11 *Let N and M^m be smooth manifolds and let Q be a closed subpolyhedron of PN. If* $\dim Q < m$, *then for every generic smooth map $f\colon N \to M$ the manifold $\hat{\Sigma}_f$ is disjoint from Q.*

Proof Let us fix some triangulation of Q. Then Q is the union of its open simplexes $\mathring{\sigma}_i$, which are smooth submanifolds of PN. This union is countable (and even finite if N is compact). Hence by Theorem C.9 the set S of maps $f\colon N \to M$ such that $\hat{\Gamma}_f|_{PN}$ is transverse to each $L_i := \mathring{\sigma}_i \times M$ is massive, and in particular dense in $C^\infty(N, M)$. Since each $\dim L_i \leq \dim Q \times M < 2m$ and $\dim P(N \times M) - \dim PN = 2m$, the image of $\hat{\Gamma}_f|_{PN}$ is disjoint from each L_i and hence from $Q \times M$ for each $f \in S$. In fact, $f \in S$ if and only if the image of $\hat{\Gamma}_f|_{PN}$ is disjoint from $Q \times M$. We have $Q \times M = PN \times M \cap R$, where R is the preimage of Q under the projection $P(N \times M) \setminus N \times PM \to PN$. Hence $f \in S$ if and only if $\hat{\Gamma}_f^{-1}(PN \times M) = \hat{\Sigma}_f$ is disjoint from $\hat{\Gamma}_f^{-1}(R) = Q$.

It remains to show that S is open in $C^\infty(N, M)$. Since $Q \times M$ is a closed subset of $P(N \times M) \setminus N \times PM$, the set of maps $PN \to P(N \times M) \setminus N \times PM$ whose image is disjoint from $Q \times M$ is an open subset of $C^\infty\big(PN,\ P(N \times M) \setminus N \times PM\big)$. On the other hand, the map $C^\infty(N, M) \to C^\infty\big(PN,\ P(N \times M) \setminus N \times PM\big)$ given by $f \mapsto \hat{\Gamma}_f|_{PN}$ is continuous by the proof of Theorem C.9. Hence S is open. □

Theorem C.12 *Let $f\colon N^n \to M^m$ be a smooth map between smooth manifolds, where N is compact, $m \geq n$, such that f is i-transverse for all i. Let P^p be a closed subpolyhedron of N contained in Σ_f and suppose that $k \geq p + j$, where $j = j(p)$ is the maximal number such that $p \leq (j + 1)n - jm - j^2$. Then for every generic smooth map $g\colon N \to \mathbb{R}^k$ the set $\Sigma_{f \times g}$ is disjoint from P, where $f \times g\colon N \to M \times \mathbb{R}^k$ is the joint map.*

Let us note that $j(p) \leq 1$ for all p due to $p \leq \dim \Sigma_f = 2n - m - 1$. Thus $j(p) = 1$ for each $p > 3n - 2m - 4$.

Proof The projection $\hat{\Sigma}_{f \times g} \to \Sigma_{f \times g}$ is surjective and is a restriction of the projection $\hat{\Sigma}_f \to \Sigma_f$. Hence $\Sigma_{f \times g}$ is disjoint from P if and only if $\hat{\Sigma}_{f \times g}$ is disjoint from the preimage Q of P under the projection $\hat{\Sigma}_f \to \Sigma_f$. Clearly, $\hat{\Sigma}_{f\times g} = \hat{\Sigma}_f \cap \hat{\Sigma}_g$. So Q is disjoint from $\hat{\Sigma}_{f\times g}$ if and only if it is disjoint from $\hat{\Sigma}_g$. Thus we need to show that for every generic smooth map $g\colon N \to \mathbb{R}^k$ the manifold $\hat{\Sigma}_g$ is disjoint from Q. To prove this, by Lemma C.11 it suffices to show that $\dim Q < k$.

Indeed, let Q_i be the preimage of $P_i := P \cap \Sigma^i_f$ under the same projection. We have $\dim Q_i = \dim P_i + i - 1$. For each $i \leq j$ we have $\dim Q_i \leq p + i - 1 \leq p+j-1 < k$. Since f is i-transverse, $\dim \Sigma^i_f \leq n-i(m-n+i) = (i+1)n-im-i^2$ (see Appendix C.6). In particular, $\dim \Sigma^{i+1}_f < \dim \Sigma^i_f$. Also, the definition of j implies that $p > \dim \Sigma^{j+1}_f$. Then for each $i > j$ we have $\dim P_i \leq \dim \Sigma^i_f \leq \dim \Sigma^{j+1}_f - j + 1 - i \leq p + j - i$. Hence $\dim Q_i \leq p + j - 1 < k$. Thus $\dim Q < k$. □

Acknowledgments I am grateful to P. Akhmetiev, L. Funar, A. Gorelov, M. Kazarian, M. Skopenkov and V. Vassiliev for stimulating conversations and useful remarks, the referees for useful feedback, and A. Papadopoulos for proofreading the text.

References

1. M. Adachi, *Embeddings and Immersions* (Iwanami Shoten, Tokyo, 1984) (Japanese); English transl. in Transl. Math. Monogr., vol. 124, Amer. Math. Soc., 1993
2. P.M. Akhmet'ev, Prem-mappings, triple self-interesection points of an oriented surface, and Rokhlin's signature theorem. Mat. Zametki **59**(6), 803–810, 959 (1996), English transl., Math. Notes **59**, 581–585 (1996)
3. P.M. Akhmet'ev, On isotopic and discrete realization of maps of n-dimensional sphere in Euclidean space. Mat. Sb. **187**(7), 3–34 (1996); English transl., Sb. Math. **187**, 951–980 (1996)
4. P.M. Akhmet'ev, Embeddings of compacta, stable homotopy groups of spheres, and singularity theory. Uspekhi Mat. Nauk **55**(3), 3–62 (2000); English transl., Russian Math. Surveys **55**, 405–462 (2000)
5. P.M. Akhmetiev, Pontryagin–Thom construction for approximation of mappings by embeddings. Topol. Appl. **140**, 133–149 (2004)
6. P.M. Akhmet'ev, A remark on the realization of mappings of the 3-dimensional sphere into itself. Trudy MIAN **247**, 10–14 (2004), English transl., Proc. Steklov Inst. Math **247**, 4–8 (2004)
7. P.M. Akhmetiev, S.A. Melikhov, Projected and near-projected embeddings. Zap. Nauch. Sem. POMI **498**, 75–104 (2020). Reprinted in J. Math. Sci. **255**, 155–174 (2021); arXiv:1711.03520
8. P.M. Akhmetiev, D. Repovš, A.B. Skopenkov, Obstructions to approximating maps of n-manifolds into $\mathbb{R}^{2n}$ by embeddings. Topol. Appl. **123**, 3–14 (2002)
9. H.A. Akitaya, R. Fulek, C.D. Tóth, Recognizing weak embeddings of graphs. ACM Trans. Algorithms **15**(50), 27 pp. (2019). arXiv:1709.09209
10. V.I. Arnol'd, S.M. Guseĭn-Zade, A.N. Varchenko, *Singularities of Differentiable Maps. Vol. I: The Classification of Critical Points, Caustics and Wave Fronts* (Nauka, 1982) (Russian); English transl. in Mono. in Math., vol. 82, Birkhäuser, Boston, 1985; 2nd Russian ed., MCCME, Moscow, 2004
11. T. Banchoff, Critical points and curvature for embedded polyhedra. J. Diff. Geom. **1**, 245–256 (1967)

12. T.F. Banchoff, O. Johnson, The normal Euler class and singularities of projections for polyhedral surfaces in 4-space. Topology **37**, 419–439 (1998)
13. E.D. Bloch, Polyhedral representation of discrete Morse functions. Discrete Math. **313**, 1342–1348 (2013). arXiv:1008.3724
14. J.M. Boardman, Singularities of differentiable maps. Inst. Hautes Études Sci. Publ. Math. **33**, 21–57 (1967)
15. S. Buoncristiano, C.P. Rourke, B.J. Sanderson, *A Geometric Approach to Homology Theory*, London Math. Soc. Lecture Note Series, vol. 18 (Cambridge Univ. Press, 1976)
16. J.S. Carter, M. Saito, *Surfaces in* 3*-space that do not Lift to Embeddings in* 4*-space*, Knot theory (Warsaw, 1995), vol. 42 (Banach Center Publ., Polish Acad. Sci. Inst. Math., Warsaw, 1998), pp. 29–47
17. F.T. Farrell, Right-orderable deck transformation groups. Rocky Mountain J. Math. **6**, 441–447 (1976)
18. R. Fulek, J. Kynčl, Hanani-Tutte for approximating maps of graphs, in *34th International Symposium on Computational Geometry, LIPIcs*. Leibniz Int. Proc. Inform., vol. 99, No. 39 (Schloss Dagstuhl. Leibniz-Zent. Inform., Wadern, 2018), 15 pp. arXiv:1705.05243
19. W. Fulton, R. MacPherson, A compactification of configuration spaces. Ann. Math. **139**, 183–225 (1994)
20. L. Funar, P.G. Pagotto, Braided surfaces and their characteristic maps. New York J. Math. **29**, 580–612 (2023). arXiv:2004.09174
21. C.G. Gibson, K. Wirthmüller, A.A. du Plessis, E.J.N. Looijenga, *Topological Stability of Smooth Mappings*, Lecture Notes in Math., vol. 552 (Springer, 1976)
22. C.A. Giller, Towards a classical knot theory for surfaces in $\mathbb{R}^4$. Illinois J. Math. **26**, 591–631 (1982)
23. M. Golubitsky, V. Guillemin, *Stable Mappings and their Singularities*, Grad. Texts in Math., vol. 14 (Springer, 1973). (Russian transl., Mir, 1977)
24. A.A. Gorelov, *Lifting Maps Between Graphs to Embeddings*, Essays on Topology — Dedicated to Valentin Poénaru (L. Funar and A. Papadopoulos, eds.), arXiv/2404.12287
25. M. Gromov, *Partial Differential Relations*, Ergebn. der Math., vol. 9 (Springer, Berlin, 1986); Russian transl. in Mir, Moscow, 1990
26. M.W. Hirsch, *Differential Topology*, Grad. Texts in Math., vol. 33 (Springer, 1976)
27. K. Houston, Topology of differentiable mappings, in *Handbook of Global Analysis*, ed. by D. Krupka, D. Saunders (Elsevier, 2008), pp. 493–532, 1213
28. M. Kazarian, Morin maps and their characteristic classes. Preprint. Available at https://nanopdf.com/download/morin-maps-and-their-characteristic-classes_pdf
29. M.A. Kervaire, J.W. Milnor, Groups of homotopy spheres. I. Ann. Math. **77**, 504–537 (1963)
30. W.L. Marar, D. Mond, Multiple point schemes for corank 1 maps. J. Lond. Math. Soc. **39**, 553–567 (1989)
31. J. N. Mather, Stability of C^∞ mappings. IV. Classification of stable germs by R-algebras. Inst. Hautes Études Sci. Publ. Math. **37**, 223–248 (1969)
32. J.N. Mather, Stability of C^∞ mappings. VI: The nice dimensions, in *Proc. of Liverpool Singularities Symp. I*, Lecture Notes in Math., vol. 192 (Springer, 1971), pp. 207–253
33. C. McCrory, Cobordism operations and singularities of maps. Bull. Amer. Math. Soc. **82**, 281–283 (1976)
34. S.A. Melikhov, Sphere eversions and realization of mappings. Proc. Steklov Inst. Math. **247**, 143–163 (2004); arXiv:math.GT/0305158; Russian transl., Tr. Mat. Inst. Steklova **247**, 159–181 (2004)
35. S.A. Melikhov, The van Kampen obstruction and its relatives. Tr. Mat. Inst. Steklova **266**, 149–183 (2009). Reprinted in: Proc. Steklov Inst. Math. **266**, 142–176 (2009); arXiv:math.GT/0612082
36. S.A. Melikhov, Transverse fundamental group and projected embeddings. Proc. Steklov Inst. Math. **290**, 155–165 (2015); arXiv:1505.00505; Russian transl., Tr. Mat. Inst. Steklova **290**, 166–177 (2015)
37. S.A. Melikhov, Combinatorics of combinatorial topology. arXiv:1208.6309v2

38. S.A. Melikhov, A triple-point Whitney trick. J. Topol. Anal. **12**(4), 1041–1046 (2020). arXiv:2210.04016
39. S.A. Melikhov, Lifting generic maps to embeddings. Triangulation and smoothing. Stud. Sci. Math. Hung. **59**(2), 124–141 (2022). arXiv:2011.01402
40. S.A. Melikhov, J. Zając, Contractible polyhedra in products of trees and absolute retracts in products of dendrites. Proc. Amer. Math. Soc. **141**, 2519–2535 (2013). arXiv:1102.0696
41. B. Morin, Formes canoniques des singularités d'une application différentiable. C. R. Acad. Sci. Paris **260**, 5662–5665 (1965); Russian transl. in Osobennosti differenciruemyh otobrazhenij, ed. by V.I. Arnold (Mir, Moscow, 1968), pp. 153–158
42. J.R. Munkres, *Elementary Differential Topology*, Ann. Math. Studies, vol. 54 (Princeton Univ. Press, 1966). Russian transl. in J. Milnor, J. Stasheff, Harakteristicheskie klassy (Mir, Moscow, 1979), pp. 270–359
43. P. Petersen V, Fatness of covers. J. Reine Angew. Math. **403**, 154–165 (1990)
44. V. Poénaru, Homotopie régulière et isotopie. Publ. Math. Orsay, *Univ. Paris XI, U.E.R. Math.*, Orsay (1974). No. 106 (23), 54 pp.; Available at http://sites.mathdoc.fr/PMO/feuilleter.php?id=PMO_1974
45. I.R. Porteous, Simple singularities of maps, in *Proc. Liverpool Singularities Symposium, I (1969/70)*, Lecture Notes in Math., vol. 192 (Springer, 1971), pp. 286–307
46. F. Ronga, "La classe duale aux points doubles" d'une application. Compos. Math. **27**, 223–232 (1973)
47. F. Ronga, Desingularisation of the triple points and of the stationary points of a map. Compos. Math. **53**, 211–223 (1984)
48. C. Rourke, B. Sanderson, The compression theorem I. Geom. Topol. **5**, 399–429 (2001). arXiv:math.GT/9712235
49. C. Rourke, B. Sanderson, The compression theorem III: applications. Alg. Geom. Topol. **3**, 857–872 (2003). arXiv:math.GT/0301356
50. S. Satoh, Lifting a generic surface in 3-space to an embedded surface in 4-space. Topol. Appl. **106**, 103–113 (2000)
51. M. Shiota, Thom's conjecture on triangulations of maps. Topology **39**, 383–399 (2000)
52. K. Siekłucki, Realization of mappings. Fund. Math. **65**, 325–343 (1969)
53. D.P. Sinha, Manifold-theoretic compactifications of configuration spaces. Selecta Math. **10**(3), 391–428 (2004). arXiv:math.GT/0306385
54. M. Skopenkov, On approximability by embeddings of cycles in the plane. Topol. Appl. **134**, 1–22 (2003). arXiv:0808.1187
55. A. Szűcs, The Gromov–Eliashberg proof of Haefliger's theorem. Stud. Sci. Math. Hungar. **17**, 303–318 (1982)
56. A. Szűcs, Multiple points of singular maps. Math. Proc. Camb. Philos. Soc. **100**, 331–346 (1986)
57. A. Szűcs, Topology of $\Sigma^{1,1}$-singular maps. Math. Proc. Camb. Philos. Soc. **121**, 465–477 (1997)
58. A. Szűcs, On the cobordism group of Morin maps. Acta Math. Hungar. **80**, 191–209 (1998)
59. A. Verona, *Stratified Mappings – Structure and Triangulability*, Lecture Notes in Math., vol. 1102 (Springer, Berlin, 1984)
60. C.T.C. Wall, Lectures on C^∞-stability and classification, in *Proceedings of Liverpool Singularities — Symposium I (1969/70)*, Lecture Notes in Math., vol. 192 (Springer, Berlin, 1971), pp. 178–206
61. C.T.C. Wall, Geometric properties of generic differentiable manifolds, in *Geometry and Topology — Rio de Janeiro, Brazil 1976*, Lecture Notes in Math., vol. 597 (Springer, Berlin, 1977), pp. 707–774
62. E.C. Zeeman, *Seminar on Combinatorial Topology* (Inst. Hautes Etudes Sci. Publ. Math., Paris, 1966)

Chapter 14
"Barcodes" for Continuous Maps and a Brief Introduction to Alternative Morse Theory

Dan Burghelea

Dedicated to Valentin Poénaru for his 90th anniversary

Abstract This chapter reviews the description of "barcodes" for a continuous real-valued map $f : X \to \mathbb{R}$ and explains how to recover the Morse complex of a Morse function from them. In this presentation, the barcodes appear as the support of two vector-space valued maps, one defined on the Euclidean plane $\mathbb{R}^2$ and the other on the "above diagonal" half plane $\mathbb{R}^2_+$.

Keywords Morse theory · Barcodes · Algebraic topology

AMS Codes 46M20, 57R19, 55E05, 55N35

14.1 Introduction

Classical Morse Theory (C.M.T) considers smooth functions $f : M \to \mathbb{R}$ on smooth manifolds M whose all critical points are non-degenerate (generic smooth functions) and, under some conditions, relates them to the homology of the underlying manifold. This is done by providing a family of chain complexes, each associated with some additional data, but all isomorphic[1] and therefore called *Morse complex* which calculate the homology of the manifold. The Morse complex explains the relations between the number c_r of critical points of the index r and the

[1] When the chain complex is a complex of κ-vector spaces, as it will be the case in this chapter, the unicity will follow from calculations; when the chain complex is of modules over an arbitrary ring this was established in [7].

D. Burghelea (✉)
Department of Mathematics, The Ohio State University, Columbus, OH, USA
e-mail: burghele@math.ohio-state.edu

L. Funar, A. Papadopoulos (eds.), *Essays on Topology*,
https://doi.org/10.1007/978-3-031-81414-3_14

dimension β_r of the r-homology vector spaces (the r-Betti numbers) and detects existence of instantons (e.g. isolated trajectories between rest points) for a vector field which admits the function f as Lyapunov function. The Morse complex is also equipped with an $\mathbb{R}$-filtration by sub complexes which calculate the homology of the piece $f^{-1}((-\infty, t])$. This filtration is locally constant in t outside the critical values of f. The elementary Morse theory, as summarily reviewed in Sect. 14.3, has extensions to various type of smooth infinite-dimensional manifolds and smooth Whitney stratified spaces and to the case when the set of non degenerate critical points consists of submanifolds rather than points.

Alternative Morse Theory (A.M.T) begins with a continuous map $f : X \to \mathbb{R}$ and refines the set $CR(f)$ of critical values[2] of f into a collection of four types of intervals which, in topological persistence homology theory[3] are referred to as *barcodes*.

In our work they appear as the points in the support of two types of vector spaces-valued maps $\hat{\delta}^f_r$ and $\hat{\gamma}^f_r$ defined on $\mathbb{R}^2$ and $\mathbb{R}^2 \setminus \Delta$, where Δ denotes the diagonal in $\mathbb{R}^2$, associated with each $r \in \mathbb{Z}_{\geq 0}$ and field κ. The multiplicity of each barcode is the dimension of the corresponding vector space, possibly infinite. Since in this chapter only the restriction of the map $\hat{\gamma}^f_r$ to the $\mathbb{R}^2_+ := \{x, y \in \mathbb{R}^2 \mid x < y\}$ will appear, one denotes this restriction by ${}^+\hat{\gamma}^f_r$. The restriction to the below diagonal half plane, $\mathbb{R}^2_- := \{x, y \in \mathbb{R}^2 \mid x < y\}$, denoted in [3] by ${}^-\hat{\gamma}^f_r$, can be derived from ${}^+\hat{\gamma}_r$ by the formula ${}^-\hat{\gamma}^f_r(a, b) = {}^+\hat{\gamma}^{-f}_r(-b, -a)$.

These maps (actually the points of their supports with their multiplicity) permit to define a chain complex of κ-vector spaces. This chain complex is also equipped with an $\mathbb{R}$-filtration determined by the critical values and when considered for a Morse function is isomorphic to the Morse complex with its filtration. In particular, as in the case of a Morse function f, from this complex, determined by barcodes, one can recognize the number of critical points corresponding to each critical value, the homology of the manifold, of the sub-levels $f^{-1}((-\infty, t])$, of the levels $f^{-1}(t)$, etc. and one detects presence of instantons for a gradient-like vector field; all these in a considerably larger class of situations than in the C.M.T.

A key merit of the theory is that the numerical invariants which appear are *computer friendly*, i.e. when considered for nice spaces (for example finite simplicial complexes and simplicial maps) they can be calculated by implementable algorithms, cf. [1].

The purpose of this chapter is to provide the definition of the maps $\hat{\delta}^f_r$, ${}^+\hat{\gamma}^f_r$, ${}^+\hat{\mu}^f_r$, ${}^+\hat{\omega}^f_r$ and ${}^+\hat{\lambda}^f_r$ for an arbitrary continuous map, relate them to the homology of the underlying space when f is tame , and present the arguments to establish the isomorphism of the chain complex defined by $\hat{\delta}^f_r$, ${}^+\hat{\gamma}^f_r$ and ${}^+\hat{\mu}^f_r$ with the Morse complex when associated with a Morse function.[4] However in this chapter we do this only in the case that the involved chain complexes are of vector spaces

[2] The values t for which the homology of the level $f^{-1}(t)$ changes.

[3] Proposed in [8] further extended in [6] and [4].

[4] For a Morse function on a compact manifold all ω^f_r, λ^f_r, μ^f_r vanish.

of finite dimension, in particular for the subcomplexes describing the $\mathbb{R}$-filtration. The concluded isomorphism is not canonical. The existence of an isomorphism compatible with the filtration is algebraically more subtle and will be discussed in subsequent work.

The chapter begins with a few observations about chain complexes in Sect. 14.2, and a brief review of elementary Morse theory in Sect. 14.3. Section 14.4 provides the definitions of the maps $\hat{\delta}^f_r$, $^+\hat{\gamma}^f_r$, $^+\hat{\mu}^f_r$, $^+\hat{\omega}^f_r$ and $^+\hat{\lambda}^f_r$ and of the associated chain complex $(C^{\delta,\gamma,\mu}_*, \partial^{\delta,\gamma,\mu}_*)$ and of its $\mathbb{R}$-filtration $(C^{\delta,\gamma,\mu}(t)_*.\partial^{\delta,\gamma,\mu}_*)$ and formulates the precise relation between the homology and the vector space-valued maps, cf. Theorem 14.4.5.

Section 14.5 collects results needed to establish the non canonical isomorphism of $(C^{\delta,\gamma,\mu}(t)_*.\partial^{\delta,\gamma,\mu}_*)$ with the t-subcomplex of the Morse complex. The theory can be extended to A.M-N.T (M-N.T stands for Morse-Novikov theory and A. for Alternative, cf. [2]).

Note In 1961–1962, as a second year undergraduate student I met Valentin Poénaru at that time a young and charismatic researcher at the Mathematical Institute of the Romanian Academy in Bucharest. In view of my interest in topology he has invited me to attend his seminar in differential topology; at that time the seminar discussed Morse Theory. Despite our rather short intersection in Bucharest (he soon left Romania), directly, through his seminar, or may be indirectly, by his reputation, he has much encouraged my wish to become a topologist. I am pleased to dedicate to him this paper, containing a few considerations about, or at least related to Morse theory, a topic in topology, which has remained directly or indirectly in the back of much of my mathematics and turns out to have more and more relevance outside topology.

14.2 Chain Complexes of κ-Vector Spaces

For a chain complex of κ-vector spaces

$$(C_*, \partial_*) \equiv \{ \cdots \longrightarrow C_{n+1} \xrightarrow{\partial_{n+1}} C_n \xrightarrow{\partial_n} C_{n-1} \longrightarrow \cdots , \ \partial_{k-1} \cdot \partial_k = 0\},,$$

a Hodge decomposition is provided by a decomposition of each C_n,

$$C_n = C_n^- \oplus H_n \oplus C_n^+$$

s.t. $\partial_n : C_n \to C_{n-1}$ is described as the matrix

$$\partial_n = \begin{bmatrix} 0 & 0 & 0 \\ 0 & 0 & 0 \\ \underline{\partial_n} & 0 & 0 \end{bmatrix}$$

with $\underline{\partial}_n$ an isomorphism; clearly $H_n(C_*, \partial_*) = H_n$. Such a decomposition exists and is unique up to a non-canonical isomorphism. When C_n are finite-dimensional, one let $c_n = \dim C_n$, $\beta_n = \dim H_n$ and $\rho_n = \operatorname{rank} \partial_n$; and these numbers are related by

$$c_n = \beta_n + \rho_n + \rho_{n-1}.$$

The existence of Hodge decompositions and the equality above implies :

Observation 14.2.1 *Two of these three sets of numerical invariants determine up to a non-canonical isomorphism a chain complex of finite-dimensional vector spaces.*

The proof is straightforward. For more details, if necessary, see [2] section 8.

An $\mathbb{R}$-filtration of (C_*, ∂_*) consists of a family of subcomplexes $(C_*(t), \partial_*) \subset (C_*, \partial_*)$, i.e.$\partial_*(C_*(t) \subset C_{*-1}(t)$, indexed by $t \in \mathbb{R}$ s.t. $(C_*(t), \partial_*) \subseteq (C_*(t'), \partial_*)$ for $t < t'$ and $\bigcup_t (C_*(t), \partial_*) = (C_*, \partial_*)$. The complex $(C_*(t), \partial_*)$ is referred to as the t-filtration or the t-subcomplex of (C_*, ∂_*).

14.3 Classical Morse Theory

Morse theory for finite-dimensional smooth manifolds considers proper smooth maps bounded from below with all critical points non degenerate.[5]

To such a function and to an arbitrary field κ one associates a collection of chain complexes of finite dimensional κ-vector spaces equipped with an $\mathbb{R}$-filtration, locally constant for $t \in \mathbb{R} \setminus CR(f)$, which calculates the homology of M and of $f^{-1}((-\infty, t])$, whose components C_n, resp. $C_n(t)$, are the κ-vector spaces generated by the critical points of index n, resp. the critical points of index n of critical values smaller or equal to t. The boundary maps $\partial_n : C_n \to C_{n-1}$ depend on additional data. but different additional data provide isomorphic complexes. Any such complex will be named *Morse complex of* f.

The additional data contains a smooth vector field X which has f as good Lyapunov function, in Milnor's terminology a *gradient like vector field*, cf. [9]; precisely, $X(f)(x) < 0$ iff $x \in M \setminus Cr(f)$ with $Cr(f) = \{x \in M \mid df(x) = 0\}$ and for any $x \in Cr(f)$ one can find a chart, called *Morse chart*, $\varphi_x : (U_x, x) \to (\mathbb{R}^n, 0)$, U_x open neighborhood of x in M, s.t.

$$f \cdot \varphi_x^{-1}(x_1, \cdots x_n) = -1/2 \sum_{i \leq k} x_i^2 + 1/2 \sum_{i \geq k+1} x_i^2$$

$$\varphi_x^* X = \sum_{i \leq k} x_i \partial_{x_i} - \sum_{i \geq k+1} x_i \partial_{x_i}.$$

[5] A critical point $x \in Cr(f) := \{x \in M \mid df(x) = 0\}$ is non-degenerate if the Hessian of f at x, i.e. $\partial^2 f / \partial x_i, \partial x_j(x)$, in a coordinate system (and then in any) is a non-degenerate quadratic form; the index of x is the number of negative eigenvalues of the Hessian.

Equivalently, $X = -grad_g f$ for g a complete Riemannian metric on M which in the neighborhood of each critical point is flat.

For a smooth vector field X on a smooth manifold M^n call *rest point* a point $x \in M$ s.t. $X(x) = 0$ and *trajectory* a smooth map $\gamma : \mathbb{R} \supset U \to M$ which satisfies $d\gamma(t)/dt = X(\gamma(t))$.

Denote by γ_y the maximal trajectory with $\gamma_y(0) = y$, $y \in M$, and define

$$W_x^{\mp} := \{y \in M \mid \lim_{t \to \mp\infty} \gamma_y(t) = x\},$$

the unstable (−) resp. stable (+) set of the rest point x. If the vector field is as above, then these sets are actually submanifolds diffeomorphic to the Euclidean space of dimension $index(x)$ resp. $n - index(x)$.

An *additional data* consists of a gradient-like vector field X which has the unstable manifolds transversal to the stable manifolds and has f as a good Lyapunov function plus a collection $\mathcal{O} = \{o_x, x \in Cr(f)\}$ of orientations o_x for any unstable manifold W_x^-. Gradient-like vector fields exist, and any vector field with hyperbolic rest points having f as Lyapunov function can be perturbed arbitrary little on arbitrary small neighborhood of the rest points, but still remaining the same in a smaller neighborhoods of these points, to get the stable and unstable manifolds transversal, cf. [10] or [5].

If x is a critical point of index k and y a critical point of index $(k - 1)$ then $W_x^- \cap W_y^+$ is a union of components γ, each submanifold of dimension one called *instanton* (= isolated trajectory between rest points).

The orientation o_x compared to the orientation o_y followed by the orientation from x to y along the instanton γ provides a sign, $\epsilon(\gamma) = \pm 1$, and then one defines the *algebraic cardinality* $I(x, y) = \sum_\gamma \epsilon(\gamma)$ of the set of instantons from x to y.

The additional data $(X, \mathcal{O})$ provides the vector spaces $C_k^{X,\mathcal{O}}$ generated by rest points of X, the same as the critical points of f of index k, hence equipped with a base. The subspace $C_k^{f,X,\mathcal{O}}(t)$ is generated by the rest points x with $f(x) \leq t$ and the linear map $\partial_k^{X,\mathcal{O}} : C_k^{X,\mathcal{O}} \to C_{k-1}^{X,\mathcal{O}}$ is given by the matrix with entries $I(x, y)$.

The main theorem of elementary Morse theory claims that if f is a proper smooth function bounded from below with all critical points non degenerate and $(X, \mathcal{O})$ is an additional data, then these vector spaces and linear maps described above define a chain complex $(C_*^{X,\mathcal{O}}, \partial_*^{X,\mathcal{O}})$ with $\mathbb{R}$-filtration $(C_*^{X,\mathcal{O}}(t), \partial_*^{X,\mathcal{O}})$ which calculates the homology of X and of $f^{-1}((-\infty, t])$. The t-filtration subcomplex has all vector spaces of finite dimension. This implies the famous Morse inequalities, the independence of additional data up to a non-canonical isomorphism, i.e. the complexes derived from different additional data are isomorphic, cf. [7], as well as the fact that rank $\partial_k \neq 0$ implies existence of instantons for any vector field having f as Lyapunov function.

We refer to any of these complexes $(C_*^{X,\mathcal{O}}, \partial_*^{X,\mathcal{O}})$ and the sub complexes $(C_*^{f,X,\mathcal{O}}(t), \partial_*^{X,\mathcal{O}})$ as Morse complexes. The A.M.T as proposed has an extension to A.M-N.T (Alternative Morse-Novikov theory) but this is not discussed in this paper.

14.4 The Maps $\hat{\delta}^f_r$, ${}^+\hat{\gamma}^f_r$ and ${}^+\mu^f_r, {}^+\lambda^f_r, {}^+\omega^f_r$ and the Associated Chain Complex $(C^{\delta,\gamma,\mu}_*, \partial^{\delta,\gamma,\mu}_*)$

The Maps $\hat{\delta}^f_r, {}^+\hat{\mu}^f_r, {}^+\hat{\gamma}^f_r, {}^+\hat{\lambda}^f_r, {}^+\hat{\omega}^f_r$

Let $f : X \to \mathbb{R}$, be a continuous real-valued map and let H_* denote the singular homology[6] with coefficients in a fixed field κ.

For any $a \in \mathbb{R}$ denote by

$$\begin{aligned} X^f_a &:= f^{-1}((-\infty, a]),\ X^a_f := f^{-1}([a, \infty)) \\ X^f_{<a} &:= f^{-1}((-\infty, a)),\ X^{>a}_f := f^{-1}((a, \infty)). \end{aligned} \tag{14.1}$$

The real number $a \in \mathbb{R}$ is called *regular value* if

$$H_*(X^f_a, X^f_{<a}) \oplus H_*(X^a_f, X^{>a}_f) = 0$$

and *critical value* if it is not regular. Denote the set of critical values by $CR(f)$.

For any $a \in \mathbb{R}$, denote by

$$\begin{aligned} \mathbb{I}^f_a(r) &:= img(H_r(X^f_a) \to H_r(X)),\ \mathbb{I}^a_f(r) := img(H_r(X^a_f) \to H_r(X)) \\ \mathbb{I}^f_{<a}(r) &:= img(H_r(X^f_{<a}) \to H_r(X)),\ \mathbb{I}^{>a}_f(r) := img(H_r(X^{>a}_f) \to H_r(X)) \end{aligned} \tag{14.2}$$

with the arrows representing the inclusion induced linear maps.

Note that $H_r(X_{<a}) = \varinjlim_{\epsilon\to 0} H_r(X_{a-\epsilon})$ and then $\mathbb{I}_{<a}(r) = \varinjlim_{\epsilon\to 0} \mathbb{I}_{a-\epsilon}(r)$. Similarly $H_r(X^{>a}) = \varinjlim_{\epsilon\to 0} H_r(X^{a+\epsilon})$ and then $\mathbb{I}^{>a}(r) = \varinjlim_{\epsilon\to 0} \mathbb{I}^{a+\epsilon}(r)$.

For any $a, b \in \mathbb{R}$ let

$$\begin{aligned} \mathbb{F}^f_r(a, b) &:= \mathbb{I}^f_a(r) \cap \mathbb{I}^b_f(r), \\ \mathbb{F}^f_r(< a, b) &:= \mathbb{I}^f_{<a}(r) \cap \mathbb{I}^b_f(r), \\ \mathbb{F}^f_r(a, > b) &:= \mathbb{I}^f_a(r) \cap \mathbb{I}^{>b}_f(r), \end{aligned} \tag{14.3}$$

and define:

$$\boxed{\hat{\delta}^f_r(a, b) := \mathbb{F}_r(a, b)/(\mathbb{F}_r(< a, b) + \mathbb{F}_r(a, > b))}. \tag{14.4}$$

[6] Or any other homology theory which satisfies the Eilenberg–Steenrod axioms and commutes with the direct limits.

Let

$$\delta_r^f(a,b) := \dim \hat{\delta}_r^f(a,b) \in \mathbb{Z}_{\geq 0} \cup \infty.$$

We also introduce

$$\begin{aligned} \mathbb{I}_{\infty}^f(r) &:= \cup_{a\in\mathbb{R}} \mathbb{I}_a^f(r) = H_r(X), \\ \mathbb{I}_f^{-\infty}(r) &:= \cup_{a\in\mathbb{R}} \mathbb{I}_f^a(r) = H_r(X), \\ \mathbb{I}_{-\infty}^f(r) &:= \cap_{a\in\mathbb{R}} \mathbb{I}_a^f(r), \\ \mathbb{I}_f^{\infty}(r) &:= \cap_{a\in\mathbb{R}} \mathbb{I}_f^a(r) \end{aligned} \tag{14.5}$$

and then define

$$\boxed{{}^{+}\hat{\mu}_r^f(a) := (\mathbb{I}_a^f(r) \cap \mathbb{I}_f^{\infty}(r))/(\mathbb{I}_{<a}^f(r) \cap \mathbb{I}_f^{\infty}(r))} \tag{14.6}$$

and

$${}^{+}\mu_r^f(a) := \dim{}^{+}\hat{\mu}_r^f(a) \in \mathbb{Z}_{\geq 0} \cup \infty.$$

Observation 14.4.1 *The assignment $\hat{\delta}_r$ defines the map*
$\hat{\delta} : \mathbb{R}^2 \rightsquigarrow \kappa$*-Vector* spaces
and the assignment ${}^{+}\hat{\mu}_r$ *defines the map*
${}^{+}\hat{\mu}_r : \mathbb{R} \rightsquigarrow \kappa$*-Vector* spaces,
which in view of the definition of regular / critical values have their supports contained in $CR(f) \times CR(f)$ *and* $CR(f)$ *resp. If* f *is bounded from above, hence* $\varprojlim_{x\to\infty} \mathbb{I}_f^x(r) = 0$*, then* ${}^{+}\hat{\mu}_r^f(a) = 0$ *for any* $a \in \mathbb{R}$.

For $a, b \in \mathbb{R}$ let

$$\begin{aligned} \mathbb{T}_r^f(a,b) &:= \ker(H_r(X_a^f) \to H_r(X_b^f)) \text{ if } \quad a \leq b \\ \mathbb{T}_r^f(<a,b) &:= \ker(H_r(X_{<a}^f) \to H_r(X_b^f)) \text{ if } \quad a \leq b \\ \mathbb{T}_r^f(a,<b) &:= \ker(H_r(X_a^f) \to H_r(X_{<b}^f)) \text{ if } \quad a < b \end{aligned} \tag{14.7}$$

and define:

$$\boxed{{}^{+}\hat{\gamma}_r^f(a,b) := \mathbb{T}_r(a,b)/(\iota\mathbb{T}_r(<a,b) + \mathbb{T}_r(a,<b))}, \tag{14.8}$$

with

$${}^{+}\gamma_r^f(a,b) := \dim \hat{\gamma}_r^f(a,b) \in \mathbb{Z}_{\geq 0} \cup \infty,$$

where $\iota : \mathbb{T}_r(< a, b) \to \mathbb{T}_r(a, b)$ is the inclusion induced linear map. When we want to specify the source and the target we write ${}^c\iota_a^b : \mathbb{T}_r(a, c) \to \mathbb{T}_r(b, c), a < b < c$ for the inclusion induced linear map ι. The map ι in the definition of ${}^+\hat{\gamma}_r^f$ above is actually ${}^b\iota_{<a}^a$.

For $y < a$, introduce

$$\boxed{{}^+\hat{\lambda}_r^f(a) := \cap_{y<a} img({}^a\iota_y^{<a} : \mathbb{T}_r(y, a) \to \mathbb{T}_r(< a, a))} \tag{14.9}$$

and

$${}^+\lambda_r^f(a) := \dim {}^+\hat{\lambda}_r^f(a) \in \mathbb{Z}_{\geq 0} \cup \infty.$$

Note that ${}^a\iota_y^{<a}(\mathbb{T}_r(y, < a)) = 0$,[7] and therefore

$$\begin{aligned} & img({}^a\iota_y^{<a} : \mathbb{T}_r(y, a)/\mathbb{T}_r(y, < a) \to \mathbb{T}_r(< a, a)) \\ & = img({}^a\iota_y^{<a} : \mathbb{T}_r(y, a) \to \mathbb{T}_r(< a, a)). \end{aligned}$$

For $x > a$ observe that $\mathbb{T}_r((a, x)/{}^x\iota_{<a}^a \mathbb{T}_r(< a, x) \subseteq coker(H_r(X_a) \to H_r(X_a))$ and then introduce

$$\boxed{{}^+\hat{\omega}_r^f(a) := \cap_{x>a}(\mathbb{T}_r(a, x)/{}^x\iota_{<a}^a \mathbb{T}_r(< a, x))} \tag{14.10}$$

and

$${}^+\omega_r^f(a) := \dim {}^+\hat{\omega}_r^f(a) \in \mathbb{Z}_{\geq 0} \cup \infty.$$

Observation 14.4.2 *The assignment ${}^+\hat{\gamma}_r^f$ defines the map*
${}^+\hat{\gamma}_r^f : \mathbb{R}_+^2 \rightsquigarrow \kappa$*-Vector* spaces
and the assignments ${}^+\hat{\lambda}_r^f$ and ${}^+\hat{\omega}_r^f$ define the maps
${}^+\hat{\lambda}_r^f : \mathbb{R} \rightsquigarrow \kappa$*-Vector* spaces *and*
${}^+\hat{\omega}_r^f : \mathbb{R} \rightsquigarrow \kappa$*-Vector* spaces
which in view of the definition of regular/critical values have their supports contained in $CR(f) \times CR(f)$ and $CR(f)$.

If f is bounded from below then ${}^+\hat{\lambda}_r^f(a) = 0$. If f is homologically tame, cf. Definition 14.4.4 below, then ${}^+\hat{\omega}_r^f(a) = 0$.

[7] Note that $\mathbb{T}_r(y, < a) \subseteq \mathbb{T}_r(y, a)$.

Relationship with Barcodes

The point $(a, b) \in supp\delta_r^f$ corresponds to what in [1] is called an r-closed barcode $[a, b]$ of multiplicity $\dim \hat{\delta}_r^f(a, b)$ when $a \leq b$ and to an $(r-1)$-open barcode (b, a) when $a > b$with the multiplicity $\dim \hat{\delta}_r(a, b)$.[8] Similarly the point (a, b) with $a < b$ in $supp^+\gamma_r^f$ corresponds to what in [1] is called an r-closed-open bar code $[a, b)$ of multiplicity $\dim {}^+\hat{\gamma}_r^f(a, b)$.

The Chain Complexes $(C_*^{\delta,\gamma,\mu}, \partial_*^{\delta,\gamma,\mu})$ and $(C_*^{\delta,\gamma,\mu}(t), \partial_*^{\delta,\gamma,\mu})$

Define

$$
\begin{aligned}
C_n^- &:= \bigoplus_{\{(a,b)\in\mathbb{R}^2 | a<b\}} {}^+\hat{\gamma}_{n-1}^f(a, b) \\
H_n &:= \bigoplus_{\{(a,b)\in\mathbb{R}^2\}} \hat{\delta}_n^f(a, b) \oplus \bigoplus_{\{a\in\mathbb{R}\}} {}^+\hat{\mu}_n(a) \\
C_n^+ &:= \bigoplus_{\{(a,b)\in\mathbb{R}^2 | a<b\}} {}^+\hat{\gamma}_n^f(a, b) \\
C_n &:= C_n^- \oplus H_n \oplus C_n^+
\end{aligned}
\tag{14.11}
$$

and $\partial_n : C_n \to C_{n-1}$ by the matrix

$$
\partial_n = \begin{bmatrix} 0 & 0 & 0 \\ 0 & 0 & 0 \\ id & 0 & 0 \end{bmatrix}
$$

with the t-filtration provided by

$$
\begin{aligned}
C_n(t) := & \bigoplus_{\{(a,b)\in CR(f)\times CR(f) | a<b\leq t\}} {}^+\hat{\gamma}_{n-1}(a, b) \oplus \\
& \bigoplus_{\{(a,b)\in CR(f)\times CR(f) | a\leq t\}} \hat{\delta}_n^f(a, b) \oplus \bigoplus_{\{(a,b)\in CR(f)\times CR(f) | a\leq t<b\}} {}^+\hat{\gamma}_n^f(a, b) \oplus \\
& \bigoplus_{\{a\in CR(f) | a\leq t\}} {}^+\hat{\mu}_r(a) \oplus \\
& \bigoplus_{\{(a,b)\in CR(f)\times CR(f) | a<b\leq t\}} {}^+\hat{\gamma}_n^f(a, b).
\end{aligned}
\tag{14.12}
$$

[8] This is because a and b should denote the ends of an interval.

Clearly $\partial_n(C_n(t)) \subset C_{n-1}(t)$. This subcomplex has a Hodge decomposition with components

$$
\begin{aligned}
C_n^-(t) &= \bigoplus_{\{(a,b)\in CR(f)\times CR(f)\mid a<b\leq t\}} {}^+\hat{\gamma}_{n-1}(a,b)\\
H_n(t) &= \bigoplus_{\{(a,b)\in CR(f)\times CR(f)\mid a\leq t\}} \hat{\delta}_n^f(a,b) \oplus \bigoplus_{\{a\in\mathbb{R}\mid a\leq t\}} {}^+\hat{\mu}_r(a)\oplus\\
&\qquad \bigoplus_{\{(a,b)\in CR(f)\times CR(f)\mid a\leq t<b\}} {}^+\hat{\gamma}_n(a,b)\\
C_n^+(t) &= \bigoplus_{\{(a,b)\in CR(f)\times CR(f)\mid a<b\leq t\}} {}^+\hat{\gamma}_n^f(a,b).
\end{aligned}
\tag{14.13}
$$

From now on, one denotes the above complex (14.11) by $(C_*{}^{\delta,\gamma,\mu}, \partial_*^{\delta,\gamma,\mu})$ and the t-filtration subcomplex (14.13) by $(C_*^{\delta,\gamma,\mu}(t), \partial_*^{\delta,\gamma,\mu})$.

The Case Where $H_r(X_a, X_{<a})$ Is Finite-Dimensional

Consider the obvious linear maps

1. $\mathbb{F}_r(a,b)/\mathbb{F}_r(<a,b) \rightarrowtail \mathbb{I}_a(r)/\mathbb{I}_{<a}(r) \twoheadleftarrow coker(H_r(X_{<a}) \to H_r(X_a)) \rightarrowtail H_r(X_a, X_{<a})$ with the last arrow induced from the long exact sequence of the pair $(X_a, X_{<a})$,
2. $\mathbb{T}_r(a,b)/\iota\mathbb{T}_r(<a,b) \rightarrowtail coker(H_r(X_{<a}) \to H_r(X_a)) \rightarrowtail H_r(X_a, X_{<a})$ with the last arrow induced from the long exact sequence of the pair $(X_a, X_{<a})$ and the first induced from the inclusions $\mathbb{T}_r(a,b) \subseteq H_r(X_a)$ and $\mathbb{T}_r(<a,b) \subseteq H_r(X_{<a})$,
3. $\mathbb{T}_{r-1}(a,b)/\mathbb{T}_{r-1}(a,<b) \twoheadleftarrow coker(H_r(X_{<b}, X_a) \to H_r(X_b, X_a)) \rightarrowtail H_r(X_b, X_{<b})$

with the last arrow induced from the long exact sequence of the triple $(X_a \subseteq X_{<b} \subset X_b)$ and the first derived from the diagram

$$
\begin{array}{ccccc}
H_r(X_{<b}, X_a) & \longrightarrow & H_r(X_b, X_a) & \twoheadrightarrow & coker(H_r(X_{<b}, X_a) \to H_r(X_b, X_a))\\
\downarrow & & \downarrow & & \downarrow\\
\mathbb{T}_{r-1}(a,<b) & \longrightarrow & \mathbb{T}_{r-1}(a,b) & \twoheadrightarrow & \mathbb{T}_{r-1}(a,b)/\mathbb{T}_{r-1}(a,<b)
\end{array}
$$

The arrow $\rightarrowtail$ resp. $\twoheadrightarrow$ indicates injective resp. surjective linear map.

As an immediate consequence one has

Observation 14.4.3

(i) $\dim H_r(X_a^f, X_{<a}^f) \geq \dim \mathbb{F}_r(a,b)/\mathbb{F}_r(<a,b) \geq \dim \mathbb{F}_r(a \times [b,b') \geq \delta_r^f(a,b)$ *is a consequence of 1. above.*

(ii) $\dim H_r(X_a^f, X_{<a}^f) \geq \dim coker(H_r(X_{<a}) \to H_r(X_a)) \geq \dim \mathbb{T}_r(a,x)/\iota\mathbb{T}_r(<a,x) \geq {}^+\omega_r^f(a)$ *is a consequences of 2. above.*

(iii) $\dim H_r(X^f_a, X^f_{<a}) \geq \dim \mathbb{T}^f_r(a,b)/\iota\mathbb{T}_r(<a,b) \geq \dim \mathbb{T}_r(a \times (b',b]) \geq \dim {}^+\hat{\gamma}_r(a,b)$ *is a consequence of 2. above.*

(iv) $\dim H_r(X^f_b, X^f_{<b}) \geq \dim \mathbb{T}^f_{r-1}(a,b)/\mathbb{T}_{r-1}(a,<b) \geq \dim \mathbb{T}_{r-1}((a',a] \times b) \geq \dim {}^+\hat{\gamma}_{r-1}(a,b)$ *is a consequence of 3. above.*

(v) $\dim H_r(X_a, X_{<a}) \geq \dim coker(H_r(X_{<a} \to H_r(X_a)) \geq \dim \mathbb{I}_a(r)/\mathbb{I}_{<a}(r)$ *is a consequence of 1. above and* $\dim H_r(X_a, X_{<a}) \geq \dim \mathbb{T}_{r-1}(<a,a) \geq \dim {}^+\hat{\lambda}_{r-1}(a)$ *follows from (14.9).*

Homological Tameness

Definition 14.4.4

1. The continuous map f is called *homologically tame* if for any $t \in \mathbb{R}$ there exists $\epsilon(t) > 0$ s.t. the inclusion induced linear map $H_r(X_t) \to H_r(X_{t+\epsilon})$ and $H_r(X^t) \to H_r(X^{t-\epsilon})$ are isomorphisms for any $\epsilon, 0 < \epsilon < \epsilon(t)$.
2. The continuous map f is called *homologically finite* (better said left-homologically finite) if for any $t \in \mathbb{R}$, $\dim H_r(X_t, X_{<t}) < \infty$.

All real-valued maps on finite- or infinite-dimensional manifolds considered in C.M.T are tame and homologically finite.

By Observation 14.4.3, if f is homologically finite then the maps δ^f_r, ${}^+\gamma^f_r$, ${}^+\mu_r$, ${}^+\lambda^f_r$ and ${}^+\omega_r$ are $\mathbb{Z}_{\geq 0}$-valued functions with the same support as $\hat{\delta}^f_r$, ${}^+\hat{\gamma}^f_r$, ${}^+\hat{\mu}_r$, ${}^+\hat{\lambda}_r$ and ${}^+\hat{\omega}^f_r$.

The tameness implies that the set $CR(f)$ is a discrete subset of $\mathbb{R}$ and therefore can be totally ordered as a sequence $\{\cdots < c_i < c_{i+1} < \cdots\}$, possibly infinite in both directions, and the filtration of $(C^{\delta,\gamma,\mu}_*, \partial^{\delta,\gamma,\mu}_*)$ is locally constant in $t \in \mathbb{R} \setminus CR(f)$. If f is in addition bounded from below then the sequence of critical values is bounded from below and each of the subcomplexes $(C^{\delta,\gamma,\mu}_*(t), \partial^{\delta,\gamma,\mu}_*)$ is a chain complex of finite-dimensional vector spaces, hence the considerations in Sect. 14.2 apply.

The following theorem, whose proof is presented in the next section, establishes the relation between the maps $\hat{\delta}^f_r, {}^+\hat{\mu}^f_r, {}^+\hat{\gamma}^f_r, {}^+\hat{\lambda}^f_r$ and the homology of the pair $(X_a, X_{<a})$, and of the spaces X_a and X.

Theorem 14.4.5 *Suppose f is homologically finite and tame. Then*

1. For $s \in \mathbb{R} \setminus CR(f)$, one has $H_r(X_s, X_{<s}) = 0$ and for $a \in CR(f)$

$$H_r(X_a, X_{<a}) = \bigoplus_{s \in CR(f)} \hat{\delta}^f_r(a,s) \oplus^+ \hat{\mu}^f_r(a) \oplus$$

$$\bigoplus_{s \in (-\infty,a) \cap CR(f)} {}^+\hat{\gamma}^f_{r-1}(s,a) \oplus {}^+\hat{\lambda}^f_{r-1}(a) \oplus$$

$$\oplus \bigoplus_{s \in (a,\infty) \cap CR(f)} {}^+\hat{\gamma}^f_r(a,s)$$

2. *If in addition f is bounded from below, then for any $t \in \mathbb{R}$,*

   (a) $H_r(X_t) = \bigoplus_{(a,b)|a\leq t} \hat{\delta}^f_r(a,b) \oplus \bigoplus_{(a,b)|a\leq t<b} {}^+\hat{\gamma}^f_r(a,b) \oplus \bigoplus_{a\leq t} {}^+\hat{\mu}^f_r(a),$

   (b) $H_r(X) = \bigoplus_{(a,b)} \hat{\delta}^f_r(a,b) + \bigoplus_a {}^+\hat{\mu}_r(a)$

   with $a, b \in CR(f)$.

3. *If in addition f is bounded from below and from above, in particular if X is compact, then*

   (a) $H_r(X_t) = \bigoplus_{(a,b)|a\leq t} \hat{\delta}^f_r(a,b) \oplus \bigoplus_{(a,b)|a\leq t<b} {}^+\hat{\gamma}^f_r(a,b),$

   (b) $H_r(X) = \bigoplus_{(a,b)} \hat{\delta}^f_r(a,b)$

   for $t \in \mathbb{R}$ and $a, b \in CR(f)$.

Note that since f is bounded from below, a situation we meet in C.M.T, the sums in item 2(a) contain only finitely many non-vanishing terms.

The proof of this theorem is sketched in Sect. 14.5.

If $f : M \to \mathbb{R}$ is a proper smooth function on a finite-dimensional manifold[9] with all critical points non-degenerate, denote by $c_r(f, a)$ the number of critical points of index r with $f(x) = a$. Then one has:

Proposition 14.4.6 (Morse Lemma)

$$c_r(f, a) = \dim H_r(M_a, M_{<a}).$$

Proposition (14.4.6) remains true if M is an infinite-dimensional smooth manifold and f satisfies Palais–Smale condition.[10] Theorem 14.4.5 and Proposition 14.4.6 imply the main result of this paper, Theorem 14.4.7.

Theorem 14.4.7 *Suppose f is a proper Morse function bounded from below on a smooth manifold or a smooth function on a smooth infinite-dimensional manifold which satisfies Palais–Smale condition whose critical points are non degenerate. Then for any t the chain complexes $(C^{f,X,\mathcal{O}}_*(t), \partial^{X,\mathcal{O}}_*)$ and the chain complex $(C^{\delta,\gamma,\mu}(f)_*(t), \partial^{\delta,\gamma,\mu}_*)$ are isomorphic.*

Proof Since both complexes are complexes of finite-dimensional vector spaces, in view of Observation 14.2.1, it suffices to check that the dimension of the r-components and the dimension of r-homology vector spaces are the same.

Indeed,

(a) $\dim C^{f,X,\mathcal{O}}_r(t)$, which by Proposition 14.4.6 is equal to $\dim(\oplus_{a\leq t} H_r(X_a, X_{<a}))$, as implied by C.M.T., is equal to $\dim(C^-_r(t) \oplus H_r(t) \oplus C_r(t)^+) = \dim C^{\delta,\gamma,\mu}_r(f)(t)$, cf. (14.13) as calculated by Theorem 14.4.5, item 1,

---

[9] Or smooth function which satisfies Palais-Smale condition on an infinite dimensional manifold.

[10] Any sequence of points $x_i \in M$ s.t. $df(x_i) \to 0$ contains a subsequence convergent to a critical point.

(b) the dimension of the homology of the chain complex $(C_*^{f,X,\mathcal{O}}(t), \partial_*^{X,\mathcal{O}})$, which by C.M.T is the dimension of $H_r(M_t)$, is equal to the dimension of the homology of the complex $(C_*^{\delta,\gamma,\mu}(t), \partial_*^{\delta,\gamma,\mu})$, as given by Theorem 14.4.5, item 2. ■

One expects that the chain complexes $(C_*^{\delta,\gamma,\mu}, \partial_*^{\delta,\gamma,\mu})$ and $(C_r^{X,\mathcal{O}}, \partial^{X,\mathcal{O}})$ are isomorphic by an isomorphism which preserves the $\mathbb{R}$-filtration; this issue will be addressed later, being algebraically more involved.

In particular, all conclusions about relationships between the rest points and instantons (for a vector field which admits a Morse function as Lyapunov) derived via homology, can be derived from bar codes and in considerably more general situations, involving more general spaces and more general flows.

14.5 About Proofs

Since in this section we will use direct and inverse limits it will be useful to have in mind the following facts.

1. (A1) For a direct system of short exact sequences of vector spaces the direct limit remains an exact sequence.
2. (A2) For an inverse system of short exact sequences

$$0 \longrightarrow \{A_\alpha, {}^A\iota_\alpha^{\alpha'}\} \longrightarrow \{B_\alpha, {}^B\iota_\alpha^{\alpha'}\} \longrightarrow \{C_\alpha, {}^C\iota_\alpha^{\alpha'}\} \longrightarrow 0 \; ,$$

by passing to the inverse limit one obtains the exact sequence

$$0 \longrightarrow \varprojlim_\alpha A_\alpha \longrightarrow \varprojlim_\alpha B_\alpha \longrightarrow \varprojlim_\alpha C_\alpha \longrightarrow \varprojlim{}'_\alpha A_\alpha \longrightarrow \cdots$$

with $\varprojlim'_\alpha A_\alpha = 0$ if the system $\{A_\alpha, {}^A\iota_\alpha^{\alpha'}\}$ satisfies Mittag–Leffler condition, in particular if ${}^A\iota_\alpha^{\alpha'}$ are surjective or if ${}^A\iota_\alpha^{\alpha'}$ are injective and A_α are subspaces of a finite-dimensional vector space A.

For any $a, \alpha, \beta \in \mathbb{R}, \alpha < \beta$ introduce

$$\boxed{\mathbb{F}_r^f(a \times [\alpha, \beta)) := \mathbb{F}_r(a, \alpha)/(\mathbb{F}_r(< a, \alpha) + \mathbb{F}_r(a, \beta)).} \tag{14.14}$$

As shown in [2] one has the following Proposition.

Proposition 14.5.1 *For any $a, b_1, b_2, b_3 \in \mathbb{R}$ with $b_1 < b_2 < b_3$ one has the obviously induced linear maps ι and π which provide the short exact sequence*

$$0 \longrightarrow \mathbb{F}_r(a \times [b_2, b_3)) \xrightarrow{i} \mathbb{F}_r(a \times [b_1, b_3)) \xrightarrow{\pi} \mathbb{F}_r(a \times [b_1, b_2)) \longrightarrow 0.$$

Unless obvious to the reader, details on the description of ι and π and on the proof of this proposition can be found in [2].

In view of this Proposition (14.5.1), for any $x', x, y, y',\ x' < x < y < y'$ one has the commutative diagrams

$$\begin{array}{ccc} \mathbb{F}_r(a \times [x', y)) & \twoheadleftarrow & \mathbb{F}_r(a \times [x', y')) \\ \uparrow & & \uparrow \\ \mathbb{F}_r(a \times [x, y)) & \twoheadleftarrow & \mathbb{F}_r(a \times [x, y')) \end{array},$$

both cartesian and co-cartesian with the horizontal arrows surjective and the vertical arrows injective. This implies (exercise left to the reader) the commutation of the direct and inverse limits stated as Observation 14.5.2, cf. ([3]).

Observation 14.5.2 *The diagram above induces a canonical isomorphism*

$$\underset{x \to -\infty}{\underrightarrow{\lim}}\ \underset{y \to \infty}{\underleftarrow{\lim}}\ \mathbb{F}_r(a \times [x, y)) \to \underset{y \to \infty}{\underleftarrow{\lim}}\ \underset{x \to -\infty}{\underrightarrow{\lim}}\ \mathbb{F}_r(a \times [x, y)). \tag{14.15}$$

Similarly, for any $a, b \in \mathbb{R}$ and $a \leq \beta < \gamma$ and $\alpha < \beta < b$, introduce

$$\boxed{\begin{array}{l} \mathbb{T}_r(a \times (\beta, \gamma]) := \mathbb{T}_r(a, \gamma)/(\iota\mathbb{T}_r(< a, \gamma) + \mathbb{T}_r(a, \beta)) \\ \mathbb{T}_r((\alpha, \beta] \times b) := \mathbb{T}_r(\beta, b)/(\iota\mathbb{T}_r(\alpha, b) + \mathbb{T}_r(\beta, < b)) \end{array}} \tag{14.16}$$

and one can verify as in [2] the following, proposition:

Proposition 14.5.3

1. *For any* $a, b_1, b_2, b_3 \in \mathbb{R},\ a \leq b_1 < b_2 < b_3$, *one has the obviously induced linear maps* ι *and* π *which provide the short exact sequence*

$$0 \longrightarrow \mathbb{T}_r(a \times (b_1, b_2]) \xrightarrow{\iota} \mathbb{T}_r(a \times (b_1, b_3]) \xrightarrow{\pi} \mathbb{T}_r(a \times (b_2, b_3])) \longrightarrow 0.$$

2. *For any* $a_1, a_2, a_3, b \in \mathbb{R},\ a_1 < a_2 < a_3 < b$, *one has the obviously induced linear maps* ι *and* π *which provide the short exact sequence*

$$0 \longrightarrow \mathbb{T}_r((a_1, a_2] \times b) \xrightarrow{\iota} \mathbb{T}_r((a_1, a_3] \times b) \xrightarrow{\pi} \mathbb{T}_r((a_2, a_3] \times b)) \longrightarrow 0.$$

Details on the description of ι and π and the verifications of this proposition can be found in [2] or [1].

In view of this Proposition (14.5.3), for $a < y' < y < x' < x$ resp. $x' < x < y < y' < b$, one has the commutative diagrams

$$\begin{array}{ccccccc} \mathbb{T}_r(a \times (y, x]) & \twoheadleftarrow & \mathbb{T}_r(a \times (y', x]) & \quad & \mathbb{T}_r((x, y'] \times b) & \twoheadleftarrow & \mathbb{T}_r((x', y'] \times b) \\ \uparrow & & \uparrow & & \uparrow & & \uparrow \\ \mathbb{T}_r(a \times (y, x']) & \twoheadleftarrow & \mathbb{T}_r(a \times (y', x']) & & \mathbb{T}_r((x, y] \times b) & \twoheadleftarrow & \mathbb{T}_r((x', y] \times b) \end{array}$$

both cartesian and co-cartesian, whose horizontal arrows are surjections and the vertical arrows are injections. As before they imply the commutation of the direct and inverse limit stated below, cf. [3].

Observation 14.5.4 *The diagrams above induce the canonical isomorphisms*

$$\varprojlim_{x \to a} \varinjlim_{y \to \infty} \mathbb{T}_r(a \times (x, y]) \to \varinjlim_{y \to \infty} \varprojlim_{x \to a} \mathbb{T}_r(a \times (x, y]), \tag{14.17}$$

$$\varinjlim_{y \to b} \varprojlim_{x \to -\infty} \mathbb{T}_r((x, y] \times b) \to \varprojlim_{x \to -\infty} \varinjlim_{y \to b} \mathbb{T}_r((x, y] \times b). \tag{14.18}$$

If one defines $\mathbb{F}_r(a \times (-\infty, \infty))$ as either one of the two limits (14.15), in view of (14.14), one concludes that

$$\begin{aligned} &\mathbb{F}_r(a \times (-\infty, \infty)) \\ &\quad := \varprojlim_{y \to \infty} \varinjlim_{x \to -\infty} \mathbb{F}_r(a \times [x, y)) = \varprojlim_{y \to \infty} \mathbb{I}_a(r)/(\mathbb{I}_{<a}(r) + \mathbb{I}_a(r) \cap \mathbb{I}^y(r)). \end{aligned}$$

By passing to inverse limit ($y \to \infty$), in view of A2 and of the finite dimensionality of $\mathbb{I}_a(r)/\mathbb{I}_{<a}(r)$, the exact sequence

$$0 \longrightarrow (\mathbb{I}_a(r) \cap \mathbb{I}^y(r))/(\mathbb{I}_{<a}(r) \cap \mathbb{I}^y(r)) \longrightarrow \mathbb{I}_a(r)/\mathbb{I}_{<a}(r) \longrightarrow \mathbb{I}_a(r)/(\mathbb{I}_{<a}(r) + \mathbb{I}_a(r) \cap \mathbb{I}^y(r)) \longrightarrow 0$$

remains exact. One obtains

$$\begin{aligned} 0 \to \varprojlim_{y \to \infty} (\mathbb{I}_a(r) \cap \mathbb{I}^y(r))/(\mathbb{I}_{<a}(r) \cap \mathbb{I}^y(r)) \to \mathbb{I}_a(r)/\mathbb{I}_{<a}(r) \\ \to \varprojlim_{y \to \infty} \mathbb{I}_a(r)/(\mathbb{I}_{<a}(r) + \mathbb{I}_a(r) \cap \mathbb{I}^y(r)) \to 0. \end{aligned}$$

Since

$$\varprojlim_{y \to \infty} (\mathbb{I}_a(r) \cap \mathbb{I}^y(r)/(\mathbb{I}_a(r) \cap \mathbb{I}^y(r)) \simeq^+ \hat{\mu}_r(a)$$

and

$$\varprojlim_{y\to\infty} \mathbb{I}_a(r)/(\mathbb{I}_{<a}(r) + \mathbb{I}_a(r) \cap \mathbb{I}^y(r)) = \mathbb{F}_r(a \times (-\infty, \infty)$$

one obtains

$$\mathbb{I}_a(r)/\mathbb{I}_{<a}(r) \simeq \mathbb{F}_r(a \times (-\infty, \infty)) \oplus {}^+\hat{\mu}_r(a). \tag{14.19}$$

Similarly, if one defines $\mathbb{T}_r((-\infty, a) \times a)$ using either one of the limits in (14.18), for example

$$\mathbb{T}_r((-\infty, a) \times a) = \varinjlim_{y\to a} \varprojlim_{x\to-\infty} \mathbb{T}_r((x, y] \times a),$$

when f is homologically finite, one obtains

$$\mathbb{T}_r(< a, a) = \mathbb{T}_r((-\infty, a) \times a) \oplus {}^+\lambda_r^f(a). \tag{14.20}$$

Indeed, by passing to the inverse limit $(x \to -\infty)$ in the short exact sequence

$$0 \longrightarrow {}^a\iota_x^y\mathbb{T}_r(x, a)/{}^{<a}\iota_x^y\mathbb{T}_r(x, < a) \longrightarrow \mathbb{T}_r(y, a)/\mathbb{T}_r(y, < a) \longrightarrow \mathbb{T}_r(y, a)/(\mathbb{T}_r(y, < a) + {}^a\iota_x^y\mathbb{T}_r(x, a)) \longrightarrow 0,$$

since $\dim \mathbb{T}_r(y, a)/\mathbb{T}_r(y, < a)$ is bounded,[11] in view of A2 one obtains

$$0 \longrightarrow {}^+\lambda_r^f(a) \longrightarrow \mathbb{T}_r(y, a)/\mathbb{T}_r(y, < a) \longrightarrow \varprojlim_{x\to-\infty} \mathbb{T}_r(y, a)/(\mathbb{T}_r(y, < a) + \iota\mathbb{T}_r(x, a)) \longrightarrow 0,$$

and the by passing to the direct limit $(y \to a)$ in view of A1 one obtains (14.20).

Similarly, one defines $\mathbb{T}_r(a \times (a, \infty))$ using either one of the limits in (14.17), for example

$$\mathbb{T}_r(a \times (x, y]) = \varinjlim_{y\to\infty} \varprojlim_{x\to a} \mathbb{T}_r(a \times (x, y]).$$

By passing to the inverse limit $(x \to a)$ in the short exact sequence

$$0 \longrightarrow \mathbb{T}_r(a, x)/\iota\mathbb{T}_r(< a, x) \longrightarrow \mathbb{T}_r(a, y)/\iota\mathbb{T}_r(< a, y) \longrightarrow \mathbb{T}_r(a, y)/(\iota\mathbb{T}_r(< a, y) + \mathbb{T}_r(x, a)) \longrightarrow 0,$$

because $\dim \mathbb{T}_r(a, x)/\mathbb{T}_r(< a, x) < \dim H_r(X_a, X_{<a}) < \infty$, one obtains in view of A2

$$0 \longrightarrow {}^+\omega_r^f(a) \longrightarrow \mathbb{T}_r(a, y)/\mathbb{T}_r((< a, y) \longrightarrow \varprojlim_{x\to a} \mathbb{T}_r(a, y)/(\mathbb{T}_r(< a, y) + \iota\mathbb{T}_r(x, a)) \longrightarrow 0,$$

---

[11] By Observation 14.4.3 item 3 $\dim \mathbb{T}_r(y, a)/\mathbb{T}_r(y, < a)$ is finite and bounded by $\dim H_r(X_a, X_{<a})$.

Then, by passing to the direct limit ($y \to \infty$), in view of A1,one obtains

$$0 \longrightarrow {}^{+}\omega_r^f(a) \longrightarrow \mathbb{T}_r(a,\infty)/\iota\mathbb{T}_r((<a,\infty) \longrightarrow \mathbb{T}_r(a \times (a,\infty)) \longrightarrow 0,$$

hence

$$\mathbb{T}_r(a,\infty)/\iota\mathbb{T}_r((<a,\infty) \simeq {}^{+}\omega_r^f(a) \oplus \mathbb{T}_r(a \times (a,\infty)). \tag{14.21}$$

The commutative diagram

$$\begin{array}{ccccccccc}
 & & 0 & & 0 & & 0 & & \\
 & & \uparrow & & \uparrow & & \uparrow & & \\
0 & \longrightarrow & \mathbb{I}_{<a}(r) & \longrightarrow & \mathbb{I}_a(r) & \longrightarrow & \mathbb{I}_a(r)/\mathbb{I}_{<a}(r) & \longrightarrow & 0 \\
 & & \uparrow & & \uparrow & & \uparrow & & \\
 & & H_r(X_{<a}) & \longrightarrow & H_r(X_a) & \longrightarrow & coker(H_r(X_{<a}) \to H_r(X_a)) & \longrightarrow & 0 \\
 & & \uparrow & & \uparrow & & \uparrow & & \\
 & & \mathbb{T}_r(<a,\infty) & \longrightarrow & \mathbb{T}_r(a,\infty) & \longrightarrow & \mathbb{T}_r(a,\infty)/\iota\mathbb{T}_r(<a,\infty) & \longrightarrow & 0 \\
 & & \uparrow & & \uparrow & & \uparrow & & \\
 & & 0 & & 0 & & 0 & &
\end{array}$$

whose all rows and columns are exact sequences implies the isomorphism

$$coker(H_r(X_{<a}) \to H_r(X_a)) \simeq \mathbb{I}_a(r)/\mathbb{I}_{<a}(r) \oplus \mathbb{T}_r(a,\infty)/\iota\mathbb{T}_r(<a,\infty). \tag{14.22}$$

The exact homology sequence of the pair $(X_\alpha, X_{<a})$ induces the isomorphism

$$H_r(X_a, X_{<a}) = coker(H_r(X_{<a}) \to H_r(X_a)) \oplus T_{(r-1)}(<a, a). \tag{14.23}$$

Consequently for f homologically finite (14.23), (14.22) followed by (14.19), and (14.20) imply

$$\begin{aligned} H_r(X_a, X_{<a}) \simeq\ & \mathbb{F}_r(a \times (-\infty,\infty)) \oplus \mathbb{T}_r(a \times (a,\infty)) \\ & \oplus \mathbb{T}_{r-1}((-\infty,a) \times a) \oplus {}^{+}\hat{\mu}_r(a) \oplus {}^{+}\omega_r^f(a) \oplus {}^{+}\lambda_{r-1}^f(a) \end{aligned} \tag{14.24}$$

Proposition 14.5.5 *Suppose* $\dim H_r(X_a, X_{<a}) < \infty$. *Then:*

1. *There exists a finite set of real numbers* $\{x_1, x_2, \cdots x_r\}$ *depending on* a, *with* $-\infty < x_1 < x_2 \cdots < x_r < \infty$ *s.t.*
   (a) $\hat{\delta}_r^f(a,x) = 0$ *of* $x \neq \{x_1, \cdots x_r\}$, *and if in addition* f *is tame, then*
   (b) $\mathbb{F}_r(a \times [x,y)) \simeq \bigoplus_{\{i \mid x \le x_i < y\}} \hat{\delta}_r^f(a, x_i)$.

2. *There exists a finite set of real numbers* $\{y_1, y_2, \cdots y_r\}$ *depending on a with* $a < y_1 < y_2 \cdots y_r < \infty$ *s.t.*

   (a) ${}^{+}\hat{\gamma}^f_r(a, y) = 0$ *of* $y \neq \{y_1, \cdots y_r\}$ *and if in addition* f *is tame then*
   (b) $\mathbb{T}_r(a \times (x, y]) \simeq \bigoplus_{\{i | x < y_i \leq y\}} {}^{+}\hat{\gamma}^f_r(a, y_i)$.

3. *There exists a finite set of real numbers* $\{x_1, x_2, \cdots x_k\}$ *depending on a with* $-\infty < x_1 < x_2 \cdots x_k < a$ *s.t.*

   (a) $\hat{\delta}^f_{r-1}(x, a) = 0$ *of* $x \neq \{x_1, \cdots x_k\}$ *and if in addition* f *is tame then*
   (b) $\mathbb{T}_r((x, y] \times a) \simeq \bigoplus_{\{i | -\infty < x_i \leq y\}} {}^{+}\hat{\gamma}^f_r(x_i, a)$.

The proof is presented in [2] but the for the reader's convenience it is also sketched below.

Note that in view of Proposition 14.5.1, the $\mathbb{Z}$-valued function $\dim \mathbb{F}_r(a \times [x, y))$, $x < y$, increases when y increases to ∞ or x decreases to $-\infty$ and the other variable remains constant.

Similarly, in view of Proposition 14.5.3, the function $\dim \mathbb{T}_r(a \times (x, y])$, $a < x < y$ increases, when y increases to ∞ or x decreases to a and the other variable x, resp. y, remains constant and the function $\dim \mathbb{T}_r((x, y] \times b)$, $x < y < b$ increases when y increases to b or when x decreases to $-\infty$ and the other variable remains constant.

Note also that if f is tame then the inclusion $\mathbb{I}^y_f \subseteq \mathbb{I}^{y-\epsilon}_f$ is an isomorphism for ϵ small enough and, in view of Proposition 14.5.1, the function $\dim \mathbb{F}_r(a \times [x, y))$ is continuous from the left in x and in y when y, resp. x, remains constant.

Similarly if f is tame then the inclusion $\mathbb{T}_r(x, b) \subseteq T_r(x + \epsilon, b)$ is an isomorphism for ϵ small enough, and in view of Proposition 14.5.3, the function $\dim \mathbb{T}_r((x, y] \times b)$ is continuous from the right in y and in x when x, resp. y, remains constant.

Also if f is tame then the inclusion $\mathbb{T}_r(a, y) \subseteq T_r(a, y + \epsilon)$ is an isomorphism for ϵ small enough, in view of Proposition 14.5.3, the function $\dim \mathbb{T}_r(a \times (x, y])$ is continuous from the right in both variables x and y when y, resp. x, remains constant.

Then the jumps at y for the first two functions and at x for the third function, when they appear, in view of Propositions 14.5.1 and 14.5.3, are given by

$$\begin{aligned}
&\lim_{\epsilon \to 0} (\dim \mathbb{F}_r(a \times [x, y + \epsilon)) - \dim \mathbb{F}_r(a \times [x, y))) \\
&\quad = \lim_{\epsilon \to 0} \dim \mathbb{F}_r(a \times [y, y + \epsilon) = \delta^f_r(a, y), \\
&\lim_{\epsilon \to 0} (\dim \mathbb{T}_r(a \times (x, y) - \dim \mathbb{T}_r(a \times (x, y - \epsilon])) \\
&\quad = \lim_{\epsilon \to 0} \dim \mathbb{T}_r(a \times (y - \epsilon, y]) =^{+} \gamma^f_r(a, y), \\
&\lim_{\epsilon \to 0} (\dim \mathbb{T}_r((x - \epsilon, y] \times a) - \dim \mathbb{T}_r((x, y] \times a)) \\
&\quad = \lim_{\epsilon \to 0} \dim \mathbb{T}_r((x - \epsilon, x] \times a) = {}^{+}\gamma^f_r(x, a).
\end{aligned} \tag{14.25}$$

In view of Propositions (14.5.1) and (14.5.3) and of the tameness of f, observe that:

1. $dim\mathbb{F}_r(a \times [x, y)) = 0$ for x very large towards ∞, or for $y < -K$ with K large enough towards ∞,
2. $dim\mathbb{T}_r(a \times (x, y]) = 0$ for x large enough towards ∞ or for x closed enough to a
3. $dim\mathbb{T}_r((x, y] \times a) = 0$ for y closed enough to a or for $x < -K$ with K large enough towards ∞.

and then

$$F_r(a \times (-\infty, \infty)) = \bigoplus_{t \in \mathbb{R}} \hat{\delta}_r(a, t), \tag{14.26}$$

$$\mathbb{T}_r(a \times (a, \beta]) = \bigoplus_{t | a < t \leq \beta} {}^{+}\hat{\gamma}_r^f(a, t), a < \beta \in \mathbb{R}, \tag{14.27}$$

$$\mathbb{T}_r(a \times (\beta, \infty)) = \varinjlim_{y \to \infty} \mathbb{T}_r(a \times (\beta, y]) = \oplus_{s > \beta} {}^{+}\hat{\gamma}_r^f(a, s), \tag{14.28}$$

$$T_{r-1}((-\infty, a) \times a) = \bigoplus_{t \in (-\infty, a)} \hat{\gamma}_{r-1}(t, a). \tag{14.29}$$

Proof of Theorem 14.4.5 First recall that:

1. If f is tame then ${}^{+}\hat{\omega}_r^f(a) = 0$,
2. If f is bounded from below then ${}^{+}\hat{\lambda}_r^f(a) = 0$,
3. If f is bounded from above then ${}^{+}\hat{\mu}_r^f(a) = 0$.

Then Item 1. in Theorem 14.4.5 follows on the nose from (14.24), (14.26),(14.28), (14.29) and (1.) above.

Note that since f is tame and bounded from below, then for any a there are only finitely many critical values smaller than a, say $c_1, c_2, \cdots c_k \leq a$, hence a finite numbers of values s, $s \leq a$, such that the vector space $\mathbb{I}_s / \mathbb{I}_{<s}$ is not trivial and all these vector spaces are of finite dimension. Then $\mathbb{I}_a(r) = \oplus_{s \leq a} \mathbb{I}_s / \mathbb{I}_{<s}$ and then, in view of (14.19) and (14.26),

$$\mathbb{I}_a(r) = \oplus_{s \leq a} \mathbb{I}_s(r) / \mathbb{I}_{<s}(r) = \bigoplus_{s \leq a, t \in \mathbb{R}} \hat{\delta}_r^f(s, t) \oplus \bigoplus_{s \leq a} {}^{+}\hat{\mu}_r^f(s). \tag{14.30}$$

To check Item 2. in Theorem 14.4.5 note that $H_r(X_a) = \mathbb{I}_a(r) \oplus \mathbb{T}_r(a, \infty)$ and in order to calculate $\mathbb{T}_r(a, \infty)$ observe the following:

(i) Since f is tame in view of (14.28) one has

$$\mathbb{T}_r(a \times (\beta, \infty)) = \varinjlim_{y \to \infty} \mathbb{T}_r(a \times (\beta, y]) = \oplus_{s>\beta} \, {}^{+}\hat{\gamma}_r^f(a, s).$$

(ii) Since f is also bounded from below, for any a the set of critical values smaller or equal to a is finite. Then let $c_1 < c_2 < \cdots c_k \leq a$ be these critical values and observe that in view of Proposition 14.5.5 item 2 the set $\bigcup_{1\leq i\leq k} supp\, {}^{+}\hat{\gamma}_r \cap (c_i \times (c_i, \infty)) \subset \mathbb{R}^2_+$ is finite. Then $\mathbb{T}_r(a, \infty) = \bigoplus_{1\leq i\leq k} \mathbb{T}_r(c_i \times (c_i, \infty))$[12] and in view of (14.28) we have

$$\begin{aligned} \mathbb{T}_r(a, \infty) &= \bigoplus_{1\leq i\leq k} \mathbb{T}_r(c_i \times (c_i, \infty)) = \bigoplus_{1\leq i\leq k} (\bigoplus_{c_i<s} {}^{+}\hat{\gamma}_r^f(c_i, s) \\ &= \bigoplus_{(t,s)|t\leq a} {}^{+}\hat{\gamma}_r^f(t, s). \end{aligned} \tag{14.31}$$

Combining (14.30) and (14.31) one completes the proof of part (a) in Item 2. of Theorem 14.4.5. To prove part (b), one passes to the direct limit when $a \to \infty$, replaces X by X_b and ∞ by b, and one repeats the arguments.

Item 3. in Theorem 14.4.5 follows from Item 2. in Theorem 14.4.5 and 2. and 3. above. ■

Proof of Proposition 14.4.6 For each critical point $x_1, x_2, \cdots x_p$ located on the level $f^{-1}(a)$, choose a Morse chart for f and choose ϵ small enough such that the discs of radius ϵ (in these charts) provide closed disjoint neighborhoods of the critical points. Denote each such neighborhood by $D(i)$ with $x_i \in D(i)$ and let $D = \sqcup D(i)$. The set of critical points $\{x_1, x_2, \cdots x_p\}$ is contained in the interior of D. Let r be the number of critical points of index r; clearly $r \leq p$. Let $M' := M \setminus \{x_1, \cdots x_p\}$ and observe that M'_a is a manifold with boundary whose interior is $M_{<a}$, hence $H_r(M'_a, M_{<a}) = 0$ for any r. Then $H_r(M_a, M_{<a}) = H_r(M_a, M'_a)$ which by excision is isomorphic to $H_r(D_a, D_a \setminus \{x_{i_1}, \cdots x_{i_r}\}), x_{i_1}, \cdots x_{i_r}$ the critical points of index r, $D_a = f^{-1}(-\infty, a]) \cap D$, which by Morse lemma is isomorphic to the direct sum of r copies of $H_r(D^r, \partial D^r) = \kappa$.

[12] Because $\mathbb{T}_r(a \times (a, \infty)) = \mathbb{T}_r(a, \infty)/\mathbb{T}_r(< a, \infty)$.

References

1. D. Burghelea, *New Topological Invariants for Real- and Angle-Valued Maps; an Alternative to Morse-Novikov Theory* (Word scientific Publishing Co. Pte. Ltd, Singapore, 2017)
2. D. Burghelea, Alternative to Morse–Novikov theory for a closed one form (I). Eur. J. Math. **6**(3), 713–730 (2020)
3. D. Burghelea, Level persistence and bar codes for continuous maps. A measure theoretic perspective (in preparation)
4. D. Burghelea, T.K. Dey, Persistence for circle valued maps. Discrete Comput. Geom. **50**, 69–98 (2013). arXiv:1104.5646
5. D. Burghelea, L. Friedlander, Th. Kappeler, On the space of trajectories of a generic vector field. An. Univ. Vest Timis. ser. mat. -inform. **XLVIII**(1, 2), 45–126 (2010)
6. G. Carlsson, V. de Silva, D. Morozov, Zigzag persistent homology and real valued functions, in *Proc. of the 25th Annual Symposium on Computational Geometry* (2009), pp. 247–256
7. O. Cornea, A. Ranicki, Rigidity and gluing for Morse and Novikov complexes. J. Eur. Math.Soc. **5**, 343–394 (2003)
8. H. Edelsbrunner, D. Letscher, A. Zomorodian, Topological persistence and simplification. Discrete Comput. Geom. **28**, 511–533 (2002)
9. J. Milnor, *Lectures on the h-Cobordism Theorem*. Notes by L. Siebenmann and J. Sondow Princeton Univ. Press. 1965 Mathematical Surveys and Monographs, vol. 108 (American Mathematical Society, Providence, 2004)
10. S. Smale, On gradient dynamical systems. Ann. Math. **74**, 199–206 (1961)

Chapter 15
A Note on the Quasi Simple Filtration of Profinite Groups

Francesco G. Russo

To Professor Valentin Poénaru for his 90th Birthday

Abstract We illustrate some intuitions of Poénaru (of more than 60 years ago) in connection with the property of quasi simple filtration, which was formulated later by Brick and Mihalik for finitely presented groups. This property adapts certain topological decompositions, which are proper to Low Dimensional Topology and Riemannian Geometry, to the context of Algebraic Topology, mostly to fundamental groups and universal coverings. We sketch some steps in the evolution of the property of quasi simple filtration, showing that there are significant margins of generalization to algebraic fundamental groups in theory of profinite groups.

Keywords Geometric simple connectivity · Profinite groups · Algebraic fundamental groups · Etale

Mathematics Subject Classification (2020) Primary: 20F05, 55M05; Secondary: 57M35, 57M40

15.1 Some Observations on the Concept of Exhaustibility of Dehn

Given a connected locally finite CW-complex X, Siebenmann [59] studied the notion of *simple connectivity at infinity* for X in 1968, requiring that for each compact set C in X there is a compact set D in X such that loops in $X \setminus D$ are homotopically trivial in $X \setminus C$. Siebenmann [59] showed that Euclidean spaces can

F. G. Russo (✉)
School of Science and Technology, Mathematics Division, University of Camerino, Camerino, Italy
e-mail: francesco.russo@unicam.it

L. Funar, A. Papadopoulos (eds.), *Essays on Topology*,
https://doi.org/10.1007/978-3-031-81414-3_15

be characterized by this property among contractible open topological n-manifolds for $n \geq 5$. The cases of lower dimensions were treated later by Wall [63] and Freedman [12], involving different techniques. Also Casson, Gersten and Stallings [15, 60, 61] worked on the notion of simple connectivity at infinity, because of its special role among Euclidean spaces, but a successive formulation was more appropriate for group theory. We may introduce the notion of simple connectivity at infinity for an abstract group G in the following way: G is simply connected at infinity if some finite complex X with fundamental group $\pi_1(X)$ equal to G has universal covering $\widetilde{X}$ which is simply connected at infinity.

Since long time topological groups have been involved in Abstract Harmonic Analysis, Mathematical Physics and Differential Geometry; these structures are algebraic, but possess interesting geometric properties. We must be more formal and introduce what we mean by "abstract group" and "discrete group", since this will be useful later on. Given a nonempty set G having a multiplication map

$$\cdot : (x, y) \in G \times G \mapsto \cdot(x, y) = xy \in G, \tag{15.1.1}$$

which is nothing else than a binary operation in G, and a neutral element $1 \in G$ with respect to such multiplicative map, that is, $\cdot\,(x, 1) = \cdot(1, x) = x$ for all $x \in G$, we may require the existence of an inversion map

$$(\,)^{-1} : x \in G \mapsto (x)^{-1} = x^{-1} \in G, \tag{15.1.2}$$

which allows us to think at the inverse of an element in G, that is, at some $y = (x)^{-1} \in G$ such that the following compatibility condition is satisfied

$$1 = \cdot(x, y) = \cdot(y, x). \tag{15.1.3}$$

Asking that $\cdot$ is associative means a compatibility condition of the type

$$\cdot\,(\cdot(x, y), z) = \cdot(x, \cdot(y, z)) \tag{15.1.4}$$

for all $x, y, z \in G$. The existence of an inverse (for each element of G) means that the map $(\,)^{-1}$ is surjective. Also the commutativity can be formulated in similar terms, requiring that $\cdot\,(x, y) = \cdot(y, x)$ for all $x, y \in G$. In this way we may reformulate the well known algebraic axioms of group G. In addition, if we endow G with a topology $\mathcal{T}(G)$ and the product space $G \times G$ with the product topology of $\mathcal{T}(G)$, the pair $(G, \mathcal{T}(G))$ is a *topological group* if the multiplication map and the inversion map are continuous. Briefly, one refers to a topological group G assuming a construction of this type and omitting $\mathcal{T}(G)$.

A topological group having a compact topology $\mathcal{T}(G)$ is called *compact group*, and, following [19, 22, 58], we always refer to compact groups as topological groups with a compact Hausdorff topology. Note that compact groups which are totally disconnected are called *profinite groups*, see [22, Theorem 1.34]. It is also good to mention that profinite groups can be always realized as projective limits of finite groups as per [22, Lemma 1.26, Propositions 1.31, 1.33]. Moreover the topology which is given in profinite groups turns out to be the topology in which each open subgroup has finite index, see [22, Chapter 1] and [58, Chapters 1 and 2].

Of course, finite groups are compact groups with the discrete topology, and, in particular, they are profinite groups. We say that a topological group G is an *abstract group* if it is a *discrete group*, that is, if G is a topological group with the discrete topology. Some authors refer to abstract groups just as sets with the axioms of group without the presence of a topology a priori, but this is clearly equivalent to requiring from the beginning (along with the axioms of group) a discrete topology: this always makes multiplication and inversion continuous. The additive group of the integers $\mathbb{Z}$ (with the induced topology from the usual topology of the real line) is an example of infinite noncompact discrete abelian group. The orthogonal group $\mathrm{O}(n)$ and the special orthogonal group $\mathrm{SO}(n)$ are examples of compact nonabelian groups which are connected and nonprofinite for all $n \geq 2$, according to [22, Examples 1.6]. Looking only at discrete groups, there are three relevant classes (each class properly contained in the next one) which we are going to discuss:

RELEVANT IMPLICATIONS FOR DISCRETE GROUPS (15.1.5)

$$\text{finitely presented groups} \Longrightarrow \text{finitely generated groups} \Longrightarrow \text{discrete groups}$$

Once the meanings of "abstract group" and "discrete groups" are clarified, we note that Davis [8] produced examples of proper actions of Coxeter (discrete) groups on contractible manifolds showing that universal coverings of closed aspherical manifolds are not necessarily homeomorphic to Euclidean spaces. Then several authors started to investigate new topological properties, which were more general than the notion of simple connectivity at infinity of Siebenmann and appropriate to characterize contractible universal coverings of compact manifolds. The contributions of Gabai and others [4, 13] are placed along this line of research, exploring the so-called "tameness conditions" in Low Dimensional Topology. The contributions of Poénaru [42–44] dealt with different concepts of exhaustibility in the context of open manifolds. Later Brick and Mihalik [2, 3] formulated a corresponding version in group theoretical terms (always looking at discrete groups) and their concept is reported below. This was used also by Gersten and Stallings [60].

Definition 15.1.1 (See [3], Section 1, p. 208) A piecewise linear space X is *quasi simply filtered* (briefly, QSF) if for any compact $C \subset X$ there exists a simply connected polyhedron K and a piecewise linear map $f : K \to X$ such that $C \subset f(K)$ and $f|_{f^{-1}(C)} : f^{-1}(C) \to C$ is a piecewise linear homeomorphism. The finitely presented group G is quasi simply filtered if there exists a polyhedron P such that $\pi_1(P) = G$ and whose universal covering $\widetilde{P}$ is quasi simply filtered.

Due to the difficulty of checking whether finitely presented groups satisfy Definition 15.1.1 or not, several authors introduced alternatives, as the following concept of Funar and Otera [14, 35, 36].

Definition 15.1.2 (See [14], Definitions 2.2 and 2.3) A polyhedron is *weakly geometrically simply connected* (briefly, WGSC) if it admits an exhaustion by compact and simply connected subpolyhedra. A finitely presented group G is weakly geometrically simply connected if there exists a polyhedron P such that $\pi_1(P) = G$ and whose universal covering $\widetilde{P}$ is weakly geometrically simply connected.

Definition 15.1.2 generalizes the notion of *geometrically simply connected*(or GSC) for a polyhedron, where one requires only the presence of a handlebody decomposition without handles of index one. In fact, the handlebody decomposition without handles of index one is a special exhaustion by compact simply connected subpolyhedra, see for details [49, 53].

RELEVANT IMPLICATIONS FOR TOPOLOGICAL SPACES (15.1.6)

$$\text{GSC} \Longrightarrow \text{WGSC} \Longrightarrow \text{QSF}$$

One can see that finitely presented groups which are simply connected at infinity are in particular weakly geometrically simply connected. In fact [14, Corollary 3.1] shows that Definitions 15.1.1 and 15.1.2 are equivalent for finitely presented groups. The following results must be mentioned here, because they show that QSF has a good behaviour with respect to subgroups and extensions.

Theorem 15.1.3 (See [3], Sections 3 and 4) *Let G be a finitely presented group.*

(i) *G is quasi simply filtered if and only if so is a finite index subgroup of G.*
(ii) *Automatic, CAT(0), combable, hyperbolic and one-relator groups are quasi simply filtered.*
(iii) *Assume that* $1 \to A \to G \to B \to 1$ *is a short exact sequence of infinite finitely presented groups. Then G is quasi simply filtered.*

As noted in several works of Otera and others [30–38], the main open questions concerning Definitions 15.1.1 and 15.1.2 are the following:

Question 15.1.4 (See [48, 49]) Is it true that all finitely presented groups are QSF? In other words, are there examples of finitely presented groups which are not QSF?

RELEVANT IMPLICATIONS FOR FINITELY PRESENTED GROUPS (15.1.7)

$$\text{WGSC} \Longleftrightarrow \text{QSF}$$

In a series of three contributions [48, 49], Poénaru worked on the positive answer of Question 15.1.4, constructing first a representation theory for quasi simply filtered groups and then showing the existence of certain universal representations in connection with the property of being quasi simply filtered. Roughly speaking, this is the main idea of answering Question 15.1.4. On the other hand, the methods and the techniques, which are involved in [48, 49], are laborious and require knowledge of several areas of Geometric Group Theory, Differential Geometry and Low Dimensional Topology. Let's report just two observations on the relevance of Question 15.1.4.

Remark 15.1.5 As noted in [30, Remark 7], a positive answer to Question 15.1.4 implies the 3-dimensional Universal Covering Conjecture and hence works as alternative topological argument to Perelman's technique on the Ricci flow in [39–41].

A second consequence is the following:

Remark 15.1.6 The QSF would be the first nontrivial geometric property satisfied by all finitely presented groups, in contrast with most of the geometric properties, introduced by Gromov [16].

It is impossible to get further into the details of Remarks 15.1.5 and 15.1.6 here. We shall just inform the reader to look at the works of Lasheras and others [5–7] for properties of similar nature to the quasi simple filtration. Further properties have been studied also by Mihalik and others [26–29] in connection with "tameness conditions" of more algebraic nature.

Remark 15.1.7 We have already mentioned the scientific production of Otera and others on the topic, but it is also appropriate to recall here the works of Poénaru [42–54] which have been the origin of most of the ideas which we described in Definitions 15.1.1, 15.1.2, Theorem 15.1.3, Question 15.1.4 and Remarks 15.1.5 and 15.1.6. It is also good to note that Dehn's work has been fundamental in several parts of [42, 43] and especially in [43], where the intuitions of Dehn [10, 11] are very clear in the arguments of the main proofs.

15.2 What Happens to the Profinite Groups?

As we had to formalize the notion of "abstract group" and "discrete group", it is appropriate here to make clear the meaning of "generation" of a discrete group and of a topological nondiscrete group. The two notions may be very different.

Definition 15.2.1 (See [9], Chapter III) A discrete group G is *finitely generated*, if there is a finite subset $X \subset G$ such that, for any $g \in G$, there exists a finite sequence $x_1, x_2, \ldots, x_n$ of elements in $X \cup X^{-1}$ with $g = x_1 x_2 \ldots x_n$.

In Definition 15.2.1, X^{-1} denotes the set of inverses of the elements of X.

Remark 15.2.2 We know that a finitely generated discrete group G with finite generating set X is always epimorphic image of the discrete free group $F(X)$ on X, see [9, Definition 7.A.1, Remark 7.A.3]. Moreover one usually refers to the positive integer $|X|$ as *rank* of $F(X)$; in fact $|X|$ denotes the minimal number of generators which are necessary to write an arbitrary element of $F(X)$, as indicated by Definition 15.2.1. One can also see from classical results in [9] that any finitely generated group G on X is isomorphic to the quotient group $F(X)/N$, where N is the kernel of some epimorphism from $F(X)$ to G. If we think at $G \simeq F(X)/N$ and if additionally N is generated by a finite set R, we say that G is *finitely presented* on $|X|$ generators with relators R, writing briefly $G = \langle X \mid R \rangle$.

A classical example of finitely presented group is a finitely presented abelian discrete group A on $|X| = n$ generators. We know from [22, Appendix 1] that such group decomposes as

$$A = \mathbb{Z}^m \oplus \mathbb{Z}(p_1^{\alpha_1}) \oplus \ldots \oplus \mathbb{Z}(p_k^{\alpha_k}) \tag{15.2.1}$$

into the direct sum of finitely many copies $\mathbb{Z}^m$ of $\mathbb{Z}$ (with $m \geq 0$ integer) and finitely many cyclic groups $\mathbb{Z}(p_i^{\alpha_i})$ of prime power order $p_i^{\alpha_i}$ for some p_i primes, $\alpha_i \geq 0$ integers, $i = 1, \ldots, k$ and $n = m + k$. In this situation $F(X) = \mathbb{Z}^n$ is the free abelian group on n generators and

$$\mathbb{Z}(p_i^{\alpha_i}) = \langle a_i \mid p_i^{\alpha_i} a_i = 0 \rangle = \mathbb{Z}/p_i^{\alpha_i}\mathbb{Z}, \tag{15.2.2}$$

where $p_i^{\alpha_i}\mathbb{Z} = \{p_i^{\alpha_i} g \mid g \in \mathbb{Z}\}$ is a normal subgroup of $\mathbb{Z}$ and

$$\pi_i : g \in \mathbb{Z} \longmapsto \pi_i(g) = g + p_i^{\alpha_i}\mathbb{Z} \in \mathbb{Z}(p_i^{\alpha_i}) \tag{15.2.3}$$

is the canonical epimorphism of $\mathbb{Z}$ onto the integers modulo $p_i^{\alpha_i}$. Then we may easily check that

$$\begin{aligned} \pi : & (a_1, \ldots, a_m, a_{m+1}, \ldots, a_n) \in \mathbb{Z}^n \\ & \mapsto (a_1, \ldots, a_m, a_{m+1} + p_{m+1}^{\alpha_{m+1}}\mathbb{Z}, \ldots, a_n + p_n^{\alpha_n}\mathbb{Z}) \in A \end{aligned} \tag{15.2.4}$$

is epimorphism of groups and its kernel N determines the relations $p_i^{\alpha_i} a_i = 0$ for all $1 \leq i \leq k$ and consequently the finite presentation for A with n generators (of rank n) and $k \leq n$ relations (plus the obvious additional relation which is due to the fact that A is abelian) is the following:

$$\begin{aligned} A = \langle & a_1, \ldots, a_m, a_{m+1}, \ldots, a_n \mid p_1^{\alpha_1} a_1 = \ldots = p_k^{\alpha_k} a_k = 0, [a_i, a_j] \\ & = 0 \; \forall i, j \in \{1, 2, \ldots, n\} \rangle. \end{aligned} \tag{15.2.5}$$

The situation of finitely presented abelian discrete groups fits naturally the ideas in Definitions 15.1.1 and 15.1.2, but for profinite groups we need a different notion of finite presentation.

Definition 15.2.3 (See [58], §2.4) A profinite group G is *topologically finitely generated* if there exists a finite dense subset X in G, that is, $\overline{\langle X \rangle} = G$ for some finite subset X of G. In particular, G is called *procyclic* if $X = \{x\}$. We also say that a subset X of a profinite group G *converges to* 1 if every open subgroup of G contains all but a finite number of the elements in X. In particular, if X generates topologically G and converges to 1, we say that X is *a set of topological generators of G converging to* 1.

Important differences between Definitions 15.2.1 and 15.2.3 appear already with one generator, but we shall recall some classical constructions, in order to appreciate these differences.

Definition 15.2.4 ([22], Definition 1.25) Let $j, k \in \mathbb{N}$ and $\varphi_{jk} : G_k \to G_j$ be homomorphisms of finite groups with $j \le k$, where $\{G_k : k \in \mathbb{N}\}$ is a collection of finite groups. We call $\{(G_k, \varphi_{jk}) : j \le k \text{ and } j, k \in \mathbb{N}\}$ an *inverse system* (or *projective system*) of finite groups, if

(a) φ_{jj} is the identity map of G_j for all $j \in \mathbb{N}$,
(b) $\varphi_{jl} = \varphi_{jk} \circ \varphi_{kl}$ for all $j, k, l \in \mathbb{N}$ with $j \le k \le l$.

In other words, Definition 15.2.4 ensures the existence of the following commutative diagram

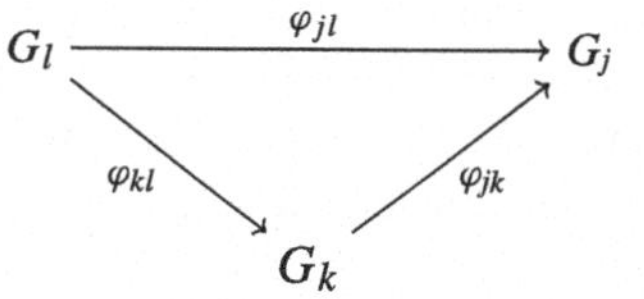

(15.2.6)

Furthermore, projective limits of Definition 15.2.4 may be embedded into the cartesian product P of the finite groups G_k. In fact P can be endowed with the product topology the groups G_k and is a profinite group as well. This allows us to have a better perspective, since we may think at profinite groups always as closed subgroups of cartesian products of finite groups.

Proposition 15.2.5 (See [22], Lemma 1.26) *Given a projective system of finite groups G_k and $P = \prod_{k \in \mathbb{N}} G_k$ cartesian product of the finite groups G_k (with the product topology), the set*

$$G = \left\{(g_j)_{j \in \mathbb{N}} \in P \ : \ \varphi_{jk}(g_k) = g_j \text{ whenever } j \le k \text{ and } j, k \in \mathbb{N}\right\}, \qquad (15.2.7)$$

is a closed subgroup of P. Moreover,

(a) *if* $\mathrm{inc} : G \to P$ *is the embedding of G into P and* $\mathrm{pr}_j : P \to G_j$ *the projection of P onto G_j, then $\varphi_j = \mathrm{pr}_j \circ \mathrm{inc} : G \to G_j$ is a continuous homomorphism and $\varphi_j = \varphi_{jk} \circ \varphi_k$ is satisfied for all $j \le k$ with $j, k \in \mathbb{N}$;*
(b) *both P and G are profinite groups.*

Motivated by Proposition 15.2.5 and [22, Definition 1.27], the group G in Proposition 15.2.5 is called *inverse limit*, or *projective limit*, of the system of finite groups G_k and we write

$$G = \varprojlim_{k \in \mathbb{N}} G_k. \qquad (15.2.8)$$

Note that either we define compact totally disconnected groups to be profinite groups, or we may define projective limits of finite groups to be profinite groups: the equivalence is shown in [22, Theorem 1.34]. Of course, there is no reason to focus on a set of indices $I = \mathbb{N}$ in the present discussion: one can rephrase Definition 15.2.4 and Proposition 15.2.5 for an arbitrary set of indices I which is just a directed set, but for what we are going to see, it is enough to concentrate on $I = \mathbb{N}$. It should be also mentioned that projective limits allow us to describe locally compact abelian groups which aren't necessarily compact but still preserve interesting topological properties (see [20, 21] for new classes of locally compact groups which have been studied in literature).

Example 15.2.6 The p-adic integers (p prime) are largely illustrated in [19, 22, 58]. Define

$$\varphi_n \ : \ u+p^{n+1}\mathbb{Z} \in \mathbb{Z}(p^{n+1}) \longmapsto \varphi_n(u+p^{n+1}\mathbb{Z}) = u+p^n\mathbb{Z} \in \mathbb{Z}(p^n), \qquad (15.2.9)$$

which is a surjective homomorphism of finite p-groups. We may remove the notation with double indices as in Definition 15.2.4, since it can be simplified in the present circumstances. The p-adic integers $\mathbb{Z}_p$ may be realized as the projective limit of $\{(\mathbb{Z}(p^n), \varphi_n) \ : \ n \in \mathbb{N}\}$, considering

$$\mathbb{Z}_p = \varprojlim_{n\in\mathbb{N}} \mathbb{Z}(p^n). \qquad (15.2.10)$$

This group is actually a projective limit of finite p-groups (i.e.: a *pro-p-group*). Moreover it is compact abelian, totally disconnected and is the pro-p-completion of the usual additive group $\mathbb{Z}$ of the integers, see [58, Example 2.1.6]. Note also that $\mathbb{Z}_p$ is the projective limit of cyclic groups, so one can refer briefly to this fact, saying that $\mathbb{Z}_p$ is a *procyclic group*, see [58, §2.7]. If $\mathbb{P}$ denotes the set of all primes, we mention that

$$\widehat{\mathbb{Z}} = \varprojlim_{n\in\mathbb{N}} \mathbb{Z}(n) = \prod_{p\in\mathbb{P}} \mathbb{Z}_p \qquad (15.2.11)$$

is the profinite completion of $\mathbb{Z}$, see [58, Examples 2.1.6, 2.3.11, Proposition 3.2.2] and [19, Chapter 4]. We may summarise the information of the structure of projective limit with the following commutative diagram:

$$\begin{array}{ccccccccccc}
\cdots \longleftarrow & \frac{\mathbb{Z}}{p^{n-1}\mathbb{Z}} & \xleftarrow{\varphi_{n-1}} & \frac{\mathbb{Z}}{p^{n}\mathbb{Z}} & \xleftarrow{\varphi_{n}} & \frac{\mathbb{Z}}{p^{n+1}\mathbb{Z}} & \longleftarrow & \cdots & \longleftarrow & \mathbb{Z}_p & \\
& \uparrow{\scriptstyle \mathrm{pr}_{|}} & & \uparrow{\scriptstyle \mathrm{pr}_{|}} & & \uparrow{\scriptstyle \mathrm{pr}_{|}} & & & & \uparrow{\scriptstyle \mathrm{pr}} & \\
\cdots = & \widehat{\mathbb{Z}} & = & \widehat{\mathbb{Z}} & = & \widehat{\mathbb{Z}} & = & \cdots & = & \widehat{\mathbb{Z}} & \qquad (15.2.12)
\end{array}$$

We have the natural projection $\mathrm{pr} : \widehat{\mathbb{Z}} \to \mathbb{Z}_p$ which restricts gradually to $\mathrm{pr}_{|} : \widehat{\mathbb{Z}} \to \mathbb{Z}/p^n\mathbb{Z}$ onto each term $\mathbb{Z}/p^n\mathbb{Z}$ of the projective limit which realizes $\mathbb{Z}_p$.

Ribes and Zalesskii [58, Definition 2.4.5] use the symbol $d(G)$ to denote the *minimal number of topological generators converging to* 1 of a profinite group G. Note that a profinite group is procyclic if and only if it is the projective limit of finite cyclic groups. Hence $d(\mathbb{Z}_p) = 1$ and $d(\mathbb{Z}_p \times \mathbb{Z}_p) = 2$ in Example 15.2.6, but also $d(\widehat{\mathbb{Z}}) = 1$ and $d(\mathbb{Z}_p \times \mathbb{Z}_q) = 1$ when p and q are different primes, see [58, Example 2.4.2]. In the category of profinite groups there are *free profinite groups* $F_{\mathcal{C}}(X)$ on a finite set X of topological generators [58, Chapter 3], so $d(X)$ represents the *rank* of $F_{\mathcal{C}}(X)$, that is, the minimal number of topological generators converging to 1 which are necessary to obtain an element of $F_{\mathcal{C}}(X)$.

Remark 15.2.7 We shall mention that the construction of $F_{\mathcal{C}}(X)$ is quite elaborate, since it involves projective limits and other techniques which disappear in the discrete case, see [58, Propositions 3.3.2 and 3.3.6]. Note also that $\mathcal{C}$ will always denote the class of finite groups, but in principle one can consider $\mathcal{C}$ to be any nonempty class of finite groups satisfying certain reasonable conditions of stability (for instance, under subgroups, finite direct products, quotients, see [58, §2.1]). In fact [58, Chapters 2 and 3] deal with pro-$\mathcal{C}$-groups, but here we work only with profinite groups and eventually with pro-p-groups, just thinking of projective limits of finite groups in the first case and of projective limits of finite p-groups in the second case.

Definition 15.2.8 (See [58], §7.8) Given a topologically finitely generated profinite group G on a finite set of topological generators X, there exists a continuous epimorphism $\varepsilon : F_{\mathcal{C}}(X) \to G$ of the free profinite group $F_{\mathcal{C}}(X)$ onto G such that $G \simeq F_{\mathcal{C}}(X)/\ker \varepsilon$. If there exists also a finite set R of topological generators of the closed normal subgroup $\ker \varepsilon$ of $F_{\mathcal{C}}(X)$, called *finite set of defining relators* associated with X, we say that G has a *profinite presentation* with finite generating set X and finite relators R. Equivalently, we say that the profinite group G is *profinitely presented*, or *topologically finitely presented*.

Of course, Example 15.2.6 shows that there are profinitely presented groups which are not finitely presented, but for finite groups the two notions are equivalent.

Example 15.2.9 According to Definition 15.2.8 and to [58, §3.5, §5.6, §5.7], one can check that the direct sum $\mathbb{Z}_p^r$ of $r \geq 0$ copies of $\mathbb{Z}_p$ is a profinitely presented abelian group on r topological generators. In particular, $d(\mathbb{Z}_p^r) = r$, that is, $\mathbb{Z}_p^r = \overline{\langle x_1, \ldots, x_r \rangle}$ and the defining relators induce the equations $[x_i, x_j] = 1$ for all $i, j \in \{1, \ldots, r\}$. It should be mentioned from [58, Chapter 3] that $\mathbb{Z}_p^r$ is the free abelian pro-p-group on r topological generators. Of course, $\mathbb{Z}_p$ is infinitely generated if we put the discrete topology on it.

We shall now summarize [58, Examples 9.2.11 and 9.2.12] here, because they help to understand the nature of generalization which we are going to make for Definitions 15.1.1 and 15.1.2.

Remark 15.2.10 Consider a proper nonsingular connected algebraic curve C of genus t over the complex field $\mathbb{C}$. As a topological space C is a compact oriented manifold of dimension two and is in fact a sphere with t handles added, that is, we may think at C as

$$C = S^2 \sharp \underbrace{T \sharp \ldots \sharp T}_{t\text{-times}}, \tag{15.2.13}$$

where

$$S^2 = \{(x, y, z) \in \mathbb{R}^3 \ : \ x^2 + y^2 + z^2 = 1\} \tag{15.2.14}$$

is the usual sphere of $\mathbb{R}^3$,

$$T = \left\{ (x, y, z) \in \mathbb{R}^3 \ : \ \left(R - \sqrt{x^2 + y^2}\right)^2 + z^2 = r^2 \right\} \tag{15.2.15}$$

the torus of $\mathbb{R}^3$ (with center at the origin, axis of symmetry in z, internal radius R and tubular section r) and $\sharp$ denotes the well known operator of connected sum, see [23, Theorem 11.3].

The *algebraic fundamental group* $\widehat{\pi}_1(C)$ (or *étale*) of C was introduced by Grothendieck, Raynaud and others [55] as the profinite completion of the usual fundamental group $\pi_1(C)$, see [55, Exp.10, Théorème 2.6, Théorème 2.9]. Basically Raynaud and others [55] consider the discrete group $\pi_1(C)$ and this turns out to be a surface group (in the usual sense of [23, Chapters 11 and 26]), hence with $2t$ generators $a_1, b_1, \ldots, a_t, b_t$ subject to one relation $[a_1, b_1][a_2, b_2] \ldots [a_t, b_t] = 1$, so we have a presentation in the usual sense of Definition 15.2.1.

From [55, Exp.10, Théorème 2.6, Théorème 2.9], $\widehat{\pi}_1(C)$ has exactly the same presentation, but now this turns out to be a profinite presentation in the sense of Definition 15.2.8, that is, an appropriate quotient of a free profinite group, that is,

$$\widehat{\pi}_1(C) = \overline{\langle a_1, b_1, \ldots, a_t, b_t \ : \ [a_1, b_1][a_2, b_2] \ldots [a_t, b_t] = 1 \rangle}. \tag{15.2.16}$$

The Seifert-Van Kampen's Theorem [23, Theorem 23.3] (in its original formulation) allows us to write $\pi_1(C)$ via the free amalgamated product of a free group $F(a_1, b_1)$ of rank 2 and of a free group $F(a_2, b_2, \ldots, a_t, b_t)$ of rank $2t - 2$ with an amalgamated infinite cyclic subgroup $\mathbb{Z}$ for $t > 1$, that is, we have

$$\begin{aligned} \pi_1(C) &= \langle a_1, b_1, \ldots, a_t, b_t \ : \ [a_1, b_1][a_2, b_2] \ldots [a_t, b_t] = 1 \rangle \\ &= \langle a_1, b_1 \rangle \coprod_{[b_1, a_1] = [a_2, b_2] = \ldots = [a_t, b_t]} \langle a_2, b_2, \ldots, a_t, b_t \rangle \\ &\simeq F(a_1, b_1) \coprod_{\mathbb{Z}} F(a_2, b_2, \ldots, a_t, b_t). \end{aligned} \tag{15.2.17}$$

Now it happens exactly the same for $\widehat{\pi}_1(C)$ when we move in the category of profinite groups:

$$\begin{aligned}\widehat{\pi}_1(C) &= \overline{\langle a_1, b_1 \rangle} \coprod_{[b_1,a_1]=[a_2,b_2]=\ldots=[a_t,b_t]} \overline{\langle a_2, b_2, \ldots, a_t, b_t \rangle} \\ &\simeq F_{\mathcal{C}}(a_1, b_1) \coprod_{\widehat{\mathbb{Z}}} F_{\mathcal{C}}(a_2, b_2, \ldots, a_t, b_t), \end{aligned} \tag{15.2.18}$$

that is, also $\widehat{\pi}_1(C)$ may be written in terms of a profinite proper free amalgamated product of a free profinite group $F_{\mathcal{C}}(a_1, b_1)$ of rank 2 and of a free profinite group $F_{\mathcal{C}}(a_2, b_2, \ldots, a_t, b_t)$ of rank $2t - 2$ with a procyclic amalgamated subgroup $\widehat{\mathbb{Z}}$ for $t > 1$. We omit to give more details on the notion of free amalgamated product of two profinite groups: it is quite technical and adapts to the category of profinite groups the logic of construction of amalgamated product for abstract groups, see [58, §9.2 and §9.3].

We want to give two examples, which show the degree of generalization which one is going to do in the present context. One is offered here, but another is offered in the successive section.

Example 15.2.11 With the notations of Remark 15.2.10, assume that we have a proper nonsingular connected algebraic curve C_0 over $\mathbb{C}$ such that $\pi_1(C_0) = \mathbb{Z}^6$. This can be realized for instance by $C_0 = S^2 \sharp T \sharp T \sharp T$. On the basis of Example 15.2.6, we get

$$\widehat{\pi}_1(C_0) = \widehat{\mathbb{Z}} \times \widehat{\mathbb{Z}} \times \widehat{\mathbb{Z}} \times \widehat{\mathbb{Z}} \times \widehat{\mathbb{Z}} \times \widehat{\mathbb{Z}} \tag{15.2.19}$$

which has an extremely complicated structure both at an algebraic level and at a topological level. In particular, $\pi_1(C_0)$ is a locally compact discrete abelian group, which embeds naturally into $\widehat{\pi}_1(C_0)$, which is its profinite completion (in fact is also its compactification). However both $\pi_1(C_0)$ and $\widehat{\pi}_1(C_0)$ are totally disconnected, so in this process of compactification via profinite completions there are no chance to look at conditions of connectivity at the level of topological groups. On the other hand, C_0 is connected at a topological level and its universal covering is $\widetilde{C}_0 = \mathbb{R}^6$ and by the original works of Siebenmann [59] this Euclidean space can be characterized among contractible open topological 6-manifolds by the simple connectivity at infinity. Consequently, one can think on the evidence of this example to introduce general notions of simple connectivity at infinity for profinitely presented groups, namely properties such as WGSC and QSF.

We have all that we need, in order to formulate the following new concept:

Definition 15.2.12 Consider a profinitely presented group G as per Definition 15.2.8. We say that G is *profinitely quasi simply filtered* or PQSF (respec. *profinitely weakly geometrically simply connected* or PWGSC), if there exists a proper nonsingular connected algebraic complex curve C such that $\widehat{\pi}_1(C) = G$ and the universal covering $\widetilde{C}$ of C is QSF (respec. WGSC).

Of course, Definition 15.2.12 specializes to Definition 15.1.1 if we replace the algebraic fundamental group with the usual fundamental group. A series of natural questions arise on the model of what has been made in [3, 14, 35, 36]. For instance, we may reformulate Definition 15.1.2 for profinite groups and check whether it is equivalent to Definition 15.2.12, or not, for profinite groups.

Theorem 15.2.13 *If G is PWGSC, then G is PQSF.*

Proof Assume then that G is PWGSC. Thus there exists a proper nonsingular connected algebraic complex curve C such that the universal covering $\widetilde{C}$ is WGSC. It is known that for general topological spaces the implication from WGSC to QSF is always valid. Hence $\widetilde{C}$ is QSF, so the result follows by Definition 15.2.12.

The converse of Theorem 15.2.13 holds in a series of situations, when the dimension is appropriate and we can infer from previous results that QSF implies WGSC. For instance, this happens very often; for $n \neq 4$ it is known that an open n-manifold is WGSC if and only if it is GSC, while in dimension 4 there are examples of open 4-manifolds which are WGSC but not GSC. On the other hand, the argument of the proof of [14, Theorem 1.1] does not work for PWGSC groups and PQSF groups, so different techniques should be used.

15.3 Some Open Problems from the Perspective of the Profinite Groups

There are a series of open questions and conjectures which arise naturally, after the discussions of the previous sections. Firstly, one expects a positive answer below:

Question 15.3.1 Is it true the converse of Theorem 15.2.13?

On the other hand, one has to be careful, since the usual techniques, which are proper to the context of nonprofinite groups, cannot be applied in the proof of Theorem 15.2.13.It should be also mentioned that there exists a Profinite Graph Theory in [57], hence it might be interesting to look at the counterpart of Cayley graphs for profinite groups in [57], as alternative for the proof of Theorem 15.2.13 and eventually of its viceversa.

Secondly, it is interesting to investigate consequences at the level of the structure of profinite groups for the notions in Definition 15.2.12. In fact the structure of the algebraic fundamental group can be investigated with the methods of profinite group theory in Algebraic Geometry, as indicated by [1, 18, 24, 25, 62]. Let's give just an idea of some complications which may arise.

Remark 15.3.2 The algebraic fundamental group of a scheme over a field of prime characteristic $p > 0$ is a very mysterious object. In other words, we can replace the choice of C in Definition 15.2.12 with a "scheme" in the sense of Grothendieck, see [55]. The nonsingular algebraic curves on fields of characteristic zero (such as $\mathbb{C}$) describe a very special situation.

In fact even in the case of smooth affine curves, the algebraic fundamental group over a field of prime characteristic is not yet completely understood. Some portions are understood (but not all) thanks to Grothendieck-Riemann Existence Theorem [55, XIII, Corollary 2.12]. The main difficulty arises from the wildly ramified covers. For instance, if C is a smooth affine curve over an algebraically closed field of characteristic $p > 0$ and X is the smooth completion of C (of genus t), the structure of $\widehat{\pi}_1(C)$, which is both the algebraic fundamental group of C and a profinite group, is not completely understood in connection with X. Recent works can be found in [1, 18, 24, 25].

Question 15.3.3 Formulate Definition 15.2.12, involving affine curves over algebraically closed fields of prime characteristic, instead of proper nonsingular connected algebraic complex curves. Then investigate the corresponding PWGSC and PQSF.

When C is a smooth affine curve over an algebraically closed field of characteristic p, Abhyankar's Conjecture, which was proved by Raynaud [56] and Harbater [17], classifies the finite quotients of $\widehat{\pi}_1(C)$. In particular, $\widehat{\pi}_1(C)$ is not finitely generated, hence it is not determined by the set of its finite quotients. The structure of $\widehat{\pi}_1(C)$ is currently investigated by several authors, who are studying its subgroups. Then it is appropriate to look at alternative properties, in order to determine $\widehat{\pi}_1(C)$ and PWGSC and PQSF can help in this.

We end with the example that was mentioned in order to explain Remark 15.2.6 and in addition to Example 15.2.11. This is to illustrate some rich constructions which appear already if we want to focus on pro-p-groups instead of profinite groups.

Example 15.3.4 According to [58, Example 9.2.12], a Demushkin group G is a pro-p-group (with p odd prime) having the following profinite presentation

$$G = \overline{\langle a_1, b_1, \ldots, a_t, b_t \ : \ a_1^{p^n}[a_1, b_1][a_2, b_2]\ldots[a_t, b_t] = 1\rangle} \tag{15.3.1}$$

$$= \overline{\langle a_1, b_1\rangle} \coprod_{[b_1,a_1]a_1^{-p^n}=[a_2,b_2]=\ldots=[a_t,b_t]} \overline{\langle a_2, b_2, \ldots, a_t, b_t\rangle}$$

with n natural number or eventually ∞ (the latter case just means that $a_1^{p^n} = 1$). This group is a maximal pro-p-quotient of the algebraic fundamental group in Remark 15.2.10, that is, it is a maximal pro-p-quotient of the algebraic fundamental group of a proper nonsingular connected algebraic complex curve of genus t. Demushkin groups are actively investigated since they are at the boundary of Algebraic Geometry and Profinite Galois Theory and we have reported this example just as an evidence of the complexity of the profinite groups, already when we deal with curves which are defined over fields of characteristic zero.

References

1. L. Bary-Soroker, M. Kumar, Subgroup structure of fundamental groups in positive characteristic. Commun. Algebra **41**, 3705–3719 (2013)
2. S. Brick, M. Mihalik, Group extensions are quasi simply filtrated. Bull. Austral. Math. Soc. **50**, 21–27 (1994)
3. S. Brick, M. Mihalik, The QSF property for groups and spaces. Math. Z. **220**, 207–217 (1995)
4. D. Calegari, D. Gabai, Shrinkwrapping and the taming of hyperbolic 3-manifolds. J. Am. Math. Soc. **19**, 385–446 (2006)
5. M. Cárdenas, F. Lasheras, A. Quintero, D. Repovš, One-relator groups and proper 3-realizability. Rev. Mat. Iberoam. **25**, 739–756 (2009)
6. M. Cárdenas, F. Lasheras, A. Quintero, R. Roy, Quasi-isometries and proper homotopy: the quasi-isometry invariance of proper 3-realizability of groups. Proc. Am. Math. Soc. **147**, 1797–1804 (2019)
7. M. Cárdenas, F. Lasheras, A. Quintero, R. Roy, A topological equivalence relation for finitely presented groups. J. Pure Appl. Algebra **224** (2020). Article ID 106300
8. M. Davis, Groups generated by reflections and aspherical manifolds not covered by Euclidean space. Ann. Math. (2) **117**, 293–324 (1983)
9. P. de la Harpe, *Topics in Geometric Group Theory* (University of Chicago Press, Chicago, 2000)
10. M. Dehn, Über unendliche diskontinuierliche Gruppen. Math. Ann. **71**, 116–144 (1912)
11. M. Dehn, *Papers on Group Theory and Topology [Translated and Introduced by John Stillwell]* (Springer, New York, 1987)
12. M. Freedman, The topology of four-dimensional manifolds. J. Diff. Geom. **17**, 357–453 (1982)
13. M. Freedman, D. Gabai, Covering a nontaming knot by the unlink. Algebr. Geom. Topol. **7**, 1561–1578 (2007)
14. L. Funar, D.E. Otera, On the WGSC and QSF tameness conditions for finitely presented groups. Groups Geom. Dyn. **4**, 549–596 (2010)
15. S. Gersten, J. Stallings, Casson's idea about 3-manifolds whose universal cover is $\mathbb{R}^3$. Int. J. Algebra Comput. **1**, 395–406 (1991)
16. M. Gromov, *Asymptotic Invariants of Infinite Groups, Geometric Group Theory, Vol. 2* (Sussex, 1991). London Math. Soc. Lecture Note Ser., vol. 182 (Cambridge Univ. Press, Cambridge, 1993), pp.1–295
17. D. Harbater, Abhyankar's conjecture on Galois groups over curves. Invent. Math. **117**, 1–25 (1994)
18. D. Harbater, A. Obus, R. Pries, K. Stevenson, Abhyankar's conjectures in Galois theory: current status and future directions. Bull. Am. Math. Soc., New Ser. **55**, 239–287 (2018)
19. W. Herfort, K.H. Hofmann, F.G. Russo, *Periodic Locally Compact Groups* (de Grutyer, Berlin, 2018)
20. W. Herfort, K.H. Hofmann, F.G. Russo, A short note on coproducts of abelian pro-Lie groups. Monatsh. Math. **204**, 887–892 (2024)
21. W. Herfort, K.H. Hofmann, F.G. Russo, Neat locally compact abelian p-groups. Math. Ann. **90**, 3471–3511 (2024)
22. K.H. Hofmann, S. Morris, *The Structure of Compact Groups* (de Gruyter, Berlin, 2006)
23. C. Kosniowski, *A First Course in Algebraic Topology* (Cambridge University Press, Cambridge, 1980)

24. M. Kumar, Fundamental group of affine curves in positive characteristic. J. Algebra **399**, 323–342 (2014)
25. M. Kumar, Embedding problems for open subgroups of the fundamental group. Ann. Inst. Fourier (Grenoble) **67**, 2623–2649 (2017)
26. M.L. Mihalik, Solvable groups that are simply connected at ∞. Mat. Z. **195**, 79–87 (1987)
27. M.L. Mihalik, Compactifying coverings of 3-manifolds. Comment. Math. Helv. **71**, 362–372 (1996)
28. M.L. Mihalik, Group extensions and tame pairs. Trans. Am. Math. Soc. **351**, 1095–1107 (1999)
29. M.L. Mihalik, S. Tschantz, Tame combing of groups. Trans. Am. Math. Soc. **349**, 4251–4264 (1997)
30. D.E. Otera, On Poénaru inverse-representations. Quaest. Math. **40**, 277–293 (2017)
31. D.E. Otera, V. Poénaru, Easy representations and the QSF property for groups. Bull. Belg. Math. Soc. Simon Stevin **19**, 385–398 (2012)
32. D.E. Otera, V. Poénaru, Finitely presented groups and the Whitehead nightmare. Groups Geom. Dyn. **11**, 291–310 (2017)
33. D.E. Otera, V. Poénaru, Tame combings and easy groups. Forum Math. **29**, 665–680 (2017)
34. D.E. Otera, V. Poénaru, Topics in geometric group theory I, in *Handbook of Group Actions V*, vol. 48, ed. by L. Ji et al. Adv. Lect. Math. (International Press, Somerville; Higher Education Press, Beijing, 2020), pp. 301–346
35. D.E. Otera, F.G. Russo, On the WGSC property in some classes of groups. Mediterr. J. Math. **6**, 501–508 (2009)
36. D.E. Otera, F.G. Russo, On topological filtrations of groups. Period. Math. Hung. **72**, 218–223 (2016)
37. D.E. Otera, V. Poénaru, C. Tanasi, On geometric simple connectivity. Bull. Math. Soc. Sci. Math. Roum. (Nouv. Ser.) **53**, 157–176 (2010)
38. D.E. Otera, F.G. Russo, C. Tanasi, On 3-dimensional WGSC inverse-representations of groups. Acta Math. Hung. **151**, 379–390 (2017)
39. G. Perelman, The entropy formula for the Ricci flow and its geometric applications. Preprint (2002). ArXiv:math.D6/0303109
40. G. Perelman, Ricci flow with surgery on three-manifolds. Preprint (2003). ArXiv:math.D6/0303109
41. G. Perelman, Finite extinction time for the solutions of the Ricci flow on certain three-manifolds. Preprint (2003). ArXiv:math.D6/0307245
42. V. Poénaru, On the equivalence relation forced by the singularities of a non degenerate simplicial map. Duke Math. J. **63**, 421–429 (1991)
43. V. Poénaru, Killing handles of index one stably and π_1^∞. Duke Math. J. **63**, 431–447 (1991)
44. V. Poénaru, Almost convex groups, Lipschitz combing, and π_1^∞ for universal covering spaces of closed 3-manifolds. J. Diff. Geom. **35**, 103–130 (1992)
45. V. Poénaru, Geometry á la Gromov for the fundamental group of a closed 3-manifold M^3 and the simple connectivity at infinity of $\tilde{M}^3$. Topology **33**, 181–196 (1994)
46. V. Poénaru, Geometric simple connectivity and low-dimensional topology. Proceed. Steklov Inst. Math. **247**, 195–208 (2004)
47. V. Poénaru, Equivariant, locally finite inverse representations with uniformly bounded zipping length, for arbitrary finitely presented groups. Geom. Ded. **167**, 91–121 (2013)
48. V. Poénaru, All finitely presented groups are QSF. arXiv:1409.7325. Preprint (2014)
49. V. Poénaru, Geometric simple connectivity and finitely presented groups. Preprint (2014). arXiv:1404.4283
50. V. Poénaru, Topics in geometric group theory II, in *Handbook of Group Actions V*, ed. by L. Ji, et al. Adv. Lect. Math., vol. 48 (International Press, Somerville; Higher Education Press, Beijing, 2020), pp. 347–398
51. V. Poénaru, On geometric group theory, in *Topology and Geometry. A Collection of Essays Dedicated to Vladimir G. Turaev*, ed. by A. Papadopoulos. IRMA Lect. Math. Theor. Phys., vol. 33 (European Mathematical Society, Berlin, 2021), pp. 399–432

52. V. Poénaru, C. Tanasi, Representations of the Whitehead manifold Wh^3 and Julia sets. Ann. Toulouse **IV**, 665–694 (1995)
53. V. Poénaru, C. Tanasi, Equivariant, almost-arborescent representations of open simply-connected 3-manifolds; a finiteness result, in *Memoirs AMS*, vol. 169 (American Mathematical Society, New York, 2004)
54. V. Poénaru, C. Tanasi, On the 1-handles of the product $V^3 \times B^n$ for a simply connected open 3-manifold V^3. Topol. Appl. **160**, 748–765 (2013)
55. M. Raynaud, Séminaire de géométrie algébrique du Bois Marie 1960/61 (SGA 1), dirigé par Alexander Grothendieck, in *Lecture Notes in Mathematics*, vol. 224 (Springer XXII, Berlin, 1971)
56. M. Raynaud, Revêtements de la droite affine en caractéristique $p > 0$. Invent. Math. **116**, 425–462 (1994)
57. L. Ribes, *Profinite Graphs and Groups* (Springer, Berlin, 2017)
58. L. Ribes, P. Zalesskii, *Profinite Groups* (Springer, Berlin, 2000)
59. L.C. Siebenmann, On detecting Euclidean space homotopically among topological manifolds. Invent. Math. **6**, 245–261 (1968)
60. J. Stallings, The piecewise-linear structure of Euclidean space. Proc. Camb. Philos. Soc. **58**, 481–488 (1962)
61. J. Stallings, Brick's quasi-simple filtrations for groups and 3-manifolds, in *Geom. Group Theory*, vol. 1 (London Math. Soc., London, 1993), pp. 188–203
62. R. Vakil, K. Wickelgren, Universal covering spaces and fundamental groups in algebraic geometry as schemes. J. Theor. Nom. Bordeaux **23**, 489–526 (2011)
63. C.T.C. Wall, On simply-connected 4-manifolds. J. Lond. Math. Soc. **39**, 141–149 (1964)

Chapter 16
Lifting Maps Between Graphs to Embeddings

Alexey Gorelov

Abstract In this chapter, we study conditions for the existence of an embedding $\widetilde{f}\colon P \to Q \times \mathbb{R}$ such that $f = \mathrm{pr}_Q \circ \widetilde{f}$, where $f\colon P \to Q$ is a piecewise linear map between polyhedra. Our focus is on non-degenerate maps between graphs, where non-degeneracy means that the preimages of points are finite sets.

We introduce combinatorial techniques and establish necessary and sufficient conditions for the general case. Using these results, we demonstrate that the problem of the existence of a lifting reduces to testing the satisfiability of a 3-CNF formula. Additionally, we construct a counterexample to a result by V. Poénaru on lifting of smooth immersions to embeddings.

Furthermore, by establishing connections between the stated problem and the approximability by embeddings, we deduce that, in the case of generic maps from a tree to a segment, a weaker condition becomes sufficient for the existence of a lifting.

Keywords Graph · Polyhedron · Simplicial complex · Embedding · Computational topology

2020 Mathematics Subject Classification Primary 05C10, 57M15, 57Q35; Secondary 57N35, 05E45

16.1 Introduction

16.1.1 Background

In this chapter, we consider the following problem: given compact one-dimensional polyhedra P and Q, and a piecewise linear map $f\colon P \to Q$, what conditions must

A. Gorelov (✉)
Université Grenoble Alpes, CNRS, Institut Fourier, Grenoble, France
e-mail: algor512@gmail.com

L. Funar, A. Papadopoulos (eds.), *Essays on Topology*,
https://doi.org/10.1007/978-3-031-81414-3_16

be satisfied for the existence of a piecewise linear embedding $\widetilde{f}\colon P \to Q \times \mathbb{R}$ such that $f = \mathrm{pr}_Q \circ \widetilde{f}$? If such an embedding exists, we refer to it as a *(codimension one) lifting of f to an embedding*, and we say that f *lifts to an embedding*. This problem can also be formulated in other contexts, such as for smooth maps between manifolds or continuous maps between topological spaces. Additionally, it is possible to consider liftings of codimension k, namely embeddings $\widetilde{f}\colon P \to Q \times \mathbb{R}^k$ such that $f = \mathrm{pr}_Q \circ \widetilde{f}$. In this case, maps that admit such lifting are sometimes referred to as k-prems (short for k-projected embedding), see [1, 15, 28].

In this chapter, our focus is on non-degenerate maps. A piecewise linear map $f\colon P \to Q$ is called *non-degenerate* if the set $f^{-1}(q)$ is finite for every point $q \in Q$. Additionally, we call a simplicial map $f\colon K \to L$ *non-degenerate* if the corresponding piecewise linear map $|f|\colon |K| \to |L|$ is non-degenerate. It is easy to see that a simplicial map $f\colon K \to L$ between finite simplicial complexes is non-degenerate if and only if it is injective on each simplex; that is, a k-simplex $A \in K$ maps to a k-simplex $f(A) \in L$.

In the one-dimensional case, non-degenerate simplicial maps are synonymous with graph homomorphisms. This establishes an interesting connection between the problem we are addressing in this chapter and certain problems in graph theory. In fact, a graph homomorphism $f\colon G \to H$ is known as an H-colouring of the graph G. Each lifting of the map f to an embedding introduces orders on sets of vertices of the same colour, which, in turn, determine the orders on the edges. These constructions, which involve colourings along with orders, are actively studied in graph theory, see e.g. [7]. For example, so-called track layouts can be viewed as liftings of graph homomorphisms to complete graphs.[1]

It should be highlighted that the one-dimensional piecewise linear version of the lifting to an embedding problem plays a special role among other versions of this problem. For instance, the problem of liftability to an embedding of smooth immersions between manifolds can be reduced to the one-dimensional piecewise linear version of the problem [5, 21]. Also, as we will show below in Theorem 16.3, the multidimensional piecewise linear version of the problem reduces to the one-dimensional case.

16.1.2 Structure

Let us now outline the general structure of the chapter.

The second section introduces several combinatorial concepts that are utilized throughout the rest of the chapter. In particular, we prove the following necessary conditions for the existence of a lifting to an embedding:

[1] It is worth noting that the author has not come across research on track layouts for other types of colouring.

Theorem 16.1 *Let $f: K \to L$ be a non-degenerate simplicial map between finite simplicial complexes such that $|f|: |K| \to |L|$ lifts to an embedding. Then the following holds:*

(1) all the covering maps $p_n: |K_f^{(n)}| \to |\widetilde{K}_f^{(n)}|$, $n > 1$ are trivial;
(2) there are no n-obstructors for f for any $n > 1$.

Here $K_f^{(n)}$ and $\widetilde{K}_f^{(n)}$ denote simplicial models for the ordered and unordered configuration spaces of n points with the same image under f, and an n-obstructor is a path in $K_f^{(n)}$ that realises a cyclic permutation of points; the reader may refer to the next section for the formal definitions.

Further, we prove two necessary and sufficient conditions for the existence of a lifting.

Theorem 16.2 *Let $f: K \to L$ be a non-degenerate simplicial map between finite simplicial complexes. Then the piecewise linear map $|f|: |K| \to |L|$ lifts to an embedding if and only if there is a collection of mutually compatible linear orders on the sets $f^{-1}(v)$, $v \in V(L)$.*

Furthermore, there is a bijection between such collections of linear orders and the isotopy classes of liftings.

Informally, the compatibility here means that the orders on the vertices induce orders on the simplices; again, we postpone the formal definition until the second section.

Theorem 16.3 *Let $f: K \to L$ be a non-degenerate simplicial map between finite simplicial complexes. Assuming that the covering map $p_2: |K_f^{(2)}| \to |\widetilde{K}_f^{(2)}|$ is trivial, there is a 3-CNF formula Γ_f,[2] such that the piecewise linear map $|f|: |K| \to |L|$ lifts to an embedding if and only if Γ_f is satisfiable. Furthermore, there is a bijection between the assignments that satisfy Γ_f and the isotopy classes of liftings.*

We will explicitly describe the formula Γ_f just before we state Theorem 16.4 in the second section.

In the third section, we explore connections between the problem of lifting of maps between graphs and the problem of lifting of smooth immersions to embeddings. In particular we show that the assumptions in a result stated by V. Poénaru in [21] need to be strengthened by including the condition of the satisfiability of Γ_f. The main result in this section demonstrates that without this stronger assumption, the main result on the existence of liftings of immersions stated in [21] cannot hold:

Theorem 16.4 *Each 3-CNF formula Γ of a specific form, as described in the statement of Theorem 16.4 on page 329, can be realised as Γ_f of a map $f: G \to H$*

[2] A boolean formula Γ is said to be in 3-CNF if it can be expressed as $\Gamma = \bigwedge_i (\alpha_i \vee \beta_i \vee \gamma_i)$, where each of α_i, β_i, and γ_i is either a variable or its negation.

between graphs. Moreover, there exists a generic immersion $g\colon S \looparrowright B$ *of a surface* S *with boundary into a handlebody* B*, such that* H *is the graph of multiple points of* g*, and* f *is the restriction of* g *on* $f^{-1}(H)$.

Therefore, our Theorem 16.4 can be considered as a correction of the main lemma used in the proof of Théorème 1 in [21], and hence of the theorem itself.

In the last section, we establish a connection between the problem of the existence of liftings of maps between graphs and the problem of approximation by embeddings. By using this connection, we deduce the main result of the section:

Theorem 16.5 *For the generic simplicial maps from a tree to a segment the non-existence of 2-obstructors is a necessary and sufficient condition for the existence of a lifting.*

16.1.3 Conventions

Throughout this chapter, we adhere to the following notation and conventions.

We use the letters K and L to represent finite simplicial complexes, with $|K|$ and $|L|$ denoting their geometric realisations. All maps between simplicial complexes are assumed to be simplicial. Additionally, for a simplicial map $f\colon K \to L$, we denote the induced piecewise linear map between $|K|$ and $|L|$ by $|f|$.

Given a simplicial complex or a graph K, we use $V(K)$ and $E(K)$ to denote the set of vertices and the set of edges of K, respectively. Moreover, $\mathrm{sk}_n K$ denotes the n-skeleton of K—a simplicial complex consisting of all simplices of K of dimension lower than or equal to n.

As previously noted, non-degenerate simplicial maps between one-dimensional simplicial complexes are essentially graph homomorphisms. Occasionally, particularly in examples, it will be more convenient to consider multigraph homomorphisms.

Let G and H be multigraphs.[3] A pair (f_V, f_E) of maps $f_V\colon V(G) \to V(H)$ and $f_E\colon E(G) \to E(H)$ is called a *multigraph homomorphism* $f\colon G \to H$ if, for any edge $e \in E(G)$, the map f_V takes the endpoints of e to the endpoints of $f_E(e)$.

As a graph, a multigraph G has a geometric realisation denoted by $|G|$. Specifically, $|G|$ is the union of $|V(G)|$ points p_v representing the vertices $v \in V(G)$, and $|E(G)|$ segments I_e representing the edges $e \in E(G)$, where the segments are glued to the points according to the incidence relation.[4] Clearly, $|G|$ is a one-dimensional polyhedron uniquely defined up to piecewise linear homomorphism.

Furthermore, a multigraph homomorphism $f\colon G \to H$ induces a piecewise linear map $|f|$ between the geometric realisations of G and H, which maps $p_v \in |G|$ to $p_{f_V(v)} \in |H|$ and homeomorphically maps I_e to $I_{f_E(e)}$. Thus, by taking

[3] Graphs that can contain loops and multiple edges.

[4] Certainly, loops are represented by loops (or segments with both endpoints identical) in $|G|$.

triangulations of $|G|$ and $|H|$ in which $|f|$ is simplicial, we can obtain a simplicial map. Therefore, whenever multigraph homomorphisms are used in this text, they can be replaced by simplicial maps after subdividing the involved multigraphs. However, to simplify the notation, we choose to use multigraph homomorphisms in some places of the text.

16.2 Combinatorics of Liftings to Embeddings

In this section, we will discuss combinatorial techniques that are useful for studying the existence of liftings to embeddings of maps between graphs.

Following our conventions, let K and L be simplicial complexes, and $f: K \to L$ be a non-degenerate simplicial map between them. Let us denote by $K_f^{(n)}$ a simplicial complex whose vertices are n-tuples of distinct vertices of K that map to the same vertex by f. Additionally, a set $\{(v_1^1, \dots, v_n^1), \dots, (v_1^{k+1}, \dots, v_n^{k+1})\}$ of vertices forms a k-simplex if for each $j = 1, \dots, n$ the set $\{v_j^1, \dots, v_j^{k+1}\}$ forms a k-simplex A_j in K. Note that $A_j \cap A_s = \varnothing$ for all $j \neq s$, and all the simplices A_j have the same image under f.

The symmetric group S_n acts naturally on $K_f^{(n)}$ by permuting the points in n-tuples. Thus, we can define the unordered version of $K_f^{(n)}$, which we denote by $\widetilde{K}_f^{(n)} = K_f^{(n)}/S_n$. Clearly, S_n also acts on the geometric realisation $|K_f^{(n)}|$ of $K_f^{(n)}$. This action is properly discontinuous, so it induces a covering map $p_n: |K_f^{(n)}| \to |\widetilde{K}_f^{(n)}|$ that forgets the order of points. Moreover, it is clear that p_n is a principal S_n-bundle.[5]

Definition (n-Obstructor) An *n-obstructor* for $f: |K| \to |L|$ is a vertex $(x_1, x_2, \dots, x_n) \in K_f^{(n)}$ and a path from $(x_1, x_2, \dots, x_n)$ to $(x_n, x_1, x_2, \dots, x_{n-1})$ in $K_f^{(n)}$.

The reader may refer to Example 16.1 below to see an example of 2-obstructor.

Lemma 16.1 *For any non-degenerate simplicial map $f: K \to L$ and $N > 1$, the following statements are equivalent:*

(1) all the covering maps $p_k: |K_f^{(k)}| \to |\widetilde{K}_f^{(k)}|$, $1 < k \le N$ are trivial;
(2) there are no k-obstructors for f for $1 < k \le N$.

Proof The implication (1) $\Rightarrow$ (2) is straightforward. Indeed, assume there is a k-obstructor that defines a path $\gamma: I \to |K_f^{(k)}|$. Then $p_k \circ \gamma$ is a loop in $|\widetilde{K}_f^{(k)}|$, but γ is not a loop in $|K_f^{(k)}|$, implying that p_k is not trivial.

[5] For a short introduction to this topic the reader may refer to [19].

Now, let us prove the implication (2) $\Rightarrow$ (1). Assume there are no k-obstructors for any $1 < k \leq N$. Let $|K_f^{(k)}| = \bigsqcup_i C_i$, where $\{C_i\}_i$ are connected components of $|K_f^{(k)}|$.[6]

Clearly, the action of S_k on $|K_f^{(k)}|$ induces an action of S_k on $\{C_i\}_i$. Let us prove that this action is free. Suppose it is not. Then there exists $\sigma \neq \mathrm{id} \in S_k$ and a connected component C such that $\sigma(C) = C$. Let p be a prime divisor of $\mathrm{ord}(\sigma)$, and define $\sigma' = \sigma^{\mathrm{ord}(\sigma)/p}$. Clearly, $\sigma'(C) = \sigma^{\mathrm{ord}(\sigma)/p}(C) = C$. Moreover, since $\mathrm{ord}(\sigma') = p$, the cycle decomposition of σ' consists of cycles of length one, and one or more disjoint cycles of length p. Let $(s_1\ s_2\ \ldots\ s_p)$ be one of them.

Now consider any vertex $x \in C$. As $\sigma'(x) \in C$, there exists a path γ in C connecting x and $\sigma'(x)$. In fact, it defines k paths in K

$$\begin{array}{llll}
\gamma_1\colon x_1 \to y_1^1 & \to \ldots \to y_1^l & \to x_{\sigma'(1)} \\
\gamma_2\colon x_2 \to y_2^1 & \to \ldots \to y_2^l & \to x_{\sigma'(2)} \\
\vdots & \vdots & \vdots \\
\gamma_k\colon x_k \to y_k^1 & \to \ldots \to y_k^l & \to x_{\sigma'(k)}
\end{array}$$

such that $y_i^k \neq y_j^k$ for $i \neq j$, and $f(y_i^k) = f(y_j^k)$. By taking the paths $\gamma_{s_1}, \ldots, \gamma_{s_p}$, we obtain a path in $K_f^{(p)}$. It can be seen that this path defines a p-obstructor.

Therefore, the action of S_n on $\{C_i\}_i$ is free. Let $\{O_\alpha\}_\alpha$ be the set of orbits of the action. For each orbit O_α, fix an arbitrary connected component $C_\alpha \in O_\alpha$, and let $T = \bigsqcup_\alpha C_\alpha$.

For every point $x \in |\widetilde{K}_f^{(k)}|$, the preimages $p_k^{-1}(x)$ form an orbit of the action of S_n on $|K_f^{(k)}|$. Thus, the connected components containing these preimages also form an orbit. Moreover, since the action of S_n on them is free, two distinct preimages cannot lie in the same connected component. The construction of T ensures that exactly one of the points in $p_k^{-1}(x)$ lies in some of $C_\alpha \subset T$. Therefore, the restriction $p_k\big|_T$ is bijective, and we can define a map $s\colon |\widetilde{K}_f^{(k)}| \to |K_f^{(k)}|$ by putting $s(x) = \left(p_k\big|_T\right)^{-1}(x)$.

Note that p_k is an open map; indeed, for any open set $U \subset |K_f^{(k)}|$ we have $p_k^{-1}(p_k(U)) = \bigcup_{\sigma \in S_k} \sigma(U)$, and all the sets $\sigma(U)$ are open. Thus, for an open $U \subseteq |K_f^{(k)}|$, we have $s^{-1}(U) = s^{-1}(U \cap T) = p_k(U \cap T)$ which is open since p_k is open, and both U and T are open.

Therefore, s is a continuous map defining a section of a principal S_k-bundle p_k, which implies that p_k is trivial by [19, Proposition 2.1]. $\square$

[6] Since the polyhedra are locally path-connected spaces, connectedness and path-connectedness are equivalent in this case.

Now, we will state necessary conditions for the existence of a lifting. It is worth mentioning that the necessity of (2), for the case of generic immersions between manifolds and in a slightly different formulation, is proven in [21, Lemme 1.1]. The necessity of (1) for $n = 2$ was discussed in [2, §3.1] for maps to $\mathbb{R}^N$ and proven in [10, Corollary 3] for generic immersions of surfaces into $\mathbb{R}^3$.

Theorem 16.1 *Let $f: K \to L$ be a non-degenerate simplicial map such that $|f|: |K| \to |L|$ lifts to an embedding. Then the following holds:*

(1) all the covering maps $p_n: |K_f^{(n)}| \to |\widetilde{K}_f^{(n)}|$, $n > 1$ are trivial;
(2) there are no n-obstructors for f for any $n > 1$.

Proof Let us begin by noting that the points of $|K_f^{(n)}|$ and $|\widetilde{K}_f^{(n)}|$ can be considered as ordered and unordered n-tuples of distinct points of $|K|$ with the same image.

Suppose there is a lifting $\widetilde{|f|}: |K| \to |L| \times \mathbb{R}$. Let $h = \mathrm{pr}_R \widetilde{|f|}: |K| \to \mathbb{R}$. This induces a continuous map that takes $(x_1, \ldots, x_n)$ to $(h(x_1), \ldots, h(x_n))$. Since $\widetilde{|f|}$ is an embedding, all $h(x_i)$'s are distinct. Thus, it actually maps $|K_f^{(n)}|$ to the configuration space $\mathrm{Conf}_n(\mathbb{R})$ of n distinct points of $\mathbb{R}$. Moreover, it is equivariant with respect to the actions of S_n on both spaces.[7] It is easy to see that $\mathrm{Conf}_n(\mathbb{R})$ equivariantly deformation retracts to the discrete space $\mathrm{Conf}_n(\{1, 2, \ldots, n\})$ of all $n!$ permutations of the numbers $\{1, \ldots, n\}$ with the natural action of S_n. By taking an equivariant isomorphism that sends the n-tuple $(1, 2, \ldots, n)$ to $\mathrm{id} \in S_n$, we can identify $\mathrm{Conf}_n(\{1, 2, \ldots, n\})$ with S_n.

Therefore, we have an equivariant map $H: |K_f^{(n)}| \to S_n$. Now define $g: |K_f^{(n)}| \to S_n \times |\widetilde{K}_f^{(n)}|$ as $g(x) = (H(x), p_n(x))$. Clearly, g is a morphism between the principal S_n-bundles p_n and the trivial bundle $\mathrm{pr}_{|\widetilde{K}_f^{(n)}|}: S_n \times |\widetilde{K}_f^{(n)}| \to |\widetilde{K}_f^{(n)}|$, thus, it is an isomorphism by [19, Proposition 2.1]. Therefore, p_n is trivial.

The non-existence of n-obstructors then follows from Lemma 16.1. □

Example 16.1 This example, along with the description of a 2-obstructor, is presented in [2, §3.1]. The map $f: G \to H$ first appeared in the proof of Theorem 2.1 in [26].

Consider the graphs G and H and the simplicial map $f: G \to H$ as in Fig. 16.1. As shown in the figure, there is a 2-obstructor for f given by the path in $G_f^{(2)}$ defined by the two paths in G:

$$\gamma_1: a_3 \to b_4 \to c_4 \to d_3 \to c_3 \to b_3 \to a_1 \to b_1 \to c_1 \to d_1 \to c_1 \to$$
$$\to b_1 \to a_1 \to b_3 \to c_3 \to b_2 \to a_2$$
$$\gamma_2: a_2 \to b_2 \to c_2 \to d_2 \to c_2 \to b_2 \to a_2 \to b_2 \to c_3 \to d_3 \to c_4 \to$$
$$\to b_4 \to a_3 \to b_4 \to c_4 \to b_4 \to a_3$$

[7] $F: X \to Y$ is equivariant with respect to actions of a group G on X and Y if $F(g(x)) = g(F(x))$ for any $x \in X$ and $g \in G$.

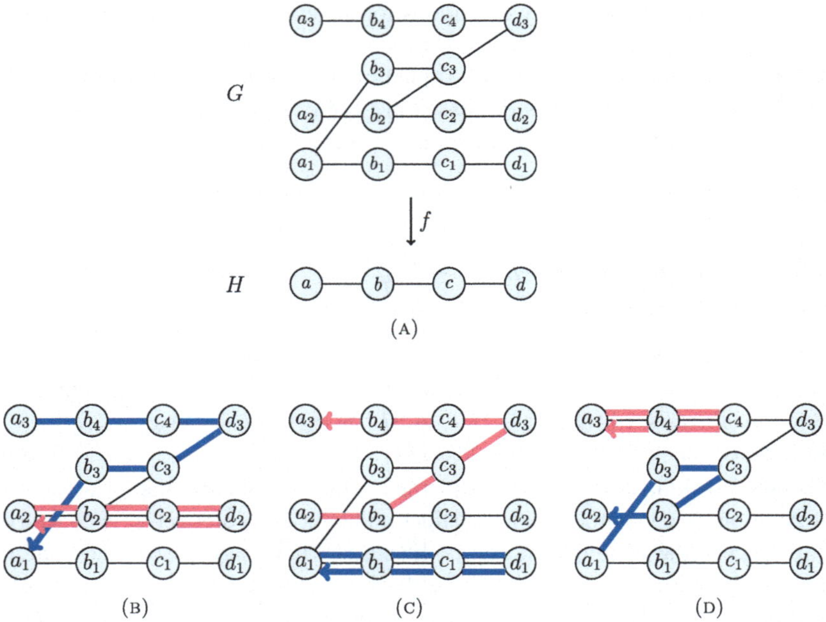

Fig. 16.1 Siekłucki's example and the 2-obstructor for it shown as a pair of paths in G (in three stages). (**a**) The map $f: G \to H$. (**b**) The first part. (**c**) The middle part. (**d**) The end part

Therefore, f does not lift to an embedding.

Theorem 16.1 provides only the necessary conditions for the existence of a lifting (see Examples 16.2 and 16.3). The next two theorems, however, provide conditions that are both necessary and sufficient.

Definition (Admissible Collection of Linear Orders) Let $f: K \to L$ be a non-degenerate simplicial map between finite simplicial complexes. For each simplex $C \in L$, let K_C be the set of simplices $A \in K$ that map to C under f. Note that since f is non-degenerate, all the simplices in K_C have the same dimension as C. Furthermore, let $V(L)$ be the set of vertices of L.

We call a collection $\{(\prec_v, K_v)\}_{v \in V(L)}$ of linear orders on the sets $K_v = f^{-1}(v)$ *admissible* if it induces linear orders on all the sets K_C, $C \in L$.

Specifically, given a pair of simplices $A, B \in K_C$, let $V_f(A, B) \subset V(A) \times V(B)$ represent all pairs of vertices (v, w) satisfying $f(v) = f(w)$. Then it is required that either $v \preceq_{f(v)} w$ holds for all $(v, w) \in V_f(A, B)$, or $v \succeq_{f(v)} w$ holds for all $(v, w) \in V_f(A, B)$.

Theorem 16.2 *Let $f: K \to L$ be a non-degenerate simplicial map between finite simplicial complexes. Then the piecewise linear map $|f|: |K| \to |L|$ lifts to an embedding if and only if there exists an admissible collection of linear orders on the sets K_v, $v \in V(L)$.*

Furthermore, there exists a bijection between the admissible collections of linear orders and the isotopy classes of liftings.

The reader may refer to [10, Proposition 4], [21, Lemme 1.4], and [16, Theorem 5] for variants of the stated theorem.

Proof Let us first, by using a lifting of $|f|$, construct an admissible collection of linear orders. Assume there is a lifting $|\widetilde{f}|\colon |K| \to |L| \times \mathbb{R}$. It defines a collection of linear orders as follows: for vertices $v, w \in V(K)$ such that $f(v) = f(w)$, we define $v \prec_f w$ if $h(v) < h(w)$, where $h = \mathrm{pr}_{\mathbb{R}} \circ |\widetilde{f}|$. Let us prove that this collection is admissible.

Suppose it is not. It implies that there exist simplices $A, B \in K$ with the same image $f(A) = f(B)$ and two pairs of vertices $f(v) = f(w)$, $f(v') = f(w')$, where $v, v' \in A$, $w, w' \in B$, such that $v \prec_f w$ but $v' \succ_f w'$.

Clearly, we have $f(\{v, v'\}) = f(\{w, w'\}) = \{f(v), f(w)\}$ where $\{a, b\}$ denotes an edge with endpoints a and b. Let $d\colon [0, 1] \to \mathbb{R}$ be the map defined as $d(t) = h((1-t)v + tv') - h((1-t)w + tw')$. Since $v \prec_f w$ and $v' \succ_f w'$, we have $d(0) = h(v) - h(w) < 0$ and $d(1) > 0$. Therefore, there exists $t' \in [0, 1]$ such that $d(t') = 0$. However, since $|f|$ is linear on the simplices of K, this implies that $|\widetilde{f}|((1-t')v + t'v') = |\widetilde{f}|((1-t')w + t'w')$, which contradicts the fact that $|\widetilde{f}|$ is an embedding.

Now suppose we have an admissible collection of linear orders $\{(\prec_v, K_v)\}_{v \in V(L)}$. Let us construct a lifting $|\widetilde{f}|$ of $|f|$ based on this collection. First, we define a map $h\colon V(K) \to \mathbb{N}$ as follows: for each K_v, we let $h\big|_{K_v}$ be the order isomorphism $K_v \to \{1, \ldots, |K_v|\}$ that takes the k-th vertex in K_v (in ascending order with respect to $\prec_v$) to k. By extending the map h linearly to the entire $|K|$, we obtain the map $h\colon |K| \to \mathbb{R}$. Take $|\widetilde{f}| = |f| \times h$. Now, let us prove that $|\widetilde{f}|$ is an embedding.

Suppose it is not. Then there exist points $v, w \in |K|$ such that $|\widetilde{f}|(v) = |\widetilde{f}|(w)$. Since v and w cannot be vertices of K by the definition of h, they lie in the interior of some simplices A and B of K, respectively, with $\dim A = \dim B > 0$.[8] The last equality follows from the fact that f is non-degenerate and hence injective on simplices, so $f(A) = f(B)$ and the simplices A, B, and $f(A)$ have the same dimension. Let $a_0, \ldots, a_n$ denote the vertices of A, and $b_0, \ldots, b_n$ denote the corresponding vertices of B, such that $f(a_i) = f(b_i)$ for each i.

Now, let $v = \sum_{i=0}^{n} \alpha_i a_i$ where the α_i's are the barycentric coordinates of v in A. Since $|f|(v) = |f|(w)$, we have $w = \sum_{i=0}^{n} \alpha_i b_i$. Therefore, we have $h(v) = \sum_{i=0}^{n} \alpha_i h(a_i) = \sum_{i=0}^{n} \alpha_i h(b_i) = h(w)$.

Because h was defined using the admissible collection of linear orders, we have either $h(a_i) \leq h(b_i)$ or $h(a_i) \geq h(b_i)$ for all i. Moreover, since $A \neq B$, at least one of the inequalities is strict. This implies that either $h(v) < h(w)$ or $h(v) > h(w)$, which contradicts the assumption that $|\widetilde{f}|(v) = |\widetilde{f}|(w)$, where $|\widetilde{f}| = |f| \times h$.

[8] In this context, we define the interior of a simplex Δ as $|\Delta| \setminus \bigcup_A |A|$, where A ranges over all proper faces of Δ.

To establish the bijection between admissible collections of linear orders and isotopy classes of liftings, we observe that each lifting $\widetilde{|f|}$ is isotopic to the lifting $\widetilde{|f|}'$ constructed in the previous step using the admissible collection of linear orders induced by $\widetilde{|f|}$. The isotopy $F\colon |K| \times [0, 1] \to |L| \times \mathbb{R}$ is defined as $F(x, t) = \big(|f|(x), (1-t)h(x) + th'(x)\big)$ where $h = \mathrm{pr}_{\mathbb{R}} \widetilde{|f|}$ and $h' = \mathrm{pr}_{\mathbb{R}} \widetilde{|f|}'$. The proof that this is an actual isotopy is left to the reader. □

The next theorem demonstrates that the problem of determining the existence of a lifting for a piecewise linear non-degenerate map between polyhedra can be reduced to the case of graphs.

Theorem 16.3 *Let $f\colon K \to L$ be a non-degenerate simplicial map. Then the corresponding piecewise linear map $|f|\colon |K| \to |L|$ lifts to an embedding if and only if $\left|f|_{\mathrm{sk}_1 K}\right| : |\mathrm{sk}_1 K| \to |\mathrm{sk}_1 L|$ lifts to an embedding. Here,* $\mathrm{sk}_1 K$ *and* $\mathrm{sk}_1 L$ *denote the one-dimensional skeletons of K and L, respectively. Moreover, there is a bijection between the isotopy classes of liftings of $|f|$ and of $\left|f|_{|\mathrm{sk}_1 K|}\right|$.*

Proof Based on Theorem 16.2, it suffices to show that a collection $\{(\prec_v, K_v)\}_{v \in V(K)}$ of linear orders is admissible with respect to f if and only if it is admissible with respect to $f|_{\mathrm{sk}_1 K}$.

It is clear that once it is admissible with respect to f, it is admissible with respect to $f|_{\mathrm{sk}_1 K}$ as well, according to the definition.

Suppose a collection is admissible with respect to $f|_{\mathrm{sk}_1 K}$, but not admissible with respect to f. This implies that there exist simplices $A, B \in K_C$ and two pairs of vertices $f(v) = f(w)$, $f(v') = f(w')$, where $v, v' \in A$, $w, w' \in B$, such that $v \prec_{f(v)} w$ but $v' \succ_{f(v')} w'$. The edges $\{v, v'\}$, $\{w, w'\}$ lie in $\mathrm{sk}_1 K$ and have the same image, but their endpoints are in the opposite order. This leads to a contradiction. □

Note that, while studying the existence of a lifting of a piecewise linear map f, it suffices to focus attention only on the subpolyhedron of the preimages of the multiple points of f. In view of Theorem 16.3, after setting f' to be a restriction of f on the set $\{x \in \mathrm{sk}_1 K \mid |f^{-1}(f(x))| > 1\} \subseteq \mathrm{sk}_1 K$, we have that f lifts if and only if f' lifts.

For instance, in the case of generic immersions of surfaces into $\mathbb{R}^3$, a similar observation allows us to narrow down the question of liftability of an immersion to the liftability of its restriction on the preimage of the set of multiple points, see [6, Theorem 3.2]. Note that this restriction is, in fact, a map between graphs.

The next theorem demonstrates that the liftability of a non-degenerate piecewise linear map $|f|\colon |K| \to |L|$ corresponding to a simplicial map $f\colon K \to L$, can be encoded as the satisfiability of a 3-CNF boolean formula, which we will define below.

The key idea is as follows: assuming the covering map $p_2\colon |K_f^{(2)}| \to |\widetilde{K}^{(2)}|$ is trivial, each trivialization of it yields two mutually inverse collections of binary relations on the preimages of the vertices of L. Specifically, by selecting a section

$s\colon |\widetilde{K}^{(2)}| \to |K_f^{(2)}|$ of p_2, we define the relation as $x\,R\,y$ when (x, y) lies in $s(|\widetilde{K}^{(2)}|)$. This relation is antisymmetric and irreflexive; therefore, to be a (strict) order, it only needs to satisfy the transitivity property. Transitivity, in turn, can be represented by a set of conditions of the form $x\,R\,y \wedge y\,R\,z \to x\,R\,z$.[9]

More precisely, let us define a boolean formula Γ_f. Denote by $\mathcal{C}$ the set of connected components of $|K_f^{(2)}|$. The involution τ on $|K_f^{(2)}|$ induces an involution on $\mathcal{C}$. Let us assume that the covering map p_2 is trivial, therefore, the latter involution is fixed-point free.

Let $O_1, O_2, \ldots, O_n$ be the orbits of the involution on $\mathcal{C}$. For each orbit O_i, pick an element $C_i \in O_i$ in the orbit, so $O_i = \{C_i, \tau(C_i)\}$, and associate a boolean variable x_i with C_i. Next, we associate the literal $\neg x_i$ with each $\tau(C_i)$. As a result, each connected component C in $\mathcal{C}$ is associated with either a variable or its negation. We refer to the literal corresponding to C as α_C.

Let T be the set of ordered triples (C, D, E) of connected components, with $C \neq E$ and $D \neq E$, such that there exist three vertices $a, b, c \in V(K)$ satisfying $(a, b) \in C$, $(b, c) \in D$, $(a, c) \in E$. Let Γ_f be the boolean function

$$\Gamma_f(x_1, \ldots, x_n) = \bigwedge_{(C,D,E)\in T} ((\alpha_C \wedge \alpha_D) \to \alpha_E) = \bigwedge_{(C,D,E)\in T} (\neg\alpha_C \vee \neg\alpha_D \vee \alpha_E)$$

Note that, aside from reordering the clauses, literals, and renaming variables, the only source of variability in the form of Γ_f arises from the initial choice of representatives for the orbits O_i. However, it is easy to check that once Γ_f contains a clause $\neg\alpha_C \vee \neg\alpha_D \vee \alpha_E$, then it must also contain the clause $\alpha_C \vee \alpha_D \vee \neg\alpha_E$.[10] This is because, once we have $(a, b) \in C$, $(b, c) \in D$, and $(a, c) \in E$, we have $(c, b) \in \tau(D)$, $(b, a) \in \tau(C)$, and $(c, a) \in \tau(E)$.

Thus, the form of Γ_f does not actually depend on the choice of representatives, and, therefore, Γ_f is uniquely defined up to reordering the clauses and the literals within clauses, and renaming the variables.

Theorem 16.4 *Let $f\colon K \to L$ be a non-degenerate simplicial map. Then the corresponding piecewise linear map $|f|\colon |K| \to |L|$ lifts to an embedding if and only if the following conditions hold:*

(1) the covering map $p_2\colon |K_f^{(2)}| \to |\widetilde{K}_f^{(2)}|$ is trivial,
(2) Γ_f is satisfiable.

Furthermore, there exists a bijection between the assignments that satisfy Γ_f and the isotopy classes of liftings.

[9] An alternative approach to expressing transitivity in terms of the covering maps p_k's is to find a pair of "compatible" trivializations for $p_2\colon |K_f^{(2)}| \to |\widetilde{K}^{(2)}|$ and $p_3\colon |K_f^{(3)}| \to |\widetilde{K}^{(3)}|$, refer to [16, Theorem 5] for further details.

[10] After double negation elimination, that is $\neg\neg x = x$.

Proof Let us assume that there exists a lifting $|\widetilde{f}|: |K| \to |L| \times \mathbb{R}$ of $|f|$. By Theorem 16.1, the covering map p_2 is trivial. Moreover, as seen in the proof of Theorem 16.1, we can obtain a continuous equivariant map H from $|K_f^{(2)}|$ to S_2 by composing $h: |K_f^{(2)}| \to \mathrm{Conf}_2(\mathbb{R})$ defined as $h((x, y)) = \big(\mathrm{pr}_{\mathbb{R}} \circ |\widetilde{f}|(x), \mathrm{pr}_{\mathbb{R}} \circ |\widetilde{f}|(y)\big)$, with the retraction $\mathrm{Conf}_2(\mathbb{R}) \to S_2$.

Let us define the variable assignment by putting $x_i = 1$ if $H(C_i) = \mathrm{id} \in S_2$, and $x_i = 0$ otherwise. This assignment is well-defined because H is constant on the connected components of $|K_f^{(2)}|$. Moreover, under this assignment, $\alpha_C = 1$ if and only if $H(C) = \mathrm{id}$.

Let us consider an implication $(\alpha_C \wedge \alpha_D) \to \alpha_E$ in Γ_f. According to the definition of Γ_f, there exist pairs $(a, b) \in C$, $(b, c) \in D$, and $(a, c) \in E$. Once $\alpha_C = 1$ and $\alpha_D = 1$, we have $\alpha_E = 1$ by the transitivity of $\prec_{f(a)}$. Therefore, the defined assignment satisfies the disjunction, and consequently, it satisfies the entire formula Γ_f.

Let us prove the reverse implication. Assume we have an assignment $\psi: \{x_i\}_{i=1}^n \to \{0, 1\}$ that satisfies Γ_f. We denote by $\psi(\alpha_C)$ the value of α_C under the assignment ψ. Now, we define a binary relation R on $V(K_f^{(2)})$ as follows: $a\,R\,b$ if $\psi(\alpha_C) = 1$ for the connected component C containing the pair (a, b), if it exists.

It is easy to see that the relation R is antisymmetric and irreflexive. Moreover, it is transitive. Indeed, suppose that pairs (a, b) and (b, c) lie in connected components C and D, respectively, and $a\,R\,b$ and $b\,R\,c$, or, in other words, $\psi(\alpha_C) = \psi(\alpha_D) = 1$. Denote by E the connected component containing the pair (a, c). If $E = C$ or $E = D$, then $\psi(\alpha_E) = \psi(\alpha_C) = 1$ or $\psi(\alpha_E) = \psi(\alpha_D) = 1$, respectively, hence $a\,R\,c$. Otherwise, we have a disjunction $\neg\alpha_C \vee \neg\alpha_D \vee \alpha_E$ in Γ_f confirming that $\psi(\alpha_E) = 1$ and hence $a\,R\,c$. Thus, R defines a collection of linear orders on the sets K_v by putting $a \prec_{f(a)} b$ if $a\,R\,b$.

Note that if $(a, b), (c, d) \in V(K_f^{(2)})$ lie in the same connected component of $K_f^{(2)}$, then, according to the definition of R, either $a \prec_{f(a)} b$ and $c \prec_{f(c)} d$, or $a \succ_{f(a)} b$ and $c \succ_{f(c)} d$.

Next, let us prove that this collection is admissible. To do this, we need to show that, given simplices A and B with the same image, we have either $a \prec_{f(a)} b$ or $a \succ_{f(a)} b$ simultaneously for all pairs of distinct vertices $(a, b) \in V(A) \times V(B)$ with $f(a) = f(b)$. However, all such pairs lie in the same connected component of $K_f^{(2)}$, which provides the desired result.

Thus, we have constructed an admissible collection of linear orders on the sets K_v, and, by Theorem 16.2, it defines a lifting of f.

Note that, in the above proof, we established a bijective correspondence between the assignments satisfying Γ_f and the collections of admissible orders. Therefore, by Theorem 16.2, this establishes a bijection between the assignments and the isotopy classes of liftings. This observation completes the proof. □

Remark Both Theorem 16.2 and Theorem 16.4 are proven under the assumption that f is a simplicial map. However, it is easy to see that the statements of these

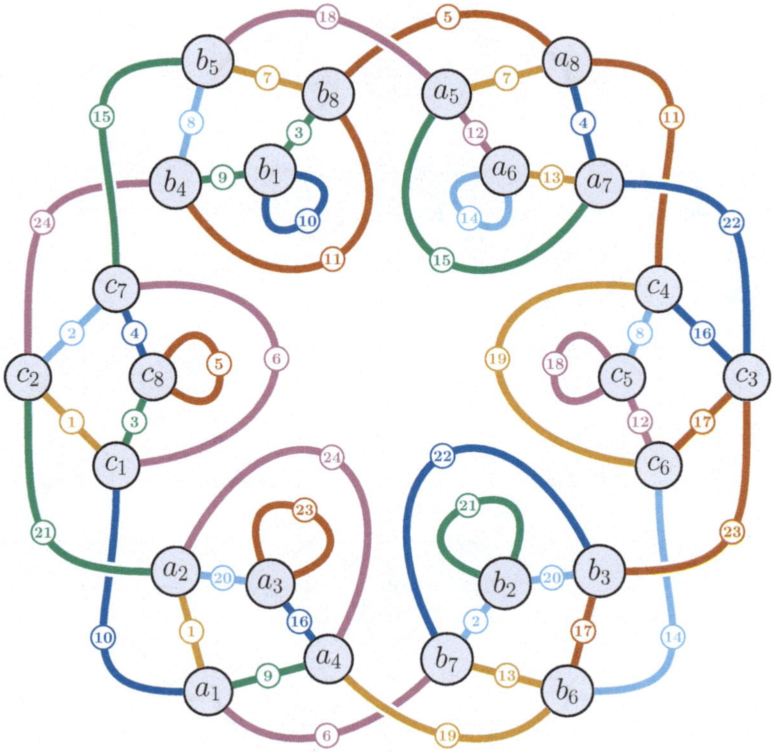

Fig. 16.2 Giller's example: graph of multiple point preimages. Vertices with the same subscript map to the same vertex, and identically labelled edges have the same image. In other words, $f(x_i) = f(y_j)$ if and only if $i = j$, and $f(e) = f(g)$ if and only if e and g share the same label

theorems remain true if we replace simplicial complexes with multigraphs, and simplicial maps with homomorphisms of multigraphs.

Example 16.2 (Giller's Example, See [10]) The provided example, known as Giller's example, illustrates an immersion of the 2-sphere S^2 into 3-dimensional Euclidean space $\mathbb{R}^3$ that cannot be lifted to an embedding. The example is discussed in the paper by Giller [10], and we refer the reader to it for details. We focus only on the graph of the multiple point preimages to understand its liftability. We denote by f the restriction of the immersion on this graph.

The multigraph of the multiple point preimages is shown in Fig. 16.2.

Additionally, Fig. 16.3 shows the corresponding graph $G_f^{(2)}$, which is the union of two circles. Note that the covering map p_2 is trivial in this case.

However, one can see that the pairs (a_4, b_4) and (b_4, c_4) lie in the outer circle of Fig. 16.3, while (a_4, c_4) lies in the inner circle. As a result, Γ_f contains clauses $(x \vee x \vee x) \wedge (\neg x \vee \neg x \vee \neg x)$, where x is a variable corresponding to the inner

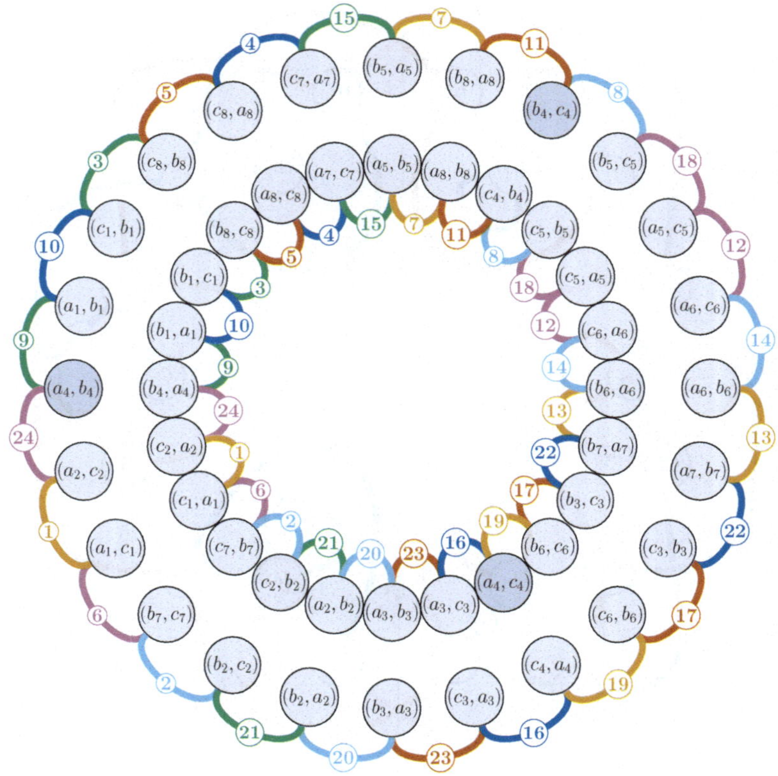

Fig. 16.3 Giller's example: the graph $G_f^{(2)}$

circle. Since this subformula is equivalent to $x \wedge \neg x$, the entire formula Γ_f is not satisfiable. Therefore, both f and the initial immersion do not lift to an embedding.

Let us now assume that for an arbitrary map f the formula Γ_f contains a clause corresponding to a triple (C, C, E). Thus, it contains a clause corresponding to a triple $(\tau(C), \tau(C), \tau(E))$ as well, and, therefore,

$$\Gamma_f = (\neg\alpha_C \vee \neg\alpha_C \vee \alpha_E) \wedge (\alpha_C \vee \alpha_C \vee \neg\alpha_E) \wedge \Gamma' = (\alpha_C = \alpha_E) \wedge \Gamma'$$

This allows us to simplify the formula Γ_f by "gluing" connected components C and E (that is, defining an equivalence relation on the points of $|K_f^{(2)}|$ that is coarser than the one defined by the partition into connected components) once we meet a triple of pairs $(a, b) \in C$, $(b, c) \in C$, and $(a, c) \in D$. This idea is implemented in [21, 22]; we will discuss it, and disprove some of the results stated in these papers, in the next section.

16.3 Liftings of Smooth Immersions

In this section, we explore connections between lifting maps between graphs and lifting generic immersions between smooth manifolds. We provide a counterexample to a result stated in [21, 22]. Additionally, we show that for each Γ that potentially could be a formula from Theorem 16.4, there exists a generic smooth immersion $f\colon X \looparrowright Y$ of a surface X with boundary into a 3-manifold Y with boundary such that the restriction f' of f onto the preimage of the multiple point set forms a map between graphs with $\Gamma_{f'} = \Gamma$.

We will use the word "generic" in the same sense in which it is used in [22]. Specifically, we say that $f\colon X \looparrowright Y$ is *generic* when

(1) $f^{-1}(\partial Y) = \partial X$,
(2) for $y \in Y \setminus \partial Y$ and $f^{-1}(y) = \{x_i\}_{i=1}^n$, the subspaces $df_{x_i}(T_{x_i}X)$ are in general position in T_yY, that is

$$\operatorname{codim} \bigcap_{i=1}^{n} df_{x_i}(T_{x_i}X) = \sum_{i=1}^{n} \operatorname{codim} df_{x_i}(T_{x_i}X),$$

(3) for $y \in \partial Y$ and $f^{-1}(y) = \{x_i\}_{i=1}^n$, the subspaces $T_y\partial Y$ and $df_{x_i}(T_{x_i}X)$ are in general position in T_yY.

Let $f\colon X \looparrowright Y$ be a generic immersion. In [22] and [21, Théorème 1], Poénaru claimed that for the existence of a lifting $\widetilde{f}\colon X \to Y \times \mathbb{R}$ to an embedding, it is necessary and sufficient that $\mu_2(f) = \nu_3(f) = 0$.

Here, the condition $\mu_2(f) = 0$ is equivalent to the triviality of the covering $P_2\colon \operatorname{Conf}_f(X,2) \to \widetilde{\operatorname{Conf}}_f(X,2)$, where $\operatorname{Conf}_f(X,2)$ represents the configuration space of ordered pairs of distinct points with the same image under f, and $\widetilde{\operatorname{Conf}}_f(X,2)$ is its unordered version. For non-degenerate piecewise linear maps between graphs K and L, this condition is equivalent to the first condition in Theorem 16.4. Indeed, both $\operatorname{Conf}_f(X,2)$ and $\widetilde{\operatorname{Conf}}_f(X,2)$ equivariantly deformation retract onto $|K_f^{(2)}|$ and $|\widetilde{K}_f^{(2)}|$, respectively. Therefore, the triviality of P_2 is equivalent to the triviality of $p_2\colon |K_f^{(2)}| \to |\widetilde{K}_f^{(2)}|$.

Now, let us consider the condition $\nu_3(f) = 0$. Following Poénaru's approach, we define a set $S_{2,3} \subseteq \operatorname{Conf}_f(X,2) \times \widetilde{\operatorname{Conf}}_f(X,3)$ that includes points $(a,b) \times \{c,d,e\}$ where $\{a,b\} \subset \{c,d,e\}$. On $S_{2,3}$, we introduce a collection $\mathbb{T}$ of all equivalence relations satisfying the following properties (denoting unordered triples as T_1 and T_2 for simplicity):

(1) if $(a,b) \times T_1 \sim (c,d) \times T_2$, then $(b,a) \times T_1 \sim (d,c) \times T_2$,
(2) if the points (a,b), (c,d) lie in the same connected component of $\operatorname{Conf}_f(X,2)$, then $(a,b) \times T_1 \sim (c,d) \times T_2$,
(3) once $(a,b) \times \{a,b,c\} \sim (b,c) \times \{a,b,c\}$, we have $(a,b) \times \{a,b,c\} \sim (b,c) \times \{a,b,c\} \sim (a,c) \times \{a,b,c\}$.

Since the intersection of two equivalence relations in $\mathbb{T}$ belongs to $\mathbb{T}$, the set $\mathbb{T}$ forms a lower subsemilattice of the lattice $\mathrm{Eq}(S_{2,3})$ consisting of all equivalence relations on the set $S_{2,3}$. Therefore, let $\sim_\nu$ denote the smallest (finest) equivalence relation in $\mathbb{T}$. We say that $\nu_3(f) = 0$ if there are no points such that $(a, b) \times \{a, b, c\} \sim_\nu (b, c) \times \{a, b, c\} \sim_\nu (c, a) \times \{a, b, c\}$.

Considering $(a, b) \times T_1 \sim (a, b) \times T_2$ for any $\sim$ from $\mathbb{T}$ and for any points $T_1, T_2 \in \widetilde{\mathrm{Conf}}_f(X, 3)$ with $\{a, b\} \subset T_1, T_2$, each equivalence relation $\sim$ from $\mathbb{T}$ defines the equivalence relation $\psi(\sim)$ on the set $\mathrm{Conf}_f(X, 2)$ by taking $(a, b)\ \psi(\sim)\ (c, d)$ if $(a, b) \times T_1 \sim (c, d) \times T_2$ for some $T_1, T_2 \in \widetilde{\mathrm{Conf}}_f(X, 3)$. One can check that $\psi\colon \mathbb{T} \hookrightarrow \mathrm{Eq}(\mathrm{Conf}_f(X, 2))$ defines an embedding of the semilattice $\mathbb{T}$ into the lattice $\mathrm{Eq}(\mathrm{Conf}_f(X, 2))$ consisting of all the equivalence relations on $\mathrm{Conf}_f(X, 2)$.

It can be verified that $\psi(\mathbb{T}) \subset \mathrm{Eq}(\mathrm{Conf}_f(X, 2))$ can be defined using the three conditions defining $\mathbb{T}$, where instead of the points of the form $(x, y) \times T$, we consider pairs (x, y). Therefore, the second condition, saying that $(a, b) \sim (c, d)$ whenever (a, b) and (c, d) lie in the same connected component of $\mathrm{Conf}_f(X, 2)$, ensures that all equivalence relations in $\psi(\mathbb{T})$ are coarser than the one defined by the partition of $\mathrm{Conf}_f(X, 2)$ into connected components.

Moreover, as demonstrated in the proof of [21, Lemme 1.2], the equivalence relation $\psi(\sim_\nu)$ (and, thus, $\sim_\nu$) can be iteratively constructed as follows: we start with the equivalence relation on $\mathrm{Conf}_f(X, 2)$ defined by the partition into connected components. We then sequentially join connected components C and D whenever we meet a triple of pairs $(a, b), (b, c) \in C$, and $(a, c) \in D$.

As discussed in the end of the preceding section, this process corresponds to simplifying the formula Γ_f by replacing terms of the form $(\neg\alpha_C \vee \neg\alpha_C \vee \alpha_D) \wedge (\alpha_C \vee \alpha_C \vee \neg\alpha_D)$ with equalities $\alpha_C = \alpha_D$. In view of Theorem 16.4, Poénaru's claim is equivalent to the fact that, for generic immersions between smooth manifolds, the part of the formula remaining after simplifications is always satisfiable.[11] The next example together with the next theorem show that the latter statement is not true.

Example 16.3 Let us consider the multigraph homomorphism shown in Fig. 16.4a. The graph $G_f^{(2)}$ (Fig. 16.4b) consists of 6 connected components that split into pairs with respect to the involution. Consequently, $p_2\colon |G_f^{(2)}| \to |\widetilde{G}_f^{(2)}|$ is trivial.

Note that for each triple (p_1, p_2, p_3), where p stands for a, b, c, or d, all the pairs (p_i, p_j), $i \neq j$ lie in different connected components of $G_f^{(2)}$. Each such triple

[11] This remark should be understood as an informal comment, since formally, the formulas Γ_f are defined only for non-degenerate simplicial maps; however, the fact that the problem of lifting to embeddings of generic immersions between manifolds reduces to the problem of lifting maps between graphs (see [21, pp. 1.21–1.22] and [5, Theorem 4.6]) makes this comment meaningful.

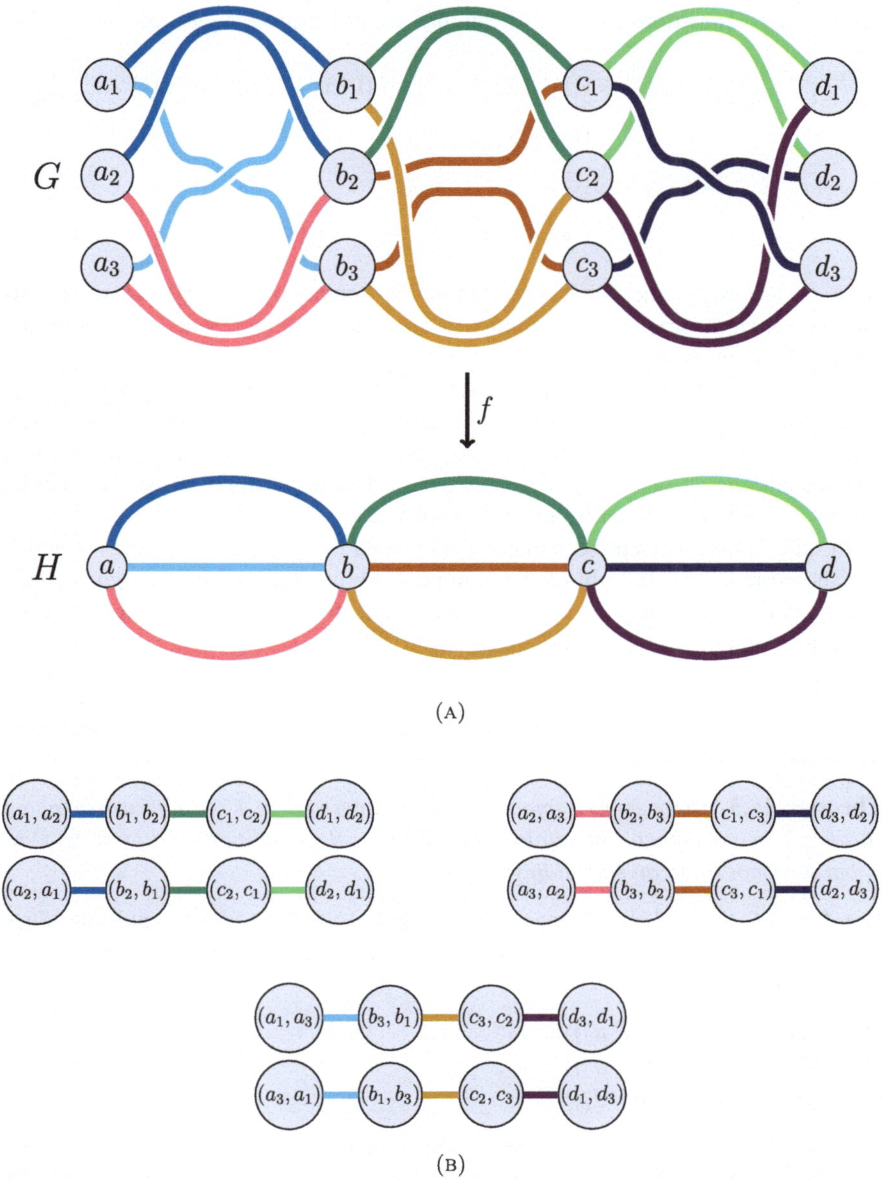

Fig. 16.4 An example of a map with "nontrivial" Γ_f. The edges of the graph G and their images in the graph H are coloured identically, the same holds for edges of $G_f^{(2)}$ and edges of G corresponding to them. Furthermore, edges of G with the same image are placed closely together, and the "upper" pairs of edges of G map to the "upper" edges of H; the same holds for the "middle" and the "lower" pairs. (**a**) The map $f : G \to H$. (**b**) The graph $G_f^{(2)}$

defines two disjunctions in Γ_f.[12] Given that, we can express Γ_f as follows:

$$\Gamma_f(x_1, x_2, x_3) = ((x_1 \wedge x_2) \to x_3) \wedge ((\neg x_1 \wedge \neg x_2) \to \neg x_3) \wedge$$
$$((x_1 \wedge x_2) \to \neg x_3) \wedge ((\neg x_1 \wedge \neg x_2) \to x_3) \wedge$$
$$((x_1 \wedge \neg x_3) \to x_2) \wedge ((\neg x_1 \wedge x_3) \to \neg x_2) \wedge$$
$$((x_1 \wedge \neg x_2) \to \neg x_3) \wedge ((\neg x_1 \wedge x_2) \to x_3)$$

Here x_1 represents the connected component containing (a_1, a_2), x_2 represents the connected component containing (a_2, a_3), and x_3 represents the connected component containing (a_1, a_3).

The first four clauses imply that either $x_1 = 1$ and $x_2 = 0$, or $x_1 = 0$ and $x_2 = 1$. In the former case, the next-to-last disjunction yields $x_3 = 0$, which contradicts $(x_1 \wedge \neg x_3) \to x_2$ as $x_2 = 0$. In the latter case, the last disjunction implies $x_3 = 1$, which contradicts $(\neg x_1 \wedge x_3) \to \neg x_2$ as $x_2 = 1$. Consequently, Γ_f is not satisfiable, and, as a result, f does not lift to an embedding.

In spite of this, it can be seen that in the formula Γ_f, there are no clauses in which literals are repeated. In light of the comments above, this means that equivalence classes under $\sim_\nu$, defined as it has been done for generic immersions before, coincide with the connected components of $|G_f^{(2)}|$.

Now, using the provided example, we will construct a counterexample to the claim of [21, Théorème 1] showing that its assumptions need to be strengthened in the sense of Theorem 16.4:

Theorem 16.5 *There exists a smooth generic immersion $j : T \looparrowright Y$ from a compact surface T to a compact 3-manifold Y that satisfies the conditions $\mu_2(j) = \nu_3(j) = 0$, but does not lift to an embedding.*

Proof Let $f : G \to H$ be the map from Example 16.3. We add three multiple edges e_1, e_2, e_3 to H, going from a to d. We also add six edges to G and define f on these edges as follows:

(1) $f(\{a_1, d_1\}) = f(\{a_2, d_2\}) = e_1$,
(2) $f(\{a_2, d_3\}) = f(\{a_3, d_2\}) = e_2$,
(3) $f(\{a_1, d_3\}) = f(\{a_3, d_1\}) = e_3$.

Let us denote the resulting graphs by the same letters G and H.

It can be observed that the covering map p_2 for the resulting map $f : G \to H$ remains trivial; moreover, the formula Γ_f also coincides with the formula of the original map.[13]

[12] Actually, each triple defines 6 disjunctions, one for each permutation of three points. However, after dropping equal disjunctions up to reordering terms, only two disjunctions remain.

[13] In fact, we have simply replaced the segments that make up the graph $G_f^{(2)}$ by joining their endpoints to form circles.

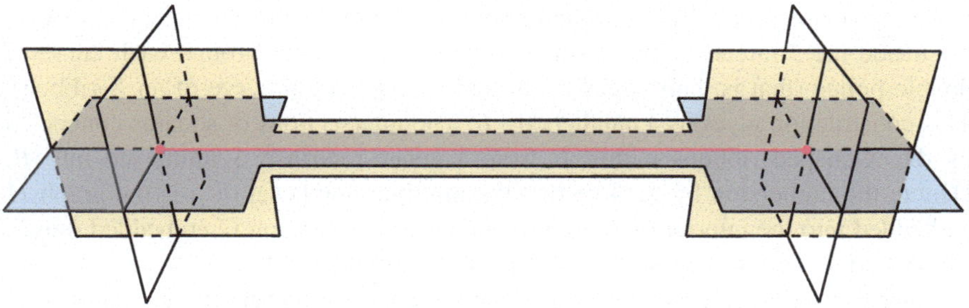

Fig. 16.5 Two triples of intersecting squares, representing triple points, and an X-ribbon between them, representing a double edge. The preimages of the pink double edge are an edge lying on the blue squares and the blue ribbon between them, and an edge lying on the orange squares and the orange ribbon

Now let us construct a surface S with boundary immersed into a handlebody B.

We represent each vertex a, b, c, d of H as an intersection of three squares with equal sides parallel to the coordinate planes, and label each square by the number from one to three. We denote the i-th square in the intersection corresponding to the vertex x by X_i. For example, the vertex $a \in V(H)$ is represented by the intersection of three squares A_1, A_2, and A_3. Furthermore, each square X_i corresponds to a vertex x_i of G. For instance, the squares A_1, A_2, and A_3 correspond to the vertices a_1, a_2, and a_3, respectively.

Thus, each vertex $x \in V(H)$ corresponds to a triple point $X_1 \cap X_2 \cap X_3$. From each such triple point, six half-edges start: two for each intersection $X_1 \cap X_2$, $X_1 \cap X_3$, and $X_2 \cap X_3$.

For each unordered pair of edges $\{x_i, y_j\}, \{x_k, y_l\} \in E(G)$ with the same image, where $f(x_i) = f(x_k) = x$ and $f(y_j) = f(y_l) = y$, we perform the following procedure.

We choose one of the half-edges from $X_i \cap X_k$ and one of the half-edges lying in $Y_j \cap Y_l$ that are currently "free" (that is, not chosen in previous steps). Then, we connect these half-edges by an X-shaped ribbon,[14] so that one plane of the X-shaped ribbon connects the side of the square X_i with the side of the square Y_j, and the other one connects the sides of the squares X_k and Y_l, see Fig. 16.5.

Note that each vertex of $\widetilde{G}_f^{(2)}$ has degree two. This ensures that there is always a free half-edge to choose. Moreover, after processing all the pairs of edges, each half-edge belongs to some X-shaped ribbon.

Observe that each X-shaped ribbon could be twisted while connecting. Additionally, when performing the steps described above, there is flexibility in choosing which half-edges to connect with each other. However, in our construction, this choice is not crucial; hence, we select free half-edges and twist the ribbons in an arbitrary way.

[14] That is, the product of two intersecting segments (forming the letter X) with another segment.

Next, we take small 3-balls around each triple point (that is, the points $X_1 \cap X_2 \cap X_3$ inside the 3-intersections of squares) and small 3-tubes around each curve of double points (that is, half-edges connected by the central curve of an X-ribbon). This construction gives us a handlebody B. The intersections of squares connected by the X-shaped ribbons within B form a generic surface S immersed into B. Denote the immersion by g. Note that the multiple points set of g is the graph H embedded into the interior of B, and its preimage is the graph G embedded into S. The boundary of S is a circle embedded into the boundary of B.

Furthermore, $g\big|_{|G|}$ acts exactly as f on the curves representing the edges of G embedded into S: g takes the curve between points x_i and y_j to the curve between points $f(x_i)$ and $f(y_j)$.

To obtain a closed 3-manifold Y, we take the double of B, which is two copies of B glued along their common boundary. The double of S, denoted as T, is then immersed into Y by the immersion j induced by $g \sqcup g \colon S \sqcup S \looparrowright B \sqcup B$. Clearly, the immersion j is in general position. The multiple point set of j is a graph $H \sqcup H$ embedded into Y, and its preimage is a graph $G \sqcup G$ embedded into T. Moreover, it is clear that $j\big|_{|G|\sqcup|G|}$ acts on the curves representing the edges of $G \sqcup G$ as $f \sqcup f$.

Let us note that the set $S_{2,3}$ in the definition of $\sim_\nu$ for the map g consists of 24 points. Each point of $S_{2,3}$ represents an ordered pair of vertices in G with the same image under g, or, equivalently, under f. In other words, the set $S_{2,3}$ coincides with the set of vertices of the graph $G_f^{(2)}$. Furthermore, it is straightforward to verify that the equivalence relation defined by the partition of the vertices of $G_f^{(2)}$ into connected components satisfies all the properties in the definition of $\sim_\nu$; therefore, these equivalence relations coincide. Given that, it becomes apparent that $\nu_3(g) = 0$. It is also evident that $\mu_2(g) = 0$ since $p_2 \colon |G_f^{(2)}| \to |\widetilde{G}_f^{(2)}|$ is trivial.

As all the multiple points of g lie in the interior of B, the set $S_{2,3}$ and the relation $\sim_\nu$ for j coincide with those of $g \sqcup g$. Therefore, since $\nu_3(g) = \mu_2(g) = 0$, we have $\nu_3(j) = \mu_2(j) = 0$.

If j lifts to an embedding then so does f. However, as shown in Example 16.3, the formula Γ_f is not satisfiable, and consequently, f does not lift to an embedding. □

It can be observed that Example 16.3 also provides a counterexample to [21, Lemme 1.7]. One can notice that the preimage of the cycle shown in [21, Fig. 1.3] does not necessarily have to be the two cycles as shown in [21, Fig. 1.4], as the paper claims, which is where the proof fails.

It is worth noting that the mentioned mistake only affects [21, Lemme 1.7], which states that a "labyrinth" (that is, a map f_j between graphs, obtained using a given immersion j of manifolds; see [21, Definition 3] for details) lifts if $\mu_2(f_j) = \nu_3(f_j) = 0$. However, the reduction from the considered case of immersions of manifolds to maps between graphs is itself correct. Thus, the immersion j lifts if and only if f_j lifts, or, equivalently, the covering map p_2 for f_f is trivial, and Γ_{f_j} is satisfiable. Our next goal is to show that these conditions cannot be weakened in the general case: we will demonstrate that for any suitable boolean formula Γ it is possible to construct a generic immersion j with $\Gamma_{f_j} = \Gamma$.

Notice that any Γ_f, as defined in the previous section, satisfies the following condition: once it contains a clause of the form $(\alpha \wedge \beta) \rightarrow \gamma$, it also contains a clause $(\neg\alpha \wedge \neg\beta) \rightarrow \neg\gamma$, up to double negation elimination. Additionally, let us note that the formulas that cannot be simplified by the algorithm described at the end of the previous section are those where, for each clause $(\alpha \wedge \beta) \rightarrow \gamma$, all variables constituting α, β, and γ are distinct.

The next theorem shows that any Γ of the form described above can be realised as Γ_f of a multigraph homomorphism f induced by a generic immersion of a surface into a handlebody, similar to what we have constructed in the previous theorem.

Theorem 16.6 *Let $\Gamma(x_1, \ldots, x_n)$ be a boolean formula of the form*

$$\Gamma(x_1, \ldots, x_n) = \bigwedge_{j=1}^{m} \left((\alpha_j \wedge \beta_j) \rightarrow \gamma_j\right)$$

where α_j, β_j, and α_j are variables or their negations. Assume that for every j

(1) the variables in the literals α_j, β_j, and γ_j are distinct,
(2) there exists j' such that $\alpha_{j'} = \neg\alpha_j$, $\beta_{j'} = \neg\beta_j$, and $\gamma_{j'} = \neg\gamma_j$ (after double negation elimination).

Then there exists a generic immersion $g\colon S \looparrowright B$ of a surface S with boundary into a handlebody B such that the restriction of g onto the preimage of the set of multiple points induces a homomorphism $f\colon G \rightarrow H$ of multigraphs. Additionally, the formula Γ_f is equivalent to Γ, differing only in variable renaming, the presence of duplicate clauses, and reordering of clauses and literals within clauses.

To prove the theorem, we will use the following lemma, which says, informally, that for each Γ as in Theorem 16.6, we can construct a multigraph homomorphism $f\colon G \rightarrow H$, whose Γ_f is equivalent to Γ, and f is similar to that we used in the construction of Theorem 16.5 which allows us, following this construction, to obtain the needed immersion $g\colon S \looparrowright B$.

Lemma 16.2 *Let $\Gamma(x_1, \ldots, x_n)$ be the same boolean formula as the one in Theorem 16.6.*

Then, there is a multigraph homomorphism $f\colon G \rightarrow H$, such that

(1) $p_2\colon |G_f^{(2)}| \rightarrow |\widetilde{G}_f^{(2)}|$ is trivial,
(2) Γ_f is equivalent to Γ in the sense of Theorem 16.6,
(3) the edges of H have two preimages,
(4) the vertices of H have three preimages,
(5) the vertices of G have degree 4,
(6) the vertices of H have degree 6,
(7) the vertices of the graph $G_f^{(2)}$ have degree two, or, equivalently, $G_f^{(2)}$ is a disjoint union of circles.

Proof First note that all disjunctions that constitute Γ split into pairs $(\alpha \wedge \beta) \to \gamma$, $(\neg\alpha \wedge \neg\beta) \to \neg\gamma$. After selecting only one disjunction from each pair, let us renumber them from 1 to $m/2$. Thus, we have $m/2$ disjunctions $(\alpha_1 \wedge \beta_1) \to \gamma_1, \ldots, (\alpha_{m/2} \wedge \beta_{m/2}) \to \gamma_{m/2}$.

We will construct the graphs G and H and the map $f : G \to H$ iteratively. To begin with, let G and H be empty graphs.

For each disjunction $(\alpha_j \wedge \beta_j) \to \gamma_j$, add three vertices v_1^j, v_2^j, and v_3^j to the graph G, and add one vertex v^j to the graph H, and define a map $f(v_i^j) = v^j$ on the new vertices v_j^i.

Next, we define a set S_i for each variable x_i, which, at the end of the procedure described below, will contain all the vertices of the connected component associated with the variable x_i. For simplicity, if $y = x_i$ is a variable, denote the corresponding set S_i by $S(y)$.

Initially, let all $S(y)$ be empty. Then, for each of the $m/2$ disjunctions $(\alpha_j \wedge \beta_j) \to \gamma_j$, we perform the following steps:

(1) If $\alpha_j = x$, add (v_1^j, v_2^j) to $S(x)$. Otherwise, if $\alpha_j = \neg x$, add (v_2^j, v_1^j) to $S(x)$.
(2) Similarly, if $\beta_j = y$, add (v_2^j, v_3^j) to $S(y)$. If $\beta_j = \neg y$, add (v_3^j, v_2^j) to $S(y)$,
(3) Finally, if $\gamma_j = z$, add (v_1^j, v_3^j) to $S(z)$. If $\gamma_j = \neg z$, add (v_3^j, v_1^j) to $S(z)$.

After performing these steps, for each j and for any $1 \le i \ne k \le 3$, exactly one of the pairs (v_i^j, v_k^j) and (v_k^j, v_i^j) belongs to some set S_l. Furthermore, $S_l \cap S_t = \varnothing$ for $l \ne t$, and $S_l \cap \tau(S_t) = \varnothing$ for any l and t. Here $\tau(S_t) = \{(y, x) \mid (x, y) \in S_t\}$ is the set of "reversed" pairs in S_t.

We proceed by choosing arbitrary linear orders on each set S_l. Using these orders, we will add edges to G and H as follows.

Let $S_l = \left\{(v_{i_1}^{j_1}, v_{k_1}^{j_1}), (v_{i_2}^{j_2}, v_{k_2}^{j_2}), \ldots, (v_{i_s}^{j_s}, v_{k_s}^{j_s})\right\}$. For each $r = 1, \ldots, s-1$ we add a new edge $e_{l,r}$ between v^{j_r} and $v^{j_{r+1}}$ in H.[15] We add two edges $e_{l,r}^1$ and $e_{l,r}^2$ connecting $v_{i_r}^{j_r}$ with $v_{i_{r+1}}^{j_{r+1}}$, and $v_{k_r}^{j_r}$ with $v_{k_{r+1}}^{j_{r+1}}$, respectively, to H, and set $f(e_{l,r}^1) = f(e_{l,r}^2) = e_{l,r}$.

Additionally, we add an edge $e_{l,s}$ between v^{j_s} and v^{j_1} to H, and two edges $e_{l,s}^1$ and $e_{l,s}^2$ between $v_{i_s}^{j_s}$ and $v_{i_1}^{j_1}$, and between $v_{k_s}^{j_s}$ and $v_{k_1}^{j_1}$, respectively, to G. We set $f(e_{l,s}^1) = f(e_{l,s}^2) = e_{l,s}$.

It is evident that at each step, two edges $((v_{i_r}^{j_r}, v_{k_r}^{j_r}), (v_{i_{r+1}}^{j_{r+1}}, v_{k_{r+1}}^{j_{r+1}}))$ and $((v_{k_r}^{j_r}, v_{i_r}^{j_r}), (v_{k_{r+1}}^{j_{r+1}}, v_{i_{r+1}}^{j_{r+1}}))$ are added to $G_f^{(2)}$. Thus, after processing a single set S_l, we obtain in $G_f^{(2)}$ two circles on the vertices of S_l and $\tau(S_l)$. Therefore, after processing all the sets S_l, the graph $G_f^{(2)}$ consists of a disjoint union of circles on the vertices of the sets S_l and $\tau(S_l)$.

[15] If there are edges between v^{j_r} and $v^{j_{r+1}}$ in H, we add a new one.

It is clear that $p_2\colon |G_f^{(2)}| \to |\widetilde{G}_f^{(2)}|$ is trivial because $S_l \cap \tau(S_l) = \varnothing$ for any S_l.

Additionally, if we associate each S_l to a variable x_l, it can be seen that each triple $\{v_1^j, v_2^j, v_3^j\}$ contributes two different disjunctions $(\alpha_j \wedge \beta_j) \to \gamma_j$ and $(\neg\alpha_j \wedge \neg\beta_j) \to \neg\gamma_j$ to Γ_f. Actually, it adds 6 disjunctions, one for each permutation of the triple, but four of them can be derived from the other two by permutation of their terms, so we consider them as duplicates.

Therefore, since these and only these disjunctions form the formula Γ, the formulas Γ_f and Γ become equivalent after elimination of the duplicates.

If follows from the construction that the vertices of H have 3 preimages, and the edges of H have 2 preimages. Furthermore, from the latter fact it follows that the map $\hat{f}\colon \widetilde{G}_f^{(2)} \to H$ induced by $f\colon G \to H$ is injective on the edges; thus, because each vertex v of H have 3 preimages under $\hat{f}$ (that is, unordered pairs of $f^{-1}(v)$), and each such preimage have degree two,[16] the vertices of H have degree six.

Finally, let us note that each vertex x of G is included into exactly 2 vertices of $\widetilde{G}_f^{(2)}$ (recall that they are unordered pairs of distinct vertices of G with the same image), and both of them have degree two. The construction provides that each such vertex corresponds to a pair of edges in G incident with the vertex x. Therefore, the vertices of G have degree four. □

Proof of Theorem 16.6 By using Lemma 16.2 we obtain a multigraph homomorphism $f\colon G \to H$ whose formula Γ_f is equivalent to Γ. To construct a generic immersion $g\colon S \looparrowright B$ let us follow the construction of $g\colon S \looparrowright B$ in Theorem 16.5.

To see that it is applicable in this case, let us note that each vertex v of H has degree 6, and each unordered pair of its distinct preimages form a vertex in $\widetilde{G}_f^{(2)}$ of degree 2, hence it is possible to represent v as an intersection of 3 squares, and assign a half-edge coming from the triple point of this intersection to each edge in $\widetilde{G}_f^{(2)}$ incident to the pair of preimages of v. □

It is worth noting that B can be embedded in $\mathbb{R}^3$. Thus, a natural question arises: for which examples is it possible to glue the boundary of S to a surface embedded in $\mathbb{R}^3 \setminus B$ such that we obtain a generic immersion $S \looparrowright \mathbb{R}^3$ that realises Γ?[17]

In some cases, such as the examples from Theorem 16.5 and certain examples obtained from the construction in Example 16.3 by joining the endpoints of the connected components of $G_f^{(2)}$ to obtain circles, it can be shown that it is not possible to glue the boundary of S in such a way. One can demonstrate this by checking that the boundary of S is not null-homologous in $\mathbb{R}^3 \setminus H$. With a computer, checking this is straightforward, using a sort of Wirtinger presentation for $\pi_1(\mathbb{R}^3 \setminus H, \mathbb{Z})$ as described in [4, Proposition 9.1.9].

[16] Recall that $G_f^{(2)}$ is a disjoint union of circles, so $\widetilde{G}_f^{(2)}$ is as well.

[17] Note that this question is different from the [5, Proposition 4.19] and [13], where it is proven that an appropriate graph H can be realised as the multiple point set of a generic immersion $S \looparrowright \mathbb{R}^3$.

16.4 Approximation by Embeddings

In this section, we will establish a connection between the existence of a lifting to an embedding and the concept of approximation by embeddings for generic maps. We will define the notion of generic maps in our context and show that the condition of the triviality of the covering map p_2 becomes sufficient for the existence of a lifting in the special case of generic maps from trees to segments.

Before we proceed, let us introduce some definitions.

Definition (Approximability by Embeddings) Let $f: G \to H$ be a simplicial map between graphs, and let $i: |H| \hookrightarrow S$ be a continuous embedding of $|H|$ into a surface S with a metric d. We say that f is *approximable by embeddings* if for every $\varepsilon > 0$, there exists a continuous embedding $j: |G| \hookrightarrow S$ such that $d(j(x), i(|f|(x))) < \varepsilon$ for every $x \in |G|$.

It is worth noting that the approximability of f generally depends on the embedding i. However, as we will demonstrate later, it does not depend on the choice of a metric on S. Moreover, as we will show later, the approximability of f is equivalent to the existence of a compatible with f embedding of $|G|$ into a neighborhood of $i(|H|)$ in S, and, therefore, depends only on the geometry of this neighborhood.

Definition (Ribbon Graph, See [8, Definition 1.5] or [18, Definition 2.4]) *A ribbon graph* $\mathcal{H}$ is a surface with boundary represented as the union of two sets of discs: a set $V(\mathcal{H})$ of vertices or 0-handles, and a set $E(\mathcal{H})$ of edges or 1-handles, satisfying the following conditions:

(1) $D_v \cap D_w = \varnothing$ and $D_e \cap D_g = \varnothing$ for different vertices $v, w \in V(\mathcal{H})$ and different edges $e, g \in E(\mathcal{H})$,
(2) for each edge D_e there are exactly two vertices D_v and D_w that intersect D_e; moreover, $D_e \cap D_v$ and $D_e \cap D_w$ are two disjoint arcs lying in the boundary of D_e, and in the boundaries of D_v and D_w, respectively.

By [12, Theorem 1.3], an embedded graph $f: |H| \to S$ lies in a triangulation of S; that is, there are a homeomorphism $h_s: S \to |K_s|$, a piecewise linear homeomorphism $g_s: |H| \to |H_s|$, where H_s subdivides H, and a simplicial embedding $f_s: H_s \hookrightarrow K_s$, such that $f = h_s^{-1} \circ |f_s| \circ g_s$.[18] Recall that the surface S admits a unique piecewise linear structure.[19] Thus, $|f_s|$ is uniquely defined by f up to piecewise linear homeomorphisms of $|H_s|$ and $|K_s|$.

[18] In fact, by applying [12, Theorem 1.3] we obtain not a simplicial complex, but a polyhedron with a graph embedded as a 1-subpolyhedron; however, it is easy to obtain a piecewise linear homeomorphism $g_s: |H| \to |H_s|$ and a simplicial embedding $f_s: H_s \to K_s$ by taking suitable triangulations of these polyhedra.

[19] See, for example, [20, Theorem 5, Chapter 8].

Therefore, given an embedding $f\colon |H| \to S$, let the surface $\mathcal{H}$ be a regular neighborhood of $f(|H|)$ in S, decomposed into discs around the images of the vertices of H (these discs form the vertices of the ribbon graph $\mathcal{H}$), and strips around the edges of H (these strips, similarly, form the edges of $\mathcal{H}$).[20]

Explicitly, assume we have a simplicial embedding $f_s\colon H_s \hookrightarrow K_s$. Take the second derived subdivisions H_s'' and K_s'' of H_s and K_s, respectively, and denote by f'' the induced simplicial embedding $f''\colon H_s'' \hookrightarrow K_s''$. Denote by $N(H_s'', K_s'')$ the simplicial neighborhood of $f''(H_s'')$ in K_s'', that is, the simplicial complex consisting of the simplices that meet $f(H_s'')$, with their faces. Hence $|N(H_s'', K_s'')|$ is a regular neighborhood of $f(H)$ in S. As discussed earlier, we define $\mathcal{H}$ to be $|N(H_s'', K_s'')|$.

To obtain a decomposition of $\mathcal{H}$ into edges and vertices, for each vertex $v \in V(H)$, let us consider its image under the composition $F = |f''| \circ g\colon |H| \to |H_s''| \hookrightarrow |K_s''|$, where g is a homeomorphism $|H| \cong |H_s''|$, and take the star of $F(v)$ in $N(H_s'', K_s'')$ as the disc $D_v \in V(\mathcal{H})$.[21]

It can be seen that these discs split $\mathcal{H}$ into a set of strips, with each strip intersecting $f(|H|)$ along exactly one curve, which is a portion of the image of one of the edges of H. Thus, each edge e of the graph H corresponds to a strip D_e containing a portion of its image. Let us take these strips as edges of the ribbon graph $\mathcal{H}$.

Since a regular neighborhood of a subpolyhedron is uniquely defined up to piecewise linear homeomorphism, see [24, Theorem 3.8], a ribbon graph of a graph embedded into a surface is also uniquely defined.

Furthermore, it follows that there is a natural bijection between the vertices and the edges of a graph H embedded into a surface S, and the vertices and the edges of an induced ribbon graph $\mathcal{H}$. Thus, taking this into account, we will label the vertices and the edges of H and $\mathcal{H}$ by the same letters; for example, a vertex $v \in V(H)$ corresponds to a vertex $D_v \in V(\mathcal{H})$, and an edge $e = \{v, w\} \in E(H)$ corresponds to an edge $D_e = D_{\{v,w\}} \in E(\mathcal{H})$.

It is also important to note that, although our definition of a ribbon graph induced by an embedding differs from the definition of the normal neighborhood in [18], the ribbon graph $\mathcal{H}$ in our construction is also a normal neighborhood in the sense of [18, Definition 2.4]; indeed, all conditions from the latter definition, except the second one, are automatically satisfied. However, it is easy to see that for each edge e incident to a vertex v of H, its image $f(|e|)$ intersects the boundary of D_v at exactly one point. Therefore, it follows that the condition (2) from [18, Definition 2.4] is also satisfied. This allows us to use results from [18], using ribbon graphs in place of normal neighborhoods.

Definition ([18, Definition 2.7]) Let $f\colon G \to H$ be a simplicial map between graphs, and $\mathcal{H}$ be a ribbon graph corresponding to an embedding of $|H|$ into a surface.

[20] For the definitions of regular neighborhoods, derived subdivisions, and simplicial neighborhoods, the reader may refer to [24].

[21] Note that the image $F(v)$ is a vertex of K_s'' since H_s'' subdivides H.

We say that f is *R-approximable by an embedding* if there exists a continuous embedding $j: |G| \hookrightarrow \operatorname{Int} \mathcal{H}$ called an *R-approximation*, satisfying the following conditions:

(1) for a vertex $v \in V(G)$, $j(v) \in D_{f(v)} \in V(\mathcal{H})$,
(2) if $f(\{v, w\})$ is a vertex of H, then $j(|\{v, w\}|) \subset D_{f(\{v,w\})}$, where $|\{v, w\}|$ is a segment representing the edge $\{v, w\}$ in $|G|$,
(3) if $f(\{v, w\})$ is an edge of H, then $j(|\{v, w\}|)$ is a simple arc intersecting $D_{f(v)} \cap D_{\{f(v), f(w)\}})$ at a single point p and $D_{f(w)} \cap D_{\{f(v), f(w)\}}$ at a single point q.[22]

If j is a general position map $j: |G| \to \operatorname{Int} \mathcal{H}$ satisfying these conditions, we call j a *generic R-approximation* of f.

Lemma 16.3 ([18, Propositions 2.9]) *Let $f: G \to H$ be a simplicial map between graphs, and let $i: |H| \hookrightarrow S$ be an embedding of H into a surface S with a metric. Then the following statements are equivalent:*

(1) f is approximable by embeddings.
(2) f is R-approximable by an embedding.

From this point onwards, we focus on stable maps between graphs. The definition provided below is a one-dimensional simplicial version of the definition of piecewise linear stable maps from [16, Appendix B].

Definition Let $f: G \to H$ be a simplicial map between graphs.

We call a vertex $v \in G$ *regular* with respect to f if f is bijective on the star $\operatorname{st}(v) = \bigcup_{e \in E(G), v \in e} e$.

The map f is called *stable* if, for any vertex $w \in V(H)$,

(1) if $\deg(w) \neq 2$, all the vertices in $f^{-1}(w)$ are regular,
(2) if $\deg(w) = 2$, $f^{-1}(w)$ contains at most one non-regular vertex.

If a simplicial map f is stable, it must also be non-degenerate. Indeed, if $f(e) = v$, where e is an edge and v is a vertex, the endpoints of e are non-regular with respect to f, implying that f is not stable.

Now we are ready to state the theorem.

Theorem 16.7 *Let $f: G \to J$ be a stable simplicial map from a graph G to a path graph J (that is, a triangulation of a segment). Let $i: |J| \hookrightarrow \mathbb{R}^2$ be an embedding. Then the following statements are equivalent:*

(1) f is approximable by embeddings with respect to $i: |J| \to \mathbb{R}^2$.
(2) $|f|$ lifts to an embedding.

Let us start by proving a simple auxiliary lemma.

[22] Thus, the points $x = j^{-1}(p)$ and $y = j^{-1}(q)$ decompose $|\{v, w\}|$ into three subsegments $|\{v, x\}|$, $|\{x, y\}|$, and $|\{y, w\}|$, where $j(|\{v, x\}|) \subset D_{f(v)}$, $j(|\{x, y\}|) \subset D_{\{f(v), f(w)\}}$, and $j(|\{y, w\}|) \subset D_{f(w)}$.

Lemma 16.4 *Let S be a rectangular region $[C, D] \times [H, L]$ in $\mathbb{R}^2$. Let α and β denote the sides $\{C\} \times [H, L]$ and $\{D\} \times [H, L]$, respectively. Let γ be a simple curve connecting a point $a \in \alpha$ and a point $b \in \beta$, such that $\gamma \cap (\alpha \sqcup \beta) = \{a, b\}$. Assume Φ is a path-connected subset of S not intersecting γ, such that $\Phi \cap \alpha$ and $\Phi \cap \beta$ are finite sets of points.*

Then, for all $p \in (\Phi \cap \alpha) \sqcup (\Phi \cap \beta)$, either $p_y < s(p)_y$ or $p_y > s(p)_y$ holds simultaneously. Here $s(p) = a$ if $p \in \alpha$, $s(p) = b$ if $p \in \beta$, and q_y denotes the y-coordinate of a point q.

Proof Suppose there exists a pair of points $p, q \in (\Phi\cap\alpha)\sqcup(\Phi\cap\beta)$ such that $p_y < s(p)_y$ and $q_y > s(q)_y$, or $p_y > s(p)_y$ and $q_y < s(q)_y$. Since Φ is path-connected and does not intersect γ, there must be a path ϕ in Φ (and thus in $S\setminus\gamma$) connecting p and q. However, according to [20, Theorem 8, Chapter 2], $S\setminus\gamma$ consists of two path-connected components with p and q lying in different components. This contradicts the existence of the path ϕ. □

Proof of Theorem 16.7 First, let us number the vertices of J according to an arbitrary orientation on J and denote them by $v_0, \dots, v_N$.

Observe that any embedding $i\colon |J| \hookrightarrow \mathbb{R}^2$ induces the same ribbon graph $\mathcal{J}$ up to a piecewise linear homeomorphism.[23] Thus, it follows from Lemma 16.3 that the approximability of f by embeddings does not depend on the embedding i.

Therefore, for convenience, we can assume that i embeds $|J|$ into $\mathbb{R}^2$ with the Euclidean distance d_2 as a line segment with $i(v_k) = \left(2k + \frac{1}{2}, 0\right)$, and $\mathcal{J}$ is a ribbon $[0, 2N + 1] \times [-1, 1]$ in $\mathbb{R}^2$, where $D_{v_k} = [2k, 2k + 1] \times [-1, 1]$ and $D_{\{v_k,v_{k+1}\}} = [2k + 1, 2(k + 1)] \times [-1, 1]$.[24]

Now suppose that $|f|$ lifts to an embedding $\widetilde{|f|}\colon |G| \to |J| \times \mathbb{R}$. Let M be the maximum of $|h(p)|$, where $h(p) = \mathrm{pr}_{\mathbb{R}} \widetilde{|f|}(p)$. For any $\varepsilon > 0$, let $F\colon |G| \to \mathbb{R}^2$ be the composition $S_\varepsilon \circ \widetilde{|f|}$ where $S_\varepsilon(q, y) = \left(i(q)_x, \frac{\varepsilon y}{2M}\right)$; here $i(q)_x$ stands for the x-coordinate of $i(q)$. Therefore, putting $q = f(p)$, we have

$$d_2\left(F(p), i(q)\right) = d_2\left(\left(i(q)_x, \frac{\varepsilon h(p)}{2M}\right), (i(q)_x, 0)\right) = \left|\frac{\varepsilon h(p)}{2M}\right| < \varepsilon.$$

Thus, f is approximable by embeddings.

Now let us prove the reverse implication. Suppose f is approximable by embeddings. By Lemma 16.3, there exists an R-approximation $j\colon |G| \hookrightarrow \operatorname{Int}\mathcal{J}$. We are going to define orders on the sets $f^{-1}(v_k)$ and $f^{-1}(\{v_k, v_{k+1}\})$. It will be more convenient to start with the sets $f^{-1}(\{v_k, v_{k+1}\})$.

Consider the images of the edges from $f^{-1}(\{v_k, v_{k+1}\})$ in the rectangular regions $D_{\{v_k,v_{k+1}\}}$. It follows from the definition of R-approximation that $j(|e|) \cap D_{\{v_k,v_{k+1}\}}$

[23] Given two distinct ribbon graphs $\mathcal{J}$ and $\mathcal{J}'$, we can take arbitrary homeomorphisms between arcs $D_v \cap D_e$ and $D'_v \cap D'_e$ and then extend them to a homeomorphism $\mathcal{J} \cong \mathcal{J}'$ that preserves the handle decompositions of $\mathcal{J}$ and $\mathcal{J}'$.

[24] $\mathcal{J}$ collapses onto $i(|J|)$, hence it is a regular neighborhood by [24, Corollary 3.30].

for an edge $e \in f^{-1}(\{v_k, v_{k+1}\})$ is a simple curve γ_e going from the left side of $D_{\{v_k,v_{k+1}\}}$ to its right side, and intersecting the boundary of $D_{\{v_k,v_{k+1}\}}$ only at the endpoints. Hence, by applying Lemma 16.4, for each pair of edges $e, g \in f^{-1}(\{v_k, v_{k+1}\})$, the endpoints of γ_e have both either larger or smaller y-coordinates than the endpoints of γ_g lying in the corresponding sides of the rectangle. Therefore, we can define a linear order by putting $e \prec_{k,k+1} g$ if both endpoints of γ_e have smaller y-coordinates than the corresponding endpoints of γ_g.

Now let us proceed to the sets $f^{-1}(v_k)$. The situation with them, although slightly more complex, is essentially the same. In what follows, we put $\gamma_w = j(|\operatorname{st} w|) \cap D_{f(w)}$ for a vertex $w \in V(G)$.

First note that, since f is stable, all the vertices in $f^{-1}(v_0)$ and $f^{-1}(v_N)$ have degree one, and for any $w \in f^{-1}(v_0)$ the set γ_w is a simple curve connecting $j(w)$ with a point on the right side of ∂D_{v_0}, and this point is the only intersection of γ_w with the boundary of D_{v_0}. Therefore, we can order the vertices $\{w_i\}$ in $f^{-1}(v_0)$ with respect to the order of the y-coordinates of the points $\gamma_{w_i} \cap \partial D_{v_0}$, which gives us a linear order $\prec_0$ on $f^{-1}(v_0)$.

The same holds true for vertices from the set $f^{-1}(v_N)$, with the only difference being that the curves γ_w for $w \in f^{-1}(v_N)$ connect $j(w)$'s with points on the left side of D_{v_N}. As for $f^{-1}(v_0)$, we can order the vertices in $f^{-1}(v_N)$ according to the order of the y-coordinates of the points $\gamma_w \cap \partial D_{v_N}$.

Now consider the set $f^{-1}(v_k)$, where $0 < k < N$. Since f is stable, for any pair of vertices $w, s \in f^{-1}(v_k)$, one of the sets γ_w and γ_s is a simple curve connecting a point on the left side of D_{v_k} and a point on its right side, and intersecting the boundary of D_{v_k} only at these two points. At the same time, the other set is clearly path-connected, and it intersects the boundary of D_{v_k} in a finite number of points, these points will be referred to as "endpoints". Therefore, by applying Lemma 16.4 we get that the y-coordinates of the endpoints of γ_w are all larger or all smaller than the y-coordinates of the endpoints of γ_s once we compare only the pairs of points lying on the same side of D_{v_k}. Hence we can define a linear order $\prec_k$ by letting $w \prec_k s$ if all the endpoints of γ_w lying on the left and right sides of D_{v_k} have the smaller y-coordinates than all the endpoint of γ_s lying on the same sides.

Finally, we are going to prove that $\{\prec_k,\ 0 \le k \le N\}$ forms an admissible collection of linear orders, and therefore, $|f|$ lifts to an embedding according to Theorem 16.2. Indeed, take a pair of edges $e = \{w_e, s_e\}, g = \{w_g, s_g\}$ of G such that $f(w_e) = f(w_g) = v_k$ and $f(s_e) = f(s_g) = v_{k+1}$. Without loss of generality, assume $e \prec_{k,k+1} g$. This implies that the "left" endpoint of γ_e, lying on the left side of $D_{\{v_k,v_{k+1}\}}$, has a smaller y-coordinate than the "left" endpoint of γ_g. However, the "left" endpoints of γ_e and γ_g are endpoints of γ_{w_e} and γ_{w_g}, lying on the same side of D_{v_k}; hence, $w_e \preceq_k w_g$. Repeating this reasoning for the "right" endpoints of γ_e and γ_g, we conclude that $s_e \preceq_{k+1} s_g$. Therefore, the orders $\prec_k$ form an admissible collection of linear orders. □

The next example shows that the assumption that f is stable in Theorem 16.7 is needed.

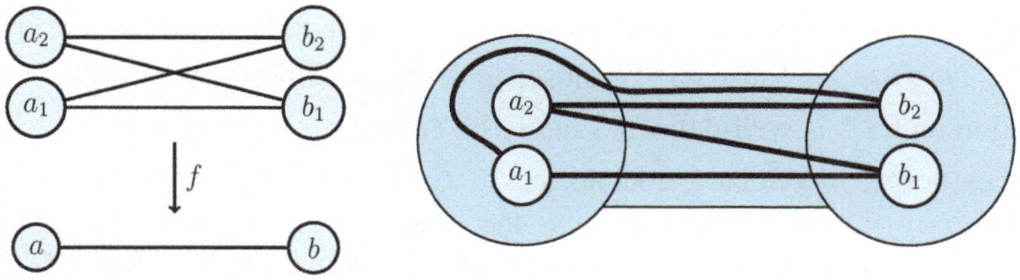

Fig. 16.6 $f: G \to I \hookrightarrow \mathbb{R}^2$ and its R-approximation

Example 16.4 Let f be a map $G \to J$, where G triangulates the circle S^1, that is shown on the left-hand side of Fig. 16.6. It is R-approximable by embeddings, see the right-hand side of Fig. 16.6, therefore, it is approximable by embeddings. However, it does not lift to an embedding, since there is a 2-obstructor for f defined as the path $(a_1, a_2) \to (b_1, b_2) \to (a_2, a_1)$ in $G_f^{(2)}$.

Now we use some known results on the problem of existence of approximation by embeddings to obtain sufficient conditions for the existence of a lifting to an embedding in the case of stable maps from trees to segments.

Definition Let $f: G \to H$ be a simplicial map between graphs, let $i: |H| \hookrightarrow S$ be an embedding of H into a surface S, and let $\mathcal{H}$ be a corresponding ribbon graph. We call a generic R-approximation $j: |G| \to \operatorname{Int} \mathcal{H}$ a $\mathbb{Z}_2$*-approximation* if for any pair of non-adjacent edges $e, g \in E(G)$ (i.e., $e \cap g = \varnothing$) the set $j(|e|) \cap j(|g|)$ is finite and consists of an even number of points.

f is called $\mathbb{Z}_2$*-approximable* if there exists a $\mathbb{Z}_2$-approximation of f.

The following theorem is proved in [9] (in general case) and in [27] (for the case when T is a trivalent tree):

Theorem 16.8 ([9, Corollary 3] and [27, Theorem 1.5]) *Let $f: T \to J$ be a simplicial map from a tree T to a path graph J, and assume that J is embedded into $\mathbb{R}^2$. If f is $\mathbb{Z}_2$-approximable then f is approximable by embeddings.*

The variant of van Kampen obstruction for $\mathbb{Z}_2$-approximability of f by embeddings described below is taken from [27] and [23].

Assume we have a simplicial map $f: G \to H$ between graphs G and H, and H is embedded into a surface S. Let $G^{\square}$ be the two-dimensional cubical complex $\bigcup\{e \times g \mid e, g \in E(G): e \cap g = \varnothing\}$. Clearly, the transposition $\tau \in S_2$ acts on $|G^{\square}|$ by $\tau(p \times q) = q \times p$. Let $\widetilde{G}^{\square}$ be a two-dimensional cell complex $G^{\square}/S_2$, whose cells are images of the cubes of $G^{\square}$ under the quotient map. Therefore, 2-cells of $\widetilde{G}^{\square}$ might be associated with unordered pairs of disjoint edges of G.

Take any generic R-approximation $j: |G| \to \mathcal{H}$ of f, where $\mathcal{H}$ stands for a ribbon graph corresponding to the embedding $|H| \hookrightarrow S$; by "generic" here we mean that j is injective on $|V(G)|$, and for any pair of edges $e, g \in E(G)$ the set $j(|e|) \cap j(|g|)$ consists of a finite number of points.

Let $G_f^\square \subset G^\square$ be the subcomplex $\bigcup\{e \times g \mid e, g \in E(G) : e \cap g = \varnothing, f(e) \cap f(g) = \varnothing\}$. Since $|G_f^\square|$ is closed under the involution on $|G^\square|$, the induced involution on $|G_f^\square|$ is well-defined. Let $\widetilde{G}_f^\square$ be $G_f^\square/S_2$.

Let us define a cellular 2-cochain $c_v \in C^2(\widetilde{G}^\square; \mathbb{Z}_2)$ by putting $c_v(e \cdot g) = |j(|e|) \cap j(|g|)| \bmod 2$, where $e \cdot g$ is the image of $e \times g$ under the projection $G^\square \to \widetilde{G}^\square$. Observe that c_v is in fact a relative cochain lying in $C^2(\widetilde{G}^\square, \widetilde{G}_f^\square; \mathbb{Z}_2)$. Indeed, for edges $e, g \in E(G)$ with $f(e) \cap f(g) = \varnothing$, the curves $j(|e|)$ and $j(|g|)$ lie in different discs with respect to the chosen generic R-approximation and therefore do not intersect. Let $v_f = [c_v] \in H^2(|\widetilde{G}^\square|, |\widetilde{G}_f^\square|; \mathbb{Z}_2)$ be the corresponding cohomology class.

It can be shown that v_f is well-defined by following [25, Lemmas 3.2–3.5]. These lemmas consider the case of the classical van Kampen obstruction, but it is easy to adapt them to our case.[25]

Theorem 16.9 ([27, Proposition 3.1]) *Suppose $f : G \to H$ is a simplicial map between graphs G and H, and H is embedded into an oriented surface S. Then $v_f = 0$ if and only if f is $\mathbb{Z}_2$-approximable.*

Now let us introduce a van Kampen-like obstruction for the existence of a lifting. We start with a cochain-free definition of the corresponding cohomology class inspired by [3] and [14]. Next, we present an explicit cochain representing it, defined in a way similar to the cochain c_v defining v_f.

Recall that f induces the principal S_2-bundle $p_2 : |G_f^{(2)}| \to |\widetilde{G}_f^{(2)}|$. Let $g : |\widetilde{G}_f^{(2)}| \to \mathbb{R}P^\infty$ be a map classifying it.[26] Let w_1 be the generator of $H^1(\mathbb{R}P^\infty; \mathbb{Z}_2) \cong \mathbb{Z}_2$. We define $w_f \in H^1(|\widetilde{G}_f^{(2)}|, \mathbb{Z}_2)$ as the pullback $g^* w_1$.[27]

It follows from the definition that w_f is the first Stiefel–Whitney class of the one-dimensional real vector bundle associated with the covering p_2.[28] Indeed, this vector bundle is a pullback under g of the canonical line bundle over $\mathbb{R}P^\infty$, and the latter bundle is associated with the universal bundle $S^\infty \to \mathbb{R}P^\infty$. Therefore, $w_f = 0$ if and only if the bundle associated with p_2 is orientable, that is, admits a non-vanishing section. Hence $w_f = 0$ if and only if p_2 is trivial.

Now let $\widetilde{f} : |G| \to |H| \times \mathbb{R}$ be a generic lifting of f; as before, the word "generic" means that $\widetilde{f}$ is injective on $V(G)$, and for any pair of edges $e, g \in E(G)$, the set $\widetilde{f}(|e|) \cap \widetilde{f}(|g|)$ consists of a finite number of points. Let c_w be a cochain in $C^1(\widetilde{G}_f^{(2)}; \mathbb{Z}_2)$ that takes the value $|\widetilde{f}(|e|) \cap \widetilde{f}(|g|)| \bmod 2$ on an edge $e \cdot g \in E(\widetilde{G}_f^{(2)})$,

[25] Note that v_f coincides with the classical van Kampen obstruction for graphs when $f : G \to \{p\} \hookrightarrow \mathbb{R}^2$.

[26] The reader may refer to [19], [17], or [11] for the needed information on universal bundles and classifying maps.

[27] See [17, Theorem 7.1] for a computation of the ring $H^\bullet(\mathbb{R}P^\infty; Z_2)$.

[28] See [11, Theorem 3.1] or [17, Theorem 7.1].

where by $e \cdot g$ we denote the edge of $\widetilde{G}_f^{(2)}$ corresponding to an unordered pair of edges e and g.

Lemma $w_f = [c_w]$.

Proof Let $F\colon |G| \to |H| \times \mathbb{R} \times \mathbb{R}$ be an embedding such that $\widetilde{f} = \mathrm{pr}_{|H|\times\mathbb{R}} \circ F$, and let $\rho\colon |\widetilde{G}_f^{(2)}| \to \mathbb{R}P^1$ be a map that takes an unordered pair $\{p, q\}$ of points of $|G|$ to an equivalence class of lines parallel to a vector going from the point $\mathrm{pr}_{\mathbb{R}\times\mathbb{R}}(F(q))$ to the point $\mathrm{pr}_{\mathbb{R}\times\mathbb{R}}(F(p))$.

Clearly, we can choose F so that ρ is a local homeomorphism at each $\{p, q\} \in |\widetilde{G}_f^{(2)}|$ where $\widetilde{f}(p) = \widetilde{f}(q)$.

It can be seen that p_2 is the pullback of the universal S_2-bundle under the composition $i \circ \rho$, where i is the inclusion $\mathbb{R}P^1 \hookrightarrow \mathbb{R}P^\infty$; indeed, it follows from the universal property of pullbacks that there is a morphism between principal S_2-bundles $(i \circ \rho)^*(S^\infty \to \mathbb{R}P^\infty)$ and p_2, which must be an isomorphism by [19, Proposition 2.1].

As the embedding $i\colon \mathbb{R}P^1 \hookrightarrow \mathbb{R}P^\infty$ induces an isomorphism $H^1(\mathbb{R}P^\infty; \mathbb{Z}_2) \cong H^1(\mathbb{R}P^1; \mathbb{Z}_2)$,[29] $w_f = \rho^* w_1$, where w_1 is the generator of $H^1(\mathbb{R}P^1; \mathbb{Z}_2)$.

Note that $\rho^{-1}([0 \times \mathbb{R}]) = \left\{\{p, q\} \mid \widetilde{f}(p) = \widetilde{f}(q)\right\} \subset |\widetilde{G}_f^{(2)}|$. Let us consider a (simplicial) cycle C. Clearly, there is an Euler cycle going through the edges belonging to C.[30] Let ψ_C be a non-degenerate piecewise linear map $\psi_C\colon S^1 \to |\widetilde{G}_f^{(2)}|$ realising it. According to the definition of c_w,

$$c_w(C) = \left|\psi_C^{-1}\left(\left\{\{p, q\} \mid \widetilde{f}(p) = \widetilde{f}(q)\right\}\right)\right| \bmod 2 = \left|(\rho \circ \psi_C)^{-1}([0 \times \mathbb{R}])\right| \bmod 2.$$

Since ρ is a local homeomorphism at the points $\{p, q\}$ where $\widetilde{f}(p) = \widetilde{f}(q)$, and ψ_C is a local homeomorphism at the preimages of the interior points of the edges of $\widetilde{G}_f^{(2)}$, the last number is, in fact, the sum of the local degrees modulo 2 at the preimages of $[0 \times \mathbb{R}]$. Therefore, $[c_w]([C])$ is the degree modulo 2 of the map $\rho \circ \psi_C$.

On the other hand, $w_f([C]) = w_1([\rho \circ \psi_C(S^1)])$ is also equal to the degree modulo 2 of the map $\rho \circ \psi_C$. Thus, $w_f = [c_w]$.[31] □

Now, we will prove the following theorem:

Theorem 16.10 *Let $f\colon G \to J$ be a stable simplicial map from a graph to a path graph. Then $w_f = 0$ if and only if $v_f = 0$.*

---

[29] The induced map of (cellular) cochain complexes is essentially truncating the cochains of dimension two and higher. Since all differentials in both complexes are zero, we obtain the desired isomorphism.

[30] Recall that an Euler cycle is a cycle that visits every edge exactly once. In our case, it visits each edge in C exactly once and does not include other edges.

[31] Note that $H^1(X; \mathbb{Z}_2) \cong \mathrm{Hom}_{\mathbb{Z}_2}(H_1(X; \mathbb{Z}_2), \mathbb{Z}_2)$ by the universal coefficient theorem.

Sketch of a Geometrical Proof Take a small neighborhood of $|\widetilde{G}_f^{(2)}|$ in $|\widetilde{G}^{\square}|$ homeomorphic to a one-dimensional disc bundle over $|\widetilde{G}_f^{(2)}|$. After taking suitable triangulations, we can observe that the total space of the bundle is the closure of the complement of $|\widetilde{G}_f^{\square}|$ in $|\widetilde{G}^{\square}|$ (see Fig. 16.7). Therefore, the cohomology class v_f belongs to the second cohomology group of the Thom space of this bundle. Furthermore, it can be shown that the Thom isomorphism maps w_f to v_f. □

Cochain-Level Proof Let $v_0, \ldots, v_N$ be the vertices of J. By replacing the triangulations of G and J with their subdivisions, we can assume that if $f^{-1}(v_i)$ contains a non-regular vertex, then all the preimages of v_{i-1} and v_{i+1} are regular.

Let $|\widetilde{f}|\colon |G| \to |J| \times \mathbb{R}$ be a generic lifting of f. By choosing S_ε as in the proof of Theorem 16.7, we obtain a generic approximation of f.

Note that the cellular cochains in $C^\bullet(\widetilde{G}^{\square}, \widetilde{G}_f^{\square}; \mathbb{Z}_2)$ and the simplicial cochains in $C^\bullet(\widetilde{G}_f^{(2)}; \mathbb{Z}_2)$ could be interpreted as, respectively, cubical cochains in $C^\bullet(G^{\square}, G_f^{\square}; \mathbb{Z}_2)$ and simplicial cochains in $C^\bullet(G_f^{(2)}; \mathbb{Z}_2)$, equivariant (as functions) with respect to the involution. Thus, let $c_v \in C^2(G^{\square}, G_f^{\square}; \mathbb{Z}_2)$ be the equivariant cubical 2-cochain defining v_f and corresponding to the generic approximation, and let $c_w \in C^\bullet(G_f^{(2)}; \mathbb{Z}_2)$ be the equivariant simplicial 1-cochain defining w_f and corresponding to $\widetilde{f}$.

Note that $G_f^{(2)}$ can be naturally embedded into $G^{\square}$ as the diagonals of the squares $(a, b) \times (c, d)$ where $f(a) = f(c)$ and $f(b) = f(d)$. It follows from the definition of c_w and c_v that c_v takes the value 1 on exactly those squares that contain a diagonal lying in $G_f^{(2)}$ on which c_w takes 1. The opposite is also true: c_w takes the value 1 on the diagonals of those squares on which c_v takes the value 1.

It can be observed that $|G_f^{(2)}|$ lies in the complement of $|G_f^{\square}|$ in $|G^{\square}|$. Furthermore, as a subcomplex of $G^{\square}$, $G_f^{\square}$ precisely consists of those squares and edges that do not intersect $|G_f^{(2)}|$.

Consider two regular vertices $a, b \in V(G)$ with the same image $f(a) = f(b) = v_i$ lying in the interior of J (i.e., $0 < i < N$). Let a be incident to edges e_1 and e_2, and b be incident to edges g_1 and h_1, where $f(g_1) = f(e_1)$ and $f(h_1) = f(e_2)$. The star of $a \times b$ in $G^{\square}$ consists of a large square divided into four smaller squares $\{e_1, e_2\} \times \{g_1, h_1\}$.[32] Observe also that $G_f^{(2)}$ forms a diagonal of the larger square composed of diagonals of $e_1 \times g_1$ and $e_2 \times h_1$.

Now, let us consider the case where b is a non-regular vertex (the case where a is non-regular is symmetrical). We can assume that b is incident to edges $g_1, \ldots, g_k, h_1, \ldots, h_l$, and a is incident to edges e_1 and e_2 such that $f(g_1) = f(g_2) = \cdots = f(g_k) = f(e_1)$ and $f(h_1) = f(h_2) = \cdots = f(h_l) = f(e_2)$. As illustrated in Fig. 16.7, the star of $a \times b$ in $G^{\square}$ is a $(k + l)$-page book, with the spine consisting of two segments, namely $e_1 \times b$ and $e_2 \times b$. Each page corresponds

[32] The star of a vertex v in a cubical complex C is the union of cubes of C that have v as a vertex.

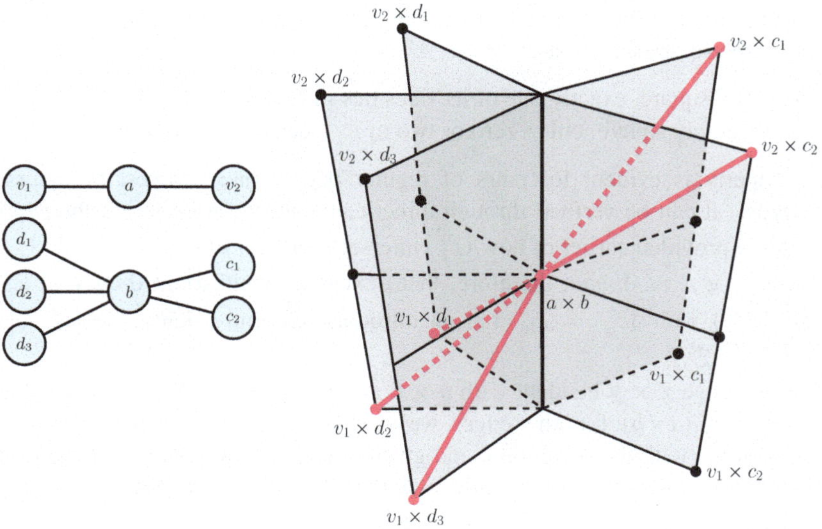

Fig. 16.7 The neighborhood of $a \times b$ in $G^{\square}$. The graph $G_f^{(2)}$ is represented by the pink lines

to an edge $v \in \{g_1, \ldots, g_k, h_1, \ldots, h_l\}$ that is incident to b and consists of two half-pages, namely $e_1 \times v$ and $e_2 \times v$. Regarding $G_f^{(2)}$ around $a \times b$, it consists of $k+l$ diagonals of half-pages, one in each page. Moreover, in the pages $\{e_1, e_2\} \times g_i$, those are diagonals of the half-pages $e_1 \times g_i$, while in the pages $\{e_1, e_2\} \times h_i$, those are diagonals of the half-pages $e_2 \times h_i$.

If $a, b \in f^{-1}(v_0)$ or $a, b \in f^{-1}(v_N)$, both a and b are regular degree one vertices. Therefore, the star of $a \times b$ in $G^{\square}$ is a single square formed by the edges incident to them, and $G_f^{(2)}$ is a diagonal of that square.

Hence, the closure of the complement of $G_f^{\square}$ in $G^{\square}$ can be seen as a collection of books connected with strips running along the graph $G_f^{(2)}$. Note that $G^{\square}$ does not contain any edges where both endpoints belong to $G_f^{(2)}$.

For each $a \times b \in V(G_f^{(2)})$, we choose a subset $E(a \times b)$ of edges incident with $a \times b$ in $G^{\square}$ as follows. When $a \times b$ is formed by a pair of regular degree one vertices, we simply select one of the two edges incident with $a \times b$, no matter which one, and put it in $E(a \times b)$. In the case where a is a regular degree two vertex, and b is a regular or non-regular vertex, we define $E(a \times b)$ as either $\{a \times h_1, \ldots, a \times h_l, e_1 \times b\}$ or $\{a \times g_1, \ldots, a \times g_k, e_2 \times b\}$, using the notation provided earlier. The selection of $E(b \times a)$ is done symmetrically with respect to the involution, meaning that $E(b \times a) = \tau(E(a \times b)) = \{y \times x \mid x \times y \in E(a \times b)\}$.

We refer to the squares of $G^{\square}$ that contain the edges of $G_f^{(2)}$ (on the diagonal) as full squares, while the squares that do not contain such edges are referred to as empty squares.

An important observation is that for any vertex $a \times b \in V(G_f^{(2)})$, the following holds for the squares of its star:

(1) for a full square, exactly one of its sides lies in $E(a \times b)$,
(2) for an empty square, either zero or two of its sides lie in $E(a \times b)$.

This property is evident for pairs of regular degree one vertices. For pairs of other types, it can be verified through direct calculations, considering the explicit description provided earlier of how $G_f^{(2)}$ intersects the star of $a \times b$.

Suppose $w_f = 0$, and, therefore, there exists an equivariant 0-cocycle $c'_w \in C^0(G_f^{(2)}; \mathbb{Z}_2)$ with $\delta c'_w = c_w$. Let us introduce an equivariant 1-cocycle $c'_v \in C^1(G^{\square}, G_f^{\square}; \mathbb{Z}_2)$.

For any edge $x \times y$ incident with $a \times b$ and belonging to $E(a \times b)$, we define $c'_v(x \times y) = c'_w(a \times b)$; for other edges, we set $c'_v(x \times y) = 0$. Since the sets $E(a \times b)$ are compatible with the involution, and given that c'_w is equivariant, it follows that c'_v is equivariant as well. Additionally, note that c'_v assigns zero values to edges not intersecting $G_f^{(2)}$. Therefore, c'_v is well-defined.

Based on the properties of the sets $E(a \times b)$ discussed above, we observe the following:

(1) On empty squares, the coboundary $\delta c'_v$ evaluates to zero. This is because each empty square contains an even number of edges on which c'_v takes the same value.
(2) On full squares, the value of $\delta c'_v$ is equal to the value of c_w on the diagonal. Indeed, a full square contains one edge from each $E(a \times b)$ and $E(a' \times b')$, where $a \times b$ and $a' \times b'$ are the endpoints of the diagonal belonging to $G_f^{(2)}$. Consequently, $\delta c'_v$ sums up the values of c'_w on $a \times b$ and $a' \times b'$. Since $\delta c'_w = c_w$, their sum is equal to the value of c_w on the diagonal.

Thus, we have $\delta c'_v = c_v$, implying that $v_f = 0$.

Conversely, assume that $v_f = 0$. Let $c'_v \in C^1(G^{\square}, G_f^{\square}; \mathbb{Z}_2)$ be an equivariant 1-cochain such that $\delta c'_v = c_v$. We define an equivariant cochain $c'_w \in C^0(G_f^{(2)}; \mathbb{Z}_2)$ on the vertices $a \times b$ of $G_f^{(2)}$ as follows:

(1) If a and b are regular vertices of degree one, the value of c'_w on $a \times b$ is the sum of the values of c'_v on the two edges incident to $a \times b$.
(2) Otherwise, it is the sum of the values of c'_v on the two "spine edges", which are the edges forming the spine of the star of $a \times b$ (as we have discussed above, geometrically, this star is a book). In the case when both a and b are regular degree two vertices, it is important to carefully choose the "spine edges" for $a \times b$, so that our choice is compatible with the involution. This means that if $x \times y$ and $x' \times y'$ are the "spine edges" for $a \times b$, then the edges $y \times x$ and $y' \times x'$ must be the "spine edges" for $b \times a$.

As mentioned earlier, the values of c_v on squares coincide with the values of c_w on their diagonals. Thus, to establish $\delta c'_w = c_w$, it suffices to show that $c'_w(a \times b) + c'_w(a' \times b')$ equals the sum of the values of c'_v on the edges of the full square containing a diagonal $\{a \times b, a' \times b'\} \in E(G_f^{(2)})$.

We claim that for a vertex $a \times b$ of $G_f^{(2)}$ and a full square S in its star, the value of c'_w on $a \times b$ equals the sum of the values of c'_v on the two edges of S incident to $a \times b$. If a and b are regular vertices of degree one, this follows from the definition of c'_w.

Otherwise, S contains exactly one of the two spine edges of $a \times b$ as a side. Let us denote this edge by e_+, and let the other side of S incident to $a \times b$ be denoted by g. Together with an empty square S', S forms a full page in the book of $a \times b$. The other spine edge e_- and g are the sides of S' incident to $a \times b$. The other two sides of S' do not intersect $G_f^{(2)}$, thus, c'_v takes the value zero on them. Consequently, since $0 = c_v(S') = \delta c'_v(S') = c'_v(e_-) + c'_v(g)$, the cochain c'_v takes the same value on the edges g and e_-. Therefore, $c'_w(a \times b) = c'_v(e_+) + c'_v(e_-) = c'_v(e_+) + c'_v(g)$, where e_+ and g are the sides of S incident to $a \times b$.

Therefore, for every full square S containing a diagonal $\{a \times b, a' \times b'\} \in G_f^{(2)}$, we conclude that $c'_w(a \times b) + c'_w(a' \times b')$ equals the sum of the values of c'_v on the sides of S. Thus, we have $\delta c'_w = c_w$ and $w_f = 0$. □

Theorem 16.11 *Let $f : T \to J$ be a stable simplicial map from a tree to a path graph. Then the following statements are equivalent:*

(1) There are no 2-obstructors for f;
(2) $|f|$ lifts to an embedding.

Proof If f lifts to an embedding, the non-existence of 2-obstructors follows from Theorem 16.1.

Now assume that there are no 2-obstructors for f. Therefore, it follows from Lemma 16.1 that p_2 is trivial, and, thus, $w_f = 0$.

Applying Theorem 16.10, we conclude that $v_f = 0$. Hence, Theorem 16.9 says that f is $\mathbb{Z}_2$-approximable with respect to an embedding $|J| \hookrightarrow \mathbb{R}^2$. Since T is a tree, by Theorem 16.8, f is approximable by embeddings. Finally, f lifts to an embedding by applying Theorem 16.7. □

The proven result has the potential for various generalizations. For instance, we believe that the non-existence of 2-obstructors condition remains sufficient for the existence of a lifting even for an arbitrary, not necessarily stable, non-degenerate map from a graph to a segment.

Furthermore, instead of considering maps to a segment, we can consider maps to trees. However, the following example shows that the stated 2-obstructor condition is not sufficient for maps from arbitrary graphs to trees, while the question of its sufficiency for maps from trees to trees remains open.

Example 16.5 Let T be a tripod with 4 vertices O, a, b, and c, where O is the central vertex. Let $|f|: S^1 \to |T|$ be a "walking around" $|T|$.

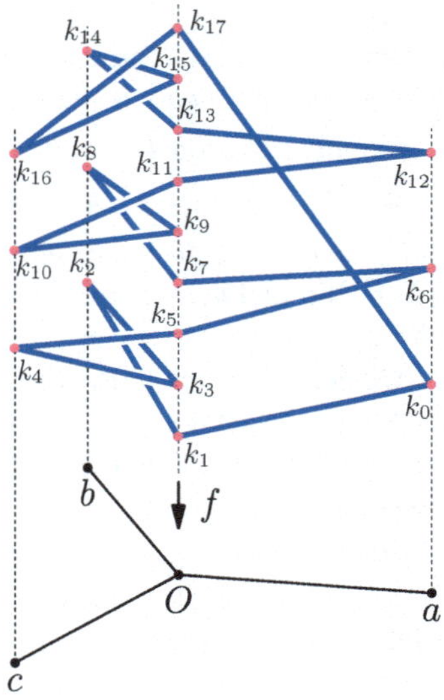

Fig. 16.8 The 3-winding map $f\colon C_{18} \to T$

Specifically, let C_6 be a hexagon triangulating S^1, whose six vertices v_i numbered in the counterclockwise order starting from zero, and let $g\colon C_6 \to T$ be simplicial map defined by

$$f(v_0) = a,\ f(v_1) = O,$$
$$f(v_2) = b,\ f(v_3) = O,$$
$$f(v_4) = c,\ f(v_5) = O.$$

Now, consider the composition $f\colon C_{18} \xrightarrow{w} C_6 \xrightarrow{g} T$ illustrated in Fig. 16.8, where C_{18} represents an 18-gon. Here the first map is a simplicial analogue of the standard 3-winding $w\colon S^1 \to S^1$ defined as $z \mapsto z^3$ assuming that S^1 is embedded into the complex plane as the unit circle centered at zero. Explicitly, $w(k_i) = v_{(i \bmod 6)}$ where $k_0, \dots, k_{17}$ are the vertices of C_{18} numbered in the counterclockwise order.

Note that f has a 3-obstructor because w does, and, therefore, it does not lift to an embedding.

We claim that f does not have any 2-obstructors. To demonstrate this, let us look at the graph $(C_{18})^{(2)}_f$.

In this graph, a vertex corresponding to a pair (k_i, k_j) of distinct vertices from $f^{-1}(O)$ has degree two when $w(k_i) = w(k_j)$, meaning that it is also a vertex of $(C_{18})^{(2)}_w$, and degree one otherwise.

Consider a vertex (k_t, k_s) of $(C_{18})^{(2)}_f$ where $f(k_t) = f(k_s) \neq O$. This is a degree-four vertex. Moreover, two of the four vertices adjacent to it have degree one, precisely those that do not lie in $(C_{18})^{(2)}_w$.

Consequently, by removing the vertices of degree one from $(C_{18})^{(2)}_f$, we get $(C_{18})^{(2)}_w$. Since the latter graph does not have a 2-obstructor path, the same holds true for the former graph.

Acknowledgments I express my strong gratitude to Sergey Melikhov for introducing me to this subject and for the invaluable discussions and patient guidance he provided throughout my research. I am also thankful to Olga Frolkina for her careful reading and critique of the draft version of this text, and to Francis Lazarus and Martin Deraux for many fruitful comments on the final version.

Last but not least, I am very grateful to Petr Akhmetiev and the other participants of the Geometric Topology Seminar at Steklov Mathematical Institute for the valuable comments and ideas expressed during the discussion of some of the results presented in the present chapter.

References

1. P. Akhmetiev, Prem-mappings, triple self-intersection points of oriented surfaces, and the Rokhlin signature theorem. Math. Notes **59**, 581–585 (1996)
2. P. Akhmetiev, D. Repovš, A. Skopenkov, Obstructions to approximating maps of n-manifolds into $\mathbb{R}^{2n}$ by embeddings. Topol. Appl. **123**, 3–14 (2002). https://core.ac.uk/download/pdf/82701277.pdf
3. M. Bestvina, M. Kapovich, B. Kleiner, Van Kampen's embedding obstruction for discrete groups. Invent. Math. **150**(2), 219–235 (2002). https://arxiv.org/abs/math/0010141
4. R. Brown, *Topology and Groupoids* (Booksurge LLC, Charleston, 2006). www.groupoids.org.uk
5. J.S. Carter, M. Saito, *Knotted Surfaces and Their Diagrams* (American Mathematical Soc., Providence, 1998)
6. J.S. Carter, M. Saito, Surfaces in 3-space that do not lift to embeddings in 4-space. Banach Center Publications **42**(1), 29–47 (1998)
7. V. Dujmovic, A. Por, D.R. Wood, Track layouts of graphs. Discret. Math. Theor. Comput. Sci. **6** (2004). https://arxiv.org/abs/cs/0407033
8. J.A. Ellis-Monaghan, I. Moffatt, *Graphs on Surfaces: Dualities, Polynomials, and Knots*. Springer Briefs in Mathematics (Springer, Berlin, 2013)
9. R. Fulek, J. Kynčl, Hanani-Hutte for approximating maps of graphs (2018). https://arxiv.org/abs/1705.05243
10. C. Giller, Towards a classical knot theory for surfaces in $\mathbb{R}^4$. Il. J. Math. **26**(4), 591–631 (1982)
11. A. Hatcher, Vector bundles and K-theory, Hatcher, A. (2017). https://pi.math.cornell.edu/~hatcher/VBKT/VB.pdf
12. P. Hoffman, B. Richter, Embedding graphs in surfaces. J. Comb. Theory Ser. B **36**(1), 65–84 (1984)
13. G.-S. Li, On self-intersections of immersed surfaces. Proc. Am. Math. Soc. **126**(12), 3721–3726 (1998)
14. S. Melikhov, The van Kampen obstruction and its relatives. Proc. Steklov Inst. Math. **266**, 142–176 (2009). https://arxiv.org/abs/math/0612082
15. S. Melikhov, Lifting generic maps to embeddings. I: triangulation and smoothing (2020). https://arxiv.org/abs/2011.01402
16. S. Melikhov, Lifting generic maps to embeddings: II. The double point obstruction (2020). https://arxiv.org/abs/1711.03518

17. J.W. Milnor, J.D. Stasheff, *Characteristic classes*. TEXromancers (2022). https://aareyanmanzoor.github.io/assets/books/characteristic-classes.pdf
18. P. Minc, Embedding simplicial arcs into the plane. Topol. Proc. **22**, 305–340 (1997)
19. S.A. Mitchell, *Notes on Principal Bundles and Classifying Spaces*. Lecture Notes. University of Washington (2001). https://sites.math.washington.edu/~mitchell/Notes/prin.pdf
20. E.E. Moise, *Geometric Topology in Dimensions 2 and 3*, vol. 47 (Springer Science & Business Media, Berlin, 2013)
21. V. Poénaru, Homotopie régulière et isotopie. Rend. Sem. Mat. Univers. Politec. Torino **37**(2), 15–56 (1979). Publications mathématiques Orsay (1974). http://sites.mathdoc.fr/PMO/afficher_notice.php?id=PMO_1974_A26
22. V. Poénaru, Some invariants of generic immersions and their geometric applications. Bull. Am. Math. Soc. **81**(6), 1079–1082 (1975)
23. D. Repovš, A. Skopenkov, A deleted product criterion for approximability of maps by embeddings. Topol. Appl. **87**(1), 1–19 (1998)
24. C.P. Rourke, B.J. Sanderson, *Introduction to Piecewise-Linear Topology* (Springer, Berlin, 1972)
25. A. Shapiro, Obstructions to the imbedding of a complex in a Euclidean space: I. The first obstruction. Ann. Math. **66**, 256–269 (1957)
26. K. Siekłucki, Realization of mappings. Fund. Math. **65**(3), 325–343 (1969) (eng)
27. M. Skopenkov, On approximability by embeddings of cycles in the plane (2019). https://arxiv.org/abs/0808.1187
28. A. Szücs, On the cobordism groups of immersions and embeddings. Math. Proc. Camb. Philos. Soc. **109**(2), 343–349 (1991)

Chapter 17
A Note on the Discriminant Loci of a Hypersurface of Bi-degree (*m*, *n*)

Susumu Tanabé

Hommage à Valentin Poénaru

Abstract We present a topological method for the study of the discriminant loci of an algebraic variety defined in a product of projective spaces. Our approach is distinguished by the efficient use of the notion of groupoid in the description of the monodromy. As an example, we deal here with the discriminant loci of a hypersurface of bi-degree (m, n).

Keywords Discriminant loci · Braid monodromy · Groupoid

AMS Subject Classification 20F36 (primary); 14D05, 32S40 (secondary)

17.1 Groupoid in the Research on Monodromy

In this chapter, we present a topological method for the study of the discriminant loci of an algebraic variety. Given the remarkable property of the discriminant loci of a variety defined in a product of projective spaces, mentioned in the Remark 17.2.2 below, we hope that our approach may contribute to the elucidation of certain geometric and analytic properties hidden deep down in the phenomenon of monodromy. As an objective of considerable importance yet realizable, we may

Partially supported by Max Planck Institut für Mathematik, Research fund of the Galatasaray University project number 1123 "Analysis and topology of algebraic functions and period integrals."

S. Tanabé (✉)
Department of Discrete Mathematics, Moscow Institute of Physics and Technology (National Research University), Dolgoprudny, Moscow Region, Russian Federation
e-mail: tanabe.s@mipt.ru

L. Funar, A. Papadopoulos (eds.), *Essays on Topology*,
https://doi.org/10.1007/978-3-031-81414-3_17

mention the application to homological mirror symmetry [10, 19, Corollary 3.5, Theorem 3.6]. In order to simplify the exposition, we restrict ourselves here to the case of a bi-degree hypersurface (m, n) [13], especially to the case $n = 2$.

We hope that the examples we present in this chapter and in other papers in preparation [13, 20, 21] illustrate the thesis promoted by A. Grothendieck [8] on the advantage of groupoids in the study of the fundamental group of the complement to the discriminant loci.

> Même pour $g = 0$ (donc quand les groupes de Teichmüller correspondants sont des groupes de tresses "bien connus"), les générateurs et relations connus à ce jour dont j'ai eu connaissance, me semblent inutilisables tels quels, car ils ne présentent pas les caractères d'invariance et de symétrie indispensables pour que l'action de $\mathbf{\Gamma} = \mathbf{Gal}(\bar{\mathbf{Q}}/\mathbf{Q})$ soit directement lisible sur cette présentation. Ceci est lié notamment au fait que les gens s'obstinent encore, en calculant avec des groupes fondamentaux, à fixer un seul point base, plutôt que d'en choisir astucieusement tout un paquet qui soit invariant par les symétries de la situation. Dans certaines situations (comme des théorèmes de descente à la Van Kampen pour groupes fondmentaux) il est bien plus élégant, voir indispensable pour y comprendre quelque chose, de travailler avec des groupoïdes fondamentaux par rapport à un paquet de points base convenable, et il en est certainement ainsi pour la tour de Teichmüller.

The same thesis is described in a slightly modified form in [4, 3. Application of fundamental groupoid].

In particular, in [20] we propose the search for the global monodromy of period integrals associated with an affine hypersurface (or a complete intersection) (Calabi–Yau manifold) using their analytic continuation along paths that avoid discriminant loci. We can, therefore, interpret our trial as a kind of sequel in the spirit of the above thesis, where the focus shifts from the absolute Galois group $\mathbf{\Gamma} = \mathbf{Gal}(\bar{\mathbf{Q}}/\mathbf{Q})$ towards the monodromy of the affine algebraic variety.

Our program proposed in [20] has been realized for the bi-degree curve $(2, 2)$ defined in $\mathbb{P}^1 \times \mathbb{P}^1$ [19].

An analysis of the connection with regular and irregular singularities [16] deserves special mention. This example eloquently illustrates how the full significance of the above thesis can be evaluated in the context of global research of differential equations.

It should be pointed out that the content of the current chapter serves as a preliminary step in the realization of a collaboration project with Michael Lönne [13]. The author would like to thank Mutlu Koçar for the assistance in preparing the figures and Athanase Papadopoulos for his careful reading of the text.

17.2 Discriminant Loci of an (m, n) Bi-degree Hypersurface

This section is dedicated to the analysis of the moduli space of a hypersurface Y which is the mirror symmetry variety of a Calabi–Yau manifold of generic bi-degree (m, n) in $\mathbb{P}^{m-1} \times \mathbb{P}^{n-1}$, in the sense of V. V. Batyrev [2], [5, 4.1]. According to

Batyrev, the Newton polyhedron $\Delta(Q)$ of the Laurent polynomial that defines the affine part of Y

$$Q(z, w) = \frac{P(z, w)}{z_1 \dots z_{m-1} w_1 \dots w_{n-1}}$$

is the polar dual of the convex envelope of a product of two canonical simplices of dimensions $m-1$ and $n-1$.

The action of an algebraic torus on the variables $(z_1, \dots, z_{m-1}, w_1, \dots, w_{n-1})$ transforms the Laurent polynomial $Q(z, w)$ with general coefficients into $F_{m,n}(z, w) =$

$$1+z_1+z_2+\cdots+z_{m-1}+\frac{x}{z_1 \cdots z_{m-1}}+w_1+\cdots+w_{n-1}+\frac{y}{w_1 \cdots w_{n-1}} \tag{17.2.1}$$

which depends on two deformation parameters (x, y).

The discriminant loci S of the affine variety

$$X^{(m,n)}_{x,y} = \{(z, w) \in (\mathbb{C}^*)^{m+n-2};\ F_{m,n}(z, w) = 0\} \tag{17.2.2}$$

are presented as the union of a curve

$$C = \{(x, y);\ D(x, y) = 0\} \tag{17.2.3}$$

and of the two coordinate axes.

$$S = \{(x, y) \in \mathbb{C}^2;\ xyD(x, y) = 0\}. \tag{17.2.4}$$

By the definition of discriminant loci, the affine variety $X^{(m,n)}_{x,y}$ (17.2.2) is singular if and only if $(x, y) \in S$.

The curve C (17.2.3) admits the following Horn–Kapranov uniformisation [18]

$$(x, y) = ((\frac{s}{ms+n})^m, (\frac{1}{ms+n})^n),\ s \in \mathbb{C} \tag{17.2.5}$$

in such a way that

$$D\big((\frac{s}{ms+n})^m, (\frac{1}{ms+n})^n\big) = 0,\ \forall s \in \mathbb{C}.$$

The derivation of the uniformisation equation (17.2.5) is based on the computation of the singular loci of the period integrals of the manifold (17.2.2). It is known [3] that the period integrals

$$f(x, y) = \int_\gamma \frac{dz \wedge dw}{dF_{m,n}(z, w)},$$

for an arbitrary cycle $\gamma \in H_{m+n-3}(X^{(m,n)}_{x,y})$ satisfy the system of Horn-type hypergeometric equations,

$$\begin{aligned}&\left(\theta_x^m - x(m\theta_x + n\theta_y + 1)\cdots(m\theta_x + n\theta_y + m)\right) f(x,y) = 0,\\ &\left(\theta_y^n - y(m\theta_x + n\theta_y + 1)\cdots(m\theta_x + n\theta_y + n)\right) f(x,y) = 0,\end{aligned} \tag{17.2.6}$$

for $m \geq n \geq 2$,

$$\theta_x = x\frac{\partial}{\partial x}, \quad \theta_y = y\frac{\partial}{\partial y}.$$

The dimension of the complex vector space of solutions (17.2.6), called holonomic rank, is equal to $mn = |\chi(X^{(m,n)}_{x,y})|$. An explicit basis for the solution space of (17.2.6) can be given by the Mellin–Barnes integrals,

$$f^k_{i,j}(x,y) = \int_\Delta \frac{\Gamma(s)^{m-i}\Gamma(t)^{n-j}\Gamma(1-(ms+nt))}{\Gamma(1-s)^i\Gamma(1-t)^j} x^{-s}y^{-t}e^{-\pi\sqrt{-1}(is+jt)}ds\wedge dt, \tag{17.2.7}$$

where $(i,j) \in [0; m-1] \times [0; n-1]$ and Δ is a semigroup formed by the poles of the integrand of (17.2.7),

$$\Delta = \{(s,t) \in \mathbb{C}^2; s \in \mathbb{Z}_{\leq 0}, t \in \mathbb{Z}_{\leq 0}\}.$$

Thus the branching loci of the solutions of the differential equations (17.2.6) admit the representation (17.2.5), [18].

The global monodromy group of integrals (17.2.7) is obtained as a representation of $\pi_1(\mathbb{C}^2 \setminus S)$. The subject of our chapter is a part of the research program on this global monodromy group.

From here on, we use the notation $\omega_m = e^{2\pi\sqrt{-1}/m}$, $\omega_n^j = e^{2\pi j\sqrt{-1}/n}$ as well as $e(\lambda) = e^{2\pi\sqrt{-1}\lambda}$, that is, $\omega_m^i = e(i/m)$, $\omega_n^j = e(j/n)$. We shall also use the notation $i \in [r_1; r_2] \Leftrightarrow i \in \{r_1, \cdots, r_2\}$ for two integers $r_1 \leq r_2$.

In order to study the fundamental group $\pi_1(\mathbb{C}^2 \setminus S)$, we start with the following lemma.

Lemma 17.2.1

(a) *The curve* $C = \{(x,y); D(x,y) = 0\}$ *is irreducible.*
(b) *The curve* C *is nodal. In other words, it has only singular points where the two branches intersect transversely.*
(c) $\pi_1(\mathbb{C}^2 \setminus C) \cong \mathbb{Z}$.

Proof

(a) Thanks to (17.2.5), the branch $(x, y_j(x))$, $j = 1, \cdots, m$ of the curve C admits the following Puiseux expansion,

$$y_j(x) = (\frac{1 - m\omega_m^j x^{1/m}}{n})^n.$$

Since each branch $y_j(x)$ belongs to the same conjugacy class, and since the irreducibility of a curve is equivalent to the uniqueness of the conjugacy class of its branches, we obtain the statement.

(b) It suffices to show that the curve with the parametric presentation $C' = \{(x, y); x = (1-t)^n, y = t^m\}$ has only double singular points. Consider the polynomials $P_{k,\ell}(t) = (1 - \omega_n^k t)^m - (1 - \omega_n^\ell t)^m$ for $0 \le k \ne \ell \le n-1$.

We need to show that (1) $P_{0,\ell}(t) = 0$, $\ell \ne 0$ has no multiple solutions. (2) The system of equations $P_{0,\ell}(t) = P_{k,\ell}(t) = 0$ has no non-zero solution for $0 < k \ne \ell \le m-1$.

For the presence of a double solution of the equation $P_{0,\ell}(t) = 0$, the existence of a solution of the following system is necessary: $P_{0,\ell}(t) = P'_{0,\ell}(t) = 0$. If t_0 is a solution of the system, then it must satisfy $\omega_n^\ell P_{0,\ell}(t_0) + (1 - \omega_n^\ell t_0)P'_{0,\ell}(t_0)/m = 0$, i. e. $(1 - t_0)^{m-1}(\omega_n^\ell - 1) = 0$. Since $P_{0,\ell}(1) = -(1 - \omega_n^\ell)^m \ne 0$, we have (1).

The non-zero solutions of $P_{0,\ell}(t) = 0$ are of the form $\{\frac{1-\omega_m^i}{1-\omega_m^i\omega_n^\ell}\}$, $1 \le i \le m-1$, $(\frac{\ell}{n} + \frac{i}{m}) \not\equiv 0 \; mod \; 1$, yet those of $P_{k,\ell}(t) = 0$ are $\{\frac{1-\omega_m^j}{\omega_n^k - \omega_m^j\omega_n^\ell}\}$, $1 \le j \le m-1$, $(\frac{-k+\ell}{n} + \frac{j}{m}) \not\equiv 0 \; mod \; 1$. We recall here that $t = 0$ corresponds to an $(m-1)$–flex point of the curve C'. The argument $Arg\left(\frac{1-\omega_m^j}{\omega_n^k - \omega_m^j\omega_n^\ell} / \frac{1-\omega_m^i}{1-\omega_m^i\omega_n^\ell}\right) = -\frac{k}{n}\pi$ is never equal to 0 or 2π for $1 \le k \le n-1$. This shows (2).

From [15], the following condition (called "negativity condition at infinity": abbreviated as **NC**) is sufficient for a nodal curve C to have an abelian group $\pi_1(\mathbb{C}^2 \setminus C)$.

(**NC**) The restriction of the algebraic function $y(x)$ satisfying $D(x, y) = 0$ in the neighbourhood of infinity (the real axis being excluded) consists of n analytic branches $y_j(x)$ of the form $g_j(\tau(x))$ where $g_j(\tau)$ is a meromorphic and univalent function in the neighbourhood of the point $\tau = 0$ for $\tau(x)$: a univalent branch of the function $x^{-1/n}$. We have $\lim_{\tau \to 0}(g_j(\tau) - g_k(\tau)) \ne 0$ for $k \ne j$, i.e., it has either a pole or a finite limit.

It follows from (17.2.5) and from a linear change of variables that the function $g_j(\tau) = (\frac{1}{n})^n(1 - \frac{m\omega_m^j}{\tau})^n$ satisfies **NC**.

By Hurewicz's theorem, we have $\pi_1(\mathbb{C}^2 \setminus C)/Z = H_1(\mathbb{C}^2 \setminus C)$, but the above discussion shows that the commutator Z cancels out. It is a well known fact that $H_1(\mathbb{C}^2 \setminus C) \cong \mathbb{Z}$ for the irreducible curve C. □

Thus, $\pi_1(\mathbb{C}^2 \setminus S)$ is generated by three generators $\gamma_1, \gamma_2, \gamma_C$ where γ_1 (resp. γ_2, γ_C) corresponds to the loop that makes a turn around the curve $x = 0$ (resp. $y = 0$, C).

The uniformisation (17.2.5) indicates that the degree mn covering of the discriminant loci (17.2.4) is homotopic to the complex arrangement of $(mn + 2)$ lines.

$$\mathcal{A}_{m,n} := \{(s,t) \in \mathbb{C}^2; st \prod_{0\leq i\leq m-1, 0\leq j\leq n-1} (\omega_m^i s + \omega_n^j t + 1) = 0\}. \tag{17.2.8}$$

The following exact sequence is a consequence of the Reidemeister–Schreier theorem [23].

$$1 \to \pi_1(\mathbb{C}^2 \setminus \mathcal{A}_{m,n}) \to \pi_1(\mathbb{C}^2 \setminus S) \to^h (\mathbb{Z}/m\mathbb{Z}) \times (\mathbb{Z}/m) \to 1 \tag{17.2.9}$$

with $h(\gamma_C) = h(\gamma_1^m) = h(\gamma_2^n) = 1$.

Remark 17.2.2 Indeed, this is an example of more general phenomena observed on the discriminant loci of a Calabi–Yau hypersurface $X_{\mathbf{n}}$ of multi-degree $\mathbf{n} = (n_i)_{i=1}^K$ defined in the product of projective spaces $\prod_{i=1}^K \mathbb{P}^{n_i - 1}$. The computation of [17, section 8] indicates that the degree $\prod_{i=1}^K n_i$ covering of the discriminant loci of such a hypersurface admits the presentation as an arrangement of hyperplanes

$$\mathcal{A}_{\mathbf{n}} := \{(t_1, \cdots, t_K) \in \mathbb{C}^K; \prod_{i=1}^K t_i \cdot \prod_{(j_1,\cdots,j_K)\in\prod_{i=1}^K(\mathbb{Z}/n_i\mathbb{Z})} (\sum_{\ell=1}^K \omega_{n_\ell}^{j_\ell} t_\ell + 1) = 0\} \tag{17.2.10}$$

which consists of $K + \prod_{i=1}^K n_i$ hyperplanes. The proof can be given with the aid of the formulas presented in [18, theorem 2.6].

For the case where $n_i = 2, \forall i \in [1; K]$, T. Terasoma [22] uses the covering (17.2.10) in order to obtain a complete description of the fundamental group of all smooth hypersurfaces of multi-degree $\mathbf{n} = (2, \cdots, 2)$.

17.3 Discriminant Loci for the Case $n = 2$

17.3.1 Rotation of Punctures and Annular Braid

In this section, we shall study the fundamental group $\pi_1(\mathbb{C}^2 \setminus \mathcal{A}_{m,2})$ by means of the braid monodromy method for $m \geq 2$.

First of all, we calculate branches $s(t)$ of the line arrangement $\mathcal{A}_{m,2}$ (17.2.8) as complex linear functions in t. We set

$$T_{j,\ell}(t) = -\omega^j(1 + (-1)^\ell t), \tag{17.3.1}$$

for $j \in [0; m-1]$, $\ell \in [0; 1]$ and $\omega = \omega_m = e(1/m)$. From here on, we shall always consider the indices (j, ℓ) in $\mathbb{Z}/m \times \mathbb{Z}/2$.

We follow the change of the section of $\mathcal{A}_{m,2}$ in terms of the value of $t \in \mathbb{C}$,

$$\mathbb{S}(t) := \{0\} \bigcup_{[j,\ell]\in[0;m-1]\times[0;1]} \{T_{j,\ell}(t)\}. \tag{17.3.2}$$

Further we shall call elements of $\mathbb{S}(t)$ **punctures**, while a fixed value of t will be called **point**.

Thus we have

$$\mathbb{S}^*(t) := \mathbb{S}(t) \setminus \{0\}. \tag{17.3.3}$$

Remark 17.3.1 The discussion in Sect. 17.3.2 below indicates certain difficulties that arise if we try to generalise the argument to the case $n \geq 3$. Namely, in the cases $n \geq 5$, we encounter so-called entanglement of branches $s(t)$ of the arrangement $\mathcal{A}_{m,n}$. See [12, 6.1 Parallel transport in the model family]. If this difficulty does not require a drastic change of our view on the topology of arrangement $\mathcal{A}_{m,n}$, we would be entitled to make a statement similar to Conjecture 17.3.9 formulated at the end of the present chapter.

Lemma 17.3.2 *The branches* (17.3.1) *meet at the coincidence points* $\mathbb{D} \cup \{0\}$ *for*

$$\mathbb{D} := \{\pm 1\} \cup \{\mathbf{i} \tan \frac{j\pi}{m}\}_{j\in[0;m-1]\setminus\{\frac{m}{2}\}}, \tag{17.3.4}$$

according to the following types. Further, we shall use the notation

$$|\bar{\mathbf{j}}| = j_2 - j_1, \ |\mathbf{j}| = j_1 + j_2, \tag{17.3.5}$$

for $\mathbf{j} = (j_1, j_2)$.

(1) *Cases* $|\bar{\ell}| = 0$. *At* $t = (-1)^{\ell+1}$, *the confluence of the following m solutions happens:*

$$T_{0,\ell}((-1)^{\ell+1}) = T_{1,\ell}((-1)^{\ell+1}) = \cdots = T_{m-1,\ell}((-1)^{\ell+1}), \tag{17.3.6}$$

for $\ell \in [0; 1]$.

(2) *Cases* $|\bar{\ell}| = 1$. *At* $t = \mathbf{i} \tan \frac{|\bar{\mathbf{j}}|\pi}{m}$, *m-tuple of double solutions confluence happen:*

$$T_{j_1,0}(\mathbf{i} \tan \frac{|\bar{\mathbf{j}}|\pi}{m}) = T_{j_2,1}(\mathbf{i} \tan \frac{|\bar{\mathbf{j}}|\pi}{m}) = -e(\frac{|\mathbf{j}|}{2m})(\cos \frac{|\bar{\mathbf{j}}|}{m}\pi)^{-1}, \tag{17.3.7}$$

if $|\frac{|\bar{\mathbf{j}}|}{m}| \neq \frac{1}{2}$.

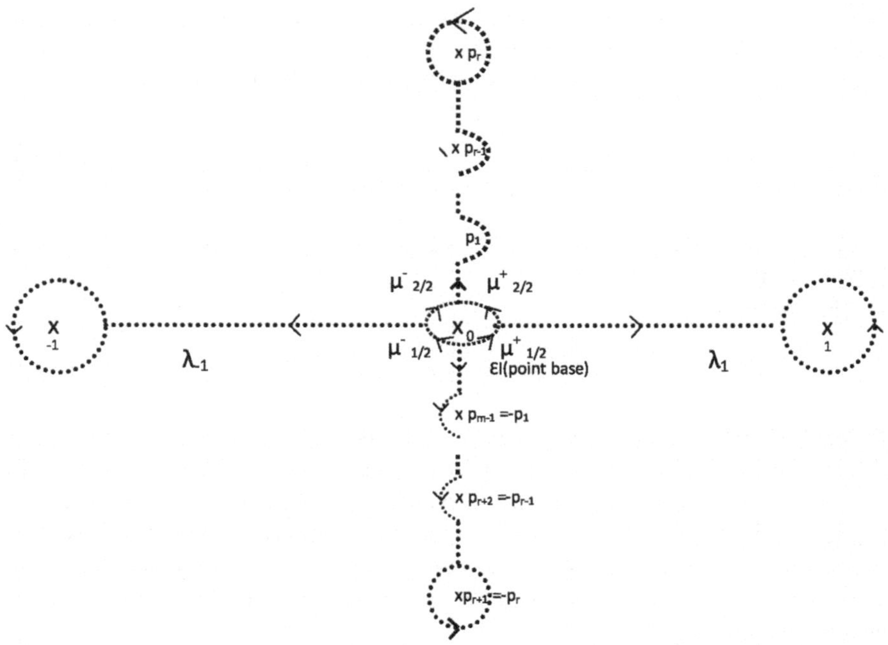

Fig. 17.1 Branching points $\mathbb{D}$ with paths starting from the base point $t = -\epsilon\mathbf{i}$

We denote by $p_k \in \mathbb{D}$ a coincidence point on the pure imaginary axis,

$$p_k = \mathbf{i}\tan\frac{k\pi}{m}, \tag{17.3.8}$$

for $k \in [0; m-1] \setminus \{\frac{m}{2}\}$. It follows that

$$\tan\frac{k\pi}{m} > 0 \ \ if \ \ k \in [1; r], \ \ \tan\frac{k\pi}{m} < 0 \, if \ \ k \in [r+1; m-1], \tag{17.3.9}$$

for $r = \lfloor\frac{m-1}{2}\rfloor$, the integer part of $\frac{m-1}{2}$.

To follow the position change of punctures (17.3.2), we fix the starting point of the motion of t on $\mathbb{C}^* \setminus \mathbb{D}$. In view of a symmetry observable in Fig. 17.1 of $\mathbb{D}$, we choose $t = -\epsilon\mathbf{i}$ as the base point for $\epsilon > 0$ small enough.

At the base point $t = -\epsilon\mathbf{i}$, the reference fibre $\mathbb{S}(-\epsilon\mathbf{i})$ consists of $2m$ punctures located on a circle with centre 0 (Fig. 17.2).

In order to describe the motion of $\mathbb{S}(t)$ along a curve starting from $t = -\epsilon\mathbf{i}$, we will make use of the notion of **rotating punctures** [14, Definition 1.4].

Definition 17.3.3 For an index set $\mathbb{J}_\mu = \{(j_k, \ell_k)\}_{k=1}^{\mu} \subset [0, m-1] \times [0; 1]$ and $t \in \mathbb{C} \backslash \mathbb{D}$, we consider a polygon $P_{\mathbb{J}_\mu}$ obtained as convex hull of μ punctures $A_{\mathbb{J}_\mu} := \{T_{j,\ell}(t)\}_{(j,\ell)\in\mathbb{J}_\mu}$. Let us assume that the punctures $\{T_{j_k,\ell_k}(t)\}_{k=1}^{\mu}$ are arranged on the boundary $\partial P_{\mathbb{J}_\mu}$ according to the anti-clockwise ordering.

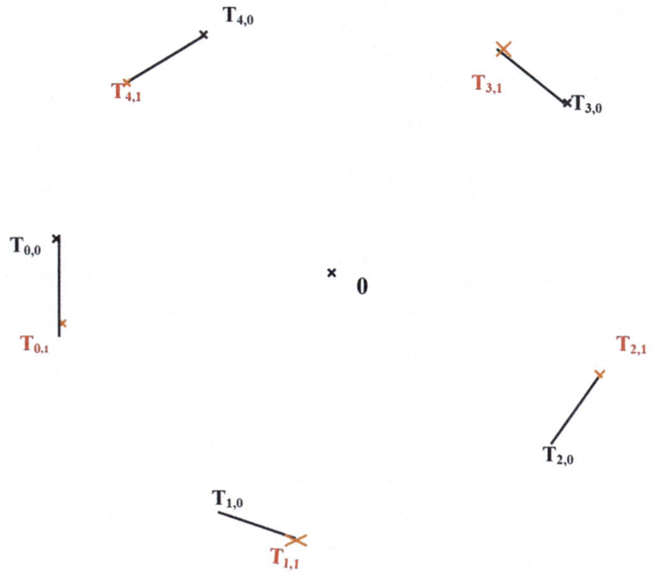

Fig. 17.2 $\mathbb{S}(-\epsilon\mathbf{i})$ for $m = 5$

In other words, $\exists x \in (P_{\mathbb{J}_\mu})^{int}$ such that

$$Arg(T_{j_k,\ell_k}(t) - x) < Arg(T_{j_{k+1},\ell_{k+1}}(t) - x)$$

$\forall (j_k, \ell_k) \in \mathbb{J}_\mu$ where $\mu + 1 \equiv 1$.

We define the rotation $R_{A_{\mathbb{J}_\mu}}$ of the set of solutions $A_{\mathbb{J}_\mu}$ as the motion of solutions like following,

$$R_{A_{\mathbb{J}_\mu}}(T_{j_k,\ell_k}(t)) = T_{j_{k+1},\ell_{k+1}}(t). \tag{17.3.10}$$

Definition (17.3.10) yields the power of rotation,

$$R^p_{A_{\mathbb{J}_\mu}}(T_{j_k,\ell_k}(t)) = T_{j_{\overline{k+p}},\ell_{\overline{k+p}}}(t), \tag{17.3.11}$$

where $\overline{k+p} \in [1; \mu]$ and $k + p \equiv \overline{k+p}$ $(mod\ \mu)$ (Fig. 17.3).

Remark 17.3.4 If the motion of t along a curve $\lambda(u) \subset \mathbb{C}^* \setminus \mathbb{D}$, $u \in [0, 1]$, such that $\lambda(0) = -\epsilon\mathbf{i}$, stirs a rotation $R_{A_{\mathbb{J}}}$, we obtain a $(2m + 1)$-braid, in the sense of [1], located in $\mathbb{C} \times [0, 1]$ traced by the solutions $\mathbb{S}(\lambda(u))$, $u \in [0, 1]$. Since $\{0\} \in \mathbb{S}(t)$, $\forall t \in \mathbb{C}$, the braid in question shall be considered as an **annular braid**, denoted by CB_{2m} (cylindrical braids), in the sense of [11]. This is a braid of $2m$ strings that correspond to $2m$ non-zero punctures that make turns around the axis $\{0\} \times [0, 1] \subset \mathbb{C} \times [0, 1]$. See Fig. 17.5.

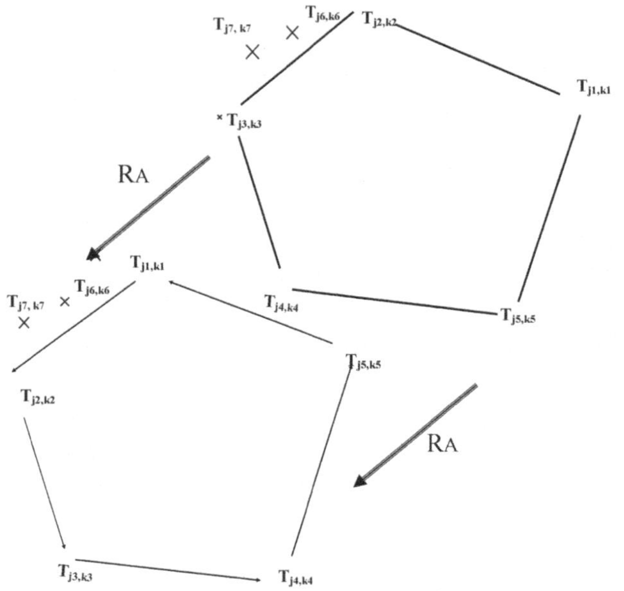

Fig. 17.3 Rotation $R_{A_{\mathbb{J}_\mu}}$ with $\mu = 5$, $A = A_{\mathbb{J}_5} = \{T_{j_\ell,k_\ell}\}_{\ell=1}^{5}$

As each solution of $\mathbb{S}(t)$ is a linear function in the complex variable t, it is legitimate to extend the notion of rotation to that of fractional rotation.

Definition 17.3.5 Let $R_{A_\mathbb{J}}$ be the rotation defined in Definition 17.3.3. For $\nu = p/q \in [0, 1] \cap \mathbb{Q}$, we define $R^{\nu}_{A_\mathbb{J}}$ as motion of solutions that satisfies $(R^{\nu}_{A_\mathbb{J}})^q = R^{p}_{A_\mathbb{J}}$ in such a way that

$$Arg\left(\frac{R^{\nu}_{A_\mathbb{J}}(T_{j_k,\ell_k}(t)) - x}{T_{j_k,\ell_k}(t) - x}\right) = \nu Arg\left(\frac{T_{j_{k+1},\ell_{k+1}}(t) - x}{T_{j_k,\ell_k}(t) - x}\right),$$

$\exists x \in (P_{\mathbb{J}_\mu})^{int}$, $\forall k \in \mathbb{Z}/|\mathbb{J}|\mathbb{Z}$. This definition can be extended to any rational number ν if we use the convention $(R_{A_\mathbb{J}})^{\nu} = (R_{A_\mathbb{J}})^{\lfloor \nu \rfloor} \cdot (R_{A_\mathbb{J}})^{\kappa}$ for $\kappa = \nu - \lfloor \nu \rfloor$.

In this chapter, actually we look at only the case where $\nu = \pm\frac{1}{2}$, while in [21] we shall deal with rotations of any rational power for algebraic functions.

17.3.2 Braid Monodromy Induced by Path Groupoids

Let us consider a continuous curve $\lambda_{P_i,P_{i+1}}$ on $\mathbb{C}^* \setminus \mathbb{D}$ starting from P_i that ends the point P_{i+1}.

It is easy to see that by composition

$$\lambda_{P_i,P_{i+2}} = \lambda_{P_i,P_{i+1}} \cdot \lambda_{P_{i+1},P_{i+2}}, \tag{17.3.12}$$

these continuous curves define a groupoid $\pi_1(\mathbb{C}^* \setminus \mathbb{D}, B)$ [4, Example 3, Example 5]. If we accept the whole space $\mathbb{C}^* \setminus \mathbb{D}$ as set B of base points, we get the fundamental groupoid where a path connecting two points is an element of a homotopy class.

We denote by ρ the standard homomorphism from the path groupoid to that of monodromy transformations acting on $\mathbb{S}^*(\bullet)$. For a path λ connecting t_1 to t_2 in $\mathbb{C}^* \setminus \mathbb{D}$, we find a diffeomorphism $\rho(\lambda)(\mathbb{S}^*(t_1)) = \mathbb{S}^*(t_2)$, in such a way that

$$\rho : \text{path } \lambda \mapsto \text{ diffeomorphism } \rho(\lambda). \tag{17.3.13}$$

By construction, the family of diffeomorphisms $\rho(\bullet)$ induced by the paths gives rise to a groupoid of monodromy transformations acting on the fibre $\mathbb{S}^*(\bullet)$ that corresponds to the composition of paths (17.3.12),

$$\rho(\lambda_{P_i,P_{i+2}}) = \rho(\lambda_{P_i,P_{i+1}}) \cdot \rho(\lambda_{P_{i+1},P_{i+2}}).$$

Such a continuous family of differomorphisms induced by the path λ has been used to describe the Picard–Lefschetz transformation of the vanishing cycles brought about by a Dehn twist [9, 1.1].

The homomorphism (17.3.13) is defined up to isotopy. Precisely, if two continuous paths on $\mathbb{C}^* \setminus \mathbb{D}$,

$$s \in [0, 1] \longrightarrow \lambda_\ell = \{t_\ell(s)\}, \quad \ell = 1, 2, \tag{17.3.14}$$

are homotopically equivalent, then each of the $2m$ trace curves $\mathbb{S}^*(\lambda_1) = \bigcup_{s\in[0,1]} \mathbb{S}^*(t_1(s))$ is isotopic to one of the trace curves $\mathbb{S}^*(\lambda_2) = \bigcup_{s\in[0,1]} \mathbb{S}^*(t_2(s))$. In this case, we say that $\rho(\lambda_1)$ is isotopic to $\rho(\lambda_2)$.

Remark 17.3.6 It is well known that the diffeomorphism ρ acting $\mathbb{S}^*(t)$, $t \notin \mathbb{D}$, can be identified with an element of the annular braid group CB_{2m} (Remark 17.3.4). The unique difference from the classical case [6, corollaire 2.8], concerning the diffeomorphism of a disc with $(2m + 1)$ perforated punctures, consists in the fact that the origin does not move in our case.

Now we pass to a concrete description of the monodromy transformation (17.3.13) in terms of the underlying path (Fig. 17.4).

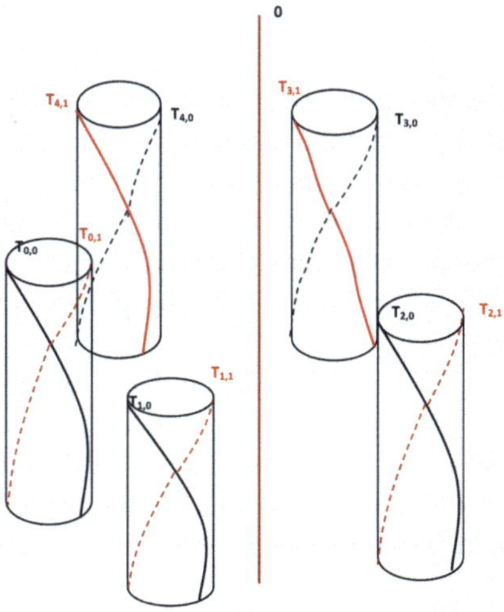

Fig. 17.4 Rotation $\rho(\mu^+_{1/2}\cdot\mu^+_{2/2}) = \prod_{j=0}^{m-1} R_{A^{(0,1)}_{j,j}}$, $m = 5$

To make the formulation simpler, we introduce the union of non-zero solutions in groups as follows,

$$A^{(0,1)}_{j,j}(t) = \{T_{j,0}(t), T_{j,1}(t)\}, \quad j \in [0; m-1], \tag{17.3.15}$$

$$A^{(1)}_{[0;m-1]} = \{T_{j,1}\}_{j=0}^{m-1}, \quad resp.\ A^{(0)}_{[0;m-1]} = \{T_{j,0}\}_{j=0}^{m-1}. \tag{17.3.16}$$

Here the punctures are arranged in such a way that $Arg(T_{j,\delta}) < Arg(T_{j+1,\delta})$, $(j, \delta) \in [0; m-1] \times [0; 1]$. For each fixed $\delta \in [0; 1]$, the convex hull of $A^{(\delta)}_{[0;m-1]}$ contains the origin. See Fig. 17.5.

For each $j \in [0; m-1]$, we look at the following pair of punctures,

$$A^{(0,1)}_{j,j+h} = \{T_{j,0}, T_{j+h,1}\}, \quad h \in [0; m-1]. \tag{17.3.17}$$

Here we use the convention $j + h \in \mathbb{Z}/m\mathbb{Z}$. In particular, we have

$$A^{(0,1)}_{j,j-k} = \{T_{j,0}, T_{m+j-k,1}\}, \quad k \in [1; m-r-1]. \tag{17.3.18}$$

The origin remains outside the convex hull of the punctures $A^{(0,1)}_{j,j-k}$ for each fixed (j, k).

We also introduce the following rotation,

$$R_{A^{(0,1)}_{[+\ell]}} = \prod_{j=0}^{m-1} R_{A^{(0,1)}_{j,j+\ell}}, \tag{17.3.19}$$

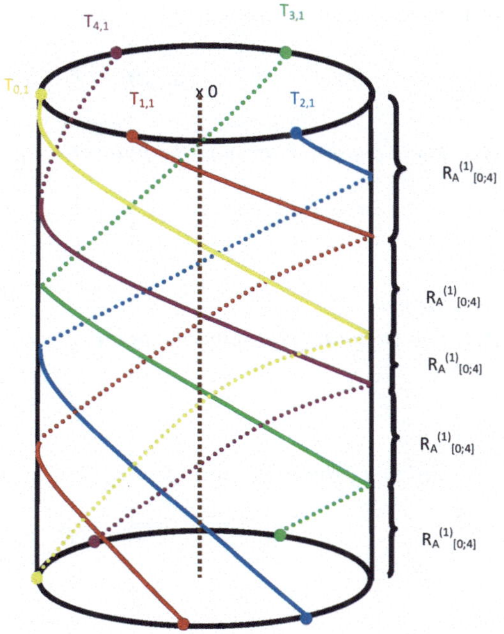

Fig. 17.5 Rotation $(R_{A^{(1)}_{[0;4]}})^5$

that can be regarded as an annular braid (Remark 17.3.4) for $\ell \in [0; r]$. The composition of (17.3.19), $\ell \in [0; k-1]$, produces a concatenation of rotations,

$$R_{A^{(0,1)}_{[\delta;k-1]}} := R_{A^{(0,1)}_{[\delta]}} \cdots R_{A^{(0,1)}_{[+(k-1)]}}, \tag{17.3.20}$$

for $\delta \in [0; 1]$.

In a similar manner, we introduce

$$R_{A^{(0,1)}_{[-\ell]}} := \prod_{j=0}^{m-1} R_{A^{(0,1)}_{j,j-\ell}}, \tag{17.3.21}$$

for $\ell \in [1, m-r-1]$.

The composition of rotations (17.3.21), $\ell \in [1, k-1]$, yields a concatenation,

$$R_{A^{(0,1)}_{-[1;k-1]}} := R_{A^{(0,1)}_{[-1]}} \cdots R_{A^{(0,1)}_{[-(k-1)]}}. \tag{17.3.22}$$

In the following theorem that is the central statement of this chapter, we use the isomorphism between the diffeomorphism group $\mathbb{S}^*(-\epsilon\mathbf{i})$ and the annular braid group CB_{2m} mentioned in Remark 17.3.6.

Theorem 17.3.7 *The following list contains all generators of the group of monodromy transformations* (17.3.13) *for the line arrangement* $\mathcal{A}_{m,2}$.

(A) *The monodromy around the origin*

$$\rho(\mu^{+}_{1/2} \cdot \mu^{+}_{2/2} \cdot (\mu^{-}_{2/2})^{-1} \cdot (\mu^{-}_{1/2})^{-1}) = (R_{A^{(0,1)}_{[0]}})^2. \tag{17.3.23}$$

(B) *The monodromy around the points* $t = \eta$,

$$\rho(\mu^{\eta}_{1/2} \cdot \lambda_\eta \cdot (\mu^{\eta}_{1/2})^{-1}) = (R_{A^{(\frac{\eta+1}{2})}_{[0;m-1]}})^m, \tag{17.3.24}$$

for $\eta = \pm 1$.

(C) *The monodromy around the point* $t = p_{m-k}, k \in [1; m-r-1]$,

$$(R_{A^{(0,1)}_{[-k]}})^2, \tag{17.3.25}$$

and that around the point $t = p_k, k \in [1; r]$, (17.3.8),

$$(R_{A^{(0,1)}_{[+k]}})^2. \tag{17.3.26}$$

Proof

(A) As t moves on the arc $\mu^{-}_{1/2}$ connecting $-\epsilon\mathbf{i}$ to $-\epsilon$ (see Fig. 17.1), each of m pairs of solutions (17.3.15) achieves a turn with angle $-\pi/2$ with respect to the point $-\omega^j$ that can be represented as

$$\rho(\mu^{-}_{1/2}) = (R_{A^{(0,1)}_{[0]}})^{-1/2}, \tag{17.3.27}$$

thanks to the commutation of rotations $R_{A^{(0,1)}_{j}}$, $j \in [0; m-1]$.

In a similar manner, we have

$$\rho(\mu^{-}_{1/2} \vec{\cdot} \mu^{-}_{2/2}) = (R_{A^{(0,1)}_{[0]}})^{-1}. \tag{17.3.28}$$

We compose the paths from left to right.

$$\mu^{-}_{1/2} \vec{\cdot} \mu^{-}_{2/2} = \{-\mathbf{i}e^{-\mathbf{i}\pi u}; u \in [0, \frac{2}{2}]\}. \tag{17.3.29}$$

Analogously we have

$$\rho(\mu^{+}_{\ell/2}) = (R_{A^{(0,1)}_{[0]}})^{1/2}, \tag{17.3.30}$$

for

$$\mu^+_{\ell/2} = \{-\mathbf{i}e^{\mathbf{i}\pi u}; u \in [\frac{\ell-1}{2}, \frac{\ell}{2}]\}, \quad \ell = 1, 2.$$

Here we remark that the full turn $\{-\mathbf{i}e(u); u \in [0, 1]\}$ leads to a braid monodromy

$$(R_{A^{(0,1)}_{[0]}})^2.$$

(B) Now we shall examine the braid monodromy induced by the movement on the path λ_1 : firstly t varies on the segment $[\epsilon, 1-\epsilon]$, then it makes a positive turn around $t = 1$ along the circle $\{1 - \epsilon e(u); u \in [0, 1]\}$, and finally it comes back to ϵ in following the segment $[1-\epsilon, \epsilon]$.

By virtue of Lemma 17.3.2 (1), the movement along λ_1 (resp. λ_{-1}) brings on the rotation

$$\rho(\lambda_1) = (R_{A^{(1)}_{[0;m-1]}})^m, \quad resp. \; \rho(\lambda_{-1}) = (R_{A^{(0)}_{[0;m-1]}})^m \tag{17.3.31}$$

for (17.3.16).

Consequently, the movement along the loop $\mu^+_{1/2} \cdot \lambda_1 \cdot (\mu^+_{1/2})^{-1}$ causes the braid monodromy

$$\rho(\mu^+_{1/2} \cdot \lambda_1 \cdot (\mu^+_{1/2})^{-1}) = (R_{A^{(0,1)}_{[0]}})^{1/2}(R_{A^{(1)}_{[0;m-1]}})^m(R_{A^{(0,1)}_{[0]}})^{-1/2}, \tag{17.3.32}$$

by virtue of (17.3.30), (17.3.31). In a similar manner, we have

$$\rho(\mu^-_{1/2} \cdot \lambda_{-1} \cdot (\mu^-_{1/2})^{-1}) = (R_{A^{(0,1)}_{[0]}})^{-1/2}(R_{A^{(0)}_{[0;m-1]}})^m(R_{A^{(0,1)}_{[0]}})^{1/2}. \tag{17.3.33}$$

We can easily see that the rotations $(R_{A^{(0,1)}_{[0]}})^{\pm\frac{1}{2}}$, and $R_{A^{(0)}_{[0;m-1]}}$ commute. To see this, let us consider a tube B_j with small enough diameter that contains $A^{(0,1)}_{j,j}$ (17.3.15) and we set

$$B_{[0;m-1]} = \cup_{j=0}^{m-1} B_j. \tag{17.3.34}$$

The rotation $R_{B_{[0;m-1]}}$ is isotopic to the rotation $R_{A^{(0)}_{[0;m-1]}}$. As the rotation $R_{A^{(0,1)}_{j,j}}$ takes place inside the tube B_j, it commutes with $R_{A^{(0)}_{[0;m-1]}}$.

(C) Now we examine the braid monodromy induced by the movement from $t = -\epsilon \mathbf{i}$ to the point of coincidence,

$$p_{m-k} = \mathbf{i}\tan(\frac{m-k}{m}) = -\mathbf{i}\tan(\frac{k}{m}), \quad k \in [1; r+1] \tag{17.3.35}$$

with

$$r := \lfloor \frac{m-1}{2} \rfloor.$$

The motion starting from $t = -\epsilon\mathbf{i}$ and arriving at $t = \epsilon\mathbf{i} + p_{m-1}$, in following first a straight line, then the circle $p_{m-1} + \epsilon e(u)$, $u \in [0; 1]$, induces the rotation

$$(R_{A^{(0,1)}_{[-1]}})^2 = \prod_{j=0}^{m-1} (R_{A^{(0,1)}_{j,j-1}})^2, \tag{17.3.36}$$

as at the point $t = p_{m-1}$, two punctures $A^{(1)}_{j,j-1}$ (17.3.18) meet.

In Fig. 17.6, we see that the rotations $R_{A^{(0,1)}_{j,j-1}}$, $j \in [1; m-1]$, commute between another.

To examine the monodromy caused at the points $t = p_{m-k}$, $k \in [2; r+1]$, we shall first follow a half turn, first in positive then in negative direction, near the points $t = p_{m-\ell}$, $\ell \in [1; k-1]$ as it is shown on Fig. 17.1, where the return movement is omitted.

At the point $t = p_{m-\ell}$, $\ell \in [1; k-1]$, during a positive half turn $p_{m-\ell} + \epsilon e(u)$, $u \in [0; 1/2]$, the solutions $\mathbb{S}^*(t)$ (17.3.2) achieve the rotation $R_{A^{(0,1)}_{[-\ell]}}$ (17.3.21).

Fig. 17.6
$\mathbb{S}(\epsilon\mathbf{i} + p_{m-1}), m = 5$

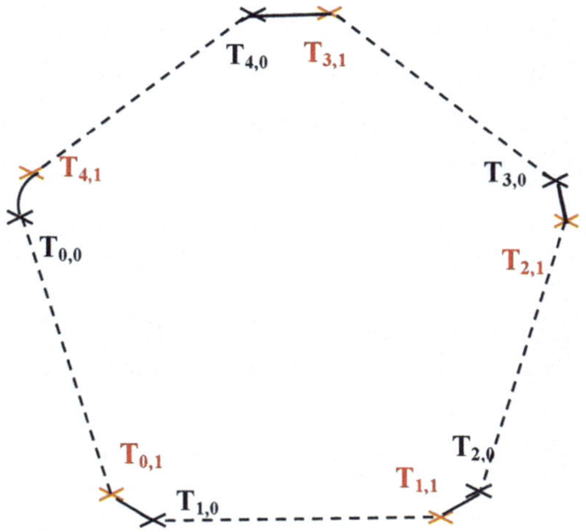

The positive turn around all points $t = p_{m-\ell}, \ell \in [1; k-1]$ gives rise to the composition $R_{A^{(0,1)}_{-[1;k-1]}}$ (17.3.22).

The full turn around $t = p_{m-k}$ (17.3.22) produces the rotation

$$(R_{A^{(0,1)}_{[-k]}})^2 = \prod_{j=0}^{m-1} (R_{A^{(0,1)}_{j,j-k}})^2, \tag{17.3.37}$$

by virtue of the definition in formula (17.3.21) applied to the case $\ell = k$.

The negative turn around all points $t = p_{m-\ell}, \ell \in [1; k-1]$, induces $R^{-1}_{A^{(0,1)}_{-[1;k-1]}}$.

Thus the loop starting from the base point $t = -\epsilon \mathbf{i}$ that makes a positive turn around $t = p_{m-k}$, following the path on Fig. 17.1, brings about the braid monodromy

$$R_{A^{(0,1)}_{-[1;k-1]}} \cdot (R_{A^{(0,1)}_{[-k]}})^2 \cdot R^{-1}_{A^{(0,1)}_{-[1;k-1]}}. \tag{17.3.38}$$

Figure 17.7 shows that the rotation (17.3.38) is isotopic to the rotation $(R_{A^{(0,1)}_{[-k]}})^2$ for the case $k = 2$.

By an inductive manner, we show the reduction of (17.3.38) to (17.3.25) for k general. The argument relies on the following observation.

During the rotation (17.3.38), the strings $T_{j-h,\delta}, (h, \delta) \in [1; k-1] \times [0; 1]$, remain either in front of or on the back of $A^{(0,1)}_{j,j-k}$ (17.3.18) for each $j \in [0; m-1]$. That is to say, in (17.3.38) the braid $R_{A^{(0,1)}_{j,j-k}}$ is not conjugate with the braids affected by $T_{j-h,\delta}, (h, \delta) \in [1; k-1] \times [0; 1]$. Other strings $T_{j-\ell,\delta}, (\ell, \delta) \in [k+1; m-1] \times [0; 1]$, are located so far from the strings $T_{j-h,\delta}, (h, \delta) \in [0; k] \times [0; 1]$, that they do not entangle with the latters. Making use of the widely spread notation [11], we can summarize this relation by the formula

$$\sigma_i(\sigma_j)^2(\sigma_i)^{-1} = (\sigma_j)^2, \ \ |i - j| \neq 1, 2m-1,$$

for $\sigma_i, \sigma_j \in CB_{2m}$.

Relying on a similar argument, we can also reduce (17.3.33) to the rotation $(R_{A^{(0)}_{[0;m-1]}})^m$.

As for the braid monodromy caused by the full turn around the point $t = p_k, k \in [1; r]$, its description is parallel to that for the cases $k \in [r+1; m-1]$ given in (17.3.38).

The rotation $\rho(\mu^+_{1/2} \vec{\cdot} \mu^+_{2/2})$ (17.3.30) caused by the half turn around the origin is followed by the rotations induced by the half turns $R_{A^{(0,1)}_{[+\ell]}}$ (17.3.19) around the points $t = p_\ell, \ell \in [1; k-1]$. Their composition is denoted by like $R_{A^{(0,1)}_{[1;k-1]}}$ (17.3.20).

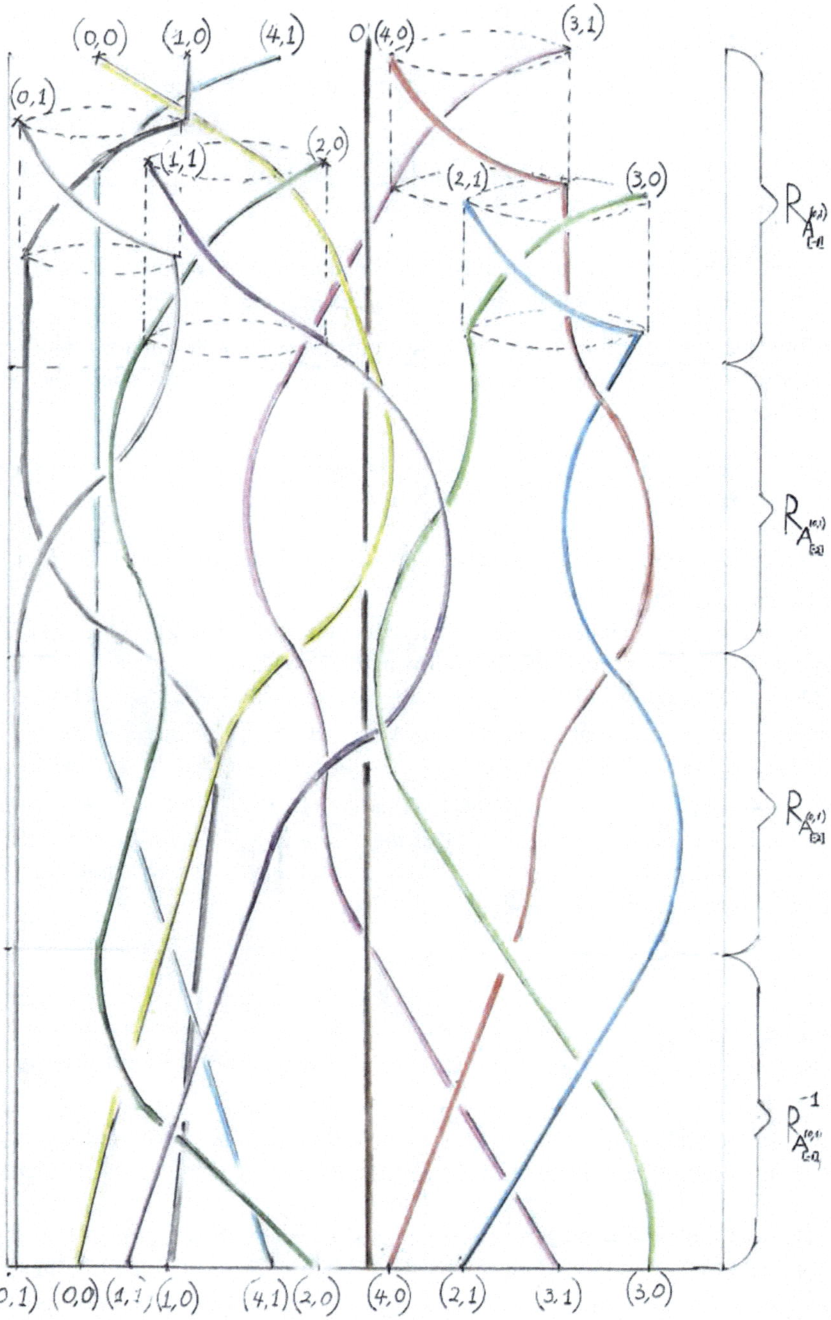

Fig. 17.7 $R_{A^{(0,1)}_{[-1]}} \cdot (R_{A^{(0,1)}_{[-2]}})^2 \cdot R^{-1}_{A^{(0,1)}_{[-1]}}$ for $m = 5$

The full turn around $t = p_k$, (17.3.19) produces the rotation

$$(R_{A^{(0,1)}_{[+k]}})^2 = \prod_{j=0}^{m-1} (R_{A^{(0,1)}_{j,j+k}})^2. \tag{17.3.39}$$

Thus the loop starting from the base point $t = -\epsilon \mathbf{i}$ that makes a positive turn around $t = p_k$, $k \in [1; r]$, after having followed the path of Fig. 17.1, brings on the braid monodromy

$$R_{A^{(0,1)}_{[0;k-1]}} \cdot (R_{A^{(0,1)}_{[+k]}})^2 \cdot R^{-1}_{A^{(0,1)}_{[0;k-1]}} \tag{17.3.40}$$

for $k \in [1; r]$.

To see that the rotation

$$(R_{A^{(0,1)}_{[0]}}) \cdot (R_{A^{(0,1)}_{[+k]}})^2 \cdot (R_{A^{(0,1)}_{[0]}})^{-1} \tag{17.3.41}$$

is isotopic to (17.3.26), we can refer to an argument similar to the reduction of (17.3.38) to (17.3.25).

Figures 17.8 and 17.9 illustrate the fact that the rotation $(R_{A^{(0,1)}_{[0]}}) \cdot (R_{A^{(0,1)}_{[+k]}})^2 \cdot (R_{A^{(0,1)}_{[0]}})^{-1}$ is isotopic to $(R_{A^{(0,1)}_{[+k]}})^2$ for $k = 1$. □

To establish a description of the fundamental group $\pi_1(\mathbb{C}^2 \setminus \mathcal{A}_{m,2})$, we consider the loops

$$T = \langle \tau_{j,0}, \tau_{j,1} \rangle_{j=0}^{m-1} \cup \{\tau_0\} \tag{17.3.42}$$

that form a system of distinguished paths [7, Definition 2], [9, Definition 1.2.3],[12, Definition 1.2] of $\mathbb{S}(-\epsilon \mathbf{i})$. The loop $\tau_{j,\delta}$ starting from a fixed point makes a positive turn around the puncture $T_{j,\delta}(-\epsilon \mathbf{i})$, $(j, \delta) \in [0; m-1] \times [0; 1]$ (Fig. 17.10).

Now let us consider the monodromy action ρ induced by the loop $\tau_{j,\delta}$ $\rho_*(\tau_{j,\delta})$ (Fig. 17.11).

In this way, we can deduce the following corollary from Theorem 17.3.7.

Corollary 17.3.8 *We get the following description of the fundamental group of the complement of the arrangement $\mathcal{A}_{m,2}$ in terms of the generators T* (17.3.42),

$$\pi_1(\mathbb{C}^2 \setminus \mathcal{A}_{m,2}) = \{T; \tau_{j,\delta} = \rho_*(\tau_{j,\delta}), (j, \delta) \in [0; m-1] \times [0; 1]\}. \tag{17.3.43}$$

Here ρ denotes the diffeomorphism of $\mathbb{S}^*(-\epsilon \mathbf{i})$ brought on by one of the rotations (17.3.23), (17.3.24), (17.3.25), (17.3.26) *from the list of Theorem 17.3.7.*

The proof is nothing but a direct application of Lemma 1.14 from [12] (Zariski–Van Kampen Theorem) to our situation.

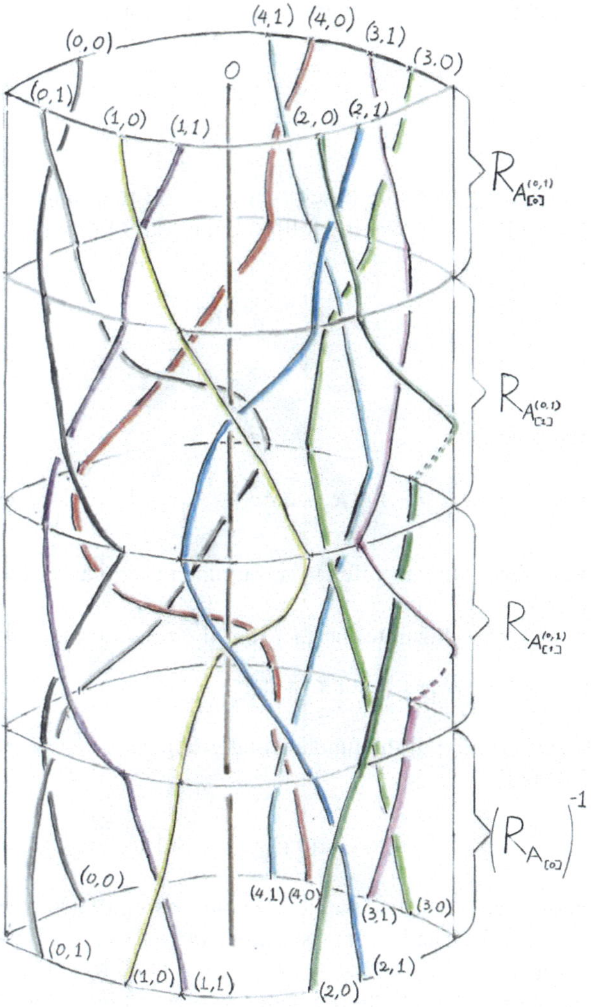

Fig. 17.8 Rotation $(R_{A^{(0,1)}_{[0]}}) \cdot (R_{A^{(0,1)}_{[1]}})^2 \cdot (R_{A^{(0,1)}_{[0]}})^{-1}$ for $m = 5$

In guise of a conclusion, we propose the following conjecture in view of a regularity that we observe in the expression of the rotations (17.3.23), (17.3.24), (17.3.25), (17.3.26). Namely, Theorem 17.3.7 implies that the global monodromy for the arrangement $\mathcal{A}_{m,2}$ is completely determined by the local monodromy.

To formulate the conjecture, we introduce some notation, effectual for general value of n.

Fig. 17.9 Rotation $(R_{A^{(0,1)}_{[1]}})^2$ for $m = 5$

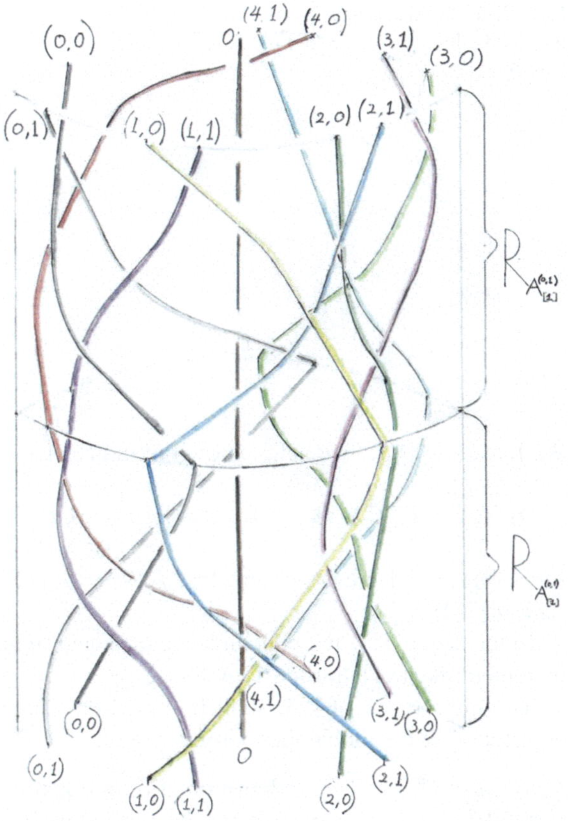

Fig. 17.10 Distinguished paths for $m = 3$

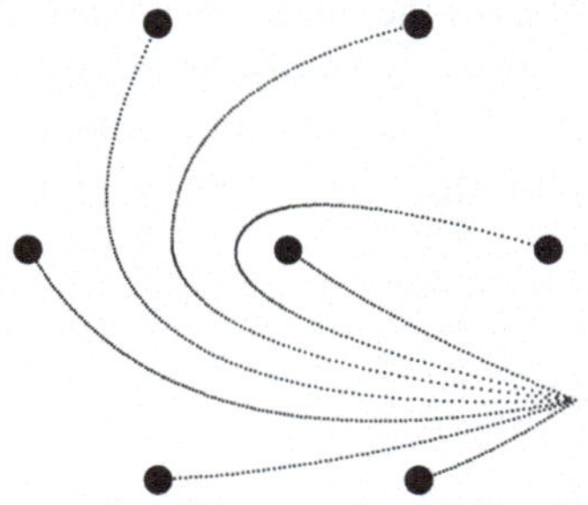

The solutions $s(t)$ of (17.2.8) can be presented as

$$T_{\mathbf{k}_\alpha}(t) = -\omega_m^{j_\alpha}(1 + \omega_n^{\ell_\alpha} t) \tag{17.3.44}$$

for $\mathbf{k}_\alpha = (j_\alpha, \ell_\alpha) \in [0; m-1] \times [0; n-1]$.

We denote the set of coincidence points $\mathbb{D}_{m,n}$ by

$$\mathbb{D}_{m,n} = \{t \in \mathbb{C};\, T_{\mathbf{k}_\alpha}(t) = T_{\mathbf{k}_\beta}(t), \exists \mathbf{k}_\alpha \neq \mathbf{k}_\beta \in [0; m-1]\times[0; n-1]\}. \tag{17.3.45}$$

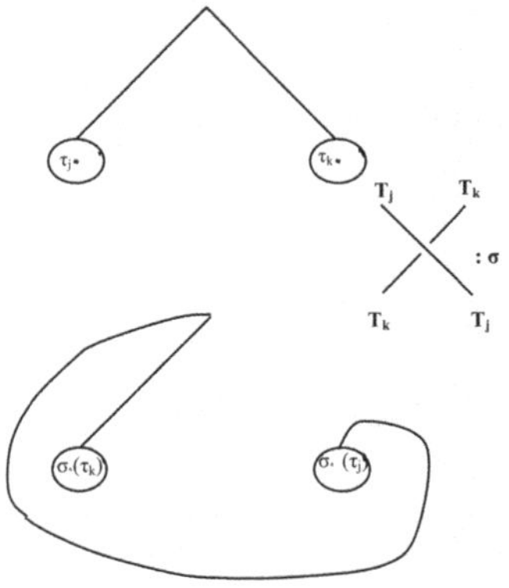

Fig. 17.11 Braid action σ induced by loops τ_j, τ_ℓ : $\sigma_*(\tau_j) = \tau_j \tau_\ell {\tau_j}^{-1}$

For fixed $\xi \in \mathbb{D}_{m,n}$, we define the set of indices $\{\Lambda_\mu(\xi)\}_{\mu=1}^{\nu_\xi}$ such that

$$T_{\mathbf{k}_\alpha}(\xi) = T_{\mathbf{k}_\beta}(\xi), \forall \mathbf{k}_\alpha \neq \mathbf{k}_\beta \in \Lambda_\mu(\xi) \subset [0; m-1] \times [0; n-1] \qquad (17.3.46)$$

for $\mu \in [1; \nu_\xi]$. Here we assume that $T_{\mathbf{k}}(\xi) \neq T_{\mathbf{h}}(\xi)$, $\forall \mathbf{k} \in \Lambda_\mu(\xi)$, $\forall \mathbf{h} \notin \Lambda_\mu(\xi)$ for each $\mu \in [1; \nu_\xi]$.

In the set $\Lambda_\mu(\xi)$, the punctures are arranged in accordance with their argument increment like in Definition 17.3.3.

The main result of [13] shall be based on Conjecture 17.3.9 on an analogous description of the global monodromy group.

Conjecture 17.3.9 The global monodromy transformation group for the arrangement $\mathcal{A}_{m,n}$, $n \geq 2$, is completely determined by its local monodromy transformations of the linear functions (17.3.44).

In other words, as the parameter t moves along a loop starting from the point $p \in \mathbb{C} \setminus \mathbb{D}_{m,n}$ that turns around ξ and no other points of $\mathbb{D}_{m,n}$ (17.3.45), the punctures $\{T_{\mathbf{k}}(t)\}_{\mathbf{k} \in \Lambda_\mu(\xi)}$, $\mu \in [1; \nu_\xi]$, (17.3.46) undergo the monodromy transformation

$$(R_{A_{\Lambda_\mu(\xi)}})^{|\Lambda_\mu(\xi)|}.$$

In particular, we remark that

$$|\Lambda_1(0)| = \cdots = |\Lambda_m(0)| = n$$

$\forall (m, n) \in (\mathbb{Z}_{\geq 2})^2$.

References

1. E. Artin, Theory of braids. Ann. Math. **48**, 101–126 (1947)
2. V.V. Batyrev, Dual polyhedra and mirror symmetry for Calabi-Yau hypersurfaces in toric varieties. J. Algebraic Geometry **3**, 493–535 (1994)
3. P. Berglund, Ph. Candelas, X. de la Ossa, A. Font, T. Hübsch, D. Jancic, F. Quevedo, Periods for Calabi-Yau variety and Landau-Ginzburg vacua. Nucl. Phys. **B 419**, 352–403 (1994)
4. R. Brown, From groups to groupoids: a brief survey. Bull. Lond. Math. Soc. **19**, 113–134 (1987)
5. D. Cox, S.Katz, *Mirror Symmetry and Algebraic Geometry*. Math Surveys and Monographs, vol. 68 (AMS, Providence, 1999)
6. A. Fathi, F. Laudenbach, V. Poénaru, Travaux de Thurston sur les surfaces. Astérisque **66** (1979)
7. A.M. Gabrielov, Intersection matrices for some singularities. Funct. Anal. Appl. **7**(3), 182–193 (1973)
8. A. Grothendieck, in *Esquisse d'un programme*. Geometric Galois Actions. 1, vol. 242. ed. by L. Schneps, P. Lochak (1997). LMS Lecture Note Series
9. S.M. Gusein-Zade, The monodromy groups of isolated singularities of hypersurfaces. Russ. Math. Surv. **32**(2), 23–69 (1977)
10. R.P. Horja, Hypergeometric functions and mirror symmetry in toric varieties. math.AG/9912109
11. R.P. Kent, D. Peifer, A geometric and algebraic description of annular braid groups. Int. J. Algebra Comput. **12**, 85–97 (2002)
12. M. Lönne, Braid monodromy of hypersurface singularities. Habilitationsschrift, Universität Hannover (2003)
13. M. Lönne, S. Tanabé, On discriminant loci of a family of bi-degree (m,n) affine hypersurfaces, in preparation
14. D. Margalit, J. McCammond, Geometric representations for the pure braid group. J. Knot Theory Ramif. **13**(1), 1–20 (2009)
15. S. Y. Orevkov, The fundamental group of the complement of a plane algebraic curve. Mat. Sb. (N.S.) **137**(179), 260–270 (1988)
16. E. Paul, J.P. Ramis, Dynamics on wild character varieties. SIGMA **11**, 068, 21 pp. (2015)
17. J. Stienstra, GKZ hypergeometric structures, in *Arithmetic and Geometry Around Hypergeometric Functions* (Istanbul, 2005) (Birkhäuser, Basel, 2007), pp. 313–372
18. S. Tanabé, On Horn-Kapranov uniformisation of the discriminantal loci. Adv. Stud. Pure Math. **46**, 223–249 (2007)
19. S. Tanabé, On monodromy representation of period integrals associated to an algebraic curve with bi-degree (2, 2). Analele Ştiinţifice ale Universităţii Constanţa, Seria Matematica **25**(1), 207–231 (2017)
20. S. Tanabé, Monodromy of period integrals associated to hypersurfaces in toric variety via Mellin-Barnes integrals, in preparation
21. S. Tanabé, M. Koçar, Braid monodromy for trinomial algebraic equation in terms of Mellin-Barnes integral representations, submitted
22. T. Terasoma, Fundamental group of non-singular locus of Lauricella F_C, arXiv:1803.06609
23. H. Zieschang, E. Vogt, H.-D. Coldewey, *Surfaces and Planar Discontinuous Groups*. Lecture Notes in Math., vol. 835 (1980)

Chapter 18
Welded Graphs, Wirtinger Groups and Knotted Punctured Spheres

Benjamin Audoux, Jean-Baptiste Meilhan, and Akira Yasuhara

Abstract We develop a general diagrammatic theory of welded graphs, and provide an extension of Satoh's Tube map from welded graphs to ribbon surface-links. As a topological application, we obtain a complete link-homotopy classification of so-called *knotted punctured spheres* in 4-space, by means of the 4-dimensional Milnor invariants introduced previously by the authors. On the algebraic side, we show that the theory of welded graphs can be reinterpreted as a theory of Wirtinger group presentations, up to a natural set of transformations; a group admitting such a presentation arises as the fundamental group of the exterior of the surface-link obtained from the given welded graph by the extended Tube map. Finally, we address the injectivity question for the Tube map, identifying a new family of local moves on welded links, called Υ moves, under which the (non extended) Tube map is invariant.

Keywords Surface-links in 4-space · Welded knotted objects · Gauss diagrams · Wirtinger group · Tube map

2020 Mathematics Subject Classification 57K45, 57K12, 20F05, 20F34, 57M15

B. Audoux
Aix Marseille Univ, CNRS, Centrale Marseille, I2M, Marseille, France
e-mail: benjamin.audoux@univ-amu.fr

J.-B. Meilhan (✉)
Univ. Grenoble Alpes, CNRS, Institut Fourier, Grenoble, France
e-mail: jean-baptiste.meilhan@univ-grenoble-alpes.fr

A. Yasuhara
Faculty of Commerce, Waseda University, Shinjuku-ku, Tokyo, Japan
e-mail: yasuhara@waseda.jp

L. Funar, A. Papadopoulos (eds.), *Essays on Topology*,
https://doi.org/10.1007/978-3-031-81414-3_18

18.1 Introduction

A *surface-link* in 4-space is the image of an embedding of a compact oriented surface in $\mathbb{R}^4$ up to ambient isotopy. The first examples of non trivial surface-links were spun links, described in the mid-twenties by Artin [1]; these are obtained as the trace of the 2π-rotation of a tangle in a half 3-dimensional (3D) subspace of $\mathbb{R}^4$ around the boundary of this subspace. Such links happen to be *ribbon* surface-links, meaning that they bound immersed 3D handlebodies in $\mathbb{R}^4$ with only *ribbon singularities*, locally shaped as the 3D thickening of an arc passing transversally through a 3-ball. Not all surface-links are ribbon, but they form an important subclass which has been extensively studied in the late sixties [31–34], when the systematic study of surface-links began. In several ways, ribbon surface-links are easier to handle than general surface-links. Indeed, they admit *simple projections*, which are projections in $\mathbb{R}^3$ with only double points—while a generic projection of a surface-link contains in general triple and branch points; for embedded spheres, being ribbon is actually equivalent to having a simple projection [31]. Ribbon surface-links also admit 3D *ribbon presentations*, see e.g. [21], which are certain ribbon-immersed disks in 3-space. But despite their simplicity, ribbon surface-links can have a central role in the study of surface-links, such as in [6] where 2-dimensional string links are classified up to link-homotopy using the fact that any such surface-link is link-homotopic to a ribbon one (a result which is generalized by Proposition 18.4.3, see below).

In 2000, ribbon surface-links received renewed interest when Satoh extended Yajima's Tube map [30]—a construction which "inflates" classical knot diagrams into ribbon surface-link diagrams—to *welded knotted objects*, and showed that this yields a surjective map from welded knots and knotoids to ribbon tori and spheres [27]; he also showed that the Tube map coincides with Artin's spun map on classical knots. Welded knotted objects were first introduced by Fenn–Rimanyi–Rourke [14] for the 4-dimensional study of motions of unknotted circles in $\mathbb{R}^3$, and can be seen as a quotient of the virtual (or, equivalently, Gauss) diagram theory. Via Satoh's Tube map, welded knotted objects provide a 2-dimensional description of ribbon surface-links, which has been used effectively in several recent works [5, 6, 23, 24].

However, welded links and string links cannot describe all type of ribbon surface-links, as it restricts to surfaces with connected components of genus 0 or 1, with 0 or 2 boundary components. To overcome this issue, the present chapter develops a theory of *welded graphs* which generalize welded knots, links and string links. We observe that the Tube map extends naturally to welded graphs (Proposition 18.2.12), thus allowing for the study of the whole class of ribbon surface-links, of any topological type. As a concrete topological application, we obtain the following classification result.

Theorem (Theorem 18.4.4) *Knotted punctured spheres are classified, up to link-homotopy, by non repeated* 4*-dimensional Milnor invariants.*

In this statement, knotted punctured spheres are properly and smoothly embedded unions of punctured spheres whose boundary is a fixed unlink (or slice link, see Remark 18.4.6), forming a vast class of surface-links which contains Le Dimet's linked disks [20] and 2-string links studied in [5, 6]. Link-homotopy is Milnor's classical notion of continuous deformation leaving distinct components disjoint at all time [25], and the classification is given using the 4-dimensional Milnor invariants of surface-links developed in [7], where this result was first announced; see Sect. 18.4 for details.

In a different direction, we use welded graphs to address the injectivity problem for the Tube map on welded knotted objects. An important open problem is indeed to determine the kernel of this map. The Tube map is known to be injective for welded braids [11] and to have a non-trivial kernel in the link case [17, 29], but the general case remains widely open. As developed in this chapter, one advantage of the notion of welded graphs is that it naturally encompasses all the known moves on welded objects for which the Tube map is invariant. Actually, using this setting, we identify a large family of local moves on welded (string) links, the Υ *moves* (Definition 18.2.7), under which the Tube map is invariant:

Proposition (Proposition 18.2.14) *The Tube map is invariant under* Υ *moves.*

These Υ moves do not seem to be induced by usual welded moves, which leads us to conjecture that they provide general obstructions for the injectivity of the Tube map on welded (string) links.

Welded graphs already appeared in the literature [9, 19], in closely related forms. Our approach is in comparison deliberately more combinatorial, given in terms of oriented graphs with edges labeled by elements in the free group generated by the vertices. This highlights a striking relationship between welded objects and *Wirtinger presentations*, which are finite presentations where each relation identifies a generator with a conjugate of another one. A celebrated result of Wirtinger provides such a presentation for the fundamental group of the complement of a (string) link described by a diagram, and it is well-known that this procedure extends to welded knots and (string) links. One of the main claims of the present chapter is that the theory of welded graphs can actually be recasted as a theory of Wirtinger presentations up to some natural transformations which preserve the associated group; see Theorem 18.3.2 for a precise statement. Moreover, this correspondence is naturally compatible with the extended Tube map, providing a nice connection between combinatorial, algebraic and topological objects, illustrated by the following result:

Proposition (Proposition 18.3.3) *For any welded graph* Γ, *the associated Wirtinger group* $G(\Gamma)$ *is isomorphic to the fundamental group of the exterior of the surface-link* $\mathrm{Tube}(\Gamma)$.

More generally, these Wirtinger presentations associated with welded graphs, yield a whole *peripheral system*, which also corresponds to a topological peripheral system for the associated ribbon surface-links. As a matter of fact, the above-mentioned 4-dimensional Milnor invariants are extracted from this peripheral system, and the above-stated link-homotopy classification of knotted punctured spheres actually follows from a more general diagrammatic result. Specifically, we introduce a notion of *self-virtualization* equivalence (sv) for welded graphs, which is the welded counterpart of link-homotopy for ribbon surface-links, and we prove the following, where *reduced* refers to a quotient of the peripheral system (see Sect. 18.3.3):

Theorem (Theorem 18.3.22) *Two welded graphs are sv-equivalent if and only if they have equivalent reduced peripheral systems.*

The chapter is organized as follows. In Sect. 18.2, welded graphs are introduced, and the relationship with welded links and string links is clarified; the connection with ribbon surface-links via an extension of Satoh's Tube map is also presented. In Sect. 18.3, we introduce Wirtinger presentations and their associated peripheral systems, and the relation to welded graphs is discussed. We also introduce the notion of self-virtualization, and prove our classification result for welded graphs up to self-virtualization. In Sect. 18.4 we return to topology; we prove that every knotted punctured spheres is link-homotopic to a ribbon one, which leads to the classification of these objects up to link-homotopy. Finally, in Sect. 18.5, we give the proof of Theorem 18.2.9 ; further technical results are also shown in Sect 18.6.

Notation For every element a of a group G, the inverse of a can be denoted by $\overline{a}$ or a^{-1}; and for every $a, b \in G$, we set $a^b = \overline{b}ab$ and $[a, b] = \overline{b}ab\overline{a}$. We denote by $F(a_1, \ldots, a_r)$ the free group on the generators $a_1, \ldots, a_r$.

18.2 Welded Graphs

18.2.1 Welded Knotted Objects

Let us start with a quick review of the "usual" welded theory. This theory was initially given in terms of virtual diagrams [18], but we will consider here the Gauss diagram point of view; see for instance [5] for a correspondence between the two approaches.

Gauss diagrams were first introduced as a pictural form for the Gauss code representation of knots. They are abstract and signed arrows whose endpoints are embedded in a compact oriented 1-dimensional manifold that shall be denoted by M hereinafter; see Fig. 18.1 for examples.

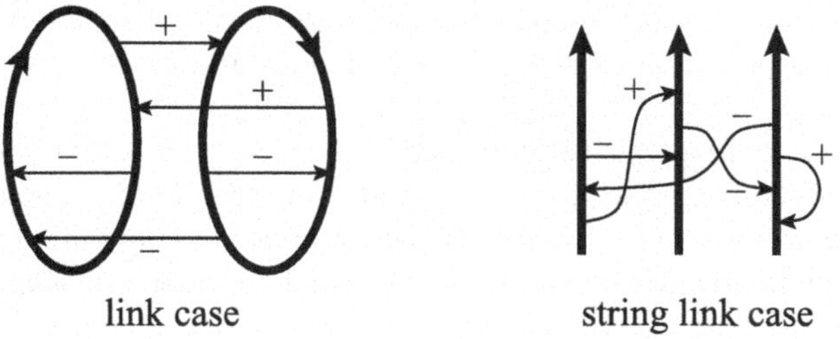

Fig. 18.1 Some Gauss diagrams

Fig. 18.2 Exchanging arrow endpoints

Welded knotted objects are defined as equivalence classes of Gauss diagrams up to the following *welded moves* (here, and in subsequent figures, ε and η are two arbitrary signs):

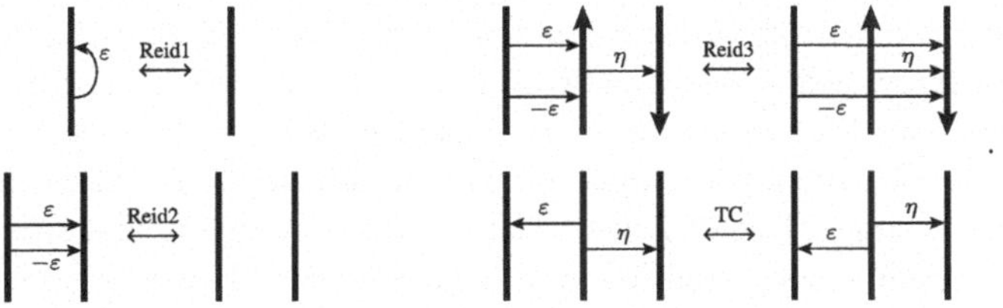

We stress that a TC move allows one to freely exchange the relative position of two adjacent arrow tails on M; TC actually stands for *Tails Commute*. Figure 18.2 represents the *exchange moves*, which are consequences of Reid2 and Reid3 moves (see [5, Fig. 4.1]), and which similarly allow to exchange the relative positions of two adjacent arrow endpoints, up to the addition of a pair of oppositely-signed arrows.

Depending on the topological nature of the 1-dimensional manifold M, one can define various types of welded knotted objects:

- *welded knots* correspond to the case where M is a single copy of S^1, and *welded links* to the case where M is a finite union of (ordered) disjoint copies of S^1;
- *welded string links* correspond to the case where M is a finite union of (ordered) disjoint intervals;
- *welded knotoids* are the quotient of 1-component welded string links, by the move which deletes or creates an arrow whose tail is adjacent to an endpoint of the interval.

We next give some notation and observations that will be useful in Sect. 18.2.3, and that shall also smoothly drag the above definitions towards the notion of welded graphs.

Observe that, up to ambient isotopy and TC moves, the relevant information for an arrow tail is not its precise position, but rather the sub-arc of M, bounded by two arrow heads, where this tail is located. This means that the information carried by an arrow can be summarized by the position of its head, together with a label indicating the sub-arc of M where the tail is located; the arrow sign can in turn be given by an overlining, or a "-1" exponent, on this label if the sign is negative. More generally, a bunch of arrows with adjacent heads can be described by a word obtained by concatenating the corresponding labels. Moreover, observe that, thanks to the TC move, bunches of adjacent arrow tails can be thought of as gathered in a single point. Such a point corresponds to the sub-arc of M containing only these tails and comprised between two arrow heads. This allows us to think of Gauss diagrams as a collection of such "tails localization" points, connected by oriented arcs decorated with words describing the sequence of arrow heads met on this portion of the Gauss diagram.

These rough observations shall be formalized in the next subsection, through the notion of welded graph.

Finally, given an arrow A of a Gauss diagram G, and assigning a letter to every sub-arc of M comprised between two heads and containing no head in its interior, we define the *head (resp. tail) conjugation of* A *by a word* w, as the insertion of the bunch of arrow heads associated with w on one side of the head (resp. tail) of A, and of the bunch of arrow heads associated with $\overline{w}$ on the other side. Notice that inserting such extra arrows splits some of the sub-arcs of M, hence changes the alphabet. For example, Fig. 18.2 can be summarized by saying that an endpoint of some arrow A_1 can be pushed through the adjacent head of an arrow A_2, at the cost of conjugating the other endpoint of A_1 by the one-letter word corresponding to A_2.

18.2.2 Definition of Welded Graphs

Definition 18.2.1 A *w-graph* is an oriented graph Γ with edges decorated by elements of the free group generated by the vertices of Γ, and with a subset of marked vertices.[1] In other words, a w-graph is the data of

- a set V_Γ of *vertices*,
- a set E_Γ of *edges*, made of triplets (a, b, w) where $a, b \in V_\Gamma$, and $w \in F(V_\Gamma)$ is the edge decoration,
- a (possibly empty) subset $V_\Gamma^\circ \subset V_\Gamma$ of *marked vertices*.

[1] These marked vertices should be thought of as "fixed" endpoints as in string links, and unmarked univalent vertices as "free" endpoints as in knotoids.

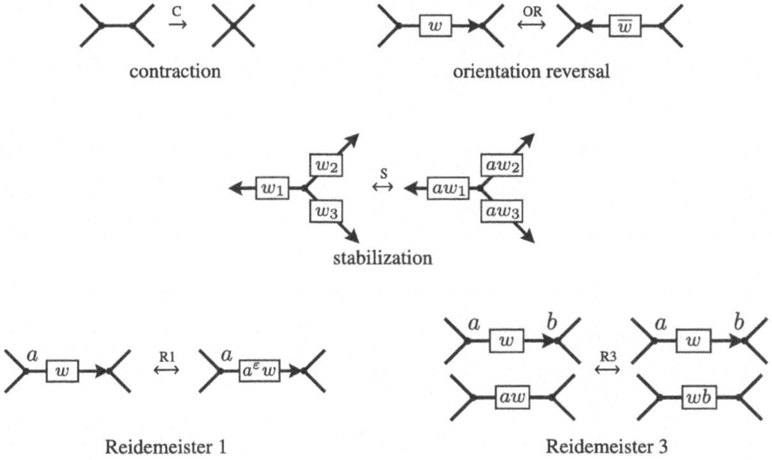

Fig. 18.3 Moves on w-graphs. Here, and in the next figure, all depicted trivalent vertices represent vertices of any valency

The connected components of the graph are ordered, and so are the marked vertices on each connected component. An edge decorated by the unit element 1 of $F(V_\Gamma)$ shall be called an *empty edge*.

In figures, we always represent marked vertices by white dots $\circ$, while unmarked vertices are represented by black dots $\bullet$.

We say that a w-graph is of *type* $\big((m_1, b_1), \ldots, (m_\ell, b_\ell)\big)$ if the underlying graph Γ has ℓ (ordered) connected components, and the i-th component has exactly m_i marked vertices and first Betti number b_i; we simply say that it has type (m, b) if $m_1 = m_2 = \cdots = m_\ell = m$ and $b_1 = b_2 = \cdots = b_\ell = b$.

We define the following moves on w-graphs, illustrated in Fig. 18.3:

- *contraction* (C), which removes an empty edge $(a, b, 1)$ from E_Γ and replaces V_Γ by $V_\Gamma/a \sim b$, where a and b are identified, under the assumption that $b \notin V_\Gamma^\circ$;
- *orientation reversal* (OR), which replaces an edge (a, b, w) by $(b, a, \overline{w})$;
- *stabilization* (S), which adds a common prefix a to all the edges connected to a given vertex b, under the assumption that all the edges are oriented outwards b, that $b \notin V_\Gamma^\circ$ and that all edge decorations of Γ are in $F\big(V \setminus \{b\}\big)$;
- *Reidemeister I* (R1), which replaces an edge (a, b, w) by $(a, b, a^\varepsilon w)$ for some $\varepsilon \in \{\pm 1\}$;
- *Reidemeister III* (R3), which replaces by wb the decoration aw of an edge $e \in E_\Gamma$, under the assumption that $(a, b, w) \in E_\Gamma \setminus \{e\}$.

Remark 18.2.2 The contraction move induces its inverse move, the *expansion* move (E), which duplicates a vertex v into v_1 and v_2, and creates a new empty edge between them. All the edges which were adjacent to v, have to be distributed between v_1 and v_2, and any occurrence of v in edge decorations can be freely replaced by v_1 or v_2.

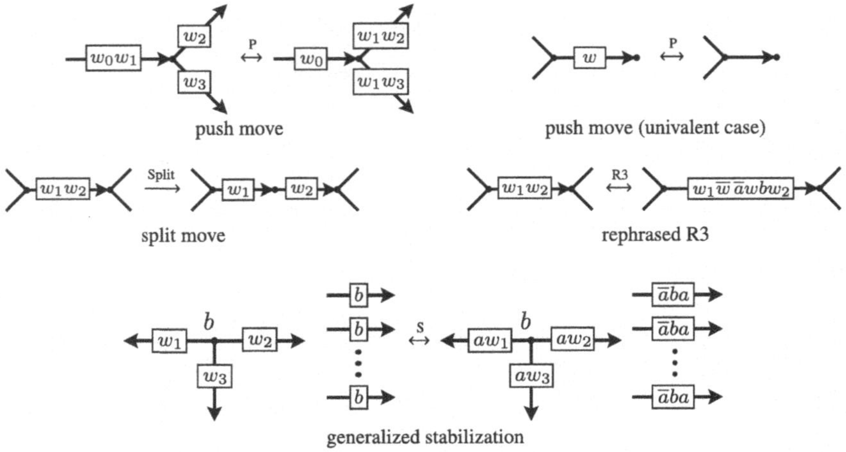

Fig. 18.4 Extra moves on w-graphs

Note that the number of marked vertices, as well as the first Betti number, are preserved by all the above moves.

Definition 18.2.3 A *welded graph* is an element of the set

$$w\text{Graph} := \{w\text{-graphs}\}/\text{C,OR,S,R1,R3},$$

and its *type* is defined as the type of any of its representatives.

More admissible moves can be obtained by combining moves C, OR, S, R1 and R3. We gather below some of these extra moves, illustrated in Fig. 18.4:

- *push move* (P), which pushes, according to the edges orientations, a word through an unmarked vertex that does not appear in any edge decoration; note that, in the case where the involved vertex is univalent, the push move simply erases the word on the incident edge;
- *split move* (Split), which splits an edge in two and separates the decoration between them. Note that it induces its inverse move, which deletes a bivalent vertex that does not appear in any edge decoration, merges the two adjacent edges and concatenates their edge decorations.

Also:

- the Reidemeister III move can be rephrased as follows: given an edge (a, b, w), insert the word $\overline{w}\,\overline{a}wb$ (or any cyclic permutation of it, or of its inverse) within the decoration of any other edge;
- in a stabilization move, the condition that the central vertex b does not appear in any edge decoration can also be relaxed, at the cost of conjugating by the prefix a every occurrence of b in the edge decorations.

We postpone to Sect 18.6 the proof that each of these extra moves are indeed consequences of moves C, OR, S and R3.

18.2.3 Relation to Welded Links and String Links

As mentioned earlier, welded graphs are closely related to welded links and string links. This is made explicit in Theorem 18.2.9 below, which is the main result of this subsection.

18.2.3.1 Statement

We define two maps

$$\psi_L : \{\text{welded links}\} \to \{\text{welded graphs of type } (0, 1)\}$$
$$\psi_{SL} : \{\text{welded string links}\} \to \{\text{welded graphs of type } (2, 0)\}$$

as follows. With a Gauss diagram D for a welded link, we associate a w-graph Γ_D. The vertices of Γ_D are all unmarked, and are in bijection with the arrow tails of D. The edges correspond to the oriented pieces of circle bounded by two "adjacent" arrow tails, which are such that they contain no arrow tail in their interior. When running according to its orientation, each such piece of circle contains an ordered (possibly empty) sequence $h_1 \cdots h_k$ of arrow heads; the decoration of the corresponding edge of Γ_D is obtained by replacing each h_i by $t_i^{\varepsilon_i}$, where t_i and ε_i are respectively the tail and the sign of the arrow which contains h_i. Finally, in order to reduce the number of vertices, empty edges can be contracted. See the left-hand side of Fig. 18.5 for an example.

For a Gauss diagram D of a welded string link, we define Γ_D in a similar way; each component of the w-graph Γ_D now contains two marked vertices, which are in bijection with the endpoints of the intervals, and the edges correspond to the oriented pieces of interval between every pair of adjacent arrow tails and/or interval endpoints. See the right-hand side of Fig. 18.5.

It is a straightforward exercise to check that this assignment $D \mapsto \Gamma_D$ is well defined on welded links and string links, hence induces the requested maps ψ_L and ψ_{SL}.

Moreover, as Fig. 18.5 illustrates, the w-graphs obtained from the above procedure, have a very specific shape. This motivates the following definitions.

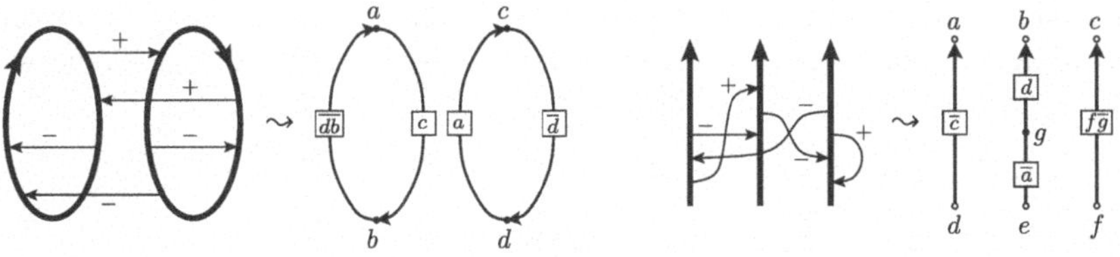

Fig. 18.5 From Gauss diagrams to w-graphs

Definition 18.2.4 A w-graph of type $(0, 1)$ is called *cyclic* if each component is homeomorphic to a circle, decomposed into bivalent unmarked vertices and oriented edges, such that each vertex has one inward and one outward oriented incident edge; see the left-hand side of Fig. 18.5.[2]

A w-graph of type $(2, 0)$ is called *linear* if each component is homeomorphic to an interval, decomposed into two univalent marked vertices, bivalent unmarked vertices and oriented edges, such that each unmarked vertex has one inward and one outward oriented incident edge; see the right-hand side of Fig. 18.5.

Lemma 18.2.5 *The maps ψ_L and ψ_{SL} are surjective.*

Proof Since the link case is strictly similar, we only give the argument in the string link case. We first observe that, up to push and contraction moves, any w-graph of type $(2, 0)$ is equivalent to a linear w-graph. Indeed, each component of a w-graph of type $(2, 0)$ is a tree with two marked vertices; we consider the shortest path of edges between these two marked vertices. Next we pick some edge e which is not in this path, but which has an endpoint a on it (and a tree T attached to its other endpoint). If a is not used in any edge decoration, we can simply apply a push move to turn e into an empty edge, and then a contraction move to delete it; otherwise, we need to first push a inside the shortest path using an expansion move, which creates an empty edge, so that we can remove e as in the former case:[3]

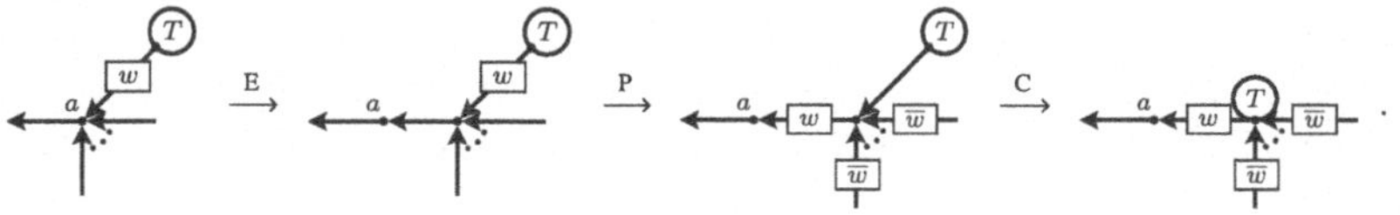

Applying this procedure recursively, we can delete all edges which are not in the shortest path. Then it only remains to apply OR moves to orient all edges coherently. Now, it remains to note that a linear w-graph Γ admits a preimage for the map ψ_{SL}. The process is summarized in the figure below:

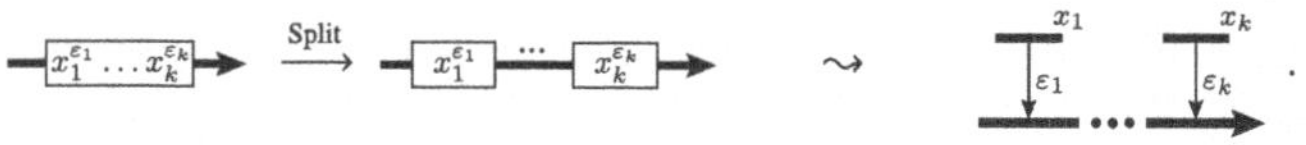

First, using Split moves, we split each word label on a linear w-graph into letter labels so that each edge is decorated by the, possibly overlined, name of a vertex. Next, we replace each (non empty) edge decoration by the head of an arrow, whose tail is attached in a neighborhood of the vertex corresponding to the letter decoration; the sign of this arrow being $-$ if the letter is overlined, and $+$ otherwise.

[2] We note that cyclic w-graphs are reminiscent of Harary and Kauffman's arc-graphs [16].

[3] Another issue would be that the vertex we are willing to push through is a marked vertex, but this can be safely avoided by pulling first the initial and final vertices away by using expansion moves.

Note that the relative position of the tails corresponding to a same letter label is irrelevant thanks to the TC move. □

Remark 18.2.6 As roughly outlined at the end of Sect. 18.2.1, and formalized by the map ψ_{SL}, word labelings on the edges of a welded graph can locally be thought of as some Gauss diagram arrows. In subsequent figures, we shall use implicitly this correspondence. Concretely, this means that we shall sometimes represent bunches of adjacent arrow heads in Gauss diagrams by some edge decoration words, whose signed letters correspond to the arrows signs and tails. A typical example follows in the next definition.

We now show that these maps are not injective. More precisely, we prove that the injectivity defect is controlled by the following moves.

Definition 18.2.7

- On welded links and string links, we define the Υ move as the following local move:

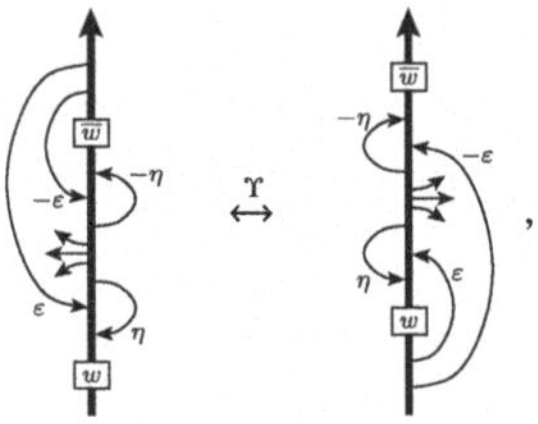

  where the three open arrows represent any number of tails whose associated heads are located outside of the picture.[4]
- On welded links, we define the *global reversal* (GR) move, as the move which reverses the orientation of a component, and reverses the sign of all arrows whose head is attached to this component.

Remark 18.2.8 We stress that the Υ move actually is a *family* of moves on welded (string) links. Indeed, there are multiple versions of the Υ move, depending on the word w, the signs ε and η, and the number of arrow tails (schematically represented by the three open arrows). See for example section 18.6.3 where we consider the case $w = 1$, $\eta = -\varepsilon$ and involving only the four self-arrows shown in the above figure (and no arrow tail).

The following is the main result of this section. As we shall see in Sect. 18.2.4, it is also relevant from the topological point of view.

[4] Since the two diagrams are identical outside the represented portion, this means that the two bunches are "identical" for both diagrams; it also means that they are not connected to any of the arrows associated with w.

Theorem 18.2.9

- *The map* ψ_L *induces a bijection between welded graphs of type* $(0, 1)$ *and welded links up to* Υ *and global reversal moves.*
- *The map* ψ_{SL} *induces a bijection between welded graphs of type* $(2, 0)$ *and welded string links up to* Υ *moves.*

In particular, the set of welded links and string links, up to Υ *and global reversal moves, embed into the set of welded graphs.*

For ease of reading, the proof of Theorem 18.2.9 is postponed to Sect. 18.5.

18.2.4 Relation to Ribbon Surface-Links

One important fact about welded links and string links, is that they are closely related to ribbon surface-links whose components are, respectively, tori and annuli, in the sense that there exists a surjective map from the former to the latter. Welded graphs are similarly related to all kind of ribbon surface-links, thus providing a more flexible combinatorial theory for the study of general ribbon surface-links. In what follows, X^4 denotes a given ambient 4-manifold in $\{S^4, \mathbb{R}^4, B^4\}$.

Definition 18.2.10

- A *ribbon handlebody* is an immersion in X^4 of a union H of oriented solid handlebodies up to ambient isotopy, such that
  - the singular set is a finite union of *ribbon disks*; here a ribbon disk is a disk D of transverse double points with one preimage D_{con}, called *contractible*, embedded in the interior of H, and one preimage D_{ess}, called *essential*, properly embedded in H;
  - if $X^4 = B^4$, $H \cap \partial B^4$ is a disjoint union of embedded disks.
- A *ribbon surface-link* is an oriented surface-link which bounds[5] a ribbon handlebody; the latter is called a *ribbon filling* of the ribbon surface-link.

We note that a given ribbon surface-link, may admit several different ribbon fillings.

By definition, the essential preimages of the ribbon disks of a ribbon handlebody H, cut H into 3-dimensional pieces, that we call *chambers*. Each chamber has the topology of a 3-ball with a finite number of handles attached, and a finite number of disks removed from its boundary. These removed disks correspond to essential preimages, and each of them is shared by two chambers (possibly twice the same). Moreover, each contractible preimage of a ribbon disk of H embeds into the interior of one of these chambers.

We say that a w-graph is *reduced* if all empty edges are loops, all non empty edges are decorated by one-letter words, and all marked vertices are univalent. We

[5] In case $X^4 = B^4$, the boundary has to be understood as the closure of $\partial H \setminus (H \cap \partial B^4)$.

can define a map

$$C : \{\text{ribbon handlebodies}\} \to \{\text{reduced } w\text{-graphs}\}/\text{OR}$$

as follows. If H is a ribbon handlebody, then $C(H)$ has:

- one unmarked vertex for each chamber and, if $X^4 = B^4$, one marked vertex for each connected component of $H \cap \partial B^4$;
- as many empty loops attached to each vertex as the genus of the corresponding chamber;
- one arbitrarily-oriented edge e_D for each ribbon disk D: the edge connects the two vertices corresponding to the two chambers that are adjacent to D_{ess}, and it is decorated by the letter corresponding to the (vertex associated with the) chamber containing D_{con}; this letter is overlined if and only if, for any given $x_0 \in D$, the concatenation of (the push-forward of) a positive basis of $T_{x_0^{\text{con}}} H$, where x_0^{con} is the preimage of x_0 in D_{con}, with (the push-forward of) a normal vector for D_{ess} at x_0^{ess}, where x_0^{ess} is the preimage of x_0 in D_{ess}, pointing into the chamber corresponding to the target vertex of e_D, is a negative basis of X^4.

Lemma 18.2.11 *The map C is a bijection.*

Proof This follows from the exact same argument as in the proof of [3, Prop. 3.7], which deals with the string link case. Roughly speaking, the proof is based on the fact that, given embedded 4-balls in S^4 with a finite number of paired embedded 3-balls on their boundary, there is a unique way, up to ambient isotopy, to connect these 3-balls with disjoint 1-handles. These 1-handles are indeed characterized by their 1-dimensional framed core; and since the framing takes values in S^2, the unit sphere of the normal bundle of the core, which is simply connected, all framing are isotopic. □

Next, consider the map

$$\partial : \{\text{ribbon handlebodies}\} \to \{\text{ribbon surface-links}\},$$

which sends a ribbon handlebody to its boundary.

Proposition 18.2.12 *The composite $\partial \circ C^{-1}$, induces a well-defined, surjective map*

$$\text{Tube} : \{\textit{welded graphs}\} \to \{\textit{ribbon surface-links}\},$$

sending a type $((m_1, b_1), \ldots, (m_\ell, b_\ell))$ welded graph to an embedded $\sqcup_{i=1}^{\ell} \Sigma_i$, where Σ_i is a genus b_i surface with m_i disjoint disks removed.

We shall sometimes call this map the *generalized Tube map*, since it naturally coincides with the known Tube map for welded (string) links, see Remark 18.2.13 below.

Proof Denote by $\widetilde{T}$ the composite $\partial \circ C^{-1}$. Note that, thanks to contraction and split moves, every w-graph is canonically equivalent to a reduced w-graph. Hence $\widetilde{T}$ naturally extends to all w-graphs. Let us check now that $\widetilde{T}$ is invariant under C, S, R1 and R3 moves.

Invariance under a contraction move is clear, and since the moves are local, invariance under R1 and R3 moves follow from the same arguments as for the original Tube map; see for instance the explicit ribbon filling changes argument given in the proof of [2, Prop. 2.5]. Let us now consider the stabilization move, which inserts a letter x_0 on the edges incident to a vertex v. Via $\widetilde{T}$, the letter x_0 corresponds to a chamber C_0, and v corresponds locally to a punctured sphere S filled by a 3-ball C; up to ambient isotopy, C and C_0 can be assumed to stand close to each other. Now, we can create[6] two oppositely-signed ribbon disks between C and C_0, such that the boundaries of the essential preimages are both parallel to a given puncture of S—see the left-hand side of the figure below, which represents the chamber C. Then we push one of these ribbon disks through C, until it splits into ribbon disks parallel to every other punctures of S:

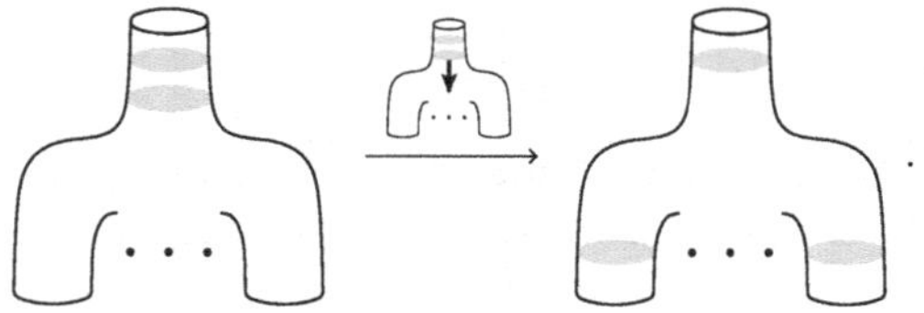

The result is the image by $\widetilde{T}$, of the stabilization move. This completes the proof that $\widetilde{T}$ is invariant under all moves on welded graphs, so that the map $\widetilde{T}$ induces the desired map Tube.

Surjectivity of Tube is clear, since any ribbon surface-link L admits at least one ribbon filling R_L by definition, $C(R_L)$ provides then a welded graph Γ_L such that $\text{Tube}(\Gamma_L) = L$. The latter part of the statement, concerning the types of welded graphs, just follows by construction. □

Remark 18.2.13 As the name suggests, the definition of our generalized Tube map follows closely the original construction, as given in [2, 3]. Specifically, by construction we have that $\text{Tube} \circ \psi_L$ and $\text{Tube} \circ \psi_{SL}$, coincide with the original Tube maps for welded links and string links, respectively.

As a matter of fact, the generalized Tube map can be defined directly, in terms of broken surface diagrams, as in [27] (we shall not recall here the language of broken surface diagrams, but refer instead the reader to [12]). This is illustrated by the following local pictures:

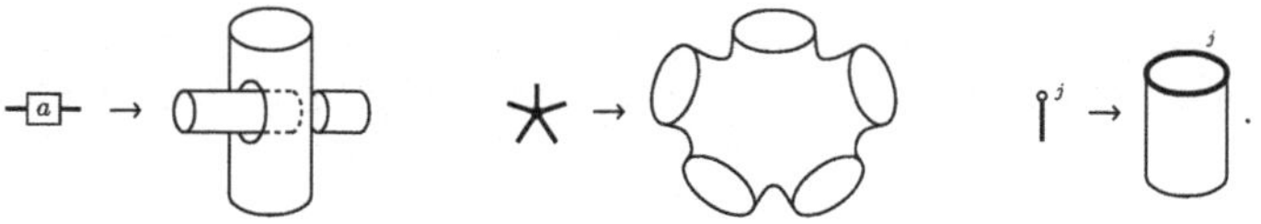

[6] See, for instance the local ribbon filling change given in [2, Fig. 16].

Here, the j-labeled bold circle on the rightmost picture denotes the j-th component of a prescribed unlink in ∂B^4. In general, for an arbitrary welded graph Γ, we obtain a diagram for Tube(Γ) by connecting the above pieces of broken surface diagrams with arbitrary embedded annuli, as prescribed by the welded graph Γ.

The Tube map, as defined by Satoh on welded knotoids and knots [27], and extended later to welded links and string links [5], served as a tool for studying ribbon 2-knots, tori and 2-string links. By Proposition 18.2.12, welded graphs now provide a general setting for studying the whole class of ribbon surface-links.

In contrast, injectivity of the Tube map has remained an intriguing question. In [11], the Tube map has been proved to be injective on welded braids, but Global Reversal moves were proved to be in its kernel for links [17, 29], and the question remains widely open for welded string links. In view of Theorem 18.2.9, we have the following.

Corollary 18.2.14 *The Tube map is invariant under* Υ *moves.*

We conjecture that, except for some very specific cases (Sect. 18.6.3), the Υ moves are not induced by the usual welded moves, and hence provide a new large class of non trivial moves in the kernel of the Tube map for welded links and string links. As a matter of fact, welded graphs intrinsically contain all the known moves under which the Tube map is invariant.

18.3 Wirtinger Groups, Peripheral Systems and Reduced Quotient

18.3.1 Wirtinger Groups

The fundamental group of the complement is an essential invariant in the topological study of embedded manifolds in codimension 2. In the case of knots and links, any diagram provides a so-called Wirtinger presentation for this group. This diagrammatical point of view on the fundamental group generalizes to other situations, such as surface-links [12] or welded knot theory [18]. The goal of the present section is to show that the theory of welded graphs can in fact be interpreted as a theory of Wirtinger presentations.

Definition 18.3.1 A *Wirtinger presentation* is a group presentation

$$\langle x_1, \ldots, x_k \mid R_1, \ldots, R_s \rangle$$

with a finite number of generators, and a finite number of relations of the form $R = \overline{x}_j x_i^w$, where x_i and x_j are two generators and w is an element of the free group $F(x_1, \ldots, x_k)$. A *Wirtinger group* is a group admitting a Wirtinger presentation.

One way to describe a Wirtinger presentation is by a decorated oriented graph with one vertex per generator, and one w-decorated edge from x_i to x_j for each relation $\overline{x}_j x_i^w$. Such a decorated graph is nothing but a w-graph, with no marked vertex. Note that performing an R1 move on this graph does not change the Wirtinger presentation since in $F(x_1, \ldots, x_k)$, we have $\overline{x}_j x_i^{x_i w} = \overline{x}_j x_i^{\overline{x}_i w} = \overline{x}_j x_i^w$. More generally, we have the following.

Proposition 18.3.2 *Welded graphs with no marked vertex are in one-to-one correspondence with Wirtinger presentations up to the following operations:*

- $\langle x_1, \ldots, x_k \mid R_1, \ldots, R_s, \overline{x}_{i_2} x_{i_1} \rangle$
- $\leftrightarrow \langle x_1, \ldots, x_{i_2-1}, x_{i_2+1}, \ldots, x_k \mid R_{1|x_{i_2} \to x_{i_1}}, \ldots, R_{s|x_{i_2} \to x_{i_1}} \rangle$;
- $\langle x_1, \ldots, x_k \mid R_1, \ldots, R_s, \overline{x}_{i_1} x_{i_2}^w \rangle \leftrightarrow \langle x_1, \ldots, x_k \mid R_1, \ldots, R_s, \overline{x}_{i_2} x_{i_1}^{\overline{w}} \rangle$;
- $\langle x_1, \ldots, x_k \mid R_1, \ldots, R_s \rangle \leftrightarrow \langle x_1, \ldots, x_k \mid R_{1|x_{i_1} \to x_{i_1}^\alpha}, \ldots, R_{s|x_{i_1} \to x_{i_1}^\alpha} \rangle$;
- $\langle x_1, \ldots, x_k \mid R_1, \ldots, R_s, \overline{x}_{i_2} x_{i_1}^w, \overline{x}_{i_4} x_{i_3}^{x_{i_1} w} \rangle$
- $\leftrightarrow \langle x_1, \ldots, x_k \mid R_1, \ldots, R_s, \overline{x}_{i_2} x_{i_1}^w, \overline{x}_{i_4} x_{i_3}^{w x_{i_2}} \rangle$

where $R_{|a \to b}$ denotes the word obtained from the word R by replacing all occurrences of a by b, and where α is any generator or inverse of a generator.

Proof It is straightforwardly checked that the first operation corresponds to a contraction move, the second one to an orientation reversal, the third one to a generalized stabilization, and the last one to a Reidemeister 3 move. □

Observe that, up to isomorphism, the four operations in Proposition 18.3.2, preserve the underlying group. For any welded graph Γ, there is hence an associated Wirtinger group, which we denote by $G(\Gamma)$, given by the presentation associated with Γ (after having unmarked all marked vertices). As a matter of fact, this group is the fundamental group of the complement of the associated ribbon surface-link via the generalized Tube map:

Proposition 18.3.3 *For every welded graph Γ, the group $G(\Gamma)$ is isomorphic to $\pi_1(X^4 \setminus \mathrm{Tube}(\Gamma))$.*

Proof Let Γ be a w-graph representative of a given welded graph. As noted in the proof of Proposition 18.2.12, this representative Γ can be assumed to be reduced, so that we can consider the associated ribbon handlebody $H = C(\Gamma)$. Recall that H is cut into chambers by ribbon disks, and that $\mathrm{Tube}(\Gamma) = \partial H$. We observe that the desired presentation for $\pi_1(X^4 \setminus \mathrm{Tube}(\Gamma))$ can be given in terms of the ribbon disks and the chambers of H. This follows from the three following facts:

- H retracts to a 1-dimensional graph in X^4, so that $X^4 \setminus H$ is connected and simply connected;
- since H is 3-dimensional, and since ribbon disks are 2-dimensional, a generic loop in $X^4 \setminus \mathrm{Tube}(\Gamma)$ avoids all ribbon disks and intersects H in a finite number of transverse points. Moreover, two loops intersecting the same chamber once with same sign are homotopic. Hence $\pi_1(X^4 \setminus \mathrm{Tube}(\Gamma))$ is generated by a set

of loops which is in bijection with chambers, a chamber c corresponding to the class of a loop intersecting H exactly once, in the interior of c;
- a generic homotopy of loops in $X^4 \setminus \mathrm{Tube}(\Gamma)$ intersects the ribbon disks in a finite number of transverse points. Each intersection corresponds to a Wirtinger relation between the two generators associated with the chambers on either side of the intersected ribbon disk, one being equal to the conjugate of the other by the generator associated with the chamber containing the contractible preimage of the ribbon disk.

Since ribbon disks and chambers of H are in bijection with edges and vertices of Γ, this presentation is nothing but the Wirtinger presentation of $G(\Gamma)$. □

In the rest of this section, we shall see that a Wirtinger presentation carries more information than just the associated group.

18.3.2 Peripheral System

The fundamental group of a codimension 2 embedding complement can be endowed with the data of so-called peripheral elements, which are given by loops staying essentially close to the embedded manifold. In the case of links, these elements are meridians and longitudes, and they form, together with the group, the peripheral system. Implicitly used by Dehn in 1914 to distinguish the right and the left-handed trefoils, peripheral systems were only formally introduced by Fox [15] in the early fifties and proved to be a complete invariant for links by Waldhausen [28] in 1968.

This notion of peripheral system has a combinatorial counterpart for welded graphs, which we introduce now. Let Γ be a w-graph, with connected components $\Gamma_1, \ldots, \Gamma_\ell$. Denote by b_i the first Betti number of Γ_i and by m_i its number of marked vertices.

Notation 18.3.4 A combinatorial path between two vertices v_1 and v_2 on Γ, is a finite sequence of adjacent edges starting at v_1 and ending at v_2. Note that, on each edge, the orientation given by the w-graph and the orientation given by the path may differ. With any combinatorial path $\mathfrak{p} = (e_1, \ldots, e_k)$, one can associate the element $w_\mathfrak{p} \in G(\Gamma)$ defined by the concatenated word $w_1^{\varepsilon_1} \ldots w_k^{\varepsilon_k}$, where w_i is the label of e_i and $\varepsilon_i \in \{\pm 1\}$ is 1 if and only if the orientations of e_i as an edge of Γ and as a path segment of $\mathfrak{p}$ coincide.

A combinatorial path corresponds in particular to a continuous path between two vertices and, reciprocally, any continuous path γ between two vertices is homotopic relative to the endpoints, to a combinatorial path $\mathfrak{p}_\gamma$. Note that two path representatives for γ differ only by some back and forth along some edges, so that the associated element in $G(\Gamma)$ is well-defined. We shall hence use simply the notation w_γ for the element $w_{\mathfrak{p}_\gamma} \in G(\Gamma)$.

Remark 18.3.5 It follows from iterated uses of Wirtinger relations that, if γ is a path between two vertices $v_1, v_2 \in \Gamma$, then $w_\gamma v_2 = v_1 w_\gamma$ in $G(\Gamma)$. This implies, in particular, that, for every $i \in \{1, \dots, \ell\}$, any two vertices of Γ_i, seen as elements of $G(\Gamma)$, are conjugate, and that if γ is a loop based at some vertex v, then v and w_γ commute.

Definition 18.3.6 *Meridians* of Γ are defined as the generators of $G(\Gamma)$, that is, the vertices of Γ. Vertices on the i-th component of Γ are sometimes more precisely called *i-th meridians*.

A *basing* for Γ is the data $\big((\mu_i)_i, \{l_1^i, \cdots, l_{b_i}^i\}_i, (a_1^i, \cdots, a_{m_i}^i)_i\big)$ where, for every $i \in \{1, \dots, \ell\}$:

- μ_i is a choice of a meridian of Γ_i;
- $l_1^i, \dots, l_{b_i}^i$ are b_i loops on Γ_i based at μ_i, that generate $\pi_1(\Gamma_i, \mu_i)$;
- $a_1^i, \dots, a_{m_i}^i$ are m_i paths that connect μ_i to each marked vertex of Γ_i.

Note that, since connected components and marked points of Γ are ordered, meridians and a_j^i paths are also ordered, whereas l_j^i loops are not.

Remark 18.3.7 If Γ has a marked vertex on each component, then there is a canonical choice for the chosen meridians, namely the minimal marked vertex on each component.

Definition 18.3.8 Let Γ be endowed with a basing $\big((\mu_i)_i, \{l_j^i\}_{i,j}, (a_j^i)_{i,j}\big)$.

For every $i \in \{1, \dots, \ell\}$, we define

- *preferred i-th loop-longitudes*, as the elements $\mu_i^{-|\lambda_j^i|_i}\lambda_j^i \in G(\Gamma)$, for $j \in \{1, \dots, b_i\}$;
- *preferred i-th arc-longitudes*, as the elements $\mu_i^{-|\alpha_j^i|_i}\alpha_j^i \in G(\Gamma)$, for $j \in \{1, \dots, m_i\}$;

where λ_j^i, resp. α_j^i, denote the element $w_{l_j^i}$, resp. $w_{a_j^i}$, of $G(\Gamma)$ and where $|w|_i$ is the algebraic number of letters in the word $w \in G(\Gamma)$ that correspond to vertices in Γ_i. A *longitude* is either a (preferred) loop or an arc-longitude.

As noted above, meridians and arc-longitudes are gathered in ordered tuples, whereas loop-longitudes are elements of unordered sets.

Proposition 18.3.9 *Let Γ' be a w-graph obtained from Γ by a welded move. There is a canonical isomorphism $\varphi : G(\Gamma) \to G(\Gamma')$, which sends any basing of Γ to a basing for Γ'.*

Proof The statement is clear for contraction, orientation reversal, stabilization and Reidemeister 3 moves. A Reidemeister 1 move applied on an edge e introduces some letter $v^{\pm 1}$ in the longitude word w_γ associated with any path γ passing through e, and changes $|w_\gamma|_i$ by ± 1. Using Remark 18.3.5, $v^{\pm 1}$ can be replaced by a $\mu_i^{\pm 1}$ factor at the beginning of the longitude word w_γ, so that it compensates the change of $|w_\gamma|_i$. □

Definition 18.3.10 A *peripheral system* for a welded graph Γ is the data

$$\big(G(\Gamma), (\mu_i)_i, \{w_{l^i_j}\}_{i,j}, (w_{a^i_j})_{i,j}\big),$$

where $\big((\mu_i)_i, \{l^i_1, \cdots, l^i_{b_i}\}_i, (a^i_1, \cdots, a^i_{m_i})_i\big)$ is a basing for Γ. Two peripheral systems for Γ are *equivalent* if they are related by a finite sequence of the following operations:

- replacing $\big(G, (\mu_i)_i, \{\lambda^i_j\}_{i,j}, (\alpha^i_j)_{i,j}\big)$ by $\big(G', (\mu'_i)_i, \{\lambda'^i_j\}_{i,j}, (\alpha'^i_j)_{i,j}\big)$, when there exists an isomorphism $\varphi : G \to G'$, some permutations $\sigma_i \in \mathfrak{S}_{b_i}$ and some elements $w_1, \ldots, w_\ell \in G'$ such that:
  - $\mu'_i = \varphi(\mu_i)^{w_i}$ for all $i \in \{1, \ldots, \ell\}$;
  - $\lambda'^i_{\sigma_i(j)} = \varphi(\lambda^i_j)^{w_i}$ for all $i \in \{1, \ldots, \ell\}$ and $j \in \{1, \ldots, b_i\}$;
  - $\alpha'^i_j = \overline{w}_i \varphi(\alpha^i_j)$ for all $i \in \{1, \ldots, \ell\}$ and $j \in \{1, \ldots, m_i\}$;
- replacing a loop-longitude by its inverse;
- precomposing an i-th arc or loop-longitude by some other i-th loop-longitude.

We emphasize that peripheral systems depend on a choice of a basing, which corresponds to the choice of a vertex v_i for each connected component Γ_i, together with a choice of a generating set for $\pi_1(\Gamma_i, v_i)$ and of paths from v_i to each marked vertex. But, since $\pi_1(\Gamma_i, v_i)$ is a free group, if follows from Nielsen transformations that equivalence classes are precisely defined to encompass any change of such choices. By Proposition 18.3.9, we thus have:

Corollary 18.3.11 *Peripheral systems up to equivalence, are well-defined for welded graphs.*

From now on, and by abuse of notation, peripheral systems shall be identified with their equivalence classes.

Remark 18.3.12 For a ribbon surface-link S, given with a ribbon filling, meridians are loops that meet the ribbon filling only once, and longitudes are defined by pushing curves on S off the surface. In light of the proof of Proposition 18.3.3, longitudes can be characterized by their intersections with the ribbon filling. But curves on $S = \text{Tube}(\Gamma)$ induce curves on Γ, and the intersections of the curves push-outs with the ribbon filling, can be read out of the curves on Γ. These precisely coincide with the longitudes of Γ. It follows that the Tube map preserves the whole peripheral system, in the sense that the peripheral system of a welded graph Γ is equal to the peripheral system of $\text{Tube}(\Gamma)$.

Remark 18.3.13 Suppose that a welded graph has only simply connected components, each having at least one marked vertex. This is for example the case for type $(2, 0)$ welded graphs, which are intimately related to welded string links by Theorem 18.2.9. In this situation, there is a preferred representative in the equivalence class of peripheral systems, corresponding to meridians being the minimal marked vertices on each component (see Remark 18.3.7). Since there are no loop-longitudes in this case, the peripheral system reduces to the data of the associated group, together with $\left(\sum_{i=1}^{\ell} m_i\right) - \ell$ elements, corresponding to paths from the minimal marked vertices to each of the other marked vertices on the same component.

In light of Remarks 18.3.12 and 18.3.13, it is clear that, for a string link L, the notion of peripheral system for L and the one for $\psi_{SL}(L)$ coincide. Since by Waldhausen's theorem, peripheral systems are classifying for classical knots, links and string links; this implies the following result.

Corollary 18.3.14 *Classical string links embed into welded graphs.*

By Theorem 18.2.9, this means that, should Υ moves be non trivial, they would not be able to change the isotopy class of a classical string link.

Remark 18.3.15 For links, the situation is slightly more intricate. Indeed, classical peripheral systems provide an orientation for each circular component, whereas welded graph peripheral systems do not. For instance, if a link L_2 is obtained from another link L_1 by mirror image and reversing all orientations, then $\psi_L(L_1) = \psi_L(L_2)$ even if L_2 and L_1 are not isotopic. It is thus an interesting question to determine when reversing some longitudes in the peripheral system of a classical links, corresponds to the peripheral system of a distinct classical link.

18.3.3 Reduced Quotients and Self-virtualization

By definition, $G(\Gamma)$ has as many generators as vertices in Γ. Nevertheless, since any two meridians on a same component are conjugate, $G(\Gamma)$ is normally generated by any choice of one generator per component (as, for example, specified by the choice of a basing). In this situation, the following classical notion, due to Milnor [25], has proved useful:

Definition 18.3.16 Let G be a group normally generated by elements $x_1, \cdots, x_p$. We denote by

$$\mathrm{R}G := G/\left\{[x_i, x_i^g] \;\middle|\; i \in \{1, \cdots p\};\, g \in G\right\},$$

the *reduced quotient of* G, which is the largest quotient where each generator commutes with any of its conjugates.

This leads to the following definition.

Definition 18.3.17 We define a *reduced peripheral system* for a welded graph Γ as the data

$$\left(\mathrm{R}G(\Gamma), ([\mu_i])_i, \{[\lambda^i_j].N_i\}_{i,j}, ([\alpha^i_j].N_i)_{i,j}\right)$$

where

- $\big(G(\Gamma), (\mu_i)_i, \{\lambda^i_j\}_{i,j}, (\alpha^i_j)_{i,j}\big)$ is a peripheral system for Γ,
- $\mathrm{R}G(\Gamma)$ is the reduced quotient of $G(\Gamma)$, and $[x]$ denotes images in $\mathrm{R}G(\Gamma)$ of $x \in G(\Gamma)$,
- $w.N_i$ is the coset of w in $\mathrm{R}G(\Gamma)$ relatively to the normal subgroup N_i of $\mathrm{R}G(\Gamma)$ generated by $[\mu_i]$.

Two reduced peripheral systems for Γ are *equivalent* if they arise from equivalent peripheral systems for Γ.

It should be noted that, since two i-th meridians are always conjugate, the reduced quotient $\mathrm{R}G(\Gamma)$ does not depend on the choice of one meridian per component. Equivalence classes of reduced peripheral systems are clearly invariant under welded moves.

Reduced quotients were introduced by Milnor in [25] for the study of links up to link-homotopies, which are homotopies that keep distinct components disjoint. In the welded context, the combinatorial counterpart of link-homotopy is the notion of sv-equivalence, defined in [5] for welded string links, and that we generalize now to welded graphs:

Definition 18.3.18 Two welded graphs differ by a *self-virtualization* (SV) if they differ by the replacement of an empty edge e by an edge decorated by a letter v or $\overline{v}$, where v is a vertex that belongs to the same component as e. Two welded graphs are *sv-equivalent* if they are related by a sequence of self-virtualizations and welded moves.

Remark 18.3.19 In terms of Gauss diagrams, an SV move is the insertion or deletion of a self-arrow, that is, an arrow which has both endpoints on a same component. In terms of virtual diagrams, this translates as replacing a self-crossing by a virtual one, or vice versa.

Proposition 18.3.20 *Reduced peripheral systems are invariant under self-virtualization.*

Proof An empty edge e yields a relation $\overline{v}_2 v_1$ in the group, where v_1 and v_2 are the two endpoints of e. Adding a decoration v on e turns the relation into $\overline{v}_2\overline{v}v_1v = \overline{v}_2[v_1, v]v_1$, which is equal to $\overline{v}_2 v_1$ in the reduced quotient since v and v_2 belong to the same component and are hence conjugate one to another (see Remark 18.3.5). The reduced group is hence invariant. Moreover, longitudes passing through e are modified from w_1w_2 to $w_1vw_2 = w_1w_2v^{w_2}$, hence define the same coset. □

Proposition 18.3.21 (Chen–Milnor presentation) *Let* $\big(G(\Gamma), (\mu_i)_i, \{\lambda^i_j\}_{i,j}, (\alpha^i_j)_{i,j}\big)$ *be a peripheral system for a welded graph* Γ. *There exists* $w^i_j \in$

$F(\mu_1, \ldots, \mu_\ell)$ for $i \in \{1, \ldots, \ell\}$ and $j \in \{1, \ldots, b_i\}$ such that $w^i_j = [\lambda^i_j]$ in $RG(\Gamma)$ and

$$RG(\Gamma) \simeq R\Big\langle \mu_1, \ldots, \mu_\ell \;\Big|\; [\mu_i, w^i_j] \text{ for all } i \in \{1, \ldots, \ell\} \text{ and } j \in \{1, \ldots, b_i\} \Big\rangle.$$

Proof The core of the proof is to show that Γ is sv-equivalent to a welded graph with a unique unmarked vertex (and possibly several marked vertices) per component. This is done component per component. Consider some component Γ_i, and pick an unmarked vertex: our goal is to eliminate all other unmarked vertices of Γ_i by contraction moves. First we remove from the edge labels, all letters corresponding to i-th meridians, using Split and self-virtualization moves. Having done so, we can successively apply generalized stabilization moves to those edges that we wish to contract, to turn them into empty edges: since no i-th meridian appear in the label that we are sweeping out, those generalized stabilization moves will only modify decorations on other components; note that only the labels of other components are affected in the process, so that previously-treated components remain of the desired form.

It remains to observe that, for the resulting welded graph with a unique unmarked vertex per component, the associated Wirtinger presentation provides the desired presentation for $RG(\Gamma)$. □

We stress that the proof of Proposition 18.3.21 provides not only a presentation for $RG(\Gamma)$, but also a "normal form" for a welded graph Γ endowed with a basing (hence with a reduced peripheral system).

Specifically, if $\big((\mu_i)_i, \{l^i_1, \cdots, l^i_{b_i}\}_{i,j}, (a^i_1, \cdots, a^i_{m_i})_i\big)$ is a basing for Γ, then the i-th component of this normal form has one unmarked vertex μ_i, b_i loops based at μ_i and decorated by $l^i_1, \ldots, l^i_{b_i}$, and m_i edges connecting μ_i to each marked vertex, and decorated by $a^i_1, \ldots, a^i_{m_i}$:

This yields the following classification of welded graphs up to self-equivalence by the reduced peripheral system:

Theorem 18.3.22 *Two welded graphs are sv-equivalent if and only if they have equivalent reduced peripheral systems.*

Proof The proof essentially follows the proof of [4, Thm. 2.1], which addresses the link case. Consider two welded graphs Γ_1 and Γ_2 with equivalent reduced peripheral systems, and suppose that Γ_1 and Γ_2 are given in normal forms as above.

The first step is to modify Γ_2 into a welded graph having the "same" reduced peripheral system as Γ_1; by this we mean that the isomorphism φ of Definition 18.3.10 sends directly meridians to meridians and longitudes to longitudes. As stated in this same definition, equivalence of reduced peripheral systems is generated by three elementary operations. The first one corresponds to conjugation, by a same word w, of a meridian and the associated loop-longitudes, and precomposition by $\overline{w}$ of the associated arc-longitudes; at the level of w-graphs, this can be realized by generalized stabilizations. The second operation reverses a loop-longitude; this can be realized by an orientation reversal move on w-graphs. The last operation precomposes a longitude by a loop-longitude from the same component; this can be realized on w-graphs by the following sequence of moves:

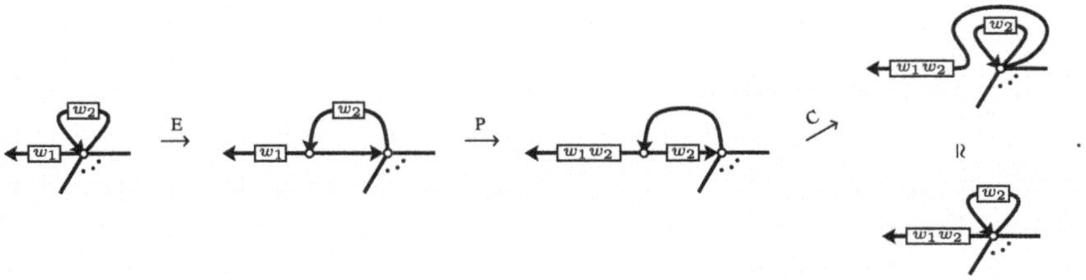

Now, Γ_1 and Γ_2 only differ by the representative words for the longitudes. Thanks to Proposition 18.3.21, we know that, for a given i-th longitude, the representative words differ by a sequence of insertion/deletion of

- conjugates of μ_i,
- commutators $[\mu_j, w_j]$ for some $j \in \{1, \dots, \ell\}$, where w_j is a j-th loop-longitude,
- commutators $[\mu_j, \mu_j^w]$ for some $j \in \{1, \dots, \ell\}$, where w is any word.

All these operations are handled in a local way in the proof of [4, Thm. 2.1], and the argument can be transposed directly to the welded graph case. □

Remark 18.3.23 Theorem 18.3.22 can be seen as a generalization of the main theorem of [4], which deals with the link case. It provides an equivalence between reduced peripheral systems and welded graphs up to sv-equivalence. Other quotients of the peripheral system, such as q-nilpotent [13, Thm. 2] or k-reduced [8, Thm. 4.3] quotients have also been investigated and have led to similar correspondences for welded links up to, respectively, w_q and self w_k-concordance. The notion of (self) w_q-concordance refers, on the one hand, to a combinatorial notion of concordance for welded objects which generalizes the classical notion of concordance for topological objects and, on the other hand, the notion of (self) w_q-equivalence based on the *arrow calculus* developed by the second and third authors [23]. This arrow calculus relies on surgery operations on welded diagrams, along certain oriented unitrivalent trees, filtered by the size of the associated tree.

It turns out that arrow calculus can be translated in terms of welded graphs. A degree q surgery corresponds to inserting a length q iterated commutator in some edge decoration, and a degree k self-surgery likewise corresponds to inserting an iterated commutator which involves at least $k + 1$ times the same component.

The diagrammatic formulation of concordance can also be extended to all welded graphs in a straightforward way. All the arguments extend to this setting, so that two welded graphs are w_q-concordant (resp. self w_k-concordant) if and only if they have equivalent q-nilpotent (resp. k-reduced) peripheral systems.

Note that, all together, this implies that, up to sv-equivalence, w_q or self w_k-concordance, the sets of welded links and string links embed in welded graphs. As a matter of fact, and compared with Theorem 18.2.9, it can indeed be directly checked by hand that Υ moves are trivial up to sv-equivalence, w_q-equivalence or self w_k-equivalence.

18.3.4 Milnor Invariants

The peripheral system and its reduced quotient are strong invariants, but they are also very difficult to compute and compare. In [26], Milnor introduced numerical invariants extracted from the nilpotent peripheral system, which are algorithmically computable, and which are still powerful link invariants. We define below similar invariants for welded graphs. A comprehensive account of Milnor invariants can be found in [7, Sec. 4], see also the survey [22].

For simplicity of exposition, here, we will restrict ourselves to the so-called "non repeated" Milnor invariants, that are extracted from the reduced peripheral system. Moreover, with our main topological result in sight (see Sect. 18.4), we shall further restrict to the following class of welded graphs.

Definition 18.3.24 A *welded forest*, is a welded graph of type $\big((m_1, 0), \dots, (m_\ell, 0)\big)$ such that $m_1, \dots, m_\ell \neq 0$. In other words, each connected component is a rooted tree: simply connected and with a minimal marked vertex.

Let Γ be a welded forest. Following Remark 18.3.7, Γ has a canonical set of meridians $\mu_1, \dots, \mu_\ell$, which corresponds to the minimal marked vertex on each component. Moreover, Γ has no non-trivial loop-longitude, so that by Proposition 18.3.21, we have that $\mathrm{R}G(\Gamma)$ is isomorphic to the reduced free group $\mathrm{R}F(\mu_1, \dots, \mu_\ell)$. Hence, following Remark 18.3.13, the reduced peripheral system of Γ is simply the data of $\big(\sum_{i=1}^{\ell} m_i\big) - \ell$ elements of $\mathrm{R}F(\mu_1, \dots, \mu_\ell)$, representing all arc-longitudes. In what follows, we shall simply denote by $\alpha^i_j \in F(\mu_1, \dots, \mu_\ell)$, the element representing the arc-longitude running to the j-th marked point on the i-th component of Γ ($i \in \{1, \cdots, \ell\}$, $j \in \{2, \cdots, m_i\}$).

Next we recall the *reduced Magnus expansion*

$$\begin{array}{ccc} \mathrm{R}F(\mu_1, \dots, \mu_\ell) & \to & \mathbb{Z}\langle\langle X_1, \dots, X_\ell \rangle\rangle/\mathcal{R} \\ \mu_i^{\varepsilon} & \mapsto & 1 + \varepsilon X_i \end{array},$$

taking values in the quotient of the ring $\mathbb{Z}\langle\langle X_1, \dots, X_\ell \rangle\rangle$ of formal power series in ℓ non-commuting variables, by the ideal $\mathcal{R}$ generated by monomials containing at least twice the same variable. The reduced Magnus expansion is known to be injective, see for instance [35].

Definition 18.3.25 For every $i \in \{1, \dots, \ell\}$, $j \in \{2, \dots, m_i\}$, and for every sequence $I = i_1 \dots i_r$ of distinct integers in $\{1, \dots, \ell\} \setminus \{i\}$, we define $\mu_\Gamma(I; i, j)$ as the coefficient of $X_{i_1} \cdots X_{i_r}$ in $E(\alpha^i_j)$. These coefficients are called the *non repeated Milnor invariants* of Γ.

Remark 18.3.26 The well-definedness of these non repeated Milnor invariants $\mu_\Gamma(I; i, j)$ is clear. Indeed, as discussed above, the welded forest Γ has a preferred set of meridians (Remark 18.3.13), and the $\left(\sum_{i=1}^{\ell} m_i\right) - \ell$ arc-longitudes forming its reduced peripheral system, induce well-defined elements α^i_j in $\mathrm{R}G(\Gamma) = \mathrm{R}F(\mu_1, \dots, \mu_\ell)$ (see Notation 18.3.4). Their invariance under self-virtualization follows from Proposition 18.3.20.

We have the following classification result.

Proposition 18.3.27 *Two welded forests are sv-equivalent if and only they have same non repeated Milnor invariants.*

Proof This is a direct consequence of Theorem 18.3.22, combined with the injectivity of the reduced Magnus expansion [35]. □

18.4 Application to Knotted Punctured Spheres up to Link-Homotopy

We can now get back to 4-dimensional topology, and deal with the case of knotted punctured spheres.

Definition 18.4.1 Let ℓ be some positive integer and $(m_1, \dots, m_\ell) \in \mathbb{N}$. A union of ℓ *knotted punctured spheres* is the image of a proper embedding in B^4 of $\Sigma = \sqcup_{i=1}^{\ell} S_i$, where S_i is a 2-sphere with m_i disjoint disks removed and such that $\partial\Sigma$ is sent to a fixed (oriented and ordered) unlink U in $\partial B^4 = S^3$, up to isotopy fixing the boundary ; we say it has type $(m_1, \dots, m_\ell)$.

As noted in the introduction, this large class of surface-links contains in particular Le Dimet's linked disks [20] (case $m_i = 1$ for all i), and 2-string links [5, 6] (case $m_i = 2$ for all i). We stress that the boundary components of each punctured sphere are ordered, so that there is a preferred meridian for each component of a union S of knotted punctured spheres, defined as a loop in $\partial B^4 \setminus S$ which is a meridian of the first boundary component.

Our purpose is to classify unions of knotted punctured spheres up to link-homotopy.

Definition 18.4.2 Two unions of knotted punctured spheres are *link-homotopic* if they are images of embeddings which are homotopic through proper immersions that keep distinct components disjoint, and that send $\partial\Sigma$ to U.

Our first result in this direction, is a generalization of [6, Thm. 3.5].

Proposition 18.4.3 *Any union of knotted punctured spheres is link-homotopic to a ribbon one.*

Proof We can use the very same strategy as used in [6, Sec. 3.2] in the 2-string link case. Pick some union S of knotted punctured spheres. The strategy goes in three steps, which can be roughly summarized as follows. First, using the fact that ∂S is an unlink, one can shrink and stretch a neighborhood of this boundary; this allows us to regard S as a 2-link (*i.e.* smoothly embedded disjoint copies of the 2-sphere) with thin, unknotted and unlinked tubes attached. Next, by a theorem of Bartels and Teichner [10, Thm. 1], there is a link-homotopy which takes this 2-link to a disjoint union of unknotted spheres. The proof then amounts to showing that, in the process of this link-homotopy, only ribbon-type linking will be created among the attached tubes, so that the final union of knotted punctured spheres is indeed ribbon. This last step uses broken surface diagrams of immersed surfaces, and consists in a systematic study of all Roseman moves and singular Roseman moves on these diagrams, which are local moves that generate link-homotopy of surface-links [6, Prop. 2.4]; the arguments in [6] being local, they apply verbatim to our more general situation. □

In [7, Sec. 5.2.3], Milnor link-homotopy invariants for surface-links were defined; they come in three flavors: Milnor maps, Milnor loop-invariants and Milnor arc-invariants. Milnor maps and Milnor loop-invariants are extracted from the loop-longitudes; in the case of a union S of ℓ knotted punctured spheres, the latter are given by curves parallel to the boundary components, which are copies of the unknot, so that Milnor maps and Milnor loop-invariants of S are trivial. Milnor arc-invariants are defined by pushing in $B^4 \setminus S$, and closing in a canonical way, the $\left(\sum_{i=1}^{\ell} m_i\right) - \ell$ boundary-to-boundary arcs given by the basing. These define a collection of elements of the reduced fundamental group of $B^4 \setminus S$, which by [6, Thm. 5.15] is isomorphic to the reduced free group generated by ℓ elements $m_1, \cdots, m_\ell$, represented by the preferred meridians of each component of S. Taking the coefficients in the reduced Magnus expansion of these elements, defines non-repeated Milnor invariants of S. We refer to [7] for details.

We can now prove the classification result announced in [7, Thm. 6.10].

Theorem 18.4.4 *Unions of knotted punctured spheres are classified, up to link-homotopy, by non repeated Milnor invariants.*

Proof By [7, Prop. 5.23], non repeated Milnor invariants are invariant under link-homotopy. Now, let S_1 and S_2 be two unions of knotted punctured spheres with the same non repeated Milnor invariants. By Proposition 18.4.3, we may assume that S_1 and S_2 are ribbon, so that by Proposition 18.2.12, they can be described as Tube(Γ_1) and Tube(Γ_2) for two welded forests Γ_1 and Γ_2. It follows from Remark 18.3.12 that the non-repeated Milnor arc-invariants of S_i coincide with those of Γ_i ($i = 1, 2$). Hence Γ_1 and Γ_2 have same non-repeated Milnor invariants, and Proposition 18.3.27 implies that Γ_1 and Γ_2 are therefore sv-equivalent. Now, in the same way as it was observed in [5] in the welded string link case, an SV move on a welded graph can be realized as a link-homotopy on its image under Tube; more precisely, the image by Tube of an SV move is a "self-circle crossing change" as illustrated in [5, Fig. 6]. Thus S_1 and S_2 are link-homotopic, and the proof is complete. □

Note that, in particular, Theorem 18.4.4 implies that linked disks (that is, unions of once-punctured knotted spheres) are all link-homotopically trivial, since linked disks admit no non-trivial longitudes. The following consequence, using results of [24], might be well-known to the experts.

Corollary 18.4.5 *Up to link-homotopy, a slice link bounds a unique union of slice disks.*

Proof Let L be a slice link in S^3, and let D_1 and D_2 be two unions of slice disks for L in the 4-ball. Pick in $S^3 \times [0, 1]$ a concordance C from L to the unlink U, and let $\overline{C}$ be the mirror image of C. By [24, Lemma 1.1], the stacking product $C \cdot \overline{C}$ is link-homotopic to the product $L \times [0, 1]$. Since, for $i = 1, 2$, a collar neighborhood of the boundary of D_i is isotopic to $L \times [0, 1]$, we have that D_i is link-homotopic to a product $C \cdot S_i$, where S_i is a union of slice disks for the unlink U. But as noted above, by Theorem 18.4.4 we have that S_1 and S_2 are link-homotopic. It follows that D_1 and D_2 are link-homotopic, as desired. □

Remark 18.4.6 The trick used in the proof of Corollary 18.4.5 above, relying on [24, Lemma 1.1], applies more generally to all knotted punctured spheres. This allows to extend the classification of Theorem 18.4.4 to a larger class of surface-links, which are images of proper embeddings in B^4 of a union of punctured spheres $\sqcup_{i=1}^{\ell} S_i$, such that the boundary is sent to a fixed *slice* link in $\partial B^4 = S^3$.

For the classification given in Theorem 18.4.4, we considered link-homotopies sending the boundary to the fixed unlink U at all time. This condition can be relaxed to define *weak link-homotopies*, as homotopies through proper immersions that may not preserve the image of the boundary.

Corollary 18.4.7 *Any two unions of knotted punctured spheres with same type are weakly link-homotopic.*

Proof Let us call *union of unknotted punctured spheres*, a union of punctured spheres which admits a ribbon filling with no ribbon disk. It follows from the same arguments as in the proof of Lemma 18.2.11, that such a union of unknotted punctured spheres is unique. We denote by S_0 these unknotted punctured spheres. Let us prove that any union S of ℓ knotted punctured spheres is weakly link-homotopic to S_0.

The strategy is to provide a union of ℓ knotted punctured spheres S' which has, on one hand, the same non repeated Milnor invariants as S, so that it is link-homotopic to S by Theorem 18.4.4, and which is on the other hand "nice enough" to be unknotted by an ambient isotopy that does not fix ∂B^4. For that, we consider the arc-longitudes of S, seen as words in the reduced free group on ℓ generators, and build an ℓ-component welded string link L realizing theses words as its longitudes: see for example [24, Fig. 2.1] or [5, Rem.4.23]. The image of L by the Tube map is an $\left(\sum_{i=1}^{\ell} m_i\right)$-component 2-string link which is ribbon. By [5, Theorem 2.30], Tube(L) is link-homotopic to a *monotone* ribbon 2-string link, the latter being the 4-dimensional trace of an ambient isotopy $(f_t)_{t\in[0,1]}$ of an unlink in S^3. On the unlink U forming the "upper" boundary of this monotone ribbon 2-string link, we glue a copy of S_0: this yields the desired union of knotted punctured spheres S'. By construction, it has the same Milnor invariants as S, and performing backward $(f_t)_{t\in[0,1]}$ on a collar neigborhood of ∂B^4 does provide an ambient isotopy which unknot S' into S_0. □

18.5 Proof of Theorem 18.2.9

We prepare below three technical lemmas that will be used in the proof of Theorem 18.2.9.

Lemma 18.5.1 *Two Gauss diagrams which differ only locally as*

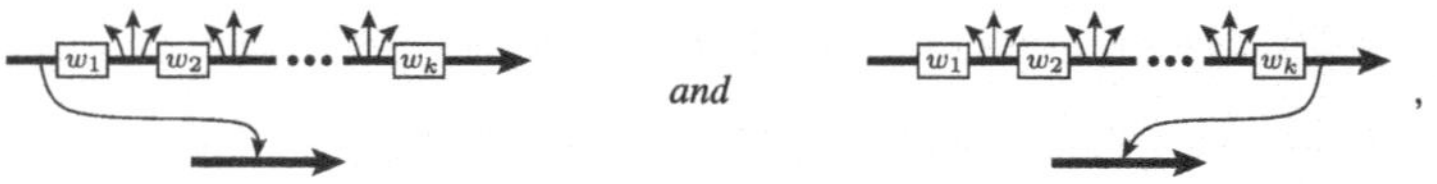

where each triad of open arrows stands for any bunch of tails, are equivalent up to welded moves, if the concatenated word $w_1 \cdots w_k$ *is trivial as a free group element.*

Proof The leftmost tail in the left-hand diagram, can be pushed to the right at the cost of conjugating its head by the word $w_1 \cdots w_k$ by the exchange moves of Fig. 18.2. But since this word is trivial as a free group element, all the conjugating arrows can be removed pairwise using Reid2 moves. □

Lemma 18.5.2 *Up to welded moves, two Gauss diagrams which differ only locally as*

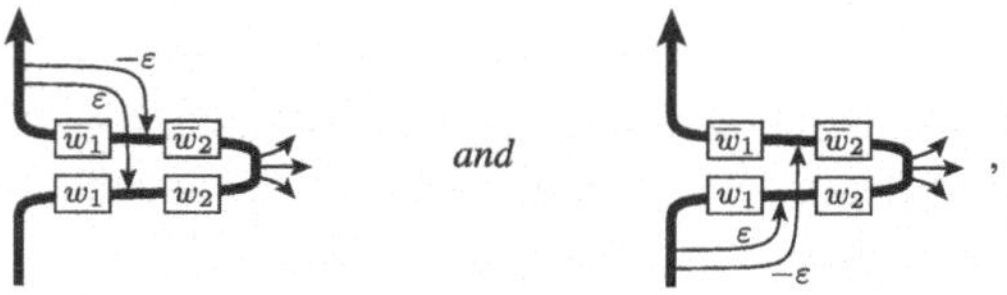

where none of the right-pointing tails are connected to w_1, *are equivalent.*

Proof We start from the left-hand diagram. We denote by a_ε and $a_{-\varepsilon}$ the depicted ε and $(-\varepsilon)$-signed arrows, and by x_t and x_b the depicted vertical top and bottom sub-arcs of the diagram. Then $\overline{w}_1$ can be pushed through the head of $a_{-\varepsilon}$, so that $a_{-\varepsilon}$ can be removed with a Reid1 move. Pushing $\overline{w}_1$ through $a_{-\varepsilon}$ in this way, this has the effect of conjugating all tails of arrows in $\overline{w}_1$ by x_t, and we denote this by replacing the label $\overline{w}_1$ by $\overline{w}_1^{x_t}$. Next, w_1 can be pushed through the head of a_ε, at the cost of conjugating w_1 by x_t.[7] We thus obtain the local diagram

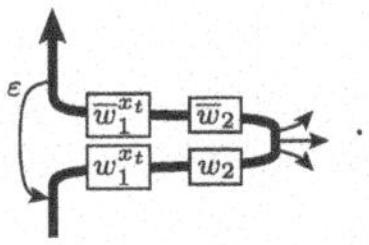

Now, observe that each letter in w_1 corresponds to two arrows, one from w_1 and one from $\overline{w}_1$, which can be assumed to have adjacent tails by move TC. When conjugated by x_t, these tails are now separated by two x_t-tailed arrows with opposite signs which can be removed using a Reid2 move. By this observation, we have that $w_1^{x_t}$ and $\overline{w}_1^{x_t}$ indeed represent inverse elements in the free group: we are therefore in the situation of Lemma 18.5.1. We apply hence Lemma 18.5.1, not only to the tail of a_ε, but also to all x_t-tailed arrows conjugating $w_1^{x_t}$ and $\overline{w}_1^{x_t}$: all these tails are pulled in this way down to the bottom interval x_b. After this deformation, the arrow a_ε can be deleted thanks to a Reid1 move, while all conjugating x_t-tailed arrows are turned into conjugating x_b-tailed arrows, since their tails now sit on x_b. As a result, we obtain the following local diagram:

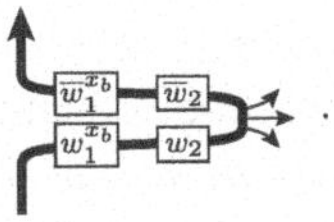

We now turn to the right-hand diagram in the statement. We can perform a similar sequence of moves as above, except that, at the final step, we only apply Lemma 18.5.1 to the arrow $a_{-\varepsilon}$ (that is, we leave all conjugating x_b-tailed arrows on x_b). This yields the very same diagram as above, which concludes the proof. □

Lemma 18.5.3 *Up to* Υ *and welded moves, two Gauss diagrams which differ only locally as:*

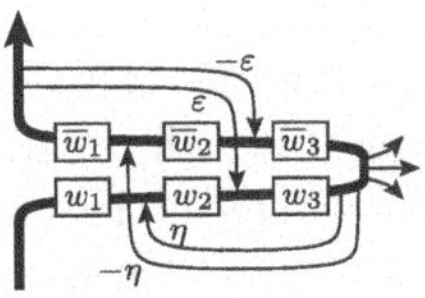

and

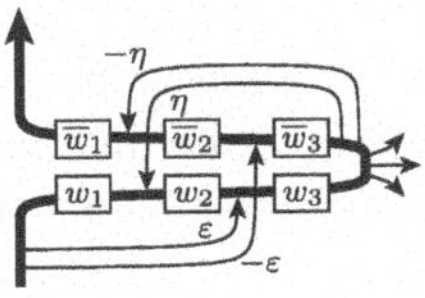

,

where none of the right-pointing tails are connected to w_1, w_2 *or* w_3, *are equivalent.*

[7] Conjugation is also by x_t since a_ε and $a_{-\varepsilon}$ have opposite signs and we are pushing w_1 and $\overline{w}_1$ in opposite directions according to the diagram orientation.

Proof We denote by a_α the depicted α-signed arrow (for $\alpha \in \{\pm\varepsilon, \pm\eta\}$), and by x_t and x_b the depicted vertical sub-arcs of the Gauss diagram. We also denote by x_r the rightmost sub-arc of the diagram, which contains the tails of arrows a_η and $a_{-\eta}$.

We first note that in the case where $w_2 = w_3 = \emptyset$, the local deformation of the statement is nothing but the Υ move; hence the case $w_2 = w_3 = \emptyset$ is immediate.

Now let us show how the case where $w_3 = \emptyset$, reduces to this $w_2 = w_3 = \emptyset$ case. This is done by pushing, in the left-hand side diagram, w_2 and $\overline{w}_2$ across the head of a_η and $a_{-\eta}$, respectively. As in the proof of Lemma 18.5.2, this deformation is achieved at the cost of conjugating these words w_2 and $\overline{w}_2$ by x_r. We can then apply the $w_2 = w_3 = \emptyset$ case, and then push back the x_r-conjugated word w_2, resp. $\overline{w}_2$, across the head of a_η, resp. $a_{-\eta}$. Being the reverse process of the initial deformation, this simply unconjugates these words, and yields the right-hand side diagram.

Finally, we observe that the general case similarly reduces to the $w_3 = \emptyset$ case. The argument is the same as above, except for one additional subtlety. Starting with the left-hand diagram, we push w_3 and $\overline{w}_3$ across the head of a_ε and $a_{-\varepsilon}$, respectively. We can then apply the $w_3 = \emptyset$ case of the statement. The point is that the initial deformation was achieved at the cost of conjugating the words w_3 and $\overline{w}_3$ by x_t (and not x_r as above). Hence we cannot immediately push back these conjugated words to obtain the desired move. But each of the x_t-conjugating arrows introduced at this initial stage, has its tail on the top vertical sub-arc x_t, and, by assumption, its head is located away from the depicted part of the diagram. Hence Lemma 18.5.1 can be applied to each of these arrows, whose tail can be moved down to the bottom vertical part x_b. Indeed, the concatenation of the words separating x_t and x_b is equivalent to the empty word, so that the lemma applies. After applying Lemma 18.5.1 in this way, w_3 and $\overline{w}_3$ can be unconjugated by pushing them back in place, and the proof is complete. □

Before proceeding with the proof of Theorem 18.2.9, we introduce some terminology.

Each connected component of a w-graph of type $(2, 0)$ has two marked vertices, which are ordered according to the orientation: the first one is called *initial vertex* and the second one is called *final vertex*. There is also a unique shortest path from the initial to the final vertex: every edge which is not on this path is called an *extra edge*. Note that the w-graph is linear if it has no extra edge.

Every vertex v of a w-graph of type $(2, 0)$ which is not the final vertex, has a unique shortest path to the final vertex: the first edge on this path is called the *out edge* of v. For vertices distinct from the initial vertex, the *in edge* is similarly defined as the first edge on the shortest path to the initial vertex. Note that the in and out edges of a vertex are distinct if and only if the vertex is on the shortest path from the initial to the final vertex.

In the same way, each connected component of a w-graph of type $(0, 1)$ has a unique shortest non contractible path: edges which are not on this path are also called *extra* edges. Observe that the w-graph is cyclic if it has no extra edge.

Proof of Theorem 18.2.9 Let us consider the string link case. First, we observe that the map ψ_{SL} is invariant under Υ moves. Indeed, on the one hand we have by definition that

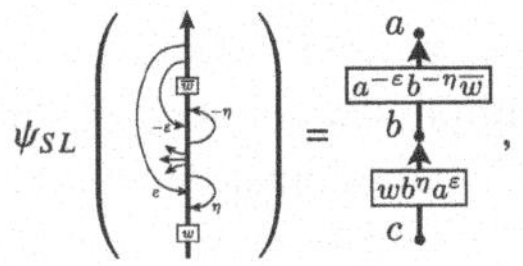

and applying a sequence of expansion, contraction and push moves to the resulting w-graph, shows that

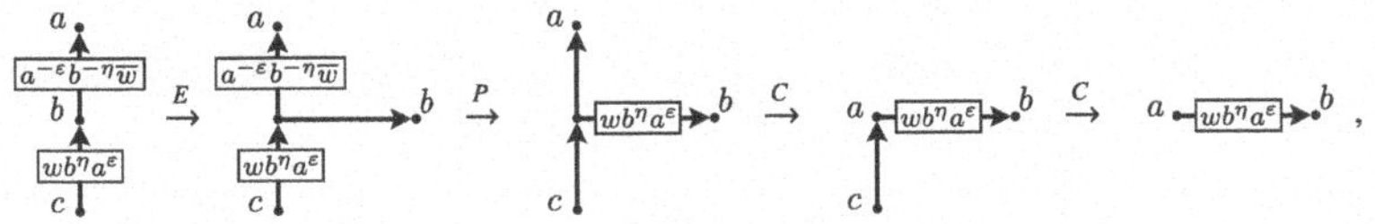

On the other hand, applying the ψ_{SL} map and a similar sequence of expansion, contraction and push moves, gives:

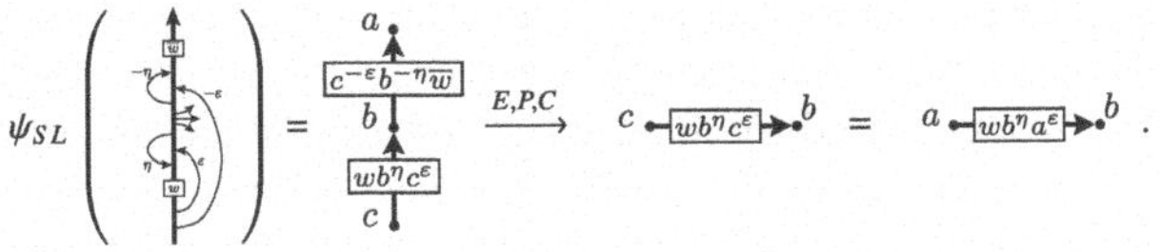

Therefore ψ_{SL} induces a well-defined map

$$\overline{\psi}_{SL} : \{\text{welded string links up to } \Upsilon \text{ moves}\} \to \{\text{welded graphs of type } (2,0)\}.$$

We now define an inverse map $\overline{\xi}_{SL}$ for $\overline{\psi}_{SL}$. Let Γ be a w-graph of type $(2, 0)$. The description of $\overline{\xi}_{SL}(\Gamma)$ will be given in three steps, summarized in Fig. 18.6. The first step provides an algorithm for turning Γ into a linear w-graph, by a canonical sequence of push and contraction moves; the second step specifies the edge labels on this linear w-graph, induced by those of Γ; the third step associates a welded string link to the resulting welded graph.

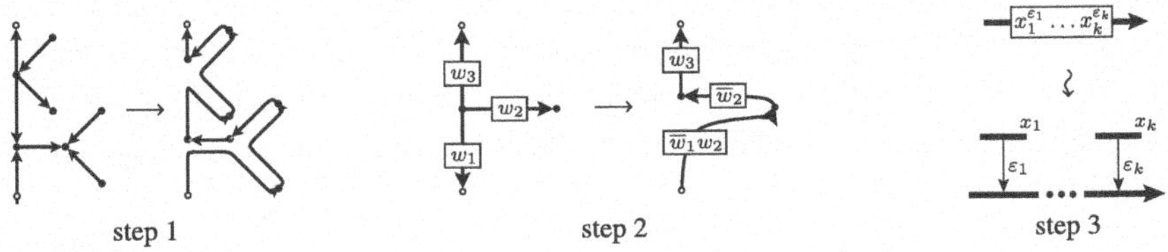

Fig. 18.6 Definition of $\overline{\xi}_{SL}$

Step 1 For every vertex of Γ, we fix an arbitrary ordering on the adjacent edges such that the out edge is last and, if it exists, the in edge is penultimate. The first edge of a vertex is called the *leading* edge of the vertex, and the second one is called the *co-leading* edge. Using these orderings clock-wisely, we obtain a planar way of embedding Γ in $\mathbb{R}^2$; we can then "contour" these planar trees (see step 1 in Fig. 18.6) to get a union I of intervals, connecting initial vertices to final ones. We orient these intervals from the initial to the final vertex. Each vertex of Γ corresponds to at least one, possibly more, points on I: we keep only the point which is closest to the final vertex. This turns I into a linear graph, whose vertices are in one-to-one correspondence with those of Γ, and whose edges are unions of edges of Γ.

Step 2 Each edge e of I is a union of edges $e_1, \cdots, e_k$ of Γ with the same or the reversed orientation. Let w_i be the decoration of e_i in Γ. Then we decorate e by the concatenation $w_1^{\varepsilon_1} \cdots w_k^{\varepsilon_k}$ where $e_i \in \{\pm 1\}$ is 1 if and only if the orientations of e and e_i agree; see step 2 in Fig. 18.6 for an illustration.

At this point, we obtain a linear w-graph Γ_l which is equivalent to Γ. Indeed, Γ_l is the result of iteratively contracting, as outlined in the proof of Lemma 18.2.5, all extra edges with a univalent unmarked endpoint, starting with the one which is closest to the final vertex of the corresponding component of I.

Step 3 We apply, on the linear w-graph Γ_l, the exact same procedure as in the latter half of the proof of Lemma 18.2.5: we split each edge by S moves to obtain a linear w-graph labeled by letters, and we replace each of these letter labels by an arrow as illustrated in step 3 of Fig. 18.6. The result is the desired welded string link $\overline{\xi}_{SL}(\Gamma)$; as observed in the proof of Lemma 18.2.5, this is indeed a preimage for Γ_l by the map ψ_{SL}.

We now have to prove that $\overline{\xi}_{SL}(\Gamma)$ is well defined, that is, $\overline{\xi}_{SL}(\Gamma)$ is invariant under all moves on w-graphs, and does not depend on the choice of planar embedding for Γ (that is, on the chosen orderings at each vertex).

First, it is easily checked that OR moves do not change the resulting Gauss diagram, so that we can freely assume that all edges are oriented toward the final vertex. Invariance under S moves (hence under push moves) is also easy to check, as they introduce pairs of arrows which can be deleted by R2 moves.

Contraction moves turn out to be the most intricate moves to check. Contraction/expansion of a leading edge, which we shall call *leading contraction/expansion* below, has no impact, so that invariance is directly checked. But contraction of a non leading edge may change drastically the position of some vertex in Γ_l.

First, let us observe that contraction of an arbitrary (non leading) edge e can always be realized by a sequence of expansion and contractions on either leading edges, or co-leading edges at a trivalent vertex; this is illustrated in the following figure, where T is any tree attached to the edge e:

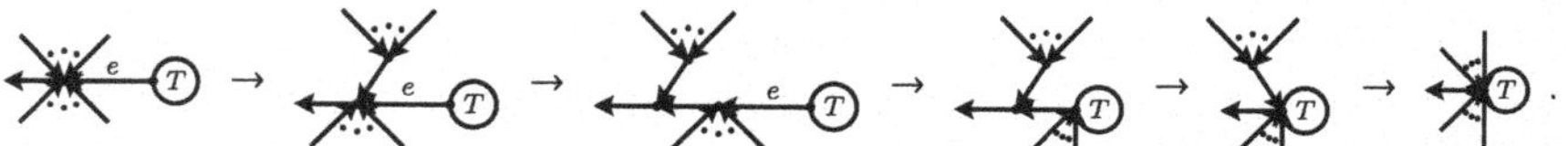

There, the first step is a leading expansion, followed by an expansion on a co-leading edge at a trivalent vertex; this turns edge e into a leading edge, so that contraction can be performed; it then only remains to perform the inverse of the first two operations.

Hence we are left with proving the case of a contraction of a co-leading edge e at a trivalent vertex. Using push moves and leading contractions, the tree attached via the leading edge can be assumed to be linear, that is, we consider a w-graph of the form:

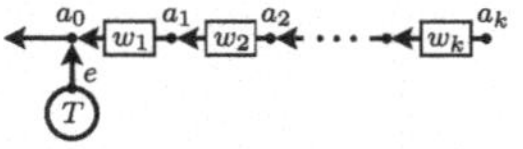

where as above, T is any tree, attached to the edge e that we would like to contract. By recursively applying push moves and leading expansions, this w-graph can be turned into the following:

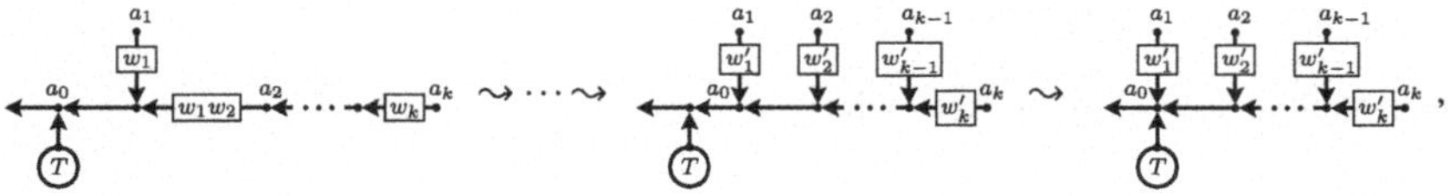

where the final step is achieved by a leading contraction. Furthermore, we may freely assume that for each $i = 1, \cdots, k$, the letter a_i appears at most once in the word w'_i. This is done by using iteratively the following trick during the above process:

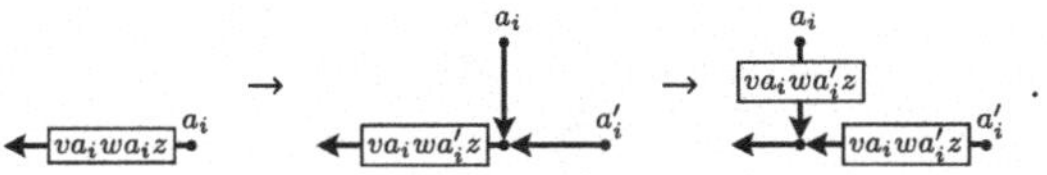

Here, the first operation is a leading expansion (which allows for replacing each label a_i by either a_i or a'_i), followed by a push move.

Now, we show the invariance under the contraction of e by induction on k. For $k = 1$, there are two cases:

- Suppose that a_1 does not appear in w_1. The figure below gives the image by $\overline{\xi}_{SL}$ of the w-graph before (left) and after (right) contraction on the edge e:

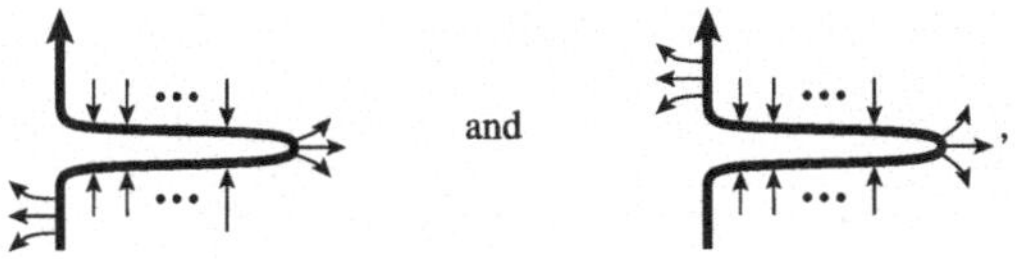

where each pair of opposed vertical arrows have adjacent tails and opposite signs. Note that the bunch of right-pointing arrows represent all occurrences of a_1 in the edge labelings, so that we know by assumption that none of these tails is connected to the vertical arrow heads of the figure. Starting from the left-hand side figure above, we aim at moving all left-pointing arrow tails from the bottom to the top vertical strand. This is achieved by Lemma 18.5.1 for any such arrow that is not connected to the vertical arrow heads; left-pointing arrows that are connected to the vertical arrow heads, come by pairs with opposite signs, so that Lemma 18.5.2 can be applied to perform the desired move.

- Suppose that a_1 appears once in w_1. Concretely this means that, in the above left-hand figure, a pair of vertical arrows corresponds to two of the right-pointing arrows:

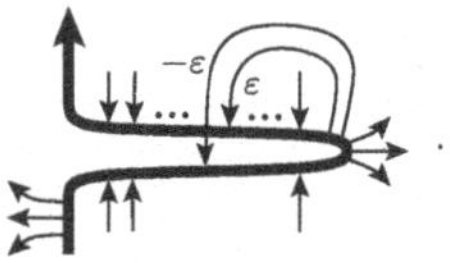

This case is handled as the previous one, except that we might use Lemma 18.5.3 instead of Lemma 18.5.2.

The inductive step is illustrated below:

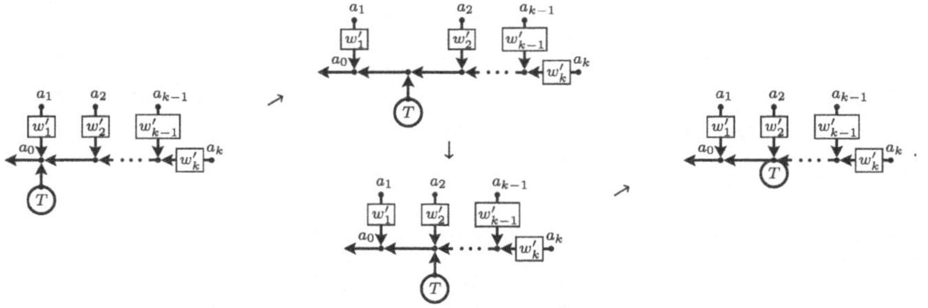

The first operation is an expansion in the "k=1 case"; the second step is a leading contraction, and the final operation is a contraction using the induction hypothesis. Finally, we undo all the previous steps to get back to the initial configuration with the desired co-leading edge contracted.

Invariance under a change of ordering of the edges at some vertex is now easily handled using push, expansion and contraction moves. First, we note that, up to expansion and contraction moves, the graph can be assumed to be uni-trivalent. Then the procedure is schematically illustrated as follows:

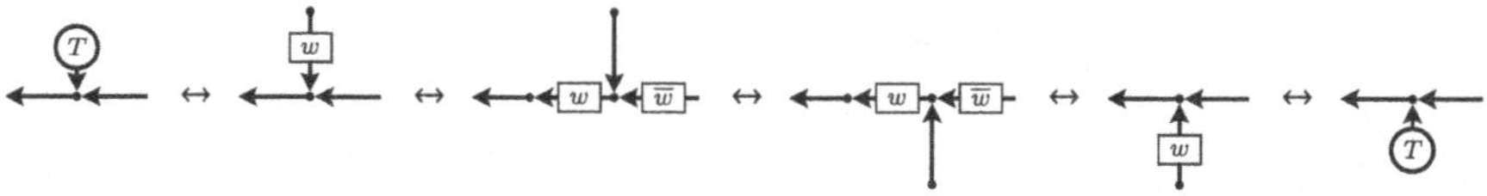

There, T denotes a tree that we would like to "move down" by reversing the cyclic order at the trivalent vertex. With contraction and push moves, we first replace T by a linear graph T_L: this is schematically denoted by a w-labeled edge in the above figure. Next we push down all the decorations of T_L, so that the vertical edges are empty and can be contracted. The reversed procedure can then be performed with the cyclic order reversed.

We next check the invariance under an R1 move. At the level of the images by $\overline{\xi}_{SL}$, such a move creates two or one arrow(s), depending on whether the edge where the move is performed, is an extra edge or not. The head h and tail t of such an arrow are separated by tails and sequences of heads such that the concatenation of the corresponding words is trivial as a free group element. Lemma 18.5.1 can hence be applied to pull t next to h. The arrow can then be removed using a Reid1 move.

For R3 moves, consider in some welded graph Γ, an edge between two vertices a and b, and decorated by a word w, and suppose that Γ contains some other edge decorated by wb. Let Γ' be the result of the R3 move that replaces the latter label by $wa^w = aw$. By construction, the linear welded graph Γ_l associated with Γ (step 2 of the definition of $\overline{\xi}_{SL}(\Gamma)$) also contains a w-labeled edge between vertices a and b. In the welded string link $\overline{\xi}_{SL}(\Gamma)$, this yields bunches of tails corresponding, respectively, to all occurrences of a and b in the edge decorations, separated by a sequence of heads which correspond to the word w. Pulling one tail from the b-bunch through the w-sequence of heads will conjugate the associated head by w. The resulting welded string link is precisely $\overline{\xi}_{SL}(\Gamma')$.

This concludes the proof that the map $\overline{\xi}_{SL}$ is well-defined. Moreover, it clearly satisfies $\overline{\xi}_{SL} \circ \overline{\psi}_{SL} = \text{Id}$. This shows that $\overline{\psi}_{SL}$ is injective, hence bijective, and the proof is complete in the string link case.

The case of welded links is handled in the same way, except that extra edges are contracted in order to get a cyclic w-graph instead of a linear one. There is however one extra issue: the resulting cyclic w-graph has no canonical orientation. This affects the 3-step procedure defining $\overline{\xi}_L$, since in step 3 we need to choose an arbitrary orientation on each circle component; this is precisely why one needs to introduce the Global Reversal move in the link case, which identifies two Gauss diagrams that differ by this choice of orientation. A more subtle point is that reversing the orientation on some component also affects step 1, as follows:

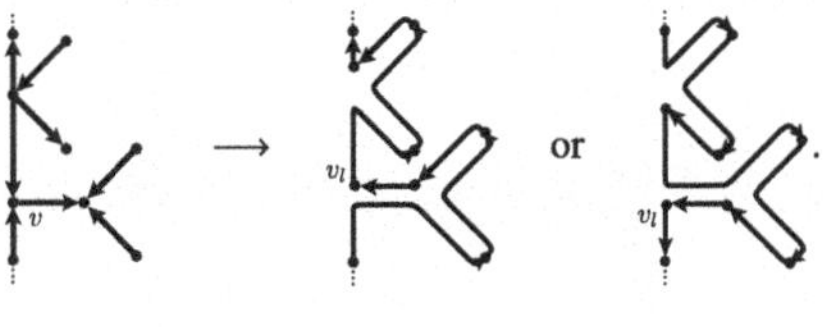

As the figure illustrates, in the linearization algorithm given in step 1, the chosen orientation is used at each vertex v of a w-graph, to define the corresponding vertex v_l in the linear w-graph. This vertex v_l represents, in the resulting Gauss diagrams, a bunch of arrow tails: Lemma 18.5.1 ensures that the Gauss diagrams resulting from the above two situations, are actually equivalent. This concludes the proof of Theorem 18.2.9. □

18.6 Extra Moves on Welded Graphs

In this section, we explain how the moves of Fig. 18.4 on welded graphs, follow from the C, OR, S and R3 moves of Fig. 18.3, as claimed at the end of Sect. 18.2. We also show how one version the ϒ move (in the sense of Remark 18.2.8) is implied by usual welded moves.

18.6.1 Push and Split Moves

We first observe that the extra push (P) and split (Split) moves are consequences of the C, OR and S moves. For this purpose, we first note that, up to OR moves, the S move can be generalized to the case where some edge is oriented toward the vertex where the move is performed; in that case, the prefix is then reversed and added at the end of the corresponding decoration:

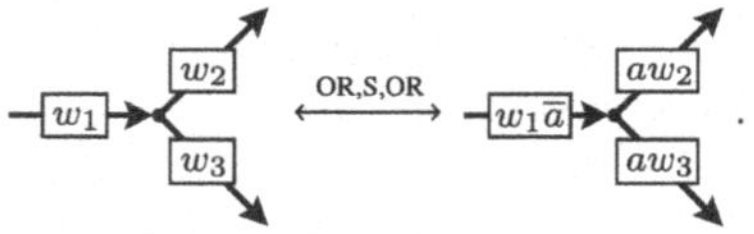

Applying this variant of the S move iteratively, provides a P move:

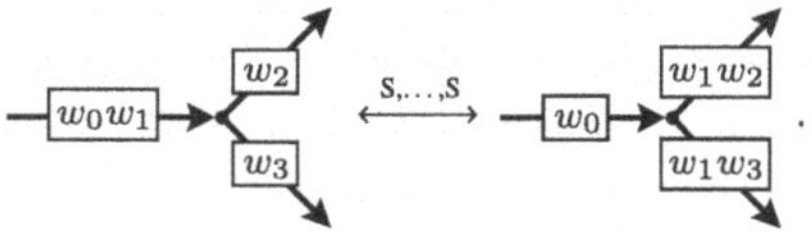

Moreover, combining this variant of the S move with a C move, allows to "split" edge decorations, thus providing a Split move:

18.6.2 Rephrased R3 and Generalized Stabilization Move

Next, we show that the Split, C, S and R3 moves on welded graphs, realize a rephrased R3 move. More precisely, we illustrate below how some edge label w_1w_2

can be replaced by $w_1\overline{w}\,\overline{a}wbw_2$, where (a, b, w) is some other edge in the graph:

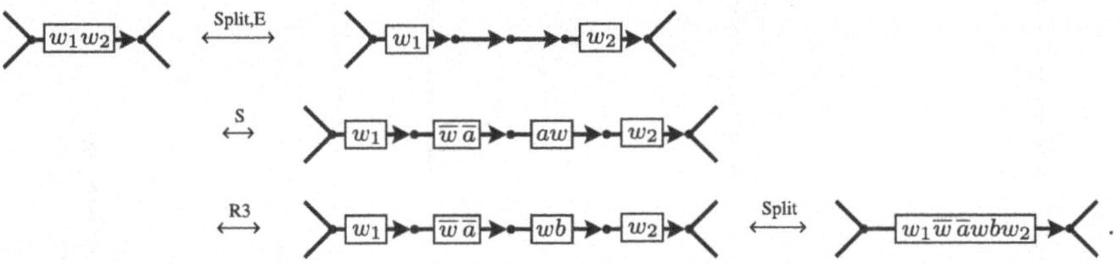

The other versions of this move, that insert any cyclic permutation of $\overline{w}\,\overline{a}wb$, or any cyclic permutation of it, are proved in a similar way.

Finally, the figure below illustrates how the generalized stabilization move, is achieved by a sequence of C,S and R3 moves:

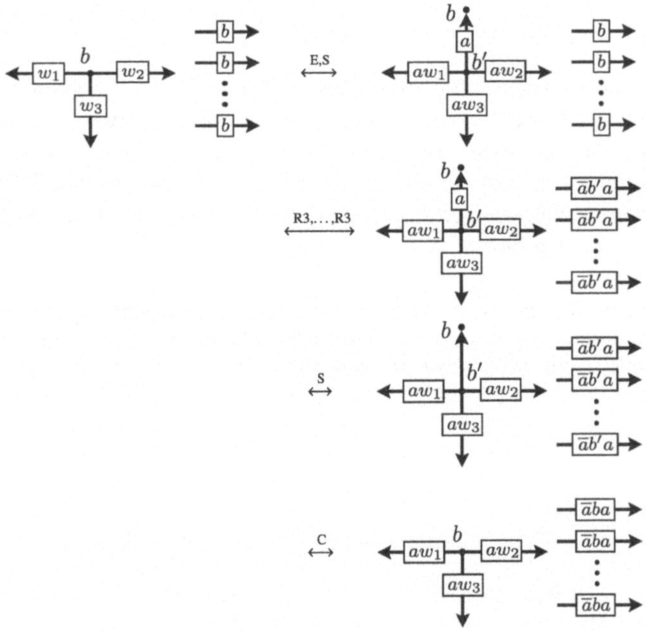

18.6.3 A Trivial ϒ Move

We finally show that, in the special case where $w = 1$, $\eta = -\varepsilon$ and involving only the four self-arrows shown in the figure of Definition 18.2.7, the ϒ move follows from Reid1,2,3 moves. This is illustrated in the following figure:

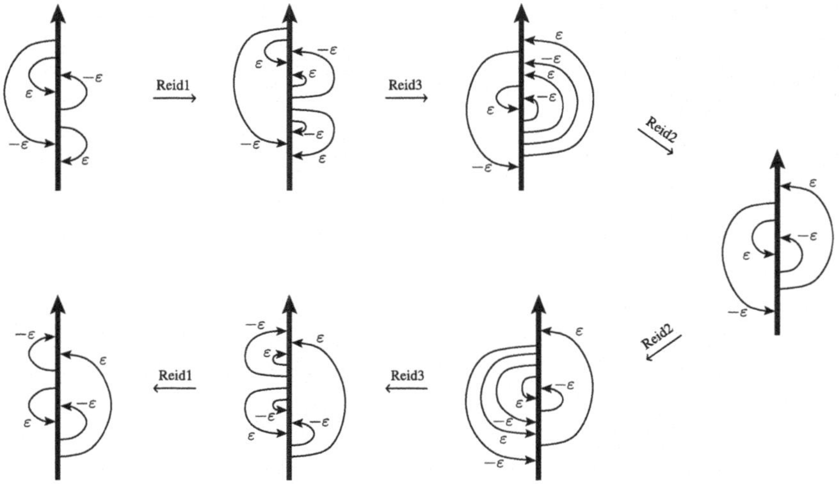

The strategy is to connect the two sides of the move to a common symmetric configuration. For both, the deformation starts by inserting two (self) arrows with Reid1 moves, then combines two Reid3 moves with TC moves to rearrange the relative positions of the arrow tails, and deletes a pair of arrows with a Reid2 move.

To the authors' knowledge, this is the only version of the Υ move that follows from the usual welded moves.

Acknowledgments The first, resp. second, author is partially supported by the project SyTriQ (ANR-20-CE40-0004), resp. the project AlMaRe (ANR-19-CE40-0001-01), of the ANR. The third author is supported by the JSPS KAKENHI grant 21K03237.

References

1. E. Artin, Zur isotopie zweidimensionalen flächen im R_4. Abh. Math. Sem. Univ. Hamburg **4**, 174–177 (1926)
2. B. Audoux, On the welded tube map, in *Knot Theory and Its Applications*, vol. 670. Contemp. Math. (Amer. Math. Soc., Providence, 2016), pp. 261–284
3. B. Audoux, Applications de modèles combinatoires issus de la topologie. Habilitation, Aix–Marseille University (2018)
4. B. Audoux, J.-B. Meilhan, Characterization of the reduced peripheral system of links. J. Inst. Math. Jussieu 1–19 (Published online 2024)
5. B. Audoux, P. Bellingeri, J.-B. Meilhan, E. Wagner, Homotopy classification of ribbon tubes and welded string links. Ann. Sc. Norm. Super. Pisa Cl. Sci. **17**(1), 713–761 (2017)
6. B. Audoux, J.-B. Meilhan, E. Wagner, On codimension two embeddings up to link-homotopy. J. Topol. **10**(4), 1107–1123 (2017)
7. B. Audoux, J.-B. Meilhan, A. Yasuhara, Milnor concordance invariant for surface-links and beyond (2021). arXiv:2109.14578
8. B. Audoux, J.-B. Meilhan, A. Yasuhara, *k*-Reduced groups and Milnor invariants. Math. Res. Lett. (to appear)

9. D. Bar-Natan, Z. Dancso, Finite type invariants of w-knotted objects II: tangles, foams and the Kashiwara-Vergne problem. Math. Ann. **367**(3–4), 1517–1586 (2017)
10. A. Bartels, P. Teichner, All two dimensions links are null homotopic. Geom. Topol. **3**, 235–252 (1999)
11. T.E. Brendle, A. Hatcher, Configuration spaces of rings and wickets. Comment. Math. Helv. **88**(1), 131–162 (2013)
12. J.S. Carter, S. Kamada, M. Saito, *Surfaces in* 4*-space*, vol. 142. Encyclopaedia of Mathematical Sciences (Springer, Berlin, 2004). Low-Dimensional Topology, III
13. B. Colombari, A diagrammatical characterization of Milnor invariants (2022). arXiv:2201.01499
14. R. Fenn, R. Rimányi, C. Rourke, The braid-permutation group. Topology **36**(1), 123–135 (1997)
15. R.H. Fox, A quick trip through knot theory, in *Topology of 3-Manifolds and Related Topics (Proc. The Univ. of Georgia Institute, 1961)* (Prentice-Hall, Inc., Englewood Cliffs, 1961), pp. 120–167
16. F. Harary, L.H. Kauffman, Knots and graphs. I. Arc graphs and colorings. Adv. Appl. Math. **22**(3), 312–337 (1999)
17. A. Ichimori, T. Kanenobu, Ribbon torus knots presented by virtual knots with up to four crossings. J. Knot Theory Ramif. **21**(13), 1240005 (2012)
18. L.H. Kauffman, Virtual knot theory. Eur. J. Combin. **20**(7), 663–690 (1999)
19. L.H. Kauffman, J.a. Faria Martins, Invariants of welded virtual knots via crossed module invariants of knotted surfaces. Compos. Math. **144**(4), 1046–1080 (2008)
20. J.-Y. Le Dimet, Cobordisme d'enlacements de disques. Mém. Soc. Math. France (N.S.) **32**, ii+92 (1988)
21. Y. Marumoto, Stable equivalence of ribbon presentations. J. Knot Theory Ramif. **1**, 241–251 (1992)
22. J.-B. Meilhan, Linking number and Milnor invariants, in *Encyclopedia of Knot Theory* (CRC Press, Boca Raton, 2021), pp. 817–829
23. J.-B. Meilhan, A. Yasuhara, Arrow calculus for welded and classical links. Algebr. Geom. Topol. **19**(1), 397–456 (2019)
24. J.-B. Meilhan, A. Yasuhara, Link concordances as surfaces in 4–space and the 4-dimensional Milnor invariants. Indiana Univ. Math. J. **71**, 2647–2669 (2022)
25. J. Milnor, Link groups. Ann. Math. (2) **59**, 177–195 (1954)
26. J. Milnor, Isotopy of links. Algebraic geometry and topology, in *A Symposium in Honor of S. Lefschetz* (Princeton University Press, Princeton, 1957), pp. 280–306
27. S. Satoh, Virtual knot presentation of ribbon torus-knots. J. Knot Theory Ramif. **9**(4), 531–542 (2000)
28. F. Waldhausen, On irreducible 3-manifolds which are sufficiently large. Ann. Math. (2) **87**, 56–88 (1968)
29. B. Winter, The classification of spun torus knots. J. Knot Theory Ramif. **18**(9), 1287–1298 (2009)
30. T. Yajima, On the fundamental groups of knotted 2-manifolds in the 4-space. J. Math. Osaka City Univ. **13**, 63–71 (1962)
31. T. Yajima, On simply knotted spheres in R^4. Osaka J. Math. **1**, 133–152 (1964)
32. T. Yanagawa, On ribbon 2-knot: the 3-manifold bounded by the 2-knots. Osaka J. Math. **6**, 447–464 (1969)
33. T. Yanagawa, On ribbon 2-knots II: the second homotopy group of the complementary domain. Osaka J. Math. **6**, 465–473 (1969)
34. T. Yanagawa, On Ribbon 2-knots III: on the unknotting Ribbon 2-knots in S^4. Osaka J. Math. **7**, 165–172 (1970)
35. E. Yurasovskaya, Homotopy String Links Over Surfaces. PhD, University of British Columbia (2008)

Chapter 19
On Geometric Representation of $\mathbb{L}$-Homology Classes

Friedrich Hegenbarth and Dušan D. Repovš

Abstract In this chapter we give a geometric representation of $H_n(B; \mathbb{L})$ classes, where $\mathbb{L}$ is the 4-periodic surgery spectrum, by establishing a relationship between the normal cobordism classes $\mathcal{N}_n^H(B, \partial)$ and the n-th $\mathbb{L}$-homology of B, representing the elements of $H_n(B; \mathbb{L})$ by normal degree one maps with a reference map to B. More precisely, we prove that for every $n \geq 6$ and every finite complex B, there exists a map $\Gamma : H_n(B; \mathbb{L}) \longrightarrow \mathcal{N}_n^H(B, \partial)$.

Keywords Generalized manifold · Cell-like map · Normal degree one map · Steenrod $\mathbb{L}$-homology · Poincaré duality complex · Periodic surgery spectrum $\mathbb{L}$ · Geometric representation · $\mathbb{L}$-homology class

2020 Mathematics Subject Classification Primary: 55R20, 57P10, 57R65 57R67; Secondary: 55M05, 55N99, 57P05, 57P99

19.1 Introduction

Giving a meaning to algebraic objects (e.g., homology classes of generalized homology theories) in terms of geometric objects, is a fundamental task in algebraic topology. The relation between singular homology and classical cobordism theory is a well-known example.

In this paper we shall consider normal degree one maps $X^n \to M^n$ with a reference map $q : M^n \to B$, where X^n is either a generalized or a topological n-manifold, M^n is a topological n-manifold, and B is a finite complex. We shall

F. Hegenbarth (✉)
Dipartimento di Matematica "Federigo Enriques", Università degli studi di Milano, Milano, Italy
e-mail: friedrich.hegenbarth@unimi.it

D. D. Repovš
Faculty of Education and Faculty of Mathematics and Physics, University of Ljubljana & Institute of Mathematics, Physics and Mechanics, Ljubljana, Slovenia
e-mail: dusan.repovs@guest.arnes.si

L. Funar, A. Papadopoulos (eds.), *Essays on Topology*,
https://doi.org/10.1007/978-3-031-81414-3_19

denote normal cobordism classes of such objects $X^n \to M^n$ with $\mathcal{N}_n^H(B)$ (resp., $\mathcal{N}_n(B)$). We emphasize that $\mathcal{N}_n^H(B)$ (resp., $\mathcal{N}_n(B)$) should not be confused with the structure set in the case when B is a topological n-manifold (or a PD_n-complex). It is easy to see that in our definition,

$$\mathcal{N}_n(B) \cong \Omega_n(B \times G/TOP),$$

where $\Omega_n(\cdot)$ denotes n-dimensional cobordisms.

A *generalized n-manifold X^n (with boundary)* is an n-dimensional Euclidean neighborhood retract (ENR) with the local homology

$$H_*(X^n, X^n \setminus \{x\}; \mathbb{Z}) \cong H_*(\mathbb{R}^n, \mathbb{R}^n \setminus \{0\}; \mathbb{Z}), \text{ for every } x \in X^n.$$

The *boundary* ∂X^n of X^n is defined by

$$\partial X^n = \{x \in X^n : H_n(X^n, X^n \setminus \{x\}; \mathbb{Z}) \cong 0\}.$$

Then ∂X^n is a generalized $(n-1)$-manifold without boundary (see Mitchell [5]). Also, X^n (resp. ∂X^n) is an n-dimensional (resp. $(n-1)$-dimensional) Poincaré duality space. We shall consider generalized manifolds with boundary as pairs of $(n+1)$-dimensional (resp. n-dimensional) generalized manifolds $(W^{n+1}, \partial W^{n+1})$.

For every $n \geq 6$, Quinn [8] has constructed an obstruction $i(X^n) \in \mathbb{Z}$, which vanishes if there exists a cell-like map $M^n \to X^n$ of M^n onto X^n, where M^n is a closed topological n-manifold (for more on cell-like maps see the survey Mitchell-Repovš [6]). The construction is done locally, in particular $i(U) = i(X^n)$, for every open subset $U \subset X^n$. Because of its nice properties, one considers $I(X^n) = 1 + 8i(X^n)$, which we shall hereafter call the *Quinn index* of X^n. In particular, $I(W^{n+1}) = I(\partial W^{n+1})$, which follows from the product property of $I(X^n)$. For more on this topic see the monograph Cavicchioli-Hegenbarth-Repovš [2].

Our main result (to be proved in Sect. 19.3) establishes a relationship between $\mathcal{N}_n^H(B)$ and the n-th $\mathbb{L}$-homology of a finite complex B, i.e. $H_n(B; \mathbb{L})$, where $\mathbb{L}$ denotes the 4-periodic surgery spectrum, i.e.

$$\mathbb{L}_0 \cong \mathbb{Z} \times G/TOP$$

(see Ranicki [9]). Namely, we represent the elements of $H_n(B; \mathbb{L})$ by normal degree one maps with a reference map to B, i.e. as the objects of the following type

$$\begin{array}{ccc} X^n & \longrightarrow & N^n \\ & & \Big\downarrow q \\ & & B \end{array} \tag{19.1}$$

where X^n is a generalized n-manifold with boundary and N^n is a topological n-manifold with boundary, and normal cobordism classes of such objects are denoted by $\mathcal{N}_n^H(B, \partial)$ (resp. $\mathcal{N}_n(B, \partial)$).

Theorem 19.1.1 *For every $n \geq 6$ and every finite complex B, there exists a map*

$$\Gamma : H_n(B; \mathbb{L}) \longrightarrow \mathcal{N}_n^H(B, \partial).$$

19.2 Geometric Representation of the Elements of $H_n(B; \mathbb{L})$

Let B be a finite complex, $n \geq 6$, and let $H_n(B; \mathbb{L})$ be the Steenrod homology of B with respect to the 4-periodic surgery spectrum $\mathbb{L}$. The Steenrod homology behaves well on the category of all compact metric spaces—see Kahn-Kaminker-Schochet [4]. The spectrum $\mathbb{L}$ is algebraically defined. It is an Ω-spectrum, i.e. it is a sequence of Δ-sets $\mathbb{L}_q$, where $\mathbb{L}_q$ is homotopy equivalent to $\Omega\mathbb{L}_{q-1}$ with $\mathbb{L}_0 \cong \mathbb{Z} \times G/TOP$ as Δ-sets. Each $\mathbb{L}_q$ consists of a sequence $\mathbb{L}_q < j >$.

We shall follow Ranicki [9, Chapter 12] to define elements of $H_n(B; \mathbb{L})$. To this end, we have to embed $B \subset S^m$, where m is sufficiently large, and consider the dual complexes of B and S^m, which will be denoted respectively as $\bar{B}$ and Σ^m. However, we shall keep the notation B and S^m also for the duals.

Let (V^m, W^m) be a pair of simplicial complexes homotopy equivalent to the pair $(S^m, S^m \setminus B)$. An element $x \in H_n(B; \mathbb{L})$ is then given by a simplicial map

$$u : (V^m, W^m) \longrightarrow (\mathbb{L}_{n-m}, *),$$

which sends W^m to 0 and satisfies a certain cycle condition. Clearly, u is well-defined by x, up to some equivalence (i.e. boundary) condition. Moreover, if $\sigma \in B$, then

$$u(\sigma) \in \mathbb{L}_{n-m} < m - |\sigma| >,$$

where $|\sigma|$ denotes the dimension of σ, i.e. $m - |\sigma|$ is the dimension of its dual cell $D(\sigma, V^m)$ in V^m.

As described in Nicas [7, pp. 25–26], the Ω-spectrum property implies the equivalence

$$\mathbb{L}_q < j > \longrightarrow (\Omega\mathbb{L}_{q-1}) < j - 1 > \cong \mathbb{L}_{q-1} < j + 1 > .$$

By iteration, one obtains the equivalence

$$\mathbb{L}_0 < n - |\sigma| > \longrightarrow \mathbb{L}_{n-m} < m - |\sigma| >$$

as Δ-sets. Note that

$$\mathbb{L}_0 < j > \cong \mathbb{Z} \times (G/TOP < j >),$$

where $G/TOP < j >$ denotes the singular complex of j-simplices of G/TOP.

The face maps

$$\partial_0, ..., \partial_j : \mathbb{L}_0 < j > \longrightarrow \mathbb{L}_0 < j - 1 >$$

leave the $\mathbb{Z}$-components invariant. Since we are using dual complexes, the face maps can be written as follows

$$\mathbb{L}_{n-m} < m - j > \longrightarrow \mathbb{L}_{n-m} < m - j - 1 > .$$

Lemma 19.2.1 *Suppose that*

$$u : (V^m, W^m) \longrightarrow (\mathbb{L}_{n-m}, *)$$

represents the element $x \in H_n(B; \mathbb{L})$ *as above, where* $(S^m, S^m \setminus B)$ *is homotopy equivalent to the simplicial model for* (V^m, W^m)*. Then under the equivalence*

$$\mathbb{L}_{n-m} < m - j > \cong \mathbb{L}_0 < n - j > \cong \mathbb{Z} \times (G/TOP < n - j >),$$

$u(\sigma), u(\tau) \in \mathbb{L}_{n-m}$ *determine the same* $\mathbb{Z}$*-component.*

Proof The subface $\sigma \prec \tau \in B$ is a certain composition of face maps denoted by ∂, with the dual δ,

$$D(\tau, V^m) \prec D(\sigma, V^m).$$

The assertion of Lemma 19.2.1 now follows from the commutativity of the following diagram

$$\begin{array}{ccccc}
u(\tau) \in \mathbb{L}_{n-m} < m - |\tau| > & \longrightarrow & \mathbb{L}_0 < n - |\tau| > & & \\
\uparrow \delta & & & \searrow^{pr} & \\
 & & & & \mathbb{Z} \\
 & & & \nearrow_{pr} & \\
u(\sigma) \in \mathbb{L}_{n-m} < m - |\sigma| > & \longrightarrow & \mathbb{L}_0 < n - |\sigma| > & &
\end{array} \tag{19.2}$$

□

Corollary 19.2.2 *Consider any pair of simplices $\sigma, \tau \in B$ of the simplicial complex B such that $\sigma \cap \tau \neq \emptyset$. Then $u(\sigma), u(\tau) \in \mathbb{L}_{n-m}$ determine the same $\mathbb{Z}$-component.*

□

Moreover, Lemma 19.2.1 also implies the following corollary.

Corollary 19.2.3 *Suppose that*

$$u, u' : (V^m, W^m) \longrightarrow (\mathbb{L}_{n-m}, *)$$

represent the same element $x \in H_n(B; \mathbb{L})$, where $(S^m, S^m \setminus B)$ is homotopy equivalent to the simplicial model for (V^m, W^m). Then for every simplex $\sigma \in B$, the elements $u(\sigma), u'(\sigma) \in \mathbb{L}_{n-m}$ determine the same $\mathbb{Z}$-component.

Proof By hypothesis, $u \sim u'$, so there is a cobordism map

$$v : \Delta^1 \times (V^m, W^m) \longrightarrow (\mathbb{L}_{n-m}, *),$$

i.e.

$$v(\Delta^1 \times \sigma) \in \mathbb{L}_{n-m} < m - |\sigma| - 1 >,$$

where

$$v(\partial_0 \Delta^1 \times \sigma) = u(\sigma) \quad \text{and} \quad v(\partial_1 \Delta^1 \times \sigma) = u'(\sigma).$$

Therefore under the face maps, we get the following diagram

$$\begin{array}{ccccc}
\mathbb{L}_{n-m} < m - |\sigma| > & & & & \\
& \searrow^{\delta_0} & & & \\
& & \mathbb{L}_{n-m} < m - |\Delta^1 \times \sigma| > & \longrightarrow & \mathbb{L}_0 < n - |\Delta^1 \times \sigma| > \\
& \nearrow_{\delta_1} & & & \downarrow {\scriptstyle pr} \\
\mathbb{L}_{n-m} < m - |\sigma| > & & & & \mathbb{Z}
\end{array} \qquad (19.3)$$

which proves the assertion of Corollary 19.2.3. □

19.3 Proof of Theorem 19.1.1

It follows from Lemma 19.2.1 and Corollaries 19.2.2 and 19.2.3 that the $\mathbb{Z}$-components depend only on $x \in H_n(B; \mathbb{L})$, and that $u(\sigma), u(\sigma') \in \mathbb{L}_{n-m}$ define the same element if σ and σ' can be connected by a chain $\sigma_1, ..., \sigma_r$ of simplices such that

$$\sigma_j \cap \sigma_{j+1} \neq \emptyset, \text{ for every } j \in \{1, ..., r-1\}.$$

This leads to the following theorem.

Theorem 19.3.1 *Suppose that B is connected. Then the construction described above defines a map*

$$I : H_n(B; \mathbb{L}) \longrightarrow L_0(\mathbb{Z}) = 1 + 8\mathbb{Z}.$$

Remark 19.3.2 The $\mathbb{Z}$-component coming from the identification

$$\mathbb{L}_0 \cong \mathbb{Z} \times G/TOP$$

is the 0-dimensional signature invariant defined by Quinn [8, Section 3.2].

The second component of the identification

$$\mathbb{L}_{n-m} < m - |\sigma| > \cong \mathbb{L}_0 < n - |\sigma| > \cong \mathbb{L}_0(\mathbb{Z}) \times (G/TOP < n - |\sigma| >)$$

associates to $x \in H_n(B; \mathbb{L})$, represented by the map, where (V^m, W^m) is a pair of simplicial complexes homotopy equivalent to the pair $(S^m, S^m \setminus B)$,

$$u : (V^m, W^m) \longrightarrow (\mathbb{L}_{n-m}, *),$$

a family of adic normal degree one maps given by $u(\sigma)$

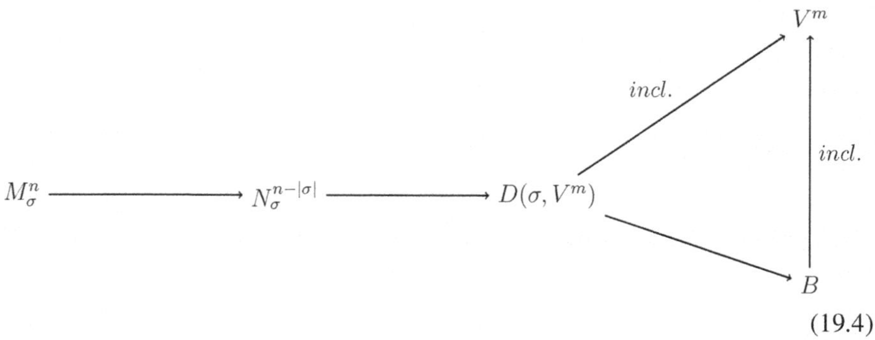

(19.4)

therefore it defines a map

$$\Gamma : H_n(B; \mathbb{L}) \longrightarrow \mathcal{N}_n^H(B, \partial).$$

Similarly, one obtains a map

$$\Gamma^+ : H_n(B; \mathbb{L}^+) \longrightarrow \mathcal{N}_n(B, \partial).$$

Clearly, the following diagram commutes

$$\begin{array}{ccc} H_n(B;\mathbb{L}^+) & \xrightarrow{\quad\Gamma^+\quad} & \mathcal{N}_n(B,\partial) \\ \Big\downarrow & & \Big\downarrow \\ H_n(B;\mathbb{L}) & \xrightarrow{\quad\Gamma\quad} & \mathcal{N}_n^H(B,\partial) \end{array} \tag{19.5}$$

Summary 19.3.3 *For every $n \geq 6$, one can construct the maps*

$$I : H_n(B; \mathbb{L}) \longrightarrow L_0(\mathbb{Z}) = 1 + 8\mathbb{Z}$$

and

$$\Gamma : H_n(B; \mathbb{L}) \longrightarrow \mathcal{N}_n^H(B, \partial)$$

via the simplicial equivalence

$$\mathbb{L}_{n-m} < m - j > \cong \mathbb{L}_0 < n - j > \cong \mathbb{L}_0(\mathbb{Z}) \times (G/TOP < n - j >).$$

This completes the proof of Theorem 19.1.1. □

Acknowledgments This research was supported by the Slovenian Research and Innovation Agency grants P1-0292, J1-4031, J1-4001, N1-0278, N1-0114, and N1-0083. We thank John Bryant, Wolfgang Lück, Washington Mio, Shmuel Weinberger, and Michael Weiss for updating us regarding the doubts about the validity of Ferry-Pedersen [3, Theorem 16.6], which in the last three decades was applied by many authors (for details see e.g. Bryant-Ferry-Mio-Weinberger [1]), and which we also wanted to apply in our original plan. We thank the editors and referees for comments and suggestions.

References

1. J. Bryant, S. Ferry, W. Mio, S. Weinberger, Erratum: "Topology of Homology Manifolds." Ann. of Math. **200**(2), 799–801 (2024)
2. A. Cavicchioli, F. Hegenbarth, D. Repovš, *Higher-Dimensional Generalized Manifolds: Surgery and Constructions*. EMS Series of Lectures in Mathematics, vol. 23 (European Math. Soc., Zürich, 2016) MR 3558558
3. S.C. Ferry, E.K. Pedersen, Epsilon surgery theory, in *Novikov Conjectures, Index Theorems and Rigidity*, vol. 2 (Oberwolfach, 1993). London Math. Soc. Lecture Note Ser., vol. 227 (Cambridge University Press, Cambridge, 1995), pp. 167–226. MR 1388311
4. D.S. Kahn, J. Kaminker, C. Schochet, Generalized homology theories on compact metric spaces. Michigan Math. J. **24**(2), 203–224 (1977). MR 0474274
5. W.J.R. Mitchell, Defining the boundary of a homology manifold. Proc. Am. Math. Soc. **110**(2), 509–513 (1990). MR 90m:57015
6. W.J.R. Mitchell, D. Repovš, The topology of cell-like mappings. Rend. Fac. Sci. Univ. Cagliari, Suppl. **58**, 265–300 (1988). MR 1122860
7. A. Nicas, *Induction Theorems for Groups of Homotopy Manifold Structures*. Memoirs of the American Mathematical Society, vol. 39, no. 267 (1982), vi+108 pp. MR 0668807
8. F.S. Quinn, Resolutions of homology manifolds and the topological characterization of manifolds. Invent. Math. **72**(2), 267–284 (1983) MR 0700771. Corrigendum, Invent. Math. **85**(3), 653 (1986). MR 0848688
9. A.A. Ranicki, *Algebraic L-Theory and Topological Manifolds*. Cambridge Tracts in Math., vol. 102 (Cambridge University Press, Cambridge, 1992) MR 1211640

Chapter 20
On Elementary Invariants of Genus One Knots and Seifert Surfaces

Christine Lescop

À Po, en souvenir de sympathiques dîners partagés
et au plaisir d'en partager d'autres !

Abstract This elementary chapter introduces easy-to-manage invariants of genus one knots in homology 3-spheres. To prove their invariance, we investigate properties of an invariant of 3-dimensional genus two homology handlebodies called the *Alexander form*. The Alexander form of a 3-manifold E with boundary contains all Reidemeister torsions of link exteriors obtained by attaching 2-handles along the boundary of E. It is a useful tool for studying Alexander polynomials and Reidemeister torsions. We extract invariants of genus one Seifert surfaces from the Alexander form of their exteriors.

Keywords Knots · 3-manifolds · Seifert surfaces · Homology 3–spheres · Alexander polynomial · Reidemeister torsion · Finite type invariants

MSC 2020 57K10, 57K31, 57K14, 57K16, 57K20

20.1 Introduction

Preliminary Conventions and Definitions In this chapter, we consider manifolds of dimension at most three, and all manifolds are oriented. We orient boundaries of manifolds with the *outward normal first* convention. Since any topological manifold of dimension at most three has a unique smooth structure up to diffeomorphism, we work with smooth (oriented) manifolds. An *isotopy* of such a manifold M is a smooth map from $\Psi\colon [0,1] \times M$ to M such that $\Psi(0,.)$ is the identity map and $\Psi(t,.)$ is a diffeomorphism for any $t \in [0,1]$. Two submanifolds P and Q of M are

---

C. Lescop (✉)
Institut Fourier, CNRS, Université Grenoble Alpes, Saint-Martin-d'Hères, France
e-mail: Christine.Lescop@univ-grenoble-alpes.fr

L. Funar, A. Papadopoulos (eds.), *Essays on Topology*,
https://doi.org/10.1007/978-3-031-81414-3_20

said to be *isotopic* if there exists an isotopy of M such that $\Psi(\{1\} \times P) = Q$. We say that a function of submanifolds of a manifold M is an *invariant* if it maps any two isotopic submanifolds to the same value. A *knot* (resp. a *link*) in a 3-manifold is an isotopy class of (images of) embeddings of the circle (resp. a disjoint union of circles) into the manifold. In this chapter, a *Seifert surface* is a connected compact (oriented) surface with one boundary component. For $\Lambda = \mathbb{Z}$, $\mathbb{Z}/2\mathbb{Z}$, or $\mathbb{Q}$, a Λ-*sphere* is a compact oriented 3-manifold with the same homology with coefficients in Λ as the standard 3-sphere S^3. $\mathbb{Z}$-spheres (resp. $\mathbb{Q}$-spheres) are also called *integer (resp. rational) homology 3-spheres*. We omit the 3 and call them integer or rational homology spheres. Unless otherwise mentioned, homology coefficients are in $\mathbb{Z}$. The cardinality or *order* of the H_1 of a $\mathbb{Q}$-sphere R is denoted by $|H_1(R)|$.

20.1.1 An Overview

Here is a short and vague introduction for readers familiar with Reidemeister torsions and Alexander polynomials. We provide complete definitions and background material in the following subsections.

We introduce an invariant $w_{\mathcal{S}}$ of genus one knots in homology 3-spheres in Theorem 20.1.2. To my knowledge, it is a new invariant. It is a simple combination of coefficients of Alexander polynomials of curves of a genus one Seifert surface. Then we introduce a second independent simple invariant w_{SL} of genus one knots in $\mathbb{Z}$-spheres in Sect. 20.1.3. As we show in Theorem 20.1.12, this second invariant w_{SL} is a combination of $w_{\mathcal{S}}$ and a (degree 3) knot invariant w_3, which appears in a surgery formula for a (degree 2) invariant of $\mathbb{Q}$-spheres. To prove the invariance of $w_{\mathcal{S}}$, we show that $w_{\mathcal{S}}$ is a combination of w_3 and a fourth invariant $W_{\mathcal{D}}$ of genus one knots, which is also new to my knowledge.

Let us outline the construction of $W_{\mathcal{D}}$. Let K be a knot that bounds a genus one Seifert surface Σ in a $\mathbb{Q}$-sphere R. Assume that $[-1, 1] \times \Sigma$ is embedded in R so that $\Sigma = \{0\} \times \Sigma$. Let $E[K]$ be the manifold obtained from the *exterior* $E = R \setminus \left(]-1, 1[\times \mathring{\Sigma}\right)$ of Σ by adding a 2-handle along K. We extract a tautological invariant $W_s(\mathcal{D}_2(\Sigma))$ of Σ from a normalized Reidemeister torsion $\mathcal{D}(\Sigma)$ of $E[K]$ in a canonical way. Then we prove that $W_s(\mathcal{D}_2(\Sigma))$ depends only on K, as outlined in Sect. 20.1.6. Our fourth invariant $W_{\mathcal{D}}(K)$ is $W_s(\mathcal{D}_2(\Sigma))$.

To study these invariants and their relationships, we use and investigate *Alexander forms* of 3-manifolds. Alexander forms are generalizations of abelian Reidemeister torsions from link complements to more general 3-manifolds. We define them from the homology of the maximal free abelian covering spaces in Sect. 20.1.4. Alexander forms contain Reidemeister torsions (and Alexander polynomials) of link exteriors obtained by attaching 2-handles.

A *genus g handlebody* is a manifold obtained from the unit ball B^3 of $\mathbb{R}^3$ by attaching g one-handles ($D^2 \times [-1, 1]$ along $D^2 \times \partial[-1, 1]$) to the boundary ∂B^3 of B^3. For $\Lambda = \mathbb{Z}$ or $\mathbb{Q}$, a (*genus g*) Λ-*handlebody* is a compact oriented 3-manifold with the same homology with coefficients in Λ as a (genus g) handlebody. A (genus g) $\mathbb{Q}$-*handlebody* is also called a (genus g) *rational homology handlebody*.

The exterior of a genus one Seifert surface in a $\mathbb{Q}$-sphere is an example of a genus two rational homology handlebody. We exhibit structural properties of the Alexander forms of genus two rational homology handlebodies in Theorem 20.1.28, which is of independent interest.

20.1.2 A First Simple Invariant of Genus One Knots

The simplest (nontrivial) link invariant is the linking number. It applies to 2-component links (J, K) in rational homology 3-spheres. In such manifolds, every null-homologous knot bounds a Seifert surface.[1] The *linking number* $lk(J, K)$ of two disjoint oriented knots J and K is the algebraic intersection of J with a transverse Seifert surface Σ_K of K when K is null-homologous. In general, $lk(J, K)$ is the algebraic intersection of J with a rational chain bounded by K. The *Alexander polynomial* $\Delta(K)$ is a simple invariant of knots in $\mathbb{Q}$-spheres discovered in 1928 by Alexander. It is the order of the first homology group of the maximal free abelian covering space of the knot complement (see Definition 20.2.1). Let us give an alternative definition of $\Delta(K)$, for a null-homologous knot K in a $\mathbb{Q}$-sphere R, using a Seifert surface Σ_K of K, as in [2, Proposition 2.3.13, page 27], for example. Embed $[-1, 0] \times \Sigma_K$ in R by an orientation-preserving embedding that naturally identifies $\{0\} \times \Sigma_K$ with Σ_K. Set $\Sigma_{K-} = \{-1\} \times \Sigma_K$. For a curve v, let v_- denote the curve of $\{-1\} \times v$. The *Seifert form* associated with $H_1(\Sigma_K)$ is the form

$$\begin{aligned} V \colon H_1(\Sigma_K) \otimes_s H_1(\Sigma_K) &\to \mathbb{Z} \\ [u] \otimes_s [v] &\mapsto lk(u, v_-). \end{aligned}$$

Then we have

$$\Delta(K) = |H_1(R)| \det\left(t^{1/2}V - t^{-1/2}({}^T V)\right)$$

where V also denotes a *Seifert matrix* of the form V, and ${}^T V$ is its transpose. We extract the (even simpler) invariant

$$\lambda'(K) = \frac{\Delta''(K)(1)}{2|H_1(R)|}$$

from $\Delta(K)$. When K has a genus one Seifert surface Σ_K, $\lambda'(K)$ is simply the determinant of an associated Seifert matrix V. (We will give definitions of Δ and

[1] To be precise, we should take an embedding image representing the knot here, but it would make the text heavier.

λ' for non-necessarily null-homologous knots in $\mathbb{Q}$-spheres in Definitions 20.2.1 and 20.2.4. See also Lemma 20.2.5.)

Let Σ be a genus one Seifert surface. A *symplectic basis* of $H_1(\Sigma)$ is a basis (α, β) of $H_1(\Sigma)$ such that the *algebraic intersection number* $\langle\alpha, \beta\rangle_\Sigma$ is one. We say that an element u of $H_1(\Sigma)$ is *primitive* if there exists an element v of $H_1(\Sigma)$ such that $\langle u, v\rangle_\Sigma = 1$. Any primitive element of $H_1(\Sigma)$ can be represented by a non-separating simple closed curve of Σ. We will prove the following easy lemma in Sect. 20.4.3.

Lemma 20.1.1 *Two homologous non-separating simple closed curves of a genus one oriented surface Σ with one boundary component are isotopic in Σ.*

Thus, for such a genus one Seifert surface, any primitive element u of $H_1(\Sigma)$ is represented by a connected curve of Σ, whose isotopy class in Σ is determined by u. We also denote this curve associated to u by u. We say that a Seifert surface Σ in a rational homology sphere R is *null-homologous* if the map induced by the inclusion from $H_1(\Sigma)$ to $H_1(R)$ maps $H_1(\Sigma)$ to 0.

We are now ready to define a simple invariant of genus one knots in integer homology 3-spheres and state one of the main theorems of this chapter.

Theorem 20.1.2 *Let $\Sigma = \Sigma_K$ be a genus one Seifert surface of a knot K in a $\mathbb{Q}$-sphere R, let (α, β) be a symplectic basis of $H_1(\Sigma)$. Let α, β, and $\gamma = -\alpha - \beta$ denote both the corresponding primitive elements of $H_1(\Sigma)$ and the associated curves in Σ and in the ambient manifold R. Define (a, b, c) so that*

$$lk(\alpha, \alpha_-) = \frac{b+c}{2}, \quad lk(\beta, \beta_-) = \frac{c+a}{2}, \quad \textit{and} \quad lk(\gamma, \gamma_-) = \frac{a+b}{2}.$$

Set

$$\lambda'(a, b, c) = \frac{ab + ac + bc + 1}{4}$$

and

$$w_\delta(a, b, c) = \frac{a+b}{2}\frac{b+c}{2}\frac{c+a}{2} - \lambda'(a, b, c)(a + b + c).$$

Then $\lambda'(K) = \lambda'(a, b, c)$ and

$$w_\delta(\Sigma) = 4a\lambda'(\alpha) + 4b\lambda'(\beta) + 4c\lambda'(\gamma) + w_\delta(a, b, c)$$

is an invariant of Σ.[2] *Furthermore, if Σ is null-homologous, then $w_\delta(\Sigma)$ depends only on $K = \partial\Sigma$. We denote it by $w_\delta(K)$.*

[2] It is independent of our choice of the symplectic basis (α, β) of $H_1(\Sigma)$.

Remark 20.1.3 Note that (β, γ) and (γ, α) are symplectic bases of $H_1(\Sigma)$. The expression of w_δ is invariant under a cyclic permutation of (a, α), (b, β), and (c, γ). We can equivalently define a to be

$$a = lk(\beta, \beta_-) + lk(\gamma, \gamma_-) - lk(\alpha, \alpha_-)$$

or

$$a = -lk(\beta, \gamma_-) - lk(\gamma, \beta_-),$$

where $lk(\beta, \gamma_-) - lk(\gamma, \beta_-) = \langle \beta, \gamma \rangle_\Sigma = 1$. Cyclic permutations lead to similar equalities for b and c. These equalities imply that a, b, and c are odd integers when Σ is null-homologous.

Remark 20.1.4 The Alexander polynomial $\Delta(K)$ of a genus one knot is $\Delta(K) = 1 + \lambda'(K)(t - 2 + t^{-1})$.

Consider the *pretzel knots* $K(a, b, c)$ of Fig. 20.1. In this figure, for an integer a, the box $\boxed{a}$ denotes a two-strand braid with $|a|$ vertical juxtapositions of the motive ⤫, if a is positive, and $|a|$ vertical juxtapositions of ⤬ otherwise. The surface $\Sigma(a, b, c)$ embedded in $(\mathbb{R}^3 \subset S^3)$ as in Fig. 20.1 is a Seifert surface of $K(a, b, c) = \partial\Sigma(a, b, c)$.

Example 20.1.5 The knot $K(-1, -1, 1)$ is the trivial knot, and we have $\lambda'(K(-1, 1, -1)) = 0$. The pictured curves α, β, and γ also represent trivial knots. In particular, we have

$$\lambda'(\alpha) = \lambda'(\beta) = \lambda'(\gamma) = 0,$$

and $w_\delta\,(K(a, b, c)) = w_\delta\,(a, b, c)$.

For the left-handed trefoil $K(1, 1, 1)$, we have $w_\delta(K(1, 1, 1)) = 1 - 3 = -2$. For the right-handed trefoil $K(-1, -1, -1)$, we have $w_\delta(K(-1, -1, -1)) = 2$. (The *signature* of $(V +^T V)$, which is another simple classical knot invariant, also distinguishes these knots.)

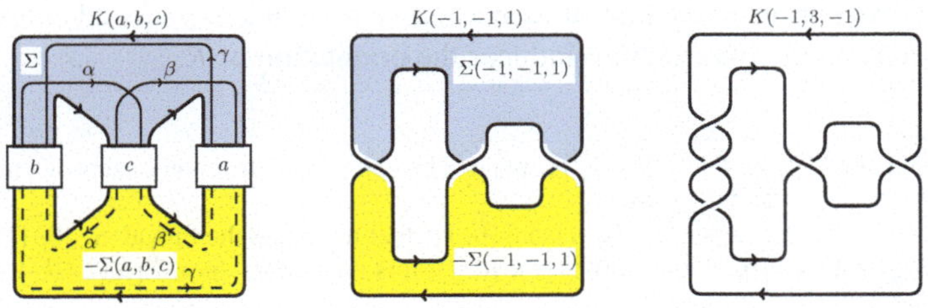

Fig. 20.1 The surface $\Sigma(a, b, c)$ with its curves α, β, and γ, $\Sigma(-1, -1, 1)$, and the figure-eight knot $K(-1, 3, -1)$

In [5], Ozsváth and Szabó introduced a powerful *categorification* $\widehat{HFK}$ of the Alexander polynomial, which detects the genus of knots. (The Alexander polynomial is a graded Euler characteristic of $\widehat{HFK}$.) Let k be an integer greater than 1. According to a theorem of Hedden and Watson [1, Theorem 1], the pretzel knots of the family $\big(K(2n+1, 2k+1, -2k-1)\big)_{n\in\mathbb{N}}$ have the same Ozsváth–Szabó Heegaard–Floer homologies $\widehat{HFK}$ and HFK.[3] We have $w_\delta\big(K(2n+1, 2k+1, -2k-1)\big) = (k+k^2)(2n+1)$. So w_δ distinguishes these knots, which have identical Ozsváth–Szabó Heegaard–Floer homology, identical Alexander polynomial, and identical vanishing signature. (The Jones polynomial also distinguishes them.)

The manifold obtained from a manifold M by reversing its orientation is denoted by $(-M)$.

Proposition 20.1.6 *Let K be a knot that bounds a null-homologous genus one Seifert surface in a $\mathbb{Q}$-sphere R. Then we have*

$$w_\delta(K \subset R) = w_\delta((-K) \subset R) \quad \textit{and} \quad w_\delta(K \subset (-R)) = -w_\delta(K \subset R).$$

Proof Changing the orientation of K exchanges the roles of α and β. Since the expression of w_δ is invariant under this exchange, the invariant w_δ is independent of the orientation of K.[4]

Reversing the orientation of the ambient rational homology sphere R multiplies the linking numbers by (-1). Therefore this operation, which changes R to $(-R)$, multiplies a, b, and c by (-1). □

The *mirror image* of the isotopy class of a knot embedding $K : S^1 \hookrightarrow \mathbb{R}^3$ is the class of the composition of K by a reflection of $\mathbb{R}^3$. It is denoted by $\overline{K}$. We have $w_\delta(\overline{K}) = -w_\delta(K)$. Therefore, w_δ vanishes for *amphicheiral knots*, which are knots isotopic to their mirror images. The figure-eight knot $K(-1, 3, -1) = K(1, -3, 1)$ is an example of an amphicheiral knot such that $\lambda'(K(-1, 3, -1)) = -1$.

20.1.3 A Second Simple Invariant of Genus One Knots

The intersection of two oriented surfaces in a $\mathbb{Q}$-sphere R is oriented so that the triple (positive normal to the first surface, positive normal to the second one, oriented tangent vector to the intersection) induces the orientation of R.

---

[3] The pretzel knot $K(1, 2k+1, -2k-1)$ is obtained by connecting the components of a trivial two-component link 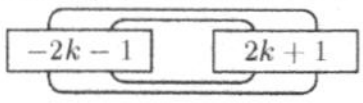

by a band $[0,1]^2$ that intersects the link along $[0,1] \times \partial[0,1]$ and replacing $[0,1] \times \partial[0,1]$ with $(\partial[0,1]) \times [0,1]$. It is nontrivial, since $\lambda'(K(1, 2k+1, -2k-1)) = -k - k^2$.

[4] For a triple $K(a, b, c)$ of pairwise distinct odd integers greater than 2, the knot $K(a, b, c)$ is *non-invertible*, i.e., it is not isotopic to $-K(a, b, c) = K(b, a, c)$. In [7], Hale Trotter proved the existence of non-invertible knots by proving that these knots are non-invertible. Unfortunately, the invariant w_δ cannot detect non-invertibility.

The definition of our second simple invariant of genus one knots is based on the following Sato–Levine invariant [6].

Definition 20.1.7 Let J and K be two disjoint null-homologous knots in a rational homology sphere R, such that the linking number $lk(J, K)$ is zero. Let Σ_J and Σ_K be two transverse surfaces in R respectively bounded by J and K, such that $\Sigma_J \cap \Sigma_K$ is disjoint from J and K. Let $\Sigma_{J,\parallel}$ be obtained from Σ_J by a slight normal push. Define the *Sato–Levine invariant* $\lambda'(J, K)$ of (J, K) to be

$$\lambda'(J, K) = -lk(\Sigma_J \cap \Sigma_K, \Sigma_{J,\parallel} \cap \Sigma_K).$$

See Definition 20.2.12 and Lemma 20.2.11 for more general definitions of $\lambda'(J, K)$.

Let $H(a, b, c)$ be the genus two handlebody $[-1, 0] \times \Sigma(a, b, c)$, where we identify $\Sigma(a, b, c)$ with $\{0\} \times \Sigma(a, b, c)$.

The boundary $\partial H(a, b, c)$ of $H(a, b, c)$ is the union of $\Sigma(a, b, c)$ and $(-\Sigma(a, b, c)_-)$, where

$$\Sigma(a, b, c)_- = (\{-1\} \times \Sigma(a, b, c)) \cup_{\{-1\}\times\partial\Sigma(a,b,c)} ([-1, 0] \times K(a, b, c))\,.$$

We think of $\partial H(a, b, c)$ as being obtained from the two copies $\Sigma(a, b, c)$ and $\Sigma(a, b, c)_-$ of $\Sigma(a, b, c)$ by inflating between these two glued copies whose boundaries are identified and sealed as in Fig. 20.2. (In Fig. 20.2, the tube  represents $|a|$ vertical juxtapositions of the motive  if a is positive, and $|a|$ vertical juxtapositions of  otherwise.) In particular, $H_0 = H(a, b, c)$ does not depend on (a, b, c), but the inclusion from $\{0\} \times \Sigma(a, b, c)$ to H_0 does.

Let R denote a $\mathbb{Q}$-sphere. Consider the three curves A, B, and C of the right-hand side of Fig. 20.2. Let $\phi\colon H_0 \hookrightarrow R$ be an embedding that maps A, B, and C to null-homologous curves in the *exterior* $E = R \setminus \phi(\mathring{H}_0)$ of $\phi(H_0)$. Any null-homologous genus one Seifert Σ in R may be written as $\phi(\Sigma(a, b, c))$ for such an embedding and for three odd integers a, b, and c (which can be chosen as in Theorem 20.1.2 or Remark 20.1.3) so that this hypothesis is fulfilled. Note that $lk(\phi(A), \phi(B)) = 0$ and that

$$\lambda'(\phi(A), \phi(B)) = \lambda'(\phi(B), \phi(C)) = \lambda'(\phi(C), \phi(A)).$$

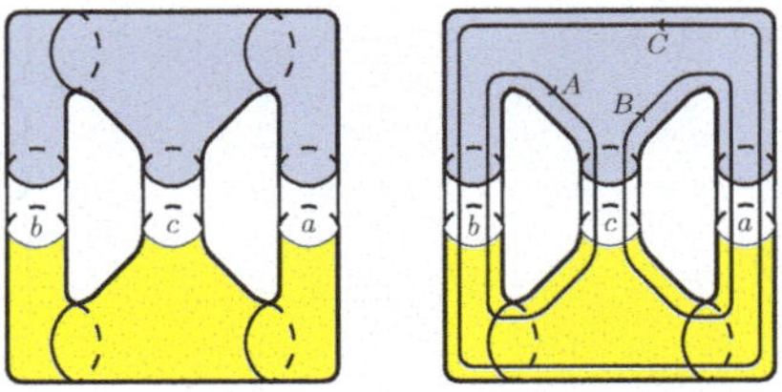

Fig. 20.2 The handlebody $H_0 = H(a, b, c)$ and the curves A, B, and C

Theorem 20.1.8 *Let $\phi\colon H_0 \hookrightarrow R$ be an embedding of H_0 into a $\mathbb{Q}$-sphere R that maps A and B to null-homologous curves in the exterior of $\phi(H_0)$. With the notation of Theorem 20.1.2, set*

$$w_{SL}\Big(\phi\big(\Sigma(a,b,c)\big)\Big) = \lambda'\big(\phi(A), \phi(B)\big) - \frac{1}{12} w_\delta(a,b,c).$$

Then $w_{SL}\big(\phi\big(\Sigma(a,b,c)\big)\big)$ is a topological invariant of the genus (at most) one knot $\partial\phi(\Sigma(a,b,c)) \subset R$. We denote it by $w_{SL}\big(\partial\phi\big(\Sigma(a,b,c)\big)\big)$.

The invariant w_{SL} is a fortiori an invariant of null-homologous genus one Seifert surfaces in $\mathbb{Q}$-spheres. It is generalized to all genus one Seifert surfaces in $\mathbb{Q}$-spheres in Theorem 20.1.33.

Example 20.1.9 Let ϕ_0 denote the trivial embedding of Fig. 20.2 of H_0 into $\mathbb{R}^3 \subset S^3$. For the knots $K(a,b,c) = \phi_0\big(K(a,b,c)\big)$, the Sato–Levine invariant $\lambda'\big(\phi_0(A), \phi_0(B)\big)$ is zero since $\phi_0(A)$ bounds a disk disjoint from $\phi(B)$. Thus $w_{SL}\big(K(a,b,c)\big) = -\frac{1}{12} w_\delta(a,b,c)$.

Let B_a be a ball B_a of $\mathbb{R}^3$ in a neighborhood of the box $\boxed{a}$ of Fig. 20.1 that intersects $\phi_0\big(\Sigma(a,b,c)\big)$ as the twisted band $\phi_0\big([0,1]^2\big)$ $=\!\boxed{a}$ so that $(\partial B_a) \cap \phi_0([0,1]^2) = \phi_0([0,1] \times \partial[0,1])$. Change the trivial embedding ϕ_0 of $\Sigma(a,b,c)$ to an embedding ϕ_J such that

- ϕ_0 and ϕ_J coincide outside the ball B_a,
- ϕ_J maps the band $[0,1]^2$ to the ball B_a so that $\phi_J([0,1] \times \partial[0,1]) = \phi_0([0,1] \times \partial[0,1])$,
- the self-linking number of $\phi_J(\beta)$ is still $\frac{c+a}{2}$, and
- the knot $\phi_J(\beta)$ is a knot J.

Figure 20.3 shows an example of a knot $\phi_J(K(-1,3,-1))$ obtained in this way, where J is the figure-eight knot.

This operation changes $K(a,b,c)$ to a knot $\phi_J\big(K(a,b,c)\big)$ with the same Alexander polynomial as $K(a,b,c)$ such that

$$w_\delta\Big(\phi_J\big(K(a,b,c)\big)\Big) = w_\delta\big(K(a,b,c)\big) + 4(b+c)\lambda'(J).$$

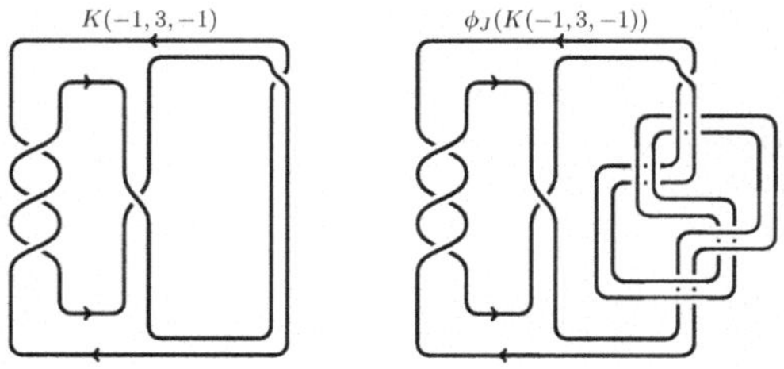

Fig. 20.3 A modification of the figure-eight knot $K(-1,3,-1)$

In the meantime, we have

$$w_{SL}\Big(\phi_J\big(K(a,b,c)\big)\Big) = w_{SL}\big(K(a,b,c)\big)$$

since $\phi_J(A)$ still bounds a disk disjoint from $\phi_J(B)$. In particular, when $4(b+c)\lambda'(J) \neq 0$, which is easily realizable, w_δ distinguishes $\phi_J\big(K(a,b,c)\big)$ from $K(a,b,c)$ while w_{SL} does not. (In the case of Fig. 20.3, $(b+c)=2$ and $\lambda'(J) = -1$.) Thus w_δ and w_{SL} are independent knot invariants.

Remark 20.1.10 To compute $\lambda'\big(\phi(A),\phi(B)\big)$, it suffices to

- choose a surface Σ_A bounded by $\phi(A)$ in the exterior of $\phi(B)$,[5]
- pick a curve c of $H_1(\Sigma_A)$, such that, $\langle x, c\rangle_{\Sigma_A} = lk\big(x,\phi(B)\big)$ for any curve x of Σ_A, and
- compute $\lambda'\big(\phi(A),\phi(B)\big) = lk(c, c_{\parallel})$, where $c_{\parallel}$ is a curve parallel to c on Σ_A.

Fix the curve $\phi(A)$ of ∂E. The kernel of the map induced by the inclusion from $H_1(\partial E)$ to $H_1(E)$ is $\mathbb{Z}[\phi(A)] \oplus \mathbb{Z}[\phi(B)]$. So, the homology class of $\phi(B)$ in $\partial E \setminus \phi(A)$ is well determined up to multiplication by ± 1 and addition of curves parallel to $\phi(A)$. In particular, $\lambda'(\phi(A),\phi(B))$ is an invariant of the curve $\phi(A)$ of ∂E. (It is similarly an invariant of the curve $\phi(B)$.) The homology class of $\phi(A)$ on ∂E is characterized by the properties that $\phi(A)$ bounds in E and that $\phi(A)$ is homologous to $\phi(\alpha)$ in $\phi(H_0)$. Performing a Dehn twist about α on $\Sigma(a,b,c)$ does not change the homology class of $\phi(A)$ on ∂E. But it changes $w_\delta(a,b,c)$ as soon as $b(b^2-1) \neq 0$, as Lemma 20.3.19 shows. Therefore, $\lambda'\big(\phi(A),\phi(B)\big)$ is not an invariant of the homology class of $\phi(A)$ in $H_1(\partial E)$. Theorem 20.1.8 can be used to reduce the computation of its variation to the computation of the variation of $w_\delta(a,b,c)$.

Theorem 20.1.8 is a direct consequence of Theorem 20.1.2 and of the following proposition [4, Proposition 7.3].

Proposition 20.1.11 *There exists an invariant w_3 of knots in rational homology 3-spheres that satisfies the following property. For any embedding ϕ of H_0 in a rational homology sphere R such that $\phi(A)$ and $\phi(B)$ are null-homologous in the exterior $R \setminus (\phi(\mathring{H}_0))$ of $\phi(H)$, we have*

$$w_3\Big(\phi(K(a,b,c))\Big) = \frac{3}{2}\lambda'(\phi(A),\phi(B)) - \frac{1}{4}w_\delta(a,b,c) - \frac{a}{2}\lambda'(\phi(A)) - \frac{b}{2}\lambda'(\phi(B)) - \frac{c}{2}\lambda'(\phi(C)).$$

[5] The *exterior* of a knot J in R is the complement of an open tubular neighborhood of J in R.

For an integer q, let $R(K; 1/q)$ denote the manifold obtained by $1/q$-surgery on a null-homologous knot K in a $\mathbb{Q}$-sphere R. In [4, Theorem 7.1], I proved a surgery formula

$$\lambda_2(R(K; 1/q)) = \lambda_2''(K)q^2 + w_3(K)q + \lambda_2(R)$$

for an invariant λ_2 of rational homology 3-spheres. The invariant w_3 that governs the linear term of this formula satisfies the property of Proposition 20.1.11, according to [4, Proposition 7.3]. It also satisfies $w_3\big(K \subset (-R)\big) = -w_3\big(K \subset R\big)$. When the ambient rational homology 3-sphere R is S^3, the invariant w_3 is the degree 3 knot invariant that changes sign under mirror image and maps the chord diagram with three diameters to (-1). It is a coefficient of the Jones polynomial in a suitable normalization.

For knots that bound null-homologous Seifert surfaces, Theorems 20.1.2 and 20.1.8 stated above can be summarized as follows.

Theorem 20.1.12 *If K is a knot in a $\mathbb{Q}$-sphere R such that K bounds a null-homologous Seifert surface in R, then the invariant $w_3(K)$ is the sum of the following independent knot invariants described in Theorems 20.1.2 and 20.1.8.*

$$w_3(K) = \frac{3}{2}w_{SL}(K) - \frac{1}{8}w_{\delta}(K).$$

This chapter and the discovery of Theorem 20.1.2 grew up from an attempt to understand surgery formulas for finite type invariants of $\mathbb{Q}$-spheres and finite type invariants of knots, such as w_3, from finite type invariants of curves on Seifert surfaces. (For a null-homologous knot K in a $\mathbb{Q}$-sphere R, the "degree one" Casson invariant λ_C satisfies the surgery formula $\lambda_C(R(K; 1/q)) = \lambda'(K)q + \lambda_C(R)$, and the knot invariant $\lambda_2''(K)$ of the quadratic term of the surgery formula above for the "degree two" invariant λ_2 is a polynomial in the entries of a Seifert matrix. See [4, Theorem 7.1].)

20.1.4 Definition of Alexander Forms and Reidemeister Torsions

We will prove Theorems 20.1.12 and 20.1.2 by studying a Reidemeister torsion associated with a genus one Seifert surface. In this subsection, we define the needed Reidemeister torsions in a way appropriate for our study from the Alexander forms described below.

Alexander forms were introduced in [3, Section 3].[6] They help study Alexander polynomials with several variables, and the coefficients ζ of [2], in particular. Below, we recall their definition and some of their properties. We develop more examples and more properties in Sect. 20.2.

Throughout this section, X denotes a connected compact oriented 3-manifold with a nonempty boundary, equipped with a basepoint $\star$. Define the *genus* $g(X)$ of X to be

$$g(X) = 1 - \chi(X) \left(= 1 - \frac{1}{2}\chi(\partial X)\right).$$

Set $g = g(X)$ and assume $g > 0$.

Let $\tilde{X}$ denote the maximal free abelian covering space of X, and let $p_X \colon \tilde{X} \to X$ be the associated covering map. Define the *group ring* Λ_X of the covering group $H_1(X)/\text{Torsion}$ of $\tilde{X}$ to be the $\mathbb{Z}$-module

$$\Lambda_X = \mathbb{Z}\left[\frac{H_1(X)}{\text{Torsion}(H_1(X))}\right] = \bigoplus_{x \in \frac{H_1(X)}{\text{Torsion}}} \mathbb{Z}\exp(x),$$

equipped with the $\mathbb{Z}$-bilinear multiplication law that maps $\big(\exp(x), \exp(y)\big)$ to $\exp(x+y)$ for any $x, y \in H_1(X)/\text{Torsion}$. We use the notation $\exp(x)$ or e^x to denote an element x of $H_1(X)/\text{Torsion}$ viewed as a generator of Λ_X to remind this multiplication law. Similarly define the group rings $\Lambda_X^{\mathbb{Q}} = \mathbb{Z}[H_1(X;\mathbb{Q})]$ and $\Lambda_X^{\frac{1}{2}} = \mathbb{Z}\big[\frac{1}{2}H_1(X)/\text{Torsion}(H_1(X))\big]$, in which Λ_X embeds. These group rings are equipped with the linear involution

$$\sum_x \lambda_x \exp(x) \mapsto \left(\overline{\sum_x \lambda_x \exp(x)} = \sum_x \lambda_x \exp(-x)\right).$$

The units of Λ_X, are its elements of the form $\exp\,(x \in H_1(X)/\text{Torsion})$ called the *positive units*, and its elements of the form $-\exp\,(x \in H_1(X)/\text{Torsion})$.

The Alexander form of X is the following invariant of the Λ_X-module

$$\mathcal{H}_X = H_1\big(\tilde{X}, p_X^{-1}(\star)\big).$$

Definition 20.1.13 Let $\mathcal{P}$ be a presentation of $\mathcal{H}_X$ over Λ_X with $(r+g)$ generators $\gamma_1, \ldots, \gamma_{r+g}$, and r relators $\rho_1, \ldots, \rho_r$, which are Λ_X-linear combinations of the γ_i.[7] Define the element $\hat{\gamma}$ of $\bigwedge^{r+g}\left(\bigoplus_{i=1}^{r+g} \Lambda_X \gamma_i\right)$ to be $\hat{\gamma} = \gamma_1 \wedge \cdots \wedge \gamma_{r+g}$.

[6] Alexander forms were called Alexander functions in [3, Section 3].

[7] Such a presentation exists. It may be obtained from a cellular decomposition of X with one 0-cell, $(r+g)$ 1-cells, r 2-cells and no other cells.

Similarly define $\hat{\rho} = \rho_1 \wedge \cdots \wedge \rho_r \in \bigwedge^r \left(\bigoplus_{i=1}^{r+g} \Lambda_X \gamma_i\right)$. The *Alexander form* $\mathcal{A}_{X,\mathcal{P}}$ of $(X, \mathcal{P})$ is the Λ_X-linear form

$$\mathcal{A}_{X,\mathcal{P}} \colon \bigwedge^g \mathcal{H}_X \longrightarrow \Lambda_X$$

that maps an element $\hat{u} = u_1 \wedge \cdots \wedge u_g$ of $\bigwedge^g \mathcal{H}_X$ to the element $\mathcal{A}_{X,\mathcal{P}}(\hat{u})$ of Λ_X defined by the equality

$$\hat{\rho} \wedge \hat{u} = \mathcal{A}_{X,\mathcal{P}}(\hat{u})\hat{\gamma} \tag{20.1}$$

where the u_i are represented as combinations of the γ_j, and the exterior products in Eq. (20.1) belong to $\bigwedge^{r+g} \left(\bigoplus_{i=1}^{r+g} \Lambda_X \gamma_i\right)$.

As proved in [3, Section 3.1], if $\mathcal{P}$ and $\mathcal{Q}$ are two presentations of $\mathcal{H}_X$ as above, then there exists a unit $\pm\exp(u)$ of Λ_X such that $\mathcal{A}_{X,\mathcal{Q}} = \pm\exp(u)\mathcal{A}_{X,\mathcal{P}}$. We will often fix $\mathcal{P}$, omit $\mathcal{P}$ from the notation, keep in mind that $\mathcal{A}_X = \mathcal{A}_{X,\mathcal{P}}$ is defined up to a multiplication by a unit of Λ_X, and call $\mathcal{A}_X$ the *Alexander form* of X.

The Alexander form $\mathcal{A}$ is a topological invariant of the pairs $(X, \star)$ up to homotopy equivalence of pairs (connected compact oriented 3-manifold with boundary and with non-negative genus, basepoint). Homotopy equivalences provide identifications between the involved Λs and $\mathcal{H}$s. Changing the location of the basepoint $\star$ of X does not affect the homotopy equivalence class of $(X, \star)$. Thus $\mathcal{A}$ is an invariant of X. But we need a reference basepoint, which we may choose anywhere we want in X. So, we fix a basepoint $\star$ of X and a preferred lift $\star_0$ of $\star$ in $\tilde{X}$.

When c is an oriented curve in X, c also denotes the homology class it carries and the associated variable in $\exp(c) \in \Lambda_X$. If furthermore c is based at $\star$, then c also denotes its own class in $\pi_1(X, \star)$. We denote the class in $\mathcal{H}_X$ of the preferred lift of c starting at $\star_0$ in $\tilde{X}$ by $\tilde{c}$.

Example 20.1.14 Let E_0 be the closure of the complement of the handlebody H_0 of Fig. 20.4 trivially embedded in S^3, as the figure suggests. We equip ∂H_0 with the based curves u_α, u_β, and u_γ of the figure. Their product $u_\alpha u_\beta u_\gamma$ is trivial in $\pi_1(\partial H_0, \star)$.

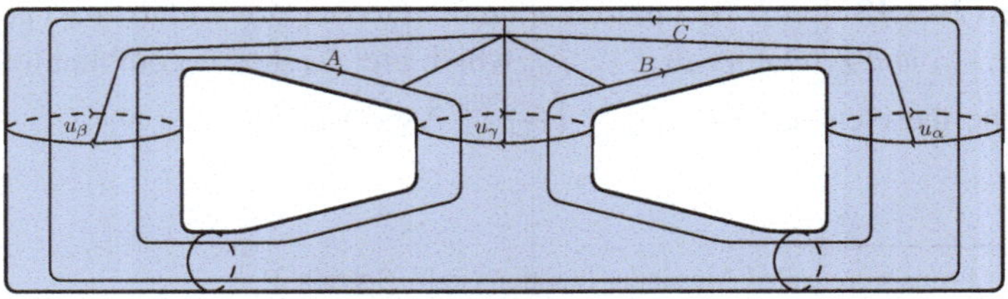

Fig. 20.4 The based curves u_α, u_β, u_γ, A, B, and C on ∂H_0

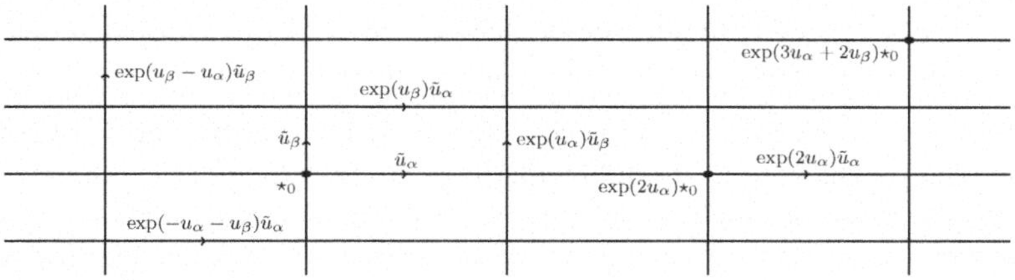

Fig. 20.5 The one-skeleton of $\tilde{E}_0$

Since $H_1(E_0) = \mathbb{Z}u_\alpha \oplus \mathbb{Z}u_\beta$, we have $\Lambda_{E_0} = \mathbb{Z}[e^{u_\alpha}, e^{-u_\alpha}, e^{u_\beta}, e^{-u_\beta}]$. We can also use the following symmetric expression

$$\Lambda_{E_0} = \frac{\mathbb{Z}[e^{u_\alpha}, e^{-u_\alpha}, e^{u_\beta}, e^{-u_\beta}, e^{u_\gamma}, e^{-u_\gamma}]}{(e^{u_\alpha}e^{u_\beta}e^{u_\gamma} = 1)}.$$

Since E_0 retracts on the wedge of the two circles u_α and u_β, $\tilde{E}_0$ is diffeomorphic to a regular neighborhood in $\mathbb{R}^3$ of the grid $\big(\mathbb{R} \times \mathbb{Z} \times \{0\}\big) \cup \big(\mathbb{Z} \times \mathbb{R} \times \{0\}\big)$ of $\mathbb{R}^3$ in Fig. 20.5. In this grid, the preimage of $\star$ is $\mathbb{Z} \times \mathbb{Z} \times \{0\}$.

In particular, we have $\mathcal{H}_{E_0} = \Lambda_{E_0}\tilde{u}_\alpha \oplus \Lambda_{E_0}\tilde{u}_\beta$. The associated presentation $\mathcal{P}_0$ of $\mathcal{H}_{E_0}$ with two generators and no relator provides the Alexander form $\mathcal{A}_{E_0,\mathcal{P}_0}$ such that $\mathcal{A}_{E_0,\mathcal{P}_0}(\tilde{u}_\alpha \wedge \tilde{u}_\beta) = 1$. Since A bounds a disk in E_0, the linear form $\mathcal{A}_{E_0,\mathcal{P}_0}(\tilde{A} \wedge .)$ vanishes.

Note that the following *Fox calculus rules* apply in $\mathcal{H}_E$. For any two based curves γ and δ of E, we have $\widetilde{\gamma\delta} = \tilde{\gamma} + \exp(\gamma)\tilde{\delta}$ and $\widetilde{\gamma^{-1}} = -\exp(-\gamma)\tilde{\gamma}$. See Fig. 20.5.

As a consequence, we get the following lemma.

Lemma 20.1.15 *Let E be a compact 3-manifold whose boundary is the genus 2 surface $(-\partial H_0)$ of Fig. 20.4. The classes of the based curves u_α, u_β, and u_γ of Fig. 20.4 satisfy the following relations.*

$$u_\alpha u_\beta u_\gamma = 1 \ \text{ in } \ \pi_1(E),$$

$$\tilde{u}_\alpha + \exp(u_\alpha)\tilde{u}_\beta + \exp(-u_\gamma)\tilde{u}_\gamma = 0 \ \text{ in } \ \mathcal{H}_E,$$

$$\tilde{u}_\beta \wedge \tilde{u}_\gamma = \exp(u_\gamma)\tilde{u}_\alpha \wedge \tilde{u}_\beta \ \text{ and } \ \tilde{u}_\gamma \wedge \tilde{u}_\alpha = \exp(-u_\beta)\tilde{u}_\alpha \wedge \tilde{u}_\beta \ \text{ in } \ \mathcal{H}_E \wedge \mathcal{H}_E.$$

□

In particular, back to Example 20.1.14, we have $\mathcal{A}_{E_0,\mathcal{P}_0}(\tilde{u}_\beta \wedge \tilde{u}_\gamma) = \exp(u_\gamma)$ and $\mathcal{A}_{E_0,\mathcal{P}_0}(\tilde{u}_\gamma \wedge \tilde{u}_\alpha) = \exp(-u_\beta)$.

Notation 20.1.16 The choice of the preferred lift $\star_0$ of $\star$ in $\tilde{X}$ provides a natural isomorphism from $H_0(p_X^{-1}(\star)) = \Lambda_X[\star_0]$ to Λ_X. The boundary map from $\mathcal{H}_X$ to

$H_0(p_X^{-1}(\star))$ composed with this isomorphism is denoted by ∂. The map ∂ maps the class $\tilde{x}$ of the lift of a based curve x that starts at $\star_0$ to $\partial(\tilde{x}) = \exp(x) - 1$.

When X is a link exterior, its genus $g(X)$ is 1. In this case, the *Reidemeister torsion* of X is the ratio of the following proposition, which is [3, Property 1, p. 640]. We use this proposition as a definition for the Reidemeister torsion.

The sign $\doteq$ stands for "equal up to multiplication by a positive unit (of Λ_X or $\Lambda_X^{\frac{1}{2}}$)".

Proposition 20.1.17 *Let X be a link exterior. Let $\tau(X)$ be its Reidemeister torsion described in [8]. For a given normalization $\mathcal{A}_{X,\mathcal{P}}$ of $\mathcal{A}_X$, the linear form $\mathcal{A}_{X,\mathcal{P}}$ is proportional to ∂. For any element u of $\mathcal{H}_X$, we have*

$$\mathcal{A}_X(u) \doteq \pm\tau(X)\partial(u).$$

Let c be a simple closed curve in the boundary ∂X of a 3-manifold X. Let $[0,1]\times c$ be a tubular neighborhood of c in ∂X. Then the manifold obtained by *attaching a 2-handle* along c is obtained by gluing a solid cylinder $[0,1]\times D^2$ along $[0,1]\times c$ so that $[0,1]\times\partial D^2$ is identified to $[0,1]\times c$, and by smoothing the angles. We denote it by $X[c]$.

Alexander forms are useful to study Reidemeister torsions since $\mathcal{A}_X$ recovers the Reidemeister torsions of the link complements obtained from X by attaching 2-handles along ∂X as above. For example, if c is based and if $X[c]$ is a link complement, then the genus of X is 2 and, up to multiplication by a unit, $\mathcal{A}_{X[c]}(.)$ is obtained from $\pm\,\mathcal{A}_X(\tilde{c}\wedge .)$ by composition by the quotient map $\Lambda_X \to \Lambda_{X[c]}$ induced by the inclusion.

20.1.5 Invariants of Genus One Surfaces in Rational Homology Spheres

Assume that the manifold X of the previous subsection is the exterior of a genus one surface $\Sigma = \phi\big(\Sigma(a,b,c)\big)$ for an embedding $\phi\colon H_0 \hookrightarrow R$ of H_0 into a $\mathbb{Q}$-sphere R. Let $E[K]$ be the 3-manifold obtained from this exterior $E = R\setminus\phi\big(\mathring{H}_0\big)$ by attaching a 2-handle along ($\partial\Sigma = K$). Then E, $E[K]$, and the Reidemeister torsion of $E[K]$ are invariants of Σ. We use the Alexander form of E to normalize the Reidemeister torsion of $E[K]$ as follows.

Notation 20.1.18 Let ε denote the $\mathbb{Z}$-linear *augmentation morphism*

$$\begin{aligned}\varepsilon\colon \Lambda_E &\longrightarrow \mathbb{Z}\\ \exp\left(x \in H_1(E)/\text{Torsion}\right) &\mapsto 1.\end{aligned}$$

An *orientation* of a rational vector space is an equivalence class of bases under the equivalence relation that identifies two bases related by a linear transformation with a positive determinant. For example, symplectic bases orient $H_1(\Sigma;\mathbb{Q})$. The natural isomorphism

$$\begin{aligned} f\colon H_1(\Sigma;\mathbb{Q}) &\to H_1(E;\mathbb{Q}) \\ z &\mapsto z - z_- \end{aligned}$$

carries the orientation of $H_1(\Sigma;\mathbb{Q})$ to an orientation of $H_1(E;\mathbb{Q})$ induced by the orientation of Σ.

Let (α,β) be a symplectic basis of Σ. View ∂E as $\Sigma \cup_K (-\Sigma_-)$. Assume that the basepoint of E is on $K \subset \partial E$ and that α and β are attached to this basepoint of E. So α and β are based curves, and their copies α_- and β_- in Σ_- are similarly based. The basis $(\alpha-\alpha_-, \beta-\beta_-)$ of $H_1(E;\mathbb{Q})$ defines the above orientation of $H_1(E;\mathbb{Q})$ and allows us to fix the sign of $\mathcal{A}_E$ so that $\varepsilon\big(\mathcal{A}_E(\tilde{\alpha}-\tilde{\alpha}_- \wedge \tilde{\beta}-\tilde{\beta}_-)\big) = |H_1(R)|$. So $\mathcal{A}_E$ is defined up to multiplication by a positive unit.

Recall that K is oriented as the boundary of Σ. We fix the sign of the Reidemeister torsion of $E[K]$ so that

$$\mathcal{A}_E(\tilde{K} \wedge u) \doteq \tau(E[K])\partial(u)$$

for any element u of $\mathcal{H}_E = \mathcal{H}_{E[K]}$. Finally, thanks to the well-known symmetry [8] of the Reidemeister torsion up to units, we can multiply a representative of $\tau(E[K])$ by a unit of $\Lambda_E^{\frac{1}{2}}$ to obtain a well-defined element $\mathcal{D}(\Sigma)$ of $\Lambda_E^{\frac{1}{2}}$ such that $\overline{\mathcal{D}(\Sigma)} = \mathcal{D}(\Sigma)$.

Note that we used twice the orientation of Σ to determine the sign of $\mathcal{D}(\Sigma)$. So, we have $\mathcal{D}(\Sigma) = \mathcal{D}(-\Sigma)$. We have proved the following proposition.

Proposition 20.1.19 *For any oriented genus one surface Σ in a rational homology sphere, the Reidemeister torsion $\mathcal{D}(\Sigma)$ normalized as above is an invariant of (the isotopy class of the image of the embedding of) Σ, with values in the group ring $\Lambda_E^{\frac{1}{2}}$.*

Let $f_*\colon \mathbb{Z}[H_1(\Sigma;\mathbb{Q})] \to \Lambda^{\mathbb{Q}}_{E[K]}$ be the isomorphism induced by the isomorphism $f\colon H_1(\Sigma;\mathbb{Q}) \to H_1(E;\mathbb{Q})$ such that $f(z) = z - z_-$. Then $f_*^{-1}(\mathcal{D}(\Sigma))$ is an invariant of Σ valued in $\mathbb{Z}[H_1(\Sigma;\mathbb{Q})]$.

We are going to extract numerical invariants from the invariant $\mathcal{D}(\Sigma)$ in Theorem 20.1.23 and Proposition 20.1.34.

Definition 20.1.20 Let V be a rational vector space. The *symmetric algebra* $S(V)$ of V is the quotient of the tensor algebra of V by the two-sided ideal generated by the elements of the form $u \otimes v - v \otimes u$. It is a graded algebra $S(V) = \oplus_{n\in\mathbb{N}} S_n(V)$ whose degree n part $S_n(V)$ is the *symmetrized* n^{th} *tensor power* $\otimes_s^n V$ of V. We set $\mathbb{Q}[[V]] = \prod_{n\in\mathbb{N}} S_n(V)$.

In particular, $S_2(V)$ is the symmetrized tensor product $V \otimes_s V$. (It is the quotient of $V \otimes V$ by the vector space generated by the differences $(u \otimes v - v \otimes u)$ for pairs $(u, v) \in V^2$.) We often denote products $u \otimes_s v$ of $V \otimes_s V$ by uv, and forget the signs $\otimes_s$ in $S_n(V)$ in the same way.

A morphism $\phi\colon V \to W$ of rational vector spaces induces a natural algebra morphism $\phi_*\colon \mathbb{Q}[[V]] \to \mathbb{Q}[[W]]$. When $V = \bigoplus_{i=1}^k \mathbb{Q}v_i$, the algebra $\mathbb{Q}[[V]]$ is isomorphic to the algebra $\mathbb{Q}[[v_1, \dots, v_k]]$ of formal power series in the commuting variables v_i, for $i = 1, \dots, k$. Note the following lemma.

Lemma 20.1.21 *For any finite-dimensional rational vector space V, the linear map*

$$\begin{aligned} \mathbb{Z}[V] &\to \mathbb{Q}[[V]] \\ \exp(v) &\mapsto \textstyle\sum_{n\in\mathbb{N}} \frac{1}{n!}v^n, \end{aligned}$$

is an injective algebra morphism.

Proof It is easy to see that the above linear map is an algebra morphism. When $V = \mathbb{Q}$, it is injective because the vectors $(a_1, a_2, \dots, a_n)$, $(a_1^2, a_2^2, \dots, a_n^2)$, …, $(a_1^n, a_2^n, \dots, a_n^n)$ are independent in $\mathbb{Q}^n$ for n pairwise distinct a_i. Let u_1, …, u_n be pairwise distinct vectors in V. Then there exists a linear form ψ on V such that $\psi(u_1)$, …, $\psi(u_n)$ are pairwise distinct. (Choose ψ outside the finitely many hyperplanes of $\mathrm{Hom}(V; \mathbb{Q})$ of linear forms that send $(u_j - u_i)$ to zero for pairs $\{u_i, u_j\}$ of distinct elements.) Then the previous case implies that $\big(\exp(u_1), \dots, \exp(u_n)\big)$ is a free system in $\mathbb{Q}[[V]]$. □

Lemma 20.1.21 allows us to consider $\Lambda_X^{\mathbb{Q}}$ as a subring of $\mathbb{Q}[[H_1(X; \mathbb{Q})]]$ by writing $\exp(v)$ as $\sum_{n\in\mathbb{N}} \frac{1}{n!}v^n$. The elements of $\Lambda_X^{\mathbb{Q}}$ that are (infinite) sums of tensors of even (resp. odd) degree in $\mathbb{Q}[[V]]$ are called *even* (resp. *odd*).

With these conventions, $\mathcal{D}(\Sigma)$ is an even element of $\mathbb{Q}[[H_1(E; \mathbb{Q})]]$, whose degree 0 part is $|H_1(R)|\lambda'(K)$, according to Lemma 20.3.8. We determine the whole series $\mathcal{D}(\Sigma)$ in Sect. 20.3.1. See Proposition 20.3.1. Its degree 2 part $\mathcal{D}_2(\Sigma)$ belongs to $H_1(E; \mathbb{Q}) \otimes_s H_1(E; \mathbb{Q})$.

Notation 20.1.22 Let V_s denote the symmetrized *Seifert form*

$$\begin{aligned} V_s\colon H_1(\Sigma; \mathbb{Q}) \otimes_s H_1(\Sigma; \mathbb{Q}) &\to \mathbb{Q} \\ u \otimes_s v &\mapsto lk(u, v_-) + lk(u_-, v). \end{aligned}$$

If ϕ maps the curves A and B of Fig. 20.2 to 0 in $H_1(E; \mathbb{Q})$, then $V_s(\alpha \otimes_s \alpha) = b + c$, $V_s(\beta \otimes_s \beta) = c + a$, and $V_s(\alpha \otimes_s \beta) = -c$. The form V_s allows us to extract the numerical invariant $V_s\left(f_*^{-1}\mathcal{D}_2(\Sigma)\right)$ of Σ.

Theorem 20.1.23 *Let Σ be a genus one Seifert surface in a $\mathbb{Q}$-sphere R. Set $W_s = W_s(\Sigma) = V_s \circ (f^{-1} \otimes_s f^{-1})$. Then $V_s\left(f_*^{-1}\mathcal{D}_2(\Sigma)\right) = W_s\left(\mathcal{D}_2(\Sigma)\right)$ is a topological invariant of Σ. Let $w_{SL}(\Sigma)$ be the invariant defined in Theorem 20.1.8 when Σ is* *null-homologous, and in Theorem 20.1.33, in general. We have*

$$W_s\left(\mathcal{D}_2(\Sigma)\right) = |H_1(R)| \left(\left(4\lambda'(\partial\Sigma) - 1\right) w_{SL}(\Sigma) + \frac{1}{4}\lambda'(\partial\Sigma) w_8(\Sigma) \right).$$

The stated invariance of $W_s\left(\mathcal{D}_2(\Sigma)\right)$ is tautological. We compute $W_s\left(\mathcal{D}_2(\Sigma)\right)$ in Sect. 20.3.2, where Theorem 20.1.23 is proved.

If K bounds a null-homologous genus one Seifert in a $\mathbb{Q}$-sphere, then $\lambda'(K) \in \mathbb{Z}$. In particular, we have $7\lambda'(K) \neq 1$. Therefore, Theorem 20.1.23 and Proposition 20.1.11 have the following immediate corollary.

Corollary 20.1.24 *For any null-homologous genus one Seifert surface Σ in a $\mathbb{Q}$-sphere R, we have*

$$w_{SL}(\Sigma) = \frac{1}{7\lambda'(\partial\Sigma) - 1} \left(2\lambda'(\partial\Sigma) w_3(\partial\Sigma) + \frac{W_s\left(\mathcal{D}_2(\Sigma)\right)}{|H_1(R)|} \right)$$

and

$$w_8(\Sigma) = \frac{4}{7\lambda'(\partial\Sigma) - 1} \left(\left(2 - 8\lambda'(\partial\Sigma)\right) w_3(\partial\Sigma) + 3\frac{W_s\left(\mathcal{D}_2(\Sigma)\right)}{|H_1(R)|} \right).$$

20.1.6 Sketch of the Proof of the Invariance of w_8

According to Corollary 20.1.24, to prove Theorem 20.1.2 for null-homologous genus one Seifert surfaces, it suffices to prove that $W_s\left(\mathcal{D}_2(\Sigma)\right)$ does not depend on the surface Σ with boundary K, as stated in Theorem 20.1.27 below. In Sects. 20.4.1 and 20.4.2, we will prove the following two lemmas, which imply Theorem 20.1.27.

Say that two oriented surfaces Σ and Σ' with boundary K in a $\mathbb{Q}$-sphere R are *K-cobordant* if there exists a diffeomorphism ψ of R isotopic to the identity of R such that $\psi(K) = K$ and $\psi(\Sigma) \cap \Sigma' = K$. Since the latter condition implies $\Sigma \cap \psi^{-1}(\Sigma') = K$, this definition is symmetric with respect to the exchange of Σ and Σ'.

Lemma 20.1.25 *Let Σ and Σ' be two genus one Seifert surfaces of a knot K in a rational homology 3-sphere R. If Σ and Σ' are K-cobordant, then*

$$W_s(\Sigma')\left(\mathcal{D}_2(\Sigma')\right) = W_s(\Sigma)\left(\mathcal{D}_2(\Sigma)\right).$$

Lemma 20.1.26 *Let R be a $\mathbb{Q}$-sphere. For any two genus one Seifert surfaces Σ and Σ' of K in R, there exists a sequence $(\Sigma_i)_{i=1,\ldots,k}$ of genus one Seifert surfaces of K such that $\Sigma = \Sigma_1$, $\Sigma' = \Sigma_k$, and, for any $i = 1, \ldots, k-1$, Σ_i and Σ_{i+1} are K-cobordant.*

Theorem 20.1.27 *Let K be a knot in a $\mathbb{Q}$-sphere. If K bounds a genus one Seifert surface Σ, then the invariant $W_s(\mathcal{D}_2(\Sigma))$ of Theorem 20.1.23 does not depend on such a genus one Seifert surface Σ. It is an invariant of the knot K.* □

20.1.7 Alexander Forms of Genus Two Homology Handlebodies

Our proof of Theorem 20.1.23 relies on a study of Alexander forms of genus two rational homology handlebodies. This study produces the following structure theorem (20.1.28), which is of independent interest. Theorem 20.1.28 involves the normalized multivariable Alexander polynomial Δ of two-component links. The polynomial of such a link is obtained from the Reidemeister torsion of its exterior by a multiplication by a unit and a change of variables. We can use Theorem 20.1.28 as a definition for the specific needed Alexander polynomial. See Definition 20.2.6 for a general definition of the normalized multivariable Alexander polynomial.

In rings $\mathbb{Q}[[V]]$ of formal power series, for a positive integer n, we use the notation $O(n)$ to denote a series of tensors (or monomials) of degree at least n. Set $\mathrm{Sh}(u) = \exp\left(\frac{u}{2}\right) - \exp\left(-\frac{u}{2}\right)$ $(= 2\sinh(u))$.

Theorem 20.1.28 *Let E be a genus two rational homology handlebody. Identify ∂E with the surface $(-\partial H_0)$ of Figs. 20.2 and 20.4 in such a way that $H_1(E;\mathbb{Q}) = \mathbb{Q}u_\alpha \oplus \mathbb{Q}u_\beta$. Let R be the $\mathbb{Q}$-sphere obtained by attaching 2-handles along u_α and u_β, and by gluing a 3-ball to the remaining boundary, which is a 2-dimensional sphere.*[8] *Set*

$$\delta_{\Delta,E}(\alpha,\beta) = -\lambda'(A)u_\beta u_\gamma - \lambda'(B)u_\gamma u_\alpha - \lambda'(C)u_\alpha u_\beta$$

and $\ell = lk(A, B)$. Let $\mathcal{A}_{E,\mathcal{P}}$ be a representative of the Alexander form of E. There exist a unique unit $\mathcal{U}$ of $\Lambda_E^{\mathbb{Q}} \subset \mathbb{Q}[[u_\alpha, u_\beta, u_\gamma]]/(u_\alpha + u_\beta + u_\gamma = 0)$ and a unique rational number $\lambda'(E; u_\alpha, u_\beta, u_\gamma)$ such that $\mathcal{A}_E = \mathcal{U}\mathcal{A}_{E,\mathcal{P}}$ satisfies the following equality.[9]

$$\mathcal{A}_E(\tilde{u}_\alpha \wedge \tilde{u}_\beta) = |H_1(R)| \left(1 + \delta_{\Delta,E}(\alpha,\beta) + \frac{1}{2}\lambda'(E; u_\alpha, u_\beta, u_\gamma)u_\alpha u_\beta u_\gamma\right) + O(4).$$

This equality defines $\lambda'(E; u_\alpha, u_\beta, u_\gamma)$.[10] *If $\ell = 0$, then we have $\lambda'(E; u_\alpha, u_\beta, u_\gamma) = \lambda'(A, B)$.*

[8] We could state this theorem without the $\mathbb{Q}$-sphere R. We have $|H_1(R)| = |H_1(E)/\mathbb{Z}u_\alpha \oplus \mathbb{Z}u_\beta|$ and $A = \ell u_\alpha$ in $H_1(E;\mathbb{Q})/\mathbb{Q}[u_\beta]$. The coefficients $\lambda'(A)$, $\lambda'(B)$, and $\lambda'(C)$ could be respectively defined from the Reidemeister torsions of $E[u_\alpha]$, $E[u_\beta]$, and $E[u_\gamma]$ valued in quotients of $\mathbb{Z}[H_1(E;\mathbb{Q})]$. See Lemma 20.2.5.

[9] Lemma 20.2.13 constrains the unit $\mathcal{U}$ to live in a smaller ring.

[10] See Corollary 20.2.24 for an alternative definition of $\lambda'(E; u_\alpha, u_\beta, u_\gamma)$.

With this normalization of $\mathcal{A}_E$, the multivariable Alexander polynomial $\Delta(A, B)$ of (A, B) satisfies the following equations.

$$Sh(u_\gamma)\Delta(A, B) = \exp\left(-\frac{\ell}{2}u_\gamma\right)\overline{\mathcal{A}_E(\tilde{u}_\alpha \wedge \tilde{u}_\beta)} - \exp\left(\frac{\ell}{2}u_\gamma\right)\mathcal{A}_E(\tilde{u}_\alpha \wedge \tilde{u}_\beta)$$

and

$$\Delta(A, B) = \exp\left(\frac{\ell-1}{2}u_\gamma\right)\mathcal{A}_E(\tilde{u}_\beta \wedge \tilde{A}) = \exp\left(\frac{\ell-1}{2}u_\gamma - u_\alpha - B\right)\mathcal{A}_E(\tilde{u}_\alpha \wedge \tilde{B}).$$

Theorem 20.1.28 shows the structure of the restriction of $\mathcal{A}_E$ to the exterior products of classes of ∂E, as stated in Corollary 20.1.31 below. In Sect. 20.2.4, we prove Theorem 20.1.28 and give other corollaries of Theorem 20.1.28. See Corollaries 20.2.22 and 20.2.24.

Note the following easy lemma.

Lemma 20.1.29 *We have $Au_\beta A^{-1}u_\beta^{-1} = B^{-1}u_\alpha^{-1}Bu_\alpha$ in $\pi_1(\partial E)$, and*

$$\partial\tilde{A}\tilde{u}_\beta - \partial\tilde{u}_\beta\tilde{A} = \exp(-u_\alpha - B)\left(\partial\tilde{B}\tilde{u}_\alpha - \partial\tilde{u}_\alpha\tilde{B}\right)$$

in $\mathcal{H}_E$. In particular, if A and B vanish in $H_1(E;\mathbb{Q})$, then we have $\partial\tilde{u}_\beta\tilde{A} = \exp(-u_\alpha)\partial\tilde{u}_\alpha\tilde{B}$ in $\mathcal{H}_E$. □

Lemma 20.1.29 implies that $\mathcal{A}_E(\tilde{A} \wedge \tilde{B}) = 0$ when A and B are rationally null-homologous in E. In general, it shows how the restriction of $\mathcal{A}_E$ to exterior products of classes of curves of ∂E is determined by $\mathcal{A}_E(\tilde{u}_\alpha \wedge \tilde{u}_\beta)$, $\mathcal{A}_E(\tilde{u}_\alpha \wedge \tilde{A})$, and $\mathcal{A}_E(\tilde{u}_\beta \wedge \tilde{A})$. The following lemma shows how $\mathcal{A}_E(\tilde{u}_\alpha \wedge \tilde{A})$ is determined by $\mathcal{A}_E(\tilde{u}_\beta \wedge \tilde{A})$ and $\mathcal{A}_E(\tilde{u}_\alpha \wedge \tilde{u}_\beta)$.

Lemma 20.1.30 *For any three elements x, y, and z of $\mathcal{H}_E$, we have*

$$\partial x\mathcal{A}_E(y \wedge z) + \partial y\mathcal{A}_E(z \wedge x) + \partial z\mathcal{A}_E(x \wedge y) = 0.$$

In particular, if z is trivial in $H_1(E;\mathbb{Q})$, then we have $\partial x\mathcal{A}_E(y \wedge z) = \partial y\mathcal{A}_E(x \wedge z)$.

Proof This is a particular case of Property 2 of [3, Section 3.2]. □

So, Theorem 20.1.28 has the following easy corollary.

Corollary 20.1.31 *Under the hypotheses of Theorem 20.1.28, we have*

$$Sh(u_\gamma)\mathcal{A}_E(\tilde{u}_\beta \wedge \tilde{A}) = \exp\left(\frac{1-2\ell}{2}u_\gamma\right)\overline{\mathcal{A}_E(\tilde{u}_\alpha \wedge \tilde{u}_\beta)} - \exp\left(\frac{1}{2}u_\gamma\right)\mathcal{A}_E(\tilde{u}_\alpha \wedge \tilde{u}_\beta),$$

and the restriction of $\mathcal{A}_E$ to the exterior products of classes of based curves of ∂E is determined by ℓ and the series $\mathcal{A}_E(\tilde{u}_\alpha \wedge \tilde{u}_\beta)$, which is determined by any determinant $\mathcal{A}_{E,\mathcal{P}}(\tilde{u}_\alpha \wedge \tilde{u}_\beta)$. □

In Sect. 20.3.3, we prove that w_δ is an invariant of genus one Seifert surfaces Σ in $\mathbb{Q}$-spheres, as announced in Theorem 20.1.2. We furthermore prove that w_δ lifts to an invariant δ_Σ valued in $H_1(\Sigma;\mathbb{Q}) \otimes_s H_1(\Sigma;\mathbb{Q})$, via the form V_s of Notation 20.1.22, as stated in the following theorem.

Theorem 20.1.32 *Let Σ be a genus one Seifert surface in a $\mathbb{Q}$-sphere R, and let (α,β) be a symplectic basis of $H_1(\Sigma)$. Set $\gamma = -\alpha - \beta$. Represent α, β, and γ by simple curves of Σ. Define $a = -lk(\beta,\gamma_-) - lk(\beta_-,\gamma)$, $b = -lk(\alpha,\gamma_-) - lk(\alpha_-,\gamma)$, and $c = -lk(\alpha,\beta_-) - lk(\alpha_-,\beta)$. Define the following elements of $H_1(\Sigma;\mathbb{Q}) \otimes_s H_1(\Sigma;\mathbb{Q})$.*

$$\begin{aligned}\delta_{2,\Sigma}(\alpha,\beta) = &-\tfrac{1}{24}(a^2+2ab+2ac+3)\alpha\otimes_s\alpha - \tfrac{1}{24}(b^2+2bc+2ba+3)\beta\otimes_s\beta\\ &-\tfrac{1}{24}(c^2+2ca+2cb+3)\gamma\otimes_s\gamma\end{aligned}$$

and

$$\delta_{\Delta,\Sigma}(\alpha,\beta) = f_*^{-1}(\delta_{\Delta,E}(\alpha,\beta)) = -\lambda'(\alpha)\beta\otimes_s\gamma - \lambda'(\beta)\gamma\otimes_s\alpha - \lambda'(\gamma)\alpha\otimes_s\beta.$$

Then the element

$$\delta_\Sigma = \delta_\Sigma(\alpha,\beta) = \delta_{2,\Sigma}(\alpha,\beta) + 4\delta_{\Delta,\Sigma}(\alpha,\beta)$$

of $H_1(\Sigma;\mathbb{Q}) \otimes_s H_1(\Sigma;\mathbb{Q})$ is an invariant of Σ such that $w_\delta(\Sigma) = V_s(\delta_\Sigma)$.

We will prove the following theorem in the end of Sect. 20.3.3, with the help of Theorem 20.1.32. It extends the invariant w_{SL} of Theorem 20.1.8 to all genus one Seifert surfaces of knots in $\mathbb{Q}$-spheres.

Theorem 20.1.33 *Let Σ be a genus one Seifert surface in a $\mathbb{Q}$-sphere R. Identify Σ to some $\phi\left(\Sigma(\check{a},\check{b},\check{c}) \subset H_0\right)$ for an embedding $\phi\colon H_0 \hookrightarrow R$ and a triple $(\check{a},\check{b},\check{c})$ of odd integers. Let $E = R \setminus \phi\big(\mathring{H}_0\big)$ be the exterior of Σ. Let $\alpha = \phi(\alpha(\check{a},\check{b},\check{c}))$ and $\beta = \phi(\beta(\check{a},\check{b},\check{c}))$ be the corresponding curves of Σ, as in Fig. 20.1. Define $a = lk(\beta,\alpha_- + \beta_-) + lk(\beta_-,\alpha+\beta)$, $b = lk(\alpha,\alpha_- + \beta_-) + lk(\alpha_-,\alpha+\beta)$, and $c = -lk(\alpha,\beta_-) - lk(\alpha_-,\beta)$. Recall the definitions of $\lambda'(E;u_\alpha,u_\beta,u_\gamma)$ from Theorem 20.1.28 and of $w_\delta(a,b,c)$ from Theorem 20.1.2. Then*

$$w_{SL}(\Sigma) = \lambda'(E;u_\alpha,u_\beta,u_\gamma) - \frac{1}{12}w_\delta(a,b,c)$$

is an invariant of Σ.

20.1.8 More Numerical Invariants of Genus One Surfaces

The torsion $f_*^{-1}(\mathcal{D}(\Sigma))$ is obviously an invariant of the surface Σ. So is its degree $(2k)$ part $f_*^{-1}(\mathcal{D}_{2k}(\Sigma))$, which is valued in the symmetric tensor product $\bigotimes_s^{2k} H_1(\Sigma;\mathbb{Q})$. We can endow this symmetric product with the canonical linear form $V_{s,2k}\colon \bigotimes_s^{2k} H_1(\Sigma;\mathbb{Q}) \to \mathbb{Q}$ that maps $(u_1 \otimes u_2 \otimes \cdots \otimes u_{2k})$ to the sum over the partitions of $\{1,2,\ldots,2k\}$ into k pairs $\{n_{2i-1}, n_{2i}\}$ of the products $\prod_{i=1}^k V_s(u_{n_{2i-1}} \otimes u_{n_{2i}})$. (There are $\frac{(2k)!}{k!2^k}$ such partitions, and $V_{s,2} = V_s$.) We get the following tautological result.

Proposition 20.1.34 *For any genus one Seifert surface in a rational homology sphere, $f_*^{-1}(\mathcal{D}(\Sigma))$ is an invariant of Σ valued in $\mathbb{Z}[H_1(\Sigma;\mathbb{Q})]$. The extracted series*

$$\left(V_{s,2k}\left(f_*^{-1}(\mathcal{D}_{2k}(\Sigma))\right)\right)_{k\in\mathbb{N}}$$

is a series of numerical invariants of Σ.

20.1.9 Questions

How does w_δ compare to other knot invariants? Is w_δ a function of the Jones polynomial for knots in S^3? Or does w_δ distinguish knots that are not distinguished by the Jones polynomial? Do the products by $(7\lambda'(\partial\Sigma) - 1)$ of the formulas of Corollary 20.1.24 hold when Σ is not null-homologous? What are the properties of the series of invariants of genus one Seifert surfaces of Proposition 20.1.34?

20.2 More on Alexander Forms

Proposition 20.1.17 allowed us to define Reidemeister torsions from Alexander forms. In this section, we also use Alexander forms to define normalized Alexander polynomials.

20.2.1 Alexander Polynomials of Knots from Alexander Forms

Definition 20.2.1 Let J be a knot in a $\mathbb{Q}$-sphere R. Let $O(J)$ be the minimal positive integer such that $O(J)J$ is null-homologous. (When J is null-homologous, we have $O(J) = 1$.) Let $X(J)$ be the exterior of J. Let m be a based meridian of J. Set

$$t^{\frac{1}{2O(J)}} = \exp\left(\frac{1}{2O(J)}m\right) \in \Lambda_{X(J)}^{\frac{1}{2}}.$$

The *normalized Alexander polynomial* $\Delta(J)$ of J is the Laurent polynomial in $\mathbb{Z}\left[t^{1/(2O(J))}, t^{-1/(2O(J))}\right]$ determined by the following conditions

$$\Delta(J) \doteq \pm\frac{\exp\left(\frac{1}{O(J)}m\right)-1}{\exp(m)-1}\mathcal{A}_{X(J)}(\tilde{m}),$$

$$\Delta(J)\left(t^{1/(2O(J))}=1\right)=\frac{|H_1(R)|}{O(J)} \text{ and } \Delta(J)(t^{-1/(2O(J))})=\Delta(t^{1/(2O(J))}).$$

See [2, 2.1.2 page 22 and 2.2.1 page 23].

Remark 20.2.2 Let g be a based curve that represents a generator of $H_1(X(J))/\text{Torsion}$. Thanks to the Poincaré duality, the algebraic intersection of such a curve with an integral chain whose boundary is $O(J)J$ is ± 1. So, we can assume that $m = O(J)g$ in $H_1(X(J))/\text{Torsion}$. As in Proposition 20.1.17, we have $\mathcal{A}_{X(J)}(\tilde{m})(\exp(g)-1) = \mathcal{A}_{X(J)}(\tilde{g})(\exp(m)-1)$. So, we have $\Delta(J) \doteq \pm\mathcal{A}_{X(J)}(\tilde{g})$. Since $\mathcal{H}_{X(J)} = H_1(\widetilde{X(J)}) \oplus \Lambda_{X(J)}[\tilde{g}]$, the polynomial $\mathcal{A}_{X(J)}(\tilde{g})$ is the *order* of $H_1(\widetilde{X(J)})$, i.e., it is the determinant of a square matrix of presentation of $H_1(\widetilde{X(J)})$ up to multiplication by a unit. We also have that

$$\left|H_1(R)\right| = \left|\varepsilon\left(\mathcal{A}_{X(J)}(\tilde{m})\right)\right| = O(J)\left|\varepsilon\left(\mathcal{A}_{X(J)}(\tilde{g})\right)\right| = O(J)\left|\text{Torsion}\left(H_1(R\setminus J)\right)\right|.$$

Example 20.2.3 The exterior $X(A)$ of the knot $\phi(A)$ of the right-hand side of Fig. 20.2 is obtained from $E = R \setminus \phi\left(\mathring{H}_0\right)$ by attaching a 2-handle along the curve u_α of Fig. 20.4. So, we have $X(A) = E[u_\alpha]$ and $H_1(X(A);\mathbb{Q}) = H_1(E;\mathbb{Q})/\mathbb{Q}[u_\alpha]$. Let $\mathcal{A}_{E,\mathcal{P}}$ be a representative of $\mathcal{A}_E$. The Alexander form $\mathcal{A}_{X(A)}$ is obtained from $\mathcal{A}_{E,\mathcal{P}}(\tilde{u}_\alpha \wedge .)$ by mapping u_α to zero (and therefore $\exp(u_\alpha)$ to 1). We write

$$\mathcal{A}_{X(A)}(u) \doteq \pm\text{ev}(u_\alpha = 0)\left(\mathcal{A}_{E,\mathcal{P}}(\tilde{u}_\alpha \wedge u)\right).$$

Since $[-u_\beta]$ is a meridian of A (pushed inside $\phi(\mathring{H}_0)$), we have

$$\Delta\left(\phi(A)\right)(t) \doteq \text{sign}\left(\varepsilon\left(\mathcal{A}_{E,\mathcal{P}}(\tilde{u}_\alpha \wedge \tilde{u}_\beta)\right)\right)\text{ev}(u_\alpha = 0)$$
$$\times\left(\frac{\text{Sh}\left(\frac{1}{O(\phi(A))}u_\beta\right)}{\text{Sh}(u_\beta)}\mathcal{A}_{E,\mathcal{P}}(\tilde{u}_\alpha \wedge \tilde{u}_\beta)\right)_{t=\exp(u_\gamma=-u_\beta)}.$$

Once Theorem 20.1.28 is proved, we will know that the above expression is even with the normalization of $\mathcal{A}_E$ of Theorem 20.1.28 instead of $\mathcal{A}_{E,\mathcal{P}}$, and we will replace $(\mathcal{A}_{E,\mathcal{P}}, \doteq)$ by $(\mathcal{A}_E, =)$.

Definition 20.2.4 Let J be a knot in a $\mathbb{Q}$-sphere R. We define $\lambda'(J)$ from the Alexander polynomial $\Delta(J)$ of Definition 20.2.1 to be

$$\lambda'(J) = \frac{O(J)\Delta''(J)(1)}{2|H_1(R)|} + \frac{1}{24}\left(1 - \frac{1}{O(J)^2}\right),$$

where the second derivative of Δ is taken with respect to the variable $t = \exp(m)$.

Lemma 20.2.5 below justifies the above definition. It provides an alternative definition for $\lambda'(J)$.

Lemma 20.2.5 *Let J be a knot in a $\mathbb{Q}$-sphere R. Let m be a based meridian of J. Let $\mathcal{A}_{X(J),\mathcal{P}}$ be a representative of the Alexander form of $X(J)$. There exists a unit $\mathcal{U}$ of $\Lambda^{\frac{1}{2}}_{X(J)} \subset \mathbb{Q}[[m]]$ such that*

$$\Delta(J) = \frac{Sh\left(\frac{1}{O(J)}m\right)}{Sh(m)}\mathcal{U}\mathcal{A}_{X(J),\mathcal{P}}(\tilde{m})$$

and

$$\mathcal{U}\mathcal{A}_{X(J),\mathcal{P}}(\tilde{m}) = |H_1(R)|\left(1 + \lambda'(J)m^2\right) + O(4).$$

If R is a $\frac{\mathbb{Z}}{2\mathbb{Z}}$–sphere or if J is null-homologous, then $\mathcal{U}$ and $\Delta(J)$ are in $\Lambda_{X(J)}$.

Proof According to Remark 20.2.2, there exists a based curve g such that $m = O(J)g$ in $H_1(X(J))/\text{Torsion}$. Then we have $H_1(X(J))/\text{Torsion} = \mathbb{Z}g$. Without loss of generality, we choose $\mathcal{A}_{X(J),\mathcal{P}}$ so that $\varepsilon\left(\mathcal{A}_{X(J),\mathcal{P}}(\tilde{g})\right) > 0$. By Definition 20.2.1, we have

$$\Delta(J) \doteq \pm\frac{\partial\tilde{g}}{\partial\tilde{m}}\mathcal{A}_{X(J),\mathcal{P}}(\tilde{m}).$$

So $\Delta(J) \doteq \pm\mathcal{A}_{X(J),\mathcal{P}}(\tilde{g})$, and there exists $k_1 \in \mathbb{Z}$ such that $\Delta(J) = \exp\left(\frac{1}{2}k_1 g\right)\mathcal{A}_{X(J),\mathcal{P}}(\tilde{g})$. Since $\mathcal{A}_{X(J),\mathcal{P}}(\tilde{g}) \in \Lambda_{X(J)}$, we have $\mathcal{A}_{X(J),\mathcal{P}}(\tilde{g}) = |H_1(R)|/O(J) + kg + O(2)$ in $\mathbb{Q}[[g]]$ for some k in $\mathbb{Z}$. Since $\Delta(J)$ is even, the degree one part of $\exp(k_1 g/2)\mathcal{A}_{X(J),\mathcal{P}}(\tilde{g})$ is zero. So, we have $k_1 = -2\frac{kO(J)}{|H_1(R)|} \in \mathbb{Z}$, and k_1 is even if $\frac{|H_1(R)|}{O(J)}$ is odd, which holds as soon as R is a $\frac{\mathbb{Z}}{2\mathbb{Z}}$-sphere. In this case, $\Delta(J)$ belongs to $\Lambda_{X(J)}$. Set

$$\mathcal{U} = \exp\left(-\frac{O(J)-1}{2}g\right)\exp\left(\frac{1}{2}k_1 g\right) \quad \text{and} \quad \mathcal{A}_{X(J)} = \mathcal{U}\mathcal{A}_{X(J),\mathcal{P}}.$$

Then

$$\begin{aligned}\mathcal{A}_{X(J)}(\tilde{m}) &= \exp\left(-\frac{O(J)-1}{2}g\right)\frac{\partial \tilde{m}}{\partial \tilde{g}}\exp\left(\frac{1}{2}k_1 g\right)\mathcal{A}_{X(J),\mathcal{P}}(\tilde{g})\\ &= \frac{\mathrm{Sh}(O(J)g)}{\mathrm{Sh}(g)}\Delta(J)\end{aligned}$$

is even, and we have

$$\mathcal{A}_{X(J)}(\tilde{m}) = |H_1(R)|(1+cm^2) + O(4).$$

Since $\mathrm{Sh}(x) = \exp\left(\frac{x}{2}\right) - \exp\left(-\frac{x}{2}\right) = x\left(1 + \frac{x^2}{24} + O(4)\right)$, we get

$$\begin{aligned}\Delta(J) &= \frac{\mathrm{Sh}\left(\frac{1}{O(J)}m\right)}{\mathrm{Sh}(m)}\mathcal{A}_{X(J)}(\tilde{m})\\ &= \frac{|H_1(R)|}{O(J)}\left(1 - \frac{1}{24}\left(1 - \frac{1}{O(J)^2}\right)m^2\right)\left(1+cm^2\right) + O(4).\end{aligned}$$

On the other hand, we have

$$\Delta(J) = \frac{|H_1(R)|}{O(J)} + \frac{\Delta''(J)(1)}{2}m^2 + O(4).$$

So, we get

$$c = \frac{1}{24}\left(1 - \frac{1}{O(J)^2}\right) + \frac{O(J)\Delta''(J)(1)}{2|H_1(R)|} = \lambda'(J).$$

Since $(O(J)-1)$ is even when R is a $\frac{\mathbb{Z}}{2\mathbb{Z}}$–sphere, $\mathcal{U}$ belongs to $\Lambda_{X(J)}$ in this case. When J is null-homologous, $O(J)$ equals one, and the expression $\Delta(J) = |H_1(R)|\det\left(t^{1/2}V - t^{-1/2}({}^TV)\right)$ shows that $\Delta(J)$ is in $\Lambda_{X(J)}$. So, we have $k_1 \in 2\mathbb{Z}$ and $\mathcal{U} \in \Lambda_{X(J)}$, too. □

20.2.2 Normalized Alexander Polynomials of Links

Definition 20.2.6 Let k be an integer such that $k \geq 2$. Set $\underline{k} = \{1, \ldots, k\}$. Let $L = (K_i)_{i\in\underline{k}}$ be a link of a $\mathbb{Q}$-sphere R with k components. For $i \in \underline{k} = \{1, \ldots, k\}$, let m_i be a meridian of the component K_i of L, and set $O_i = O(K_i)$. The Reidemeister torsion $\tau(X(L))$ of the exterior $X(L)$ of L is an element of

$$\mathbb{Z}\left[e^{\frac{1}{O_1}m_1}, e^{-\frac{1}{O_1}m_1}, e^{\frac{1}{O_2}m_2}, e^{-\frac{1}{O_2}m_2}, \ldots, e^{\frac{1}{O_k}m_k}, e^{-\frac{1}{O_k}m_k}\right].$$

It is defined up to multiplication by a unit $\exp\big(x \in H_1\big(X(L)\big)/\text{Torsion}\big)^{11}$. The multivariable Alexander polynomial $\Delta(L)$ of L is an element of $\mathbb{Z}\big[t_1^{1/(2O_1)}, t_1^{-1/(2O_1)}, \ldots, t_k^{1/(2O_k)}, t_k^{-1/(2O_k)}\big]$. It is obtained from $\tau(X(L))$, by a multiplication by a unit of $\Lambda^{\frac{1}{2}}_{X(L)}$, and by the change of variables

$$t_i^{\frac{1}{2O_i}} = \exp\left(\frac{1}{2O_i}m_i\right).$$

It satisfies the following property

$$\Delta(L)\big(t_1^{\frac{1}{2O_1}}, t_2^{\frac{1}{2O_2}}, \ldots, t_k^{\frac{1}{2O_k}}\big) = \pm\Delta(L)\Big(t_1^{-\frac{1}{2O_1}}, t_2^{-\frac{1}{2O_2}}, \ldots, t_k^{-\frac{1}{2O_k}}\Big),$$

which determines it up to sign. To fix the sign of Δ, as Turaev did in [8], we use Alexander forms as follows.

Let B be a 3-ball in $X(L)$ that intersects the boundary $\partial N(K_i)$ of the tubular neighborhood $N(K_i)$ of its component K_i along a disk D_i, for each component K_i of L. So $B \cup \cup_{i=1}^k N(K_i)$ is a handlebody in R, and $X(L)$ is obtained from the exterior E of this handlebody by attaching the 2-cells $D_1, \ldots, D_{k-1}$ to ∂E and by thickening them. Let ℓ_i be a longitude of K_i, we have $\langle m_i, \ell_i\rangle_{\partial N(K_i)} = 1$. Attach m_i and ℓ_i to a basepoint in $\partial E \cap B$ by a path on $\partial E \cap (\partial N(K_i) \cup B)$ so that the boundary δ_i of $\pm D_i$ reads $m_i \ell_i m_i^{-1} \ell_i^{-1}$ up to conjugation in $\pi_1(\partial E)$. Then $\tilde{\delta}_i = (1-\exp(\ell_i))\tilde{m}_i + (\exp(m_i)-1)\tilde{\ell}_i$ in $\mathcal{H}_E$, up to multiplication by a positive unit of Λ_E. According to [3, Property 8, p. 647], for any j and r in $\underline{k}$, the *multivariable Alexander polynomial* $\Delta(L)$ *of* L satisfies

$$\Delta(L) \doteq \text{sign}\Big(\varepsilon\big(\mathcal{A}_E(\hat{m})\big)\Big)\frac{\mathcal{A}_E\left(\hat{\delta}\left(\frac{\tilde{m}_j}{\tilde{\delta}_r}\right)\right)}{\partial \tilde{m}_j},$$

where $\hat{m} = \tilde{m}_1 \wedge \cdots \wedge \tilde{m}_k$, $\hat{\delta} = \tilde{\delta}_1 \wedge \cdots \wedge \tilde{\delta}_k$, $\hat{\delta}\left(\frac{\tilde{m}_j}{\tilde{\delta}_r}\right) = \tilde{\delta}_1 \wedge \cdots \wedge \tilde{\delta}_{r-1} \wedge \tilde{m}_j \wedge \tilde{\delta}_{r+1} \wedge \cdots \wedge \tilde{\delta}_k$, and we keep the change of variables $t_i^{1/(2O_i)} = \exp\big(\frac{1}{2O_i}m_i\big)$ understood. See also [2, 2.2.1 page 23].

[11] Here, we see $H_1(X(L))/\text{Torsion}$ as a submodule of $H_1(X(L);\mathbb{Q}) = \oplus_{i=1}^k \mathbb{Q}m_i$. Let $N(K_i)$ denote a tubular neighborhood of K_i. Let Σ_i be a surface embedded in the exterior $R \setminus \mathring{N}(K_i)$ of K_i whose boundary $\partial\Sigma_i$ is on $\partial N(K_i)$ and is homologous to $O_i K_i$ in $N(K_i)$. Every element x of $H_1(X(L);\mathbb{Q}) = \oplus_{i=1}^k \mathbb{Q}m_i$ may be written as $x = \sum_{i=1}^k (\langle x, \Sigma_i\rangle / O_i) m_i$.

Example 20.2.7 For a general two-component link $L = (K_1, K_2)$ as above, we have

$$\begin{aligned}
&\Delta(L)\,(t_1 = \exp(m_1), t_2 = \exp(m_k)) \\
&\doteq \operatorname{sign}\Big(\varepsilon\big(\mathcal{A}_E(\tilde{m}_1 \wedge \tilde{m}_2)\big)\Big) \frac{\mathcal{A}_E\Big((1-\exp(\ell_1))\tilde{m}_1 + (\exp(m_1)-1)\tilde{\ell}_1 \wedge \tilde{m}_1\Big)}{\partial \tilde{m}_1} \\
&\doteq \operatorname{sign}\Big(\varepsilon\big(\mathcal{A}_E(\tilde{m}_1 \wedge \tilde{m}_2)\big)\Big)\mathcal{A}_E\left(\tilde{\ell}_1 \wedge \tilde{m}_1\right) \\
&\doteq \operatorname{sign}\Big(\varepsilon\big(\mathcal{A}_E(\tilde{m}_1 \wedge \tilde{m}_2)\big)\Big)\mathcal{A}_E\left(\tilde{m}_2 \wedge \tilde{\ell}_2\right).
\end{aligned}$$

Since $\ell_1 = lk(K_1, K_2)m_2$ in $H_1(X(L))/\text{Torsion}$, we have $\varepsilon(\Delta(L)) = -lk(K_1, K_2)|H_1(R)|$. Observe that $|\varepsilon(\Delta(L))|$ is the order of the quotient of $H_1(X(L))$ by the image of the H_1 of one of the boundary components. In particular, for our genus one knot K and the two-component link whose exterior is the manifold $E[K]$ obtained from E by adding a 2-handle along K, $|\varepsilon(\mathcal{D}(\Sigma))|$ is

$$\left|\varepsilon\Big(\mathcal{A}_E(\tilde{\alpha} \wedge \tilde{\beta})\Big)\right| = \left|\varepsilon\Big(\mathcal{A}_E(\tilde{\alpha}_- \wedge \tilde{\beta}_-)\Big)\right| = \Big|H_1(R)\Big|\Big|\det(V)\Big| = \Big|H_1(R)\Big|\Big|\lambda'(\phi(K))\Big|$$

where V is the Seifert matrix of Σ.

Remark 20.2.8 The condition that δ_i is conjugate to $m_i\ell_i m_i^{-1}\ell_i^{-1}$ is satisfied when m_i and ℓ_i are represented by two simple closed curves that intersect at one transverse point and when they are attached to the basepoint by the same path from the intersection point to the basepoint of ∂E. Figure 20.6 shows the expression of δ_i in the π_1 of the regular neighborhood of $m_i \cup \ell_i$ with respect to the shown basepoints and paths.

Be aware that δ_i might no longer be conjugate to $m_i\ell_i m_i^{-1}\ell_i^{-1}$ if one of the curves is attached to the basepoint by another path. For example, replace ℓ_i by $\check{\ell}_i$ so that $\check{\ell}_i = \delta_i\ell_i\delta_i^{-1}$ in $\pi_1(E)$, then $\check{\delta}_i = m_i\check{\ell}_i m_i^{-1}\check{\ell}_i^{-1}$ and

$$\tilde{\check{\delta}}_i = \tilde{\delta}_i + \big(\exp(m_i) - 1\big)\big(1 - \exp(\ell_i)\big)\tilde{\delta}_i.$$

So $\tilde{\check{\delta}}_i$ and $\tilde{\delta}_i$ are not obtained from one another by a multiplication by a positive unit of Λ_E when ℓ_i is nonzero in $H_1\big(X(L); \mathbb{Q}\big)$, and δ_i and $\check{\delta}_i$ are not conjugate in this case.

Example 20.2.9 For the links $(\phi(A), \phi(B))$, $(\phi(\alpha), \phi(\beta_-))$, and $(\phi(\alpha_-), \phi(\beta))$, the meridian of $\phi(A)$, $\phi(\alpha)$, or $\phi(\alpha_-)$ is u_β^{-1}, and the meridian of $\phi(B)$, $\phi(\beta_-)$, or

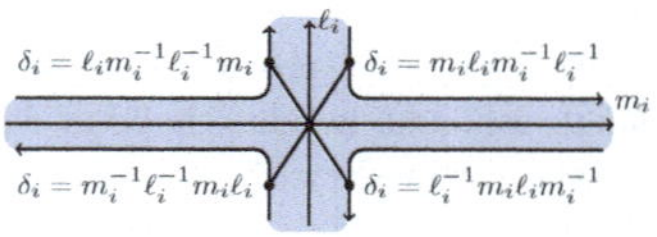

Fig. 20.6 A piece of the genus one surface obtained by plumbing tubular neighborhoods of m_i and ℓ_i

$\phi(\beta)$ is u_α. If $\varepsilon\left(\mathcal{A}_E(u_\beta^{-1} \wedge u_\alpha)\right) > 0$, then we have

$$\begin{aligned}
&\Delta(\phi(A), \phi(B))(t_A = \exp(m_A = -u_\beta), t_B = \exp(m_B = u_\alpha)) \doteq \mathcal{A}_E(\tilde{u}_\alpha \wedge \tilde{B}),\\
&\Delta(\phi(\alpha), \phi(\beta_-)) \doteq \mathcal{A}_E\left(\tilde{\alpha} \wedge \widetilde{u_\beta^{-1}}\right),\\
&\Delta(\phi(\alpha_-), \phi(\beta)) \doteq \mathcal{A}_E(\tilde{u}_\alpha \wedge \tilde{\beta}).
\end{aligned}$$

Definition 20.2.10 Let (K_1, K_2) be a two-component link in a $\mathbb{Q}$-sphere R. The coefficient $\zeta(K_1, K_2)$ is the coefficient of $m_1 m_2$ in the multivariable Alexander polynomial Δ of Definition 20.2.6 viewed as a series in $\mathbb{Q}[[m_1, m_2]]$.

The following lemma will allow us to generalize the description of $\lambda'(K_1, K_2)$ of Definition 20.1.7 in Definition 20.2.12.

Lemma 20.2.11 *Let (K_1, K_2) be a 2-component link in a $\mathbb{Q}$-sphere R such that $lk(K_1, K_2) = 0$. For $i = 1, 2$, let $N(K_i)$ denote a tubular neighborhood of K_i, and let Σ_i be a surface embedded in $R \setminus \left(N(K_{3-i}) \sqcup \mathring{N}(K_i)\right)$ such that the boundary $\partial\Sigma_i$ of Σ_i consists of r_i parallel copies of a primitive curve λ_i on $\partial N(K_i)$, with $r_i > 0$, and $\partial\Sigma_i$ is homologous to $O_i K_i$ in $N(K_i)$, where $O_i = O(K_i)$. Assume that Σ_1 and Σ_2 are transverse. Let $\Sigma_{1,\parallel}$ be obtained from Σ_1 by a slight normal push. Then the Alexander polynomial $\Delta(K_1, K_2)$ of (K_1, K_2) satisfies the following equation in $\mathbb{Q}[[m_1, m_2]]$*

$$\Delta(K_1, K_2)\left(\exp\left(\frac{1}{2O_1}m_1\right), \exp\left(\frac{1}{2O_2}m_2\right)\right) = \zeta(K_1, K_2)m_1 m_2 + O(4),$$

where

$$\zeta(K_1, K_2) = -\frac{|H_1(R)|}{r_1 r_2 O_1 O_2} lk(\Sigma_1 \cap \Sigma_2, \Sigma_{1,\parallel} \cap \Sigma_2).$$

Proof The degree 0 part of $\Delta(K_1, K_2)$ is $-lk(K_1, K_2)|H_1(R)|$, as in Example 20.2.7. According to the surgery formula [2, 1.4.8 page 12, T2 page 13], the invariant λ of the manifold $\chi = R\left((K_1; 0), (K_2; 0)\right)$ obtained from R by 0-surgery on (K_1, K_2) is

$$\lambda(\chi) = \zeta(K_1, K_2).$$

Define the quadratic form $q\colon H_2(\chi) \wedge H_2(\chi) \to \mathbb{Q}$ that maps the exterior product of the homology classes of two transverse surfaces F_1 and F_2 in χ to $lk(F_1 \cap F_2, F_{1,\parallel} \cap F_2)$. Let $([A_1], [A_2])$ be a basis of $H_2(\chi) = \mathbb{Z}[A_1] \oplus \mathbb{Z}[A_2]$. Then [2, T5.2, page 13] implies

$$\lambda(\chi) = -|\text{Torsion}(H_1(\chi))| q([A_1] \wedge [A_2]).$$

For $i \in \{1, 2\}$, let T_i denote the solid torus in χ with meridian λ_i added during the 0–surgery along K_i. Let $X = X(K_1, K_2)$ be the exterior of (K_1, K_2). There is a canonical isomorphism from $H_2(\chi)$ to $H_2(X, \partial X)$. It maps the class of a surface F in χ transverse to T_i to the class $[\check{F} = F \cap X]$. We can assume that the above basis $([A_1], [A_2])$ of $H_2(\chi)$ is represented by surfaces A_1 and A_2 intersecting the T_i as meridian disks of the T_i and we do. For $i \in \{1, 2\}$, set $\check{A}_i = A_i \cap X$ and let $\hat{\Sigma}_i$ be the closed surface of χ obtained from Σ_i by gluing meridian disks of T_i along $\partial \Sigma_i$. Set

$$r = \left| \frac{\mathbb{Z}[\lambda_1] \oplus \mathbb{Z}[\lambda_2]}{\mathbb{Z}[\partial \check{A}_1] \oplus \mathbb{Z}[\partial \check{A}_2]} \right|.$$

We have

$$q([A_1] \wedge [A_2]) = \frac{r^2}{r_1^2 r_2^2} q([\hat{\Sigma}_1] \wedge [\hat{\Sigma}_2]).$$

For $i \in \{1, 2\}$, let μ_i be a curve on the boundary of the solid torus T_i such that $\langle \mu_i, \lambda_i \rangle_{\partial T_i} = 1$. The Mayer–Vietoris sequence associated with the decomposition of χ as the union of X and $T_1 \cup T_2$ leads to an exact sequence

$$\frac{\mathbb{Z}[\lambda_1] \oplus \mathbb{Z}[\lambda_2]}{\mathbb{Z}[\partial \check{A}_1] \oplus \mathbb{Z}[\partial \check{A}_2]} \oplus \mathbb{Z}\mu_1 \oplus \mathbb{Z}\mu_2 \hookrightarrow H_1(X) \oplus H_1(T_1) \oplus H_1(T_2) \to H_1(\chi) \to 0,$$

which implies $\left|\text{Torsion}\big(H_1(X)\big)\right| = \left|\text{Torsion}\big(H_1(\chi)\big)\right| r$. We obtain

$$\zeta(K_1, K_2) = -\frac{\left|\text{Torsion}\big(H_1(X)\big)\right|}{r} \frac{r^2}{r_1^2 r_2^2} q([\hat{\Sigma}_1] \wedge [\hat{\Sigma}_2]).$$

Since $H_1(R) = \frac{H_1(X)}{\mathbb{Z}m_1 \oplus \mathbb{Z}m_2}$, we have $|H_1(R)| = \left|\det\big(\langle m_i, \check{A}_j \rangle_R\big)\right| \left|\text{Torsion}\big(H_1(X)\big)\right|$. Thus

$$\begin{aligned} |H_1(R)| &= \frac{r\left|\langle m_1, \Sigma_1 \rangle_R \langle m_2, \Sigma_2 \rangle_R\right|}{r_1 r_2} \left|\text{Torsion}\big(H_1(X)\big)\right| \\ &= \frac{r}{r_1 r_2} O_1 O_2 \left|\text{Torsion}\big(H_1(X)\big)\right|. \end{aligned}$$

Therefore, we obtain the desired formula for $\zeta(K_1, K_2)$. □

Definition 20.2.12 Under the assumptions of Definition 20.2.10, we set $\lambda'(K_1, K_2) = \frac{\zeta(K_1, K_2)}{|H_1(R)|}$.

20.2.3 A Few Lemmas about the Structure of Alexander Forms

In this subsection, E is a rational homology genus two handlebody as in Theorem 20.1.28. We start the proof of this structure theorem for the Alexander form of such rational homology genus two handlebodies by proving preliminary lemmas. Let $\mathcal{A}_{E,n}(\tilde{u}_\alpha \wedge \tilde{u}_\beta)$ denote the degree n part of $\mathcal{A}_E(\tilde{u}_\alpha \wedge \tilde{u}_\beta) = \sum_{n\in\mathbb{N}} \mathcal{A}_{E,n}(\tilde{u}_\alpha \wedge \tilde{u}_\beta)$.

Lemma 20.2.13 *Under the hypotheses of Theorem 20.1.28, let $\mathcal{A}_{E,\mathcal{P}}$ be a representative of the Alexander form of E. There exists a unique unit $\mathcal{U}$ of $\Lambda_E^{\mathbb{Q}} \subset \frac{\mathbb{Q}[[u_\alpha,u_\beta,u_\gamma]]}{u_\alpha+u_\beta+u_\gamma=0}$ such that $\mathcal{A}_E = \mathcal{U}\mathcal{A}_{E,\mathcal{P}}$ satisfies*

$$\mathcal{A}_E(\tilde{u}_\alpha \wedge \tilde{u}_\beta) = |H_1(R)| + \sum_{n\in\mathbb{N},n\geq 2} \mathcal{A}_{E,n}(u_\alpha \wedge u_\beta)$$

where

$$\mathcal{A}_{E,2}(\tilde{u}_\alpha \wedge \tilde{u}_\beta) = |H_1(R)|\delta_{\Delta,E}(\alpha,\beta)$$

and $\mathcal{A}_{E,3}(\tilde{u}_\alpha \wedge \tilde{u}_\beta) = \frac{|H_1(R)|}{2}\lambda'(E;u_\alpha,u_\beta,u_\gamma)u_\alpha u_\beta u_\gamma$ for some $\lambda'(E;u_\alpha,u_\beta,u_\gamma) \in \mathbb{Q}$.

Furthermore, when A and B are null-homologous in R, $\mathcal{U}$ is a unit of Λ_E. In particular, when A and B are null-homologous, $\mathcal{A}_E(v \wedge w) \in \Lambda_E$ for any $v \wedge w \in \mathcal{H}_E \wedge \mathcal{H}_E$. In general, $\mathcal{U}$ is a unit of

$$\begin{aligned}
&\Lambda(E;u_\alpha,u_\beta,u_\gamma)\\
&\quad= \mathbb{Z}\left[\exp\left(\frac{u_\alpha}{2\mathcal{O}(B)}\right), \exp\left(-\frac{u_\alpha}{2\mathcal{O}(B)}\right), \exp\left(\frac{u_\beta}{2\mathcal{O}(A)}\right), \exp\left(-\frac{u_\beta}{2\mathcal{O}(A)}\right)\right]\\
&\quad\cap \mathbb{Z}\left[e^{\frac{u_\beta}{2\mathcal{O}(C)}}, e^{-\frac{u_\beta}{2\mathcal{O}(C)}}, e^{\frac{u_\gamma}{2\mathcal{O}(B)}}, e^{-\frac{u_\gamma}{2\mathcal{O}(B)}}\right] \cap \mathbb{Z}\left[e^{\frac{u_\gamma}{2\mathcal{O}(A)}}, e^{-\frac{u_\gamma}{2\mathcal{O}(A)}}, e^{\frac{u_\alpha}{2\mathcal{O}(C)}}, e^{-\frac{u_\alpha}{2\mathcal{O}(C)}}\right].
\end{aligned}$$

If $|H_1(R)|$ is odd (or equivalently if R is a $\mathbb{Z}/2\mathbb{Z}$-sphere), then $\mathcal{U}$ is a unit of

$$\begin{aligned}
&\Lambda_o(E;u_\alpha,u_\beta,u_\gamma)\\
&\quad= \mathbb{Z}\left[\exp\left(\frac{u_\alpha}{\mathcal{O}(B)}\right), \exp\left(-\frac{u_\alpha}{\mathcal{O}(B)}\right), \exp\left(\frac{u_\beta}{\mathcal{O}(A)}\right), \exp\left(-\frac{u_\beta}{\mathcal{O}(A)}\right)\right]\\
&\quad\cap \mathbb{Z}\left[e^{\frac{u_\beta}{\mathcal{O}(C)}}, e^{-\frac{u_\beta}{\mathcal{O}(C)}}, e^{\frac{u_\gamma}{\mathcal{O}(B)}}, e^{-\frac{u_\gamma}{\mathcal{O}(B)}}\right] \cap \mathbb{Z}\left[e^{\frac{u_\gamma}{\mathcal{O}(A)}}, e^{-\frac{u_\gamma}{\mathcal{O}(A)}}, e^{\frac{u_\alpha}{\mathcal{O}(C)}}, e^{-\frac{u_\alpha}{\mathcal{O}(C)}}\right].
\end{aligned}$$

Proof Recall from Example 20.2.9 that u_β^{-1} is a meridian of A and that u_α is a meridian of B, when A and B are slightly pushed inside $\phi(\mathring{H}_0)$. There is no loss of generality in assuming that $\varepsilon\big(\mathcal{A}_{E,\mathcal{P}}(\tilde{u}_\alpha \wedge \tilde{u}_\beta)\big) > 0$, and we do. According to Lemma 20.2.5, there exists $(k_a, k_b) \in \mathbb{Z}^2$ such that

$$|H_1(R)|\left(1+\lambda'(A)u_\beta^2\right) = \exp\left(\frac{k_a u_\beta}{2O(A)}\right)\mathrm{ev}(u_\alpha=0)\left(\mathcal{A}_{E,\mathcal{P}}(\tilde{u}_\alpha \wedge \tilde{u}_\beta)\right) + O(4)$$

and

$$|H_1(R)|\left(1+\lambda'(B)u_\alpha^2\right) = \exp\left(\frac{k_b u_\alpha}{2O(B)}\right)\mathrm{ev}(u_\beta=0)\left(\mathcal{A}_{E,\mathcal{P}}(\tilde{u}_\alpha \wedge \tilde{u}_\beta)\right) + O(4)$$

as in Example 20.2.3. Set $\mathcal{U} = \exp(\frac{k_b}{2O(B)}u_\alpha + \frac{k_a}{2O(A)}u_\beta)$ and $\mathcal{A}_E = \mathcal{U}\mathcal{A}_{E,\mathcal{P}}$. Then

$$\mathcal{A}_E(\tilde{u}_\alpha \wedge \tilde{u}_\beta) = |H_1(R)| + \sum_{n\in\mathbb{N}, n\geq 2} \mathcal{A}_{E,n}(\tilde{u}_\alpha \wedge \tilde{u}_\beta)$$

where

$$\mathcal{A}_{E,2}(\tilde{u}_\alpha \wedge \tilde{u}_\beta) = |H_1(R)|\left(\lambda'(A)u_\beta^2 + \lambda'(B)u_\alpha^2 + c(a,b)u_\alpha u_\beta\right).$$

The coefficients of u_α^n and u_β^n in the degree n part $\mathcal{A}_{E,n}(\tilde{u}_\alpha \wedge \tilde{u}_\beta)$ of $\big(\mathcal{A}_E(\tilde{u}_\alpha \wedge \tilde{u}_\beta) \in \mathbb{Q}[[u_\alpha, u_\beta]]\big)$ are determined by the Alexander polynomials of A and B. They vanish for odd n. Since $\tilde{u}_\alpha \wedge \tilde{u}_\beta = \exp(u_\beta)\tilde{u}_\gamma \wedge \tilde{u}_\alpha$ in $\wedge^2\mathcal{H}_E$ according to Lemma 20.1.15, and because of the order 3 symmetry of the setting, $\tilde{\mathcal{A}}_E = \exp(u_\beta)\mathcal{A}_E$ satisfies

$$\tilde{\mathcal{A}}_E(\tilde{u}_\gamma \wedge \tilde{u}_\alpha) = \mathcal{A}_E(\tilde{u}_\alpha \wedge \tilde{u}_\beta) = |H_1(R)|\left(1+\lambda'(C)u_\alpha^2 + \lambda'(A)u_\gamma^2 + c(c,a)u_\gamma u_\alpha\right)$$

in $\mathbb{Q}[[u_\alpha, u_\gamma]] = \frac{\mathbb{Q}[[u_\alpha,u_\beta,u_\gamma]]}{u_\alpha+u_\beta+u_\gamma=0}$. So, we get

$$\begin{aligned}
\frac{\mathcal{A}_{E,2}(\tilde{u}_\alpha \wedge \tilde{u}_\beta)}{|H_1(R)|} &= \lambda'(A)u_\beta(-u_\alpha - u_\gamma) + \lambda'(B)u_\alpha(-u_\beta - u_\gamma) + c(a,b)u_\alpha u_\beta \\
&= \lambda'(C)u_\alpha(-u_\beta - u_\gamma) + \lambda'(A)u_\gamma(-u_\alpha - u_\beta) + c(c,a)u_\gamma u_\alpha \\
&= -\lambda'(A)u_\beta u_\gamma - \lambda'(B)u_\gamma u_\alpha - \lambda'(C)u_\alpha u_\beta \\
&= \delta_{\Delta,E}(\alpha,\beta).
\end{aligned}$$

Furthermore, $\mathcal{U} \in \mathbb{Z}\left[e^{\frac{u_\alpha}{2O(B)}}, e^{-\frac{u_\alpha}{2O(B)}}, e^{\frac{u_\beta}{2O(A)}}, e^{-\frac{u_\beta}{2O(A)}}\right]$, and

$$\tilde{\mathcal{U}} = \exp(u_\beta)\mathcal{U} \in \mathbb{Z}\left[e^{\frac{u_\alpha}{2O(C)}}, e^{-\frac{u_\alpha}{2O(C)}}, e^{\frac{u_\gamma}{2O(A)}}, e^{-\frac{u_\gamma}{2O(A)}}\right].$$

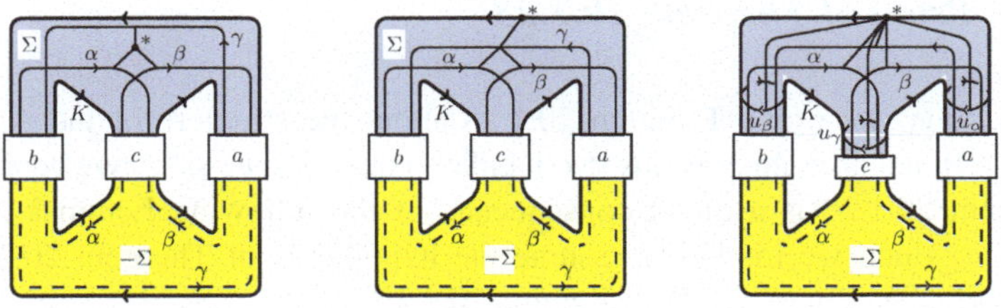

Fig. 20.7 The based curves $\alpha = \alpha(a, b, c)$ and $\beta = \beta(a, b, c)$ on the surface $\Sigma(a, b, c)$, with their basepoint moved to $K = K(a, b, c)$

So $\mathcal{U} \in \mathbb{Z}\left[e^{\frac{u_\alpha}{2O(C)}}, e^{-\frac{u_\alpha}{2O(C)}}, e^{\frac{u_\gamma}{2O(A)}}, e^{-\frac{u_\gamma}{2O(A)}}\right]$. Similarly $\mathcal{U} \in \mathbb{Z}\left[e^{\frac{u_\beta}{2O(C)}}, e^{-\frac{u_\beta}{2O(C)}}, e^{\frac{u_\gamma}{2O(B)}}, e^{-\frac{u_\gamma}{2O(B)}}\right]$. We conclude that $\mathcal{U} \in \Lambda(E; u_\alpha, u_\beta, u_\gamma)$. When Σ is null-homologous, or when $|H_1(R)|$ is odd, we can remove the 2 from the above denominators, and $\mathcal{U} \in \Lambda_o(E; u_\alpha, u_\beta, u_\gamma)$, thanks to Lemma 20.2.5.

Since $\mathcal{A}_{E,3}$ vanishes when at least one of the variables u_α, u_β or u_γ is zero, the lemma is proved. □

Remark 20.2.14 We fix the normalization of $\mathcal{A}_E$ of Lemma 20.2.13, but we keep in mind that the normalization breaks the order 3 symmetry. Indeed, according to Lemma 20.1.15,

$$\mathcal{A}_E(\tilde{u}_\beta \wedge \tilde{u}_\gamma) = \exp(u_\gamma)\mathcal{A}_E(\tilde{u}_\alpha \wedge \tilde{u}_\beta)$$

and $\mathcal{A}_E(\tilde{u}_\gamma \wedge \tilde{u}_\alpha) = \exp(-u_\beta)\mathcal{A}_E(\tilde{u}_\alpha \wedge \tilde{u}_\beta)$.

Lemma 20.2.15 *Let (a, b, c) be a triple of odd relative integers. Set $a_- = \frac{a-1}{2}$, $a_+ = \frac{a+1}{2}$, $b_- = \frac{b-1}{2}$, $b_+ = \frac{b+1}{2}$, $c_- = \frac{c-1}{2}$, and $c_+ = \frac{b+1}{2}$. The homotopy classes of the based curves $\alpha(a, b, c)$ and $\beta(a, b, c)$ of Fig. 20.7 may be written as*

$$\alpha(a, b, c) = u_\gamma^{(c-1)/2} A u_\beta^{-(b+1)/2} \quad \text{and} \quad \beta(a, b, c) = u_\alpha^{(a-1)/2} B u_\gamma^{-(c+1)/2}$$

in $\pi_1(E)$ with respect to the based curves of Fig. 20.4, and we have

$$\tilde{\alpha}(a, b, c) = \frac{\exp(c_- u_\gamma) - 1}{\partial \tilde{u}_\gamma} \tilde{u}_\gamma + \exp\left(c_- u_\gamma\right)\left(\tilde{A} + \exp(A)\frac{\exp(-b_+ u_\beta) - 1}{\partial \tilde{u}_\beta}\tilde{u}_\beta\right)$$

and

$$\tilde{\beta}(a, b, c) = \frac{\exp(a_- u_\alpha) - 1}{\partial \tilde{u}_\alpha} \tilde{u}_\alpha + \exp\left(a_- u_\alpha\right)\left(\tilde{B} + \exp(B)\frac{\exp(-c_+ u_\gamma) - 1}{\partial \tilde{u}_\gamma}\tilde{u}_\gamma\right)$$

in $\mathcal{H}_E$. □

20.2.4 Proof of Theorem 20.1.28

In this section, we prove Theorem 20.1.28 about the structure of the Alexander forms of rational homology genus two handlebodies, and we state two corollaries.

Theorem 20.1.28 is a direct consequence of the following Lemmas 20.2.16–20.2.20, in which we implicitly assume the hypotheses of Theorem 20.1.28 and use the normalization of $\mathcal{A}_E$ from Lemma 20.2.13.

Lemma 20.2.16 *We have*

$$\mathcal{A}_E(\tilde{u}_\beta \wedge \tilde{A}) = \exp(-u_\alpha - B)\,\mathcal{A}_E(\tilde{u}_\alpha \wedge \tilde{B}).$$

Proof Define $\delta \in \mathcal{H}_E$ to be $\delta = \partial\tilde{A}\tilde{u}_\beta - \partial\tilde{u}_\beta\tilde{A}$. Then $\delta = \exp(-u_\alpha - B)\left(\partial\tilde{B}\tilde{u}_\alpha - \partial\tilde{u}_\alpha\tilde{B}\right)$ in $\mathcal{H}_E$, according to Lemma 20.1.29. Since $\partial\delta = 0$, Lemma 20.1.30 implies $\partial\tilde{u}_\alpha\mathcal{A}_E(\tilde{u}_\beta \wedge \delta) = \partial\tilde{u}_\beta\mathcal{A}_E(\tilde{u}_\alpha \wedge \delta)$. So, we obtain $\partial\tilde{u}_\alpha\mathcal{A}_E(\tilde{u}_\beta \wedge \partial\tilde{u}_\beta\tilde{A}) = \exp(-u_\alpha - B)\,\partial\tilde{u}_\beta\mathcal{A}_E(\tilde{u}_\alpha \wedge \partial\tilde{u}_\alpha\tilde{B})$. □

Lemma 20.2.17 *There exists $v \in H_1(E;\mathbb{Q})$ such that*

$$\Delta(A, B) = \exp\left(\frac{\ell - 1}{2}u_\gamma + v\right)\mathcal{A}_E(\tilde{u}_\beta \wedge \tilde{A}).$$

If $\ell \neq 0$, then $v = 0$.

Let (a, b, c) be a triple of odd integers. Set $c_- = \frac{c-1}{2}$. Recall the curves $\alpha(a, b, c)$ and $\beta(a, b, c)$ of Fig. 20.7 and Lemma 20.2.15. If $c_- \neq \ell$, then we have

$$\Delta(\alpha(a, b, c), \beta_-(a, b, c)) = \exp\left(\frac{\ell}{2}u_\gamma\right)\frac{Sh(c_-u_\gamma)}{Sh(u_\gamma)}\mathcal{A}_E(\tilde{u}_\alpha \wedge \tilde{u}_\beta)$$
$$+ \exp\left(\frac{c_-}{2}u_\gamma - v\right)\Delta(A, B).$$

Proof The first assertion follows from Example 20.2.9 and Lemma 20.2.16. Let us compute the Alexander polynomial $\Delta(\alpha(a, b, c), \beta_-(a, b, c))$, for infinitely many triples (a, b, c) of odd integers. Again, Example 20.2.9 implies

$$\Delta(\alpha(a, b, c), \beta_-(a, b, c)) \doteq \mathcal{A}_E(\tilde{u}_\beta \wedge \tilde{\alpha}(a, b, c)).$$

The expression of $\tilde{\alpha}(a, b, c)$ in Lemma 20.2.15 implies

$$\mathcal{A}_E(\tilde{u}_\beta \wedge \tilde{\alpha}(a, b, c)) = \frac{\exp(c_-u_\gamma) - 1}{\partial\tilde{u}_\gamma}\mathcal{A}_E(\tilde{u}_\beta \wedge \tilde{u}_\gamma) + \exp\left(c_-u_\gamma\right)\mathcal{A}_E(\tilde{u}_\beta \wedge \tilde{A}).$$

In particular, $\Delta(\alpha(a,b,c),\beta_-(a,b,c))$ depends only on the arbitrary integer c_-. We denote it by $\Delta(c_-)$. There exists $w(c_-) \in H_1(E;\mathbb{Q})$ such that

$$\Delta(c_-) = \exp\left(\frac{2\ell - c - 1}{4}u_\gamma + w(c_-)\right)\mathcal{A}_E\big(\tilde{u}_\beta \wedge \tilde{\alpha}(a,b,c)\big).$$

Then Lemma 20.1.15 implies

$$\Delta(c_-) = \exp\big(w(c_-)\big)\exp\left(\frac{\ell}{2}u_\gamma\right)\frac{\mathrm{Sh}(c_-u_\gamma)}{\mathrm{Sh}(u_\gamma)}\mathcal{A}_E(\tilde{u}_\alpha \wedge \tilde{u}_\beta)$$
$$+ \exp\big(w(c_-)\big)\exp\left(\frac{c_-}{2}u_\gamma - v\right)\Delta(A,B).$$

Since $A = \ell u_\alpha + q u_\beta$ in $H_1(E;\mathbb{Q})$, for some $q \in \mathbb{Q}$, the degree 0 part $\mathcal{A}_{E,0}(\tilde{u}_\beta \wedge \tilde{A})$ of $\mathcal{A}_E(\tilde{u}_\beta \wedge \tilde{A})$ (and of $\Delta(A,B)$) is $-|H_1(R)|\ell$. The degree 1 part of $\Delta(c_-)$, which must be zero, is

$$|H_1(R)|\,(c_- - \ell)\,w(c_-) + |H_1(R)|\left(c_-\frac{\ell}{2}u_\gamma - \ell\big(\frac{c_-}{2}u_\gamma - v\big)\right)$$
$$= |H_1(R)|\Big((c_- - \ell)\,w(c_-) + \ell v\Big).$$

In particular $(c_- - \ell)\,w(c_-) = -\ell v$.

For any integer c_-, the degree 3 part $\delta_3(c_-)$ of $\frac{\Delta(c_-)}{|H_1(R)|}$, which must be zero, may be expressed as follows.

$$\begin{aligned}\delta_3(c_-) = {} & \tfrac{1}{6}w(c_-)^3\,(c_- - \ell) + \tfrac{1}{2}w(c_-)^2\ell v \\ & + w(c_-)c_-\left(\frac{\ell^2u_\gamma^2}{8} + \frac{c_-^2-1}{24}u_\gamma^2 + \frac{\mathcal{A}_{E,2}(\tilde{u}_\alpha\wedge\tilde{u}_\beta)}{|H_1(R)|}\right) \\ & + w(c_-)\left(-\tfrac{\ell}{2}(\tfrac{c_-}{2}u_\gamma - v)^2 + \tfrac{\Delta_2(A,B)}{|H_1(R)|}\right) \\ & + c_-\left(\frac{\mathcal{A}_{E,3}(\tilde{u}_\alpha\wedge\tilde{u}_\beta)}{|H_1(R)|} + \tfrac{\ell}{2}u_\gamma\left(\frac{c_-^2-1}{24}u_\gamma^2 + \frac{\mathcal{A}_{E,2}(\tilde{u}_\alpha\wedge\tilde{u}_\beta)}{|H_1(R)|}\right) + \frac{\ell^3u_\gamma^3}{48}\right) \\ & - \tfrac{\ell}{6}(\tfrac{c_-}{2}u_\gamma - v)^3 + (\tfrac{c_-}{2}u_\gamma - v)\tfrac{\Delta_2(A,B)}{|H_1(R)|}.\end{aligned}$$

Multiply $\delta_3(c_-)$ by $(c_- - \ell)^2$ and replace $(c_- - \ell)\,w(c_-)$ by $(-\ell v)$ in this equation to get a polynomial in the variable c_-. This polynomial is identically zero. We may write it as $(c_- - \ell)^2\,\delta_3(c_-) = \frac{1}{3}(\ell v)^3 + (c_- - \ell)Q(c_-)$ for some polynomial Q. This implies that $\ell v = 0$. Therefore, we have $v = 0$ as soon as $\ell \neq 0$.

Furthermore, for any integer c_- such that $c_- \neq \ell$, we have $w(c_-) = 0$. This proves the last assertion of the lemma. □

Lemma 20.2.18 *If $\ell = 0$ and $\Delta(A, B) \neq 0$, then the element $v \in H_1(E; \mathbb{Q})$ of Lemma 20.2.17 is zero.*

Proof Let c_- be a nonzero integer. Let $k \in \mathbb{N}$ be the smallest integer such that $\Delta_{2k}(A, B) \neq 0$. For any $j < k$, we have $\mathcal{A}_{E,2j+1}(\tilde{u}_\alpha \wedge \tilde{u}_\beta) = 0$. Indeed, if this holds for any $j' < j$, then the null degree $(2j + 1)$ part of $\Delta(c_-) = \Delta(\alpha(a, b, c), \beta_-(a, b, c))$ is $c_-\mathcal{A}_{E,2j+1}(\tilde{u}_\alpha \wedge \tilde{u}_\beta)$ according to Lemma 20.2.17. So, the null degree $(2k + 1)$ part of $\Delta(c_-)$ is

$$c_- \left(\mathcal{A}_{E,2k+1}(\tilde{u}_\alpha \wedge \tilde{u}_\beta) + \frac{1}{2} u_\gamma \Delta_{2k}(A, B) \right) - v \Delta_{2k}(A, B) = 0$$

for any nonzero integer c_-. Thus $\mathcal{A}_{E,2k+1}(\tilde{u}_\alpha \wedge \tilde{u}_\beta) = -\frac{1}{2} u_\gamma \Delta_{2k}(A, B)$ and $v = 0$. □

Lemma 20.2.19 *We have*

$$Sh(u_\gamma)\Delta(A, B) = \exp\left(-\frac{\ell}{2}u_\gamma\right) \overline{\mathcal{A}_E(\tilde{u}_\alpha \wedge \tilde{u}_\beta)} - \exp\left(\frac{\ell}{2}u_\gamma\right) \mathcal{A}_E(\tilde{u}_\alpha \wedge \tilde{u}_\beta).$$

Proof According to Lemmas 20.2.17 and 20.2.18, for any integer $c_- \neq \ell$, we have

$$\Delta(c_-) = \Delta\big(\alpha(a, b, c), \beta_-(a, b, c)\big) = \exp\left(\frac{\ell}{2}u_\gamma\right) \frac{\mathrm{Sh}(c_- u_\gamma)}{\mathrm{Sh}(u_\gamma)} \mathcal{A}_E(\tilde{u}_\alpha \wedge \tilde{u}_\beta)$$
$$+ \exp\left(\frac{c_-}{2}u_\gamma\right) \Delta(A, B).$$

Since $\Delta(c_-)$ is even, we also have

$$\Delta(c_-) = \exp\left(-\frac{\ell}{2}u_\gamma\right) \frac{\mathrm{Sh}(c_- u_\gamma)}{\mathrm{Sh}(u_\gamma)} \overline{\mathcal{A}_E(\tilde{u}_\alpha \wedge \tilde{u}_\beta)} + \exp\left(-\frac{c_-}{2}u_\gamma\right) \Delta(A, B).$$

□

Lemma 20.2.20 *If $\ell = 0$, then the coefficient $\lambda'(E; u_\alpha, u_\beta, u_\gamma)$ of Lemma 20.2.13 and Theorem 20.1.28 is $\lambda'(A, B)$.*

Proof In this case, the equations

$$\begin{aligned} \mathrm{Sh}(u_\gamma)\Delta(A, B) &= \overline{\mathcal{A}_E(\tilde{u}_\alpha \wedge \tilde{u}_\beta)} - \mathcal{A}_E(\tilde{u}_\alpha \wedge \tilde{u}_\beta) = -2\mathcal{A}_{E,3}(\tilde{u}_\alpha \wedge \tilde{u}_\beta) + O(4) \\ &= -|H_1(R)|\lambda'(E; u_\alpha, u_\beta, u_\gamma)u_\alpha u_\beta u_\gamma + O(4) \end{aligned}$$

imply $\Delta(A, B) = -|H_1(R)|\lambda'(E; u_\alpha, u_\beta, u_\gamma)u_\alpha u_\beta + O(3)$, where u_α is a meridian of B and $(-u_\beta)$ is a meridian of A (when A and B are slightly pushed inside $\phi(\mathring{H}_0)$). So, Lemma 20.2.20 follows from Definition 20.2.12, which is consistent with Definition 20.1.7, thanks to Lemma 20.2.11. □

The proof of Theorem 20.1.28 is finished. □

Proposition 20.2.21 *Let $\mathcal{A}_{E,e}(\tilde{u}_\alpha \wedge \tilde{u}_\beta)$ and $\mathcal{A}_{E,o}(\tilde{u}_\alpha \wedge \tilde{u}_\beta)$ respectively denote the even and the odd part of $\mathcal{A}_{E,e}(\tilde{u}_\alpha \wedge \tilde{u}_\beta)$. Then there exists an even element $\Gamma = \Gamma(E; u_\alpha, u_\beta, u_\gamma)$ of $\frac{\mathbb{Q}[[u_\alpha,u_\beta,u_\gamma]]}{u_\alpha+u_\beta+u_\gamma=0}$ such that*

$$Sh(u_\alpha)Sh(u_\beta)Sh(u_\gamma)\Gamma = \overline{\mathcal{A}_E(\tilde{u}_\alpha \wedge \tilde{u}_\beta)} - \mathcal{A}_E(\tilde{u}_\alpha \wedge \tilde{u}_\beta) = -2\mathcal{A}_{E,o}(\tilde{u}_\alpha \wedge \tilde{u}_\beta),$$

whose degree 0 part is $\Gamma_0 = -|H_1(R)|\lambda'(E; u_\alpha, u_\beta, u_\gamma)$. If A and B are null-homologous, then $\Gamma \in \Lambda_E$. Set

$$\cosh(u) = \frac{1}{2}\left(\exp\left(\frac{u}{2}\right) + \exp\left(-\frac{u}{2}\right)\right).$$

The multivariable Alexander polynomial $\Delta(A, B)$ of (A, B) satisfies the following equation

$$\Delta(A, B) = -\frac{Sh(\ell u_\gamma)}{Sh(u_\gamma)}\mathcal{A}_{E,e}(\tilde{u}_\alpha \wedge \tilde{u}_\beta) + Sh(u_\alpha)Sh(u_\beta)\cosh(\ell u_\gamma)\Gamma.$$

Proof Lemma 20.2.13 implies that $\left(\overline{\mathcal{A}_E(\tilde{u}_\alpha \wedge \tilde{u}_\beta)} - \mathcal{A}_E(\tilde{u}_\alpha \wedge \tilde{u}_\beta)\right)$ is a polynomial in

$$\mathbb{Z}\left[\exp\left(\frac{u_\alpha}{2O(B)}\right), \exp\left(-\frac{u_\alpha}{2O(B)}\right), \exp\left(\frac{u_\beta}{2O(A)}\right), \exp\left(-\frac{u_\beta}{2O(A)}\right)\right].$$

According to Example 20.2.3, it is a multiple of $\left(\exp(\frac{u_\alpha}{2O(B)}) - 1\right)$. It is similarly a multiple of $\left(\exp(\frac{u_\beta}{2O(A)}) - 1\right)$ and $\left(\exp(\frac{u_\gamma}{2O(A)}) - 1\right)$.

If A and B are null-homologous, then $\left(\overline{\mathcal{A}_E(\tilde{u}_\alpha \wedge \tilde{u}_\beta)} - \mathcal{A}_E(\tilde{u}_\alpha \wedge \tilde{u}_\beta)\right)$ is a polynomial of $\mathbb{Z}\left[e^{u_\alpha}, e^{-u_\alpha}, e^{u_\beta}, e^{-u_\beta}\right]$, and it is a multiple of $\left((e^{u_\alpha} - 1)(e^{u_\beta} - 1)(e^{u_\gamma} - 1) = \mathrm{Sh}(u_\alpha)\mathrm{Sh}(u_\beta)\mathrm{Sh}(u_\gamma)\right)$. Use Lemma 20.2.19 to conclude the proof. □

We can now state the following corollary of Theorem 20.1.28.

Corollary 20.2.22 *Under the hypotheses of Theorem 20.1.28, if $lk(A, B) = 0$, then*

$$\Delta(A, B) = Sh(u_\alpha)Sh(u_\beta)\Gamma,$$

with the series Γ of Proposition 20.2.21. Similarly, if $lk(B, C) = 0$, then $\Delta(B, C) = Sh(u_\beta)Sh(u_\gamma)\Gamma$, and, if $lk(C, A) = 0$, then $\Delta(C, A) = Sh(u_\gamma)Sh(u_\alpha)\Gamma$.

Assume that A and B are rationally null-homologous in E. Then we have $lk(A, B) = lk(B, C) = lk(C, A) = 0$, $\partial\tilde{u}_\beta\tilde{A} = \exp(-u_\alpha)\partial\tilde{u}_\alpha\tilde{B}$ in $\mathcal{H}_E$, and, for any $x \in \mathcal{H}_E$,

$$\begin{aligned}\mathcal{A}_E\left(x \wedge \tilde{A}\right) &= \exp(u_\gamma)\partial x \partial\tilde{u}_\alpha\Gamma\\ \mathcal{A}_E\left(x \wedge \tilde{B}\right) &= \exp(-u_\beta)\partial x \partial\tilde{u}_\beta\Gamma\\ \mathcal{A}_E\left(x \wedge \tilde{C}\right) &= \partial x \partial\tilde{u}_\gamma\Gamma,\end{aligned}$$

with the normalization of $\mathcal{A}_E$ of Theorem 20.1.28.

Proof of Corollary 20.2.22 When A is zero in $H_1(E;\mathbb{Q})$, we have $\partial\tilde{A} = 0$. Then Lemma 20.1.30 implies that $\mathcal{A}_E(. \wedge \tilde{A})$ is proportional to ∂. So, the expression of $\mathcal{A}_E\big(\tilde{u}_\beta \wedge \tilde{A}\big)$ of Corollary 20.1.31 shows the first equality. According to Lemma 20.1.29, we have $\partial\tilde{u}_\beta\tilde{A} = \exp(-u_\alpha)\,\partial\tilde{u}_\alpha\tilde{B}$. Therefore, the first equality implies the second one. Since $BAC = 1$ in $\pi_1(E)$, the third equality follows from Lemma 20.1.15. □

Remark 20.2.23 For any pair (A', B') of disjoint curves on ∂E that are obtained from A and B by Dehn twists about the curves u_α, u_β and u_γ of Fig. 20.4 (deprived of the paths to the basepoint), we have

$$\mathrm{Sh}(u_\gamma)\Delta(A', B') = \exp\left(-\frac{lk(A', B')}{2}u_\gamma\right)\overline{\mathcal{A}_E(\tilde{u}_\alpha \wedge \tilde{u}_\beta)} - \exp\left(\frac{lk(A', B')}{2}u_\gamma\right)\mathcal{A}_E(\tilde{u}_\alpha \wedge \tilde{u}_\beta)$$

as in Theorem 20.1.28. The expressions for $\mathcal{A}_E(\tilde{u}_\beta \wedge A')$ and $\mathcal{A}_E(\tilde{u}_\alpha \wedge B')$ are also valid for such curves. (Note that these curves coincide with A and B on the top half of Fig. 20.4, which contains all the basepoints.) Let us move the basepoint on $\partial\Sigma(a, b, c)$ as in Fig. 20.7 without loss. So, the curves α_- and β_- are based curves of Σ_- obtained from α and β by the natural diffeomorphism from Σ to Σ_-. For any triple (a, b, c),

$$\big(\alpha(a, b, c), \check{\beta}_-(a, b, c) = u_\alpha\beta_-u_\alpha^{-1}\big) \text{ and } \big(\check{\alpha}_-(a, b, c) = u_\beta^{-1}\alpha_-u_\beta, \beta(a, b, c)\big)$$

are examples of pairs (A', B') as above.

The *one-variable Alexander polynomial* of the link (A, B) whose component meridians are $m(A)$ and $m(B)$, is obtained from the two-variable Alexander polynomial, by setting

$$t^{\frac{1}{2O(A)O(B)}} = \exp\left(\frac{1}{2O(A)O(B)}m(A)\right) = \exp\left(\frac{1}{2O(A)O(B)}m(B)\right).$$

The following corollary of Theorem 20.1.28 and Proposition 20.2.21 allows us to compute

$$\mathcal{A}_{E,3}(\tilde{u}_\alpha \wedge \tilde{u}_\beta) = \frac{|H_1(R)|}{2}\lambda'(E; u_\alpha, u_\beta, u_\gamma)u_\alpha u_\beta u_\gamma$$

from the one-variable Alexander polynomials of A, B, C, and (A, B). (Here, (A, B) can be replaced with any (A', B') as in Remark 20.2.23.)

Corollary 20.2.24 *Recall $\ell = lk(A, B)$. The degree two part $\Delta_2(A, B)$ of $\Delta(A, B)$ is*

$$\Delta_2(A, B) = |H_1(R)|\ell\frac{1-\ell^2}{24}u_\gamma^2 - \ell\mathcal{A}_{E,2}(\tilde{u}_\alpha \wedge \tilde{u}_\beta) - |H_1(R)|\lambda'(E; u_\alpha, u_\beta, u_\gamma)u_\alpha u_\beta,$$

and we have

$$2\mathcal{A}_{E,3}(\tilde{u}_\alpha \wedge \tilde{u}_\beta) = |H_1(R)|\ell\frac{1-\ell^2}{24}u_\gamma^3 - u_\gamma\big(\ell\mathcal{A}_{E,2}(\tilde{u}_\alpha \wedge \tilde{u}_\beta) + \Delta_2(A, B)\big).$$

Proof Proposition 20.2.21 implies $\Delta(A, B) = -\frac{\mathrm{Sh}(\ell u_\gamma)}{\mathrm{Sh}(u_\gamma)}\mathcal{A}_{E,e}(\tilde{u}_\alpha \wedge \tilde{u}_\beta) + \mathrm{Sh}(u_\alpha)\mathrm{Sh}(u_\beta)\cosh(\ell u_\gamma)\Gamma$. The first equation follows easily. The second equation follows from the first one. □

20.3 The Alexander Series of a Genus One Seifert Surface

In Sect. 20.1.5, we associated a Reidemeister torsion $\tau(E[K])$ to a genus one Seifert surface Σ. In this section, we study this torsion. We view it as a series $\mathcal{D}(\Sigma)$, which is described in Propositions 20.3.1 and 20.3.6.

20.3.1 Formulas for the Alexander Series of a Genus One Seifert Surface

In this subsection, we first prove the following proposition.

Proposition 20.3.1 *Let $\phi\colon H_0 \hookrightarrow R$ be an embedding of the handlebody H_0 in a $\mathbb{Q}$-sphere R. Let $(\check{a}, \check{b}, \check{c})$ be a triple of odd integers. Set $\Sigma = \phi(\Sigma(\check{a}, \check{b}, \check{c}))$. Let $\alpha = \phi(\alpha(\check{a}, \check{b}, \check{c}))$ and $\beta = \phi(\beta(\check{a}, \check{b}, \check{c}))$ be the corresponding curves of Σ, as in Fig. 20.1. Define $a_- = -lk(\beta, \gamma_-)$, $b_- = -lk(\gamma, \alpha_-)$, $c_- = -lk(\alpha, \beta_-)$, $a_+ = -lk(\beta_-, \gamma)$, $b_+ = -lk(\gamma_-, \alpha)$, $c_+ = -lk(\alpha_-, \beta)$, $a = a_+ + a_-$, $b = b_+ + b_-$, and $c = c_+ + c_-$. We have $c_- = \frac{c-1}{2}$ and $c_+ = \frac{c+1}{2}$. Set*

$$\begin{aligned}\mathcal{E}(a, b, c) = \;& \exp\big(\tfrac{a_-}{2}u_\alpha\big)Sh(u_\beta)Sh\left(\alpha = c_- u_\gamma - b_+ u_\beta\right)\\ & - \exp\big(\tfrac{b_+}{2}u_\beta\big)Sh(u_\alpha)Sh\left(\beta = a_- u_\alpha - c_+ u_\gamma\right).\end{aligned}$$

Then we have

$$\begin{aligned}\mathcal{E}(a, b, c) = \;& \exp\left(\tfrac{a_- u_\alpha + b_- u_\beta - c_- u_\gamma}{2}\right) - \exp\left(\tfrac{a_+ u_\alpha + b_+ u_\beta - c_+ u_\gamma}{2}\right)\\ & + \exp\left(\tfrac{b_- u_\beta + c_- u_\gamma - a_- u_\alpha}{2}\right) - \exp\left(\tfrac{b_+ u_\beta + c_+ u_\gamma - a_+ u_\alpha}{2}\right)\\ & + \exp\left(\tfrac{c_- u_\gamma + a_- u_\alpha - b_- u_\beta}{2}\right) - \exp\left(\tfrac{c_+ u_\gamma + a_+ u_\alpha - b_+ u_\beta}{2}\right).\end{aligned}$$

Furthermore, with the normalization of $\mathcal{A}_E$ of Theorem 20.1.28, we have

$$Sh(u_\alpha)Sh(u_\beta)Sh(u_\gamma)\mathcal{D}(\Sigma) = \mathcal{E}(a,b,c)\overline{\mathcal{A}_E(\tilde{u}_\alpha \wedge \tilde{u}_\beta)} - \overline{\mathcal{E}(a,b,c)}\mathcal{A}_E(\tilde{u}_\alpha \wedge \tilde{u}_\beta).$$

and

$$\mathcal{D}(\Sigma) = \exp\left(\frac{1}{2}\left(-a_+u_\alpha + b_+u_\beta - c_+u_\gamma\right)\right)\mathcal{A}_E(\tilde{\alpha} \wedge \tilde{\beta}).$$

Remark 20.3.2 When R is a $\mathbb{Z}$-sphere, we can deduce that $\mathcal{D}(\Sigma)$ is in Λ_E if and only if $(a-b)$ and $(a-c)$ are in $4\mathbb{Z}$.

In this subsection, we work under the hypotheses of Proposition 20.3.1. Note that any genus one Seifert surface Σ may be written as $\phi(\Sigma(\check{a}, \check{b}, \check{c}))$ for some (and actually any) triple $(\check{a}, \check{b}, \check{c})$ of odd integers. The exterior of Σ is $E = R \setminus \phi\big(\mathring{H}_0\big)$.

We use the normalization of the Alexander form $\mathcal{A}_E$ of Theorem 20.1.28. The following lemma gives a first expression of the Reidemeister torsion $\tau(E[K])$ of $E[K]$ introduced in Sect. 20.1.5.

Lemma 20.3.3 *We have*

$$\tau(E[K]) \doteq \mathcal{A}_E\left(\tilde{\alpha} \wedge \tilde{\beta}\right).$$

Proof According to our choice of sign in Sect. 20.1.5, we have $\tau(E[K]) \doteq \frac{\mathcal{A}_E(\tilde{K} \wedge x)}{\partial x}$ for any $x \in \mathcal{H}_E$ such that $\partial x \neq 0$. Since $K = \alpha^{-1}\beta\alpha\beta^{-1} = \alpha^{-1}\left(\beta\alpha\beta^{-1}\alpha^{-1}\right)\alpha$, we have $\tilde{K} = \exp(-\alpha)\big((\partial\tilde{\beta})\tilde{\alpha} - (\partial\tilde{\alpha})\tilde{\beta}\big)$ in $\mathcal{H}_E$. Therefore, we get the announced result since α and β cannot be both rationally null-homologous in E (because they intersect homologically). □

Lemma 20.3.4 *We have*

$$\partial\tilde{u}_\alpha\partial\tilde{u}_\beta\mathcal{A}_E(\tilde{\alpha}\wedge\tilde{\beta}) = \partial\tilde{\alpha}\partial\tilde{u}_\beta\mathcal{A}_E(\tilde{u}_\alpha\wedge\tilde{\beta}) - \partial\tilde{u}_\alpha\partial\tilde{\beta}\mathcal{A}_E(\tilde{u}_\beta\wedge\tilde{\alpha}) - \partial\tilde{\alpha}\partial\tilde{\beta}\mathcal{A}_E(\tilde{u}_\alpha\wedge\tilde{u}_\beta).$$

Proof Lemma 20.1.30 implies

$$\partial\tilde{u}_\beta\mathcal{A}_E(\tilde{\alpha} \wedge \tilde{\beta}) = \partial\tilde{\alpha}\mathcal{A}_E(\tilde{u}_\beta \wedge \tilde{\beta}) - \partial\tilde{\beta}\mathcal{A}_E(\tilde{u}_\beta \wedge \tilde{\alpha})$$

and

$$\partial\tilde{u}_\alpha\mathcal{A}_E(\tilde{u}_\beta \wedge \tilde{\beta}) = \partial\tilde{u}_\beta\mathcal{A}_E(\tilde{u}_\alpha \wedge \tilde{\beta}) - \partial\tilde{\beta}\mathcal{A}_E(\tilde{u}_\alpha \wedge \tilde{u}_\beta).$$

□

Proof of Proposition 20.3.1 View $(-u_\beta)$ as a meridian of α_- (pushed inside $\phi(\mathring{H}_0)$) and u_α as a meridian of β_-. Thus

$$\alpha = -lk(\alpha,\alpha_-)u_\beta + lk(\alpha,\beta_-)u_\alpha = lk(\alpha,\beta_-)(u_\alpha+u_\beta) + lk(\alpha,\gamma_-)u_\beta$$
$$= c_- u_\gamma - b_+ u_\beta$$

in $H_1(E;\mathbb{Q})$ (as in Lemma 20.2.15, when A and B vanish in $H_1(E;\mathbb{Q})$). To prove that $\mathcal{E}(a,b,c)$ has the given symmetric six-term expression, we expand its defining expression and group the terms that contribute to

$$\exp\left(\frac{1}{2}\left(\varepsilon_a a_- u_\alpha + \varepsilon_b b_+ u_\beta + \varepsilon_c (c_- \text{ or } c_+) u_\gamma\right) + \text{monomial independent of } a, b, c\right)$$

for every $(\varepsilon_a,\varepsilon_b,\varepsilon_c)$ in $\{-1,1\}^3$.

Let $\mathcal{R} = \mathcal{E}(a,b,c)\overline{\mathcal{A}_E(\tilde{u}_\alpha\wedge\tilde{u}_\beta)} - \overline{\mathcal{E}(a,b,c)}\mathcal{A}_E(\tilde{u}_\alpha\wedge\tilde{u}_\beta)$ be the right-hand side of the third equation of the statement. Set

$$d = \exp\left(\frac{1}{2}\left(-a_+u_\alpha + b_+u_\beta - c_+u_\gamma\right)\right)\mathrm{Sh}(u_\alpha)\mathrm{Sh}(u_\beta)\mathrm{Sh}(u_\gamma)\mathcal{A}_E(\tilde{\alpha}\wedge\tilde{\beta}).$$

Recall Lemma 20.3.3. Since $\mathcal{R}$ is odd, it suffices to prove that $d = \mathcal{R}$ to finish the proof of Proposition 20.3.1. According to Remark 20.2.23 and Corollary 20.1.31, since $lk(\alpha,\beta_-) = -c_-$, we have

$$\mathrm{Sh}(u_\gamma)\mathcal{A}_E(\tilde{u}_\beta\wedge\tilde{\alpha}) = \exp\left(\frac{1+2c_-}{2}u_\gamma\right)\overline{\mathcal{A}_E(\tilde{u}_\alpha\wedge\tilde{u}_\beta)} - \exp\left(\frac{u_\gamma}{2}\right)\mathcal{A}_E(\tilde{u}_\alpha\wedge\tilde{u}_\beta).$$

Similarly, $lk(\alpha_-,\beta) = -c_+$. So, Remark 20.2.23 and Theorem 20.1.28 imply

$$\mathrm{Sh}(u_\gamma)\mathcal{A}_E(\tilde{u}_\alpha\wedge\tilde{\beta}) = \exp\left(\frac{1+2c_+}{2}u_\gamma + u_\alpha + \beta\right)\overline{\mathcal{A}_E(\tilde{u}_\alpha\wedge\tilde{u}_\beta)}$$
$$- \exp\left(\frac{u_\gamma}{2} + u_\alpha + \beta\right)\mathcal{A}_E(\tilde{u}_\alpha\wedge\tilde{u}_\beta).$$

We have $d = \exp\left(\frac{1}{2}\left(-c_-u_\gamma + u_\beta - \alpha - \beta\right)\right)\partial\tilde{u}_\alpha\partial\tilde{u}_\beta\mathrm{Sh}(u_\gamma)\mathcal{A}_E(\tilde{\alpha}\wedge\tilde{\beta})$, and Lemma 20.3.4 implies $d = e\overline{\mathcal{A}_E(\tilde{u}_\alpha\wedge\tilde{u}_\beta)} + e'\mathcal{A}_E(\tilde{u}_\alpha\wedge\tilde{u}_\beta)$ with

$$\begin{aligned} e &= \mathrm{Sh}(\alpha)\mathrm{Sh}(u_\beta)\exp\left(u_\alpha + u_\beta + (1+2c_+-c_-)\tfrac{u_\gamma}{2} + \tfrac{1}{2}\beta\right) \\ &\quad -\mathrm{Sh}(u_\alpha)\mathrm{Sh}(\beta)\exp\left(\tfrac{1}{2}\left(u_\alpha + u_\beta + (1+c_-)u_\gamma - \alpha\right)\right) \\ &= \mathrm{Sh}(\alpha)\mathrm{Sh}(u_\beta)\exp\left(\tfrac{a_-}{2}u_\alpha\right) - \mathrm{Sh}(u_\alpha)\mathrm{Sh}(\beta)\exp\left(\tfrac{b_+}{2}u_\beta\right) \\ &= \mathcal{E}(a,b,c) \end{aligned}$$

and

$$\begin{aligned}
e' = &-\mathrm{Sh}(\alpha = c_-u_\gamma - b_+u_\beta)\mathrm{Sh}(u_\beta)\exp\left(\tfrac{1}{2}\left(a_-u_\alpha - 2c_+u_\gamma\right)\right)\\
&+\mathrm{Sh}(u_\alpha)\mathrm{Sh}(\beta = a_-u_\alpha - c_+u_\gamma)\exp\left(\tfrac{1}{2}\left(b_+u_\beta - 2c_-u_\gamma\right)\right)\\
&-\mathrm{Sh}(u_\gamma)\mathrm{Sh}(\alpha)\mathrm{Sh}(\beta)\exp\left(\tfrac{1}{2}\left(-c_-u_\gamma + u_\beta\right)\right)\\
= &-\exp\left(\tfrac{1}{2}\left(a_+u_\alpha + (1-b_+)u_\beta - c_+u_\gamma\right)\right)\mathrm{Sh}(u_\beta)\\
&+\exp\left(\tfrac{1}{2}\left(a_-u_\alpha + b_+u_\beta - (2c_+ + c_-)u_\gamma\right)\right)\mathrm{Sh}(u_\beta)\\
&+\exp\left(\tfrac{1}{2}\left(a_-u_\alpha + b_+u_\beta - (2c_- + c_+)u_\gamma\right)\right)\mathrm{Sh}(u_\alpha)\\
&-\exp\left(\tfrac{1}{2}\left(-a_-u_\alpha + b_+u_\beta + (1-c_-)u_\gamma\right)\right)\mathrm{Sh}(u_\alpha)\\
&-\exp\left(\tfrac{1}{2}\left(a_-u_\alpha + (1-b_+)u_\beta - c_+u_\gamma\right)\right)\mathrm{Sh}(u_\gamma)\\
&-\exp\left(\tfrac{1}{2}\left(-a_+u_\alpha + b_+u_\beta - c_-u_\gamma\right)\right)\mathrm{Sh}(u_\gamma)\\
&+\exp\left(\tfrac{1}{2}\left(-a_-u_\alpha + (1-b_+)u_\beta + c_+u_\gamma\right)\right)\mathrm{Sh}(u_\gamma)\\
&+\exp\left(\tfrac{1}{2}\left(a_-u_\alpha + (b_+ + 1)u_\beta - (2c_- + c_+)u_\gamma\right)\right)\mathrm{Sh}(u_\gamma).
\end{aligned}$$

The sum of the second, third, and eighth lines is

$$\begin{aligned}
&\exp\left(\frac{1}{2}\left(a_-u_\alpha + b_+u_\beta - (2c_- + c_+)u_\gamma\right)\right)\\
&\quad\times\left(\exp\left(-\frac{u_\gamma}{2}\right)\mathrm{Sh}(u_\beta) + \mathrm{Sh}(u_\alpha) + \exp\left(\frac{u_\beta}{2}\right)\mathrm{Sh}(u_\gamma)\right).
\end{aligned}$$

It vanishes thanks to Lemma 20.1.15. The sum of the other lines is

$$\begin{aligned}
e' = &\quad\exp\left(\tfrac{a_+u_\alpha - b_+u_\beta - c_+u_\gamma}{2}\right) - \exp\left(\tfrac{a_-u_\alpha - b_-u_\beta - c_-u_\gamma}{2}\right)\\
&+\exp\left(\tfrac{b_+u_\beta - c_+u_\gamma - a_+u_\alpha}{2}\right) - \exp\left(\tfrac{b_-u_\beta - c_-u_\gamma - a_-u_\alpha}{2}\right)\\
&+\exp\left(\tfrac{c_+u_\gamma - a_+u_\alpha - b_+u_\beta}{2}\right) - \exp\left(\tfrac{c_-u_\gamma - a_-u_\alpha - b_-u_\beta}{2}\right)\\
= &-\overline{\mathcal{E}(a,b,c)}.
\end{aligned}$$

□

Proposition 20.3.5 *Recall the surface $\Sigma(a,b,c)$ of Fig. 20.1 embedded in S^3. The Alexander series $\mathcal{D}(a,b,c) = \mathcal{D}(\Sigma(a,b,c))$ is*

$$\begin{aligned}
\mathcal{D}(a,b,c) = \exp\left(\frac{a_-}{2}u_\alpha\right)&\frac{Sh(b_+u_\beta)}{Sh(u_\beta)}\frac{Sh(c_+u_\gamma)}{Sh(u_\gamma)}\\
&+\exp\left(\frac{b_+}{2}u_\beta\right)\frac{Sh(c_-u_\gamma)}{Sh(u_\gamma)}\frac{Sh(a_-u_\alpha)}{Sh(u_\alpha)}\\
&\qquad+\exp\left(\frac{c_+}{2}u_\gamma + \frac{1}{2}u_\beta\right)\frac{Sh(a_-u_\alpha)}{Sh(u_\alpha)}\frac{Sh(b_+u_\beta)}{Sh(u_\beta)}.
\end{aligned}$$

It satisfies the following equalities:

$$\begin{aligned} Sh(u_\alpha)Sh(u_\beta)Sh(u_\gamma)\mathcal{D}(a,b,c) = &\quad \mathcal{E}(a,b,c) - \overline{\mathcal{E}(a,b,c)} \\ = &\quad Sh\left(a_- u_\alpha + b_- u_\beta - c_- u_\gamma\right) \\ &+ Sh\left(b_- u_\beta + c_- u_\gamma - a_- u_\alpha\right) \\ &+ Sh\left(c_- u_\gamma + a_- u_\alpha - b_- u_\beta\right) \\ &- Sh\left(a_+ u_\alpha + b_+ u_\beta - c_+ u_\gamma\right) \\ &- Sh\left(b_+ u_\beta + c_+ u_\gamma - a_+ u_\alpha\right) \\ &- Sh\left(c_+ u_\gamma + a_+ u_\alpha - b_+ u_\beta\right). \end{aligned}$$

Proof Recall that $\mathcal{A}_{E_0}(\tilde{u}_\alpha \wedge \tilde{u}_\beta) = 1$ from Example 20.1.14. Then Proposition 20.3.1 implies that $\mathrm{Sh}(u_\alpha)\mathrm{Sh}(u_\beta)\mathrm{Sh}(u_\gamma)\mathcal{D}(\Sigma(a,b,c)) = \mathcal{E}(a,b,c) - \overline{\mathcal{E}(a,b,c)}$. So, it suffices to check that the first given expression $\tilde{\mathcal{D}}$ of $\mathcal{D}(a,b,c)$ satisfies this equation. To check it, we compute

$$\begin{aligned} \mathrm{Sh}(u_\alpha)\mathrm{Sh}(u_\beta)\mathrm{Sh}(u_\gamma)\tilde{\mathcal{D}} = &\ \exp\left(\tfrac{1}{2}a_- u_\alpha\right)\mathrm{Sh}(u_\alpha)\mathrm{Sh}(b_+ u_\beta)\mathrm{Sh}(c_+ u_\gamma) \\ &+ \exp\left(\tfrac{1}{2}b_+ u_\beta\right)\mathrm{Sh}(a_- u_\alpha)\mathrm{Sh}(u_\beta)\mathrm{Sh}(c_- u_\gamma) \\ &+ \exp\left(\tfrac{1}{2}c_+ u_\gamma + \tfrac{1}{2}u_\beta\right)\mathrm{Sh}(a_- u_\alpha)\mathrm{Sh}(b_+ u_\beta)\mathrm{Sh}(u_\gamma). \end{aligned}$$

Again, we expand this expression and group the terms that contribute to

$$\exp\left(\frac{1}{2}\left(\varepsilon_a a_- u_\alpha + \varepsilon_b b_+ u_\beta + \varepsilon_c (c_- \text{ or } c_+) u_\gamma\right) + \text{monomial independent of } a, b, c\right)$$

for every $(\varepsilon_a, \varepsilon_b, \varepsilon_c)$ in $\{-1,1\}^3$. There are no terms for which $\varepsilon_a = \varepsilon_b = \varepsilon_c = -1$. The six terms for which $\varepsilon_a = \varepsilon_b = \varepsilon_c = 1$ cancel. The six terms for which exactly one $\varepsilon_.$ is 1 correspond directly to terms in the right-hand side of the last equality of the statement. □

The following proposition is an immediate corollary of Propositions 20.3.1 and 20.3.5.

Proposition 20.3.6 *Under the hypotheses of Proposition 20.3.1, we have*

$$\mathcal{D}(\Sigma) = \mathcal{D}(a,b,c)\mathcal{A}_E\left(\tilde{u}_\alpha \wedge \tilde{u}_\beta\right) + \mathcal{E}(a,b,c)\Gamma$$

where $\mathcal{D}(a,b,c)$ is the even polynomial of Proposition 20.3.5, $\mathcal{E}(a,b,c)$ is defined in Proposition 20.3.1, and $\Gamma = \Gamma(E; u_\alpha, u_\beta, u_\gamma)$ is defined in Proposition 20.2.21.

Proof Proposition 20.3.1 implies

$$\begin{aligned} \mathrm{Sh}(u_\alpha)\mathrm{Sh}(u_\beta)\mathrm{Sh}(u_\gamma)\mathcal{D}(\Sigma) = &\left(\mathcal{E}(a,b,c) - \overline{\mathcal{E}(a,b,c)}\right)\mathcal{A}_E(\tilde{u}_\alpha \wedge \tilde{u}_\beta) \\ &+ \mathcal{E}(a,b,c)\left(\overline{\mathcal{A}_E(\tilde{u}_\alpha \wedge \tilde{u}_\beta)} - \mathcal{A}_E(\tilde{u}_\alpha \wedge \tilde{u}_\beta)\right). \end{aligned}$$

□

20.3.2 The Low Degree Terms of $\mathcal{D}(\Sigma)$

In this section, we compute $\mathcal{D}(\Sigma)$ up to order two under the hypotheses of Proposition 20.3.1 to prove Theorem 20.1.23. The vector space

$$H_1(\Sigma;\mathbb{Q}) \otimes_s H_1(\Sigma;\mathbb{Q}) = \mathbb{Q}(\alpha \otimes_s \alpha) \oplus \mathbb{Q}(\beta \otimes_s \beta) \oplus \mathbb{Q}(\gamma \otimes_s \gamma)$$

is equipped with the canonical symmetric non-degenerate bilinear form

$$\begin{aligned} \mathcal{B}\colon (H_1(\Sigma;\mathbb{Q}) \otimes_s H_1(\Sigma;\mathbb{Q}))^2 &\to \mathbb{Q} \\ (u \otimes_s v, x \otimes_s y) &\mapsto \langle u, x\rangle_\Sigma \langle v, y\rangle_\Sigma + \langle u, y\rangle_\Sigma \langle v, x\rangle_\Sigma, \end{aligned}$$

which maps $(\alpha \otimes_s \alpha, \beta \otimes_s \beta)$ to 2. The tensor

$$V_s^* = \frac{a}{2}\alpha \otimes_s \alpha + \frac{b}{2}\beta \otimes_s \beta + \frac{c}{2}\gamma \otimes_s \gamma$$

is the dual of the symmetrized Seifert form V_s introduced in Sect. 20.1.5, with respect to $\mathcal{B}$. $V_s(u \otimes_s v) = \mathcal{B}(u \otimes_s v, V_s^*)$. So V_s^* is a canonical element of $H_1(\Sigma;\mathbb{Q}) \otimes_s H_1(\Sigma;\mathbb{Q})$.

Lemma 20.3.7 *Recall the isomorphism f of Proposition 20.1.19. Set*

$$P = (f \otimes_s f)(V_s^*) = \frac{a}{2}u_\alpha^2 + \frac{b}{2}u_\beta^2 + \frac{c}{2}u_\gamma^2.$$

Then P is a canonical element of $H_1(E;\mathbb{Q}) \otimes_s H_1(E;\mathbb{Q})$. Furthermore, we have

$$P = u_\alpha \otimes_s \beta - u_\beta \otimes_s \alpha \quad \textit{and} \quad \mathcal{E}(a,b,c) = -P + O(3).$$

Proof We have

$$\begin{aligned} &u_\alpha \otimes_s (\beta = a_- u_\alpha - c_+ u_\gamma) - u_\beta \otimes_s (\alpha = c_- u_\gamma - b_+ u_\beta) \\ &= P + \frac{1}{2}\left(-u_\alpha^2 - u_\gamma u_\alpha + u_\beta u_\gamma + u_\beta^2\right) = P. \end{aligned}$$

According to Proposition 20.3.1, we have $\mathcal{E}(a,b,c) = u_\beta \otimes_s \alpha - u_\alpha \otimes_s \beta + O(3)$. □

Lemma 20.3.8 *We have*

$$\mathcal{D}(a,b,c) = \lambda'(a,b,c) + O(2)$$

and the degree 0 part of $\mathcal{D}(\Sigma)$ is $|H_1(R)|\lambda'(K)$.

Proof The first expression of $\mathcal{D}(a,b,c)$ in Proposition 20.3.5 shows that

$$\varepsilon(\mathcal{D}(a,b,c)) = b_+c_+ + c_-a_- + a_-b_+ = \lambda'(a,b,c)$$

while the second one shows that $\mathcal{D}(a,b,c)$ is even. Then Proposition 20.3.6, Theorem 20.1.28, and Lemma 20.3.7 imply that the degree 0 part of $\mathcal{D}(\Sigma)$ is $|H_1(R)|\lambda'(K)$. □

Note the following easy lemma.

Lemma 20.3.9 *Recall the polynomial invariant P from Lemma 20.3.7, and the polynomial $\delta_{2,\Sigma}(\alpha,\beta)$ of Theorem 20.1.32. Set $\delta_{2,E}(\alpha,\beta) = f_*(\delta_{2,\Sigma}(\alpha,\beta))$. Then we have*

$$\begin{aligned}
\delta_{2,E}(\alpha,\beta) &= -\tfrac{1}{24}(a^2+2ab+2ac+3)u_\alpha^2 - \tfrac{1}{24}(b^2+2bc+2ba+3)u_\beta^2 \\
&\quad \tfrac{1}{24}(c^2+2ca+2cb+3)u_\gamma^2 \\
&= \tfrac{a^2u_\alpha^2+b^2u_\beta^2+c^2u_\gamma^2}{24} - \tfrac{a+b+c}{6}P - \tfrac{u_\alpha^2+u_\beta^2+u_\gamma^2}{8} \\
&= \tfrac{1}{24}\left((a+b)^2+8\lambda'(a,b,c)+4\right)u_\alpha u_\beta \\
&\quad +\tfrac{1}{24}\left((b+c)^2+8\lambda'(a,b,c)+4\right)u_\beta u_\gamma \\
&\quad +\tfrac{1}{24}\left((c+a)^2+8\lambda'(a,b,c)+4\right)u_\gamma u_\alpha.
\end{aligned}$$

□

Lemma 20.3.10 *With the notation of Lemma 20.3.9, the degree two part $\mathcal{D}_2(a,b,c)$ of the polynomial $\mathcal{D}(a,b,c)$ of Proposition 20.3.5 is*

$$\begin{aligned}
\mathcal{D}_2(a,b,c) &= \tfrac{(a^2-1)}{16\times 24}(3+3bc+a(b+c))u_\alpha^2 + \tfrac{(b^2-1)}{16\times 24}(3+3ca+b(c+a))u_\beta^2 \\
&\quad + \tfrac{(c^2-1)}{16\times 24}(3+3ab+c(a+b))u_\gamma^2 \\
&= -\tfrac{1}{12}w_S(a,b,c)P + \tfrac{\lambda'(a,b,c)}{4}\delta_{2,E}(\alpha,\beta),
\end{aligned}$$

where $w_S(a,b,c)$ is defined in Theorem 20.1.2.

Proof Recall

$$\begin{aligned}
\mathcal{D}(a,b,c) &= \exp\left(\tfrac{1}{2}a_-u_\alpha\right)\frac{\mathrm{Sh}(b_+u_\beta)}{\mathrm{Sh}(u_\beta)}\frac{\mathrm{Sh}(c_+u_\gamma)}{\mathrm{Sh}(u_\gamma)} + \exp\left(\tfrac{1}{2}b_+u_\beta\right)\frac{\mathrm{Sh}(c_-u_\gamma)}{\mathrm{Sh}(u_\gamma)}\frac{\mathrm{Sh}(a_-u_\alpha)}{\mathrm{Sh}(u_\alpha)} \\
&\quad + \exp\left(\tfrac{1}{2}c_+u_\gamma + \tfrac{1}{2}u_\beta\right)\frac{\mathrm{Sh}(a_-u_\alpha)}{\mathrm{Sh}(u_\alpha)}\frac{\mathrm{Sh}(b_+u_\beta)}{\mathrm{Sh}(u_\beta)},
\end{aligned}$$

$\mathrm{Sh}(u) = \exp\left(\frac{u}{2}\right) - \exp\left(-\frac{u}{2}\right) = u\left(1+\frac{u^2}{24}+O(4)\right)$ and

$$\frac{\mathrm{Sh}\,(a_-u_\alpha)}{\mathrm{Sh}(u_\alpha)} = a_-\left(1+\frac{a_-^2-1}{24}u_\alpha^2\right) + O(4).$$

Since the degree 2 part of $\exp\left(\frac{1}{2}a_-u_\alpha\right)\frac{\mathrm{Sh}(b_+u_\beta)}{\mathrm{Sh}(u_\beta)}\frac{\mathrm{Sh}(c_+u_\gamma)}{\mathrm{Sh}(u_\gamma)}$ is

$$b_+c_+\left(\frac{1}{8}a_-^2u_\alpha^2+\frac{b_+^2-1}{24}u_\beta^2+\frac{c_+^2-1}{24}u_\gamma^2\right),$$

we get

$$\begin{aligned}\mathcal{D}_2(a,b,c) &= \tfrac{1}{8}\left(a_-^2b_+c_+u_\alpha^2+b_+^2c_-a_-u_\beta^2+c_+^2a_-b_+u_\gamma^2\right)\\ &\quad+\left(\tfrac{1}{8}a_-b_+u_\beta^2+2c_+a_-b_+u_\beta u_\gamma\right)\\ &\quad+\tfrac{a_-(b_++c_-)}{24}(a_-^2-1)u_\alpha^2+\tfrac{b_+(c_++a_-)}{24}(b_+^2-1)u_\beta^2\\ &\quad+\tfrac{b_+c_+(c_+^2-1)+a_-c_-(c_-^2-1)}{24}u_\gamma^2\\ &= \tfrac{1}{8}\left(a_-^2b_+c_+u_\alpha^2+b_+^2c_-a_-u_\beta^2+c_+^2a_-b_+u_\gamma^2\right)\\ &\quad+\left(\tfrac{1}{8}a_-b_+u_\beta^2+c_+a_-b_+(u_\alpha^2-u_\beta^2-u_\gamma^2)\right)\\ &\quad+\tfrac{a_-a_+(b+c)}{48}(a_--1)u_\alpha^2+\tfrac{b_+b_-(c+a)}{48}(b_++1)u_\beta^2\\ &\quad+\tfrac{c_-c_+(b_+(c_++1)+a_-(c_--1))}{24}u_\gamma^2.\end{aligned}$$

Thanks to the symmetry of $\mathcal{D}_2(a,b,c)=d_au_\alpha^2+d_bu_\beta^2+d_cu_\gamma^2$, it suffices to compute

$$\begin{aligned}d_a &= \tfrac{a_-a_+}{48}\left(6b_+c_++(b+c)(a_--1)\right)=\tfrac{a^2-1}{8\times 48}\left(3(b+1)(c+1)+(b+c)(a-3)\right)\\ &= \tfrac{a^2-1}{8\times 48}\left(3+3bc+a(b+c)\right)\end{aligned}$$

to find the first expression of $\mathcal{D}_2(a,b,c)$.

For the second one, recall $16w_8(a,b,c)=2(a+b)(b+c)(c+a)-4(ab+bc+ca+1)(a+b+c)$. We have

$$\frac{\lambda'(a,b,c)}{4}\delta_{2,E}(\alpha,\beta)-\frac{1}{12}w_8(a,b,c)P=c_au_\alpha^2+c_bu_\beta^2+c_cu_\gamma^2$$

where

$$\begin{aligned}c_a &= -\tfrac{1}{16\times 24}(ab+ac+bc+1)(a^2+2ab+2ac+3)\\ &\quad+\tfrac{1}{16\times 24}\left(2a(a^2(b+c)+b^2(a+c)+c^2(a+b))+8a^2bc+4a(a+b+c)\right)\\ &= \tfrac{(a^2-1)}{16\times 24}(3+3bc+a(b+c)).\end{aligned}$$

□

Proposition 20.3.11 *Let Σ be a surface as in Proposition 20.3.1. Set $\delta_E(\alpha,\beta)=\delta_{2,E}(\alpha,\beta)+4\delta_{\Delta,E}(\alpha,\beta)$, where $\delta_{2,E}(\alpha,\beta)$ and $\delta_{\Delta,E}(\alpha,\beta)$ are respectively defined in Lemma 20.3.9 and in Theorem 20.1.28. The degree two part $\mathcal{D}_2(\Sigma)$ of $\mathcal{D}(\Sigma)$ satisfies*

$$\frac{1}{|H_1(R)|}\mathcal{D}_2(\Sigma)=w_{SL}(\Sigma)P+\frac{\lambda'(\partial\Sigma)}{4}\delta_E(\alpha,\beta),$$

where w_{SL} and P are respectively defined in Theorem 20.1.33 and Lemma 20.3.7.

Proof According to Proposition 20.3.6, $\mathcal{D}(\Sigma) = \mathcal{D}(a,b,c)\mathcal{A}_E\left(\tilde{u}_\alpha \wedge \tilde{u}_\beta\right) + \mathcal{E}(a,b,c)\Gamma$. Lemmas 20.3.8 and 20.3.10 imply

$$\mathcal{D}(a,b,c) = \lambda'(a,b,c) - \frac{1}{12}w_\delta(a,b,c)P + \frac{\lambda'(a,b,c)}{4}\delta_{2,E}(\alpha,\beta) + O(4)$$

where $\lambda'(a,b,c) = \lambda'(\phi(K))$. Lemma 20.3.7 implies $\mathcal{E}(a,b,c) = -P + O(3)$. Theorem 20.1.28 implies $\mathcal{A}_E\left(\tilde{u}_\alpha \wedge \tilde{u}_\beta\right) = |H_1(R)|\left(1 + \delta_{\Delta,E}(\alpha,\beta)\right) + O(3)$, and Proposition 20.2.21 implies $\Gamma = -|H_1(R)|\lambda'(E; u_\alpha, u_\beta, u_\gamma) + O(2)$. Therefore, we have

$$\begin{aligned}\frac{1}{|H_1(R)|}\mathcal{D}_2(\Sigma) = &\frac{\lambda'(a,b,c)}{4}\left(\delta_{2,E}(\alpha,\beta) + 4\delta_{\Delta,E}(\alpha,\beta)\right)\\ &- \frac{1}{12}w_\delta(a,b,c)P + \lambda'(E; u_\alpha, u_\beta, u_\gamma)P,\end{aligned}$$

where $w_{SL}(\Sigma)$ is defined to be $w_{SL}(\Sigma) = \lambda'(E; u_\alpha, u_\beta, u_\gamma) - \frac{1}{12}w_\delta(a,b,c)$ in Theorem 20.1.33. □

Recall

$$\begin{aligned}W_s = V_s \circ (f^{-1} \otimes_s f^{-1})\colon \otimes_s^2 H_1(E = R \setminus (\mathring{H}))/\text{Torsion} &\to \mathbb{Q}\\ u_\alpha^2 &\mapsto b + c\\ u_\alpha u_\beta &\mapsto -c\end{aligned}$$

from the statement of Theorem 20.1.23. Note the following easy lemma.

Lemma 20.3.12 *The polynomial $w_\delta(a,b,c)$ of Theorem 20.1.2 equals*

$$w_\delta(a,b,c) = -\frac{a^2(b+c) + b^2(c+a) + c^2(a+b)}{8} - \frac{abc}{2} - \frac{a+b+c}{4}$$

and we have $W_s(\delta_{2,E}(\alpha,\beta)) = w_\delta(a,b,c)$, $W_s(\delta_{\Delta,E}(\alpha,\beta)) = a\lambda'(\phi(A)) + b\lambda'(\phi(B)) + c\lambda'(\phi(C))$, $W_s(\delta_E(\alpha,\beta)) = w_\delta(\Sigma)$, and $W_s(P) = 4\lambda'(\phi(K)) - 1$.

Proof Thanks to Lemma 20.3.9, we have

$$\begin{aligned}W_s(\delta_{2,E}(\alpha,\beta)) = &\frac{a^2(b+c) + b^2(c+a) + c^2(a+b)}{24}\\ &- \frac{a+b+c}{6}(4\lambda'(\phi(K)) - 1) - \frac{a+b+c}{4}.\end{aligned}$$

□

Theorem 20.1.23 is a direct consequence of Proposition 20.3.11 and Lemma 20.3.12. □

20.3.3 The Invariant δ of a Genus One Seifert Surface

In this section, we prove that $\left(\delta_\Sigma(\alpha,\beta) = f_*^{-1}\left(\delta_E(\alpha,\beta)\right)\right)$ is an invariant of Σ such that $w_\delta(\Sigma) = V_s(\delta_\Sigma)$, as stated in Theorem 20.1.32.

Remark 20.3.13 Corollary 20.1.24 implies that $w_{SL}(\Sigma)$ is an invariant of Σ when Σ is null-homologous. Since P is also an invariant of Σ, Proposition 20.3.11 implies Theorem 20.1.32 when Σ is null-homologous and $\lambda'(K) \neq 0$. The proof below shows that Theorem 20.1.32 holds without these restrictive assumptions.

First note the following lemma.

Lemma 20.3.14 *The polynomial $\delta_E(\alpha,\beta)$ of Proposition 20.3.11 depends only on the homology classes of α and β on Σ.*

Proof Since a, b, and c are determined by the Seifert form, they depend only on the homology classes of α and β on Σ. So do $\lambda'(\phi(A))$, $\lambda'(\phi(B))$, and $\lambda'(\phi(C))$ according to Lemma 20.1.1. □

The group of diffeomorphisms of Σ up to isotopy acts on $H_1(\Sigma) = \mathbb{Z}[\alpha] \oplus \mathbb{Z}[\beta]$ as $SL_2(\mathbb{Z})$. It is generated by the Dehn twists about the curves α and β of Fig. 20.1. Thus it is sufficient to prove that $\delta_\Sigma(\alpha,\beta)$ is invariant under a Dehn twist about α and a Dehn twist about β in order to prove Theorem 20.1.32. Let τ_α be the Dehn twist about a curve parallel to α that changes β to $\beta' = \beta\alpha$. Thanks to the order 3 symmetry of the setting, it suffices to prove that $\delta_\Sigma(\alpha,\beta)$ is invariant under τ_α.

The twist τ_α changes β_- to $\beta'_- = \beta_-\alpha_-$. It leaves α and α_- unchanged. Therefore, it leaves the curve $u_\alpha = \alpha\alpha_-^{-1}$ of Figs. 20.4 and 20.7 unchanged ($u'_\alpha = u_\alpha$), and it changes $u_\beta = \beta_-^{-1}\beta$ to $u'_\beta = \alpha_-^{-1}u_\beta\alpha$. Recall $\breve{\alpha}_- = u_\beta^{-1}\alpha_- u_\beta$ from Remark 20.2.23. Then we have

$$u'_\beta = u_\beta\breve{\alpha}_-^{-1}\alpha.$$

Lemma 20.3.15 *We have*

$$\delta_{\Delta,E}(\alpha',\beta') - \delta_{\Delta,E}(\alpha,\beta) = \frac{1}{8}(b^2-1)u_\alpha u_\gamma.$$

Proof Since $\tilde{u}'_\beta = \tilde{u}_\beta + \exp(u_\beta - \alpha_-)(\tilde{\alpha} - \tilde{\breve{\alpha}}_-)$, we have

$$\mathcal{A}_E(\tilde{u}'_\alpha \wedge \tilde{u}'_\beta) = \mathcal{A}_E(\tilde{u}_\alpha \wedge \tilde{u}_\beta) + \exp(u_\beta - \alpha_-)\mathcal{A}_E(\tilde{u}_\alpha \wedge \tilde{\alpha} - \tilde{\breve{\alpha}}_-)$$

where $\partial\tilde{u}_\beta\mathcal{A}_E(\tilde{u}_\alpha \wedge \tilde{\alpha}) = \partial\tilde{u}_\alpha\mathcal{A}_E(\tilde{u}_\beta \wedge \tilde{\alpha}) + \partial\tilde{\alpha}\mathcal{A}_E(\tilde{u}_\alpha \wedge \tilde{u}_\beta)$ according to Lemma 20.1.30. Using the similar equality for $\breve{\alpha}_-$ and the expressions of $\mathcal{A}_E(\tilde{u}_\beta \wedge$

$\tilde{\alpha}$) and $\mathcal{A}_E(\tilde{u}_\beta \wedge \tilde{\tilde{\alpha}}_-)$ given by Corollary 20.1.31 and Remark 20.2.23, we get

$$\begin{aligned}\mathrm{Sh}(u_\gamma)\partial\tilde{u}_\beta\mathcal{A}_E(\tilde{u}'_\alpha \wedge \tilde{u}'_\beta) = \\ \mathrm{Sh}(u_\gamma)\partial\tilde{u}_\beta\mathcal{A}_E(\tilde{u}_\alpha \wedge \tilde{u}_\beta) \\ +\exp(u_\beta - \alpha_-)\mathrm{Sh}(u_\gamma)\exp(\alpha_-)\left(\exp(u_\alpha)-1\right)\mathcal{A}_E(\tilde{u}_\alpha \wedge \tilde{u}_\beta) \\ +\partial\tilde{u}_\alpha\exp(u_\beta - \alpha_-)\left(\exp\left(\tfrac{1+2c_-}{2}u_\gamma\right)\exp\left(\tfrac{1+2c_+}{2}u_\gamma\right)\right)\overline{\mathcal{A}_E(\tilde{u}_\alpha \wedge \tilde{u}_\beta)}.\end{aligned}$$

So, we obtain

$$\begin{aligned}\partial\tilde{u}_\beta\mathcal{A}_E(\tilde{u}'_\alpha \wedge \tilde{u}'_\beta) &= \left(\partial\tilde{u}_\beta + \exp(u_\beta)\partial\tilde{u}_\alpha\right)\mathcal{A}_E(\tilde{u}_\alpha \wedge \tilde{u}_\beta) \\ &\quad -\partial\tilde{u}_\alpha\exp(u_\beta - \alpha_-)\exp(c_+u_\gamma)\overline{\mathcal{A}_E(\tilde{u}_\alpha \wedge \tilde{u}_\beta)} \\ &= (\exp(-u_\gamma)-1)\mathcal{A}_E(\tilde{u}_\alpha \wedge \tilde{u}_\beta) \\ &\quad -\partial\tilde{u}_\alpha\exp((1+b_-)u_\beta)\overline{\mathcal{A}_E(\tilde{u}_\alpha \wedge \tilde{u}_\beta)}.\end{aligned}$$

Set $\mathcal{A}'_E = \exp(b_-u_\alpha)\mathcal{A}_E$. Then we have

$$\begin{aligned}\mathrm{Sh}(u_\beta)\mathcal{A}'_E(\tilde{u}'_\alpha \wedge \tilde{u}'_\beta) &= -\mathrm{Sh}(u_\gamma)\exp\left(\left(\tfrac{1}{2}+b_-\right)u_\alpha\right)\mathcal{A}_E(\tilde{u}_\alpha \wedge \tilde{u}_\beta) \\ &\quad -\mathrm{Sh}(u_\alpha)\exp\left(\left(-\tfrac{1}{2}-b_-\right)u_\gamma\right)\overline{\mathcal{A}_E(\tilde{u}_\alpha \wedge \tilde{u}_\beta)} \\ &= -\mathrm{Sh}(u_\gamma)\exp\left(\tfrac{b}{2}u_\alpha\right)\mathcal{A}_E(\tilde{u}_\alpha \wedge \tilde{u}_\beta) \\ &\quad -\mathrm{Sh}(u_\alpha)\exp\left(-\tfrac{b}{2}u_\gamma\right)\overline{\mathcal{A}_E(\tilde{u}_\alpha \wedge \tilde{u}_\beta)},\end{aligned}$$

and the degree two part of $\mathrm{Sh}(u_\beta)\mathcal{A}'_E(\tilde{u}'_\alpha \wedge \tilde{u}'_\beta)$ is zero. Therefore, the degree one part of $\mathcal{A}'_E(\tilde{u}'_\alpha \wedge \tilde{u}'_\beta)$ is zero, and $\delta' = \delta_{\Delta,E}(\alpha', \beta')$ is the degree two part of $\frac{1}{|H_1(R)|}\mathcal{A}'_E(\tilde{u}'_\alpha \wedge \tilde{u}'_\beta)$. Set $\delta = \delta_{\Delta,E}(\alpha, \beta)$. We have

$$\begin{aligned}u_\beta\left(1+\tfrac{u_\beta^2}{24}\right)(1+\delta') &= -u_\gamma\left(1+\tfrac{u_\gamma^2}{24}\right)\left(1+\tfrac{b^2}{8}u_\alpha^2\right)(1+\delta) \\ &\quad -u_\alpha\left(1+\tfrac{u_\alpha^2}{24}\right)\left(1+\tfrac{b^2}{8}u_\gamma^2\right)(1+\delta)+O(4) \\ &= u_\beta\left(1+\delta+\tfrac{u_\gamma^2-u_\alpha u_\gamma+u_\alpha^2}{24}+\tfrac{1}{8}b^2u_\alpha u_\gamma\right)+O(4).\end{aligned}$$

So $\delta' = \delta + \frac{1}{8}b^2u_\alpha u_\gamma + \frac{u_\gamma^2-u_\alpha u_\gamma+u_\alpha^2-(u_\alpha+u_\gamma)^2}{24} = \delta + \frac{1}{8}(b^2-1)u_\alpha u_\gamma$. □

Lemma 20.3.16 *The Dehn twist τ_α about α changes (a, b, c) to*

$$(a' = a + 2b, b' = 2b + c, c' = -b)$$

and the element $(u_\alpha, u_\beta, u_\gamma)$ of $H_1(\partial E)^3$ to $(u'_\alpha = u_\alpha, u'_\beta = u_\beta + u_\alpha = -u_\gamma, u'_\gamma = u_\gamma - u_\alpha)$.

Proof In $H_1(\partial E)$, we have $\alpha_- = c_+ u_\gamma - b_- u_\beta$, $\alpha' = \alpha$, $\beta' = \beta + \alpha$, and $\gamma' = \gamma - \alpha$. We also have $lk(\beta_-, \gamma) = -a_+$, $lk(\gamma_-, \alpha) = -b_+ = -\frac{b+1}{2}$, and $lk(\alpha_-, \beta) = -c_+$. Since $lk(\gamma'_-, \alpha) = -b_+ - \frac{b+c}{2}$, we obtain $b' = 2b + c$. Similarly $lk(\alpha_-, \beta') = -c'_+ = -c_+ + \frac{b+c}{2}$, and $c' = -b$. Since $lk(\beta'_-, \gamma') = -a_+ - \frac{b+c}{2} - b_- + c_-$, we get $a' = a + 2b$. □

Lemma 20.3.17 *We have*

$$\delta_{2,E}(\alpha', \beta') - \delta_{2,E}(\alpha, \beta) = \frac{1}{2}(1 - b^2) u_\alpha u_\gamma$$

Proof Set $\delta''_2 = \delta_{2,E}(\alpha', \beta') - \delta_{2,E}(\alpha, \beta)$ and recall

$$\delta_{2,E}(\alpha, \beta) = \frac{a^2 u_\alpha^2 + b^2 u_\beta^2 + c^2 u_\gamma^2}{24} - \frac{a+b+c}{6} P - \frac{u_\alpha^2 + u_\beta^2 + u_\gamma^2}{8}$$

from Lemma 20.3.9. We have

$$\begin{aligned}
\delta''_2 &= -\tfrac{b}{3} P - \tfrac{1}{8}\left(u_\alpha^2 - u_\beta^2 + u_\gamma^2 - (u_\beta^2 - u_\alpha^2 - u_\gamma^2)\right) \\
&\quad + \tfrac{1}{24}\left((4b^2 + 4ab) u_\alpha^2 + (2b + c)^2 u_\gamma^2 - b^2 u_\beta^2 + b^2 (u_\gamma - u_\alpha)^2 - c^2 u_\gamma^2\right) \\
&= -\tfrac{1}{24}(4ab u_\alpha^2 + 4b^2 u_\beta^2 + 4bc u_\gamma^2) - \tfrac{1}{24}\left(6u_\alpha^2 - 6u_\beta^2 + 6u_\gamma^2\right) \\
&\quad + \tfrac{1}{24}\left((4b^2 + 4ab) u_\alpha^2 + (4b^2 + 4bc + c^2) u_\gamma^2 - b^2 u_\beta^2\right) \\
&\quad + \tfrac{1}{24}\left(b^2 (2u_\gamma^2 + 2u_\alpha^2 - u_\beta^2) - c^2 u_\gamma^2\right) \\
&= \tfrac{1}{24}(6b^2 - 6)(u_\alpha^2 - u_\beta^2 + u_\gamma^2) = \tfrac{1}{2}(1 - b^2) u_\alpha u_\gamma.
\end{aligned}$$

□

Lemma 20.3.18 *Under the hypotheses of Proposition 20.3.11, the polynomial $\delta_E(\alpha, \beta)$ is an invariant of Σ.*

Proof This is a direct consequence of Lemmas 20.3.15 and 20.3.17. □

Proof of Theorem 20.1.32 Lemma 20.3.18 implies that $\delta_\Sigma(\alpha, \beta) = f_*^{-1}\left(\delta_E(\alpha, \beta)\right)$ is an invariant of Σ. □

Proof of Theorem 20.1.33 Since $\mathcal{D}_2(\Sigma)$, the polynomial P of Lemma 20.3.7, and $\delta_E(\alpha, \beta)$ are invariants of Σ, and since $P \neq 0$, Proposition 20.3.11 implies that $w_{SL}(\Sigma)$ is an invariant of Σ, too. □

Lemma 20.3.19 *With the notation of Lemma 20.3.16, we have*

$$w_\delta(a', b', c') - w_\delta(a, b, c) = \frac{1}{2} b(b^2 - 1).$$

Proof This follows from Lemmas 20.3.17 and 20.3.12. □

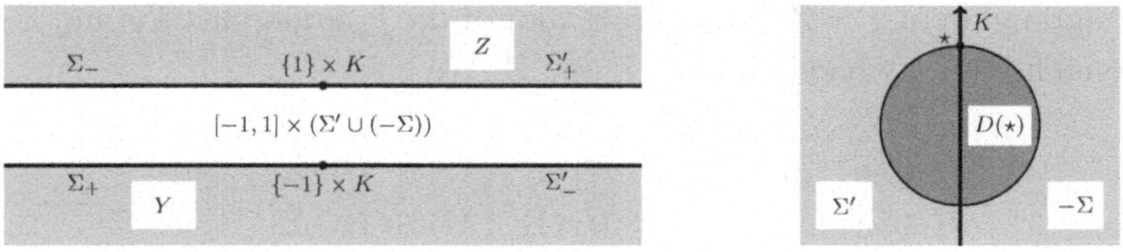

Fig. 20.8 The manifolds Y and Z, and the disk $D(\star)$ in $\left(\Sigma' \cup (-\Sigma)\right)$

20.4 Proof of Invariance of $W_s\left(\mathcal{D}_2(\Sigma)\right)$

20.4.1 Invariance of $W_s\left(\mathcal{D}_2(\Sigma)\right)$ under Cobordism

In this subsection, we prove that $W_s\left(\mathcal{D}_2(.)\right)$ is invariant under K-cobordism of genus one Seifert surfaces, as stated in Lemma 20.1.25.

Let Σ and Σ' be two null-homologous genus one Seifert surfaces of a knot K in a rational homology 3-sphere R. Assume $\Sigma \cap \Sigma' = K$. Also assume that $\Sigma' \cup_K (-\Sigma)$ is smooth, and consider a tubular neighborhood $[-1,1] \times \left(\Sigma' \cup_K (-\Sigma)\right)$ of $\Sigma' \cup_K (-\Sigma)$ in R. Set $\Sigma'_+ = \{1\} \times \Sigma'$, $\Sigma'_- = \{-1\} \times \Sigma'$, $(-\Sigma_+) = \{-1\} \times (-\Sigma)$, and $\Sigma_- = \{1\} \times (\Sigma)$. The 3-manifold $R \setminus \left(]-1,1[\times \left(\Sigma' \cup_K (-\Sigma)\right)\right)$ has two connected components Y and Z with $\partial Y = \Sigma'_- \cup (-\Sigma_+)$ and $\partial Z = \Sigma_- \cup (-\Sigma'_+)$. (This notation for Σ'_- and Σ_- is compatible with the previous notation except in a tubular neighborhood of K.)

The *Lagrangian* of a 3-manifold X with boundary is the kernel of the map induced by the inclusion from $H_1(\partial X; \mathbb{Q})$ to $H_1(X; \mathbb{Q})$.

Let $\star$ be a basepoint of K. Let $D(\star)$ be a disk of $\left(\Sigma' \cup_K (-\Sigma)\right)$ such that $\star$ is on $\partial D(\star)$ and the intersection $D(\star) \cap K$ is an interval, as in the right part of Fig. 20.8. Let $I_K \subset K$ be the interval $K \setminus (K \cap \mathring{D}(\star))$.

Define $Y * Z$ to be $Y \cup Z \cup ([-1,1] \times D(\star))$. Then $\partial(Y * Z)$ separates R into two genus 4 rational homology handlebodies $Y * Z$ and $E_{YZ} \cong [-1,1] \times \left(\Sigma' \cup_{I_K} (-\Sigma)\right)$. For an element u of $H_1(\Sigma)$, u_+ denotes the class of $\{-1\} \times u$ (the curve u is pushed in the direction of the positive normal to Σ, which is the negative normal to $(-\Sigma)$). Let $\mathcal{L}_\Sigma$ be the image of the injective map

$$\begin{array}{lcl} H_1(\Sigma; \mathbb{Q}) & \to & H_1(\partial(Y * Z); \mathbb{Q}) \\ u & \mapsto & [u_+] - [u_-], \end{array}$$

and let $\mathcal{L}_{\Sigma'}$ be the image of the similar map $(u' \mapsto [u'_+] - [u'_-])$ from $H_1(\Sigma'; \mathbb{Q})$ to $H_1(\partial(Y * Z); \mathbb{Q})$. Then the Lagrangian of E_{YZ} is the direct sum $\mathcal{L}_\Sigma \oplus \mathcal{L}_{\Sigma'}$.

The Lagrangian of $Y * Z$ is the direct sum of the Lagrangians $\mathcal{L}_Y$ and $\mathcal{L}_Z$ of the two rational homology handlebodies Y and Z. We have

$$\begin{aligned}
H_1(\partial Y;\mathbb{Q}) &= H_1(\Sigma'_-;\mathbb{Q}) \oplus H_1(\Sigma_+;\mathbb{Q}),\\
H_1(\partial Z;\mathbb{Q}) &= H_1(\Sigma_-;\mathbb{Q}) \oplus H_1(\Sigma'_+;\mathbb{Q}),\\
H_1(\partial(Y * Z);\mathbb{Q}) &= H_1(\partial Y;\mathbb{Q}) \oplus H_1(\partial Z;\mathbb{Q})\\
H_1(Y;\mathbb{Q}) &= H_1(\partial Y;\mathbb{Q})/\mathcal{L}_Y,\\
H_1(Z;\mathbb{Q}) &= H_1(\partial Z;\mathbb{Q})/\mathcal{L}_Z,
\end{aligned}$$

and

$$H_1(Y * Z;\mathbb{Q}) = H_1(Y;\mathbb{Q}) \oplus H_1(Z;\mathbb{Q}) = H_1\big(\partial(Y * Z);\mathbb{Q}\big)/(\mathcal{L}_Y \oplus \mathcal{L}_Z).$$

Construct the link exterior $Y[K]$ (resp. $Z[K]$) by attaching a 2-handle along $\{-1\} \times K$ to Y (resp. $\{1\} \times K$ to Z). Choose a symplectic basis (α, β) of $H_1(\Sigma)$ and a symplectic basis (α', β') of $H_1(\Sigma')$. Normalize the Alexander forms $\mathcal{A}_Y$ and $\mathcal{A}_Z$ of Y and Z, with respect to these bases, so that the even normalized Reidemester torsions of $Y[K]$ and $Z[K]$ are

$$\mathcal{D}(Y[K]) = \mathcal{A}_Y(\tilde{\alpha}_+ \wedge \tilde{\beta}_+) \in \Lambda_Y^{\mathbb{Q}} = \mathbb{Z}[H_1(Y;\mathbb{Q})]$$

and

$$\mathcal{D}(Z[K]) = \mathcal{A}_Z(\tilde{\alpha}'_+ \wedge \tilde{\beta}'_+) \in \Lambda_Z^{\mathbb{Q}} = \mathbb{Z}[H_1(Z;\mathbb{Q})],$$

respectively, as in Lemma 20.3.3. As shown in Remark 20.4.4 below, there is no canonical sign for the Reidemeister torsions of $Y[K]$ and $Z[K]$. We fix arbitrary signs so that the above equalities hold.

In the definition of $Y * Z$ as $Y \cup Z \cup ([-1, 1] \times D(\star))$, shrink $[-1, 1] \times \partial D(\star)$ to $\partial D(\star)$ so that the basepoint $\star$ is in $Y \cap Z$. Note that $E = R \setminus (]1, 1[\times(\Sigma \setminus K))$ is obtained from $Y * Z$ by identifying Σ'_- with Σ'_+. The manifold $E' = R \setminus \big(]1, 1[\times(\Sigma' \setminus K)\big)$ is similarly obtained from $Y * Z$ by identifying Σ_- with Σ_+. Let $i_E \colon Y * Z \hookrightarrow E$ and $i_{E'} \colon Y * Z \hookrightarrow E'$ denote the corresponding inclusion maps (whose homotopy classes are well-defined). Then we have the following lemma.

Lemma 20.4.1 *We can normalize $\mathcal{A}_{Y*Z}$ so that it has the following properties.*

- *The Alexander form $\mathcal{A}_{Y*Z}$ is valued in $\Lambda_{Y*Z}^{\mathbb{Q}} = \mathbb{Z}[H_1(Y;\mathbb{Q}) \oplus H_1(Z;\mathbb{Q})]$, which contains $\Lambda_Y^{\mathbb{Q}}$ and $\Lambda_Z^{\mathbb{Q}}$.*
- *If u_Y, v_Y are two based curves of ∂Y, and if u_Z, v_Z are two based curves of ∂Z, then*

$$\mathcal{A}_{Y*Z}(\tilde{u}_Y \wedge \tilde{v}_Y \wedge \tilde{u}_Z \wedge \tilde{v}_Z) = \mathcal{A}_Y(\tilde{u}_Y \wedge \tilde{v}_Y)\mathcal{A}_Z(\tilde{u}_Z \wedge \tilde{v}_Z).$$

- *If u, v, and w are three based curves of ∂Y (resp. of ∂Z) in $\mathcal{H}_{Y*Z}$, then $\mathcal{A}_{Y*Z}(\tilde{u} \wedge \tilde{v} \wedge \tilde{w} \wedge .)$ is zero.*
- *There exists $\eta \in \{-1, 1\}$ such that*

$$\varepsilon\left(\mathcal{A}_{Y*Z}(\tilde{\alpha}'_+ - \tilde{\alpha}'_- \wedge \tilde{\beta}'_+ - \beta'_- \wedge \tilde{\alpha}_+ - \tilde{\alpha}_- \wedge \tilde{\beta}_+ - \beta_-)\right) = \eta |H_1(R)|.$$

Let $ev(u'_\alpha = u'_\beta = 0)$ (resp. $ev(u_\alpha = u_\beta = 0)$) be the map from $\Lambda^{\mathbb{Q}}_{Y*Z}$ to $\Lambda^{\mathbb{Q}}_E$ (resp. $\Lambda^{\mathbb{Q}}_{E'}$) induced by i_E (resp. by $i_{E'}$). Then

$$\mathcal{D}(\Sigma) = \eta ev(u'_\alpha = u'_\beta = 0)\left(\mathcal{A}_Y(\tilde{\alpha}_+ \wedge \tilde{\beta}_+)\mathcal{A}_Z(\tilde{\alpha}'_+ \wedge \tilde{\beta}'_+)\right)$$

and

$$\mathcal{D}(\Sigma') = \eta ev(u_\alpha = u_\beta = 0)\left(\mathcal{A}_Y(\tilde{\alpha}_+ \wedge \tilde{\beta}_+)\mathcal{A}_Z(\tilde{\alpha}'_+ \wedge \tilde{\beta}'_+)\right).$$

Proof The first four properties are easy to prove. The Alexander form $\mathcal{A}_{E'}$ of $E' = R\setminus(]-1, 1[\times (\Sigma' \setminus K))$ is normalized so that $\varepsilon(\mathcal{A}_{E'}(\tilde{\alpha}' - \tilde{\alpha}'_- \wedge \tilde{\beta}' - \tilde{\beta}'_-)) = |H_1(R)|$ as in Sect. 20.1.5. Since E' is obtained from $Y * Z$ by attaching two 2-handles along $u_\alpha = \alpha_+\alpha_-^{-1}$ and $u_\beta = \beta_-^{-1}\beta_+$, we have

$$\mathcal{A}_{E'} \doteq \eta \mathrm{ev}(u_\alpha = u_\beta = 0)\left(\mathcal{A}_{Y*Z}(\tilde{\alpha}_+ - \tilde{\alpha}_- \wedge \tilde{\beta}_+ - \tilde{\beta}_- \wedge .)\right)$$

where $\mathrm{ev}(u_\alpha = u_\beta = 0)$ sends the homology classes of u_α and u_β to zero. Thus, we get

$$\begin{aligned}\mathcal{D}(\Sigma') &\doteq \eta \mathrm{ev}(u_\alpha = u_\beta = 0)\left(\mathcal{A}_{Y*Z}(\tilde{\alpha}_+ - \tilde{\alpha}_- \wedge \tilde{\beta}_+ - \tilde{\beta}_- \wedge \tilde{\alpha}'_+ \wedge \tilde{\beta}'_+)\right)\\ &\doteq \eta \mathrm{ev}(u_\alpha = u_\beta = 0)\left(\mathcal{A}_{Y*Z}(\tilde{\alpha}_+ \wedge \tilde{\beta}_+ \wedge \tilde{\alpha}'_+ \wedge \tilde{\beta}'_+)\right)\\ &\doteq \eta \mathrm{ev}(u_\alpha = u_\beta = 0)\left(\mathcal{A}_Y(\tilde{\alpha}_+ \wedge \tilde{\beta}_+)\mathcal{A}_Z(\tilde{\alpha}'_+ \wedge \tilde{\beta}'_+)\right).\end{aligned}$$

Since $\mathcal{A}_Y(\tilde{\alpha}_+ \wedge \tilde{\beta}_+)\mathcal{A}_Z(\tilde{\alpha}'_+ \wedge \tilde{\beta}'_+)$ is even, we can replace $\doteq$ by $=$. We similarly get $\mathcal{D}(\Sigma) = \eta \mathrm{ev}(u'_\alpha = u'_\beta = 0)(\mathcal{A}_Y(\tilde{\alpha}_+ \wedge \tilde{\beta}_+)\mathcal{A}_Z(\tilde{\alpha}'_+ \wedge \tilde{\beta}'_+))$. □

Since $\mathcal{A}_Y(\tilde{\alpha}_+ \wedge \tilde{\beta}_+)$ and $\mathcal{A}_Z(\tilde{\alpha}'_+ \wedge \tilde{\beta}'_+)$ are both even, the following lemma allows us to conclude the proof of Lemma 20.1.25.

Lemma 20.4.2 *With the notation of Theorem 20.1.23 for W_s, for any element S of*

$$\left(H_1(\partial Y; \mathbb{Q}) \otimes_s H_1(\partial Y; \mathbb{Q})\right) \oplus \left(H_1(\partial Z; \mathbb{Q}) \otimes_s H_1(\partial Z; \mathbb{Q})\right),$$

we have

$$W_s(\Sigma')\big(\mathrm{ev}(u_\alpha = u_\beta = 0)(S)\big) = W_s(\Sigma)\big(\mathrm{ev}(u'_\alpha = u'_\beta = 0)(S)\big).$$

Proof It suffices to prove the lemma for an element $S = y \otimes y$ where $y \in H_1(\partial Y)$ or $y \in H_1(\partial Z)$. Let $y \in H_1(\partial Y) = H_1(\Sigma'_-) \oplus H_1(\Sigma_+)$. Since $H_1(\partial(Y * Z); \mathbb{Q}) = \mathcal{L}_Y \oplus \mathcal{L}_Z \oplus \mathcal{L}_\Sigma \oplus \mathcal{L}_{\Sigma'}$, there exists a unique $(u, u') \in H_1(\Sigma; \mathbb{Q}) \times H_1(\Sigma'; \mathbb{Q})$ such that

$$y = u_+ - u_- + u'_+ - u'_-$$

in $H_1(\partial(Y*Z); \mathbb{Q})/(\mathcal{L}_Y \oplus \mathcal{L}_Z)$. Since $i_{E*}(y) = u_+ - u_-$, we have $u = f^{-1}(i_{E*}(y))$, and

$$W_s(\Sigma)\left(\mathrm{ev}(u'_\alpha = u'_\beta = 0)(S) = i_{E*}(y) \otimes i_{E*}(y)\right) = 2lk(u_+, u_-).$$

We similarly get

$$W_s(\Sigma')\left(\mathrm{ev}(u_\alpha = u_\beta = 0)(S)\right) = 2lk(u'_+, u'_-).$$

Since $(u'_+ - u_-) \in \mathcal{L}_Z$, we have $lk(u'_+, u'_-) = lk(u_-, u'_-) = lk(u, u')$, and $lk(u_+, u_-) = lk(u_+, u'_+) = lk(u, u')$. For $y \in H_1(\partial Z; \mathbb{Q})$ such that $y = u_+ - u_- + u'_+ - u'_-$ in $H_1(Y * Z; \mathbb{Q})$, we similarly obtain

$$W_s(\Sigma)\Big(\mathrm{ev}(u'_\alpha = u'_\beta = 0)(y \otimes y)\Big) = W_s(\Sigma')\Big(\mathrm{ev}(u_\alpha = u_\beta = 0)(y \otimes y)\Big) = lk(u, u').$$

□

Lemma 20.1.25 is proved. □

Remark 20.4.3 The hypothesis that S belongs to $\left(\otimes_s^2 H_1(\partial Y; \mathbb{Q})\right) \oplus \left(\otimes_s^2 H_1(\partial Z; \mathbb{Q})\right)$ is necessary in Lemma 20.4.2. For example, for $S = (\alpha_+ - \alpha_-) \otimes (\alpha_+ - \alpha_-)$, $W_s(\Sigma')\left(\mathrm{ev}(u_\alpha = u_\beta = 0)(S)\right) = 0$, while $W_s(\Sigma)\left(\mathrm{ev}(u'_\alpha = u'_\beta = 0)(S)\right) = lk(\alpha_+, \alpha_-)$ is not zero, in general.

Remark 20.4.4 Fix a normalization $\mathcal{A}_Y$ of the Alexander form of Y. For two based curves u_1 and u_2 of Y, the integer $\varepsilon\left(\mathcal{A}_Y(\tilde{u}_1 \wedge \tilde{u}_2)\right)$ depends only of the homology classes of u_1 and u_2 in Y. It is equal to $\pm\left|H_1(Y)/\left(\mathbb{Z}[u_1] + \mathbb{Z}[u_2]\right)\right|$. We denote it by $\varepsilon\mathcal{A}_Y\left(u_1 \wedge u_2\right)$. Choose a symplectic basis (α, β) of $H_1(\Sigma)$ such that $\alpha \notin \mathcal{L}_Y$.

Let $\psi\colon \Sigma \to \Sigma'$ be a diffeomorphism. Then $R(Y, \psi) = Y/(\Sigma \sim_\psi \Sigma')$ is a $\mathbb{Q}$-sphere if and only if $\varepsilon\mathcal{A}_Y\left(\alpha_+ - \psi_*(\alpha)_- \wedge \beta_+ - \psi_*(\beta)_-\right) \neq 0$. Below, we omit the subscripts $+$ and $-$ because they are useless here.

We are about to prove that there exist diffeomorphisms $\psi_r\colon \Sigma \to \Sigma'$ and $\psi_s\colon \Sigma \to \Sigma'$ such that $\varepsilon\mathcal{A}_Y\big(\alpha - \psi_{r*}(\alpha) \wedge \beta - \psi_{r*}(\beta)\big)\varepsilon\mathcal{A}_Y\big(\alpha - \psi_{s*}(\alpha) \wedge \beta -$

$\psi_{s*}(\beta)\big) < 0$. So, such diffeomorphisms do not provide a canonical sign for the Reidemeister torsions of $Y[K]$.

Let (k_1, k_2, v_1, v_2) be a basis of $H_1(\partial Y, \mathbb{Q})$ such that (k_1, k_2) is a basis of $\mathcal{L}_Y$, and $\langle k_i, v_j\rangle_{\partial Y} = \delta_{ij}$. ($\delta_{ii} = 1$, and $\delta_{ij} = 0$ if $i \neq j$.) Then $\varepsilon\mathcal{A}_Y(v_1 \wedge v_2) \neq 0$, and, for any two curves u_1 and u_2 of ∂Y, we have

$$\varepsilon\mathcal{A}_Y(v_1 \wedge v_2)k_1 \wedge k_2 \wedge u_1 \wedge u_2 = \varepsilon\mathcal{A}_Y(u_1 \wedge u_2)k_1 \wedge k_2 \wedge v_1 \wedge v_2$$

in $\bigwedge^4 H_1(\partial Y, \mathbb{Q})$.

Define $\delta_Y(u_1 \wedge u_2) = \langle k_1, u_1\rangle_{\partial Y}\langle k_2, u_2\rangle_{\partial Y} - \langle k_1, u_2\rangle_{\partial Y}\langle k_2, u_1\rangle_{\partial Y}$. Then $\delta_Y(v_1 \wedge v_2) = 1$ and

$$\delta_Y(u_1 \wedge u_2) = \frac{\varepsilon\mathcal{A}_Y(u_1 \wedge u_2)}{\varepsilon\mathcal{A}_Y(v_1 \wedge v_2)}.$$

In particular, $R(Y, \psi)$ is a $\mathbb{Q}$-sphere if and only $\delta_Y\left(\alpha - \psi_*(\alpha) \wedge \beta - \psi_*(\beta)\right) \neq 0$.

If $\mathcal{L}_Y \cap H_1(\Sigma; \mathbb{Q}) \neq \{0\}$, then there exist a primitive element γ_Y of $H_1(\Sigma)$ such that $\mathcal{L}_Y \cap H_1(\Sigma; \mathbb{Q}) = \mathbb{Q}\gamma_Y$ and a primitive element γ'_Y of $H_1(\Sigma')$ such that $\mathcal{L}_Y \cap H_1(\Sigma'; \mathbb{Q}) = \mathbb{Q}\gamma'_Y$. Then we choose the above basis (k_1, k_2, v_1, v_2) so that $k_1 = \gamma_Y$ and $k_2 = \gamma'_Y$. So, we have

$$\delta_Y\Big(\alpha - \psi_*(\alpha) \wedge \beta - \psi_*(\beta)\Big) = \langle\gamma_Y, \alpha\rangle_\Sigma\langle\gamma'_Y, \psi_*(\beta)\rangle_{\Sigma'} - \langle\gamma_Y, \beta\rangle_\Sigma\langle\gamma'_Y, \psi_*(\alpha)\rangle_{\Sigma'}.$$

If $\mathcal{L}_Y \cap H_1(\Sigma; \mathbb{Q}) = \{0\}$, then there exists a symplectic isomorphism $\lambda_Y\colon H_1(\Sigma; \mathbb{Q}) \to H_1(\Sigma'; \mathbb{Q})$ such that $\mathcal{L}_Y = \mathbb{Q}(\alpha - \lambda_Y(\alpha)) \oplus \mathbb{Q}(\beta - \lambda_Y(\beta))$. Choose the above basis (k_1, k_2, v_1, v_2) so that $k_1 = \alpha - \lambda_Y(\alpha)$ and $k_2 = \beta - \lambda_Y(\beta)$. Thus, we have

$$\begin{aligned}\delta_Y\Big(\alpha - \psi_*(\alpha) \wedge \beta - \psi_*(\beta)\Big) &= \langle\lambda_Y(\alpha), \psi_*(\alpha)\rangle_{\Sigma'}\langle\lambda_Y(\beta), \psi_*(\beta)\rangle_{\Sigma'} \\ &\quad + \big(1 - \langle\lambda_Y(\alpha), \psi_*(\beta)\rangle_{\Sigma'}\big)\big(1 + \langle\lambda_Y(\beta), \psi_*(\alpha)\rangle_{\Sigma'}\big) \\ &= 2 - \langle\lambda_Y(\alpha), \psi_*(\beta)\rangle_{\Sigma'} + \langle\lambda_Y(\beta), \psi_*(\alpha)\rangle_{\Sigma'}.\end{aligned}$$

Choose a symplectic basis $(\psi_0(\alpha), \psi_0(\beta))$ of $H_1(\Sigma')$ such that $\langle\psi_0(\alpha), \gamma'_Y\rangle_{\Sigma'} \neq 0$ if $\mathcal{L}_Y \cap H_1(\Sigma; \mathbb{Q}) \neq \{0\}$, and $\langle\lambda_Y(\alpha), \psi_0(\alpha)\rangle_{\Sigma'} \neq 0$ if $\mathcal{L}_Y \cap H_1(\Sigma; \mathbb{Q}) = \{0\}$. This condition can be rephrased as $\psi_0(\alpha) \notin \mathcal{L}_Y \oplus \mathbb{Q}\alpha$.

Let ψ_k be the symplectic map that transforms (α, β) to $(\psi_k(\alpha) = \psi_0(\alpha), \psi_k(\beta) = \psi_0(\beta) + k\psi_0(\alpha))$. Such a map is induced by a diffeomorphism still denoted by ψ_k from Σ to Σ'. Thus, there exists $R \in \mathbb{N}$ such that for any $r > R$, we have

$$\delta_Y\Big(\alpha - \psi_r(\alpha) \wedge \beta - \psi_r(\beta)\Big)\delta_Y\Big(\alpha - \psi_{-r}(\alpha) \wedge \beta - \psi_{-r}(\beta)\Big) < 0.$$

We can simultaneously get the same results for Z, starting with a symplectic basis (α, β) such that $\alpha \notin \mathcal{L}_Y \cup \mathcal{L}_Z$ and a symplectic basis $(\psi_0(\alpha), \psi_0(\beta))$ of $H_1(\Sigma')$ such that $\psi_0(\alpha) \notin (\mathcal{L}_Z \oplus \mathbb{Q}\alpha) \cup (\mathcal{L}_Y \oplus \mathbb{Q}\alpha)$. In particular, there exists $R_2 \in \mathbb{N}$ such that for any integer r in $\mathbb{R}\setminus]-R_2, R_2[$, $R(Y, \psi_r) = Y/(\Sigma \sim_{\psi_r} \Sigma')$ and $R(Z, \psi_r) = Z/(\Sigma \sim_{\psi_r} \Sigma')$ are $\mathbb{Q}$-spheres.

20.4.2 From a Genus One Seifert Surface to Another One

Say that two genus one Seifert surfaces Σ and Σ' of K in a $\mathbb{Q}$-sphere R are *weakly K-cobordant* if there exists a sequence $(\Sigma_i)_{i=1,\dots,k}$ of genus one Seifert surfaces of K such that $\Sigma = \Sigma_1$, $\Sigma' = \Sigma_k$, and, for any $i = 1, \dots, k-1$, the surfaces Σ_i and Σ_{i+1} are K-cobordant. In this section, we prove that any two genus one Seifert surfaces of the same knot K in a $\mathbb{Q}$-sphere are weakly K-cobordant. This proves Lemma 20.1.26.

We start with two oriented genus one surfaces Σ and Σ' with boundary K in R. Since they induce the same framing of K, we can perform an isotopy of Σ so that $\Sigma \cap \Sigma'$ is the disjoint union of K and n circles embedded in the interiors of Σ and Σ', for some $n \in \mathbb{N}$.

We proceed by induction on $n = n(\Sigma, \Sigma')$. When $n = 0$, Σ and Σ' are K-cobordant. Otherwise, it suffices to prove that, for any pair (Σ, Σ') of surfaces as above, there exist genus one surfaces $\tilde{\Sigma}$ and $\tilde{\Sigma}'$, respectively weakly K-cobordant to Σ and Σ', such that $n(\tilde{\Sigma}, \tilde{\Sigma}') < n(\Sigma, \Sigma')$, to conclude by induction.

Let C be a connected component of $\Sigma \cap \Sigma' \setminus K$. It is oriented as in the beginning of Sect. 20.1.3. On Σ, the curve C can be of one of the three following types:

- (Σ, D): when $\pm C$ bounds a disk embedded in Σ,
- (Σ, K): when $\pm C$ cobounds an annulus ($[0, 1] \times S^1$) embedded in Σ with K
- (Σ, p): when C does not separate Σ.

Similarly, C is of one of the types (Σ', D) (when $\pm C$ bounds a disk embedded in Σ'), (Σ', K) or (Σ', p) on Σ'.

Lemma 20.4.5 *If $\Sigma \cap \Sigma' \setminus K$ contains a curve of type (Σ', D) or a curve of type (Σ, D), then Σ and Σ' are respectively K-cobordant to genus one surfaces $\tilde{\Sigma}$ and $\tilde{\Sigma}'$ such that $n(\tilde{\Sigma}, \tilde{\Sigma}') < n(\Sigma, \Sigma')$.*

Proof Assume that there is a curve C of type (Σ', D). Then there is such a curve such that the disk D_C bounded by C in $\pm\Sigma'$ does not contain another component of $\Sigma \cap \Sigma'$. We can assume that a product $[-1, 1] \times D_C$ meets Σ along $[-1, 1] \times \partial D_C$, as in Fig. 20.9, and we do. We have $\varepsilon[-1, 1] \times \partial D_C \subset \Sigma$, for some $\varepsilon \in \{-1, 1\}$. Let $\Sigma(D_C)$ be the surface obtained from Σ by replacing $\varepsilon[-1, 1] \times \partial D_C$ with $\varepsilon(\partial[-1, 1]) \times D_C$. Let $\Sigma(D_C, K)$ be the connected component of $\Sigma(D_C)$ that contains K.

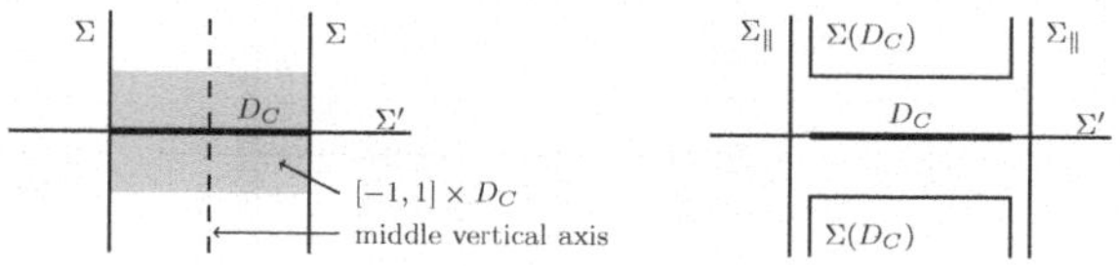

Fig. 20.9 Around the disk D_C (The 3–dimensional picture is obtained by rotating the picture around the middle vertical axis.)

The number $n\big(\Sigma(D_C, K), \Sigma'\big)$ of connected components of $\Sigma(D_C, K) \cap \Sigma'$ distinct from K satisfies $n\big(\Sigma(D_C, K), \Sigma'\big) < n(\Sigma, \Sigma')$. Let $\Sigma_\parallel$ be a surface obtained from Σ by pushing the interior of Σ in its normal direction that does not go towards D_C near C. Then $\Sigma(D_C, K) \cap \Sigma_\parallel = K$. Thus $\Sigma(D_C, K)$ is K-cobordant to Σ.

When C is of type (Σ, D), $\Sigma(D_C, K)$ is a genus one surface K-cobordant to Σ. Otherwise, $\Sigma(D_C, K)$ is a disk. This disk is K-cobordant to a genus one surface $\tilde{\Sigma}(D_C, K)$ obtained by connected sum with the boundary of a tiny solid torus near $\Sigma(D_C, K)$. We can perform such a connected sum without affecting the intersection with Σ' and $\Sigma_\parallel$. □

Lemma 20.4.6 *Assume that $\Sigma \cap \Sigma' \setminus K$ contains a curve C, which is of type (Σ, K) on Σ, and of type (Σ', K) on Σ'. Then Σ is isotopic to a surface $\tilde{\Sigma}$, and Σ' is isotopic to a surface $\tilde{\Sigma}'$ such that $n(\tilde{\Sigma}, \tilde{\Sigma}') < n(\Sigma, \Sigma')$.*

Proof Consider the surfaces $\tilde{\Sigma}$ and $\tilde{\Sigma}'$ such that $\tilde{\Sigma}$ (resp. $\tilde{\Sigma}'$) is the closure of the surface obtained from Σ (resp. from Σ') by removing the annulus between K and C in Σ from Σ (resp. in Σ' from Σ'). □

Lemma 20.4.7 *Assume that $\Sigma \cap \Sigma' \setminus K$ contains two curves C_1 and C_2 such that*

- *there exist an annulus A' in Σ' and $\varepsilon_1 \in \{-1, 1\}$ such that $\partial A' = \varepsilon_1(C_1 \cup C_2)$ and the interior of A' does not meet Σ,*
- *neither C_1 nor C_2 is of type (Σ, D).*

Then there exists a genus one surface $\tilde{\Sigma}$ K-cobordant to Σ such that $n(\tilde{\Sigma}, \Sigma') < n(\Sigma, \Sigma')$.

Proof Let $[-1, 1] \times A'$ be a collar of A' that intersects Σ' along $A' = \{0\} \times A'$, and $(\pm\Sigma)$ along $[-1, 1] \times \partial A'$. Our hypotheses guarantee that the coorientation of Σ points either towards A' on $[-1, 1] \times \partial A'$ or towards $\Sigma' \setminus A'$. In other words, there exists $\varepsilon_2 \in \{-1, 1\}$ such that the orientation of both components of $[-1, 1] \times \partial A'$ matches the orientation of $\varepsilon_2\Sigma$. Set $\check{\Sigma} = \Sigma \setminus (\varepsilon_2]-1, 1[\times \partial A')$. The surface $\check{\Sigma}$ is the disjoint union of

- an annulus and a disk with two holes, when C_1 and C_2 are both of type (Σ, p),
- a genus one surface with one boundary component and two annuli, when C_1 and C_2 are both of type (Σ, K),
- an annulus and a disk with two holes otherwise, when one of the curves C_1 and C_2 is of type (Σ, p) and the other one is of type (Σ, K).

Set $\Sigma(A') = \check{\Sigma} \cup (\varepsilon_2(\partial[-1, 1]) \times A')$. See Fig. 20.10. The surface $\Sigma(A')$ is the union of pieces of non-positive Euler characteristics glued along circles. It is an

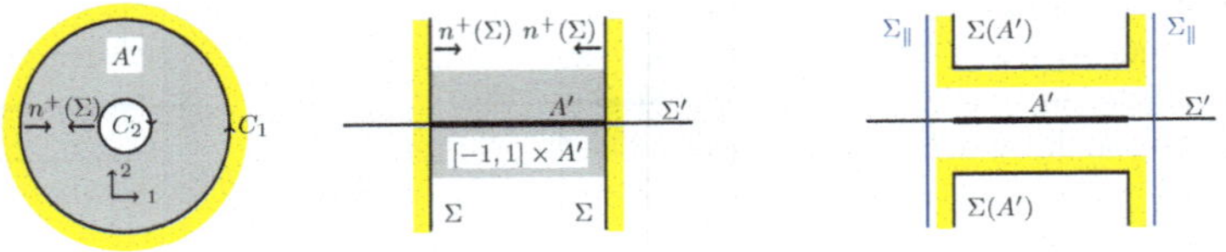

Fig. 20.10 Around the annulus A', when the positive normals to Σ point towards A' (For the second or third figure, the 3–dimensional picture is the product of this figure by a circle. The side of the negative normal to Σ or $\Sigma(A')$ is colored light.)

oriented surface whose boundary is K. Its Euler characteristic is (-1). Its connected components have non-positive Euler characteristics. Since the sum of the Euler characteristics of its components is -1, exactly one of its connected components has Euler characteristic -1, and it must be a genus one Seifert surface $\tilde{\Sigma}$ of K.

Furthermore, $\Sigma(A')$ is K-cobordant to a surface $\Sigma_{\|}$ obtained from Σ by pushing the interior of Σ in its normal direction that does not go towards A' near $\partial A'$. The component $\tilde{\Sigma}$ of $\Sigma(A')$ is K-cobordant to $\Sigma_{\|}$, too. □

Of course, we can exchange the roles of Σ and Σ' in the statement of Lemma 20.4.7 and get the lemma.

Lemma 20.4.8 *If $\Sigma \cap \Sigma' \setminus K$ contains two curves C_1 and C_2 such that*

- *there exist an annulus A in Σ and $\varepsilon \in \{-1, 1\}$ such that $\partial A = \varepsilon(C_1 \cup C_2)$ and the interior of A does not meet Σ',*
- *neither C_1 nor C_2 is of type (Σ', D).*

Then there exists a genus one surface $\tilde{\Sigma}'$ K-cobordant to Σ' such that $n(\Sigma, \tilde{\Sigma}') < n(\Sigma, \Sigma')$. □

The following lemma allows us to conclude the proof of Lemma 20.1.26 using the previous four lemmas.

Lemma 20.4.9 *Let Σ and Σ' be two genus one surfaces of a knot K in a $\mathbb{Q}$-sphere R such that $\Sigma \cap \Sigma'$ is the disjoint union of K and n circles which are embedded in the interiors of Σ and Σ', with $n \in \mathbb{N}$, and $\Sigma \cap \Sigma' \setminus K$ contains*

- *no curve of type (Σ, D) on Σ, no curve of type (Σ', D) on Σ',*
- *no curve which is both of type (Σ, K) and (Σ', K),*
- *no pair (C_1, C_2) of curves (as in Lemma 20.4.7) such that $\pm(C_1 \cup C_2)$ is the boundary of an annulus of Σ' whose interior does not meet Σ,*
- *no pair (C_1, C_2) of curves (as in Lemma 20.4.8) such that $\pm(C_1 \cup C_2)$ is the boundary of an annulus of Σ whose interior does not meet Σ'.*

Then $\Sigma \cap \Sigma' = K$.

Proof Since R is a $\mathbb{Q}$-sphere and since $(\Sigma' \cup -\Sigma)$ is an oriented 2-cycle, the algebraic intersection of any closed curve with $(\Sigma' \cup -\Sigma)$ is zero. Let s be a point near K on the negative side of $(\Sigma' \cup -\Sigma)$ (the side of the negative normal). For any

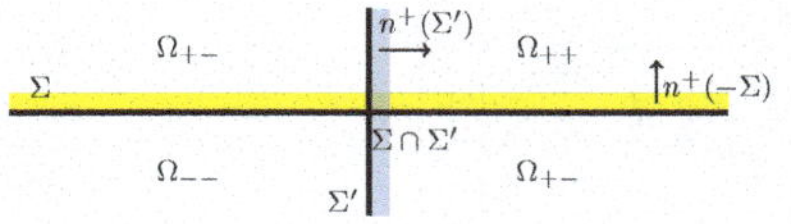

$$W(\Omega_{++}) = W(\Omega_{-+}) + 1$$
$$W(\Omega_{+-}) = W(\Omega_{-+})$$
$$W(\Omega_{--}) = W(\Omega_{-+}) - 1$$

Fig. 20.11 Weights around a component of $\Sigma \cap \Sigma' \setminus K$

point of $R \setminus (\Sigma' \cup (-\Sigma))$, define $W(p)$ as the algebraic intersection of a path from s to p transverse to $(\Sigma' \cup (-\Sigma))$ with $(\Sigma' \cup (-\Sigma))$. Then $R \setminus (\Sigma' \cup (-\Sigma))$ has several connected components Ω_i weighted by integers $W(\Omega_i)$, where $W(\Omega_i) = W(p \in \Omega_i)$. The difference $W(\Omega_j) - W(\Omega_i)$ is the algebraic intersection of a path from a point of Ω_i to a point of Ω_j with $\Sigma' \cup (-\Sigma)$ as in Fig. 20.11. The components adjacent to K are weighted by 0 and 1.

Weight the connected components of $\Sigma' \setminus (\Sigma \cap \Sigma')$ by the weight of the adjacent component of $R \setminus (\Sigma' \cup (-\Sigma))$ in the direction of the positive normal to Σ'. Similarly weight the connected components of $\Sigma \setminus (\Sigma \cap \Sigma')$ by the weight of the adjacent component of $R \setminus (\Sigma' \cup (-\Sigma))$ in the direction of the negative normal to Σ. So, the components adjacent to K get the weight one, and the weight on $\Sigma \setminus (\Sigma \cap \Sigma')$ or on $\Sigma' \setminus (\Sigma \cap \Sigma')$ varies by ± 1 across a component of $\Sigma \cap \Sigma'$. In particular, $\Sigma \cap \Sigma'$ contains an even number of (necessarily parallel) components of type (Σ', p). Half of them are oriented in one direction, and the other half are oriented in the opposite direction. So, if there are curves of $(\Sigma \cap \Sigma') \setminus K$ of type (Σ', p), there is an annulus A' between two such consecutive transverse intersection curves C_1 and C_2 of type (Σ', p) with opposite direction. This is forbidden by our assumptions. Transverse intersection curves of type (Σ, p) are similarly forbidden. □

So our reduction is finished, and Lemma 20.1.26 is proved. □

20.4.3 Curves on Genus One Seifert Surfaces

In this section, we prove that two homologous non-separating simple closed curves ξ and ξ' of a genus one oriented surface Σ with one boundary component are isotopic in Σ, as stated in Lemma 20.1.1.

Assume that these curves ξ and ξ' are transverse without loss of generality. Cut Σ along ξ to obtain a disk with two holes, as in Fig. 20.12. We view this disk with two holes as $D_e \setminus (\mathring{D}_i \sqcup \mathring{D}_0)$, for a disk D_e of $\mathbb{R}^2$ and two disjoint disks D_0 and D_i in the interior $\mathring{D}_e$ of D_e, where $\partial D_0 = -\partial\Sigma$, and ∂D_e and ∂D_i are respectively identified to two copies ξ_e and ξ_i of ξ. According to the Jordan–Schoenflies theorem, any simple closed curve of D_e bounds a disk in D_e.

If ξ' does not meet ξ, then the disk bounded by ξ' in D_e contains the disk D_i bounded by the internal copy ξ_i of ξ. So, the curve ξ' is parallel to one of the copies of ξ.

Fig. 20.12 The surface Σ cut along ξ, and a curve ξ' homologous to ξ, on the left

Otherwise, we show that we can always reduce the intersection $\xi \cap \xi'$ of ξ and ξ' by isotopy. Since the algebraic intersection number of ξ and ξ' is zero, if $\xi \cap \xi'$ is not empty, then there exists an arc p' of ξ' from ξ_i to itself on $D_e \setminus (\mathring{D}_i \sqcup \mathring{D}_0)$. Let p be the arc of $\pm\xi_i$ with boundary $-\partial p'$ such that the disk bounded by $\pm(p \cup p')$ in D_e does not contain D_i. If this disk bounded by $\pm(p \cup p')$ does not contain D_0, then there is a bigon (as in the right-hand side of Fig. 20.12) that can be eliminated, and we can reduce $\xi \cap \xi'$ by an isotopy. Otherwise, pick an arc q from the external copy ξ_e of ξ to itself. The arc q cannot meet the interior of the disk bounded by $\pm(p \cup p')$. Therefore, it cobounds a disk with an arc of ξ_e in $D_e \setminus (\mathring{D}_i \sqcup \mathring{D}_0)$. So, there is a least a bigon that can be suppressed by isotopy. □

References

1. M. Hedden, L. Watson, On the geography and botany of knot Floer homology. Selecta Math. (N.S.) **24**(2), 997–1037 (2018)
2. C. Lescop, *Global Surgery Formula for the Casson-Walker Invariant*. Annals of Mathematics Studies, vol. 140 (Princeton University Press, Princeton, 1996)
3. C. Lescop, A sum formula for the Casson-Walker invariant. Invent. Math. **133**(3), 613–681 (1998)
4. C. Lescop, Surgery formulae for finite type invariants of rational homology 3-spheres. Algebr. Geom. Topol. **9**(2), 979–1047 (2009). https://doi.org/10.2140/agt.2009.9.979
5. P. Ozsváth, Z. Szabó, Holomorphic disks and knot invariants. Adv. Math. **186**(1), 58–116 (2004)
6. N. Sato, Cobordisms of semiboundary links. Topol. Appl. **18**(2–3), 225–234 (1984)
7. H.F. Trotter, Non-invertible knots exist. Topology **2**, 275–280 (1963)
8. V.G. Turaev, Reidemeister torsion in knot theory. Uspekhi Mat. Nauk **41**(1) (247), 97–147, 240 (1986)

Chapter 21
On a Family of Hyperbolic Brunnian Links and Their Volumes

Dušan D. Repovš and Andrei Yu. Vesnin

Abstract An n-component link L is said to be *Brunnian* if it is non-trivial but every proper sublink of L is trivial. The simplest and best known example of a hyperbolic Brunnian link is the 3-component link known as "Borromean rings". For $n \geq 2$, we introduce an infinite family of n-component Brunnian links with positive integer parameters $Br(k_1, \dots, k_n)$ that generalize examples constructed by Debrunner in 1964. We are interested in hyperbolic invariants of 3-manifolds $S^3 \setminus Br(k_1, \dots, k_n)$ and we obtain upper bounds for their volumes. Our approach is based on Dehn fillings on cusped manifolds with volumes related to volumes of ideal right-angled hyperbolic antiprisms.

Keywords Hyperbolic Brunnian link · Adams move · Augmented link · Ideal right-angled antiprism

2020 Mathematics Subject Classification 57K10, 57K32, 52B10

D.R. was supported by the Slovenian Research and Innovation Agency (program P1-0292 and grants J1-4031, J1-4001, N1-0278, N1-0114, and N1-0083). A.V. was supported by the Ministry of Science and Higher Education of Russia under agreement no. 075-02-2024-1437.

D. D. Repovš (✉)
Faculty of Education, Faculty of Mathematics and Physics, University of Ljubljana & Institute of Mathematics, Physics and Mechanics, Ljubljana, Slovenia
e-mail: dusan.repovs@guest.arnes.si

A. Y. Vesnin
Sobolev Institute of Mathematics, Russian Academy of Sciences, Novosibirsk, Russia

Regional Mathematical Center, Tomsk State University, Tomsk, Russia
e-mail: vesnin@math.nsc.ru

L. Funar, A. Papadopoulos (eds.), *Essays on Topology*,
https://doi.org/10.1007/978-3-031-81414-3_21

21.1 Introduction

If a link L in S^3 is nontrivial, yet every proper sublink of L is trivial, we say that L is *Brunnian* (or that L has the *Brunnian property*). Links with this property were studied by Milnor [17] who called them *almost trivial* links. In 1961 Debrunner [9] renamed them as *Brunnian* links, in honor of Hermann Brunn whose early contributions [8] to knot theory also included examples of such links. In recent decades, Brunnian links have been under investigation from several points of views. In particular, we refer to Bai [5], Bai and Ma [6], Bai and Wang [7], and Kanenobu [13, 14] for hyperbolic and satellite properties of Brunnian links, Lei, Wu and Zhang [16] for intersecting subgroups of Brunnian link groups, and Habiro and Meilhan [12] for finite-type invariants and Milnor invariants of Brunnian links. We would also like to mention molecular Borromean rings considered by Wang and Stoddart [22] and a possible physical realization of higher-order Brunnian structures studied by Baas [4].

The simplest and best known example of a 3-component Brunnian link is the 3-component link 6^3_2 in Rolfsen's notations [18] which is also known as the "Borromean rings". We denote it by $\mathcal{B}$. It is well-known from Thurston [20, Chapter 3] that the link $\mathcal{B}$ is hyperbolic, the complement $S^3 \setminus \mathcal{B}$ can be decomposed into two copies of an ideal right-angled octahedron, and $\mathrm{vol}(S^3 \setminus \mathcal{B}) = 7.327724$, up to six digits.

We recall that a polyhedron in a hyperbolic 3-space $\mathbb{H}^3$ is said to be *ideal* if all of its vertices belong to $\partial\mathbb{H}^3$, and it is called *right-angled* if all of its dihedral angles are equal to $\pi/2$. By virtue of the Andreev theorem [3], every ideal right-angled hyperbolic polyhedron is determined by its 1-dimensional skeleton, up to an isometry of $\mathbb{H}^3$. An initial list of ideal right-angled hyperbolic polyhedra and their volumes was presented by Egorov and Vesnin [10, 21], and the upper volume bounds depending only of number of vertices were obtained by Alexandrov et al. [2].

In the present chapter, we introduce for every $n \geq 2$, an infinite family of n-component Brunnian links which generalize examples constructed in 1961 by Debrunner [9]. We are interested in hyperbolic structures on the complements of these links. By using decompositions of complements of fully augmented links into pairs of ideal right-angled polyhedra by a method from an appendix by Agol and Thurston in Lackenby [15], and the Dehn filling theorem, we provide upper bounds for volumes of such hyperbolic Brunnian links.

The chapter is organized as follows: In Sect. 21.2 we introduce a family of $(3n + 2)$-component links L_n, for every $n \geq 2$, and demonstrate that their complement $S^3 \setminus L_n$ can be decomposed into four ideal right-angled antiprisms A_{2n}. Then we apply Thurston's formula from [20, Chapter 6] for volumes of antiprisms to obtain a formula for $\mathrm{vol}(S^3 \setminus L_n)$ (see Theorem 21.2.1). Next, applying Adams moves to $2n$ vertical components of L_n, we construct links L'_n with $3n$ components such that $\mathrm{vol}(S^3 \setminus L'_n) = \mathrm{vol}(S^3 \setminus L_n)$. In Sect. 21.3, we construct by Dehn fillings on $2n$ components of L'_n, a family of n-component links $Br(k_1, \dots, k_n)$, depending on the filling parameters $k_1, \dots, k_n$ (see Theorems 21.3.1 and 21.3.4). In particular, when all k_i are equal to 1, we get the Brunnian links from Debrunner [9], whose hyperbolicity was established by Bai [5].

21.2 Hyperbolic Links and Ideal Right-Angled Antiprisms

Recall that a knot or a link $K \subset S^3$ is said to be *hyperbolic* if its complement $S^3 \setminus K$ admits a complete metric of constant curvature -1. Equivalently, the 3-manifold $S^3 \setminus K = \mathbb{H}^3/G$ is hyperbolic, where $\mathbb{H}^3$ is the hyperbolic 3-space and G is a discrete, torsion-free group of isometries, isomorphic to $\pi_1(S^3 \setminus K)$.

For $n \geq 2$, let us denote a link with $3n + 2$ components by L_n, where each component is a circle and the components are linked in the same manner as shown in Fig. 21.1 for the case $n = 3$.

By using the computer program SnapPy [19], one can see that the 8-component link L_2 is hyperbolic and has $\mathrm{vol}(S^3 \setminus L_2) = 24.092184$, up to six digits. To demonstrate hyperbolicity of L_n for arbitrary $n \geq 2$ and find the volume formula for $S^3 \setminus L_n$, we shall use the approach from Lackenby [15] (see also Futer and Purcell [11]). In the terminology of Lackenby [15], the link L_n is augmented, with $2n$ "vertical" components (colored in red in Fig. 21.1), and $S^3 \setminus L_n$ admits a decomposition into two ideal polyhedra P_n and P_n' with faces identified in pairs. The polyhedra P_n and P_n' are identical. By the construction each vertical component of L_n gives to a pair of triangles in P_n with a common vertex like a bowtie, as presented in Fig. 21.2 where common vertices are red.

At the last step, in order to obtain the polyhedron P_n, we shall compress black edges which connect two vertices of valence three, to obtain a new vertex of valence four. There are $4n$ such edges which give us $2n$ black vertices, as presented in Fig. 21.3 (to complete the construction, it is necessary to identify the vertices A

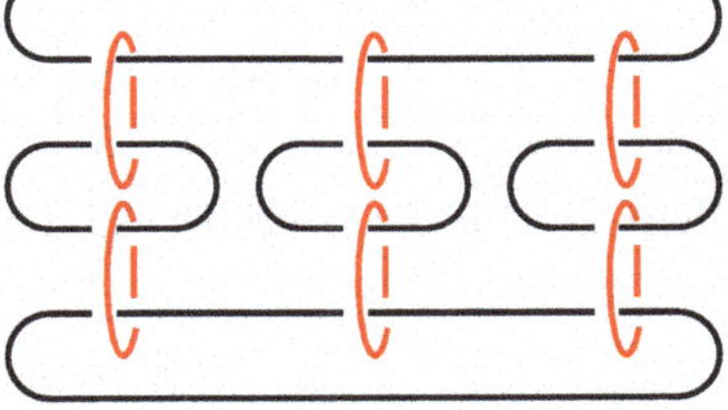

Fig. 21.1 Link L_n with $3n + 2$ components, case $n = 3$

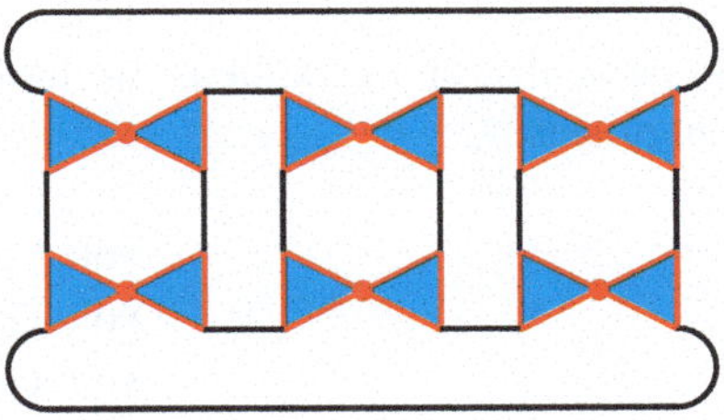

Fig. 21.2 Associating bowties to vertical components of L_n, case $n = 3$

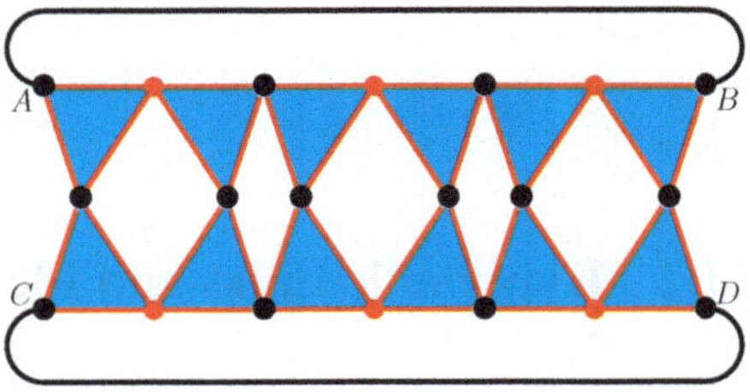

Fig. 21.3 1-Skeleton of polyhedron P_n, case $n = 3$

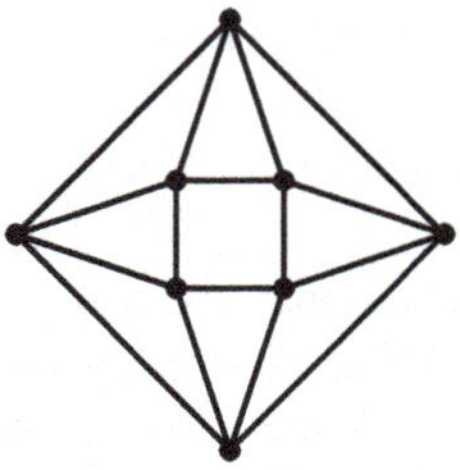

Fig. 21.4 1-Skeleton of antiprism A_4

and B, as well as the vertices C and D). As a result, P_n is an ideal right-angled polyhedron with $6n$ vertices, where $2n$ red vertices correspond to bowties and $4n$ black vertices appeared after compressing black edges. Moreover, P_n has two $2n$-gonal faces as its top and bottom, $2n$ triangles incident to the top, $2n$ triangles incident to the bottom, and $4n$ quadrilaterals on the middle level. By cutting P_n along the middle line passing through quadrilaterals, we shall see that P_n can be decomposed into two identical ideal right-angled $2n$-gonal antiprisms A_{2n}. Recall that A_n has $2n + 2$ faces, where two n-gonal faces can be considered as the top and the bottom, and $2n$ triangular faces on the lateral surface, see Fig. 21.4 for the antiprism A_4.

Volumes of right-angled antiprisms A_n, $n \geq 3$, are given by the following formula obtained by Thurston [20, Example 6.8.7], where an antiprism was named a *drum with triangles*,

$$\mathrm{vol}(A_n) = 2n\left[\Lambda\left(\frac{\pi}{4} + \frac{\pi}{2n}\right) + \Lambda\left(\frac{\pi}{4} - \frac{\pi}{2n}\right)\right]. \qquad (21.1)$$

Here, $\Lambda(\theta)$ is the Lobachevsky function defined in [20, Chapter 7] as

$$\Lambda(\theta) = -\int_0^{\theta} \log|2\sin(t)|dt.$$

In particular, up to six digits, we have the following volumes of ideal right-angled antiprisms:

$$\begin{aligned}
\mathrm{vol}(A_3) &= 3.663863,\\
\mathrm{vol}(A_4) &= 6.023046,\\
\mathrm{vol}(A_5) &= 8.137885,\\
\mathrm{vol}(A_6) &= 10.149416.
\end{aligned}$$

Fig. 21.5 Replacing a part of the projection in Theorem 21.2.2

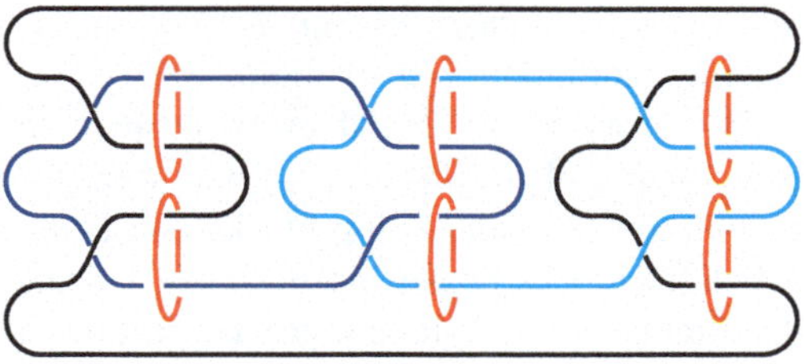

Fig. 21.6 Link L'_n with $3n$ components, case $n = 3$

Theorem 21.2.1 *For every $n \geq 2$, the following formula holds:*

$$\mathrm{vol}(S^3 \setminus L_n) = 16n \left[\Lambda\left(\frac{\pi}{4} + \frac{\pi}{4n}\right) + \Lambda\left(\frac{\pi}{4} - \frac{\pi}{4n}\right)\right].$$

Proof Indeed, from the above considerations we obtain

$$\mathrm{vol}(S^3 \setminus L_n) = 2\,\mathrm{vol}(P_n) = 4\,\mathrm{vol}(A_{2n}),$$

so the assertion follows by formula (21.1).

The following was proved by Adams [1, Corollary 5.1].

Theorem 21.2.2 (See [1]) *Let J be a link in S^3 such that $S^3 \setminus J$ is hyperbolic and J has a projection for which some part appears as in Fig. 21.5a. Let J' be the link obtained by replacing that part by the projection of J appearing in Fig. 21.5a with the one appearing in Fig. 21.5b. Then $S^3 \setminus J'$ is hyperbolic with the same volume as $S^3 \setminus J$.*

We shall call the replacement presented in Fig. 21.5 and its inverse the *Adams moves*. Let us apply Adams moves to each of $2n$ parts of related vertical components of the hyperbolic link L_n. The resulting link L'_n with $3n$ components is depicted in Fig. 21.6 for the case $n = 3$.

Theorems 21.2.1 and 21.2.2 now yield the following result.

Corollary 21.2.3 *For every $n \geq 2$, the following formula holds:*

$$\mathrm{vol}(S^3 \setminus L'_n) = 16n \left[\Lambda\left(\frac{\pi}{4} + \frac{\pi}{4n}\right) + \Lambda\left(\frac{\pi}{4} - \frac{\pi}{4n}\right)\right]. \tag{21.2}$$

21.3 A Family of Hyperbolic Brunnian Links

Suppose that $\overline{M}$ is a compact orientable 3-manifold with ∂M a collection of tori, and that the interior $M \subset \overline{M}$ admits a complete hyperbolic structure. Then M is referred to as a *cusped manifold*. For any $n \geq 2$, the complement $S^3 \setminus L'_n$ is a cusped manifold by construction.

For every $n \geq 2$ and every positive integers $k_1, \ldots, k_n$, we define an n-component link $Br(k_1, \ldots, k_n)$ with a diagram having $2n$ blocks of twist regions as follows: Consider n top blocks consisting of $2k_i + 1$ positive half-twists and n bottom blocks consisting of $2k_i + 1$ negative half-twists for $i = 1, \ldots, n$. A diagram of $Br(1, 2, 1)$ is presented in Fig. 21.7. By Rolfsen [18], $Br(k_1, \ldots, k_n)$ can be considered as a result of Dehn fillings with slopes $1/k_i$ and $-1/k_i$ for $i = 1, \ldots, n$ on $2n$ cusps corresponding to vertical components of $S^3 \setminus L'_n$ in Fig. 21.6. Thus, in total we get $2n$ twist regions with $2k_i + 1$ positive and negative half-twists, $i = 1, \ldots, n$.

If all k_i are equal to some k, then $Br(k, k, \ldots, k)$ admits a cyclic rotation symmetry of order n. The links $Br(1, 1, \ldots, 1)$ appeared already in Debrunner [9, Fig. 2] and were denoted by L_F. It was demonstrated in [9] that L_F is unsplittable, but any proper sublink of L_F is completely splittable. The 5-component link presented in Rolfsen [18, p. 69] is link $Br(1, 1, 1, 1, 1)$ in our notation. By the same argument as in [9], the following result holds.

Theorem 21.3.1 *For every n, $Br(k_1, \ldots, k_n)$ is an n-component Brunnian link.*

Proof It is easy to see that after removing one component of $Br(k_1, \ldots, k_n)$, all other components can be transformed to trivial ones by a sequence of simplifying underpassing moves from (a) to (b) shown in Fig. 21.8.

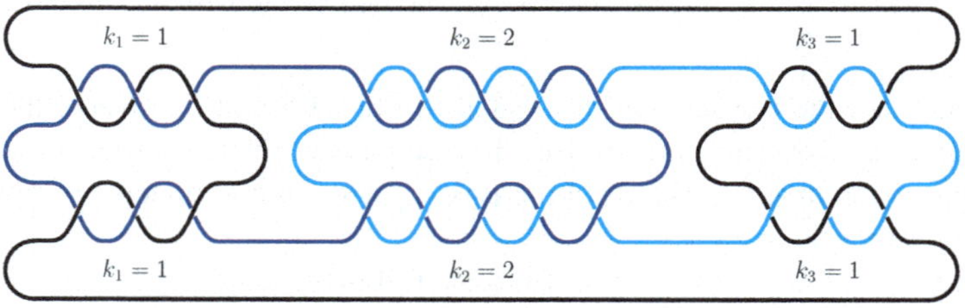

Fig. 21.7 Link $Br(1, 2, 1)$

Fig. 21.8 The simplifying underpassing move

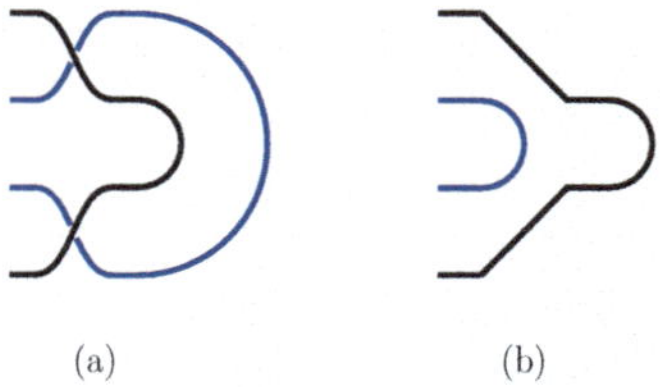

A practical method to check the hyperbolicity of Brunnian links was presented by Bai [5] who proved that for $n \geq 2$ with $k_i = 1, i = 1, \ldots, n$, the link $Br(k_1, \ldots, k_n)$ is hyperbolic [5, Theorem 1.3]. The smallest one is a 2-component link $Br(1, 1)$ with 12 crossings and $\mathrm{vol}(S^3 \setminus Br(1, 1)) = 12.528922$, up to six digits. Moreover, it can be recognized by using SnapPy [19] that $Br(1, 1) = L12n1180$, a non-alternating link with 12 components. At the same time, for $k_i \geq 3$, twisted regions in the diagram of $Br(k_1, \ldots, k_n)$ have more than 6 crossings. Therefore, the following result from Futer and Purcell [11, Theorem 1.7] can be applied.

Theorem 21.3.2 (See [11]) *Let K be a link in S^3 with a prime, twist-reduced diagram $D(K)$. Suppose that every twist region of $D(K)$ contains at least 6 crossings and that each component of K passes through at least 7 twist regions (counted with their multiplicity). Then every non-trivial Dehn filling of all the components of K is hyperbolic.*

The relation between volumes of cusped manifold and hyperbolic manifolds obtained by Dehn filling is due to Gromov and Thurston [20, Theorem 6.5.6].

Theorem 21.3.3 (See [20]) *Suppose M is a complete hyperbolic manifold of finite volume and that $N \neq M$ is a complete hyperbolic manifold obtained topologically by replacing certain cusps of M by solid tori. Then* $\mathrm{vol}(N) < \mathrm{vol}(M)$.

Theorem 21.3.4 *For hyperbolic links $Br(k_1, \ldots, k_n)$, the following upper bound holds:*

$$\mathrm{vol}(S^3 \setminus B(k_1, \ldots, k_n)) < 16n \left[\Lambda\left(\frac{\pi}{4} + \frac{\pi}{4n}\right) + \Lambda\left(\frac{\pi}{4} - \frac{\pi}{4n}\right)\right].$$

Proof Since $Br(k_1, \ldots, k_n)$ can be obtained by Dehn filling on $2n$ cusps of the hyperbolic link L'_n, the result follows by Corollary 21.2.3 and Theorem 21.3.3.

Corollary 21.3.5 *For every $n \geq 2$, the value $\beta_n = 16n\left[\Lambda\left(\frac{\pi}{4} + \frac{\pi}{4n}\right) + \Lambda\left(\frac{\pi}{4} - \frac{\pi}{4n}\right)\right]$ is the limit point for volumes of hyperbolic Brunnian links with n components.*

We conclude the chapter by the following open problems concerning hyperbolic Brunnian links.

Problem 21.3.1 What is the smallest volume hyperbolic Brunnian link with n components?

It is well known that the link "Borromean rings" is arithmetic [20, Chapter 7].

Problem 21.3.2 Which Brunnian links are arithmetic?

Also, the following interesting problem was formulated by Bai and Ma [6, Problem 7.0.8].

Problem 21.3.3 (See [6]) Let $B(n)$ be the number of Brunnian links with n or fewer crossings, $B_h(n)$ the number of hyperbolic Brunnian links with n or fewer crossings, and denote

$$\limsup_{n\to\infty} \frac{B_h(n)}{B(n)} = a, \qquad \liminf_{n\to\infty} \frac{B_h(n)}{B(n)} = b.$$

Is then $a = b$, $b = 0$, or $a < 1$?

Acknowledgments We thank the editors and referees for comments and suggestions.

References

1. C. Adams, Thrice-punctured spheres in hyperbolic 3-manifolds. Trans. Am. Math. Soc. **287**, 645–656 (1985)
2. S. Alexandrov, N. Bogachev, A. Vesnin, A. Egorov, On volumes of hyperbolic right-angled polyhedra. Sbornik: Mathematics **214**(2), 148–165 (2023). https://doi.org/10.4213/sm9740e. Preprint version is available at https://arxiv.org/abs/2111.08789
3. E.M. Andreev, On convex polyhedra of finite volume in Lobachevskii space. Math. USSR-Sbornik **12**(2), 255–259 (1970)
4. N.A. Baas, New states of matter suggested by new topological structures. Int. J. Gen. Syst. **42**(2), 137–169 (2013). https://doi.org/10.1080/03081079.2012.728403. Preprint version is available at https://arxiv.org/abs/1012.2698
5. S. Bai, *Hyperbolic Brunnian Links*. Preprint available at https://arxiv.org/abs/2104.12637
6. S. Bai, J. Ma, Satellite constructions and geometric classification of Brunnian links. J. Knot Theory Ramif. **30**(10) (2021), art. no. 2140005. https://doi.org/10.1142/S0218216521400058. Preprint version is available at https://arxiv.org/abs/1906.01253
7. S. Bai, W. Wang, New criteria and constructions of Brunnian links. J. Knot Theory Ramif. **29**(13) (2020), art. no. 2043008. https://doi.org/10.1142/S0218216520430087. Preprint version is available at https://arxiv.org/abs/2006.10290
8. H. Brunn, *Über Verkettung*, Sitzungsberichte der Bayerischen Akademie der Wissenschaften. Math.-Naturwiss. Klasse **22**, 77–99 (1892)
9. H. Debrunner, Links of Brunnian type. Duke Math. J. **28**, 17–23 (1961)
10. A. Egorov, A. Vesnin, Volume estimates for right-angled hyperbolic polyhedra. Rendiconti dell'Istituto di Matematica dell'Università di Trieste **52**, 565–576 (2020). https://doi.org/10.13137/2464-8728/30958. Preprint version is available at https://arxiv.org/abs/2010.11147
11. D. Futer, J.S. Purcell, Links with no exceptional surgeries. Commentarii Mathematici Helvetici **82**, 629–664 (2007). https://doi.org/10.4171/CMH/105
12. K. Habiro, J.-B. Meilhan, Finite type invariants and Milnor invariants for Brunnian links. Int. J. Math. **19**(6), 747–766 (2008). https://doi.org/10.1142/S0129167X08004820
13. T. Kanenobu, Satellite links with Brunnian properties. Arch. Math. **44**, 369–372 (1984)
14. T. Kanenobu, Hyperbolic links with Brunnian properties. J. Math. Soc. Jpn. **38**(2), 295–308 (1986)
15. M. Lackenby, The volume of hyperbolic alternating link complements. With an appendix by I. Agol and D. Thurston. Proc. Lond. Math. Soc. **88**, 204–224 (2004). https://doi.org/10.1112/S0024611503014291
16. F. Lei, J. Wu, Y. Zhang, On intersecting subgroups of Brunnian link groups. Algebr. Geom. Topol. **16**, 1043–1061 (2016). https://doi.org/10.2140/agt.2016.16.1043

17. J. Milnor, Link groups. Ann. Math. **59**, 177–195 (1954)
18. D. Rolfsen, *Knots and Links*. Corrected reprint of the 1976 original, Mathematics Lecture Series, vol. 7 (Publish or Perish, Houston, 1990)
19. SnapPy, Version 3.1 (May 2023). https://snappy.math.uic.edu
20. W.P. Thurston, *The Geometry and Topology of Three-Manifolds*. Collected Works of William P. Thurston with Commentary, vol. IV. With a preface by S.P. Kerckhoff, Edited by B. Farb, D. Gabai, S.P. Kerckhoff (American Mathematical Society, Providence, 2022). ISBN: 9781470463915; 9781470451646
21. A. Vesnin, A. Egorov, Ideal right-angled polyhedra in Lobachevsky space. Chebyshevskii Sbornik **21**(2), 65–83 (2020) (in Russian). https://doi.org/10.22405/2226-8383-2020-21-2-65-83. English preprint version is available at https://arxiv.org/abs/1909.11523
22. Y. Wang, J.F. Stoddart, Molecular Borromean rings: from controlled construction to potential applications. Chem **3**(1), 17–18 (2017). https://doi.org/10.1016/j.chempr.2017.06.009

Chapter 22
"All the Resources of Pure Mathematics": Mathematical Physics, and Mathematics Against Physics

Arkady Plotnitsky

> The most powerful method of advance that can be suggested at present is to employ all the resources of pure mathematics in attempts to perfect and generalise the mathematical formalism that forms the existing basis of theoretical physics, and after each success in this direction, to try to interpret the new mathematical features in terms of physical entities. —Paul A. M. Dirac [22, p. 1]

> What we will eventually need [is] a new sort of quantum topology, or something like it, not yet available to us. —Valentin Poénaru [62, p. 702]

Abstract This chapter argues for an asymmetry in the reciprocal impact of modern mathematics and modern physics on each other, an impact commonly seen more symmetrically. By modern mathematics I refer to mathematics that, from its emergence around 1800 to our own time, was defined, as abstract mathematics, by separating, abstracting, itself from natural objects and thus from physics, as well as from the objects of daily thinking. Modern physics, which emerged two centuries earlier, along with what we call modernity itself, and has continued to our own time as well, is a mathematical-experimental science, with mathematics defining this conjunction. As such it has not and could have not separated itself from mathematics. Indeed, as this chapter argues, twentieth-century physics, especially relativity and quantum theory, was able to use modern mathematical theories born from the separation of modern mathematics from physics. Reciprocally, throughout its history, beginning with the invention of calculus, modern physics has had a major influence on mathematics, including modern mathematics and, with relativity and quantum theory, on the twentieth- and twenty-first century of mathematics. This chapter argues for an asymmetry in this reciprocal impact, in favor of the impact

A. Plotnitsky (✉)
Literature, Theory, and Cultural Studies Program, Philosophy and Literature Program, Purdue University, West Lafayette, IN, USA
e-mail: plotnits@purdue.edu

L. Funar, A. Papadopoulos (eds.), *Essays on Topology*,
https://doi.org/10.1007/978-3-031-81414-3_22

of modern mathematics on modern physics, while this mutual impact is commonly seen more symmetrically, and sometimes by way of a reverse asymmetry, in favor of theoretical physics in helping to resolve some outstanding problems of pure mathematics. This chapter will challenge and qualify the latter view by arguing that theoretical physics cannot do so short of becoming mathematics, by, as modern mathematics itself has done, separating itself from physics. This chapter sees all modern physics as defined primarily by the invention of new theories, comprised of mathematized conceptual conglomerates, capable of predicting and, in some cases, representing the physical reality responsible for the phenomena considered, which are part of this reality. The invention of new mathematics, at least new for physics (this mathematics can be borrowed from already existing mathematics), is, the chapter argues, what has most fundamentally defined modern physics from its emergence until our own time.

Keywords Four-dimensional topology · Modern physics · Modern mathematics · Quantum field theory (QFT) · Quantum mechanics (QM) · Yang–Mills theory

Subject Classification 00A30, 01A20, 01A60, 01A65, 81–03, 81P05

22.1 Introduction

This chapter argues for an asymmetry in the reciprocal impact of modern mathematics and modern physics on each other, an impact commonly seen more symmetrically. By modern mathematics I refer to mathematics that, from its emergence around 1800 to our own time, was defined by separating, abstracting, itself from natural objects and thus from physics, as well as from the objects of daily thinking. Modern physics, which emerged two centuries earlier, along with what we call modernity itself, and has continued to our own time as well, is a mathematical-experimental science, with mathematics defining this conjunction. As such it has not, and could have not, separated itself from mathematics. Indeed, as this chapter argues, twentieth-century physics, especially relativity and quantum theory, was able to use modern mathematical theories born from the separation of modern mathematics from physics. Reciprocally, throughout its history, beginning with the invention of calculus, modern physics has had a major influence on mathematics, including modern mathematics and, with relativity and quantum theory (QT), on the twentieth- and twenty-first century of mathematics. This chapter argues for an asymmetry in this reciprocal impact, in favor of the impact of modern mathematics on modern physics, while this mutual impact is commonly seen more symmetrically, and sometimes even by way of a reverse asymmetry, in favor of theoretical physics in helping to resolve some outstanding problems of pure mathematics. This chapter will challenge and qualify the latter view by arguing that theoretical physics cannot do so short of becoming mathematics, by, as modern mathematics has done, separating itself from physics. In other words, while such a solution can be found by a theoretical physicist, it still must be mathematical. Conversely, a physicist or a mathematician can solve a physical problem by a mathematical means, not only by inventing new mathematical concepts from physics but also by using purely mathematical calculations, manipulations of symbols, and proofs.

This chapter will, thus, be concerned with the undeniable significance and "effectiveness" of mathematics in physics that Eugene Wigner famously saw as "unreasonable" [81]. It would be impossible to consider modern physics otherwise given its mathematical character. The present perspective on this subject (intimated in Plotnitsky [56]) is grounded in the abstract nature of this mathematics and in a radically new way, not considered by Wigner, this mathematics works in QT, the physical theory with which this chapter will be primarily concerned. Wigner, too, made his main contributions in QT, especially in his work on symmetry and group theory in quantum mechanics (QM) and quantum field theory (QFT). Ultimately, however, this chapter is about mathematics as a form of thought, still, beyond mathematics itself, best exemplified by theoretical physics.

In the present view, then, from its introduction on, modern physics is defined by the invention of new theories, comprised of mathematized conceptual conglomerates capable of predicting and, in some cases, representing the physical reality responsible for the phenomena considered. This invention, it follows, also required the invention of new mathematics, at least new for physics (this mathematics can be borrowed from already existing mathematics), thus, relegating philosophy or *philosophical* physics, which grounded the preceding history of physics from and, arguably, before Aristotle on, to the margins of *mathematical* physics. That does not mean that these margins are not relevant, because they may always invade the center, especially at the time of crises, such as that leading to the rise of QT and then QFT, which went through several crises of its own. For one thing, we don't know why this mathematics works, which is the main reason for Wigner's appeal to the unreasonableness of its effectiveness in physics. More significantly for this chapter, the new epistemology that emerged with QT is also part of a new philosophy of reality and probability (as all QT, but not all classical physics, is probabilistic).

The new mathematics of QT, capable of dealing with the experimental data of quantum physics, still had to be invented, however, as it was by Werner Heisenberg in the case of QM, grounded in this new philosophy. In this respect, Heisenberg, who sometimes referred to his QM as a new "calculus" (e.g., Heisenberg, [39, p. 109]), followed the project, just described, that defined modern physics from its emergence, with René Descartes and Galileo's use of mathematics and, more spectacularly, Sir Isaac Newton's invention of calculus, until our own time. At the same time, in contrast to classical physics or relativity, in QM, one also confronts a non-mathematizable remainder of the reality considered, a remainder that may be beyond any representation or even conception and is assumed to be in the interpretation adopted in this chapter, following Heisenberg and Niels Bohr. This remainder is also responsible for making the mathematical and epistemological character of probability, as irreducible to causality, in QT, different from the character of probability, always in principle reducible to causality, in classical physics. (The concept of causality is explained below.) Nevertheless, the use of modern mathematics, as able to predict the probabilities of quantum experiments, was the most decisive mathematical aspect of QM and QFT.

This is why, while my argument for the asymmetrical primacy of mathematics in the relationships between mathematics and physics can be made for an earlier, even the earliest known, history of these relationships, this chapter is primarily concerned with the project of *modern mathematics* and its relation to *modern physics*. As indicated above, these two designations commonly refer, as they will be here, to different historical periods. Modern mathematics, along with its avatars such as pure or abstract mathematics, refers to the mathematics that emerged around 1800 and has consolidated around 1900, extending to our own time. This mathematics was shaped and eventually defined by its divorce (a) from a *representation* of natural objects and, thus, from modern physics, uniformly defined (until QT) by mathematically idealized representations of nature and predictions enabled by these representations, and (b) from our general phenomenal intuition, no longer able to comprehend the objects of modern mathematics fully and even partially.

By contrast, modern physics refers to the rise of physics as a mathematical-experimental science, at the time of modernity itself in, roughly, the sixteenth century. This designation is fitting because modern physics was part of and shaped the rise of modernity itself, as defined by what is known as the scientific revolution, which also helped to shape the technological revolution, sometimes viewed more narrowly as the industrial revolution.[1] The rise of new mathematics—algebra, analytic geometry, and calculus—at the time is rarely referred to as a mathematical revolution, although this rise is often viewed as part of the scientific revolution. In fact, some of this mathematics, especially algebra, was modern because it was not concerned with the description of nature and thus physics.

In reflecting on the rise of modern physics with Descartes, Galileo, and others, Martin Heidegger observed: "Modern science is experimental because of its mathematical project" [38]. This is so not only or even primarily because of the role of measurement vs. Aristotle's physics as a physics of qualitative observation, important as this role is. Ancient Greek geometry, geo-*metry*, was also a science [*episteme*] of physical space, such as that of surfaces of the ground, and measurements in space. This view was restored to geometry, by a reversal, by Albert Einstein's general relativity (GR), which made the physics of gravity a geometry, by using modern mathematics, Riemannian geometry. By that time (1915), modern geometry was separate from physics, a separation amplified by the role of topology, a modern discipline, although it has an earlier genealogy, in particular, in the work of Leonhard Euler. Euler was one of the key figures, along with Jean le Rond D'Alembert and Pierre Simon de Laplace, in the history of the relationships between

[1] The emergence of modernity was shaped by several revolutions or, as they took a while and have involved continuous developments, transformations, such as political-economic (the rise of capitalism), the advent of democracy (capitalist democracy), or the rise of secular humanism and related literature and art. These revolutions, especially scientific and technological-industrial ones, were interrelated and influenced each other.

mathematics and physics in the eighteenth century, before the emergence of modern mathematics defined by its separation from physics. As Euler's work on topology indicates, the situation is more complex, in both directions, also given cases such as that of Karl Friedrich Gauss and Bernhard Riemann in the nineteenth century, and even some twentieth-century cases, such as that of David Hilbert and Hermann Weyl. I shall further comment on these complexities below. For the moment, the most essential reason for Heidegger's claim is that from its emergence on modern physics—classical physics, relativity, special (SR) and general (GR), and QT, the main types of fundamental theories (theories dealing with the ultimate constitution of nature) now, with several theories comprising each—uses mathematics to relate to and especially to predict the quantitative data found in observed phenomena.

On the other hand, René Thom would argue that this relation is already found in Aristotle, insofar as one sees his physics as in effect qualitative topology [52, 53, 72]. This argument would imply that, contrary to the view expressed by Heidegger (still valid for modern physics but missing topology), all physics that we know, that of Aristotle included, is experimental because of its mathematical project. The main difference would, then, be that, unlike Aristotle's qualitative topology of physics, the mathematics used in modern physics is quantitative, or rather *also* quantitative, which is necessary if physics needs to relate to measurements of the quantitative data observed in phenomena, rather than only to phenomena themselves. That, however, only confirms (a) that all known physics, from Aristotle on, is defined by mathematics, and (b) the more fundamentally grounding role of mathematics that underlies the role of measurement in modern physics. Measurement is mathematics, but mathematics, including in its use in physics, is much more than measurement.

As the term "phenomenon," introduced by Immanuel Kant, implies, "phenomena" themselves and the way we experience them belong to thought, even if they may appear in thought as the result of the interaction between thought (or in the first place, material bodies, specifically, brains) and nature. Accordingly, as Kant argued, phenomena constructed by thought may not correspond to real objects in nature, designated by Kant as things-in-themselves or *noumena* [44]. Kant used the term understanding [*Verstand*] exclusively to refer to phenomena, as opposed to reason [*Vernunft*], which may reach things-in-themselves or *noumena*.[2]

However, in classical physics or relativity, SR (which deals with electromagnetism) or GR (which is a non-Newtonian theory of gravity), phenomena observed may be identified with the objects considered for all practical purposes, including in allowing the mathematized concepts of the theory, ideally, to represent and, by

[2] There are further complexities to Kant's view, and technically *noumena* and things-in-themselves are not the same (e.g., [70]), but these complexities may be put aside for the present purposes.

using this representation, predict the behavior of these objects. Kant realized this as well in the case of classical physics, even if he, arguably, saw this representation, by Euclidean geometry and Newtonian mechanics, as corresponding to the ultimate nature of physical reality. He allowed, however, that this claim or assumption may only be practically justified, which would be sufficient for modern physics to function objectively [44, p. 115], [59, pp. 57–58]. All modern physics deals with (mathematically) idealized models of physical reality, which are sufficient for predicting the behavior of the systems considered within the proper scopes of these models.

This type of identification, even for practical purposes, of the phenomena observed, *quantum phenomena*, with the objects considered, *quantum objects*, becomes difficult, if possible at all, in QT, and it is strictly precluded in the set of interpretations of quantum phenomena and QT primarily considered here, to be designated as the "reality without realism" (RWR) interpretations, following [59, 60].[3] By quantum phenomena I refer to observed physical phenomena in considering which one must take into account the Planck constant, h (a quantity of a very small magnitude $6.62607015 \times 10^{-34} Joule \times Herz^{-}1$) introduced by Max Planck in 1900 in inaugurating quantum physics.[4] In these interpretations, quantum phenomena, as manifested in what is observed in observational instruments, are irreducibly different from quantum objects, which by interacting with observational instruments, are responsible for the emergence of quantum phenomena.

The (RWR) interpretation adopted in this chapter takes this difference between objects and phenomena considered in QT beyond Kant. First, the ultimate constitution of the reality considered, responsible for quantum phenomena, is assumed in this interpretation to be not only beyond representation or knowledge but also beyond conception, beyond the reach of thought altogether, while Kant sees things-in-themselves as merely beyond representations or at most beyond knowledge [44, p. 115], [59, p. 57]. I shall refer to such interpretations as *strong* RWR interpretations, vs. *weak* ones, in which case, for example, in the interpretation assumed by Heisenberg in his discovery of QM, this reality is merely beyond representation or knowledge. It can be shown that Kant's epistemology does not quite amount to the epistemology of weak RWR interpretations either [59, pp. 57–58]. (Hereafter, unless qualified, by RWR interpretations, I shall refer to strong ones.) Secondly, while quantum objects are still assumed to be beyond conceptions in the present interpretation, in this interpretation (but not on all RWR interpretations), they are

[3] While they are commonly, including here, interpreted jointly, quantum phenomena could be interpreted independently of a QT, such as QM or QFT, dealing with them. By contrast, an interpretation of a given QT entails, however implicitly, an interpretation of quantum phenomena. I shall qualify this difference when necessary.

[4] This definition may require qualifications (e.g., [59, , pp. 37–38]). These qualifications are, however, not germane for this chapter. All quantum phenomena considered here involve h, which is essential to the ultimate constitution of nature, assuming that QM and QFT are correct, because h reflects the Planck scale, the ultimate scale of this constitution in the present-day understanding of fundamental physics.

no longer assumed to be entities existing independently of our interaction with nature and thus to be part of the ultimate constitution of the reality responsible for quantum phenomena. Instead, the concept of quantum object, either in general or in any specific kind (say, an electron or a photon), is only applicable at the time of observation. Only the ultimate nature of reality responsible for observed quantum phenomena, and it follows, at the time of observation, for quantum objects, is assumed to exist independently, without us being able to represent or even conceive of it. As explained below, the main reason for this view has to do with high-energy quantum regimes. In most RWR interpretations, including that of Bohr, quantum objects belong to this independent reality.

The conception of RWR presupposes more general conceptions of reality and existence, assumed here to be primitive concepts and not given analytical definitions. By "reality" I refer to that which is assumed to exist, without making any claims concerning the *character* of this existence, claims that define realism or ontology.[5] I understand existence as a capacity to have effects on the world or our interactions with the world, effects that manifest another form of reality, viewed here as "actual," in the case of quantum physics, manifested as phenomena observed in observational instruments. The assumption that anything exists is always inferred from such effects. The concept of RWR may apply to both the ultimate reality responsible for quantum phenomena and to the ultimate (mental) reality responsible for thinking, including in mathematics [60]. It follows that, in RWR interpretations, the mathematical formalism of QT does not represent any objects in nature or the observed phenomena, described by classical physics, with the addition of SR in high-energy quantum regimens. This formalism only relates to observed phenomena by predicting, in general probabilistically, the data observed in these phenomena, which makes this formalism strictly mental, just as is mathematics itself. Mathematics need not relate to anything physical in any way and may be purely mental, a fact of which modern mathematics took advantage in liberating itself from physics, thus realizing its abstract nature, inherent in it from its lost origins on. Physics must, however, do so in one way or another.

Heisenberg found this new way to connect abstract mathematics to physics in his discovery of QM. His approach made new physics engendered by abstract mathematical schemes, either invented, as they were by Heisenberg (technically, reinvented because, his matrix variables already existed in mathematics), or borrowed from already available mathematics. The approach was extended by Paul Dirac to quantum electrodynamics (QED) and thus physics in high-energy quantum regimes. While adopting it as early as 1925, Dirac expressly advocated this approach in 1931, in confronting then emerging problems of QED. He said in the statement used as my epigraph:

[5] The terms "realist" and "ontological" may designate more diverging concepts. Still, they are close in their meaning and will be used, as adjectives, interchangeably here. I shall adopt "realism," as a noun, as a general term and refer by an "ontology," as a noun, to a given representation or conception of the reality considered.

> The most powerful method of advance that can be suggested at present is to employ all the resources of pure mathematics in attempts to perfect and generalise the mathematical formalism that forms the existing basis of theoretical physics, and after each success in this direction, to try to interpret the new mathematical features in terms of physical entities. [Dirac [22, p. 1]]

One moves from mathematics, *abstract mathematics*, to physics, rather than from physics to mathematics representing this physics. Dirac, again, already practiced this approach in his work on QM and on QED, especially in deriving Dirac's equation. So did Heisenberg in creating QM in 1925, and Heisenberg's original paper was a great inspiration for Dirac. The method did not quite work in the case that Dirac had in mind in 1931, that of divergent and thus mathematically illegitimate quantities found in QED. The problem was solved or (some, including Dirac, never accepted it as a solution) handled by renormalization around 1950, essentially by manipulating QED in its original form, at least until the introduction of the renormalization group and some stratospheric mathematics, such as the motivic Galois theory, discussed in Sect. 22.4. As, however, testified to by these, mathematically new, developments, and by the new mathematics of QFT in general, Dirac's point remains valid. Even the original approach to renormalization was not without important mathematical innovations, such as those building on Feynman's path integrals (introduced by him earlier), which stimulated mathematics beyond physics.

In all modern, mathematical-experimental, physics, I argue, the creation of new physics is also a creation of new mathematics for physics, possibly by adopting some previously existing mathematics. By contrast, the use of mathematics originating in physics in mathematics requires one to leave physics behind, to convert this mathematics into *mathematics*. This could also be done by a physicist, either for the purposes of physics or by moving to a purely mathematical investigation. Conversely, a mathematician can perform purely mathematical tasks for physical reasons. It is not a matter of who does it but of what is being done and how. Thus, in both physics, at least in QM and QFT, and in mathematics, including mathematics that originates in physics, one ultimately needs to leave physics behind, in accord with the imperative of *modern* mathematics. In QM or QFT, one needed to return to physics, which was, at least in RWR interpretations, no longer that defined, as was classical physics or relativity, by the aim of representing the ultimate nature of the reality responsible for quantum phenomena. It was instead a physics of probabilistic predictions of the data registered in quantum phenomena and only of such predictions. No other predictions are, as things stand now, possible on experimental grounds. At the same time, quantum phenomena themselves are represented by classical physics (cum SR in high-energy regimes) rather than by QM or QFT.

This chapter is, thus, an affirmation of the defining role of *mathematical thinking* in physics and in human thought in general, including of course in mathematics itself, defined as a more specifically and rigorously demarcated, especially institutionally, form of mathematical thinking, although this demarcation still cannot avoid being overflown from its exterior. I shall not attempt to define mathematical thinking, which may not be rigorously possible, given its complexity and diversity. It suffices for my purposes to refer by mathematical thinking to that part of human thinking that can give rise to mathematics as a field of knowledge and research, the field to which it is not easy and perhaps impossible to give a single definition either. I shall only state this chapter's understanding of the main fields of modern mathematics, which is sufficient for my purposes and is, I would contend, in accord with how they are generally viewed. I understand algebra as the mathematical formalization of the relationships between symbols, arithmetic as dealing specifically with numbers, geometry as the mathematical formalization of spatiality, especially in terms of measurement, and topology as the mathematization of the structure of spatial-like objects apart from measurements, through their continuity or discontinuity. The corresponding mathematical fields are algebra, number theory, geometry, and topology. Analysis deals with questions of limit and related concepts, such as continuity and differentiation. There are multiple intersections between these fields, and fields that branch off them.

In considering mathematical *thinking*, this chapter, following [59, 60], gives preference to the role of conceptual thinking (in physics, focusing on mathematical aspects of physical concepts) rather than in logical propositions or calculations. Both logic and calculations are essential to mathematics and hence mathematical thinking, including conceptual thinking there, because the structure of mathematical concepts is logical, and calculations require concepts and can lead to new concepts.[6] While, however, fully respecting these features, this chapter views creative mathematical thinking as most essentially defined by the invention of new concepts, an invention that may sometimes occur in (or through) physics, but is nevertheless still that of a *mathematical* concept.

This view is based on a specific definition or *concept* of concept and of a mathematical theory as an organized assemblage of concepts. Adopting in part the definition offered in Deleuze and Guattari [19], a concept is not merely a generalization from particulars (which is commonly assumed to define concepts) or a general or abstract idea, although a concept may contain such generalizations or ideas, specifically abstract mathematical ideas, where these abstractions may also become concepts. In sum, a concept is a multicomponent structure, defined by the organization, composition, of its components, and some of these components may be concepts in turn.

This is only a general, *abstract*, definition. What meaningfully defines a concept is the specific character of its compositional structure, defined by both the nature of each component and the character of their relations within this composition, by how these components relate to each other in the structure of the concept. Such components may be more general or more specific, and some of them are concepts in

---

[6] There are still other aspects of mathematical thinking, for example, a narrative one, the role of which, as constitutive rather than merely auxiliary, has received a considerable amount of attention during recent decades. For a representative collection of essays, see [24]. The present author is among the contributors, although the article argued, from the RWR perspective, for the limited efficacy of narrative in mathematics [57].

turn. This composition is unique in a new concept, and it makes this concept unique. The creative essence of concepts is in their compositional individuality, which is only possible by virtue of their generally compositional nature but is defined by the specificity of this composition. All musical works are compositions, but each gives us different music. A group is composed of its elements, but in accordance with very specific rules, which give it its structure and role in mathematics in group theory and beyond, and then it multiplies further to still more specific concepts of a group, a finite group, a Lie group, an algebraic group, and so forth, and then into further concepts within each of these.

Of course, every concept, no matter how innovative, has a history defined by other concepts from which this concept emerges and without which it would not be possible. A new concept is always a product of an interplay of continuities with and breaks from preceding concepts, although the balance of continuities and breaks may be different. Some concepts, such as those of a group by Évariste Galois, of a manifold [*Manningfaltigkeit*] by Riemann, or of matrix variables by Heisenberg, were more radically innovative. They still, however, had continuities with earlier concepts, such as those of Adrien-Marie Legendre in the case of Galois and those of Gauss in the case of Riemann, or concepts of the preceding quantum dispersion theory in Heisenberg.

It is fitting that this chapter appears in a volume dedicated to Valentin Poénaru, a volume to which it is a special privilege for this author to contribute. Part of the genealogy of my argument in this chapter is an intriguing remark made by Poénaru. I have discussed this remark previously, but it will appear in a new light from the perspective of this chapter, as a celebration of mathematics, to which Poénaru has made his remarkable contributions. Poénaru's remark was part of his reflections on the fundamental differences in the topology of four-dimensional manifolds, vis-à-vis those of other dimensions, a difference that Yang–Mills theory unexpectedly helped to establish, as Poénaru notes, referring to the work of several key figures involved, most especially Simon Donaldson. Poénaru says: "The tools came from a really unexpected direction, the physics of the Standard Model of elementary particles and fields,... the non-linear Yang–Mills equations (nonabelian gauge theory)," suggesting that, in approaching remaining, thus far seemingly intractable, issues in the topology of $\mathbb{R}^4$, "we will eventually need ... a new sort of quantum topology, or something like it, not yet available to us" [62, p. 698]. I shall discuss this comment in Sect. 22.5.

My question in this chapter is whether, while undoubtedly connected with the physics of the standard model, these tools as such ultimately came from this physics, even if Yang–Mills theory was invented in physics. I shall argue that, as tools, they still came from mathematics, which, as abstract mathematics, was developed apart from physics, including those that motivated and shaped Chen Ning Yang and Robert L. Mills's work. Donaldson's work had little, if anything, to do with physics, which is not to say that it cannot be used in physics. But even if this type of work has connections to physics, as some of it has this fact does not change my argument. It is possible that some physical theory will provide, similarly to the way Yang–Mills theory did for Donaldson, concepts that can be developed into a mathematical theory necessary to address the problem of $\mathbb{R}^4$. It is equally possible, however, that such a theory, with or without parallel or connections to any physical theory, will come from mathematics itself. Either way it will need to be developed as a rigorous mathematical theory apart from physics.

The next section considers and contrasts the role of modern mathematics as abstract mathematics in relativity and QT. The following two sections are devoted to QT in RWR interpretations. Section 22.3 considers QM, historically grounding this discussion in Heisenberg's thinking that led him to the discovery of QM. Section 22.4 offers a parallel discussion of QFT, historically grounding it in Dirac's thinking that led him to the discovery of QED, especially his equation for the relativistic electron. The section also reflects on Yang–Mills theory and renormalization, including the renormalization group and the role of motivic Galois theory there. Section 22.5 is a case study, illustrating my main argument by considering the role of Yang–Mills theory in the topology of four-dimensional manifolds. By using Alexander Grothendieck's comments on Riemann's insight concerning a possibly discrete character of "the reality underlying space" [64, p. 33], Sect. 22.6 offers concluding reflections on the subject that shadows this chapter throughout, that of the relationships between continuity and discontinuity in mathematics and physics. Mathematically, the unfolding landscape of this chapter is governed by, in their historical succession, Einstein's field equations in GR and summit equations of QT—Heisenberg's matrix equations and Schrödinger's equation in QM, Dirac's equation in QED, and the Yang–Mills equations in QFT, grounding the standard model of particle physics, which is where QFT stands now.

22.2 Mathematics in Relativity and Quantum Theory

Hermann Weyl emphasized the divorce of modern mathematics from ordinary thinking, more specifically in the case of the concept of the continuum, but with more general implications. While, however, assuming the independence of modern mathematics from representing nature and thus from physics, Weyl also argued that physics must use the "abstract schemata" of modern mathematics [75, p. 108]. Weyl reflected on this situation in his 1918 *The continuum* [75], followed closely by his 1918 classic *Time–space–matter* [78], both shaped by modern mathematics, by which point quantum physics (although not yet QM), as defined by discontinuity, entered his thinking [78, p. 23]. Weyl, a major figure in mathematics during the time (from the 1900s on) when modern mathematics took mathematics over, also made major contributions, (primarily) mathematical and physical, to both relativity and QT. According to Weyl: "the conceptual world of mathematics is so foreign to what the intuitive continuum presents to us that the demand for coincidence between the two must be dismissed as absurd" [75, p. 108]. "Coincidence" is, however, not the same as "relation," which may be unavoidable, at least insofar as it is difficult to think of continuity, spatial or temporal, apart from general phenomenal intuition.

Weyl's assessment extends beyond the idea of continuity, given that the conceptual world of mathematics contains discrete concepts equally foreign to our phenomenal intuition of discreteness. Nevertheless, the question of the continuum has a particular significance in this regard in the foundations of mathematics in the wake in Georg Cantor's work and a new relevance in the foundations of physics in view of QT, as Weyl noted in *Space-time-matter* [78, p. 23]. Prior to QT, all modern physics, such as classical mechanics and then relativity, was comprised of mathematically continuous theories. Indeed, from Newton on, these theories were effectively defined by differential equations, ideally representing the physical reality considered, which was difficult to do or even to assume for the formalism of QT, including differential equations used there. Mathematically, the problem of the continuum was deepened by Gödel's incompleteness theorems of 1931 and Paul Cohen's proof of the undecidability of the continuum hypothesis in the 1960s. These findings made us realize that we do not and even cannot know how a continuous line, straight or curved (which does not matter topologically), is mathematically constituted by its points. In other words, there may not be a rigorous mathematical concept of continuity (at least in terms of set theory) that accords with our phenomenal intuition of continuity, in contrast to discrete mathematical entities at least some of which, especially finite ones, are in accord with our phenomenal intuition. The undecidability of Cantor's continuum hypothesis suggests that "what the intuitive continuum presents to [our thought]," for example, as space, time, and motion, may not be rigorously realizable mathematically. One could also not be certain, especially in the wake of quantum physics, that either the intuitive or mathematical continuum corresponds to anything that actually exists in nature rather than only in human thought, at least an intuitively phenomenal continuum, if not a mathematical one, which may not exist even as a mental reality in view of the undecidability of the continuum hypothesis.

Does this mean that the ultimate constitution of nature, including, in Riemann's apt phrase, "the reality underlying space," is discrete [64, p. 33]? This is possible, and the idea has been and continues to be advocated, in particular in view of the difficulties of QFT (e.g., Hagar [37]). I shall discuss the subject below, merely stating now that quantum phenomena are always discrete relative to each other, regardless of interpretation. The question is whether they could or not be continuously connected to each other, and an answer depends on an interpretation. In RWR interpretations, the ultimate constitution of the reality considered is beyond any conception and hence cannot be assumed to be either continuous or discrete, or any combination of both, any more than anything else, if a concept other than either continuous or discrete, or a combination of both, is possible. This impossibility in principle precludes assuming any continuous connections between discrete quantum phenomena, making enigmatic the fact that they are related, and some of them are. Otherwise, there would be no probabilistic connections or correlations, which are well established experimentally, between them, or some of them (some quantum events are strictly random).

Ironically, twentieth-century physics, such as relativity and QT, was able to benefit from this divorce between modern mathematics and our ordinary phenomenal intuition, as well as, especially in QT, between mathematics and a representation of the physical world, *in its ultimate constitution*. I qualify because, this representation appears to be unavoidable and is expressly required in RWR interpretations at the level of quantum phenomena, represented by classical physics (with the addition of SR in high-energy quantum regimes handled by QFT). Weyl, aware of this use of abstract mathematics in new physics, added to his assessment: "Nevertheless, those abstract schemata supplied us by mathematics must underlie the exact sciences of domains of objects in which continua play a role" [75, p. 108]. But what kind of role?

Einstein's relativity, essentially a continuous theory, appears to have been primarily on Weyl's mind. The first German editions of both *The continuum* and *Space–time–matter*, containing Weyl's own contributions to GR, were published in 1918. QM was a few years away and the state of QT at the time, later referred to as the "old quantum theory" or, given that part of it was classical, "semi-classical" quantum theory, was unsettled. Eventually, Weyl came to play an important role in the development of QM, first, by assisting Schrödinger's work on Schrödinger's equation, the main equation of QM. Weyl then made major original contributions of his own, especially as concerns the role of symmetry and group theory in QM and QFT. His book, *The theory of groups and quantum mechanics* [76], originally published in German in 1928, only a few years after the introduction of QM, had a major impact on the subsequent development of QM and QFT. He also introduced the concept of gauge symmetry, which (invented in his unsuccessful attempt to unify GR and electromagnetism) has been central for QFT, including Yang–Mills theory. While the case would require a separate discussion, Weyl's work would illustrate the key aspects of the relationships, argued to be asymmetrical in favor of mathematics, between modern physics and modern mathematics considered in this chapter. Thus, the concept of gauge symmetry emerged, in Weyl's work, from physics and yet was developed by Weyl purely mathematically, allowing one to separate it from physics, as it was in Donaldson's work in using Yang–Mills theory in topology. Weyl's work also provides many examples in GR and, especially by using group theory, QT, of solving physical problems by a mathematical means—both by inventing new mathematical concepts and by technical work.

While, however, relativity, especially GR, and QT, especially QM and QFT, were both great beneficiaries of modern mathematics, there were crucial differences between their use of this mathematics. The primary reason for these differences was the fundamental roles of discontinuity and, correlatively, probability in QT, roles that became fully manifested in QM and QFT (1925–1927), eventually leading to RWR interpretations of them. Relativity used modern mathematics in a representational (realist) way, just as did classical physics, such as classical mechanics or classical electromagnetism, in electromagnetism with intimations of the type of relationships between modern mathematics and physics found in relativity. This is because, while the concept of an electromagnetic field allows for a mathematical representation, such as that by means of Maxwell's equations, this concept is not something corresponding to daily phenomenal intuition. Maxwell's electromagnetic theory, a classical field theory, always remained the main inspiration and model for Einstein, including in GR and his failed attempts to unify GR and electromagnetism. The mathematized physical representations beyond daily concepts was possible in both Maxwell's theory and relativity because the break, more radical in relativity, from our daily intuition and concepts need not entail a divorce from a realism or ontology defined by physical or mathematical concepts, possibly purely mathematical ones. It is possible, including in QT, to assume a purely mathematical ontology in the absence of a physical one, such as that defined by physical concepts of classical physics or relativity.

In relativity, this ontology was, again, defined by mathematized physical concepts. SR was an electromagnetic theory, with a mathematical representation of the reality considered defined by Minkowski's concept of spacetime in accord with the Lorentz invariance, corresponding to the change, vs. classical physics, of certain physical concepts, such as that of addition of velocities. Riemannian geometry proved to be a right fit for the physics of GR, the physics that emerged from the equivalence principle of inertial and gravitational mass. This mathematics allowed for the necessary correction of Newton's theory of gravity, which, while an excellent approximation, was technically incorrect even within its proper scope, given, for example, its failure to correctly predict the behavior of Mercury. GR was a field theory, on the model of Maxwell's electromagnetism, rather than Newtonian mechanics of moving solid bodies. Ontologically and epistemologically, however, GR was no different from Newtonian mechanics, or its classical field counterpart, Maxwell's electromagnetism. More remarkable (in accord with Weyl's argument above), was the fact that, in a radical departure from classical physics, the relativistic law of addition of velocities (defined by the Lorentz transformation) in special relativity, $s = \frac{v+u}{1+(vu/c)^2}$, for collinear motion ($c$ is the speed of light in a vacuum), was beyond human intuition of motion.[7] We cannot conceive of this kind of motion. Hence, this concept of motion is no longer a mathematical refinement of a daily concept of motion, of the kind the physical concept of motion in classical physics is. Relativity was the first physical theory that defeated our ability to form a visualization or any phenomenal conception of an elementary physical process. Nevertheless, SR, which deals with electromagnetism (in which SR is inherent because Maxwell's equations are relativistically invariant), and then GR still offered

[7] Photons have a special significance as they only exist as moving, in a vacuum, with c, which limits the physical speed of all processes in nature, as reflected in the Lorentz invariance, governing the formula just mentioned. Were it possible to put a clock on a photon, it would stand still, while at the same time the photon would be found in all locations of its trajectory at once within its own spatial frame of reference. Rigorously, this means that the very concepts of clock and (measuring) rod lose their meaning in the frame of reference of a moving photon (there are no other photons), as does the concept of frame of reference. Although and in part because not applicable to photons, the concept of frame of reference is crucial to SR. GR is not, globally, Lorentz invariant: it has a larger symmetry group or groups (depending on the spacetime considered). It is, however, Lorenz invariant locally or infinitesimally, and it respects (as do all known physical phenomena) c as the limit on the speed of all physical processes considered. GR is a gauge theory of the Lorentz group which acts on the tangent space. As noted, the concept of gauge invariance was introduced by Weyl in this context, in his unsuccessful attempt to unify GR and electromagnetism, which is, arguably, impossible apart from QFT.

a mathematically idealized representation of the physical reality considered, just as did classical physics. In GR, this representation is defined by the Riemannian geometry, which requires tensor calculus for establishing the field equations of the theory, such as Einstein's field equations, generally in the form

$$G_{\mu\nu} + \Lambda_{\mu\nu} = \kappa T_{\mu\nu}$$

($G_{\mu\nu}$ is the Einstein tensor; $G_{\mu\nu} = R_{\mu\nu} - \frac{1}{2} R g_{\mu\nu} R_{\mu\nu}$is the Ricci tensor, a symmetric second-degree tensor that depends only the metric tensor and its first and second derivatives; R is the scalar curvature; $T_{\mu\nu}$ is the stress energy tensor; Λ is the cosmological constant, and $\kappa = \frac{8\pi G}{c^4} \approx 2.077 \times 10^{-43} N^{-1}$ is the Einstein gravitation constant, and G is the Newtonian gravitational constant.)

Deceptively simple in appearance, this equation unfolds into a system of ten nonlinear, hyperbolic-elliptic partial differential equations. This is a far cry from anything based in our daily concepts. Infinitesimally, the spaces considered are metrically Euclidean in accordance with Riemann's concept of manifold, mathematically representing physical space. SR applies in a local or infinitesimal neighborhood of each point, which gives the corresponding local spacetime the Minkowskian metric. Physically, this metric corresponds to the fact that all reference frames that are in free fall are equivalent. The concept of spacetime was introduced by Hermann Minkowski in 1908. It was not part of special relativity as originally developed by Einstein, and was initially accepted by Einstein with some reluctance, but it proved useful in developing GR. While uncontroversial as a mathematical tool, as a physical concept, spacetime involves complexities and its status as such has been debated, unlike, or much more than, the status of Riemannian spaces of GR as representational concepts. As explained below, however, their representational status has been questioned as well.

While using mathematical and physical concepts that defeat our general phenomenal intuition, relativity, SR and GR, remains a realist and causal theory, in fact a deterministic theory, because its predictions are ideally exact. By "causality" I refer, *ontologically*, to the architecture of the physical reality and behavior of the systems associated with this architecture. This architecture is causal if the state of this reality, A, at a given point determines, by a given law (such as Newton's law of gravity), the state of this reality at any future point, X, and A is itself determined by the same law by any previous state. (This definition does not require A to be a cause of X, which may be something else, for example, the force of gravity in Newton's mechanics.) By determinism, I refer, *epistemologically*, to the possibility, ideally, of exact predictions concerning the behavior of the systems considered, which need not allow for determinism, even when they are causal. In the case of individual or small (nonchaotic) systems in classical physics, causality and determinism are coextensive. On the other hand, while classical statistical physics and chaos theory are, ontologically, causal, they are not, epistemologically, deterministic, because the complexity of the systems considered requires the recourse to probability in predicting their behavior. This recourse is, however, merely due to our insufficient knowledge of the reality considered, assumed to be causal. This assumption becomes difficult to sustain in QT (either QM or QFT) and is in principle precluded in RWR interpretations, because no ontology of the ultimate constitution of the reality responsible for quantum phenomena is possible. The recourse to probability in QT, in these interpretations, is due to this impossibility rather than to a lack, for practical reasons, of knowledge of the constitution of the ultimate reality responsible for quantum phenomena, a reality beyond conception and hence in principle (rather than merely practically) beyond any possible knowledge [4, v. 2, pp. 34, 62].

Relativity, beginning with SR, did introduce new features of measurement and changed the role of measuring instruments (rods and clocks), making space and time defined by them, rather than something assumed to exist independently and is then measured by rods and clocks, as were Newtonian absolute space and time. Nevertheless, just as classical mechanics, SR predicted the outcomes of measurements on the basis of its mathematical formalism which, in the first place, represented how these outcomes come about. This was possible because, as in classical mechanics, one could, for all practical purposes, assume that, in Bohr's words, "the phenomena concerned may be observed without disturbing them appreciably" [4, v. 1, p. 53].

This assumption is no longer possible in quantum physics *regardless of interpretation*. In RWR interpretations, QM or QFT is, in view of this impossibility, representationally disconnected from both the physical emergence of quantum phenomena, no longer represented by QM or QFT, or otherwise, and these phenomena themselves, represented by classical theory and not by QM or QFT. Accordingly, both classical physics (with the addition of SR in high-energy regimes) and QT become necessary, without being unified or even unifiable within a single theory [61]. Classical physics is necessary to describe the observed quantum phenomena, which we can "see," while QT is necessary to predict the data observed in them, the data made possible by the RWR-type reality (which is beyond representation or even conception) in its interaction with experimental technology. Classical physics, however, cannot predict these data. This incapacity led to the rise of QT, beginning with Planck's discovery of the quantum nature of radiation in 1900. From Planck on as well, QT has always been a probabilistic theory, which is, as noted, fully in accord with quantum experiments, as no other predictions are, in general, possible.[8]

As indicated above, whether a representation of physical reality is connected to our general phenomenal intuition or not, the possibility of this representation arises because classical physics and relativity are based on the idea that we can observe the phenomena considered without disturbing them sufficiently to affect them. As a result, we can, for all practical purposes, identify classical or relativistic phenomena with the corresponding objects in nature in their independent behavior and (ideally)

[8] In certain special circumstances, such as those of the so-called EPR (Einstein-Podolsky-Rosen) type experiments [28], it is possible to make conditional predictions of some quantities involved with probability equal to one, *determinstically*, but still *without assuming causality* or, in the first place, realism, thus allowing for RWR interpretations [59, pp. 227–272]. Accordingly, determinism is sometimes, conditionally, possible without causality in quantum but not is classical physics, which is always causal.

represent their behavior and predict it by using this representation, keeping in mind qualifications necessary for classical statistical physics or chaos theory, which are, while causal, still probabilistic. This identification helps realism, but it does not guarantee it, even in the case of classical mechanics, where representational idealizations are more in accord with our phenomenal experience. As noted, Kant already realized this difficulty, even though he granted Euclidean geometry and Newtonian physics the capacity to represent the ultimate nature of physical reality, at least with a practical justification, which he thought sufficient for the objective functioning of a theory [44, p. 115]. In other words, the realist nature of classical physics and relativity is ultimately an assumption, justified for all practical purposes with the scopes of these theories. This assumption is also made in this chapter and is used for handling the physics of the observable parts of measuring instruments and, via them, of quantum phenomena, keeping in mind that the data observed cannot be predicted by classical classical (cum SR in high-energy quantum regimes).

It has been questioned whether the mathematical architecture of relativity corresponds to the architecture of nature, as opposed to serving as a mathematical model for correct predictions concerning relativistic phenomena (e.g., [8]). In this case, these predictions are ideally exact, deterministic, as opposed to the probabilistic or statistical predictions of QT, even in dealing with most elementary individual quantum phenomena. This is a fundamental difference due to the impossibility, in principle, of controlling the interference of observational instruments with the quantum object under investigation, regardless of interpretation.

In contrast to relativity, QM or QFT, in RWR-type interpretations, only used (abstract) mathematics for providing probabilistic or statistical predictions concerning the outcomes of quantum experiments, without offering a representation or, in strong RWR interpretations, even a conception of the ultimate constitution of the reality responsible for these outcomes. It is natural and even unavoidable in human thinking to assume that something happens between events we observed. The sense that something happened is one of the most essential elements of human thinking and even defines it as concerns temporality. In RWR interpretations, however, the expression "something happened" is inapplicable to the ultimate constitution of the reality responsible for quantum phenomena. According to Heisenberg:

> There is no description of what happens to the system between the initial observation and the next measurement. ... The demand to "describe what happens" in the quantum-theoretical process between two successive observations is a contradiction in adjecto, since the word "describe" refers to the use of classical concepts, while these concepts cannot be applied in the space between the observations; they can only be applied at the points of observation. [Heisenberg [40, pp. 57, 145]].

The same would apply to the word "happen" or "system," or any word we use, whatever concept it may designate, including reality. This difficulty does not arise when "reality" refers to a "reality without realism" and hence is a "word" without a concept attached to it, if one can still speak of a word in this case. Heisenberg adds: "But the problems of language are really serious. We wish to speak in some way about the structure of the atoms and not only about 'facts'—the latter being, for instance, the black spots on a photographic plate or the water droplets in a cloud chamber. However, we cannot speak about the atoms in ordinary language" [40, pp. 178–179]. Nor is it possible in terms of ordinary concepts, from which ordinary language is indissociable. Heisenberg's formulation, however, allows for a mathematical representation of the reality between observations, defining quantum phenomena, which was a view adopted by Heisenberg at the time of this statement [40, pp. 59–75, 145, 167–186]; [42, pp. 71–88]; [59, pp. 75–76]. The words "happens" or "physical" (or any word) need no longer be part of this representation, which only requires mathematical symbols, possibly defined with the help of ordinary language. As Heisenberg said earlier, shortly after the introduction of QM, at the time assuming a (weak) RWR interpretation rather than a mathematical realism of his later view, mathematics is "fortunately" free from the limitations of ordinary language and concepts:

> It is not surprising that our language [or concepts] should be incapable of describing processes occurring within atoms, for . . . it was invented to describe the experiences of daily life, and these consist only of processes involving exceedingly large numbers of atoms. Furthermore, it is very difficult to modify our language so that it will be able to describe these atomic processes, for words can only describe things of which we can form mental pictures, and this ability, too, is a result of daily experience. Fortunately, mathematics is not subject to this limitation, and it has been possible to invent a mathematical scheme—the quantum theory [QM]–which seems entirely adequate for the treatment of atomic processes. [Heisenberg [39, p. 11]].

Words can do more, and they can in fact describe things of which we cannot form mental pictures, including mathematical symbols. This qualification does not, however, undermine Heisenberg's main point, amounting to the fact that our language and thought are the product of our evolutionary biological and neurological development defined by the experiences consisting only of processes dealing with huge numbers of atoms. Accordingly, there is no special reason to assume that they will be able to describe nature on the atomic or still smaller scales, or conversely, on very large, say, cosmological scales. One might instead doubt that one can do so even mathematically, as Heisenberg eventually assumed, given that mathematics, too, is only human and as such is the product of the same evolutionary development. We just might be fortunate to be capable of using it to predict quantum phenomena.

It remains the case, however, that mathematics, especially abstract mathematics, such as that used in QM and QFT, does have a great, even if not complete, degree of freedom from the limitation of daily language and thinking. As such, in physics or in mathematics itself, mathematics enables us to *relate* to things in nature and mind which are, as forms of RWR-type reality, beyond the reach of thinking, including mathematical thinking [60, pp. 83–95]. In QM and QFT it does so by enabling us to estimate probabilities or statistics of quantum events, to which Heisenberg refers here by speaking of his scheme, quantum mechanics, as "entirely adequate for the treatment of atomic processes." (Not everyone, beginning, again, with Einstein, would see this, merely probabilistically predictive, treatment itself as *entirely* adequate.) In fact, it is equally fortunate that nature allows us to use such a "mathematical scheme," because the freedom of mathematics from the limitations of common language and concepts does not a priori guarantee that any such scheme will work, and it might not work beyond the scales now handled by QFT.

In RWR interpretations, then, the ultimate reality responsible for quantum phenomena is *constructed by thought as unconstructible by thought*. While doubted or rejected by many, even most, physicists and philosophers, beginning, again, with Einstein, on philosophical grounds, this interpretation is not as speculative or far-fetched as it might appear. Nobody has ever seen, at least thus far, an electron or a photon, or any quantum object as such, even if composite and macroscopic (such as a Bose–Einstein condensate). One could only observe the traces of the interactions between them and measuring instruments, traces that define quantum phenomena but that cannot be identified with the reality responsible for them (and thus quantum objects), because quantum phenomena are also the products of the interaction between this reality and millions of atoms, say, in a cloud chamber or on a photographic plate. This fact does not preclude a realist theory representing the ultimate reality responsible for quantum phenomena, or even a realist interpretation of QM or QFT. But it does open the possibility of RWR interpretations, in which this reality is unrepresentable otherwise or even inconceivable.

The mathematical formalism of the theory, as that of Hilbert spaces over $\mathbb{C}$, is, too, unvisualizable even in finite dimensions, especially as representing anything physical, doing which is, again, not possible in RWR interpretations. All that we can *see* (observe or even imagine) are the measuring instruments and the data observed in them, as quantum phenomena, which we establish, *create*, each time uniquely, with the help of observational technology. These observations cannot, thus, be assumed to "observe" something already there as observations are in classical physics or relativity, which would also allow one to convert an observation into a measurement of an already existing property.

In RWR interpretations, the irreducible difference between quantum phenomena and the reality responsible for them (in the present view, with the concept of a quantum object only applicable at the time of observation) separates observation and measurement. In quantum physics, an observation is a unique act of creation of quantum phenomena. A measurement is a subsequent measurement of one or another property of the phenomena observed in the instruments used, but never of the quantum object considered. By the same token, QM or QFT only predicts the outcomes of measurements of the properties of quantum phenomena, defined by our observational technology and described by classical physics (cum SR in high-energy quantum regimes). This fact also makes classical physics fundamental physics (a position well earned by it but often denied to it) and an essential part of all physics, including QT, where it describes the data found in quantum phenomena observed in the instruments used, along with the observable parts of these instruments. On the other hand, it cannot predict these data, thus also irreducibly separating it or SR (added in high-energy regimes) from QT. Thus, all three theories are necessary, while remaining separate from each other. GR is a different case because it cannot, thus far, be related to quantum phenomena and hence QT at all.

RWR interpretations makes quantum physics the first truly *experimental* science insofar as one no longer merely follows what is bound to happen in any event but rather experiments with reality by creative new reality and a new course of reality in each experiment. In classical physics and relativity, one can and usually does, as, paradigmatically, in observing the motion of planets in the solar system, merely follow what happens, because the intervention of technology, beginning with our bodies, can be neglected. This also allows, at least in some cases (in relativity in all cases), for ideally exact, deterministic, predictions. In quantum physics, our observations and thus quantum phenomena are always, in principle, defined by human interventions into the course of reality, shaping this course by means of experimental technology, the role of which is irreducible, constitutive. Such interventions are also possible in classical physics and relativity, but they are not essential because after each such intervention we can still observe phenomena without measurably disturbing them. In quantum physics, they are unavoidable: our observations are creations of quantum phenomena and each time a new unique reality. All assignment of probabilities is human as well, although the statistics of actual observations is objective, which also makes QM or QFT falsifiable.

In QM or QFT, moreover, these probabilistic predictions are only possible by using *ad hoc* rules added to the formalism rather than being part of it, such as Born's rule, which relates, by using complex conjugation, complex quantities of the quantum-mechanical formalism (the so-called amplitudes) to real numbers corresponding to the probabilities of quantum events. While there are other rules that have the same function, I shall only refer to Born's rule here, the first and still most commonly used. One takes the square moduli of the eigenvalues of the operators associated with quantum variables, such as position, momentum, or energy, or equivalently, multiply these eigenvalues by their complex conjugates. Doing so gives one real numbers, corresponding, once suitably normalized, to the probability densities of observed events. In the simplest case, when ψ is a wave function for a point particle in position space, the probability density function $p(x, y, z)$ for predicting a measurement of the position of this particle (measured at time t_0) at some future time t_1 equals to $|\psi(x, y, z, t_1|^2$. Integrating over this density gives the probability or (if one repeats the experiment many times) statistics of finding the particle in a given area.

Born's rule allows us to move from complex quantities of the formalism to real quantities related to probabilities, which must be real numbers, by means of a purely mathematical procedure, the structure of which has no representational relation to physical reality any more than does the formalism itself. The only such relations are those to quantum phenomena observed in measuring instruments, relations defined by representing these outcomes by the formalism (using functions of real variables) of classical physics and not that of QM or QFT (defined by Hilbert spaces over $\mathbb{C}$). Born's rule brings physics into the purely abstract mathematics of QM or QFT. Although Born's or similar rules are connected naturally to the formalism of QM (given that complex conjugation naturally connects complex and real numbers), they are added to rather than contained in this formalism. We do not know why these rules work, but they do. But then, we don't know while the whole scheme of QM or QFT works either, insofar as it has no representational connections to any physics, either that of the ultimate nature of reality responsible for quantum phenomena or that of quantum phenomena themselves, represented by classical physics.

In departure from classical probability theory, the probabilities involved are nonadditive, or non-Kolmogorovian: the joint probability of two or more mutually exclusive alternatives in which an event might occur is not equal to the sum of the probabilities for each alternative, as in Kolmogorovian probability theory, where $p(A) + p\bar{A} = 1$ ($\bar{A}$ is an event independent of A, and p is probability). In Kolmogorovian probability theories, for any countable sequence of disjoined sets (corresponding to mutually exclusive events) $E_1, E_2, \ldots$ the probability $P(\bigcup_{t=1}^{\infty}) = \sum_{t=1}^{\infty} P(E_i)$.

In QM (in most interpretations), the probabilities of mutually exclusive events are derived by using the addition of "amplitudes". This term is adopted from classical wave theory because the equations of QM or QFT formally contain qualities that are similar to amplitudes in classical wave or field equations. These amplitudes, however, no longer serve their classical physical function, as part of a continuous representation of motion. Instead, they are related, via Born's rule, to discrete transitions events observed in measuring instruments and described by classical physics. At least in RWR interpretations, these amplitudes are not representations of anything physical, unlike in classical physics where amplitudes and (wave) superposition represent physical processes. They are no longer amplitudes of physical motions, but are formal mathematical entities, "probability (density) amplitudes." As such they are related to possible alternatives in which expected events might happen. Born's rule is then applied to the sum of these amplitudes, giving real numbers corresponding to the sum of probabilities of these events.

Thus, in RWR interpretations, the mathematical formalism of QM (accompanied by Born's rule) uses "abstract schemata" of modern mathematics, invoked by Weyl, in a radically new way [75, p. 108]. It established a new type of relationship between the mathematics of a physical theory and experimental data, and thus between mathematics and physics, or mathematics and nature, in fact requiring us to rethink the concept of physics as science.[9] As such, QM occupies a special place in the history of modern physics as mathematical-experimental science. *First*, it does so by virtue of using the mathematics, modern mathematics, born from its divorce from physics, abstracted from physics. *Second*, more radically epistemologically (vis-à-vis relativity or such classical theories as chaos theory, which also use modern mathematics), QM, at least in RWR interpretations, keeps this divorce in place as concerns a representation or even a conception of the ultimate constitution of the reality responsible for quantum phenomena. And yet, *third*,

[9] It can, for example, be done in informational terms, following quantum information theory, as discussion from the present RWR perspective by this author in [59, pp. 307–328], by arguing that Heisenberg's approach leading him to his discovery of QM was quantum-informational in spirit, while, by the same token, quantum information theory can be seen as Heisenbergian in spirit. See also [15] for a stimulating related argument, which, however, does not assume an RWR interpretation.

as an irreducibly probabilistic theory, even in dealing with the simplest possible quantum systems, QM predicts the data observed fully in accord with what is experimentally observed, as no other predictions concerning quantum phenomena are experimentally possible. All these features of QM are retained in high-energy quantum regimes and QFT, which indeed adds further radical features to QT. Prepared as it was by the preceding history of QT, especially Bohr's 1913 theory, QM was a radical transformation of theoretical physics, defined by the birth of physics out of the spirit of mathematics.

22.3 "The Heisenberg Method": The Birth of Physics Out of the Spirit of Mathematics

In confronting the experimental situation defined by quantum phenomena, Heisenberg, in a radical move that changed theoretical and even experimental physics, was compelled to conceive QM as a theory designed only for predicting the probabilities of what was observed in measuring instruments without representing the behavior of quantum objects [41]. The radical nature of this move was immediately grasped by Bohr [4, v. 1, pp. 48–52]. This was not surprising given that Bohr's 1913 atomic theory was the main precursor of Heisenberg's approach. Bohr's concept of "quantum jump" defining his theory already entailed the same epistemology, even if only as part of his "semi-classical" theory. (The rest was classical.)

Shortly before his paper was published, Heisenberg said:"What I really like in this scheme [QM] is that one can really reduce *all interactions* between atoms and the external world ... to transition probabilities" [between quantum measurements] [Heisenberg, Letter to Kronig, 5 June 1925; cited in Mehra and Rechenberg [49, v. 2, p. 242]; emphasis added]. By referring to the "interactions between atoms and the external world," this statement suggests that QM was only predicting the effects of these interactions observed in quantum phenomena, defined in observational instruments, and amenable, along with the observed parts of these instruments, to a treatment by classical physics. Establishing these effects required special instruments beyond human bodies. Human bodies are sufficient in some cases and are good models of the instruments used in classical physics, but not in quantum physics, which is irreducibly technological, or as noted already in relativity, which was also irreducibly technological, while, however, allowing SR and GR to be causal and in fact deterministic, and in the first place, realist theory. This was because the observational instruments used in relativity (rods and clocks) allowed one to observe the systems considered without disturbing them and (ideally) mathematize their independent behavior. This is no longer possible in considering quantum phenomena, which poses major difficulties for, even if does not preclude, realism. While not found in Bohr's 1913 theory, Heisenberg's view just cited was adopted by

Bohr as a defining feature of his interpretation in all its versions.[10] It is also assumed by the present interpretation.

As indicated earlier, this procedure, linked to combining and yet separating observation and measurement (vs. classical physics and relativity, both of which would make observation and measurement co-extensive) replaces measurement in the sense of measuring some preexisting properties of objects, which has defined the idea of measurement from ancient Greek geometry, geo-*metry*, on. Instead, an observation, establishing a quantum phenomenon, becomes an act of creation, each time unique, defined by a decision of thought and enabled by using an observational instrument, without measuring the properties of quantum objects or even observing them. What is observed instead are the effects of the interactions between quantum objects and observational instruments manifested in these instruments. Quantum objects are unobservable as such and, in RWR interpretations, placed beyond representation or, by the time of Bohr's ultimate interpretation (introduced in the late 1930s), even conception. In the present interpretation, moreover, the "concept" of a quantum object, while still assumed to be an RWR-type entity and hence unrepresentable or, again, unconceivable (and thus not having a specifiable concept associated with it), is only applicable at the time of an observation. The data observed could then be measured classically. Observational instruments also contain a quantum stratum through which they interact with quantum objects or with the ultimate (RWR) reality responsible for quantum phenomena, and in the present view quantum objects at the time of observation. As quantum, this stratum and this interaction are also beyond representation and ultimately conception in RWR interpretations.

It was Heisenberg's renunciation of a geometrical and continuous (and correlatively, causal) representation of quantum objects and behavior, by replacing the geometry of trajectories with the algebra of probabilities, that led him to the discovery of QM, an approach bemoaned, on all three counts—algebra, RWR epistemology, and the irreducible role of probability—by Einstein as "the Heisenberg method" [25, p. 378]. The geometry of trajectories was still retained, even if only partially in Bohr's 1913 atomic theory, in which so-called stationary states (those of constant energy of electrons) were mechanically represented by electrons' orbits around the nuclei. By contrast, the discontinuous transitions between these states, "quantum jumps," had no mechanical representation, in conflict with both classical mechanics and classical electrodynamics alike. This was a revolutionary move, one of the most radical ever in the history of physics, which implied an RWR interpretation for this part of his theory.

This part was, however, decisive for understanding the constitution of atoms, as Heisenberg came to realize, which led him to basing QM on this idea, while abandoning the classical part of Bohr's theory. As he said, writing to Wolfgang Pauli:

[10] Bohr adjusted his views, sometimes significantly, a few times, eventually arriving at his ultimate, RWR, interpretation around 1937, after a decade of debate with Einstein [59, pp. 178–191].

> It is really my conviction that an interpretation of the Rydberg formula [for wavelength of spectral lines] of atomic in terms of circular and elliptical orbits (according to classical geometry) does not have the slightest physical significance. And all my wretched efforts are devoted to killing totally the concept of an orbit—which one cannot observe anyway—and replace it by a more suitable one. [Heisenberg to Pauli, 9 July 1925; cited in Mehra and Rechenberg [49, v. 2, p. 284]

This more suitable concept was just that of the energy state of an electron, as manifested in an observational instrument, without representing the motion of the electron at all, any more than it would in a quantum jump. In Heisenberg's theory and in QM or QFT thereafter, in RWR interpretations, there are only discrete transitions, "quantum jumps," between quantum phenomena, observed in measuring instruments, while QM or QFT is a theory providing the transition probabilities between the events defining these phenomena, without providing and even precluding a description or, in strong RWR interpretations, even conception of what happens between such events.

There remains a question whether a purely mathematical representation (apart from ordinary language or concepts) of the ultimate nature of physical reality between quantum events is possible. It is not possible in Bohr's interpretation, at least in its ultimate version, or in the present one. On the other hand, Heisenberg eventually came to assume this possibility, exhibiting a Platonist tendency. I speak of a tendency, because it would be difficult to attribute to Heisenberg a full-fledged Platonism (which is not always the same as Plato's philosophy, Plato's own Platonism), that Heisenberg retained the view of all physical reality as ultimately material. He also retained certain affinities with Bohr's interpretation, which may be seen as closer to Aristotle than Plato, on which I comment below. While more apparent in his later philosophical writings, this tendency toward and an interest in Plato is found in Heisenberg's earlier thinking, including during his invention of QM, accompanied by his reading of Plato's *Timaeus* [40, pp. 59–75], [42, pp. 71–88]; [56, pp. 471–472]. (Heisenberg's father was a classicist.) Plato's ontology of reality, in his case mental reality, was mathematical or mathematical-philosophical and thus free of physical concepts and ultimately reaching beyond ordinary language and concepts, which is, to some degree, although not completely, possible for philosophical concepts and discourse. As such, this ontology served as an inspiration, even if, as purely mental, not strictly a model, for Heisenberg's mathematical sense of the ultimate nature of reality in quantum physics, in contrast to Aristotle's physics, defined by physical concepts [60, pp. 259–270].

If, however, one accepts Thom's argument, cited above, in Thom [72], that Aristotle's physics was (proto)topological and in this sense mathematical in representing his physical concepts, one can speak of two divergent mathematical trends of ancient Greek thinking, bridged by mathematics, geometry in Plato and topology in Aristotle. One might, then, with due caution, relate these two trends, respectively, Heisenberg's, more Platonist, view, supplemented by algebra, and Bohr's, more Aristotelian view, apart from Aristotle's realism, which is a crucial difference, of course.

We might not be as modern as we, or some of us, like to think, or ancient Greek thinking is more modern than we, or some of us, like to think. Its modern aspects transpire, for example, in emphasizing the creative nature of philosophical and, thus, by implication all theoretical thinking, in Plato, and the theoretical- or even mathematical-experimental nature of science [*episteme*] in Aristotle. A better argument may be that, one, yet again, deals with a long history of a complex and fluid interplay between the old and the new, defined as such at each juncture of this history. Throughout this history, however, supporting my main argument here, physics appears to have been defined by its mathematical project from Plato and Aristotle or even the pre-Socratics on. QT did not change this, but, especially with QM, as defined, mathematically and epistemologically, by Heisenberg and Bohr, it changed the nature of this project. Nevertheless, just as earlier physics, QT continues to oscillate between Plato and Aristotle, with Heisenberg moving it towards Plato and Bohr towards Aristotle. And yet, Heisenberg also used Aristotle's concept of "*potentia*" (alien to Bohr's thinking) in his later interpretation of QM, thus bringing Plato and Aristotle together and, thus, yet again, confirming the complexity of the interplay of continuities and breaks in this history, from the pre-Socratics on [40, p. 44]. Mathematics provides an unbroken, ever transforming, line or rather field of continuity in this history, with Bohr and Heisenberg moving it away from representing the ultimate nature of physical reality in QT, but thus, seemingly (but, I argue, only seemingly) paradoxically, also giving mathematics an even greater role in physics.

It is true that QM was co-discovered, as wave mechanics, by Schrödinger, whose thinking was governed by the realist imperative of finding a physical description and mathematical representation of quantum objects and processes, in this case in wave terms. He envisioned a more geometrical and realist theory of quantum phenomena and continuous (and causal) connections between them. Schrödinger, however, could not sustain his agenda, especially in dealing with the discreteness of quantum phenomena and the probabilistic nature of quantum predictions, which were and have remained since well confirmed experimental facts. To bring his theory into accord with these facts, he made several mathematical moves that were in conflict with this agenda, but were consistent with Heisenberg's agenda [59, pp. 145–166]. These were these mathematical moves that led Schrödinger to his, correct, formalism of QM. Thus, in his case too, it was ultimately mathematics that brought physics to where it needed to be.

Both formalisms were quickly proven to be mathematically equivalent (or more accurately translatable into each other), unsurprisingly in view of the circumstances just described, which fact did not affect Heisenberg's agenda did not help Schrödinger's. Indeed, QM ultimately proved to be a disappointment to Schrödinger. While accepting QM as an effective mechanism of predictive calculations, he came to repudiate it conceptually. He was not happy with its fundamentally probabilistic character either. He described QM as "a doctrine born of distress" and as "perhaps after all only a convenient calculational trick," which, he further lamented, "today ... has attained influence of unprecedented scope over our basic attitude toward nature" [66, pp. 154, 167]. At it happened, the scope of its influence in terms of its radical epistemology (primarily on Schrödinger's mind here), while then strong in Copenhagen and Göttingen, has never been especially great, because most preferred a realist theory of quantum phenomena.[11]

Heisenberg expressly used mathematics never previously used in physics, that of unbounded infinite-dimensional matrices over $\mathbb{C}$. He was famously unaware of the existence of matrix algebra and reinvented it, although unbounded infinite matrices were not previously studied in mathematics either. It is not even entirely clear that Heisenberg thought of the entities he introduced as matrices, rather than just double-indexed arrays of quantities. That they could be seen as matrices was first realized by Max Born, Heisenberg's post-doctoral supervisor, on reading Heisenberg's paper. QM was recast into Hilbert-space form (over $\mathbb{C}$), by von Neumann, who also gave the first rigorous proof of the mathematical equivalence of Heisenberg's and Schrödinger's versions, although this equivalence was more or less assumed by then. Von Neumann presented this version in his classic 1932 book [73], also in juxtaposition to yet another version of the formalism, introduced by Dirac, using Dirac's delta function [21]. The delta function was not considered mathematically legitimate at the time, because its value is zero everywhere except at zero, while its integral over the entire real line is equal to one, which is not a mathematically legitimate *function*. The delta function was, however, given a mathematical legitimacy by means of a new concept, a distribution (in effect a functional), due to Sergei Sobolev and Laurent Schwarz (who introduced the term). This is an example of both the influence of physics on mathematics and the fact that one needs to give such ideas a mathematical rigor to make them part of mathematics.[12]

One of the most unexpected and implicative new features of Heisenberg's algebra was the noncommutative nature, $PQ - QP \neq 0$, for his variables. These variables (replacing the functions of the real variables of classical mechanics), initially conceived by Heisenberg in terms of linear algebra, as matrix elements, were defined more formally as Hermitian operators in Hilbert spaces over $\mathbb{C}$, for example, those associated, respectively, with the momentum, p, and the coordinate, q. For this reasons, these operators are also known as "observables," which is a misleading term because they are never physically observed, but only enable QM or QFT to predict the probabilities of quantities, such as p or q, observed in quantum phenomena and described by classical physics. While it was unknown to Heisenberg, who was, as were others, worried about it, noncommutativity was a familiar feature of mathematics by then. The role of this feature proved to be momentous. Most famously, it came to represent or more accurately be associated with Heisenberg's uncertainty relations, $\Delta q \Delta p \approx h$ (where q is the coordinate,

[11] Schrödinger returned to his original ideas in the 1950s in another attempt to develop a realist and causal wave-like theory based on his equation [67]. This attempt does not appear to me to have been successful and its impact has been limited.

[12] There are alternative formalisms, such as in terms of C^* algebras or more recently monoidal categories, all thus far mathematically equivalent to the Hilbert space formalism, still used most.

p is the momentum in the corresponding direction), discovered by Heisenberg in 1927. The uncertainty relations reflect the impossibility of simultaneous exact measurements of the momentum and the coordinate associated with a quantum object but, in RWR interpretations, observed in measuring instruments.

This impossibility was understood by Bohr in terms of complementarity, his most famous concept, developed shortly after the uncertainty relations. Complementarity is defined by

(A) a mutual exclusivity of certain phenomena, entities, or conceptions; and yet
(B) the possibility of considering each one of them separately at any given point; and
(C) the necessity of considering all of them at different moments of time for a comprehensive account of the totality of phenomena that one must consider in quantum physics.

Bohr's ultimate, RWR, interpretation, developed around 1937, gave the RWR-type epistemology to complementarity, via his concept of phenomenon. Phenomena were defined strictly by what is observed in measuring instruments, under the impact of quantum objects, and described by classical physics, while the independent behavior of quantum objects and hence how these phenomena come about was placed beyond conception [4, v. 2, p. 64]. Complementarity was now applied to quantum phenomena and not quantum objects.

Complementarity implies two incompatible pictures of what is observed, as phenomena, in the instruments used. The *possible* information about a quantum object, the information to be found in observational instruments, in only obtainable in the mutually incompatible evidence obtainable under different experimental conditions [4, v. 2, p. 40]. On the other hand, once made, either measurement, say, that of the position, will provide the complete *actual* information about the system's state, as complete as possible, at this moment in time. One could never obtain the complementary information, provided by the momentum measurement, at this moment in time, because to do so one would need simultaneously to perform a complementarity experiment on it, which is impossible. But one can, by (B), always decide to perform either one or the other experiment. Each experiment establishes the only reality there is at this moment in time, and the alternative decision would establish a different reality. Rather than arbitrarily selecting one or other part of a preexisting physical reality, as is ideally possible in classical physics, our decisions concerning which experiment to perform establish the *single* reality which defines what *type* of quantity (although not its value, which can only be predicted probabilistically) can be observed or predicted and precludes the complementary alternative. That we can freely or sufficiently freely make a decision (a preferable category to that of choice) concerning which experiment to perform is in accordance with the very idea of experimentation in all physics or science [3, p. 699]. As against classical physics or relativity, however, implementing our decision will allow us to make only certain types of possible predictions and will irrevocably exclude certain other, complementary, types of predictions. Our decisions define reality and its future course, while QM enable possible predictions, unavoidably probabilistic, concerning this course.

Complementarity, thus, reflects the fact, indicated earlier, that, in accord with and bringing together two main meanings of the word "experiment" (as a test and as innovative creation), the practice of quantum physics became the first practice of physics or science that is both, and jointly, *generatively* experimental and *generatively* mathematical. It is generatively experimental as experimental physics in its conventional sense because it is no longer defined, as in classical or relativistic physics, in tracking the independent behavior of the systems considered and measuring their independent properties. It is defined by *unavoidably* creating configurations of experimental technology containing traces of its interactions with nature that reflect the fact that what happens is unavoidably defined by what kinds of experiments we perform, by how we affect physical reality by a unique act of observation as creation. I qualify by "unavoidably" because, while the phenomena observed in classical physics or relativity may be affected by experimental technology and while we do stage experiments there, in principle one can observe these phenomena without appreciably affecting what is thus observed, which also allows one to see these phenomena as ideally representing the corresponding objects. Thus, one still follows what would have happened. In quantum physics, the experiment is a creation of a new material and phenomenal configuration that enables one, by performing a measurement on the observable part of the instrument used, to define the probabilities for outcomes of future experiments, without, at least in RWR interpretations, it being possible to follow, track, how these outcomes come about.

Correlatively, the practice of theoretical physics is *generatively* mathematical because it no longer consists, as in classical physics or relativity, in offering an idealized mathematical representation of quantum objects and behavior. It consists in inventing abstract mathematical schemes enabling us to predict, probabilistically or statistically, the outcomes of quantum events. These schemes are unrelated to and thus are not helped by our general phenomenal intuition, as are those of classical mechanics or, in a more limited way, relativity. On the other hand, they are not limited by this intuition either.

Indeed, one experiments with mathematics as well. It is experimenting with abstract mathematics, divorced from our phenomenal intuition or even physical concepts that, with Heisenberg, became the way to new theories in fundamental physics. Dirac's work in QED was arguably an even more striking example of this new approach to theoretical physics, which, as noted, he expressly advocated as "the most powerful method of advance that can be suggested at present" [22]. Heisenberg was the first practitioner of this method, "the Heisenberg method," lamented by Einstein on, among others, this account. Heisenberg had to suspend, to "forget," the ways of using mathematics in classical physics and even relativity, to arrive at his new mathematical scheme. His physics was defined by finding the mathematics probabilistically predicting the data in question. He was able to use the already available equations of classical mechanics, by Bohr's correspondence principle, which states that, at the classical limit, say, for large quantum numbers when electrons are far away from nuclei, where one can use classical physics, the predictions obtained by classical and quantum theory should coincide. Heisenberg gave this principle a mathematical form by postulating that the equations used were formally the same. The variables, however, could not have been the same, because classical variables would give wrong prediction for low quantum numbers, thus suggesting that new variables were necessary for the same equation.

Heisenberg was the only person who thought in this way, a new mathematical way of thinking in physics, at the time, reversing the approach, common then, in which one would continue to use standard classical variables and modify the equations by using, often ingeniously, ad hoc, the Bohr correspondence principle differently for each case considered. "The Heisenberg method" implied a radical transformation of theoretical physics, enabled by a new, essentially mathematical, way of thinking, defined by an invention of a new predictive mathematical scheme, rather than by finding the mathematics that would represent the reality considered.

I would like to follow this birth of physics out of the spirit of mathematics, to paraphrase Friedrich Nietzsche's famous title, *The birth of tragedy out of the spirit of music* [50], a new way of using and, given its purely abstract nature (rather than a representational relation to physical reality), a new way of *inventing* mathematics in physics, or even mathematics as physics. In his 1925 paper, introducing QM [41], Heisenberg began with an observation that reflected the necessity of a departure from the classical ideal of continuous mathematical representation of individual physical processes. He said:

> In quantum theory it has not been possible to associate the electron with a point in space, *considered as a function of time*, by means of observable quantities. However, even in quantum theory it is possible to ascribe to an electron the emission of radiation" [the effect of which is observed in a measuring instrument]. [Heisenberg [41, p. 263]; emphasis added]

Referring to what happened between experiments, via a classical concept of emission, this statement would require qualifications in RWR interpretations, which preclude referring to anything that happens between observation. These considerations were, however, to come later. A measurement could associate an electron with a point in space, with QM capable of predicting the probability for finding its position in a given area. But it is not possible to do so by linking, as in classical physics, this association to a function of time (as a real variable) representing the continuous motion of this electron, which would then allow one to predict this position ideally exactly. Heisenberg's paper did not offer a treatment of electrons in stationary states, in which one could speak of the position of an electron in an atom. (A consistent treatment of stationary states was given later in Born and Jordan [7] and Born et al. [5].) An instantly repeated measurement could give the value, *the same value*, of its position, which instant repetition is, however, an idealization (e.g., Heisenberg [39, p. 56], von Neumann [73, pp. 355–356], Schrödinger [66, pp. 158–159]). Schrödinger's time-dependent equation made it possible to predict a (physical) quantum state for any variable, including for electrons in stationary states.

Heisenberg states next:

In order to characterize this radiation we first need the frequencies which appear as functions of two variables. In quantum theory these functions are

$$\nu(n, n-\alpha) = 1/h\{W(n) - W(n-\alpha)\}$$

and in classical theory in the form

$$(\nu, \alpha) = \alpha\nu(n) = \alpha/h(dW/dn).$$

[Heisenberg [41, p. 263]]

This difference, central to Bohr's 1913 theory (which introduced these functions in QT), leads to the difference between classical and quantum theories as regards the combination relations for frequencies, which correspond to the Rydberg–Ritz combination rules, reflecting "the discrepancy between the calculated orbital frequency of the electrons and the frequency of the emitted radiation" [41, p. 263]. This was a key new feature, incompatible with classical electrodynamics, of Bohr's 1913 theory. However, Heisenberg then says, "in order to complete the description of radiation [in conformity, by the correspondence principle, with the classical Fourier representation] it is necessary to have not only frequencies but also the amplitudes" [41, p. 263]. On the one hand, then, the equations of QM must formally contain amplitudes as well as frequencies. On the other hand, these amplitudes could no longer serve their classical physical function (as part of a continuous representation of motion) but were instead related to discrete transitions between stationary states, reflecting a radically different way of thinking.

As explained, in Heisenberg's theory and in QM since then, these "amplitudes" are no longer amplitudes of physical motions, but are instead formal mathematical entities, "probability (density) amplitudes," linked, via Born's rule, to the probabilities of transitions between stationary states, manifested in the spectral data observed in quantum experiments. The probability amplitude is just $\langle\lambda_i|\psi\rangle$($\lambda_i$ is an eigenvalue and ψ is the wave function) and Born's rule says that the corresponding probability is the square of the amplitude or is the amplitude multiplied by its own complex conjugate, or $P_i = \langle\lambda_i|\psi\rangle^2$. This makes the term "amplitude" purely mathematical or symbolic [4, v. 1, p. 18].

In classical physics the mathematics of (wave) superpositions represent physical processes; in QM, at least in RWR interpretations, it does not. Schrödinger's wave function, ψ, is no longer assumed to correspond to any physical waves, as Schrödinger initially hoped it would, but was instead associated with probability-densities for future predictions, the distribution of which was "wave-like" in pattern. From this viewpoint, one can discard the idea of waves in quantum theory altogether. There are no waves, but only correlated wave-like, but always discrete, patterns predictable by QM. The corresponding probabilities were derived by a form of

Born's rule, which Heisenberg postulates for this limited case.[13] Heisenberg then says:

> The amplitudes may be treated as complex vectors, each determined by six independent components, and they determine both the polarization and the phase. As the amplitudes are also functions of the two variables n and α, the corresponding part of the radiation is given by the following expressions:
>
> Quantum-theoretical:
>
> $$Re\{A(n, n-\alpha)e^{i\omega(n,n-\alpha)t}\}$$
>
> Classical:
>
> $$Re\{A_\alpha(n)e^{i\omega(n)\alpha)t}\}$$
>
> [Heisenberg [41, p. 263]]

The problem—a difficult and, "at first sight," even insurmountable problem—is that "the phase contained in A would seem to be devoid of physical significance in quantum theory, since frequencies are in general not commensurable with their harmonics" and, as a result, "a geometrical interpretation of such quantum-theoretical phase relations in analogy with those of classical theory seems at present scarcely possible" [41, pp. 263–265]. A new, quantum-theoretical, form of geometry enters, the geometry of Hilbert spaces over $\mathbb{C}$. This geometry is no longer analogous to the geometry used in classical theory (or relativity), insofar as it is only used to predict the probabilities of the outcomes of quantum experiments and not to represent how these outcomes come about. Following that of Bohr in his 1913 theory, Heisenberg's approach was based on the shift from calculating the probability of finding a moving electron in a given state to calculating the probability of an electron's transition from one state to another, without describing the physical mechanism responsible for this transition. Heisenberg's theory is more fully in harmony with this thinking because there were no longer orbits treated classically.

Heisenberg says next: "However, we shall see presently that also in quantum theory the phase has a definitive significance which is *analogous* to its significance in classical theory" [41, p. 264]; emphasis added. "Analogous" could only mean here that, rather than being analogous physically, the way the phase enters mathematically is analogous to the way the classical phase enters mathematically in classical theory, in accordance with the mathematical form of the correspondence principle, now the correspondence postulate, insofar as quantum-mechanical equations are formally the same as those of classical physics. Heisenberg only considered a toy model of an aharmonic quantum oscillator, and thus needed only a Newtonian equation for it, rather than the Hamiltonian equations required for a full-fledged

[13] Born begins his paper, introducing Born's rule, with a reference to Heisenberg but notes that his rule "has so far been applied exclusively to the calculation of stationary states and vibration amplitudes associated with transitions" [5, p. 863]. Born extends the rule to predicting any states, which belatedly (in 1954), brought him a Nobel Prize. The right formula famously occurred in a footnote added in proofs [5, p. 865], note.

theory, developed in Born and Jordan [7] and Born et al. [5]. As Heisenberg explains,

> [If one considers] a given quantity $x(t)$ [a coordinate as a function of time] in classical theory, this can be regarded as represented by a set of quantities of the form
>
> $$A_\alpha(n)e^{i\omega(n)\alpha t},$$
>
> which, depending on whether the motion is periodic or not, can be combined into a sum or integral which represents $x(t)$:
>
> $$(x,t) =_\alpha \sum_{-\infty}^{\infty} A_\alpha(n)e^{i\omega(n)\alpha t}$$
>
> or
>
> $$(x,t) =_\alpha \int_{-\infty}^{\infty} A_\alpha(n)e^{i\omega(n)\alpha t}.$$
>
> [Heisenberg [41, p. 264]]

Heisenberg next makes his most decisive and most extraordinary move. He notes that "a similar combination of the corresponding quantum-theoretical quantities seems to be impossible in a unique manner and therefore not meaningful, in view of the equal weight of the variables n and α." "However," he says, "one might readily regard the ensemble of quantities $A(n, n-\alpha)e^{i\omega(n,n-\alpha)t}$ [an infinite square matrix] as a representation of the quantity $x(t)$" [41, p. 264]. In retrospect, once one deals with the transitions between stationary states, matrices appear naturally, with rows and columns linked to these states respectively. This naturalness, however, became apparent, or one might say, became natural, only in retrospect. At the time, it was an extraordinary guess and a great conceptual invention, not unmotivated, but little prepared by the preceding history of physics.

However, it does not by itself establish an algebra of these arrangements, for which one needs to find rigorous rules for adding and multiplying these elements. Otherwise, Heisenberg cannot use these variables in the equations of his new mechanics. In essence, one needed to define the multi-, in fact infinite-, dimensional matrix algebra, which is what Heisenberg did, unaware of its existence in mathematics, as part of modern mathematics. Matrix multiplication was introduced by Arthur Cayley in 1856. For a quantum-theoretical version of the classical equation considered, which would apply (no longer as an equation of motion) to these variables, Heisenberg needed to construct the powers of such quantities, beginning with $x(t)^2$, which was actually all he needed (and noncommutativity was not needed to define it). The answer in classical theory is obvious and, for the reasons just explained, unworkable in QM, where, Heisenberg proposed:

It seems that the simplest and most natural assumption would be to replace classical [Fourier] equations ... by

$$b(n, n-\beta)e^{i\omega(n,n-\beta)t} =_{\alpha} \sum_{-\infty}^{\infty} A(n, n-\alpha)A(a-\alpha, n-\beta)e^{i\omega(n,n-\beta)t}$$

or

$$b(n, n-\beta)e^{i\omega(n,n-\beta)t} =_{\alpha} \int_{-\infty}^{\infty} A(n, n-\alpha)A(a-\alpha, n-\beta)e^{i\omega(n,n-\beta)t}.$$

[Heisenberg [41, p. 265]]

This is the main mathematical postulate, the (matrix) multiplication postulate, of Heisenberg's theory, according to him, "an almost necessary consequence of the frequency combination rules" [41, p. 265]. Although it is commutative in the case of x^2, it is in general noncommutative, expressly for position and momentum variables, and Heisenberg, without realizing it, used this noncommutativity in solving his equation, as Dirac was the first to notice [49, v. 4, p. 129].

Thus, all the key features of the formalism, eventually axiomatized by von Neumann's Hilbert-space version, were contained in Heisenberg's paper, apart from Schrödinger's equation, introduced a few months later. The key parallel mathematical innovation (when time was considered) was the complex wave function, $\psi(t)$, a vector in an infinite-dimensional complex Hilbert space, which enables one to predict the probabilities of quantum events (H is an energy operator also known as a Hamiltonian):

$$i\hbar\frac{d}{dt}|\psi(t)\rangle = \hat{H}|\psi(t)\rangle.$$

For one dimension:

$$i\hbar\frac{\partial\psi(x,t)}{\partial t} = \left[\frac{\hbar^2}{2m}\frac{\partial}{\partial x^2} + V(x,t)\right]\psi(x,t).$$

One can see why Bohr spoke of Heisenberg's discovery as inaugurating "a new era of mutual stimulation of mathematics and mechanics" [4, v. 1, p. 51]. As a mechanics or physical theory (QM was not mechanics in the usual sense either), QM was born out of the spirit of abstract mathematics, as part of modern mathematics divorced or liberated from representing nature and thus from physics, in this case specifically as matrix algebra, part of linear algebra, in infinite dimensions, over $\mathbb{C}$, a characteristically modern concept.[14] It was reconnected to

[14] It is revealing that Schrödinger, who discovered his equation differently, in his published papers recast this discovery in terms of an eigenvalue problem, as an extension of matrix algebra to functional analysis [65], [58, pp. 84–99], [59, pp. 145–166]. Functional analysis was a key development of modern mathematics, through which QM was axiomatized by von Neumann [73].

physics, only in terms of in probabilistic predictions, via Born's rule, necessary for defining probabilities as real numbers. In early days of QM, replacing real-function variables of classical physics with operator-operators in Hilbert spaces over $\mathbb{C}$ in QM, thus in effect replacing classical mathematics with modern mathematics, was called quantization. At the same time, quantization, supplemented by Born's rule, also replaced, in the case of individual systems, determinism of classical mechanics with the probabilistic character of quantum predictions. In other words, quantization was a new relationship that emerged from the Heisenberg method between modern mathematics and modern physics, as a mathematical-experimental science, here underlain by RWR epistemology. This was a broader meaning of "the Heisenberg method," as concern the relationships between mathematics and physics, underpinned by an RWR-type epistemology, revolutionary as well, anticipated as it was by Bohr's 1913 atomic theory. This meaning, in all aspects just outlined, was understood by Bohr, as a full elaboration of the sentence cited above makes clear:

> It will interest mathematical circles that the mathematical instruments created by the higher algebra play an essential part in the rational formulation of the new quantum mechanics. Thus, the general proofs of the conservation theorems in Heisenberg's theory carried out by Born and Jordan are based on the use of the theory of matrices, which go back to Cayley and were developed especially by Hermite. It is to be hoped that a new era of mutual stimulation of mechanics and mathematics commenced. To the physicists it will at first seem deplorable that in atomic problems we have apparently met with such a limitation of our usual means of visualization. This regret will, however, have to give way to thankfulness that mathematics in this field, too, presents us with the tools to prepare the way for further progress. [Bohr [4] v. 1, p. 51]

"Visualization" is used in the sense of the German *Anschaulichkeit*, which has a more general meaning of phenomenal intuition, beyond which quantum objects and behavior were placed in Heisenberg's approach. While a new era of mutual stimulation of physics and mathematics is also at stake, Bohr speaks of a "mutual stimulation of *mechanics* and mathematics" because the difference between classical and quantum physics essentially concerned individual quantum objects and their behavior. In classical physics, the theory that mathematically represents individual objects and their behavior is classical mechanics. In QM, which considers individual quantum objects and behavior, mathematics plays an essentially different role, at least in RWR interpretations, because, in Bohr's formulation in the same article, "Atomic Theory and Mechanics" (with "mechanics" used in title for the same reason), "in contrast to ordinary [classical] mechanics, the new quantum mechanics does not deal with a space–time description of the motion of atomic particles" [4, v. 1, p. 48].

As far as the physicists' attitude is concerned, the subsequent history has proven that Bohr was too optimistic. Discontent with "the limitation" in question has never subsided. Einstein, again, was in the forefront of this resistance. He did not find satisfactory or even acceptable this state of affairs as concerns physics or the

---

Hilbert, who was arguably the first to use the word "eigen," was considering the eigenvalues of integral operators as infinite matrices early in the twentieth century.

relationships between mechanics, or physics in general, and mathematics, and he never stopped seeing this limitation as anything less than deplorable. Schrödinger was quick to join Einstein, followed by most physicists and philosophers, guided by the hope that a future theory will return us to the realist form of the relationships between mathematics and physics.

Beyond being merely undesirable, the irreducible role of probability and lack of realism in "the Heisenberg method" were disqualifying features of QM as a fundamental theory for Einstein, even though he admitted that this "algebraic" method might become viable in the future, with much reluctance and hoping, in fact being convinced, otherwise [25, p. 378]. Einstein associated these, for him unacceptable, epistemological features with algebra, not without reasons, but, as explained below, not entirely accurately either. Einstein's pronouncements to the effect God "is . . . not playing at dice" are famous (e.g., Born [6, p. 88]). He was equally unhappy with the recourse to probability in dealing even with individual systems, even the simplest possible one. He was, however, most concerned with the absence of realism in considering the ultimate constitution of nature, aware, in any event, that the irreducible recourse to probability would be automatic as a result, a point made by Bohr in his debate with Einstein (e.g., Bohr [4, v. 2, p. 34]). It does not appear that Einstein contemplated realist interpretations of QM, possibly because his views were shaped by his debate with Bohr, who adopted an RWR interpretation. Einstein accepted, on practical grounds, probabilistic theories in physics and made major contributions to such theories, classical and quantum. However, he never accepted the idea, defining QM, that a fundamental physical theory, a theory dealing with the ultimate constitution of nature, could be probabilistic. In RWR interpretations, this is the only possible account of quantum phenomena, no matter how elementary. Einstein, however, saw this feature as a fault of the theory rather than, even as a possibility, as something that may be a feature of nature or our interactions with it. By contrast, according to Bohr, at stake in quantum physics was not "an arbitrary renunciation of a more detailed analysis of atomic phenomena" but "a recognition that such an analysis is *in principle* excluded," or at least may be "*in principle* excluded" [4, v. 2, p. 62]. I qualify because, in the present view, such is the case in RWR interpretations, even if Bohr himself had in mind a stronger claim that this is the case regardless of interpretation. According to Einstein:

> There is no doubt that quantum mechanics has seized hold of a beautiful element of truth and that it will be a touchstone for a future theoretical basis in that it must be deducible as a limiting case from that basis, just as electrostatics is deducible from the Maxwell equations of the electromagnetic field or as thermodynamics is deducible from statistical mechanics. I do not believe that quantum mechanics will be the starting point in the search for this basis, just as one cannot arrive at the foundations of mechanics from thermodynamics or statistical mechanics. . . .
>
> [P]erhaps the success of the Heisenberg method points to a purely algebraic method of description of nature, that is, to the elimination of continuous functions from physics. Then, however, we must give up, in principle, the space–time continuum [at the ultimate level of reality]. It is not unimaginable that human ingenuity will some day find methods which will make it possible to proceed along such a path. At present however, such a program looks like an attempt to breathe in empty space. [Einstein [25, pp. 361, 378]]

For Einstein the proper basis for his "search for a more complete conception" had to be elsewhere [25, pp. 361, 375]. He might have been right if the type of theory he wanted had to be a more geometrical theory on the model of Maxwell's electrodynamics, and then GR, grounded in Riemannian geometry. Such a theory was, he thought, unlikely to emerge from QM or QFT, and this is probably true, although there have been attempts to develop QFT along these lines. Neither theory is, however, "a *description* of nature" because, at least in RWR interpretations, neither offers a representation of the ultimate constitution of nature, which was Einstein's imperative for a fundamental theory. Accordingly, Einstein did not see QM as a viable "method" of "the description of nature" and as a "useful point of departure for future development" [26, p. 83].[15]

Einstein, who was no naïve realist, was aware that, just as GR was, a theory of the kind he wanted could only be a suitably mathematized *conceptual* idealization, the product of "a free conceptual construction" [26, p. 47]. The main problem of the Heisenberg method for Einstein was that Heisenberg's theory only provided the *algebra* of probabilities for the outcomes of quantum experiments, without a proper realist idealization, essentially geometrical in character (even if it contained algebra, as relativity did), provided by GR. Or, so it appeared before singularities entered the theory within merely a year from the discovery of the theory with Schwarzschield's 1915 work, although it took much longer to accept them as a permanent feature of the theory. Einstein never reconciled himself to the idea of singularities in relativity and thought they should ultimately be avoided. Would he have been convinced by the current evidence and Roger Penrose's 1965 proof (a decade after Einstein's death in 1955), for which Penrose was awarded the 2020

[15] There is an additional dimension to the situation. In the EPR paper [28], Einstein, by using a thought experiment, now known as the EPR experiment, argued that QM itself, regardless of interpretation, is incomplete because the EPR experiment allowed one to claim that the exact value of both conjugate variables, the position, q, and the momentum, p, could be assigned to a quantum object, while QM allowed for no such assignment (consistently with the uncertainty relations). Alternatively, if complete, EPR argued, QM would be nonlocal by allowing an instantaneous physical action at a distance, and, as such, in violation of relativity. Bohr counterargued that EPR's argument was based on the assumptions that were inconsistent (or at least ambiguous) in view of what could in fact be experimentally established, including in the EPR experiment, which would allow one to see QM as both complete within its proper scope (as complete as possible given what can actually be observed or predicted) and local [3]. While the debate concerning the subject and the EPR-Bohr exchange has continued and still does, I shall put it aside, because it does not affect my argument concerning QM in RWR interpretations, which would be rejected by Einstein in any event. Einstein acknowledged that his argument would not apply if QM were, as it is in RWR interpretations, only a statistical theory of ensembles or a probabilistic, rather than deterministic, theory of individual quantum systems, neither of which would be unacceptable to him in a fundamental physical theory (e.g., Born [6, pp. 105, 155]). He still believed, however, on the basis the EPR argument, as grounded in assumptions challenged by Bohr, that a "more complete" (i.e., realist and deterministic) theory of individual quantum behavior could be found. I have considered the subject in detail on several occasions, arguing, in accord with Bohr's view, that Einstein's argument, while subtle, is ultimately unsustainable, thus allowing one to see QM as both complete within its proper scope and local (e.g., Plotnitsky [59, pp. 227–272]).

Nobel prize in physics, that singularities of black holes were a consequence of GR, without esoteric assumptions leading to singularities in previous solutions of Einstein's equation of GR [54]? Penrose shared the prize with Andrea Ghez and Reinhard Genzel, who experimentally confirmed the existence of a massive black hole at the center of the Milky Way. It is true that black holes are ultimately quantum objects, which, however, does not help to mitigate their role in physically confirming the role of singularities with GR as such. Assuming that such a theory is possible, it is difficult to know how far we are from quantum gravity or an alternative theory of gravity, which appears necessary given the existence of black holes, or whether such a theory will bring gravity into harmony with other fundamental forces of nature. One can safely bet, however, that it will require a formidable mathematics, in which both algebra and geometry are likely to be equally germane.

While Einstein's assessment of the Heisenberg method and QM as essentially algebraic was to the point, it requires qualifications, giving it a greater complexity than Einstein's brief statement conveys. This complexity also brings geometry back into the Heisenberg method and QM. The Heisenberg method was fundamentally algebraic. It was also irreducibly probabilistic and allowed for, and was even conducive to, RWR interpretations by virtue of the algebraic features of QM, even though it could not preclude realism, in principle possible by means of other theories or by means of alternative interpretations of QM. But to be fundamentally algebraic is not the same as to be exclusively algebraic. The Heisenberg method and QM or QFT did not exclude geometry or topology, and led to new developments in both fields, for example, in noncommutative geometry. QM brought with it a new way of geometrical thinking and a new, more abstract, form of geometry, that of Hilbert spaces (over $\mathbb{C}$), to fundamental physics. Einstein would not have seen this feature as removing his main concerns, which are epistemological, and it does not. But it qualifies his claim mathematically. QM and QFT led to a new synergy of algebra and geometry in physics, a synergy by then in place in mathematics.

Some qualifications of Einstein's assessment are immediately necessary in view of the basic mathematical structure of QM and QFT. Thus, in saying that "we must give up, in principle, the space-time continuum," Einstein must have had in mind the spacetime continuum in representing the ultimate character of physical reality by a geometrical theory, such as GR, grounded in Riemannian geometry. Einstein was aware that the idea that this character may be discrete had been around for long time by then. As noted, this possibility was considered, as early as 1854, by Riemann in his Habilitation lecture, which Einstein knew. Riemann brought this possibility up, on Kantian lines, in connection with the difference between our continuous phenomenal representation of physical space and "the reality underlying space," which may be discrete [64, p. 33]. A subtler part of Riemann's argument concerned the difference between the discrete and continuous nature of the reality underlying space in the infinitely small. If it is discrete then the ground of the metric relation of the corresponding manifold is given by the mathematical concept of this manifold itself. On the other hand, if it is continuous, this ground is given by the concept of phenomena justified by experience and defined by the physical principles arising from these phenomena. Putting the question of singularities aside for the moment,

GR is a manifestation of the second case, as an example of both this grounding and a change of physical principles vis-à-vis those of Newton's theory of gravity.

QT, however, gave new reasons for the first possibility. The idea that "the reality underlying space" may be discrete has acquired additional grounds in view of the difficulties of QFT, its successes notwithstanding. As noted, the idea of this discreteness was, in view of these difficulties, advocated by, among others, Heisenberg from the 1930s on (e.g. Hagar [37, pp. 69–73]). His line of reasoning concerning it largely, albeit not entirely, faded away because of the successes of renormalization of QED in 1950 and then electroweak theory in the 1970s. Nevertheless, the difficulties of QFT (such self-interaction of elementary particles or vacuum polarization,) and, in view of the infinite values of the quantities considered, the necessity of renormalization has been seen as a problem by many physicists and philosophers, beginning, as noted, with Dirac, the founder of QED. Richard Feynman, one of the architects of the renormalizations of QED (for which he was awarded a Nobel prize) often reflected on these difficulties, including their arguably bottom line:

> It always bothers me that, according to the laws [of QT] as we understand them today, it takes a computing machine an infinite number of logical operations to figure out what goes on in no matter how tiny a region of space, and no matter how tiny a region of time. How can all that be going on in that tiny space? Why should it take an infinite amount of logic to figure out what one tiny piece of space/time is going to do" [Feynman [31, p. 57]].

If space is discrete this problem disappears, but there is no discrete mathematics thus far to handle quantum phenomena as QFT is able to do as a mathematically continuous theory, its difficulties notwithstanding. As Feynman noted on the same occasion, the problem may also disappear if physics is no longer mathematical, a possibility hypothesized by him, albeit without giving it much of a chance [Feynman [31, pp. 57–58]]. One might indeed doubt that this will happen any time soon.

The idea of the ultimately discrete nature of "the reality underlying space" or physics beyond the current QFT scales has acquired new currency in some attempts to reconcile QFT and GR, possibly by moving to a new type of theory that would account for both quantum and gravitational phenomena. Among well-known proposals, also aimed to offer alternatives to string and M-Brane theories (which are continuous), are the loop quantum gravity and causal network theory. A promising investigation along quantum-informational lines that supports to the possibility that "the reality underling space" is discrete at the Planck scale was undertaken in D'Ariano and Perinotti [16], [17]. The authors derive both Maxwell's theory and Dirac's equation (in the absence of interactions) from discrete cellular-automata architecture, and a set of fundamental principles, which do not rely on continuity, making the spacetime continuum an emergent phenomenon. These articles further suggest that the Lorentz invariance and hence SR is broken at the Planck scale, while preserving locality, defined by the assumption that physical systems can only be physically influenced by their immediate environment. These articles are also relevant here mathematically because of their use of geometric group theory, grounded in the realization that discrete mathematical objects, such as groups, defined in algebraic terms, can be considered as geometric-like objects and studied with geometric to topological techniques, developed for continuous mathematical entities. There are other theories of this type in mathematics, such as Grothendieck's étale topology, discussed later in this chapter. Still, while they have earlier precursors in algebra or even in geometry, such as finite geometries that began to be developed in the nineteenth century, these are recent developments, competing with centuries or even millennia of the dominance of continuous thinking and mathematics based on it, which led to an immense multitude and richness of theories. The very fact that these developments are modeled on continuous mathematics is a symptom of its dominance.

On the other hand, in RWR interpretations, QM or QFT does not require the assumption that the reality underlying space or the ultimate nature of physical reality is discrete, because these interpretations make a more radical assumption that the ultimate constitution of reality responsible for quantum phenomena is inconceivable and, hence, cannot be assumed to be discrete any more than continuous. If these interpretations of quantum phenomena themselves, which are strictly discrete relative to each other, are in place, it does not matter whether a theory predicting what is observed in these phenomena is mathematically continuous or discrete. Continuity is a mathematical feature of the formalism of QM and QFT in their currently standard forms, just as it is that of the formalism of classical physics or relativity. Unlike in the latter theories, however, the continuous mathematics of QM and QFT relates, via the corresponding operator algebra (cum Born's rule), to the data observed as discrete phenomena by predicting the probabilities or statistics of the occurrence of these data. At the same time, each quantum phenomenon, while discrete relative to any other, is described classically in the continuous space and time just as observed phenomena are in classical physics or relativity. I leave aside the question of whether the continuity of space or time is only a feature of human thinking rather than of nature, although that it is could be supported by extending the RWR view beyond QT to all physics (e.g., Plotnitsky [59]).

On the one hand, then, confirming Einstein's view, this relation is established via the algebra of Hilbert-space operators, which gives the Heisenberg method its fundamentally algebraic character. The continuous functions, over $\mathbb{R}$, used in classical physics or relativity (for variables such as position or momentum) are replaced by operators and their algebra in Hilbert spaces over $\mathbb{C}$. On the other hand, qualifying Einstein's view, these Hilbert spaces themselves are continuous geometrical objects, defined analogously to mathematical spaces that represent physical spaces in classical physics and relativity, with the concept of distance ("norm" in Hilbert spaces) defined algebraically in both cases.

Hilbert spaces give one the possibility, often taken advantage of by physicists, to think geometrically, by using both the rigor of algebraic formalism and our phenomenal geometrical intuition as heuristic help and guidance, in accord with "Think Geometrically, Prove Algebraically" [69, p. 277]. The principle was introduced in algebraic geometry over finite fields (an important fact in the context of the relationships between algebra and geometry, or discreteness and continuity), but is equally applicable in QM and QFT [60, pp. 76–83]. Dirac, one of the greatest algebraic virtuosi of quantum theory, was, nevertheless, fond of referring to geometrical thinking in QM and QFT (e.g., Farmelo [30]). QFT often deals with Hilbert spaces whose continuity is denser than that of regular continua such as the (real number) spacetime continuum of classical physics or relativity. Besides, continuous functions are retained, because these Hilbert spaces are those of continuous functions, which are infinite-dimensional vectors when one is dealing with continuous variables, such as "position" and "momentum," represented by operators. More accurately, these operators are abstract mathematical elements that allow one to predict the value as the position or the momentum (but never both together because of the uncertainty relations) observed in measuring instruments.

In sum, rather than only algebraic, the Heisenberg method, was equally, a new geometrical method of theoretical physics, bringing algebra and geometry together. This method was adopted and developed by Dirac in founding QED, as, in Bohr's words, "a most striking illustration of the power and fertility of the general quantum-mechanical way of description" [4, v. 2, p. 64]. Given the RWR interpretation adopted by Bohr at the time of this statement *does not describe* the ultimate nature of reality responsible for quantum phenomena, which are *described*, but by classical physics, it might be more precise to speak of the quantum-mechanical way of *thinking*.

22.4 "The Most Powerful Method of Advance": Dirac and the Mathematics of QFT

Most interpretations, including those based on the idea of quantum field, of QM or QFT (in both low or high-energy regimes) adopt one or another concept of an elementary particle. Virtually all of these concepts are defined by the feature, special to quantum physics, that, while elementary particles of the same type cannot be distinguished from each other, these types themselves are rigorously distinguishable. Thus, two electrons could be distinguished by changeable properties associated with them, such as their positions in space or time, momentums, energy, or the directions of spins. Such properties are subject to the uncertainty relations and complementarity. Thus, it is possible to locate two different electrons in separate regions in space, or, by the Pauli exclusion principle, in different states in general (which is not possible in the case of bosons, particles with an integer spin, two or more of which could be in the same state). It is not possible to distinguish them from each other by their mass, charge, or spin. In RWR interpretations, properties defining electrons or other elementary particles within each type could only be *associated* with quantum objects by means of the effects observed in measuring instruments and are not attributable to these objects independently. This, however, still allows one to maintain both the indistinguishability of particles of the same type and the distinguishability of these types themselves because both can be consistently defined by such effects.

In the present interpretation, moreover, the concept of elementary particle (or any quantum object) is only applicable at the time of observation. Hence, rigorously, one cannot speak of the same quantum object, such as the same electron, in two successive measurements. The assumption, however, that it is the same electron is a statistically permissible idealization in low energy (QM) regimes. A statistical dimension of the situation is still present because one cannot be certain that one encounters the same electron in an experiment designed to detect it after it was emitted from a source even in low-energy (QM) regimes, although the probability that it would be a different electron is low. As explained below, in high-energy (QFT) regimes, speaking of the "same" electron detected more than once even in the course of a single experiment involving several observations loses its meaning altogether. In the present interpretation, speaking of an elementary particle or any quantum object *nearly* amounts to a protocol for establishing the relationships, in general probabilistic, between quantum events—nearly but not entirely because the quantum is still assumed to exist, as something beyond conception, at the time of observation. All that is physically observed is only a property of the observed part of the instrument used but not of the quantum object considered.

The elementary character of a particle is defined by the fact that there is no experiment that allows one to associate the corresponding effects on instruments with more elementary individual quantum objects. Once such an experiment becomes conceivable or is performed, the status of a quantum object as an elementary particle could be challenged or disproven, as it happened when hadrons and mesons were discovered to be composed of quarks and gluons. This composite nature will manifest itself in a new set of effects observed in each corresponding experiment. "Elementary particles" cannot be assumed to be fundamental elementary constituents, "building blocks," of nature in RWR interpretations, because the latter preclude any assumption concerning this constitution. Nor is it possible to apply to elementary particles any specified concept of particle, any more than any other concept, such as wave or field, although the concept of quantum field could be defined as a mode of an independent RWR reality, rather than as a quantum object [59, pp. 273–306].[16]

All aspects of the concept of elementary particle just outlined are fully applicable in both low and high energy quantum regimes. In the high-energy quantum regimes, however, the concept of elementary particle acquires new features in view of the following situation, to which the mathematical architecture of QFT, beginning with

[16] Quantum field is a complex and, especially physically, still, a century after its introduction, unsettled concept, including in relation to the concept of a particle. Mathematically, although wave functions are still used in QFT, a quantum field is defined by operators, replacing wave functions, on which operators ("observables") act, in QM, by associating with every point x in a Hilbert-space used, an operator acting on this space. (For historical reasons, this procedure is sometimes referred to as a "second quantization.") The literature on the subject is massive, and I shall only mention a few works. See [47] for a basic introduction, [55] for a standard textbook, [74] for a classic treatment, [84] for the present author's sentimental favorite, [43] for a subtler realist view, and [59, pp. 273–306] for RWR interpretation.

QED, responds. In fact, with Dirac's equation, this *mathematical* architecture was discovered first and then was understood as corresponding to this *physical* situation. The case is thus a manifestation of the nature QFT as defined by the invention of abstract mathematical architectures, which, in RWR interpretations, are not tied to a representation of physical reality ultimately responsible for quantum phenomena, but only used for probabilistically predicting the data observed.

Speaking for the moment in classical terms, suppose, considering QED, that one arranges for an emission of an electron, at a given high energy, from a source and then performs an observation at a certain distance from that source by placing a photographic plate there. The probability or, if we repeat the experiment with the same initial conditions (defined by the observed classical and hence repeatable state of the emitting device), statistics of the outcomes would be properly predicted by QED. But what is the outcome? The answer is not what our classical or even quantum-mechanical intuition would expect. If one considers an electron in the low energy (QM) regime, it is impossible, because of the uncertainty relations, to predict the place of the collision exactly as is in principle possible in classical physics. The emitted electron could be found anywhere in a given area, to which such a prediction applies, or elsewhere, or not found at all. Nor can an emission of an electron be guaranteed. There is a small but nonzero probability that a collision will not be observed or that the observed trace is not that of the emitted electron. Once, however, one moves to high-energy regimes, beginning with those governed by QED, in subsequent observations one can find not only an electron (or nothing), but also other particles: a positron, a photon, an electron-positron pair. That is, in RWR interpretations, one can register the events or phenomena that we associate with such entities.

QED predicts which among such events can occur, and with what probability, and in RWR interpretations, QFT does so without representing or even allowing one to conceive of how such events come about. The corresponding Hilbert-space machinery becomes more complex, in the case of Dirac's equation making the wave function a four-component Hilbert-space vector, as opposed to a one-component or, if one considers spin, two-component vector in QM. These four components represent the fact that Dirac's equation is an equation for both the (free) electron and the (free) positron, including their spins. They can transform into each other or other particles, such as photons, in the corresponding high-energy processes. In RWR interpretations, these transformations are only manifested in observational instruments, as nothing can be said or even thought of concerning what happens between observations. By the same token, one can no longer speak of the same electron, positron, and so forth as detected in two successive measurements. It is in responding to this situation that the interpretation adopted here assumes the concept of a quantum object, including that of an elementary particle, to be an idealization (still of the RWR type and hence beyond conception) that only applies at the time of observation, rather than something existing independently [59, pp. 273–306].

In the elegant simplicity of its symbolic form, reproduced on the plate in Westminster Abbey commemorating Dirac, Dirac's equation is:

$$i\gamma \cdot \delta\psi = m\psi.$$

Similarly to that of Einstein's field equations in GR, this simplicity is deceptive. As Feynman noted, in commenting on "Dirac's matrices," but making a more general point: "Why should the fundamental laws of Nature be so that one cannot explain them to a high-school student—but only to a quite advanced graduate student? And we claim they are simple? In what sense are they simple? Because we can write them in one line. But it takes 8 years of college education to understand the symbols. Are there any simple ideas in these laws?" (A letter to T. A. Welton, February 10, 1947; cited in Schweber [68, p. 408]). The "one line" of Dirac's equation encodes a complex mathematical machinery. As introduced by Dirac, the equation was:

$$\left(\beta mc^2 + \sum_{k=1}^{3} \alpha_k p_k c\right) \psi(x,t) = i\hbar \frac{\partial \psi(x,t)}{\partial t}$$

$$\alpha_i^2 = \beta^2 = I_4 \text{ (I_4 is the idenitity matrix)}$$

$$\alpha_i \beta + \beta \alpha_i = 0$$

$$\alpha_i \alpha_j + \alpha_j \alpha_i = 0$$

This equation unfolds into four coupled linear first-order partial differential equations for the four quantities that make up the wave function. $\boldsymbol{P}$ is the momentum operator. The wave function $\psi(t,x)$ takes value in a Hilbert space $X = C^4$ (Dirac's spinors are elements of X). For each t, (t,x) is an element of $H = L^2(R^3) \otimes C^4$. This architecture allows one to predict the probabilities of high-energy (QED) events associated with the relativistic electron, just as Schrödinger's equation does for the nonrelativistic electron in QM.

An analogous, if more complex, situation is found in the case of Yang–Mills equations. In this case, one begins with quantum fields from the start. I consider, first, the general mathematics of Yang–Mills theory, in its fiber-bundle version, which transforms these initially (mathematized) *physical concepts* into purely *mathematical concepts*. These concepts handle Yang–Mills theory in mathematics and now, but not initially, commonly in physics.

In this language, Yang–Mills theory establishes that under transformations of a local gauge (changes of local trivialization of the principal fiber bundle) the quantum field considered transforms exactly in the way a connection (which is a gauge field) transforms on a principal fiber bundle. The gauge field strength is the curvature F_A of the connection. The energy of the fields is now given (up to a constant) by the Yang–Mills action functional: $\int_X \|F_A\|^2 dvol_g$. By the principle of least action, the equations of motion are now given by the Euler–Lagrange equations of this functional, the Yang–Mills equation: $d_A \star F_A = 0$. ($\star$ is the Hodge star operator acting on two forms.) Solutions of these equations are termed Yang–Mills connections or instantons. Just as the symbolic form of Dirac's equation ($i\gamma \cdot \delta\psi = m\psi$), this equation is a paragon of mathematical simplicity and elegance and is just as deceptive. Furthermore, the theory just sketched still needed to be quantized and thus also made probabilistic in accord with the experimental data observed, which is probabilistic, as is all quantum data. This was done by means of Feynman path integrals (not entirely rigorously defined mathematically), which integrate complex amplitudes, by representing wave functions by the so-called propagators (which specify the probability amplitudes for observing particles in two respective positions in a given time). One still needs Born's rule to get the probabilities of what is observed. Donaldson's theorem, proven by using the moduli space of instantons on four-dimensional manifolds, used the classical rather than quantized version of Yang–Mills theory, confirming that it had little to do with physics but only with the mathematics just sketched as pure mathematics.

While, however, in accord with my asymmetrical argument, modern mathematics, not the least, differential topology, drives the physics of Yang-Mills theory, none of the language and concepts of their contemporary mathematics are found in Yang and Mills's original paper. It was much independent of this mathematics, to which there are no references in the paper. While already prominent in mathematics itself, these concepts were not as routine in theoretical physics at the time (in 1954) as they became since, and in any event Yang and Mills did not use them. Atiyah's and then Donaldson's work dates nearly three decades later, in an entirely different mathematical landscape, which also defined contemporary work in QFT and string theory. Yang and Mills's paper built on the preceding works on isospin (isotopic spin) in QFT, in which the topology of QFT was cast. (Isospin is a "flavor" quantum number, which was eventually related to up- and down-quark content of a hadron or meson particle). Yang and Mills proceed as follows:

> Let ψ be a two-component wave function describing a field with isotopic spin 1/2. Under an isotopic gauge transformation it transforms by
>
> $$\psi = S\psi',$$
>
> where S is a 2×2 unitary matrix with determinant unity. In accordance with the previous section, we require, in analogy with the electromagnetic case, that all derivatives of ψ appear on the following combination:
>
> $$(\partial_\mu - i\epsilon B_\mu)\psi.$$
>
> B_μ are 2×2 matrices such that for $\mu = 1, 2$, and 3, B_μ is Hermitian and B_4 is anti-Hermitian. Invariance requires that
>
> $$(\partial_\mu - i\epsilon B'_\mu)\psi' = (\partial_\mu - i\epsilon B_\mu)\psi. \qquad (22.2)$$

Combining (1) and (22.2), we obtain the isotopic gauge transformation of B_μ

$$B'_\mu = S^{-1} B_\mu S + \frac{i}{\epsilon} S_{-1} \frac{\partial S}{\partial x_\mu}. \tag{22.3}$$

The last term is similar to the gradient term in the gauge transformation of electromagnetic potentials. In analogy to the procedure of obtaining gauge invariant field strengths in the electromagnetic case, we define now

$$F_{\mu\nu} = \frac{\partial B_\mu}{\partial x \nu} - \frac{\partial B_n u}{\partial x_\mu} + i\epsilon(B_\mu B_\nu - B_\nu B_\mu). \tag{22.4}$$

One easily shows from (22.3) that

$$F'_{\mu\nu} = S^{-1} F_{\mu\nu} S. \tag{22.5}$$

under an isotopic gauge transformation. Other simple functions of B than (22.4) do not lead to such a simple transformation property. The above lines of thought can be applied to any field ψ with arbitrary isotopic spin. [82, pp. 192–193]

They then develop a nonabelian gauge invariance theory, the Yang–Mill theory, and quantize it, which enables them to handle isotopic spin, a purely quantum entity. This is a major achievement. But could one say that Yang and Mills "developed (essentially independently, of the mathematical literature) the theory of principal bundles and connections in order to explain the concept of gauge symmetry and gauge invariance as it applies to physical theories"? Admittedly, coming from *Wikipedia*, this statement requires caution, but the sentiment is hardly uncommon in this and similar cases [The Yang–Mills equations, *Wikipedia*]. I don't think that one can really say this. This language is a reflection of new mathematical concepts, in part, but only in part, prompted by Yang–Mills theory, but developed by mathematics to serve mathematical needs. Of course, this mathematics can then be returned to physics. This fact, however, does not undermine my point that, while physics needs mathematics, mathematics does not need physics, even though it may be helped by mathematical *ideas* invented in physics.

The QCD (quantum chromodynamics) Lagrangian, which is that of the SU(3) Yang-Mills theory, governing the interaction of quarks and gluons, including quark confinement, reflects a great mathematical complexity of the theory

$$\mathcal{L} = \sum_q (\bar{\psi}_{qi} i \gamma_\mu [\delta_{ij} \partial_\mu + ig(G^\alpha_\mu t_\alpha] \psi_q j - m_q \bar{psi}_q i \psi_q i) - \frac{1}{4} G^\alpha_{\mu\nu} G^{\mu\nu}_\alpha$$

$G^{\mu\nu}_\alpha$ is a color field tensor;
G^μ_α is a four potential of the gluon field ($\alpha = 1, ...8$);
t_α are 3×3 Gell-Mann matrices, generators of the SU (3) color group;
$f^{\alpha\beta\gamma}$ are structure constant of the SU(3) color group;
ψ_i is the Dirac spinor of the quark field (i is a color);
$g = \sqrt{4\pi\alpha_s} (\hbar = c = 1)$ is a color charge (strong coupling constant).

Note the appearance of the Dirac spinor, ubiquitous in QFT. This might still look elegant or even simple to the practitioners of QCD, a very elite group. Once, however, one begins calculating, unfolding its symbols, as, in David Gross and Frank Wilczek's or Hugh D. Politzer's remarkable papers on asymptotic freedom [33, 34, 63], which brough them their Nobel prize, this elegance retreats. One enters the calculational nightmares for which QCD and the standard model are famous but from which new theories, perhaps quantum gravity, may yet be born, just as it happened with Heisenberg's invention of QM and Dirac's invention of QED, which eventually brought physics to Yang–Mills theory.

Bohr, as noted, saw "Dirac's ingenious quantum theory of the electron" as "a most striking illustration of the power and fertility of the general quantum-mechanical way of description," or better, way of thinking [4, v. 2, p. 64]. In Bohr's ultimate, RWR, interpretation in place by this point, the word "description" only applied to quantum phenomena, described by classical physics, without allowing for description or even a conception of the ultimate constitution of reality responsible for these phenomena. This also precluded any mathematical representation of this constitution, in contrast to Heisenberg's view at the time of his comment. By defying the possibility to maintain the identity of a quantum object, specifically an elementary particle, within a single experiment, QFT brought with it a new concept of multiplicity into quantum physics, consistent with RWR interpretations. This feature of QFT compelled Heisenberg, who made major contributions to QFT, to see Dirac's discovery as an event of a major, even unique, significance. He described it "as perhaps the biggest change of all the big changes in physics of our [twentieth] century. It was a discovery of utmost importance because it changed our whole picture of matter. ... It was one of the most spectacular consequences of Dirac's discovery that the old concept of the elementary particle collapsed completely" [42, pp. 31–33]. It was experimenting with abstract mathematics that led Dirac to this discovery, just as it led Heisenberg to the discovery of QM.

Group theory, central for physics, for example, in view of Noether's theorems, is a crucial part of this mathematical technology, because while quantum data can only be handled probabilistically, they also have a complex ordering, by obeying symmetries, including local symmetries. The latter have been especially important in QFT, not the least in leading to discoveries of new particles, such as quarks and gluons inside the nucleus, eventually establishing the standard model of particle physics. When QFT predictions concern the effects associated with elementary particles of a given type, such as electrons or quarks (there are six types of quarks), the mathematics used involves an irreducible representation of the corresponding group. This is how Murrey Gell-Mann and Georges Zweig discovered quarks. They were considering the SU(3) flavor symmetry group for hadrons, and noticed that no hadrons were associated with the irreducible representations or, as I call them, "Galois atoms" of this group [60]. To assume that there were elementary particles associated with these representations implied, however, that hadrons were not elementary particles but composites of new particles, named quarks by Gell-Mann.

Galois atoms, as a *concept* (although not the term), were introduced by Wigner [80], whose work played a major role in the history of symmetries and group representations in QT, including as mathematically different from those in classical physics and relativity. This difference was in part due to the role of discrete symmetries, such as mirror symmetries, not associated with conservation laws as some continuous symmetries are in accordance with Noether's theorems. Wigner's work was central for understanding discrete symmetries in QT. This is yet another testimony to both the role of abstract mathematics in QT to the immense reach of Galois's concept of group, far beyond the initial problem in algebra this concept was developed to address. Galois theory is a paradigmatic early concept of abstract algebra that has played an important role in numerous developments of modern mathematics. Eventually, Galois theory also entered QFT in the context of renormalization.

While the role of Galois theory in renormalization has been rarely considered beyond those (not very many) who work on it, the subject of renormalization in general in QFT has been extensively covered in physical and philosophical literature. I shall accordingly only summarize essential points most relevant in the present context. It was realized by the early 1930s that the computations provided by QED were reliable only to a first order of perturbation theory. Perturbation theory is a set of approximation methods using mathematical perturbation for describing a complex quantum system by using a simple one with a known solution and adding to it a "perturbing" Hamiltonian representing a small disturbance of that simple system. These computations led to the appearance of infinities when one attempted to use the formalism for calculations that would provide closer approximations matching the experimental data. These difficulties were eventually handled through the renormalization procedure, which became a crucial part of QFT and remains so, because QFT as currently used still contains these divergencies, specifically certain divergent and, hence, mathematically illegitimate integrals. Roughly, the renormalization procedure manipulates such infinite integrals by, at a certain stage of calculations, replacing them with finite integrals through artificial cut-offs. These cut-offs have no proper mathematical justification and are performed by putting in, by hand, experimentally obtained numbers that make these integrals finite, which removes the infinities (which happen to cancel each other) from the final results of calculations (e.g., Teller [71, pp. 149–168]). Yang–Mills theory is not renormalizable in dimensions higher than four, which does not affect the standard model, which is four-dimensional. The renormalization of the corresponding Yang–Mills theories was crucial for both the electroweak unification and QCD, which comprise the standard model.

At some point in this history a new concept, that of the renormalization group, was introduced, with the work of Kenneth G. Wilson being arguably most significant in the development of its role. (The concept also played an important role and was in part developed in solid state physics.) The renormalization group is a mathematical technique, rather than a group in its technical sense (it may be seen as a semigroup), especially as part of the so-called effective QFTs, that enable one to properly investigate the changes of a quantum system as represented ("viewed") at different scales. It reflects the changes in the underlying force laws of QFT, changes due to the fact that the energy scale at which physical processes occur varies, with energy-momentum and resolution distance scales obeying the uncertainty relations. Actual computations are difficult and massively assisted by, and in fact impossible apart from, the use of computers. Conceptually, one evaluates the transformations in the formalism in accordance with Feynman's diagrams, which are, however, only heuristic guides to such calculations in most views, including Feynman's own (e.g., Plotnitsky [59, pp. 300–302]).

Quite unexpectedly, Galois theory was brought in to recast the mathematics involved in a new form. The formal mathematical complexities are nearly prohibitive even for those working in QFT by virtue of using some stratospheric techniques, such as Grothendieck's theory of motives. These complexities cannot be considered here. My point is the remarkable fact itself of the role of a "Galois group" in renormalization theory. The idea was initially suggested by Pierre Cartier, who designated the corresponding Galois group a "cosmic Galois group," in view, one suspects, of its potential significance in the extension of QFT enabling one to handle gravity [9]. It is sufficient to consider the basic statement of how a specific ("motivic") Galois group enters in Alain Connes and Matilde Marcolli's article. The article and their related work also provide necessary formal definitions for a mathematically informed reader, as the mathematics is formidable and requires advance levels of training.[17] They say:

> The divergences of quantum field theory are a *highly structured phenomenon* [even beyond the renormalization group, which already structures them]. More precisely, they provide data that define an action of a specific "*motivic Galois group*" U^* on the set of physical theories.
>
> In particular, this exhibits the renormalization group as the action of a one parameter subgroup $\mathbb{G}a \subset U^*$ of the above Galois group. . . .
>
> The natural appearance of the "motivic Galois group" U^* in the context of renormalization confirms a suggestion made by Cartier in [9], that in the Connes–Kreimer theory of perturbative renormalization one should find a hidden "cosmic Galois group" closely related in structure to the Grothendieck–Teichmüller group. The question of relations between the work of Connes–Kreimer, motivic Galois theory, and deformation quantization was further emphasized by Kontsevich in [45]. . . .
>
> The "motivic Galois group" U^* acts on the set of dimensionless coupling constants of physical theories, through the map of the corresponding group G to formal diffeomorphisms constructed in Connes and Kreimer [12].
>
> This also realizes the hope formulated in Connes [10] of relating concretely the renormalization group to a Galois group. . . .
>
> These facts altogether [all together?] indicate that the divergences of Quantum Field Theory, far from just being an unwanted nuisance, are a clear sign of the presence of totally unexpected symmetries of geometric origin. [Connes and Marcolli [13, pp. 4073–4075]; emphasis added].

[17] See Connes and Marcolli [14, pp. 95–136] for a comprehensive exposition, which also explores the role of Riemann's differential geometry in this context. It is a long work, again, prohibitively difficult technically.

This conclusion counters the common sentiments, some more strongly negative, including by Dirac, the founder of QED, toward the divergencies of QFT and renormalization from the 1930s on. These sentiments were tempered but far from vanquished by the successes of QFT enabled by renormalization, beginning with calculating the Lamb shift in the 1950s by renormalizing QED by Sin-Itiro Tomonaga, Julian Schwinger, and Feynman, who were awarded the Nobel prize for this work. Even those who were less worried about renormalization, such as Schwinger or Freeman Dyson, would not think of it as a reflection of deeper aspects of the mathematics of QFT, moreover, making "the divergences of quantum field theory ... a *highly structured phenomenon*" and, thus, making them more properly mathematical. For mathematics is a creation of more rigorously structured, conceptually and logically, compositions, sometimes elevating this creation to the organization of structures, compositions of compositions, such as sequences of algebraic fields in Galois theory in algebra, or sequences QFTs in the motivic Galois theory of QFT. I see this elevation as the Galois principle, defined below. The use of mathematics in physics is the use these compositions in dealing with the phenomena considered, in QFT, in RWR interpretations, by predicting the data observed. Born's rule is still needed.

Connes and Marcolli's conclusion also suggests that the use of continuous mathematics in QFT may not be a problem, reflected in the necessity of renormalization, that would *require* this mathematics to be replaced with discrete mathematics, making QFT a mathematically discrete theory at least at the deeper scales, ultimately the Planck scale. Instead, the continuous mathematics of QFT, including the very necessity of renormalization, *might* be a manifestation of how nature or (a necessary qualification in RWR interpretations) our interactions with nature work in high-energy quantum regimes. Discrete alternatives cannot be excluded. As noted, advocating them is understandable and has been around for a long time, along with attempts to develop such alternatives. This is not an easy task, and most such attempts thus far tend to be based on transferring the methods of continuous mathematics to the discrete domain. This is not a criticism. Geometric group theory and Grothendieck's étale-cohomology theory show the effectiveness of this approach. As in dealing with other outstanding problems of fundamental physics, it is difficult to predict how this situation will be resolved, if it will ever be. On the other hand, "the symmetries of geometric origin" invoked by Connes and Marcolli do not appear to be in conflict with to RWR interpretations of QFT. These interpretations may, however, also apply to a discrete or finite theory that may replace QFT in its present form of a continuous theory, dealing with the irreducibly discrete phenomena.

The understanding of the physical meaning and implications of the motivic Galois theory is a complex task, which, as far as I know, have been barely pursued. What remains remarkable is an unexpected and potentially far-reaching role of Galois theory in the mathematics of QFT, including in linking algebra and geometry there, just as they are sometimes linked by Galois theory, for example, in the mathematics of Riemann surfaces and then Grothendieck's topos theory and étale cohomology, which led Grothendieck to motivic cohomologies. The latter would require a separate discussion beyond my scope.[18] The point that I would like to make is that purely mathematical aspects of Galois theory suggest that its role in QFT and renormalization group may not be so surprising, especially in retrospect, admittedly a suspect way to make things look expected. In any event, one can formulate a more general principle, the Galois principle, at work here.

First, one can think of the standard Galois theory in algebra as related to the "scales" corresponding to the fields extending the original field, by a fundamental theorem of Galois theory. The theorem states that given a finite Galois extension K/k, there is a bijection between the set of subfields $k \subset E \subset K$ and the subgroups $H \subset K$. Then, E is given by the set of invariants of K under the action of $H, E = K^H = a \in K : ga = a, g \in H$. If H is a normal subgroup then $G/H \cong Gal(E/k)$ and conversely, if E/k is a normal field extension, then the corresponding subgroups $Gal(K/k)$ is a normal group. In the case of renormalization group, one deals with a transformation of theories, QFTs, rather than fields, making the corresponding Galois theory metatheoretical. The principle itself, however, remains in place, in both cases, referring to structuring structures, composing compositions.*The Galois principle states that a given set of domains (fields, manifolds, or theories) is transformed by means of an algebraic structure, such as a group or semi-group (or some other structure) so that each domain is associated with a substructure of this structure, such as a subgroup by the Galois group of a field.*

The connection between Galois theory and Riemann surfaces was established long before topos theory, and found, for example, in Weyl's 1913 book on Riemann surfaces, known to J-P. Serre and possibly Grothendieck:

> The group of cover transformations, regarded as an abstract group, expresses purely and completely everything in the relation between the normal covering surface $\tilde{F}$ and the base surface $\boldsymbol{F}$ which has the character of analysis situs [topology]. This group is also called the Galois group of $\tilde{\boldsymbol{F}}$. It is in fact the analog of the Galois group of a normal algebraic field (of finite degree) over a base field. [79, p. 58]

Grothendieck's key insight (in part via Serre's work), leading him to étale cohomology, as a version of a Weil cohomology, was to generalize, in terms of category theory, the concept of "open set," beyond a subset of the algebraic variety (which gives only a very weak topology, Zariski's topology, too weak to develop a proper cohomology theory over finite fields). This was possible because the concept of sheaf and of the cohomology of sheaves could be defined by any category, rather than only that of open sets of a given space. Étale cohomology is defined by this type of replacement, specifically by using the category of étale mappings of an algebraic variety, which become "open subsets" of the finite unbranched covering spaces of the variety, a vast and radical generalization of Riemann's concept of a covering space. Part of the genealogy of this generalization was the fact that the fundamental group of a topological space, say, again, a Riemann surface, could be defined in two ways. First, it can be defined more geometrically, as the group of equivalence

[18] See Marcolli [48] for an introduction to motives in the set of contexts under discussion here.

classes of the sets of all loops at a given point, with the equivalence relation given by homotopy. Alternatively, a fundamental group can be defined as a group of transpositions of covering spaces. In this definition, it is analogous to the Galois group of the algebraic closure of a field, thus confirming the genealogy of étale cohomology in both Riemann's and Galois's work. Étale mappings give a sufficient number of open sets to define adequate cohomology groups for some coefficients for algebraic varieties over finite fields. For complex varieties, one recovers the standard cohomology groups with coefficients in any constructible sheaf.

The category of étale mappings is a topos, a concept that is, for now, the most abstract algebraic form of spatiality in mathematics. It was seen as such by Grothendieck himself, who compared his contribution to the idea of space in mathematics to that of Einstein in GR in physics and that of Schrödinger in QM [35, pp. 68]. A reference to Schrödinger is not self-evident. It may, however, be plausibly seen as referring to the role of Hilbert spaces over $\mathbb{C}$ in QM, in Schrödinger's case infinite-dimensional ones, through which one relates to events in actual physical space represented as 3D real manifold. I shall further comment on this point in the next section. Time is a different matter, which will have to be put aside here.[19] I might add that Grothendieck's theory of étale cohomologies, based on Galois theory, was also one of the *motivations* for his concept of motive, although the Grothendieck–Teichmüller theory preceded étale cohomologies (e.g., A'Campo et al. [1]).

22.5 On Reasonable and Unreasonable Effectiveness, and Ineffectiveness, of Physics in Mathematics: Yang–Mills Theory and Differential Topology

The influence of twentieth-century physics, including QFT, on contemporary mathematics is a vast subject, which would require a full-length study to be considered with a reasonable comprehensiveness.[20] Only a few aspects of this influence will be addressed here. Indeed, my argument here concerns a contrary, if perhaps (in musical terms, preserving a certain harmony) contrapuntal relationships: a limited nature of this influence, significant or even momentous as an impetus of QM and QFT may have been in mathematics. This point amplifies my more general argument that in twentieth and twenty-first century physics, as a mathematical

[19] The subject was discussed from the RWR perspective by this author elsewhere [61].

[20] A useful collection of essays dealing with the traffic between them at the contemporary stage "one century after Hilbert" (to cite its title, Foundations of mathematics and physics one century after Hilbert), some of which address the role of contemporary physics in mathematics, is Kouneiher [46]. The contributors include several prominent figures, some cited in this chapter, and many essays contain useful references. Epistemological issues, however, receive little attention. Most contributions adopt manifestly realist or Platonist positions.

science, and in mathematics itself, mathematics always leaves physics behind. In QT in RWR interpretations, mathematics leaves behind its representational function, thus following the project of modern mathematics, which leaves all physics behind altogether, including when it converts what it takes from physics into mathematics. My subject in this section is a more recent history of this situation, still only to be addressed by way of an example, a case study. This example is the role of Yang-Mills theory in the topology of four-dimensional manifolds, including $\mathbb{R}^4$, a unique object. Not only is four the only dimension, n, for which $\mathbb{R}^n$ has an exotic smooth structure, but $\mathbb{R}^4$ has an uncountable number of exotic smooth structures.

As noted, Yang–Mills theory is grounded in nonabelian local gauge symmetries. The concepts of local and specifically gauge symmetries have been central to field theories, classical and quantum, beginning with Maxwell's electrodynamics, the first known theory defined by a gauge symmetry. Arguably, the single defining event (if there can be one) of this history in the twentieth-century was Weyl's idea that invariance under the change of a scale or "gauge" (*Eichinvariaz*) might form a local symmetry of GR. Weyl's attempt, based on this idea, to unify GR and electromagnetism did not work, but the idea received a new lease on life in QT, courtesy of Weyl himself, Vladimir Fock, Fritz London, Wolfgang Pauli and others. In QT, gauge theories involve local phase changes in quantum fields, like $\hat{\psi} \rightarrow e^{(i\pi(x)t)}\psi$.

Initially, these theories were abelian, including in quantum electrodynamics, which has the symmetry group U(1) and one spinor gauge field, the electromagnetic four-potential, with the photon as its gauge boson. As discussed above, however, in 1954, in dealing with problems of strong interactions, Yang and Mills introduced, as a major novelty, nonabelian gauge theories, by generalizing the gauge invariance of electromagnetism, based on the action of SU(2) symmetry group on the isospin doublet of proton and neutron. Eventually the theory was applied to the weak forces, with the group SU(2), but as a local symmetry group, while the isospin SU(2) symmetry is global. A global symmetry keeps a given property invariant for a transformation that is applied simultaneously at all points of spacetime. A local symmetry keeps a property invariant when a different symmetry transformation applies at each point of spacetime. (A global symmetry is also a local symmetry.) The standard model is a non-abelian local gauge theory governed by the tensor product of three symmetry groups U(1)⊗SU(2)⊗SU(3) and broken symmetries. It contains twelve gauge bosons: the photon, three weak bosons, and eight gluons.

Yang-Mills theory (especially in the 1970s, after the renormalization of the electroweak theory gave the theory its centrality in QFT and the standard model) stimulated new mathematics, just as (in descending temporal order) QED, QM, and GR did earlier in the twentieth century or classical physics, especially through partial differential equations, from throughout the history of modern mathematics. Beyond, however, a more general point, argued for here, that *stimulating* new mathematics is *not new mathematics*, there is a difference in the case of QT. Classical physics is a mathematical idealization of objects and processes that we observe in daily life, such as the motion of solid bodies. As such, it is a form of mathematical abstraction but only in the following sense: only some, mathematizable, properties of such objects and processes are considered, while their other properties are disregarded. In GR or already SR this is no longer the case, although the representational, realist role of differential equations is retained. Mathematicians, however, eventually began to study differential equations apart from their role in physics, including those that have no connection to physics at all, thus bringing this field in accord with modern mathematics. That took a while, and had to wait until the twentieth century. Of course, even when it was connected to physics, the study of differential equations was a mathematical field, developed by some of the most important figures in the history of mathematics, such as Euler (even before the rise of modern mathematics), Laplace, Cauchy, and Poincaré. This, however, is not the same as expressly making this field modern mathematics. As Gromov remarked, in prefacing his book *Partial differential relations*, which had a major impact in several areas of topology and geometry, apart from connections to physics:

> The classical theory of partial differential equations is rooted in physics, where equations (are assumed to) describe the laws of nature. Law abiding functions, which satisfy such an equation, are very rare in the space of all admissible functions (regardless of a particular topology in a function space).
>
> Moreover, some additional (like initial or boundary) conditions often insure the uniqueness of solutions. The existence of these is usually established with some a priori estimates which locate a possible solution in a given function space.
>
> We deal in this book with a completely different class of partial differential equations (and more general relations) which arise in differential geometry rather than in physics. Our equations are, for the most part, undetermined (or, at least, behave like those) and their solutions are rather dense in spaces of functions. [Gromov [32],"Foreword"]

This is in 1986, two centuries into modern mathematics. Gromov's parenthesis "are assumed to" is in accord with the present philosophical position, in which this can only be an assumption, especially as concerns the differential equations of QM and QFT, such as Schrödinger's, Dirac's, or the Yang–Mills, which are not equations of motion in RWR interpretations. (I am not ascribing this position to Gromov.) The work of Atiyah was especially notable for its use of Yang-Mills theory in making it part of mathematics, in accord with the view expressed by Gromov here, although Atiyah himself takes a different view of this situation, in part by disparaging the role of algebra in QT or even in mathematics in general.[21] Then Donaldson, Atiyah's

[21] Atiyah goes as far as stating that "algebra is the offer made by the devil to mathematicians," which even makes one "stop thinking" [2, p. 7] (cited in [53, p. 15]). It is not easy to distinguish in this unconditional way algebra and geometry, as this chapter shows and as Atiyah acknowledged, still holding on to his negative sentiment toward algebra. Intriguingly, Weyl made a similar, remark in 1939: "In these days the angel of topology and the devil of abstract algebra fight for the soul of each individual mathematical domain" [77, p. 500]. Made in a more genial vein, this comment is penetrating in suggesting the interplay on both in each area of mathematics. It may be that the devil of algebra even for Atiyah and, certainly for Weyl, is only that of algebra as the "machine" (Atiyah's word) of calculation, algebra deprived of geometry, or topology. The algebra advocated in this chapter embraces and sometimes creates geometry and topology, just as the geometry and topology advocated here embrace and sometimes create algebra. It would be difficult to find greater manifestations of this interplay than Weyl's or Atiyah's own work, including that on the

student, showed that the differential classification of smooth four-dimensional manifolds was very different from their classification (up to homeomorphism) as topological manifolds. A crucial difference, which made Donaldson's work unexpected, was that, while Atiyah studied mathematically the solution of the Yang–Mills equations, Donaldson used these equations for approaching purely mathematical problems. (There was important work along these lines by others, such as Karen Uhlenbeck and Clifford Taubes.) Soon thereafter Michael Freedman used Donaldson's work to demonstrate the existence of the exotic $\mathbb{R}^4$. This is one of the most remarkable results in differential topology, although the existence of similarly exotic manifolds, such as seven-dimensional spheres, Milnor spheres, was known earlier. Gauge theory became part of mathematics as such, albeit not without some further help from physics. In the 1990s Edward Witten and Nathan Seiberg develop new gauge-field methods, based on supersymmetry theory, that led to important topological invariants, Seiberg–Witten invariants, which can be used to prove results similar and even stronger than those of Donaldson's theory, more easily. Technical details related to so-called moduli spaces are important but are secondary here. Seiberg–Witten theory still uses physical concepts, such as that of monopole. In any event, the local gauge theory, introduced by Yang and Mills independently of any investigation in pure mathematics, has become a major part of pure mathematics.

In other words, it was mathematics that made this theory properly mathematical, with some physicists, such as Witten, contributing to this process, primarily in terms of concepts, intuitions, and partial proofs, rather than full mathematical proofs and other features necessary in mathematics as a disciplinary field. Physics may provide mathematics with a new concept or a new mathematical problem, and a new concept is always a new problem or set of problems, keeping the concept alive as gives rise to new concepts and, and as problems [60, pp. 52–76]. Once, however, a concept enters mathematics it must live its new mathematical life, equally defined by logical and other aspects required by mathematics. It leaves physics behind. The combination of modern mathematics and the twentieth-century physics gave this situation a new dimension insofar as new mathematical concepts invented in physics may be abstract, rather than representational, concepts with a history in modern mathematics, divorced from physics. Both QM and QFT, including Yang–Mills theory, or, on realist lines, GR, are examples of using such concepts.

Accordingly, rather than following technical aspects of the mathematics developed from Yang–Mills theory, I would like to reflect further on this point, via an intriguing, if cryptic, comment by Poénaru, a pure mathematician. He is known for, among other things, his co-discovery of a Mazur manifold (a contractible, compact, smooth four-dimensional manifold with boundary which is not diffeomorphic to the standard four-dimensional ball). I shall attempt to give this comment a suitable sense, without claiming that this sense is Poénaru's own. Poénaru reflects on the

mathematics of gauge theory by Weyl and the mathematics of Yang–Mills theory by Atiyah. As this chapter argues, the relationships between continuity and discontinuity in these mathematical fields, or in present-day and likely future fundamental physics, are no less complex.

fundamental differences in the topology of four-dimensional manifolds, vis-à-vis those in all other dimensions, a difference that Yang–Mills theory unexpectedly helped to establish. "The tools," he says, "came from a really unexpected direction, the physics of the Standard Model of elementary particles and fields, ... the non-linear Yang–Mills equations (nonabelian gauge theory)" [62, pp. 698]. He then notes:

> If X^n is a smooth manifold homeomorphic to $\mathbb{R}^4$ then, if also $n = 4$, it was known since a while that X^n is diffeomorphic to the standard $\mathbb{R}^n$. This basic fact has been proved by Moise for $n \leq 3$ and by Stallings for $n \geq 5$.
>
> Moreover it was also known, essentially via the work of Kervaire and Milnor that, in dimensions $n = 4$, the difference between DIFF and TOP (which can be there in dimensions $n \geq 5$) is completely controlled by the discrete invariants of algebraic topology, let us say by some "discrete quantum numbers." (Poénaru, [62], p. 698)

An invocation of "quantum numbers" is unusual, although the language of "quantization" was used in differential topology in dealing with operators adopted from QT, such as Dirac's operators. Quantum numbers used in QT are always discrete, which was one of the key aspects of QT from its onset, even before QM, which gave this discreteness a more rigorous mathematical treatment. Quantum numbers represent values of conserved quantities of quantum systems. (In RWR interpretations, the dynamics, as well as quantum numbers, only refer to what is observed in observational experiments, and not to the independent properties of quantum objects.) They correspond to eigenvalues of operators that commute with the Hamiltonian and hence can be (ideally) exactly known at the same time as the energy of the system. A specification of all quantum numbers of a quantum object fully characterizes its basic state, and these numbers can in principle be measured together. QM or QFT is different from classical mechanics, where the values of all quantities characterizing the system are changing continuously or are constant. This is why this difference is sometimes referred to as the quantization of these quantities, technically, only some of them because some quantum numbers, such as spin, have no classical counterparts. There is no classical physics of spin. One can, however, classically observe and measure, and predict, as an effect of the interaction between a quantum object and the instrument used, either the value of the spin (divided by $\hbar = h/2\pi$) or its direction. The value is constant for any given type of particle, and in certain circumstances, this direction can be predicted with probability one, as in the EPR type experiments [28], without requiring causality or realism and hence consistently with RWR interpretations [59, pp. 227–272]. Quantum numbers take values in discrete sets of integers or half-integers. They are indeed discrete invariants.

While Poénaru's comparison between the discrete invariants of algebraic topology and "discrete quantum numbers" is peculiar, it may be more than merely a metaphor. It signals a potentially important point:

> All the facts above, which were already well-known by, let us say 1970 or before, fail to be true when $n = 4$. This is so because we have here the following surprising result.

> **Theorem E (Freedman-Donaldson).** There exist open manifolds smooth X^4 s.t.[such that]
> $X^4 = \mathbb{R}^4$ TOP but $X^4 \neq \mathbb{R}^4$ DIFF
> It is here that one can really perceive that there is a big mysterious chasm between dimension four and the other dimensions. In dimension four the discrete invariants of algebraic topology are of no use when one tries to compare DIFF and TOP. ...
> The theorem in question is the first sign that the difference between DIFF and TOP, which is non-existent at category level in dimensions ≤ 3, and completely amenable to the discrete techniques of algebraic topology in dimensions ≥ 5, is a big wide-open gap in dimension exactly four, totally un-understandable by algebraic topology. I think that understanding this gap is an important problem, and I tend to view the smooth 4-dimensional Schoenflies issue [which is to describe, up to diffeomorphism, the smooth compact bounded n-manifolds Δ^n gotten by splitting S^n along a smooth embedding $S^{n-1} \subset S^n$] as a first small step in that direction. But what we will eventually need is a new sort of quantum topology, or something like it, not yet available to us. This should allow us to make good sense of the gap DIFF/TOP in dimension four, as well let us say, as we understand that kind of issue today in dimensions higher than five, via the sophisticated methods of differential and algebraic topology, methods which seem powerless for our gap. [Poénaru [62, pp. 699, 702]]

My main interest is the suggestion that "what we will eventually need is a new sort of quantum topology, or something like it, not yet available to us." Poénaru does not explain what sort of quantum topology or what he means by quantum topology. One might surmise, given the role of Donaldson's use of Yang–Mills theory, that Poénaru had in mind some mathematics coming from QFT, possibly from topological quantum field theory, which he mentions, via, in addition to Yang–Mills theory, Seiberg–Witten equations [62, p. 698]. This field has made major contributions to mathematics. Some have even argued that these areas of theoretical physics, especially extended to string and M-Brane theory, were more significant in mathematics than, thus far, in physics. Yang–Mills theory is different because it emerged more directly in physics itself and in response to the problems of QFT at the time, eventually helping to solve some of them, thus leading to major advances in physics. It was only made into or became part of a purely mathematical theory, entirely separated by physics, by Atiyah, Donaldson, and others later. On the other hand, the Seiberg–Witten theory depends on physical concepts, such as monopoles, and as such retains connections and has implications for physics.

There are, thus, two issues at stake. The first is a possible use QFT or other forms of theoretical physics, perhaps some form of string or M-brain theory, as a source of the mathematical concepts or theories for dealing with the topological problem in question. The second is the purely mathematical nature of the mathematical concepts and theories necessary, as concerns the relationships between continuity (in the topology and geometry of manifolds) and discontinuity (of the algebra of discrete invariants, such as cohomological ones) in topology, regardless of where this mathematics originates. This new mathematics may then also become important for physics, but that is not at issue at the moment.

As concerns the first issue, even if some form of QFT or its extension, or any form of QT, will be such a source, these concepts or theories will still need to be made *mathematics* proper to solve the problem in question. It is instructive to considers from this perspective a reflection on these relations by Atiyah. This reflection is a few decades old but is not out of date and provides plenty of food for thought. Atiyah was actively involved in the debates concerning these relations and had strongly positive views of the role of physics in inventing new mathematics. At the same time, he was aware that such inventions require more rigorous treatment, especially rigorous proofs, in mathematics itself. He says:

> In the last quarter of the 20th century, ... there has been a tremendous incursion of new ideas from physics into mathematics. This is perhaps one of the most remarkable stories of the whole century. It requires perhaps a whole lecture on its own but basically, quantum field theory and string theory have been applied in remarkable ways to get new results, ideas and techniques in many parts of mathematics. By this I mean that the physicists have been able to predict that certain things will be true in mathematics based on their understanding of the physical theory. Of course, that is not a rigorous proof, but it is backed by a very powerful amount of intuition, special cases, and analogies. These results predicted by the physicists have time and again been checked by the mathematicians and found to be fundamentally correct, even though it is quite hard to produce proofs, and many of them have not yet been fully proved....
>
> One gets another beautiful story with explicit results, going by the name of mirror symmetry. All this comes from quantum field theory in 1+1 dimensions.
>
> If we move up one dimension, where we have 2-space and 1-time, this is where the theory of Vaughan–Jones of invariants of knots comes in. This has had an elegant explanation or interpretation, in quantum field theory terms.
>
> Also coming out of this is what are called ‘quantum groups’. Now, the nicest thing about quantum groups is their name. They are definitely not groups! If you were to ask me for the definition of a quantum group, I would need another half hour. They are complicated objects, but there is no question that they have a deep relationship with quantum theory. They emerged out of physics, and they are being applied by hard-nosed algebraists who actually use them for definite computations.
>
> If we move up one step further, to fully four-dimensional theory (three-plus-one dimensions), that is where Donaldson’s theory of four-dimensional manifolds fits in, and where quantum field theory has had a major impact. In particular, it led Seiberg and Witten to produce their alternative theory, which is based on physical intuition and gives marvellous results mathematically as well. All of these are particular examples. There are many more.
>
> Then there is string theory and this is already passé! M-[Brane] theory is what we should talk about now, and that is a rich theory, again with a large number of mathematical aspects to it. Results coming out of it are still being digested, and will keep mathematicians busy for a long time to come. [Atiyah [2, pp. 11–13]]

“Quantum groups” are indeed not groups (roughly, they are noncommutative, such as Hopf, algebras with added structure). But then, they are not exactly quantum either. They are called quantum because of noncommutativity, and they have been used in QFT and string and M-Brane theories. Nevertheless, there is a conflation here of the term “quantum,” pertinent to Poénaru’s point under discussion. The essence of the quantum in physics is the combination of *physical discreteness* and *probability*, moreover, at least in RWR interpretations, in the absence, in contrast to classical physics, when the latter uses probability, of the underlying causality. (This causality amounts to determinism in dealing with the ultimate individual constituents of the systems considered.) At the same time, quantum probabilities are predicted, via the law of amplitudes and Born’s rule, by *continuous theories*, over $\mathbb{C}$, such as QM and QFT. Mathematical theories mentioned by Atiyah are not really concerned with probability, never even mentioned by Atiyah in the article just cited. Discreteness is mentioned but only in the purely mathematical context, such as that of discrete groups.

But that just confirms my point. While prompted by physics, the theories Atiyah invokes have been developed mathematically as they must have been to become the mathematical theories they are now. Just as much, in fact more, mathematics in these areas has no connections to physics or makes such connections essentially irrelevant. Naturally, it is not only a matter of a rigorous mathematical proof, to which one cannot reduce mathematics, and Atiyah is right both to qualify the absence of such proofs and not to limit mathematics to them. The decades following Atiyah's article have proven that the "Results coming out of [M-Brane theory] ... kept mathematicians busy for a long time to come." But that had often little to do with physics. As noted, Donaldson's work using Yang–Mills theory, has no physics in it. His first paper on the subject does not even mention the word "quantum" or "physics," or "discreteness" or "probability" [23]. This was to be expected, for one thing, because, as noted, he used the differential-geometrical Yang–Mills theory rather than the quantized one. Of course, those physicists (or mathematicians) who would use such mathematically developed theories in physics would have to be concerned with quantum discreteness and probability if they want to use them to predict quantum data.

Atiyah's reflection echoes, perhaps deliberately, Robert MacPherson's (1995) statement from a widely circulated letter in organizing a special program on QFT at the Princeton Institute for Advanced Study:

> The goal is to create and convey an understanding, in terms congenial to mathematicians, of some fundamental notions of physics, such as quantum field theory. The emphasis will be on developing the intuition stemming from functional integrals. One way to define the goals of the program is by negation, excluding certain important subjects commonly pursued by mathematicians whose work is motivated by physics. In this spirit, it is not planned to treat except peripherally the magnificent new applications of field theory, such as Seiberg–Witten equations to Donaldson theory. Nor is the plan to consider fundamental new constructions within mathematics that were inspired by physics, such as quantum groups or vertex operator algebras. *Nor is the aim to discuss how to provide mathematical rigor for physical theories. Rather, the goal is to develop the sort of intuition common among physicists for those who are used to thought processes stemming from geometry and algebra.* [Dijkgraaf [20]; emphasis added]

Ludwig Faddeev, my former, (strongly) mathematically oriented, professor of QM and QFT at the University of St. Petersburg, who cited the passage, suggested the emphasized sentence as a definition of "mathematical physics," more commonly defined as application of mathematical methods in physics, a view questioned by Faddeev [29, p. 555]. It seems to me that what this sentence proposes is closer to something like "physical mathematics": mathematics emerging from physics and its intuitions.[22] I have no quarrel with this goal. In fact, it is fully consistent with my

[22] It is revealing that in discussing QFT on this occasion, Faddeev never mentions either discreteness (or discontinuity) or probability either. Would not a new sense of these concepts, as

argument here, because, valuable as they may be, such intuitions and concepts they give rise to must still be developed into rigorous mathematical concepts and theories. This process is, however, likely to require new strictly *mathematical* intuitions as well. As I have argued here, it is doubtful, for example, that valuable mathematically as they might have been the concepts found in Yang and Mills's paper would have been sufficient for Atiyah's and Donaldson's work on four-dimensional manifolds. Their mathematics was developed via intuitions and techniques of differential topology and geometry far beyond anything found in Yang and Mills's paper.

As it happened, the lectures in the program, as represented in two volumes at a formidable length of 1500 pages, deal primarily with *mathematical* intuitions common among physicists, primarily among string theorists, who dominated theoretical physics at the time (in the late 1990s), especially at the Institute or at Princeton University. In other words, these physicists in this work were already in effect mathematicians, used more "thought processes from geometry and algebra," even if they didn't always offer rigorous mathematics, especially proofs. This is quite different from what one finds in Yang and Mills's paper, or in the works of founding figures, such as Heisenberg and Dirac. Their work, mathematically defined as it was, combined these mathematical intuitions with essentially physical ones. In fact, in conflict with the above cited statements, Witten's and related works is dominant presence in the lectures. MacPherson's sense of "intuitions common to physicists" was somewhat narrow and in effect related to intuitions of string theorists, most of whose intuitions were mathematical, to begin with.

This is not a criticism, not only because the volume contains magnificent mathematics, but also because this mathematics confirms my point of the importance of mathematics beyond physics even in physics. Also, in fairness, while bypassing certain key aspects of QFT as a physical theory, physical intuitions are not entirely absent among these investigations. They are instead different from those of preceding physics. String and M-Brane theorists have been criticized for losing the connections to physics and making these theories pure mathematics instead, although most physicists working in this field do not strictly practice it as mathematics either, in particular as concerns offering rigorous proof. Awarding the Fields medal to Witten in 1990 became controversial because he never proved any theorem. This also led to a reverse criticism to the effect that string theorists are not really doing mathematics either. I shall not enter these debates here. Neither criticism is entirely fair. Mathematics is not only about proofs, and string or M-Brane theorists, such as Witten, introduced important mathematical concepts. There are also mathematicians contributing to these theories, including by supplying such proofs. Accordingly, as concerns mathematics, the situation is mixed. As concerns physics, while the connections of string and M-Brane theories to experimental data

---

discussed here (also crucial for Feynman, whom Faddeev mentions as an influence and inspiration), also be a new intuition from which mathematics could benefit and in fact had benefited, especially, in the case of discreteness and discontinuity, in areas quite remote from physics? This question may also be addressed to many, even most contributions to [20], including the one by Faddeev.

are indirect, sometimes highly indirect, they are not entirely absent, and they did lead to new physical intuitions, even if, thus far, mostly hypothetical ones. But that's not something to be dismissed either.

The perspective on theoretical physics offered here, underlain by an RWR interpretation of quantum phenomena and QT, offers a different way of thinking about the nature of theoretical physics, at least if the latter is quantum, even if without removing the uncertainty regarding the future of fundamental physics, including whether it will be quantum. On the other hand, one might welcome this uncertainty, because it may lead to surprising discoveries. As I have argued here, in the wake of Heisenberg's creation of QM, the practice of theoretical physics no longer consists, as in classical physics or relativity, in offering an idealized mathematical representation of physical objects and behavior, but by inventing abstract mathematical schemes capable of predicting the outcomes of experimental events. It is a new form of mathematical physics, no longer placing mathematics in the service of representing nature, if still relating this mathematics to physical phenomena and thus to nature by means of probabilistic predictions.

This mathematics itself is the first part of my twofold argument concerning modern mathematics as abstract mathematics divorced from physics even in physics itself, insofar as it is, at least in RWR interpretations, divorced from a representation of the ultimate reality responsible for quantum phenomena, and only relates to these phenomena in terms of probabilistic predictions. Heisenberg's invention of QM, Dirac's of QED, and Yang and Mills's of their theory are my main cases illustrating this part. The second part of my argument is that, in accord with the project of modern mathematics, even if it may originate in physics, a mathematical concept or theory has to leave physics (including that responsible for the emergence of this concept or a theory) behind and thus to become modern mathematics. Donaldson's work is an example of that second case: while it uses Yang–Mills theory, which originated in QFT, this work is not really different from the work in differential topology that has no connection to physics, and it brings Yang–Mills theory into the fold of modern mathematics. Atiyah's work on Yang–Mills theory has already done so. Thus, Donaldson already built on a purely mathematical theory, not even using the quantized version of the theory, thus making the physics of Yang–Mills theory essentially irrelevant. By contrast, Seiberg and Witten's work could eventually correspond to the first part, assuming that it will ever prove viable physically and allow for an RWR interpretation. It is true that, while serving as a major stimulus for mathematics, half a century of major efforts by thousands of talented, some exceptionally talented, physicists have not offered anywhere enough to demonstrate a physical viability of string or M-Brane theories. This fact has led to questioning whether this line of research has any potential at all to be viable for physics. Apart, however, from the not always fair nature of this questioning, this failure may be due to the difficulty of the task of creating a viable quantum or quantum-scale gravity theory (possibly unifying all fundamental forces). No other approach to do so has done any better thus far. It is difficult to predict what, if anything, will succeed.

It is, I also argue, equally difficult to know whether the way to solve the $\mathbb{R}^4$ problem will come from physics, such as QFT, or mathematics itself, and whether there will be anything "quantum" about this new topology. There may, however, be deeper conceptual reasons behind Poénaru's appeal to some new "quantum topology," which exceed the connections to QT. Because Poénaru does not elaborate, I cannot be certain that this is what he had in mind, but I shall suggest these reasons because they are, in my view, important. As this chapter argues, QT, from its inception on, but especially in QM and QFT, has been about the relationships between continuity and discontinuity and has given these relationships an entirely new form. For that reason, it might offer some ideas that will help to handle the complexity of the case of continuous manifolds and discontinuous algebraic invariants in algebraic topology in dimension four. Of course, as emphasized throughout this chapter, in QFT (or QM), these relationships are between mathematically continuous structures and discontinuous quantum phenomena, including those associated with quantum numbers. They are, moreover, probabilistic and as such involve physical events, which is not the case when it comes to algebraic invariants, "quantum numbers," used in topology. Only the *relationship* between the continuous and the discontinuous is shared.

As noted, it is possible that one will need some form of discrete mathematics to address the problems, still mathematically and physically formidable of QFT, or problems of extending quantum theory to gravity, possibly bringing all forces of nature together or at least consistently. That may also include the discrete mathematics of space, or in the first place the assumption that space, or to return to Riemann's prescient phrase, "the reality underlying space," is discrete, the possibility entertained by Riemann (without anything akin to quantum theory available at the time), which, he argued, would make physics different [64, p. 33]. It is far from clear at present whether such an alternative will prove to be workable, which leaves open the possibility that one will need to continue to handle quantum discreteness by means of continuous mathematics, moreover of two essentially different types, that of QT for predicting them and that of classical physics for describing them.

In any event, at issue at the moment is a need for an adequate *mathematical* theory of the relationships between the *continuous* topology of four-manifolds, including $\mathbb{R}^4$ itself (defined by the difference between their topological and differential structure) and *discrete* algebraic invariants, because the methods that handle these relationships in dimension five and higher, are not workable. (The issue is, again, nonexistent in lower dimensions). The relationships between the continuous mathematics and discrete quantum numbers, manifested in the discreteness of quantum phenomena, in QT, may provide a suitable parallel and possibly new mathematical ideas, either from the current QFT or some new future theory still necessary in physics as well in view of the quantum gravity problem. But, first of all, if one assumes an RWR interpretation, in QT, the continuous mathematics, over $\mathbb{C}$, which would be necessary for the DIFF $\neq$TOP problem, has no inherent mathematical connections to discrete quantum numbers or any discrete data, which are observed in physical phenomena, described by a totally different mathematics, over $\mathbb{R}$, of classical physics. The ultimate nature of physical reality responsible for these discrete data is beyond representation or even conception and hence, cannot be either continuous or discrete, at least if our theory of this reality is quantum, but possibly in general [59, p. viii]. The only relationships between these discrete data are, again, established by observations and measurements, independent of QT, and transition probabilities, provided by QT, with the help of Born's rule, added to QT. The situation is thus entirely different from that of the DIFF ≠TOP problem in dimension four, unless—a tantalizing possibility!—the theory that would handle this problem would prove to be probabilistic and thus involve some concept of event. Any such theory, whatever its origin, would, again, still have to be developed as a properly mathematical theory.

It is true that, as Poénaru must have been aware of, dimension four has a very special apace in physics in view of SR and GR, and hence in QT, even apart from quantum gravity, because all high-energy measurement requires the use of SR (and even QM has to be and is consistent with SR). This is also, mathematically, the dimension of Yang–Mills theory, which, as noted earlier, is not renormalizable in higher dimensions, or many topological QFTs. According, these theories use the Minkowskian spacetime, which is topologically $\mathbb{R}^4$, but not geometrically so, given that the Lorentz metric is different from the Euclidean metric. As also noted earlier, however, while the concept is uncontroversial mathematically, the status of spacetime in SR or spacetimes (there are many) in GR as a physical concept is a separate matter, because in any actual observations space and time remain separate, 3D space and time. We do not observe spacetime but only space and time. Technically, we observed rods and clocks as instruments that register the data which we associate with space and time, an association that involves further complexities, put aside here [61]. It is difficult to see the relationships between this physics and the mathematical DIFF ≠TOP problem. These connections are more likely to come, if they are to come at all, from the mathematical structure of some "quantum topology" of and its relations to "quantum numbers" in a QT that handles this physics.

22.6 Conclusion: The Discrete, the Continuous, and the Unthinkable with Grothendieck

The argument of this chapter would question or at least qualify the "prediction" made by Grothendieck concerning the "expected renewal" ("if it must yet come") of our understanding of the ultimate nature of space, or by implication physical reality, in reflecting on Riemann's insight that "the reality underlying space" may be discrete. On the other hand, this argument allows for Riemann's view or (they are not the same) Grothendieck's conjecture concerning the nature of this reality.[23] According to Grothendieck:

[23] I have considered Grothendieck's argument previously [60, 61]. While, however, building on these earlier reflections, the discussion to follow modifies them in view of the argument of this chapter.

> It must be already fifteen or twenty years ago that, leafing through the modest volume constituting the complete works of Riemann, I was struck by a remark of his "in passing". He pointed out that it could well be that the ultimate structure of space is discrete, while the continuous representations that we make of it constitute perhaps a simplification (perhaps excessive, in the long run ...) of a *more complex reality*; That for the human mind, "the continuous" was easier to grasp than "the discontinuous", and that it serves us, therefore, as an "approximation" to apprehend the discontinuous.
>
> This is a remark of a surprising penetration in the mouth of a mathematician, at a time when the Euclidean model of physical space had never yet been questioned; in the strictly logical sense, it is rather the discontinuous which traditionally served as a mode of technical approach to the continuous.
>
> Mathematical developments of recent decades have, moreover, shown a much more intimate symbiosis between continuous and discontinuous structures than was imagined, even in the first half of this century.
>
> In any case finding a "satisfactory" model (or, if necessary, a set of such models, "satisfactorily connecting" to each other) of "continuous", "discrete" or of "mixed" nature - such work will surely involve a great conceptual imagination, and a consummate flair for apprehending and unveiling new type mathematical structures.
>
> This kind of imagination or "flair" seems rare to me, not only among physicists (where Einstein and Schrödinger seem to have been among the rare exceptions), but even among mathematicians (and here I speak with full knowledge).
>
> To summarize I predict that the expected renewal (if it must yet come) will come from a mathematician in soul well informed about the great problems of physics, rather than from a physicist. But above all, it will take a man with "philosophical openness" to grasp the crux of the problem. This is by no means a technical one but rather a fundamental problem of natural philosophy [35, pp. 67–68]; cited in Connes [11, pp. 164, 195], emphasis added.[24]

One could not quite say that "the Euclidean model of physical space had never yet been questioned" at the time of Riemann's lecture in 1854, given that non-Euclidean geometry and, with it, the view that physical space may not be Euclidean had been around for two decades by then.[25] Also, "Riemann's remark" was not really "'in passing.'" It was germane to his argument (e.g., Plotnitsky [60, pp. 118–132]). One must, however, keep in mind Grothendieck's quotation marks and the fact that this is a recollection, occasioning Grothendieck's own view, rather than an analysis of Riemann's argument.

Grothendieck's insight itself, which reverses the conventional relationships between continuity and discontinuity, is profound. His greatest work in algebraic geometry was grounded in extending the cohomological architecture of algebraic topology, developed for continuous manifolds, to discrete algebraic varieties, via his two key concepts, a scheme and a topos, which he saw as his most important con-

[24] Connes, who cites the passage, does not consider it despite its relevance to his article, entitled "Geometry and the quantum." He merely adds Riemann's remarks, presumably to provide his reader with the context of Grothendieck's comment. Connes's argument itself does bear on the relationships between the continuous and the discontinuous in mathematics and fundamental physics. Grothendieck's passage is also cited in Papadopoulos [51, pp. 59–60] in his informative discussion of the genealogy of topological aspects of Riemann's thinking, without, however, discussing the passage itself either.

[25] It is conceivable that by "the Euclidean model" Grothendieck means a continuous model. This possibility was suggested to the present author by Athanase Papadopoulos.

tributions, including in redefining the concept of space in mathematics. In the next paragraph, he compares his contribution to the idea of space in mathematics, defined by these concepts, to that of Einstein and Schrödinger [35, p. 68]. Grothendieck may be right in this claim, especially given topos theory as arguably the most abstract concept of spatiality in mathematics thus far, with a remarkable genealogy. This genealogy was initiated by Riemann's concept of covering space and by Galois theory, applied to the fundamental group, and absorbed a formidable array of concepts of modern mathematics. Unless Grothendieck refers to Schrödinger's work on GR (which, while possible, seems unlikely), a reference to Schrödinger may appear strange. In fact, it makes, or (I am not sure that Grothendieck had this sense in mind) can be given, a sense. QM established a new type of relationship, fundamentally probabilistic in nature, between purely mathematical spaces, Hilbert spaces over $\mathbb{C}$, in Schrödinger's case infinite-dimensional ones, and the (three-dimensional) physical space, mathematically represented as a real manifold. Technically, it was Heisenberg who first defined this architecture of QM, while Schrödinger initially thought that the wave function would represent quantum processes in space and time and, as noted, never accepted the epistemological architecture of QM advocated by Heisenberg and Bohr.

More significant are implications of Grothendieck's reversal of the standard relationships between the discrete and the continuous, the possibility of which in physics bears especially on QFT. As considered in this chapter, in QFT the continuity of space or time leads to difficulties, not found in QM. These difficulties have made the view that "the reality underlying space" might be discrete to have periodically reemerged from the rise of QFT to our own time, even apart from the problem of quantum gravity in considering which and hypothetical approaches to it this view is more common. On the other hand, QFT is a viable theory in handling all other fundamental forces of nature, even if, not in the way many would prefer, especially in view of the necessity of renormalization. There is, thus far, no effective discrete theory for doing so.

QFT allows one to make or rather give a clear sense to the point stated by Grothendieck for "the human mind" (where this point is less certain and is not claimed by Riemann) to the effect that " 'the continuous' ... serves ... as an 'approximation' to apprehend the discontinuous." As discussed earlier, some believe that continuous (high-energy) QFTs may only be an approximation of discontinuous ones, which should apply at the higher energy scale, ultimately at the Planck scale. Even if this will not prove to be the case and we will continue to use a mathematically continuous theory, such as QFT, this view is consistent and reasonable on physical grounds, and is in accord with Riemann's argument, which is physical, concerning the discrete reality underlying space [60, pp. 118–131]. Riemann's statements concerning the subject there clearly show that the way Riemann sees the situation or at least what he says is different from what Grothendieck says in his comments (again, keeping in mind that one deals with Grothendieck's recollection rather than a reading of Riemann). It is conceivable that Riemann thought along the lines stated by Grothendieck. Riemann's actual point in the Habilitation lecture was, however, that the reality underlying space or, by implication, the ultimate constitution of physical reality, may be either discrete or continuous (in the immeasurably small), and that the determination of this reality would be different depending of which it is. This difference is as follows: "in a discrete manifold the principle of metric relations is already contained in the concept of the manifold, but in a continuous one it must come from something else," namely from physics [64, p. 33]. "The continuous representations that we make of [the discrete structure of space] constitute perhaps a simplification (perhaps excessive, in the long run ...) of a *more complex reality* [emphasis added]," Grothendieck says, thus suggesting that the discrete may be more complex than the continuous. This is an intriguing possibility, although Grothendieck does not explain why this may be the case. On the other hand, in RWR interpretations (assuming that the ultimate nature of physics is quantum), the ultimate constitution of reality is beyond the reach of thought and hence is neither continuous nor discontinuous, any more than conforming to any concept we can form.

As such, this reality may indeed be "a more complex reality,' more complex than any we can imagine. The assumption that a reality is beyond thought allows this reality to be unimaginably (literally speaking!) complex. Indeed, it is likely that, limited by our evolutionary biological and neurological nature, we can imagine very little of how nature ultimately is, if the word "to be" applies, and the same may be true about how thought itself is, including in dealing with mathematical reality. The DIFF ≠TOP problem in dimension four might be a manifestation of that richness, preventing us from handling it in the way we do in dimension five and higher. It is, again, not clear that QT may help to deal with this problem. That Yang–Mills theory helped to establish it does not guarantee that any mathematics coming from QT or any theoretical physics can solve it. But QT makes us think differently about it. Over a hundred years ago, Henri Lebesgue, one of the founders of twentieth-century integration and measure theory, observed (in commenting on the paradoxes of set theory, shaking the foundations of mathematics then) that the fact that we cannot imagine or mathematically define objects, such as "sets," that are neither finite nor infinite, does not mean that such objects do not exist [36, pp. 261–273], [18, p. 258]. Lebesgue did not specify in what type of domain, material or mental, such entities might exist. The statement is, however, a profound reflection of the limit of our thought concerning the nature of reality, either material or mental but also on our thought capacity to conceive of such limits. Similarly, the fact that we cannot conceive of entities that are neither continuous nor discontinuous does not mean that such entities do not exist, including in nature, a possibility brought about by QT. It does not, however, follow either that such entities do exist.

Lebesgue's insight was a response to the problem of the continuum and the debates concerning it shaped by Cantor's work and especially his continuum hypothesis that, as discussed earlier, was given the deeper and more radical understanding, defining our view now, by Gödel's incompleteness theorems of 1931 and then Cohen's proof of the undecidability of the continuum hypothesis in the 1960s. The undecidability of Cantor's continuum hypothesis suggests that the concept of continuum may not be realizable mathematically and does not exist in nature, for example, as space, time, or motion. It need not follow that the ultimate constitution of physical reality, including, "the reality underlying space," is discrete. Quantum phenomena are irreducibly discrete relative to each other, but not in the constitution of each, given that their observable parts are described by continuous mathematics, that of classical physics and SR. In RWR interpretations, the reality ultimately responsible for quantum phenomena is beyond thought and hence cannot be assumed to be either continuous or discrete, or any combination of both. Only this reality itself is assumed to exist.

There is no evidence that Riemann ever contemplated this type of reality, which is not surprising given that QT was half a century away, with no indication that anything like it could come about. Nor does it appear that Grothendieck had ever considered this type of reality either, even if without accepting it. His invocation of Einstein and Schrödinger, rather than Bohr or Heisenberg, suggests that he would have been unlikely to accept this type of conception, assumed by Bohr or Heisenberg. Nor does Grothendieck appear to have accepted or contemplated an RWR-type interpretation of mathematical reality, although some of his ideas suggest such an interpretation [60, pp. 175–179].

One might also doubt, although one cannot exclude, Grothendieck's prediction: "I predict that the expected renewal (if it must yet come) will come from a mathematician in soul well informed about the great problems of physics, rather than from a physicist." History suggests otherwise. Things were more mixed early with Descartes and Newton, both of whom were physicists as well as mathematicians. It is also true that Riemann and Gauss, and a few other mathematicians made major contributions to physics. Among major examples would be Euler, D'Alembert, Laplace, or Hamilton, or in the case of expressly modern mathematics, Hilbert, von Neumann, and Weyl, and quite a few others, some (admittedly only a limited sample) mentioned in this chapter, throughout the twentieth- and twenty-first century. These contributions were, however, primarily mathematical, unlike Gauss's and Riemann's work in physics as such. Both of course also made mathematical contributions to physics, but their work I am referring to now was physics, which, it is also true, had in turn a shaping influence on their mathematics, certainly in the case of Riemann, who worked on several expressly physical projects.

Still when it comes to breakthroughs and renewals, especially by the time physics began to use modern mathematics, they mostly came from physicists: Galileo, Newton, Maxwell, Boltzmann, Planck, Einstein, Bohr, Heisenberg, and Dirac, to name only some of the key figures. It took Einstein and his GR to give a rigorous physical content to Riemann's insights by bringing together the physics of gravitation and Riemannian geometry, even though Riemann, "the genius of Riemann, solitary and uncomprehended," was credited by Einstein [27, p. 190]. Einstein's characterization is not quite accurate. Riemann and his genius were neither entirely solitary nor entirely uncomprehended by his contemporaries, Gauss, for one, unlike say, Galois and his genius, although even Galois's case has complexities in this regard. In any event, it was Einstein who discovered the physics in question and realized that it needed a certain new type of mathematics to be handled, even though he was helped by his friend Marcel Grossman, a mathematician, in finding and mastering this mathematics. QT is a different story. While some of the key ideas that existed in mathematics earlier were used, and while mathematicians, such as Weyl and von Neumann, importantly contributed, primarily mathematically, to QT, the credit for the *creation* of QT goes strictly to physicists, beginning with Planck, Einstein, and Bohr in the old, pre-quantum-mechanical, quantum theory, and then Heisenberg, Schrödinger, Dirac, and others, virtually exclusively physicists.

As my argument in this chapter would suggest, given the history of modern physics (which is my Bayesian prior), solving the main outstanding problems of fundamental physics, whatever degree of "renewal" this requires, is more likely to come from a physicist or several physicists, admittedly, greatly informed mathematically, than from a mathematician. On the other hand, one might be hesitant to predict the origins of mathematics necessary to deal with these problems, that is, whether this mathematics will emerge in physics, as in Heisenberg's discovery of QM, or in mathematics already in place, as in Einstein's discovery of GR by using Riemannian geometry. One might, again, conversely, get help from physics in dealing with the difficult problems of mathematics, as in Donaldson's use of the Yang-Mills theory. I have not argued here otherwise. I have argued, however, that this help is limited because one would still need to give this mathematics the full measure of *mathematics*. In both cases, thus, which is my main argument here, mathematics ultimately carries the day more than physics does, at least modern physics, which is a mathematical-experimental science, with mathematics defining this conjunction.

The view just outlined may differently justify Grothendieck's claim. It is, again, of some interest that, while giving a potential primacy to the discontinuous, he invokes, as examples of the necessary "kind of imagination and 'flair'," Einstein and Schrödinger. Both were thinkers of continuity, who tried to avoid the discontinuous in physical reality at all cost, even if with imagination and flair. Would one not need instead the kind of imagination and flair for the discontinuous found in Heisenberg and Bohr? A safer bet, which Grothendieck might have shared, would be that it will still be *mathematics* that will move physics forward, as it has done throughout the history of modern physics as mathematical-experimental science, with mathematics, again, defining this conjunction. In this sense, a physicist who would be able to do so would be a mathematician, "a mathematician in soul," thus confirming Grothendieck's prediction, even if not in the way Grothendieck had in mind. Grothendieck was, however, right to think that "above all, it will take a man with 'philosophical openness' to grasp the crux of the problem. This is by no means a technical one but rather a fundamental problem of natural philosophy." It might also be a woman like Emmy Noether of Noether's theorems fame and a key figure in the genealogy of Grothendieck's mathematics, in which she appears by name in the important concept of a Noetherian scheme. When it comes to discontinuity, Heisenberg, especially at the time of his invention of QM, and Bohr would, again, be better examples, certainly in their philosophical openness to the discontinuous, in a manifested juxtaposition to Einstein and Schrödinger, as thinkers of continuity. It is not a matter of avoiding continuity and geometry, in the first case, or discontinuity and algebra, in the second, which was not the aim of these figures and is, arguably, impossible. It is a matter of a relative balance and the vision of the ultimate nature of reality and the means by which we relate to it.

Heisenberg's thinking leading him to his invention QM established an entirely new type of relationship between continuity and discontinuity, which also redefined the relationships between mathematics and physics, bringing about "the new era" of their mutual stimulation [4, v. 1, p. 51]. First, these relationships (combining both kinds just mentioned) were between the *mathematical continuity* of the formalism of QM and the *physical discontinuity* of quantum phenomena, always discrete relative to each other. Secondly, epistemologically more radically, he abandoned the assumption that QM must represent the continuous (and causal) connections between these discrete phenomena and even the assumption of the existence of such connections. These assumptions grounded all preceding physical theories, such as classical physics and relativity, and made these theories realist. Instead, the ultimate reality responsible for the existence of quantum phenomena was assumed by Heisenberg to be beyond representation and then in Bohr's strong RWR interpretation beyond conception and thus *un*-thinkable.

Heisenberg's approach, however, retained a realist description of quantum phenomena, along with observable parts of the instruments used, while assuming, necessarily, that these instruments also have quantum parts (which are, as quantum, thus allowing for RWR interpretations) enabling their interaction with quantum objects. This description was by classical physics. This view, fully developed by Bohr as part of in his interpretation (and adopted by the present interpretation), contains both pairs—"reality without realism" and "reality amenable to realism," and "QM and classical physics," defined by two fundamentally different types of mathematics—without, it follows, any physical (mechanical) connections between the components of each pair. The only connection between them is that by probabilistic predictions concerning the second entity in each pair, enabled by the first, by means the formalism of QM cum Born's rule. This connection would not, however, be possible without observational technology built by experimentalists (or sometimes existing in nature, but converted into observational technology by us), enabling the creation of quantum phenomena. Nor of course, indeed in the first place, would it be possible without the ultimate constitution of the reality responsible, in interacting with this technology, for the emergence of these phenomena. No would it be possible without our bodies and brains (ultimately the products of this reality as well) capable of observing this phenomena and of creating the mathematics, such as that of QM, or QFTM and classical physics, of dealing with them.

Accordingly, while in RWR interpretations this reality is beyond human thought, it is still assumed to be part of nature, in accord with Lebesgue's observation that something that cannot be thought by us can, nevertheless, exist in nature or thought, even mathematical thought, as Gödel's theorems confirmed [60]. As Lebesgue also understood, however, it is equally impossible to be certain that any such reality does exist. Assuming this reality, even if only with a practical justification (no other justification is claimed in this chapter or is possible in RWR interpretations) is a philosophical wager. It was, however, this wager that led Heisenberg to QM, defined by a new mathematics combining both continuity and discreteness, and correlatively geometry and algebra, and enabling QM to predict quantum phenomena, irreducibly discrete relative to each other. These predictions were probabilistic as they needed to be on experimental grounds. As a result, with quantum physics, modern physics acquired a new type of relationship among the discrete, the continuous, and the unthinkable.

It was, however, the mathematical project that had defined modern physics, as mathematical-experimental science, from Descartes and Galileo but was transformed by Heisenberg that brought about this new situation in physics, with implications far beyond physics, beginning with mathematics itself. Quantum physics is a precious gift of nature and thought to physics, mathematics, and philosophy, as well as culture, for its emergence was also a transformative event of culture, just as were the emergence of modern physics and modern mathematics.

Acknowledgments I am grateful to Athanase Papadopoulos for inviting this contribution and for invaluable exchanges on the subjects considered in this chapter. My thinking concerning these subjects also benefited greatly from my discussions with G. Mauro D'Ariano, whom I would like to thank as well.

References

1. N. A'Campo, L. Ji, A. Papadopoulos, On Grothendieck's tame topology, in *Handbook of Teichmüller Theory*, ed. by A. Papadopoulos, vol VI (European Mathematical Society, Zürich, 2016), pp. 521–536
2. M. Atiyah, Special article. Mathematics in the 20th century. Bull. Lond. Math. Soc. **34**, 1–15 (2002)
3. N. Bohr, Can quantum-mechanical description of physical reality be considered complete? Phys. Rev. **48**, 696–702 (1935)
4. N. Bohr, *The Philosophical Writings of Niels Bohr*, vol. 3 (Ox Bow Press, Woodbridge, 1987)
5. M. Born, Quantenmechanik der Stoßvorgänge. Z. Phys. **38**, 863–867 (1926)
6. M. Born, *The Einstein-Born Letters* (Trans. I Born) (Walker, New York, 2005)
7. M. Born, P. Jordan, Zur Quantenmechanik Z. Phys. **34**, 858–888 (1925)
8. J. Butterfield, C.J. Isham, Spacetime and the philosophical challenge of quantum gravity, in *Physics Meets Philosophy at the Planck Scale: Contemporary Theories of Quantum Gravity*, ed. by C. Callender, N. Huggett (Cambridge University Press, Cambridge, 2001), pp. 33–89
9. P. Cartier, A mad day's work: from Grothendieck to Connes and Kontsevitch, the evolution of concepts of space and symmetry. Bull. Am. Math. **38**(4), 389–408 (2001)
10. A. Connes, Symmétries Galoisiennes et renormalization, in *Poincaré Seminar 2002: Vacuum Energy, Renormalization*, ed. by B. Duplantier, V. Rivasseau (Birkhäuser, Boston/Cambridge, 2003), pp. 174–204
11. A. Connes, Geometry and the quantum, in *Mathematics and Physics One Century After Hilbert*, ed. by J. Kouneiher (Springer Nature, Switzerland AG, 2018), pp. 159–196
12. A. Connes, D. Kreimer, Renormalization in quantum field theory and the Riemann-Hilbert problem. II. Commun. Math. Phys. **216**(1), 215–241 (2001)
13. A. Connes, M. Marcolli, Renormalization and motivic Galois theory. Int. Math. Res. Not. **76**, 4073–4092 (2004)
14. A. Connes, M. Marcolli, *Noncommutative Geometry, Quantum Fields, and Motives* (American Mathematical Society, Hindustan Book Agency, Providence, 2007)

15. G.M. D'Ariano, Physics without physics. Int. J. Theor. Phys. **56**(1), 97–128 (2017)
16. G.M. D'Ariano, P. Perinotti, Derivation of the Dirac equation from principles of information processing. Phys. Rev. A **90**, 062106 (2014)
17. G.M. D'Ariano, P. Perinotti, Quantum cellular automata and free quantum field theory. Front. Phys. **12**(1), 1–11 (2017)
18. J.F. Dauben, *Georg Cantor: His Mathematics and Philosophy of the Infinite* (Princeton University Press, Princeton, 1990)
19. G. Deleuze, F. Guattari, *What Is philosophy?* (Trans. H. Tomlinson, G. Burchell) (Columbia University Press, New York, 1996)
20. V. Dijkgraaf, *Quantum Fields and Strings: A Course for Mathematicians*, vols. I, II (AMS, IAS, Providence, 1999)
21. P.A.M. Dirac, *The Principles of Quantum Mechanics* (Clarendon, Oxford, 1930)
22. P.A.M. Dirac, Quantized singularities in the electromagnetic field. Proc R. Soc. Lond. A **133** 821, 60–72, 60–72 (1931). https://doi.org/10.1098/rspa.1931.0130
23. S. Donaldson, An application of gauge theory to four dimensional topology. J. Differ. Geom. **18**, 279–315 (1983)
24. A. Doxiadis, B. Mazur (eds.), *Circles Disturbed: The Interplay of Mathematics and Narrative* (Princeton University Press, Princeton, 2012)
25. A. Einstein, Physics and reality. J. Franklin Inst. **221**, 349–382 (1936)
26. A. Einstein, *Autobiographical Notes* (Trans. P. A. Schilpp) (Open Court, La Salle, 1949)
27. A. Einstein, The problem of space, ether, and the field in physics (1934), in *Beyond Geometry: Classic Papers from Riemann to Einstein*, ed. by P. Pesic (Dover Mineola, New York, 2007), pp. 173–188
28. A. Einstein, B. Podolsky, N. Rosen, Can quantum-mechanical description of physical reality be considered complete (1935)? in *Quantum Theory and Measurement*, ed. by J.A.Wheeler, W.H. Zurek (Princeton University Press, Princeton, 1983), pp. 138–141
29. L. Faddeev, *Fifty Years of Mathematical Physics: Selected Works of Ludwig Faddeev* (World Scientific, Singapore, 2016)
30. G. Farmelo, Dirac's hidden geometry. Nature **437**, 323 (2005)
31. R. Feynman, *The Character of Physical Law* (MIT Press, Cambridge, 1994)
32. M. Gromov, *Partial Differential Relations* (Springer, Berlin-Heidelberg, 1986)
33. D.J. Gross, F. Wilczek, Asymptotically free gauge Theories I. Phys. Rev. D **8**, 3633–3651 (1973). https://doi.org/10.1103/PhysRevD.8.3633
34. D.J. Gross, F. Wilczek, F. Asymptotically free gauge theories II. Phys. Rev. D **9**, 980–993 (1974). https://doi.org/10.1103/PhysRevD.9.980
35. A. Grothendieck, *Récoltes et semailles, I, II. Réflexions et témoignage sur un passé de mathématicien* (Galimard, Paris, 2022)
36. J. Hadamard, Cinq lettres sur la théorie des ensembles. Bull. S. M. F. t. **33**, 261–273, 261–273 (1905)
37. A. Hagar, *Discrete or Continuous: The Quest for the Fundamental Length in Modern Physics* (Cambridge University Press, Cambridge, 2016)
38. M. Heidegger, *What Is a Thing?* (Trans. V. Deutsch) (Gateway, South Bend, 1967)
39. W. Heisenberg, *The Physical Principles of the Quantum Theory* (Trans. K. Eckhart, F. C. Hoyt) (Dover, Mineola, 1930/1949)
40. W. Heisenberg, *Physics and Philosophy: The Revolution in Modern Science* (Harper and Row, New York, 1962)
41. W. Heisenberg, Quantum-theoretical re-interpretation of kinematical and mechanical relations (1925), in *Sources of Quantum Mechanics*, ed. by B.L. Van der Waerden (Dover, Mineola, 1968), pp. 261–277
42. W. Heisenberg, *Encounters with Einstein, and Other Essays on People, Places, and Particles* (Princeton University Press, Princeton, 1989)
43. G. Jaeger, Exchange forces in particle physics. Found. Phys. **51**(1), 1–31 (2021). https://doi.org/10.1007/s10701-021--00425-0

44. I. Kant, *Critique of Pure reason* (Trans. P. Guyer, A. W. Wood) (Cambridge University Press, Cambridge, 1997)
45. M. Kontsevich, Operads and motives in deformation quantization. Lett. Math. Phys. **48**(1), 35–72 (1999)
46. J. Kouneiher (ed), *Foundations of Mathematics and Physics One Century After Hilbert* (Springer Nature, Switzerland AG, 2018)
47. M. Kuhlmann, Quantum field theory. *The Stanford Encyclopedia of Philosophy*, ed. by E.N. Zalta, Fall 2021 edn. https://plato.stanford.edu/
48. M. Marcolli, *Feynman Motives* (World Scientific, Singapore, 2009)
49. J. Mehra, H. Rechenberg, *The Historical Development of Quantum Theory*, vol. 6 (Springer, Berlin,2001)
50. F. Nietzsche, *The Birth of Tragedy and the Case of Wagner* (Trans. W. Kaufmann) (Vintage, New York, 1967)
51. A. Papadopoulos, Physics in Riemann's mathematical papers, in *From Riemann to Differential Geometry and Relativity*, ed. by L. Ji, A.Papadopoulos, S.Yamada (Springer/Nature, Switzerland AG, 2017), pp. 151–207
52. A. Papadopoulos, Topology and biology: from Aristotle to Thom, in *Geometry in History*, ed. by S.G. Dani, A. Papadopoulos (Springer/Nature, Switzerland AG, 2019), pp. 89–128
53. A. Papadopoulos, René Thom: from mathematics to philosophy. in *Handbook of the History and Philosophy of Mathematical Practice*, ed. by B. Sriraman (Springer/Nature, Switzerland AG, 2024)
54. R. Penrose, Gravitational collapse and space-time singularities. Phys. Rev. Lett **14**(3), 57–59 (1965). https://doi.org/10.1103/PhysRevLett.14.57
55. M.E. Peskin, D.V. Schroeder, *An Introduction to Quantum Field Theory* (CRS Press, London, 1995)
56. A. Plotnitsky, On the reasonable and unreasonable effectiveness of mathematics in classical and quantum physics. Found. Phys. **41**, 466–491 (2011). https://doi.org/10.1007/s10701-010-9442-2
57. A. Plotnitsky, The disaster of the diagonal, in *Circles Disturbed: The Interplay of Mathematics and Narrative*, ed. by A. Doxiadis, B. Mazur (Princeton University Press, Princeton, 2012)
58. A. Plotnitsky, *The Principles of Quantum Theory, from Planck's Quanta to the Higgs Boson: The Nature of Quantum Reality and the Spirit of Copenhagen* (Springer/Nature, New York, 2016)
59. A. Plotnitsky, *Reality Without Realism: Matter, Thought, and Technology in Quantum Physics* (Springer/Nature, Heidelberg, 2021)
60. A. Plotnitsky, *Logos and Alogon: Thinkable and Unthinkable in Mathematics, from the Pythagoreans to the Moderns* (Springer/Nature, Heidelberg, 2022)
61. A. Plotnitsky, t is not time: reality, causality, and the arrow of events in quantum theory. Eur. Phys. J. 1–22 (2023)
62. V.A. Poénaru, Glimpse into the problems of the fourth dimension. in *Geometry in History*, ed. by S.G. Dani, A. Papadopoulos (Springer/Nature, Switzerland AG, 2019), pp. 677–703
63. H.D. Politzer, Reliable perturbative results for strong interactions. PRL **30**(26), 1346–1349 (1973). https://doi.org/10.1103/PhysRevLett.30.1346
64. B. Riemann, On the hypotheses that lie at the foundations of geometry (1854), in *Beyond Geometry: Classic Papers from Riemann to Einstein*, ed. by P. Pesic (Dover, Mineola, 2007), pp. 23–40
65. E. Schrödinger, *Collected Papers on Wave Mechanics* (Trans. J. F. Shearer) (Chelsea, New York, 1982)
66. E. Schrödinger, The present situation in quantum mechanics, in *Quantum Theory and Measurement*, ed. by J.A. Wheeler, W.H. Zurek (Princeton University Press, Princeton, 1983), pp. 152–167
67. E. Schrödinger, *Interpretation of Quantum Mechanics: Dublin Seminars (1949–1955) and Other Unpublished Essays* (Ox Bow Press, Woodbridge, 1995)

68. S.S. Schweber, *QED and the Men Who Made It: Dyson, Feynman, Schwinger, and Tomonaga* (Princeton University Press, Princeton, 1994)
69. J. Silverman, J. Tate, *Rational Points on Elliptic Curves* (Springer, Heidelberg/New York, 2015)
70. N.F. Stang, Kant's transcendental idealism, in *The Stanford Encyclopedia of Philosophy*, ed. by E.N. Zalta, U. Nodelman, Winter 2023 edn. https://plato.stanford.edu/archives/win2023/entries/kant-transcendental-idealism/
71. P. Teller, *An Interpretive Introduction to Quantum Field Theory* (Princeton University Press, Princeton, 1997)
72. R. Thom, Les intuitions topologiques primordiales de l'aristotélisme. Revue Thomiste **88**(3), 393–409 (1988)
73. J. Von Neumann, *Mathematical Foundations of Quantum Mechanics* (Trans. R. T. Beyer) (Princeton University Press, Princeton, 1932/1983)
74. S. Weinberg, *The Quantum Theory of Fields, Volume 1: Foundations* (Cambridge University Press, Cambridge, 2005)
75. H. Weyl, *The Continuum: A Critical Examination of the Foundation of Analysis* (Trans. S. Pollard, T. Bole) (Dover, Mineola, 1928/1994)
76. H. Weyl, *Theory of Groups and Quantum Mechanics* (Trans. H. P. Robertson) (Dover, Mineola, 1931/1984)
77. H. Weyl, Invariants. Duke Math. J. **5**, 489–502 (1939)
78. H. Weyl, *Space–Time–Matter* (Trans. H. L. Brose) (Dover, Mineola, 1952)
79. H. Weyl, *The Concept of a Riemann Surface* (Trans. G. L. MacLane) (Dover, Mineola, 2013)
80. E.P. Wigner, On unitary representations of the inhomogeneous Lorentz group. Ann. Math. **40**, 149–204 (1939)
81. E.P. Wigner, The unreasonable effectiveness of mathematics in the natural science. Commun. Pure Appl. Math. **13**, 1–14 (1960)
82. C.N. Yang, R.L. Mills, Conservation of isotopic spin and isotopic gauge invariance. Phys. Rev. **96**(1), 191–195 (1954)
83. Wikipedia, Yang–Mills Equations. https://en.wikipedia.org/wiki/Yang-Mills-equations
84. A. Zee, *Quantum Field Theory in a Nutshell* (Princeton University Press, Princeton, 2010)

Chapter 23
The Case for Absolute Geometry

Victor Pambuccian

Abstract This is a plea for a neglected area of geometry: absolute geometry. Among the theorems shown to hold in absolute geometry is Urquhart's theorem, which, having been previously referred to as the most "elementary" theorem of Euclidean geometry, may henceforth be considered to be the most "elementary" theorem of absolute geometry. Prominently among the open problems mentioned is that of the existence in non-elliptic metric planes in which every segment has a midpoint of generalized centers of mass for any system of n material points with equal weights.

Keywords Foundations of geometry · Absolute geometry · Metric planes

AMS Classification 01A60, 03A05, 03B30, 51-03, 51F05, 51F10, 51P05

23.1 Introduction

23.1.1 Continuity

When someone claims today to be doing research in geometry, chances are the geometry in question is differential geometry, topological geometry, algebraic geometry, or finite geometry. The first two kinds of geometry take continuity for granted, find it to be a fundamental component of geometry. The third kind assumes a rich algebraic structure to be an essential aspect of geometry, while the fourth disparages continuity in favor of a world with only finitely many geometric objects, a world made impossible, as shown in [32], by a wide range of reasonable geometric assumptions. None of them is in the tradition of synthetic geometry, in which

V. Pambuccian (✉)
School of Mathematical and Natural Sciences, Arizona State University - West Campus, Phoenix, AZ, USA
e-mail: pamb@asu.edu

L. Funar, A. Papadopoulos (eds.), *Essays on Topology*,
https://doi.org/10.1007/978-3-031-81414-3_23

only purely hypothetical-deductive reasoning on geometric objects connected by geometric relations is allowed.

Continuity, however, and the unique models it engenders, is far from being an intuitive or intellectual necessity. It cannot be invoked by an appeal to a claim of geometry mimicking reality, not now and not when mathematicians made continuity a topic of foundational research, as can be seen from Hilbert's remarks of 1898:

> a homogeneous continuum, that would allow indefinite divisibility and would thus achieve the infinitely small, cannot be encountered anywhere in nature. The infinite divisibility of the continuum is an operation existing only in thought, only an idea, which is refuted by our observations of nature and by the experience drawn from physics and chemistry.[1] (Transl. V.P.)

The more recent insights of loop quantum gravity reinforce Hilbert's views of 1898, for they imply that spacetime exhibits a granular structure [39].

And Richard Dedekind himself, who gave us the first rigorous construction of the real numbers, thought that many of the properties of a space deemed really existing would remain valid in the absence of continuity:

> The assumption of this property of the line is nothing else than an axiom by which we attribute to the line its continuity, by which we find continuity in the line. If space has at all a real existence it is not necessary for it to be continuous; many of its properties would remain the same even were it discontinuous. And if we knew for certain that space was discontinuous there would be nothing to prevent us, in case we so desired, from filling up its gaps, in thought, and thus making it continuous [...][2] (Transl. W. W. Beman)

That one should not confuse the game of geometry with nature itself, had been noted already by Giordano Bruno in *La cena de le ceneri*, in 1584: "Altro è giocare con la geometria, altro è verificare con la natura." In a similar vein, Albert Einstein stated on January 27, 1921, during his Ceremonial lecture in the Prussian Academy of Sciences that: "As far as the laws of mathematics refer to reality, they are not certain, and as far as they are certain, they do not refer to reality. [Insofern sich die Sätze der Mathematik auf die Wirklichkeit beziehen, sind sie nicht sicher, und insofern sie sicher sind, beziehen sie sich nicht auf die Wirklichkeit.]"

Many results in geometry are first proved in a setting that includes very generous assumptions. That less is needed is usually seen at a later stage. Several results that

---

[1] ein homogenes Kontinuum, das die fortgesetzte Teilbarkeit zuliesse und somit das Unendlich-Kleine realiseren würde, [wird] in der Wirklichkeit nirgends angetroffen. Die unendliche Teilbarkeit eines Kontinuums ist nur eine in Gedanken vorhandene Operation, nur eine Idee, die durch unsere Beobachtungen der Natur und die Erfahrungen der Physik und Chemie widerlegt wird. [18, p. 705].

[2] Die Annahme dieser Eigenschaft der Linie ist nichts als ein Axiom, durch welches wir erst der Linie ihre Stetigkeit zuerkennen, durch welches wir die Stetigkeit in die Linie hineindenken. Hat überhaupt der Raum eine reale Existenz, so braucht er doch *nicht* notwendig stetig zu sein; unzählige seiner Eigenschaften würden dieselben bleiben, wenn er auch unstetig wäre. Und wüßten wir gewiß, daß der Raum unstetig wäre, so könnte uns doch wieder nichts hindern, falls es uns beliebte, ihn durch Ausfüllung seiner Lücken in Gedanken zu einem stetigen zu machen [...] [9, p. 18–19].

appeared to require the differential geometric apparatus turned out to be provable, in a more general form, inside synthetic differential geometry, in the style of H. Busemann and A. D. Alexandrov. Those still assume that space is continuous, but the fact that "many of its properties would remain the same even were it discontinuous" had been noticed already by M. Pasch and later by D. Hilbert. With the realization, going back to Th. Skolem in 1923, that —"if we are interested in producing an axiomatic system, we can *only* use first-order logic" [14, p. 472])— the axiomatic foundation of geometry lost interest in categorical axiomatizations, that have only one model, that of a continuous geometry, given that these cannot be axiomatized in first-order logic.

The question whether a geometric statement remains valid in the absence of continuity is thus not simply a question of following an Ockham's razor philosophy of weakening the assumptions, but has a deeper philosophical meaning. Whether a statement is valid inside a first-order theory or not amounts to knowing that the statement depends only on geometric assumptions, without the admixture of set theory, which is indispensable for the introduction of continuity, of the real numbers, into geometry. If a geometric intuition is to be considered independently of those of time and quantity, which give rise to algebra and set theory, then basic geometric results should be true in first-order theories. It is worth knowing whether a geometric result has an elementary core or whether it completely collapses in the absence of continuity. While it may seem that some theories require the construction of the real numbers, this is not always the case, as is shown in [3], where results in general relativity are expressed inside a first-order theory.

23.1.2 The Parallel Postulate

Another aspect of space that one ought to question as an assumption in the proof of a geometric statement is the validity of the Euclidean parallel postulate. Questioning the need to make such an assumption leads to what János Bolyai called *absolute geometry*, the theory consisting of those statements that are true in both Euclidean and hyperbolic geometry. While many elementary results for which only Euclidean proofs were known turned out to be true in absolute geometry, it is still very hard to prove any result in an absolute setting. To present absolute geometry, we will use the language $L_{B\equiv}$ introduced by Alfred Tarski, in which the only variables are for *points*, and in which there are two predicate symbols, for *betweenness* and *equidistance*, the former being a ternary one, B, with $B(abc)$ to be read as "b lies between a and c (and b may be a or c, and $a = b = c$ is also allowed)" and the latter a quaternary one $\equiv$, with $ab \equiv cd$ to be read as "b is as distant from a as d is from c" (or "ab is congruent to cd"), The axioms of plane asbolute geometry, a theory to be referred to as $\mathscr{A}$, are:

A 1 $ab \equiv pq \wedge ab \equiv rs \rightarrow pq \equiv rs,$

A 2 $ab \equiv cc \rightarrow a = b$

A 3 $(\forall qabc)(\exists x)\, B(qax) \wedge ax \equiv bc$

A 4 $B(abc) \wedge B(a'b'c') \wedge ab \equiv a'b' \wedge bc \equiv b'c' \wedge ad \equiv a'd' \wedge bd \equiv b'd' \rightarrow (dc \equiv c'd' \vee a = b)$

A 5 $B(aba) \rightarrow a = b$

A 6 $(\forall abcde)(\exists f)\, B(cea) \wedge B(dba) \rightarrow B(efd) \wedge B(bfc)$

A 7 $(\exists abc) \neg(B(abc) \vee B(bca) \vee B(cab))$

A 8 $u \neq v \wedge au \equiv av \wedge bu \equiv bv \wedge cu \equiv cv \rightarrow (B(abc) \vee B(bca) \vee B(cab))$

Here A3 is a segment transport axiom, stating that any segment can be transported on any ray, A4 denotes the five-segment axiom, which is a form of the side-angle-side triangle congruence axiom in the set-up in which there are no angles and thus no angle-congruence, A6 denotes the Pasch axiom, A7 denotes a lower-dimension axiom, stating that the dimension is at least 2, and A8 is an upper-dimension axiom stating that the dimension is at most 2.

The models of $\mathscr{A}$ are called *Hilbert planes* [17] and they have been described algebraically by W. Pejas [36]. They are submodels of *projective-metric coordinate planes* $\mathfrak{P}(K, k)$ over an ordered Pythagorean field K (one in which $x^{2+1} \in K$ for all $x \in K$), with k an element of K, to be referred to as the *orthogonality constant* (or the *metric constant*). $\mathfrak{P}(K, k)$ consists of a set of points and lines—the former to be denoted by (x, y, z) the latter by $[u, v, w]$ (determined up to multiplication by a non-zero scalar, not all coordinates (elements of K) being allowed to be 0)—endowed with a notion of incidence—point (x, y, z) being incident with line $[u, v, w]$ if and only if $xu+yv+zw = 0$—and an orthogonality of lines, under which lines $[u, v, w]$ and $[u', v', w']$ are orthogonal if and only if $uu' + vv' + kww' = 0$. By the *affine-metric plane* $\mathfrak{A}(K, k)$ we mean the projective plane $\mathfrak{P}(K, k)$ from which the line $[0, 0, 1]$, as well as all the points on it have been removed. All its points can be normalized to $(x, y, 1)$, for which we will write (x, y). The order of K induces in the usual manner an order on the lines of $\mathfrak{A}(K, k)$.

If we denote by R the ring of *finitely bounded* elements, i.e., $R = \{x \in K : (\exists n \in \mathbf{N})\, |x| < n\}$ and by P the ideal of *infinitely small* elements of K, i.e., $P = \{0\} \cup \{x \in K : x^{-1} \notin R\}$, then all Hilbert planes are isomorphic to a plane of the following three types:

Type 1 $E = \{(a, b) : a, b \in M\} \subset \mathfrak{A}(K, 0)$, *where M is an R-module $\neq (0)$;*

Type 2 $E = \{(a, b) : a, b \in M\} \subset \mathfrak{A}(K, k)$ *with $k \neq 0$, where M is an R-module $\neq (0)$ included in $\{a \in K \mid ka^2 \in P\}$, that satisfies the condition*

$$a \in M \Rightarrow ka^2 + 1 \in K^2;$$

Type 3 $E = \{\mathbf{x} : Q(\mathbf{x}) > 0, Q(\mathbf{x}) \notin J\} \subset \mathfrak{A}(K, k)$ *with $k < 0$, where $J \subseteq P$ is a prime ideal of R that satisfies the condition*

$$ka^2 + 1 > 0, ka^2 + 1 \notin J \Rightarrow ka^2 + 1 \in K^2,$$

with K satisfying

$$\{a \in K : ka^2 \in R \setminus P\} \neq \emptyset.$$

The meaning of k in this context can be best described by mentioning that its sign is the same as that of $\alpha + \beta + \gamma - \pi$, where α, β, and γ are the measures of the three angles of a triangle.

Thus, a Hilbert plane is either (i) a part of an ordered Euclidean plane with free mobility (Type 1), thus a plane of Dehn-type, as these were first described in [10], or (ii) an infinitely small neighborhood of the origin in a plane with an arbitrary non-zero orthogonality constant, or (iii) a generalized Beltrami–Cayley–Klein model, consisting of the interior of a circle, with, possibly, an infinitely small collar around its circumference removed.

Let **NE** stand for the statement that the sum of the angles of a triangle is $\leq 180°$, i.e., that the metric constant k is ≤ 0. It can be expressed in several ways. We choose the statement that the midline in a triangle is less than or equal to half of the opposite side. Formally, with L denoting the defined notion of collinearity, with $L(xyz)$ defined by $B(xyz) \vee B(yzx) \vee B(zxy)$ and M denoting the midpoint predicate, with $M(abc)$ defined by $B(abc) \wedge ba \equiv bc$:

NE $\neg L(oab) \wedge M(oma) \wedge M(onb) \wedge M(apb) \wedge B(aqb) \wedge aq \equiv mn \rightarrow B(aqp)$.

23.1.3 Why Statements True in $\mathscr{A}$ Ought to be Proved Inside $\mathscr{A}$

Very few theorems are known to be true in $\mathscr{A}$ and proofs inside $\mathscr{A}$ tend to be incomparably longer and to require much more ingenuity than proofs inside Euclidean geometry. While some proofs developed in the Euclidean setting carry over to the absolute setting virtually without change, such as the proofs in [34] of Fagnano's problem or Descube's 1880 proof [33] of the Steiner–Lehmus theorem, others, such as the results proved in [16, 20, 23, 27–30], require a completely different approach. An even higher level of complexity is required for proofs inside a more austere version of absolute geometry put forward by Johannes Hjelmslev [19], as can be seen from the proof that the medians of a triangle lie in a pencil in [6, §4.10, Satz 14, p. 74].

One could ask: Why the trouble, if one believes the world around us to be Euclidean and continuous? Even in that case, the usual proof [8, 1.4, p. 10] that the medians are concurrent is correct but profoundly misleading, as it depicts the fact of the concurrence of the medians as a consequence of uniquely Euclidean properties, which is simply false.

To see the issue under another angle, imagine a philosopher P who proves that all humans have a certain property π, and uses in the proof in an essential way the fact that humans have flat nails. It turns out, on closer inspection, that all mammals have property π. Is P's proof not embarrassing, even open to ridicule? P being anthropocentric, in tune with the audience, could claim a philosopher couldn't care less about the other mammals, and thus finds nothing embarrassing with the proof.

We will present in this chapter two theorems that allow insight into the truth of universal statements inside $\mathscr{A}$+**NE** and inside a weaker theory of absolute geometry without having found actual proofs inside those theories. This will allow us to show that the existence of Gergonne's point or of Nagel's point can be proved in $\mathscr{A}$+**NE**, that M. L. Urquhart's "most simple theorem" is true in $\mathscr{A}$, that the existence of centers of gravity for finite sets of points with equal masses concentrated in each point requires even less than $\mathscr{A}$ + **NE**, and that Federico Commandino's theorem on the concurrence of the lines joining a vertex of a tetrahedron with the center of gravity of the opposite face holds in a very general setting.

23.2 Universal Statements True in the Euclidean and the Hyperbolic Plane Over the Reals Are True in $\mathscr{A}$ + NE

Let $\mathbb{E}$ and $\mathbb{H}$ denote the Euclidean plane over the field of real numbers and the Beltrami–Cayley–Klein model of the hyberbolic plane over the real numbers (there is noting special about this model; we chose it just to fix ideas).

Theorem 23.1 *Let σ be a universal sentence in $L_{B\equiv}$, such that σ holds in both $\mathbb{E}$ and $\mathbb{H}$. Then $\mathscr{A}$ +* **NE** $\vdash \sigma$.

Proof Suppose $\mathscr{A} \nvdash \sigma$. Then there is a model $\mathfrak{M}$ of $\mathscr{A}$ in which $\neg\sigma$ holds. If $\mathfrak{M}$ is a Hilbert plane of Type 1, then $\neg\sigma$, as an existential statement, holds in any extension of $\mathfrak{M}$, in particular in $\mathfrak{M}'$, the Cartesian plane over $\overline{K}$, the real closure of K. However, $\mathbb{E}$ and $\mathfrak{M}'$ are elementarily equivalent, so $\neg\sigma$ would have to hold in $\mathbb{E}$ as well, a contradiction. If $\mathfrak{M}$ were to be a Hilbert plane of Type 3, then $\neg\sigma$, as an existential statement, would hold in any extension of $\mathfrak{M}$, in particular in $\mathfrak{M}''$, the Beltrami–Cayley–Klein plane over $\overline{K}$, the real closure of K (the negative orthogonality constant k can be normalized to -1, given that $\overline{K}$ is Euclidean). Since $\mathfrak{M}''$ and $\mathbb{H}$ are elementarily equivalent, $\neg\sigma$ would have to hold in $\mathbb{H}$ as well, a contradiction. Since σ holds in Hilbert planes of Type 3, it must hold in those of Type 2 with $k < 0$, given that the algebraic translation of σ is a function of the coordinates of the points involved and k and thus is the same in both cases.

We believe that more is true. Namely that:

If σ is a universal sentence in $L_{B\equiv}$, of the form $\alpha \to \beta$, where β is expressed in $L_{L\equiv}$, with α satisfiable in both $\mathbb{E}$ and $\mathbb{H}$, such that σ holds in both $\mathbb{E}$ and $\mathbb{H}$, then $\mathscr{A} \vdash \sigma$.

However, this is currently an open problem.

Corollary 23.1 *The following two statements hold in $\mathscr{A}$ + **NE**:*

(i) The lines joining the vertices of a triangle with the points of tangency of the incircle with the opposite sides are concurrent (the Gergonne point).
(ii) The lines going through the vertices of a triangle and dividing its perimeter in two are concurrent (the Nagel point).

Proof That these statements hold in $\mathbb{E}$ is well-known. They also hold, with the same proof based on Ceva's theorem, in $\mathbb{H}$, given that there is a hyperbolic version of Ceva's theorem (see [22, Th. 34.7, p. 458]). The result follows by applying Theorem 23.1.

In 1968, in an obituary [13], it was revealed that M. L. Urquhart considered the following statement to be "the most elementary theorem of Euclidean geometry" :

If a, e, and f are three non-collinear points, $B(aec)$, $B(afd)$, $B(dbe)$, $B(cbf)$, and $ae + eb = af + fb$, then $ac + cb = ad + db$.

Several proofs of it are known in Euclidean geometry ([41, p. 159–160], [35], [42]) and one [11] for its validity in $\mathbb{H}$. This allows, using Theorem 23.1, to conclude that Urquhart's theorem holds in $\mathscr{A}$+**NE**. However, we have the stronger statement:

Corollary 23.2 *Urquhart's theorem holds in $\mathscr{A}$*

Proof The proof of Urquhart's theorem presented in [41, p. 159–160] has only one step which might not hold in $\mathscr{A}$. It is the statement that the internal angle bisectors of $\widehat{cbd}$ and of the supplement of the angle $\widehat{bdf}$ intersect. According to [31], this is the case in all Hilbert planes satisfying the *Lotschnittaxiom* (and all Hilbert planes of Type 2 satisfy the *Lotschnittaxiom*), provided that the measure of the supplement of the angle $\widehat{cbd}$ is not infinitely small. One can easily achieve the conditions required for the intersection of the two internal angle bisectors mentioned above, by choosing, in a Hilbert plane with orthogonality constant x_5, $a = (0, 0)$, $f = (x_1, 0)$, $b = (x_2, x_3)$, and the line ae having slope x_4, with x_1, x_2, x_3, x_4, x_5 each independent of all the previous ones—i.e., for $i \in \{0, 1, 2, 3, 4\}$, $x_{i+1} \notin \mathbb{Q}_i$, where $\mathbb{Q}_{i+1} = \overline{\mathbb{Q}_i(x_{i+1})}$, with $\mathbb{Q}_0 = \mathbb{Q}$, where by $\overline{K}$ we have denoted the real closure of K, and the order in $\mathbb{Q}_i$ is chosen conveniently. Since the equation $ac + cb = ad + db$ holds for these values for a, f, b, which determine the values of e, c, d, the equation must hold universally, given the fact that x_1, x_2, x_3, x_4, x_5 are independent "transcendentals", so the fact that x_5 is negative or positive makes no difference for the validity of the equation. This is the case for Hilbert planes of Type 3 as well.

23.3 Centers of Mass for Finite Sets of Points

A very general notion of *center of mass* for a finite set of points with "weights" attached to them can be obtained if we consider finite systems of *material points* (A_i, μ_i), where the A_i are points and μ_i are some weights from some additively written commutative semigroup S, define the operation of *union* $\mathfrak{A} \cup \mathfrak{B}$ of two such systems $\mathfrak{A} = \{(A_i, \mu_i)\}$ and $\mathfrak{B} = \{(B_j, \nu_j)\}$, to be the system $\mathfrak{C} = \{(C_k, \pi_k)\}$, where the C_k are the A_i or the B_j, and the π_k are μ_i or ν_j in case $A_i \neq B_j$, and $\mu_i + \nu_j$ in case $A_i = B_j$.

A *center of mass* Γ is a map mapping a finite system of material points $\mathfrak{A}$ to a material point (G, γ), subject to the following axioms:

(I) $\Gamma((A, \mu)) = (A, \mu)$
(II) $\Gamma((A, \mu), (B, \mu)) = (M, \nu)$, where M is the midpoint of AB and ν depends on A, B, and μ
(III) $\Gamma((A, \mu), (B, \nu)) = (C, \pi)$, where C is a point of AB, and π depends on A, B, μ, and ν
(IV) $\Gamma(\mathfrak{A} \cup \mathfrak{B}) = \Gamma(\Gamma(\mathfrak{A}) \cup \Gamma(\mathfrak{B}))$

Note that (II) makes no specification of the value of ν as a function of A, B, and μ. In the Euclidean setting, one would have expected ν to be 2μ, but here the only thing that is stipulated is the position of the center of mass. In (III), again, there is no requirement on the value of π in relation with A, B, μ and ν; only the position of the center of mass is specified, and even that is kept vague. One would have expected that C would be required to lie *between* A and B, but we make no such stipulation.

The question is now: What is the most general geometry which allows the existence of centers of mass? We can broaden the field of candidates as an answer to this question by not insisting that the center of mass be an actual point of the geometry in question, but rather that it be an *ideal point* of that geometry. Such centers of mass will be referred to as *generalized centers of mass*. This can be understood as follows. Suppose in our geometry every segment AB has a midpoint, to be denoted by $m(A, B)$. We will think of any two distinct lines g and h as having an ideal point in common. The ideal point may be a point of our geometry, in case the two lines intersect, or one that is not a point of our geometry. Given a third line k, suppose we have a way to decide whether k passes through the same ideal point that is common to g and h. We then say that g, h, and k *lie in a pencil*. Suppose also that in our geometry for every point P and three lines g, h, and k lying in a pencil, with P not incident with any of the lines g, h, and k, there exists a line p, such that P is incident with p and g, h, and p lie in a pencil. In such a geometry, any system of two material points with equal masses has a center of mass, namely the midpoint of the segment formed by the two points. Also any system of three material points with equal masses will have a generalized center of mass provided that the medians of any triangle lie in a pencil. Suppose our geometry satisfies this condition as well. Then, a system consisting of four distinct material points (A_i, μ), with $i \in \{1, 2, 3, 4\}$, will have a generalized center of mass if and only if the lines $\langle m(A_1, A_2), m(A_3, A_4)\rangle$, $\langle m(A_1, A_3), m(A_2, A_4)\rangle$, $\langle m(A_1, A_4), m(A_2, A_3)\rangle$ lie in a pencil, and so are the first two lines and any line joining A_i with the possibly imaginary point determined by the pencil formed by the medians of the three points among $\{A_1, A_2, A_3, A_4\}$ with index different from i (for $i \in \{1, 2, 3, 4\}$). The generalized center of mass of a system of four material points, with equal weights attached to each point, can thus be understood as being determined by two lines (all other lines on which the generalized center of mass should lie on have to lie in a pencil with those two lines). In the same manner, one can say that a system of five material points $\{(A_i, \mu) : i \in \{1, 2, 3, 4, 5\}\}$ has a generalized center of mass if any three of the lines joining A_i with the generalized center of mass of the four points among A_1, A_2, A_3, A_4, A_5 with index different from i (for $i \in \{1, 2, 3, 4, 5\}$) lie in a pencil, and any two of these lines also lie in a pencil with any line joining the midpoint of A_jA_k with the generalized center of mass of $A_lA_mA_n$, where $\{j, k, l, m, n\} = \{1, 2, 3, 4, 5\}$. In this manner one can proceed inductively and define what it means for a system of n material points, with equal weights attached to each point, to have a generalized center of mass, by asking that the lines joining any of the n points with the generalized center of mass of the remaining $n-1$ points, and the lines joining the midpoint of any segment formed by the n points with the generalized center of mass of the remaining $n-2$ points go through some possibly ideal point.

We will now introduce the most general geometry in which generalized centers of mass exist.

23.4 Non-elliptic Metric Planes

Metric planes, as presented in [6] (for different axiomatization see [25], [26]), are defined in the language of groups. A certain subfamily of metric planes are the structures that we propose as the right geometric candidate for the existence of generalized centers of mass for finite sets of material points. The idea behind metric planes is to single out some key properties that the composition of line reflections have, that are, on the one hand, common to the three classical geometries, and on the other hand are strong enough to allow the embedding of the geometries associated with those reflections into Pappian projective planes. The main idea is to notice that if a and b are line reflections, and if $ab = ba$ (multiplication being here composition), then the lines defined by the reflections a and b must be perpendicular, and conversely. Since perpendicular lines intersect, and the composition of the reflections in them amounts to the reflection in their point of intersection, we can choose to refer to the product ab, in case $ab = ba$, as to a *point reflection*. In a similar manner, the geometric notion of point-line incidence can now be expressed in the language of line reflections. The point P determined by the point reflection ab, where a and b denote line reflections with $ab = ba$, is said to be incident with the line determined by the line reflection g if $Pg = gP$.

The axiom system consists of a *fundamental assumption*, that G be a group (written multiplicatively) generated by an invariant set S of involutory elements, as well as five axioms (see below). The elements of S will be denoted by lowercase Latin letters, and will be referred to as *line reflections* (or simply *lines*). Involutory products of two elements in S will be denoted by uppercase Latin letters, and will be referred to as *point reflections* (or simply *points*). For any two elements α and β in G, $\alpha \mid \beta$ denotes the fact that $\alpha \cdot \beta$ is involutory (in case both α and β are line reflections, then we also write $\alpha \perp \beta$ for $\alpha \mid \beta$). The notation $\alpha_1, \ldots, \alpha_n \mid \beta_1, \ldots, \beta_m$ stands for the conjunction of all $\alpha_i \mid \beta_j$ with $i \in \{1, \ldots, n\}$ and $j \in \{1, \ldots, m\}$. The relation $P \mid g$ (or $g \mid P$) may be read as "point P is incident with line g", and $g \mid h$ as "line g is orthogonal to line h." The axioms for metric planes state that

M 1 *For all points P and Q there is a line g such that $P, Q \mid g$.*

M 2 *If, for points P and Q and lines g and h, we have $P, Q \mid g, h$, then $P = Q$ or $g = h$.*

M 3 *If the lines a, b, and c and the point P are such that $a, b, c \mid P$, then there is a line d such that $abc = d$.*

M 4 *If the lines a, b, c, and g are such that $a, b, c \perp g$, then there is a line d such that $abc = d$.*

M 5 *There are lines g, h, j such that $g \perp h$, but none of $j \perp g$, $j \perp h$, or $j \mid gh$ holds.*

By **M1** and **M2** there is a unique line joining two distinct points P and Q. We will denote it by $\langle P, Q \rangle$. It is, in general, not true that a midpoint exists for every pair of points. In this context, we say that M is a midpoint for A and B if $A^M = B$. Here we have denoted by α^β the element $\beta^{-1}\alpha\beta$ (so A^B denotes the reflection of A in B, P^g the reflection of P in g, and g^P the reflection of g in P). Let **M** denote the axiom asking for the existence of a midpoint for any pair of points. It is, in general, not true that a midpoint, should it exist, is unique. In fact, to ensure uniqueness (see [6, pp. 53–54]), it is necessary and sufficient to assume that the metric plane is not elliptic. This means that it satisfies the axiom (here 1 stands for the neutral element of G)

∼P *For all lines a, b, c, we have $abc \neq 1$.*

An alternate requirement which is equivalent to **∼P**, is that, for any line g and any point P not incident with g, there exists a unique line h incident with P and perpendicular to g (if $P|g$, then the perpendicular through P to g is always unique, it is Pg). This requirement is obviously false in the case of plane elliptic geometry, in which, for every line there is a point from which one can draw infinitely many perpendiculars to that line (that there are infinitely many perpendiculars from a *pole* of a line to that line follows from [6, Satz 20, §6,12, p. 121], which states that there are no finite elliptic planes, so every line must be incident with infinitely many points). Double elliptic geometry is already excluded by the axioms for metric planes, as there is a unique line joining two distinct points in metric planes.

Let $\mathscr{M}$ stand for the theory axiomatized by the fundamental assumption as well as **M1-M5, M**. Particularly simple axiom systems for theories synonymous with $\mathscr{M} + \sim$**P** have been proposed in [40] and [5].

By Bachmann [6, Satz 15, p. 43], given any point P not on all the lines g, h, k, which are lying in a pencil, there is a unique line p, with $P \mid p$ and p, g, h lying in a pencil. This means that metric planes do indeed have the property that we needed so that generalized centers of mass can be defined and their position determined for systems consisting of four or more material points.

To get *Euclidean planes*, in other words to have both a Euclidean metric, expressed by $\mathbf{R}^*$, and the Euclidean parallel postulate, expressed by $\mathbf{V}^*$, valid, one adds two axioms to $\mathscr{M}$, namely:

R* *If* $a, b \perp c$ *and* $a \perp d$, *then* $b \perp d$.

V* *For any two lines* a *and* b, *there is a point* C *with* $a, b \mid C$ *or a line* c *with* $a, b \perp c$.

Two lines a and b that are in the situation described by $\mathbf{V}^*$—i.e., such that there is a point C with $a, b \mid C$ or a line c such that $a, b \perp c$—are said to be *connectible*. To obtain *hyperbolic planes*, we need to add to $\mathscr{M}$ the axioms $\sim \mathbf{V}^*$ and

H *If* $a, b, c \mid P$ *and if none of the pairs* (a, g), (b, g), *and* (c, g) *is connectible, then* $a = b$ *or* $b = c$ *or* $c = a$.

By Bachmann [6, p. 219], $\sim\mathbf{V}^*$ implies $\sim\mathbf{P}$, and by Bachmann [6, p. 202], $\sim\mathbf{P}$ also follows in the presence of $\mathbf{R}^*$ and $\mathbf{V}^*$.

Let σ_n denote the universal statement saying that a generalized center of mass exists for any system of n material points with equal weights. The statement is rather long and has been described in detail in the previous section.

By Galperin [15], σ_n holds in $\mathbb{H}$, given that centers of mass exist in the hyperbolic plane.

We can now show that

Theorem 23.2 $\mathscr{M} + (\mathbf{R}^* \vee \mathbf{H}) \vdash \sigma_n$.

Proof First, let us show that $\mathscr{M} + (\sim\mathbf{V}^* \wedge \mathbf{H}) \vdash \sigma_n$. Suppose that were not the case, so there would exist a model of $\mathscr{M} + (\sim\mathbf{V}^* \wedge \mathbf{H})$ in which σ_n does not hold. Such a model would have to be, by Bachmann [6, Satz 2 & 3, p. 233], isomorphic to a Beltrami–Cayley–Klein model $\mathfrak{M}$ of hyperbolic geometry over an ordered field K. Let $\overline{K}$ denote the real closure of K. The Beltrami–Cayley–Klein model over $\overline{K}$, to be referred to as $\mathfrak{H}$, is elementarily equivalent to $\mathbb{H}$, and, given that $\neg\sigma_n$ is an existential statement that holds in $\mathfrak{M}$, it must hold in $\mathfrak{H}$ as well, given that $\mathfrak{M}$ is a submodel of $\mathfrak{H}$, and thus, by the elementary equivalence of $\mathfrak{H}$ and $\mathbb{H}$, $\neg\sigma_n$ holds in $\mathbb{H}$ as well, a contradiction. That generalized centers of mass exist in the presence of $\mathbf{R}^*$ and $\mathbf{V}^*$ follows from the fact that the incidence structure of Euclidean planes is that of Pappian affine planes, in which generalized centers of mass are known to exist. Now the result follows from the fact that we have both $\mathscr{M} + (\sim\mathbf{V}^* \wedge \mathbf{H}) \vdash \sigma_n$ and $\mathscr{M} + (\mathbf{V}^* \wedge \mathbf{R}^*) \vdash \sigma_n$.

There is every reason to believe that $\mathscr{M} + \sim\mathbf{P} \vdash \sigma_n$.

Similarly, one could ask for the minimal axiom system in which Federico Commandino's theorem (see [43]) on the center of mass of a tetrahedron with equal weights attached holds (Fig. 23.1). In our "imaginary point" language, it asks for the lines joining the vertices of the tetrahedron with the generalized centers of mass of the opposite sides to go through a possibly imaginary point, the generalized center

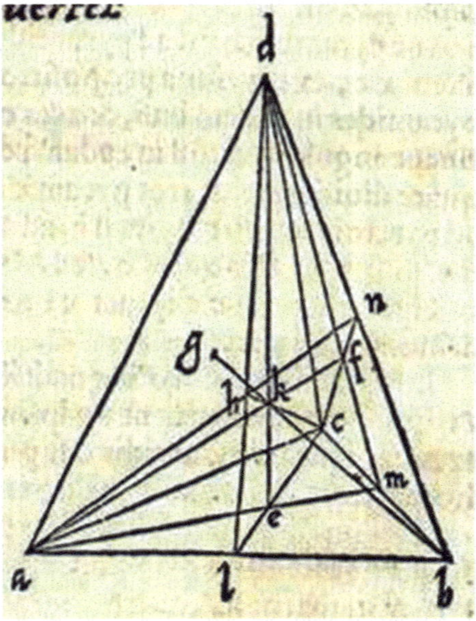

Fig. 23.1 From Federico Commandino's *De Centro Gravitatis Solidorum*, 1565

of mass of the tetrahedron. An axiom system for three-dimensional metric spaces, in terms of plane reflections, has been proposed in [1]. One could, in analogy to our tactic in the two-dimensional case, get the axioms for Euclidean and hyperbolic spaces and then use model-theoretic results, based on the fact that Commandino's theorem holds not only in three-dimensional Euclidean sapce, but also in three-dimensional hyperolic space (as shown in [15]), to get a result. However, it is very likely that Commandino's theorem holds in the geometry defined by Ahrens' axiom system, if one adds to it the existence and uniqueness of midpoints (or in the axiom system in [21], in which midpoints exist).

23.5 Conclusion

We have seen that several results which are usually proved inside Euclidean geometry hold in a much more general setting. However, we did not provide actual proofs for any of these results. There are many more results that very likely hold in $\mathscr{A}$. Here we list only three examples of problems for which the answer is unknown in $\mathscr{A}$: (i) If two triangles have pairwise congruent internal angle bisectors, then they are congruent (this was shown for the Euclidean case in [24]); (ii) given two points A and B, does there exist a point P, such that $\widehat{APB}$ is 120°?; (iii) if $\widehat{UVW}$ is an acute angle, then we know how to construct four points P such that $\widehat{APB} \equiv \widehat{UVW}$ (lay off $\widehat{UVW}$ from A, such that one leg becomes the ray $\overrightarrow{AB}$, while the other is ϱ and reflect A in the perpendicular dropped from B to ϱ; since $\widehat{UVW}$ can be laid in two ways, there are two solutions from P from this construction; doing the same thing in B rather than A provides other two solutions); are there any other points? are there infinitely many points P with $\widehat{APB} \equiv \widehat{UVW}$? One can also ask whether there is an absolute version for the following result of Dmitriev and Dynkin [12], [7], and [2]: If P is an arbitrary point in the interior of a convex n-gon $A_1A_2 \dots A_n$ and we denote $A_{n+1} = A_1$, then

$$\min_{k=1,\dots,n} \widehat{PA_kA_{k+1}} \leq \frac{\pi}{2} - \frac{\pi}{n}. \tag{23.1}$$

Equality occurs if and only if $A_1A_2 \dots A_n$ is a regular n-gon and P is its center.

The absolute version of (23.1) would be (here we denote, in addition to $A_{n+1} = A_1$, $A_{n+2} = A_2$)

$$\min_{k=1,\dots,n} \widehat{PA_kA_{k+1}} \leq \frac{\sum_{i=1}^{n} \widehat{A_iA_{i+1}A_{i+2}}}{2n}. \tag{23.2}$$

In general, there are numerous problems that one can assign to students in a Euclidean geometry class. On the other hand, a book on plane absolute geometry would have very few problems for which we know a synthetic solution. Elementary geometry has been largely abandoned from school mathematics curricula or preserved only in a computational form, as analytic geometry. If we want to avoid a repeat of Valentin Poénaru's experience in school, "I really hated the silly maths which the school tried to teach me" [37, p. 706]—an experience which is not restricted to occupied countries, as the remarks in [38] make clear—then emphasizing absolute geometry and number theory offers a way out. They are both as elementary as it gets and they are not amenable to algorithmic solutions.

This will certainly not happen in the real world, except in radically alternative schools, whose aim is to foster creativity, not the satisfaction that comes from following procedural steps, in that desperate attempt to overcome what Günther Anders [4] called "the shame of not being a thing."

References

1. J.H. Ahrens, Begründung der absoluten Geometrie des Raumes aus dem Spiegelungsbegriff. Math. Z. **71**, 154–185 (1959)
2. J. Akiyama, M. Nakajima, A generalization of a problem posed in Mathematics Olympiad. Fac. Sci. Tokai Univ. **27**, 59–61 (1992)
3. H. Andéka, J.X. Madarász, I. Németi, G. Székely, Axiomatizing relativistic dynamics without conservation postulates. Stud. Logic **89**(2), 163–186 (2008)
4. G. Anders, *Die Antiquiertheit des Menschen* (Band I. C. H. Beck, Mnchen, 1980)
5. C. Augat, H. Hähl, An axiom system for absolute plane geometry in terms of collinearity and congruence. Beitr. Algebra Geom. **66** (2025) https://doi.org/10.1007/s13366-024-00758-7
6. F. Bachmann, *Aufbau der Geometrie aus dem Spiegelungsbegriff*. Ergänzte Auflage, vol. 2 (Springer, Berlin, 1973)

7. Á. Besenyei, The Brocard angle and a geometrical gem from Dmitriev and Dynkin. Am. Math. Month. **122**(5), 495–499 (2015)
8. H.S.M. Coxeter, *Introduction to Geometry*, 2nd edn. (Wiley, New York, 1969)
9. R. Dedekind, *Stetigkeit und irrationale Zahlen* (Vieweg, Braunschweig, 1872)
10. M. Dehn, Die Legendre'schen Sätze über die Winkelsumme im Dreieck. Math. Ann. **53**, 404–439 (1900)
11. O. Demirel, E. Soytürk Seyrantepe, The theorems of Urquhart and Steiner–Lehmus in the Poincaré ball model of hyperbolic geometry. Mat. Vesnik **63**(4), 263–274 (2011)
12. N. Dmitriev, E. Dynkin, On the characteristic roots of stochastic matrices. C. R. Acad. Sci. URSS **49**, 159–162 (1945)
13. D. Elliott, M.L. Urquhart. J. Aust. Math. Soc. **8**, 129–133 (1968)
14. J. Ferreirós, The road to modern logic — an interpretation. Bull. Symbol. Logic **7**, 441–484 (2001)
15. G.A. Galperin, A concept of the mass center of a system of material points in the constant curvature spaces. Commun. Math. Phys. **154**(1), 63–84 (1993)
16. D. Harutyunyan, A. Nazaryan, V. Pambuccian, The Hajja-Martini inequality in a weak absolute geometry. J. Geom. **110**(2), artice no 24, 16pp. (2019)
17. D. Hilbert, *Grndlagen der Geometrie* (B. G. Teubner, Stuttgart, 1977)
18. D. Hilbert, *David Hilbert's Lectures on the Foundations of Arithmetic and Logic, 1917–1933*, ed. by W. Ewald, W. Sieg. In collaboration with U. Majer and D. Schlimm (Springer, Berlin, 2013)
19. J. Hjelmslev, Om Geometriens almene Grundlag, in *Den 11te skandinaviske matematikerkongress i Trondheim, 22–25 August 1949, 1–12* (J.G. Tanum, Trondheim, 1952)
20. I. Hociotă, V. Pambuccian, Acute triangulation of a triangle in a general setting revisited. J. Geom. **102**, 81–84 (2012)
21. H. Lenz, Zur Axiomatik der absoluten Geometrie des Raumes. Arch. Math. **19**, 205–213 (1968)
22. G.E. Martin, *The Foundations of Geometry and the non-Euclidean Plane* (Springer, New York, 1982)
23. A. Nazaryan, Equilateral triangles have minimal area and perimeter among all triangles containing a given circle in Hilbert planes. J. Geom. **114**, 25 (2023)
24. V. Oxman, A purely geometric proof of the uniqueness of a triangle with prescribed angle bisectors. Forum Geom. **8**, 197–200 (2008)
25. V. Pambuccian, Constructive axiomatization of non-elliptic metric planes. Bull. Polish Acad. Sci. Math. **51**, 49–57 (2003)
26. V. Pambuccian, Orthogonality as single primitive notion for metric planes. With an appendix by H. and R. Struve. Beitr. Algebra Geom. **48**, 399–409 (2007)
27. V. Pambuccian, Acute triangulation of a triangle in a general setting. Can. Math. Bull. **53**, 534–541 (2010)
28. V. Pambuccian, Absolute geometry proofs of two geometric inequalities of Chisini. J. Geom. **108**, 265–270 (2017)
29. V. Pambuccian, C. Schacht, Euler's inequality in absolute geometry. J. Geom. **109**, 8, 1–11 (2018)
30. V. Pambuccian, Existence of special rainbow triangles in weak geometries. Georgian Math. J. **26**(4), 489–498 (2019)
31. V. Pambuccian, C. Schacht, Lippmann's axiom and Lebesgue's axiom are equivalent to the Lotschnittaxiom. Beitr. Algebra Geom. **60**(4), 733–748 (2019)
32. V. Pambuccian, R. Struve, Axiom systems implying infinity in the foundations of geometry. Beitr. Algebra Geom. **62**, 109–120 (2021)
33. V. Pambuccian, H. Struve, R. Struve, The Steiner–Lehmus theorem and "triangles with congruent medians are isosceles" hold in weak geometries. Beitr. Algebra Geom. **57**, 483–497 (2016)
34. V. Pambuccian, H. Struve, R. Struve, A comprehensive analysis of the axiomatic frameworks in which Fagnano's theorem holds. J. Geom. **116**(1), 6 (2025)

35. D. Pedoe, The most "elementary" theorem of Euclidean geometry. Math. Mag. **49**(1), 40–42 (1976)
36. W. Pejas, Die Modelle des Hilbertschen Axiomensystems der absoluten Geometrie. Math. Ann. **143**, 212–235 (1961)
37. V. Poénaru, Memories from my former life: the making of a mathematician, in *Geometry in History*, ed. by S. Dani, A. Papadopoulos pp. 705–734 Springer, Cham (2019)
38. V. Poénaru, Some disillusioned thoughts about school and teaching, in *Handbook of the History and Philosophy of Mathematical Practice*, ed. by B. Sriraman (Springer, Cham, 2024)
39. C. Rovelli, *Quantum Gravity* (Cambridge University Press, Cambridge, 2004)
40. E. Schröder, Zur projektiven Einbettung normaler absoluter Ebenen. Mitt. Math. Ges. Hamb. **39**, 77–98 (2019)
41. G. Szekeres, Kinematic geometry: an axiomatic system for Minkowski space-time. J. Austral. Math. Soc. **8**, 134–160 (1968)
42. G. Wanner, The Cramer-Castillon problem and Urquhart's 'most elementary' theorem. Elem. Math. **61**, 58–64 (2006)
43. G. Wanner, F. Commandino, de centro gravitatis solidorum, 1565. Elem. Math. **73**(4), 179—181 (2018)

Chapter 24
The Reconstruction of Theaetetus' Theory of Ratios of Magnitudes

Stelios Negrepontis and Dimitrios Protopapas

Abstract In the present chapter we obtain the reconstruction of Theaetotus' theory of ratios of magnitudes based, according to Aristotle's *Topics* 158b, on the definition of proportion in terms of equal anthyphairesis. Our reconstruction is built on the anthyphairetic interpretation of the notoriously difficult *Theaetetus* 147d6-e1 passage on Theaetetus' mathematical discovery of quadratic incommensurabilities, itself based on the traces it has left on Plato's philosophical definition of Knowledge in his dialogues *Theaetetus*, *Sophist* and *Meno*. Contrary to earlier reconstructions by Becker, van der Waerden and Knorr, our reconstruction reveals a theory that (a) applies only to the restricted class of pairs of magnitudes whose anthyphairesis is finite or eventually periodic, and (b) avoids the anachronistic use of Eudoxus' definition 4 of Book V of Euclid's *Elements*.

Keywords Theaetetus · Plato · Eudoxus · Plato's *Theaetetus* · *Sophist* · *Meno* · Euclid's *Elements* · Theory of ratios of magnitudes · Periodic anthyphairesis · Logos criterion · True opinion plus logos · Name plus logos · Becker · van der Waerden · Knorr · Burnyeat · Incommensurability · Pythagoreans

Subject Classification Primary 00A30, Secondary 01A20

24.1 Introduction

Aristotle, in *Topics* 158b29-35, informs us of a theory of ratios of magnitudes based on an anthyphairetic definition of proportion that predates Eudoxus' theory. Since Becker [5] this theory was attributed to Theaetetus (Sect. 24.2). The importance

S. Negrepontis (✉) · D. Protopapas
Mathematics Department, Athens University, Athens, Greece

L. Funar, A. Papadopoulos (eds.), *Essays on Topology*,
https://doi.org/10.1007/978-3-031-81414-3_24

of this early theory cannot be overestimated, as it is the first step towards the construction of the real numbers, a project that was brought to fruition only in modern times by Dedekind [10] in 1872, following closely Eudoxus' theory of ratios of magnitudes, the theory of ratios of magnitudes that replaced Theaetetus' theory. Attempts to a satisfactory understanding of the theory have been hampered by the realization that Eudoxus' condition (Book V, Definition 4 of the *Elements*) appears to be necessary for the reconstruction (Becker [5], van der Waerden [48], Knorr [18]). This apparent anachronism has led some researchers to believe either that Theaetetus' theory was not rigorous, treating the analogue of the crucial Proposition V.9 of the *Elements* as self-evident (Knorr), or even that such theory never really existed (Acerbi [1], Saito [42]) (Sect. 24.3).

Previous historians of Mathematics along with Platonic scholars have failed to realize the opportunities offered to modern readers of Plato for a double understanding of Theaetetus' Mathematics and Plato's Philosophy by the plan outlined in the *Theaetetus*. Plato sets the fundamental question on philosophical knowledge in the beginning of the dialogue, where (a) philosophical knowledge is correlated with Theaetetus' mathematical discovery of a general theorem of incommensurability, as reported in the *Theaetetus* 147d3-e1, urging in fact Theaetetus to approach the philosophical question on knowledge by imitating his mathematical discovery, and (b) philosophical knowledge is described (i) as True Opinion plus Logos or, equivalently, (ii) as Name plus Logos (Sect. 24.4). Some of these opportunities offered by Plato, but ignored by modern scholars, as philosophical imitations of Theaetetus' mathematical discovery consist in the divisions of the Angler, the Noble Sophistry and the Sophist in the *Sophist*, where each is described as providing knowledge by Name plus Logos, and in the *Meno*, which provides knowledge of the diameter to the side of a square, described as True Opinion plus Logos. Both the entities described as Plato's Name plus Logos in the *Sophist* and as True Opinion plus Logos in the *Meno* reveal lucidly that Logos, the fundamental concept of Plato's philosophy, is an imitation of the Logos Criterion for anthyphairetic periodicity, namely the condition that the ratio of two successive anthyphairetic remainders is equal to the ratio of two subsequent successive remainders (Sect. 24.2).

We conclude that

(a) contrary to the arithmetical reconstruction of Theaetetus' proof regarding quadratic incommensurabilities provided by Euclid in Proposition X.9 of the *Elements*, which both van der Waerden [48] and Knorr [18] suggest (Sect. 24.9), Theaetetus' theorem reported in passage 147d3-e1 states that if lines a and b satisfy $a^2 = Nb^2$ for non-square number N, or more generally $Aa^2 = Cb^2$

for natural numbers A, C, such that AC is not square, then the anthyphairesis of a to b satisfies the Logos Criterion for eventual periodicity (and hence, by Proposition X.2, a and b are incommensurable) marking the transition from the Pythagorean method of finitization by the preservation of Gnomons to the Logos Criterion and the birth of Theaetetus' theory of ratios of magnitudes (Sects. 24.6 and 24.7);

(b) contrary to Burnyeat's [8] claim (Sect. 24.9), Plato had a strong motive to exhibit the method of proof of Theaetetus, since Plato's Logos is the philosophical imitation of Theaetetus' Logos Criterion, revealing the anthyphairetic nature and structure of an intelligible Being, and thus forming the basis of Plato's own philosophy (Sect. 24.5);

(c) contrary to the insistent interpretation of Logos by Plato scholars, as "account", "discourse", "rational definition","rational explanation" and "reason" (Sect. 24.9), we suggest that the meaning of Plato's Logos is the philosophical analogue of Theaetetus' Logos Criterion for anthyphairetic periodicity, resulting in geometric incommensurability and philosophical Knowledge and Oneness of the intelligible Beings (Sect. 24.5);

(d) contrary to claims made by Acerbi and Saito, Theaetetus' theory of proportion for magnitudes is certainly existent, since essential use of this theory is made in Theaetetus' theorem on quadratic incommensurabilities (Sect. 24.7), in the Theaetetean Book X of the *Elements*, including the palindromicity of anthyphairesis of a to b if $a^2 = Nb^2$ for a non-square number N and the solution to Pell's problem in relation with the Platonic dialogue *Statesman* [33] and through its imitation in Plato's philosophy in the *Sophist*, the *Meno* and other dialogues (Sect. 24.5); and,

(e) contrary to the reconstructions put forward by Becker [5], van der Waerden [48] and Knorr [18] (Sect. 24.3), Theaetetus' theory of ratios of magnitudes is NOT about the class of all ratios of magnitudes, as Eudoxus' theory of ratios, presented in Book V of the *Elements*, BUT is strictly about the limited class of all ratios with finite or eventually periodic anthyphairesis (Sect. 24.8).

Our thesis is supported by the facts that

(a) all of Theaetetus' work is strictly about periodic anthyphairesis, in fact

[1] the eventual periodicity of the anthyphairesis of lines a to b, satisfying $Ma^2 = Nb^2$, for MN not square number (the statement is suggested in the *Theaetetus* 147d3-e1 passage, proof of the theorem in Sect. 24.7 using tools only from a restored Book II of the *Elements*, plus a pigeon-hole argument, Negrepontis, Farmaki, [29]);

[2] the palindromic periodicity of the anthyphairesis of the surds $\sqrt{N}$ for any non-square number N (the statement of theorem [2] is suggested in Plato's *Statesman*, cf. Negrepontis [24, 32]; Negrepontis et al. [33];

[3] the eventual periodic anthyphairesis of lines a to b, satisfying more general quadratic expressions, including the Application of Areas in defect (Sect. 24.6 using tools only from a restored Book II of the *Elements*, Negrepontis, Farmaki, [29]), and the use of this result to show that 12 classes of alogoi lines, including the minor, despite being alogoi, are determined by an eventually periodic anthyphairesis (*Scholia in Euclidem* X.135, X.185, cf. Negrepontis and Farmaki [28], Negrepontis et al. [34]); and

[4] the Pell theorem: If N is a non-square number, then find natural numbers x, y such that $y^2 = Nx^2 + 1$ (resulting from the palindromic periodicity theorem, Negrepontis et al. [33]);

(b) Plato's use of Theaetetus's work is exclusively about ratios with periodic anthyphairesis; and

(c) the *Anonyma Scholia to the Elements* X.2, X.135 and X.185 confirm that Theaetetus' theory of ratios was exclusively about ratios in periodic anthyphairesis, and that Eudoxus' more general theory of ratios (presented in Book V of the *Elements*) was met with strong disapproval by Platonic circles. For a more detailed analysis of Theaetetus' work and its effect to Plato's philosophy the reader is referred to Section 2 of Negrepontis, Protopapas, [31].

Once it is made clear that Theaetetus' theory of ratios is strictly and only for ratios with finite or eventually periodic anthyphairesis (certainly not because the definition of equal anthyphairesis does not allow it, but because ratios with non-periodic anthyphairesis are not wanted), then we will have no problem in reconstructing the theory, all the while avoiding the use of the anachronistic Eudoxus' condition of Definition 4 of Book V of the *Elements* (Sect. 24.10).

We provide further support towards the confirmation of our reconstruction in a separate publication (Negrepontis and Protopapas, [31]), by showing that the proofs of all the propositions in the Theaetetean Book X of the *Elements* do not need Eudoxus' theory of ratios which relies on Definition 4 of Book V of the *Elements*, but only the limited Theaetetus' theory of ratios applied solely to ratios with terms in eventually periodic anthyphairesis.

24.2 Aristotle, in *Topics* 158b, Reports of a Pre-Eudoxean Theory of Ratios for Magnitudes Based Upon Equality of Anthyphairesis and While the Originator of the Theory Is not Mentioned, It Is Generally Attributed to Theaetetus

24.2.1 A Pre-Eudoxean Theory of Proportions for Magnitudes Based on the Equality of Anthyphairesis, Is Reported by Aristotle in the Topics *158b29-35*

Book V of Euclid's *Elements* is devoted to the presentation of Eudoxus' theory of ratios of magnitudes, on which Books X, XII, and XIII of the *Elements* are based. However, Aristotle in *Topics* 158b29-35 reports on a clearly preceding theory of ratios of magnitudes, based on the definition of proportion in terms of equality of antanairesis/ anthyphairesis.

> ἔοικε δὲ καὶ ἐν τοῖς μαθήμασιν
> ἔνια δι' ὁρισμοῦ ἔλλειψιν οὐ ῥᾳδίως γράφεσθαι,
> οἷον ὅτι ἡ παρὰ τὴν πλευρὰν τέμνουσα τὸ ἐπίπεδον
> ὁμοίως διαιρεῖ τήν τε γραμμὴν καὶ τὸ χωρίον.
> τοῦ δὲ ὁρισμοῦ ῥηθέντος εὐθέως φανερὸν τὸ λεγόμενον·
> τὴν γὰρ αὐτὴν ἀνταναίρεσιν ἔχει τὰ χωρία καὶ αἱ γραμμαί·
> ἔστι δ' ὁρισμὸς τοῦ αὐτοῦ λόγου οὗτος.
> It also appears that in mathematics,
> it is not easy to construct proofs of some things because of the deficiency of a definition,
> for instance that the line cutting a plane figure parallel to its side
> divides the area and the line similarly.
> But once the definition has been stated, the proposition is evident at once.
> For the areas and the lines have the same reciprocal subtraction (antanairesis, anthyphairesis);
> and this is the definition of "same ratio".
> [Translation by Smith [2]].

That the term "antanairesis" in Aristotle's *Topics* coincides with the term "anthyphairesis" in Euclid's *Elements* (Propositions VII.1 & 2, X.2) is confirmed by Alexander Aphrodisieus' *Commentary to Aristotle's Topics* 545, 16–17.

We can safely reconstruct the definition that Aristotle has in mind as follows.

Let a, b be magnitudes, $a > b$, with finite or infinite anthyphairesis.

$a = k_0 b + c_1, c_1 < b,$
$b = k_1 c_1 + c_2, c_2 < c_1,$
$c_1 = k_2 c_2 + c_3, c_3 < c_2,$
...
$c_{n-1} = k_n c_n + c_{n+1}, c_{n+1} < c_n,$
...

We set $Anth(a, b) = [k_0, k_1, k_2, \ldots, k_n, \ldots]$.

Two dyads of magnitudes $a, b, a > b$ and $c, d, c > d$ have *equal anthyphairesis* if $Anth(a, b) = Anth(c, d)$. The definition of proportion that Aristotle has in mind in his *Topics* passage is then the following:

Definition (Aristotle, *Topics* 158b29-35) If $a, b, a > b$, and $c, d, c > d$, are two dyads of magnitudes, we say that $a/b = c/d$ if $Anth(a, b) = Anth(c, d)$.

The Proposition that Aristotle has in mind is a simpler version of Proposition VI.1 of the *Elements*, and may be stated in modern notation as follows:

Proposition (*Elements*, VI.1) *If a, b, c are line segments, then* $(ac)/(bc) = a/b$.

Proof Let the anthyphairesis of a to b be the following:
$a = k_0 b + c_1, c_1 < b,$
$b = k_1 c_1 + c_2, c_2 < c_1,$
$c_1 = k_2 c_2 + c_3, c_3 < c_2,$
...
$c_{n-1} = k_n c_n + c_{n+1}, c_{n+1} < c_n,$
...
Then
$ac = k_0 bc + c_1 c, c_1 < b,$
$bc = k_1 c_1 c + c_2 c, c_2 < c_1,$
$c_1 c = k_2 c_2 c + c_3 c, c_3 < c_2,$
...
$c_{n-1} c = k_n c_n c + c_{n+1} c, c_{n+1} < c_n,$
...
It is clear that $Anth(a, b) = Anth(ac, bc)$, hence $(ac)/(bc) = a/b$.

24.2.2 Although Aristotle Does not Mention the Originator of This Theory, Nevertheless Modern Historians of Greek Mathematics Attribute the Theory of Proportion to Theaetetus, the Great Geometer of Plato's Academy

Aristotle does not mention the discoverer of this theory of proportion for magnitudes. Becker, van der Waerden and Knorr regard Theaetetus as the originator of this theory of proportion. In Knorr's own words:

> ...an attractive thesis by O. Becker [1933]: that in the period before Eudoxus an attempt had been made to produce a general theory of proportions, valid for commensurable and incommensurable magnitudes alike, via a proportion-concept based on anthyphairesis. [Footnote 43] [18, pp. 125–126].
> ...a proportion theory based on anthyphairesis was indeed the keystone of Theaetetus' theory of irrational lines. [18, p. 126].
> ...we argue that Theaetetus introduced the anthyphairetic theory of proportion as the necessary auxiliary to his study of irrationals, not as a formal remedy of the 'crisis' in proportion theory occasioned by the discovery of incommensurability. [18, p. 290].
> [Footnote 43] O. Becker, 'Voreudoxische Proportionenlehre', QS 1933, 2:B, pp. 311-333. This thesis was also proposed by O. Toeplitz, H. Zeuthen and E. Dijksterhuis. For a

discussion, see B. L. van der Waerden, SA 1954, pp. 175-179 and our chapter VIII, Sections II-III. [18, p. 130].

24.2.3 *Theaetetus' Logos Criterion for the Eventual Periodicity of an Anthyphairesis*

As we shall see in effect, fundamental within Theaetetus' theory of ratios of magnitudes is the Logos Criterion for anthyphairetic periodicity, an almost immediate consequence of Theaetetus' definition of proportion.

Proposition 24.2.3.1 (Logos Criterion for the Periodicity of an Anthyphairesis) Let a, b be two line segments, with $a > b$, and anthyphairesis given as follows:
$a = k_0 b + e_1$, with $b > e_1$,
$b = k_1 e_1 + e_2$, with $e_1 > e_2$,
...
$e_{m-1} = k_m e_m + e_{m+1}$, with $e_m > e_{m+1}$,
$e_m = k_{m+1} e_{m+1} + e_{m+2}$, with $e_{m+1} > e_{m+2}$,
...
$e_{n-1} = k_n e_n + e_{n+1}$, with $e_n > e_{n+1}$,
$e_n = k_{n+1} e_{n+1} + e_{n+2}$, with $e_{n+1} > e_{n+2}$,
...
and assume that there are indices $m < n$ such that $e_n/e_{n+1} = e_m/e_{m+1}$ ("Logos Criterion"). Then the anthyphairesis of a to b is eventually periodic, in fact $Anth(a, b) = [k_0, k_1, \ldots, k_m, period(k_{m+1}, k_{m+2}, \ldots, k_n)]$.

Proof By the Theaetetean definition of analogy, provided to us by Aristotle in the *Topics* 158b29-35 passage, $Anth(e_n, e_{n+1}) = Anth(e_m, e_{m+1})$. But $Anth(a, b) =$
$[k_0, k_1, \ldots, k_m, Anth(e_m, e_{m+1})] =$
$[k_0, k_1, \ldots, k_m, k_{m+1}, k_{m+2}, \ldots, k_n, Anth(e_n, e_{n+1})] =$
$[k_0, k_1, \ldots, k_m, k_{m+1}, k_{m+2}, \ldots, k_n, Anth(e_m, e_{m+1})] =$
$[k_0, k_1, \ldots, k_m, k_{m+1}, k_{m+2}, \ldots, k_n, k_{m+1}, k_{m+2}, \ldots, k_n, Anth(e_n, e_{n+1})] =$
$\ldots =$
$[k_0, k_1, \ldots, k_m, period(k_{m+1}, k_{m+2}, \ldots, k_n)]$.

24.3 The Problematic Reliance on Eudoxus' Condition (Definition V.4 of the *Elements*) in the Reconstructions of Theaetetus' Theory of Ratios of Magnitudes by Becker and by Knorr

Becker assumes that Theaetetus' theory of ratios of magnitudes, based on the definition of proportion in terms of equal anthyphairesis, is a theory for the same class of ratios of magnitudes as is Eudoxus' theory. Therefore, his reconstruction of

Theaetetus' proof of the analogue of the crucial Proposition V.9 of the *Elements*, is inevitably in need of Eudoxus' condition (Definition V.4); van der Waerden essentially agrees with this reconstruction:

> O. Becker has advanced an ingenious hypothesis for this. From the proportionality
> (1) $a : b = c : d$,
> one deduces first the equality of the areas
> (2) $ad = bc$
> then interchanges b and c, and finally returns to the proportionality
> (3) $a : c = b : d$
> It is this last step which involves the proposition on rectangles, that Aristotle talked about:
> (4) $a : c = ad : cd$,
> since
> $a : c = ad : cd = cb : cd = b : d$
> follows from (4) and (2).
> The fragments from Aristotle, which have been cited, make it very plausible, that (3) was indeed derived from (2) in this manner.
> The deduction of (2) from (1) involves another proposition. For, from (1) follows
> $ad : bd = a : b = c : d = bc : bd$;
> Now it remains to prove: *P. If in a proportion the consequents are equal then the antecedents* (ad and bc) *are equal as well.*
> The proof of P. requires use of the so-called "lemma of Archimedes", that is formulated as follows: *Q. If A and B are comparable magnitudes (e.g. both line segments), and if A is less than B, then a certain multiple nA exceeds B.* [48, pp. 177–178].

Note that what van der Waerden calls "lemma of Archimedes" is Definition 4 of Book V of the *Elements*, and we refer to it as "Eudoxus' condition".

> All becomes clear now. Evidently, Theaetetus began his book with an exposition of the theory of proportions, based on the antanairesis-definition. Following his usual procedure, he started with lemmas which would be needed later on; among these is R = X 1. In propositions X 2, 3 he established the theory of the infinite or finite antanairesis, thus obtaining at the same time a criterium for the commensurability of two line segments or two areas. It is probable that the next thing was the theory of proportions, based on the antanairesis-definition to which Aristotle refers. Euclid omitted this part, because he had already given another theory of proportions in Book V. [48, pp. 178–179].

Knorr [18], although not fully accepting Becker's reconstruction, nevertheless does not question Becker's assumption that Theaetetus' theory is about the same class of ratios of magnitudes as is Eudoxus' theory, and that therefore a rigorous proof of the analogue of Proposition V.9 would need Eudoxus' condition.

> The simple elegance of the Euclidean proof in contrast to the cumbersome anthyphairetic version scarcely needs comment. Becker and the adherents to his thesis on the pre-Eudoxean anthyphairetic theory wish to maintain that this theorem was proved within that theory and made necessary for the first time the use of the convergence-principle. I do not fully accept this view. For the algebra in (b) [the anthyphairetic proof] is messy, inevitably so; but by Greek methods of algebra it would be horrendous, and I have no doubt in this case prohibitively difficult to effect formally. I find it significant that Theorem 6 [if for magnitudes A, B, C, we have $A/C = B/C$, then $A = B$ (*Elements* V.9)] is absent from Book VII. This leads me to believe that in the initial stages in the theory of proportions, both for integers and for magnitudes (via *anthyphairesis*), this deceptively easy-looking step was omitted. But in the course of the complete formulation of the theory, and probably in context of Theorem 7 [Given four lines A, B, C, D; $A/B = C/D$ if and only if

> $AD = BC$ (*Elements* VII.19 and VI.16)] (as Becker maintains), the necessity of Theorem 6 was detected. The anthyphairetic verification must certainly have been abandoned, however, after it was discovered that certain principles worked out after an attempt at a proof - i.e. X,1 and V, Def. 4 - could be turned around into an alternative definition of proportion (V, Def. 5). Thus, the stage was reached that at which the whole theory of proportions could now be put into Eudoxean-Euclidean form. [18, p. 340].

Therefore, if we assume that Theaetetus' theory was about essentially all ratios of magnitudes, we are forced to accept either an anachronistic use of Eudoxus' condition or some sloppy, non-rigorous approach by Theaetetus. Both are highly implausible. Thus, the previous reconstructions of Theaetetus' theory have a serious problem and, consequently, a reconstruction that is rigorous and avoids any use of Eudoxus' condition will have a serious advantage.
A satisfactory reconstruction of Theaetetus' theory of ratios must not make any use of Eudoxus' condition (V, Definition 4).

24.4 Our Strategy to Examine the Central Role that the Imitation of Theaetetus' Theorem, Reported in the *Theaetetus* 147d-148b, Plays in Plato's Philosophy, Proves Fruitful Towards the Understanding of Both the Method of Theaetetus' Proof and the Nature of Plato's Philosophy

24.4.1 Theaetetus' *Account of the Close Connection Between Theaetetus' Mathematical Discovery and Plato's Philosophy*

We have no ancient mathematical sources at our disposal, informing us on the method used by Theaetetus to prove his quadratic incommensurabilities. On the other hand, we have Plato's dialogues intact and, as it has been shown by one of us (Negrepontis [22–24]), Plato's dialogues contain traces of these Mathematics at the very core of their philosophy. In particular, in his trilogy *Theaetetus-Sophist-Statesman*, Plato guides us to the connection between his philosophical theory of knowledge of intelligible Beings and Theaetetus' mathematical method of proving quadratic incommensurabilities.
Step 1. The philosophical problem on knowledge, Theaetetus *145d-147c*

In the beginning of the dialogue *Theaetetus*, Socrates sets the basic unifying problem that runs through the whole trilogy: How can we obtain the knowledge of the fundamental entities in Plato's philosophy, the intelligible Beings, also referred to as Platonic Ideas?

Step 2. Theaetetus' realization that his recent mathematical discovery is similar to the philosophical problem on knowledge, Theaetetus *147c-e*

After some inconclusive discussion, Theaetetus recognizes that Socrates' philosophical question appears to be related to a mathematical discovery that he recently made (with a friend of his). In *Theaetetus* 147d-e, Theaetetus describes his discovery but Plato's language is obscure and in need of careful and reasoned interpretation. Among earlier historians of Mathematics, including van der Waerden [48] and Knorr [18] and Platonic scholars there is general agreement:

(a) that Theodorus proved that if $a^2 = Nb^2$ for a non-square number N, $3 \leq N \leq 17$ (or, possibly, $3 \leq N < 17$), then a, b are incommensurable lines, and
(b) that Theaetetus discovered and proved a general incommensurability theorem, whose minimal content is the following:

Theaetetus' Theorem (*weak form*).

If a, b are line segments, satisfying $a^2 = Nb^2$ with N a non-square number, or more generally satisfying $a^2/b^2 = C/A$ with A, C natural numbers such that AC is a non-square number, then a, b are lines incommensurable/irrational to each other.

The exact statement and the method of proof of the theorem remains to be determined.

Step 3. It is urged by Socrates that philosophical knowledge can be achieved by imitating Theaetetus' mathematical discovery, Theaetetus *148c9-d7*

Shortly after, at *Theaetetus* 148c-d, Socrates exhorts Theaetetus to "try to imitate" his mathematical discovery in order to answer the philosophical question about the knowledge of the Ideas.

ΣΩ. Θάρρει τοίνυν περὶ σαυτῷ καὶ τὶ οἴου Θεόδωρον λέγειν,
προθυμήθητι δὲ παντὶ τρόπῳ
τῶν τε ἄλλων πέρι καὶ ἐπιστήμης
λαβεῖν λόγον τί ποτε τυγχάνει ὄν.;
ΘΕΑΙ. Προθυμίας μὲν ἕνεκα, ὦ Σώκρατες, φανεῖται. 148c9-d3
Socrates. You may be reassured then, about Theodorus' account of you,
and set your mind on finding a definition of knowledge, as of anything else,
with all the zeal at your command.
Theaetetus. If it depends on my zeal, Socrates, the truth will come to light.
[translation by Cornford [9]]
ΣΩ. Ἴθι δή—καλῶς γὰρ ἄρτι ὑφηγήσω—
πειρῶ μιμούμενος (148d4-5) τὴν περὶ τῶν δυνάμεων ἀπόκρισιν,
ὥσπερ ταύτας πολλὰς οὔσας ἑνὶ εἴδει περιέλαβες,
οὕτω καὶ τὰς πολλὰς ἐπιστήμας ἑνὶ λόγῳ προσειπεῖν. 148d4-7
Socrates. Forward, then. On the way you have just shown so well.
Take as a model your answer about the roots:
just as you found a single character to embrace all that multitude,
so now try to find a single formula that applies to the many kinds of knowledge.
[translation by Cornford [9]]
socrates: Come on then. You've just given us a good start.
Try mimicking your answer about powers:
just as there were many of them, and yet you covered them all with a single form,
try now to apply a single account to the many knowledges there are.
[translation by Christopher Rowe [40]]

Step 4. It is further stated by Socrates that philosophical knowledge consists of Name plus Logos, equivalently of True Opinion plus Logos, Theaetetus *201d-202c*

Later in the *Theaetetus* 201d-202c two equivalent definitions are given for the Knowledge of an intelligible Being (the first described with greater clarity in the *Sophist* 218c4-5):

> δεῖ δὲ ἀεὶ παντὸς πέρι τὸ πρᾶγμα αὐτὸ μᾶλλον διὰ λόγων ἢ
> τοὔνομα μόνον συνωμολογῆσθαι χωρὶς λόγου. Sophist 218c4-5
> however, we ought always in every instance to come to agreement about the thing itself by "Logos"
> rather than about the mere name without "Logos".
> [translation by Fowler [36], modifications by the authors]

Definition (1) The Knowledge (episteme) of an intelligible Being is not Name only but Logos, as well.

Equivalently,

> ὅταν μὲν οὖν ἄνευ λόγου τὴν ἀληθῆ δόξαν τινός τις λάβῃ,
> ἀληθεύειν μὲν αὐτοῦ τὴν ψυχὴν περὶ αὐτό, γιγνώσκειν δ' οὔ·
> τὸν γὰρ μὴ δυνάμενον δοῦναί τε καὶ δέξασθαι λόγον ἀνεπιστήμονα εἶναι περὶ τούτου·
> προσλαβόντα δὲ λόγον δυνατόν τε ταῦτα πάντα γεγονέναι καὶ τελείως πρὸς ἐπιστήμην
> ἔχειν Theaetetus 202b8-c5
> When therefore a man acquires without "Logos" the true opinion about anything,
> his mind has the truth about it, but has no knowledge;
> for he who cannot give and receive the "Logos" of a thing is without knowledge of it;
> but when he has acquired also the "Logos" he may possibly have become all that I have said and may now be perfect in knowledge.
> [translation by Fowler [36], modifications by the authors]

Definition (2) The Knowledge of an intelligible Being is True Opinion about this true Being plus Logos.

The attempts to understand these definitions of intelligible knowledge in the *Theaetetus* are not successful. Schematically we have the following

Interplay between Theaetetus' Mathematics and Plato's Philosophy in Plato's *Theaetetus*

philosophical question on Knowledge by Socrates	*geometrical/mathematical discovery by Theaetetus*
Step 1 145d4-e9 Question set	
145e-147c Inconsequential Discussion on the matter	
147c4 Hint that knowledge is infinite	
	Step 2 147c7-d1 Realization by Theaetetus of a connection between the question on Philosophical Knowledge and his mathematical discovery

philosophical question on Knowledge by Socrates	*geometrical/mathematical discovery by Theaetetus*
	147d3-148b2 Account of Theaetetus' discovery
Step 3 148c9-d7 "peiro mimoumenos" The urge to imitate Theaetetus' discovery in order to asnwer the philosophical question	
Step 4 201d-202c Knowledge as Name plus Logos, equivalently as True Opinion plus Logos	
206c-210d [end of dialogue] Despite attempts, failure to understand the true meaning of Knowledge and Logos	

24.4.2 *Outline of Our Strategy on the Basis of* Theaetetus' *Account*

Combining Steps 3 and 4 we conclude that philosophical knowledge on the one hand is described as Name plus Logos, equivalently as True Opinion plus Logos, on the other hand as imitation of Theaetetus' mathematical discovery.

The philosophical knowledge of the line a whose square doubles the area of the square with side b is given in the *Meno* as True Opinion plus Logos. In Sect. 24.5.1, we find that the philosophical terms True Opinion, Logos and Knowledge have a quite specific geometrical meaning:
True Opinion consists in the first two steps of the anthyphairesis of the diameter a to the side b of a square,
Logos consists in the Logos Criterion for the periodicity of the anthyphairesis of the diameter a to the side b of a square, and
Knowledge of the diameter a to the side b consists in the infinite periodic anthyphairesis where the sequence of successive quotients is $[1, 2, 2, 2, \ldots]$.

The philosophical knowledge of each of the intelligible Beings Angler, Noble Sophistry and Sophist is given in the *Sophist*, the sequel of the *Theaetetus* in the Platonic trilogy *Theaetetus-Sophist-Statesman* devoted to knowledge, as Name plus Logos. In Sect. 24.5.2, we find that the philosophical terms Name, Logos and Knowledge have a quite specific geometrical meaning:
Knowledge of the Angler consists in its definition in terms of a philosophical imitation of an anthyphairesis,
whose anthyphairetic remainders are the *Names*, including the Angler itself, and which possesses *Logos*, the philosophical imitation of Theaetetus' Logos Criterion for periodic anthyphairesis.

Similar definitions are given for the Noble Sophistry and especially for the Sophist. Since philosophical knowledge has also been described as the imitation of Theaetetus' theorem, we expect that in fact Theaetetus proved not only the incommensurability of each power, but also its anthyphairetic periodicity.

In Sect. 24.5.3 we arrive at the same conclusion by carefully examining the philosophical wording of the description of Theaetetus' mathematical discovery in the *Theaetetus* 147d3-e1 passage, and arguing that "the powers" in the 147d6-e1 passage should be rendered as a distributive plural, an interpretation that suggests that the method of proof of the incommensurabilities of the powers by Theodorus and Theaetetus was anthyphairetic. In particular, the philosophical statement that Theaetetus proved that the infinite multitude of parts in which each power is divided by anthyphairesis can be collected into One strongly suggests that every power is a self-similar One, a Oneness that is conceivable only under periodic anthyphairesis. We are thus led again to the conclusion that Theaetetus proved that the anthyphairesis of every power with respect to the one foot line is periodic, or more generally

Theaetetus' Theorem (*strong form*).
If a, b are line segments, satisfying $a^2 = Nb^2$ with N non-square number, or more generally satisfying $a^2/b^2 = C/A$ with A, C natural numbers such that AC is a non-square number, then the anthyphairesis of a to b is eventually periodic, hence a, b are lines incommensurable/irrational to each other.

In conclusion, our approach proves to be quite fruitful. It yields both

(a) that Theaetetus' theorem, reported in the *Theaetetus* 147d3-b2, states that a quadratic equation of the form $a^2 = Nb^2$ with N non-square number, or more generally satisfying $a^2/b^2 = C/A$ with A, C natural numbers such that AC is a non-square number, possesses an eventually periodic anthyphairesis, established in terms of Theaetetus' Logos Criterion, and
(b) that the intelligible Being/Idea of Plato is the philosophical analogue of a dyad of kinds satisfying *Logos*, the philosophical analogue of the Logos Criterion of Theaetetus for anthyphairetic periodicity.
In addition, it makes clear
(c) that the role of Theaetetus' theory of ratios of magnitudes is fundamental, both for Theaetetus' theorem, proved with an appeal to the Logos Criterion for anthyphairetic periodicity, and for Plato's account of philosophical Knowledge in terms of Logos, the philosophical imitation of Theaetetus' Logos Criterion.

24.5 Plato's Imitation of Theaetetus' Discovery in the *Sophist* Reveals (a) the Anthyphairetic Nature of Plato's Philosophy, in Particular the Crucial Concept of Logos as the Philosophical Imitation of the Logos Criterion for Anthyphairetic Periodicity, and (b) the Anthyphairetic Method Used by Theaetetus in the Proof of Quadratic Incommensurabilities, in Particular the Role it had in the Birth of Theaetetus' Theory of Proportions of Magnitudes

24.5.1 The Philosophic Imitation of Theaetetus' Theorem in Plato's Sophist *218b5-221c4*

The anthyphairetic nature of the Divisions in the *Sophist* and in the *Statesman*, in particular of the Logos in these Divisions, have been examined elsewhere (Negrepontis [22–24, 26]; Negrepontis et al. [33]), and the reader is referred there for the detailed arguments. In the present subsection we go through the simpler Division in the *Sophist*, that of the Angler.

24.5.1.1 The Knowledge of the Intelligible Being Angler by Name and Logos in the *Sophist* 218b-221c

The method of "Name and Logos" (cf. *Theaetetus* 201e2-202b5, *Sophist* 218c1-5, 221a7-b2, 268c5-d5), is exemplified at the start of the *Sophist* 218b-221c by the definition of the Being Angler. The Angler is certainly a lowly paradigm of an intelligible Being, but even for lowly entities, there is an intelligible Being, cf. *Parmenides* 130b7-e3. In the diagram below, we reproduce the binary division process which leads to the Angler.

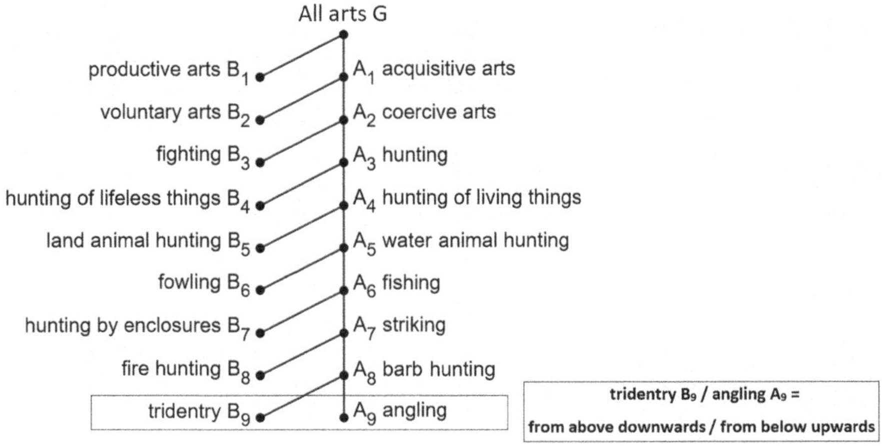

24.5.1.2 The Logos Criterion of the Angler

The description of the Logos Criterion of the Angler is strongly suggested by the *Sophist* 220e2-221c3 passage (which we have, for the sake of convenience, divided into two parts [A] and [B]):

> [A] *Stranger* Then of striking which belongs to barb-hunting [A_8],
> that part which proceeds downward from above ('anothen eis to kato'),
> is called, because tridents are chiefly used in it, tridentry [B_9],
> I suppose....
> *Stranger* The kind that is characterized by the opposite sort of blow,
> which is practiced with a hook and strikes, ...
> and proceeds from below upwards ('katothen eis tounantion ano'),
> being pulled up by twigs and rods.
> By what name, Theaetetus, shall we say this ought to be called?
> *Theaetetus* I think our search is now ended and we have found the very thing we set before us a while ago as necessary to find.
> *Stranger*: Now, then, you and I are only agreed about the name of angling, [A_9]
> 220e2-221b1.
> [B] but we have acquired also a satisfactory 'Logos' of the thing itself.
> For (gar)
> of art as a whole, half was acquisitive,
> and of the acquisitive, half was coercive,
> and of the coercive, half was hunting,
> and of hunting, half was animal hunting,
> and of animal hunting, half was water hunting, and of water hunting [A_5]
> the whole part from below (to katothen tmema holon) was fishing, [A_6]
> and of fishing, half was striking,
> and of striking half was barb-hunting, [A_8]
> and of this [A_8]
> the part in which the blow is pulled
> from below upwards (to peri ten katothen ano) was angling. [A_9]
> 221b1-c3.
> [translation by Fowler [36], with modifications by Negrepontis S.].

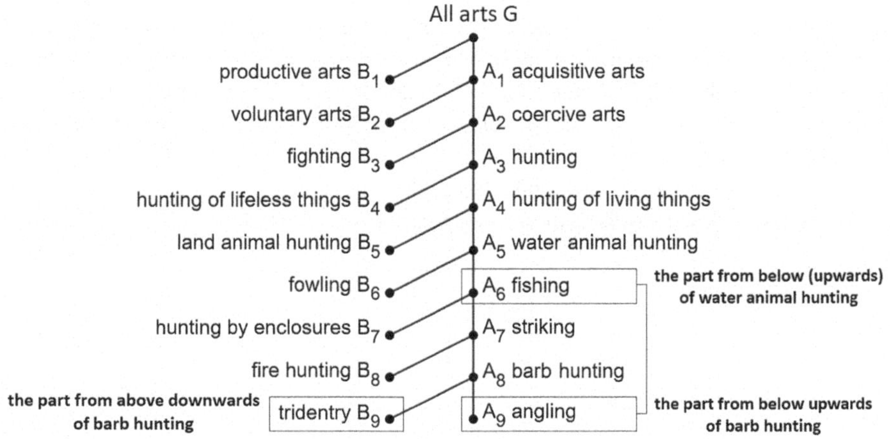

In [A], the opposing relation of Tridentry to Angling is carefully explained: all Fishing with a hook is divided into

Tridentry (=Fishing with a trident), which is described as
Fishing with a hook with an art that proceeds from above downwards, and
Angling (=Fishing with a rod), which is described as
Fishing with a hook with an art that proceeds from below upwards.

But now, in [B], it is claimed that "the Logos" of the Angling has been found, too. The justification — the proof — that we have indeed found the Logos, too, is contained in the remainder [B], since this remaining part [B] starts with a "for" ("gar"), and this justification can be seen to consist of:

(i) an accurate recounting of all the division steps, abbreviated in the sense that of the two species into which each genus is divided, only the one that contains the Angler is mentioned, while the species opposite to it is omitted;
(ii) a reminder that the last species, the angling, is characterized as the part of its genus that proceeds "from below upwards"; and,
(iii) the ONLY new information (since (i) and (ii) are repetitions of things already contained in the Division and in [A]), which concerns the species of fishing, three steps before angling, and which informs us for the first time that this species is "the whole part from below" of its genus.

Since this is an abbreviated account, there is no explicit information on the species opposite to "fishing"—namely "fowling"—but since "fishing" was described not simply as "the part from below" of its genus, but emphatically as "the whole part from below", it follows that the opposite species—"fowling"—must be characterized as "the (whole) part from above" of the same genus. In fact, there can be no other justification for the presence of the term "whole" in the description of "fishing" with a view to arguing that we have obtained "Logos", except to indicate and imply this description for its opposite species, "fowling".

We recall that due to the word "for" ("gar"), part [B] is explicitly a justification of the claim that we have succeeded in finding the "Logos" of the Angling. We may then ask: what is the "Logos" of the Angler that reasonably results from such a justification? There can really be only one answer: the "Logos" we are looking for is the equality of the "philosophic ratio" of Tridentry to Angling, namely the equality of the ratio "of from above downwards to from below upwards", to the ratio "of Fowling to Fishing". Thus:

fowling B_6/fishing A_6 = tridentry B_9/angling A_9 =
from above downwards/from below upwards.

Since the species Tridentry and Angling form a pair of opposite species, and the species Fowling and Fishing form another pair of opposite species in the Division Scheme for the Angler, the resulting "Logos" bears a most uncanny similarity to the Logos Criterion for the periodicity of the anthyphairesis of geometric magnitudes, and of geometric powers in particular.

24.5.1.3 The Name and Logos of the Angler

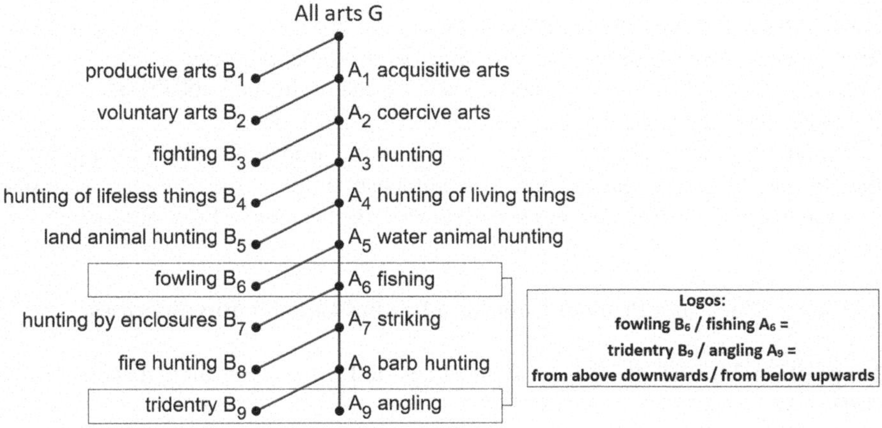

Thus the structure of Plato's intelligible Being, such as the Angler (but also of the Sophist) in the *Sophist*, described as Name and Logos, appears to be a dyad whose anthyphairesis satisfies Theaetetus' Logos Crirerion, and is thus in periodic anthyphairesis. It follows that the meaning of "Logos" in Plato's description of the knowledge of an intelligible Being as Name and Logos or as True Opinion and Logos, is exactly the "Logos Criterion for anthyphairetic periodicity".

Since philosophic knowledge (a) is provided by periodic anthyphairesis, and (b) is an imitation of Theaetetus' theorem, it follows that we expect that the proof of Theaetetus' theorem is provided by periodic anthyphairesis and the Logos Criterion.

24.5.2 *The Interpretation of the Knowledge in the* Meno *80d-86a and 97a-98b as True Opinion Plus Logos (or Causal Logismos) and as Recollection is the Knowledge of the Anthyphairesis of the Diameter to the Side of a Square, Achieved by Theaetetus' Logos Criterion*

In the present subsection we outline our interpretation, developed in Negrepontis [26], that in the *Meno* 80d-86e and 97a-98b, for a line a whose square is double the square of a given line b,
Knowledge is the full infinite anthyphairesis of a to b,
True Opinion consists of the first two stages of this anthyphairesis,
Logos (or Causal Logismos) is the Logos Criterion of this anthyphairesis, and
Recollection is the repetition of the ratio in the Logos Criterion.
The model for the whole argument in the *Meno* is Theaetetus' streamlined proof of the original Pythagorean proof of the incommensurability of the diameter to the side of a square given in 24.5.2.2 below, with the help of the simple Proposition 24.5.2.1.

24.5.2.1 The Cross Product Condition Implies Equal Anthyphaireses

Proposition 24.5.2.1 If a, b, c, d are lines and $ad = bc$, then $Anth(a, b) = Anth(c, d)$.

Proof Let $Anth(a, b) = [k_0, k_1, \ldots]$. We proceed by induction.
Thus $a = k_0 b + e_1$, with $e_1 < b$. Then $cb = ad = k_0 bd + e_1 d$.

By Proposition I.44 of the *Elements* (by this reference we always mean and make use of the second part of the proof, following the application of Proposition I.42), there is a line segment f_1, such that $e_1 d = bf_1$. Since $e_1 < b$, it is clear that $f_1 < d$.

Thus $cb = k_0 bd + e_1 d = k_0 bd + bf_1$; then $c = I_0 d + f_1$ with $f_1 < d$.

Thus $Anth(a, b) = [k_0, k_1, \ldots] = [k_0, Anth(b, e_1)]$, while $Anth(c, d) = [I_0, Anth(d, f_1)]$.

Thus, we have that $e_1 d = bf_1$ and we must prove that $Anth(b, e_1) = Anth(d, f_1)$.

We continue as in the first step, and we finish the proof by induction.

Note For a number of reasons, including

(a) the association of incommensurability with the Infinite, suggesting anthyphairetic proof using Proposition X.2, in the Pythagorean stories concerning Hippasus,
(b) the introduction and knowledge by the Pythagoreans of the side and diameter numbers, which are the anthyphairetic convergents of the anthyphairesis of the diameter to the side of a square, and
(c) the anthyphairetic interpretation of Zeno's paradoxes and arguments,

we conclude that the fundamental Pythagorean discovery of the incommensurability of the diameter to the side of a square was proved by an anthyphairetic method, employing the method of Application of Areas. The reader is referred to Negrepontis et al. [32] for a detailed account.

24.5.2.2 Anthyphairesis of the Diameter of a Square with Respect to its Side

Proposition 24.5.2.2 (Pythagorean) If a, b are lines such that $a^2 = 2b^2$, then a is the diameter of the square with side b, and $Anth(a, b) = [1,\text{period}(2)]$.

Proof (Theaetetus)
[Part I]
By the (isosceles) Pythagorean theorem, line a is constructed as the diameter of the square with given side b. Clearly $a > b$. Since $a^2 = 2b^2$, it follows that $a < 2b$ (if we had $a \geq 2b$, then $a^2 \geq 4b^2 > 2b^2 = a^2$, contradiction). Thus $b < a < 2b$. Hence,

$a = b + c_1, c_1 < b.$

Claim. $b > 2c_1, 3b > 2a$ (if we had $b \leq 2c_1 = 2a - 2b$, then $3b \leq 2a$, hence $9b^2 \leq 4a^2 = 8b^2$, contradiction). Hence,

$b = 2c_1 + c_2.$

$b > 2(a - b), 3b > 2a.$

[Part II, Logos Criterion]
Then $c_1 = a - b$, hence $c_2 = b - 2c_1 = b - 2(a - b) = 3b - 2a$, hence
$bc_2 = b(3b - 2a) = 3b^2 - 2ab$, and
$c_1^2 = (a - b)(a - b) = a^2 + b^2 - 2ab = 3b^2 - 2ab.$
Hence $bc_2 = c_1^2$. Since $b > c_1$, and

$bc_2 = c_1^2,$

clearly $c_2 < c_1$, thus the second step is indeed anthyphairetic

$b = 2c_1 + c_2, c_2 < c_1.$

By $bc_2 = c_1^2$ and Proposition 24.5.2.1, we have
$Anth(b, c_1) = Anth(c_1, c_2)$, namely $b/c_1 = c_1/c_2$, according to the Theaetetean definition.
Then $Anth(c_1, c_2) = Anth(b, c_1) = [2, Anth(c_1, c_2)]$. Hence,
$Anth(a, b) =$
$[1, 2, Anth(c_1, c_2)] =$
$[1, 2, Anth(b, c_1)] =$
$[1, 2, 2, Anth(c_1, c_2)] =$
$\ldots =$
$[1, period(2)]$.

24.5.2.3 True Opinion in the *Meno* Corresponds to Part I of the Proof in 5.2.2

The first anthyphairetic relation $a = b + c_1$, $c_1 < b$ is established in the *Meno* 82e2- 83e10, in the equivalent form $b < a < 2b$ (with $b = 2$); the second partial anthyphairetic relation $b = 2c_1 + c_2$ is established in the *Meno* 83e1-10 in the equivalent form $2a < 3b$ with $b = 2$); and the fact that Plato calls these two anthyphairetic relations True Opinion can be seen from *Meno* 83e11-84d2, 97a9-c5 and *Symposium* 210a4-e6.

24.5.2.4 Logos in the *Meno* Corresponds to Part II, Logos Criterion of the Proof in 5.2.2

From the statement that *Knowledge is True Opinion plus Something* (*Meno* 97c6-d3) and the anthyphairetic interpretation of True Opinion, we expect

[1a] that the *Knowledge* of line a with respect to the line b will be the full infinite anthyphairesis of a to b, and

[1b] that the *Something* needed will be (a philosophic analogue of) the *Logos Criterion* of Proposition 24.5.2.2. And, indeed, Plato finally states in the *Meno* 86b6-c3, 97d4-98a4 that Knowledge is True Opinion plus *Causal Logismos* (*Logismos Aitias*) or *Logos*.

We find support for this heuristic idea [1] in that
[2] Plato in the *Meno* 97d4-98a4 describes the Logos as a Bond, exactly as in the *Sophist*, where in the Division of the Sophist the Logos (clearly the imitation of Logos Criterion) is also described as bond 251d8-9, 253a4-c1, 268c5-d5.
Furthermore, [3] in the *Timaeus* 31c2-32a7 we find the following revealing passage:

> And *the fairest (kallistos) of bonds (desmon) is that which most perfectly collects into one itself and the things which it binds together (sundoumena)*; and to effect this in the fairest manner is *the natural property of proportion (analogio)*. For whenever the middle term of any three numbers, *or volumes (ogkon) or powers (dunameon)*, is such that as the first term is to it, so is it to the last term, and again, conversely, as the last term is to the middle, so is the middle to the first, then the middle term becomes, in turn, the first and the last, while the first and last become in turn middle terms, and the necessary consequence will be that all the terms are *the same (ta auta)*, and being *the same (ta auta) they all are one (hen panta estai)*. *Timaeus* 31c2-32a7 [translated by W.R.M. Lamb [38] with modifications by the authors]

Thus the Logos Criterion in Theaetetus' proof (of Proposition 24.5.2.2) of the Pythagorean incommensurability qualifies as "the fairest bond", a proportion in three terms, such that all three terms become "the same", namely are "collected into One", precisely as with the powers in the *Theaetetus* 147d-e.
[4] Plato describes Knowledge in the *Meno* as *Recollection*, which as explained by Proclus is recollection of "logoi", and this is seen as a clear reference to the nature of the Logos Criterion as a repetition of ratios.

We are thus justified in identifying *Meno*'s Logos or Causal Logismos as Theaetetus' Logos Criterion.

24.5.2.5 Note

The nature of Plato's intelligible Being as a philosophic imitation of a dyad in the philosophical analogue of periodic anthyphairesis by means of the Logos Criterion is revealed in other Platonic dialogues, as well, such as the *Republic*, the *Parmenides* and the *Timaeus*. (Negrepontis [26] for the *Republic*, [27] for the *Parmenides*, Negrepontis and Kalisperi [30] for the *Timaeus*).

24.5.3 The Direct Anthyphairetic Interpretation of the Theaetetus *147d3-e1 Passage*

We outline here the translation and interpretation of the *Theaetetus* 147d3-e1 given in Negrepontis [24] and Negrepontis et al. [33]. The interpretation, independent of those given in Sects. 24.6.1 and 24.6.2, fully confirms that the *Theaetetus* 147d3-e1 passage should be understood in terms of periodic anthyphairesis.
On the basis of arguments provided from the *Theaetetus* text itself, given in Negrepontis [24], pp. 341–343, we are led to a distributive reading for the plural "powers" of the *Theaetetus* 147d7-8 crucial statement.

> ἐπειδὴ ἄπειροι τὸ πλῆθος αἱ δυνάμεις ἐφαίνοντο
> Well, since each power was seen to be infinite in multitude

The distributive reading is fully confirmed by the two existing ancient anonymous commentators. The distributive reading of 147d7-8 concerning Theodorus' lesson, forces the whole sentence 147d6-e1 into a distributive reading of the plural "powers". Taking into account the arguments presented in the paper cited, the passage 147d3-148b3 must be rendered as follows:

> Since, in the aftermath of Theodorus' lesson,
> every power (αἱ δυνάμεις) was shown (ἐφαίνοντο)
> to be [divided into] an infinite multitude of parts (ἄπειροι τὸ πλῆθος),
> Theaetetus and his companion had the idea and in fact succeeded in collecting the infinite multitude of every power into One (συλλαβεῖν εἰς ἕν).
> [Negrepontis [24], p. 343]

Thus, our rendering of the *Theaetetus* 147d3-e1 is the following:

> Theaet.: Theodorus was proving for us via diagrams something about powers,
> in particular about the three-foot-power and the five-foot-power
> –demonstrating that these are not commensurable in length with the one-footpower–
> and selecting each power individually in this way up to the seventeen-foot power;
> but in this one he somehow stopped.
> Well, since each power was seen to be infinite in multitude
> (ἐπειδὴ ἄπειροι τὸ πλῆθος αἱ δυνάμεις ἐφαίνοντο),

> it occurred to us to do something on these lines:
> to try to collect [the infinite multitude] into one (συλλαβεῖν εἰς ἕν),
> by which (ὅτῳ) we could refer to each of all these powers (πάσας ταύτας τὰς δυνάμεις).

We have yet to make mathematical sense of the still enigmatic meaning of "collection into one". The rendering of the *Theaetetus* 147d3-e1 given in Negrepontis [24] strongly suggests an anthyphairetic method. The collection of the anthyphairetic infinity of a power (the effect, the conclusion of Theaetetus' proposition) is possible only because the square of the power is commensurable to the one foot square (the cause, the assumption of Theaetetus' proposition). Accordingly, we are guided to the following

Proposition (Provisional Statement of Theaetetus' Theorem) *If a is a power, namely a is a line incommensurable with respect to the one foot line b, and a^2 is commensurable to b^2, then the infinity of the anthyphairesis of a to b, which arises because of the incommensurability of a to b, can be "collected" into one.*

We now must make some mathematical sense of the philosophical "collection" of the infinitely many parts of each power into one. The reason that the method is called Collection in One is explained in the *Parmenides*: the periodic anthyphairesis of a to b results in the equalization of a and b, in the sense either that the (finite) number of the parts of a is equal to the number of the parts of b (*Parm.* 144d4-e3, with a = One, b = Being); or, equivalently, that the (finite) number of times that Being is not-Being is equal to the number of the not-Beings (*Soph.* 257a4-6, with a = Being, and b = not-Being), both following closely Zeno's *Fragment* B3.

Thus, Collection, meaning equalisation of all parts produced by the infinite anthyphairetic Division, is a consequence of anthyphairetic periodicity. The Oneness of a Platonic Idea is thus not the Oneness of a partless entity, such as the One of the first hypothesis in the *Parmenides* or the geometric point, but the Oneness of a self-similar One with parts, in fact an infinite multitude of parts, the One of the second hypothesis in the *Parmenides*, such that each part is equalized to each other [Negrepontis [24, p. 373] and [27]].

We have thus arrived, by a different argument, at the same conclusion as will be reached by the end of Sect. 24.7 of the purely mathematical statement of Theaetetus' mathematical discovery described in the *Theaetetus* 147d3-e1:

Proposition (Theaetetus) *If a, b are lines and N is a non-square number, such that $a^2 = Nb^2$, then the anthyphairesis of a to b is eventually periodic; hence a, b are incommensurable lines.*

24.6 The Anthyphairetic Reconstruction of Theaetetus' Proof Regarding Quadratic Incommensurabilities (*Theaetetus* 147d3-e1), Part I: The Computation of Every Step in the Anthyphairetical Process, via the Pythagorean Application of Areas, Constructed by Anthyphairetic Substitution

Plato's method of the acquisition of Knowledge of intelligible Beings, described as True Opinion & Logos in the *Meno*, described equivalently as Name & Logos in the *Sophist* and the *Statesman*, or as Division and Collection, according to our interpretation (outlined in Sect. 24.5), is the philosophical imitation of an anthyphairetic division followed by Logos, having the form e.g. for the Angler $B_6/A_6 = B_9/A_9$, for the diameter a to the side b, $b/c_1 = c_1/c_2$. But Plato's method is also described in the *Theaetetus* 148c-d as the philosophical imitation of Theaetetus' mathematical method of proof of his theorem on quadratic incommensurabilities.

Putting the two together, we certainly expect that Theaetetus' method of proof of his theorem on quadratic incommensurabilities will be anthyphairetic, in fact through the use of anthyphairetic substitution. We present our anthyphairetic reconstruction of the proof of Theaetetus' theorem in the present Sect. 24.6 and in the next Sect. 24.7 (cf. also Negrepontis et al. [34]).

In Sect. 24.6, starting with an arithmetical quadratic equation, we present the construction of the anthyphairetic process, employing essentially the Pythagorean tools connected with the Application of Areas, the same tools that, as we have argued in the paper just mentioned, were used for the proof of the incommensurabilities by the Pythagoreans and by Theodorus. This is a process technically demanding, but not too different from the special cases worked out by Theodorus. The great novel idea that completes the proof will be given in Sect. 24.7.

24.6.1 The Pythagorean Application of Areas in Book II of the Elements

Emblematic within Book II of the *Elements* is Proposition II.4. It has long been understood that the proper/original place of the Pythagorean Theorem is not at its current position as Proposition I.47, proved with the help of Propositions I.37 and 1.41, but as Proposition II.4/5, proved with the help of Proposition II.4. This is strongly suggested by a variety of reasons, but mainly by the proof of the isosceles Pythagorean theorem described in Plato's *Meno* 82a-85b.

Furthermore, the method of Application of Areas appears in the *Elements* in general form, as Propositions VI.27 & 28 (in defect) and VI.29 (in excess), where use of Eudoxus' theory of ratios of magnitudes is made. On the other hand, Eudoxus' theory is presented in Book V and it has also long been understood that the proper place of the original Pythagorean method, with no use of ratios, belongs to Book II of the *Elements*.

Application of Areas is the process by which, given an area and a line a, one can construct a parallelogram on a (usually a rectangle) whose area is equal to the given area. The process involves the construction of a suitable line b, such that the parallelogram with sides a and b has area equal to the given area. The Pythagoreans used two Propositions on Applications of Areas, in defect & in excess, which can be proved by using the Pythagorean theorem and Propositions II.5 & 6 respectively, and are variations of Proposition II.4, precisely suited for these proofs. The Pythagorean Application of Areas in defect takes the position II.5/6 and in excess the position II.6/7. Thus, the first part of Book II of the *Elements*, partially completed with the addition of the Pythagorean theorem and the Pythagorean Application of Areas has the following content:

Proposition (*Elements* II.1, 2, 3) *If* $a_1, a_2, ..., a_n$ *and* b *are lines, then* $(a_1 + a_2 + \ldots + a_n)b = a_1b + a_2b + \ldots + a_nb$.

The distributive rule of the "multiplication" of magnitudes.

Proposition (*Elements* II.4) *If* a, b *are lines, then* $(a + b)^2 = a^2 + b(b + 2a)$.

[Proof with the use of Gnomons.]

Proposition II.4 is the fundamental proposition of the Pythagorean Book II of the *Elements*, the emblematic proposition for the hypothesis of Geometric Algebra.

Proposition (Pythagorean Theorem, II.4/5) *If an orthogonal triangle has sides* a, b *around the right angle and hypotenuse* c, *then* $a^2 + b^2 = c^2$.

[Proof from Proposition II.4.]

Propositions II.5 & 6 are each equivalent to Proposition II.4, given in a completion of the square form. With the Pythagorean theorem and the completion of squares II.5 & II.6 available, it is quite simple to prove the two fundamental forms of the Pythagorean Application of Areas:

Proposition (*Elements* II.5) *If* a, x *are lines, with* $a/2 > x$, *then* $(a/2)^2 - (a/2 - x)^2 = x(a - x)$.

Proposition (Pythagorean Application of Areas in Defect, II.5/6) *Given lines* a, m, *with* $a/2 > m$, *to construct a line* x, *such that* $x(a - x) = m^2$.

[Proof from Pythagorean theorem and Proposition II.5.]

Proposition (*Elements* II.6) *If* a, x *are lines, then* $(a/2 + x)^2 = (a/2)^2 + x(a + x)$.

Proposition (Pythagorean Application of Areas in Excess, II.6/7) *Given lines* a, m, *to construct a line* x, *such that* $x(a + x) = m^2$.

[Proof from Pythagorean theorem and Proposition II.6: construct the triangle with sides m, $a/2$ about a right angle and set y for the hypotenuse of the triangle.
By the Pythagorean theorem $y^2 = m^2 + (a/2)^2$. Certainly $y > a/2$. Set $x = y - (a/2)$.
Apply II.6: $y^2 = (a/2 + x)^2 = (a/2)^2 + x(a + x)$. Hence $x(a + x) = m^2$.]

The close relation of the construction of a mean proportional of two lines [II.14] with the Pythagorean Application of Areas in Defect [II.5/6] is seen by the following

Proposition 24.6.1.1 Let a be a given line.

(a) [II.5/6] if a line m is given, such that $a/2 > m$, then construct line x, such that $x(a-x) = m^2$;
(b) [II.14] if a line x is given, such that $a/2 > x$, then construct line m, such that $x(a-x) = m^2$.

Proof By the Pythagorean theorem II.4/5 and Proposition II.5, applied in that order for (a) and in reverse order for (b).

Note In all circumstances, the Propositions of Euclid's *Elements* are either stated in modern notation by the authors or the translation by Heath [15] is quoted.

24.6.2 The Use of the Pythagorean Application of Areas for the Generation of the Infinite Anthyphairesis

24.6.2.1 For the Computation of the Infinite Anthyphairesis of a to b, Satisfying an Equation of the form $Aa^2 = Cb^2$, It Is Necessary to Deal with the Computation of the Anthyphairesis of a to b Satisfying an Equation of the Form $A_1b^2 = B_1bc + C_1c^2$

The computation of the anthyphairesis of a to b, in case $Aa^2 = Cb^2$, with A, C natural numbers reduces to the computation of the anthyphairesis of b to c, $Anth(a, b) = [k, Anth(b, c)]$, satisfying a quadratic equation in excess $A_1b^2 = B_1bc + C_1c^2$.

It is easy to see that the proof of the anthyphairesis of a, b, $a > b$, with $Aa^2 = Cb^2$ for non-square natural numbers A, C, leads necessarily, from the second step on, to the proof of the incommensurability of the quadratic equations in excess of the form $A_1a^2 = B_1ab + C_1b^2$, for natural numbers A_1, B_1, C_1, with non-square discriminant $B_1^2 + 4A_1C_1 = 4AC$.

Proposition 24.6.2.1 If a, b are lines, with $a > b$, and $Aa^2 = Cb^2$, with A, C natural numbers with $A < C$, and non-square AC [in particular if $a^2 = Nb^2$ for some non-square number N], then there are a number k, a line c, $c < b$, and natural numbers A_1, B_1, C_1, such that $Anth(a, b) = [k, Anth(b, c)]$, and $A_1b^2 = B_1bc + C_1c^2$.

Proof Let J be a natural number such that $J^2 < 4AC < (J+1)^2$, and k the integer part of $J/(2A)$. Then $a = kb + c$, $c < b$, $2Ak \leq J$, $Ak^2 < C$. Set $A_1 = C - Ak^2$, $B_1 = 2Ak$, $C_1 = A$, and verify that $A_1b^2 = B_1bc + C_1c^2$, and $Anth(a, b) = [k, Anth(b, c)]$.

Thus, even if we are interested mainly or only in the incommensurability of the case $a^2 = Nb^2$, we are forced to consider the more general quadratic equation $Aa^2 = Bab + Cb^2$.

24.6.2.2 The Anthyphairetic Division $a = kb + c$ of Line a to Line b, Satisfying a Quadratic Equation $Aa^2 = Bab + Cb^2$ Is Computed by Means of the Pythagorean Method of Application of Areas; and the Anthyphairetic Substitution Results in a Quadratic Equation $A_1b^2 = B_1bc + C_1c^2$

Proposition 24.6.2.2

(a) An arithmetical quadratic equation in excess $Aa^2 = Bab + Cb^2$, where a, b are lines, with $a > b$ and A, B, C are natural numbers with non-square discriminant $B^2 + 4AC$, is transformed, upon setting $x = Aa - Bb$, into the Application of Areas in excess $x(Bb + x) = ACb^2$.

(b) The solution of the Application of Areas in excess yields, upon setting J the natural number such that $J^2 < B^2 + 4AC < (J + 1)^2$, and k the integer part of $(B + J)/(2A)$, the anthyphairetic relation $a = kb + c$, with $c < b$.

(c) $B < 2Ak \leq J + B$, $Ak^2 < Bk + C$.

(d) Upon setting $A_1 = Bk + C - Ak^2$, $B_1 = 2Ak - B$, $C_1 = A$, we obtain again an arithmetical quadratic equation in excess $A_1b^2 = B_1bc + C_1c^2$.

Proof

(a) is routine.

(b) The solution of the Application of Areas, according to Proposition II.6/7, yields

[b1] $2y = Bb + 2x$
[b2] $2y = Jb + c_1$, with $c_1 < b$. From [b1] and [b2] we get
[b3] $2x = (J - B)b + c_1$, with $c_1 < b$. Since $x = Aa - Bb$, we get
[b4] $2Aa = 2Bb + 2x = (B + J)b + c_1$, with $c_1 < b$.
[b5] $a = [(B + J)/(2A)]\,b + c_1/(2A)$
[b6] $a = kb + (Db + c_1)/(2A)$, with $D < 2A$.
Setting $(Db + c_1)/(2A) = c$, and noting that $c < b$, we conclude that
[b7] $a = kb + c$, with $c < b$.

Thus k is the integer part of the arithmetical ratio $(J + B)/(2A)$
[This is the general algebraic step, obtained from
the solution of the Application of Areas in Excess].

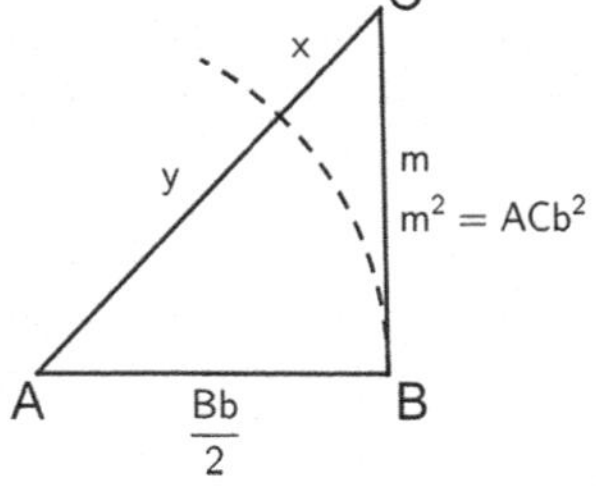

(c) Clearly $a < (k+1)b \leq 2kb$ and $Bb < Aa$, hence $Bb < Aa < 2Akb$, hence $B < 2Ak$. By the definition of k, $2Ak \leq J + B$, hence $2Ak - B \leq J$. By the definition of J, we have $(2Ak - B)^2 \leq J^2 < B^2 + 4AC$, hence by II.7, we have $4A^2k^2 + B^2 < B^2 + 4AC + 4ABk$, $4A^2k^2 < 4AC + 4ABk$, hence $Ak^2 < Bk + C$.
(d) By substitution $A = (kb+c)^2 = B(kb+c)b + Cb^2$,
hence $Ak^2b^2 + Ac^2 + 2Akbc = Bkb^2 + Bbc + Cb^2$,
hence $(Bk + C - Ak^2)b^2 = (2Ak - B)bc + Ac^2$,
hence $A_1b^2 = B_1bc + C_1c^2$.

It is next proved that the quadratic equation in defect $Aa^2 + Cb^2 = Bab$ is necessarily transformed into a quadratic equation in excess, after a finite number of successive steps of anthyphairetic substitution.

Proposition 24.6.2.3 An arithmetical quadratic equation either in defect $Aa^2 + Cb^2 = Bab$, or of the form $Aa^2 + Bab = Cb^2$, or of the form $Aa^2 = Cb^2$ with lines a, b incommensurable, under repeated anthyphairetic substitution eventually turns into an arithmetical quadratic equation in excess.

Proof Let $Aa^2 + Cb^2 = Bab$ be a quadratic equation in defect, with a, b incommensurable, $a > b$. Then the discriminant $B^2 - 4AC$ is a non-square number. We set $x = Bb - Aa$, and obtain the Application of Areas in defect II.5/6, $x(Bb - x) = ACb^2$. We set J the natural number such that $J^2 < B^2 - 4AC < (J+1)^2$, and k the integer part of $(B + J)/(2A)$. The solution of the Application of Areas in defect yields, in analogy to the corresponding step for a quadratic equation in excess (Proposition 24.6.2.2), the anthyphairetic relation $a = kb + c$, with $c < b$. [This is the general algebraic step.]

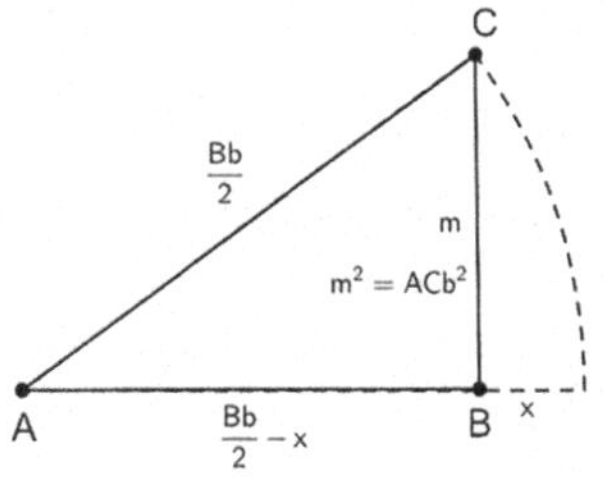

Next we employ the anthyphairetic substitution of a by $kb + c$ in the initial equation $Aa^2 + Cb^2 = Bab$, and we have $A(kb+c)^2 + Cb^2 = B(kb+c)b$; by Propositions II.1, 4 we have $(Ak^2 + C)b^2 + 2Akbc + Ac^2 = Bkb^2 + Bbc$.

Claim $Ak^2 + C \neq Bk$. [If $Ak^2 + C = Bk$, then $2Akbc + Ac^2 = Bbc$, then $2Akb + Ac = Bb$, hence $Ac = (B - 2Ak)b$, hence b, c commensurable, hence a, b commensurable, impossible.]
We then distinguish the following cases:

Case 1.	Case 2.	Case 3.	Case 4.	Case 5.
$Bk > Ak^2 + C$, $2Ak > B$	$Bk > Ak^2 + C$, $2Ak < B$	$Bk > Ak^2 + C$, $2Ak = B$	$Bk < Ak^2 + C$, $2Ak < B$	$Bk < Ak^2 + C$, $2Ak \geq B$
$A_1 = Bk - Ak^2 - C$, $B_1 = 2Ak - B$, $C_1 = A$ are natural numbers	$A_1 = Bk - Ak^2 - C$, $B_1 = B - 2Ak$, $C_1 = A$ are natural numbers	$A_1 = Bk - Ak^2 - C$, $[B_1 = 0]$, $C_1 = A$ are natural numbers	$A_1 = Ak^2 + C - Bk$, $B_1 = B - 2Ak$, $C_1 = A$ are natural numbers	$A_1 = Ak^2 + C - Bk$, $B_1 = 2Ak - B$, $C_1 = A$ are natural numbers

The anthyphairetic substitution $a = kb + c$ in equation $Aa^2 + Cb^2 = Bab$ yields:

	$A_1b^2 + B_1bc = C_1c^2$	$A_1b^2 = C_1c^2$	$A_1b^2 + C_1c^2 = B_1bc$	Impossible
			Application of areas in defect with $B_1 < B$	
	The next step: $b = mc + d, d < c$ $A_2 = C_1 - A_1m^2 - B_1m$ $B_2 = B_1 + 2A_1m$ $C_2 = A_1$	The next step: $b = mc + d, d < c$ $A_2 = C_1 - A_1m^2$ $B_2 = 2A_1m$ $C_2 = A_1$		
$A_1b^2 = B_1bc + C_1c^2$ Application of Areas **in Excess**	$A_2c^2 = B_2cd + C_2d^2$ Application of Areas **in Excess**	$A_2c^2 = B_2cd + C_2d^2$ Application of Areas **in Excess**	Eventually Application of Areas **in Excess**	

The (many but manageable) details are left to the reader.

24.7 The Anthyphairetic Reconstruction of Theaetetus' Proof Regarding Quadratic Incommensurabilities (*Theaetetus*, 147d3-e1), Part II: The Novel Idea of Theaetetus, A Pigeonhole Argument, Results in the Preservation of the Application of Areas/Gnomons, Hence Both in the Periodicity of the Anthyphairesis and in the Birth of the Theory of Proportions of Magnitudes

We present the basic new idea of Theaetetus, *a pigeonhole argument* based on the *invariance of the discriminant under anthyphairetic substitution* (Sect. 24.7.2). The argument implies that there are two stages in the anthyphairesis of an arithmetical quadratic equation, say the mth and afterwards the nth stage, so that the mth, say e_m, and the (m+1)th, e_{m+1}, anthyphairetic remainder satisfy the same arithmetical quadratic equation as the nth, e_n, and (n+1)th, e_{n+1}, remainders. We would then have $Ae_m^2 = Be_me_{m+1} + Ce_m^2$ and $Ae_n^2 = Be_ne_{n+1} + Ce_n^2$ (for some natural numbers A, B, C). This is a generalization of the Pythagorean *Invariance of Gnomons*, the condition by which the Pythagoreans referred to the appearance of the same coefficients in each arithmetical quadratic equation that arises at every anthyphairetic step, after some point in the process. As it is not difficult to see (shown in Sect. 24.7.1, below) this repetition implies that $Anth(e_m, e_{m+1}) = Anth(e_n, e_{n+1})$. It is precisely by this condition and by some natural inductive argument, that the anthyphairesis of the initial line segments a to b is eventually periodic.
At this point, we compare Plato's imitation, involving Logos, in effect the Logos Criterion for anthyphairetic periodicity, e.g. $B_6/A_6 = B_9/A_9$ for the Angler (in Sect. 24.5) and the condition on equal anthyphairesis, which arose naturally from our reconstruction. Through this comparison we conclude that Theaetetus reached at the definition of proportion of magnitudes in terms of equal anthyphairesis, by interpreting the condition $Anth(e_m, e_{m+1}) = Anth(e_n, e_{n+1})$ within the anthyphairesis as a means to denote the proportion $e_m/e_{m+1} = e_n/e_{n+1}$. This moment marks *the birth of the Logos Criterion for anthyphairetic periodicity* (Subsection 24.2.3), Plato's dialectical meaning of his term *Logos* (in the *Sophist* as in Sect. 24.5 and in his whole work), and *the birth of Theaetetus' theory of ratios of magnitudes*. Thus, Theaetetus' theory of ratios of magnitudes is, from its initial conception, inseparably connected with periodic anthyphairesis.

24.7.1 The Transition from the Preservation of Gnomons to the Equality of the Anthyphaireses

Proposition (***Elements* I.44**) *Given lines a, b, c, to construct a line d, such that* $ab = cd$.

The propositions that follow depend on Proposition I.44. Furthermore, note that Proposition I.44 is in the second part of Book I of the *Elements*, which may be considered preparatory for Book II; and Book II culminates with the proof of incommensurability at Proposition II.8/9.

Lemma 24.7.1.1 *If* $Aa^2 = Bab + Cb^2$ *and* $Aa^2 = Bad + Cd^2$, *then* $b = d$.

Proof By hypothesis $Bab + Cb^2 = Bad + Cd^2$. Suppose b is different from d, say $b > d$. Then $Bab > Bad$ and $Cb^2 > Cd^2$, hence $Bab + Cb^2 > Bad + Cd^2$, a contradiction.

Proposition 24.7.1.2 (This Proposition's Condition Is Referred to as *the Preservation of Quadratic Equations/Gnomons*) If $Aa^2 = Bab + Cb^2$ and $Ac^2 = Bcd + Cd^2$, then $ad = bc$.

Proof By Proposition I.44, there is a line e, such that $bc = ae$. From the assumption we have $Aa^2c = Babc + Cb^2c$, whence $Aa^2c = Baae + Cbae$, whence $Aac = Bae + Cbe$, whence $Aacc = Baec + Cbec$, whence $Ac^2a = Bcea + Ce^2a$, whence $Ac^2 = Bce + Ce^2$. By Lemma 24.7.1.1, $d = e$. Hence $ad = bc$.

Proposition 24.7.1.3 If $Aa^2 = Bab + Cb^2$ and $Ac^2 = Bcd + Cd^2$, then $Anth(a, b) = Anth(c, d)$.

Proof If $Aa^2 = Bab + Cb^2$, and $Ac^2 = Bcd + Cd^2$, then $ad = bc$ (Proposition 24.7.1.2); and, if $ad = bc$, then $Anth(a, b) = Anth(c, d)$ (Proposition 24.5.2.1)

24.7.2 Theaetetus' Fundamental New Idea: The Pigeon-Hole Argument: Based Upon Our Interpretation, the Mathematical Idea Theaetetus Had, According to Plato's Theaetetus *147d7, Was a Pigeonhole Principle, Drawn from the Invariance of the Discriminant of the Quadratic Equation Under Anthyphairetic Substitution*

Propositions 24.7.1.2 and 24.7.1.3 are natural extensions of the Pythagorean and Theodorus' approach and they do not qualify as the major new idea, described in the *Theaetetus* 147d7. The cases worked out by Theodorus, $a^2 = Nb^2$ for non-square N up to $N = 17$, lead to the formation of the discriminant and do indicate the reappearance of the same discriminant in every step of the solution of the required Application of Areas in Excess, suggesting the invariance of the discriminant under anthyphairetic substitution.

The fundamental new idea that Theaetetus must have had as a result of Theodorus' lesson, can only be the pigeon-hole principle that is needed for the proof of the general periodicity theorem. Theaetetus must have noticed that in all the cases of the proofs of incommensurability by Theodorus the discriminant of the quadratic equation in excess remains constant.

As noted in detail in Negrepontis et al. [33], Subsection 6.5, Plato's appeal, in the *Statesman* 272d-e, for the necessity of transferring from the Kronos era to the Zeus era in his related myth, appears to be a philosophical version of a pigeonhole argument, employed exactly at the point where it is needed in the proof of Theaetetus' palindromic periodicity theorem, a theorem deeper than the simple periodicity theorem presently considered and which must have been proved at a later stage.

24.7.2.1 The Preservation of an Arithmetic Quadratic Equation's Discriminant Under Anthyphairetic Substitution

Proposition 24.7.2.1 If lines a, b, with $a > b$, satisfy the quadratic equation in excess $Aa^2 = Bab + Cb^2$, where A, B, C are natural numbers with non-square discriminant $B^2 + 4AC$, $a = kb + c$, with $c < b$, and $A_1 = Bk + C - Ak^2$, $B_1 = 2Ak - B$, $C_1 = A$, as in Proposition 24.6.2.2, then the discriminant of the produced quadratic equation in excess $A_1b^2 = B_1bc + C_1c^2$ is equal to the discriminant of the given quadratic equation in excess.

Proof The proof is routine: $B_1^2 + 4A_1C_1 = (2Ak - B)^2 + 4(Bk + C - Ak^2)A =$ (by Proposition II.7)
$(4A^2k^2 + B^2 - 4ABk) + (4ABk + 4AC - 4A^2k^2) = B^2 + 4AC.$

Note here that the discriminant of every quadratic equation has a clear geometric content, namely it is directly related to the square of the hypotenuse in the fundamental right-angled triangle that we employ for the solution of each quadratic equation.

24.7.2.2 Theaetetus' General Periodicity Theorem

Proposition 24.7.2.2 If the lines $a, b, a > b$, satisfy either $Aa^2 = Cb^2$ with non-square discriminant $4AC$, or a quadratic equation in Excess $Aa^2 = Bab + Cb^2$ with non-square discriminant $B^2 + 4AC$ (including the case $Aa^2 = Cb^2$) or a quadratic equation in Defect $Aa^2 + Cb^2 = Bab$ with non-square discriminant $B^2 - 4AC$, then the anthyphairesis of a to b is eventually periodic.

Proof We denote the anthyphairesis of a to b as follows:
$a = k_0b + e_1$, with $b > e_1$,
$b = k_1e_1 + e_2$, with $e_1 > e_2$,
...
$e_{m-1} = k_me_m + e_{m+1}$, with $e_m > e_{m+1}$,
...
$e_{n-1} = k_ne_n + e_{n+1}$, with $e_n > e_{n+1}$,
...

A quadratic equation in defect turns into a quadratic equation in Excess after a finite number of anthyphairetic substitutions, according to Proposition 24.6.2.3. The discriminant is kept invariant in all anthyphairetic substitutions by Proposition 24.7.2.1.

A pigeonhole argument, based on the invariance of the discriminant, yields two equations in excess in steps, say $m < n$, with the same integer coefficients:

$A_m e_m^2 = B_m e_m e_{m+1} + C_m e_{m+1}^2$ and $A_n e_n^2 = B_n e_n e_{n+1} + C_n e_{n+1}^2$, with $A_m = A_n$, $B_m = B_n$, $C_m = C_n$.

By Proposition 24.7.1.2, $e_m e_{n+1} = e_{m+1} e_n$.

By Proposition 24.5.2.1, $Anth(e_m, e_{m+1}) = Anth(e_n, e_{n+1})$.

It now follows by the Theaetetean definition of proportion that $e_m/e_{m+1} = e_n/e_{n+1}$ and therefore, by proposition 24.2.3.1, the anthyphairesis of a to b is eventually periodic.

Note We believe that the birth of Theaetetus' theory of ratios of magnitudes occurred precisely at the moment when Theaetetus rephrased the established condition of the Pythagoreans regarding the preservation of Application of Areas/Gnomons, implying anthyphairetic periodicity and appearing at the conclusion of his theorem, in terms of equality of anthyphaireses and then in terms of proportion, equality of ratios, according to the definition provided to us by Aristotle in *Topics* 158b29-35. Thus anthyphairetic periodicity is built in the initial conception of the theory.

24.8 The Theory of Theaetetus Is *Only* About Ratios of Magnitudes with Eventually Periodic Anthyphairesis

24.8.1 The Birth of Theaetetus' Theory of Ratios of Magnitudes

According to our reconstruction in Sect. 24.7 of Theaetetus' theorem, reported in Plato's *Theaetetus* 147d-148b, the birth of Theaetetus' theory of ratios of magnitudes occurred at the exact point where the older, originally Pythagorean finitization of the infinite anthyphairesis by the preservation of Gnomons/Application of Areas was replaced by the definition of proportion in terms of the equality of anthyphairesis and the resulting Logos Criterion for the periodicity and the finitization of the knowledge of a geometric anthyphairesis. Thus, the initial role of Theaetetus' theory of ratios was to rephrase the Pythagorean finitizer, the preservation of the Gnomons, which strictly speaking applied only for the infinite anthyphairesis of the diameter to the side of a square with a more general finitizer, the Logos Criterion, that applied to every eventually periodic anthyphairesis.

24.8.2 Logos in Plato's Philosophy

According to our interpretation of Plato's Logos, as used in the *Sophist* and in the *Meno* in Sect. 24.5 but in fact throughout Plato's philosophy, Plato's Logos is the philosophical imitation of the Logos Criterion, the fundamental tool of acquiring knowledge of a Platonic intelligible Being, Plato's Idea. The intelligible Being therefore, has the structure of a dyad of kinds in imitation of a dyad of lines in geometric periodic anthyphairesis. Thus, Platonic Logos is the philosophic finitizer and provider of Knowledge of the infinite anthyphairetic structure of a Platonic Idea.

Thus Theaetetus' theory of ratios of magnitudes was conceived as a tool to study eventually periodic anthyphairesis, and Plato used Theaetetus' theory in order to describe his philosophical theory of intelligible Beings. Theaetetus' theory of ratios turned out to be crucial for Plato's philosophical system, because this theory was initially conceived by the Eleatics Parmenides and Zeno who had no theory of ratios of magnitudes and no Logos to rely upon. Zeno's finitizer of the philosophical infinite anthyphairesis, described in *Fragments* B3 & B1, and mentioned in the Platonic dialogues *Parmenides* Second Hypothesis and *Sophist*, is the precursor of Platonic Logos (shown in Negrepontis [27]).

If the basic role of Theaetetus' theory of ratios is to be an effective finitizer and producer of knowledge and Oneness, by applying it to eventually periodic anthyphairesis, equivalently to quadratic equations in the form of Pythagorean Application of Areas with arithmetical coefficients, then Theaetetus' theory cannot be a theory for all ratios of magnitudes, but only for ratios of magnitudes a and b, such that the anthyphairesis of a to b is eventually periodic.

24.8.3 Definition 3 in the Theaetetean Book X of the Elements

That Theaetetus' theory of ratios is only for ratios in eventually periodic anthyphairesis may be seen from Definition X.3 of the *Elements* on rational and alogos/irrational lines, and from the remarkable *Anonymon Scholion* X.2 *to the Elements*. Theaetetus' theory of ratios is applied to Book X of the *Elements*, which is generally considered to be due to Theaetetus, and concerns the theory of "alogoi"/irrational lines.

Definition (X.3 of the *Elements*) Let then the assigned straight line be called rational (rhete), and those straight lines which are commensurable with it, whether in length and in square, or in square only, rational (rhetai), but those that are incommensurable with it irrational (alogoi).

Thus, given an assumed line r, a line a possesses ratio according to Definition X.3 with respect to r if a and r are commensurable in square, namely if a^2, r^2 are commensurable squares. But according to Theaetetus' theorem, the anthyphairesis of a to r is eventually periodic. Thus the condition that a line a possesses ratio with respect to a line r in Definition X.3 clearly refers to Theaetetus' theory of ratios and is a condition closely related and in fact stronger than eventual anthyphairesis. (Two *Anonyma Scholia to the Elements*, *Scholion* V.22 and *Scholion* V.24 refer to this definition).

24.8.4 The Comparison of Theaetetus' to Eudoxus' Theory of Ratios of Magnitudes in the **Anonymon Scholion** *X.2* **to Euclid's Elements** *and Proclus'* **Commentary to Euclid** *285,5-286,11*

The situation is quite different for Eudoxus' theory of ratios of magnitudes, presented in Book V of the *Elements*.

Definition (V.4 of the *Elements*) Magnitudes are said to have a ratio to one another which can, when multiplied, exceed one another.

Thus a Eudoxean ratio of magnitudes is defined for those magnitudes a and b, such that Eudoxus' condition (mentioned also in Sect. 24.3) is satisfied, namely $a < nb$ and $b < ma$ for some numbers m and n.

The remarkable *Scholion* X.2 *to the Elements* compares the two definitions, Eudoxean and Theaetetean, and criticizes the Eudoxean theory, given in Book V of the *Elements*, as being so general that it annihilates the difference between rational and alogoi lines, defined in the Theaetetean Book X of the *Elements* (we omit the references to commensurable lines in the *Scholion*, to make clearer the opposition between the two theories).

Scholion X.2 *to the Elements*
Τὰ [...] μαθήματα φανταστικῶς νοοῦμεν,
We obtain knowledge of the magnitudes by imagination.
[...]
διὸ καὶ τὰ [μαθήματα] εἰς ἄπειρον διαιρεῖται,
therefore the magnitudes are divided ad infinitum
[...]
ἡ δὲ φαντασία πλῆθος ἄπειρον ἔχει· διὸ τὰ φανταστὰ ἄπειρα
Imagination possesses an infinite multitude; thus entities of imagination are infinite.
καὶ τὰ μεγέθη οὖν ὡς φανταστὰ ἄπειρα καὶ ἡ τομὴ αὐτῶν.
And the magnitudes as being entities of imagination are infinite and infinitely divisible

The unusual term "phantasia" refers to the divisibility ad infinitum of the magnitudes. We conjecture that the origin of the term is precisely in the sentence 147d7-8 of the *Theaetetus*, "since the powers 'ephainonto' infinite in multitude", and was probably concocted by Proclus. There is a long passage in Proclus' *Commentary to Euclid*, 285,5-286,11 about imagination and the infinite (cf. also *Commentaries to Euclid* 50-58, where imagination is also called "pathetikos nous", passive intellect):

> ἐν τῇ φαντασίᾳ τὸ ἄπειρον ὑφίστασθαι μόνον
> οὐ νοούσης τὸ ἄπειρον τῆς φαντασίας.
> in the imagination the infinite only exists
> without the imagination's knowing the infinite
> 285,5-7 ὥσπερ γὰρ τὸ σκότος τῷ μὴ ὁρᾶν ἡ ὄψις γινώσκει,
> οὕτως ἡ φαντασία τὸ ἄπειρον τῷ μὴ νοεῖν.
> Just as the sight recognizes the darkness by not seeing,
> so the imagination (φαντασία) recognizes the infinite (ἄπειρον) by not understanding it.
> ἅμα γὰρ νοεῖ
> καὶ μορφὴν ἐπάγει τῷ νοουμένῳ καὶ πέρας,
> καὶ τῇ νοήσει
> τὴν τοῦ φαντάσματος ἵστησι διέξοδον,
> καὶ διέξεισιν αὐτὸ
> καὶ περιλαμβάνει. 285,7-10
> For when the imagination knows,
> it simultaneously assigns to the object of its knowledge a Form and a Finite,
> and in knowing
> brings to an end the movement through the imagined object, and
> goes through it, and
> comprehends it. [translation by Morrow [21]]

But when the finitizer (peras), namely the Logos, is added, then the knowledge of the infinite is obtained, and the infinite is comprehended (perilambanei) (into One). We thus understand that the meaning of the *imagination* is the *(anthyphairetic) infinite without Logos and knowledge*.

Scholion X.2 in the sequel compares Theaetetus' and Eudoxus' theories of ratios:

The *Scholion* is strongly critical of Eudoxus' theory, exactly because it obliterates the difference between rational and irrational lines, literally between those lines that possess Logos and those that do not.

Thus, Theaetetus' theory is clearly the superior one both for mathematical use and of course for Plato's philosophy, exactly because it is limited only to ratios that provide knowledge of the infinite, namely to ratios with eventually periodic anthyphairesis.

Scholion X.2, lines 7–22

<table>
<tr><th>possessing Logos
in the sense of Eudoxus' theory</th><th>possessing Logos
in the sense of Theaetetus' theory</th></tr>
<tr><td>Εἰ πάντα τὰ μεγέθη τὰ πεπερασμένα δύναται
πολλαπλασιαζόμενα ἀλλήλων ὑπερέχειν
(τοῦτο δὲ ἦν τὸ λόγον ἔχειν,
ὡς ἐν τῷ πέμπτῳ μεμαθήκαμεν),
if all finite magnitudes have the power by
being multiplied to exceed each other
(for such was the condition that a magnitude
possesses logos,
as we were taught
in Book Five of the Elements)</td><td></td></tr>
<tr><td>τίς μηχανὴ
τὴν τῶν ἀλόγων
ἐπεισφέρειν διαφοράν;
how on earth (tis mechane)
[Eudoxus' theory] can ever distinguish
between the rational and the alogon?</td><td></td></tr>
<tr><td></td><td>ἢ ὅτι τὸ μέτρον
[...]
θέσει δὲ ἐν τοῖς μεγέθεσι
διὰ τὴν ἐπ' ἄπειρον τομήν;
πρὸς γὰρ
πῆχυν ἢ σπιθαμὴν
ἤ τι τοιοῦτον γνώριμον μέτρον
τὸ ῥητὸν καὶ τὸ ἄρρητον γνωρίζομεν.</td></tr>
<tr><td>possessing Logos
in the sense of Eudoxus' theory</td><td>possessing Logos
in the sense of Theaetetus' theory</td></tr>
<tr><td></td><td>while we obtain knowledge of
the rational and the irrational
with respect to a known assumed line
[definition X.3, Theaetetus' theory]</td></tr>
<tr><td colspan="2">καὶ μὴν τὸ λόγον ἔχειν
to possess logos is meant</td></tr>
<tr><td>ἄλλως μὲν
ἐπὶ τῶν μεγεθῶν λέγεται
τῶν πεπερασμένων καὶ ὁμογενῶν,
differently
for magnitudes finite and homogeneous</td><td></td></tr>
<tr><td></td><td>[...]
ἄλλως
ἐπὶ τῶν ῥητῶν προσαγορευομένων.
differently
for rational magnitudes</td></tr>
</table>

ὅπου μὲν γὰρ ὁ *λόγος μόνον* καὶ ἡ σχέσις θεωρεῖται τῶν πεπερασμένων μεγεθῶν κατὰ τὸ μεῖζον καὶ ἔλαττον, for [in Eudoxus' theory] *logos* and the relation of the finite magnitudes is considered only according to the greater and the smaller [*definition V.4 of the* Elements]	
	[...] ὅπου δὲ πρὸς τὸ ἐγκείμενον μέτρον *τὴν τῶν ῥητῶν ἡμῖν πρὸς τὰ ἄλογα* *διαφορὰν παρέσχετο.* while [in definition X.3 of *Theaetetus' theory*] the assumed measure *does provide the difference* *between the rational and the irrational lines.*

[*Scholion* X.2 rendered into English by the authors]

24.8.5 Conclusion

From Sects. 24.8.1–24.8.4, we have no doubt that, contrary to the reconstructions put forward by Becker [5], van der Waerden [48], and Knorr [18] (Sect. 24.3), Theaetetus' theory of ratios of magnitudes is NOT about the class of all ratios of magnitudes, as Eudoxus' theory of ratios, presented in Book V of the *Elements*, BUT is strictly about the limited class of all ratios with finite or eventually periodic anthyphairesis.

24.9 Our Critique of the Failure of Earlier Historians of Greek Mathematics and Scholars of Plato to Realize the Fundamental Role that the Philosophical Imitation of Geometrical Periodic Anthyphairesis Plays in Plato's Philosophy, in Particular in the Dialogues *Theaetetus*, *Sophist* and *Meno*, and to Reach an Essential Understanding of Both Greek Mathematics and Plato's Philosophy

Bertand Russell writes:

> It is noteworthy that modern Platonists, with few exceptions, are ignorant of mathematics in spite of the immense importance that Plato attached to arithmetic and geometry, and the immense influence that they had on his philosophy. [41, p. 131]

Paraphrasing Bertrand Russell, we may say that both modern Platonists and historians of Greek Mathematics did not realize the immense importance that Plato attached to periodic anthyphairesis, and the immense influence that it had on his philosophy. As a result Plato's philosophy was ill understood and some fundamental achievements of Greek Mathematics had lost the only source that could save them from oblivion.

Euclid with his proof of Proposition X.9 of the *Elements* may have contributed negatively towards a true understanding of the method that Theaetetus used for the proof of his general theorem, reported by Plato in the *Theaetetus* 147d-148b. Quite independently from the corrections that are needed to make the proof rigorous, our main concern is that Euclid gives an arithmetical proof, that is close to the methods used by Archytas, rather than those of Theaetetus (in 24.9.1). Van der Waerden and Knorr might be misdirected by Euclid's proof of X.9 and *Scholion* X.62 towards the translations (in 24.9.2, 24.9.5) and the arithmetical proofs (24.9.3, 24.9.6) they have suggested.

Historians of Greek Mathematics and scholars on Plato did not take into account the revealing close connection between Theaetetus' proof of incommensurabilities, reported in the *Theaetetus* 147d-148b, and the philosophical descriptions of the Knowledge of intelligible Beings, as imitations of Theaetetus' proof (148c-d), and as Name plus Logos, equivalently as True Opinion plus Logos (201–202), realized in the *Sophist* and in the *Meno*, along with other Platonic dialogues.

Burnyeat cannot be right, either in the translation of the *Theaetetus* 147d-e (in 24.9.2, 24.9.5), or in the claim that Plato had no motive to indicate Theaetetus' method (in 24.9.4, 24.9.6). In our opinion the most serious failure of the Platonists is the misunderstanding of the anthyphairetic nature of Plato's Logos (in 24.9.7).

24.9.1 Theaetetus' Theorem Receives a Defective But Correctible Proof in Proposition X.9 of the Elements, Which Is Arithmetical, Probably due to Archytas

Theaetetus' theorem described in the *Theaetetus* 147d3-e1 passage has been given a problematic arithmetical proof rendered in different ways by various authors, and this difference has wide implications both for the history of Greek Mathematics and for Plato's philosophy. We will discuss the Euclidean defective arithmetical proof of the theorem of Theaetetus given in the *Elements* as Proposition X.9, provide a completion/ correction of the proof, and indicate that it is an Archytan type proof.

The *Anonymous Scholion to Euclid's Elements* X.62 correlates Proposition X.9 with the difficult *Theaetetus* 147d3-e1 passage stating:

Τὸ θεώρημα τοῦτο Θεαιτήτειόν ἐστιν εὕρημα,
καὶ μέμνηται αὐτοῦ ὁ Πλάτων ἐν Θεαιτήτῳ
This theorem is a Theaetetean find,
and Plato recalls it in the *Theaetetus*

Proposition (*Elements* X.9, Translated by Heath [15] with Modifications by the Authors)

[1] The squares on straight lines commensurable in length have to one another the ratio which a square number has to a square number; and
[2] squares which have to one another the ratio which a square number has to a square number also have their sides commensurable in length.
[3] But the squares on straight lines incommensurable in length do not have to one another the ratio which a square number has to a square number; and
[4] squares [of straight lines] which do not have to one another the ratio which a square number has to a square number also do not have their sides commensurable in length either.

We are only interested in [4] which may be stated as follows:

> If a, b are lines, such that $a^2/b^2 = C/A$ with AC non-square number, then a, b are incommensurable.

***Proof of statement [4] in the* Elements** Finally, let the square on a not have to the square on b the ratio which a square number has to a square number. I say that a is incommensurable in length with b. For, if a is commensurable with b, then [by [1]] the square a^2 on a has to the square b^2 on b the ratio which a square number m^2 has to a square number n^2. But it does not, therefore a is not commensurable in length with b.

The arithmetical proof is problematic in that it is incomplete on two counts. Indeed, since we have assumed that a and b are commensurable, it follows that there are numbers m, n and a line c, such that $a = mc$, $b = nc$. Then obviously $a^2 = m^2c^2$, $b^2 = n^2c^2$.

The First Difficulty in the Euclidean Proof

We certainly want to conclude that $(m^2c^2)/(n^2c^2) = m^2/n^2$, a mixed proportion of the type $x/y = p/q$, where x, y are magnitudes and p, q are natural numbers. Euclid never supplies the required definition of a mixed proportion, but states and proves Propositions X.5 and 6, propositions involving a mixed proportion, as if he had such a definition. Thus Euclid's proofs are problematic. As van der Waerden notes:

> Indeed, Euclid's proofs contain logical errors for which we cannot hold Theaetetus responsible. Moreover these proofs depend on the theory of proportions developed in Book V, following Eudoxus, which was not yet known to Theaetetus. [48, p. 175]

On the other hand, it is quite natural to supply an adequate definition, most probably the one that Theaetetus himself had given:

Definition 24.9.1.1 If a, b are magnitudes, and m, n are numbers, then the proportion $a/b = m/n$ holds if $Anth(a, b) = Anth(m, n)$.

Note Earlier definitions by Knorr [18, pp. 253–254] and Taisbak [45, pp. 21–25] provide an arithmetical definition of a mixed ratio and correct the proof of Proposition X.5.

With this definition it is "immediately obvious" (εὐθέως φανερὸν) as Aristotle would say (*Topics* 158b33), that $(m^2c^2)/(n^2c^2) = m^2/n^2$, and hence that $a^2/b^2 = m^2/n^2$. Hence $m^2/n^2 = C/A$.

Note that this is a completely arithmetical relation.

The Second Difficulty in the Euclidean Proof

The left-hand side is a ratio of the form square number to square number, while the right-hand side is not. Euclid appears to regard it as an obvious contradiction, but it certainly is not. In order to show that this relation is impossible, we need to appeal to

Proposition (*Elements*, VIII.8) *If there are k mean proportionals between two numbers, then there are k mean proportionals between any two numbers which have the same ratios with the original numbers.*

Indeed, since the terms of the ratio m^2/n^2 possess the mean proportional mn, and since $m^2/n^2 = C/A$, it follows from Proposition VIII.8 that the numbers C, A possess also a mean proportional M, namely $AC = M^2$. But AC is not a square number, hence we now have a contradiction.

The problematic nature of the Euclidean proof of Proposition X.9 has been pointed out by van der Waerden, Knorr and Mazur:

> In Book X of the Elements, two proofs occur, so completely different in style from the rest of the book, that they have to be considered as later additions, viz. the proofs of X 9, 10. For, whereas Book X as a whole excels in its strictly logical structure and its extremely brief and elegant proofs, these two proofs not only contain superfluous parts, but also gaps. [van der Waerden [48, p. 159]].

> Further, Euclid does not offer a construction of roots in his demonstration of X,9, a theorem related to those in discussion in the *Theaetetus-passage*. [Knorr [18, p. 71]].
> THEOREM 7. Two lines are commensurable in length if and only if the squares constructed on them have the ratio which a square integer has to a square integer. (*Elements* X,9) [Knorr [18, p. 225]].
> Turning now to Theorem 7 we observe that although it has been stated in its Euclidean form (X,9), it is not in this form serviceable as a criterion of incommensurability. In addition, a result of the following type is needed:
> THEOREM 8. The least terms of a ratio of square integers are also square integers. [Knorr [18, p. 232]].

> Specifically, you will find—in various places—the claim that Theaetetus' theorem is proven in Proposition 9 of Book X of Euclid's *Elements*. It doesn't serve any purpose here to list the places where you find this incorrect assertion, except to say that it is incorrect, and it remains a thriving delusion since at least one important article published as late as 2005 repeats it. It is an especially strange delusion since nothing subtle is going on here.
> Even a cursory glance at Proposition 9 of Book X will convince you that what is being demonstrated there—if you take it in a modern perspective—is an utter triviality. Proposition 9 of Book X stands, though, for an important issue in ancient thought if taken on its own terms, but it won't prove irrationality of anything for us, let alone irrationality of all the numbers that Theaetetus proves. One might imagine that Heath's commentary on

> this— which is perfectly clear, and says exactly what is indeed proved in Proposition 9 — would dispel the misconception that Theaetetus' theorem about the irrationality of surds is contained in this proposition, but it seems that this has held on with some tenacity. [Mazur [20, p. 236]]

The corrected proof of Proposition X.9 is arithmetical, and thus not, according to our argument in Sects. 24.5–24.7, the original Theaetetean proof, certainly not by Theaetetus. We note that Proposition VIII.8 of the *Elements* is attributed to Archytas, and thus the completed/corrected arithmetical proof of X.9 must be attributed to Archytas or his students. Since Book X is certainly attributed to Theaetetus, it follows that Euclid modified the original work.

24.9.2 The Translation of the **Theaetetus** *147d3-e1 Passage by Historians of Greek Mathematics van der Waerden and Knorr and by the Platonic Scholar Burnyeat, and Our Critique of Their Translations*

We present in this section the standard translations of the Theaetetus 147d3-e1 passage by van der Waerden, Knorr and Burnyeat and we defer our critique of their translations to Sect. 24.9.5.

24.9.2.1 Translation by van der Waerden

> Here our Theodorus drew (or wrote) something about sides of squares (περὶ δυνάμεών τι ἔγραφε) and showed (ἀποφαίνων) that those of three or five feet are not commensurable in length with those of one foot, and in this manner he took up one after another up to the one of 17 feet; here something stopped him (or: here he stopped). [48, p. 142].
> Now it occurred to us (Theaetetus and another young man), since the number of roots appeared to be infinite, to try to collect them under one name, by which we could henceforth call all the roots. [48, p. 166].

24.9.2.2 Translation by Knorr

> Theaet.: Theodorus was proving for us via diagrams something about powers,
> in particular about the three-foot-power and the five-foot-power –
> demonstrating that these are not commensurable in length with the one-foot-power
> – and selecting each power individually in this way up to the seventeen-foot-power;
> but in this one for some reason he encountered difficulty.
> Now, this is what occurred to us:
> that, since we recognized the powers to be unlimited in number,
> we might try to collect them under a single name,
> by which we would designate all these powers. [18, pp. 62–63].

24.9.2.3 Translation by Burnyeat

THEAETETUS. Theodorus here was drawing diagrams to show us something
about powers namely that a square of three square feet and one of five square feet
aren't commensurable, in respect of length of side, with a square of one square foot;
and so on, selecting each case individually, up to seventeen square feet.
At that point he somehow got tied up.
Well, since the powers seemed to be unlimited in number,
it occurred to us to do something on these lines:
to try to collect the powers under one term
by which we could refer to them all. [8, p. 493].

24.9.3 The Arithmetical Reconstructions of the Proof of the Theorem of Theaetetus on Quadratic Incommensurabilities by van der Waerden and by Knorr

We will briefly describe the method that van der Waerden and Knorr suggested, in mutually substantial agreement, as the method that Theaetetus used for his proof of the quadratic incommensurabilities.

24.9.3.1 Reconstruction by van der Waerden

What is for us more important than these trivial definitions, is the proposition given at the end, very briefly but nevertheless clearly:
Line segments, which produce a square, whose area is an integer, but not a square number, are incommensurable with the unit of length.
...
How would the proof of such a proposition be made with classical methods? Indirectly of course. Supposing that two lines, which produce squares of areas n and 1, had commensurable lengths, one would try to show that n is a square number.
The proof necessarily breaks up into a geometrical and a number-theoretical part. From the assumed commensurability of the sides, one first deduces geometrically that the ratio of the sides is equal to that of two numbers p and q, and that the ratio of the areas of the squares which they produce is equal to that of p^2 and q^2, so that
(1) $n : 1 = p^2 : q^2$,
or, equivalently,
(2) $p^2 = nq^2$.
Then it has to be shown, number-theoretically, that it follows from (1) and (2) that n is a square number. [48, pp. 166–167].

It is Zeuthen's judgment [Zeuthen, 1910, p. 395], that the greatest merit of Theaetetus is however not to be found in the geometric analysis of the irrationality problem, but in the arithmetic part, in the proof therefore that (2) can hold only if n is a square. According to this Judgment, Theaetetus had discovered and proved several propositions from the arithmetical Books VII and VIII which were needed for this proof.

I do not share Zeuthen's view on this. For, as we have seen before, Book VII is of older date and forms the foundation of the Pythagorean theory of numbers. But, once one has at his disposal the propositions of Book VII, the number-theoretical part of Theaetetus' proof has no further difficulties. One can, for instance, reason as follows: q and p in (1) can of course be taken to be relatively prime. Then it follows from VII 27 that p^2 and q^2 (and p^3 and q^3 as well) are relatively prime, so that, by VII 21, they are the least of the numbers in that ratio. But in the left member of (1) occur n and 1 which are also relatively prime and hence also the least in this ratio. It follows that $n = p^2$ and $1 = q^2$, so that n is a square. The case of cubes proceeds analogously.

In my opinion the merit of Theaetetus lies therefore not in his contribution to the theory of numbers, but in his study of incommensurable line segments which produce commensurable squares. [48, p. 168].

24.9.3.2 Reconstruction by Knorr

THEOREM 5. If a square contains an integral number of unit areas, and if that number is a square integer, then the side of the square will be commensurable with the unit-line; but if that number is non-square, then the side will be incommensurable with the unit. (*Theaetetus*, 148A) [18, p. 225].
[*Theorem 5.* If a, b are lines and N is a non-square number such that $a^2 = Nb^2$, then a, b are incommensurable]
THEOREM 7. Two lines are commensurable in length if and only if the squares constructed on them have the ratio which a square integer has to a square integer. (*Elements* X,9) [18, p. 225].
[*Theorem 7.* If a, b are lines, and A, C are natural numbers, such that AC is not a square number, and $Aa^2 = Cb^2$, then a, b are incommensurable]
The theorems admit of completely straightforward demonstrations, as soon as the relevant general theorems on relative primes are available, as we shall show. Portions of that theory were already implicit in Archytas' proof of the theorem on epimoric ratios, namely, the examination of the techniques of reducing the terms of ratios. Hence, by the first or second decade of the fourth century, the number theory from which the proofs of Theorems 5 and 6 follow as easy deductions was in preparation. We are thus justified in accepting the most obvious interpretation of Plato's account: that Theaetetus himself instituted the proofs of the theorems which he is made to enunciate in the dialogue. [18, p. 226].

24.9.3.3 A Rendering of the Proof of the Theorem of Theaetetus on Quadratic Incommensurabilities Suggested by Zeuthen, van der Waerden and Knorr

Proposition 24.9.3.1 If $Aa^2 = Cb^2$ and AC is not a square number, then a, b are incommensurable.

Proof Suppose a, b are commensurable.
Then $a = mc$, $b = nc$ for some natural numbers m, n, and line c.
We may assume by the arithmetical Proposition VII.21 that m, n are relatively prime.

Then $Am^2 = Cn^2$.
By Proposition VII.19, $m^2/n^2 = C/A$.
By Proposition VII.27, m^2, n^2 are relatively prime.
By Proposition VII.20, there is a natural number k, such that $C = km^2$, $A = kn^2$.
Then $AC = k^2m^2n^2$, a square number, contradiction.

24.9.4 The Strict Neutrality Thesis of Burnyeat on the Method of Proof Employed by Theaetetus

Although most historians of Greek Mathematics lean towards an arithmetical method of proof for Theaetetus' incommensurabilities, nevertheless Burnyeat [8], rejected all reconstructions proposed till then [1978], whether arithmetical or anthyphairetic, on the ground that Plato's *Theaetetus* passage, our only ancient source for Theaetetus' discovery, gives no information on the method followed, either by Theodorus originally or by Theaetetus afterwards; in fact, Burnyeat expressed the view that

> Plato has no motive to indicate to the reader whether he has in mind any definite method of proof. [8, p. 505].

Indeed, Burnyeat writes:

> In the face of this impasse the question that needs to be asked is whether Plato has any reason to leave a hint, let alone so indeterminate and ambiguous a hint, as to the mathematical methods used in Theodorus' lesson. As every reader of the dialogue knows, the mathematical scene illustrates a point about definition and examples. When Theaetetus is first asked what knowledge is, he replies by giving examples of knowledge: geometry and the other mathematical sciences he is learning with Theodorus, cobblery and other crafts-each and all of these are knowledge (146cd). Socrates puts him right with an analogy: his answer is like that of someone who, on being asked what clay is, replies, "There is potters' clay, brickmakers' clay, and so on, each and all of which are clay," giving a list of clays instead of the simple, straightforward answer, "It is earth mixed with liquid" (146d-147c). It is at this point that Theaetetus says, "It looks easy now, Socrates, when you put it like that" (147c), and proceeds to tell his story. Theodorus' part in the story does not depend on whether or not he could continue past 17; his role is to provide examples of incommensurability. His case-by-case proof of their incommensurability is mentioned because, if one is not going by a general definition or rule of the kind the boys devised, it is only via construction and proof that examples of incommensurability are forthcoming: construction to obtain a length such as $\sqrt{3}$, which is not marked on any ruler, and proof to show that, divide how you will, you can find no unit to measure without remainder both it and a 1-foot line. Beyond that, *Plato has no motive to indicate to the reader whether he has in mind any definite method of proof* or any particular cause for Theodorus to stop at 17. This is not to deny, of course, the legitimacy of speculating about what methods would be available to Theodorus or other fifth-century mathematicians for proving various cases of incommensurability. But this must be an independent inquiry; there is no good reason to expect that the answer is to be squeezed out of one ambiguous sentence in Plato's dialogue. [8, p. 505] .

We will discuss Burnyeat's neutrality thesis in Sect. 24.9.6, below.

24.9.5 Critique of the Translations by van der Waerden, Knorr and Burnyeat

These translations are problematical in at least two counts.

The first difficulty is with the almost universal rendering

> since the number of roots appeared to be infinite (24.9.2.1, van der Waerden)
> since we recognized the powers to be unlimited in number (24.9.2.2, Knorr)
> since the powers seemed to be unlimited in number (24.9.2.3, Burnyeat)

of the sentence

> *ἐπειδὴ ἄπειροι τὸ πλῆθος αἱ δυνάμεις ἐφαίνοντο,* 147d7-8.

In Negrepontis [24] and Negrepontis et al. [33], arguments have been presented, according to which the only conceivable way to make sense of the sentence 147d7-8 is to read the plural "the powers" in a *distributive* sense:

> since *each power* was seen to be infinite in multitude

It is then strongly suggested that *each power* was seen to be *infinite* in multitude, as a result of Theodorus' lesson, because Theodorus was subjecting each power to anthyphairetic division, and thus proving, by means of Proposition X.2 of the *Elements*, that each power, of those he examined, was incommensurable to the one foot line.

The second difficulty is the almost universal trivialized rendering

> to try to collect them under one name,
> by which we could henceforth call all the roots. (24.9.2.1, van der Waerden)
> we might try to collect them under a single name,
> by which we would designate all these powers. (24.9.2.2, Knorr)
> to try to collect the powers under one term
> by which we could refer to them all. (24.9.2.3, Burnyeat)

of the sentence

> *πειραθῆναι συλλαβεῖν εἰς ἕν,*
> *ὅτῳ πάσας ταύτας προσαγορεύσομεν τὰς δυνάμεις.* 147d8-e1

In the two papers quoted above, the corresponding distributive plural gives the following rendering:

> to try to collect [the infinite multitude] into one (συλλαβεῖν εἰς ἕν),
> by which (ὅτῳ) we could refer to *each of all these powers* (πάσας ταύτας τὰς δυνάμεις).

Thus Theaetetus' mathematical discovery consists in first finding a general method to prove that every power, namely every line a such that $a^2 = Nb^2$, for some non-square number N, is divided into an infinite multitude of parts, namely has infinite anthyphairesis with respect to b, hence is incommensurable to b, but in addition that all these infinitely many parts of each power can be "collected into one".

Here Plato is employing philosophical language to describe a mathematical theorem, in fact his philosophical method of Name and Logos by which he defined the Sophist, is equivalently described as Division and Collection; as we outlined in

Sect. 24.5.3, Platonic Collection into One refers to the fundamental One-and-Many property of intelligible Beings, whose ultimate cause is definitely anthyphairetic periodicity.

24.9.6 Critique of the Arithmetical Non-anthyphairetic Reconstructions of Theaetetus' Theorem Described in the Theaetetus *147d3-e1 by van der Waerden and Knorr, and of the Neutrality Thesis by Burnyeat*

24.9.6.1 The Arithmetical Reconstructions of Theaetetus' Theorem Suggested by van der Waerden and Knorr are Unable to Explain the Relevance of Theaetetus' Theorem to Plato's Philosophy

We find difficult to understand how Zeuthen and van der Waerden, on the one hand have suggested that the method of Theodorus was anthyphairetic, employing Proposition X.2 of the *Elements* (cf. van der Waerden [48, pp. 144–146]), and on the other hand have suggested that the method of Theaetetus was arithmetical, employing Proposition VII.27 of the *Elements*, even though according to *Theaetetus* 147d3-e1, young Theaetetus had conceived his proof following the lesson by Theodorus. Perhaps Zeuthen, van der Waerden and other historians were misled by the proof of X.9 presented by Euclid in the *Elements*. In fact, Knorr writes the following:

> The essential content of this passage, as it might appear in the survey literature on the history of mathematics [Footnote 3] is this: Plato distinguishes two theorems - that of Theodorus, to the effect that the square roots of 3, 5, ... , 17 are irrational; and that of Theaetetus, that the square root of any non-square integer is irrational.
> [...]
> In fact, this is the sense given to it in a scholium to the ninth proposition of the tenth book of the Elements ("two lines are commensurable in length if and only if the squares generated on them are in the ratio of square numbers.") The scholiast calls this "the discovery of Theaetetus." It would appear that his source for this claim is precisely our passage from the Theaetetus. For he cites it and interprets it as asserting merely a restricted form of the Euclidean theorem. Now, accepting this natural reading of the passage,
> [...] [18, pp. 64–65]
> [Footnote 3] Accounts of this basic type appear in the histories by W. W. R. Ball [4], J. Tropfke [46], J. E. Hofmann [16], F. Lasserre [19], and C. Boyer [7], [18, p. 97]

Regardless of the reason which led to van der Waerden's misconception, and despite the fact that he did realize the logical errors and gaps in Euclid's proof, he did not consider that the original method of proof by Theaetetus could have been anything other than arithmetical. Knorr [18] is at least self-consistent in adopting throughout, for the Pythagoreans, Theodorus and Theaetetus, arithmetical methods.

Van der Waerden writes:

> In the dialogue Plato gives an example of a mathematical investigation made by Theaetetus. The example is rather dragged in; it has to serve as an introduction to a philosophical discussion, but it does not fit very well. [48, p. 166].

It is clear from this statement that van der Waerden does not realize the close connection between Theaetetus' mathematical discovery and Plato's philosophy. The theorem of Theaetetus, reported in the *Theaetetus* 147d-e, does establish the incommensurability of quadratic powers, but it accomplishes not only this but something much more, it provides complete knowledge of the power; by proving anthyphairetic periodicity, by Proposition X.2, we immediately have incommensurability, but by obtaining at a finite stage the period of the anthyphairesis we also have complete knowledge of an infinite entity at a finite stage. Thus the mathematical example is not "rather dragged", and it is not the case that "it does not fit very well", but on the contrary it reveals the anthyphairetic basis of Plato's philosophy. The same remarks apply for Knorr's somewhat different, but also arithmetical and non-anthyphairetic reconstruction.

24.9.6.2 The Neutrality Thesis of Burnyeat Fails to Take into Account (a) the Passage *Theaetetus* 148c-d on the Urge to Imitate Theaetetus' Method, Indicating Great Interest by Plato of Theaetetus' Method, and (b) the Anthyphairetic Nature of the Actual Imitation, as Outlined in Sects. 24.5–24.7

Burnyeat's translation of the *Theaetetus*' 147d-e (given in Sect. 24.9.2.3) is almost identical to the translation by van der Waerden, Knorr, and is thus subject to the same critique that theirs is. Burnyeat's thesis (presented in 24.9.4) that Plato has no motive to provide any information of Theaetetus' method is most certainly wrong. For how would we reconcile Plato's lack of motive or interest to provide some hint of the method of proof by Theaetetus with Plato's avowed urge and intention to imitate this method of Theaetetus on the subject of the Knowledge of the intelligible Beings (*Theaetetus* 148d4-5), a subject of supreme importance for Plato's philosophy? Moreover, Burnyeat's neutrality thesis (24.9.4) fails to take into account the anthyphairetic nature of the imitation, realized in the *Sophist* and the *Statesman*.

24.9.7 The Problem with the Philosophical Interpretations of Plato's Logos

The most fundamental concept in Plato's philosophy is that of "Logos". As Plato writes in the *Sophist* 260a-b

> if we were deprived of this [Logos], we would be deprived of philosophy, which would be the greatest calamity

As we have indicated in Sect. 24.5, employing passages from the *Sophist* and the *Meno*, the standard interpretations of Logos, as "account", "discourse", "rational definition", "rational explanation" and "reason" are not the correct ones. Plato's Logos is an imitation of Theaetetus' Logos Criterion for the periodicity of a geometric anthyphairesis, providing not only incommensurability, but complete Knowledge of its infinite structure and Equalization of all its parts, collecting all in a One, and thus making it an ideal candidate for an intelligible Being.
Theaetetus 202c2-5

> for he who cannot give and receive Logos (logon) of a thing is without knowledge (anepistemona) of it; but when he has acquired Logos (logon) he may possibly have become all that I have said and may now be perfect in knowledge (epistemen).

Note Logos is rendered in the classical translation by Fowler [36] as rational explanation.
Sophist 260a5-b2, *Logos* is necessary for the Knowledge of intelligible Beings.

> Stranger Our object was to establish that the Logos (logon) is one of the kinds of the Beings (ton onton). For if we were deprived of this [Logos], we would be deprived of philosophy, which would be the greatest calamity; moreover, we must at the present come to an agreement (diomologesasthai) about the nature (ti pot' estin) of Logos (λόγον), and if we were robbed of it [Logos], there would be no true Being (med' einai) and we could no longer be able to provide Logos (legein), and we should be robbed of it [Logos], if we agreed that there is no mixture (meixin) of anything with anything.

Note The classical translation by Fowler [36] renders all six references to "Logos" as "discourse".
Statesman 286a4-7, *Logos* is necessary for the Knowledge of intelligible Beings.

> We must therefore study so as to have the power (dunaton einai) to give and receive (dounai kai dexasthai) the Logos (logon) of each of them; for the immaterial [intelligible] Beings, which are the noblest and greatest, can be made known (saphos deiknutai) by Logos (logoi) only, and it is for their [the intelligibles'] sake (heneka) that all Logoi are provided (legomena).

Note The classical translation by Fowler and Lamb [37] renders "logon" as "rational definition", "logoi" as "reason", and "legomena" as "all we are saying".
Republic 534b3-6

> "And do you not also give the name dialectician (dialectikon) to the man who is able to exact an account (logon) of the true Being of each thing? And will you not say that the one who is unable to do this, in so far as he is incapable of rendering Logos (logon) to himself and others, does not possess intelligence about the matter?"

Note Logos is rendered in the classical translation by Shorey [39] as "account".
The renderings of the crucial term "logos" in the *Meno* 86b7 and 86c2 by Platonic scholars as "argument" and "word", respectively.

> There are other things about the argument that I would not confidently affirm, but that we shall be better men, more courageous and less idle, if we think we ought to inquire into what we do not know, instead of thinking that because we cannot find what we do not know we ought not to seek it — that I would do battle for, so far as possible, in word and deed. [86b6-c2, in Halper [14, p. 239]]
> For he goes on asserting, forcefully and without any ambiguity, that in all that was said there was only one point (81 d 5 - e 1; 86 b 2 - 4) for which he would fight strenuously, in word and deed (kai logdi kai ergoi), whenever he was able to do so. [Klein [17, p. 182]]
> As for the other points, I wouldn't absolutely insist on the argument. But I would fight, both in word and deed, for the following point: that we would be better, more manly, and less lazy if we believed that we ought to inquire into what we do not know, than if we believed that we cannot discover what we do not know and so have no duty to inquire. [86b6–c2, in Scott [43, p. 121]]

Logos is rendered as argument and word.

> Socrates: Yes, I think so too, Meno. I wouldn't support every aspect of the argument with particular vigour, but there's one proposition that I'd defend to the death, if I could, by argument and by action: that as long as we think we should search for what we don't know we'll be better people– less fainthearted and less lazy– than if we were to think that we had no chance of discovering what we don't know and that there's no point in even searching for it. [Waterfield [49, p. 124]]

Logos is rendered as argument.

> So I seem to myself, Meno. All the other points I have made in support of the argument are not such as I can confidently assert; but that, if we are convinced we should inquire after what we do not know, we should get better and braver and less lazy than if we believe that we are neither able nor obliged to inquire after things we do not know – this is something for which I am determined to fight, so far as I am able, both in word and deed. [86b6-c2, in Ebert [12, p. 197]]
> Socrates: I think so too, Meno. In defense of the argument I would not affirm the other points very strongly, but that we would be better, more manly, and less lazy by believing that one should search for what one doesn't know than if we believed that we cannot discover what we do not know and should not even search for it – that is something over which I would fiercely contend, if I were able, in both word and deed. [Sedley and Long, [44, p. 23]]
> As far as the other points (ta alla) are concerned, I wouldn't altogether take a stand on the argument. But that we will be better and more manly and less idle if we think one should inquire into that which one doesn't know than if we think it isn't possible to discover what we don't know and that we don't need to inquire into it: that is something I would certainly fight for to the end, if I could, in both word and deed. [85d12-86c2, in Fine [13, p. 150]]
> I would not confidently assert the other things said in defense of this account, but that we would be better and braver and less idle if we believe that one ought to inquire concerning those things he fails to know than if one believes it is not possible to discover nor necessary to inquire concerning those things one fails to know (εἰ οἰοίμεθα ἃ μὴ ἐπιστάμεθα μηδὲ δυνατὸν εἶναι εὑρεῖν μηδὲ δεῖν ζητεῖν), I would fight for in both word and deed as far as I am able. (Meno 86B6-C2; emphasis added) [Benson [6, p. 62]]

Nehamas [35] expresses the opinion that the meaning that Plato has in mind with Logos must come from a study of the Divisions in the *Sophist* and the *Statesman*.

> The answer [to the meaning of Logos in Plato] can only be given using the long and complicated divisions through which the Sophist and the Politicus attempt to define their subject matter. Both dialogues try to define a species of art (techne) or science (episteme), that of the sophist and that of the statesman (cf. Pol, 258b2-7). In both dialogues, the art in question is defined by being located within a determinate network of other arts and sciences. And the process by means of which the definition is reached provides, at the same time, both an explanation of the aporetic ending of the Theaetetus and an Illustration of the obscure opening of the Philebus. Plato, therefore, does try to answer the question of Theaetetus. But his answer is not cryptically contained within the dialogue itself, either negatively (as Cornford argued) or positively (as Fine suggested). His answer is given in the two dialogues that follow the Theaetetus. [35, p. 36]

24.10 The Reconstruction of Theaetetus' Theory of Ratios of Magnitudes Only for Ratios with no Recourse to Eudoxus' Condition (Definition 4 in Book V of the *Elements*)

The detailed reconstruction of the proof of Theaetetus' theorem reveals that Theaetetus' theory of magnitudes is *essentially a theory of ratios of lines*, designed principally for the needs of periodic anthyphairesis. Thus the Logos Criterion for anthyphairetic periodicity is essentially just a restatement in more suggestive and convenient language of the more heavy and burdensome Pythagorean preservation of Application of Areas/Gnomons. As shown in Sect. 24.8, Theaetetus' theory is for the limited class of those ratios of lines with finite or eventually periodic anthyphairesis.

Exactly because of the limited nature of the theory, it is possible to prove, with the help of the crucial Proposition 24.10.1.4, all the basic properties of this theory *with no recourse to Eudoxus' condition* (as was the case with the previous reconstructions). The extension of the theory to areas and surfaces is achieved with the help of the analogue of the key Proposition VI.1 of the *Elements*. The fact that this theory is really a theory not for general magnitudes, but principally for lines, makes necessary the many special cases in the proof of its Propositions. This feature is pointed out by Aristotle in *Analytics Posterior* 74a, and the vehicle of passing from lines to areas and surfaces is exactly the proposition (analogue to Proposition VI.1 of the *Elements*) singled out by Aristotle in the *Topics* 158b to inform us about this theory.

24.10.1 The Introduction of the Generalized Side and Diameter Numbers

24.10.1.1 The Generalized Side-Diameter Numbers p_n, q_n

Definition 24.10.1.1 Let there be an infinite anthyphairesis of a to b:
$a = k_0 b + e_1, e_1 < b$
...
$e_{m-2} = k_{m-1} e_{m-1} + e_m, e_m < e_{m-1}$
$e_{m-1} = k_m e_m + e_{m+1}, e_m < e_{m+1}$
$e_m = k_{m+1} e_{m+1} + e_{m+2}, e_{m+1} < e_{m+2}$
...
$e_{n-2} = k_{n-1} e_{n-1} + e_n, e_n < e_{n-1}$
$e_{n-1} = k_n e_n + e_{n+1}, e_{n+1} < e_n$
....
We set
$p_1 = 1, q_1 = k_0$, and
$p_n = k_{n-1} p_{n-1} + p_{n-2}, q_n = k_{n-1} q_{n-1} + q_{n-2}$ for every n.

Note The (classical) side and diameter numbers are identified as the generalized side and diameter numbers of the anthyphairesis of the diagonal (diameter) to side of a square.
[Indeed the anthyphairesis of diameter to side of a square is
$k_0 = 1, k_n = 2$ for all $n = 1, 2, \ldots$.
Then the generalized side-diameter numbers are recursively defined by
$p_1 = 1, q_1 = k_0 = 1$, and,
$p_{n+2} = 2p_{n+1} + p_n, q_{n+2} = 2q_{n+1} + q_n$.
Then
$p_{n+2} = 2p_{n+1} + p_n = p_{n+1} + p_{n+1} + p_n = p_{n+1} + (p_n + q_n) + p_n = p_{n+1} + q_{n+1}$,
and
$q_{n+2} = 2q_{n+1} + q_n = q_{n+1} + q_{n+1} + q_n = q_{n+1} + 2p_n + q_n + q_n = q_{n+1} + 2(p_n + q_n) = 2p_{n+1} + q_{n+1}$,
and the recursive definition of the side and diameter numbers is derived
$p_{n+2} = p_{n+1} + q_{n+1}$,
$q_{n+2} = 2p_{n+1} + q_{n+1}$].
A major difference between the classical side and diameter numbers and the generalized ones is that the recursive rule defining the generalized side and diameter numbers is far inferior to the classical rule, in that at the stage n they are defined in terms of the previous two stages, $n - 1$ and $n - 2$, and not just in terms of the immediately previous one $n - 1$.

24.10.1.2 The Generalized Side and Diameter Numbers Are Useful in Expressing Anthyphairetic Remainders in Terms of the Initial Magnitudes *a* and *b*

Proposition 24.10.1.2 [the anthyphairetic remainders in terms of the generalized side and diameter numbers] With the notation of Definition 24.10.1.1, we have $e_n = (-1)^n(q_n b - p_n a)$ for all $n = 1, 2, \ldots$.

Proof $e_1 = a - k_0 b = (-1)^1(q_1 b - p_1 a)$.
Assume inductively that:
$e_{n-2} = (-1)^{n-2}(q_{n-2}b - p_{n-2}a)$, and $e_{n-1} = (-1)^{n-1}(q_{n-1}b - p_{n-1}a)$.
Then:
$e_n =$
$e_{n-2} - k_{n-1}e_{n-1} =$
$(-1)^{n-2}(q_{n-2}b - p_{n-2}a) - (-1)^{n-1}k_{n-1}(q_{n-1}b - p_{n-1}a) =$
$(-1)^n[q_{n-2}b - p_{n-2}a + k_{n-1}(q_{n-1}b - p_{n-1}a)] =$
$(-1)^n[(q_{n-2} + k_{n-1}q_{n-1})b - (p_{n-2} + k_{n-1}p_{n-1})a] =$
$(-1)^n(q_n b - p_n a)$.

Proposition 24.10.1.3 If $Aa^2 = Bab+Cb^2$, and $ad = bc$, then $Ac^2 = Bcd+Cd^2$.

Proof From the assumption we have $Aa^2 = Bab + Cb^2$,
whence $Aa^2c = Babc + Cb^2c$,
whence $Aa^2c = Babc + Cbad$,
whence $Aac = Bbc + Cbd$,
whence $Aac^2 = Bbc^2 + Cbdc$,
whence $Aac^2 = Bcad + Cad^2$,
whence $Ac^2 = Bcd + Cd^2$.

Proposition 24.10.1.4

1. For every finite ordered sequence of natural numbers $k_0, k_1, \ldots, k_n$, there are lines a and b, $a > b$, such that $Anth(a, b) = [period(k_0, k_1, \ldots, k_n)]$, and $p_{n+1}a^2 = (q_{n+1} - p_n)ab + q_n b^2$.
2. If lines c and d satisfy $Anth(c, d) = [period(k_0, k_1, \ldots, k_n)]$, then $p_{n+1}c^2 = (q_{n+1} - p_n)cd + q_n d^2$.

Proof

1. The first $n + 1$ steps of the anthyphairesis of a to b are:
   $a = k_0 b + e_1, e_1 < b$
   $b = k_1 e_1 + e_2, e_2 < e_1$
   $e_1 = k_2 e_2 + e_3, e_3 < e_2$
   ...
   $e_{n-2} = k_{n-1}e_{n-1} + e_n, e_n < e_{n-1}$
   $e_{n-1} = k_n e_n + e_{n+1}, e_{n+1} < e_n$.

We also impose the cross product condition $ae_{n+1} = be_n$. The cross product condition implies the periodicity of the anthyphairesis; indeed, by Proposition 24.7.1.3, $Anth(a, b) = Anth(e_n, e_{n+1})$, and thus by induction:

$Anth(a, b) = [k_0, k_1, \ldots, k_n, Anth(e_n, e_{n+1})]$
$= [k_0, k_1, \ldots, k_n, Anth(a, b)]$
$= [k_0, k_1, \ldots, k_n, k_0, k_1, \ldots, k_n, Anth(e_n, e_{n+1})]$
$= \ldots$
$= [period(k_0, k_1, \ldots, k_n)]$.

The cross product relation further insures that a, b satisfy an arithmetical quadratic equality in excess; indeed, by Proposition 24.10.1.2, $e_n = (-1)^n(q_n b - p_n a)$, and $e_{n+1} = (-1)^{n+1}(q_{n+1}b - p_{n+1}a)$, hence, $ae_{n+1} = (-1)^{n+1}a(q_{n+1}b - p_{n+1}a)$, and

$be_n = (-1)^n b(q_n b - p_n a) = (-1)^{n+1} b(p_n a - q_n b)$.

Thus, the cross product condition takes the form: $a(q_{n+1}b - p_{n+1}a) = b(p_n a - q_n b)$, hence $p_{n+1}a^2 = (q_{n+1} - p_n)ab + q_n b^2$.

2. Since $Anth(a, b) = Anth(c, d)$, we have that the anthyphairesis of c to d is purely periodic. Since the generalized side and diameter numbers depend solely on the sequence of quotients $k_0, k_1, \ldots, k_n$ it follows that the generalized side r_i and diameter s_i numbers of c to d coincide with the generalized side p_i and diameter q_i numbers of a to b, respectively, namely $r_i = p_i$, $s_i = q_i$ for $i = 1, 2, \ldots, n, n+1$; hence, denoting by f_i the ith anthyphairetic remainder of c to d, it follows that: $cf_{n+1} = df_n$, hence, by Proposition 24.10.1.3, c, d satisfy the same arithmetical quadratic equation, therefore:
   $p_{n+1}c^2 = (q_{n+1} - p_n)cd + q_n d^2$.

From this result by completing the anthyphairesis of both a to b and c to d upwards we obtain, without difficulty, the general case. The details are left to the reader.

24.10.2 Theaetetus' Theory of Ratios of Magnitudes

Proposition 24.10.1.4 makes possible the proof of the fundamental Proposition 24.10.1.4: in Theaetetus' theory, for lines a, b, c, d, the condition of proportion $a/b = c/d$ is equivalent to the equality of the cross products $ad = bc$, with no recourse to Eudoxus' condition. In turn, following essentially the old idea of Becker [5], we can prove not only the Alternando property for lines, but also the crucial analogue of Proposition V.8 of the *Elements*.

Definition 24.10.2.1 A line a possesses *Theaetetean ratio* with respect to line b if the anthyphairesis of a to b is finite or eventually periodic.

Definition 24.10.2.2 (of Proportion) Two Theaetetean ratios of lines a/b and c/d are equal in the Theaetetean sense, $a/b = c/d$, if $Anth(a, b) = Anth(c, d)$.

Proposition 24.10.2.3 (Analogue of the Transitive Property, Proposition V.11 for Lines, the Analogous Proposition for Numbers Is not Explicitly Stated in Book VII, But Is Implicitly Used) Let a, b and c, d and e, f be pairs of lines possessing Theaetetean ratios, namely with finite or eventually periodic anthyphairesis. Then if $a/b = c/d$, and $c/d = e/f$, then $a/b = e/f$;

Proof By definition 24.10.2.1, either $Anth(a, b) = [k_0, k_1, \ldots, k_n, period(k_{n+1}, k_{n+2}, \ldots, k_m)]$ if it is eventually periodic, or $Anth(a, b) = [k_0, k_1, \ldots, k_n]$ if it is finite. Also, by definition 24.10.2.2, $a/b = c/d$ means that $Anth(a, b) = Anth(c, d)$. By definition 24.10.2.2 again, $c/d = e/f$ means that $Anth(c, d) = Anth(e, f)$. Thus $Anth(a, b) = Anth(e, f)$ and this, by the same definition, means that $a/b = e/f$.

Proposition 24.10.2.4 (Analogue of Proposition VI.16 for lines, Proposition VII.19 for Numbers and a Fundamental Proposition) Let a, b and c, d be pairs of lines possessing Theaetetean ratios, namely with either finite or eventually periodic anthyphairesis. Then $a/b = c/d$ (namely $Anth(a, b) = Anth(c, d)$) if and only if $ad = bc$.

Proof Let a, b and c, d be two pairs of lines with either finite or eventually periodic anthyphairesis. If $ad = bc$, then by Proposition 24.5.2.1, $Anth(a, b) = Anth(c, d)$, namely $a/b = c/d$ (definition 24.10.2.2). If $a/b = c/d$, then by Proposition 24.10.1.4, a, b and c, d satisfy the same arithmetical quadratic equation, say $Aa^2 = Bab + Cb^2$ and $Ac^2 = Bcd + Cd^2$. By Proposition 24.7.1.2, $ad = bc$.

Proposition 24.10.2.5 (Analogue of Proposition V.9 for Lines, the Analogous Proposition for Numbers Is not Explicitly Stated in Book VII, But Is Implicitly Used) If lines a, b and a, c have finite or eventually periodic anthyphairesis, and $a/b = a/c$, then $b = c$.

Proof By definition $a/b = a/c$ means that $Anth(a, b) = Anth(a, c)$. By Proposition 24.10.2.4, $ac = ab$, hence $b = c$.

Note For the restrictive class of lines a, b and c, d with finite or eventually periodic anthyphairesis, the proofs of the Fundamental Proposition 24.10.2.4 and 24.10.2.5 are provided without any use of Eudoxus' condition, Definition V.4 of the *Elements*, thus avoiding the anachronism that plagues the earlier reconstructions. This by itself is a strong argument in favor of our reconstruction.

Regarding the next proposition we will prove, the *Alternando*, van der Waerden writes:

> But one property, viz. the interchange of the means, causes difficulty. And now it is curious that, according to Aristotle, it was indeed exactly the proof of this proposition, which at first led to difficulties. "Formerly", says Aristotle in Anal. Post. 15, "this proposition was proved separately for numbers, for line segments, for solids and for periods of time. But after the introduction of the general concept which includes numbers as well as lines, solids and periods of time" (viz. the concept of magnitude), "the proposition could be proved in general".
> The proof for numbers can be found in Book VII (VII 13).What was the old proof for line segments?

> O. Becker has advanced an ingenious hypothesis for this. From the proportionality
> (1) $a : b = c : d$,
> one deduces first the equality of the areas
> (2) $ad = bc$,
> then interchanges b and c, and finally returns to the proportionality
> (3) $a : c = b : d$. [48, p. 177].

van der Waerden refers to the following remarkable Aristotle's *Analytics Posterior* 74a17-25 passage:

> καὶ τὸ ἀνάλογον ὅτι καὶ ἐναλλάξ,
> *ᾗ ἀριθμοὶ καὶ ᾗ γραμμαὶ καὶ ᾗ στερεὰ καὶ ᾗ χρόνοι,*
> ὥσπερ ἐδείκνυτό ποτε χωρίς,
> ἐνδεχόμενόν γε κατὰ πάντων μιᾷ ἀποδείξει δειχθῆναι·
> ἀλλὰ διὰ τὸ μὴ εἶναι ὠνομασμένον τι ταῦτα πάντα ἓν,
> *ἀριθμοί μήκη χρόνοι στερεά,*
> καὶ εἴδει διαφέρειν ἀλλήλων,
> χωρὶς ἐλαμβάνετο.νῦν δὲ καθόλου δείκνυται·
> *οὐ γὰρ ᾗ γραμμαὶ ἢ ᾗ ἀριθμοὶ ὑπῆρχεν,*
> ἀλλ' ᾗ τοδί, ὃ καθόλου ὑποτίθενται ὑπάρχειν.
> Again, the law that proportionals alternate might be supposed to apply
> *to numbers qua numbers, and similarly to lines, solids and periods of time*;
> as indeed it used to be demonstrated of these subjects separately.
> It could, of course, have been proved of them all by a single demonstration,
> but since there was no single term to denote the common quality of
> *numbers, lengths, time and solids,*
> and they differ in species from one another, they were treated separately;
> but now the law is proved universally;
> for the property did not belong to them
> *qua lines or qua numbers,*
> but qua possessing this special quality which they are assumed to possess universally.
> [translation by Tredennick [3]]

We will follow Becker's idea for the proof of

Proposition 24.10.2.6 (*Alternando*, Analogue of Proposition V.16 for Lines, Proposition VII.13 for Numbers) If $a/b = c/d$ and $Anth(a, c)$, $Anth(b, d)$ are finite or eventually periodic, then $a/c = b/d$.

Proof By Proposition 24.10.2.4, $ad = bc$. Since $Anth(a, c)$ is finite or eventually periodic, then by Proposition 24.10.2.4, $a/c = b/d$.

Proposition 24.10.2.7 (*Ex Equali*, Analogue of Proposition V.22 for Lines, Proposition VII.14 for Numbers) If a, b, c, d, e, f are lines and $a/b = d/e$ and $b/c = e/f$ and $Anth(a, c)$, $Anth(d, f)$ are finite or eventually periodic, then $a/c = d/f$.

Proof By the Fundamental Proposition (24.10.2.4), we have:
$ae = bd$, and
$bf = ce$.
Then $abf = ace = bcd$,
hence $af = cd$.
By the Fundamental Proposition (24.10.2.4), $a/c = d/f$.

Proposition 24.10.2.8 (Perturbed Proportion, Analogue of Proposition V.23 for Lines) If a, b, c, d, e, f are lines and $a/b = e/f$ and $b/c = d/e$, and $Anth(a, c)$, $Anth(d, f)$ are finite or eventually periodic, then $a/c = d/f$.

Proof By the Fundamental Proposition (24.10.2.4), we have
$af = be$ and $be = cd$.
By transitivity (24.10.2.3), $af = cd$.
By the Fundamental Proposition (24.10.2.4), we have $a/c = d/f$.

Proposition 24.10.2.9 (Part (a) is the Analogue of Proposition V.12 for Lines and Proposition VII.12 for Numbers While Part (b) Is the Analogue of Proposition V.19 for Lines and Proposition VII.11 for Numbers) If a, b, c, d are lines and $Anth(a, b) = Anth(c, d)$ is finite or eventually periodic anthyphairesis, then:

(a) $Anth(a+c, b+d)$ is finite or eventually periodic, and $(a+c)/(b+d) = a/b$, and
(b) if $a > c$ and $b > d$, then $Anth(a-c, b-d)$ is finite or eventually periodic, and $(a-c)/(b-d) = a/b$.

Proof

(a) By proposition 24.5.2.1, and since $Anth(a, b) = Anth(c, d)$, we have: $ad = bc$.
Then: $ad + ab = bc + ab$,
$a(b + d) = b(a + c)$.
By proposition 24.5.2.1, we have that $Anth(a, b) = Anth(a + c, b + d)$ and then,
by definition 24.10.2.2 $(a + c)/(b + d) = a/b$.
(b) By proposition 24.10.2.4, and since $Anth(a, b) = Anth(c, d)$, we have: $ad = bc$.
Then: $ab + ad = ab + bc$,
$ab - bc = ab - ad$,
$b(a - c) = a(b - d)$.
By proposition 24.5.2.1, we have that $Anth(a, b) = Anth(a - c, b - d)$ and then,
by definition 24.10.2.2 $(a - c)/(b - d) = a/b$.

Proposition 24.10.2.10 (Part (a) Is the Analogue of Proposition V.18 for Lines and Part (b) Is the Analogue of Proposition V.17 for Lines, But There Is No Analogue for Numbers) If a, b, c, d are lines and $Anth(a, b) = Anth(c, d)$ is finite or eventually periodic anthyphairesis, then:
(a) $Anth(a+b, b) = Anth(c+d, d)$ is finite or eventually periodic, and $(a+b)/b = (c + d)/d$, and
(b) if $a - b > b$ and $c - d > d$, then $Anth(a - b, b) = Anth(c - d, d)$ is finite or eventually periodic, and $(a - b)/b = (c - d)/d$.

Proof
(a) Consider the case where $Anth(a, b) = Anth(c, d)$ and both are finite.
Let their anthyphairesis be denoted by $[k_0, k_1, \ldots, k_n]$. Then, it is easy to demonstrate, that the anthyphairesis of $a + b$ to b and of $c + d$ to d, differs from the anthyphairesis of a to b and c to d, respectively, only in the first quotient.
Essentially, instead of k_0 the first quotient now becomes $k_0 + 1$.
Then $Anth(a + b, b) = Anth(c + d, d) = [k_0 + 1, k_1, \ldots, k_n]$ and, by definition 24.10.2.2,
$(a + b)/b = (c + d)/d$. Similarly, we can demonstrate that the proposition is true when the anthyphairesis of a to b is eventually periodic.
(b) Consider the case where $Anth(a, b) = Anth(c, d)$ and both are finite.
Let their anthyphairesis be denoted by $[k_0, k_1, \ldots, k_n]$.
Then, it is easy to demonstrate, that the anthyphairesis of $a - b$ to b and of $c - d$ to d, differs from the anthyphairesis of a to b and c to d, respectively, only in the first quotient.
Essentially, instead of k_0 the first quotient now becomes $k_0 - 1$.
Then $Anth(a - b, b) = Anth(c - d, d) = [k_0 - 1, k_1, \ldots, k_n]$ and, by definition 24.10.2.2,
$(a - b)/b = (c - d)/d$. Similarly, we can demonstrate that the proposition is true when the anthyphairesis of a to b is eventually periodic.

Proposition 24.10.2.11 (*Topics* Proposition, Analogue of VI.1 for Lines, Propositions VII.17, 18 for Numbers) If a, b, c are lines, with $Anth(a, b)$ eventually periodic, then $(ac)/(bc) = a/b$.

Proof As already noted, we have that $Anth(a, b) = Anth(ac, bc)$, hence $(ac)/(bc) = a/b$.

Proposition 24.10.2.12 (Analogue of Proposition V.9 for Rectilinear Areas) If A, B, C are rectilinear areas, with $A/B = A/C$, and $Anth(A, C)$ is finite or eventually periodic, then $B = C$.

Proof By the definition of $A/B = A/C$, we have $Anth(A, B) = Anth(A, C)$.
Let r be an assumed line.
By Proposition I.45 of the *Elements*, there are lines a, b, c, such that $A = ar$, $B = br$, $C = cr$.
By Proposition 24.10.2.11, $A/B = (ar)/(br) = a/b$, $A/C = (ar)/(cr) = a/c$.
Hence, by transitivity (24.10.2.3), $a/b = a/c$ and by the fundamental Proposition 24.10.2.4, $ac = ab$, hence $b = c$. Hence $B = br = cr = C$.

Proposition 24.10.2.13 (*Alternando*, Analogue of Proposition V.16 for Rectilinear Areas) If A, B, C, D are rectilinear areas, with $A/B = C/D$, and $Anth(A, C)$, $Anth(B, D)$ are finite or eventually periodic, then $A/C = B/D$.

Proof By the definition of $A/B = C/D$, we have $Anth(A, B) = Anth(C, D)$.
Let r be an assumed line.
By Proposition I.45 of the *Elements*, there are lines a, b, c, d, such that
$A = ar$, $B = br$, $C = cr$, $D = dr$.

By Proposition 24.10.2.11, $A/B = (ar)/(br) = a/b$, $C/D = (cr)/(dr) = c/d$. Hence, by transitivity (24.10.2.3), $a/b = c/d$ and by the fundamental Proposition 24.10.2.4, $ad = bc$. By Proposition 24.10.2.6 (Alternando for lines) $a/c = b/d$ and by Proposition 24.10.2.11 and transitivity, $A/C = B/D$.

Proposition 24.10.2.14 (Analogue to V.22, Part (a) Is the *Ex Equali* for Rectilinear Areas and Part (b) Is the *Ex Equali* for a Mixed Proportion)
(a) If A, B, C, D, E, F are rectilinear areas, with $A/B = D/E$, and $B/C = E/F$, and $Anth(A, C)$, $Anth(D, F)$ are finite or eventually periodic, then $A/C = D/F$.
(b) If A, B, C are rectilinear areas, d, e, f are lines, with $A/B = d/e$, $B/C = e/f$, and $Anth(A, C)$, $Anth(d, f)$ are finite or eventually periodic, then $A/C = d/f$.

Proposition 24.10.2.15 (Analogue to V.23, Part (a) Is the Perturbed Proportion for Rectilinear Areas and Part (b) Is the Perturbed Proportion for a Mixed Proportion)
(a) If A, B, C, D, E, F are rectilinear areas, with $A/B = E/F$, and $B/C = D/E$, and $Anth(A, C)$, $Anth(D, F)$ are finite or eventually periodic, then $A/C = D/F$.
(b) If A, B, C are rectilinear areas, d, e, f are lines, with $A/B = e/f$, $B/C = d/e$, and $Anth(A, C)$, $Anth(d, f)$ are finite or eventually periodic, then $A/C = d/f$.

Proofs of Propositions 24.10.2.14 and 24.10.2.15 are analogous to the proof of Proposition 24.10.2.12 and are left to the reader.
Note The fragmentation of the proof of a property in separate cases for lines, areas, volumes, as described by Aristotle in the *Analytics Posterior* passage quoted above, is confirmed, not only for the *Alternando* property, but also for the other properties of Theaetetus' theory of proportion.

References

1. F. Acerbi, Drowning by multiples remarks on the fifth book of Euclid's Elements, with special emphasis on Prop. 8. Arch. Hist. Exact Sci. **57**, 175–242 (2003)
2. Aristotle, *Topics Books I and VIII, with Excerpts from Related Texts*. Translated with a Commentary by R. Smith (Clarendon Press, Oxford, 1997)
3. Aristotle, *Postetior Analytics*. Translation by H. Tredennick, The Loeb Classical Library (Harvard University Press, Cambridge, 1960)
4. W.W.R. Ball, *A Short Account of the History of Mathematics* (Dover Publications, New York, 1960). (reprint of the 4th ed., 1908)
5. O. Becker, Eine Voreudoxische Proportionenlehre und ihre Spuren bei Aristoteles und Euklid. Quellen und Studien zur Geschichte der Mathematik, Astronomie und Physik **2**, 311–333 (1933)
6. H.H. Benson, *Clitophon's Challenge: Dialectic in Plato's Meno, Phaedo and Republic* (Oxford University Press, New York, 2015)
7. C.B. Boyer, *A History of Mathematics* (Wiley, New York, 1968)
8. M.F. Burnyeat, The philosophical sense of Theaetetus' mathematics. Isis **69**, 489–513 (1978)
9. F.M. Cornford, *Plato's Theory of Knowledge: The Theaetetus and the Sophist of Plato* (The Liberal Arts Press, New York, 1957)
10. R. Dedekind, *Stetigkeit und irrationale Zahlen* (Friedrich Vieweg, Braunschweig, 1872)

11. E.J. Dijksterhuis, *De elementen van Euclides: De ontwikkeling der griekscche wiskunde voor Euclides. Boek i der elementen* (P. Noordhoff, Groningen, 1929)
12. T. Ebert, The theory of recollection in Plato's Meno": against a myth of platonic scholarship, in *Gorgias-Menon. Selected Papers from the Seventh Symposium Platonicum*, ed. by M. Erler, L. Brisson (Academia Verlag, Sankt Augustin, 2007), pp. 184–198
13. G. Fine, *The Possibility of Inquiry: Meno's Paradox from Socrates to Sextus* (Oxford University Press, Oxford, 2014)
14. E.C. Halper, A lesson from the Meno, in *Gorgias-Menon. Selected Papers from the Seventh Symposium Platonicum*, ed. by M. Erler, L. Brisson (Academia Verlag, Sankt Augustin, 2007), pp. 234–242
15. T.L. Heath, *The Thirteen Books of Euclid's Elements*, vol. 3 (Cambridge Univeristy Press, Cambridge, 1908)
16. J.E. Hofmann, *Geschichte der Mathematik*, vol. 3 (De Gruyter, Berlin, 1953)
17. J. Klein, *A Commentary on Plato's Meno* (University of Chicago Press, Chicago, 1965)
18. W.R. Knorr, *The Evolution of the Euclidean Elements: A Study of the Theory of Incommensurable Magnitudes and its Significance for Early Greek Geometry* (Reidel, Dordrecht, 1975)
19. F. Lasserre, *The Birth of Mathematics in the Age of Plato* (World Publishing Company, Cleveland/New York, 1966). (repr. of 1964 edition)
20. B. Mazur, How did Theaetetus prove his theorem? in *The Envisioned Life: Essays in Honor of Eva Brann*, ed. by E.T.H. Brann, P. Kalkavage, E. Salem (Paul Dry Books, Philadelphia, 2007), pp. 227–250
21. G.R. Morrow, *Proclus: A Commentary on the First Book of Euclid's Elements* (Princeton University Press, Princeton, 1970)
22. S. Negrepontis, Η Ανθυφαιρετική Φύση της Διαλεκτικής του Πλάτωνος (in Greek), in *Διεπιστημονική Προσέγγιση των Μαθηματικών και της Διδασκαλίας τους* edited by Φ. Καλαβάση, Μ. Μεϊμάρη (Gutenberg, Athens, 2000), pp. 15–77
23. S. Negrepontis, Plato's theory of knowledge of forms by division and collection in the Sophistes is a philosophic analogue of periodic anthyphairesis (and modern continued fractions) (2012). arXiv: 1207.2950
24. S. Negrepontis, The anthyphairetic revolutions of the Platonic ideas, in *Revolutions and Continuity in Greek Mathematics*, ed. by M. Sialaros (De Gruyter, Berlin, 2018), pp. 335–381
25. S. Negrepontis, Plato on geometry and the geometers, in *Geometry in History*, ed. by S.G. Dani, A. Papadopoulos (Springer, Berlin, 2019), pp. 1–88
26. S. Negrepontis, The anthyphairetic interpretation of knowledge as recollection in Plato's *Meno* 80d-86e and 97a-98b, in *Handbook of History and Philosophy of Mathematical Practice*, ed. by B. Sriraman (Springer, Cham, 2024), pp. 471–519
27. S. Negrepontis, The periodic anthyphairetic nature of the one of the second hypothesis in Plato's *Parmenides*, in *Proceedings of the Conference Mathématiques et Musique: des Grecs a Euler, Strasbourg, 10–11 September 2015*, ed. by X. Hascher, A. Papadopoulos (Hermann, Paris, TA)
28. S. Negrepontis, V. Farmaki, *History of Greek Mathematics*, vol. I (In Greek) (Ekkremes, Athens, 2019)
29. S. Negrepontis, V. Farmaki, Pythagorean book II of the *Elements* restored and Pythagorean incommensurabilities reconstructed, in *Essays on Geometry: Celebrating the 65th birthday of Athanase Papadopoulos*, ed. by A.M. Uludağ, A. Zeytin (Springer, Cham, 2025), pp. 279–338
30. S. Negrepontis, D. Kalisperi, The mystery of Plato's receptacle in the *Timaeus* resolved, in *Handbook of History and Philosophy of Mathematical Practice*, ed. by B. Sriraman (Springer, Cham, 2024), pp. 521–598
31. S. Negrepontis, D. Protopapas, The restoration of book X of the *Elements* to its original Theaetetean form. Ganita Bharati Bullet. Ind. Soc. History Math. **45**(2) (2023)
32. S. Negrepontis, V. Farmaki, M. Brokou, The central role of incommensurability in pre-Euclidean Greek mathematics and philosophy. Ganita Bharati, Bullet. Ind. Soc. History Math. **42**(1–2), 1–34 (2020)

33. S. Negrepontis, V. Farmaki, M. Brokou, The Pell equation in the Pythagoreans, Theaetetus, and Hindu mathematics, in *Handbook of History and Philosophy of Mathematical Practice*, ed. by B. Sriraman (Springer, Cham, 2024), pp. 915–1021
34. S. Negrepontis, V. Farmaki, D. Kalisperi, Pre-Eudoxean geometric algebra. Ganita Bharati Bullet. Ind. Soc. History Math. **44**(2), 107–152 (2022)
35. A. Nehamas, Episteme and logos in Plato's later thought. Archiv fur Geschichte der Philosophie **66**, 11–36 (1984)
36. Plato, *Theaetetus. Sophist.* Translation by H.N. Fowler. The Loeb Classical Library (Harvard University Press, Cambridge, 1921)
37. Plato, *Statesman. Philebus. Ion.* Translation by H.N. Fowler and W.R.M. Lamb. The Loeb Classical Library (Harvard University Press, Cambridge, 1925)
38. Plato, *Timaeus.* Translation by W.R.M. Lamb. The Loeb Classical Library (Harvard University Press, Cambridge, 1925)
39. Plato, *Republic, Books 6-10.* Translation by P. Shorey. The Loeb Classical Library (Harvard University Press, Cambridge, 1935)
40. Plato *Theaetetus and Sophist.* Translation by Ch. Rowe. Cambridge Texts in the History of Philosophy, reprint edn. (Cambridge University Press, Cambridge, 2015)
41. B. Russell, *History of Western Philosophy* (George Allen & Unwin, London, 1946)
42. K. Saito, Phantom theories of pre-eudoxean proportion. Sci. Context **16**, 331–347 (2003)
43. D. Scott, *Plato's Meno* (Cambridge University Press, Cambridge, 2005)
44. D. Sedley, A. Long, *Plato: Meno and Phaedo* (Cambridge University Press, Cambridge, 2011)
45. C. Taisbak, *Coloured Quadrangles: A Guide to the Tenth Book of Euclid's Elements* (Museum Tusculanum Press, Copenhagen, 1982)
46. J. Tropfke, *Geschichte der Elementar-Mathematik*, vol. 7, 2nd edn. (Berlin/Leipzig, De Gruyter, 1921–1924)
47. B.L. van der Waerden, Arithmetik. Math. Ann. **120**, 687–693 (1947/1949)
48. B.L. van der Waerden, *Science Awakening* (Kluwer Academic Publishers, Groningen, 1954). [Translated from the Dutch edition of 1950 by A. Dresden]
49. R. Waterfield, *Meno and Other Dialogues. Charmides, Laches, Lysis, Meno* (Oxford University Press, Oxford, 2005)
50. H.G. Zeuthen, Sur la constitution des livres arithmétiques des Eléments d'Euclide et leur rapport a la question de irrationalité. Oversigt det kgl. *Dansk Videnskabernes Selskabs Forhandlinger* **5**, 295–437 (1910)
51. H.G. Zeuthen, Hvorledes Mathematiken i Tiden fra Platon til Euklid blev rationel Videnskab. DVSSk **1**(8), 199–369 (1917)

Chapter 25
Mathematical Works

Valentin Poénaru

Abstract This chapter is an autobiographical perspective on the mathematical work of Valentin Poénaru, from 1955 until 2000.

Keywords Biography · Valentin Poénaru

AMS Codes 01A60, 01A70, 5503

I was born in Bucharest (Romania) in October 1932 and studied mathematics at the University of Bucharest from 1951 to 1955. I started publishing research articles towards the end of my second year of study, and soon began an intense collaboration with my classmates Ciprian Foias (who much later became our colleague at Orsay, before moving to Bloomington Indiana, USA) and Georges Gussi (who remained in Romania). All our joint articles [1–10], concerned linear and especially non-linear partial differential equations, in the way that, at the time, they were practiced by J. L. Lions (one of our correspondents) and others.

But since, on the one hand, my own turn of mind was rather geometrical and, on the other, I realized that in analysis I couldn't really compete with my friend Foias, I quickly turned to topology.

There, in Bucharest, I was in almost total mathematical solitude and, perhaps because of this, I started thinking about the Poincaré hypothesis (the leitmotif of my career ever since).

In 1955 I was appointed assistant at the University of Bucharest, a position from which I was expelled (as an "enemy of the people"), along with some forty other colleagues, at the beginning of October 1958. It was a relatively late turn of the

These report was written in 2000.

V. Poénaru (✉)
IHES, Bures-sur-Yvette, France
e-mail: valpoe@hotmail.com

L. Funar, A. Papadopoulos (eds.), *Essays on Topology*,
https://doi.org/10.1007/978-3-031-81414-3_25

screw, following the events of Budapest in 1956. Meanwhile, in April 1958, it turned out that my first, admittedly naive, attempt to prove the Poincaré hypothesis had collapsed.

Between October 1958 and December 1959, I was "full-time unemployed", and as this situation was becoming rather untenable, I accepted a very modest job (janitor) at the Romanian Mathematical Society. But there was mathematics to keep my spirits up, and towards the end of 1959, a good theorem emerged from the rubble of the Poincaré abortive attempt:

There exists a compact manifold of dimension four M^4, smooth, contractible, such that its boundary ∂M^4 is not simply connected, $\pi_1(\partial M^4) \neq 0$ so $M^4 \neq B^4$ *but* $M^4 \times I \overset{\text{DIFF}}{=\!=} B^5$.

One of the many spin-offs of this result is *the existence of a smooth involution on S^4 whose set of fixed points is a non-simply connected 3-dimensional homology sphere*. With the help of Jean Leray (and also Georges de Rham) I managed to get this work published in France [11]. What's more, Barry Mazur, with whom I have since developed the strongest of friendships, independently discovered the same result, which now bears both our names. There is a huge literature on the subject. Finding non-trivial factorizations of the hypercube of dimension five earned me a change of status: in May 1960 I became a research associate at the Institute of Mathematics of the Romanian Academy. At that time, I proved a second theorem [12, 14, 15] well outdated by subsequent developments since, but which, at the time, was an achievement:

Let Δ^3 be a homotopy ball of dimension three. If we multiply it with B^2 and do the connected sum with enough $S^2 \times B^3$, we find:

$$\Delta^3 \times B^2 \# S^2 \times B^3 \# S^2 \times B^3 \# \cdots \# S^2 \times B^3 \overset{\text{DIFF}}{=\!=} B^5 \# S^2 \times B^3 \# S^2 \times B^3 \# \cdots \# S^2 \times B^3.$$

This 5-dimensional (and therefore crude) result can be seen as a very vague precursor to a 4-dimensional result, and to the very useful and infinitely harder Theorem 2.

In August 1962, I was invited to give a half-hour talk at the ICM (Stockholm). I managed to get there and leave Romania for good. In this escapade, I was greatly helped by colleagues and/or friends from France, Switzerland, the United States and even Romania, to whom I remain highly grateful. (I won't go into the details here.)

From September 1962 to October 1963, I was hosted by the IHÉS, and also by the Universities of Geneva and Lausanne. I took a refugee status, in France, and passed my Doctorat d'État at the University of Paris, in March 1963, with Charles Ehresmann as thesis director. This thesis consists essentially of the results cited above.

In the meantime, for the purposes of my research on the Poincaré hypothesis, I became interested in Smale's theory of immersions. In [13] I take the step (codimension > 0) $\Rightarrow$ (codimension $= 0$), which has been important for subsequent developments. The collaboration with André Haefliger [16] which on the one hand gives the theory of immersions its definitive form and on the other extends it, also, to the PL category, is considered a classic article in the subject, and the theory of "immersions with folds" that I developed [17, 18] has been widely used and cited by Gromov and Eliashberg. The articles [20, 26, 34] are also part of the same work group.

In October 1963, I emigrated to the United States, where I was successively a researcher at Harvard, at IAS Princeton, again at Harvard (Visiting Professor) and then at Northeastern (Full Professor). In 1964, I did another abortive attempt at the Poincaré hypothesis. Rightly or wrongly, this determined me to work in almost complete silence, as far as this subject was concerned, for the next twenty years. I will come back to this later.

In October 1967, I returned to France as an associate professor at Orsay. It was Jean Cerf, with considerable help from Henri Cartan, who brought me to Orsay with the idea that I would stay there for good, just as he did for Larry Siebenmann and Harold Rosenberg.

In 1969, I became a naturalized French citizen and a Professor at Orsay, with a career reconstitution three years later, and, apart from two leaves to the CNRS, I stayed there for the rest of my career.

It was from my research into the Poincaré hypothesis that I became interested in the theory of singularities [19, 22, 24, 25, 27, 28]. My main contribution to the subject includes the theory of *C^∞ singularities in the presence of the symmetry of a compact Lie group*—structural stability à la John Mather, versal deployment, etc.; the monograph [28] presents the set of results.

At the time, around 1974, I was in need of at least a partial change of ideas, which led me to physics. The equivariant versal deployment had already got me thinking about symmetry breaks, à la Salam–Weinberg–Higgs. And there I was lucky enough to find myself at the right time (1974), in the right place (Paris), where there was Louis Michel (at the IHÉS), Maurice Kleman (at Orsay) and Gérard Toulouse (at the ÉNS) who were interested in the same subject, from the condensed matter point of view. The idea was already in the air that homotopy groups should be used to classify defects in condensed matter. During this period, I had a very fruitful collaboration with G. Toulouse and a few other French physicists. [26, 29–31, 33, 36, 40]

I can try to summarize the main results of our contribution:

(i) We linked the algebraic structure of homotopy groups to the dynamics of defects; in particular, we showed that there exists an obstruction to the crossing of defect lines linked to the non-commutativity of the fundamental group of the space of the "order parameter" (this concerned a type of material that was still hypothetical at the time, but has since been synthesized: "biaxial nematics"; these are liquid crystals.

(ii) Partly prompted by the physicist Alexander Polyakov (currently at Princeton), we generalized (i) to higher dimensions, demonstrating a confinement mechanism linked to supersymmetry. The Lie super-bracket involved here is the Whitehead product $\pi_p \times \pi_q \to \pi_{p+q-1}$ [29].

(iii) We have shown how the π_3 of the order parameter space can create "textured" defects, thus explaining the "double rings" discovered in the laboratory by Yves Bouligand.

Much has been done on the subject over the years, but our work was among the very first.

Around the same time (1976–1977), Albert Fathi, François Laudenbach and I decided to produce a text with complete proofs for the theorems announced by William Thurston concerning measured foliations, diffeomorphisms of surfaces, ergodic properties of these objects and the compactification of Teichmüller space. The seminar we organized on this subject at Orsay resulted in the volume [35], unanimously acclaimed by the specialists. The result was an invaluable service to the mathematical community.

Back to physics. After a few years of intense collaboration, my own scientific interests (which, as far as physics was concerned, turned towards quantum field theory) and those of Gérard Toulouse (which turned towards the statistical mechanics of neural systems) took different paths. So, around 1977, I met Geoff Chew (Berkeley), whose ideas won me over; it has to be said that his "bootstrap" had also won over a whole generation of particle physicists. Chew and I, together with other collaborators, developed the "topological bootstrap" with particular reference to discrete quantum numbers, spinorology and the fundamental symmetries C, P, T [37–39, 41–50]. I won't try to summarize these works. It is worth adding here that while my collaboration with Toulouse was respectable and respected, that with Geoff Chew, with his not always orthodox ideas, was controversial. I don't myself have a definitive judgement on this question, but I do know that thanks to this collaboration, I have learned an enormous amount of physics, and also inherited a whole mathematical problematics that was interesting, but probably still too difficult, at the time we were involved in this research, see [51]. I should also add that, directly or indirectly, some of my friend Geoff Chew's ideas formed part of the fabric of string theory, of which several key players, such as John H. Schwarz and David Gross, were students. On this subject, I would like to mention the articles by Steve Weinberg, John H. Schwarz and David Gross, in the volume in tribute to G. F. Chew, "A passion for Physics", World Scientific (1985).

All this takes us up to around 1980, and I pick up the thread of the Poincaré hypothesis, which continued to occupy most of my time and energy, even if this remained mostly underground, with very few external manifestations, for a very long time.

In parallel with the Poincaré hypothesis, I have been thinking about other low-dimensional topology problems. For example, I formulated the following Conjecture around 1960:

Let $N^5(K^2)$ be a regular 5-dimensional neighborhood of a finite contractible simplicial 2-complex K^2. Then

$$N^5(K^2) \overset{\text{DIFF}}{=\!=} B^5$$

In 1970, I succeeded in proving the "honeycomb representation theorem" [21] and [57] (improved version), which was to become the first step in my program. Around 1972, my project took shape in the form of two (conjectural) stages of roughly equal and, in fact, largely similar difficulty:

(i) "The Coherence Theorem" (which corresponds to (theorem 2) + (conjecture 4)) and
(ii) the implication (Coherence theorem) $\Rightarrow (\Delta^3 = B^3)$ (which corresponds to Theorem 3).

Around 1973, I became convinced (as did Dave Gabai) that no traditional tricks, in fact no finite process, could ever overcome (i) and (ii); we'd only ever succeed in shifting the difficulties to another place. As I had always been obsessed by Barry Mazur's approach to the Schoenflies problem, I decided to try and iterate, indefinitely, the long procedure that proves the existence of the honeycomb representation, in an attempt to chase away difficulties ad infinitum. But it wasn't until around 1980 that this infinitely complicated process began to converge. It was at this time that I dropped all other research activities, so that I could devote myself to the Poincaré hypothesis, and to nothing else (1980–1989).

But before I can continue the story, I must digress for a moment. Still motivated by Poincaré, I had formulated a certain conjecture about knots, around 1974; a written trace can be found in [32] where, incidentally, my own version of Casson's handles can also be found, see also [23]. This conjecture soon became known as "Poénaru's Conjecture" and generated a great deal of interest and work. It is too technical for me to set out here, and low-dimensional just mention one corollary, which was also my motivation.

Let M^4 be a compact 4-dimensional smooth manifold such that $H_*(M^4) = H_*(pt)$ and admitting a handle decomposition

$$M^4 = B^4 + h^2 + H^3,$$

where h^2 is an index 2 *handle and H^3 a handle of index* 3. *Then $M^4 \overset{\text{DIFF}}{=\!=} B^4$.*

The "Poénaru Conjecture" resisted all the efforts of specialists until Dave Gabai succeeded in proving it, with a completely new technology ("Foliations and the Topology of 3-manifolds" III, Jour. of Diff. Geom 26 (1987), p. 479–536). This was the beginning of my mathematical relationship with Dave, which is so important to me.

In 1989, I announced the proof of Poincaré's Conjecture [52]; this turned out to be a little premature as there was still a bit missing. Dave is extremely interested in my work, and has already gone through some very substantial parts of it. I should add, too, that over the years we've spent hundreds of intense hours working together on Poincaré (and later on the $\pi_1^\infty(M^3)$). He told me that over the last ten years, he has devoted around a quarter of his total time available for mathematics to these programs.

On the other hand, still around 1989–1990, the first partial results concerning the $\pi_1^\infty(\widetilde{M^3})$ are starting to come out too, [54–56]. As they are much easier than the Poincaré work, by some factor I dare not name, they attracted much more attention, and many people became interested in them. Also, for [56], I was once again in a friendly competition with Andrew Casson.

In the spring of 1995, Gabai and M. Freedman found a gap in the paper that at the time was called the "Po V", part of my approach to Poincaré. (This is, essentially, the "gluing" between Theorem 3 and what, at the time, was a weaker form of Theorem 2.) To fill this gap, I developed a two-stage strategy: the following theorem ("Po V-A" [64]), to begin with:

Let Δ^4 be a compact smooth manifold of dimension four such that $\partial\Delta^4$ is a homology sphere and such that for any compact $K \subset \operatorname{int}(\Delta^4)$ we can find a geometrically simply connected compact smooth submanifold M^4 such that $K \subset M^4 \subset \Delta^4$. Then the open manifold

$$\operatorname{int}(\delta^4 \sharp (S^2 \times B^2) \sharp (S^2 \times B^2) \sharp \cdots)$$

obtained by making connected sums along the boundary of infinitely many copies of $S^2 \times B^2$, is geometrically simply connected at infinity.

This corresponds to that part of the proof of Theorem 2 which was completed in November 1998, and which is discussed in Part 2. The second (and final) step in filling the hole is the "Po V-B", i.e. the proof of Conjecture 4. This work, joint with Dave, has yet to be completed.

More recently (September 1999) I managed to finish the proof of the following theorem:[1]

Theorem 25.1 *Let M^3 be a 3-dimensional closed manifold. Then the universal covering $\widetilde{M^3}$ is simply connected at infinity, $\pi_1^\infty(\widetilde{M^3}) = 0$).*

The proof of this theorem, announced in [67] and [65], is contained in [68] and [70]. The history of the problem is long, and low-dimensional just point out here that the analog of Theorem 1 in all dimensions $n \geq 4$ is known to be false (M. Davis 1982), even when $\pi_1(M^n)$ is a "generic" ("tame") group, as are, for example, hyperbolic groups in Gromov's sense. Theorem 1 is equally relevant to the theory of low-dimensional varieties and the theory of discrete groups. My proof was the culmination of 20 years' work, punctuated by preliminary or partial results [54–56, 58–60, 69]. Article [65] gives a panoramic view of the entire program. The demonstration techniques for Theorem 1 are inspired by B below, but unlike B for

[1] Footnote by Poénaru in 2024: I remind you that this notice dates from the year 2000. The part of the text between Theorem 1 and Theorem 3 inclusive is overtaken and corrected by what is explained in my text *Mathematical Adventures*, published in the present volume. In it, I explain that the proof of Theorem 1 had a flaw (which Dave and I found), and that this theorem has been completely overtaken by my QSF Theorem. Indeed, after a dozen years' work, from the debris of the proof of Theorem 1 emerged the proof of my QSF theorem, of which Theorem 25.1 is only a small special case. All this is recounted in the text *Mathematical Adventures*.

which dimension four is crucial, here we're working in very high dimensions, to arrive, only at the end, at a 3-dimensional conclusion.

For the rest, Δ^3 will be a 3-dimensional homotopy ball, # will mean connected sum along the edges and "geometrically simply connected" will mean the possibility of finding a handle decomposition (in the DIFF category), without index one handles. That said, my main results to date concerning the Poincaré hypothesis can be summed up in the following two theorems.

Theorem 25.2 *For all Δ^3, the smooth, open 4-dimensional manifold*

$$X^4 = \text{int}((\Delta^3 \times I)\#(S^2 \times B^2)\#(S^2 \times B^2)\# \cdots) \tag{25.1}$$

is geometrically simply connected.

The proof is in the articles [53, 57, 61, 62, 64] . After a number of intermediate versions, very carefully scrutinized and criticized by Dave Gabai, the final [64] step wasn't finished until November 1998.

Theorem 25.3 (The Stable Poincaré Conjecture) *Let Δ^3 be such that $\Delta^3 \times I$ is geometrically simply connected. Then so is Δ^3, and therefore $\Delta^3 = B^3$.*

Overviews of this work, as it appeared at various times, can be found in [52, 63, 66] and, also, in an article by Dave Gabai: "Valentin Poénaru's program for the Poincaré Conjecture", in the volume "Geometry, Topology and Physics for Raoul Bott" S.T. Yau (ed.), Int. Press (1995).

The whole program represents around 45 years of work, but things didn't really get off the ground until the 70s, as mentioned above. This is closely related to Thurston's geometrization (or uniformization) conjecture. My techniques are mainly based on dimension four (to be discussed later) and almost never on dimension three itself.

Once we have Theorems 25.2 and 25.3 above, what is still missing to complete the proof of the Poincaré hypothesis is the demonstration of the following Conjecture:

Conjecture 25.4 With X^4 defined as in (1) above, we have the following implication:

(2) If X^4 is geometrically simply connected, then $\Delta^3 \times I$ is, also, geometrically simply connected.

From 1995 onwards, I worked, more or less full-time, in collaboration with Dave Gabai, on the proof of this conjecture (let us say, it was my main research project). With Dave, we did a long job on it and, provided there is no hidden flaw in it, we are close to the goal.

We'll call a compact, differential submanifold of codimension zero $\Delta^n_{Sch} \subset S^n$, which is such that $\partial\Delta^n_{Sch} = S^{n-1}$, a Schoenflies ball. In a seminal work, Barry Mazur demonstrated in 1959 that such a Δ^n_{Sch} is always topologically standard $\Delta^n_{Sch} \overset{\text{TOP}}{=\!=} S^n$, and shortly afterwards Steve Smale proved that this result is also true in the DIFF context, provided that $n \neq 4$. (Actually, for $n = 5$, Kervaire–Milnor's work also comes into play.) The case, still open, $n = 4$ is one of the major problems of differential topology in dimension four.

From the beginning, Dave Gabai and I knew that our efforts concerning Conjecture 4 above had an impact on 4-dimensional Schoenflies. But it was only towards the beginning of the 2000s that our ideas on this crystallized sufficiently so as to be able to formulate a concrete and detailed program (with promising partial results) concerning the following conjecture:

Conjecture 25.5 Any Δ^4_{Sch} that is geometrically simply connected is differentially standard, $\Delta^4_{Sch} \overset{\text{DIFF}}{=\!=} S^4$.

This led to a new research project, again in collaboration with Dave Gabai. Above, I was a little too quick to present Barry's contribution, which shows, in fact, that

(3) $\Delta^4_{Sch} \setminus \{\text{one boundary point}\} \overset{\text{DIFF}}{=\!=} B^4 \setminus \{\text{one boundary point}\}$.

To be more precise, Barry proves this for n arbitrary, and by Tychonov compactification, this gives the TOP equality written above; (3) is the starting point of my approach with Dave.

There were a few other mathematicians who invested time and energy in studying my work. There is, for example, François Laudenbach (who spent six months full-time reading and verifying the honeycomb representation theorem [57] and gave talks on it at one of the IHÉS Fridays), Louis Funar (who generalized [55] a lot, who provided me with one of the lemmas for Theorem 1 and who, together with one of Dave's former PhD students, Siddhartha Gadgil (Stony Brook), worked on variations on the theme of our "Po V-B"). There were also Frank Quinn (Virginia Tech) et Serguei Matveev (Cheliabinsk). But Dave Gabai's role over the years has been crucial for me. He knows my work in depth better than anyone else, he has lectured on it in various places and, above all, he is able to handle all the intricacies of the machine and make it work, all by himself.

I would like to end this reprot by adding the following points:

I had a number of PhD students, including André Gramain, Jean Barge, Jean Lannes, François Latour, Pierre Vogel and Louis Funar. Over the years, other future mathematicians have followed my "cours de troisième cycle" cycle teaching at Harvard and Orsay, including John Mather, Barry Simon, John Hamal Hubbard, Pierre de la Harpe, Lucien Guillou, Alexis Marin, Michel Boileau, Francis Bonahon and others.

Since I have been at Orsay, I have given undergraduate and postgraduate courses, which I think have been good, and every year I have also given a postgraduate course, trying never to talk about the same subject twice; reference [51] reflects one of these courses. Of course, I have practiced such teaching at foreign universities, too.

I didn't hesitate to take on a number of administrative tasks: I was Director of the Mathematics Laboratory at Orsay in the early 70s and, later, a member of the Scientific Council of the same university. On several occasions, I have been a member of the CNU and the CNRS National Committee (each time, appointed). I was a member of the Collège de France Commission for the Cours Peccot (Lichnerowicz Commission) and a member of the Commission des Thèses des Universités parisiennes (Lelong Commission). Since 1998, I have been a member of the Board of Directors of the association "Les Amis de l'IHÉS", and I am still secretary of the executive committee, which grew out of this Board.

I have been a guest speaker at a large number of scientific events, which I won't attempt to list, but I mention a conference on singularities, at the same time as Hironaka, and to physics conferences, at the same time as Pierre Gilles de Gennes, Kenneth Wilson, Philip Anderson or Gerard 't Hooft, and three conferences devoted entirely to my work on Poincaré: one in Pasadena 1993 (USA) and two in Trento 1995, 1996 (Italy).

References

1. V. Poénaru, C. Foias, G. Gussi, On the Cauchy problem for elliptic partial differential equations (en roumain). Bull. Sti. **VII**, 97–103 (1955) (Bucarest)
2. V. Poénaru, C. Foias, G. Gussi, On the boundary value problem for linear differential operators of order two (en roumain). Bull. Sti. **II**, 699–721 (1955) (Bucarest)
3. V. Poénaru, C. Foias, G. Gussi, Une méthode directe dans l'étude du problème de Cauchy pour les équations aux dérivées partielles, hyperboliques, du second ordre, en deux variables. Rev. Math. pures appl. (Bucarest) **1**, 61–98 (1956)
4. V. Poénaru, C. Foias, G. Gussi, Une méthode directe dans l'étude des équations aux dérivées partielles, hyperboliques, quasi-linéaires, en deux variables. Math. Nachr. (Berlin) **15**, 89–116 (1956)
5. V. Poénaru, C. Foias, G. Gussi, A direct method for the Cauchy problem for quasi-linear hyperbolic equations with two variables (en russe). Dokl. Akad. Nauk USSR (Moscou) **12**, 381–382 (1957)
6. V. Poénaru, C. Foias, G. Gussi, On the basic approximation theorem for semi-groups of linear operators. Proc. Natl. Acad. Sci. (U.S.A.) **53**, 616–618 (1957)
7. V. Poénaru, C. Foias, G. Gussi, L'étude de l'équation $\frac{du}{dt} = A(t)u$ pour certaines classes d'opérateurs non bornés de l'espace de Hilbert. Trans. Am. Math. Soc. (U.S.A.) **86**, 335–347 (1957)
8. V. Poénaru, C. Foias, G. Gussi, Verallgemeinerte Lösungen für p-parabolische quasilineare Systeme. Math. Nachr. (Berlin) **71**, 1–8 (1958)
9. V. Poénaru, C. Foias, G. Gussi, Generalized solutions for quasilinear differential equations in Banach space (en russe). Dokl. Akad. Nauk USSR (Moscou) **19**, 884–887 (1958)
10. V. Poénaru, C. Foias, G. Gussi, Sur les solutions généralisées de certaines équations linéaires et quasi-linéaires dans l'espace de Banach. Rev. Math. pures appl. (Bucarest) **3**, 283–304 (1958)
11. V. Poénaru, Les décompositions de l'hypercube en produit topologique. Bull. Soc. Math. France **8**, 113–129 (1960)
12. V. Poénaru, Sur quelques propriétés des variétés simplement connexes à trois dimensions. Rend. Mat. (Rome) **20**, 225–269 (1961)
13. V. Poénaru, Sur la théorie des immersions. Topology (Oxford) **1**, 81–100 (1962)
14. V. Poénaru, Produits cartésiens de variétés différentiables sur un disque. Compte-rendu du Congrès international des Mathématiciens, Stockholm (1962), pp. 481–489
15. V. Poénaru, Sur les variétés tridimensionnelles ayant el type d'homotopie de S3 (Thèse). Topologie et géométrie différentielle **6**, 1–67 (1964)

16. V. Poénaru, A. Haefliger, La classification des immersions combinatoires, Publications Mathématiques de l'I.H.E.S., vol. 23 (1964), pp. 75–91
17. V. Poénaru, On regular homotopy and isotopy. Preprint Institute for Advanced Study, Princeton (U.S.A.) (1965), pp. 1–165
18. V. Poénaru, On regular homotopy in codimension 1. Ann. Math. **83**, 257–265 (1966)
19. V. Poénaru, Un théorème des fonctions implicites pour les espaces d'applications Co, Publications mathématiques de l'I.H.E.S. **38**, 93–124 (1970)
20. V. Poénaru, Homotopy theory and differentiable singularities, dans le volume "Manifolds", Amsterdam (1970), pp. 106–132
21. V. Poénaru, Sur al structure des sphères d'homotopie lisses en dimension 3, lère partie. Preprint Pub- lication mathématique d'Orsay (1971), pp. 1–425 (chap. I–V)
22. V. Poénaru, *Analyse différentielle*. Lectures Notes n°, vol. 371 (Springer, Berlin, 1974), 28 pp.
23. V. Poénaru, A remark on simply collected 3-manifolds. Bull. Am. Math. Soc. **80**, 1203–1204 (1974)
24. V. Poénaru, Lectures on the singularities of Co mappings. Cours IME 1974, dans el volume "Singularities of analytic spaces" CIME (1975), pp. 83–145
25. V. Poénaru, Stability of equivariant smooth mappings. B.A.M.S. **81**, 1125–1126 (1975)
26. V. Poénaru, Some invariants of generic immersions and geometric applications. B.A.M.S. **81**, 1079–1082 (1975)
27. V. Poénaru, Versal unfoldings of G-invariant functions. B.A.M.S. **82**, 86–88 (1976)
28. V. Poénaru, *Singularités* C^0 *en présence de symétrie*. Lectures notes n°, vol. 510 (Springer, Berlin, 1976), 174 pp.
29. V. Poénaru, G. Toulouse, The crossing of defects ni ordered media and the topology of -3 manifolds. J. Phys. **7**, 887–895 (1977)
30. V. Poénaru, G. Toulouse, Topological solitons and graded Lie algebras. J. Math. Phys. **20**, 13–19 (1979)
31. V. Poénaru, Y. Bouligand, B. Derrida, Y. Pomeau, G. Toulouse, Distorsions with double topological character: the case of cholesterics. J. Phys. **35**, 863–867 (1978)
32. V. Poénaru, Some remarks on low-dimensional topology and immersion theory, dans le volume "Differential topology" CIME 1976 (1979), pp. 131–138
33. V. Poénaru, Elementary algebraic topology related to the theory of defects and textures, dans el volume "Matière mal-condensée", Les Houches 1978 (1979), pp. 265–318
34. V. Poénaru, Homotopie régulière et isotopie. Rend. Mat. Torino (1979), pp. 15–56
35. V. Poénaru, A. Fathi, F. Laudenbach, et al., Travaux de Thurston sur les surfaces. Astérisque **66–67** (1979), 284 pp.
36. V. Poénaru, Superalgebras and confinement ni condensed matter physics, dans el volume "Recent developments ni Gauge theories" (Cargèse 1979), édité par t'Hooft et al., Plenum Publications (1980), pp. 303–312
37. V. Poénaru, G.F. Chew, Topological bootstrap prediction of three colored, eight-flavoured quarks. Phys. Rev. Lett. **54**(4), 229–231 (1980)
38. V. Poénaru, G.F. Chew, J. Finkelstein, R. McMurry, A topological theory of electric charge. Phys. Lett. **100B**(1), 53–56 (1981)
39. V. Poénaru, G.F. Chew, Topological bootstrap theory of hadrons. Zeit. Phys. C. Part. Fields **11**, 59–93 (1981)
40. V. Poénaru, Some aspects of the theory of defects of ordered media and gauge fields related to foliations. Commun. Math. Phys. **80**, 127–136 (1981)
41. G.F. Chew, J. Finkelstein, R. McMurry, V. Poénaru, Topological theory of electromagnetism. Phys. Rev. D **24**(8), 2287–2304 (1981)
42. V. Poénaru, G.F. Chew, J. Finkelstein, Topological theory of electroweak vector bosons. Phys. Rev. D. **24**(10), 2764–2765 (1981)
43. V. Poénaru, G.F. Chew, Topological representation of leptons. Zeit. Phys. C. Part. Fields **14**, 233–239 (1982)
44. V. Poénaru, G.F. Chew, J. Finkelstein, B. Nicolescu, Topological compositeness of quarks, leptons and electroweak bosons. Zeit. Phys. C. Part. Fields **41**, 289–291 (1982)

45. V. Poénaru, G.F. Chew, J. Finkelstein, A single sheet classical surface for electroweak hadronic currents. Zeit. Phys. C. Part. Fields. **14**, 363–367 (1982)
46. V. Poénaru, Topological bootstrap theory of strong interactions (Exposé général pour la VIe session d'Études biennales de Physique nucléaire, Aussois 1981). Rendiconti Torino **39**, 1–14 (1981)
47. V. Poénaru, G.F. Chew, Parity asymetry from diquarks. Phys. Rev. D. **30**(7), 1579–1584 (1984)
48. V. Poénaru, G.F. Chew, D. Issler, B. Nicolescu, *GeV* Partons and *TeV* Hexons from a topological viewpoint, dans le volume "XIXe Rencontre de Moriond", Ed. Frontières (1984), pp. 143–165
49. V. Poénaru, G.F. Chew, Single-surface basis for topological particle theory. Phys. Rev. D. **32**(10), 2683–2686 (1985)
50. V. Poénaru, B. Nicolescu, From Baryonium ot Hexons, dans el volume "A Passion for Physics," Chew jubilee, Berkeley 1984 (World Scientific, Singapore, 1985), pp. 195–221
51. V. Poénaru, C. Tanasi, Introduzione alla Geometria e alla Topologia dei Campi di Yang-Mills. Suppl. ai Rendiconti del Circolo Matematico di Palermo, s. **II**(13), 1–55 (1986)
52. V. Poénaru, Infinite Processes and the 3-dimensional Poincaré Conjecture: an outline of the outline of the proof, Research announcement. Prépublications d'Orsay 89–06 (1989)
53. V. Poénaru, Infinite Processes and the 3-dimensional Poincaré Conjecture, I : the collapsible pseudo-spine representation theorem. Topology **31**(3), 625–656 (1992)
54. V. Poénaru, On the equivalence relation forced by the singularities of a non-degenerate simplicial map. Duke Math. J. **63**(2), 421–429 (1991)
55. V. Poénaru, Killing handles of index one stably and π_1^∞. Duke Math. J. **63**(2), 431–447 (1991)
56. V. Poénaru, Almost convex groups, Lipschitz combing and n i for universal covering spaces of closed 3-manifolds. J. Diff. Geom. **35**, 103–130 (1992)
57. V. Poénaru, Infinite processes and the 3-dimensional Poincaré Conjecture, II: the Honeycomb representation theorem. Prépublications d'Orsay 93–14 (1993)
58. V. Poénaru, Geometry "à la Gromov" for the fundamental group of a closed 3-manifold M3 and the simple connectivity at infinity of M3. Topology **3**(1), 181–196 (1994)
59. V. Poénaru, C. Tanasi, Hausdorff combing of groups and n i for universal covering spaces of closed 3-manifolds. Ann. Sc. Norm. Pisa **20**(3), 387–414 (1993)
60. V. Poénaru, C. Tanasi, k-Weakly almost convex groups and simple connectivity at infinity for universal covering spaces of closed 3-manifolds. Geom. Dedic. **48**, 57–81 (1993)
61. V. Poénaru, Infinite processes and the 3-dimensional Poincaré Conjecture, III : the Algorithm. Prépublications d'Orsay 92–10 (1992)
62. V. Poénaru, Processus infinis et conjecture de Poincaré en dimension trois IV: Le théorème de non sauvagerie lisse (The smooth tameness theorem), Part A. Orsay Preprint, 92–83 (1992); Part B, Orsay Preprint, 95–33 (1995)
63. V. Poénaru, Three lectures on higher-dimensional topological methods in three-dimensional topology. Suppl. Rend. Palermo, 203–217 (1997)
64. V. Poénaru, Geometric simply connectivity in smooth four-dimensional. Topology Prépublications d'Orsay 98–63 (1998). A longer detailed version of this paper is in the process of being typed (I am very grateful to the Harvard Math. Department for funding this operation.)
65. V. Poénaru, π_1^∞ and infinite simple homotopy type ni dimension three (H. Nencka, ed.) Madeira 98, Contemporary Math. 223 AMS (1999), pp. 1–28
66. V. Poénaru, A Program for the Poincaré Conjecture, and some of its ramifications, in the volume "Low-dimensional topology" (World Scientific) A. Banyaga et al. edit. (1999)
67. V. Poénaru, $\pi_1^\infty \tilde{M}^3 = 0$ (A short outline of the proof). Prépublications d'Orsay 99–73 (1999)
68. V. Poénaru, Universal covering spaces of closed 3-manifolds are simply connected at infinity (with an appendix by L. Funar). Prépublications d'Orsay 2000-20 (2000)
69. V. Poénaru, C. Tanasi, Representations of the Whitehead manifold Wh^3 and Julia sets. Ann. Toulouse **4**(3), 655–694 (1995)
70. V. Poénaru, C. Tanasi, Equivariant, Almost-Arborescent Representations of Open simply-connected 3-manifolds; A Finiteness result. Prépublications d'Orsay 99–24 (1999)

Chapter 26
Mathematical Adventures (A Letter to Barry Mazur)

Valentin Poénaru

Abstract This chapter outlines the evolution and the key ideas behind Poénaru's approach to the 3-dimensional Poincaré Conjecture and the Schoenflies Conjecture in dimension 4.

Keywords Geometric simple connectivity · QSF · Simplicial map · Representation · Whitehead manifold · Homotopy 4-disk

AMS Codes 01A60, 01A70, 5503

September 2016

Dear Barry,

In recent years, a number of distinct mathematical trends seem to have come to full fruition, while some other trends are, for me at the time being, still a big open battle field. So they are the subject of a very intense research activity, right now.

And I felt an irresistible urge to write about all these things. Also it did me a lot of good to write it down right now; I got to feel that writing can act like a sort of liberation.

The result is the present text, in the form of a long letter addressed to you. It is certainly in no way a mathematical paper, although it has certainly a lot to do with mathematics, with my own mathematics, essentially. This story will start about 1993, at a time when I thought that I had already finished my proof of the Poincaré Conjecture. As I have just told you, this text is supposed to be largely nontechnical, and I will rather stress the human and emotional aspects. Nevertheless a minimum of technicalities will still be unavoidable, otherwise it would become gibberish. So let me start by describing to you how my Poincaré Conjecture program looked like, as of 1993.

V. Poénaru (✉)
IHES, Bures-sur-Yvette, France
e-mail: valpoe@hotmail.com

L. Funar, A. Papadopoulos (eds.), *Essays on Topology*,
https://doi.org/10.1007/978-3-031-81414-3_26

There were, to begin with, four papers, I will call them Po I to Po IV, (and here you are free to decide whether the "Po" actually stands for Po, i.e. for me, or for Poincaré), the principal result of which was the following theorem:

(1) *Let* Δ^3 *be the Poincaré* 3-*ball. Then, the smooth open four-manifold* $\mathrm{int}(\Delta^3 \times I)$ *is "geometrically simply connected at long distance".*

Although this long distance terminology was suggested by you, in the early days, during some long joggings around the Yvette basin, I will remind you now what it all means. To begin with, a smooth manifold is said to be geometrically simply-connected (GSC) if the 1-handles cancel with the 2-handles. A smooth open manifold X is said to be geometrically simply connected at long distance if for any compact subset K of X, there is a compact smooth codimension-zero GSC submanifold of X, sandwiched between K and X itself.

These papers Po I to Po IV had been, by then, very carefully checked by David Gabai, whom I will call Dave from now on. A bit later, I will tell you about the paper Po VI, which proves that if the smooth compact bounded 4-manifold $\Delta^3 \times I$ is GSC, then Δ^3 itself is the 3-ball (which, of course is, exactly, the three-dimensional Poincaré Conjecture). This is now just a silly little comment, but in my approach it is more convenient to deal with the compact bounded Δ^3, the homotopy 3-ball, rather than with the closed Σ^3, the homotopy 3-sphere. This, of course, makes no difference for the final outcome.

But the present story really starts with the paper Po V, which was sitting in the middle of the whole program. What Po V was supposed to prove was the following implication, using the conclusion (1) of Po IV, as a premice

(2) *If* $\mathrm{int}(\Delta^3 \times I)$ *is GSC at long distance, then* $\Delta^3 \times I$ *itself is GSC.*

So how was I supposed to go about, proving that? Here I have to go back to the papers Po I and Po II which, among other things, showed that Δ^3 can be "REPRESENTED". This is an old concoction of mine which will occur big in this present text, and which I will have to describe for you, at least approximately. Just a word of caution, right now: my REPRESENTATIONS should not be mixed up with the well-known group representations, although as you will see later when we will move to group theory, the two notions are in a certain sense dual to each other.

So let us start with some object K, which we assume to be a simplicial complex of dimension three, that is really the meaningful case for the whole theory, essentially because there is a rich structure for the double points set of the nondegenerate maps of 2d complexes into 3d complexes, which one wants to make use of. A REPRESENTATION of K is a sort of GSC resolution of it, more precisely it is

(3) *A nondegenerate simplicial map* f *of some REPRESENTATION space* X *into* K, *where* X *is a GSC complex of dimension two or three, not necessarily locally finite, where* f *is "zippable" and where* f *is also "essentially surjective", a notion which needs to be defined, of course, since* X *might be of dimension two. But I will skip that definition now.*

Zippable means that the double points of f are commandeered by the singularities, the points where f fails to be an immersion. More specifically, a nondegenerate map is zippable if when one kills all its singularities by zipping (i.e. by folding maps), then it becomes an embedding, i.e. the double points get killed too. Clearly, there is here a notion of strategy floating around too, and this plays a big role.

Turning finally to the GSC requirement, although GSC stems from Morse theory, it makes good sense for general cell-complexes too. Later on, when we will move to group theory, we will meet it again in the context of complexes which are not even necessarily locally finite.

So, there is a REPRESENTATION of Δ^3, meaning, in the present context, the following thing, on the lines of (3) above

(4) *A nondegenerate map f from a finite 2-dimensional complex X into Δ^3.*

And at this point I will say a few more words about those papers Po I to Po IV. The paper Po I shows that Δ^3 can be REPRESENTED, like in (4), while Po II and Po III develop a sort of calculus with REPRESENTATIONS, changing one REPRESENTATION into another one; Dave and I called it the honeycomb calculus. One essentially looks at a given REPRESENTATION with some magnifying glasses, seeing then finer and finer details; it is a sort of inverse block-transformation. This has to be iterated indefinitely in the next papers, in particular in Po IV, so as to achieve (1) above, which corresponds to some sort of invariant limit situation of the infinite iterative process. It was useful for me, at that point, to think of the renormalization group. Po II is really very long and messy, and since Laudenbach at some point had strong doubts about it, we spent a whole semester going carefully through it and reorganizing it more neatly. He got then so familiar with Po II, that he even lectured on it, at IHES, afterwards.

Once we are given a REPRESENTATION like in (3) or (4) above, there is the idea of thickening it into some higher dimension. For the purposes of the three-dimensional Poincaré Conjecture, the appropriate dimension for such a thickening turns out to be exactly four. As you shall see, my whole Poincaré program hardly brushes dimension three; all the big action goes on in four dimensions. But then, in the context of geometric group theory, to which we will come later, that appropriate dimension for the thickening is a very high one. But then also, the present high dimensions have little in common with those high dimensions from the work of Smale or Stallings, which everybody knows about. We do not ask now quite the same things from them.

Since a very long time ago, I knew that in order to show that the four-manifold $\Delta^3 \times I$ was GSC, it sufficed to find a REPRESENTATION (4) which has a certain combinatorial property, which I will call COHERENCE. I will only try to give you here the general flavor of what this means. The map f in (4) has double points, which are to be thought of, both as a subset M_2 of X and also as a subset M^2 of $X \times X$, with an obvious map of the following form, in between:

(5) *A map π from M^2 to M_2.*

Generically, this is something like a $\mathbb{Z}/2\mathbb{Z}$ principal fibration. COHERENCE certainly implies that this fibration is trivial, i.e. not a covering space, and it implies much more too. Since COHERENCE involves also a strategy, it is in a certain sense a non-commutative version of the triviality of this fibration (5). And that mere triviality does not imply COHERENCE. There is, also, another funny little thing here. The triviality of the principal fibration (5) clearly gets inherited by any reasonable piece of the map π. But not so for COHERENCE, which is, really, a non-localizable global property.

Once all these things have been said, I can go back now to Po V which, making now a bit more precise what I have said before, was supposed to deduce from Po IV that Δ^3 admits COHERENT REPRESENTATIONS, which in turn would have then implied that $\Delta^3 \times I$ is GSC. This, together with Po VI, would have then also clinched my proof of the Poincaré Conjecture. So far so good.

In 1993, Alberto Tognoli, whom I am sure you have met, as I am also quite sure that you know his work on the Nash structures and in real-algebraic geometry, started to organize a nice meeting at Levico, in the middle of the Italian Alps, all devoted to my Poincaré program, for the Summer of 1994. Dave thought that Po V was a good topic for that meeting. He was in Paris during the Winter-Spring of 1993–1994, and we started checking together, very carefully, Po V. Mike Freedman was around too, and he joined us occasionally. All this was happening at IHES.

But before I go on with my story, I have to open a lengthy parenthesis. Everybody knows that I have spent a lot of time and energy, actually a sizable portion of my life, on the three-dimensional Poincaré Conjecture. Some people know that I have also been involved in other types of research too. For instance, I have certainly been involved rather big in the prehistory of Gromov's h-principle, in singularity theory and, of course, in Physics. Concerning my career as a physicist, if I may call it that, I did together with some physicists friends, some respectable (and respected) work on the defects of ordered media in condensed matter physics. We connected, for instance, the dynamics of the defects in question to supersymmetry, the kind of supersymmetry or rather super-algebras coming from the Whitehead product for the homotopy groups and the generic lack of commutativity for the fundamental group, and this uncovered a new, until then unknown phenomenon. We also connected that dynamics to the same piece of mathematics which is involved in the, for the time being rather mythical but still very intriguing, magnetic monopoles, the magnetic monopoles of t'Hooft and Polyakov. This again, explained a number of things which seemed paradoxical at the time.

But I also did much less orthodox, not that respectable nor respected work, together with Geoff Chew from Berkeley, a very fascinating personality indeed, to whom I became very close. This piece of work, on the so-called topological bootstrap of the elementary particles, was pooh-poohed by many, but I do not regret my involvement with this highly unorthodox piece of research. It certainly was great fun, and then I also learned a lot of Physics, so to say directly from the horse's mouth, in the process. And I do intend to come back to Physics one day, I even have some more concrete plans concerning that, but as you will soon see, my hands are too full with other things, right now and for the foreseeable future too.

Now, something which only very few people know is that, since the early seventies, I have also spent much time and energy on the four-dimensional DIFF Schoenflies problem. As part of your own big breakthrough in 1958 or so, you had shown among other things, that any smooth four-dimensional Schoenflies ball was homeomorphic to the standard ball, and that the homeomorphism in question is actually a diffeomorphism, except maybe at a boundary point. Since about the very early sixties it was also known, as a result of combining your work with Smale's h-cobordism theorem and with the Kervaire–Milnor theory, that in all dimensions other than four, things behaved OK, i.e. in dimensions other than four, the smooth Schoenflies balls are always standard. But I was obsessed with the issue of smoothing that four-dimensional boundary point, and I went big at it with all kind of techniques, like Whitney's theory of differentiable functions on arbitrary closed sets. But nothing seemed to work.

So, let's go back to Dave and myself checking Po V, with occasional visits from Mike Freedman. And one day he came with what turned out to be a very crucial piece of advice. Said Mike: You guys, while trying to show that $\Delta^3 \times I$ is GSC, why don't you go with your techniques on the smooth 4d Schoenflies ball too.

For me, this was as if some scales fell off my eyes. I had thought so hard, both about the Poincaré Conjecture and about the smooth four-dimensional Schoenflies ball, but I had never tried to join the two projects together. In particular, I had never before thought about the GSC issues involved in the Schoenflies problem, although I had been so much involved with them for the $\Delta^3 \times I$. From then on, I did start to think jointly about the two projects. This became like a leitmotive, as far as I was concerned.

At this point there occurred a very big mishap for me. We found a big irreparable gap in the paper Po V, and I will not try to describe to you now how I felt then. We still went to Levico, Frank Quinn and quite a number of other people joined us too, and Dave and myself lectured on the papers Po I to Po IV. People were happy enough, so that a second Levico meeting was organized, essentially with the same crowd, for the Summer 1995, on my Poincaré project again.

There is a very beautiful mountain lake in Levico and, during that Summer 1994, when my schedule allowed it, I took long walks around the lake, pondering on the dead Po V. And during these walks I devised a big two-proged project, of how to get back that lost COHERENCE, the still missing link for my proof of the Poincaré conjecture, which I so badly needed. As you shall see, it took me about twelve years to fulfill this project and to get, at the same time, the geometric simple connectivity of the four-dimensional smooth Schoenflies ball too.

I have hardly talked about Po VI, so far, and this is the good time to tell you very briefly what it contains. I will start with a definition. Let LK be a framed link, inside the boundary of the 4d ball. There is then a canonical smooth open four-manifold, let's call it $V(LK)$, floating around. This is the result of the following three successive operations: Use LK to add the obvious 2-handles to the 4-ball, then make a connected sum along the boundary with infinitely many copies of $S^2 \times D^2$ and then, finally, remove any boundary point. I will assume from now on that we deal here with the null-framing for our LK.

Assume now that you have the COHERENCE theorem (which was still missing at this time in my story), which states that for any Δ^3 there is a COHERENT REPRESENTATION. Incidentally, as you shall soon see, to get that COHERENCE, it would actually suffice to know that the 4-manifold $\Delta^3 \times I$ is GSC. Then, assuming the COHERENCE THEOREM (or, equivalently, that $\Delta^3 \times I$ is GSC), for a particularly doctored COHERENT representation, there is a zipping strategy, which produces an LK like above, with the following two features:

(6-A) *The 4-manifolds* $(\Delta^3 \times I) \# (S^2 \times D^2)$*'s and* $\{B^4$ *with some 2-handles added along LK*$\}$ *are diffeomorphic.*

This is an easy one, if one has the COHERENCE theorem; but what follows next is highly nontrivial.

(6-B) *That doctoring mentioned above is such that one can continue the construction of the LK with an infinite process which produces a new way of compactifying V(LK), different from the obvious one, and I will call this the strange compactification. And this strange compactification is such that from it one can read that the fundamental group at infinity of V(LK) is pro-free.*

From here on there is an easy chain of arguments, climbing down from 4d to 3d, and showing successively that

(i) π_1 of the complement of LK in $S^3 = \partial B^4$ is free;
(ii) the $(\Delta^3 \times I) \# (S^2 \times D^2)$'s occuring in (6-A) is diffeomorphic to $B^4 \# (S^2 \times D^2)$'s. And from this, by simply going to the boundary we get that
(iii) Δ^3 is the standard 3-ball, hence end of the proof of the Poincaré Conjecture, provided, of course, that one has the COHERENCE theorem, to begin with.

Po VI was the topic of Levico two. Dave has looked a bit at it and he liked it. But then we got involved in the new papers Po V-A and Po V-B, the replacements of the dead Po V and also, a bit later various events occurred too, I will tell you about them in due time, so we never had the opportunity to go back to that Po VI together again. But Dave often advised me to return to Po VI and try to get more out of it; he thought it had nice interesting potentialities. Frank Quinn also liked what he saw of Po VI.

I will go back in time now, to the middle seventies, when Milen and I were living in Orsay. In the evenings, I used to walk through the small streets, and think about mathematics. And in one of these walks, a sort of joke, or a mathematical pun, occurred to me. Well, COHERENCE implies GSC for $\Delta^3 \times I$, but then what about the converse statement? At that time, this seemed to me like a funny, non-serious, jesting question, but actually, it turns out that this converse is true too. If $\Delta^3 \times I$ is GSC, then Δ^3 admits a COHERENT REPRESENTATION. Coherence is a funny item, at face value it is 2d and combinatorial, but it also looms big in 4d, in the DIFF context. On the other hand, it is irrelevant in dimensions larger than four, where the difference between coherence and noncoherence vanishes into thin air.

Now, during my long walks around the Levico lake, I realized that I better turn things upside down and instead of proving first the COHERENCE, so as to get my GSC, I better prove the GSC business first, so as to achieve COHERENCE, afterwards, as a consequence. The point here is also that, without COHERENCE, the infinite process in Po VI cannot be triggered at all. The COHERENCE theorem really sits in the very center of my Poincaré program, and the GSC is just a means for achieving it; it is a mere tool.

There is another item which I want to talk to you about, now. In all those papers, from Po IV to Po VI, there occurs the following kind of object, crucial for the whole program. These objects are noncompact smooth four-manifolds X, with the following general kind of structure. The interior of X is the standard Euclidean four-space, while the boundary is a framed link, let's say right now with finitely many components. When I talk here about "standard" $\mathbb{R}^4$'s, I have to remind you that, according to Freedman and to Donaldson's Yang–Mills theory (or to Seiberg–Witten, if one wishes rather to invoke that, and then things are even easier), there are infinitely many smooth exotic $\mathbb{R}^4$'s, there are actually uncountably many of them. And I call "sort of links" the objects X introduced above.

Clearly, any classical link can be changed into a sort of link, and these particular sort of links will be called smoothly tame. They can be nicely compactified into a smooth standard 4-ball. There is another example of sort of link, or rather of sort of knots, namely the Casson Handles. You may remember that Mike Freedman's breakthrough for proving the 4-dimensional Poincaré Conjecture in the category TOP, was to show that all the Casson Handles, and a priori there are uncountably many of them (their moduli space, if we may call it that, is a Cantor set), are all topologically trivial, hence topologically tame too. But because exotic $\mathbb{R}^4$'s do exist, some Casson Handles (maybe all of them?) have to be smoothly wild; they do not admit the nice smooth compactification described above. To me, there is here, for the sort of links, a very fascinating distinction between smoothly tame and smoothly wild; this is completely foreign to the three-dimensional world of classical links and knots. I would like to dig more into this some day, and possibly connect it to quantum field theory. Remember that classical links are very much related to the quantum entanglement brought about by the quantum mechanics of several particle systems. So, what kind of quantum entanglement might be connected with the sort of links and with their strange wild versus tame phenomenology? An Italian physicist friend even suggested to me one day how this might be connected with the quantum mechanics of black holes. But this is just pure speculation, of course.

As I just told you, sort of links pop up all over the place, in the papers Po IV, Po V-A and B, and in Po VI too. The main issue, each time, is to show that those particular sort of links which do occur, are smoothly tame. And this is invariably proved using an infinite process. Now, since the three-dimensional sphere is compact, this may sound funny. But then, since wild Artin–Fox arcs do not exist in four dimensions, if one wants to show that a given sort of link is smoothly tame, it suffices to extend its boundary, from a framed link to a copy of the Euclidean 3-space. And this kind of thing, an infinite process can do. Let us be a bit more precise here. The sort of links which appear in my work are invariably themselves created by an infinite process, and the issue is whether this can be extended to an infinite process capable of producing the desired boundary enlargement. You see, also, that the big and still very mysterious issue of the gap between the categories DIFF versus TOP in dimension four, and exactly in dimension four (where contrary to what happens in dimensions higher than four, where the difference between DIFF and TOP is perfectly well understood, the discrete invariants of algebraic topology are, now in 4d, totally mute), looms big in my Poincaré program too, although it is, at face value, only 3d homotopy spheres which are involved at this point.

Of course, independently of all this, it would be nice to find quantum topological kind of invariants, capable of detecting smoothly wild sort of links (among topologically tame ones, let's say).

In my present story, I will still have to come back later to the two-pronged plan for replacing the dead Po V, which I devised in Levico. But some other things, which happened before that, will have to be told first. Besides Poincaré and Schoenflies, there was also a third problem which has kept me busy for a very long time, namely trying to prove that all universal covering spaces of closed 3-manifolds are simply-connected at infinity.

As explained above, the very first little step in my Poincaré approach was to REPRESENT Δ^3. But then, other 3-dimensional objects may be REPRESENTED too; and there is here a little remark. If K can be REPRESENTED, like in (3), then it has to be simply-connected to begin with. And now, in order to deal with that fundamental group at infinity, I started to REPRESENT the universal covering spaces of closed 3-manifolds. This turned out to be quite an efficient tool, and about 1990 I managed to prove a number of theorems with the following gist. If the fundamental group of some closed three-manifold satisfies any of a long list of nice geometric conditions, like Gromov's hyperbolicity, Cannon's almost convexity, being automatic, Thurston's combability, and so on, then the universal covering space is simply-connected at infinity, in other words, the fundamental group at infinity of the universal covering space in question is zero, which is the same thing as saying that the π_1^∞ of the fundamental group itself is zero. These kind of things, as such, are specific for dimension three; they, generally speaking, fail dismally in dimensions four and more. Also, Andrew Casson, completely independently, proved similar things, and this is neither the first nor the last time when we two came up with similar thoughts.

I should also tell you here a little fact. Assume that you give me an irreducible closed 3-manifold. Then, saying that its universal covering space is simply connected at infinity is equivalent to the statement that the covering space in question is the Euclidean 3-space.

Now, while thinking about those universal coverings, I very naturally thought about the simply-connected wild open 3-manifolds too, and so I started to REPRESENT Whitehead's classical manifold. I remind you that Whitehead's manifold is the classical example of an open contractible 3-manifold, coming with $\pi_1^\infty \neq 0$. Then I got very worried, because something seemed to go drastically wrong, and I was beginning to fear that my whole theory was in shambles. I realized then that I had stumbled over chaotic behaviour. And being not very knowledgeable about that, I rushed to IHES, hoping to find Dennis Sullivan, and get some advice from him. Dennis was out of town, and I bumped into Hamal Hubbard instead. I told him about my problems, and very nicely he explained to me that I had rediscovered the Julia sets. We spent the whole day together, and he taught me the basics of quadratic maps, of Julia sets and of the Mandelbrot set, everything which I needed to know about them, for my own work.

Incidentally, the whole issue of the dynamical systems connected with the Whitehead manifold, is still largely unchartered territory, to the best of my knowledge. You may remember Smale's solenoid, that typical hyperbolic dynamics, coming with its rational zeta function. There is another, analogous but trickier and more subtle, non-hyperbolic dynamical system, coming with the Whitehead manifold and which certainly deserves to be studied. What about its zeta function, for instance? But the kind of dynamics which I have just mentioned now is, a priori at least, something quite different from that Julia sets creating dynamical chaotic behaviour which comes with the previously mentioned REPRESENTATIONS of the Whitehead manifold and with their Whitehead nightmare, a notion to be described, later on, in this letter. I certainly do not claim any explicit relation between the two things.

Finally I can come back now to the two-pronged Levico project. The first of the two prongs became a paper called Po V-A, the main result of which was a purely four-dimensional theorem. Let M be a smooth compact 4-manifold with connected boundary. There is then, to begin with, a big stabilization of it, a canonically attached smooth open 4-manifold, which I will call $\mathrm{Stab}\,(M)$, gotten as follows

(7) $\mathrm{Stab}\,(M)$ *is gotten by the following two steps: add to* M *a connected sum along the boundary, of infinitely many* $S^2 \times D^2$*'s. Then delete all the boundary points.*

I can state now the theorem of Po V-A:

"If int(M) *is GSC at long distance, then* $\mathrm{Stab}\,(M)$ *is GSC, provided that the boundary of* M *satisfies some mild homological requirement."*

Together with Po IV, this implies that $\mathrm{Stab}\,(\Delta^3 \times I) = \mathrm{int}\,(\Delta^3 \times I \,\#\, \infty \,\#\, (S^2 \times D^2))$ is GSC. But proving this theorem was a serious big battle, and I still remember walking through the streets of Paris and thinking that if I managed to prove this one, then I did not need to feel ashamed in front of anybody, because I would have proved a theorem as difficult as any.

I discussed big Po V-A with Dave. He was coming regularly to Paris and I was flying regularly to Los Angeles, in order to go to CalTech, where he was then. After very lengthy discussions he got finally convinced that Po V-A was correct. Then, due to a very generous grant from Harvard, which you got for me, the paper got typed. But it was more like a book, rather than like a regular math paper, since it had over 350 pages. It saw the light of day as an IHES preprint, in 2001.

There was then the second prong of the Levico lake plans, the Po V-B, which is an even trickier business than Po V-A, but that was still in the limbo for quite a while. Its aim was to deduce, starting from the Po V-A which I have just described, i.e. from the fact that $\mathrm{Stab}\,(\Delta^3 \times I)$ is GSC, that $\Delta^3 \times I$ itself was GSC and, of course, the issue of Δ^4, the smooth Schoenflies ball, being GSC was now in the air too.

But, in the meanwhile, I had also started to work on another big project, namely to prove, in full generality, that the universal covering spaces of all closed 3-manifolds were simply connected at infinity. Remember that as I have already told you above, in the early nineties, I had proved this under the additional assumption that the fundamental group of the manifold itself satisfies any of a list of nice geometric conditions. But now I wanted to do the same thing without any additional assumptions.

When Dave and I worked together, we always liked to have, apart from the main topic of the day, some additional other mathematical topics, which we discussed while relaxing. And so, while working on Po V-A, we often talked about the not yet existing Po V-B, about those universal covering spaces, or about the Schoenflies ball. Was it GSC?, and provided it was GSC, was it standard?

Time was passing by, and the first of October 2001, I was due for retirement from Orsay, i.e. from the French university. But that Spring of 2001, I was still teaching. And since essentially since ever, my teaching was all concentrated in the Spring term. My teaching load had always been very light, but being a bit paranoic about this issue, I was always worried that the teaching load might increase in the future, which it actually never did. Anyway, that Spring, my total teaching load consisted of one very small math course for second year chemistry students, a grand total of about 24 hours in all, plus a graduate course. I had given graduate courses every year, on a large variety of topics which I will not bother to enumerate here. In the late 1990s I tended to give quite advanced courses on topics like enough of Freedman and of Donaldson's Yang–Mills (or of the Seiberg–Witten equations and their moduli spaces, depending on the years) to be able to show that exotic R^4's existed, or on measured foliations and Teichmüller theory including the Thurston compactification of Teichmüller space, or on pieces of my own work. But for this last graduate course, my colleagues had asked me to do something more elementary, like Smale's h-cobordism theorem, or immersion theory.

Of course, I did them both, except that now that immersion stuff appeared as a paragraph in Gromov's h-principle theory. Of course also, as an introduction to Gromov's trick to prove his theorem, I told the students about Thom's version of Smale's trick for immersions. If I would have to do all this again today, then I would rather use Eliashberg and Mishenko's holonomic approximation theorem, but that did not exist then. I remember also, that one day, in order to allow the students to relax, after we had been ploughing through differential relations in higher jet spaces, I told them: let's do something lighter and more fun today, I will give you a really fast proof of the Poincaré Conjecture in high dimensions, something quite different from the h-cobordism theorem, and we shall see how high dimensions we actually get. So, we did very fast some very elementary engulfing, introducing all the necessary background on the way. I had not prepared in advance, and we got something like the generalized Poincaré Conjecture in dimensions higher than nine. This last graduate course of mine, at Orsay, was a big success. The other years, I used to have three or four students and/or people who attended, now they were more than thirty. And they liked it, I even got some big applauses in the end. Most of my students were bright young people from École Normale, and several of them became, later on, completely accomplished mathematicians. But one student was different. He was not only a bit older than the others, he was an English businessman, with his own small computer software company in London, who studied mathematics just for fun. I met him several times, afterwards, in Paris, in London, or in New-York. A very nice person, indeed, and we stayed friends.

But then I also had to give that small undergraduate course. This took place in another part of the Orsay Campus and, whether it rained or snowed, I always preferred to walk there, rather than drive. But then I had to think about something during those walks, so as not to get bored. And I developed some sort of elaborate mathematical joke. Together with Dave, we were banging our heads against the walls, about that GSC issue, both for $\Delta^3 \times I$ and for the Schoenflies ball. My game now was to take the various concepts which Dave and I were working with, turn them upside down, change their names and functions, and sometimes change their dimensions too. And guess what popped up, like from a magician's hat: all of a sudden, I had a sketchy program, not for proving that Δ^4 was GSC, but for proving that if it was GSC then it was standard too. This was a funny idea, but I got hooked on it. And I soon realized that a certain 2d lemma was necessary for the whole scheme to work. This mathematical joke occurred sometimes towards the end of February 2001, and I spent a good part of that Spring trying to prove the lemma. During a visit to Trento, where Tognoli had invited me to give some talks, I thought that I might have an idea for the proof, but I was still very shaky about it. I might add here, as a parenthesis, that in Trento, the town with that counter-reformation council in the 1550s, there is also a very nice University, and that Levico, that place I mentioned before, is a beautiful vacation resort, next to Trento. All that part of the world is like a fairy-land, with beautiful mountains, lakes, castles and rich vineyard.

A bit later, that same year 2001, I flew to Caltech, as I was doing periodically, to work on Po V-B with Dave. And I already mentioned that our habit was that, after intense discussions on the main topic of the day, we would take a break, and change of math topic, for relaxation. And so, during one of those breaks, I told Dave about my little 2d lemma, but not yet about the rest of the joke. The next day, he came with a counter-example, and I remember that I went home quite depressed. But, during the night, I realized two things. Firstly, that what I really needed was not that 2d statement, but something else, a rather more elaborate 2d and 4d cobordism business. And then, Dave's counterexample offered a suggestion for how to go about for proving that.

So, the next day, I told Dave about the new prospective lemma. And Dave asked me: what the D. do you care about these kind of items, which seem so far removed from your usual quests? Then I told him about the whole jesting project. And I never saw him so excited. The ceilings, in the Caltech rooms are pretty low, and Dave made such a jump that he almost hit one, and he said something to the following effect: Is it possible that so close to the ideas I have been playing and toiling with so much lately, there was another set of ideas, both so close and so strangely different too, and that I was never aware of them! And then we sat down and put together what we have always called since then, The Pasadena Lemma. From that day on, Dave and I got embarked on a big joint project, let us call it the GSC Schoenflies project, in addition to our other various projects, namely to show that any GSC smooth Schoenflies ball is standard. And the Pasadena lemma was supposed to be a key ingredient for this.

This is a good time for another longish parenthesis. Quite many years before I had to go into retirement, I was already very upset about that. I felt that from retirement to the tomb there is only one little step; not that I am afraid about that tomb and, as you know, I do not believe neither in God, nor in life after death. But I hated the idea of retirement. And so I started to ask big for help, for some other positions, for afterwards, outside of France of course, where that was no longer possible; I was by then sixty-nine, and that was that. Dave, who was still at Caltech, went to the University president, with the OK of the math department, telling him that he wanted to have me hired as a professor. But the president vetoed that idea, because of my age. Then Dave had another brilliant idea, he got for me a very nice lush NSF grant, which allowed me, for several years, to come to Princeton University, where Dave was moving then, to spend something like three or four months a year there. This came with a very good salary, allowing me not only to come for those extended visits to Princeton, but also to pay for Hannibal's computer graphics education. That was quite a good investment, as Hannibal himself told me, at the time. Then, Alberto Tognoli also got for me a three year visiting professorship in Trento, and that additional salary allowed us to repair the roof of our house.

I am mentioning these financial details, because they are far from being irrelevant for us. You see, a normal French university professor starts his or her career as a civil servant, with the pension clock starting to click, at the age of about twenty, when he or she enters École Normale Spérieure or Ecole Polytechnique. I have landed here in France, coming from the outer space, entering the system only at the age of about thirty-five, and so my clock clicked only for a shorter time. And there is a certain numerical parameter entering into the computation of the pension, for almost everybody it is 80%, while in my case it is only 60%. So, the additional American and Italian incomes were more than welcome. Unfortunately, that could not last indefinitely.

So, for a number of years I was commuting between Princeton and Trento, with a stop in Paris in between, to see my doctors and to arrange for the American visa. Milen joined me quite a lot too and, incidentally, since she did not have a US salary, like I had, she did not need a US visa either. All this was quite fun. But there was also a very sad thing concerning the Trento episode. Alberto Tognoli, who in the beginning had big projects for research together with me, very soon got a big nervous breakdown, from which he never recovered. And the reason was not some metaphysical issue, but a very sad personal story. Because of various reasons which I will not go into here, his three children refused to see him any longer. As Alberto told me, they only communicated through their lawyers. So, he essentially copped out of everything, and I hardly saw him during my Trento period. I had to make there my way, all by myself. And by the time my appointment in Trento ended, Alberto had already retired, moved out of town, and soon afterwards he got a very nasty cancer and died. It was really a very sad affair, it certainly saddened me a lot.

Let's go back to maths now. In the very early 2000s I thought I had completed my project of showing that all universal covering spaces of closed three-dimensional manifolds were simply-connected at infinity. But then, after serious checking together with Dave, we found a very big gap. And, while in Trento, I devised a very nice way to repair it. The main step in my whole approach had been to somehow manage to replace some big infinite quotient-space projection, of a very untractable type, by a still infinite, but now much more manageable inclusion map. And the big difficulty was to actually construct that inclusion map in question. While in Trento, I got a new, now correct idea of how to go about that. Here is the general flavor of the whole business. Originally, in the flawed approach, one started by adding some 2d gutters to some singular 3d object, then everything was smoothened by a thickening in some high dimension. Now, in the new approach, that same singular 3d object gets smoothened into very high dimensions, in the very beginning. Then, making use of the additional dimensions (and that word "additional", meaning in addition to the original three, is crucial here), some indentations, let's call them ditches, get drilled. This drilling is actually an infinite process, the convergence of which has to be very carefully watched and controlled. It looked to me that this did the trick, and I flew then to Princeton, where I was planning to work out my new scheme, while continuing to go on with Po V-B, of course.

I forget now whether all this had happened end of 2002 or beginning 2003. The point is that, when I got to Princeton, I got the big shock of my life. Perelman had just announced his proof of the Poincaré Conjecture, actually of the full Geometrization Conjecture, of which the Poincaré Conjecture is only a piece.

For some reason, I had always felt very sure of myself when it came to prove the Poincaré Conjecture by geometric methods, and rightly or wrongly, I was never worried about competition from that direction. But I had always been worried that some nonlinear PDEs might do the thing. And how right I was to worry about that. You may ask why did I not go myself into nonlinear PDE, the answer of course is that human life and energy are limited, and so one has to make strategical choices, about how to use one's available time and energy.

It was my turn now to get a nervous breakdown, with a big fit of depression, which lasted for about three weeks. In the meanwhile, I had been invited to give a lecture at an AMS meeting at NYU. And I was in such low spirits that I had decided to decline that invitation. Fortunately, Dave shook me and he convinced me that I could not do such a thing. So I went to NYU and everything went well. But it is funny how I managed to pull myself out of that state of black depression. I heard that Casson had announced a sequence of four or five successive lectures somewhere in Texas, with a mysterious title, which made everybody think that he was announcing his own proof of the Poincaré Conjecture; it was well known that he was working on it, on the whole geometrization conjecture too, with big computer help. So, I said to myself that since there was now more than one player in the game, I could join in too. Then I decided to concentrate on finishing Po V-B, telling to myself that as long as I got to the finish, it did not matter that I was not the first one to get there. I also decided to throw then and there to the garbage my new universal covering spaces project, since that simple connectivity at infinity for the universal covering spaces of closed 3-manifolds was now an easy consequence of the geometrization of 3-dimensional manifolds which Perelman had proved.

Incidentally, that Casson rumor turned out to be a dud. Dave, who took the plane to go to listen to those lectures, came back completely disappointed. He thought that Casson had had some ideas, then realized in the last minute that there was a serious gap, and so he just improvised some lectures, which were certainly not very inspiring. But by then, by the time the dust had settled, I had also gotten out of the state of depression.

So, back I was to work. Frank Quinn was also visiting Princeton University at that time, and the three of us, Dave, Quinn and myself, we had intense days of work, on issues like Po V-B, or GSC Schoenflies and, at the same time, we started to learn things about the Ricci flow. The three of us worked very intensely, and I still remember how Dave, who is an immensely energetic person, got so tired one day that he fell asleep, in the middle of one of our talks. Frank and I let him rest, and we quietly moved to another office, where we went on with our business.

When I was much younger, I did some sports at high level (essentially mountain-climbing, as you well know), and I understand very well what it means to be in top physical shape. During this Princeton period, as a result of all that intense activity, I felt now in top mathematical shape. And this allowed me to do something which I had wished to do since long, but had never quite managed before. This was to understand how the Po V-B mechanism, or as much as I could understand of it then, the mechanism which from the fact that int $(\Delta^3 \times I \# \infty \# (S^2 \times D^2))$ is GSC (which is the result of Po V-A, remember, and see here (7) too) derived that $\Delta^3 \times I$ itself is GSC, how that same kind of mechanism could also be used to show that the Schoenflies ball is GSC. This is really very technical and not appropriate for the present letter. So I will only give you here a hint, in a Scientific American style. Just like int $(\Delta^3 \times I \# \infty \# (S^2 \times D^2))$ gets used for $\Delta^3 \times I$, so your old classical result that the smooth four-dimensional Schoenflies ball, with one boundary point removed, is standard, gets used for the Schoenflies ball itself. But the details of this are quite subtle and complicated.

By now, it was my turn to buck up Dave, who had also been completely shaken by the Perelman news. He told me that he felt like somebody living at Hiroshima, with an atomic bomb exploding on top of him. Everything he had ever done, he said to me, had now become superseded. Dave also said that at least me, Po, I may be able to save something from the wreckage, since most of what I had done was four-dimensional. But then, he also managed to pull himself together, essentially by doing some new mathematics, which was not covered by the geometrization theorem. And this really bucked him up. By then, a common friend, his name is not mentioned in this letter and it will not be now either, advised Dave to step back and leave that piece of mathematics he was so happy about, to a younger man, a PhD student who was also working on something like it. "You are already established said the friend, and that young man is not". I was scandalized, and I convinced Dave to stick to that piece of mathematics and live happily with it. These were times loaded with emotion, indeed.

My combined $\Delta^3 \times I$ and Schoenflies ideas were developing and getting more precise, but by then I had also lost my audience as far as my Poincaré Conjecture program was concerned. I will come back to this a bit later.

Perelman was supposed to appear soon in Princeton, and we all (and this includes by now quite a large crowd), needed to know more about the Ricci flow, in preparation for that visit. Our preparation started with some very good lectures by Richard Hamilton at Rutgers, then Simon Brendle, who looked like he was by far the youngest person in the Princeton Math department, gave us a very serious little course on the Ricci flow, and finally Perelman himself arrived.

His first lecture was supposed to be for everybody, and so it took place in the largest lecture room available at the Princeton University Math department, with a very large audience. Neither the name of Poincaré nor that of Thurston were ever mentioned in that first lecture, but I think it was great. The big hero was Boltzmann who, for more than one reason, is one of my great heroes too. Perelman mostly talked about statistical mechanics, about Boltzmann's entropy, about his own concept of entropy, inspired by string theory, and he explained what that entropy could do for the Ricci flow. This involved a very spectacular step; the heat equation had to be played in reverse, something which is reputedly impossible to do. But in his version of that equation, he had a magical non-linear term which just, just did the trick. It was essentially the only ingredient capable of doing that. I remind you that what I am trying to describe here, is the main step for showing that the Ricci flow is gradient, a big discovery of Perelman's. And, by the way, the renormalization flow, an even more formidable flow, rather mythical for the time being, is also believed, or conjectured, to be gradient. I loved all of that, quite a lot. But, sitting next to me, was Armand Borel, an old friend, who clearly disliked it big. He kept whispering into my ears, in French of course, "Il se fout de notre gueule".[1]

When the lecture ended, the Chang couple, who had organized the whole visit, suggested a nice elegant restaurant dinner. But Perelman refused very politely, saying that since it was Passover, he could not join. Then Dave immediately said to him: Please come to my house, my wife is preparing a nice Setter party. Perelman gladly accepted, then Dave turned to me and asked me whether I wished to join too. I answered: With great pleasure. Then Dave turned to Frank Quinn and asked him the same question. The Virginian, with his Southern manners, answered: I would be honored. So we all had our Setter at the Gabai's home. I will not describe to you now all the details of the Setter, since you know quite well how that goes. I must say that, in such occasions, I put a kippa on my head, and then I do like all the others do. And Quinn did the same too. So, around a big table, each of us got a copy of that little book, written both in Hebrew and in English, and then we took turns reading aloud, successively, the various paragraphs of it. Those who could, and that means Dave, his wife Dana and his mother-in-law, did it in Hebrew. Those of us who could not, and this means Perelman, Quinn and myself, we did it in English. And after the ceremonial part, came a very nice dinner party. I was sitting next to Perelman, who

[1] Editors' note: A polite translation would be: "He's laughing in our faces".

told me about his dreary life in Russia. He obviously could have emigrated since long, but I understood that his mother could not have been moved abroad. Next day he gave three more advanced lectures, for the cognoscendi, and those were excellent too. A large part of whatever knowledge I have about the Ricci flow, comes from Hamilton, from Brendle and from Perelman himself. And of course I have also read some of the things which friends of mine like Gérard Besson or John Morgan have written on the topic.

But by now nobody wanted to look any longer at my Poincaré papers, Po V-A and B, nor at Po VI. The general consensus was that Perelman's work was very hard, but that mine was even more so, and then of course it did not do the geometrization of three-manifolds either. So nobody wanted to invest time and energy in those papers any longer. I managed to finish, completely alone, in 2006, Po V-B, which proves that both $\Delta^3 \times I$ and the 4d smooth Schoenflies ball are GSC. So, finally I got back the COHERENCE theorem which had gotten lost in 1994, and hence I had now my own Poincaré proof too. A forty page long announcement, typed at IHÉS, the help of which was constantly essential for me, and which went to the ArXiv site, was the only public sign that I had finished the job. But nobody was interested to look into it any longer. And the very long Po V-B paper remains a hand-written manuscript, it never got typed, nor read by anybody.

By then, I had also started to work on a completely different project. Sometimes in the early years of 2000, I do not remember exactly when, I had a PhD student, Daniele Otera, who was writing a thesis on geometric group theory; later he became one of my collaborators, you have met him at Autrans. Incidentally, since forever, me and Daniele we have always spoken Italian together. And, one day, he told me that there was a group theoretical property, called QSF, discovered by Brick, Mihalik and John Stallings which, said Daniele, should interest me, and indeed it did.

I have said already that I do not want to write a technical text, and so I will not go here into a formal, explicit definition of QSF. The name stands, anyway, for "quasi-simple-filtration", and please do not blame me for it, I have not chosen it. But I will give you some hints of what it is all about. To begin with, QSF is a property which a locally compact space which is simply-connected might or might not have. You may think of it as the "correct" version of simple connectivity at infinity, and I will soon explain that word "correct". In point of fact, for open 3-manifolds, QSF is just equivalent to simple connectivity at infinity. The QSF concept has been inspired by the already mentioned old work of mine and of Casson's on those universal covering spaces. The QSF is, actually, a more singular version of a notion which I have introduced, at that early occasion, the so-called Dehn-exhaustibility. But there is a virtue which the QSF notion has, and which singles it out, it is group-theoretical, not just topological. What this means is that, if you give me two compact spaces with the same fundamental group, then the universal covering space of the first is QSF IFF the same is true for the second one. So, it makes sense to define QSF, not only for locally compact spaces, but for finitely presented groups too; when it comes to groups, that QSF definition which I have actually not given you explicitly, is presentation independent. And, while many nice groups fail to be simply connected at infinity, nobody has ever seen a finitely presented group which is not QSF. This should explain my earlier term "correct". QSF is also closely connected with GSC but, apart from a trivial implication from GSC to QSF, that precisely complete connection is a theorem, not just part of the definition; I will come back to this, a bit later. And the connection with the fundamental group at infinity, in dimension three, is a theorem too.

Incidentally, you have probably already noticed a funny little thing; more often than not, when I move around in the mathematical world, I seem to stumble on that GSC concept. And before digging more into that, I feel like giving you a fast refresher into the basics of GSC. Clearly, GSC implies simple-connectivity (SC), but what about the converse? Here is, to begin with, the complete answer to that question in the case of smooth compact manifolds.

If the dimensions are five or more, then it was shown by Smale that the answer is yes; this is part of the proof of his h-cobordism theorem, of course. If the dimension is four, then that converse is NOT always true. This was proved by Casson. Interestingly enough, he started with those 4-manifolds which you and I have constructed in the very early days, more than fifty years ago, and then he developed a very cute little group-theoretical argument, worth knowing by us all. When one is in dimension three, then saying that SC implies GSC becomes equivalent to the Poincaré Conjecture, so the answer is now YES but that is, by this time, Perelman's work, of course.

The trichotomy above extends for open manifolds, provided one adds the appropriate conditions at infinity. Finally, it remains the case of the noncompact manifolds with non-empty boundary. There we have a total mess without any clean general results. But once all these things have been said, I feel like reminding you of something concerning that first line: for compact manifolds of dimension at least five, SC implies GSC. There was a little drama here, in 1960, when this was proved, involving Smale and Stallings, and this is certainly part of the history of mathematics; it certainly deserves being told, but this is not the proper place for that.

What I have told you, so far, is just the story of SC versus GSC in the manifolds context, and there things get smoothened out, we have even learned that if one goes very high-dimensional, then abstractly speaking, angles get drastically rounded up, so to say. But there is now also the context of cell-complexes, for that same issue. There clearly only the two-dimensional skeleta are concerned. So we can restrict now happily to 2d complexes. And here our whole story suddenly takes a much more dramatic twist, irrespective of whether the complex is finite or not. The difference between SC and GSC corresponds roughly now to the immense gap which separates mere contractibility from collapsibility.

Finally, I can come back to my own relation to GSC. Both $\Delta^3 \times I$ and the Schoenflies ball are 4-dimensional and compact, but a priori it is not said that one has to get involved with the GSC issue, when dealing with them. Both by chance and by choice it so happens that this was my approach to them. By contrast with this, in my approach to the QSF property for finitely presented groups, I had to deal exactly with GSC, but now in the context of those smooth noncompact high-dimensional manifolds with non-empty boundary always in contexts which are outside of the well-chartered territory. And then also, in all cases, the REPRESENTATION space X is always a cell complex, not a manifold.

And after this general prentice, I will go back to the main story. Once I got aware, via Daniele Otera, that there was this notion of QSF, I also realized that my work on the simple connectivity at infinity of the 3d universal covering spaces, in the completely general case, that work which I had decided to throw away to the garbage, while in Princeton, in the Spring of 2003, actually contained techniques capable, potentially at least, of showing that all finitely presented groups were QSF. Incidentally, using my earlier π_1^∞ work from the very early 1990s, it could be shown that all the Gromov hyperbolic groups, all the automatic groups, and all the others, were QSF. Of course, at the time when I did that earlier work, the QSF concept did not yet exist. And then also, in those early days I was so obsessed with the 3d coverings, that I did not see the rest of the world. I only saw some trees, but I missed the forest; I did not realize that while I thought that I was doing 3d topology, I was actually doing group theory. Notwithstanding all this, the old techniques did imply what I just said above, concerning the Gromov groups and the others. This actually did not need any reference to the 3-manifolds or to their covering spaces, all those were actually irrelevant, and they look to me today rather as a red herring, as far as these purely group-theoretical issues are concerned.

But then, saying that all finitely presented groups are QSF, is sticking one's neck really very far out. On the one hand, the statement itself is very highly non-trivial. I can tell you, almost quantitatively, how non-trivial it is. Consider those groups G which are fundamental groups of closed 3-manifolds; and in the framework of Gromov's theory of random groups, such G's are very rare events. If you want to show that such a G is QSF which it actually is, then there are only two ways for achieving this. Either you use the full geometrization theorem of 3-manifolds, which Perelman has proved, or you invoke my claimed general result, which is, of course, totally independent of Perelman's work. I say here "general", because what I claim is that all finitely presented groups are QSF.

And the general consensus, let's say the standard philosophy, is that any property which is valid for ALL finitely presented groups has to be then trivial too, so my claim goes completely against that standardly accepted philosophy.

I will tell you now a few more things concerning my work on this subject. Any finitely presented group can of course, be presented as the fundamental group of some finite simplicial complex. Traditionally, one takes here a 2d complex, but I have to start by choosing very carefully some compact singular 3-manifold, call it M, having the chosen G as its fundamental group. You see, while I am dealing with completely general finitely presented groups, not using from the group structure anything beyond mere discrete symmetry and the existence of a compact fundamental domain, I have to be very choosy when it comes to the finite presentations of those groups.

Once I have my presentation M for G, I take its universal covering space and then REPRESENT it, like in (3). So, I work now with a nondegenerate map of the following form

(8) *A nondegenerate map f from a REPRESENTATION space X which is GSC, to the universal covering space of M, call it $\widetilde{M}$. Our f here is zippable, of course.*

I will call such an (8) a REPRESENTATION of G. Actually, G and $\widetilde{M}$ are quasi-isometric objects, so one can think of the map f as going into G, just the other way around with respect to what the mundane group representations do.

A little parenthesis here: for our G, I use both presentations and REPRESENTATIONS; the two words do not mean the same thing and should not be mixed up.

Now, our X is certainly noncompact, actually violently so, as you shall soon see, it is in fact immensely huge, and so, generally speaking, our REPRESENTATION from Formula (8) might exhibit the following phenomenon, which, years ago already, I have called the Whitehead nightmare: The map f which, in general, thoroughly explores every nook and hook of $\widetilde{M}$, a bit like Feynman's path integral, might come back and hit infinitely often a same compact spot K of $\widetilde{M}$. This kind of nightmare is well present in other places too, not just for my REPRESENTATION of the group G. The Casson Handles, the gropes of Freedman and Quinn, and of course the classical Whitehead manifold, which inspired me for the name, to begin with, all exhibit the nightmare in question. Incidentally, that Whitehead manifold was ever present in my mind, a bit like a scarecrow signalizing things which I had to avoid, while working on Po VI, during my work in group theory, and of course also when I found that its REPRESENTATIONS exhibit that dynamical chaos, related to the Julia sets.

In order to prove that all finitely presented G's are QSF, one has to come to grips, frontally, with Whitehead's nightmare, and deal with it. This is what I have actually done. Thickening the REPRESENTATION (8) into some very high dimension is to be used now, and the really interesting geometric action will happen in those dimensions which are in addition to the initial three and distinct from them.

But let us stop here for an instant, and notice something. When working on the Poincaré Conjecture, I have used REPRESENTATIONS, where the REPRESENTATION space, the source of the map f was a 2-dimensional finite complex. And one of the major difficulties then was to find a COHERENT REPRESENTATION, something of paramount importance in dimension exactly four. For QSF, I use again REPRESENTATIONS, but the representation space is this time non-compact, and the major difficulty now is that the set of double points of the map f might not be a closed subset of X, because of Whitehead's nightmare. This property of the double points set not being closed becomes very manifest when you work with a two-dimensional REPRESENTATION space X, something which can become necessary too. Anyway, the real action takes place now in very high dimensions, where the COHERENCE issue has vanished into thin air. But in both cases, Poincaré and QSF, the main hero is the double point set of the REPRESENTATION map, but then also, in the script, the role which that hero plays is completely different in those two cases.

Going now back to the REPRESENTATION (8), notice that there is, in its context, the canonical free action of G on $\widetilde{M}$. It makes sense then to consider an equivariant situation, where G also acts freely on X too, with the map f respecting the two actions. Actually, constructing for our given G a REPRESENTATION which is equivariant is a highly nontrivial thing, even when the presentation M of G is non-singular, and when one then actually deals with fundamental groups of smooth 3-manifolds. Incidentally, notice that, in contrast with this, a mundane group representation is automatically equivariant. And at an early stage, I realized that I very much needed to prove the following lemma:

"Any finitely presented group G admits a locally finite REPRESENTATION with the following additional two properties:

(i) *It is equivariant;*
(ii) *It has uniformly bounded zipping length."*

That local finiteness is not an idle issue here; the natural tendency of X is not to be locally finite. In point of fact, if one wants to get a REPRESENTATION for G with all the three features from the lemma above, then the representation space X needs to be immensely huge, something which only makes Whitehead's nightmare be even more dramatic, of course.

The central piece of my proof for QSF is a certain compactification theorem, for which that uniformly bounded zipping length is very much required. Then, equivariance is necessary, all over the place, for the whole process to work, particularly in the grand final argument, which derives the main theorem from the compactification result just mentioned. And I still remember that walk, under the starry sky, one very cold winter evening, when that grand final argument in question dawned on me.

While I was completely involved with these matters, I went to Grenoble to give a lecture and also to discuss my work with Louis Funar, whom by now you know; he was the main organizer of the Autrans meeting. And let me tell you first a bit about him. He has been my PhD student way back in 1990. Funar was one out of a batch of very bright Romanian graduate students who landed at Orsay immediately after Ceausescu's fall. He and I had already corresponded a bit before, and so he chose me as his thesis advisor, when he came to Orsay; his other comrades chose Michel Raynaud, Arnaud Beauville, Larry Siebenmann, Pierre Pansu or others. Most of them are by now completely accomplished mathematicians, with very good University positions in the Western world. Hardly any of them returned to Romania.

As a teenager, Louis has been one of those very bright kids winning international mathematical olympiads. And such young guys no longer went to the regular high school, they were put together in some sort of college, like in Cambridge, Oxford or Harvard, where they were trained in maths and Science by professors from the University. That means an immensely vast mathematical culture, but Louis also has a very good humanistic culture, he did not need the high school for that.

Louis has done big work in several mathematical areas, like for instance quantum topology; and, being my student he used the tools of quantum topology, Hecke algebras and other things, in order to study the ends of wild 3-manifolds. Then, to Dave's great amusement, at some point Louis ganged up with a former PhD student of Dave's, Siddharta Gadgil, now professor in Bombay, and together they wrote quite a number of papers on arguments related to GSC. But Funar is also a big specialist in geometric group theory, and it is him who introduced Daniele Otera to QSF. Since I was travelling a lot, at the time when Daniele was writing his thesis, he moved from Paris to Grenoble where Louis took good care of him, and during my visit to Grenoble which I am talking about now, Daniele was there too.

Louis had spent years trying, unsuccessfully, to construct a finitely presented group which was not QSF, so he was the person to whom I wanted to talk about my projects. His first reaction to my ideas about showing that all finitely presented groups were QSF, was about like this: If you would tell me that you can prove that all fundamental groups of closed manifolds which are $K(\pi, 1)$ are QSF, then this would be a very great theorem. Saying that all G's are QSF is much more than that, it would be an earthquake, since it would go against everything which we believe in. He was skeptical about my lemma too, he thought that only groups with algorithmically solvable Word problem could have uniformly bounded zipping length.

But then, very few years later, I finally did manage to prove that lemma. Funnily enough, although the statement talks about some uniform boundedness, an infinite process is needed for the whole proof. That is also how, an immensely huge, nevertheless locally finite and/or locally compact representation space X got created. And I wrote a paper with the proof of that lemma, had it typed at IHES, and submitted it to the today Blue Journal, while working on the next two papers with the complete QSF proof. After quite a while, Etienne Ghys, the then main editor of the Blue Journal, wrote back to me to tell me that my paper was rejected. He also sent me three referee reports, and a personal letter too. The reports did not say, neither that the paper was wrong, nor that it was uninteresting, but they just said that it was crazy. Ghys added in his letter that he thought that the paper was good, that he was fed up with his referees, and that he was planning to resign from the editorial board, which he actually did.

So, I completely re-wrote the paper, had it typed again, and sent it now to Geometriae Dedicata. The first referee report which I got from them, was a long six or seven page long letter. The referee wrote that the result was important, but that he was not convinced that the proof was correct, and he asked a lot of questions too. He also complained that the paper was not reader-friendly. I re-re-rewrote it, had it typed again and this kind of process was iterated and re-iterated, during a number of years, I do not even remember how many years, nor how many times. But the final result was a much better paper, or at least so I think; that referee (or referees) did actually a good job. Then, by October 2012, a few days after my 80th birthday, and while still in Fiji, I got the news that my paper was finally accepted for publication. This was quite a nice birthday gift.

But this paper which proves my lemma is only the first part of the QSF trilogy. And parts two and three were finally written, in their completely final form, only by the early Spring of 2013. When I say "written" I mean here hand-written, since I cannot, yet at least, type in TeX. In the meanwhile I have talked a lot to Louis about this work, and by now he starts believing that I might actually have it. He even has big plans for how to reorganize the whole subject of geometric group theory, once one knows that all finitely presented groups are QSF. Parts II and III have by now been typed, and without some friendly help, I do not know how I would have managed.

But the trilogy looks now more like a book than like a regular paper. Parts II, III are by now online at the ArXiv site and they have been submitted to a journal too.

There is also a certain CODA to the trilogy, and I will tell you now a bit about it. As already said earlier, in order to show that all finitely presented G's are QSF, one has to battle with the Whitehead nightmare. And, at this point I will introduce a definition. A finitely presented group G will be called EASY, if it admits a REPRESENTATION without the nightmare in question. It is easy to show that any G which is easy is also QSF. Notice that the two "easy" used in the last sentence, do not have quite the same meaning. And, from the present vantage-point, what my old 3d universal covering space papers had actually proved was that the Gromov hyperbolic groups, the automatic groups, and the others, were all easy, they can avoid the Whitehead nightmare. Similarly, Perelman's work implies that all fundamental groups of closed 3-manifolds are easy too. At this point, what the CODA to the trilogy is supposed to prove is that, as a consequence of the fact that all finitely presented groups are QSF, they are easy too. But, so far, this is just a conjecture. That connection between GSC and QSF, which I mentioned earlier, has to be used now; to be slightly more specific, one should use here a theorem of Funar and Otera saying that if G is QSF, then G is the fundamental group of a smooth high-dimensional manifold M, of dimension much higher than three, of course, the universal covering space of which is GSC. So, my Conjecture, which should be the CODA to the trilogy, is that being easy is a universal property of all finitely presented groups; I feel like calling this fact "the impertinent conjecture", you can see why. But keep in mind that the "easy" used here is not everybody's "easy". Incidentally, the converse of the Funar–Otera result quoted above is trivially true, so that their theorem offers an alternative definition for QSF, albeit not a presentation-invariant one, unlike the original definition, which I did not give you explicitly. Anyway, by now you should know what QSF means, for groups at least.

These days, Daniele Otera and myself are actively involved in trying to deduce the impertinent conjecture from my QSF theorem.

But, so far, we have only managed to prove something weaker, which I feel like mentioning here. Years ago, Funar and Otera introduced the notion "weakly geometrically simply-connected" (WGSC), which means the existence of an infinite exhaustion by finite complexes which are simply-connected.

What Daniele and myself proved is that if one modifies the definition of REPRESENTATIONS, by substituting WGSC to GSC, then the Whitehead nightmare can be avoided.

But the actual impertinent conjecture seems quite difficult. The GSC concept is quite subtle, with mystery still becoming around it.

Let me put all this a bit differently too. Once one has managed to master the Whitehead nightmare, well enough so as to show that all G's are QSF, then one should be able to prove next that for every individual finitely presented group there is actually a strategy for avoiding the nightmare in question. But then, before one can get rid of the Whitehead nightmare, one has to start by living big with it. That is what the trilogy actually does.

Before we leave this group-theoretical adventure, here are some inter-related philosophical questions. How is it really possible to have a property for finitely presented groups, which is, at the same time, both highly nontrivial and also completely universal? And, then also, where are the "difficult" group-like objects hiding? or, if you wish, is there a framework where the standard philosophy can be restored?

I have an idea that there is here, waiting to be discovered, a much larger category of objects, connected maybe to the (relatively) recently discovered quasi-crystals or to the Penrose aperiodic tilings, vastly bigger than the category of finitely presented groups and encompassing it. This should be something like the quasi (or almost)-periodic functions, beyond the periodic ones. I believe that I have to learn Alain Connes' non-commutative geometry, in order to give flesh and bones to my ideas here. But this has to wait; I have other things cooking in the oven right now.

At the same time as this big group-theoretical quest was evolving, I had also a completely different mathematical adventure going on too, which I will tell you briefly about, now.

Ever since we got the Pasadena Lemma, Dave and I started working together on the GSC Schoenflies project the aim of which, I remind you, is to show that each smooth GSC 4-dimensional Schoenflies ball is standard. And at some time in the Fall of 2008 we were both staring at the blackboard, looking at some tricky potential counter-example, which we were trying to understand. And then, really simultaneously, something struck us both. The so-called Po Conjecture was staring there at us, smiling like a Cheshire cat.

Here I have to explain you something. Way back, in the early seventies, when I was trying hard to find my way with the Poincaré Conjecture, I had developed a little strategy, which depended on a conjectural statement, which although never published, caught the attention of many people at the time, and which came to be known as the Po Conjecture, and here it is really meant Po and not Poincaré. But soon my plan which was supposed to use it collapsed, and then I lost interest in it too. Not so the others.

At face value, the Po Conjecture was an issue in knot theory, which depended on a parameter, itself an odd number, call it k. For $k = 1$, the thing was trivial, then Laudenbach proved it for $k = 3$. Next, somebody did it for $k = 5$, then somebody else again claimed it for $k = 7$, but his proof was very long and nobody had the courage to read it. In the meanwhile, a big new mathematical star had appeared at the horizon, that was our friend David Gabai. And Dave developed a brand new technology in 3-dimensional topology, based on a class of very special tight foliations which he had introduced, and he started writing a big three-part paper in which he proved all sort of interesting things, using that technology. And the third of those papers was the proof of the Po Conjecture, now for all values of k at once.

And then we two met, in the early 1980s, at IHÉS. And I told Dave something which hardly anybody knew then, namely that the Po Conjecture also had another facet, it was not only a statement about knots, but it was also, at the same time, a strong result in DIFF four-dimensional topology. We talked about all this, and eventually these things found their place in the final version of his paper. One can find there also the so-called generalized Po Conjecture, a statement which we two believe to be beyond the present state of the art, in mathematics. Dave also knew then that I was working on the Poincaré Conjecture, but I was not yet ready to start talking about it in any detail.

And I started doing that, a few years later, at a big conference in Luminy, a beautiful resort, in the Calanques, close to Marseille; Dave was there in the audience. And, straight after that conference, Dave came back with me to Paris and then he stayed with me for a while. I say with me, because Milen and the children were in Norway at the time. Our big friendship and mathematical collaboration started then.

So, now in 2008, that Po Conjecture seemed to reappear in the middle of our Schoenflies project. During 2009 we developed a big program for our GSC Schoenflies thing. There was still a Homological Conjecture which needed to be proved; we knew how to do the free part, but not the torsion part. Anyway, things looked very promising, at the end of 2009. But then sometime in February 2010, I got an email with very bad news from Dave. Our Pasadena lemma was irreparably wrong; that e-mail contained a very explicit counterexample which killed it completely. You may remember that in the Fall of 2010, I have sent you a little text, "A tale of maths and physics", where this very dramatic e-mail is mentioned, and where I afterwards tell things about the Casimir effect, about Boltzmann and Ramanujan.

The Summer of 2010 is when Alexandra and Tony got married. And there, in Fiji, I started developing a completely new strategy for something which we call now "The new, revised, Pasadena lemma". Like the old dead lemma, this is again a mixture of 2d and of 4d and, most importantly it does the same good for us as the old dead lemma was supposed to have done. But it was now not only much much trickier than the old one, it was much more integrated into the whole Schoenflies approach too. A rather beautifully looking scheme came out, but it had to rely on several conjectural statements, some 4d topology and some homology too. In the Spring of 2011, Dave came for a more extended visit to Paris, and we discussed this at length. Everything looked good, except that all those conjectural statements looked more and more dubious, as we were discussing them in more detail. And so we got quite discouraged, and we had to start from scratch again.

Then, in 2012, I flew again to Fiji, with another scheme in my mind for proving the new revised Pasadena lemma. All the time we had been badly pestered by the possible existence of some kind of bad cycles. And Dave flew from Princeton to Paris so that we could discuss things together. We got convinced that there was really no way to conjure away our bad cycles, they were unavoidably there, present.

But we found a rather interesting procedure for cutting them, using the GSC property of the Schoenflies ball which we were looking at. Even with that, the end of the road was still not in view. And since 2012 Dave and myself have been continuously bottling with that "new revised Pasadena lemma" and also with a related statement, "the Mock–Pasadena lemma", supposedly a crutch for the real thing.

This piece of work, trying to show that any GSC 4d Schoenflies ball is standard is an active field of investigation, for Dave and myself, right now.

Changing now the topic, towards the end of 2014 and the beginning of 2015 I decided to take again a look at that very long old Po V-B manuscript from 2006, the one which was never typed and which nobody ever read either. It was, by then, completely out of my own mind too.

Then I realized that, for some reasons which maybe were understandable in those early years, when I wrote PO V-B, I had sort of blinders, at that time, failing to see the big forest which was looming behind those nearest trees. That made that my own re-reading, of that old paper of mine, the old Po V-B, was a very painful and rather heavy affair.

The fact is that, years ago, while I was writing that Po V-B I did not know where I would get, until towards the end, and the written paper keeps the traces of the various unnecessary detours and blind alleys. Now I knew exactly what the correct scheme was, and so life was much easier. But once I understood well the whole idea, I thought things over again, and rather than following blindly the old Po V-B, I started by disentangling the Schoenflies part from my Poincaré papers, completely, and thereby I managed to vastly simplify and also shorten the arguments for Schoenflies, which I extracted from Po V-B. I found a lot of simplifying tricks too.

So, by June 2016, I had a completely clean proof that all 4d Smooth Schoenflies balls are GSC, a proof without any connections with nor even references too, to the Poincaré Conjecture. The final paper, is by now typed, at today, of course. It is still a long one, but a considerably shorter than the typed Po V-B might have been, had that existed.

I will end here my long letter to you, and as you can see, I do have some things to do, for the next years to come.

A small final thought. I have sent you last Fall my "Memories from my former life", which end about thirty years before the present story, told here, starts. In the meanwhile, I had met Milen and we got married, we got our two children Hannibal and Alexandra, and then of course too, a lot of mathematics, the prehistory of the present story, also happened.

All this became part of me, and so I am certainly not quite the same person any longer, as the main hero of those "Memories ...". But then, in those Memories themselves, the story starts when my age was from eleven to about thirteen or fourteen, and then it ends, rather abruptly, just before I became thirty. Clearly, already then, a certain change of person occurred, I was no longer the same me, at the end of that particular episode. Finally, it is hard for me, today, not to compare the two texts. Each of the two stories concerns a span of time of approximately twenty years. And the thing which really strikes me is that the density and intensity of time and energy spent on maths is considerably higher in this second story, the one I just told you now, than in the first one. This is probably a normal thing.

You may remember that you have gotten already an earlier, incomplete version of this long letter in September 2013. Now the story is more complete and brought up to date.

But I do not only feel awe, awe in front of all this big mathematics, I feel Angst too. I am also filled with apprehension, fearing that there might be some hidden gap, crack or mistake, somewhere. And then, I am anxiously waiting to see the general reaction to the QSF saga. It is a critical time, indeed.

Much love for both of you, from Po

POST-SCRIPT 2024. Eventually, some years ago, Parts II and III of the TRILOGY have been refused by the referees. They were considered too hard to read and understand, and, moreover, not enough self-contained. And so, during a three or four years work, I have produced new versions of them, re-written from scratch. They are now more reader-friendly and more self-contained, but also longer than the older versions. They are now available online, on the CNRS website "HAL".

Here is the reference for the re-written TRILOGY II:

Finitely presented groups and geometric simple connectivity, (2023); part (A) hal 03954340 v1, part (B) hal 03954350 v1, part (C) hal 03954353 v1.

For the re-written TRILOGY III:

A proof that all finitely presented groups are QSF, (2023) , hal 03954357 v1.

There is also a shorter, overview version of the QSF TRILOGY, which should be helpful. This consists of the following two papers:

D.E. Otera and V. Poénaru, *Topics in geometric group theory I*, in Handbook of group actions V (ed. L. Ji, A. Papadopoulos and S.-T. Yau), Int. Press p.301–346 (2020), and *Topics in geometric group theory II*, in the same volume of the Handbook of group actions, p. 347–398 (2020).

Quite recently, there is a sequel to these things. This is a very advanced project paper, joint work of Louis Funar, Daniele Otera and myself, concerning a proof that all finitely presented groups are geometrically simply connected (and this GSC property is much stronger than QSF), and also that they can avoid the Whitehead nightmare (and this means that they are "easy".) This is announced in my paper *On the Whitehead nightmare and some related topics*, in EMS Surveys Math. Sci, a volume for the 80th anniversary of Dennis Sullivan (2023). This project, about which I am very optimistic, uses the technology of the QSF TRILOGY, appropriately twisted. With respect to that TRILOGY, two big difficulties have to be overcome now:

(i) Getting rid of infinite foaminess (infinitely generated second homotopy group), while staying GSC, and
(ii) Bringing things from infinity to finite distance, staying GSC.

A provisional written version exists already.

Concerning the rather esoteric ideas concerning a new, much larger category, encompassing the finitely presented groups, ideas expressed in the 2016 letter, there is a bit more to be said now. In particular, I have discovered a precise connection between a certain kind of violation on GSC and non-commutative geometry, which should be relevant for the topic. There are here two papers of mine, on this subject: *On geometric group theory*, in: Topology and groups, essays dedicated to V. Turaev, in IRMA Lectures in Mathematics and Theoretical Physics (ed. A. Papadopoulos), p. 349–432 (2021), and *Classical differential topology and non-*

commutative geometry, in Surveys in geometry I (ed. A. Papadopoulos), Springer, p. 309–341 (2022).

Concerning now the 4-dimensional DIFF Schoenflies, I have now a long paper on line, proving that the 4d DIFF Schoenflies ball is GSC:

All smooth four-dimensional Schoenflies balls are geometrically simply connected, arXiv: 1609.05094 (2016).

There is also a shorter over-view version available: *All 4-dimensional smooth Schoenflies balls are geometrically simply-connected*, in Surveys in Geometry I (ed. A. Papadopoulos), Springer p. 369–307 (2022).

Finally, Dave and myself are still working on the big project of showing that 4d DIFF Schoenflies balls which are GSC, are standard. And, as such, the ideas concerning this topic in the 2016 letter are no longer actual, as such.

Chapter 27
Adventures in Physics

Valentin Poénaru

To the memory of my friend Gérard Toulouse

Abstract This chapter is a focus on the joint work of Valentin Poénaru and Gérard Toulouse on the dynamics of defects of ordered media.

Keywords Homotopy groups · Non-commutativity · Defects

AMS Code 00A79

I am a mathematician, and my personal field of research is geometric topology, more precisely low-dimensional manifolds and geometric group theory. In my life, I have worked big on the Poincaré Conjecture and related topics.

During the years, my mathematical interests were quite varied. I have also worked on the singularities of mappings, under the effect of a symmetry group, where I happened to stumble on the Higgs potential too, but I did not understand its meaning, then. But there I quite naturally met the topic of symmetry breaking, something which fits quite naturally with the things I will tell you about now.

In the early seventies, I was professor at the Mathematics Department of the University of Paris-Sud (Orsay). Our neighbors and friends, the Williamses, an Anglo-Belgian couple, were working at the Condensed Matter Physics Department of the same university. Claudine Williams was organizing monthly lectures on general scientific topics, for the members of her department. Around 1973, I gave there an introductory lecture on topology, for the members of Claudine's department. As soon as I finished my lecture, two members of the audience approached me, saying that the kind of things I had talked about was exactly what they needed. The two were Gérard Toulouse and Maurice Kléman. I understood

V. Poénaru (✉)
IHES, Bures-sur-Yvette, France
e-mail: valpoe@hotmail.com

L. Funar, A. Papadopoulos (eds.), *Essays on Topology*,
https://doi.org/10.1007/978-3-031-81414-3_27

from them that topology was found out to help understand the defects of ordered media. At that time only two groups in the world had started to work on this topic. In Paris, to be more precise, in Orsay and Bures-sur-Yvette, there were Toulouse, Kléman and Louis Michel from IHÉS. But then there was also the Moscow group of Volovik and Mineev, working in the team of Dzialoshinsky. Incidentally, a bit later, a younger physicist, Misha Monastyrsky joined them. Some years later, Misha started coming to Bures/Orsay. We met and we became friends.

This is how things were in those old days. But some thirty years later, the situation changed drastically. With the work of Ed Witten, Barry Simon (who, incidentally, had been my student in a topology course at Harvard, in the early I960s) and others, with topological quantum field theory, a topic which fascinates me, and with other things like that, topology became a big hot topic in physics.

But let us go back to those early 1970s. Gérard and me started talking a lot together. He taught me physics and I taught him topology. And since at the time I had some problems with my own mathematical work, I was welcoming a temporary change of horizon. Gérard was a charming personality, full of enthusiasm, of inspiration and of clever ideas. He was very good at communicating these things too. Among the many physicists which I have met in my life, he was the one with whom I could interact best.

Concerning our joint work, it was already known at the time that defects of ordered media were connected with the homotopy groups of the manifold of internal states, a space which somehow encodes symmetry breaking. In particular, the existence of defect lines was connected with the fundamental group of that manifold of internal states, and similarly the defect points had to do with the second homotopy group.

But Gérard and me, we were looking at the dynamics of defects, and we discovered that the non-commutativity of that fundamental group, when present, prevented the defect lines from crossing [1]. Many physicists outside the topic of condensed matter became interested in our result, partly, I think, because non-commutativity is so fundamental in quantum theory. Actually, the famous Russian field theorist Sasha Polyakov suggested to us to see what happens with our little theory when one tries to extend it to higher dimensions; that was clearly outside of condensed matter physics.

And there we had a beautiful surprise. The totality of homotopy groups, when endowed with the Whitehead product, becomes a super-symmetric algebra. And we found that the noncommutativity of the Whitehead product is the obstruction for the crossing of higher-dimensional defects [2]. So, somehow supersymmetry is now involved. Gérard told me that he never had a discovery that gave him so much pleasure.

I will introduce here a little parenthesis which I find amusing. In those early seventies, Jean Dieudonné, the famous French mathematician, had written a book on the mathematics of Bourbaki. And when it came to topology, he had written that this topic had no connection with physics. When, a few years later he published a second edition, he corrected that statement and then he quoted Gérard, myself, and Louis Michel too.

Our third and last joint work, a piece of work done by the two of us, together with Yves Pomeau, Bernard Derrida and Yves Bouligand, concerned the following puzzling issue. In his study of biological tissues, Bouligand, who was a biologist but at the same time a distinguished physicist, had discovered the, now, so-called Bouligand rings. These were defect rings, existing in a situation when the potential obstruction for crossing was living in the center of the fundamental group of the manifold of internal states and which nevertheless stayed hooked together, refusing to cross. But non-commutativity was not the explanation.

What we discovered was that the explanation for this paradoxical situation was the occurrence in the game of a new actor, the third homotopy group of the manifold of internal states.The deep reason for the occurrence of this totally new ingredient was that now we were in the context of textures, when long range effects become important too [3]. And I think that the mathematics involved here is not without relations with the—still conjectural—magnetic monopoles of t'Hooft and Polyakov.

We have reached by now the later 1970's. I had managed to overcome those difficulties which were bogging me earlier, and so I went back to my mathematical research. Actually, I still had another, shorter incursion into physics, when I was helping Geoff Chew from Berkeley to put up a topological bootstrap theory. But those years of collaboration with Toulouse and the others were a very exciting and interesting period in my scientific career. Gérard was a very important figure in that scientific life of mine.

Geoff Chew was a very charismatic person to whom I became very close. But this is not the place to talk about him nor about our unfulfilled projects. I will rather end with a little mathematical discussion. We have just talked about the applications of homotopy groups to the theory of defects. But what about homology? Well, in some sense homotopy groups give information about local pieces of the set of defects. And I believe there is a "defect cycle" which puts together this local information into something global, and which brings homology groups into the field. Actually, in [4], under very very special assumptions, I have managed to construct this cycle and proved how it connects to surgery of defects and to the Volterra process. But I do not know how to proceed in the general case without those too special assumptions, nor whether this has any interest for the ordered media of condensed matter physics. Is there room for it in more esoteric areas of physics, like quantum field theory, quantum gravity, and all that?

27.1 A Short Appendix on Bouligand Rings

Yves Bouligand developed an experimental technique for studying the cholesteric order. And he found defect lines, called now the Bouligand rings, which stayed linked although the homotopical obstruction for their crossing lived in the center of the fundamental group of the manifold of internal states. So non-commutativity was not what kept them linked.

We are actually here in the presence of textures, ordered media where the order parameter is everywhere defined, and the region where it is not constant plays the role of defects. It is now a completely new actor, the third homotopy group of the manifold of internal states, which is the obstruction for crossing. More details are in [4].

References

1. V. Poénaru, G. Toulouse, The crossing of defects in ordered media and the topology of 3-manifolds. J. Phys. Fr. **38**(8), 887–895 (1977)
2. V. Poénaru, G. Toulouse, Topological solitons and graded Lie algebras. J. Math. Phys. **20**, 13–19 (1979)
3. Y. Bouligand, B. Derrida, V. Poénaru, Y. Pomeau, G. Toulouse, Distorsions with double topological character: the case of cholesterics. J. Phys. Fr. **39**(8), 863–867 (1978)
4. V. Poénaru, *Elementary Algebraic Topology Related to the Theory of Defects and Textures, Les Houches 1978, Session XXXI, Ill-Condensed Matter*, ed. by R. Balian, R. Maynard, G. Toulouse (North Holland, Amsterdam, 1984), pp. 263–318

Chapter 28
Dimitru Poénaru, My Father

Valentin Poénaru

Abstract This chapter is a panorama of Valentin Poénaru's life in Romania before 1962.

Keywords Biography · Valentin Poénaru · History

AMS Code 01A70

My father's friends used to call him Mitu, but me, I soon dropped the tata-mama, Romanian version of dad-mummy, and started calling my parents sherif and sherifa, a feminine version of sherif, which I had concocted myself.

Before I go on, let me say that some of the things I will tell you now might overlap with things I have already told in previous texts, in particular, in the article *Memories from My Former Life: The Making of a Mathematician*, which appeared in the volume *Geometry in History* edited by S. G. Dani and A. Papadopoulos (Springer, 2019). But I will try, as much as possible, to avoid overlaps.

My father was born in 1899, in Oltenia, a province of old Romania. Old Romania, the land before 1918, was bordered by the Carpathian mountains to the West, and by the Danube, the Black Sea and the river Prut to the East. It consisted of two provinces, Muntenia to the South-West, Oltenia was its most Western piece, and Moldova to the North-East. Do not mix this up with the Moldavian Republic, called Bessarabia by the Romanians, located between the rivers Prut and Dniester.

On the other side of the Carpathian mountains was Transylvania, part of the Kingdom of Hungary, itself a component of the Austro-Hungarian empire. Transylvania was inhabited by Hungarians and Germans, majoritarian in the towns, and Romanians, majoritarian in the countryside. When my father was born, his own father was about 55 years old. His wife, my grandmother, was German, descendant from a group of Swabs, brought by the Habsburg empress Maria-Theresa in the

V. Poénaru (✉)
IHES, Bures-sur-Yvette, France
e-mail: valpoe@hotmail.com

L. Funar, A. Papadopoulos (eds.), *Essays on Topology*,
https://doi.org/10.1007/978-3-031-81414-3_28

eighteenth century, and settled in Western Transylvania. My grandfather came from a family of old country-side nobility, and he had the idea that working was a shameful thing. So, for many years he had earned his living by selling, piece by piece, the land he owned, without ever thinking of using it otherwise. And when all that was sold out, my grandparents lived in complete poverty. What saved them was that, by the time he was 10 years old, my father passed successfully the then very hard competition for entering high-school. That earned him a scholarship, which became the income of his parents. My father often told me that he wished he had been born orphan, his parents had never helped him, they were only a weight. High-school studies in Romania of the day lasted from age 10 to 18. Then, at 14 you had three options: either *réal*, a terminology borrowed from the educational system of Germanic countries, which meant science, or *modern*, meaning modern languages, or *classic*, meaning Latin and Greek. For girls, who were totally separated for boys, the only choices possible were the last two. My father chose réal and my mother chose modern; but they had not met yet.

My father also had an older sister whom I never met, since she died of tuberculosis in the early 1920s. Her husband had already been killed in the war, In 1917. You will hear, later, about their son Costica, my cousin. I cannot spell here correctly his name: in Romanian the letters a, i, s, t also have variants with a slightly different pronunciation and also written with slightly different characters, which my computer, of course, does not have.

Until the Summer of 1916, Romania was neutral in the big war. But then, that summer 1916, the country joined the allied France, England and Russia, in the hope of grabbing Transylvania away from Hungary. Then, very soon, the German army invaded the country, starting with Oltenia. Many people fled to Moldova; my father joined them, taking his scholarship with him, away from his parents. Thus, he finished his high-school studies in Iaşy, the capital of Moldova. He wished to continue further studies, but now he was penniless. Some French officers who were there advised him to enroll as a medical student, choosing from the beginning to become a medical officer of the army, and hence earning right away a salary. I must add here that my father spoke very good French, with an accent of course, but always making sure to use all those subtleties of French grammar, which we rather tend to avoid today.

The war ended, my father was now a medical student in Bucharest, but he had always loved the starry sky, was very interested in astronomy, and so he also enrolled for one year as a student in mathematics. At the medical faculty, he met a girl he fell in love with, Elena Lupescu; her nickname was Lela. They married and I am their son.

When Mitu and Lela settled in a flat in Bucharest, his parents who were then living essentially in the street, joined them and occupied a room in the house. My grandfather died when I was less than three and I hardly remember him. My paternal grandmother, Maria, was a highly unpleasant person, essentially at permanent war with my parents and with me. Now, before I tell you more things concerning my parents, here is a little detail. When I was born, in 1932, my father announced to their friends: *habemus filius!* He had a rather good knowledge of latin. Much later in

1945, he met a Hungarian priest with whom he became friends. Their only common language was latin, and they exchanged letters in latin. As my father told me, this was not the latin of Cicero, but a simpler version, called church latin.

Once his studies were finished, my father was a military doctor, whose job was the organization of medical supplies for the army. By the end of the 1930s, he had the rank of colonel. But he also had a private medical practice, he was a dentist. And at the end of the Second World War, when he left the army, my father devoted himself completely to his medical practice, which was very successful and flourishing.

Before going on, I have to tell you about my mother's background. She was born in 1900 in Bucharest. Her own mother, Aristia, was an elementary school teacher, while her father was an accountant. He was actually on top of the profession: after many years of being the main accountant of the Romanian commercial maritime fleet, he became the personal accountant of the queen. My maternal grandmother had about 12 brothers and sisters; they all fared well, so my mother had a lot of cousins, to which she was quite close.

In decreasing order of age, the five children of my maternal grand-parents were: Christian, Lucina, my mother Elena, Ecaterina and Marian, the younger brother. Concerning now the older brother, Christian, an agronomical engineer, he was soon ostracized by hid parents, who very strongly disapproved of his wife. So, I hardly saw him, and I must confess that I never knew, nor understood exactly, why and how all this break in the family happened. Anyway, my mother was very attached to her two sisters and her younger brother, as well as to their consorts. And for my father, who had not much other family to stick to, they became his close family too. Something very similar happened much later on to me, when Milen's family became my family too.

My aunt Lucina was married to a cousin of her own mother, Petre Mincu, a very rich and successful businessman, and also estate owner. Later, in 1947 or so, when the communists had taken complete control of the country, when Lucina was already dead, because of heart problems, Petre had lost all of his fortune and he was forced to move, in exile, to a distant Transylvanian town. My father visited him often and brought him money to survive. And he was not the only person whom my father helped this way.

The younger sister Ecaterina, nickname Ica, was a high-school teacher, married to an army officer, Victor Stanescu, and they had a daughter, Liliana, six months younger than me; you will hear much more about her. The youngest in that family, Marian, a chemical engineer, was married to a Russian-Greek girl, Lealea, whose Odessa based family had fled the Russian Revolution. She had relatives in Lebanon, Turkey and other places around. The only members of all this large family whom Milen ever met were Lealea and Liliana.

When I was young, Liliana was for me like a sister and for both of us the godparents were Lucina and Petre, who treated us like their own children; they had no other. As children and youngsters, until about 1946, when life in the country changed abruptly, Liliana and me always spent the first half of the summer in the country mansion of our godparents, and the second half in their magnificent large villa, in Predeal, in the Carpathian mountains.

You may have noticed that the only young descendants of that large family tree I just described were Liliana and me. And then, as you will find out, later on, sadly enough, the tortuous ways of life completely separated me from Liliana, and we became like complete strangers. I come back now to my mother herself. She was a lab-doctor; but she never dealt with the patients, only with the microbes themselves. She worked in the Romanian version of the Paris Pasteur institute, and her special field was tetanus; she was producing the anti-tetanus vaccine for the country. I visited her relatively often in her lab, where I also met her boss, professor Condrea, who let me happily browse through his beautiful non-medical books. The point is that biology and even less, medicine, could never clinch my big thirst for the universal and the absolute, which eventually made me become a mathematician.

Now, my mother was very busy with her job and could not very much take care of me as a young child. So I had a German nanny, quite a usual thing in those days. My nanny, Litzi, was a descendant of a saxon group of Germans, brought by the king of Hungary in Eastern Transylvania, a thousand years before. This means that when I was a small child, there were two German-speaking persons around, my paternal grandmother and Litzi; they both, of course spoke normally Hochdeutsch, meaning normal German. They could also speak their respective dialects, swab and saxon, but then they could not understand each-other. The fact is that both Romanian and German became my mother languages.

So, when I was a child, I had a cosy life. But this is how Romania of my childhood and youth was: fridges were not available, we only had some sort of ice boxes; then, for heating the house, for getting hot water, or for running the kitchen oven, the only available source of energy was wood. Also, when you bought a chicken, or any other bird, it came with all its feathers on. My parents needed both cook and chambermaid; also, as an army officer, my father was entitled to an orderly, he lived in a small room in our house, he was cutting wood, plucking chicken, and all that. In this way, his life was cosyer than in a military barrack.

Actually, both my mother, her two sisters and her sister-in-law were excellent cooks; none of my two grandmothers were. My mother seldom had the time to do the cooking herself, but she taught both Litzi and the cook how to do it.

My father loved good food, particularly the one I just mentioned, and he loved good wine too. My parents, who had many good friends, often had dinner parties. According to the custom of those days, these never started before 10 p.m. As a child, and later as a teenager, I loved to listen to the very interesting conversations at my parents' dinners.

And now I open a parenthesis and come back to my cousin Costica. He was a bit more than twenty years older than me, and we liked each other a lot. Since his parents had disappeared very early, he was brought up by my father, and by our common grandmother. He was a technician, sort of half-engineer, But he had a problem. Already as a child he started stealing, and then later on he stole from the factories or companies where he worked. This had earned him several periods of jail. I must add that in the Romania I knew, stealing was much less punished than anything having to do with politics.

Costica was the beloved grandson of our common grandmother, who disliked my father and hated my mother and me. She always said I was a no-good, with intellectual tastes which were too sissified, not worthy of a real guy.

Litzi and Costica were quite good friends, and at some stage Litzi married a friend of Costica. After several years, they divorced, and she came back to live with us. Sometimes, when my parents were absent, Litzi and Costica organized dinner parties of their own. They were flying less high than the parties of my parents, but I still enjoyed them big. Costica was always in good mood and he had a very pleasant sense of humor.

My father, to whom I come back now, had a big impact on my intellectual development. We were fortunate to live in a part of the town with big parks and public gardens. And whenever he could, he loved to take long walks with me. When he left the army, at the end of 1945, these walks became even more often and longer. He told me a lot of interesting things, and I largely owe to him my interest in history. He also had a very large library, several thousands of books. About three quarters of them were in French, then in German and Romanian, and even a few in English; my mother had a bit of knowledge of that language. I found there many books on science, philosophy, literature and history. Having all that close at hand was essential for my development.

But my father also believed in education, and very soon, let's say 1940 or beginning 1941, he thought I should learn French. In those days, a lot of French young men were prisoners of war of the Germans, who treated them much better than they had done for the Poles, or later for the Russians. So, there was a young prisoner, Monsieur Caillavé, from the Basque country whom, because he was not in good health, the Germans let loose in Romania, then a vassal country of them, to be taken care of by the Vichy embassy in Bucharest. This gave him lodging and food, but since he lacked pocket money, he offered private lessons in French. He came to teach me French for about a year, then he was allowed to go back home. And now I could devour the books by Jules Verne and Alexandre Dumas, which I had immediate access to.

There is here a little detail which you have to understand. School in Romania only occupied a half-day, leaving a lot of time for free reading, for private lessons and, most importantly as far as I was concerned, for free thinking.

Then, when I was about nine years old, my father thought that since I spoke fluently German, I should also get to know classical German literature. So he got for me a teacher, Mrs. Emma Szabo, a German from Transylvania. She was rather a strange person, both a convinced member of the communist party, and also romantically nostalgic for the Nazis. She did not think there was a contradiction there. But she was a good teacher. She introduced me to the work of the Jewish German poet Heinrich Heine, who became one of my very much beloved poets. And she could also tell me a lot of little interesting stories, like the following. In the play "Don Carlos" by Schiller, one of the characters says: "Give us freedom of thought" (I spare you the German version). But during the Hitler time, this was changed into: "Give us freedom", with the "thought" dropped.

A little detail now. I had a very good life-long friend, Mircea. When we were children our respective parents got close too. And, like me, Mircea also had had a German nanny. So, Emma Szabo also taught German literature to Mircea. As you will see, this opened a whole pattern of shared private lessons. When we were about 10 or 11, we went together to be taught English, by Mrs Mary Lazarescu, a real English woman, born Mary Cobb.

About the same time too, I started to study mathematics, on my own, with great passion. At the age of about 13–14, I had already completely mastered calculus, and I really enjoyed it big. My father certainly encouraged me in this activity; after all he was very fond of culture, scholarship and learning. But at the same time, while I was in high-school, my father tried to dissuade me from making mathematics my profession. He always said this was the surest road to starvation. Both my parents would have very much liked me to become a medical doctor, like themselves. Specifically, my father thought that the most comfortable career for me was to become a dentist, like himself, and inherit hid flourishing practice. I did not like at all those ideas, and medicine was certainly not my kind of thing. Also, little did my father dream that by 1958 private medical practice was to be abolished in our country.

So, I loved very much my parents, but we did not always agree about everything. For instance, already as a youngster, I was dreaming of leaving Romania, for a more scientifically advanced place. But this also came with a big worry: how would I be able to manage alone, in a foreign land, where I knew nobody. I could not have dreamt then, that once I made a mathematical breakthrough—this happened about 1959–1960—I would play, as the French would say "dans la cour des grands", and I would get fantastic friends, like my best friend Barry Mazur from Cambridge Massachusetts, or Christopher Zeeman from Cambridge England, the Scandinavian cousin of whom would become my much beloved second wife, Milen. On this issue of leaving Romania, I again disagreed with my parents, who were very attached to their country. My father sometimes said that he would prefer to die in jail in Romania rather than living somewhere abroad.

Coming back to my education, my father thought that my totally personal activity of maths studies should be buttressed by some private lessons too. And so, he found out that there was a university professor, at the engineering school, who offered private lessons of advanced mathematics, at a high price, of course. So, professor Ionescu-Bujor came to see me, to gauge what I knew and what I needed. He found out that concerning calculus I was OK, but then, when he saw on my desk a fat book of trigonometry, full of lengthy formulae, he started laughing; he immediately taught me the magnificent Euler formula, concerning complex exponentials, which makes all of that classical trigonometry obsolete. Then, for a year or so, he taught me Cantor's theory of infinite sets, something I loved with passion and also projective geometry, a topic which, strangely and unfortunately, is no longer taught in France today. And, according to the already established pattern, professor Ionescu-Bujor also went to give lessons to Mircea. Very unsuccessfully, he tried to teach him calculus. Mircea told me that he had big trouble in understanding the infinite, and since I was in love with it, that saddened me.

Professor Ionescu-Bujor was an excellent teacher, but he was not a real mathematician, in the creative sense. And later on, as a first year university student, when I got under the protective shield of my now boss, professor Moisil, I found out that Moisil despised Ionescu-Bujor. They were both important members of the Communist Party, but they disliked each other. This did not make my life easy, but I will not tell you more about it now.

Anyway, soon I finished high-school and entered the university as a math student. Then I had big political problems with the communist system, I will not tell you again about that. And when I finished my studies I only got a very miserable job at the University.

When in 1956 I married Sanda, that turned out to be an earthquake in our family life, and it saddened a lot my father's last two years of life. My parents and Sanda's parents disliked each other at first sight and continued to do so big, forever after. And her parents strongly disapproved of me, they said I was a no-good, with a miserable career, and with no future. Actually, my salary was about one third of Sanda's income, and my father had to help me with money every month. My ex-mother on law, who particularly disliked me, always said that an actress's salary should totally go to her dresses and other toilette things. There was no sympathy at all, either, between my parents and Sanda. Actually, I realize that had I stayed married to her, I would have never achieved anything interesting in my life. Being together with her, I had very little leisure to think, I was always pushed around to do this and that; according to my ex-mother in law, that was a man's job.

At the beginning of our marriage, we lived in a room in her parents's house. But soon it was clear that we had to move out, and it was my job to find a new lodging. This was in a communist country: lodgings could not be rented, that did not exist. When something was free, you needed permission from the state to settle; that usually needed bribe. So, for six months my main activity was to be out, hunting for a place to live. But I was totally unsuccessful, although I spent some money. Then, one day, may mother found out that a young couple in her lab was leaving their lodging. My father, through his connections, managed to contact a state official in charge of these things, and bribed him. That meant an amount of money which exceeded my yearly income. But now we had our room, 4 meters by 4 meters, to live in. This was in a relatively elegant apartment, where the original owners had been squeezed in one room. Three other rooms were occupied by three other families (that included us), and living-room, bathroom and kitchen were all common. Our food and cooking was taken care of by Sanda's former German nanny, and we have never entered that very crowded kitchen. This should give you a picture of how life was in communist Romania of those days. And when I was in Moscow, in 1961, I saw even worse things.

In the fall of 1957, my father's health deteriorated a lot; he got serious liver problems. Personally, I also think that he got very depressed because he understood that private medical practice would soon get abolished. That happened a few months after he left us. Anyway, in April 1958, on Easter day, my father died. I do not have the strength, here and now, to tell you about the last months of his life and about his end. The sorrow would just crush me.

And few months later, I was kicked out of my very modest university job, I had been labeled enemy of the people. Sanda and me divorced, and I moved fast back to my mother's flat, in the former office of my father, before anybody else could grab it. The divorce was a real liberation, and I became a new man, but I have already written about that. From 1958 to 1962, the year I left Romania for good, I lived together with my mother, in the family apartment. And once I left, for 12 years, we could only communicate by phone or letters. I could not travel back, and for all that long period of time, permission to travel abroad was constantly refused to her. As a cruel joke, that permission was finally given a few days after she was no longer there. She died of heart problems in 1974. Whenever I think of her I feel like weeping.

I will end this with a couple of additions, and I start by coming back to my cousin Liliana. She would have liked to become good friends with Sanda, but that did not work out well and we started seeing each other much less often. She became a construction engineer, but never got married. I think her father was responsible for that. For many years, he prevented any young man who would have liked to see her to come close and then, later on, he blamed her big for not having a husband. And after a number of years, in about 1961, she joined the communist party and so, of course, I never told her about my plans to leave the country forever. And once that happened, she violently disapproved it. In the 1980s she came for a short trip to Paris, but when we met our relations were cold. Then, finally, about 2015 or so, we exchanged a number of phone calls. Now she was again very friendly, like long before. She even told me that my father would have been very proud to see me now. Then, very sadly, she passed away. Incidentally, I often tried to imagine what my father would think, seeing me in my new life.

I will tell you now something more joyous. I have already told you that my father was very fond of good wine. But once the communist regime had hit very hard the wine producers, the wine available in the state shops—and there was no other—was awful. Normally, according to public rumor, it was made out of car tires or hay. Actually, when it happened to be out of grapes, this was publicized with big letters. So my father decided to produce his own wine. On the black market he bought a big engine to squeeze the juice out of the grapes we bought on the official market. By then, I was about 17, and very fond of physical exercise. So it was me who handled the squeezing press. My father started reading books on how to make wine, and after a few years we managed to produce some rather good one, at least good for us then.

Finally, since all those years of the by now so distant childhood and youth were so full of nasty events, I feel I owe you also some hints about politics and history.

Although of partly German ancestry, my father never liked what the Germans did, neither under the emperor William the second, and even less under Hitler. His sympathies went with France and England. And influenced by the ideas of the American admiral Mahan, in the nineteenth century, he thought that with her powerful navy, Britain could never loose a war. Of course, he did not realize that since Mahan, the world had changed a lot, nor that the new technologies had now really changed everything.

And I will end this text by trying to give you an image of the country I grew up in. Romania had never been a real democracy, but then, around 1935, it became a real dictatorship, the royal dictatorship of King Charles. And I will tell you now my earliest childhood memory. I had just become 3 years old, and I was very proud of that. In a public garden, I was playing with my friend Radu Don, who was 5, while our two German nannies were chatting on a nearby bench. In our game, the following sentence was uttered: "A foreign para-trooper is coming to shoot the King". And, immediately, out of a bush, appeared a policeman who arrested us and wanted to drag us to the police station. With difficulty, our nannies managed to convince him, given our age, to let us free. This little incident left a big trace in my mind, for several years. And every time I saw a cop, I felt like running away.

When France collapsed, in 1940, King Charles, who had not supported Hitler, was obliged to abdicate, and Romania became a vassal State to Germany. The power was now in the hands of the Iron Guard, the most extreme antisemitic fascist party of Europe, and they declared that they will rule the country for the next 1000 years. As you shall see, they lasted only for four months. During those months they did all kinds of horrible things. They murdered jews and political opponents, totally disorganized the country and created a complete mess. But soon they got into an open conflict with the army, with which they had previously to share power. For his further plans, Hitler needed Romania to be functional, so he helped general Antonescu, his stooge, to take over the power. And the country entered the war, against Russia and the Western powers.

Thus, you may notice that during my childhood and youth, I had to live through four successive dictatorships: the royal dictatorship, the Iron Guard dictatorship, the Antonescu dictatorship, and a bit later Stalinist communist dictatorship. Not much opportunity to breath freely.

Here is also a little detail to make you get an idea of the place I grew up in. My English teacher, Mrs Mary Lazarescu, had a first husband who was shot by the Royal dictatorship in the 1930s. He was an important member of the Iron Guard. And her second husband died in a communist jail.

Finally, I come back to my family. During the war, my father was in charge of the medical depot of the army, located in Bucharest, not far from our house. This was not just a storing place, it also had workshops for repairing medical equipment. It also employed a lot of civilians: doctors, engineers and workers. Actually, one of those workers, Mr Sirbu, became a good friend, and had often dinners with us. I think my father was very much liked by the people working with him.

By March 1944 the war was going very badly for Hitler. The Russians were very close to the Romanian border and the Americans and British were deep inside Italy. Romania was the main oil producer for Hitler, and it was clear that soon it will be bombed by air. Schools in Bucharest closed down, and the 2nd of April my father took me to safety by car, to the large villa of my godparents, in Predeal. Litzi came with me, and soon we were joined by my godparents, aunt Ica, Liana, and also a lot of other people too.

Starting the 4th of August, Bucharest was bombed big, my father's depot was moved to safety to Western Transylvania, many people fled the town, and my mother joined my father. In the beginning of June, my father arranged that a military truck which was coming to his place should come and fetch me from Predeal and bring me to my parents. They were now in the little town of Pincota, which was half-Hungarian half-Romanian. We were using the living-room of the Fekete family, a Hungarian couple of about the same age as my parents.

Both my parents and myself could not resist comparing how the two communities geared their respective existences, and the comparison was certainly not in favor of the Romanians. For my parents, it was actually a cultural revolution. They understood that, contrary to what they had always been told, the Hungarians were nice, able people. We had often dinner with the Fekete couple, and they became good friends of my parents.

The 23rd of August 1944 there was a political earthquake, and we got this important piece of news in Pincota, on the radio: Young King Michael, successor of Charles, had joined, together with the old political parties, the peasant party. Liberals, socialists, and now the communists too, toppled general Antonescu whom he arrested, broke with the Germans, and joined the war on the other side.

So, all of a sudden, the front moved from Eastern Moldova to Western Transylvania, and the Romanian army was fighting now against Hitler. Of course, being in Pincota had become now dangerous, and my father's depot was moved back to Bucarest.

All this had spared a lot of suffering to Romania which never became a battlefield, like willing, nilling, Hungary and Chechoslovakia did. At the time, Stalin congratulated King Michael and bestowed on him a big Russian medal. But very soon later, he chased him away and installed communist dictatorship. There would be much more to tell you, but this is a good time to stop.

Chapter 29
Father Andrei

Valentin Poénaru

Abstract This chapter is a collection of remembrances from Po's life in Romania, and his relation with a monk called Andrei.

Keywords Biography · Valentin Poénaru · Christianity

AMS Code 01A70

Around 1951, while living in Bucharest, Romania, I got to know an orthodox monk, Brother Andrei Scrima, later to become Father Andrei. I was nineteen years old and he was about ten years older than me. And soon we became friends. A bit later I will tell you how it happened that I met Brother Andrei, but right now I will rather tell you things about him. Brother Andrei was an assistant at the Faculty of Philosophy of the University of Bucharest. Then he became a monk. His cell was located in the headquarters of the Bucharest Patriarchate, a sort of mini-Vatican of the patriarch, head of the autocephalous Orthodox Church of Romania. Brother Andrei served as a librarian for that institution. Quite often, various people, mostly academics, visited him there. His cell became a place for many interesting meetings where philosophy, metaphysics bordering theology, and other such topics were discussed. I was a frequent visitor, sometimes alone, and we became friends. Brother Andrei was member of a community inside the Orthodox Church, practicing the Iesihastos, a completely individualistic mystical activity, which in English is called the *prayer of the heart*. This prayer consists of very few words: "Lord Jesus Christ, Son of God, Have pity on me a sinner." This prayer, constantly repeated, is part of the training for the next step, a totally wordless mystical contemplation. I knew also other members of that community, like Virgil S. (a philosopher), or Father Benedict G. (a monk), or father Alexe B. (a priest). I actually visited quite often father Alexe and his family (remember that orthodox priests are married). One of his daughters,

V. Poénaru (✉)
IHES, Bures-sur-Yvette, France
e-mail: valpoe@hotmail.com

L. Funar, A. Papadopoulos (eds.), *Essays on Topology*,
https://doi.org/10.1007/978-3-031-81414-3_29

Ileana, a friend of mine, married another good friend of mine, Eugen, and their son, Matei, my godson, is now a professor of mathematics at the University of Nancy. A second daughter, Mariuca, is a medical doctor in Texas. The communist regime was, usually, relatively tolerant for the orthodox creed, but certainly not so for the other Christian churches, for judaism, or for islam. But in the middle of the 1960s, the regime cracked down on *Iesihastos*; there started persecutions and political trials. Father Benedict died in jail. But by then I was not living in Romania any longer.

Towards 1955, the life of Brother Andrei changed completely. Krishna Menon, Nehru's second in command, came for a state visit to Bucarest. Remember that Jawaharlal Nehru was the first president of India. The communist regimes courted India a lot, and so Menon was received with very high honors. But the Indian guest also insisted in visiting the patriarchy, and there the only person who could speak English was father Andrei; it so happens that he also had some rudimentary knowledge of Sanscrit. Krishna Menon was so impressed by brother Andrei that he insisted in bringing him back to India with him, to teach philosophy there. The Romanian government could not refuse the demand of such a highly honored guest. So, Brother Andrei left for India, where he served as a professor of philosophy at a university, for a couple of years. After that, he moved to an orthodox monastery in the mountains of Lebanon, which became his home for many years. By then he was no longer Brother Andrei, but Father Andrei, and he also had some big plans. The two old branches of Christianity, the Orthodox and the Catholic Churches were not on speaking terms since the XIth century. He decided to try to reconcile them, and so he managed to organize a meeting, somewhere in Greece, where the pope in Rome and the patriarch of Constantinople met. And once the two high priests were together, they decided that from now on they were friends. Forever after that, Father Andrei was received with big open doors, both at the patriarchate of Istanbul and at the Vatican. He was, by now, well-known on the world scene.

Before I can tell you how me and Father Andrei met again, I have to go back to the beginning and tell you things about myself.

I was born in an orthodox country and I got baptized; but my two beloved children, Hannibal and Alexandra, which Milen and I got much later were not. Neither my parents nor me were practicing Christians, and I continue not to be that. For reasons of tradition, my parents and me as a child went to church at Christmas and Easter, but that was it. As a young teenager, my big interest was philosophy and my favorite book—my "livre de chevet" as the French would say—was Kant's *Kritick der Reiner Vernunft* (Critique of pure reason). But then I shifted from philosophy to physics and from there to mathematics. Finally I became a mathematician. But at the time when I was eighteen or nineteen my interest in philosophy, which was still alive even after I had moved to science, triggered an interest in mysticism, and so I got to know the Iesihastos community; that is how I met Brother Andrei. And it so happened that a few years after Brother Andrei had left, I met the professor for whom he had worked at the Faculty of Philosophy, Anton Dumitriu. At the time when I was seeing Brother Andrei at the patriarchy, Anton Dumitriu was in jail, for political reasons. I had read some of his books, concerning the paradoxes of logic and other related topics, which had interested me. And now that he was free, we got to know each other, and we had many nice talks together. But, of course, my big interest was mathematics. In the Summer of 1962 I was preparing myself to go to the International Congress of Mathematicians, in Stockholm, where I was an invited speaker, and I knew that I would never come back to Romania, forever after. Father Alexe, whom I have already mentioned, was one of the relatively few people who knew that I did not intend to return; and he gave me his blessing. A number of years later, in 1967, when Milen and me settled in Bures-sur-Yvette, in the Paris region, I started to see Father Andrei again. He was doing by now a certain amount of traveling, and when he stopped in Paris, he was living in a cell inside an ecumenical center run by Dominicans, called Istina, in what was then the very beautiful park of Saint Cloud. In liturgical Russian,"Istina" means eternity. So we resumed our talks, we had dinners together, and I remember that he impressed Milen quite a lot. Father Andrei and me discussed many things, but we did not always agree. In particular, I did not always agree with his views concerning the third world, to which he had become very attached, nor concerning our Western society, for me the only one where one could happily pursue scientific research. But a topic which had always fascinated me, and which I often discussed with Father Andrei, was the creative power of the infinite. I will try to give you an example of what that magical power can do. Since about 1900 there was a famous problem in mathematics, the Schoenflies conjecture. Roughly stated, it says that in any dimension n, certain two n-dimensional objects, a priori very different from each other, are actually the same; never mind the exact details. Anyway, by about 1925 it was known that for n equal to three or less, the conjecture was true; beyond that was complete mystery. Then, In the late 1950's Barry Mazur, using the magic power of the infinite, proved the Schoenflies Conjecture for all n; and he was barely 18 at the time. I will only give you here a caricature of his argument, at high-school level. Consider an infinite alternate sum of plus ones and minus ones. Arranging the computation in two ways, one way you can get one, the other way you can get zero. Hence you have "proved" that one equals zero, which is nonsense, of course. But Barry's proof was a fantastically beautiful argument, so new and original that it took the mathematical community six months to understand and accept it. It so happens, also, that a few years later, Barry and I met and became best friends; but that is another story.

Father Andrei was very much interested in these ideas concerning the creative power of the infinite, and in the fact that using this power one can manage to harness the absurdity. But then, the township of Paris decided to convert the Park of Saint Cloud into a stadium. So the park was erased, Istina disappeared in the process, and father Andrei stopped coming. We continued to communicate by mail, between Paris and Lebanon.

In 1990, when communism was gone, Father Andrei moved back to his cell at the Bucharest patriarchy where, a number of years later, he died. But in the meanwhile he had written a couple of books, where among many other things, he talked about me, his friend. And some years later, that earned me an invitation to the Vatican, to a theological conference. I thought I would be out of place there and, very politely, of course, I refused.

- -

After this story, you might ask whether I believe in God or not. In the few remaining lines, I will try to provide sort of an answer to that question. I should start by saying that I do not go to church and that organized religion is something totally foreign to me. Furthermore, I consider that the question whether God exists or not is meaningless. For me, at least, the word "exists" refers to this world, to things accessible by our senses, by scientific experiment, or by scientific theory. This restricts to our universe, our world. Of course, there is also the scenario of the parallel worlds hovering around, the multiverse. String theory, but not only string theory, may suggest that, and I personally like the suggestion. But I do not think it bears on our present discussion. Anyway, existence is a term which should be restricted to our world, universe or multiverse. Now, at least when we look at the world around us, it is certainly beautiful: beautiful plants and animals, beautiful forests and mountains, then certainly the beauty of the starry sky, and all that. But then, behind the beauty of the sunlight, on a bright summer day, looms the majestic beauty of Maxwell's equations. The big deep beauty of the world is actually its mathematical laws. And as far as their ontological status is concerned, I think that this stays a big mystery. Of course, when it comes to biology as a science, that is still a bit like physics had been before Galileo and Descartes. Misha Gromov predicts that when biology will reach that more advanced state, then some new, not yet discovered mathematics, presumably immensely beautiful, will be involved. But that is still to come.

So we come now to the world of mathematics, which to my mind is as real and objective as our world. Behind the reality of our world stands that very beautiful mathematical reality. Mathematics is actually an infinitely large, infinitely complex and fantastically beautiful structure. Let us call it the infinite castle; and we humans have only access to a very small, finite part of it. At this point I would like to think that some entity, let us call it Transcendency, would be able to contemplate the infinite castle in its totality, all at once. Imagine grasping the totality of the integers, all at once, like we grasp some small finite piece of it. At that level, Gödel's undecidability gets a completely different outlook, and there are no more unsolved mathematical problems.

This may be just a dream of mine, but I think that, just like mathematics, Transcendency has objective reality, totally outside of our world, and that it is accessible to us only by mystical contemplation, a strictly individual and individualistic activity. This kind of activity should require training, and passing from apprentice to master. Personally I never did it, but I do believe that it can be done. And I think that I have some good reasons for that belief.

Concerning Transcendency, I take the same view as Pseudo-Dionysius the Areopagite, the Neoplatonic philosopher and Christian theologian of the fifth century, namely that Transcendency is apophatic, meaning beyond expression. There is no meaning in putting adjectives on top of it, like the various religions do for their Gods, things like almighty, merciful, and all that. Making any precise statements concerning it would be meaningless; possibly one could only use negatives, like "it is not this, nor that, nor ..." And it certainly makes no sense to talk about the Creation of the world, either. The world just is; it makes no sense to talk about its creation. It is not like a pair of shoes or a suit of clothes, which need shoe-maker and tailor. When it comes to the relation which Transcendency might have with our world, I feel sympathy with what Meister Eckhart, the great mystic of the Rhine valley, said many centuries ago: God is not part of the world, and His existence is less than that of the smallest mouse. Then there is the Christian idea which I cannot accept: that God the Father sends his Son to our humble and peripheral planet Earth, in order to save humanity. And why not just the man from Nazareth, rather than God the Son. That idea, which all religions have, that God cares a lot for us, sounds absurd to me. I cannot believe in any special role for humanity in this world. In an infinite universe, or even in a finite very large one, there should be many more intelligent species, some highly superior to us. I sympathise here with Giordano Bruno. I cannot express to you how desperately depressed and lonely I would feel, if it would turn out that we are alone in the universe.

Chapter 30
The Magic Summer of 1964

Valentin Poénaru

Abstract This chapter gives some glimpses on the early years of Po in the West along with his family.

Keywords Biography · Valentin Poénaru · Geometric simple connectivity

AMS Code 01A70

In a span of few months, in 1964, Milen and I met in Cambridge, England in the beginning of June. We fell in love with each other and on the 3rd of October we got married in Oslo. That is a telegraphic account of what happened that magic Summer. But before I can properly develop that story, I have to tell you about the two very distinct lines of events which made it possible to happen at all. And even before, here are some data: In that Summer 1964, Milen was 18 and I was 31; and the 3rd of October of this year 2021, together with Alexandra and Tony, we celebrated the 57th anniversary of our marriage, in a very nice restaurant, on a cliff on the sea, in Nice. I need to explain you those two lines because, a priori, all this was a highly unlikely event. Milen was Scandinavian and I came from Romania, in the backwoods of the Balkans. And so I will start with the first line: how did I become friends with Christopher Zeeman, the very distinguished English mathematician, later to become Sir Christopher. And then I will come to the second line, how came that an English mathematician had a cousin who was a beautiful Scandinavian girl.

Here is how I got together with Christopher.

In 1959, while living in Bucarest, I was jobless and I was making my living by giving private math lessons (which was black marketeering), because the previous

---

Written in 2021

V. Poénaru (✉)
IHES, Bures-sur-Yvette, France
e-mail: valpoe@hotmail.com

L. Funar, A. Papadopoulos (eds.), *Essays on Topology*,
https://doi.org/10.1007/978-3-031-81414-3_30

September I had been kicked out of my very modest University job, since I had been declared to be an enemy of the people. But then I made an important mathematical discovery, which was the big quantum jump in my life. What matters for our present story is that about the same time, in Cambridge, Christopher also made a mathematical breakthrough. I cannot resist telling you that story; it is both quite interesting, and it played a big role in my own life and in Milen's.

Christopher was then a young lecturer at the University of Cambridge and he gave a course in topology, our common field at the time. And he wanted his students to work independently, so he gave them mathematical papers to read and to report about. His youngest student, barely 18, was Roger Penrose, to become later a world-famous mathematical physicist, now a Nobel prize winner. Penrose vastly improved the theorem he was supposed to lecture about, coming with a completely new and original line of argument. Christopher discovered that this argument could be vastly extended and lead to a really revolutionary result. But there was a problem: the whole thing depended on something which Christopher did not know whether it was true or false, or whether anything was known about it. And so he went to a famous Cambridge professor, Henry Whitehead, and asked for advice. Whitehead laughed and said: but I have already proved this thirty years ago!

So there it was now, the big theorem of Penrose, Whitehead and Zeeman. I heard about this and Christopher also heard about my non-trivial factorization of the five-dimensional cube (that was my discovery mentioned above). And since we were each interested in the other's work, we started corresponding

Two things should be added here before I go on.

Barry Mazur, who was already a very famous mathematician, made independently the same discovery as me and that brought us together. We became best friends. It so happens that many years later it was discovered that the class of four-dimensional manifolds we had discovered, named today after us, have properties which are not true in any other dimension different from four. I will come to this now. I will open a parenthesis, and explain in more precise terms what those mathematical discoveries which I just mentioned were exactly.

The Penrose–Whitehead–Zeeman theorem concerns piecewise linear (PL) manifolds. It says that if a simply-connected n-dimensional manifold M has all of its homotopy groups, up to dimension q, with q smaller than $n/2$, equal to zero, then it can be embedded into a Euclidean space of dimension $2n-q$. The methods of proof of this theorem were an important brick for the techniques used by John Stallings, a few years later, for his proof of the Poincaré Conjecture, in a PL context, in all dimensions ≥ 7. Soon afterwards, Christopher extended this to dimensions 5 and 6, using a very clever trick.

I turn now to the discovery which, independently of each other, Barry and I we made. And we move from the context of PL manifolds to infinitely differentiable ones.

We found that there exist contractible, compact bounded 4-manifolds N, with non-simply connected boundary (hence, topologically, they are certainly not the 4-ball), such that their cartesian product with an arc is the 5-ball. This result, which triggered a certain amount of literature, was born in unsuccessful attempts to prove the 3-dimensional Poincaré conjecture. It so happens that these non-trivial factors N had another very intriguing property, discovered years later by Andrew Casson. This concerns the notion of geometric simple connectivity (GSC), about which I will say a few words now.

Trivially, GSC implies simple connectivity. Steve Smale discovered that for smooth manifolds of dimension 5 or more, the converse implication holds too, that is, $\pi_1 = 0$ implies GSC, and this was an important step in his proof of the Poincaré conjecture, in dimensions 5 and higher, in the smooth context.

Many years later came Perelman's celebrated result, from which it follows right away that in dimension three too, $\pi_1 = 0$ implies GSC.

Now Casson showed that our nontrivial factors N cannot be GSC. So, in dimension 4, and only in dimension 4, $\pi_1 = 0$ does not imply GSC. This, and the existence, unexplainable by algebraic topology, of uncountably many smooth structures, exactly in dimension four, makes that dimension 4 very very special. This is a topic which fascinates me. With this I close now my mathematical parenthesis.

Now we will change completely the register. Whitehead's work has always fascinated me and one of my latest papers, written in the Summer 2021, is called "Whitehead's Nightmare"; this is a mathematical term concocted by me, which I will not explain to you now. I will just tell you that long ago, during a walk under the starry sky, in a cold winter night, I devised that term, thinking of some strange mathematical facts. This was while I was waiting for Hannibal, then about eleven, to finish his session of orthophony, demanded by his school.

A few years after Christopher and I connected with each other, I managed to escape from Romania, and started a completely new existence. And very very soon I landed in France, to stay with Barry and Gretchen in Bures-sur Yvette. Barry was visiting IHÉS, one of the best research institutes in the world, for maths and physics, located in Bures. Christopher and his wife Rosemary were visiting there too, so we finally met and became friends. And since by now I had lost my original parental home, IHÉS has somehow become, ever since, the substitute for it. This ends the story of my first line of events, and now I will pass to the second one, the Scandinavian connection of Christopher.

We will move now many many years back into the past, to Russia in 1917, at the time of the Russian revolution. It was also the time of the First World War, but Denmark was a neutral country with solid Russian connections. And it so happens that in 1917 there were two young Danes, Rigmor Zeeman and Peter Dreyer, each working for their respective trade companies, in Saint Petersburg (re-named Petrograd during the first world war). These are very key characters in our present story, because in time they became Milen's grandparents. And Rigmor, later called Meo by everybody, always told us that she actually first met her Peter in one of the main squares of Saint Petersburg, while in that place too, the Bolshevik red guards were storming the Winter Palace, and brought Lenin to power.

Anyway, they got back to Copenhagen, got married, and starting essentially from scratch, managed to build an impressive economic empire. They were the main owners of Hôtel d'Angleterre, the best hotel in Copenhagen, and member of a league of the ten top hotels in Europe. And they were also the most important wine merchants in Danemark.

You may have noticed that my first line of events involved some whiff of adventure, but the second line involves even more. Rigmor had several brothers and sisters. And in the very early twenties, one brother, Christian, went to become a monk in a Buddhist monastery in Kamchatka, an icy peninsula coming out of Siberia towards the Pacific Ocean. But after three years of that, he decided he had enough, moved to Japan and got married to an English girl. They had two children, our Christopher and his older sister Catharine. Soon, there was a terrible earthquake in Japan, Christian's business collapsed, and the marriage started doing the same. So, Christopher's father embarked on a ship for San Francisco, presumably to get some help. During the trip he mysteriously disappeared and was never heard of ever since. The by now lonely wife moved with her two children back to England and that is how Christopher became a British mathematician. And soon he also became a beloved nephew of Milen's grandmother Meo. I may add that during all his adult life Christopher tried to find out what had happened to his father, whether he was still alive, and searched for him, sometimes in quite unbelievable places, like Buddhist monasteries in Tibet. He was thinking of the Shangri-La story.

A second brother of Rigmor, Fleming, joined the French foreign legion and then also disappeared. But twenty or so years later he sent a postcard back home: everything is fine, I am in Brazil; that is what he wrote. He had made a new life in Brazil. And, of course, Christopher went to see him there.

The third brother, Herbert, stayed home and did not move to far-away lands. He became a satyrical journalist, and during the Second World War he started writing little puns against the Germans, in a kind of Danish which he thought the Germans could not decode. But they did, and came and shot him dead in front of his family. Of course, he was connected with the anti-nazi resistance too. His memory is honored today in Denmark.

With this I end my second line of events, and we move now to 1964. By then I was living on the other side of the Atlantic, in Cambridge Massachusetts, and I was teaching at Harvard. Christopher was not only a great mathematician, but also a fantastic organizer. He got a very large amount of money from the British government, in order to put up a very big and fancy topology conference in Cambridge. All the important topologists of the day were there, in Cambridge, that year 1964. I came there, from Cambridge Masachussets, already in May, to give some lectures in Oxford and Manchester. And, of course, Barry and Gretchen came there too. As a matter of fact, they were there, together with me and Milen, from the very beginning of our love affair, through most of that magic Summer. And so they were not only my best friends, but became Milen's best friends too. And we think of them as very close members of our family. When I first got there, Christopher told me that his young Danish cousin, that was Milen, will soon come for a longish visit, in order to perfect her English. And he suggested that I should accompany her to the next May ball, in the forthcoming month of June. You have to know that these so-called May balls (happening possibly in June) are big fancy social events in the British society, taking place in Cambridge or Oxford. And Milen and I had free access to this fancy event because she was Christopher's cousin and I was a guest of the Caius and Gonville college in Cambridge.

Anyway, a few days later, about the 3rd of June, during a party at Christopher and Rosemary's house, appeared a beautiful blond blue eyed girl, that was Milen, and essentially right away, some magic thing passed between us two. But before I continue, I have to tell you a bit about Milen's background. Rigmor and Peter had three children: a daughter Trine and two sons, a younger Peter and Dines. Trine was an artist, a sculptor, and soon she got married with a fellow artist; and they had a daughter, Milen. That marriage, not very much approved of by the family, very soon collapsed, and Trine married an American, with whom she had a second daughter, Christina. The second marriage did not last long either, and finally Trine married a Norwegian lawyer, Johannes—we always called him Jo—and moved to Oslo, together with her two daughters. Together with Jo, a son got born, Gunnar. That is how Milen, although born in Denemark, mostly grew up in Norway.

Milen had actually never met her biological father (that happened only much later, when we were already together, but I will not tell about that now). Her real father figure became Jo, and more will be told here about him. But here is a little amusing detail: When the family of five met for dinner, each of the two parents spoke his or her native language (and you have to know that Danish and Norwegian are quite close) while each of the three kids addressed each parent in his or her respective language.

But now, before anything else, I should tell you that Milen was very different from all the other girls I had known before. She was always happy and smiling, never bored, enjoying things as they came. On the other hand, she was also very clever and curious about interesting things; we can have long conversations about all kinds of subjects. We both enjoyed sports, an important part of our lives, and did a lot together: skying, bicycling, running, hiking. Milen is an artist, in time she became a great painter. At the same time she has a very big knowledge concerning various religious or magic-shamanistic beliefs and their history. This is useful for her art. Being together with her has been an immense thing in my life.

Back now to 1964. After that party where we first met, Milen and I started going out together, often accompanied by Barry and Gretchen; as already said, they were my best friends and soon became Milen's too. And we fell in love with each other. I still remember the first time I took her out for dinner, at "The turk's head", a nice restaurant in Cambridge. And then, towards the middle of June, came that famous May ball, at King's College in Cambridge. I had to rent a tuxedo, and Milen borrowed a very fancy dress from Catharine, Christopher's sister. And that was quite an event: several orchestras, big dancing, of course, and fancy meal in a big beautiful garden in front of the historic building where the thing took place. We were surrounded by the classy British society, but we only cared about our own magic universe. And by then we knew that we wanted to be together forever.

By then Milen's family had heard that things were happening, and they wanted Milen to come home and tell what was going on. So, for two weeks, Milen went to Skagen, on the Northern coast of Jutland, where all her family was for holiday, while I made a short trip to IHÉS.

When we found ourselves back together in Cambridge, after this short break, we planned to get married. Barry and Gretchen took a short flight to Bucharest, to bring the news to my mother. And we started to plan to go to Oslo to talk to Milen's parents. I understand that Christopher had already written very reassuring things concerning me, to his aunt. And he took me for a long walk to explain to me the family set-up. Here is how Christopher put things: the family in Oslo was essentially a colony of the central power, in Copenhagen, namely Milen's grandparents, Rigmor and Peter, who were the pinnacle of a big family clan. So, Trine and Jo could not decide anything, the decision stood with the grand-parents. And so, towards the end of August, I flew with Milen to Oslo. I was quite nicely received by Trine and Jo, but as Christopher had warned me, they were quite non-committal concerning the marriage. Actually, two days later Milen's grandmother arrived too, to see me. I think she decided that she liked me, and here is what happened. I had actually invited Milen to come for a travel to France, and her grandmother said: why don't you go together to Paris, and when you come back we will see about the marriage. And so we flew to Paris, actually we went to Bures, where the IHÉS had put a flat at our disposal, for the occasion. And we certainly did a lot of things in Paris and in the Vallée de Chevreuse.

There is now a funny little story which I have to tell you. We gave our first dinner party. I invited, for the occasion, Léon Motchane, the director of the IHÉS, and his main secretary and future wife, Annie, and also Alexander Grothendieck, the most important mathematician of the second half of the XXth century and his wife Mireille, all friends of mine. But you have to understand that Milen had never cooked, her mother did not trust her in the kitchen, and my knowledge in the matter was restricted to hard-boiling an egg and opening a can. So the gastronomy was a disaster: we opened a can of smoked salmon, that was quite usual then, but it was rotten, then the roast beef was burned, and the only thing which saved, a bit, the day, was a big Danish ice-cream with chocolate sauce. But everybody was happy and Milen was adopted by the IHÉS. Then after some short trips to Bruxelles, where I had the only ex-Romanian relatives in the West, and to Lausanne, to visit Georges de Rham, we got back to Copenhagen, towards the end of September. There, we were received with open arms, and plans for the marriage started. That is when I also met old Peter, Milen's grandfather. I had heard that he was a big bossy guy, but actually he turned out to be a very friendly old man, full of interesting stories. Unfortunately, the next time I returned from the US to Europe, he was already dead.

But Meo, Milen's grandmother, adopted me, in time, as a new grandson. In time too Milen's step-father Jo and I became best friends. And I still remember the long interesting conversations which I had with Jo and his brother-in law Kjell, an important Norwegian diplomat, on history and world affairs. We took a ferry from Copenhagen to Oslo, where we got married in the town Hall, and then the two of us, Trine, Jo, Gunnar and Ki, all went back to Copenhagen for the big marriage feast. But a few days later I had to fly back to the US, to my new appointment, at the Institute for Advanced Study in Princeton, the American model of the European IHÉS. Milen had to wait for her immigrant visa, and she joined me in November. And then our married life started unfolding, and it has continued happily to today.

I would like to add some additional comments. When Milen and I first met, our only common language was English, and we have continued to speak English together, ever since. Soon Milen learned French too, through reading books, more exactly the six volume historical novel "Les rois maudits", initially with my help, and then also taking courses at Princeton University. After a number of years we moved to France, to Bures again, always feeling a bit like American immigrants in Europe. We are both equally fluent in English and French, let's say we are bi-lingual, but that does not include our private "mother-languages", Danish and Norwegian for her, Romanian and German for me. And our two children, Hannibal and Alexandra, born in France, are both completely fluent in English and French. And the four of us speak very good Italian too. A last linguistic comment: strictly speaking it is Trine, Milen's mother, who was first cousin of Christopher. But Christopher always referred to Milen as his cousin, and later extended that term to me too. And I continued to use it.

In the painting represented in Figure 30.1, you can see me standing in front of Milen who is pregnant with Alexandra, and with Hannibal who was a little kid. The painter is an old friend of mine, Herman Braun-Vega, from Peru. He spent virtually his entire career as an artist in Paris, and died of Parkinson's in 2019. The painting was done in 1981.

Fig. 30.1 The family Poénaru in 1981, by Herman Braun-Vega

Chapter 31
Po, A Mathematician Between Humanism and Science

Corrado Tanasi and Franco Marchese

Abstract This is a personal tribute to Po by his long time Italian collaborator and another friend.

Keywords Biography · Valentin Poénaru

Mathematics Subject Classification 01 A 80

From the many conversations held with mutual friends, the following account was born, which is a loving attempt to outline the figure of Valentin Poénaru, with all his unconventionality and, above all, with that singular ability to live in a liminal condition between humanism and science, balancing between the two cultures. This amounts to rejecting, both intellectually and practically, the harmful dichotomy that has long sought to erect a wall between the humanistic and scientific worlds. That is no small feat.

We met Poénaru over forty years ago. He was invited in 1977 to the Department of Mathematics at the University of Palermo for a series of lectures that were held for many years afterwards, soon becoming a summer tradition and allowing a friendship to form.

We were immediately struck by the name he had chosen for himself, Po, rejecting his given name. Perhaps he found the name, Valentin, annoyingly sweet; almost certainly he felt his last name, Poénaru, was too obviously linked to what he always referred to as "my former homeland," Romania, which he had excluded from his horizon with a *damnatio memoriae* that extended even to his name. And surely the

C. Tanasi (✉)
Università di Palermo, Dipartimento di Matematica e Informatica, Palermo, Italy
e-mail: corrado.tanasi@unipa.it

F. Marchese
Liceo Scientifico "S. Cannizzaro", Palermo, Italy
e-mail: paps02000l@istruzione.it

L. Funar, A. Papadopoulos (eds.), *Essays on Topology*,
https://doi.org/10.1007/978-3-031-81414-3_31

memory of Edgar Allan Poe, a genius much more troubled than he, also played a role.

As we spent more time together, the traits of his personality became increasingly clear. Above all, an insatiable and voracious curiosity that led him to explore the fields of philosophy, literature, and, most of all, his beloved history. In these academic fields, seemingly so different and distant from his chosen one, mathematics, Po acquired a culture far from amateurish, which, thanks to his prodigious memory, allowed him to speak for hours about the Punic Wars (not by chance, his son is named Hannibal) or the Bolshevik Revolution with a level of expertise that even a professional historian would have envied. But without the arrogance of someone who knows he belongs to an elite caste, which professional historians, the "academics" (we are speaking, of course, of Italy), too often display. His historical narratives were rigorous and at the same time engaging, precise and detailed, yet also as captivating as a novel.

On the other hand, nothing was ever so profoundly distant, so radically foreign to Po, as the rigidity and, in some cases, arrogance of the academic world. We never saw him wear a jacket; his dress code only included long or short pants depending on the heat, sandals or hiking boots, and often improbable t-shirts. Po was very fond of these brightly colored t-shirts, to the point of continuing to wear them with sovereign indifference even when they were significantly frayed.

Po's literary culture was vast and surprisingly varied. While it may seem normal–after all, he was primarily a mathematician!–that he was well-acquainted with a book not exactly famous like Flatland by Edwin A. Abbot, it was a mystery that he could discuss an author long since removed from the Italian canon, such as Giovanni Papini. But note, not the more well-known and more detestable Papini, but rather the imaginative and still readable one from Gog, a collection of stories from 1931 that hadn't been reprinted in Italy for decades (to read it, one would have had to buy a French translation). How and where Po came into contact with Papini? We do not know, but we imagine that the answer lies in his "former homeland," Romania, a land with a vibrant Central European tradition before the Sovietization erased everything, and a true melting pot of people, languages, and cultures.

At this point, it is time to clarify the particular relationship that Po established with languages. What we are about to say certainly applies to Italian, of which we are direct witnesses, but we suspect it also applies to French, German, and English, the language of his American years and, always, the one spoken at home with his wife Milen (a name that Po etymologically relates to "smile"; it may be a paretymology, but it is a happy paretymology: we have rarely seen such a bright smile). Thus, in relation to Italian, Po has a relaxed and pragmatic attitude. If he does not know a term, he does not try to use a synonym, nor does he make an effort to engage in a roundabout way to resolve the shortfall: he simply relies on hybridization, creating a lexical (but also syntactic) mold: one of Po's expressions that has long entered our family lexicon is: "vuoi ti dire un sicreto ?" (do you want me to tell you a secret?), pronounced with a smiling wink that makes him quite similar to a gnome who has just discovered a precious nugget of gold and followed by a hearty laugh, because the revealed "secrets" are quite trivial, relating to, say, his preference for Perrier mineral water, "petillante," of course. This reliance on lexical molding, carried out with a gentle disregard for danger, produces surreal effects when Po encounters so-called "false friends." If you ask a supermarket clerk how long one can "guardare" (look at) a certain olive oil, it may happen that the clerk is taken aback, unless they know French and can automatically connect "guardare" with the french "garder" (to preserve). The masterpiece of this imaginative linguistic procedure that makes Po's Italian flavorful and unique is embedded in the story (Po loves to tell stories and has an extraordinary narrative talent, a storytelling ability reminiscent of an ancient bard) of a mountain adventure he experienced when he was young in Romania: "la corda si è cassata, lui è tombato ed è disparuto: morto" (the rope snapped, he fell, and he disappeared: dead). One does not know whether to weep for the tragedy so vividly described or to laugh at its hilarious linguistic rendition. We cannot exclude the possibility that in what we call "poenarese" there is also a modest amount of unconscious (or self-satisfied?) playfulness: perhaps Po could speak a less "naive" Italian, as he would say, but he is not interested in doing so; he finds much more enjoyment in invention than in conformity.

To conclude these lines, few but affectionate, we return to Po's profession as a mathematician. During an interview in 1999 for an Italian studies magazine called "Chichibìo," which no longer exists, responding to the question, "In Italy, the divorce between humanistic and scientific culture has a long tradition, which is hardly something to be proud of. How have you experienced the relationship between the two cultures?" he said: "The divorce between humanistic and scientific culture is a fairly recent phenomenon, whose origin, at least in its current form, can be traced to the last century. This divorce can easily be traced back to the structure of secondary and university education in modern states, but since this education ultimately depends on the collective decisions of politicians and citizens, the entire society bears its share of responsibility. It seems to me that literary education in our universities deserves particular criticism. At the secondary education level, especially in France, the written word dominates everything; other expressive media, such as drawing, painting, or music, are considered entirely secondary, if not outright discouraged. And furthermore: a poorly understood or poorly assimilated Cartesianism means that the only thought recognized as such is that which expresses itself through precise words, leaving no room for dreams, imagination, or intuition. This brings me to the second part of the question. I believe that when I do mathematics, I do it with both halves of my brain; in fact, my entire being is involved, not just the predominantly logical and discursive part of the intellect. Therefore, it is no longer possible to think in terms of conflict between the two poles of culture, and some form of integration must occur between them, *ipso facto*. I am certainly not the only man of science convinced of this thesis; but from here to the moment when all of this will become part of education, there is still a long way to go."

Index

L. Funar, A. Papadopoulos (eds.), *Essays on Topology*,
https://doi.org/10.1007/978-3-031-81414-3

U

V

W

X

Y

Z

The manufacturer's authorised representative in the EU is Springer Nature Customer Service Centre GmbH, Europaplatz 3, 69115 Heidelberg, Germany. If you have any concerns regarding our products, please contact ProductSafety@springernature.com

Printed and bound by CPI Group (UK) Ltd, Croydon, CR0 4YY
30/09/2026
02225588-0020